Petroleum Geology of the North European Margin

Norwegian Petroleum Society

Organizing and Editing Subcommittee

Elisabeth Holter,
Conference organizer
(Norwegian Petroleum Society)

Anthony M Spencer, Chairman
(BP Petroleum Development)

Sverre O Johnsen (Norwegian Institute of Technology)

Atle Mørk (Continental Shelf Institute)

Eigill Nysæther (Saga Petroleum)

Per Songstad (Norsk Hydro)

Ånon Spinnangr (Statoil)

Petroleum Geology of the North European Margin

Proceedings of the North European Margin Symposium (NEMS '83), organized by the Norwegian Petroleum Society and held at the Norwegian Institute of Technology (NTH) in Trondheim, 9–11 May, 1983.

Edited by
A.M. Spencer et al.

Published by
Graham & Trotman Ltd.
for the Norwegian Petroleum Society

First published in 1984 by
Graham & Trotman Limited
Sterling House
66 Wilton Road
London SW1V 1DE

Sold and distributed in North America
by International Human Resource
Development Corporation (Publishers)
137 Newbury Street, Boston,
MA 02116, USA.

British Library Cataloguing in Publication Data

North European Margin Symposium
(1983: Trondheim) Petroleum geology
of the North European margin.
1. Gas, Natural in submerged lands —
Europe
2. Petroleum in submerged lands —
Europe
3. Continental shelf — Europe
I. Title II. Spencer, A.M.
III. Norwegian Petroleum Society
553.2'8'094 TN882.E9

ISBN 0-86010-486-9

Typeset in Great Britain by Tek-Art,
Croydon.
Printed and bound in Great Britain by
Page Brothers, Norwich.

CONTENTS

Foreword

This volume contains the papers presented at the North European Margin Symposium which was held at the Norwegian Institute of Technology in Trondheim from 9 to 11th May 1983. The idea for the meeting was conceived by the Geology and Geophysics Committee of the Norwegian Petroleum Society, in meetings consisting of R P F Hardman, J Freake, I Gausland, A M Spencer, R G Todd, F Aamodt, E Nysæther and G Kobjørnsen. The Trondheim meeting was planned and organized by an Organizing and Editing Subcommittee consisting of E Holter, O S Johnsen, A Mørk, E Nysæther, P Songstad, A M Spencer and Å Spinnangr.

There were 216 participants at the meeting in Trondheim and 34 papers were given, most of which were invited. The meeting was opened by Dr K S Heier, Director of the Norwegian Geological Survey and a concluding address was given by E Bergsager of the Norwegian Petroleum Directorate, who talked about future exploration in Norway.

The symposium in Trondheim was the latest in a generally biennial series of meetings* organized by the Norwegian Petroleum Society (Norsk Petroleumsforening — NPF) comprising:

Jurassic Northern North Sea Symposium (JNNSS), Stavanger (1975)
Mesozoic Northern North Sea Symposium (MNNSS), Oslo (1977)
Norwegian Sea Symposium (NSS), Tromsø (1979)
The Sedimentation of the North Sea reservoir rocks, Geilo (1980)
Norwegian Symposium on Exploration (NSE), Bergen (1981)
North European Margin Symposium (NEMS), Trondheim (1983)

In view of the current significance of the subject of the meeting, a new method of publication has been adopted by NPF in order to ensure fuller distribution and more permanent record of the articles presented. This has involved refereeing the articles and we thank all of the referees who helped with this difficult task. The NPF is most grateful to the Norwegian Institute of Technology for making available appropriate facilities.

Organizing and Editing Subcommittee

*Information on the earlier symposia and their publications can be obtained from: Norsk Petroleumsforening, Postboks 1897, Vika, Oslo 1, Norway

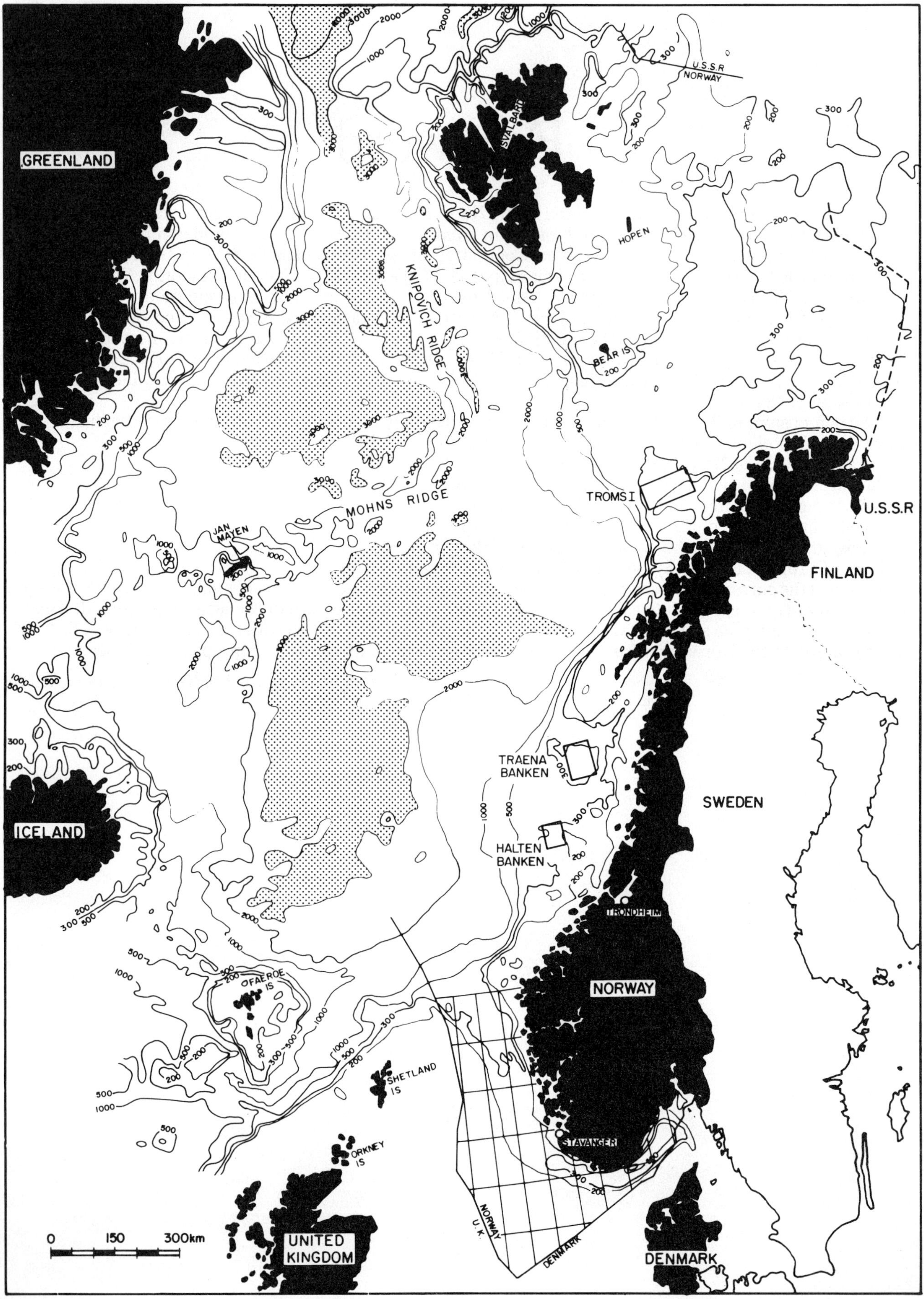
GREENLAND
SVALBARD
HOPEN
BEAR IS
KNIPOVICH RIDGE
MOHNS RIDGE
JAN MAYEN
TROMSI
U.S.S.R
NORWAY
FINLAND
TRAENA BANKEN
HALTEN BANKEN
SWEDEN
ICELAND
TRONDHEIM
NORWAY
FAEROE IS
SHETLAND IS
ORKNEY IS
STAVANGER
UNITED KINGDOM
DENMARK
NORWAY
U.K.
DENMARK
0 150 300km

Introduction

Knut S. Heier
Director General, The Geological Survey of Norway

Two hundred years of mapping and drilling have given us a reasonable model of the continents, and nearly two decades of deep sea drilling have shown at least a cartoon of the oceanic crust. The transition between continent and ocean remains, however, almost unknown, the 'new frontier' of geological exploration. Much current research in earth science is based on applying the plate tectonic model to problems as diverse as continental orogeny, palaeo-oceanography, and metallogeny. Plate tectonics, developed in large part from geophysical and geological information gathered from the world's oceans, is now playing a key role in explaining the geological events and processes that shaped the continents. The continental margins play key roles in the plate tectonic model. Moreover, they represent the largest unexplored area where we might find gas and oil.

We speak of two types of continental margin, active and passive. A third type, transform margins, of which the San Andreas fault system is an example, may occur in conjunction with either active or passive margins. An active margin is occasioned by the collision or the convergence of two crustal plates, one of which carries a continent at its edge. The active margin along the Pacific coast of South America is formed by the Nazca Plate being subducted or drawn down beneath the American plate, upon which rides the continent of South America.

This type of continental margin is characterized by deep ocean trenches, volcanoes and earthquakes. It is also the probable breeding ground for some of our more valuable sulphide ores, especially those known as massive sulphides and porphyry coppers. Therefore a study of active margins can contribute to our knowledge of tectonic forces, ore deposits and earthquakes. Through a study of the processes that take place today at the active margins we may hope to gain insight on the formation of our own stratabound sulphide deposits in the Caledonian mountain chain that formed several hundred million years ago. This insight may ultimately lead to the finding of new ore deposits in such areas.

When a spreading centre splits, a continent and a new ocean basin is formed, the two halves of the old continent move apart and their margins gradually subside. These margins, which do not coincide with plate boundaries, are known as passive margins. They contain the geologic history of rifting; they show the chemical transition between continental and oceanic crust, and they contain the oldest marine sediments of the present spreading cycle. Lying atop the deeper crust of passive margins is a thick wedge of sediments. Indeed, about half of all the marine sediments deposited in the ocean basins during the past 200 Ma is on the lower continental slopes and continental rise. Because of their thickness and their proximity to the continents, these sediments are potential sources of oil and gas.

In this symposium, the North European Margin Symposium, we shall discuss the development of the sedimentary basins at a passive margin during late Palaeozoic to Tertiary times. Their setting, structure, stratigraphy and history have been intensively studied during the last 20 years, all with the purpose of exploration for hydrocarbons. In retrospect our geological understanding of this area before 1960 was ridiculously primitive. Rocks of the Jurassic and Cretaceous were known from Norway at Andøya and from some scattered localities elsewhere without their true significance and relationship to the geology of the continental shelf being realized.

Seismic measurements carried out by the Lamont Geological Observatory during the years 1951 to 1953 in the Norwegian and Greenland seas and off the coast of Trøndelag (the landward side of the southern part of the Vøring plateau) showed a 'seismic depth' to the underlying crystalline

Petroleum Geology of the North European Margin © Norwegian Petroleum Society (Graham & Trotman, 1984) pp. 1–2.

basement rocks of approximately 5000 metres. As far as I know this was the first indication of the presence of thick layers of consolidated sediments between the crystalline basement and the Quaternary surface deposits. The results were not published until 1959.

The Geological Survey of Norway did their first aeromagnetic measurements over the continental shelf in 1962. As this work continued the different basins of sedimentary rocks were discovered and delineated.

During the three days of the Symposium a total of 35 papers were given, covering the whole range from the evolution of the margins and mechanisms of basin subsidence to the detailed geology of individual basins. A major section of the conference deals with syntheses of results obtained during the exploration for hydrocarbons on the Norwegian Continental Shelf north of 62°N.

1 Cenozoic evolution of the continental margin off Norway and western Svalbard

Olav Eldholm, Eirik Sundvor,† Annik M. Myhre* and Jan Inge Faleide**
*Department of Geology, University of Oslo, Norway
† Seismological Observatory, University of Bergen, Norway

Following the Mesozoic phases of crustal extension the region between Norway and Greenland became relatively quiet, with regional subsidence dominating in a large basinal area which extended into the North Sea and the Barents Sea. At about Palaeocene/Eocene time the mid-Atlantic rift progressed northward and sea-floor spreading started along the newly formed plate boundary. The initiation of Cenozoic rifting and the subsequent drifting phase has fundamentally defined the structural and depositional framework of the present day margin off Norway and western Svalbard. Particularly important observations are: the existence of marginal highs apparently associated with one or several episodes of volcanic activity; the remarkably different amounts of marginal subsidence, with respect to both region and time; the effects of the mid-Oligocene change in relative plate motion on the margin structures; the two-stage evolution of the Greenland Sea with its associated shear movements.

INTRODUCTION

The continental margin off Norway and western Svalbard has developed as a passive margin since late Palaeocene time, its history being closely related to the progressive northward opening of the North Atlantic—Arctic oceans. A physiographic map of the Norwegian-Greenland Sea reveals quite clearly that the margin changes character along its strike (see Fig. 1). A few observations are particularly noteworthy. The width of the continental shelf changes from a maximum of about 220 km at 66°N to only 20 km off Vesterålen and about 30 km at 79°N. The continental slope gradient also varies considerably, being steepest off the Vesterålen islands. The marginal Vøring Plateau, which covers an area of about 35 000 km^2, breaks the continental slope between 65 and 67°N. Two major regional sedimentary basin provinces, the North Sea and the Barents Sea basins, extend on to the continental margin. The present day plate boundary, the Knipovich Ridge spreading centre, lies as close as 80 km to the Svalbard coastline.

These observations suggest variety in structural style and depositional regime along the margin. It is evident that the formation of the present margin has been governed by the main events in the history of the plate tectonic evolution of the Norwegian-Greenland Sea. Therefore, the passive margin also varies in age along its strike.

The objective of this paper is to describe, in regional terms, the Cenozoic evolution of the margin. This requires a first-order plate tectonic model for the post-Palaeocene opening of the Norwegian-Greenland Sea. Moreover, the history of the margin depends upon the geological conditions that existed prior to the initiation of Cenozoic sea-floor spreading, as well as an understanding of the marginal highs which have been mapped in several areas. Then, we will proceed to describe the main margin provinces by selected regional type sections and discuss some of the features which will focus centrally in forthcoming investigations.

This paper covers the margin from the northernmost part of the Faeroe-Shetland Channel in the south to the Spitsbergen Fracture Zone in the north (Fig. 1). We have attempted to include most of the recent studies off Norway and Svalbard. A more comprehensive listing of earlier work is given by Eldholm and Talwani (1981). Obviously, an understanding of the eastern margin of the Norwegian-Greenland Sea is closely related to the geology of its conjugate part; here we refer to the accompanying papers of Håkansson and Stemmerik (1984), Larsen (1984) and Surlyk *et al.* (1984) dealing with East Greenland. The main results from the Jan Mayen Ridge have been summarized by Myhre *et al.* (in press).

Petroleum Geology of the North European Margin © Norwegian Petroleum Society (Graham & Trotman, 1984) pp. 3—18.

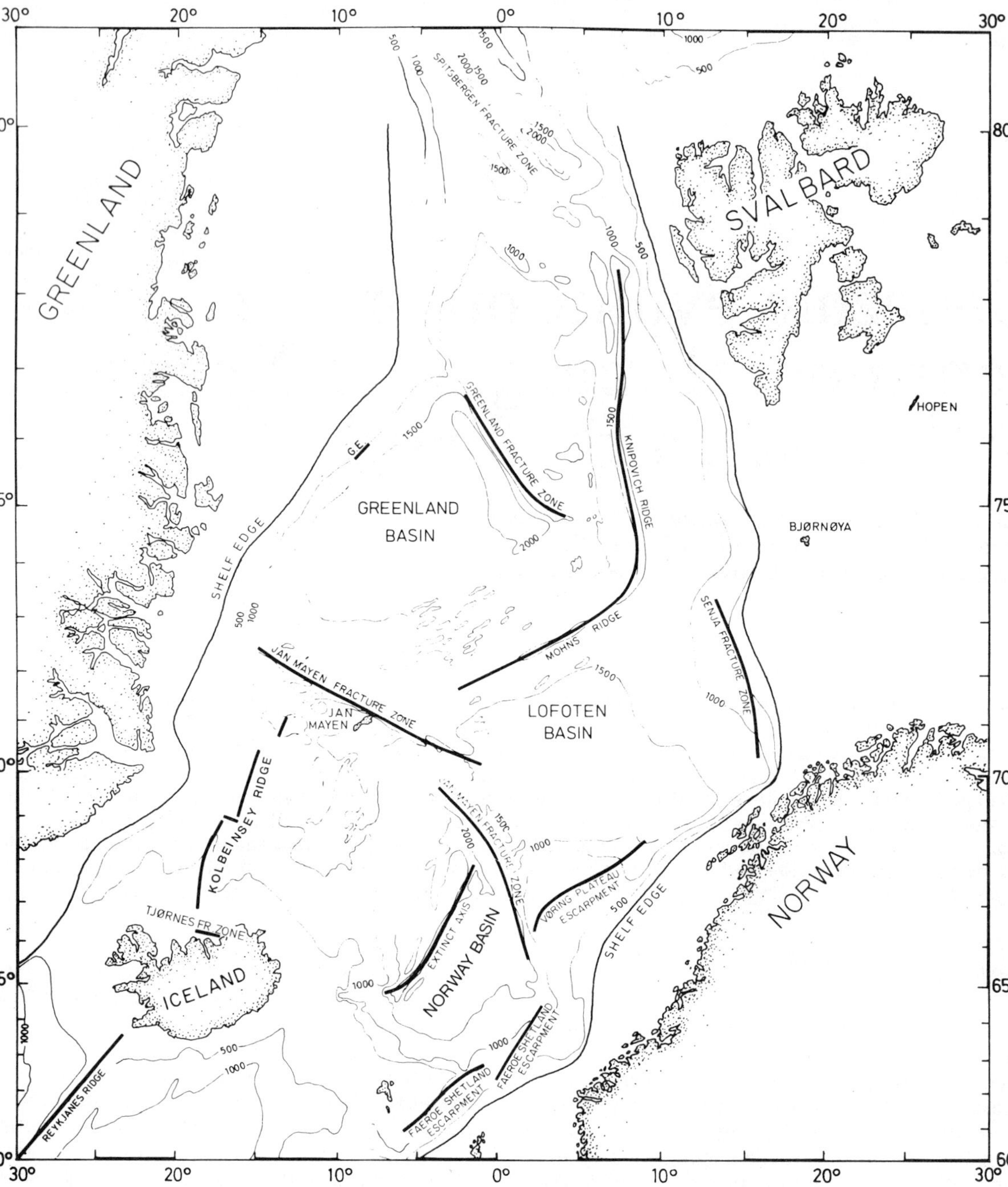

Fig. 1. Regional bathymetry and major structural features in the Norwegian-Greenland Sea. Slightly modified from Talwani and Eldholm (1977).

SITUATION PRIOR TO LATE PALAEOCENE

There appears to be general agreement that Cenozoic sea-floor spreading in the Norwegian-Greenland Sea was initiated during the time period between magnetic anomalies 25 and 24B. With reference to the LaBrecque *et al.* (1977) polarity time-scale the opening occurred at about 58 Ma, in the late Palaeocene.

Studies based primarily on seismic data tied to drill-site information indicate that the region between Norway and Greenland has been a depositional area at least since the Carboniferous (Rønnevik and Navrestad, 1977; Jørgensen and Navrestad, 1981; Rønnevik, 1981; Rønnevik *et al.*, 1983). Sedimentation started on an epeirogenic subsiding post-Caledonian surface. In late Palaeozoic times a tectonic phase created a system of rotated fault blocks with sedimentation filling in the fault-bounded basins. From late Palaeozoic to early Mesozoic times sedimentation gradually smoothed the relief, forming a large regional basin in early Jurassic times. A new main phase of extensional tectonics, Kimmerian, which started in mid and culminated in late Jurassic times, formed a system of basins and ridges (Figs. 2 and 3). During Cretaceous and into Palaeocene times the relief was again smoothed by marine sedimentation and regional subsidence in large epicontinental basins.

Pre-Cenozoic sea-floor spreading within this regional basin during Permian times has been proposed by Russel (1976) and Russel and Smythe (1978), whereas Smythe *et al.* (1983) favour pre-Albian spreading. Bott (1975, 1978) and Hanisch (1984) suggested Mesozoic spreading to the north of the Faeroe-Shetland Channel. Price

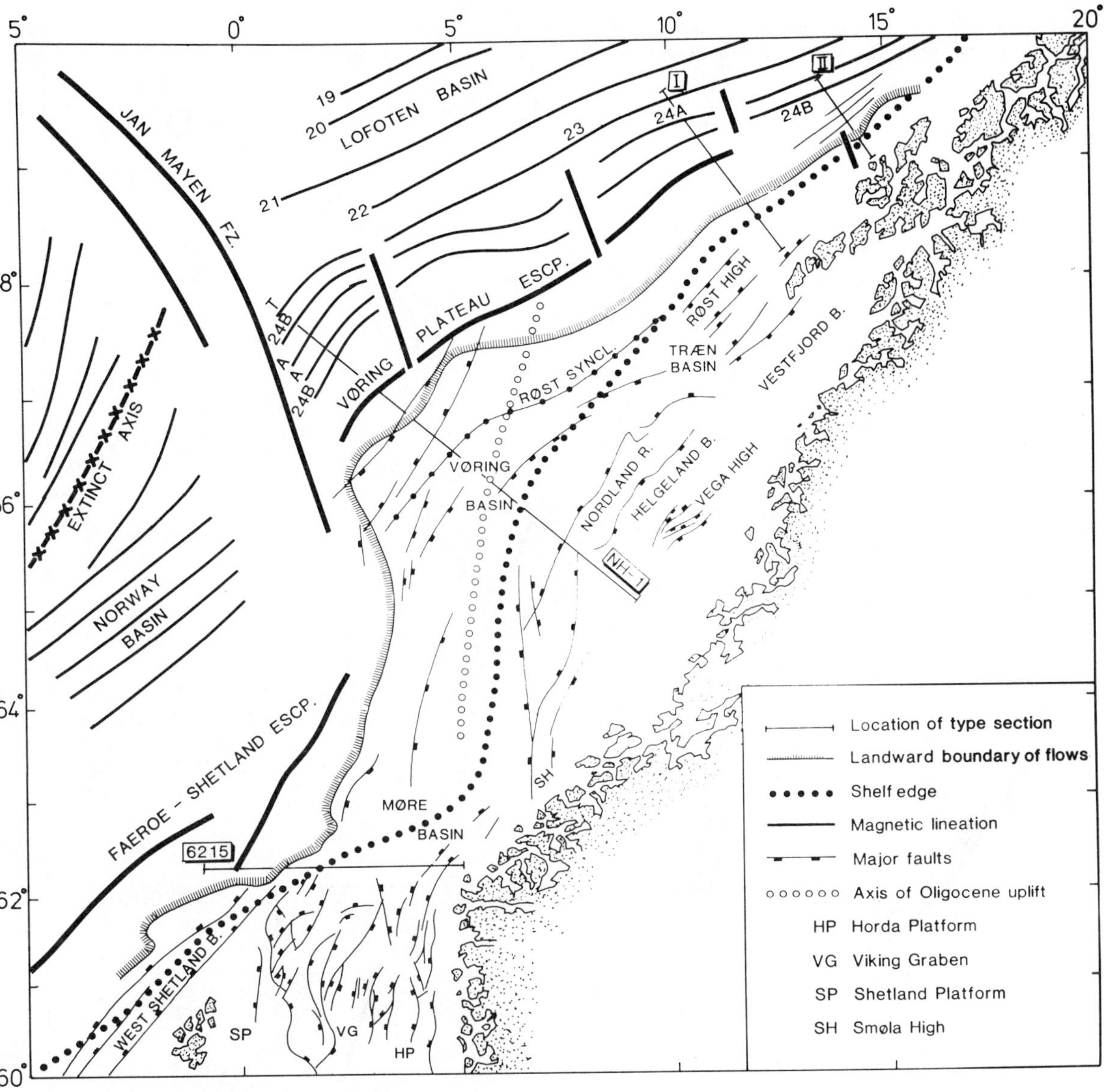

Fig. 2. Main regional features of the Norwegian margin between 60 and 70°N. Based on Talwani and Eldholm (1977), Eldholm *et al.* (1979), Jørgensen and Navrestad (1981), Hamar *et al.* (1980), Bleie *et al.* (1982), Hinz *et al.* (1982), Hagevang *et al.* (1983), Talwani *et al.* (1983), Rønnevik *et al.* (1983). The magnetic lineations in the Norway Basin have not been identified. An analysis of this problem is presented by Nunns and Peacock (1982).

and Rattey (in press) have proposed a 30-km wide zone of pre-Cenomanian Cretaceous oceanic crust beneath the inner Vøring Basin. This is based on the existence of pre-Cenomanian lavas or sills in the Møre Basin and the absence of base-Cretaceous or older reflectors in a narrow region of the inner Vøring Basin. We concur with Price and Rattey (in press) that none of the present observations necessitate a Permian ocean. However, we find it difficult to accept a narrow zone of early Cretaceous oceanic crust north of 62°N, as a zone of this kind would be expected to yield a definitive magnetic and gravimetric signature. Moreover, there is no evidence of thermal contraction in the overlying Mesozoic sediments. Finally, existence of oceanic crust should give rise to a high-impedance, identifiable reflector not deeper than the level of the adjacent top-Jurassic sediments. On the other hand, the evidence of early Cretaceous tectonism along the Norwegian margin (Rønnevik *et al.*, 1979; Hinz *et al.*, 1982; Price and Rattey, in press) may indeed reflect a northward continuation of the stress regime which eventually caused sea floor spreading in the southern Rockall Trough during the Albian—Mastrichtian interval (Roberts *et al.*, 1981). In our judgement, this event did not approach the stage of oceanic crustal generation at the Norwegian continental margin.

In earliest Tertiary times, according to our model, sea-floor spreading occurred along a plate boundary in the North Atlantic, continuing into the Labrador Sea (Kristoffersen and Talwani, 1977; Srivastava, 1978). The crust between Norway and Greenland, which had been subjected to several tensional phases in Mesozoic times without complete crustal break-up, was again being thinned with generation of oceanic crust in late Palaeocene times (Talwani and Eldholm, 1977).

PLATE TECTONIC SUMMARY

Subsequent to the start of sea-floor spreading the incipient ocean between Norway and Greenland gradually widened and deepened. However, the present distribution of oceanic ridges and fracture

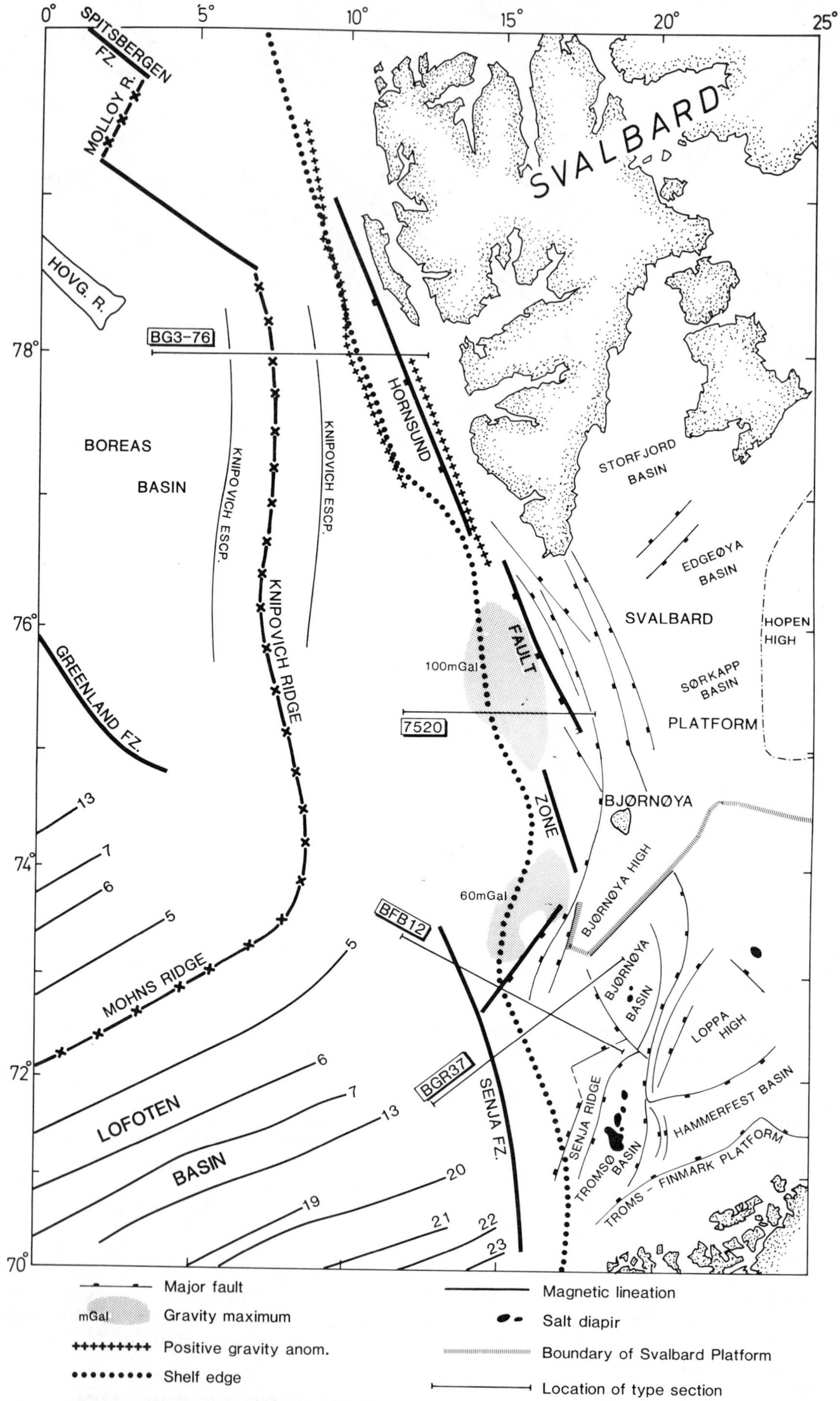

Fig. 3. Main regional features of the Norwegian margin between 70 and 80°N. Based on Talwani and Eldholm (1977), Sundvor and Eldholm (1979), Myhre *et al.* (1982), Rønnevik *et al.* (1982), Faleide *et al.* (in press a, b).

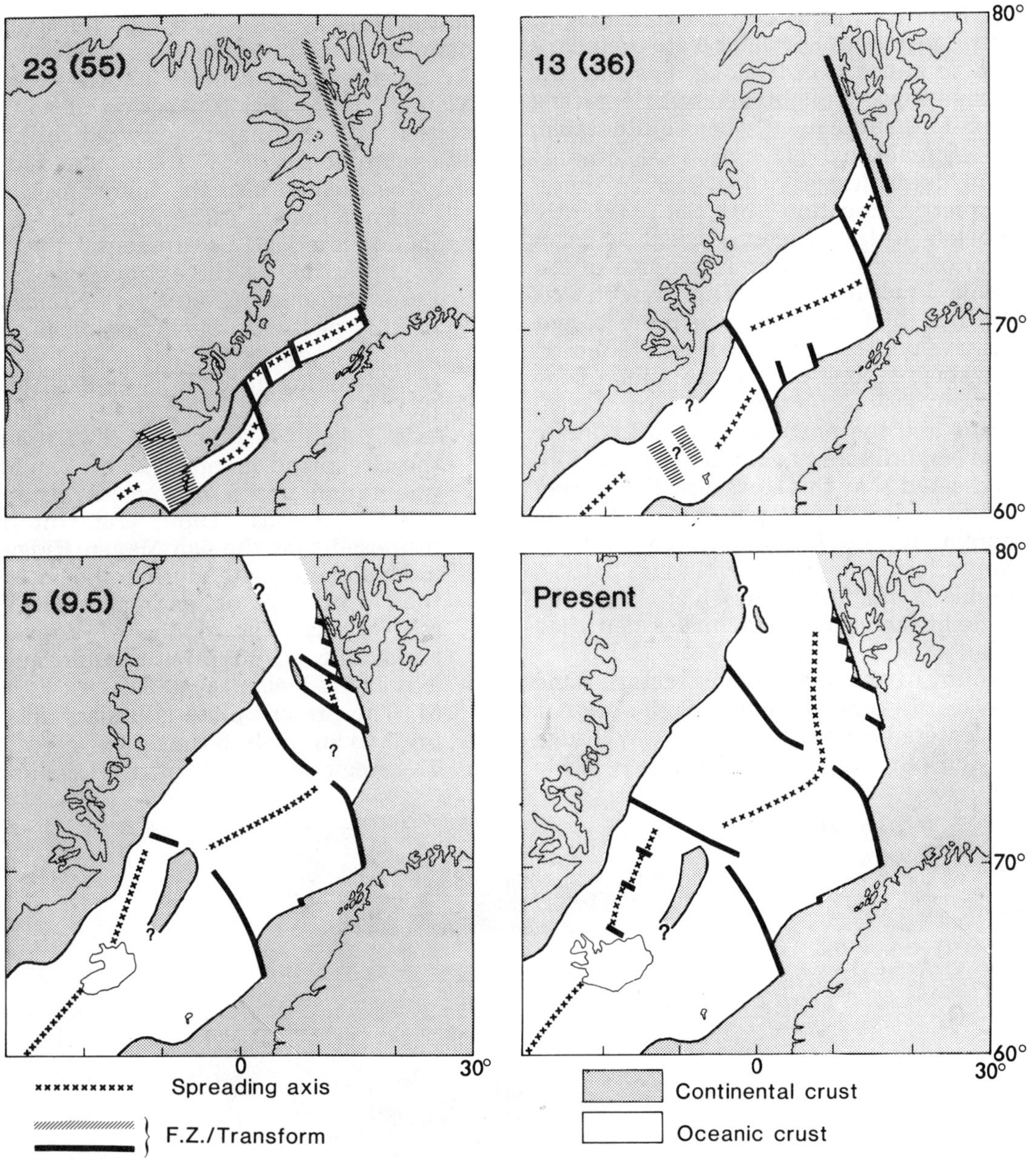

Fig. 4. Simplified model showing principal events in the plate tectonic evolution of the Norwegian-Greenland Sea. Based mainly on Talwani and Eldholm (1977) and Myhre *et al.* (1982).

zones in the Norwegian-Greenland Sea reflects a rather complex history of development. A detailed model for Cenozoic plate tectonic history was proposed by Talwani and Eldholm (1977). Later work has added new information and suggested changes in this model, particularly relating to the formation of the Norway Basin and the oldest part of the Iceland Plateau (Nunns, 1980; Vogt *et al.*, 1980; Nunns and Peacock, 1982; Unternehr, 1982) and the evolution of the Greenland Sea (Myhre *et al.*, 1982).

The Cenozoic history of the continental margin is closely related to the plate tectonic development which is characterized by two major evolutionary phases and local migration of the plate boundary (Fig. 4).

During the first main evolutionary phase, between anomalies 25/24 and 13, Greenland moved in a NNW direction relative to Eurasia. A major change in the pole of rotation at about the time of anomaly 13 caused the relative movement between Greenland and Eurasia to become almost west-northwest. These directions are clearly expressed in the azimuths of the fracture zones. The change in the relative spreading direction is associated with the cessation of spreading in the Labrador Sea when the plate geometry in the North Atlantic changed from a three-plate to a two-plate configuration, with Greenland becoming part of the North American Plate (Kristoffersen and Talwani, 1977; Srivastava, 1978). This event is highly significant for the development of the sea floor and margins north of the Greenland-Senja Fracture Zone. Prior to the time of anomaly 13, Greenland slid along Svalbard with formation of oceanic crust only in the southern Greenland Sea. Since anomaly 13 time, however, sea-floor spreading has taken place along the entire plate boundary between Svalbard and Greenland (Fig. 4).

The two major fracture-zone systems, Jan Mayen and Greenland-Senja, transecting the Norwegian-Greenland Sea separate regions which have evolved differently in terms of plate tectonics (Figs. 4 and 5). We also note that the actual location of these systems was probably related to pre-Cenozoic features and that their interaction with the continent has influenced the Cenozoic margin.

In the central region, the original plate boundary was subjected to minor offsets along its strike. At anomaly 23 time it was straightened out (Hagevang *et al.*, 1983) and has subsequently been quite stable, with largely symmetrical generation of oceanic crust along the Mohns Ridge.

The asymmetric location of the presently active Knipovich Ridge suggests recent changes in the location of the plate boundary north of the Greenland-Senja Fracture Zone. The present axis is not older than 5—6 Ma and even more recent changes may have occurred (Sundvor and Eldholm, 1979). A two-stage model for the formation of the Greenland Sea, in which the southern part opened in early Eocene and the northern part at anomaly 13 times, has been proposed by Myhre *et al.* (1982). They also indicated that the Hovgaard Ridge may be a continental fragment split off from the Svalbard margin at about anomaly 13 time. It should be borne in mind, however, that no pattern of plate-boundary migration has yet been deciphered and no linear magnetic anomalies identified in the Greenland Sea.

The area south of the Jan Mayen Fracture Zone is structurally very complex. Whereas the present spreading is centered along the Kolbeinsey Ridge north of Iceland, an extinct spreading axis exists in the Norway Basin. Between these main spreading centres there may have existed relatively short-lived spreading centres at the eastern Iceland Plateau, representing a period of adjustment associated with the major westward shift. Tentatively, the tectonic history (Talwani and Eldholm, 1977) may be divided into the following stages (Fig. 5):

(1) Pre-anomaly 23 spreading west of the Faeroe-Shetland Escarpment.

(2) Spreading at the extinct axis in the Norway Basin (anomaly 23-20 time). Nunns and Peacock (1982) postulate spreading from this axis also prior to anomaly 23.

(3) Spreading in the Norway Basin and/or Iceland Plateau, characterized by complex westward axial migration (anomaly 20-7 time). A part of the Greenland margin, i.e. the Jan Mayen Ridge, was completely separated just prior to anomaly 6 time. Nunns (1980) and Unternehr (1982) proposed that the Jan Mayen Ridge acted as an independent plate during the interval between the formation of anomalies 20-7 and 13-7, respectively. The models of Vogt *et al.* (1980) and Nunns (1980) do not require an intermediate axis at the Iceland Plateau.

(4) The present plate boundary was formed just prior to anomaly 5 time.

The regional trend of the shallow transverse

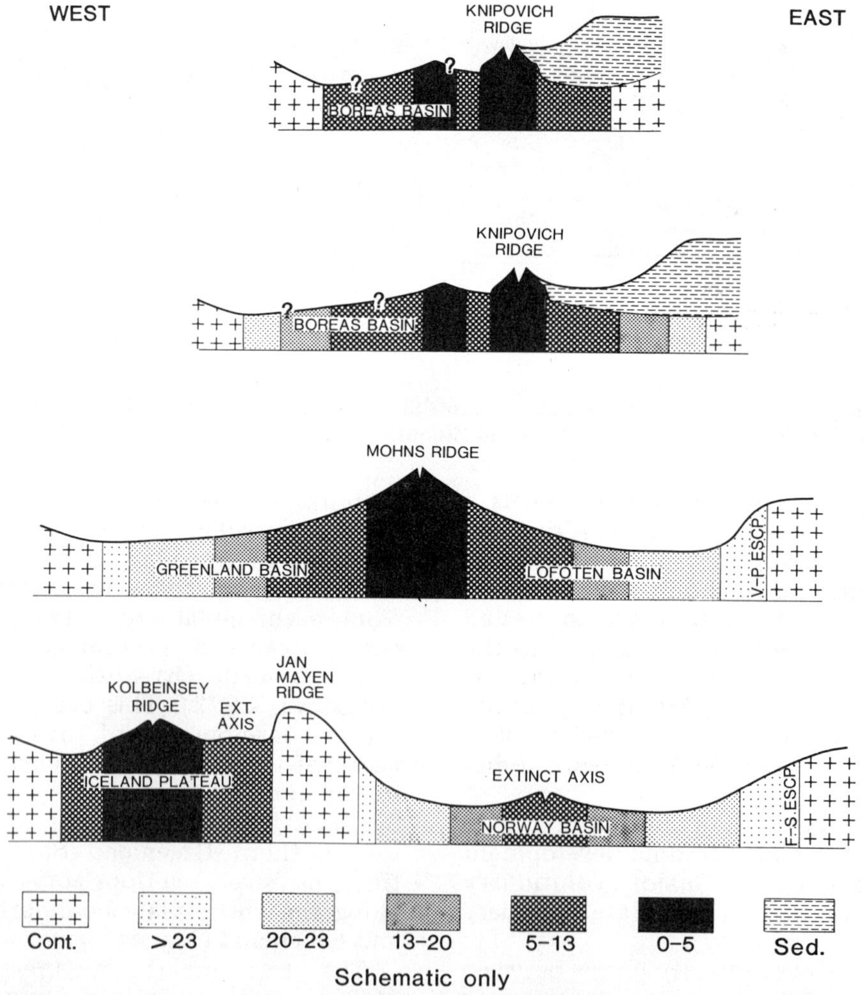

Fig. 5. Schematic evolutionary diagrams along sections across the main basins of the Norwegian-Greenland Sea. Based on the models of Talwani and Eldholm (1977) and Myhre *et al.* (1982). There is little conclusive information about the pre-Knipovich Ridge in the Greenland Sea. Alternative models have been proposed for parts of the area south of the Jan Mayen Fracture Zone.

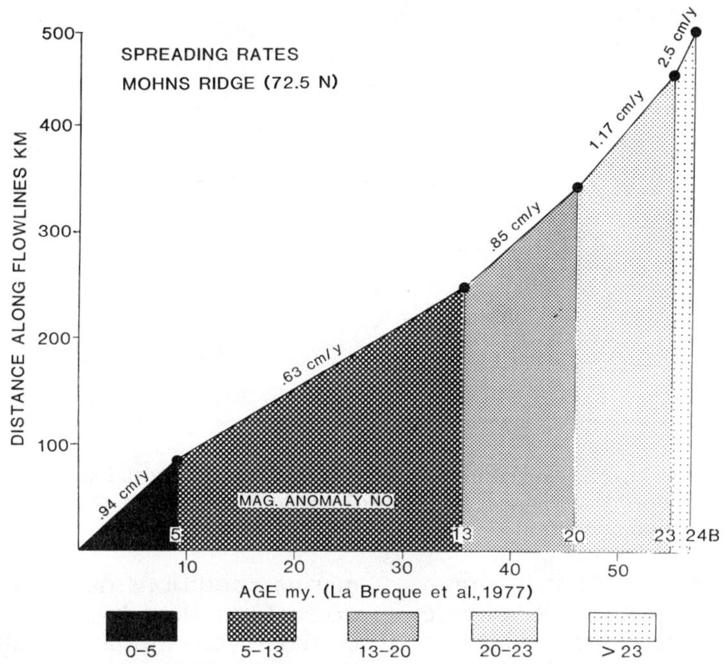

Fig. 6. Speading rates along flow-lines in the Lofoten-Greenland basins.

Greenland—Iceland—Faeroes Ridge lies along flow lines which describe the relative plate motion indicating that the presently emergent part of the plate boundary, the Iceland 'hot spot', has been in existence throughout the history of the ocean.

With the exception of the first few million years, when there was a rapid generation of oceanic crust (2.5 cm/y), the rate of sea-floor spreading has been relatively slow through Cenozoic times, with a somewhat increased rate since anomaly 5 time (Fig. 6).

MARGINAL HIGHS

During early investigations of the Norwegian margin, buried acoustic basement highs were mapped beneath the outer Vøring Plateau and the lower slope south of the Jan Mayen Fracture Zone (Talwani and Eldholm, 1972), as well as on the Greenland margin between 75.3°N and the Greenland Fracture Zone (Eldholm and Windisch, 1974). These highs were bounded landward by escarpment-like features and associated with the continent—ocean boundary. The escarpments also separated areas with typical changes in the geophysical parameters. The Deep Sea Drilling Project (DSDP) drilling proved that the acoustic basement reflector at the outer Vøring Plateau was related to basalts of early Eocene age (Talwani *et al.*, 1976).

Results from multichannel seismic surveys have, however, shown complex structures at the highs and in their environs. A main observation is the existence of zones of seaward dipping reflectors beneath the sediment—basalt interface. These zones exist beneath the Faeroe-Shetland and Vøring basement highs and near the base of the slope in the Lofoten Basin (Hinz and Weber, 1976; Hinz and Schlüter, 1978a; Talwani, 1978; Eldholm *et al.*, 1979; Talwani *et al.*, 1981). Eldholm and Sundvor (1980) noted that the dipping reflectors always occurred landward of anomaly 23, beneath the smooth reflector earlier identified as acoustic basement. These observations have made it necessary to re-examine the idea that the escarpments form the continent—ocean boundary.

Off Norway, the area in the vicinity of the Vøring Plateau Escarpment has been most thoroughly investigated. On the basis of sub-surface seismic data, Hinz *et al.* (1982) and Talwani *et al.* (1983) divided the area into the zones shown schematically in Fig. 7. Zone III is underlain by normal oceanic crust formed by sea-floor spreading. Zone II, between the escarpment and the oceanic crust, consists of an outer part characterized by a dipping reflector sequence below the early Eocene basalt and an inner part, where the sub-basalt reflectors are horizontal or sub-horizontal and rest on an underlying base reflector, K. The width of the two sub-zones varies along the margin. Zone I marks the extent of a distinct seismic horizon believed to represent flows of the same age as the smooth volcanic reflector in Zone II. The landward limit of Zone II is also shown in Fig. 2.

The series of dipping reflectors appears to be composed of individual seaward dipping divergent beds which are slightly convex upward. In this zone, Mutter *et al.* (in press) have found a seismic velocity distribution yielding an approximately normal oceanic layer 3 sequence, whereas the thickness of the layer 2 sequence is greatly expanded relative to normal oceanic crust. A possible low-velocity layer is indicated near the base of the dipping reflectors. Finally, they note that this velocity structure is similar to that of eastern Iceland.

The fact that similar seaward dipping sequences as well as marginal highs have been observed along many rifted passive margins (Hinz, 1981) indicates that they may refer to a distinct event in the history of the early margin formation. Two separate explanations have been given for the data off

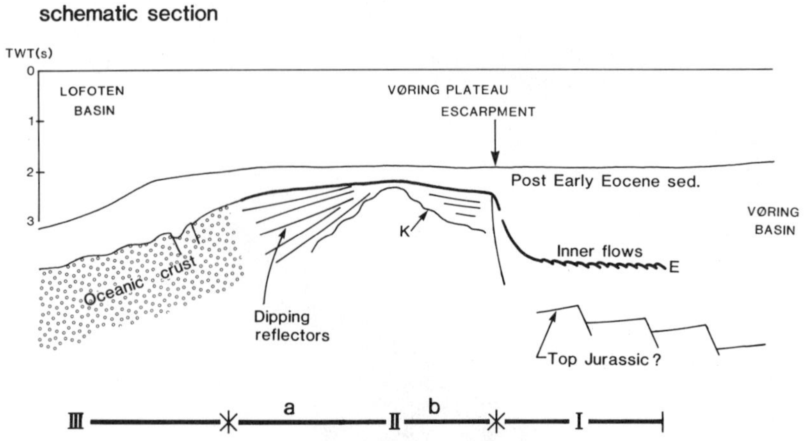

Fig. 7. Geological schematic section across the Vøring Plateau Escarpment. Zone III normal oceanic crust. Zone II sub-basalt reflectors (a: dipping, b: horizontal or subhorizontal). Zone I inner flows.

Norway (Fig. 8). Mutter *et al.* (1982; in press) advocated that the dipping reflectors are volcanic layers of oceanic crust formed during sub-aerial conditions. On the other hand, Hinz (1981) suggested that the major part of the reflectors consist of volcanic rocks extruded over a thinned, contaminated, continental crust in sub-aerial to shallow marine conditions during the late rifting stage of ocean formation. In terms of the continent—ocean boundary, the model of Mutter *et al.* (1982) favours the Vøring Plateau Escarpment as a primary feature, but Hinz (1981) requires continental crust farther seaward, towards the boundary of Zone II.

According to Mutter *et al.* (1982) reflector K is

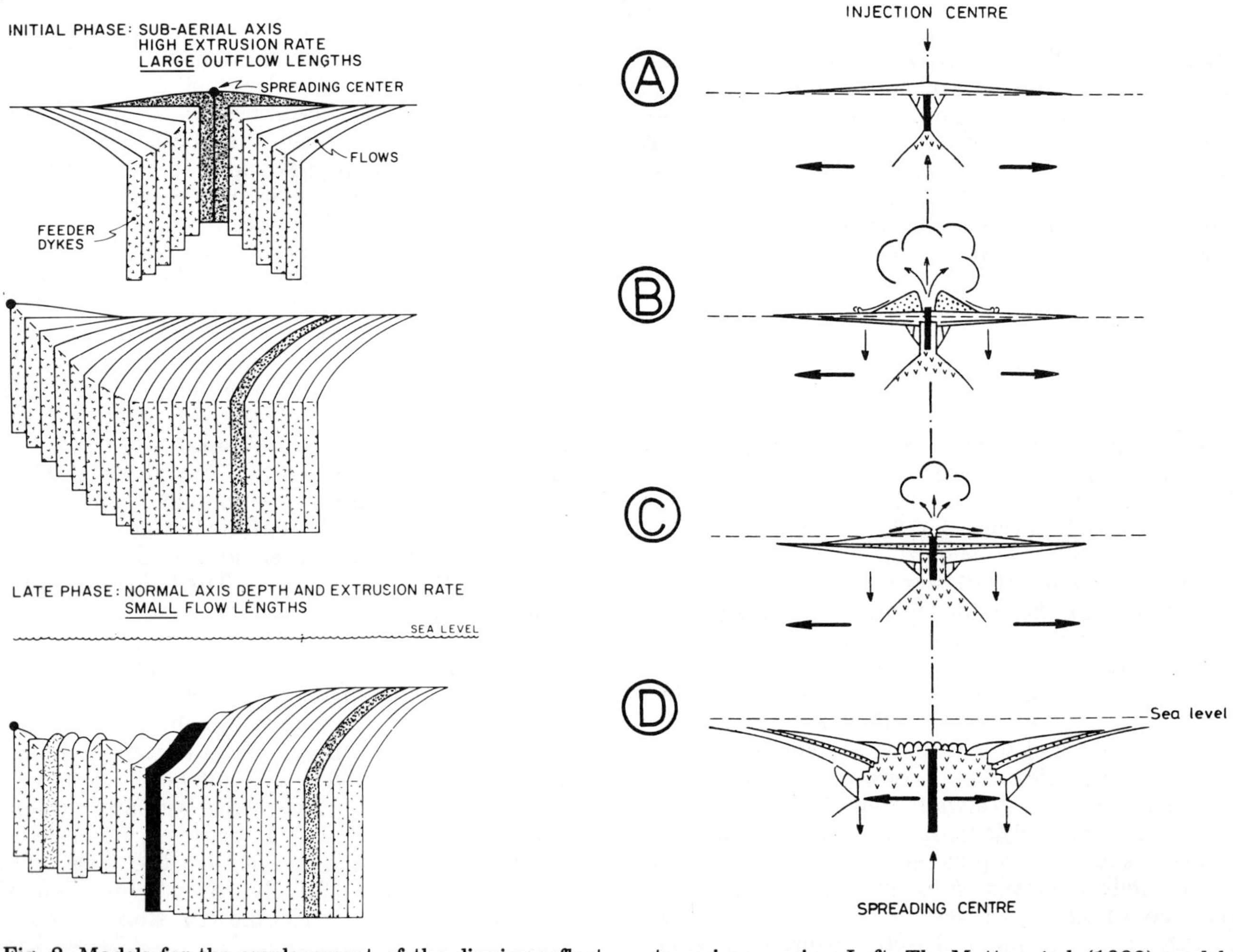

Fig. 8. Models for the emplacement of the dipping reflectors at passive margins. Left: The Mutter *et al.* (1982) model in which large flow lengths and high intrusion rates occur initially. The weight of the flows depresses the underlying oceanic crust. As time progresses the region subsides and the mode of crustal generation changes to normal sea-floor spreading with small flow lengths and normal extrusion rates. Right: The Hinz (1981) model in which the dipping reflectors largely are extruded over thinned continental crust under subaerial conditions. Normal oceanic crust is developing during stages C—D and the subsidence of the axial region causes a reversal of the dip of the volcanics towards the axis.

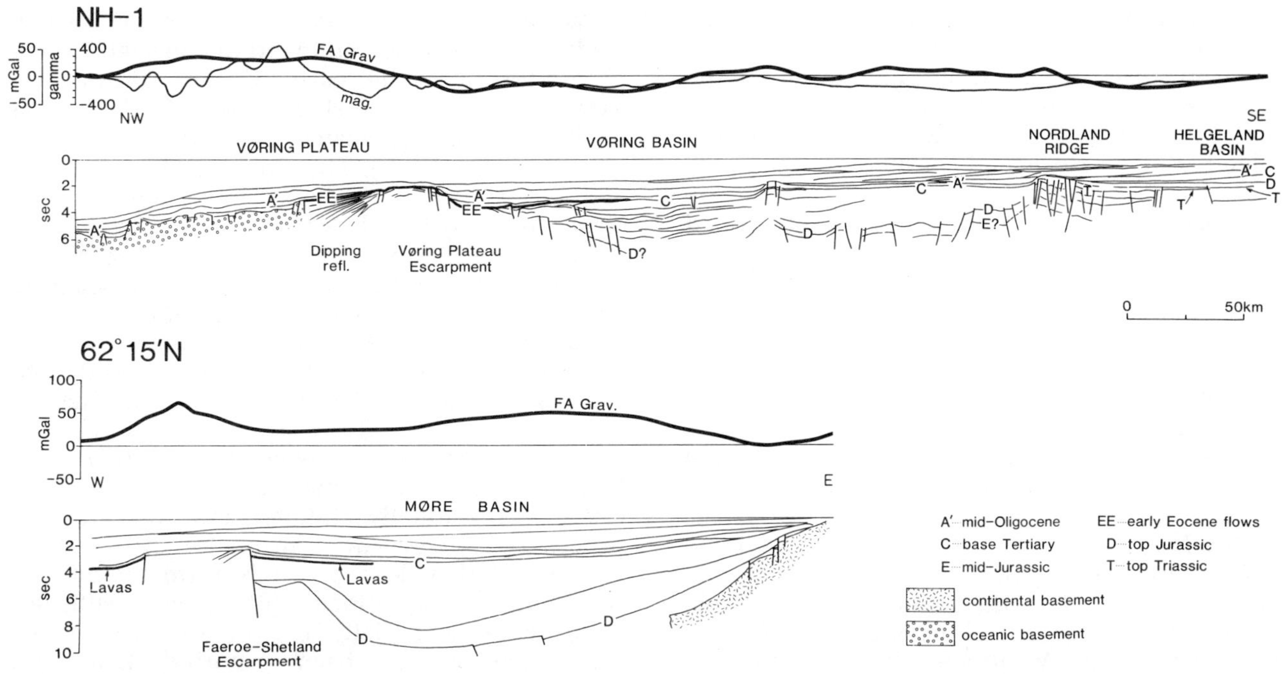

Fig. 9. Geoseismic sections across the southern Møre Basin and Nordland Basin. Profile 62°15′ from Hamar *et al.* (1980) with free-air gravity from Talwani and Eldholm (1972) Profile NH1 from Skogseid (1983). Profile locations shown in Fig. 2.

probably the dyke—lava interface, whereas Hinz (1981) suggests that it forms a late Cretaceous surface which existed before the onset of sea-floor spreading. It has also been suggested that the early Eocene flows and underlying reflectors could originate from an episodic volcanic event just prior to anomaly 23 time, which overprinted existing features on both the oceanic and continental crust (Eldholm *et al.*, 1979). Regardless of the mode of emplacement, however, the fact that magnetic lineations 24A and 24B overlie the dipping reflectors (Hagevang *et al.*, 1983) has to be explained when the type of the underlying crust is considered.

MARGIN 60—70°N

The inner part of the margin is characterized by a number of structural elements formed before mid Cretaceous times, whereas features relating to the Cenozoic passive margin formation dominate the outer part (Fig. 2).

In the south, the North Sea Basin terminates at the margin. Its northern part is characterized by extensional tectonics of Mesozoic age, with the prominent north-trending Viking Graben being bounded by the Horda Platform and its northern continuation to the east and the East Shetland Platform to the west. The West Shetland Platform forms the eastern boundary of the Faeroe-Shetland Channel. The West Shetland Basin contains sediments as old as Palaeozoic, but the westward extent of these rocks beneath the Faeroe-Shetland Channel is not known, because no seismic penetration is obtained below a zone of early Tertiary extrusives (Ridd, 1981). This region was probably faulted in Permo-Triassic times but the main structural activity took place from late Jurassic to early Cretaceous times (Cashion, 1975). After early Tertiary extrusive volcanism, the Faeroe-Shetland Channel became a major depocentre. There is still controversy about the nature of the crust underneath the western flank of the Faeroe-Shetland Channel and the Faeroe Islands, and different investigators have put the continent—ocean boundary on different sides of the Faeroes. We believe that this question is closely related to the understanding of the nature of the basement highs further north. The extrusives in the channel proper, which Ziegler (1982) related to the Thulean volcanic episode in the North Atlantic, have been interpreted as reflecting the same volcanic episode as those on the margin further north. In Fig. 2 we indicate a continuity of the Early Eocene flows all the way to the Vesterålen margin at 69°N. Some of the volcanics have been emplaced as sills, as indicated by Hamar *et al.* (1980) (Fig. 9). The fact that the southern flows do not lie far from the North Sea, where there is evidence of widespread volcanic activity dated as 52.5—55 Ma (Jacqué and Thouvenin, 1975), provides an additional clue about the age of the flows.

The southern margin consists of three regional provinces bounded by the marginal escarpments, the Jan Mayen Fracture Zone and the Norwegian coastline. These are the Møre, Nordland and Lofoten-Vesterålen margins.

The Møre Basin, comprising the main part of the Møre margin, appears to be a relatively unstructured basin except for the flank towards the coastline. The absence of structural elements may partly reflect that it has been a major depocentre since late Jurassic times, with as much as 9—10 km of post-Jurassic sediments (Hamar *et al.*, 1980) causing poor seismic resolution at depth (Fig. 9). Rønnevik *et al.* (1983) pointed out that the pre-Upper Jurassic strata are thin, reflecting uplift and erosion of a late Palaeozoic depocentre during early Mesozoic times. Rønnevik and Navrestad (1977) indicated several sub-basins below the Cretaceous sequence, which correspond to the magnetic lows mapped by Åm (1970).

The Nordland margin may be divided into two parts, separated by a structurally complex transition zone (Kristiansund-Bodø Fault Complex). To the east is a platform-like region on the central shelf, bounded by the structural lineaments along the western flank of the Nordland Ridge and the outer Smøla High (Fig. 2). This area comprises the Nordland Ridge, Helgeland Basin and the Vega High. It was structured by the Kimmerian movements (Rønnevik *et al.*, in press) but had developed as a depocentre in the late Triassic/early Jurassic times. Except in local sub-basins, the cover of younger sediments is relatively thin.

The Vøring Basin to the west was formed by block-faulting and rapid subsidence in late Jurassic and early Cretaceous times. Both the Møre and Vøring basins comprise thick Upper Cretaceous sequences (Fig. 9). Rønnevik *et al.* (1983) have mapped a pre-Tertiary rift zone, the Smøla Rift, close to the Vøring Plateau Escarpment. To the north the Nordland margin extends into the Træn and Vestfjord basins, the latter containing predominantly pre-Cretaceous sediments. The Træn Basin, which pinches out along a regional flexure towards the Lofoten-Vesterålen shelf, is characterized on the other hand by a very thick Cretaceous sequence (Jørgensen and Navrestad, 1981) and appears to be genetically related to the Møre and Vøring basins.

On the main part of the Lofoten-Vesterålen shelf, the Cenozoic section is absent and the region appears to have been a positive area for long periods prior to Cenozoic times (Eldholm *et al.*, 1979). The main shelf is underlain by a relatively thin sedimentary cover, forming a small block-faulted basin on the central shelf off Lofoten (Jørgensen and Navrestad, 1981). This basin is bounded by a basement ridge landward of the shelf edge, the Røst High (Figs. 2 and 10). A narrow Cenozoic wedge outcrops just landward of the shelf edge and progrades over the early Eocene flows, which extend almost to the shelf edge. Off Vesterålen, a marked change in structural style is observed at about 69°N (Fig. 2). The elongated basement ridge at the outer shelf terminates in a zone of complex structures, north of which the Cenozoic wedge widens considerably. This change is also reflected in the gravity field, which suggests a significant change in subsurface geology (Eldholm *et al.*, 1979).

In early Palaeocene times the incipient margin had become part of a regionally subsiding broad basin with smoothed relief, but the southern part of the Nordland Ridge was not buried before Palaeocene or early Eocene times (Jørgensen and Navrestad, 1981). During that time, the underlying crust was gradually thinned and weakened. The process of rifting of continental crust prior to formation of a deep ocean is normally associated with updoming, block-faulting and subsidence prior to drifting (Bally and Snelson, 1980). A rift unconformity has been reported from many margins. On the Norwegian margin there is little evidence of either a phase of regional crustal doming or of a rift unconformity. It appears rather that the marginal subsidence related to the initiation of sea-floor spreading is in general flexural in character. Although there is clear evidence of an earlier Mesozoic rift phase, the region experienced tectonically quiet conditions lasting for about 50 Ma in late Cretaceous times. There are some indications, however, of Tertiary faulting in areas of the margin, most often due to the reactivation of older lineaments. No evidence is seen for the typical structural pattern which would be predicted

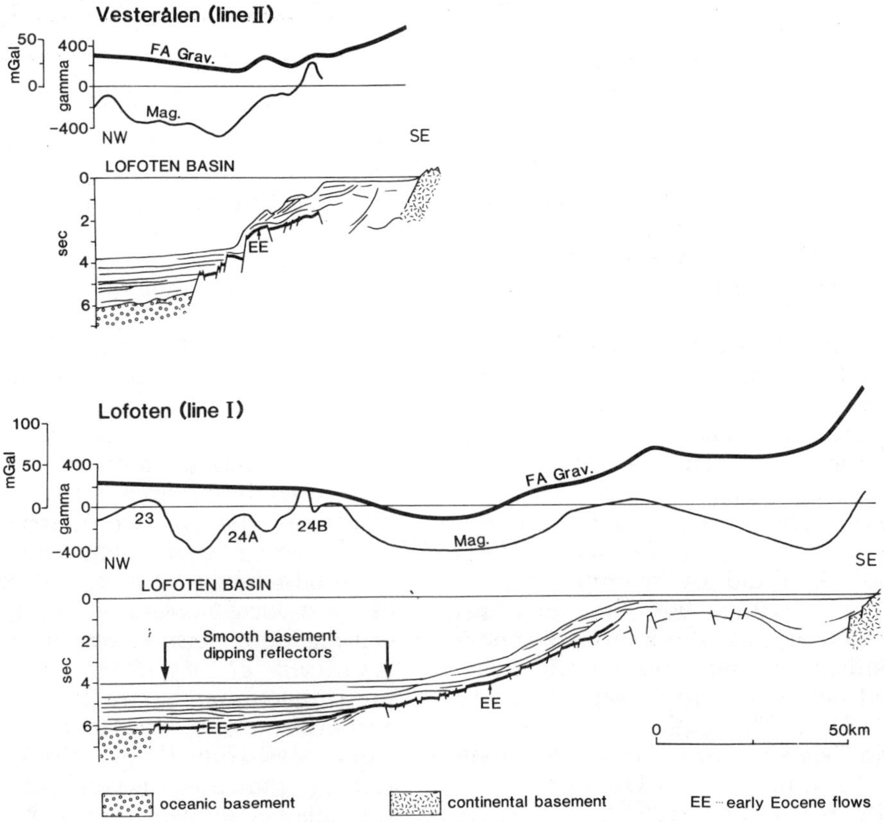

Fig. 10. Geoseismic sections across the Lofoten-Vesterålen margin (Eldholm *et al.*, 1979). Profile locations in Fig. 2.

prior to the drift phase. A Palaeocene rift unconformity has been reported from the Jan Mayen Ridge (Garde, 1978; Gairaud *et al.*, 1978), which is the conjugate part of the Møre margin. Myhre *et al.* (in press) have, however, proposed that this horizon could be an equivalent to the late Kimmerian unconformity formed in late Jurassic times.

The mode of Cenozoic margin subsidence changes from south to north. The continental slope off Lofoten-Vesterålen reveals major faulting of the early Eocene flows at the continental slope (Fig. 10). Eldholm *et al.* (1979) noted that the new plate boundary crossed the pre-opening basin diagonally and approached the adjacent craton near Vesterålen. Thus the rifting took place at the very flank of the basin, where relatively little sediment had accumulated. They suggested that the mode of marginal subsidence was in some way related to the nature of the underlying crust.

Maps of the total Cenozoic sediment thickness (Rønnevik and Navrestad, 1977; Jørgensen and Navrestad, 1981) show maximum thicknesses, about 3.5 seconds in the Møre Basin where a major depocentre has existed since late Jurassic times. To the north the cover rapidly increases in thickness away from the coast and approaches an average value of about 2 seconds on the main shelf. The variation in thickness is often associated with underlying structures. The marginal highs are covered by much thinner sections than the landward basins. Moreover, the absence of Tertiary sediments on the main Lofoten shelf is counteracted by increased thicknesses on the adjacent oceanic crust in the Lofoten Basin (Eldholm and Windisch, 1974).

Rønnevik and Navrestad (1979) mapped two regional reflectors within the main sedimentary wedge above the base-Tertiary. These are dated as the top-Palaeocene and mid-Tertiary (base-Miocene of Jørgensen and Navrestad, 1981), respectively. The two lower sequences are thickest near the coast where they form local fan-shaped maxima, whereas the upper sequence attains a maximum thickness near the shelf edge. All three units have a seismic character indicative of progradational deposition. The main margin appears to have been built out from a coastline to the east or southeast. South of Lofoten, however, the main bulk of the material has been transported from the north and northeast. Jørgensen and Navrestad (1981) have also identified a mid-Oligocene reflector A′, in the outer Møre Basin (Fig. 10). This reflector is considered to have formed by the global drop in sea level, and this subsequently caused the main depositional area to move to the shelf edge.

In the Vøring Basin a major synclinal structure, the Røst Syncline, has been mapped by Rønnevik *et al.* (1983) (Fig. 2), who believe it is the result of both Kimmerian and Tertiary movements. Furthermore, they proposed that the Nordland margin was influenced by mid Tertiary compressional tectonics, producing domal structures which trend in a more northerly direction than the older Kimmerian structures. This event has affected the Smøla Rift and the Nordland Ridge and formed an elongated area of inversion crossing the margin from 64.2°N to 68°N (Fig. 2). Hinz *et al.* (1982) note that the Palaeogene is updomed and partially truncated by mid-Oligocene and late Miocene erosion. Mutter (in press) suggests that the deformation took place throughout Cretaceous times, reaching its final stage between mid-Eocene and mid-Oligocene. He also indicates a diapiric origin rather than simple compression. Rønnevik *et al.* (1979) dated the inversion to mid-Oligocene, whereas Hinz *et al.* (1982) implied only a general Oligocene age. According to Hinz *et al.* (1982), it was caused by either igneous intrusion into the underlying Cretaceous sequence, mobilization of pre-Kimmerian evaporites or by deep crustal movements. On the other hand, Rønnevik *et al.* (1983) related these structures to shear zones which cannot, at present, be predicted from the Norwegian-Greenland Sea plate tectonic model. We suggest a possible relationship with a change in stress regime during mid-Oligocene times in the entire Norwegian-Greenland Sea region, which may have triggered reactivation of older features. We would stress that better mapping is required prior to detailed formulation of possible models. We note that the area of Eocene mud diapirs, mapped by Talwani and Eldholm (1972) and further described by Caston (1976), lies in the inversion area. It is thus possible that the movements responsible for the uplift have also triggered the mud movements.

MARGIN 70–80°N

In a regional sense this margin comprises three main parts. South of about 73°N it forms the western boundary of a large basinal region of the Barents Sea. Further north it terminates the stratigraphically elevated Svalbard Platform and the northernmost part lies off the island of Spitsbergen (Fig. 3).

Three major tectonic phases have produced the main structural elements in Fig. 3 (Rønnevik *et al.*, 1982; Faleide *et al.*, 1984 a):

(1) The Svalbardian phase in late Devonian to early Carboniferous times.

(2) The mid and late Kimmerian phases in mid Jurassic to early Cretaceous times.

(3) Cenozoic tectonism related to the progressively northward opening of the Norwegian-Greenland Sea.

The westernmost elements of the large basinal area between Norway and the Svalbard Platform are the Tromsø and Hammerfest basins and the Senja Ridge (Fig. 3).

The Tromsø Basin is very deep, and bounded in the west by the Senja Ridge. In the east it is separated from the Hammerfest Basin by an intensively tectonized hinge zone, dominated by extensional structures and tilted fault blocks. A large deep-seated salt massif piercing towards the surface lies in the central basin. Salt diapirs have been mapped from sea floor to a depth of more than 10 km. Because of the great depths, little is known about the pre-Jurassic history, i.e. when the Tromsø Basin was part of a larger regional basin. The salt layer is dated as Permian (Faleide *et al.*, in press a), but Upper Devonian evaporites may exist in a graben beneath the present Tromsø Basin (Rønnevik *et al.*, 1982). Triassic clastics were deposited during a tectonic quiet period character-

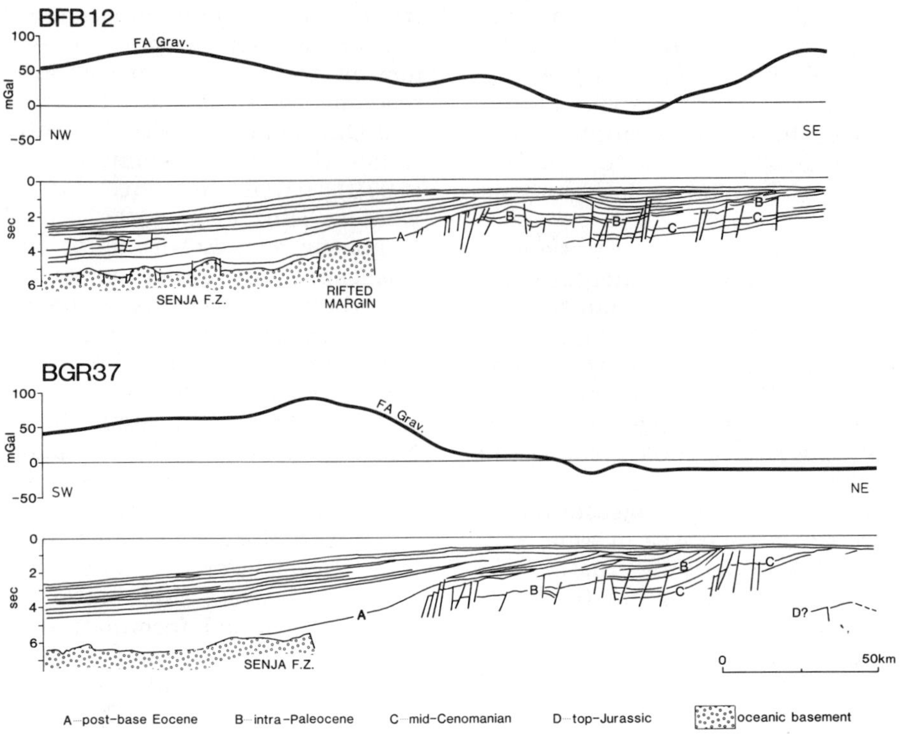

Fig. 11. Geoseismic sections across the Barents Sea margin. Profile BGR 37 crosses the Senja Fracture Zone and B FB 12 both the rifted margin linking the Senja Fracture Zone with the Hornsund Fault Zone and the oceanic part of the Senja Fracture Zone. The profiles have been described by Hinz and Schlüter (1978b). This interpretation is by Faleide and Gudlaugsson (1981). Free-air gravity from Faleide *et al.* (in press b). Profile locations in Fig. 3.

ized by regional subsidence. The development of a separate sub-basin was initiated by mid Kimmerian block-faulting followed by late Kimmerian movements. A thick sequence of Lower Cretaceous sediments accumulated contemporaneously with rapid subsidence and vigorous salt movements. In Cenomanian times the basin had been filled in, the strong differential movements receded and the Tromsφ Basin became part of the regionally subsiding area in the southwestern Barents Sea.

The Senja Ridge (Fig. 3) is a faulted sedimentary high buried by a post-Cenomanian sequence. It may be underlain by basement relief or intrusive bodies as it is associated with a prominent local gravity maximum. Faulting in a partly compressional regime has severely disturbed the sediments but no stratigraphic information is available from within the high. Rφnnevik *et al.* (1982) suggested it was inverted by late Lower Cretaceous reactivation of the Kimmerian fault system causing wrench movements along the southern boundary fault of the Bjφrnφya Basin. Hence, the Senja Ridge was part of the early Cretaceous depocentre prior to the inversion, implying large amounts of early Cretaceous sediments within the present high.

The Bjφrnφya Basin is a complex, elongated north-northwest-trending trough south of Bjφrnφya, to the southwest of the Bjφrnφya Basin, continues into a still deeper basin at the margin. The basin developed much like the Tromsφ Basin, with formation of the present basin by rapid early Cretaceous subsidence following the Kimmerian faulting. The basin infill was derived both from the uplifted Loppa High and the Bjφrnφya High (Fig. 3).

The Svalbard Platform covers the area north of the regional basin province. The platform includes several structural elements but because of many common features it can be considered as one regional province (Faleide *et al.*, 1984 a). The Bjφrnφya High represents a southwestern extension of the platform. With respect to the region to the south, the platform is characterized by shallower depths to the seismic stratigraphic levels, higher velocities in the pre-Cretaceous Mesozoic sequence and a regionally higher level of the gravity field. The platform is underlain by a thick, relatively flat-lying Jurassic—Triassic sequence. The main faulting shown in Fig. 3 had taken place prior to the end of the Permian (Rφnnevik *et al.*, 1983).

The Bjφrnφya High (Fig. 3), which appears to be underlain by a shallow basement, was developed as a positive feature not later than early Permian times.

Two major structural elements, the Senja Fracture Zone and the Hornsund Fault Zone, define the continent—ocean boundary. Although the detailed relationship has not yet been worked out, Fig. 3 suggests a connection between the trends and locations of Cenozoic margin features and older structures. Thus the late-Mesozoic structural patterns have defined the line of subsequent continental break-up.

The Senja Fracture Zone is a first-order feature in the isostatic gravity field (Grφnlie and Talwani, 1982). The prominent gravity anomaly was defined by Talwani and Eldholm (1977) as the continent—ocean boundary south of 73°N. Later multichannel seismic data show that oceanic basement extends eastward to the location of the gravity anomaly, where the basement signature disappears quite suddenly. On the landward side of the anomaly no basement reflector has yet been found. Moreover, there is basement relief at the fracture zone (Fig. 11).

North of 74°N a major north-northwest trending fault zone, the Hornsund Fault Zone, has been delineated on the central continental shelf (Figs. 3 and 12). It marks the boundary between two regions with distinctive reflection properties. Only a few scattered reflectors are observed on the landward side, whereas a typical layered seaward dipping sequence exists west of the fault. Sundvor and Eldholm (1979) and Sundvor *et al.* (1979) mapped the fault zone as a continuous feature but indicated that it was probably composed of several down-faulted blocks. Myhre *et al.* (1982) proposed that the fault zone was composed of three approximately subparallel segments of which the central segment was shifted about 20 km to the east. They also pointed out that the geophysical signature of the area landward of the fault zone was indeed similar to that of the Svalbard Platform.

A sedimentary basin extends from the fault zone to the Knipovich Ridge and appears to be predominantly underlain by oceanic crust. On the seismic lines (Fig. 12) the oceanic basement can be traced with certainty to a distance of 5—30 km from the fault zone. Except for the very young crust, which is bounded by the Knipovich Ridge escarpments (Fig. 12), the oceanic basement surface is quite smooth and does not have the typical hyperbolic character of oceanic crust, but exhibits a moderate amount of faulting. The continuity of the basement reflector indicates that the continent—ocean boundary is located close to or at the Hornsund Fault Zone.

The southern segment of the Hornsund Fault Zone and the Senja Fracture Zone are linked together by a northeast-trending fault southwest of Bjørnøya (Fig. 3). Myhre *et al.* (1982) proposed that the Hornsund Fault Zone marked a sheared margin segment in Palaeogene times. During Eocene and early Oligocene times a rifted basin opened north of the Greenland-Senja Fracture Zone (Fig. 4). At this time the Hornsund Fault Zone was a sheared margin with the connecting fault to the Senja Fracture Zone representing a rifted margin segment. At anomaly 13 time a fragment of the Svalbard margin was cut off and the central Hornsund Fault Zone segment was established. This segment has subsequently acted as a rifted margin. It is quite interesting to note how well these features are reflected in the gravity field. In particular, there are prominent positive gravity anomalies on the seaward side of the rifted margins. The difference in amplitude level is ascribed to the younger and more elevated basement high off the central segment (Figs. 3 and 12).

From focal mechanism solutions, matching of symmetric topography on either sides of the Knipovich and Nansen Ridges, and satellite imagery Savostin and Karasik (1981) predicted a Spitsbergen microplate and inferred a plate boundary across the margin at about 77°N. Unternehr (1982) suggested that Svalbard moved as an independent plate between anomalies 5 and 7. No features that

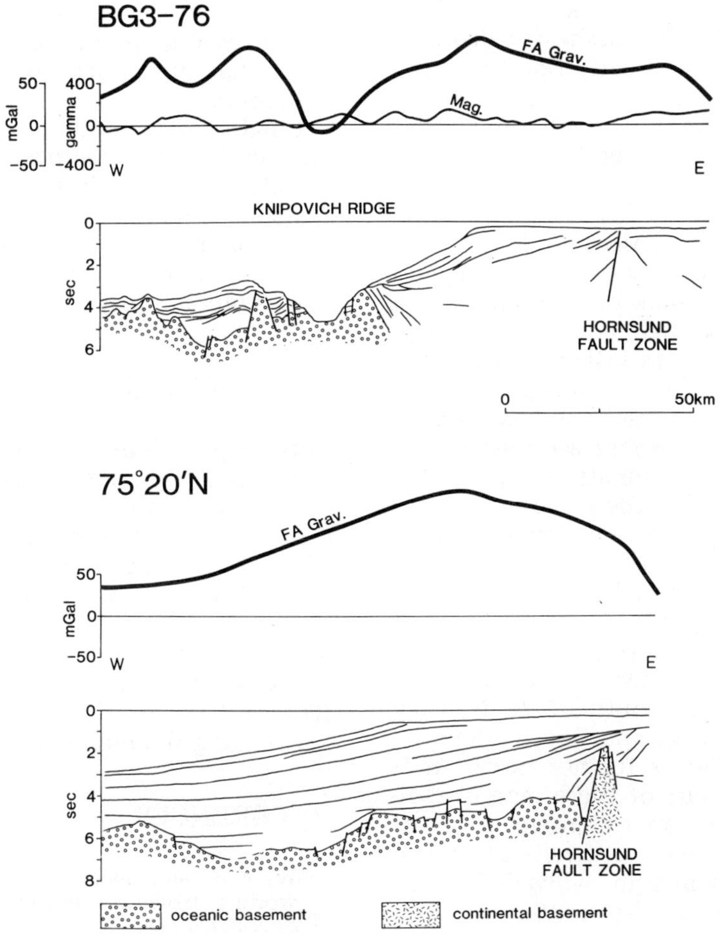

Fig. 12. Geoseismic sections across the Svalbard margin (Myhre *et al.*, 1982; Myhre, in press). Profile locations in Fig. 3.

could be related to young or recent plate boundaries have yet been observed at the margin.

The entire slope and outer shelf is covered with thick Cenozoic sediments. South of 74°N three regional reflectors, A, B and C (Fig. 11), have been mapped by Faleide *et al.* (1984 a). The deepest seismic stratigraphic level, on the margin south of the Bjørnøya High, is usually reflector C of Cenomanian age. This is an unconformity generated on a flat-lying shelf where the older relief had been levelled (Fig. 11). The B—C sequence prograded from the north, in a homogeneous environment, under fairly stable (pre-rift) tectonic conditions. Reflector B is interpreted as a base Upper Palaeocene shelf unconformity. The area was affected by block-faulting which started approximately at B time. The accompanying subsidence tilted the shelf sequences to the west and led to an entirely different depositional environment. In the A—B sequence there is evidence of initial progradation, erosional channels and local unconformities. Faleide *et al.* (in press a) interpreted this as sedimentation in a basin or over the shelf-edge with fluctuating sea level and sporadic tectonic activity. In terms of the plate tectonic history, the A—B sequence is characterized as syn-rift sediments. Later uplift truncated and intensively faulted the A—B sequence near the present ocean—continent boundary. This was followed by generation of oceanic crust, subsidence and burial by a huge post-A (post-drift) sedimentary wedge. The major part of the A sequence consists of prograded terrigenous sediments derived from an emerged Barents Sea, east of the sedimentary wedge. Rønnevik *et al.* (1982) have related the A and B reflectors to the drops in sea level in mid Oligocene and early Eocene times, respectively.

West of the Hornsund Fault Zone the maximum sediment thickness is not less than 7 km. The sediments are well stratified and single reflectors can be followed regionally, except for areas where the multiples tend to mask the reflectors. The upper sediments are composed of a series of prisms and wedges revealing a continuous outbuilding and a seaward migration of the shelf edge. Only a few reflectors have been delineated just seaward of the fault zone. This is partly due to multiples, but in some profiles the reflector patterns indicate a very rapid history of deposition where sediments have been dumped into a young immature rift. To the west the basin is dammed by the ridge mountains and in the northernmost part of the basin the young sediments overflowed the axial mountains and were deposited in the present rift valley (Fig. 12).

Myhre *et al.* (1982) proposed that the entire sequence of sediments north of 76°N had been deposited in front of the Hornsund Fault Zone since mid Oligocene times. South of 76°N, they also inferred Lower Oligocene and Eocene deposits in addition to the sequences further north. It is probable that the major part of these margin sediments were derived from an extensive cover of Palaeocene/Eocene deposits on Svalbard which have been eroded as the margin evolved. Manum and Throndsen (1978) have shown that an overburden of about 1.7 km has been removed from the exposed early Tertiary rocks on Spitsbergen.

SUMMARY

The margin off Norway and Svalbard has evolved as a passive margin since late Palaeocene times and comprises segments of late Palaeocene, Eocene and mid-Oligocene ages. South of 70°N and off the central Hornsund Fault Zone marginal basement highs developed during the initial stage of oceanic basin formation. There is no general agreement as to the crustal nature of the basement highs or regarding the possible generation of oceanic crust prior to Cenozoic times. We favour the idea that episodes of crustal thinning may have occurred but that no oceanic crust was generated prior to late Palaeocene times.

Except for the southern Barents Sea margin, there are no reflectors that can easily be defined as relating to a rift or break-up unconformity. In fact, the margin experienced a tectonically quiet period lasting at least 50 Ma prior to the Cenozoic formation of oceanic crust. Moreover, there is no evidence of regional updoming prior to rifting. In this context we must also point out that the mode of marginal subsidence changes considerably, possibly as a consequence of the pre-rift geology. These observations should be considered when making general models for passive margin evolution.

The margin reveals greatly differing amounts of Cenozoic subsidence and sedimentation along its strike. The Lofoten-Vesterålen region and the western Svalbard Platform contain little in the way of Cenozoic sediments and these areas appear to have resisted subsidence for long periods prior to Cenozoic times. On the other hand, the Møre Basin, which was a major depocentre during Mesozoic times, has continued to receive a large volume of sediments.

Finally, we would emphasize the recent observations indicating structural inversion and reactivation of older structures in Cenozoic times. In the Barents Sea such phenomena are most likely related to the events in plate tectonic development, whereas the connection with major plate tectonic events, if any, is much more obscure further south.

ACKNOWLEDGEMENT

This paper depends to a large extent on the continuing support of data acquisition and interpretation by the Universities of Bergen and Oslo provided by the Norwegian Petroleum Directorate. The long-standing close co-operation with colleagues from Bundesanstalt für Geowissenschaften und Rohstoffe, Hannover, and Lamont-Doherty Geological Observatory of Columbia University, New York, is gratefully appreciated. We thank Jakob Skogseid for invaluable assistance and Brian Sturt for reading the manuscript.

REFERENCES

Bally, A.W. and Snelson, S. 1980. Facts and principles of world petroleum occurrence: realms of subsidence. *Can. Soc. Petrol. Geol. Mem.*, 6, 9—94.

Bleie, J., Oppebøen, K.A. and Nysæther, E. 1982. *The Hydrocarbon Potential of the Northern Norwegian Shelf in the Light of Recent Drilling.* Norwegian Petroleum Society, ONS-82.

Bott, M.H.P. 1975. Structure and evolution of the Atlantic floor between northern Scotland and Iceland. *Norges Geol. Unders.* **316**, 195—200.

Bott, M.H.P. 1978. The origin and development of the continental margins between the British Isles and South-eastern Greenland. In: D.R. Bowes and B.E. Leake (eds.) Crustal Evolution in Northwestern Britain and Adjacent Regions. *Geol. J., Spec. Iss.* **10**, 377—392.

Cashion, W.W. 1975. The geology of the West-Shetland Basin. *Offshore Eur.*, **216**, 1—7.

Caston, V.N.D. 1976. Tertiary sediments of the Vøring Plateau, Norwegian Sea, recovered by Leg 38 of the Deep Sea Drilling Project. In: M. Talwani *Initial Reports Deep Sea Drilling Project,* Vol. 38, U.S. Govt. Printing Office, Washington, D.C., pp. 761—782.

Eldholm, O. and Sundvor, E. 1980. The continental margins of the Norwegian-Greenland Sea: Recent results and outstanding problems. *Phil. Trans. R. Soc. London, A,* 77—86.

Eldholm, O. and Talwani, M. 1981. The passive margins of Northern Europe and East-Greenland. Geodynamics Ser. *Am. Geophys. Union,* **6**, 30—44.

Eldholm, O. and Windisch, C.C. 1974. Sediment distribution in the Norwegian-Greenland Sea. *Geol. Soc. Am. Bull.*, **85**, 1661—1676.

Eldholm, O., Sundvor, R. and Myhre, A.M. 1979. Continental margin off Lofoten-Vesterålen. *Mar. Geophys. Res.*, **4**, 3—35.

Faleide, J.I. and Gudlaugsson, S.T. 1981. Geology of the western Barents Sea — A regional study based on marine geophysical data. Cand. real. thesis, Univ. of Oslo, 160 pp.

Faleide, J.I., Gudlaugsson, S.T. and Jacquart, G. 1984 a. Evolution of the western Barents Sea. *Mar. Pet. Geol.* (in press).

Faleide, J.I., Gudlaugsson, S.T., Johnson, B., Myhre, A.M. and Eldholm, O. 1984b. Free-air gravity anomaly maps of the Greenland Sea and the Barents Sea. *Norsk Polarinst. Skr.* **180** (in press).

Gairaud, H., Jacquart, G., Aubertin, F. and Beuzard, P. 1978. The Jan Mayen Ridge synthesis of geological knowledge and new data. *Oceanol. Acta,* **1**, 335—358.

Garde, S.S. 1978. Zür geologischen Entwicklung des Jan Mayen Ruckens nach geophysikalischen Daten. Dissertation, Univ. Clausthal, 74 pp.

Grønlie, G. and Talwani. M. 1982. The free-air gravity field of the Norwegian-Greenland Sea and adjacent areas. *Earth Evol. Ser.*, **2**, 79—103.

Hagevang, T., Eldholm, O. and Aalstad, I. 1983. Pre-23 magnetic anomalies between Jan Mayen and Greenland-Senja Fracture Zones in the Norwegian Sea. *Mar. Geophys. Res.* **5**, 345—363.

Hamar, G.P., Jacobsson, K.H., Ormaasen, D.E. and Skarpnes, O. 1980. Tectonic development of the North Sea north of central highs. In: *The Sedimentation of the North Sea Reservoir Rocks.* Norwegian Petroleum Society 11 pp.

Hanisch, J. 1984. Late Cretaceous opening of the NE Atlantic and W Spitsbergen Orogeny: A plate tectonic model. This volume, pp. 187—198.

Hinz, K. 1981. A hypothesis on terrestrial catastrophes. Wedges of very thick oceanward dipping layers beneath passive continental margins — their origin and paleo-environment significance. *Geol. Jb.*, **E22**, 3—28.

Hinz, K. and Schlüter, H-U. 1978a. Der Nordatlantik — Ergebnisse geophysikalischer Untersuchungen der Bundesanstalt für Geowissenschaften und Rohstoffe an nordatlantischen Kontinentalrändern. *Erdöl-Erdgas-Z.*, **94**, 271—280.

Hinz, K. and Schlüter, H-U. 1978b. The geological structure of the western Barents Sea. *Mar. Geol.*, **26**, 199—230.

Hinz, K. and Weber, J. 1976. Zum geologischen Aufbau des Norwegischen Kontinentalrandes und der Barents-See nach reflexionsseismischen Messungen. *Erdöl Kohle, Erdgas, Petrochemie,* 3—29.

Hinz, K., Dostman, H.J. and Hanisch, J. 1982. *Structural Framework of the Norwegian Sea.* Norwegian Petroleum Society, ONS-82.

Håkansson, E. and Stemmerik, L. 1984. Wandel Sea Basin — The North Greenland equivalent to Svalbard and the Barents Shelf. This volume, pp. 97—108.

Jacqué, M. and Thouvenin, J. 1975. Lower Tertiary tuffs and volcanic activity in the North Sea. In: A.W. Woodland (ed.) *Petroleum and the Continental Shelf of North-west Europe,* Vol. 1. Applied Science Publishers, London, pp. 455—465.

Jørgensen, F. and Navrestad, T. 1981. The geology of the Norwegian Shelf between 62°N and the Lofoten Islands. In: L.V. Illing and G.D. Hobson (eds.). *Petroleum Geology of the Continental Shelf of NW Europe,* Institute of Petroleum, London, pp. 407—413.

Kristoffersen, Y. and Talwani, M. 1977. Extinct triple junction south of Greenland and the Tertiary motion of Greenland relative to North America. *Geol. Soc. Am. Bull.*, **88**, 1037—1049.

LaBrecque, J.L., Kent, D.V. and Cande, S.C. 1977. Revised magnetic polarity time scale for late Cretaceous and Cenozoic time. *Geology,* **5**, 330—335.

Larsen, H.C. 1984. Geology of the East Greenland Shelf. This volume, pp. 329—340.

Manum, S.B. and Throndsen, T. 1978. Rank of coal and dispersed organic matter and its geological bearing in the Spitsbergen Tertiary. *Norsk Polarinst. Årbok,* **1977**, 159—177.

Mutter, J.C. Cenozoic and late Mesozoic stratigraphy and subsidence history of the Norwegian margin. *Geol. Soc. Am Bull.* (in press).

Mutter, J.C., Talwani, M. and Stoffa P.L. 1982. Origin of seaward dipping reflectors in oceanic crust off the Norwegian margin by 'Subaerial sea-floor spreading'. *Geology,* **10**, 353—357.

Mutter. J.C., Talwani, M. and Stoffa P.L. Structure of the oceanic crust adjacent to the Norwegian margin. *J. Geophys. Res.* (in press).

Myhre, A.M. 1984. The Svalbard margin: A geophysical study. Dr. scient. Thesis., Univ. of Oslo (in press).

Myhre, A.M., Eldholm, O. and Sundvor, E. 1982. The margin between Senja and Spitsbergen fracture zones: Implication from plate tectonics. *Tectonophysics,* **89**, 1—32.

Myhre, A.S., Eldholm, O. and Sundvor, E. The Jan Mayen Ridge: Present status. *Polar Res.* (in press).

Nunns, A.G. 1980. Marine Geophysical Investigations in the Norwegian-Greenland Sea Between the Latitudes of 62°N and 74°N. Ph.D. Thesis, Univ. Durham, UK, 185 pp.

Nunns, A.G. and Peacock, J.H. 1982. Correlation, Identification and Inversion of Magnetic anomalies in the Norway Basin. *Earth Evol. Ser.*, **2**, 130—138.

Price, I. and Rattey, R.P. Cretaceous tectonics of Mid-Norway: Implications for the Rockall Faeroes—Shetland rift. *J. Geol. Soc. Lond.* (in press).

Ridd, M.F. 1981. Petroleum geology west of the Shetlands. In: L.V. Illing and G.D. Hobson (eds.). *Petroleum Geology of the Continental Shelf of NW Europe.* Institute of Petroleum, London, pp. 414—425.

Roberts, D.G., Masson, D.G. and Miles, P.R. 1981. Age and structure of the southern Rockall trough: New evidence. *Earth Planet. Sci. Lett.*, **52**, 115—128.

Russel, M.J. 1976. A possible lower Permian age for the onset of ocean floor spreading in the northern North Atlantic. *Scott. J. Geol.*, **12**, 315—323.

Russel, M.J. and Smythe, D.K. 1978. Evidence for an early Permian oceanic rift in the northern North Atlantic. In: Paleorift Systems with Emphasis on the Permian Oslo Rift, I. *Nato Adv. Study Inst., Ser. C.,* **36**, 173—179.

Rønnevik, H.C. 1981. Geology of the Barents Sea. In: L.V. Illing and G.D. Hobson (eds.). *Petroleum Geology of the Continental Shelf of Northwest Europe.* Institute of Petroleum, London, pp. 395—406.

Rønnevik, H. and Navrestad, T. 1977. Geology of the Norwegian shelf between 62°N and 69°N. *GeoJournal,* **1**, 33—46.

Rønnevik, H., Beskow, B. and Jacobsen, H.P. 1982. *Structural and Stratigraphic Evolution of the Barents Sea.* ONS-82, Norwegian Petroleum Society, 29 pp.

Rønnevik, H.C., Eggen, S. and Vollset, J. 1983. Exploration of the Norwegian shelf. *Petroleum Geochemistry and Exploration of Europe.* Blackwell Scientific Press, Oxford, pp. 71—94.

Rønnevik, H. Jørgensen, F. and Motland K. 1979. *The Geology of the Northern Part of the Vøring Plateau.* Norwegian Petroleum Society NSS/12, 21 pp.

Savostin, L.A. and Karasik, A.M. 1981. Recent plate tectonics of the Arctic Basin and of northeastern Asia. *Tectonophysics,* **74**, 111—145.

Skogseid, J. 1983. A Marine Geophysical Study of Profiles between the Vøring Plateau margin and the Jan Mayen Ridge. Cand. scient. thesis, Univ. Oslo.

Smythe, D.K., Chalmers, J.A., Skuce, A.G., Dobinson, A. and Mould, A.S. 1983. Early opening history of the

North Atlantic — I. Structure and origin of the Faeroe-Shetland Escarpment. *Geophys. J.R. Astron. Soc.*, **72**, 373—398.

Srivastava, S.P. 1978. Evolution of the Labrador Sea and its bearing on the early evolution of the North Atlantic. *Geophys. J. Astron. Soc.*, **52**, 313—357.

Sundvor, E. and Eldholm, O. 1979. The western and northern margin off Svalbard. *Tectonophysics*, **59**, 239—250.

Sundvor, E., Myhre, A.M. and Eldholm, O. 1979. The Svalbard continental margin. *Norwegian Sea Symposium* (Norwegian Petroleum Society) NSS/6, pp. 1—25.

Surlyk, F. *et al.* 1984. Permian Basin of East-Greenland. This volume, pp. 303—316.

Talwani, M. 1978. Distribution of basement under the eastern North Atlantic Ocean and the Norwegian Sea. *Geol. J., Spec. Iss.*, **5**, 347—376.

Talwani, M. and Eldholm, O. 1972. The continental margin off Norway: A geophysical study. *Geol. Soc. Am. Bull.* **83**, 3575—3608.

Talwani, M. and Eldholm, O. 1977. Evolution of the Norwegian-Greenland Sea. *Geol. Soc. Am. Bull*, **88**, 969—999.

Talwani, M., Mutter, J.E. and Eldholm, O. 1981. The initiations of opening of the Norwegian Sea. *Oceanol. Acta.* 4 (supplement) 23—30.

Talwani, M., Mutter, J. and Hinz, K. 1983. Ocean continent boundary under the Norwegian continental margin. In: M.H.P. Bott *et al.* (eds.). *Structure and Development of the Greenland—Scotland Ridges, New Methods and Concepts.* Proc. NATO Adv. Res. Inst. Bressanone 1981, Plenum New York, pp. 121—131.

Talwani, M., Udintsev, G. *et al.* (eds.) 1976. In: *Initial Reports Deep Sea Drilling Project*, Vol. 38. U.S. Government Printing Office, Washington D.C., 1256 pp.

Unternehr, P. 1982. Etude Structurale et Cinématique de la Mer de Norvège et du Greenland. Evolution du microcontinent de Jan Mayen. Thesis, Univ. Bretagne, France, 227 pp.

Vogt, P.R., Johnson, G.L. and Kristjansson, L. 1980. Morphology and magnetic anomalies north of Iceland. *J. Geophys. Res.* **47**, 61—66.

Ziegler, P.A. 1982. Geological Atlas of Western and Central Europe. Shell Int. Petrol., Maatsch. B.V., 130 pp.

Åm, K. 1970. Aeromagnetic investigations of the continental shelf of Norway, Stad—Lofoten (62—69°N). *Norges Geol. Unders.*, **266**, 49—61.

2 Structural highs and basins in the western Barents Sea

Hans Rønnevik and Hans-Petter Jacobsen
Norwegian Petroleum Directorate

The Barents Sea overlies an intracratonic basin, which borders a passive shear margin in the west. Twelve major sequence boundaries have been defined. The structural development of the Barents Sea can be summarized in the following way. Upper Devonian-Lower Carboniferous graben formation as a result of a major sinistral wrench in the west and a conjugate dextral wrench through the central part of the Barents Sea. Epeirogenic subsidence of the whole of the area as one basin in Upper Carboniferous and Lower Permian. Middle Permian elevation of the western rim of the Barents Sea basin. Early Middle Kimmerian wrenching in the western and northern Barents Sea with formation of intrabasinal highs and basins. The wrenching was related to a dextral N—S system in the west and conjugate NE—SW wrenches into the Barents Sea. Mid-Cretaceous to mid-Tertiary structuring of the Senja Ridge as a result of a southward movement of the area from Svalbard to the Bjørnøya Basin.

INTRODUCTION

The regional structural setting of the western Barents Sea has been discussed previously (Rønnevik 1981, 1982). The interpretation is based on the regional geophysical surveys carried out by the Norwegian Petroleum Directorate since 1973 (Fig. 1). The present paper will discuss briefly the regional setting and then discuss in more detail the Bjørnøya Basin.

REGIONAL GEOLOGICAL SETTING

The Barents Sea overlies an intracratonic basin (Fig. 2). In the west the basin borders the oceanic Norwegian Sea with a passive shear margin that started to develop in the Palaeocene. The two major shear margin elements are related to the Hornsund Fault Zone in the north and the Senja Fracture Zone in the south. In between these two major fault zones there is another NNW—SSE shear-margin element. The shear-margin elements can be followed into transform fault zones in the oceanic crust to the west.

Petroleum Geology of the North European Margin © Norwegian Petroleum Society (Graham & Trotman, 1984) pp. 19—32.

Upper Devonian—Lower Carboniferous development

The oldest sequence that can be mapped regionally in the Barents Sea was deposited in fault-bounded basins (Fig. 3). This sequence is interpreted to be of Upper Devonian to Lower Carboniferous age (Rønnevik, 1982) (see Fig. 16). The basal sequence boundary is difficult to map north of 74° and east of 27° due to poor data quality.

The area east of 25° is dominated by a major NE—SW oriented basin (Fig. 16), which is a half graben in the south bounded by faults on both sides, down-thrown to the east. The northern half of the basin is a graben. The basin corresponds to the Nordkapp Basin that has been defined at a younger level. West of the Nordkapp Basin there is a NE—SW graben system from approximately 73° and south-westwards. The basin trend cannot be mapped west of 20°. Hence one can only surmise a possible continuation beneath the younger Tromsø Basin. To the south of the basin trend there is a marked E—W fault at about 71°30′N and a marked fault zone to the south and east of this. These fault systems were formed during the Kimmerian tectonic phase. A marked half graben bounded by a fault down-thrown to the north is situated south of Hopen. Saltflow structures are not observed in this basin.

The dominating fault direction in Upper Devonian—Lower Carboniferous is NE—SW. The faults are extensional. The east flank of the Cale-

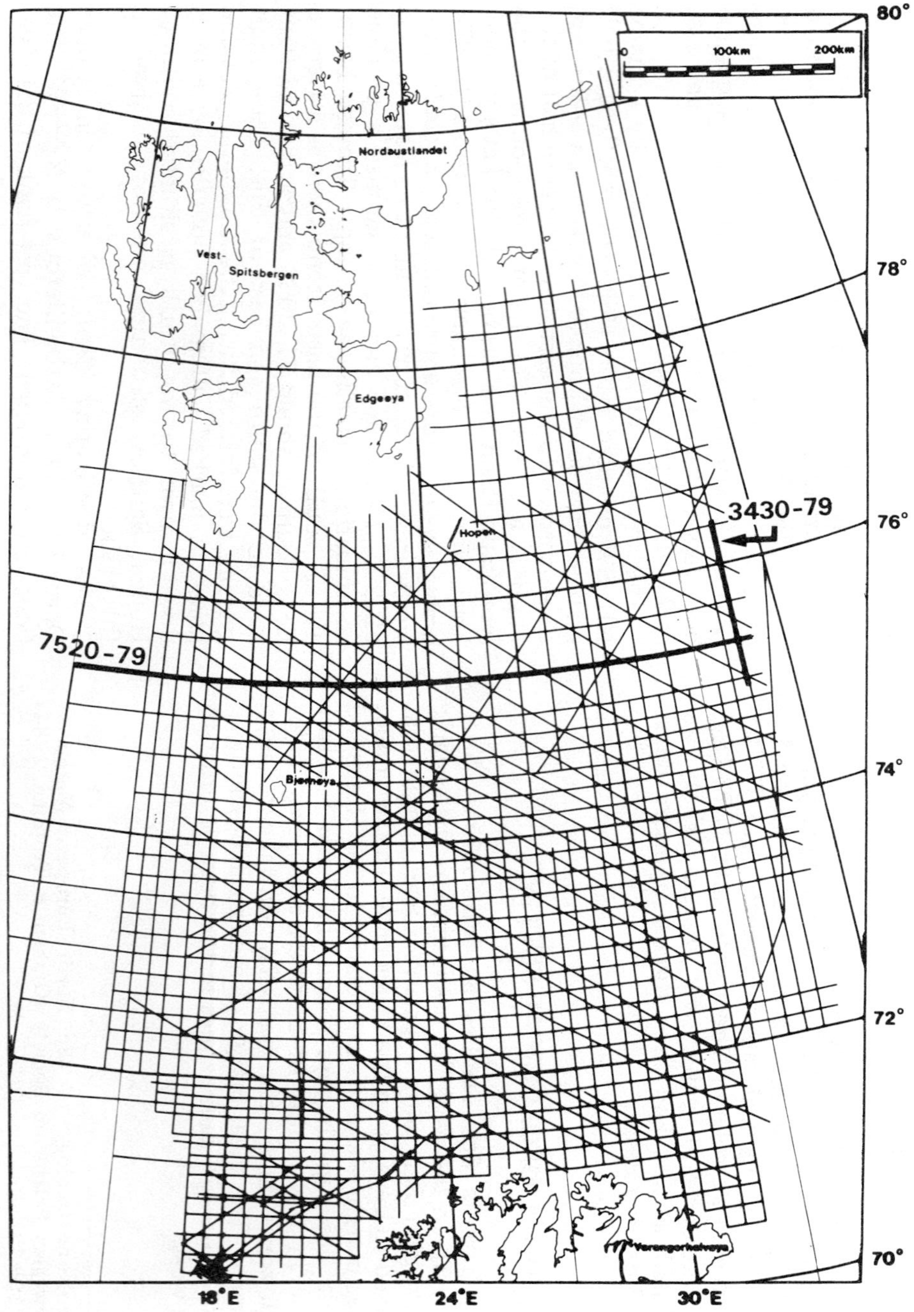

Fig. 1. Geophysical survey grid.

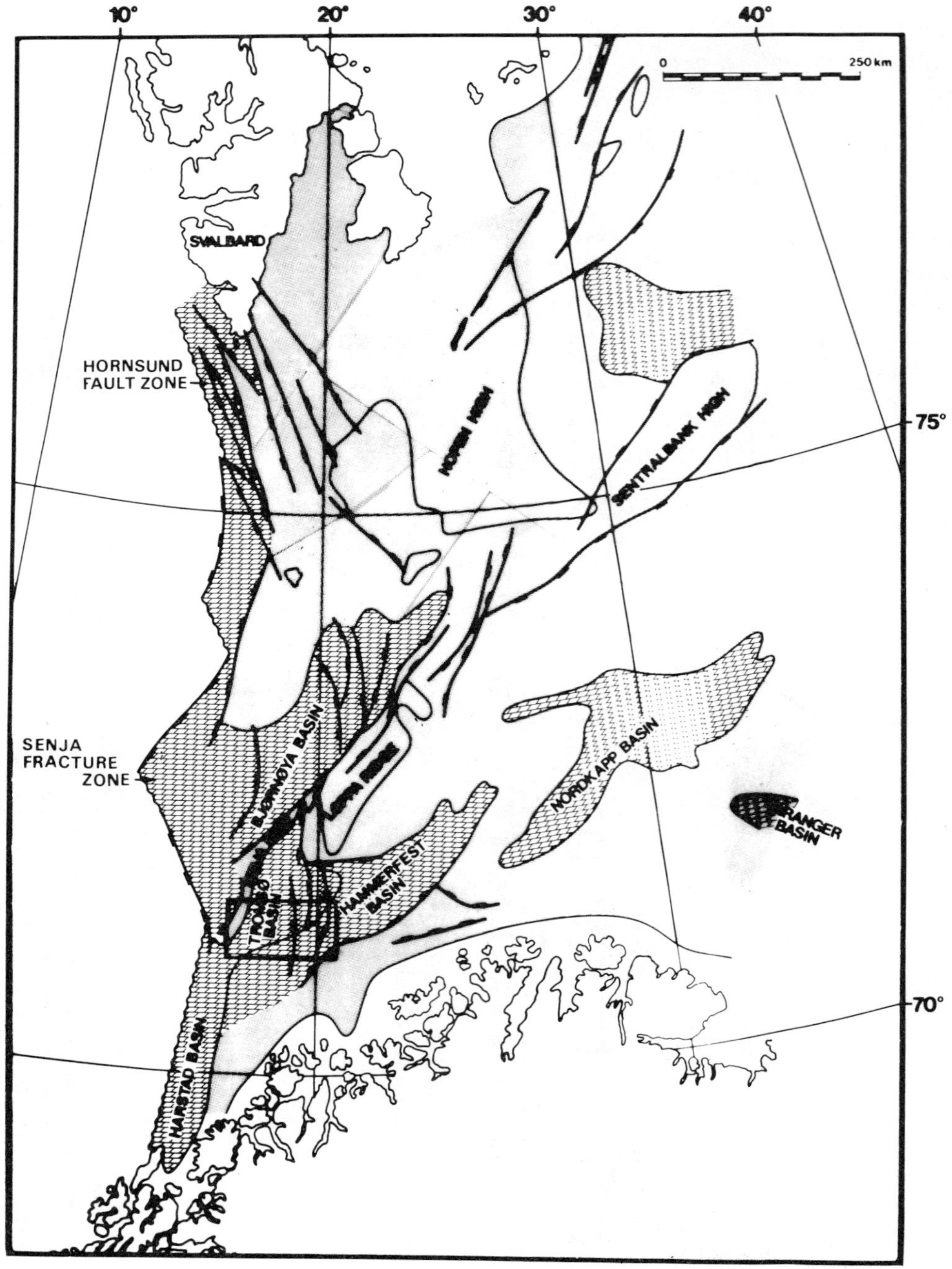

Fig. 2. Main structural elements in the Barents Sea.

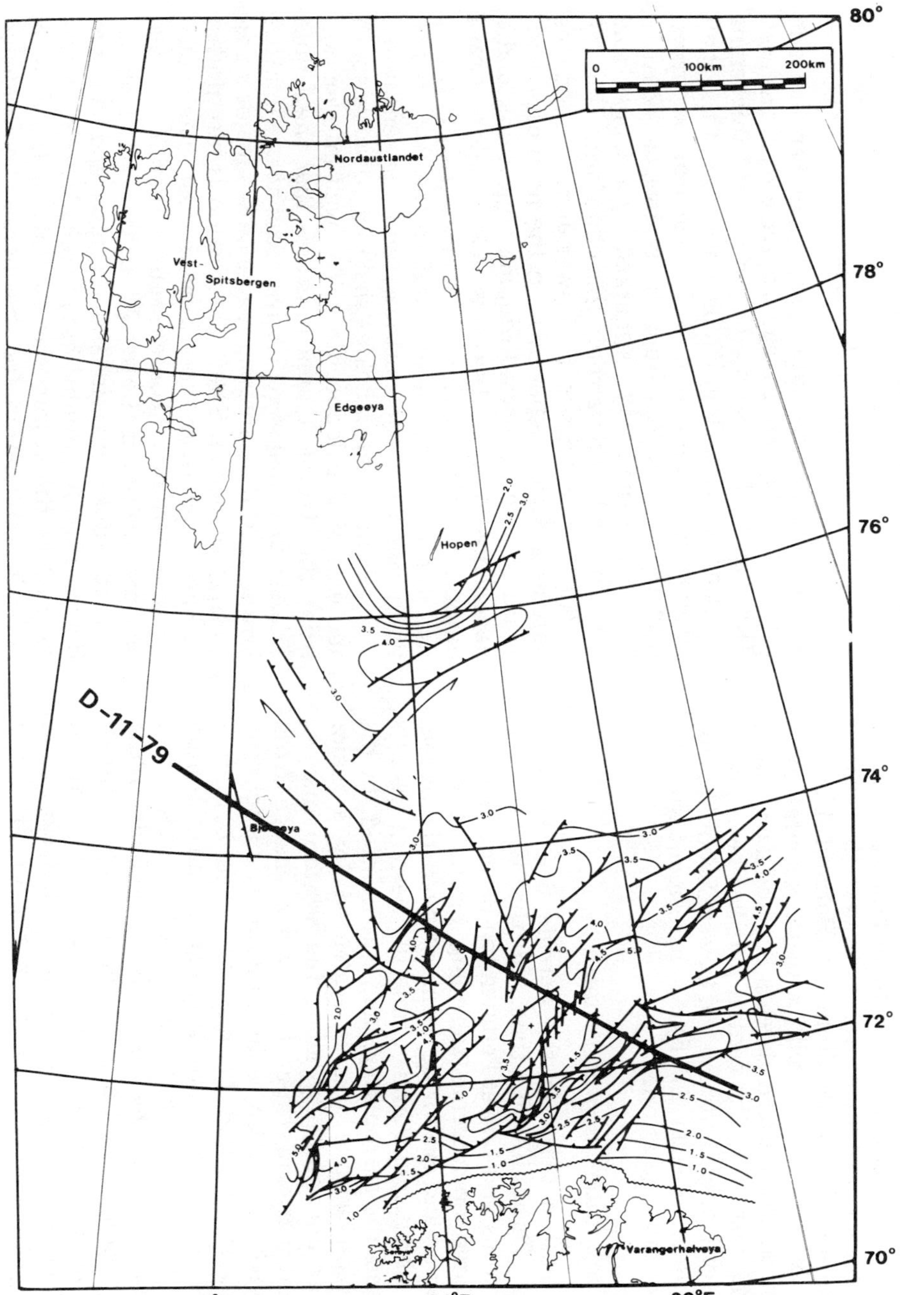

Fig. 3. Isochrone map of Horizon H (late Devonian). Two-way time in seconds.

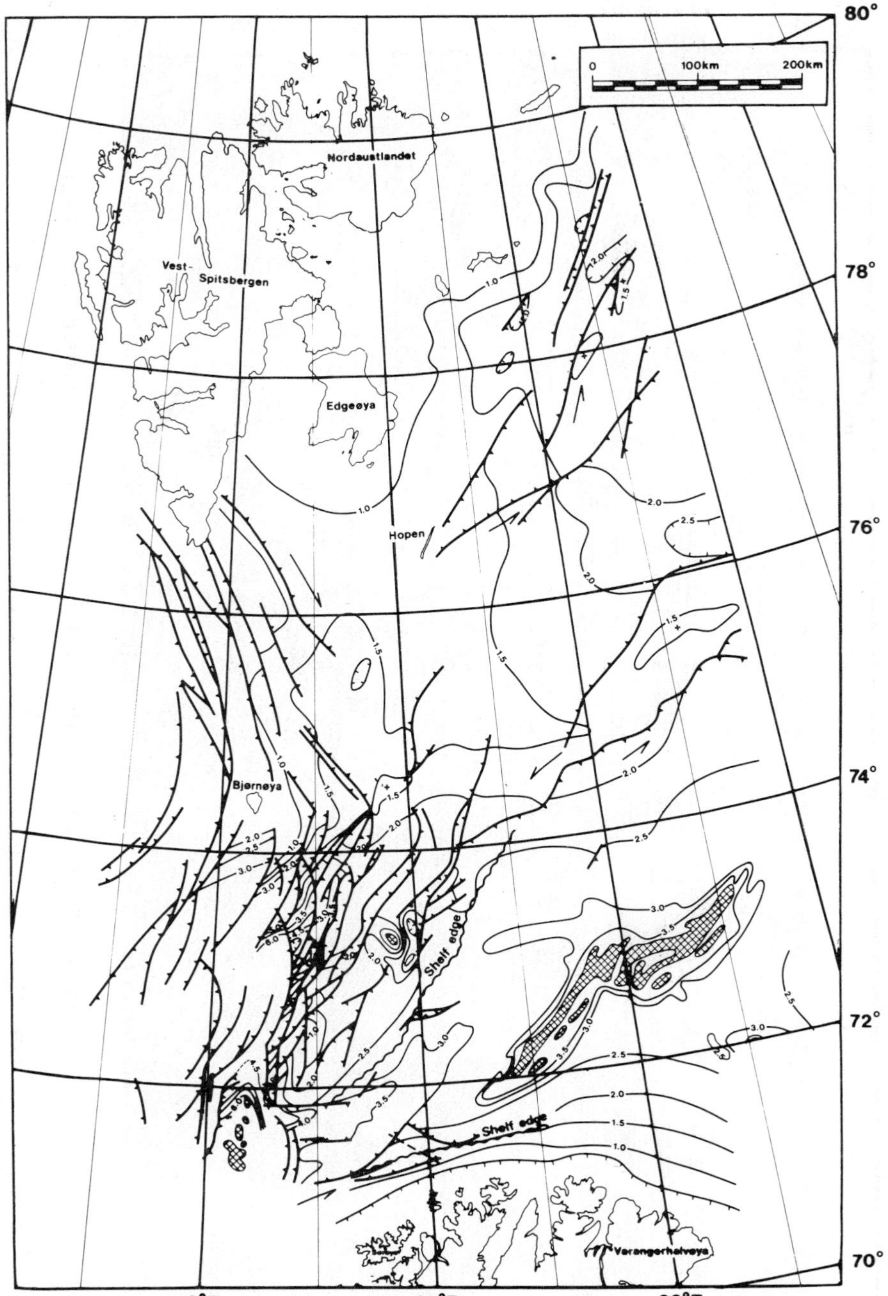

Fig. 4. Isochrone map of Horizon F_1 (mid Permian). Two-way time in seconds.

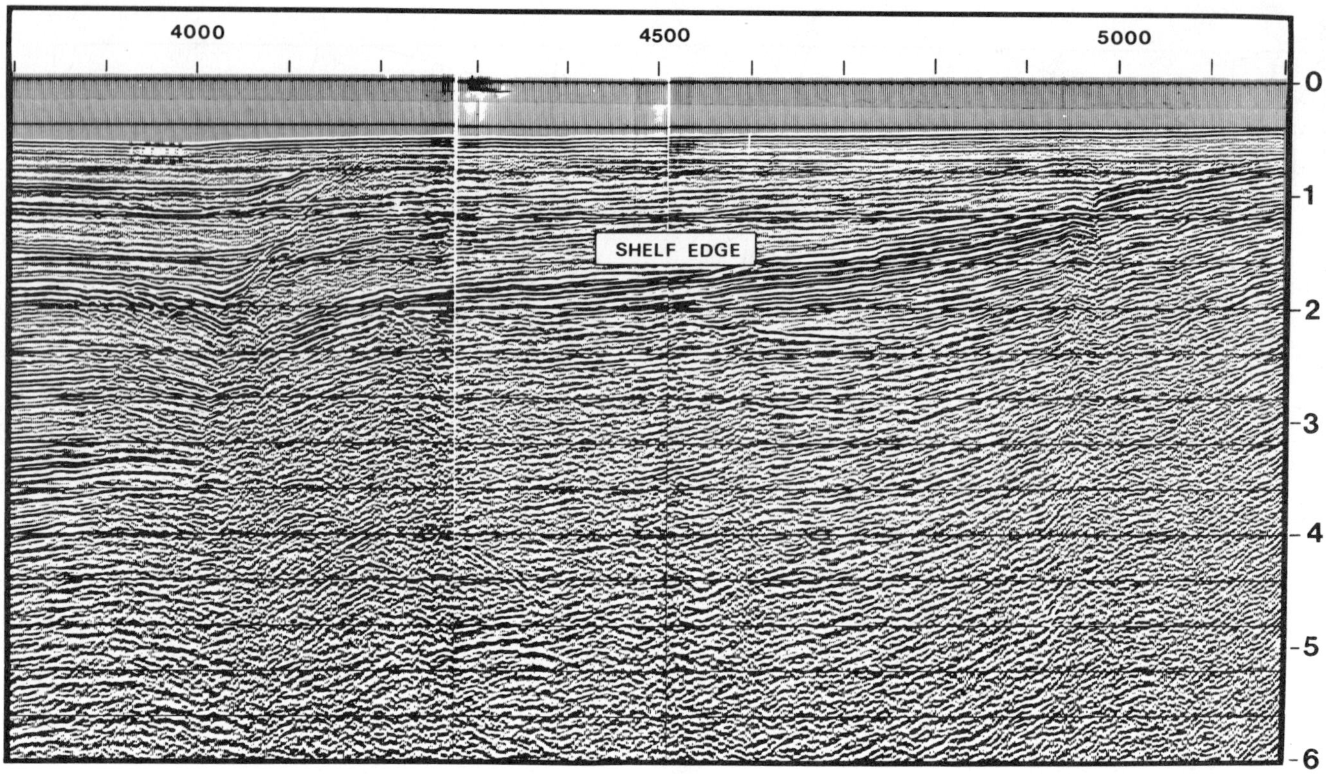

Fig. 5. Seismic section showing the marked carbonate shelf edge along the southern rim of the basin.

donian orogen straddles the Norwegian coast around 28°E. To the east of the Caledonides there is only minor faulting, suggesting that the NE—SW systems follow Caledonian structural trends.

In the central part of the Barents Sea there is an important WNW—ESE fault zone between 30°E and 35°E north of 72°N, which becomes NNW—SSE in the western part of the Barents Sea. Along this fault zone in the east there are possible compressional structures. The fault system is considered to be related to a major wrench system that crosses the Barents Sea. In the literature (Harland, 1969), a mega tectonic model with a major sinistral shear system in the western Barents Sea and a conjugate dextral system in the central part has been postulated for Upper Devonian and Lower Carboniferous. This model (Ziegler, 1978) is in agreement with the present interpretation.

Mid Carboniferous—mid Permian

The fault activity in the Barents Sea ceased in Upper Carboniferous time. In the Upper Carboniferous and Lower Permian the Barents Sea was one large epeirogenic subsiding basin, in which carbonate and evaporitic sediments were deposited.

At about mid Permian time the western Barents Sea and the surrounding land was uplifted. In the reflection seismic this is evidenced by a major angular unconformity along the western margin. In mid Permian time there was a change from carbonate to clastic sedimentation. The isochrone map over the mid Permian sequence boundary (Fig. 4) shows the southeastern area as an unfaulted stable province. This tectonic setting existed over the larger part of the Barents Sea. The faults in the northwestern part are the result of mid Kimmerian and younger tectonic overprint. At the mid Permian sequence boundary there is a marked carbonate shelf edge along the southern rim of the basin between 20 and 28° (Fig. 5). In the west this palaeoshelf can be followed northwards across the Hammerfest Basin and along the east flank of the Loppa Ridge. The existence of a shelf edge both in the Hammerfest Basin and in the area to the north and south demonstrates that the Troms Finmark Fault Zone was not active in late Palaeozoic time. East of 28° the shelf edge breaks up into isolated carbonate buildups.

Upper Permian—Lower Jurassic

The clastic sedimentation in Upper Permian and Lower Triassic had a major source area in the east. This can be demonstrated by a marked thickening towards the east (Fig. 6). The position of the shelf edge at approximate mid Triassic can be traced over the Barents Sea from the Norwegian coast to 77°N. The shelf edge is marked by a depositional thinning and westward dipping clinoforms (Fig. 7).

The Upper Triassic to Lower Jurassic sequence defines a major depocentre in the northeast, with south to southwestward dipping clinoforms, and a depocentre in the southwest with low angle eastward dipping clinoforms (Figs. 8 and 9). This indicates that the major sediment input to the Barents Sea Basin was from the northeast and the southwest. The seismic data indicates shaleout towards south and east of the depocentres. The well information from the Troms 1 area indicates that the Upper Triassic is a major regressive sequence, while the Lower to Middle Jurassic sequence is a major transgressive sequence. The transgression is interpreted to have been from west towards east. The southwestern sediment source must, therefore, have become inactive in Lower Jurassic. A relatively increased subsidence in the Tromsø and western Hammerfest Basin must have taken place; this seems to predate the Kimmerian faulting in the area. A possible explanation for the increased subsidence

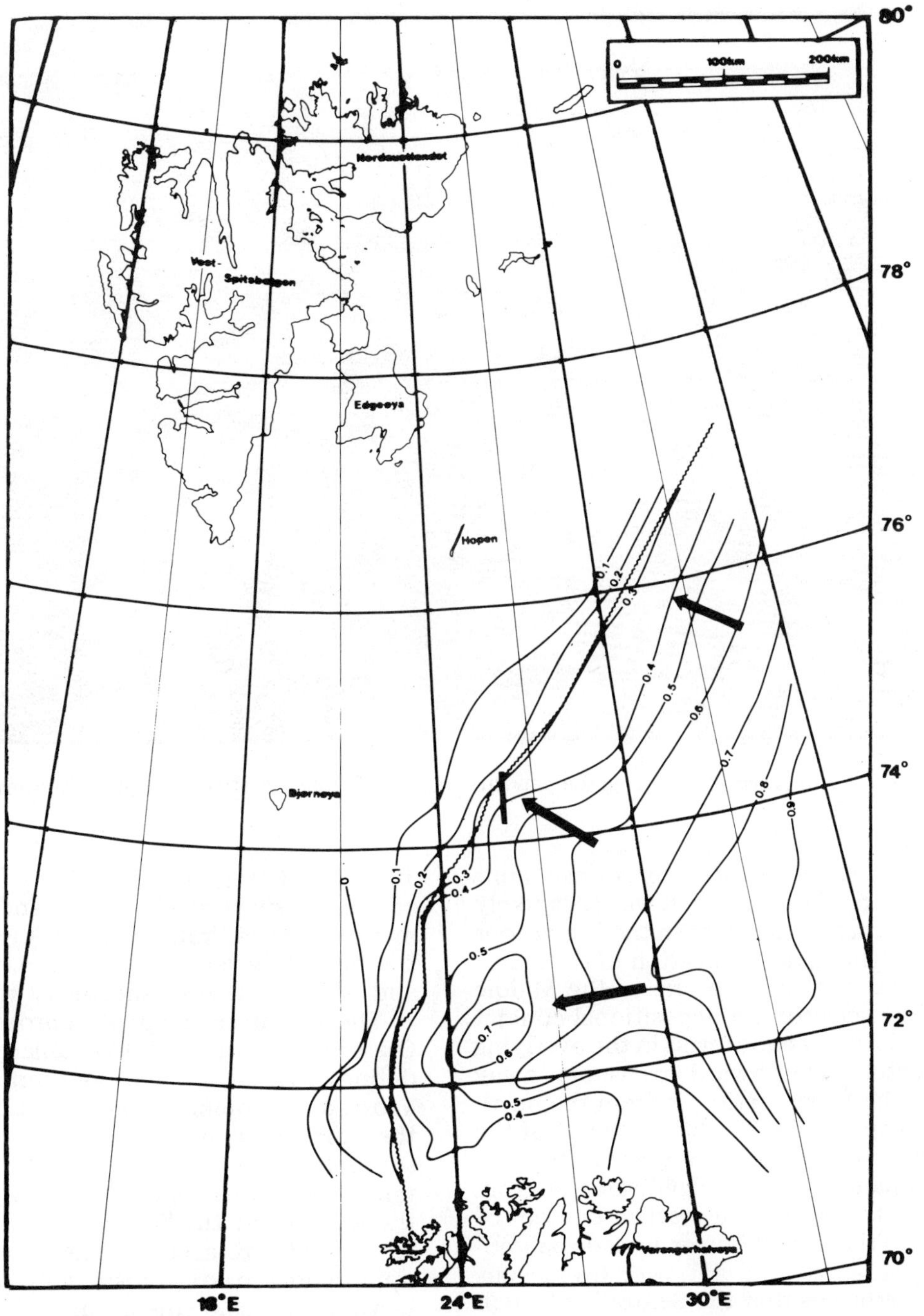

Fig. 6. Isopach map between Horizons E and E[1] (early—mid Triassic). Two-way time in seconds. Arrows show inferred transport directions. Wavy line shows shelf edge.

can be the formation or expansion of a major rim syncline in relation to salt movements in the Tromsø Basin.

The whole of the Barents Sea seems to have been one epeirogenic subsiding basin in the Upper Triassic and Lower Jurassic, although a thickening of the sequence south of West-Spitsbergen indicates a more rapidly subsiding area. There is a thickening related to salt movement in the Nordkapp Basin.

Middle Jurassic—Lower Cretaceous

The early Middle Kimmerian tectonic movements resulted in subsidence of the Hammerfest, Tromsø and Bjørnøya Basins, with the stable Loppa Ridge in between. In the Hammerfest Basin erosional scars indicate an uplift prior to the start of the subsidence.

The Hammerfest Basin is bounded towards the south by a marked fault zone called the Troms Finmark Fault Zone (Fig. 9). In the north the basin is bounded by a fault zone. East of 22°, the fault zone changes into a flexure. In the west the Hammerfest Basin is separated from the Tromsø Basin with a marked hinge zone, which can be followed northwards along the westflank of the Loppa Ridge and across the Bjørnøya Basin. The hinge zone separates the Bjørnøya basin into an eastern shallow province and a western deep province (Fig. 11).

There is a marked thickening of the Upper Jurassic and Lower Cretaceous sequence across the hinge zone. The major subsidence occurred in Lower Cretaceous (Fig. 12).

The Bjørnøya Basin is bounded in the south by a major fault zone, which continues northeastwards across the whole of the Barents Sea. Along this fault complex in the east there is a high, referred to

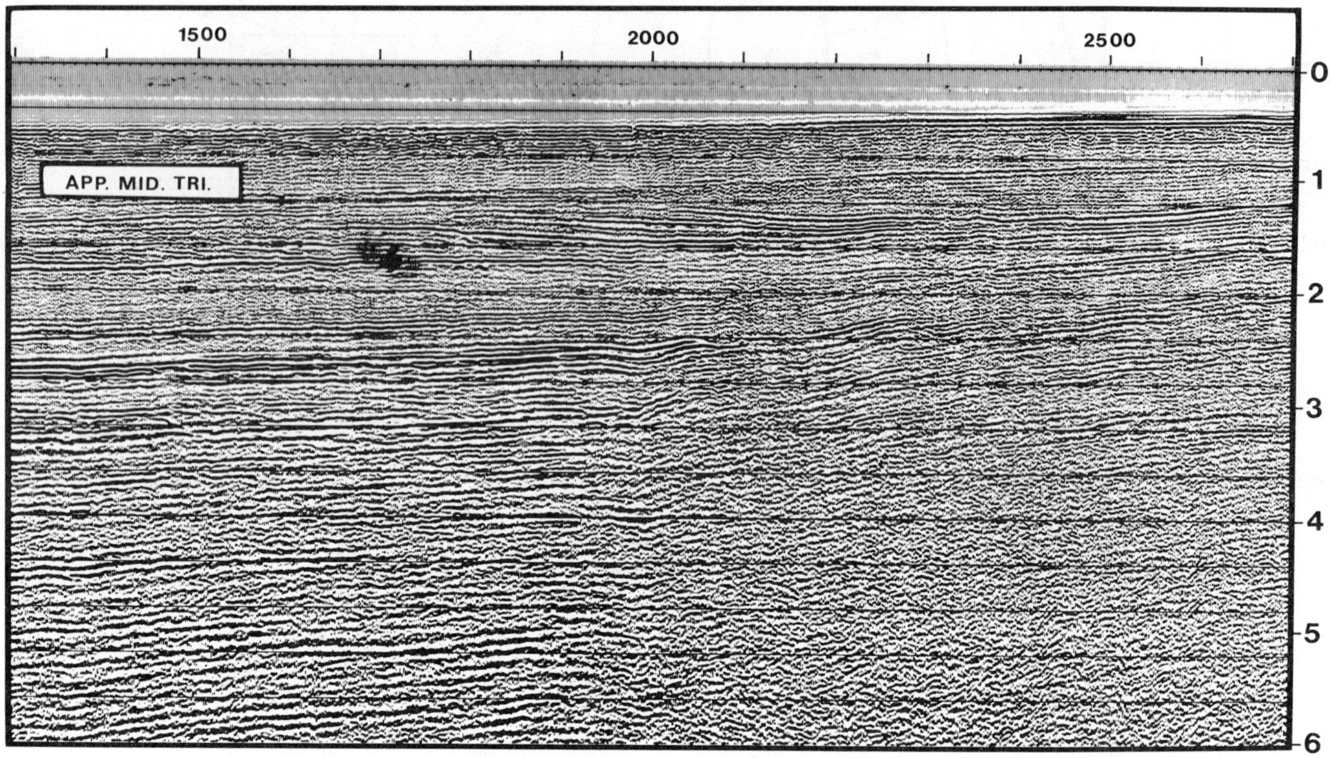

Fig. 7. Seismic section showing the depositional thinning and westward dipping clinoforms.

as the Sentralbank High, bounded by reverse faults (Figs. 9, 14 and 15). Movements have most likely taken place along these faults both in the Cretaceous and the Tertiary. The major indication of the oldest possible age is that the subcrop line of the Middle Jurassic (Fig. 9) seems to be a depositional contact east of 27°. The internal clinoforms in the overlying sequence indicate a southward transport away from the high. The progradation is from the north across the Nordkapp Basin. To the southeast of the Nordkapp basin there is a high that acted as a secondary provenance area. Dredged samples from the eastern Barents Sea show a marked content of reworked Upper Triassic and Lower Jurassic palynomorphs in Upper Jurassic and Lower Cretaceous rocks. This also indicates that the Sentralbank High acted as a source area for sediments from the Middle Jurassic onwards. The subcrop of the Middle Jurassic boundary to the west of approximately 27° is an erosional contact, affected also by younger uplifts than the Kimmerian movements.

North of the Sentralbank High there is an eastward dipping basin with Upper Jurassic to Lower Cretaceous strata. The area to the north of the fault zone was generally uplifted and eroded. The effect of the Kimmerian and younger structuring is seen on the mid Permian isochrone map (Fig. 4). Northeastward from Hopen Island a complex NE—SW-oriented fault zone can be followed, along which several domal uplifts bounded by reverse faults are mapped. Thin inliers of post Middle Jurassic strata are possible to the north of the mapped subcrop of the Middle Jurassic sequence boundary.

The structuring during the early Middle Kimmerian phases can be explained by the formation of a major dextral wrench system along the hinge zone extending from the transition between the Hammerfest Basin and Tromsø Basin through the Bjørnøya Basin, and also by the formation of three conjugate sinistral NE—SW trending fault systems marked by the Troms—Finmark Fault Zone (Fig. 9), the boundary fault of the Bjørnøya Basin (Fig. 11) and the fault zone northeastward from Hopen.

The isochrone map of approximate base Upper Cretaceous (Intra Cenomanian) (see Fig. 10) defines two basins. The southern basin overlies the Tromsø, Hammerfest and Nordkapp Basins, and the northern one overlies the Bjørnøya Basin. The southern basin trend is structured very little by mid Cretaceous and younger faulting, except in the west where the Senja Ridge came into existence from mid Cretaceous. The basin has undergone uplift and erosion, most likely during a Laramide tectonic phase. This resulted in erosion of the Upper Cretaceous prior to deposition of Palaeocene in the basin. The hiatus can be seen as an erosional relief, especially at the flank of the Nordkapp Basin.

STRUCTURAL DEVELOPMENT OF THE BJØRNØYA BASIN

The Bjørnøya Basin has been subject to several tectonic phases from Middle Jurassic; the pre Middle Jurassic basin setting is unknown. Both on Bjørnøya to the north and on the Loppa Ridge to the south there is marked hiatus between Permian and late Triassic strata. A similar hiatus can also be expected in the basin. Salt is not observed in the basin, and hence there is no indirect evidence for late Palaeozoic basins beneath the Middle Jurassic sequence boundary. However, the seismic data demonstrate that there are sediments beneath this sequence boundary. The isochrone map over top Middle Jurassic (Fig. 11) shows that the Bjørnøya Basin deepens and widens westwards. Part of the deepening is caused by major fault zones down-

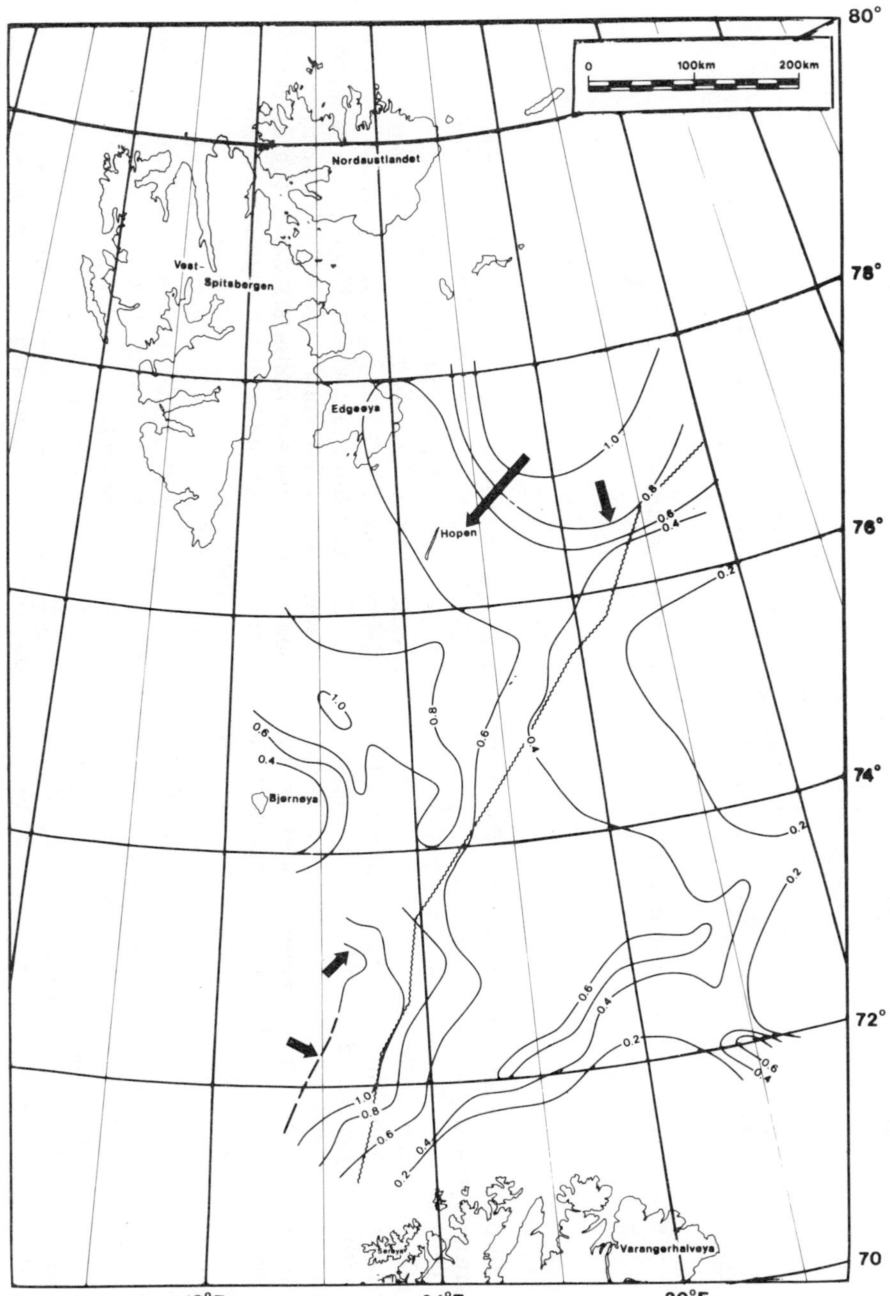

Fig. 8. Isopach map between Horizons D^{11} and D^{1} (Upper Triassic—Lower Jurassic). Two-way time in seconds. Arrows show transport direction. Wavy line shows shelf edge position at top Lower Triassic.

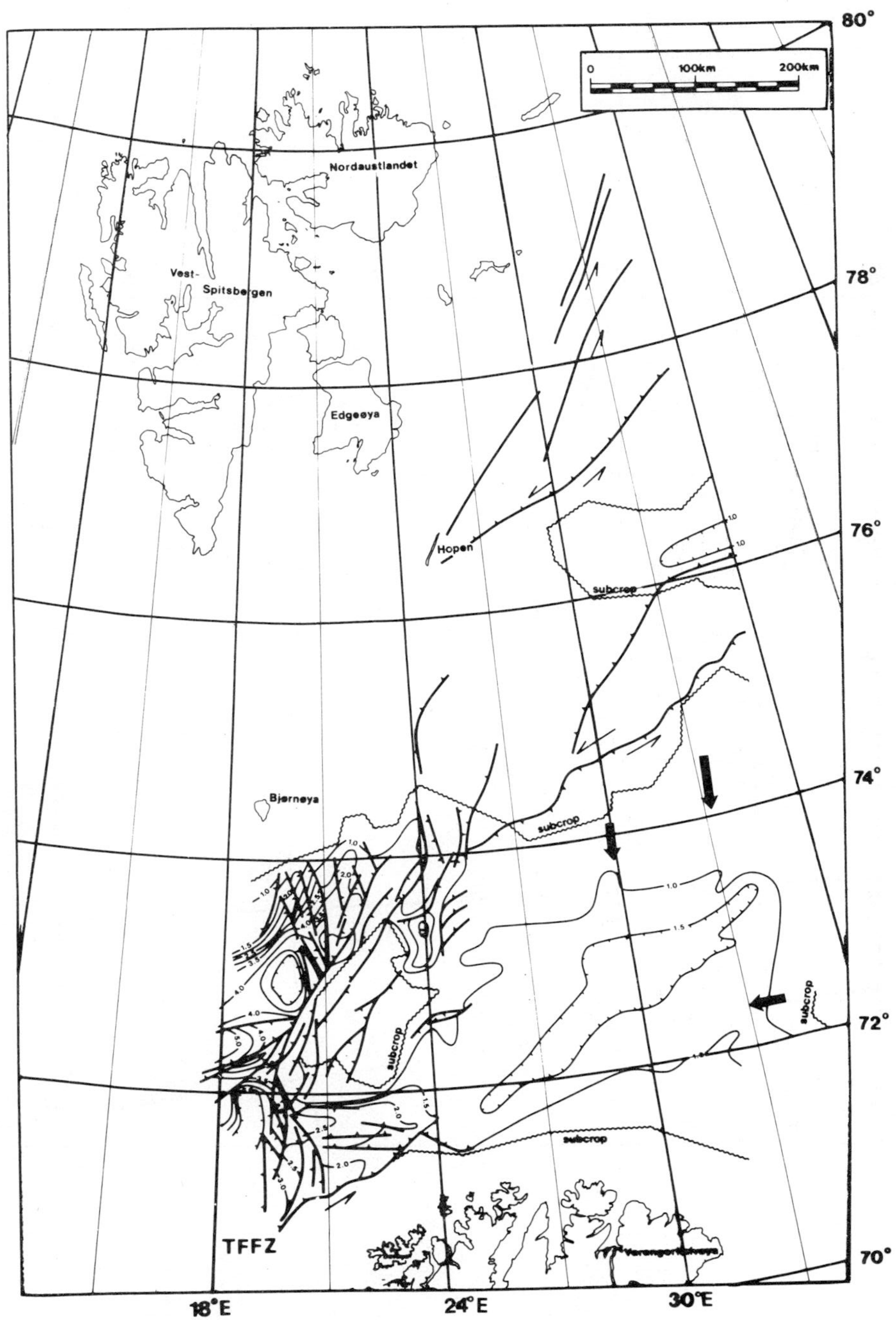

Fig. 9. Isochrone map of Horizons D^{1} (mid Jurassic). Two-way time in seconds. Arrows show transport directions.

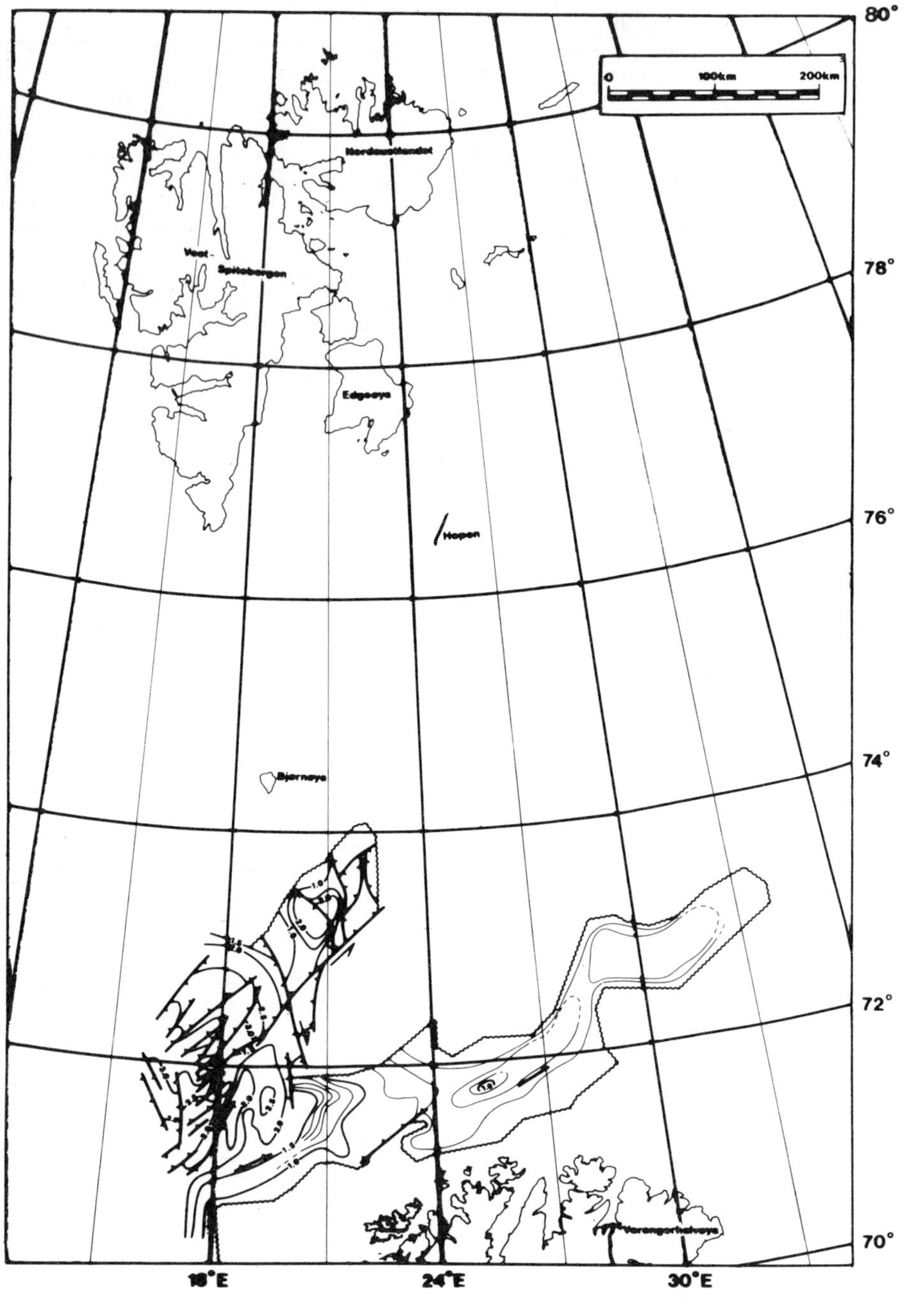

Fig. 10. Isochrone map of Horizon C (top Lower Cretaceous). Two-way time in seconds.

thrown to the west at around 20 and 21°. The fault zones are curved with the concave side opening westwards. The eastern fault is the western boundary of a marked horst, bounded to the east by an eastward dipping curved fault. This fault merges towards the south with a major NE—SW fault zone that marks the southern boundary of the Bjørnøya Basin. There is no corresponding boundary fault in the north and the curved faults might continue in a northwesterly direction. The data quality is not good enough to substantiate this hypothesis.

The geoseismic section along the axis of the Bjørnøya Basin (see Figs. 18 and 11) illustrates the general westward dip of the basin axis. In the western part of the basin, the youngest basin fill is intensely faulted by westward dipping listric faults, which are unrelated to the major Kimmerian faults. In the eastern part of the basin the Kimmerian faults became reactivated both in mid-Cretaceous and Tertiary. The Tertiary faulting was strong and resulted in the formation of new faults as seen east of shotpoints 2000 and 3000 in Fig. 18. In the west the Tertiary faulting, which took place in several phases until the Miocene, resulted in basin inversion. Due to the southern boundary fault, the Bjørnøya Basin has the character of an asymmetric basin in a N—S direction.

The profiles across the basin (Figs. 17 — 19) show that the young faulting is much more intense along the northern margin than the southern one. The geoseismic sections across the basin show that the Middle Jurassic (D′) and mid Cretaceous (C) sequence boundaries converge towards the north, but they do not meet within the presently defined

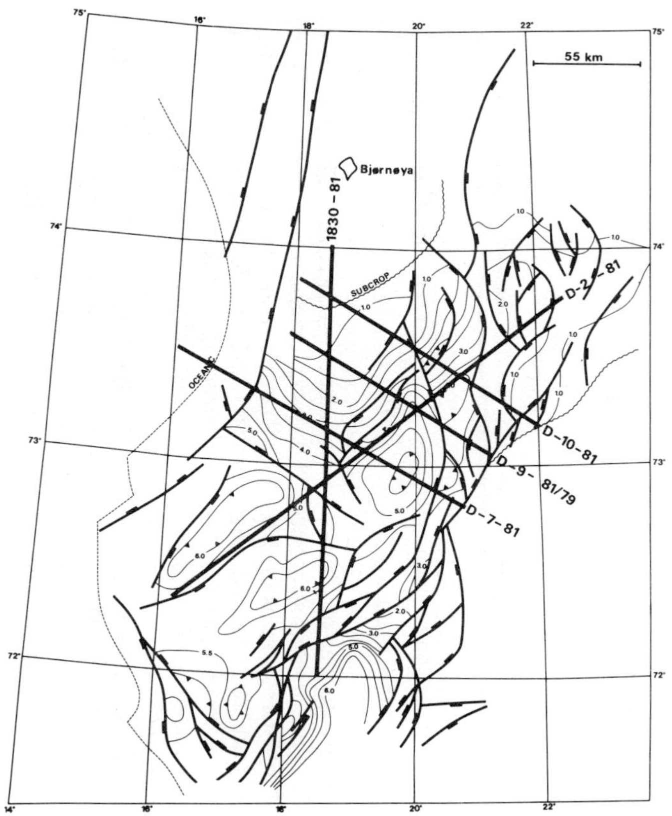

Fig. 11. Isochrone map of Horizon D^1 in the Bjørnøya Basin (Middle Jurassic). Two-way time in seconds.

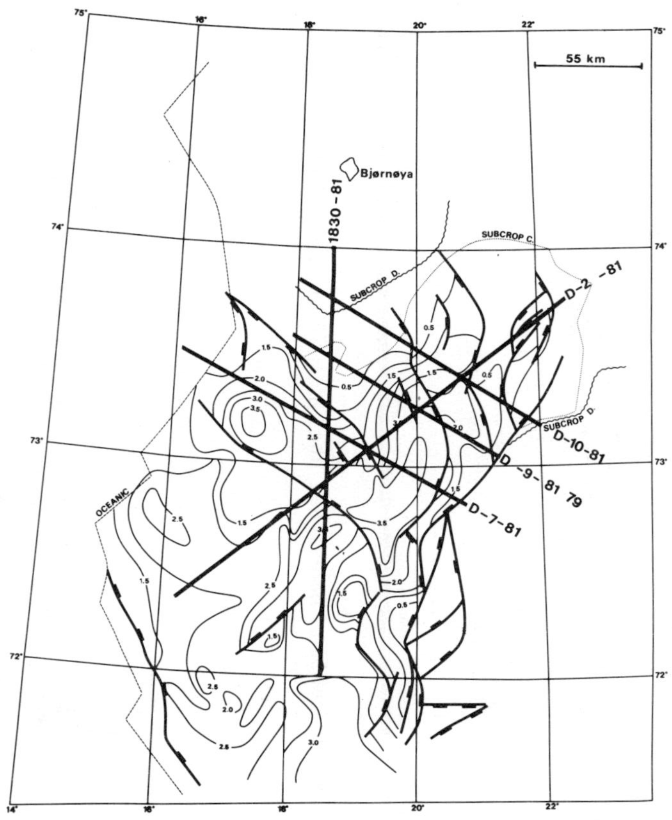

Fig. 12. Isopach map between Horizons D^1 and C (Middle Jurassic to Lower Cretaceous) in the Bjørnøya Basin. Two-way time in seconds.

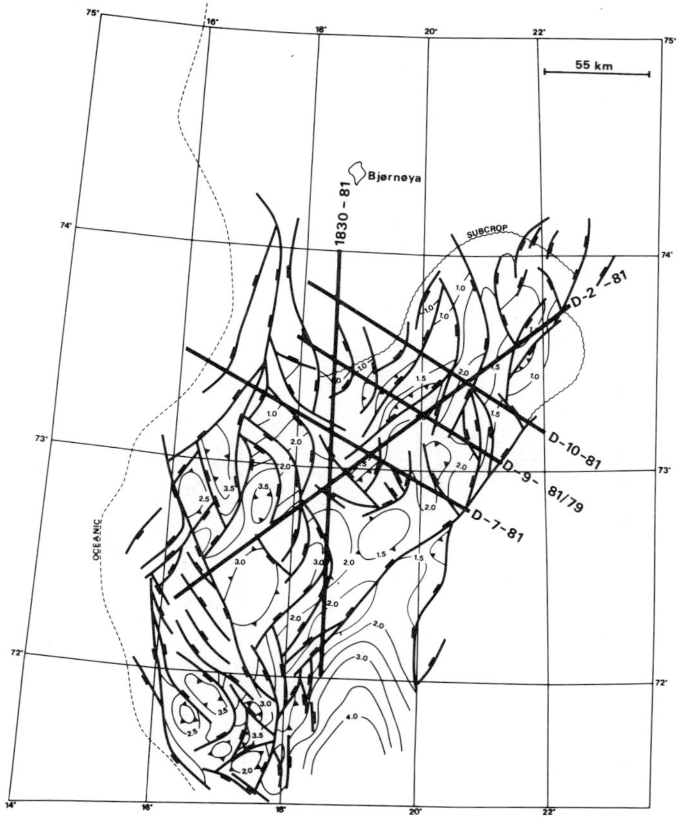

Fig. 13. Isochrone map of Horizon C (top Lower Cretaceous) in the Bjørnøya Basin. Two-way time in seconds.

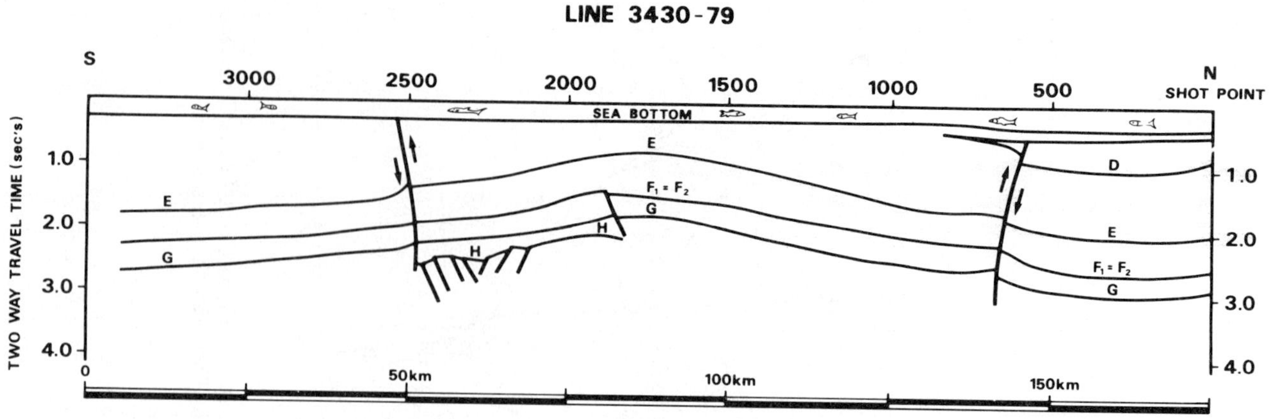

Fig. 14. Geoseismic section of Barents Sea, line 3430—79 (see Fig. 1 for location) showing the Sentralbank High, bounded by reverse faults.

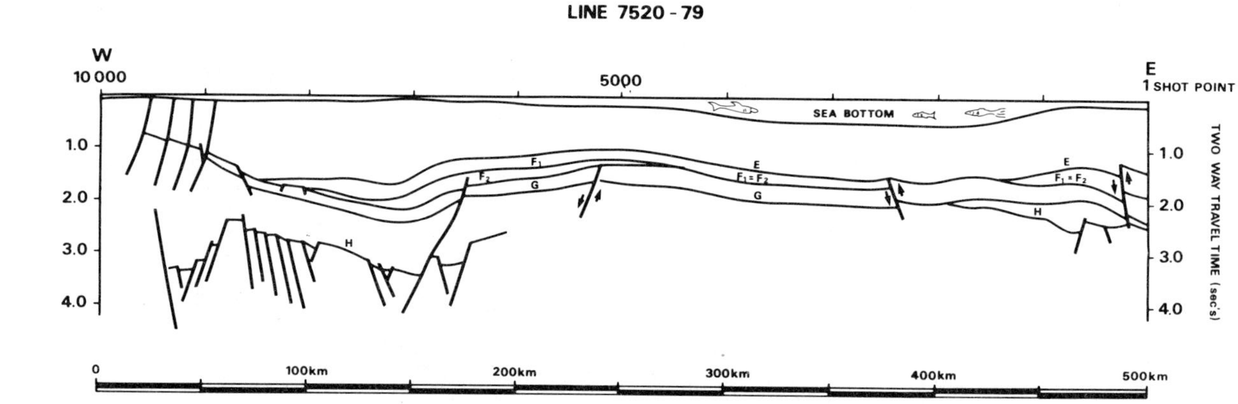

Fig. 15. Geoseismic section of Barents Sea, line 7520—79 (see Fig. 1 for location).

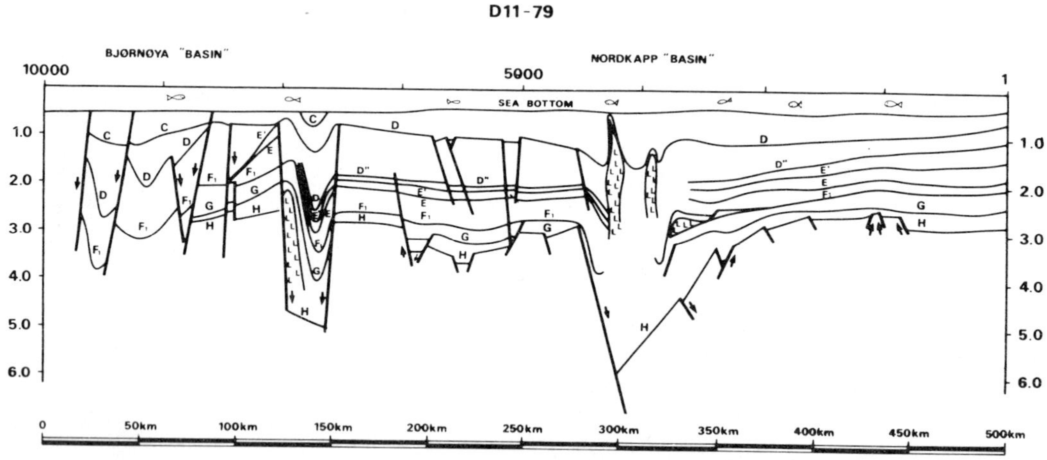

Fig. 16. Geoseismic section of Barents Sea, line D-11—79 (see Fig. 3 for location).

basin limit. The northern basin hinge was originally situated further to the north. The present subcrop is the result of a marked post mid-Cretaceous uplift.

The mid Cretaceous (C) and top Palaeocene sequence (B) boundaries also converge towards the north; they converge close to each other and might be close to the palaeo-hinge. The clinoforms in the sequence above the top Palaeocene sequence boundary indicate a transport component from the north. The area to the north of the subcrop of the mid Cretaceous has been a provenance area since the earliest Tertiary. The organic material in dredged Lower Cretaceous rocks south of Bjørnøya shows vitrinite reflectivity of about 0.6%. This maturity is an independent indication of an uplift after some degree of burial.

East of 20°, the Bjørnøya Basin has been limited to the south by a marked boundary fault since Middle Jurassic. The Loppa Ridge also limits the basin to the south here. Further west the southern limit of the basin is marked by the Senja Ridge. The N—S cross section (Fig. 19) demonstrates that

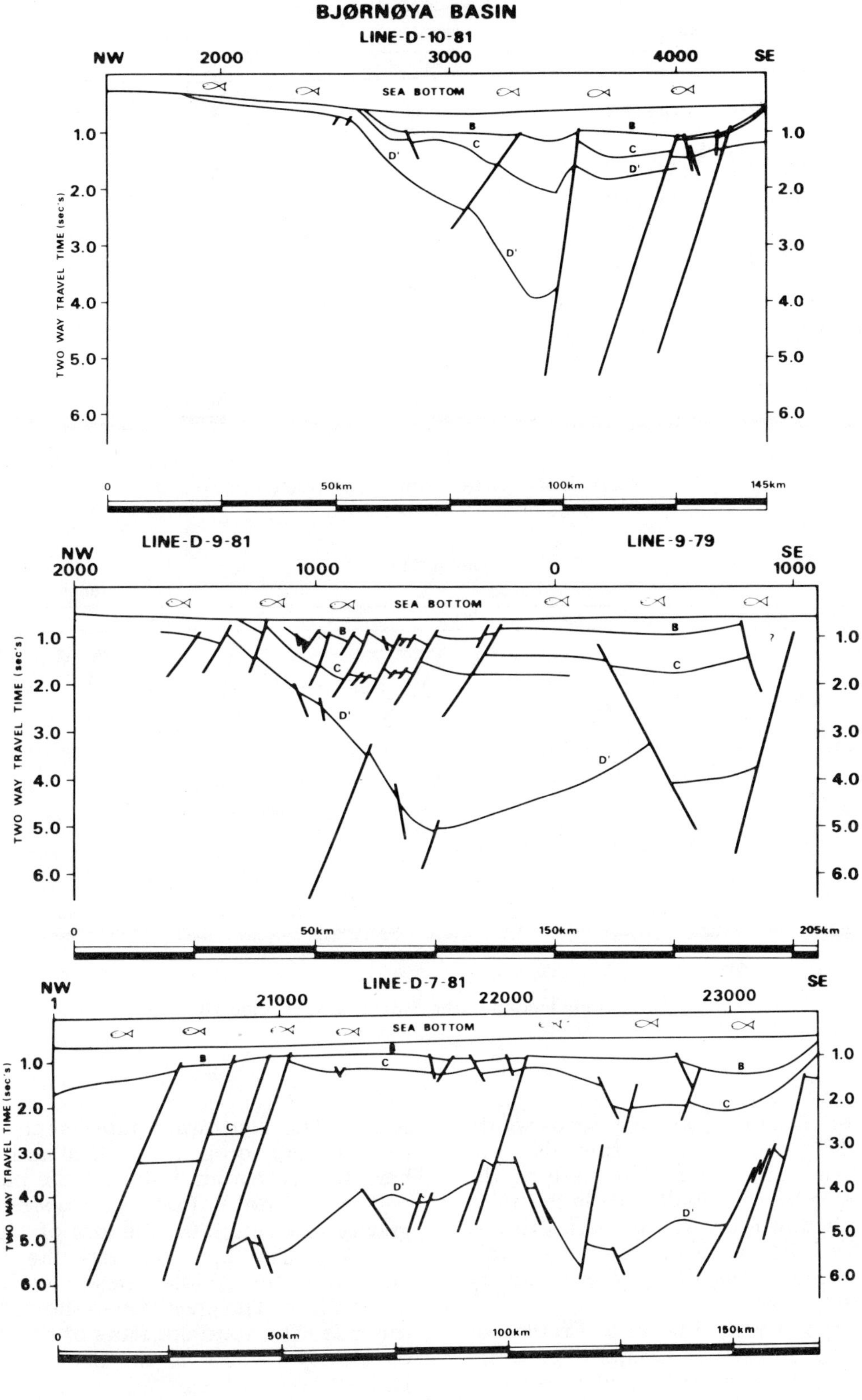

Fig. 17. Geoseismic sections in the Bjørnøya Basin: lines D-10—81, D-7—81, D-9—81/79.

the Senja Ridge is underlain by thick Upper Jurassic to Lower Cretaceous strata. The isopach map (Fig. 12) shows that the whole area to the west of the hinge zone was a depocentre in pre mid-Cretaceous time. There existed a direct continuation from the Tromsø Basin into the Bjørnøya Basin. The major N—S depocentre reaches more than 3.5 seconds in two-way time thickness. Lower Cretaceous constitutes the major part of the basin fill. In the north the map indicates a marked NW—SE basin trend limited to the west by a major down to the east fault. There is a thickness increase across the curved down to the west faults. These faults converge with the NE—SW oriented fault and continue into the hinge zone to the west of the Loppa Ridge and transition between the Tromsø Basin and the Hammerfest Basin.

The depocentre under the Senja Ridge from Middle Jurassic to mid Cretaceous demonstrates clearly that the ridge is a mid Cretaceous and

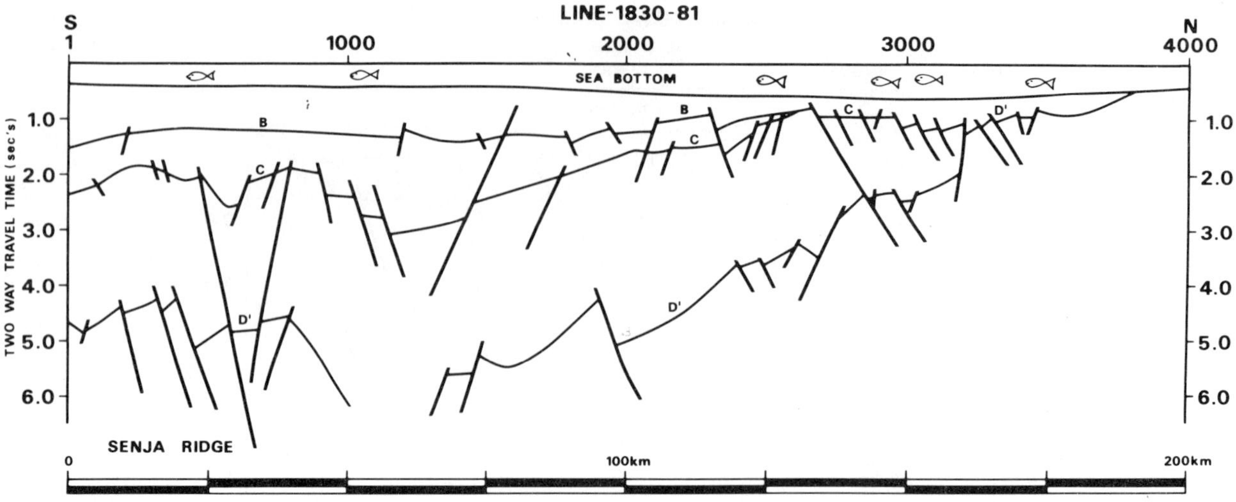

Fig. 18. Geoseismic section in the Bjørnøya Basin: line D-2—81.

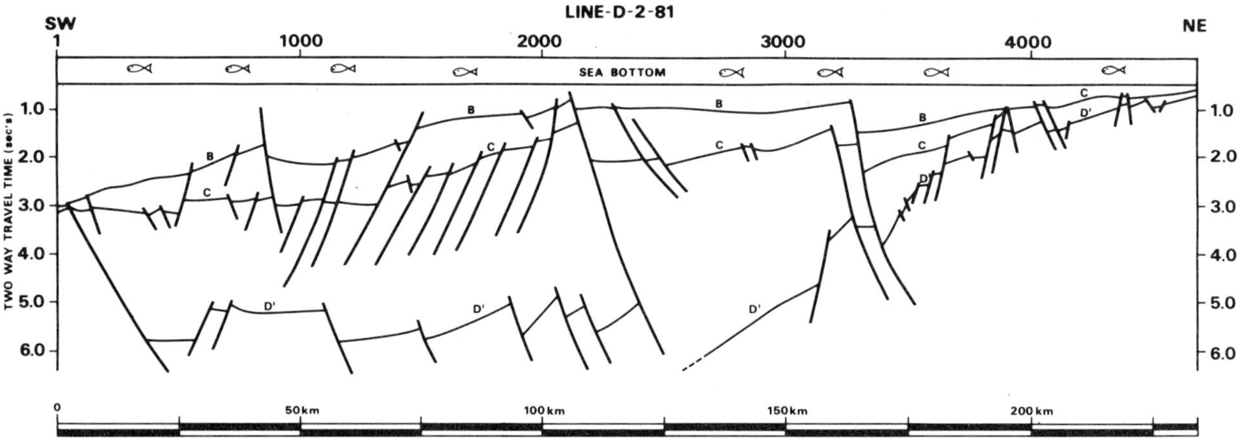

Fig. 19. Geoseismic section in the Bjørnøya Basin: line 1830—81.

younger structure. The ridge, which extends southwestwards to 18° and then southwards, was tectonically active until the middle part of the Tertiary. The major NE—SW fault-zone in the north shows faults both down-thrown to the north and to the south. This fault zone is interpreted as a mid Cretaceous extension of the Kimmerian boundary fault of the Bjørnøya Basin.

In the Bjørnøya Basin, the mid Cretaceous isochrone map (Fig. 13) shows a major depocentre west of 18°. The depocentre is limited to the east by a marked curved fault zone that extends into the Senja Ridge in the south. The map also shows a second depocentre east of 19°. The two subbasins are separated by a complex high trend, which is in the focal point of the NE—SW and NW—SE fault trends. The high is related to compressional tectonics. A similar focal point of fault trends and compressional tectonics is observed at the northeastern end of the Bjørnøya Basin and in the Senja Ridge. The compressional movement seems related to the southward movement of a stable block to the north of the Bjørnøya Basin. The major part of this southward movement is taken up as compression in the basin and the Senja Ridge. The southward movements are related to the Tertiary orogeny on Svalbard. The orogeny here was active until the middle part of Tertiary.

The Bjørnøya Basin corresponds to a major gravity low (Fig. 20), the axis of which is oriented NE—SW. The axis parallels the structural axis defined at the Middle Jurassic level, but is shifted northwards. The gravity map shows two lows along the axis. The southern flank of the basin is marked by gravity highs over the Senja Ridge and Loppa Ridge.

SUMMARY

The structural development of the Barents Sea can be summarized as follows:

(1) Upper Devonian—Lower Carboniferous graben formation as a result of a major sinistral wrench in the west and a conjugate dextral wrench through the central part of the Barents Sea.

(2) Epeirogenic subsidence of the whole of the area as one basin in Upper Carboniferous and Lower Permian.

(3) Middle Permian elevation of the western rim

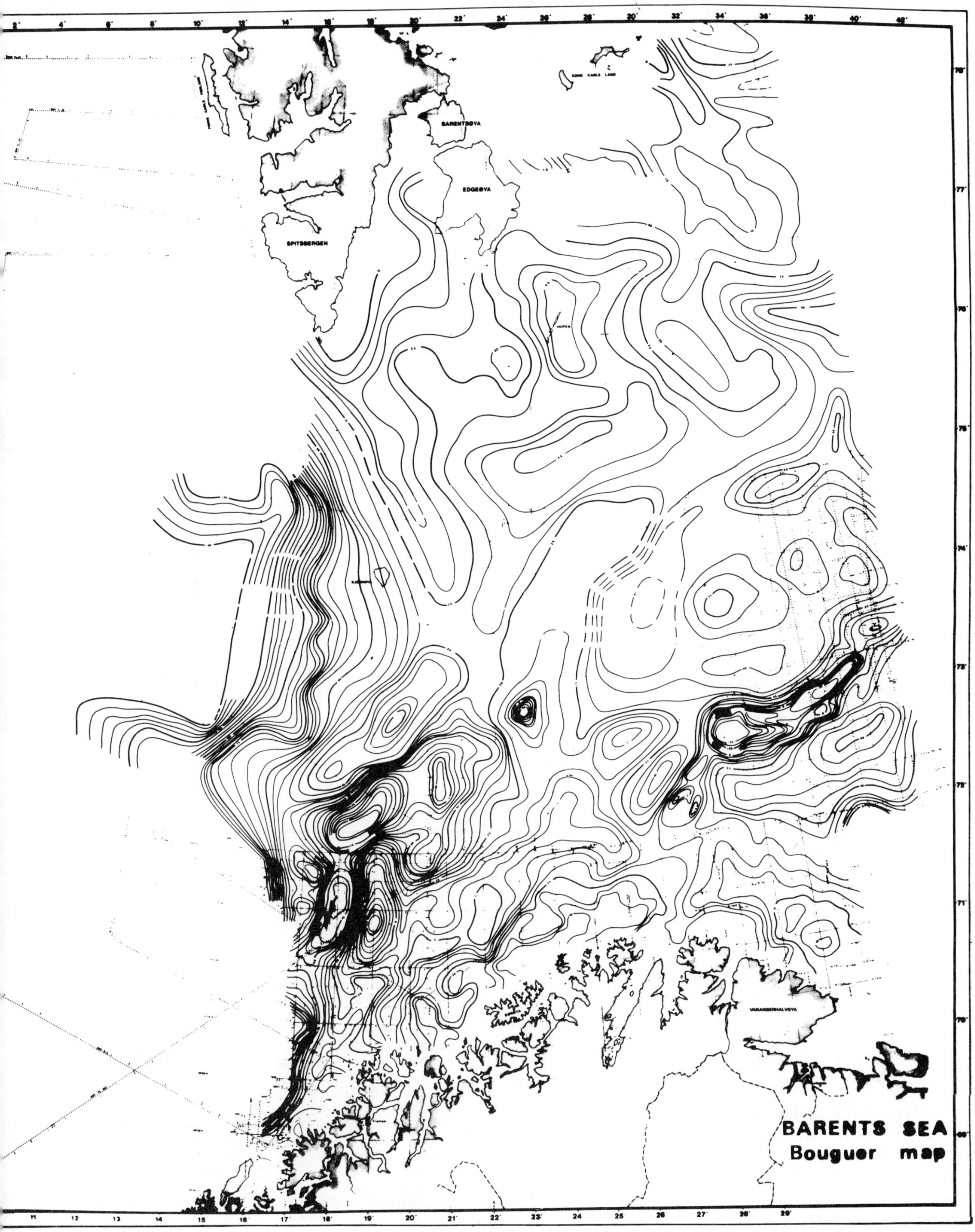

Fig. 20. Bouguer map.

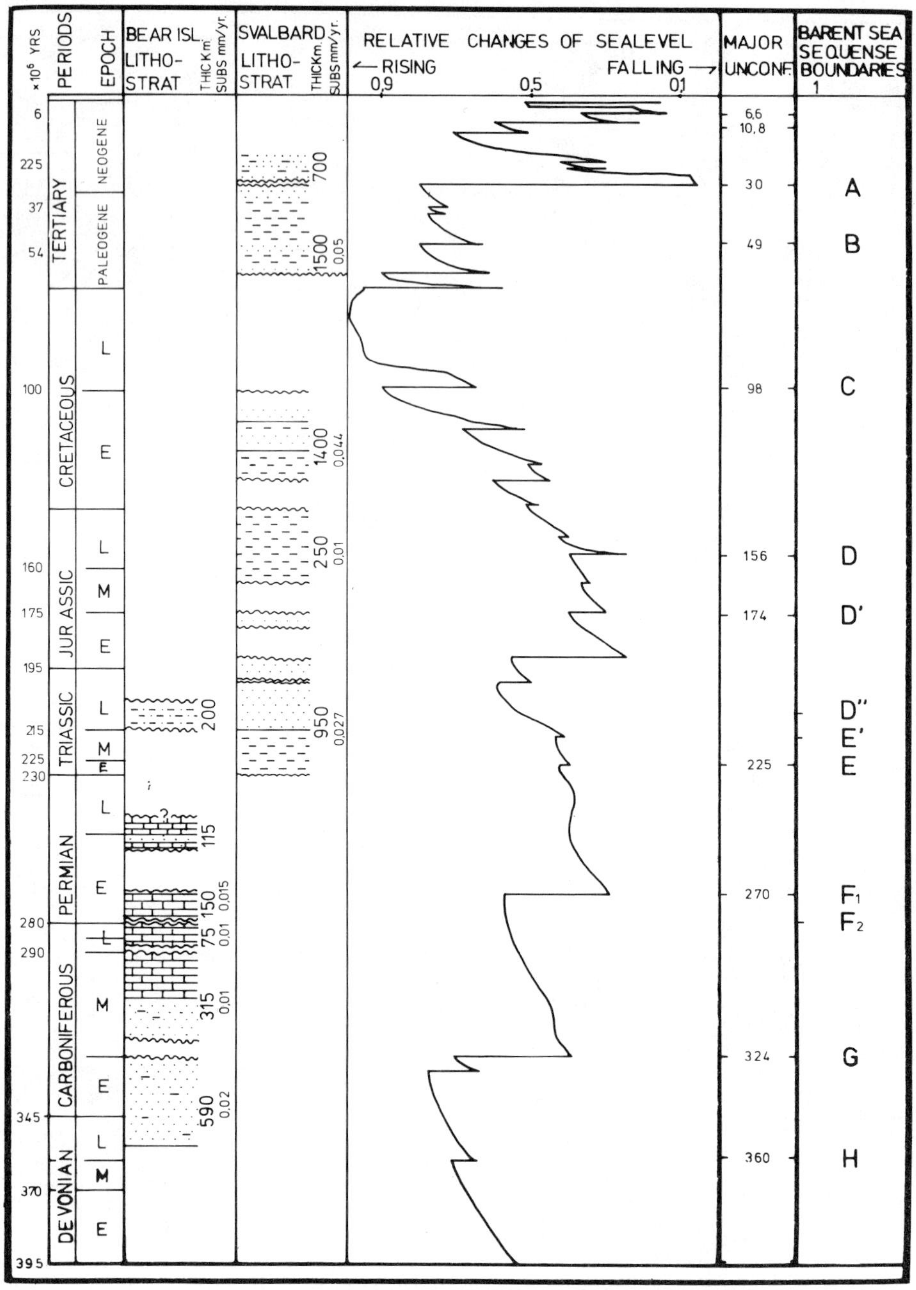

Fig. 21. Ages of sequence boundaries.

of the Barents Sea Basin.

(4) Early Middle Kimmerian wrenching in the western and northern Barents Sea with formation of intrabasinal highs and basins. The wrenching was related to a dextral N—S system in the west and conjugate NE—SW wrenches into the Barents Sea.

(5) Mid-Cretaceous to mid Tertiary structuring of the Senja Ridge as a result of a southward movement of the area from Svalbard to the Bjørnøya Basin.

REFERENCES

Gjelberg, J. and Steel, R. 1979. Middle Carboniferous sedimentation in relation to tectonic climatic and sea level changes on Bjørnøya and Spitsbergen. *Norwegian Sea Symposium.* Norwegian Petroleum Society NSS/27.

Gloppen, T.G. and Westre, S. 1982. Petroleum potential big off northern Norway. *Oil Gas J.* **80**(23), 114—127.

Harland, W.B. 1969. Contribution of Spitsbergen to understanding of tectonic evolution of North Atlantic Region. *Am. Assoc. Petrol. Geol., Mem.*, **12**, 817—851.

Mørk, A. and Worsley, D. 1979. The Triassic and Lower Jurassic succession of Svalbard. A review. *Norwegian Sea Symposium.* Norwegian Petroleum Society NSS/29.

Nalivkin, D.V. 1960. *The Geology of the USSR. A Short Outline.* Pergamon Press, London.

Rønnevik, H.C. 1981. Geology of the Barents Sea. In: L.V. Illing and G. Hobson (eds.), *Petroleum Geology of the Continental Shelf of North Western Europe. Heyden,* London, pp. 395—406.

Rønnevik, H.C. 1982. *Structural and Stratigraphic Evolution of the Barents Sea.* NPD-Contribution No. 1, Norwegian Petroleum Society, ONS-82.

Ziegler, P.A. 1978. North Western Europe: tectonics and basin development. *Geol. Mijnb.*, **57**, 589—626.

3 The Askeladden gas find – Troms I

Steinar Westre
Statoil, Harstad, Norway

The Askeladden gas field lies in the central part of block 7120/8 in the Troms I area, approximately 100 km off the north Norwegian coast. The gas field covers a large part of the block. So far two wells have been drilled in the licence, and different gas—water contacts (GWC) have been found in the two wells. Reservoir rocks in the Askeladden field are sandstone sequences of mid and early Jurassic age. The post-mid Jurassic sediments contain no reservoir rocks, and consist predominantly of claystones, occasionally with streaks of limestone, dolomite and sandstone. Based on sedimentary structures and textures in the cores, the Middle and Lower Jurassic strata in the two wells are interpreted as a generally transgressive marine clastic coastal and inner shelf sequence. Below the reservoir lies several hundred metres of sandstones interbedded with minor shale layers.

Source rock analysis from the two wells drilled in the licence shows that the best potential source rocks for oil and gas are the Upper Jurassic shales. Hauterivian to Triassic shales are also good source beds for gas. Gas composition analysis shows that the gas is relatively dry, containing approximately 5% CO_2.

INTRODUCTION

In the autumn of 1979, 20 blocks on Tromsøflaket were opened for allocation. In spring 1980 block 7119/12 was awarded to a group of companies with Statoil as operator. The results from the first well drilled in that block confirmed most of our assumptions, but the result was disappointing on one important point. The oil and gas that had probably once been in the reservoir, was there no longer. It had probably escaped by leakage along faults. All the information and experience from the first drilling season were then used in a re-evaluation of the Troms I box, and when the government, in late 1980, asked the oil companies to update their applications, Statoil chose block 7120/8 as their first priority block.

Block 7120/8 or Production Licence 064 (PL 064) was then awarded on 27 March 1981 to the following group of companies:

Den Norske Stats Oljeselskap a.s. (operator)	50%
Esso Exploration and Production Norway A/S	25%
Norsk Hydro Produksjon A/S	15%
Elf Aquitaine Norge A/S	5%
Phillips Petroleum Company Norway	5%

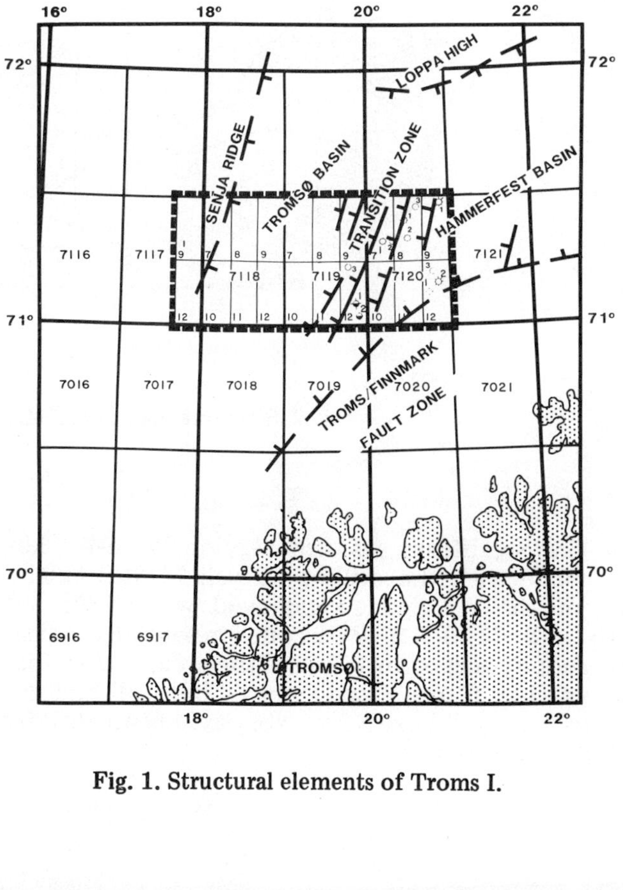

Fig. 1. Structural elements of Troms I.

Petroleum Geology of the North European Margin © Norwegian Petroleum Society (Graham & Trotman, 1984) pp. 33—39.

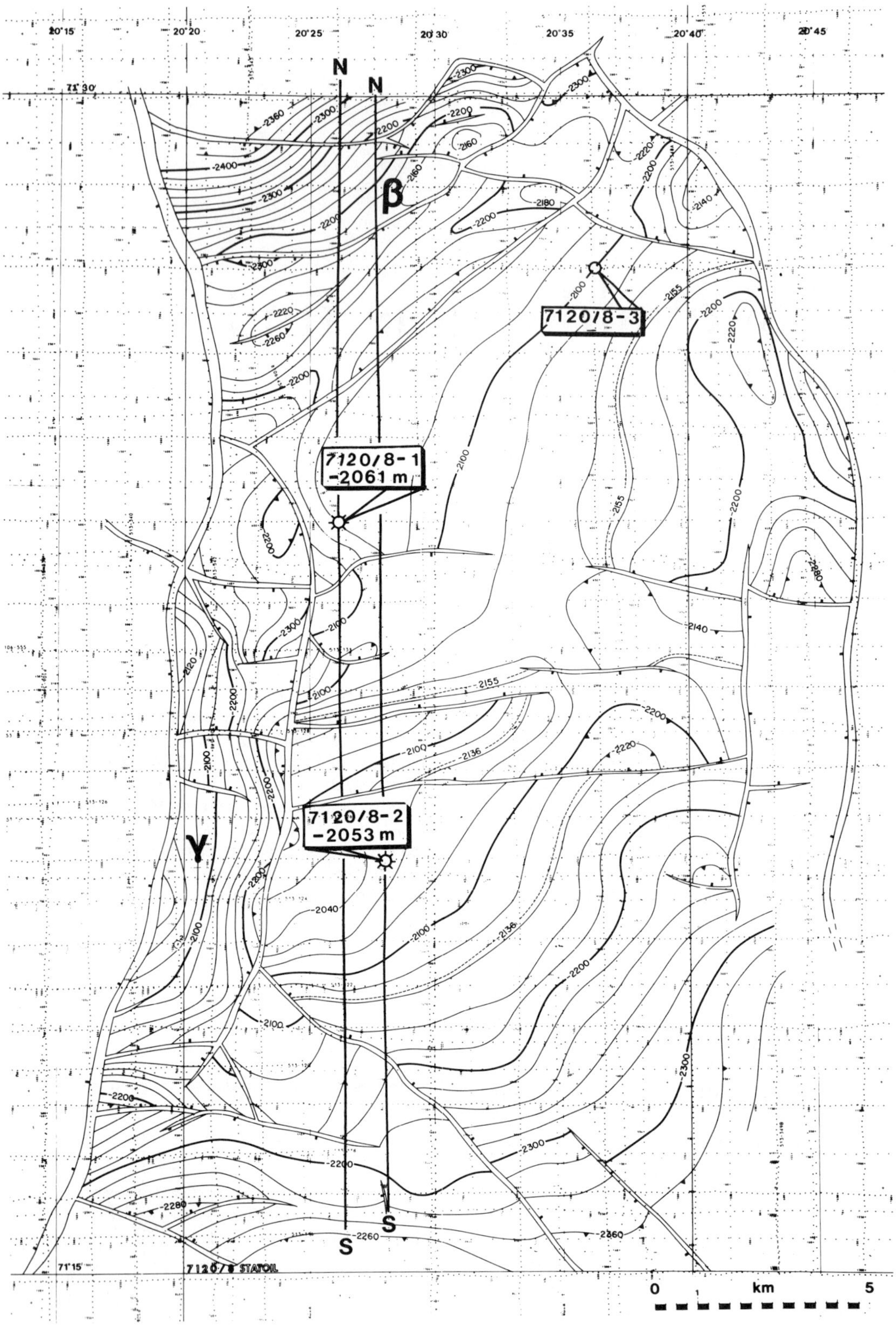

Fig. 2. Structure map of top Middle Jurassic. (Contour interval 20 m.)

The first significant gas discovery on Tromsøflaket was made with the drilling of well 7120/8-1. The size of the Askeladden gas field was confirmed with well 7120/8-2. With our present geological and geophysical knowledge of the area, it is almost certain that future discoveries of both oil and gas will be made offshore North Norway. Future drilling will give the answer.

TECTONICS

Block 7120/8 is situated in the Hammerfest Basin close to the transition zone between that basin and the Tromsø Basin (Fig. 1). The Hammerfest Basin is an E-trending basin with large amounts of Upper Palaeozoic to Cenozoic sediments. Two main fault patterns are recognized in the Hammerfest Basin, one more or less N—S and the other E—W (Fig. 2).

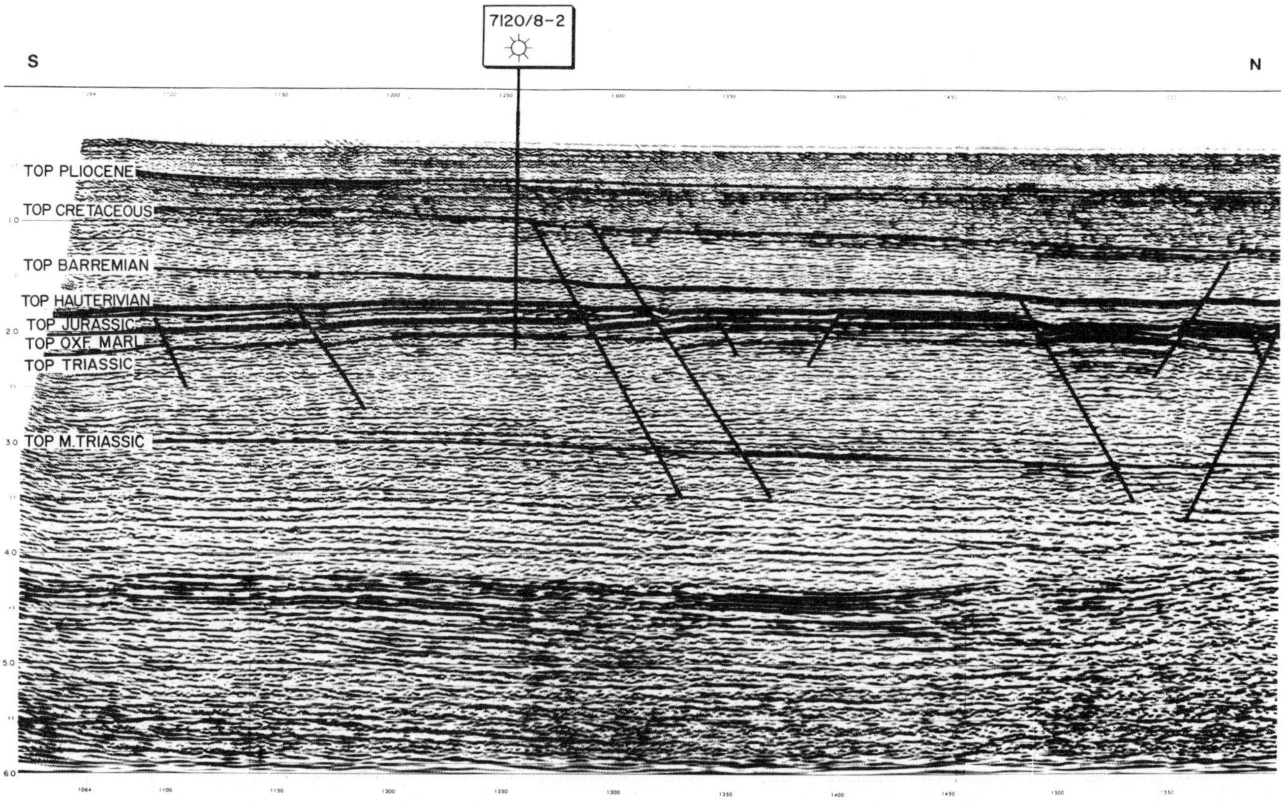

Fig. 3. Seismic line ST 8106—352 through well 7120/8-2. (See Fig. 2 for location.)

Due to poor data resolution, the tectonics of the pre-Middle Triassic section is not well known. So far the area seems relatively undisturbed by Hercynian as well as early Kimmerian tectonic events. However, near the end of mid-Jurassic times, some doming took place in the Hammerfest Basin. A low-relief anticline, accompanied by minor E-trending normal faults, was created. These faults die out downwards in Lower Jurassic/Upper Triassic strata (Fig. 3). The movements were probably initiated during the main Kimmerian orogenic event and seem to have ceased near the end of late Jurassic times.

The main tectonic phase in the area was, however, the tensional regime introduced at the Jurassic—Cretaceous boundary. This fits well with the late Kimmerian tectonic event. A series of large NNW-trending normal faults were created, probably along weakness zones in the basement. The main phase of movements seems to have ceased in pre-Cenomanian time, although a few of the larger faults have been active several times and may be traced on seismic sections through the top Cretaceous reflector (Fig. 3). Some of these faults are found to be associated with shows of 'heavy' hydrocarbons in the Tertiary section, and may thus be interpreted as being conductors of fluids and gases from deeper levels.

SEISMIC INTERPRETATION

Since 1974 Statoil has continuously updated the evaluation of the Troms I area. New interpretations have been made as seismic coverage, the quality of the seismic data, and our geological understanding of the area have improved. The latest seismic interpretation and mapping of block 7120/8 is based on a 1 × 1 km seismic grid of data shot in 1975 and 1981. The 1975 data were reprocessed in 1982. The quality of these data is good. The interpretation has been carried out on migrated sections. This has not led to any tie problems due to the gentle dip of the structures.

The following horizons have been interpreted and mapped (see Fig. 3): top Pliocene; top Cretaceous; top Cenomanian; top Barremian; top Hauterivian; top Jurassic; top Oxfordian Marl; top Middle Triassic.

LITHOSTRATIGRAPHIC CORRELATION

The gamma-ray and sonic logs from 7120/8-1 and 7120/8-2 are correlated as shown on Fig. 4. For most formation tops there are good similarities of the log response in both wells.

The main lithology in the Quaternary and Tertiary sections is grey claystone with stringers of siltstone and dolomite, except for the uppermost very sandy and gravelly part. In the Cretaceous, grey claystones are interspersed with limestone and dolomite stringers that often create good markers. For example, top Barremian is marked by an increase in both velocity and density. Top Jurassic is marked by a gamma-ray increase where the more radioactive Jurassic shales begin. The Middle Jurassic sandstone comes in at a sharp gamma-ray decrease, while top Oxfordian Marl is less easy to correlate, but is picked where the gamma-ray starts to decrease and the velocity to increase.

BIOSTRATIGRAPHY

The palaeontological datings of the two wells are generally in good accordance with the log correla-

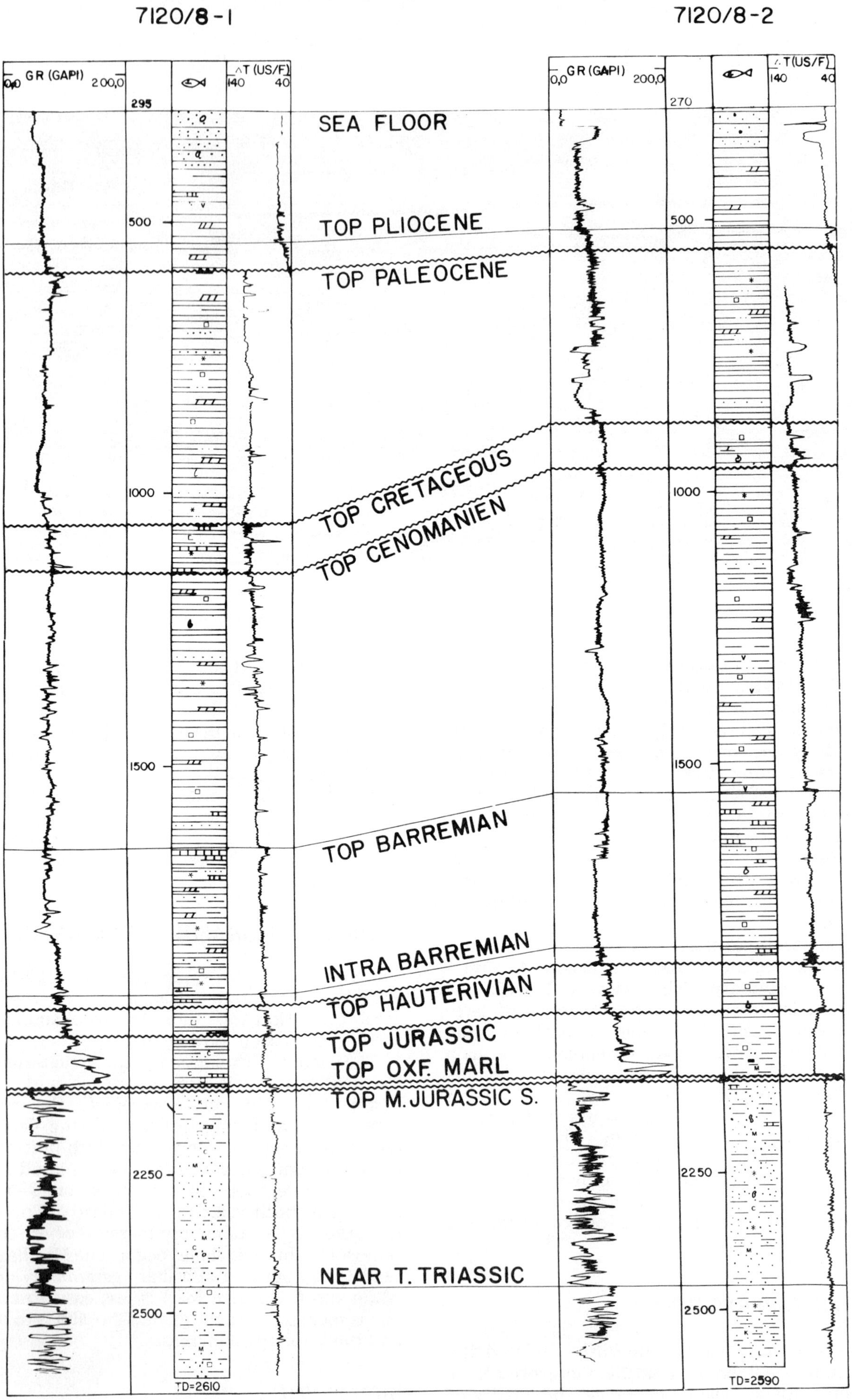

Fig. 4. Stratigraphic log correlation, Askeladden field.

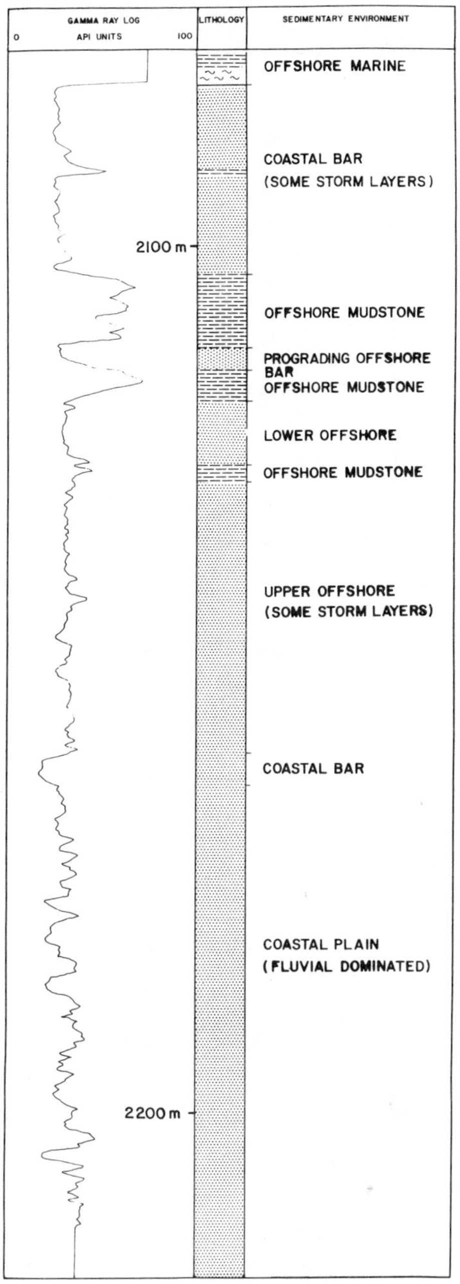

Fig. 5. Main reservoir section of well 7120/8-2.

tion. The most important exceptions are the dating of top Pliocene, which is uncertain due to reworking (the log top is also difficult to pick), and top Barremian, which could also be of Albian/Aptian age.

The biostratigraphic work on the two wells indicates that sediments of Pliocene age rest unconformably on early Eocene sediments. There is no good evidence for the presence of rocks of Miocene, Oligocene or late and Middle Eocene age. Traces of tuffaceous claystone have been reported and dated as early Eocene/late Paleocene. Sediments of late Paleocene age rest unconformably on Campanian claystones. There is no evidence of rocks of early Paleocene or Maastrichtian age. No palaeontological evidence has been recorded for rocks of Santonian to Turonian age, and the Companian claystone is therefore probably unconformable on Cenomanian claystone. A full sequence of beds from Cenomanian to Hauterivian age can be recognized. Early Cretaceous rocks of Hauterivian age rest unconformably on organic rich claystones of Ryazanian age. A sequence of organic rich claystones ranging in age from Ryazanian to early Portlandian have been recognized.

A very thin marly interval (less than 10 m) of earliest Kimmeridgian/Oxfordian age is present. The interval is probably bounded by unconformities at both the top and the base, since there is no good evidence for the presence of late Kimmeridgian sediments, and it directly overlies sands considered to be Middle Jurassic in age. The Middle and early Jurassic sequence comprises mainly sandstone. The sequence ranges in age from early Bajocian to Hettangian. The remainder of the sequence penetrated by the two wells is dated as Upper Triassic (Rhaetian to Norian).

SEDIMENTARY ENVIRONMENT

Based on sedimentary structures and textures in the cores, the Middle and Lower Jurassic in the two wells is interpreted as a generally transgressive marine clastic coastal and inner shelf sequence. The upper reservoir section in 7120/8-1 was deposited under the influence of relatively high energy, while the same section in 7120/8-2 was deposited in a more protected position, and the sediments are more strongly bioturbated (Fig. 5).

RESERVOIR CONTINUITY

So far two wells have been drilled in block 7120/8, but the different sedimentological units seem to be relatively uniform across the eastern Troms I area. If the depositional environment is taken into consideration, i.e. an overall transgressive regime with a very delicate balance between sediment input and basin subsidence, it is reasonable to expect only minor thinning or thickening of the different reservoir units across the Askeladden field. However, some variations in facies development, and thus reservoir parameters, will of course occur.

WATER ZONE

Below the reservoir lie several hundred metres of sandstones interbedded with minor shale layers. The shales are probably not very extensive laterally, and there will therefore probably be communication between the sandstone layers. Porosity and permeability are not as good as in the reservoir, but probably good enough to give acceptable water flow. The horizontal water flow is probably best because of the good and laterally extensive reservoir sandstone bodies.

RESERVOIR PROPERTIES

Originally, the reservoir rock in the Troms I area must have had excellent properties, due to the circumstances under which it was deposited. The sand was clean, well sorted and had favourable grain sizes. During burial, three main factors worked together to impair reservoir quality:

(1) Unfavourable burial history, i.e. the reservoir rock was buried to great depths early on.

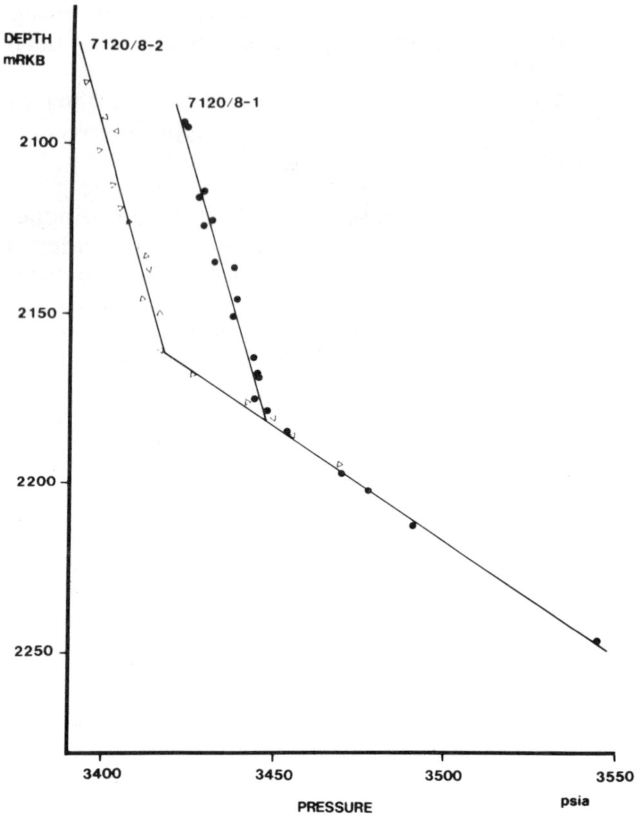

Fig. 6. RFT Pressure comparison, wells 7120/8-1 and 8-2.

(2) Lack of overpressures in the structures during most of the burial, due to lack of effective seal. Leakage was probably caused by late movements along fault planes.

(3) No hydrocarbons in the structures for the same reason as above.

The result is that the reservoir properties in parts of the Troms I area appear to have been markedly reduced. Fortunately, this seems to apply only to the deeper parts of the transition zone and probably also to the Tromsø Basin. In block 7120/8 a more favourable burial as well as tectonic history has to a great extent led to the preservation of good reservoir properties. Hence, the net pay average porosity in the Askeladden field is approximately 15%.

Evaluated log porosity compares well to measured core porosity, but is in some intervals slightly lower than the measured values. This is quite normal and is believed to be caused by overburden pressure and clay-bound water in the pores.

GAS AND WATER PRESSURE REGIMES

Gas gradients of 0.204 g/cm^3 and 0.203 g/cm^3 were obtained in the reservoirs from RFT-measurements in wells 7120/8-1 and 7120/8-2, respectively. The water gradient is calculated to be 1.101 g/cm^3 in both wells (Fig. 6). This indicates that there is pressure communication between the northern and southern compartments of the Askeladden field in the water zone, but not in the gas zone. This means that the Askeladden field is divided into a northern and a southern structure, and the point of separation is as indicated on Figs. 2 and 7.

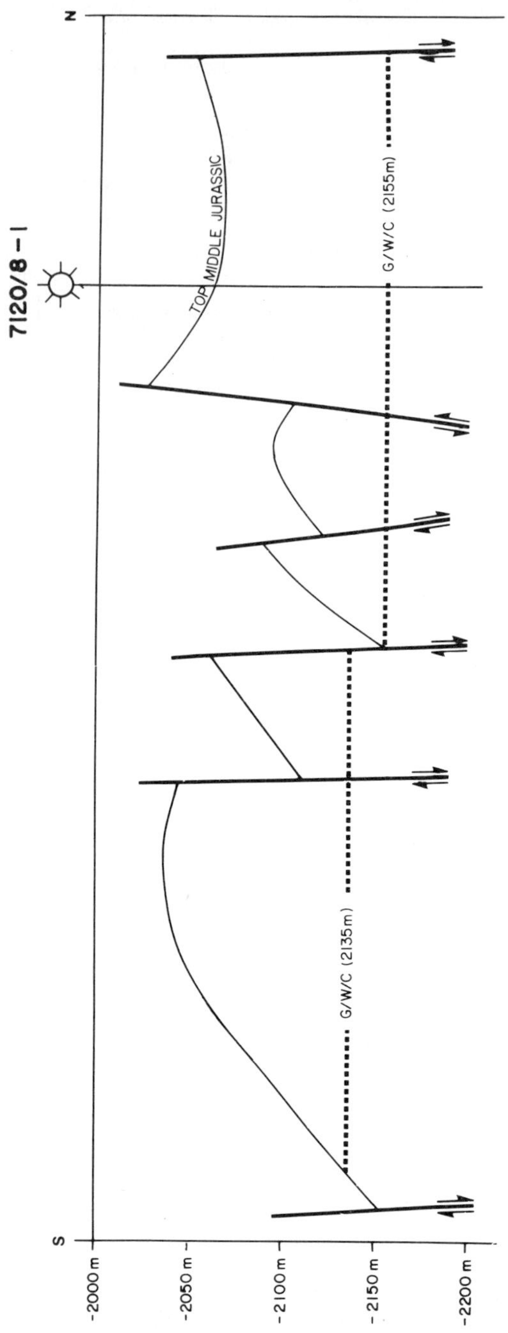

Fig. 7. Structural cross-section along line 513–344 through well 7120/8-1 (See Fig. 2 for location.)

GAS COMPOSITION

Gas composition analysis of hydrocarbons from the Drill Stem Tests performed in the two wells (Table 1) shows that the gas compositions are

TABLE 1. Test results (maximum flow)

Well 7120/8-1	
Gas rate:	953 100 Sm^3/D
Condensate rate:	50 Sm^3/D
Choke size:	25 mm
Test tubing:	127 mm
Well 7120/8-2	
Gas rate:	470 400 Sm^3/D
Condensate rate:	17.5 Sm^3/D
Choke size:	17.5 mm
Test tubing:	89 mm

almost identical in the northern and southern compartments of the Askeladden field. This was also indicated by the gas gradients established from the RFT-pressure points taken in the two wells. The gas is relatively dry and contains approximately 5% CO_2 and no H_2S.

SOURCE ROCKS

Source rock analysis from the two wells drilled in the licence shows that the best potential source rocks for oil and gas are the Upper Jurassic shales. Hauterivian to Triassic shales are also good source beds for gas.

The post-Hauterivian claystones have generally a poor potential for hydrocarbons, with mostly type III and IV kerogen. The Upper Jurassic shales have high Total Organic Carbon (TOC) values which increase downwards. The kerogen is mainly mixed type II and III. The Lower Jurassic to Upper Triassic sequence consists of sandstone with subordinate layers of claystone and coal. The claystone is partly carbonaceous. The claystones have fairly high TOC values; type III kerogen predominates in this section. The interbeds of coal contain type III kerogen and have a rich potential for gas.

Geochemical analysis indicates that the gas in the Askeladden field is derived from a gas-prone 'non-marine'-type source rock of Lower Jurassic/ Upper Triassic age. Contributions from older, deeper sources, however, cannot be ruled out, due to lack of data. In no way, however, does this gas resemble late-stage gas generated from an oil-prone source rock like the Upper Jurassic shales.

The first migration of gas into the reservoir rocks in the Askeladden field probably started in late Cretaceous or early Tertiary times.

ACKNOWLEDGEMENTS

The author wishes to express his gratitute to Statoil and the partners in PL 064, Block 7120/8, for permission to publish this paper. Special thanks are also due to the colleagues in Statoil's exploration department for valuable discussions.

REFERENCE

Gloppen, T.G. and Westre, S. 1982. Petroleum potential big off northern Norway. *Oil Gas J.* **80** (23), 114—127.

4 Nomenclature of the main structural features on the Norwegian Continental Shelf north of the 62nd parallel

Roy H. Gabrielsen, Roald Færseth†, Georg Hamar‡ and Hans Rønnevik§*
*Saga Petroleum, Høvik, Norway; †Norsk Hydro, Sandvika, Norway
‡ Statoil, Forus, Norway; § Norwegian Petroleum Directorate

A committee of nomenclature of structural features of the northern Norwegian Continental Shelf was formed in November 1982. The committee had the following aims: To present a list of names and definitions for the major tectonic features off-shore Noway north of the 62nd parallel; to propose nomenclature rules to be followed in the future for that area.

The committee has concluded this work, knowing that future adjustment will be necessary in the light of new data. Four main principles were used: The structural features should be named after a nearby geographical place, preferably an offshore name. The second part of the name should give information about the nature of the structure. The same geographical name should not be used for more than one tectonic feature. Names that have been defined or have been in common use in the literature should be given in preference to new ones.

Accordingly, the committee has tried to give accurate definitions of the various structural features, and as far as possible to avoid renaming. Only a few structures which had not been defined previously are proposed, although adjustments to existing structural features have been made, based on new knowledge of the area.

INTRODUCTION

Both the hydrocarbon potential of the Norwegian Continental Shelf north of the 62nd parallel and the interest of the academic world in the plate tectonics of the northern Norwegian Continental Margin has led to an explosion in research activity in this area in the last 15 years. Although the amount of new data collected in this period is enormous, we are still in a pioneering period in the geological understanding of the area, inasmuch as the area to be investigated covers approximately 850 000 km^2.

Nevertheless, a great many structural features have been discovered, described and named in the last few years, and the need for a systematic nomenclature and a summary of descriptions of these structural features has become increasingly clear. With this in mind, a committee was initiated in late autumn 1982, with one representative from each of the three Norwegian oil companies (Norsk Hydro, Saga Petroleum and Statoil) and a representative from the Norwegian Petroleum Directorate, to give summary definitions of the structural features on the Norwegian Continental Shelf north of the 62nd parallel.

The intention has been to define and to describe the major structural features, to summarize the present state of knowledge and to present a list of references, including the papers which contain the original description. In a few cases new names or redefinitions have been proposed. In most cases, the broadest definition of the structural elements has been used. This has been done to make subdivision of the structures possible, so that in the future a fault zone can be divided into several segments, a basin into several sub-basins etc.

Two main problems have been encountered. Firstly, at the present stage of investigation a summary will suffer from inhomogeneity in the data base: the drilled areas have been most comprehensively mapped. Secondly, the structural pattern has changed with time, so that the pattern of structural high and low areas are not the same at, e.g. top Permian, Middle Jurassic and base Cretaceous levels. Accordingly, the present proposals for nomenclature primarily relate to the structure at base Cretaceous level, although the earlier and later history of some of the structures is indicated.

Also, the magnetic and gravity anomaly signatures of the features have been noted. Even here,

Petroleum Geology of the North European Margin © Norwegian Petroleum Society (Graham & Trotman, 1984) pp. 41—60.

the data base is sparse for most of the area. The data surveyed are those presented by Åm (1970, 1975).

Nomenclature: principles and definitions

The following principles should be used in future names:

(1) Structures should be named after the nearest geographical place: island, fjord, fishing ground, etc.

(2) The Norwegian geographical name should be given priority on the Norwegian Continental Shelf, i.e. Bjørnøy Basin should be used instead of Bear Island Basin.

(3) The second part of the name should give a precise description of the nature of the structure (see 'Structural terms' below).

(4) The same geographical name should not be used for more than one structure: The multiple use of names (e.g. Vøring Basin, Vøring Plateau, and Vøring Escarpment) is unfortunate and should be avoided.

The committee recommends that the above principles are used in future. However, to avoid confusion, we do not propose to change names that have been in use in the literature for a long time, even if they are not in accordance with the above principles.

Structural terms

In the existing nomenclature of structures, the use of structural terms is to some extent arbitrary. For example, the terms 'ridge' and 'structural high' as well as 'fault zone' and 'fault system' are used side by side. A more precise use of structural terms is recommended. Thus, the broad term 'structural high' should be used when the nature of the structure is unknown, whereas the term 'ridge' should be used when the structure fits the strict definition of a 'ridge' (see below).

Thus a structure may be termed a 'structural high' at an early stage of investigation, but when more precise information is available, the correct term ('ridge', 'dome') can be applied. This principle would ensure that the name of the structure gives the reader maximum information about the nature of the feature.

The following definitions of structural terms are mainly based on Bates and Jackson (1980). It should be noted that not all the terms mentioned below are used in the present nomenclature proposal, but they are recommended for more detailed naming of structural features in the future. In general, the structural terms may be used to refer to a specific geological surface or time. Accordingly, the definitions, expecially of geomorphological features, are valid even when the features are not reflected in the present topography. Recommended Norwegian translations of the structural terms are included in parentheses.

High or structural high (Høyde or strukturell høyde): A general term for an elevated tectonic feature such as a crest, culmination, anticline, dome, horst, ridge, spur. Should be used when the precise nature of the feature is unknown.

Dome (Dom): A general term for any smoothly rounded rock-mass. The term is also applied to broadly up-archedregions. The structure is primarily not fault-bounded.

Horst (Horst): An elongate, relatively uplifted crustal unit or block that is bounded by faults along its long sides. It is a structural form and may, or may not, be expressed geomorphologically. Being a structural term, 'horst' is preferred to 'ridge', which is a geomorphological term (see below), when a feature is known to be fault-bounded.

Ridge (Rygg): (a) A general term for a long, narrow, geomorphological, elevation, usually sharp-crested with steep sides, occurring either independently or as a part of larger geomorphological element.
(b) (Marine geology): An elongate, steep-sided elevation of the ocean floor, having rough topography.

Spur (Utstikker): A subordinate ridge or lesser elevation that projects sharply from the crest or side of a hill, mountain or other high; a small hill extending from a prominent structural high.

Plateau (Platå): (a) (Geomorphological). Any comparatively flat area of great extent and elevation: an extensive region considerably elevated (generally more than 150 m in altitude) above the adjacent area or above sea level. It is commonly limited on at least one side by an abrupt descent, has a flat or nearly smooth surface but may be dissected by deep valleys and surrounded by highs. The largest part of its total surface is, however, at or near the summit level. A plateau may be of tectonic, residual or volcanic origin.
(b) (Marine geology) A broad, more or less flat-topped and ill-defined elevation of the sea-floor, generally over 150 m in elevation.

Platform (Plattform): The part of a continent that is covered by flat-lying or gently tilted strata (mainly sedimentary) and which is underlain at varying depth by a basement of rocks that were consolidated during earlier deformations. A platform is a part of a craton.

Basin (Basseng): A low area, tectonic in origin, in which sediments have accumulated (e.g. a circular centrocline, a fault-bounded intramontane feature, or a linear crustal downwarp). Such features were basins at the time of sedimentation, but are not necessarily so today.

Pull-apart-basin (Skjœr-basseng): Basin or depression associated with *en echelon* faults in a strike-slip fault system. The basins are usually rhomb-shaped and elongate parallel to the strike of the fault zone (see e.g. Rodgers, 1980).

Graben (Graben): An elongate, relatively depressed crustal unit or block that is bounded by faults on its long sides. It is a structural form that may or may not be geomorphologically expressed as a rift valley.

Rift (Rift): A long, narrow continental graben (or trough) that is bounded by normal faults; a graben of regional extent. It marks a zone along which the entire thickness of the lithosphere has been thinned or ruptured under extension.

Trough (Trau): (a) (Geomorphological) Any long, narrow depression in the Earth's surface.
(b) (Marine geology) An elongate depression of the

sea-floor which is wider and shallower than a trench, and with less steeply dipping sides. Troughs and trenches are gradational forms, a trough may develop from a trench by becoming filled with sediment.

Fault valley (Forkastningsdal): A linear depression produced by faulting; e.g. a small, narrow valley created within a major fault zone by relative depression of narrow slices, or a large graben situated between tilted block movements, or a valley produced by relative uplift on opposite sides of two parallel reverse faults.

Escarpment (Brattkant): A long, more or less continuous cliff or relatively steep slope facing in one general direction, breaking the continuity of an area by separating two levels or gently sloping surfaces. The escarpment is produced by erosion or by faulting. The term is often used synonymously with 'scarp'.

Fault zone (Fortkastningssone): A fault that is composed of more than one fracture or fault segment, frequently arranged in an *en echelon* pattern. This type of geometry is found on different scales in most faults.

Fault set (Forkastningssett): A group of faults that are parallel or subparallel, and that are related to a particular deformational episode.

Fault system (Forkastningssystem): Two or more interconnecting fault sets.

Fault complex (Forkastningskompleks): A group of faults that interconnect and intersect, having the same or different ages.

Splay fault (Splittforkastning): One of a series of minor faults at the extremities of a major fault. Splay faults are frequently associated with rifts.

Shear zone (Skjærsone): A zone of deformation characterized by complex fracturing due to shear strain. Geometrically, the zone may encompass several fracture sets (see above), folding, local unconformities and local depocentre (pull-apart-basins, see above). The main (synthetic) faults frequently are arranged in an *en echelon* fashion.

Fracture zone (Bruddsone): Elongate zone of unusually irregular topography that often separates regions of different depths on the ocean floor. Such zones commonly cross and apparently displace the mid-oceanic ridges by faulting.

STRUCTURAL ELEMENTS BETWEEN 62° AND 69°N

The Norwegian Shelf outside Møre-Lofoten is bordered to the west by the Tertiary oceanic crust of the Norwegian Sea and shows a development that is characteristic of passive continental margins.

Following the Caledonian orogeny which ended in late Silurian—early Devonian time, the uplifted mountain belt was subject to intense erosion. Coarse-grained fluvial sediments were deposited in intramontane basins in Devonian time and are preserved today in fault-bounded basins along the coast of western Norway. During the following geological periods up to mid Jurassic time there is no evidence for any major tectonic events in this area. Sediments of Triassic and Lower Jurassic age were deposited in an epeirogenic subsiding basin under conditions varying from continental to shallow marine.

From mid Jurassic time the area was subject to extensional forces which created a large number of generally normal faults with three main trends, N—S, NE—SW and ENE—WSW. This tectonic activity, combined with Cretaceous differential subsidence, created most of the structures described in this paper.

The structural picture of the area at base Cretaceous level shows a platform (the Trøndelag Platform) off the coast of Norway, which to the west and north is bordered by a basinal area. The transition between the platform and the basins is represented by a tectonically complex zone. Within each of these main structural elements, i.e. the platform, the fault complex and the basinal area, a number of sub-elements have been described.

In addition to the three main Mesozoic fault directions, a fourth (NW—SE) trend has been recognized. These faults resulted from Tertiary tectonic activity, although they might represent a reactivation in Tertiary time of older structures.

The main structural elements between 62°N and the Lofoten Islands are summarized in Figs. 1, 2, and 3.

Trøndelag Platform (Trøndelagplattformen)

Named after Trøndelag County.

References

The name was introduced by N. B. Hollander in talks given in 1981, but no written definition was given (Hollander, 1981). See also Aanstad *et al.* (1981).

Position

63°—67°N, 6°20′—14°E. The platform is bounded against the Caledonides of the Norwegian mainland by the Rana Fault Complex and the Møre Fault Zone, and is limited on the west and northwest by Kristiansund-Bodø Fault Complex. The Kristiansund-Bodø Fault Complex separates the Trøndelag Platform from the deep Møre, Trœn and Vestfjord basins to the west and north.

Description

The Trøndelag Platform represents a late Mesozoic—early Cretaceous stable area. It includes the Froan Basin, Helgeland Basin, Vega High, and Frøya High. in Lower and Middle Jurassic times the platform was part of a sedimentary basin which extended further to the west into the area of the Vøring Basin. Cretaceous faulting along the western flank of this positive structural feature defines the western edge of the Trøndelag Platform.

Age

The Trøndelag Platform was initiated in early Cretaceous time.

Helgeland Basin (Helgelandbassenget)

Named after an area in Nordland County.

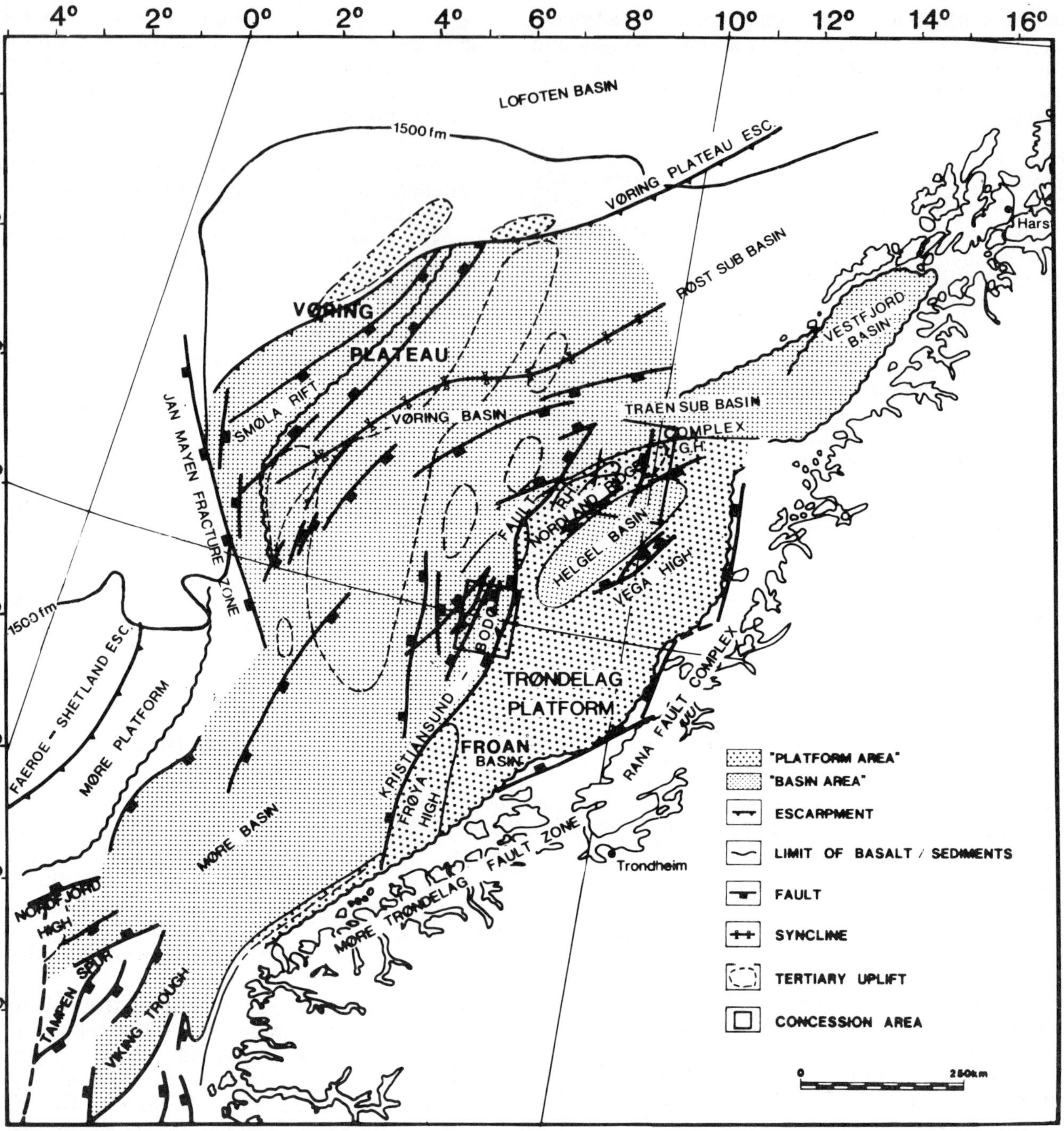

Fig. 1. Main structural elements of the Møre-Lofoten Shelf, mid Norway. G.H. = Grønøy High, R.H. = Rødøy High. Fault pattern after Rønnevik *et al.* (in press).

References

The structure was defined by Rønnevik *et al.* (1975) and later discussed by Jørgensen and Navrestad (1979, 1981).

Position

65°30′–66°30′N, 8°–11°30′E. The basin is bounded by the Vega High to the southeast and by the Nordland Ridge and the crestal part of the Trøndelag Platform to the west.

Description

The Helgeland Basin is intimately connected with the development of the Nordland Ridge, and is a depocentre on the lower part of the south easterly tilted platform.

Age

The Helgeland Basin is mainly a Cretaceous feature, although signs of subsidence relative to the Nordland Ridge may be traced from mid Jurassic time.

Vega High (Vegahøyden)

Named after an island off the Nordland coast.

References

The structure was called the Vega Ridge by Rønnevik and Navrestad (1976). See also Rønnevik and Navrestad (1977) and Aanstad *et al.* (1981).

Position

65°20′–65°45′N, 9°20′–20°30′E. The high is situated within the Trøndelag Platform, and is bounded by the Helgeland Basin to the northwest.

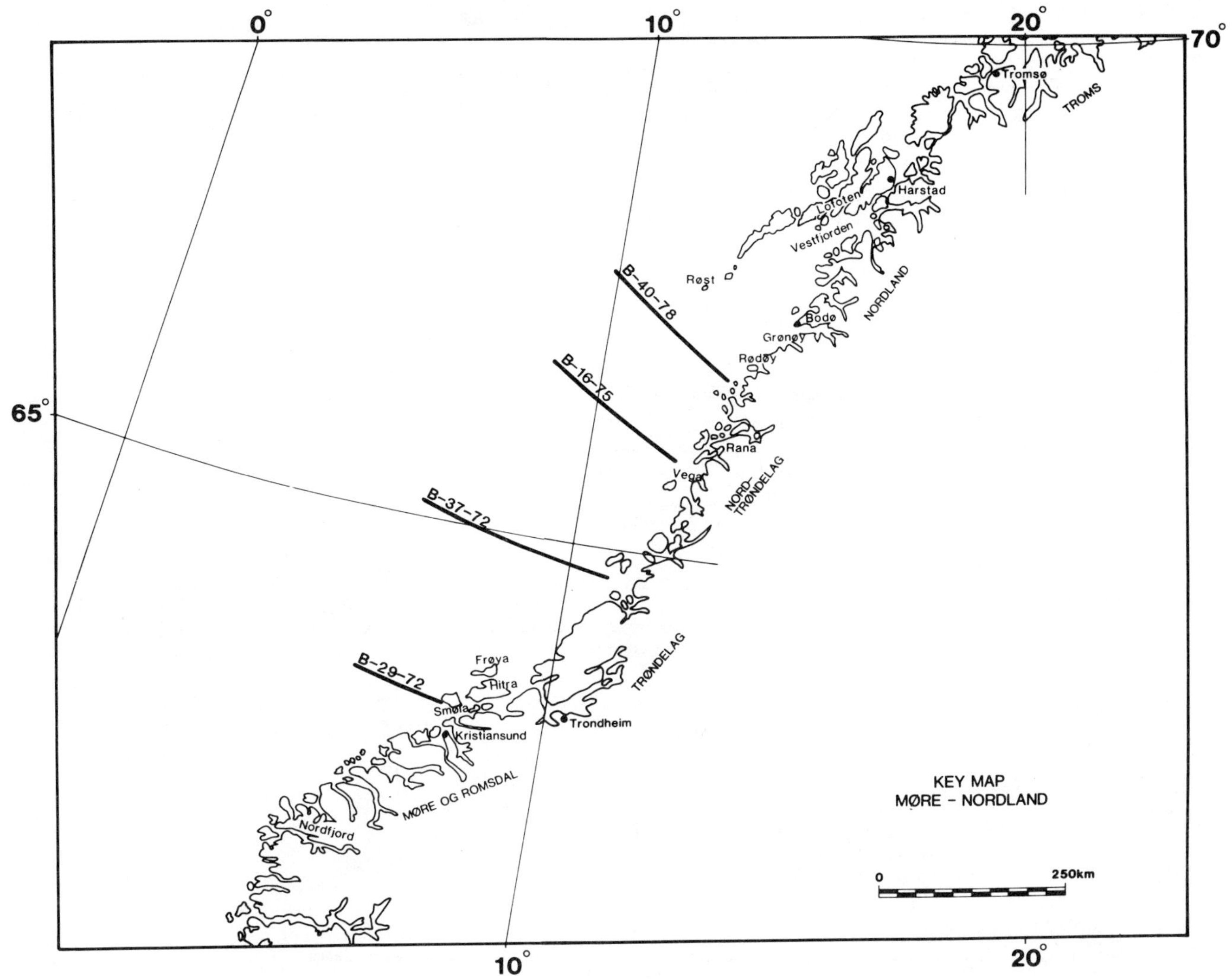

Fig. 2. Location map: Møre-Lofoten Shelf. Location of profiles in Fig. 3 are indicated.

Description

The Vega High is a NE—SW oriented domal feature situated southeast of the Helgeland Basin. The central part of the structure is characterized by a graben, interpreted as a collapse-structure.

Age

The high was initiated during early Cretaceous times, and movements took place through much of the Cretaceous period.

Comments

The structure was termed the Vega Ridge by Rønnevik and Navrestad (1976). New data indicate that the structure does not fit this definition. The graben structure situated on the top of the Vega High was termed the Vega Fault Zone by Price and Rattey (in press).

Froan Basin (Froanbassenget)

Named after an island off the coast of Sør-Trøndelag.

Reference

Bøen *et al.* (1984).

Position

63°30′—65°00′N, 7°—10°E.

Description

The Froan Basin is filled with a moderately thick Mesozoic and Tertiary sequence, and is part of the Trøndelag Platform.

Age

The basin was initiated during early Mesozoic times. Erosion took place in early Cretaceous times, followed by tilting towards the southwest during the mid Cretaceous orogenic event.

Frøya High (Frøyahøyden)

Named after a major island off Sør-Trøndelag County.

References

See Hinz (1972b). The structure was termed Smøla High by Price and Rattey (in press). The structure is also treated by Hamar and Hjelle (1984).

Position

63°—64°N, 6°10′—7°40′E.

Description

The structure is interpreted as a shallow basement high bordered by NNE-trending faults against the Møre Basin to the west and by the Hitra Basin to the east.

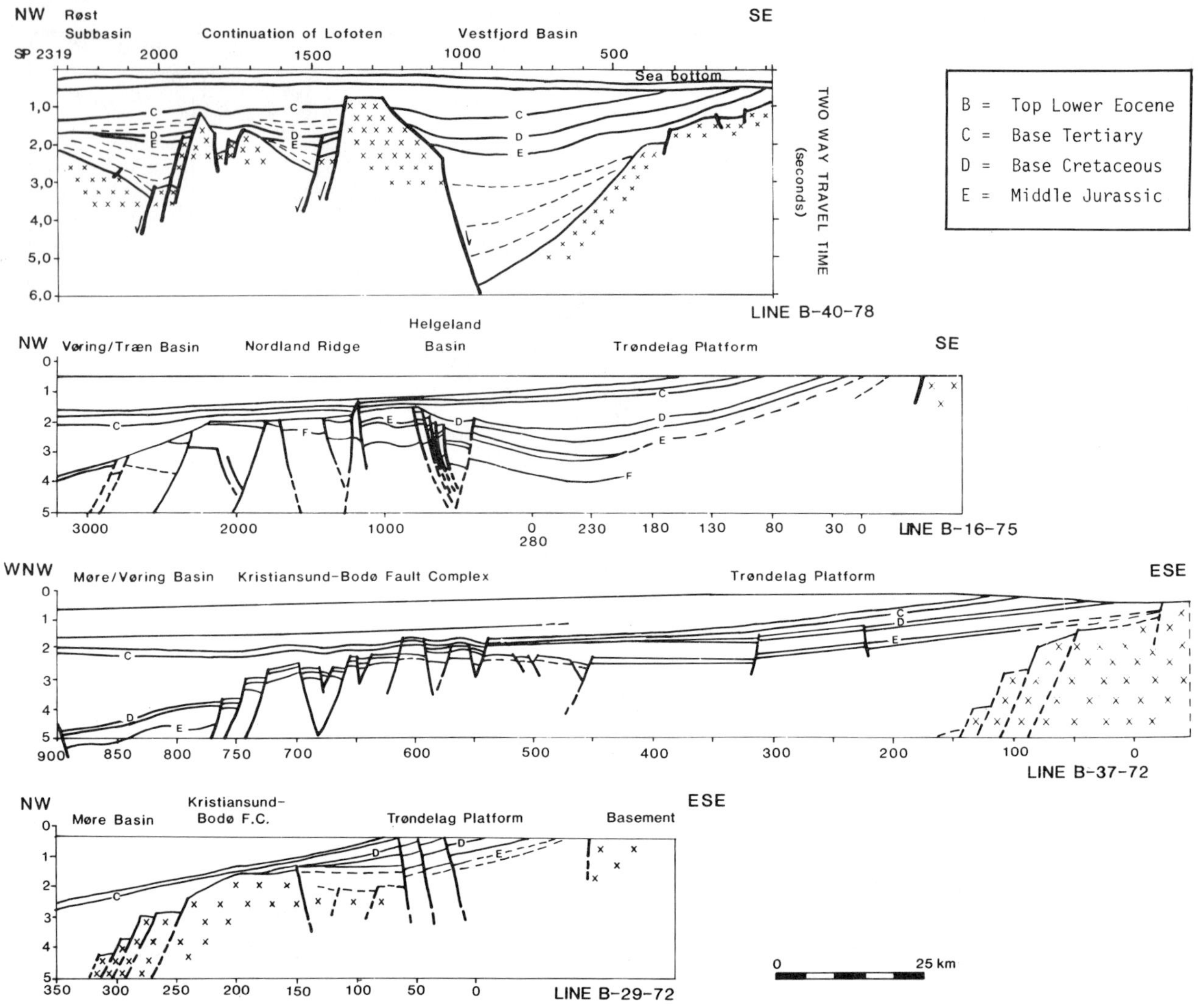

Fig. 3. Principal profiles of the Møre-Lofoten Shelf. For location of profiles, see Fig. 2. After Jørgensen and Navrestad (1981) and Gabrielsen and Robinson (1984).

Age

The structure probably formed in early Cretaceous times, but reactivation took place during Tertiary times.

Comments

The name Smøla High was proposed by Price and Rattey (in press). This is unfortunate, as the name Smøla had already been used for a basin northwest of the Kristiansund-Bodø Fault Complex (Rønnevik *et al.*, 1983).

Rana Fault Complex (Rana forkastningskompleks)

A new name after a fjord in Nordland County.

Position

63°—68°N, 7°—15°E.

Description

The Rana Fault Complex separates the post-Caledonian strata in the west from the Precambrian and Lower Palaeozoic rocks which subcrop sea-bottom to the east. The overall trend is NE—SW to NNE—SSW, parallel to the Norwegian coastline. The complex is composed of several fault sets, each characterized by faults arranged *en echelon* with downthrow to the west.

Age

A thick sequence including Triassic and possibly pre-Triassic strata occurs west of the Rana Fault Complex. East of the fault complex Cenozoic(?) rocks are found directly on metamorphic basement. This indicates either Triassic synsedimentary faulting or Triassic sedimentation across the Rana Fault Complex with later pre-Cenozoic erosion of the sediments east of this complex.

Møre-Trøndelag Fault Zone (Møre-Trøndelag forkastningssone)

Named after Møre and Trøndelag Counties.

References

The same structural feature was termed the Møre-Trøndelag Line by Price and Rattey (in press), and it coincides with the Møre-Trøndelag Zone defined from Satellite imagery on the Norwegian mainland by Gabrielsen and Ramberg (1979), which has also been described by Oftedahl (1975) and Aanstad *et al.* (1981).

Position

62°—64°N, 3°—12°E.

Description

On the Norwegian mainland, the Møre-Trøndelag Fault Zone affects Precambrian and, in part, Caledonian rocks which are dominated by an ENE—WSW structural grain. This zone includes some of the major faults known in the area, such as the Beistadfjorden Fault, the Tarva Fault, the Hitra Fault and the Verran Fault (Oftedahl, 1972, 1975). Most of the faults are normal, although Aanstad *et al.* (1981) described field evidence for small-scale strike-slip movements.

This structural trend continues offshore where it delimits major structural elements such as the Trøndelag Platform, the Horda Platform and the Møre Basin.

Age

The ENE—WSW structural grain in the northwestern part of Norway can be related to the Caledonian deformations. Slivers of Devonian (Siedlecka and Siedlecki, 1972), and Jurassic (Oftedahl, 1972, 1975) sediments occur within the fault zone together with Devonian and Permian dykes (Storetvedt, 1967; Priem *et al.*, 1968; Råheim 1974). In off-shore areas Price and Rattey (in press) have suggested dextral transcurrent movements on the ENE—WSW system in Cretaceous time, which altogether suggest a long-lived tectonic activity along the Møre-Trøndelag Fault Zone.

Kristiansund-Bodø Fault Complex (Kristiansund-Bodø forkastningskompleks)

Previously undefined. Named after two towns in Norway.

Position

64°—67°N, 6°—12°E.

Description

The Kristiansund-Bodø Fault Complex separates the Trøndelag Platform from the adjoining sedimentary basins. The complex is composed of fault sets, each characterized by faults partly arranged in *en enchelon* fashion. In the southwest (the Haltenbank area), individual faults mostly trend NNE. To the northeast, this trend meets the ENE fault trend which is dominant in the Nordland Ridge. Both fault trends parallel important structural grains on the Norwegian mainland.

The fault complex changes character from the northeast, where it defines a ridge-like feature, towards the south, where it is a complex horst-and-graben-type system (Gabrielsen and Robinson, 1984). It terminates against the Møre-Trøndelag Fault Zone. Several structural features are associated with the Kristiansund-Bodø Fault Complex (from north to south): Nordland Ridge, Grønøy High, Rødøy High and Frøya High. The fault complex was a part of the regional sedimentary basin until late Jurassic times.

Age

The Fault Complex resulted from late Mesozoic—early Cenozoic activation of a possibly older zone of weakness. The zone was also in places influenced by Tertiary tectonic activity.

Nordland Ridge (Nordlandryggen)

Named after Nordland County.

References

Originally defined by Rønnevik *et al.* (1975). See also Jørgensen and Navrestad (1979, 1981), Rønnevik *et al.* (1979, 1982), Hinz *et al.* (1982), Price and Rattey (in press), and Gabrielsen and Robinson (1984).

Position

65° 30′—66° 40′N, 7°—12° E. The structure is located within the Kristiansund-Bodø Fault Complex.

Description

The Nordland Ridge is bordered by the deep Vøring Basin to the west and northwest, the Vestfjord Basin to the north, and by the Helgeland Basin to the south and east. The structure is cut by faults of various trends, producing several high and low parts along its NE—SW-trending main axis. Two principal structures are named Rødøy and Grønøy Highs. The Nordland High was part of a major depocentre throughout late Palaeozoic and early Mesozoic times. After the Mesozoic faulting, the structure was covered by a thin Cretaceous and Tertiary sedimentary sequence.

Age

The main phases of deformation occurred in late Jurassic—early Cretaceous times.

Grønøy High (Grønnøyhøyden)

Named after a minor island off the coast of Nordland, mid-Norway.

Reference

Larsen and Skarpnes (1984).

Position.

66° 20′—66° 35′N, and 9° 50′—10° 25′E.

Description

The Grønøy High is an elongated NE-trending elevation. The high is a part of the Nordland Ridge of the Kristiansund-Bodø Fault Complex.

Age

The structure was initiated by early Mesozoic tectonic movements and has been reactivated as recently as late Tertiary (Miocene). In large parts of this high, late Jurassic to early Cretaceous and Miocene sediments are lacking.

Rødøy High (Rødøyhøyden)

Named after a minor island off the coast of Nordland.

References

This structural element is defined by Gowers and Lunde (1984), and Larsen and Skarpnes (1984).

Position

65° 45′—66° 40′N, 8°—9° 30′E, within the Kristiansund-Bodø Fault Complex.

Description

The Rødøy High is a wedge-shaped, fault-bounded structural high of eroded Palaeozoic or even older rocks. The structure, which may be considered a part of the Nordland Ridge, is bounded by NE-trending major faults separating it from the Trœn Basin to the west and the Helgeland Basin to the ENE.

Age

The Rødøy High may date from late Palaeozoic times, when it acted as an area of erosion (see Gowers and Lunde, 1984; Larsen and Skarpnes, 1984). It was covered by sediments during Triassic and Jurassic times, which were removed by subsequent early Cretaceous erosion.

Vestfjord Basin (Vestfjordbassenget)

Named after the major fjord south of Lofoten archipelago.

References

See Rønnevik and Navrestad (1977) and Jørgensen and Navrestad (1979, 1981).

Position

67°—68°N, 11°—14°30′E.

Description

The asymmetric Vestfjord Basin is bounded by sets of NE and ENE-trending normal faults. A major fault system controls the WNW border, whereas the basin becomes gradually shallower towards the southeast. The basin is filled with more than 10 km of Mesozoic and Cenozoic sediments. The basal sequence may consist of continental red beds of Permain/Carboniferous age.

Age

The major phase of subsidence started in Triassic and lasted into Middle Jurassic times.

Vøring Plateau (Vøringplatået)

References

The structure was defined by Nansen (1904) and later discussed by Åm (1970), Hinz and Moe (1971), Johnson *et al.* (1968), Hinz (1972a), Sellevoll (1975), Talwani and Eldholm (1972), Rønnevik *et al.* (1979), Eldholm and Talwani (1981), Hagevang *et al.* (1983).

Position

The approximate extent of the Vøring Plateau is shown by the 1500 m contour, extending from 66° to 68°30′N and 2°—9°E. It is terminated by the Lofoten Basin to the northwest, the Jan Mayen Fracture Zone to the south, and the Vøring Basin to the east.

Description

The Vøring Plateau is a Tertiary marginal high, which structurally is divided into an inner and an outer part by the Vøring Plateau Escarpment. In the central part of the plateau the crustal thickness is estimated to less than 20 km (Hinz, 1972). In the outer part there are indications of shallow basement, overlain by igneous rocks related to the Tertiary opening of the North Atlantic which, in turn, are covered by a thin, Tertiary sedimentary sequence.

Over the topographically highest part of the Vøring Plateau, the basement has a thin Tertiary cover. In the southeast the basement dips steeply to the southeast and becomes buried below the thick sedimentary sequence in the Vøring Basin. The inner part (which is identical to the Vøring Basin) is subdivided into two basinal elements (Smøla Rift and Røst Sub-basin).

Age

The Vøring Plateau is a structural feature which is related to the Tertiary opening of the North Atlantic. However, this area may also have existed as a structural high in Cretaceous and even pre-Cretaceous time, when the Vøring Basin developed.

Comments

Although there are three structures named 'Vøring' in the area (Vøring Plateau, Vøring Basin, and Vøring Plateau Escarpment, see below), these names are well established in the literature, and re-naming would be confusing.

Vøring Plateau Escarpment (Vøringplatåets brattkant)

References

The Vøring Plateau Escarpment has been discussed in several articles (i.e. Talwani and Eldholm, 1972, 1977; Sellevoll, 1975; Talwani *et al.* 1981; Hinz *et al.*, 1982).

Position

67°—69°N, 2°—9°E.

Description

The Vøring Plateau Escarpment is a steep, NE—SW trending structural feature, which according to Talwani and Eldholm (1972) represents a change from continental to oceanic crust and divides the Vøring Plateau into an inner and outer part. The escarpment also marks the boundary between the northwest part of the Vøring Plateau covered by thin Tertiary to Quaternary deposits and the southeast part with a thick Tertiary and pre-Tertiary sequence. The Vøring Plateau Escarpment is associated with distinct gravity and magnetic trends. The escarpment also defines the limit of a magnetic quiet zone situated landward of the escarpment.

Age

Following Talwani and Eldholm (1972), the Vøring Plateau Escarpment is a first-order structural boundary related to the opening of the Norwegian-Greenland Sea in late Cretaceous—early Tertiary times.

Jan Mayen Fracture Zone (Jan Mayen bruddsone)

Named after the island of Jan Mayen.

References

Sykes (1965) attributed an E-trending pattern of earthquakes near the island of Jan Mayen to a major fracture zone that offsets the ridge crest. The extent and importance of the fracture zone are discussed by Johnson and Heezen (1967), Talwani and Eldholm (1972, 1977) and Sellevoll (1975).

Position

70°–72°30'N, 0°–25°W.

Description

According to Talwani and Eldholm (1977), the Jan Mayen Fracture Zone consists of two distinct segments. The western segment, lying approximately between 2° and 15°W has an azimuth of 110 degrees. The eastern segment trends from N 130°E at its western end to about N 150°E at its eastern end. The more northerly trend of the eastern segment is attributed to a more northerly motion of Greenland relative to Norway in early Tertiary time than at present. Further to the southeast, the tectonically complex zone separating the Møre and Vøring Basins represents a possible structural continuation of the Jan Mayen Fracture Zone. Both segments of the Jan Mayen Fracture Zone are clearly characterized by a negative free-air anomaly, the axis of which is nearly parallel to the fracture zone (Talwani and Eldholm, 1977). Along the southeastern continuation of this trend, similarly oriented gravimetric lineaments have been mapped (Aanstad *et al.*, 1981).

Age

The Jan Mayen Fracture Zone is interpreted as a transform fault-zone related to early Tertiary opening of the North Atlantic. However, data to the southeast, within the zone connecting the Møre and Vøring Basins, indicate that the fracture zone follows an older zone of weakness (Rønnevik *et al.* 1975).

Vøring Basin (Vøringbassenget)

References

The structure was defined by Åm (1970) and Rønnevik *et al.* (1975) and was later described by Jørgensen and Navrestad (1979, 1981), Hinz *et al.* (1982), and Price and Rattey (in press).

Position

65°–67°30'N, 2°–10°E. The basin is bounded to the southwest by the Jan Mayen Fracture Zone and Møre Basin, to the east by the Kristiansund-Bodø Fault Complex, to the northeast by the Trœn Basin and to the west by the Vøring Plateau Escarpment.

Description

The Vøring Basin trends NE. It contains thick Cretaceous deposits and apparently is time equivalent to the Møre Basin to the southwest. The two basins are connected by a tectonically complex zone which may represent a continuation of the Jan Mayen Fracture Zone. The Vøring Basin is subdivided into the Smøla Rift, Røst Sub-basin, and the Trœn Sub-basin.

Age

The Vøring Basin developed during Cretaceous time.

Røst Sub-basin (Røst subbasseng)

Named after one of the minor islands of Lofoten archipelago.

References

Called 'Røst Syncline' by Rønnevik *et al.* (1983). See also Hintz *et al.* (1982, part of the northern Vøring Basin).

Position

66°45'–68°N, 5°30'–8°E.

Description

The basin is a 100-km wide depression with a NE–SW axis. It is flanked by a NE-trending fault bounded high to the southeast and a series of rotated fault blocks to the northwest.

Age

The structure was initiated in late Jurassic—early Cretaceous times, and has been reactivated in connection with early to late Alpine movements.

Smøla Rift (Smølariften)

Named after an island on the Møre coast.

References

Defined by Rønnevik *et al.* (1983). See also Jørgensen and Navrestad (1981), Hinz *et al.* (1982) (Vøring Basin) and Price and Rattey (in press) (Inner Vøring Basin). See also Bøen *et al.* (1983) for a more extensive description.

Position

65°30'–67°45'N, 2°–5°40'E.

Description

The Smøla Rift is bordered by the Vøring Plateau Escarpment to the north and northwest, and the Jan Mayen Fracture Zone to the southwest. Tertiary basalt flows are believed to cover the western part of the structure, whereas the eastern part was influenced by compression during Eocene times.

Age

The Smøla Rift was initiated in pre-Cretaceous times, although a reversal took place in Tertiary times. Note that the name Smøla has also been informally used on a sedimentary unit onshore mid-Norway (Siedlecka, 1975).

Trœn Sub-basin (Trœn subbassenget)

Named after an island off coastal Nordland.

References

See R/nnevik and Navrestad (1976), Jørgensen and Navrestad (1981) and Hinz *et al.* (1982).

Position

66°45'–67°15'N, 8°–10°15'E. Located between the Røst Sub-basin and the Nordland Ridge.

Description

An elongated basin with the longest axis oriented NE–SW. The basin is bordered by the Nordland

Ridge to the southeast. It has a maximum depth at about 66°45′N and 9°30′E, where it contains more than 6000 m of Cretaceous sediments. The Trœn Sub-basin is considered a sub-basin of the Vøring Basin.

Age

The major subsidence took place in Cretaceous times.

Comments

The Trœn Sub-basin was originally termed Trœn Basin by Rønnevik and Navrestad (1976).

Møre Basin (Mørebassenget)

Named after Møre County.

References

The structure was defined by Rønnevik *et al.* (1975) and has since been discussed by Jørgensen and Navrestad (1981), Hinz *et al.* (1982), and Price and Rattey (in press).

Position

61°30′—65°N, 0°30′—7°E, situated between the Møre Platform to the west and the Kristiansund/Bodø Fault Complex to the east.

Description

The Møre Basin is a well-defined NE-trending structural feature at base Cretaceous level. While the southeast and eastern margins of the basin are well-defined, the northwest termination is not known. However, Rønnevik *et al.* (1975) suggested the Faeroe-Shetland Escarpment as a probable northwestern border. Towards the northeast the Møre Basin is bounded by a NNW-oriented tectonic zone, which separates the Møre and Vøring Basins. This tectonic zone might represent a landward continuation of the Jan Mayen Fracture Zone. The Cretaceous sequence defining the Møre Basin may exceed 5000 m. It is dominantly a clastic sequence, but seismic data suggest the existence of igneous rocks, either as sills or lava flows.

Age

The Møre Basin is defined by its Cretaceous infill. Below the Cretaceous strata, there is evidence of a block-faulted sedimentary sequence. Seismic data and wells located in the southern part of the basin indicate that the pre-Cretaceous sequence is Jurassic in age. However, as the Jurassic deposits are relatively thin, it seems that the Møre Basin was not a major depocentre in Jurassic time.

Møre Platform (Møreplattformen)

Named after the Møre County, mid Norway.

Reference

Hamar *et al.* (1980).

Position

The Møre Platform is situated west of the Møre Basin, north of 62°10′N.

Description

The platform is an elongate structure separated from the Shetland Platform to the west by a narrow trough filled with sediments and volcanics. The platform is surrounded by Tertiary lava flows to the west, north and south. To the east, volcanic rocks included in the mid Cretaceous sedimentary pile are found in a 100—150 km broad zone. The Møre Platform shows signs of internal deformation, and strongly tilted fault blocks overlain by Palaeocene sediments are found locally. The platform is believed to be underlain by continental crust. The platform is tilted towards the north and west.

Age

The Møre Platform is a result of Tertiary movements.

Comments

The committee considers it unfortunate that the name Møre has been used for two structural elements (Møre Platform and Møre Basin). However, the names have been in use for several years and renaming is not considered appropriate.

Nordfjord High (Nordfjordhøyden)

The structure is named after a major fjord in Sogn og Fjordane County.

References

The structure was defined by Hamar *et al.* (1980) using the name Nordfjord Horst. See also Hinz *et al.* (1982).

Position

The Nordfjord High is situated in the southwestern part of the Møre Basin, between 62°—62°05′N and 1°30′—2°E.

Description

The structure is a NE-oriented basement high covered by sediments of Mesozoic and Tertiary age. The high is bordered by major faults on its northwest and southeast sides. To the south the Nordfjord High is separated from the Tampen Spur by a deep depression.

Age

The high was formed during early Mesozoic times together with the Tampen Spur, and was later, probably by end Jurassic and early Cretaceous times, separated from this by rejuvenated movements.

Faeroe-Shetland Escarpment (Fœrøy-Shetland brattkanten)

Named after the Faeroe and Shetland Islands.

Reference

The name was introduced by Talwani and Eldholm (1972).

Position

61°—64°N, 5°W—3°E.

Description

Talwani and Eldholm (1972) described a buried escarpment which they assumed to represent the

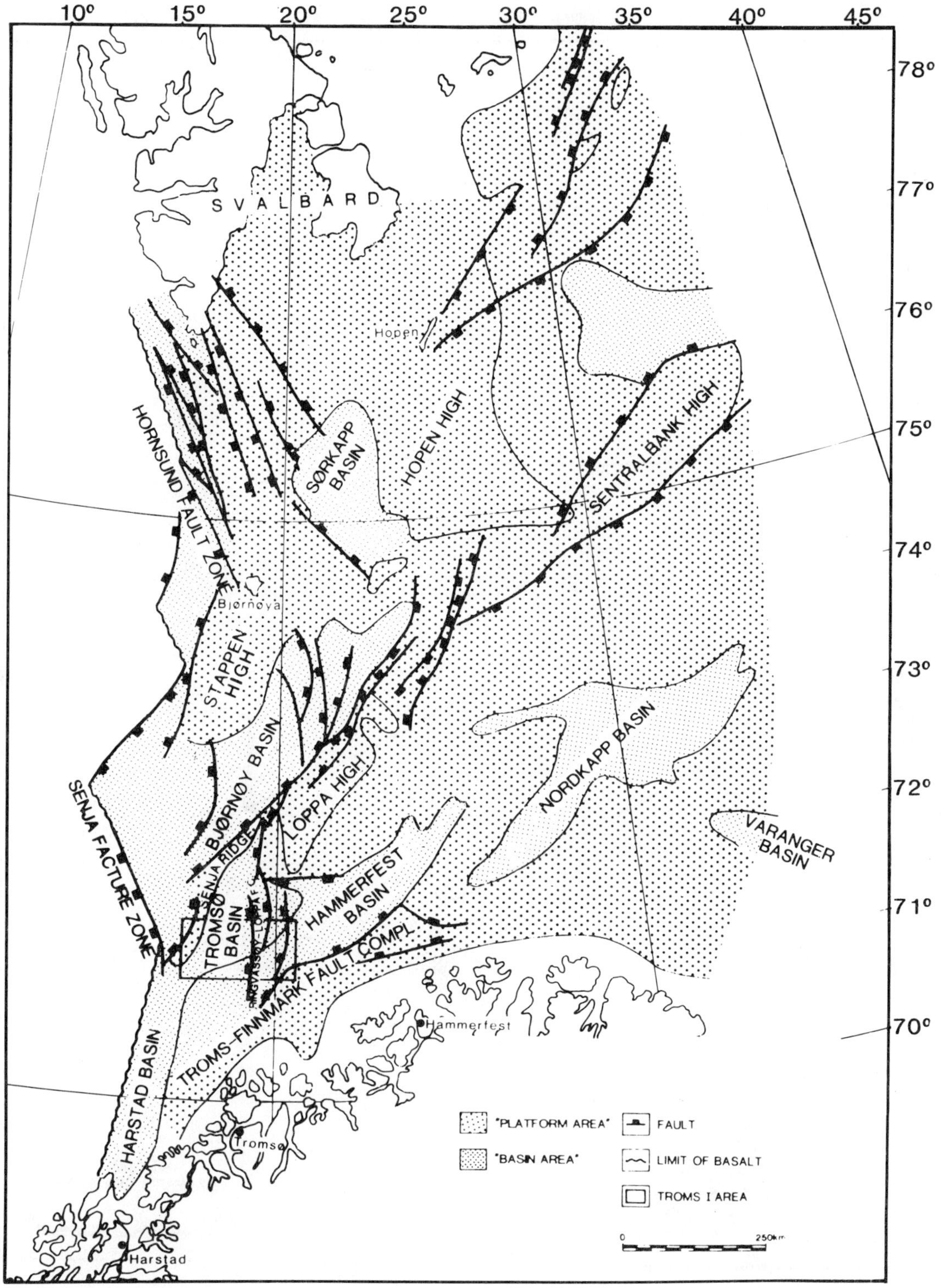

Fig. 4. Main structural elements of the Barents Sea between the Norwegian mainland and Svalbard. Fault pattern after Rønnevik *et al.* (in press).

eastern boundary of a basement high south of the Jan Mayen Fracture Zone. This feature, with an overall NE-trend, apparently consists of two segments which nearly coincide with the western edge of the Faeroe-Shetland Channel. The Faeroe-Shetland Escarpment is defined by a gravity gradient and by well-defined magnetic lineaments.

Age

Talwani and Eldholm (1972) attributed the Faeroe-Shetland Escarpment to the early Tertiary opening of the Norwegian Sea.

STRUCTURAL ELEMENTS OF THE BARENTS SEA AND ADJACENT AREAS

The Barents Sea is bordered by the Norwegian-Greenland Sea to the west, the Arctic Ocean to the north, and Novaya Zemlya and Franz Josef Land to the east and northeast, while the Norwegian mainland and the Kola Peninsula define its southern limit. Physiographically, this shallow epicontinental sea is divided into a northern and southern province by the Barents Sea Valley (e.g. Malod and Mascle, 1975; Hinz and Schlüter, 1978).

Geologically, the Barents Sea has undergone a

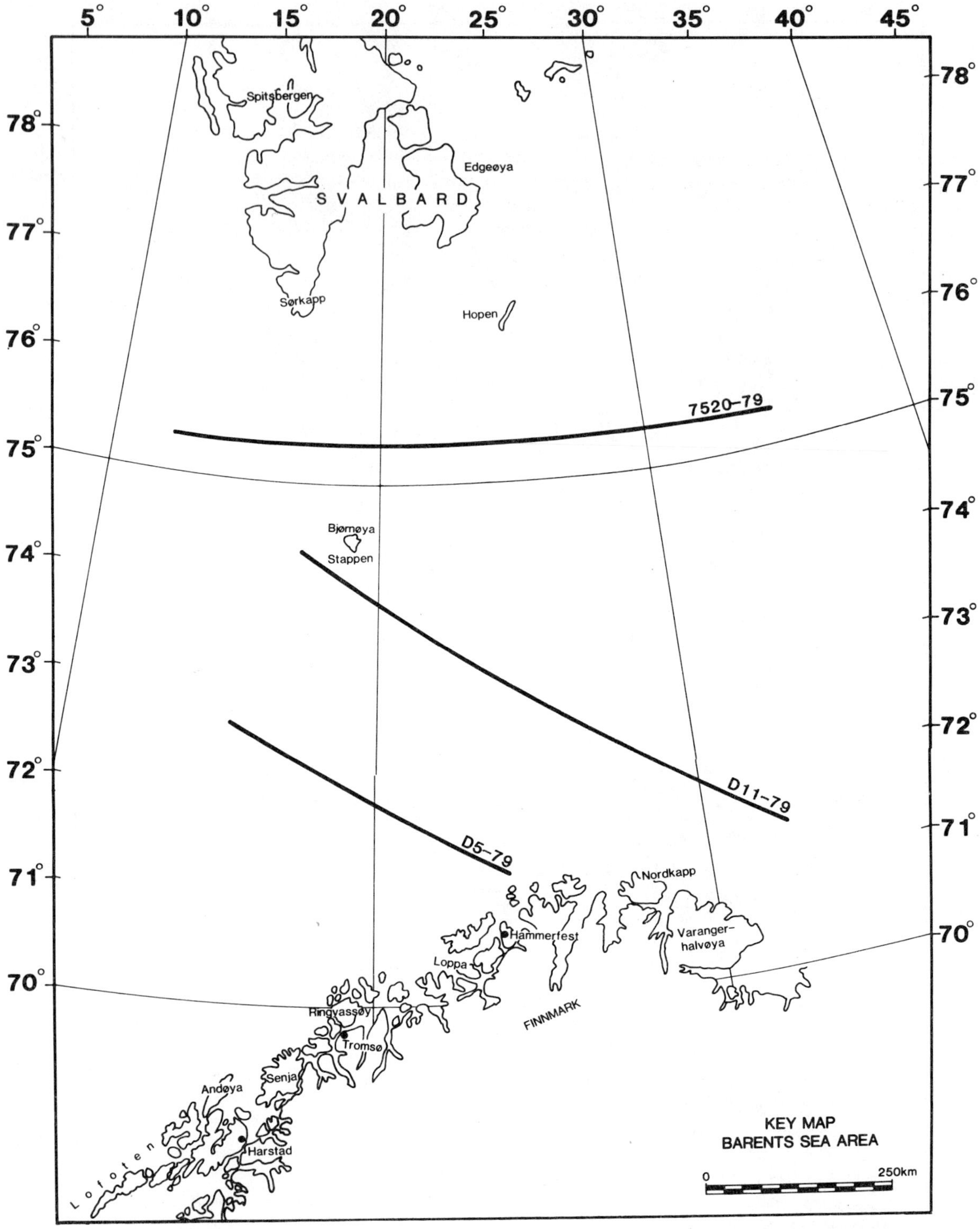

Fig. 5. Location map: Barents Sea. Locations of profiles in Fig. 6 are indicated.

development from late Palaeozoic to Cenozoic which includes early (Devonian—early Permian) sedimentation in intramontane basins in the south and simultaneous development of a carbonate-shelf to the north. Evaporites were also deposited in this period. From late Permian times a gradual transgression took place from the north, culminating with a regional transgression in late Jurassic times when dark clays were deposited in large parts of the southern Barents Sea. From early Cretaceous times, shallow marine to sublittoral conditions prevailed in the east. In Tertiary times a land area in the east supplied large amounts of sediment to the continental margin in the west (Rønnevik, 1981).

The structural history includes two phases of Caledonian movements (Sturt *et al.*, 1978), the Svalbardian phase (Harland, 1965, 1969), the Kimmerian tectonic phase (Jurassic—early Cretaceous), and influence from the opening of the North Atlantic during Palaeocene times. Precambrian structures in the basement may also have affected the tectonic development of the area (Gabrielsen, in press). In Mesozoic and Cenozoic times, halokinesis was important. The continental shelf of the western Barents Sea is bordered by two major fracture zones, namely the Senja and Hornsund Fault Zones to the west and the Troms-Finnmark Fault Complex to the south. Of these, the Senja Fracture Zone and the Hornsund Fault Zones to the west may be considered the most fundamental, because these systems mark the transition zone between oceanic and continental crust. Myhre *et al.* (1982) have indicated that the central segment of the Hornsund Fault Zone and the NNE segment southwest of Bjørnøya are rifted passive margins, whereas the other parts of the boundary are either of the sheared or sheared-rifted type. The margin is characterized by elongated positive gravity anomalies.

The Troms-Finnmark Fault Complex represents the major divide between subsided parts of the continental shelf and the Troms-Finnmark Platform, which forms a rim along the metamorphic rocks of the Norwegian mainland. The fault zone is believed to represent an old zone of weakness in the crust (Gabrielsen, in press).

The main tectonic features of the Barents Sea area are summarized in Figs. 4, 5 and 6.

Troms-Finnmark Fault Complex (Troms-Finnmark forkastningskompleks)

Named after Troms and Finnmark Counties.

References

Originally identified by Moe (1974). Also described by Syrstad *et al.* (1976) and Gabrielsen (in press).

Position

68°30′—71°30′N, 16°—27°E.

Description

The Troms-Finnmark Fault Complex is the major structural divide between the sediments of the continental shelf and a landward sedimentary platform. The fault complex consists of *en echelon* major faults with smaller faults on the platform side.

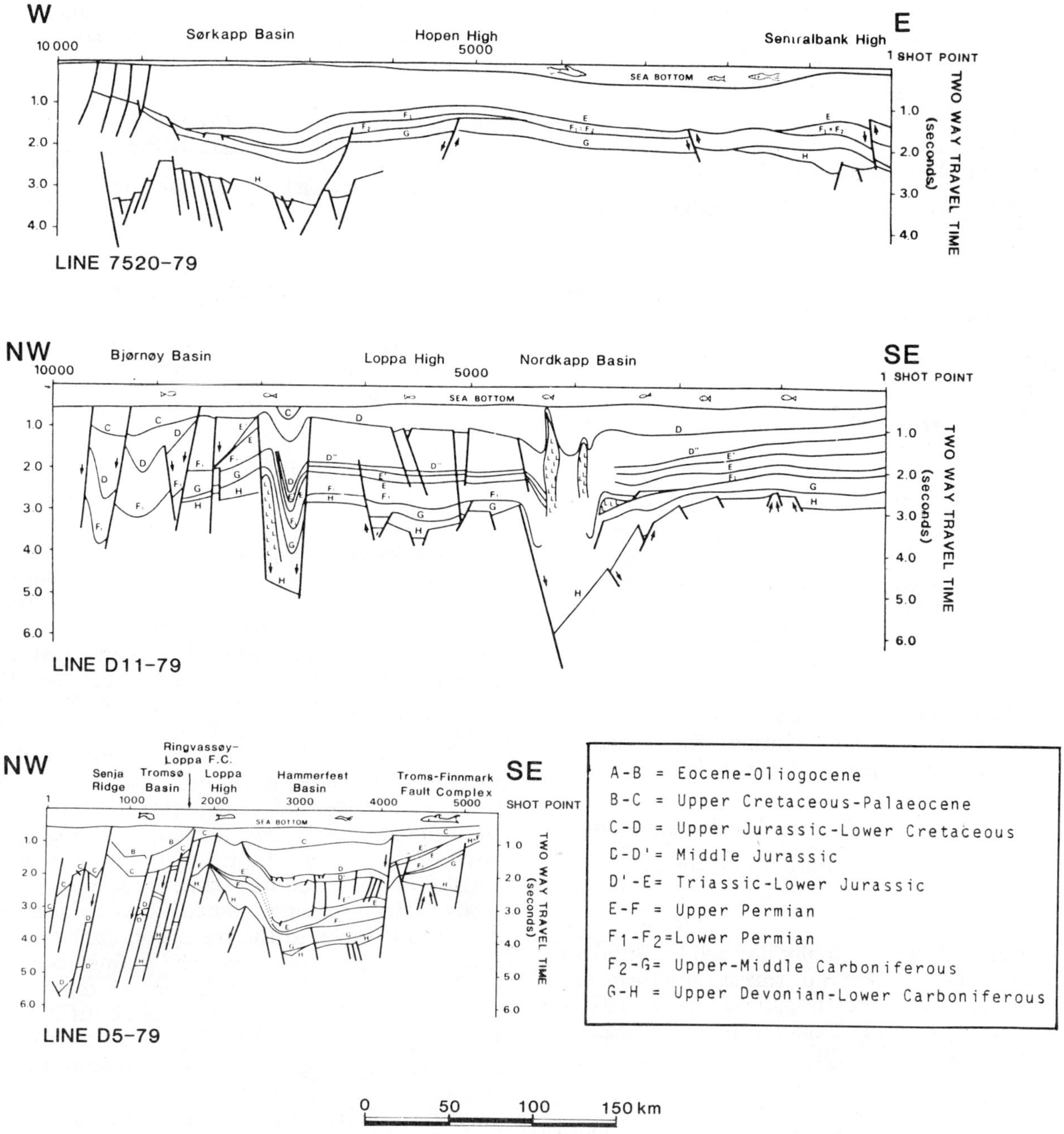

Fig. 6. Principal profiles of the Barents Sea region. Locations of profiles are indicated in Fig. 5. After Rønnevik *et al.* (1982).

The southern part of the system strikes NE, but the zone turns in a more ENE direction at about 19°E longitude. The system is less conspicuous towards the south. To the east, another abrupt change of direction occurs at 20°30′E, where the fault zone changes to a NW-trend. The Fault Complex parallels the shoreline along a large part of the coast of Troms and Finnmark, and marks the southeastern borders of the Tromsø and Hammerfest Basins. The Fault Complex is characterized by a marked increase in the gravity gradient (Syrstad *et al.* 1976).

Age

The Fault Complex is believed to be an old zone of weakness, parts of which might have been initiated as early as late Palaeozoic times. The fault complex has been periodically active since then, probably until Tertiary times.

Comments

The previous name of this structure used in literature was Troms-Finnmark Fault Zone. The term fault complex, implying 'a group of faults that interconnect and intersect, having the same or different ages', is preferred here.

Senja Fracture Zone (Senjabruddsonen)

Named after the island of Senja, Troms County.

References

Originally mapped by Talwani and Eldholm (1972). See also Talwani and Eldholm (1977), Sundvor and Eldholm (1979), Myhre *et al.* (1982).

Position

73°30′—70°20′N, 13°—16°E.

Description

The Senja Fracture Zone is a NNW-trending element, which in the south separates continental and oceanic crust. It was originally defined as an oceanic transform. On the oceanward side, the structure has been interpreted as a counterpart of the Greenland Fracture Zone (Talwani and Eldholm 1977).

Age

The fracture zone is related to the sliding of Greenland past Svalbard during the Tertiary opening of the Norwegian-Greenland Sea.

Hornsund Fault Zone (Hornsund forkastningssone)

Named after a fjord in western Spitsbergen.

References

Myhre *et al.* (1982). Originally defined by Sundvor and Eldholm (1976). Parts of this system have also been described by Sundvor and Eldholm (1979), Sundvor *et al.* (1979), Faleide and Gudlaugsson (1981) and Vogt *et al.* (1981).

Position

The Hornsund Fault Zone consists of several separate segments, from west of Svalbard to Bjørnøya, between about 74°—79°N and 9°—13°E.

Description

The Hornsund Fault Zone coincides roughly with the border between oceanic and continental crust. The zone consists of three major NNW-trending segments with westerly throw, which show varying characteristics along strike. At 77°N one single major fault with a minimum throw of 4 seconds has been mapped, whereas the segment between about 76°30′ and 75°40′N is characterized by a series of step faults. By contrast, the southernmost part consists of one major fault (Myhre *et al.*, 1982). The faults are considered to have resulted from an initial dextral shear event, followed by normal faulting.

Age

Formation of the Hornsund Fault Zone has been directly linked to the Tertiary opening of the Norwegian-Greenland Sea. This event is divided into an early Palaeocene shear phase, and a second tensional phase starting about 36 Ma (Oligocene). It should, however, be noticed that the Hornsund Fault Zone parallels the late Devonian fault trend of Svalbard.

Ringvassøy-Loppa Fault Complex (Ringvassøy-Loppa forkastningskompleks)

Named after two islands on the coast of Troms and Finnmark.

References

New definition. See also Øvrebø and Talleraas (1977), Faleide and Gudlaugsson (1981), Gloppen and Westre (1982) and Gabrielsen (in press). Previously called the Tromsø/Hammerfest Basin Transition Zone.

Position

70°45′—72°30′N, 18°30′—20°00′E.

Description

The Ringvassøy-Loppa Fault Complex trends NNE. Its southern segment is the Tromsø/Hammerfest Basin Transition Zone, which is characterized by a NNE-striking fault cross-cutting the E—W system of the Hammerfest Basin. The faults have a listric geometry with the concave part facing the Tromsø Basin and thus show the typical configuration of detached normal faults. The geometric relation to the southern part of the Troms-Finnmark Fault Complex and indications that the transition zone has had the effect of a long-lived hinge line, implies that the zone may be interpreted as a deep-seated zone of weakness. The listric fault systems characteristic of the zone may thus be secondary expressions of deep-seated basement-involved movements. The largest faults of the transition zone are the two most westerly ones, which define the eastern shoulder of the Tromsø Basin. At them the base Cretaceous reflector drops from about 2.5 seconds to more than 5 seconds (two-way time) and the faults are frequently seen to be reactivated to cut the overlying Upper Cretaceous and even Tertiary sequences. Further north, the western limit of the Loppa Ridge lies on the continuation of the Tromsø/Hammerfest Basin Transition Zone. This is considered the northernmost part of the Ringvassøy-

Loppa Fault Complex. The Ringvassøy-Loppa Fault Zone is associated with a slight positive gravity lineament trend.

Age

The Ringvassøy-Loppa Fault Zone seems to have a tectonic status similar to the Troms-Finnmark Fault Zone. It may thus involve basement and is characterized by reactivation. The last movements, associated with differential subsidence, affected the Tertiary section.

Comments

It is recommended that the name Ringvassøy-Loppa Fault Zone is applied for the structure described above. Accordingly, the Tromsø/Hammerfest Basin Transition Zone should be considered a segment of this fault complex.

Harstad Basin (Harstadbassenget)

Named after the town of Harstad, Troms County.

References

Defined by Rønnevik *et al.* (1975). Also described by Øvrebø and Talleraas (1977) and Rønnevik *et al.* (1982, in press).

Position

North of Andøya, between 69°—71°N and 16°30′—17°45′E.

Description

The basin is elongated N—S and is faultbounded to the south and east. The western boundary is defined as the transition zone to the Eocene oceanic crust of the North Atlantic. The northern continuation of the basin is presently unknown. The structure is marked by the great regional gravimetric anomaly associated with the oceanic/continental crust transition.

Age

The basin is undrilled, and the age of the sediments cannot be dated by correlation to wells on present data. However, the basin probably was initiated in Jurassic times or earlier, as indicated by analogy to Andøya (Dalland, 1975, 1979).

Tromsø Basin (Tromsøbassenget)

Named after the town of Tromsø, Troms County.

References

First defined by Rønnevik *et al.* (1975). See also Hinz and Weber (1975), Syrstad *et al.* (1976), Øvrebø and Talleraas (1977), Hinz and Schlüter (1978), Talleraas (1979), Følnstelien and Horvei (1979), Faleide and Gudlaugsson (1981), Gloppen and Westre (1982) and Rønnevik *et al.* (1982, in press).

Position

71°—72°30′N, 18°—20°E.

Description

The Tromsø Basin is bordered by the Senja Ridge to the west, the transition zone to the Hammerfest Basin to the east, and the Troms-Finnmark Fault Zone to the south and southeast. The basin is a very deep structure, typically asymmetric, and may be said to consist of several sub-basins. Tectonically it is a clear contrast to both the Senja Ridge and the Hammerfest Basin. The most spectacular structural feature is the NNE-trending central basin axis, defined by a smooth flexure and a series of salt stocks. Generally, the western part of the basin is the deepest; either this is as a result of differential salt movements (Øvrebø and Talleraas, 1977) or it is a primary, structural effect. The Tromsø Basin is associated with a gravity low (Syrstad *et al.*, 1976) which is believed to reflect large masses of salt in the deep parts of the basin (below 4 km). Local gravity highs over diapirs imply heavy cap rock material.

Age

The major development of the basin is post-mid Jurassic. However, an older (late Palaeozoic) topography, influenced by the major fault zones, should not be ruled out. The Tromsø Basin may thus have acted as a depocentre at this time, although it was not developed as a separate basin in mid Carboniferous to early Jurassic times. Fast subsidence in late Cretaceous times, combined with large-scale halokinetic movements, separated the Tromsø Basin from the Hammerfest Basin. Palaeocene reactivation of the major fault systems has increased the basin relief.

Hammerfest Basin (Hammerfestbassenget)

Named after a town in Finnmark County.

References

The basin was defined by Rønnevik *et al.* (1975), and has been discussed by Øvrebø and Talleraas (1977), Talleraas (1979), Følnstelien and Horvei (1979), Gudlaugsson and Faleide (1981), Gloppen and Westre (1982), and Rønnevik *et al.* (1982, 1983).

Position

71°—72°30′N, 20°—26°E.

Description

The Hammerfest Basin is bordered to the north by the Loppa High, to the south by the Troms-Finnmark Fault Complex and to the west by the Tromsø Basin. The basin narrows eastwards and gradually dies out in this direction, although the trend of the basin axis (ENE—WSW) may be followed eastwards into the Nordkapp Basin. A central, elongated, domal structure parallels the basin axis. This is associated with an internal, E-trending fault system. In contrast to the Tromsø Basin, the Hammerfest Basin has not been affected by halokinesis. The Hammerfest Basin is characterized by low free-air (—21 mgal) and Bouguer gravity anomalies.

Age

The main basin subsidence started in late mid Jurassic times, although the Troms-Finnmark Fault Complex may have been active even before this. Basin subsidence was renewed in late Cretaceous times.

Nordkapp Basin (Nordkappbassenget)

Named after the northernmost tip of the Norwegian mainland, at the island of Magerøy.

References

Originally defined by Hinz and Weber (1975). Also treated by Øvrebø and Talleraas (1977), Hinz and Schlüter (1978), Marty *et al.* (1979), Faleide and Gudlaugsson (1981), Rønnevik (1981) and Rønnevik *et al.* (1982, in press).

Position

71°30′—73°30′N, 25°—35°E.

Description

The Nordkapp Basin is a NE-trending, elongated structure. Defined at Upper Devonian—Lower Carboniferous level, the southwestern part of the structure is asymmetric, the northwest flank being defined by a major normal fault. Its southeastern flank consists of a fault zone influenced by an extreme basinward rotation. Via an E-trending central segment, the northeastern part of the basin has a more symmetric graben-like configuration (Rønnevik *et al.*, 1982). The upper part (Jurassic and younger) of the sedimentary pile is deformed mainly by flexuring, and the basin is heavily affected by halokinesis. The Nordkapp Basin is associated with a significant gravity low.

Age

The structure is a late Devonian—early Carboniferous rift basin. Renewed subsidence started in mid Triassic, and continued into early Tertiary times. The later movements are associated with halokinesis.

Comments

The term Nordkapp Basin (North Cape Basin) had been used by Eldholm and Talwani (1977) to include the Hammerfest Basin plus the Nordkapp Basin (present terminology) and a large part of the area south of the Sentralbank High. In view of the original maps presented by Hinz and Weber (1975), and the well established present terminology, this broader use of the term should be avoided.

Varanger Basin (Varangerbassenget)

Named after a peninsula in eastern Finnmark.

Key references

Treated by Rønnevik (1981) and Rønnevik *et al.* (in press).

Position

South of 72°20′N and east of 35°E.

Description

Due to lack of data in the area, not much is known about the structure. The Varanger Basin is affected by halokinesis, mainly salt pillows. The structure is characterized by a gravity low.

Age

The basin was initiated in Triassic times.

Comments

The Varanger Basin may be part of a larger basin. However, due to lack of data the eastern extent of the structure is unknown.

Senja Ridge (Senjaryggen)

Named after the island of Senja, Troms County.

References

Originally defined by Sundvor (1971). Also treated by Hinz and Weber (1975), Rønnevik *et al.* (1975), Syrstad *et al.* (1976), Øvrebø and Talleraas (1977), Hinz and Schlüter (1978), Faleide and Gudlaugsson (1981), Rønnevik *et al.* (1983) and Gabrielsen (in press).

Position

71°00′–72°30′N, 16°30′—18°30′E.

Description

The Senja Ridge is a tectonic high sub-parallel to the continental margin. The structure marks the western margin of the Tromsø Basin, and is defined by two major step-like fault systems, one westerly dipping (on the western side), and one easterly dipping. It is, however, difficult to give a more detailed description of the southern parts of the structure, due to poor data quality. Towards the north, the Senja Ridge is split into several elements, and it terminates south of the Bjørnøya Basin (Faleide and Gudlaugsson, 1981). In detail the Senja Ridge consists of several segments defined by swarms of NNW- and NNE-striking faults, frequently arranged in an *en echelon* pattern. These sets are locally broken by NW-trending structures, giving very complex structural segments. One of these segments is found between 71°05′ and 71°15′N, another north of 71°40′N. The Senja Ridge is situated within a magnetic quiet zone (Faleide and Gudlaugsson, 1981). However, a pronounced positive, N-trending gravity anomaly is associated with the ridge. Syrstad *et al.* (1976) indicated that the gravity anomaly may be explained by a lateral density contrast to the surrounding sediments (2.65 and 2.20 g/cm^3), and a non-magnetic basement structure with a density of 2.75 g/cm^3.

Age

The topography of the higher levels of the Senja Ridge obviously resulted from late Cretaceous to Tertiary movements, which were associated with dextral shear along the continental margin in an early phase of the ocean floor spreading (e.g. Hinz and Weber, 1975; Rønnevik *et al.*, 1975,) although this may be controversial (Spencer *et al.*, 1984). However, the strong indications of relief on the basement cannot be explained by this deformation alone, and earlier movements should be expected.

Comments

The Senja Ridge was defined originally as a continuous tectonic high which might be followed all the way from about 71°00′ to the southern tip of Spitsbergen (Øvrebø and Talleraas, 1977), although the original figure of Rønnevik *et al.* (1975) restricted the structure to the area west of the Tromsø Basin. New data from the area seem to confirm the latter interpretation as the correct one.

Loppa High (Loppahøyden)

Named after a minor island on the coast of Finnmark County.

References

Defined by Rønnevik *et al.* (1975). Recognized as 'Übergangszone—Einheit 6' and western part of 'Transitional Unit—Unit V' by Hinz and Weber (1975) and Hinz and Schlüter (1978). Also treated by Øvrebø and Talleraas (1977), Faleide and Gudlaugsson (1981), Rønnevik (1981) and Rønnevik *et al.* (1983).

Position

72° 00′—73° 30′N, 20° 00′—24° 00′E.

Description

The Loppa High is situated between the Hammerfest and Bjørnøy Basins. The high is wedge-shaped and is bordered on the south, against the Hammerfest Basin, by a E-striking zone of faults and flexures. To the northwest, the Loppa High is bounded towards the Bjørnøy Basin by a NE-trending fault zone. The eastern limit is indicated by a NW-trending fault zone southwest of a major salt structure at approximately 73°15′N, 23°20′E (Kristoffersen and Elverhøi, 1978). The Loppa High is associated with positive gravity and magnetic anomalies. This has been explained by a shallow metamorphic basement containing magnetic bodies underlying the high.

Age

The western boundary fault is believed to be an old zone of weakness (Rønnevik *et al.*, in press) whereas the present outline of the structure is a result of Kimmerian tectonism.

Comments

The original name of the structure was the Loppa Ridge. Considering the triangular shape of the structure, the more neutral term Loppa High is proposed.

Bjørnøy Basin (Bjørnøybassenget)

Named after an island between Spitsbergen and the Norwegian mainland.

References

Originally defined by Øvrebø and Talleraas (1977). Also treated by Hinz and Schlüter (1978), Marty *et al.* (1979), Faleide and Gudlaugsson (1981), Rønnevik (1981) and Rønnevik *et al.* (1982, 1983).

Position

72° 30′—74° 30′N, 18° 00′—25° 30′E.

Description

The NE-trending Bjørnøy Basin is a complex half-graben-like structure. It is fault-bounded against the Loppa Ridge towards the southeast, but the nature of its northern margin is unknown. The eastern, shallower part of the basin is heavily faulted, and indications of compression have been reported from this area (Rønnevik *et al.*, 1982). The Bjørnøy Basin is associated with a negative gravity anomaly (Faleide and Gudlaugsson, 1981).

Age

Basin subsidence probably started in late mid Jurassic and continued into Lower Tertiary times.

Stappen High (Stappenhøyden)

Named after the southernmost point of Bjørnøya.

References

Described by Faleide and Gudlaugsson (1981), who used the term Bjørnøy High. See also Rønnevik *et al.* (1982).

Position

The island of Bjørnøya marks the highest point of the structure which is situated between 73° and 75°N.

Description

The Stappen High is bounded to the west by the southern segment of the NNW-trending Hornsund Fault Zone. The eastern flank coincides with the western margin of the Bjørnøy Basin. To the north, the structure is limited by the Sørkapp Basin. Faleide and Gudlaugsson (1981) indicated that the high has a shallow core of basement rocks.

Age

The Stappen High probably developed as a positive element in late Palaeozoic times, according to Faleide and Gudlaugsson (1981).

Comments

The name 'Bjørnøya High', proposed by Faleide and Gudlaugsson (1981), is considered unfortunate inasmuch as the term Bjørnøy Basin had already been in use for a long time. As the term 'Bjørnøya High' has not yet been used widely, the name 'Stappen High' is preferred.

Sørkapp Basin (Sørkappbassenget)

Named after the southernmost tip of Spitsbergen.

References

Identified by Faleide and Gudlaugsson (1981) as the Hopen Basin. See also Rønnevik *et al.* (1983).

Position

Approximately 74°—76°N, 21°30′—23°E.

Description

The basin trends NW and is fault-bounded on the southwestern border. The lower part of the sedimentary sequences (Permian) seems to thicken towards the east, whereas the opposite is the case with the younger sedimentary pile, indicating a complicated tectonic history (Faleide and Gudlaugsson, 1981). The eastern flank of the basin coincides with a gravity low, which may be followed from the southern tip of Spitsbergen.

Age

According to Faleide and Gudlaugsson (1981), the Sørkapp Basin area was influenced by block-faulting in pre-Permian time. The border fault systems were reactivated later.

Comments

The structure was termed Hopen Basin by Faleide and Gudlaugsson (1981). This name has not been published, however, and we would suggest that the term Sørkapp Basin is used to avoid confusion with the Hopen High.

Sentralbank High (Sentralbankhøyden)

Named after a shallow part of the Barents Sea.

References

Defined by Rønnevik *et al.* (1982) and Bleie *et al.* (1982). See also Rønnevik *et al.* (1983).

Position

74°00′—76°30′N, 26°00′—35°00′E.

Description

The Sentralbank high is a NE-trending structural high of which both the northwestern and southwestern boundaries are reverse faults, probably related to wrenching. The structure is on trend with the northwestern boundary of the Loppa Ridge, and may be part of a complex NE-oriented fault zone crossing most of the Barents Sea Shelf (Rønnevik *et al.*, 1982). From the present data, no gravimetric or magnetic anomalies are known to be associated with the Sentralbank High.

Age

The major movements along the boundary faults of the Sentralbank High took place during the Kimmerian tectonic phase. The faults are, however, characterized by reactivation.

Hopen High (Hopen høyden)

Named after the island of Hopen.

References

The name was introduced by Faleide and Gudlaugsson (1981), and the structure defined by Rønnevik *et al.* (1983).

Position

74°00′—77°00′N, 25°00′—29°00′E.

Description

The Hopen High is a large, NE-trending domal structure, not bounded by faults, with an estimated depth to basement of only 2 km (Faleide and Gudlaugsson, 1981). Although the detailed outline of the Hopen High is unknown, the northwestern crest of the structure seems to be situated northwest of the island of Hopen. The structure is associated with a positive gravity anomaly.

Age

The structure is probably related to post-Kimmerian deformation.

Comments

The original definition of Faleide and Gudlaugsson (1981) identified the Hopen High as a positive gravity element. Recent mapping in the area indicates that the structure is a large structural element (see Rønnevik *et al.*, 1982).

ACKNOWLEDGEMENTS

We would like to express our gratitude to the Norwegian Petroleum Directorate, Norsk Hydro a.s., Saga Petroleum a.s., and Statoil for giving us the opportunity to fulfil this work which was initiated by KGG (Komité for geofysikk og geologi). The work has benefited from the comments of several colleagues: Svein Eggen, Jan Inge Faleide, Martin B. Gowers, Nils B. Hollander, Charlotte Robinson, Anthony M. Spencer, Bjarne Tveiten and Jan Vollset. Their advice has been of great value.

Charlotte Robinson and Astrid Øverland, both of Saga Petroleum a.s., corrected the English text, and typed numerous versions of the manuscript, respectively. The figures were drafted by Ingrid Crosby, Tom Johansen and Vidar Olafsen. We wish to thank them all for their assistance and patience.

REFERENCES

Bates, R.L. and Jackson, J.A. 1980. *Glossary of Geology*, 2nd edn. American Geological Institute, Falls Church, VA, 751 pp.

Bleie, J., Oppebøen, K.A. and Nysœther, E. 1982. The hydrocarbon potential of the northern Norwegian shelf in the light of recent drilling. *Offshore Northern Seas 1982. Technology Conference and Exhibition*, Stavanger 1982, E/5, pp. 1—39.

Bøen, F., Eggen, S. and Vollset, J. 1984. Structures and basins of the margin from 62 to 69°N and their development. This volume, pp. 253—270.

Dalland, A. 1975. The Mesozoic rocks of Andøya, Northern Norway. *Nor. Geol. Unders.*, **316**, 271—287.

Dalland, A. 1979. The sedimentary sequence of Andøya, northern Norway — Depositional and structural history. *Proceedings, Norwegian Sea Symposium*, Tromsø 1979. Norwegian Petroleum Society, NNS/26, pp. 1—31.

Eldholm, O. and Talwani, M. 1977. Sediment distribution and structural framework of the Barents Sea. *Geol. Soc. Am. Bull.*, **88**, 1015—1029.

Eldholm, O. and Talwani, M. 1981. The passive margin of Northern Europe and East-Greenland. In: Dynamics of Passive Margins. *Geodynamics Ser.*, **6**, 30—44.

Faleide, J.I. and Gudlaugsson, S.T. 1981. Geology of the Western Barents Sea — a Regional Study Based on Marine Geophysical Data. Unpubl. cand. real. thesis, Univ. of Oslo, 160 pp.

Fønstelien, S. and Horvei, A. 1979. Thoughts and considerations arising from the study of a 'prospective' feature in the Troms I area. *Proceedings, Norwegian Sea Symposium*, Tromsø 1979. Norwegian Petroleum Society, NSS/20, pp. 1—10.

Gabrielsen, R.H. Long-lived fault zones and their influence on the tectonic development of the southwestern Barents Sea. *J. Geol. Soc. Lond.* (in press).

Gabrielsen, R.H. and Ramberg, I.B. 1979. Fracture patterns in Norway from LANDSAT imagery: results and potential use. *Proceedings, Norwegian Sea Symposium*, Tromsø 1979. Norwegian Petroleum Society, NSS/23, pp. 1—28.

Gabrielsen, R.H. and Robinson, C. 1984. Tectonic inhomogenities of the Kristiansund-Bodø Fault Complex offshore mid-Norway. This volume, pp. 397—406.

Gloppen, T.G. and Westre, S. 1982. Petroleum potential big off northern Norway. *Oil Gas J.*, **82** (23), 114—127.

Gowers, M.B. and Lunde, G. 1984. The geological history of Trænabanken. This volume, pp. 237—252.

Hagevang, T., Eldholm, O. and Aalstad, I. 1983. Pre-23 magnetic anomalies between Jan Mayen and Greenland-Senja Fracture Zones in the Norwegian Sea. *Mar. Geophys. Res.*, **5**, 345—363.

Hamar, G.P. and Hjelle, K. 1984. Tectonic framework of the Møre Basin and the northern North Sea. This volume. pp. 349—358.

Hamar, G.P., Jacobsen, K.H., Ormaasen, D.E. and Skarpnes, O. 1980. Tectonic development of the North Sea north of the Central Highs. In: *The sedimentation of the North Sea Reservoir Rocks.* Norwegian Petroleum Society, 23 pp.

Harland, W.B. 1965. The tectonic evolution of the Arctic—North Atlantic region. *R. Soc. Lond. Phil. Trans., A,* **258**, 59—75.

Harland, W.B. 1969. Contribution of Spitsbergen to understanding of tectonic evolution of North Atlantic region. *Am. Assoc. Petrol. Geol. Mem.*, **12**, 817—851.

Hinz, K. 1972a. The seismic crustal structure of the Norwegian continental margin in the Vøring Plateau, in the Norwegian deep sea and the eastern flanks of the Jan Mayen Ridge between 66 and 68°N. *Int. Geol. Congr. Montreal, Ser.*, **8**. 28—37.

Hinz, K. 1972b. Der Krustenaufbau der Norwegischen Kontinentalrandes (Vøring Plateau) und der Norwegischen Tiefsee zwischen 66° und 68°N nach seismischen Untersuchungen. *Meteor Forsch. -Ergebnisse,* **10**, 1—16.

Hinz, K. and Moe, A. 1971. Crustal structure in the Norwegian Sea. *Nature Phys. Sci.*, **232**, 187—190.

Hinz, K., Dustmann, H.J. and Hanisch, J. 1982. Structural framework of the Norwegian Sea. *Offshore Northern Seas 1982. Technology Conference and Exhibition,* Stavanger 1982, E/4, pp. 1—12.

Hinz, K. and Schlüter, H.-U. 1978. The geological structure of the western Barents Sea. *Mar. Geol.*, **26**, 199—230.

Hinz, K. and Weber, J. 1975. Zum geologischen Aufbau des Norwegischen Kontinentalrandes un der Barents-See nach reflexionsseismische Messungen. Compendium 75/75, Erganzungsband der Zeutschrift *Erdöl Kohle, Erdgas, Petrochemie,* pp. 3—29.

Hollander, N.B. 1981. Vurdering av olje/gass potensialet i Haltenbanken-området i lys av det første borehullet. (Abstract) *Geolognytt,* **15**, 21.

Johnson, G.L., Ballard, J.A. and Watson, J.A. 1968. Seismic studies of the Norwegian continental margin. *Nor. Polarinst. Årb.* **1966**, 112—119.

Johnson, G.L. and Heezen, B.G. 1967. The morphology and evolution of the Norwegian-Greenland Sea. *Deep-Sea Res.*, **14**, 755—771.

Jørgensen, F. and Navrestad, T. 1979. Main structural elements and sedimentary succession on the shelf outside Nordland (Norway). *Proceedings, Norwegian Sea Symposium,* Tromsø 1979. Norwegian Petroleum Society, NSS/11, pp. 1—20.

Jørgensen, F. and Navrestad, T. 1981. The geology of the Norwegian Shelf between 62°N and the Lofoten Islands. In: L.V. Illing and G.D. Hobson (eds.), *Petroleum Geology of the Continental Shelf of Northwest Europe,* Institute of Petroleum, London, pp. 407—413.

Kristoffersen, Y. and Elverhøi, A. 1978. A diapir structure in Bjørnøyrenna. *Nor. Polarinst. Årbok.* **1977**, 189—198.

Larsen, R.M. and Skarpnes, O. 1984. Regional interpretation and hydrocarbon potential of the Trœnbanken area. This volume, pp. 217—236.

Malod, J. and Mascle, J. 1975. Structures geologiques de la marge continentale a l'ouest du Spitzberg. *Mar. Geophys. Res.*, **2**, 215—229.

Marty, A., Rabate, J., Cazes, Jacquart, G. and Auxietre, J.L. 1979. CEPM Barents Sea seismic survey. *Proceedings, Norwegian Sea Symposium.* Norwegian Petroleum Society, NSS/16, pp. 1—16.

Moe, A. 1974. Continental shelf north of 62° latitude. *Offshore Northern Seas 1974. Technology Conference and Exhibition,* Stavanger 1974, G-IV/5, 1—18.

Myhre, A., Eldholm, O. and Sundvor, E. 1982. The margin between Senja and Spitsbergen Fracture Zone: implication from plate tectonics. *Tectonophys.*, **89**, 33—50.

Nansen, F. 1904. The bathymetrical features of the North Polar Seas, with a discussion of the continental shelves and previous oscillations of the shore-line. Norw. North Polar Exp. 1893—1896. *Sci. Res.*, **4**, 1—232.

Oftedal, C. 1972. A sideritic ironstone of Jurassic age in Beistadfjorden, Trøndelag. *Nor. Geol. Tidsskr.*, **52**, 123—134.

Oftedahl, C. 1975. Middle Jurassic graben tectonics in mid-Norway. *Proceedings, Jurassic Northern North Sea Symposium,* Stavanger 1975, JNSS/21, pp. 1—13.

Price, I. and Rattey, R.P. Cretaceous tectonics of mid-Norway: Implications for the Rockall Faeroes—Shetland Rift. *J. Geol. Soc. Lond.* (in press).

Priem, H.N.A., Verschure, R.H., Boelrijk, N.A.I.M., Hebedea, E.H. and Thorkildsen, Chr. D. 1968. Rb—Sr and K—Ar age measurements on phlogopitic biotite from the ultrabasic lamprophyre dike on the island of Ytterøy, Trondheimsfjord, Norway. *Nor. Geol. Tidsskr.*, **48**, 319—321.

Rodgers, D.A. 1980. Analysis of pull-apart basin development produced by *en echelon* strike-slip faults. In: P.F. Ballance and H.G. Reading, (eds.). Sedimentation in Oblique-slip Mobile Zones, Spec. Publ., *Int. Ass. Sediment.*, **4**, 27—41.

Rønnevik, H.C. 1981. Geology of the Barents Sea. In: L.V. Illing and G.D. Hobson (eds.), *Petroleum Geology of the Continental Shelf of Northwest Europe,* Institute of Petroleum, London, pp. 395—406.

Rønnevik, H. and Navrestad, T. 1976. Geology of the Norwegian shelf between 62°N and 69°N. *Offshore North Sea 1976. Technology Conference and Exhibition,* Stavanger 1976, GIV/4, pp. 1—24.

Rønnevik, H. and Navrestad. T. 1977. Geology of the Norwegian Shelf between 62°N and 69°N. *GeoJournal,* **1**, 33—45.

Rønnevik, H.C., Bergsaker, E.I., Moe, A., Øvrebø, O., Navrestad, T. and Stangenes, J. 1975. The geology of the Norwegian continental shelf. In: A.W. Woodland (ed.), *Petroleum and the Continental Shelf of Northwest Europe, Vol. 1, Geology,* Institute of Petroleum, London, pp. 117—129.

Rønnevik, H., Beskow, B. and Jacobsen, H.P., 1982. Structural and stratigraphic evolution of the Barents Sea. *Offshore North Sea 1982. Technology Conference and Exhibition,* Stavanger 1982, 29 pp.

Rønnevik, H., Eggen, S. and Vollset, J. 1983. Exploration of the Norwegian Shelf. In: J. Brooks (ed.), *Petroleum Geochemistry and Exploration of Europe.* Blackwell, Oxford, pp. 71—93.

Rønnevik, H., Jørgensen, F. and Motland, K. 1979. The geology of the northern part of the Vøringplateau. *Proceedings, Norwegian Sea Symposium,* Tromsø 1979. Norwegian Petroleum Society, NSS/12, pp. 1—12.

Råheim, A. 1974. A post Caledonian syenite porphyry dyke in the Western Gneiss Region, Tusna, Central Norway. *Norsk Geol. Tidsskr.* **54**, 139—147.

Sellevoll, M.A. 1975. Seismic refraction measurements and continuous seismic profiling on the continental margin off Norway between 60°N and 69°N. *Nor. Geol. Unders.*, **316**, 219—235.

Siedlecka, A. 1975. Old red sandstone lithostratigraphy and sedimentation of the Outer Fosen area, Trondheim Region. *Nor. Geol. Unders.*, **321**, 1—35.

Siedlecka, A. and Siedlecki, S. 1972. A contribution to the geology of the Downtonian sedimentary rocks of Hitra. *Nor. Geol. Unders.*, **275**, 1—28.

Spencer, A.M., Home, P.C. and Berglund, L.T. 1984. Tertiary structural development of the western Barents Shelf: Troms to Svalbard. This volume, pp. 199—210.

Storetvedt, K.M. 1967. Magnetic properties of an ultrabasic dyke from the island of Ytterøy, Norway. *Nor. Geol. Tidsskr.* **47**, 171—184.

Sturt, B.A., Pringle, I.R. and Ramsay, D.M. 1978. The Finnmarkian phase of the Caledonian orogeny. *J. Geol. Soc. Lond.*, **135**, 597—610.

Sundvor, E. 1971. Seismic refraction measurements on the Norwegian continental shelf between Andøy and Fugløybanken. *Mar. Geophys. Res.*, **1**, 303—313.

Sundvor, E. and Eldholm, O. 1976. *Marine Geophysical Survey on the Continental Margin from Bear Island to Hornsund, Spitsbergen.* Sci. Rep. No. 3, Univ. of Bergen, Seismological Observatory, 28 pp.

Sundvor, E. and Eldholm, O. 1979. The western margin off Svalbard. *Tectonophysics,* **59**, 239—250.

Sundvor, E., Myhre, A. and Eldholm, O. 1979. The Svalbard continental margin. *Proceedings, Norwegian Sea Symposium,* Tromsø 1979. Norwegian Petroleum Society, NSS/6, pp. 1—25.

Sykes, L.R. 1965. The seismicity of the Arctic. *Seismol. Soc. Am. Bull.*, **55**, 501—518.

Syrstad, E., Talleraas, T. and Bergseth, S. 1976. Gravity modelling offshore Troms, northern Norway. *Offshore Northern Seas 1976. Exploration Geology and Geophysics, Technology Conference and Exhibition,* Stavanger 1976.

Talleraas, E. 1979. The Hammerfest Basin — an aulacogen. *Proceedings, Norwegian Sea Symposium.* Norwegian Petroleum Society, NSS/18, pp. 1—13.

Talwani, M. and Eldholm, O. 1972. Continental margin off Norway: A geophysical study. *Geol. Soc. Am. Bull.*, **83**, 3575—3606.

Talwani, M. and Eldholm, O. 1977. Evolution of the Norwegian-Greenland Sea. *Geol. Soc. Am. Bull.*, **88**, 969—999.

Vogt, P., Perry, R.K., Feden, R.H., Fleming, H.S. and Cherkis, N.Z. 1981. The Greenland-Norwegian Sea and Iceland environment: geology and geophysics. In: A.E.M. Nairn, M. Churkin and F.G. Stehli (eds.), *The Ocean Basins and Margins,* Vol. 5, Plenum Press, New York, pp. 493—598.

Øvrebø, O. and Talleraas, E. 1977. The structural geology of the Troms area (Barents Sea). *GeoJournal,* **1**, 47—54.

Åm. K. 1975. Aeromagnetic investigations on the continental shelf of Norway, Stad-Lofoten (62—69°). *Nor. Geol. Unders.*, **266**, 49—61.

Åm. K. 1975. Aeromagnetic basement complex mapping north of latitude 62°N, Norway. *Nor. Geol. Unders.*, **316**, 351—374.

Aanstad, K.M., Gabrielsen, R.H., Hagevang, T., Ramberg, I.B. and Torvanger, O., 1981. Correlation of offshore and onshore structural features between 62°N and 68°N, Norway. *Proceedings, Norwegian Symposium on Exploration,* Bergen 1981, NSE/11, pp. 1—25.

5 Depositional environment and diagenesis of Jurassic reservoir sandstones in the eastern part of Troms I area

S. Olaussen, A. Dalland,*[†§] *T.G. Gloppen* and E. Johannessen*[‡]
*Statoil Geologisk Laboratorium Stavanger, Norway
[†]Statoil, Bergen, Norway [‡]Statoil, LET III, Harstad, Norway

Cores from five wells within the eastern part of Troms I contain mineralogically and texturally mature sandstones (generally classified as quartz arenites) deposited on a stable platform during the Middle and Lower Jurassic. Deposition took place under an overall transgressive regime. The delicate balance between sediment input and basin subsidence is exemplified by the development of a very thick tidal flat sequence, the occurrence of phosphate, glauconite and chamosite, and the supermaturity of the sandstones and the lateral extension of the facies. Two different facies associations are distinguished: (1) a coastal plain/tidal flat facies association (Nordmela Fm); (2) a near shore and inner shelf facies association (Stø Fm, which overlies 1). The main reservoir sand bodies are found within the latter association.

Both regional metamorphic (e.g. Caledonian and Precambrian) and sedimentary rocks (e.g. Upper Palaeozoic) are suggested as the major provenance for the sandstones.

The excellent primary porosity in the mature sandstones has been variably reduced by silica and carbonate cementation. Stylolites, well-developed in tight sands in the western part of the area, indicate that pressure solution was a major process in generating silica cement. Secondary porosity is created by leaching of carbonate cement.

Diagenesis is partly controlled by the depositional environment. However post-depositional tectonics are interpreted as the major controlling factor of the porosity distribution.

INTRODUCTION

The Troms I area is situated between 71° and 71°30'N and between 17°40' and 21°E (Fig. 1). Before the 1983 drilling season nine wildcat wells were drilled in six of the allocated blocks. Six more wells are planned for 1983 and five of these are wildcats. The area was opened for drilling in spring 1980, and the first significant hydrocarbon find was made by Statoil in 1981. Gas was then found in a fault-bounded structure in block 7120/8 (the Askeladden field). To date, gas in non-commercial quantities has been found in four other wildcats on separate structures (Fig. 1). Five wells drilled by Statoil in the eastern part of Troms I have been used in this study. The wells are: 7119/12-1 (1980), 7119/12-2 and 7120/8-1 (1981), 7120/8-2 and 7120/7-1 (1982).

Five important structural features have been

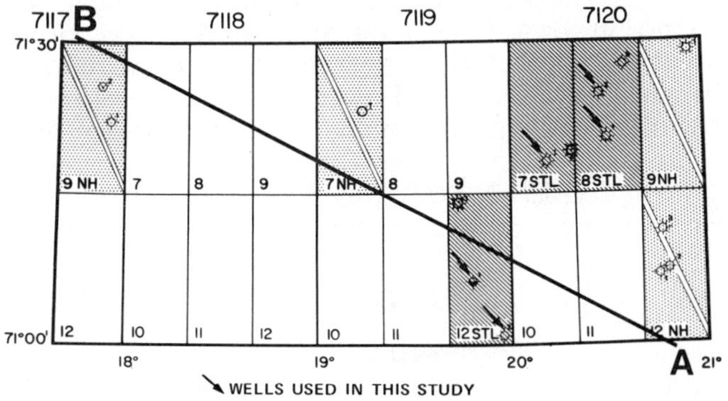

Fig. 1. Map of the Troms I area with well status. STL: Statoil operated licences, NH: Norsk Hydro operated licences.

distinguished in the Troms I area (Figs. 1 and 2. For details see Westre, 1984):

(1) The Troms-Finnmark Fault Complex, a NE-trending zone of apparently normal faults separating the Upper Palaeozoic to Cenozoic sedimentary basins from the more stable platform area to the southeast.

(2) The Hammerfest Basin is an east—west trending basin containing Upper Palaeozoic to

[§] Currently with Statoil, Bergen, Norway.

Petroleum Geology of the North European Margin © Norwegian Petroleum Society (Graham & Trotman, 1984) pp. 61—79.

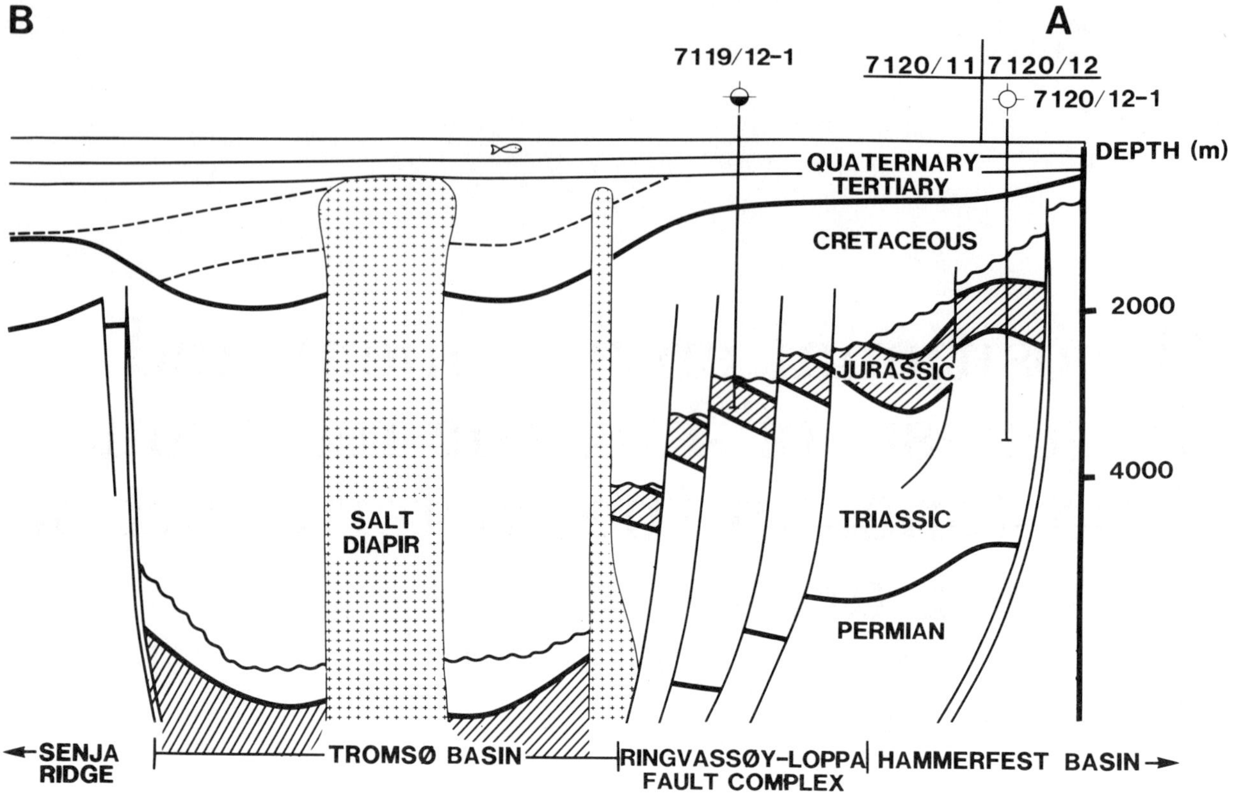

Fig. 2. Structural cross section of the Troms I area along line A—B in Fig. 1.

Cenozoic sediments, affected by minor Mesozoic tectonism.

(3) The transition zone between the Hammerfest and the Tromsϕ basins, the Ringvassϕy-Loppa Fault Complex, is intensely tectonized and dominated by extensional tectonics and tilted fault blocks.

(4) The Tromsϕ Basin, a very deep sedimentary basin, contains large amounts of Upper Palaeozoic, Mesozoic and Cenozoic sediments. The basin is dominated by salt tectonism (Fig. 2).

(5) The Senja Ridge, a structural high of post-Cenomanian age, limits the Tromsϕ Basin toward the west.

The nearest rock outcrops of comparable age are found in East Greenland and on Andϕya. Both areas display rocks of paralic and shallow marine origin. In Svalbard the Lower Jurassic is represented by marine shales and marginal marine shales and sandstones, while the Middle Jurassic section consists of a single remaniè conglomerate.

STRATIGRAPHY

The succession considered here has been named the Nordmela and Stϕ Formations by Skarpnes *et al.* (in preparation). The formations investigated (Table 1) are readily correlated in the wells.

The age of the reservoir section in the different wells is generally early to Middle Jurassic, although dating is still in progress. The dating of the section seems at present to reflect differences in the biostratigraphic schemes used by the various consultants. The accuracy of the formation dating will naturally improve as more data become available. The formation ages must therefore be regarded as preliminary.

TABLE 1. Stratigraphic data (All depths are log depth — RKB measured from drill floor.)

				Well Number Log depth (m) (RKB)				
Age	*Fm*	*Unit*		*7119/12-1*	*7119/12-2*	*7120/8-1*	*7120/8-2*	*7120/7-1*
Toarcian → Bathonian	Stϕ	S-III	Top	2658.0	1372.0	2092.0	2081.0	2408.0
			Thickness	56.5	61.5	24.5	22.5	34.0
			Base	2714.5	1433.5	2116.5	2103.5	2442.0
		S-II	Thickness	26.5	22.5	23.0	23.5	19.0
			Base	2741.0	1456.0	2139.5	2127.0	2461.0
		S-I	Thickness	70.0	61.5	48.5	43.0	58.0
			Base	2811.0	1517.5	2187.0	2170.5	2519.0
Undiff. Sinem. -- Toarcian	Nordmela	N-III	Thickness	34.0	31.5	18.0	10.0	9.0
			Base	2845.0	1549.0	2205.5	2180.5	2528.0
		N-II	Thickness	80.0	74.0	73.5	66.5	74.5
			Base	2925.0	1623.0	2279.0	2247.0	2602.5
		N-I	Thickness	80.0	96.0	51.0	57.0	48.5
			Base	3005.0	1719.0	2330.0	2294.0	2651.0

LITHOLOGY AND DEPOSITIONAL ENVIRONMENT

The succession is subdivided into two formations and six units (Table 1) reflecting two distinct facies associations: a lower, coastal plain/tidal flat sequence — The Nordmela Fm — and an upper, near shore and inner shelf sequence — the Stø Fm (the hydrocarbon-bearing sedimentary rocks in the Askeladden field).

The Nordmela Formation

Based on the depositional style and wireline log characteristics, the Nordmela Fm is subdivided into three units (N-I to N-III) which can be correlated throughout the eastern part of the Troms I area (Fig. 3). Continuous cores from this sequence are available only from block 7119/12 and most of the following interpretation is based on data from this block.

The Nordmela Fm is of early Jurassic age. Evidence for the presence of Hettangian sediments is not conclusive, and at present the most precise dating for these sedimentary rocks is undifferentiated Sinemurian—Toarcian.

The formation consists of very fine to fine-grained sandstone with increasing amounts of coarser-grained channel sandstones upwards (Fig. 3). The sandstones are frequently interbedded with thin mudstone layers. Plant roots and marine bioturbation occur together in the same bed or in alternating beds. Flaser and lenticular bedding, wave ripples, and very thin coal beds are common features. Herring-bone cross stratification is observed occasionally.

Unit N-I

This interval is about 80 m thick in well 7119/12-2, from which a nearly complete succession of cores has been studied. It is thinner (about 50 m) to the northeast, in blocks 7120/7 and —/8. The lower boundary of Unit N-I is well defined and marked by gamma log shifts from well defined box-like character to linear serrated pattern. Linear-, cylinder, minor bell and funnel serrated gamma log motifs characterize this unit.

The cores from block 7119/12 show that the unit consists of light-coloured, very fine to fine grained sandstone, frequently alternating with dark gray to black mudstone beds. Rootlets and very thin coal seams are common, as is bioturbation, mainly of marine affinity. Wave generated ripples (Fig. 4b) and flaser bedding (Fig 4a) also characterize the unit, while current ripples (sometimes 'Herring-bone', Fig. 4e) and horizontal parallel lamination are less common. The beds are often arranged into small coarsening upwards sequences, 1—3-m thick (Fig. 3).

Unit N-I is interpreted as a tidal flat/marsh deposit, mainly because of the intense intermixing of fossil plant roots and marine bioturbation combined with flaser bedding and herring-bone cross-stratification. The co-existence of marine trace fossils and rootlets in the same bed or in alternating beds throughout the succession may give the impression that the deposition took place exactly at the shoreline. This is to some extent misleading, since penetration by roots and by burrowing animals commonly occurs well after the actual time of deposition of the bed. Comparing the section with modern tidal flats, we may assume that most of the sand was deposited at the outer (higher energy) part of the tidal flat. Chenier plains may also be considered. The mudstone layers were deposited mainly in the low-energy inner tidal flat and marsh areas. A few of the mudstone belts may represent true marine sedimentation from the seaward side of the sand flat area.

Since a major part of shallow marine sandstone sequences is deposited along delta fronts, we also have to consider possible deltaic environments in order to explain the succession described above. Repeated progradation of crevasse splay deposits into an interdistributary bay could result in a succession with coarsening upwards sandstone sequences similar to those described. The sediment introduced to the bay area may have undergone sufficient reworking by waves and tides. There are a few objections to this interpretation, however. We find no delta front or prodelta deposits underneath this unit. Distributary bay deposits are nearly always mud-dominated, while Unit N-I is dominated by sandstones (70—80%). Coals are commonly well developed in interdistributary bay sequences — only thin coal streaks are found in Unit I. Grain size of the sandstone, type of sedimentary structures, and degree of bioturbation should exhibit great variation from the proximal to distal parts of the splays. The coarsening upwards units in this case show very little variation regarding these aspects. Levees and distributary channel fills cannot be identified in the succession.

The reservoir quality of the facies is generally poor, but not negligible.

Unit N-II

The base is defined at the first significant high amplitude of the gamma log above Unit N-I. The unit is similar to Unit N-I, but has a higher sandstone—mudstone ratio and a clearer separation of sandstone and mudstone beds, which gives the gamma ray log a distinctly higher amplitude than for the underlying unit. In addition, bell-shaped and cylinder-shaped serrated log appearances are more obvious in this unit. Unit N-II contains upwards fining sequences. Thin coal beds and coalified plant fragments occur more frequently, and flaser bedding is less common than in Unit N-I.

Unit N-II is about 100-m thick in 7119/12-1 and 2. As in Unit N-I, it thins towards the northeast, where it is about 70-m thick. The above description is based mainly on the study of cores from 7119/12-2.

The main difference from Unit N-I is the presence of sharp-based fining upward sandstone sequences. These appear to be channel filled or point bars, probably associated with meandering tidal channels. In 7119/12-2 the lowermost of the coarse-grained sand sequences, between 1615 and 1620 m RKB, has an opposite trend in grain size, though a sharp base with coal clasts forms the base. It is probably not a bar sequence, but rather a succession of two or more superimposed channel fills. The reason for not interpreting the channels as fluvial is the close association with marine trace fossils. These occur both in the upper part of some of the fining-upwards sequences and immediately

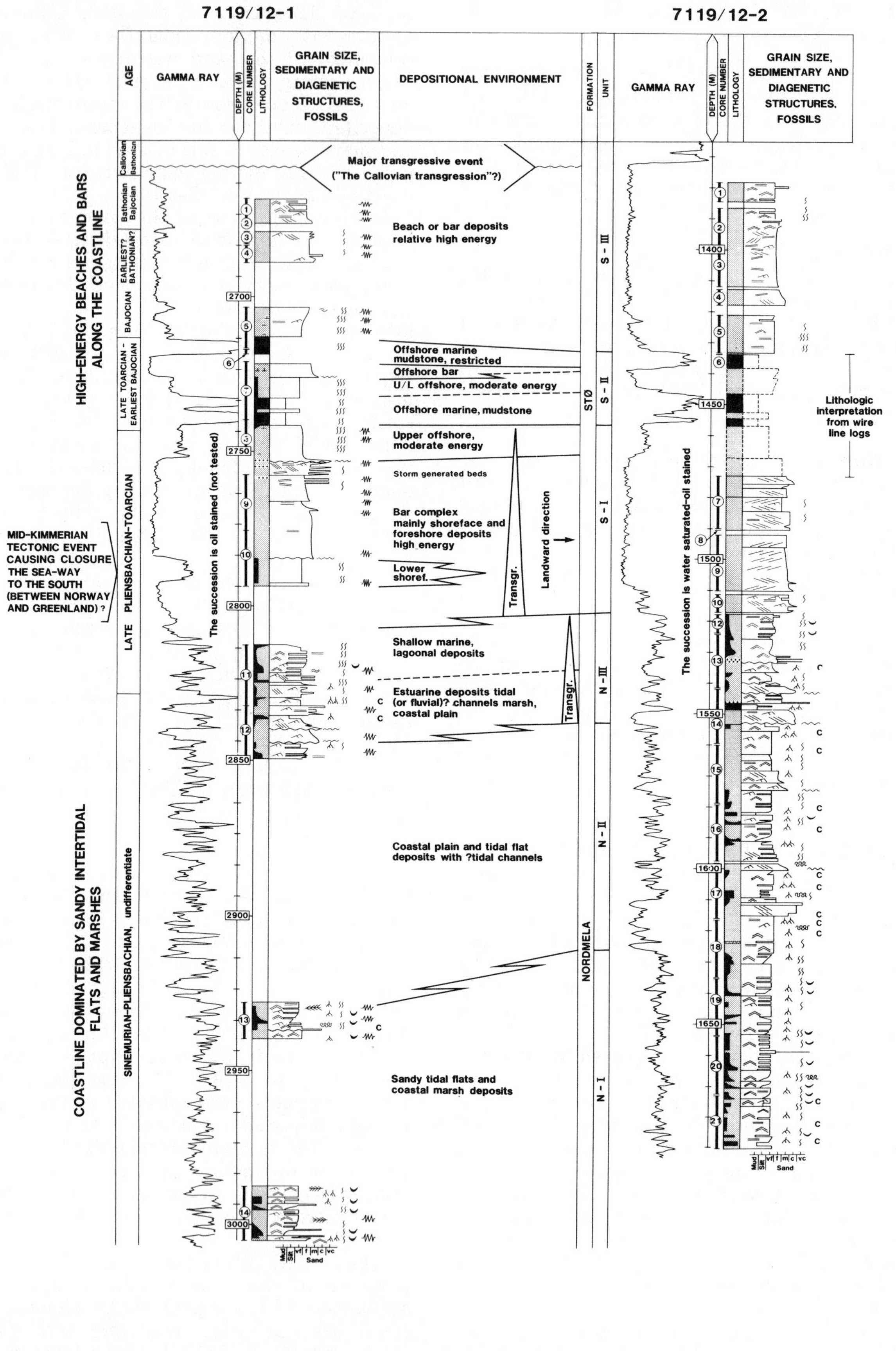

Fig. 3. Profiles of the wells, all depths are log depth, RKB. Note, due to difference in compaction the depositional

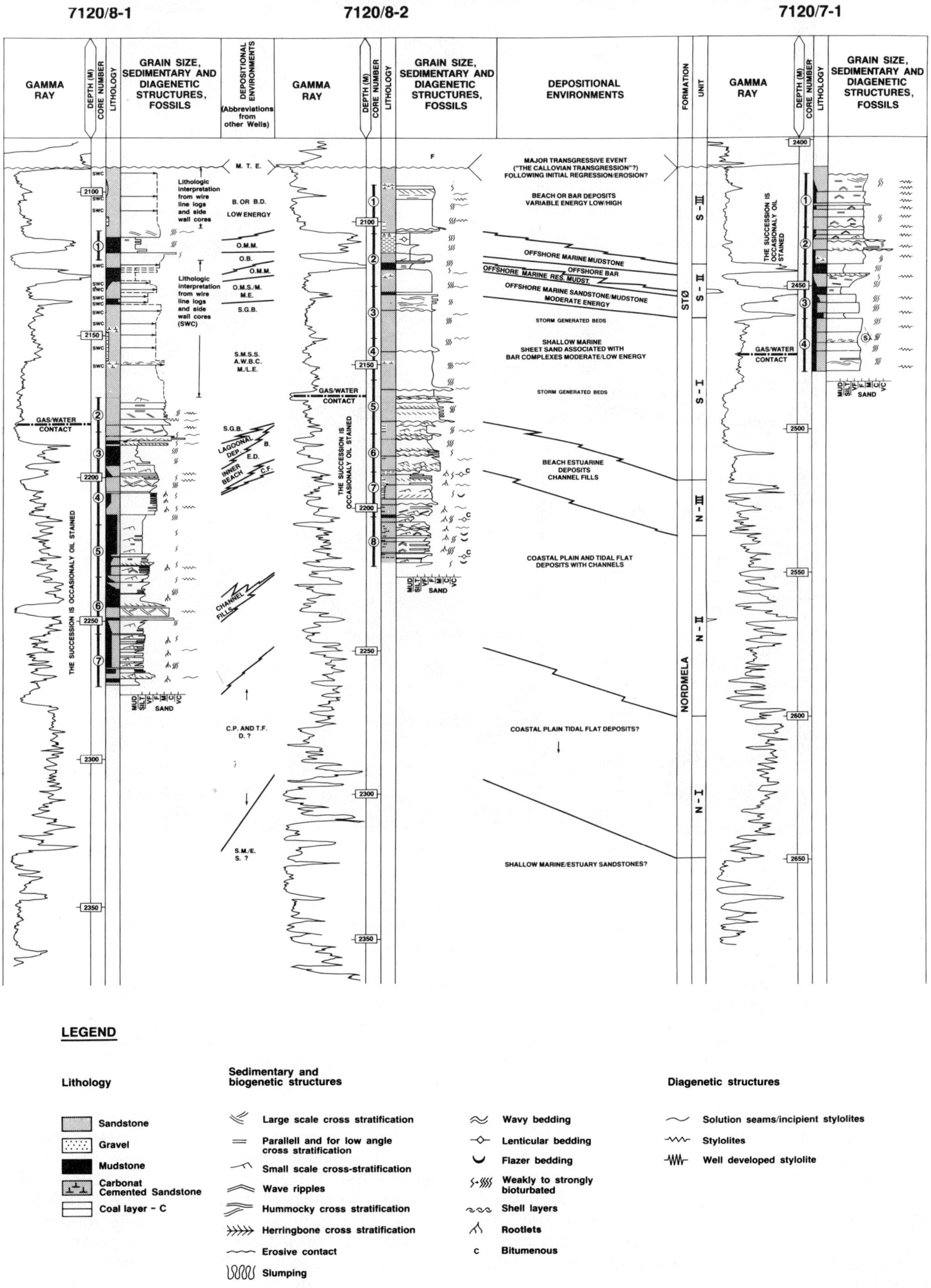

thickness of Unit S-III in the Stø Formation in 7119/12-1 and 7119/12-2 are approximately the same.

Fig. 4. Sedimentary structures of the Nordmela Fm — coastal plain—tidal flat deposits. All core sections are 15 cm wide (a) Sedimentary structures (2929 m) in 7119/12-2 all comparable with recent tidal deposits. Flaser bedding and rip up clasts with ripple lamination pass into lenticular and wavy bedding with convolute bedding on top. (b) Burrowed horizon passing up into wave ripples. Uppermost burrowed horizon (grey) is cemented by siderite. Note lack of compaction of these burrows (3000 m, 7119/12-2). (c) Small scale upward coarsening bed with rootlets capped by coaly shale (2208 m 7120/8-1). (d) Flaser and lenticular bedded sandstone with large burrows? (2249.5 m, 7120/8-1). (e) Herringbone cross strata (2932 m, 7119/12-1). (f) Low angle cross-strata with siderite layer and clast. Arrow pointing at foresets (showing a herring-bone appearance) (Unit N-III, 2199.5 m, 7120/8-1). (g) *Diplocraterion* burrow, in fine grained sandstone which is overlain by coarse grained, normal graded sandstone bed, probably a storm layer. Shore face, unit N-III (1553.2 m 7119/12-2).

beneath their bases. The development of tidal channels in Unit N-II, compared with Unit N-I where they are not observed, may be related to factors such as tidal range, proximity to estuaries and average energy. Tidal flats and marshes of today show great variation in the density and size of tidal creeks and channels. There are also strong differences in the rate of lateral migration of tidal channels from place to place.

In Unit N-II there are several differences in the depositional style between wells 7119/12 and 7120/8. Large escape burrows and branched (?crustacea) burrows (Fig. 4d) are common throughout the cored sections in the 7120/8 wells as opposed to the 7119/12 wells. In 7120/8-1 a fining upward sequence capped by non-bioturbated mud/siltstone and coal is interpreted as fluvial deposits, possibly point bar. Characteristic bell shaped, serrated gamma log traces are consistent with this interpretation. In addition, thicker mudstone sequences occur in the 7120/8 wells. This may indicate that Unit N-II in the 7120/8 block was in a more proximal position on the coastal plain where fluvial processes were more dominating and the tidal influence less important than in block 7119/12.

Reservoir quality: fair to good, with partly sheet-like sandstone bodies and isolated channel sandstones.

Unit N-III

The base is defined at the first appearance of

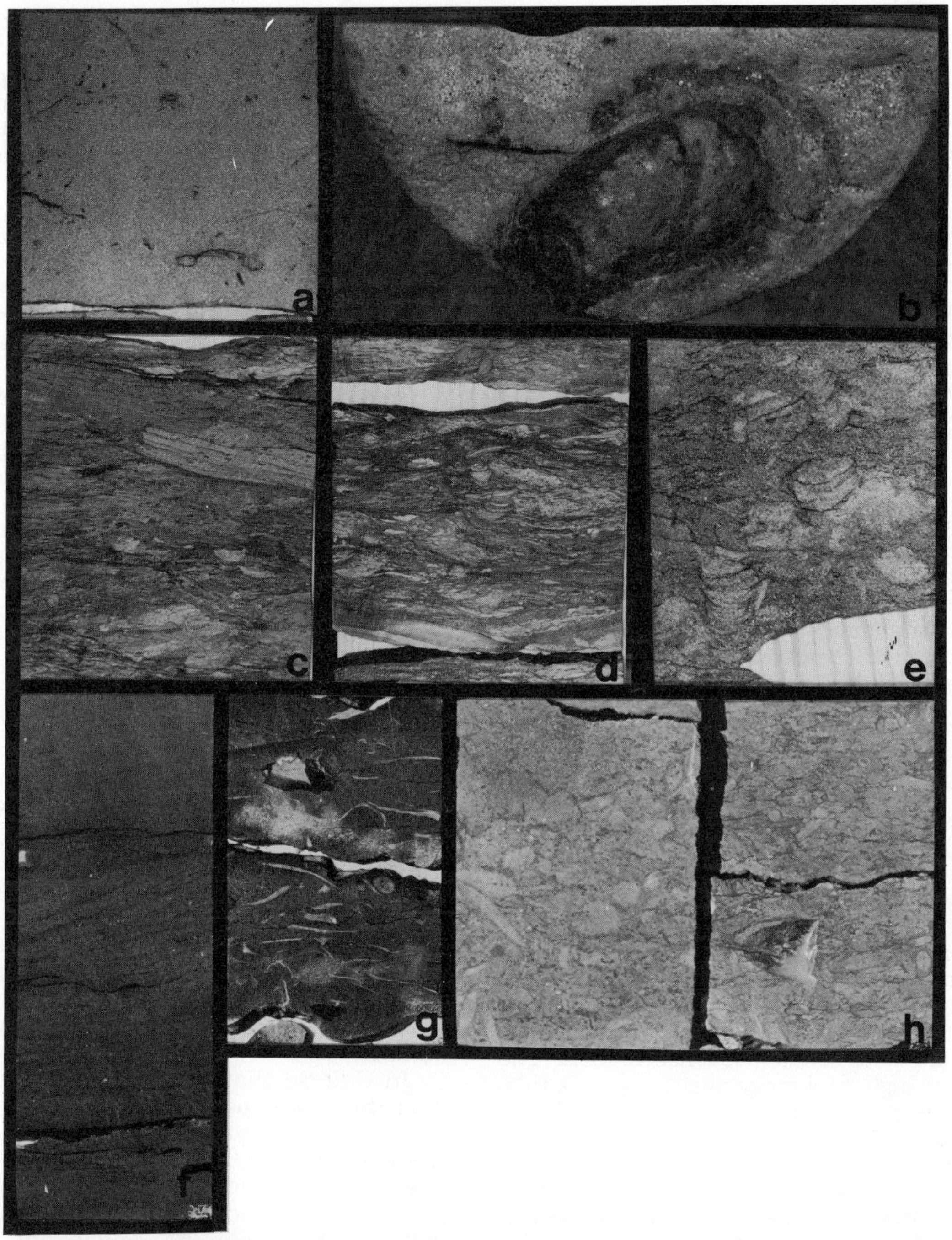

Fig. 5. Sedimentary structures from the Stø Fm. All core sections are 15 cm wide. (a, b) Very well-sorted quartz arenite with *Rhizocorallium* burrow in vertical cut, lower part of the section in a and at bedding surface in b. Offshore bar sandstone, Unit S-II (2120.4 m 7120/8-1). (c, d) *Teichichnus* and *Chondrites* burrows in offshore mudstone Unit S-II. c; elongated view of *Teichichnus* d; short axial view of a *Teichichnus* burrow. (e) Moderately well sorted quartz arenite with *Diplocraterion* burrows. Shoreface, Unit S-I (2178.1 m 7120/8-1). (f) Medium grained quartz arenite. Shale clasts and wave ripples are passing up into cross strata with tangential foreset and low angle cross-strata. Upper Shoreface, Unit S-III (2670 m, 7119/12-1). (g) Coquina sandstone with upwards decreasing abundance of shell fragments. Unit S-I (2779 m, 7119/12-1). (h) Burrowed sandstone. Left piece is ankerite cemented and shows no sign of compaction, while flattening of burrows and solution seams are common in the uncemented part (right). This association shows that the carbonate is precompactional. Unit S-III, Stø Fm (2708 m, 7119/12-1).

cylinder or bell-shaped gamma log readings. In the cores the boundary is defined by an increase in grain size, frequently overlying an erosion surface.

Cylinder-shaped gamma log with occasionally high amplitudes separate this unit from those below and above. Cores through Unit N-III exist in four of the wells (Fig. 3), all of which show a general fining upwards trend. The lower part of the unit consists of sandy sequences, on average coarser grained than the sandstones seen in the rest of the Nordmela Fm. Cross-bedding and current ripples are the most common sedimentary structures and fossil plant roots and thin coaly beds are often present (Fig. 4c). Many of the sandstone sequences, especially in block 7119/12, have erosional bases and show upwards fining trends like the channel fills within the underlying unit (Unit N-II).

The upper part of Unit N—III is finer grained, sometimes mud dominated, and commonly intensely bioturbated. Plant roots and thin coaly beds so commonly found throughout the rest of the Nordmela Fm seem to be absent in this uppermost part. Unit N-III is about 35-m thick in 7119/12-1 and -2, while it is reduced to about 25 m in the wells to the northeast (Fig. 3).

The succession of sedimentary features shown in Fig. 3 suggests that Unit III represents a transgressive event. The presence of marine trace fossils (Fig. 4g) within the sequences may (as in Unit N-II) favour a tidal channel rather than a fluvial channel origin. Other depositional facies are also present in this part of Unit N-III. Thus the basal part of 7120/8-1 is interpreted as a beach facies (Fig. 4f).

The upper, fine-grained part of Unit N-III is more thoroughly burrowed than the rest of the Nordmela Fm. It is a marine deposit, and in 7119/12-1 and -2 wave ripples indicate shallow water. The lack of plant roots and coals and a slightly different trace fossil assemblage distinguish this part from the tidal flat facies discussed earlier. A lagoonal environment is suggested.

Reservoir quality: good in the lower part, based on the same argument as for Unit II. The upper part is expected to be of poorer reservoir quality and it is laterally interfingering with mudstone.

Depositional environments

Deposition of the Nordmela Fm took place in a paralic, medium to low energy environment. Strong tidal influence during deposition is suggested and a major part of the sedimentation occurred on tidal flats. As we find little evidence of barrier bar facies associated with the tidal deposits, the tidal range is inferred to have been large, probably within the upper meso-tidal or macro-tidal range.

We expect that near-by rivers must have existed as sediment sources for such a thick sandy sequence, but the Nordmela Fm is difficult to fit into fluvial- or wave-dominated deltaic sequence models. The indications of high tidal range might suggest a tide-dominated delta, but the published models for such deltas are quite different from the succession described here.

Apparently the rivers did not form 'normal' prograding deltas, but more likely terminated in some kind of estuarine system, associated with wide tidal flat areas. This is also in accordance with the tidal nature (e.g. herring-bone structures and abundant flaser bedding, Fig. 4) of the whole sequence. In conclusion, due to the vertical and lateral facies association, a coastal plain/tidal flat facies association is suggested.

Reservoir quality of the Nordmela Fm is generally poor, but it improves upwards, due to increasing abundance of channel sandstone facies.

The Stø Formation

The formation, comprising near shore—inner shelf deposits, is subdivided into three units (S-I to S-III). A Toarcian age is suggested for the lowest unit (S-I). The middle unit is generally interpreted as having been deposited during a transgressive event and seems to be correlatable over the entire area. Near the top of this unit (S-II), an acme of the dinocyst *Nannoceratopsis gracilis* occurs in all the wells, indicating approximately the same age for this level over the whole area; this suggests a late Toarcian to early Bajocian age for unit S-II.

The uppermost unit in the Stø Fm contains very poor fossil assemblages, and is difficult to date. However, in all wells a Bajocian assemblage can be traced. A questionable Bathonian age is indicated in most wells for the top of the formation. In general there are biostratigraphic evidences for a hiatus between these two sequences. Although the hiatus is difficult to date accurately, the fossil assemblages indicate a Bajocian—Early Bathonian age for Unit S-III, while the overlying marl is suggested to be of Middle Callovian—Early Kimmeridgian age. This indicates that sediments from ?Bathonian—Early Callovian are missing.

Smooth, linear and cylinder-shaped low-reading gamma ray log patterns plus occasional linear, serrated high-reading gamma ray log breaks distinguish the Stø Fm from the underlying Nordmela Fm. The sandstones are mainly fine- to medium-grained and are classified as quartz arenites. Commonly marine macro-fossils, chamositic and glauconitic grains and authigenic phosphate occur in distinct beds. Also, burrows of marine affinity are common (i.e. *Diplocraterion*, *Rhizocorallium*, *Chondrites*, *Ophiomorpha*, *Siphonites*, *Skolithos*, *Teichicnius*, Fig. 5). Large-scale cross-stratification with low angle, high angle and hummocky appearance are distinguished within the cored sections.

Unit S-I

The 50—70-m thick sequence in all wells consists of mainly medium to fine grained sandstone containing large-scale cross-strata or massive beds interrupted by coarser grained and normal graded thin layers. In 7120/8-2 and 7120/7-1 and the uppermost part of the 7119/12 wells, the unit is more finely grained and much more burrowed, occasionally interrupted by thin wavy laminated graded bedding suggesting hummocky cross-stratification.

In Unit S-I the major part of the 7119/12 wells and the lowest part of the 7120/8 wells are interpreted as high-energy sediments in a shoreface to foreshore setting. Tidal inlet and tidal delta deposits are also considered for the lowermost section. The extensive bioturbation, common fossiliferous layers (sandy coquinas) and minor wave bedded layers in the comparable sequences in 7120/8-2 and 7120/7-1 and in the upper part of the 7119/12 wells suggest lateral and vertical transition to deposition below fair weather wave base in an open marine environment. Here the common occurrence of *Chondrites* as well as *Siphonites* burrows is consistent with an open-shelf environment. It is then natural to consider that the sediments in 7120/8-2 and 7120/7-1 were deposited further offshore or in a more protected area than those of the 7119/12 wells. The 7120/8-1 well represents a transitional facies. Unit S-I consists of shallow marine deposits, deposited in different energy or wave (flow) regimes (Fig. 3). Although the sequence is well cored, subsurface investigation is too limited to state the geometry of the sandstone bodies. However, based on the vertical and lateral succession, a shallow marine sheet sandstone associated with and/or protected by bar complexes, is interpreted as the major depositional environment.

Reservoir quality: excellent to good sheet-like reservoir sandstone bodies are expected. However, matrix content increases laterally due to position of bar complexes or coastline.

Unit S-II

The base is taken at the first significantly higher gamma reading above Unit S-I in the Stø Fm. In the cored intervals the base is defined at the first appearance of mudstone above Unit S-I.

It is possible to subdivide the unit into five subunits (Fig. 3), which are regionally extensive. Unit S-II comprises three major lithofacies: (1) totally bioturbated (e.g. *Chondrites, Teichicnus, Siphonites*), very fine grained, argillaceous quartz arenites with minor glauconite; (2) weakly bioturbated (e.g. *Rhizocorallium*, (Fig. 5a,b)) or non-burrowed, fine to medium quartz arenites and (J) burrowed (e.g. *Teichichnus, Chondrites*) to non-burrowed mud/siltstone with siderite and pyrite concretions and occasionally with glauconite and chamositic ooid-layers. These characteristics suggest an offshore marine setting. Restricted anoxic periods during deposition of the muds are indicated by non-burrowed horizons and abundant pyrite concretions.

Unit S-II is interpreted to result from a continuous transgression that left the record: (i) Barrier (?) bar complexes, with storm generated beds; (ii) marine, offshore, burrowed, very fine grained sandstones (both facies within Unit S-I); (iii) marine, offshore, mudstone and sandstone (Unit S-II). The non- and weakly-burrowed, medium grain-sized sandstone unit (Fig. 5a,b) suggests offshore bar sedimentation. The fact that the mudstone units mark the peaks of transgressions is further substantiated by their regional extension.

Reservoir quality: the mudstones probably represent barriers to vertical flow. Facies changes within the sandstone are expected, although great lateral extension is hitherto observed.

Unit S-III

The base is the first appearance of the sandstone above the upper mudstone of Unit S-II. The thickness varies from 70 m (7119/12 wells) to 20—40 m (7120/8 and -7 wells) due to erosion or non-deposition.

In all wells the lower part of this sequence consists of a strongly burrowed, very fine-grained argillaceous sandstone with minor glauconite and chamosite (Fig. 5h). Shell lags with identifiable fossils such as belemnites, including hooklets of belemnites, mussels, brachiopods, (i.e. *Lingula* and *?Discinisca*) and echinoderm spines have been observed. Minor, normal graded, non-burrowed beds, occasionally with shell lags, and lamination in a hummocky appearance, suggest storm influenced beds. These parts of the unit were deposited below normal wave base.

Generally there is an increasing grain size upwards in the lower part of the formation. In 7119/12-1, -2 and 7120/7-1, fine- to medium-grained quartz arenites with coarse- to very coarse-grained lags, comprise the remaining part of the formation. Vertical burrows (*Skolithos* and *Ophiomorpha*), low angle cross-stratification, parallel lamination and wave ripples (Fig. 5f) suggest upper shoreface deposits. In 7120/8-2 the upper cored part consists of 0.5—2 m cycles of non-burrowed sandstones, interpreted as shoreface deposits.

Unit S-III is interpreted as a shallow marine, prograding deposit. The formation has a coarsening upwards trend when including the uppermost sequence in Unit S-II, from offshore mudstone to offshore sandstone to shoreface sandstone. The coarser sandstones and shell lags are interpreted as storm generated beds. A consequence of this interpretation is a shallowing upward sequence representing a prograding marine deposit within an inner shelf, coastline system.

Reservoir quality: if we interpret this sandstone body as progradational unit within an overall transgression (relative rise of sea level) or stillstand of sea level, the lateral extension of these sandstone bodies is expected to be very large.

Deposition of the Nordmela and Stø Formations

The deposition of the Stø Fm above the Nordmela Fm indicates a change from coastal alluvial plain to high-energy coast lines.

All well logs and examination of the cores clearly show that the Stø Fm, the major part of the reservoir section in 7120/8-1, -2 and 7120/7-1, is entirely different from the underlying Nordmela Fm. The deposition of the sandstones took place in very shallow water close to the shoreline, but there was a shift in energy level, and probably also in tidal range, which led to the formation of extensive beach ridges and shelf sands. From palaeontological reports we have no evidence of a hiatus between the lower and the upper part of the succession, but the change in facies indicates at least a minor regression. It is likely that the sharp boundary observed in the wells represents a minor time break due to uplift/erosion. Consistent with this interpretation is the occurrence of phosphate both as cement and grain coatings (see Plate 1 and Fig. 7a) and glauconite in the lower part of Unit S-I in 7120/8-1. In any case, the basin configuration must have changed at this stage. If it had remained the same, we should still have had a tide-dominated sequence. Whatever geological factors that induced these changes, they must be sought outside the eastern part of the Troms I area, outside of where we have the well control. This is because there was no significant change in water depth in east Troms I. The sedimentation still took place just below sea level and formed wide-spread sheet-like sand bodies that can be correlated throughout the area. The good correlation between the wells shows that the area remained relatively flat and that no significant block faulting took place during deposition of the Stø Fm. Active faulting would have segmented the area and would have produced considerable facies differences and difficulties in correlation between the wells.

This major change in depositional environment must be explained at a more regional scale. A relatively narrow sea-way existed between Norway and Greenland during most of the Jurassic period (Sachs *et al.*, 1973). This sea-way connected the North Atlantic/North Sea to the Arctic Ocean. Comparison of fossil fauna from East Greenland and the British Isles indicates that the sea-way became closed, or at least restricted, towards the

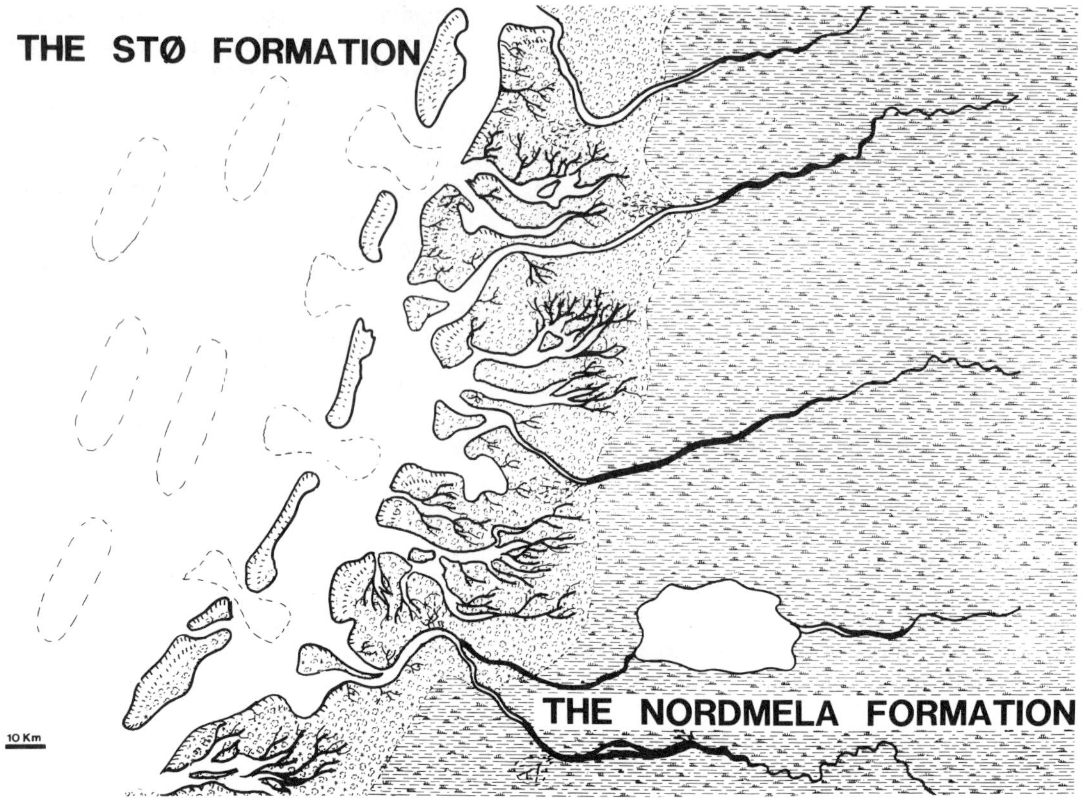

Fig. 6. A tentative reconstruction of the depositional environment of the Nordmela and the Stø Formations.

end of the Lower Jurassic and remained partly closed throughout most of the Middle Jurassic (Donovan, 1966, pp. 63–65). The closure was probably related to tectonic activity (uplift prior to and during the mid-Kimmerian phase (?)). Closing and opening of the sea-way would greatly change the pattern of seawater circulation and could have had a major influence on tidal ranges in a wide area. This means that tectonic events further south along the rift system between Norway and Greenland could have influenced the shift in facies we observe in the reservoir section in the Troms I area. Slight local tilting would perhaps also facilitate the shift from wide tidal flats to high-energy sandy beaches and bars.

If Unit I in the Stø Fm is mainly a barrier bar complex, we might expect to find lagoonal and/or non-marine deposits either at the top or at the base of the succession. This would depend on whether the sedimentation took place in a regressive or a transgressive regime. As the shallow marine sequence in the top of the Nordmela Fm was interpreted as lagoonal, we have a natural succession formed by landward migration of barrier bar systems during transgression. This could be summarized as follows:

(1) The continental side of the lagoon, represented by (?) tidal or fluvial channels and medium/high energy beach deposits of the lower part of Unit N-III.

(2) The lagoon itself, with variable bioturbation, relatively fine-grained upper part of Unit N-III. Thin bedded graded layers suggest washover beds in 7120/8-1.

(3) The transition between the lagoon and the barrier, complexes, represented by the lower part of the Stø Fm.

(4) The beach and foreshore deposits of the barrier could be the well-sorted middle part of Unit S-I (the subaerial-aeolian part of the barrier that should have been in between had little potential for being preserved).

(5) The upper part of Unit S-I would represent shoreface deposits and offshore bars parallel to the main barrier.

(6) This is capped by strongly bioturbated shallow marine sandstone, that is in turn succeeded by the marine shale of Unit S-II.

In summary, the Nordmela and Stø Formations represent two depositional systems, comprising six units with an overall transgressive frame. The two depositional systems consist of a lower part; a coastal plain/tidal flat sequence (Nordmela Fm) and an upper inner shelf/ nearshore sequence (Stø Fm). A tentative reconstruction of the depositional environment representing the two systems is presented in Fig. 6. The intersecting mudstone sequence in the middle part (Unit S-II) of the Stø Fm reflects more rapid transgressive episodes.

The exceptionally thick coastal plain/tidal flat sequence overlain by the supermature and lateral extensive shallow marine sandstones indicates quiet tectonic conditions during deposition. In addition, the shallow marine sandstone is very regular on gamma ray logs and uniform in grain-size. No well-developed coarsening upward sequences are observed, which would be expected in an progradational coastline. This amalgamation of sandstone sequences was caused by tectonic stability and a maximum balance between sediment supply and basin subsidence, which in turn gave a stationary coastline for long periods of time in the Stø Fm.

SANDSTONE COMPOSITION, DISTRIBUTION AND PROVENANCE

The methods used in the petrographic study include thin section analysis with additional cathodoluminescence, scanning electron microscopy and X-ray diffraction. Details of methods and analytical procedures are available on request to Statoil's Geological Laboratory.

The sandstones are generally classified as quartz arenites using the classification of Dott *et al.* in Pettijohn *et al.* (1972). The lower coastal plain/tidal flat succession shows a higher content of feldspar, mica, and clay minerals and is sometimes classified as subarkosic arenites/wackes and quartz wackes. The coarser grained beds in the marine sandstones have sometimes a high chert content, justifying the term sublitharenites, if the chert is regarded as a rock fragment (see below).

Silica

The monocrystalline quartz grains always show undulatory extinction. This can be divided further into slightly undulatory to strongly undulatory extinction, varying with the depth of the reservoir section. The quartz grains with a metamorphic polycrystalline texture consist either of elongated crenulated crystals or sutured crystals. Clasts of quartzite with schistosity are observed. Other polycrystalline silica grains are cherts, which consist either of fragments of microquartz or a combination of microquartz and megaquartz. Some chert fragments, especially in the Stø Fm have spherical structures, suggesting an original fossiliferous origin.

Cathodoluminescence shows at least two major colours for the quartz grains, namely brownish to red and violet (Plate 1e). In addition, abraded grains with red/brown luminescent cement shows a previous quartz cemented sandstone as source. According to Zinkernagel's scheme (1970, fig. 3) this implies a variety of sources for the quartz grains. The colours indicate plutonic and metamorphic sources and the grains with a previous cement suggest sedimentary sources.

Feldspar

Both microcline and plagioclase occur in small quantities in the upper part of the succession, while some horizons in the lowermost formation are classified as subarkosic arenites/wackes. Within the fine grained carbonate cemented sandstones, in the Stø Fm, feldspars are more common and show few signs of alteration. This contrasts with the non-cemented sections, suggesting that feldspar was originally more common in the finer grain-sized and poorer sorted sandstones.

Micas

The mica content is higher in the Nordmela Fm than in the Stø Fm. In the finer grained and more poorly sorted sandstones in the Stø Fm, mica is common. Mica also occurs as a trace mineral in the coarser grained marine sandstones of the Stø Fm. The micas are relatively small and lie in thin continuous or discontinuous wavy laminae and on stylolitic surfaces which are randomly distributed in the sandstones. White micas (muscovite/sericite) are by far the most common types. Biotite is observed only sporadically. Alteration (illitization and kaolinitization) of micas is common in the Nordmela Fm.

Accessory minerals

Other detrital constituents in minor quantities are zircon, rutile and tourmaline (zoned and partly authigenic). These are concentrated along stylolites when developed. The occurrence of carbonate, glauconite, phosphate and chamosite grains is discussed below.

Matrix

The matrix of the sandstones consists of undifferentiated material which is too fine to be identified with the petrographic microscope. X-ray diffraction and scanning electron microscopy have shown clay minerals, including authigenic and clastic kaolinite, illite, illite/smectite, chlorite, framboidal and rhombohedral pyrite, cryptocrystalline authigenic silica and bitumen as identifiable constituents. The matrix content is occasionally higher than 15% (e.g. wackes) in the coastal plain/tidal flat deposits and offshore sandstones in the middle unit of the Stø Fm. Generally, the matrix content varies with sedimentary facies in accordance with the energy of deposition.

Grain coatings

In the marine sandstones, illite, chlorite and siderite are observed as grain coatings. Illite coating of quartz grains may have given preferential quartz overgrowth in grain to grain pressure solution as quartz overgrowths are common in the tightly cemented and stylolitized parts of the succession in 7119/12-1. Phosphate and chamositic ooids are discussed below.

Discussion

The various luminescent colours of quartz and polygenetic quartz grains imply two major sources plus a minor source for the sand grains: (1) regional metamorphic rocks (e.g. Caledonian—Precambrian basement); (2) low grade or unmetamorphosed sedimentary rocks (e.g. pre-Mesozoic); (3) plutonic rocks. Chert fragments with spherical structure are very similar to those reported in Carboniferous/Permian chert rocks from Bjørnøya and Spitsbergen.

In the near shore/inner shelf deposits the matrix content and mica content of the sandstone will affect the gamma log. Heavy mineral layers are expected in the beach or bar deposits which can occasionally give high gamma responses. In addition, the higher feldspar content (K-feldspar) in the Nordmela Fm also contributes to a higher gamma reading.

The mineralogical maturity of the sandstones in general increased upwards from subarkosic to quartz-arenites.

DIAGENESIS

The major diagenetic processes altering the original reservoir properties of the sandstones are mechanical

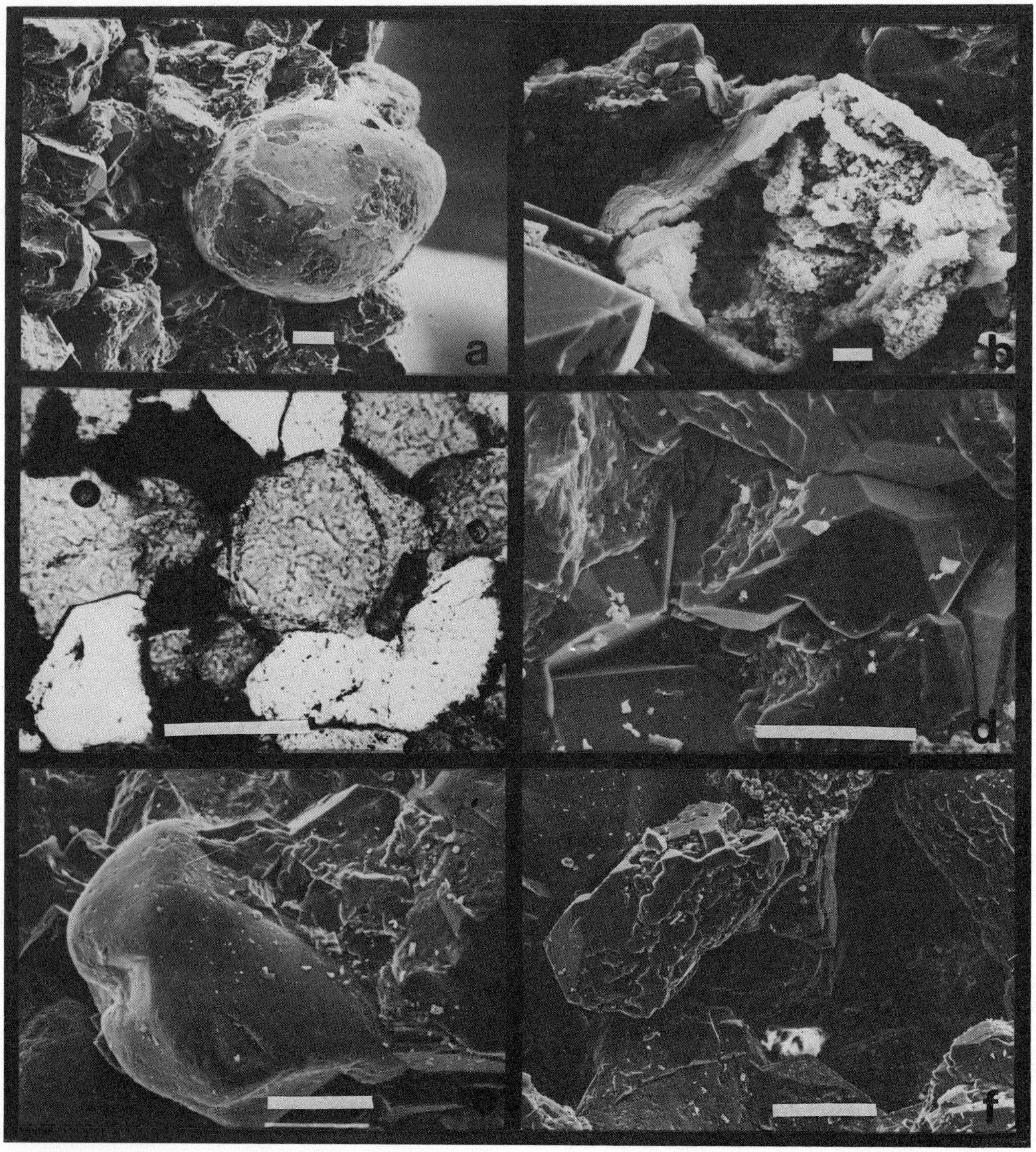

Fig. 7. (a) Quartz arenite with phosphatic grain in middle part. Note quartz overgrowth and quartz crystals growing through and displacing the phosphatic coating (SEM 2180 m, 7120/8-1). (b) Leached carbonate cement (SEM, 2179 m 7120/8-1). (c) Tight quartz cemented quartz arenite. Thin section crossed nicols, note dust rim of original grain. (2676.2 m, 7119/12-1). (d) Authigenic quartz crystals in quartz arenite. Less than 5% in measured porosity (SEM, 2676.2 m, 7119/12-1). (e) Quartz grain in ankerite cemented area showing that carbonate cement pre-dates quartz overgrowth. No quartz overgrowths exist on the grain and cement is partly leached (right part) (SEM, 2721 m, 7119/12-1). (f) Quartz grains in non-ankerite cemented area with abundant quartz overgrowth (SEM). Same rock piece as e. All scale bars: 100 μm.

compaction, quartz cementation, carbonate cementation and dissolution, kaolinitization, and illitization. Authigenic pyrite (both framboidal and rhombohedral crystals), feldspar, phosphates, iron silicates and other clay minerals are important environmental indicators but are not discussed in detail here because they have little effect on the porosity.

Mechanical compaction

Both physical compaction and chemical processes are observed in these sandstones. The undulatory nature of the quartz, plus the fracturing and crenulation of the grains in the wells, suggest that the rock has undergone considerable stress. Quartz-arenites in 7119/12-1 texturally resemble metaquartzites with common sutured and concave—convex contacts. Similar textures are less common in 7119/12-2. In the lower part of the cored interval in 7119/12-1 the quartz sandstone has a granular texture probably produced by crushing, consequently pososity is severely affected. Silt-sized quartz, produced by granulation, is squeezed in as pore filling material. This process is regarded as an important porosity-reducing agent in the deeper reservoirs.

Plastic deformation of mica grains is common in all wells, causing porosity reduction in the more argillaceous sandstone. Mechanical compaction of grains is not observed in early-cemented horizons. Crushed grains and granulation of quartz is associated with fractures.

Quartz cementation

Quartz cementation is the dominant porosity-reducing process in the sandstones. Extensive stylolitization (see Fig. 9a) of the sandstone is present at relatively shallow depth (2—2.7 km). Quartz cement occurs as authigenic overgrowths (Plate 1 and Fig. 7) sometimes defined by a dust rim on the original quartz grain (Fig. 7c).

Cathodoluminescence study of the deeper reservoir sections reveals two generations of authigenic quartz. Occasionally, a thin rim of non-luminescent quartz overgrowth is observed between the original grain and the major red/brown overgrowth (Plate 1d, e, f). Although little is known about the cause of luminescence in quartz, red colour may be caused by either high temperature diagnesis, or activation by Mn^{2+} (Nickel, 1978). However, it is reasonable to assign the different luminescent zones of quartz overgrowth to pore-waters with different chemistry and/or temperature. Normally, quartz overgrowths postdate carbonate cementation (Plate 1d; Fig. 7e); however, occasional thin quartz overgrowths predate the carbonate cement. This indicates either an earlier phase of quartz overgrowth, or carbonate cementation between two periods of quartz cementation.

The most extensive stylolitization and adjacent quartz cementation is observed in the deepest section (i.e. 7119/12-1). Here, pore-filling quartz has reduced the porosity in fine to medium grained, very well sorted quartz arenites to less than 5%. A relationship between the degree of stylolitization and porosity reduction is shown in Fig 8. Here the maximum amplitude of any one stylolite is used as an estimate of the minimum thickness of material removed along that surface (Stockdale, 1926; Mossop, 1972). Figure 8 shows that 15—20% is the maximum porosity that one can expect in these well-sorted supermature sandstones. Only stylolites in the non-burrowed marine quartz arenites were measured. The percentage removal represents a minimum compaction due to mechanical deformation. Grain-to-grain pressure solution. microstylolites and mechanical cracking of the rock are not taken into account. Also, pressure solution is greater in argillaceous sandstones (c.f. Weyl, 1959). There is also strong evidence of pressure solution in the shallowest section examined (i.e. 7119/12-2). This is indicated by sutured, stylolitic contacts between grains and thin quartz overgrowths adjacent to the point contact of the grains (Plate 1a). The porosity of the medium grained, well-sorted quartz arenites is reduced to 25—30% by mechanical compaction and quartz cementation. Generally there is a simpler linear relationship between porosity loss and burial (see Fig. 10). Some horizons in the deepest reservoir have anomalously high porosity, caused by secondary leaching. Silica cementation is considered to take place in two stages: (1) shallow lithification and (2) deeper-burial stylolitization.

Shallow lithification

Quartz cementation is best shown in the shallow-buried sandstones of 7119/12-2, where sutured grain contacts and minor quartz overgrowths occur without any evidence of stylolites or solution seams. Sutured grain contacts are interpreted to have formed by grain-to-grain pressure solution (Robin, 1978) during compaction, unhindered by high pore pressures. Overpressured zones are not detected within the Troms I area, hence the effective stress at grain contacts has been sufficient for dissolution (Robin, 1978).

The thin quartz overgrowths seen pre-dating carbonate cementation in 7119/12-2 are probably representative of the development of the first stage of the quartz authigenesis. Quartz to form these first overgrowths could have been derived either by large-scale redistribution of quartz from grain-to-grain pressure solution, or from external sources (e.g. biogenic silica). It is assumed that prior to stylolitization, the sandstones were lithified by grain-to-grain pressure solution.

Deeper-burial stylolitization

Stylolites in 7119/12-1 are common, but variable in both morphology and size (Fig. 9 left). Amplitudes of more than 5 cm have been measured. The

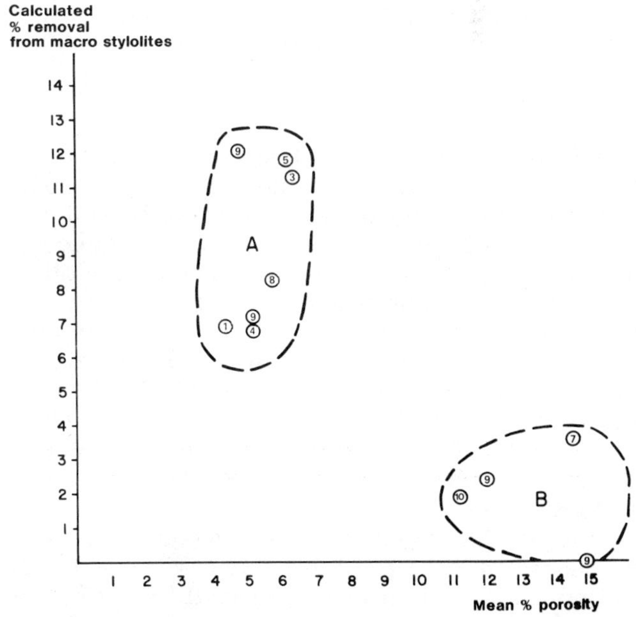

Fig. 8. Plot of percentage removal of sandstone thickness in quartz arenites by stylolites against mean porosity of different cores and core interval. Based on measurements of maximum amplitude stylolites on 362 and 140 porosity readings in non- or weakly-burrowed marine sandstones in the Stφ Fm in 7119/12-1. Numbers in circles are core numbers.

$$(\% \text{ removal} = \frac{\text{Cm — removal (Amplitude)} \times 100}{\text{Measured core interval} + \text{Cm — removal (Amplitude)}}$$

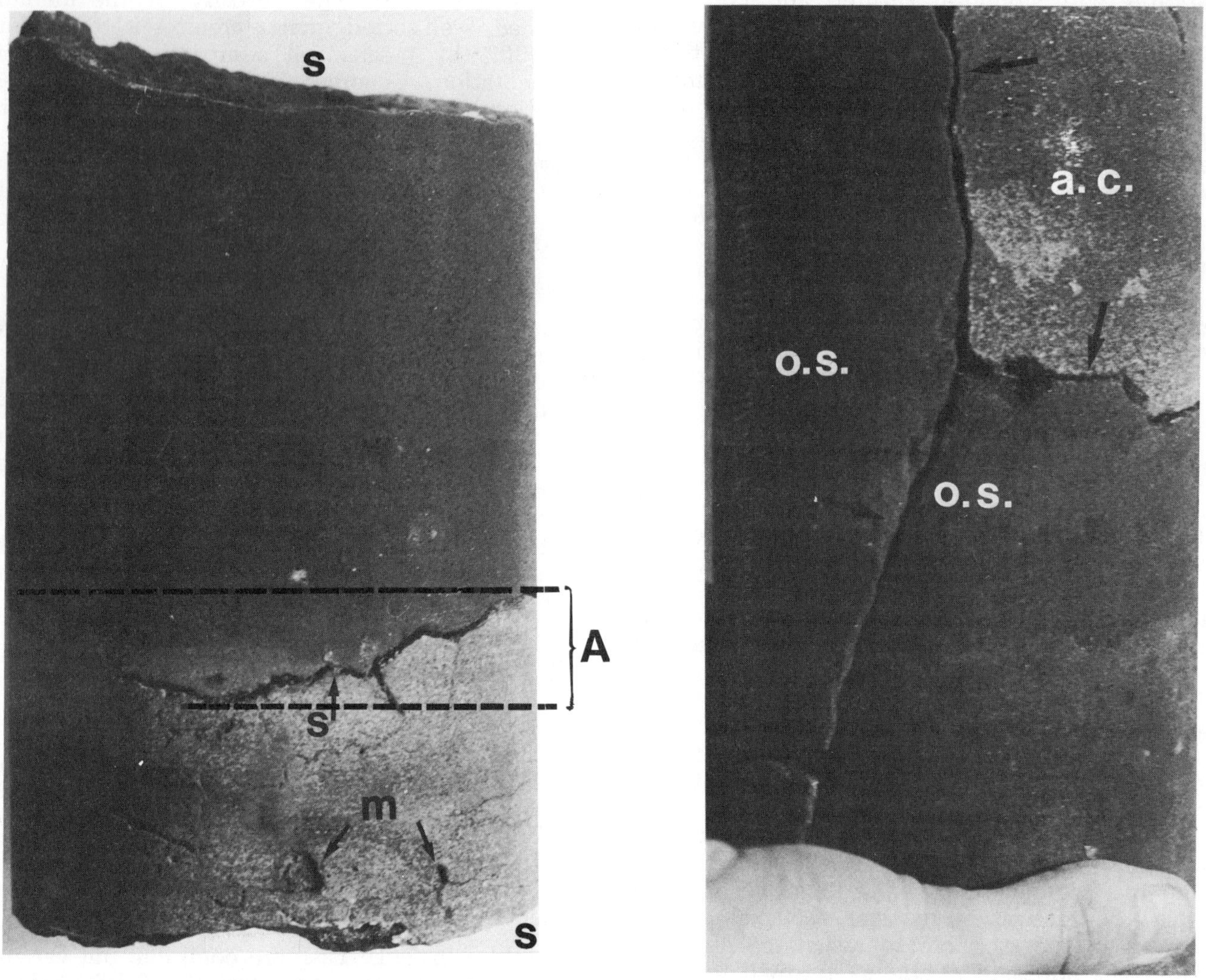

Fig. 9. Core pieces from 7119/12-1. Core width is 15 cm. (a) Stylolite amplitude showing minimum loss of the rock: A, m, moldic porosity; s, stylolite surfaces (2675 m). (b) Cracked porous quartz arenite. O.S., oil stained sandstone; a.c., ankerite cemented sandstone. Arrow shows cracks with rim of authigenic quartz crystals (2718.7 m).

stylolites are distinguished as concentrations of fine organic matter, clay minerals, mica, heavy minerals and minor diagenetic carbonate, and have formed along pre-existing thin clay-organic-rich horizons. Argillaceous zones contain most stylolites, an observation consistent with the theory that clay-rich layers in sandstones are most favourable for stylolite development (Weyl, 1959; Robin, 1978). Stylolitization is interpreted to be the main source of quartz cementation in the deeper wells in the Troms I area. High temperature and low pore pressure are thought to have favoured widespread stylolitization.

It is possible that a period of late faulting, located by seismic mapping (Gloppen and Westre, 1982), caused the sandstones to be flushed by hydrothermal solutions from the adjacent deep Tromsø Basin. These solutions may have encouraged stylolitization, and quartz cementation both by raising reservoir temperature, and by injecting silica-enriched pore waters.

In conclusion, in these five wells, a dynamic depth-related pore fill of quartz cement (Fig. 10) is a result of pressure solution, and this process is mainly related to lack of overpressure and increased temperature. Minor contributions of quartz cement from other sources are not discounted.

Carbonate cementation and dissolution

A varied development of different carbonate minerals (dolomite/ankerite, calcite and siderite) as cement occurs throughout the succession in all wells. Carbonate cement may occur as thin impermeable layers in the Stø Fm. Their lateral extent is uncertain as they have formed along localized coquinas or fossiliferous sandstone layers associated with inner shelf deposits. Lateral extension of coquinas in inner shelf deposits have been reported to extend to more than 5 km.

Calcite—dolomite/ankerite cementation

The most important carbonate cements in the reservoir sequence (marine deposits) are calcite and dolomite/ankerite. Dolomite and ankerite are mainly thought to be alteration products of calcite.

The cements occur as patchy or complete pore filling of poikilotopic and sparry crystals, and as separate single crystals distributed within the sandstones. Partly-dissolved patches of calcite (Plate 1b) are interpreted as dissolution features and are often related to occasional over-sized pores. In the deeper wells, carbonate cemented layers show little or no sign of compaction of original burrows when compared with the overlying and underlying rocks (Fig. 5h). Furthermore, stylolites which are common in the rock strata elsewhere, are uncommon in these carbonate cemented sandstones. However, fractures are common in the latter. Within carbonate cemented horizons grains have a floating texture whilst mechanical deformation and sutured contacts between the grains are the normal habit in the adjacent non-carbonate cemented layers. Fossils and ooids are commonly observed in the horizons which often show a complex generation of carbonate cement (see Fig. 13c, e, f). From these relationships, the poikilotopic calcite cement is regarded as an early pore filling cement.

Normally the poikilotopic calcite cement predates quartz overgrowth. However, in a few cases minor quartz overgrowths are observed between grain surfaces and sparry calcite cement. Textural similarity between calcite and dolomite/ankerite cements leads to the interpretation that the latter have formed by dolomitization of the former. Notably, ghosts of fossils and ooids occur. In some cases where minor ankerite/dolomite rhombs post-date quartz overgrowth, these are interpreted as being formed in a late diagenetic event. Matsumoto and Iijima (1982), in an investigation of Ca—Mg—Fe carbonates in coalfields of Japan, differentiated between authigenic carbonates in fresh water, brackish—marine and marine facies. Calcite in the marine facies was precipitated during an early diagenesis event by dissolution of calcerous shells (dehydration and recrystallization) and was contemporaneous with bacterial reduction of sulphate. Ankeritization in the marine facies was taking place from 300 to 500 m and 1000 to 2000 m burial depth as a result of migration and mixing of expelled pore waters from interbedded freshwater and brackish—marine and marine sediments. The carbonate cements in the Troms I area, may have been formed in a similar way. The occurrence of ghosts of carbonate allochems (see Fig. 13b) in the ankerite cemented sandstone is consistent with this interpretation.

Siderite

This occurs mainly in two-facies: (1) in the coastal alluvial plain/tidal flat sequences as clasts, peletoids and minor concretions, the last occasionally having a core of calcite, suggesting gradual sideritization of calcite; (2) siderite commonly occurs as concretions in the marine shales in the upper part of the reservoir sequence. Concretions in both facies show evidence of early (precompactional) formation and *in-situ* growth. The laminae continue uninterrupted through the concretions and the thicknesses of the laminae are always greater within the concretions. Burrows (*Chondrites*) are flattened in the surrounding shale, but show no sign of compaction within the concretions. The concretions in the marine shales have internal septarian cracks filled with sparry carbonate.

Siderite is commonly precipitated as an early diagenetic mineral in fresh or brackish water with

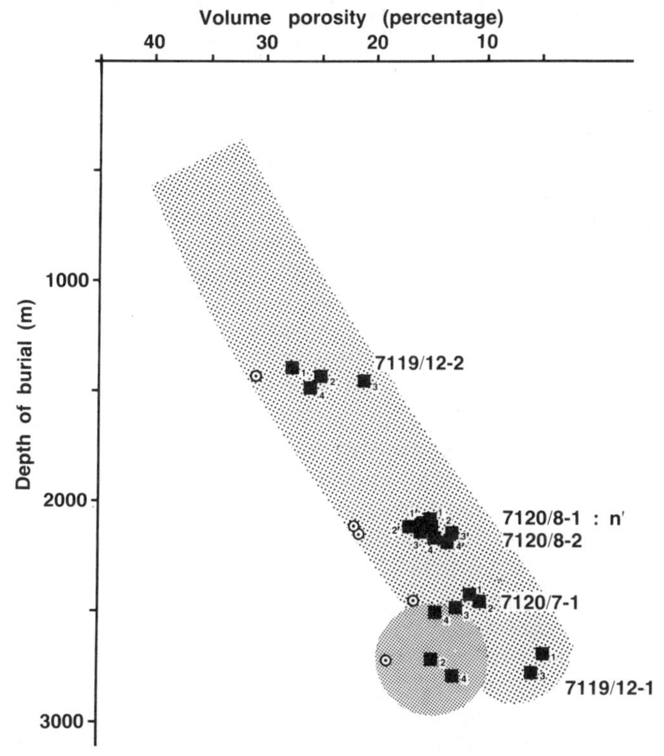

Fig. 10. Porosity versus depth of burial of the five wells investigated in the Troms I area. Four different facies in the near shore/inner shelf facies association of the Stø Fm have been chosen. 1, beach or bar deposit in unit S-III; 2, offshore bar sandstone in unit S-II; 3, offshore sandstone; and 4, bar complex sandstone in unit S-I. Note the linear trend of the porosity except for 7119/12-1. See text for discussion. Maximum porosity measured in core plugs is marked with a circle.

low sulphate concentration or below the zone of sulphate reducing bacteria (Berner, 1971). The siderite on coastal alluvial plains/tidal flats is therefore useful as an indicator of fresh water influence, while the siderite concretions in the marine shale suggest formation at a later stage of diagenesis (e.g. Matsumoto and Iijima, 1982).

Dissolution of calcite/ankerite cement and formation of secondary porosity

The strong bimodality in porosity of the quartz arenites in 7119/12-1 (Figs. 8 and 10) can be explained in three ways:

(1) Differences in textural and mineralogical composition; argillaceous zones, shale partings and locally high chert content do occur in the stylolitized sandstones. This may be a partial explanation but certainly does not explain a difference of 5—20% in medium grained quartz arenites.

(2) Locally overpressured zones may have occurred during the earlier geological events, but overpressured zones are not detected today. Due to the connection between the high porosity bar complex sandstone (Unit S-I in the Stø Fm) and the low poroxity stylolitized zones, overpressuring is highly unlikely.

(3) Carbonate cementation before burial with later dissolution of the early cement. This model is discussed below.

Secondary porosity is observed in thin sections as oversized pores, corroded grains, and partial dissolution of cement (Figs. 7b, 9a). Corroded ankerite cement is also recognized in scanning electron microscopy (Fig. 7e).

Cathodoluminescence often shows patches of 'rotten' partly dissolved bright luminescent calcite (Plate 1b). In 7119/12-1 few or no stylolites occur in the high porosity zones. However, the rocks are fractured (Fig. 9b), proving that previously a more competent rock existed, similar in appearance to the now cracked carbonate cemented sandstone (see above). The occurrence of patchy ankerite cement associated with fissures suggests dissolution (Fig. 9b). Otherwise feldspar grains or mica grains are unaltered when enclosed in patchy carbonate cement, but kaolinitized or illitized in non-cemented areas. These observations and the porosity/permeability data suggest that dissolution of carbonate cement within the reservoir has created secondary porosity. Similar textures also occur in the 7120/8 and 7120/7 wells, where they are less well developed.

The dissolution of carbonate cement during burial as observed here implies the generation of acid pore waters at depth. Thermal decarboxylation of kerogen in shales has been suggested as a mechanism for generation of acid pore water (CO_2-production), which then may enhance porosity in reservoir rocks at depth by mineral dissolution (Schmidt and McDonald, 1979). In the present case an exceptionally thick sequence of shale occupies the down-flanking basin of the structure. According to Bjorøy *et al.* (1983), this Cretaceous potential source rock is dominated by humic kerogen which is favourable for CO_2 genera-

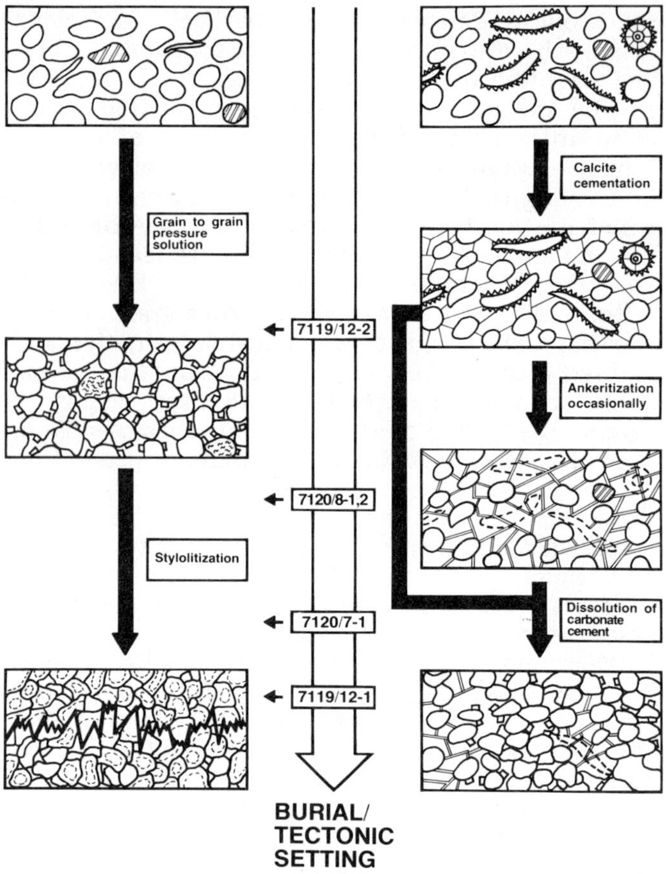

Fig. 11. Diagenetic model of the marine reservoir section (Stø Fm) of the five wells.

Fig. 12. Kaolinite as pore filling cement in 7119/12-1. (a, b) Marine sandstone: kaolinite is post-dating quartz overgrowth by overlying quartz cement (2775 m, 7119/12-1). Scale bars 100 μm and 10 μm respectively. (c, d) Coastal plain sandstone: kaolinite is predating quartz overgrowth as quartz cement is overgrowing kaolinite. Illitization of kaolinite (d) (2932 m, 7119/12-1). Scale bar: 10 μm. Note difference in crystal morphology and size of the two kaolinite occurrences.

tion, and thereby dissolution of carbonate cement. In addition, CO_2 could be generated from coal horizons in the basin. Acids other than carbonic acid may also be involved during petroleum generation (Curtis, 1983). Shales in the western part of Troms I area could then provide favourable conditions for generation of acid pore water and formation of secondary porosity in adjacent sandstones. The diagenetic model for the formation of secondary porosity in the 'high' porosity zones (middle part of S-I and S-II), as in the 7119/12-1 well, can be summarized as follows (see also right side of Fig. 11):

(1) A mineralogically mature (quartz) sand was deposited, together with some carbonate fossil fragments, minor feldspar and rock fragments.

(2) Early diagenetic $CaCO_3$ cementation, including early marine cementation.

(3) Mixing of pore water of marine and continental origin during burial in some of the horizons transformed calcite into the iron-rich dolomite/ankerite. The mixing was controlled by preferred permeability paths.

(4) Further burial generated CO_2 from decarboxylation reactions of kerogen in source rocks and produced acidic pore waters which leached the carbonates, so creating secondary porosity. Minor kaolinite was also precipitated.

This model for generation of secondary porosity is based mainly on observations from well 7119/12-1. However, we suggest that in general it is applicable also in surrounding areas.

Kaolinitization

Kaolinite is developed as a pore filling mineral in many of the sandstones. However, kaolinite is far more common in the more continually-dominated facies and tidal-flat facies. Not only abundance but also crystal size, crystal morphology and cement stratigraphy seem to be related to facies (Fig. 12). For example, in 7119/12-1 kaolinite from the

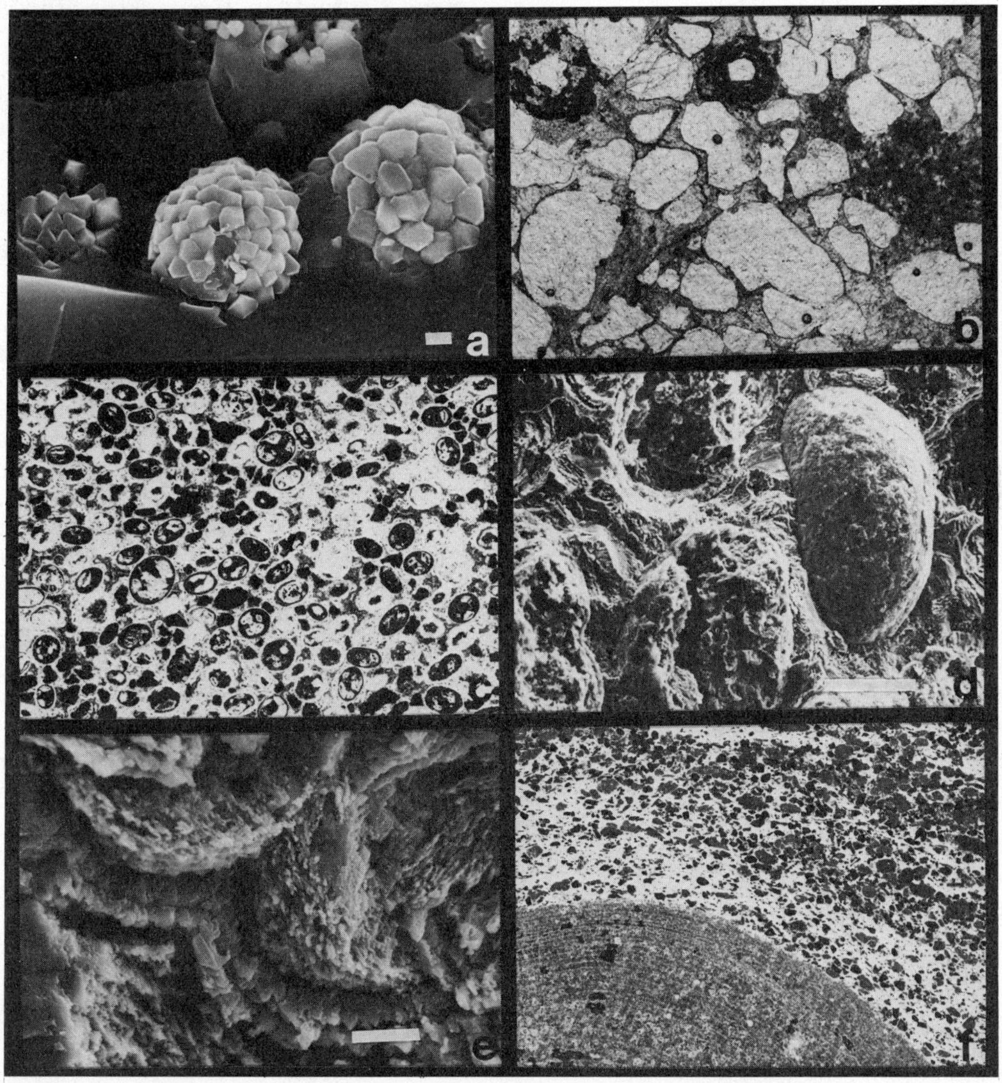

Fig. 13. Other authigenic minerals. (a) Framboidal pyrite enclosed by quartz (2667 m, 7119/12-1). Scale bars: 1 μm. (b) Ankerite cemented, medium grained sandstone with two ooid structures in upper left part and pelletoid clast (middle right part). Note floating texture of the grains (2722.3 m, 7119/12-1). (c) Chamositic ooids in sparry calcite. Negative thin section. Medium grained (2101 m, 7120/8-2). (d) SEM picture of medium grained, chamositic ooids. Note dissolution of the ooids in middle part (same as c). Scale bar: 100 μm. (e) Detail of the first appearance of the calcite fringe cement around the ooid grain suggesting an original fibrous orgin? Mg-calcite or aragonite (same as c). Scale bar: 10 μm. (f) Belemnite in lower part, with medium grained glauconite sand above (2735.30 m, 7119/12-1).

more continental facies consists of euhedral to subhedral coarse-grained crystals (Fig. 12c, d), while the marine sandstones have smaller euhedral crystals (Fig. 12a, b). In addition, the kaolinite seems to pre-date quartz authigenesis in the more continental facies while it post-dates quartz overgrowths in the more marine facies (Fig. 12).

The large amount of kaolinite in the coastal plain and tidal flat facies observed in these successions may be explained primarily by a greater flux of fresh meteoric water through a sandstone rich in feldspar and mica (e.g. Bjørlykke *et al.*, 1979). Marine porewater is alkaline and not favourable for flux-controlled kaolinite authigenesis. It is suggested by Hurst and Irwin (1982) that kaolinite in marine sandstones can grow from a stagnant porewater by a diffusion-controlled authigenesis.

Recently Curtis (1983) has suggested that acid pore water generated from compacting shales takes aluminium into solution and that simultaneous carbonate dissolution and kaolinite precipitation is possible. This model could explain the occurrence of late diagenetic kaolinite in the marine sandstones of the reservoir sequences of well 7119/12-1.

Illitization

In 7120/8-1, -2, 7120/7-1 and 7119/12-1 illite fibres post-date quartz overgrowths and occur as a decomposition product after feldspar, mica, and kaolinite (Figs. 12c, d). X-Ray diffraction analysis reveals that the 'illite fibres' are a low expandable illite/smectite, typical of fibrous 'illitic' clays formed during burial diagenesis (c.f. Hurst and Irwin, 1982). The transformation is best developed in 7119/12-1. Illitized kaolinite is mainly seen in the lower coastal plain/tidal flat unit and has a minor development in the estuarine facies, and as traces in shallow marine sandstones. Extensive pore filling illitization in the coastal plain unit is probably a product of the decomposition of kaolinite.

Illites/smectites are here interpreted as being direct precipitates from concentrated pore waters evolved during late diagenesis. The formation of authigenic illite/smectite in the less mature subarkoses in coastal plain/tidal flat units can be related to internal sources of ions (e.g. potassium from the decomposition of micas and feldspars). In sandstones from the Viking Graben illitization of kaolinite generally begins at 3—4-km depth. The occurrence of illitization at 2-km depth and deeper in the Troms I area may suggest higher diagenetic temperatures than at equivalent depths in the Viking Graben.

Other authigenic minerals

Pyrite occurs as framboidal (Fig. 13a), replacive (e.g. fossils) and rhombohedral forms. The two first forms pre-date the quartz overgrowth (Fig. 13a), while cement stratigraphy for the rhombohedral forms cannot be established. Glauconitic sandstone layers (Fig. 13f), chamositic ooid layers (Fig. 13c, and d), and phosphate-rich layers (Plate 1c; Fig. 7a) are important environmental indicators found in the marine inner shelf/near shore sequence, and represent periods with low sedimentation rates.

The occurrence of chamositic/calcerous ooids with micritic coatings and later fibrous cements (Fig. 13e) is indicative of a warm, possibly subtropical climate in Troms during the Middle Jurassic (Porrenga, 1967; Bathurst, 1975). Carbonate peletoids are locally abundant, and their elliptical shape and uniform size indicate that they are faecal pellets which were later recrystallized. Chlorite coatings in the marine sandstones are occasionally observed.

Discussion of the sandstone diagenesis

Despite the dominance of quartz cementation throughout the Troms I area, many of the diagenetic products within the sandstones in the Nordmela and Stø Formations are clearly related to original sandstone composition and depositional environment (Fig. 14). The overall transgressive sequence from coastal alluvial plain to offshore sandstone shows influence of depositional pore water composition with the following authigenic products:

(1) Abundant pore filling kaolinite and siderite in the coastal plain/tidal flat deposits, reflecting the influence of meteoric waters.

(2) Calcite cementation, glauconite, chamositic ooids, phosphate, minor chlorite coating and framboidal pyrite reflect marine pore water.

The diagenetic products generated by burial are also partly dependent on original sandstone composition (Fig. 14). Illitization occurs most commonly

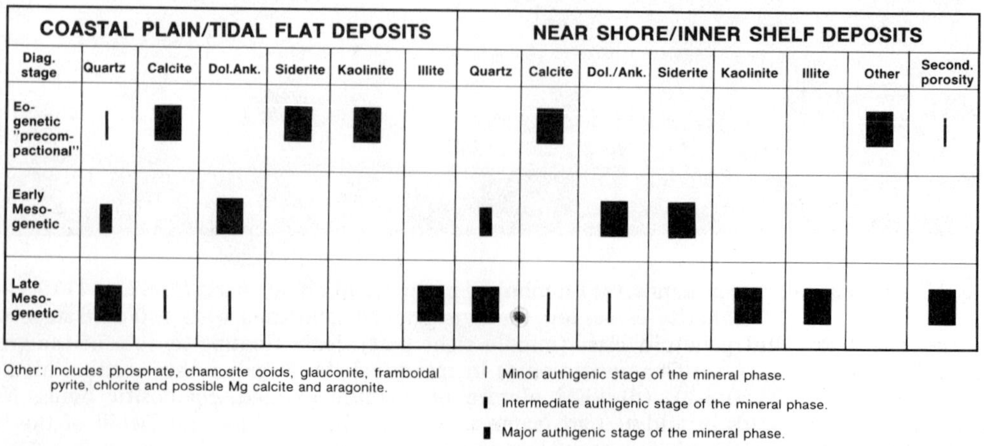

Fig. 14. Cement stratigraphy of the five wells investigated.

in the more submature sandstones where K^+ ions from feldspar and mica are available. Mixing of porewater during burial has probably altered some of the calcite cemented horizons to ankerite and siderite in the marine sequence.

The most important changes in the porosity of the reservoir sandstones were controlled by burial/ tectonic events. The main diagenetic results of these events are mechanical compaction, quartz cementation, illitization, carbonate cementation and development of secondary porosity. A summary of diagenesis for the marine succession is given in Fig. 14. In general, the diagenesis in the wells with reservoir sections buried to around 2600 m is comparable to diagenesis occurring at 3.5—4.0 km in the North Sea. Futhermore, the shallowest well at 1.4 km has a more mature diagenesis than comparable sequences in the North Sea. Two possible mechanisms are suggested for the differences between the North Sea and Troms I area:

(1) Previous deeper burial of the sections and later uplift.

(2) Deeper generated warmer water flushing the reservoir. The pathways for this water flushing may be fault controlled. This porewater may have been CO_2 -rich as a result of maturation of humic kerogen in adjacent shale, have had a low pH and high temperature favourable for carbonate dissolution and minor kaolinite precipitation. Hydrothermal waters would encourage quartz mobilization and subsequent cementation.

Diagenetic studies by Edwards (1979) of Mesozoic sandstones in the Svalbard onshore areas led him to interpret the major porosity changes as a result of tectonic movements. Although the diagenetic products were different from those in our studies (partly due to original differences in facies and sandstone composition) we would interpret the overall controls to be similar.

The present study shows that the major late diagenetic processes of the Troms I area are controlled by the post depositional tectonic development, e.g. block faulting, rate of burial and stress regimes.

ACKNOWLEDGEMENT

This paper is presented with kind permission from Statoil and the licence partners in PL 060, 064 and 077. Thanks are also due to members of the staff at Statoil's Geological Laboratory for their assistance and stimulating discussions. We are grateful to Dr A. Hurst for comments on the manuscript and to Ms Anne Beth Fløtre for typing.

REFERENCES

Bathurst, R.G.G. 1975. *Carbonate Sediments and their Diagenesis*, Developments in Sedimentology, No. 12. Elsevier, Amsterdam, 620 pp.

Berner, R.A. 1971. *Principles of Chemical Sedimentology*. McGraw-Hill, New York, 240 pp.

Bjørlykke, K., Elverhøi, A. and Malm, O.A. 1979. Diagenesis in Mesozoic sandstones from Spitsbergen and the North Sea — a comparison. *Geol. Rundschau*, **68**, 1152—1171.

Bjorøy, M., Bue, B. and Elvsborg, A. 1983. Organic geochemical analysis of the two first wells in the Troms I area (Barents Sea.) *Advances in Organic Geochemistry*. John Wiley, Chichester, pp. 16—27.

Blatt, H. 1979. Diagenetic processes in sandstones. In: P.A. Scholle, and P.R. Schluger (eds.). *Aspects of Diagenesis*, S.E.P.M. Spec. Publ. No. 26, pp. 141—157.

Blatt, H., Middleton, G. and Murray, R. 1980. *Origin of Sedimentary Rocks*, Prentice Hall, Englewood Cliffs, p. 341.

Curtis, C.D. 1983. Link between aluminium mobility and destruction of secondary porosity. *A.A.P.G. Bull.*, **67**, 380—384.

Donovan, D.T. 1966. *Stratigraphy*. Thomas Murby, London, 199 pp.

Edwards, M.B. 1979. Sandstone in Lower Cretaceous Helvetiafjellet Formation, Svalbard: bearing on reservoir potential of Barents Shelf. *A.A.P.G. Bull.*, **63**, 2193—2203.

Gabrielsen, R.H., Fœrseth, R., Hamar, G. and Rønnevik, H. 1984. Nomenclature of the main structural features on the Norwegian Continental Shelf north of 62nd parallel. This volume, pp. 41—60.

Gloppen, T.G. and Westre, S. 1982. Petroleum potential big off northern Norway. *Oil Gas J.* **80**(23), 114—127.

Hurst, A. and Irwin, H. 1981. Geological modelling of clay diagenesis in sandstones. *Clay Minerals* **17**, 5—22.

Matsumoto, R. and Iijima, A. 1981. Origin and diagenetic evolution of Ca—Mg—Fe carbonates in some coalfields of Japan. *Sedimentology* 28, 239—260.

Mossop, G.D. 1972. Origin of peripheral rim, Redwater reef, Alberta. *Can. Petrol. Geol. Bull.*, **20**, 238—280.

Nickel, E. 1978. The present status of cathode luminescence as a tool in sedimentology. *Minerals Sci. Eng.* **10**, 73—100.

Pettijohn, F.J., Potter, P.E. and Siever, R. 1972. *Sand and Sandstone*. Springer, Berlin, 618 pp.

Porrenga, D.H. 1967. *Clay Mineralogy and Geochemistry of Recent Marine Sediments in Tropical Areas*. Publ. No. 9, Fysisch-Geographisch. Lab., Univ. Amsterdam, 145 pp. 145 pp.

Robin, P-Y.F. 1978. Pressure solution at grain-to-grain contacts. *Geochim. Cosmochim. Acta*, **42**, 1383—1389.

Sachs, V.N., Basow, V.A., Dagis, A.A. Ivanova, E.F. Meledina, S.V., Mesezhnikov, M.S., Nainyayeva, T.I., Zakharov, V.A. and Shulgina, N.I. 1973. Paleozoogeography of Boreal-Realm Seas in Jurassic and Neocomian. *A.A.P.G. Bull.*, **57**, 219—229.

Schmidt, V. and McDonald, D.A. 1979. The role of secondary porosity in the course of sandstone diagenesis. In: P.A. Scholle and P.R. Schluger (eds.). *Aspects of Diagenesis*, S.E.P.M. Spec. Publ. No. 26, pp. 175—207.

Skarpnes, O., Jacobsen, W.V., Karlsson, W. and Solheim, S. Preliminary stratigraphic nomenclature of Troms and Haltenbanken. (in preparation)

Stockdale, P.B. 1926. The stratigraphic significance of solution in rock. *J. Geol.* **34**, 399—414.

Westre, S. 1984. The Askeladden gas find, Troms I. This volume, pp. 33—40.

Weyl, P.K. 1959. Pressure solution and the force of crystallization, a phenomenological theory. *J. Geophys. Res.* **64**, 2001—2025.

Zinkernagel, U. 1978. Cathodoluminescence of quartz and its application to sandstone petrology. *Contr. Sedimentol.* **8**, 1—69.

6 Porosity development in selected Jurassic sandstones from the Norwegian and North Seas, Norway – an overview

Richard E. Larese, Norman L. Haskell, Dennis R. Prezbindowski and Dan Beju
Amoco Production Company, Oklahoma, USA

Petrographic analyses of selected Jurassic sandstone core specimens from wells in the Norwegian and North Seas, Norway, indicate that porosity development has been influenced significantly by provenance and sedimentologic and diagenetic controls. In lower-energy marine sands, porosity has been reduced by the presence of detrital clay as matrix. Correspondingly, compaction of ductile argillaceous clasts (e.g. rip-ups) in several deltaic distributary channel sandstones has resulted in considerable loss of pore space as well. Locally, bioturbated intervals may be slightly porous. In these, development of argillaceous coatings on quartz surfaces by burrowing organisms has apparently restricted quartz overgrowth development.

Appreciable pore space in Troms 1 area specimens (7119/12-1) has been eliminated by pressure solution effects on both grain and stylolitic scale. Pressure solution represents an important chemical compactional mechanism in finer grained, argillaceous sands. Silica generated within these intervals migrated into adjacent coarser grained zones where it precipitated as quartz cement. Stylolitic seams are best developed along detrital carbonaceous—argillaceous (illitic) bedding laminae. Apparently, this compactional process can be dependent upon sedimentary facies characteristics.

Cementation constitutes an important mechanism for porosity reduction in several Jurassic sandstone intervals studied. Principal cementing agents include quartz, carbonate (calcite, Fe-dolomite/ankerite, siderite), and kaolinite. Where kaolinite cementation effects are pronounced, sandstone intervals are commonly microporous. Carbon and oxygen stable-isotopic analyses of carbonate cements reveal a major constructive diagenetic interaction between organic and inorganic components during early burial. This relationship with continued burial evolved towards a destructive interaction, characterized by corrosive formation waters causing secondary porosity formation.

Porosity development in selected Jurassic sandstone specimens represents a combination of 'primary' intergranular and secondary-type interstices. In some specimens, secondary porosity by dissolution of detrital and authigenic mineral constituents may account for total effective pore space. Readily identifiable voids result largely by dissolution of detrital feldspar, as discrete grains and as constituents in igneous/metamorphic rock fragments. Dissolution effects in lithic fragments, argillaceous matrix, carbonate cement/clasts, organic detritus, and authigenic feldspar also add to the pore space. At present, an accurate evaluation of the extent of carbonate decementation in Jurassic sands is problematic.

In two Jurassic cores (7119/12-1, 7/12-4), the potential for maximum porosity—permeability development lies in the nearshore marine sediments as opposed to associated deltaic and/or estuarine environments. This is largely due to the better development of marine sands in these particular wells. Diagenetic processes influenced by provenance and sedimentologic controls have significantly affected porosity—permeability characteristics of these intervals too.

INTRODUCTION

Petrographic analyses of selected Jurassic sandstone core specimens from wells in the Norwegian and North Seas (Fig. 1) suggest that porosity and permeability development has been influenced significantly by provenance and by sedimentologic and diagenetic effects. An attempt has been made to document and evaluate the impact of these factors on porosity and permeability. Approximately 180 core specimens from six wells, in addition to several miscellaneous specimens, were analysed by standard light microscopy, scanning electron microscopy, and X-ray diffractometry methods. Cementation effects in several samples were studied employing cathodoluminescene techniques. In selected specimens, carbonate cements were subjected to stable isotopic analyses ($\delta^{13}C$, $\delta^{18}O$) and organic matter to thermal maturation

Petroleum Geology of the North European Margin © Norwegian Petroleum Society (Graham & Trotman, 1984) pp. 81—95.

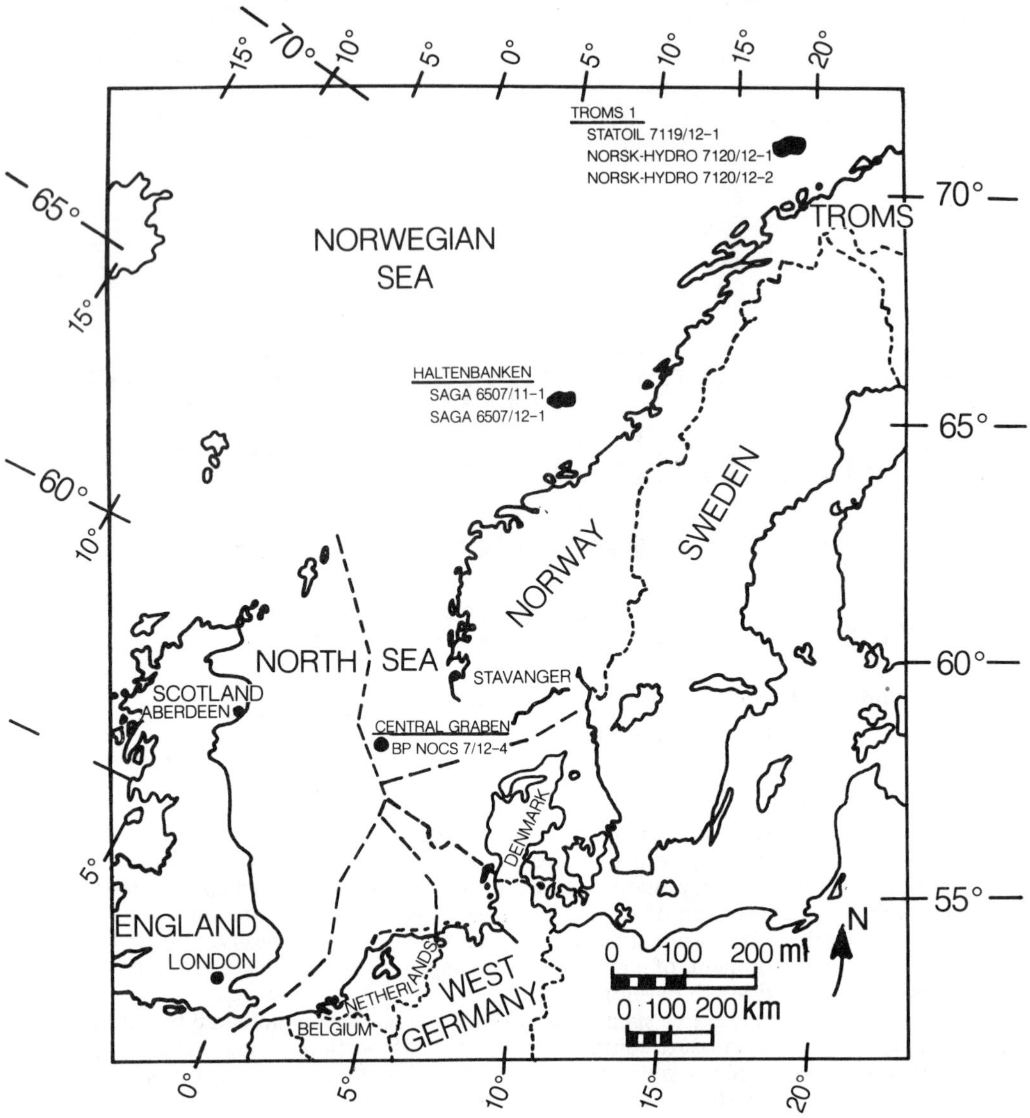

Fig. 1. Map showing location of wells from which selected Jurassic core specimens were analysed.

characterization (vitrinite reflectance). The cores studied were within these intervals:

7119/12-1	2667.0—3001.6 m	Lower—early Middle Jurassic
7120/12-1	2042.0—2058.6 m	Middle Jurassic
7120/12-2	1888.0—2052.0 m	Lower Jurassic
6507/11-1	2502.1—2543.0 m	Lower Jurassic
6507/12-1	2404.0—2715.8 m	Lower Jurassic
7/12-4	3447.5—3556.0 m	Middle—Upper Jurassic

On-site descriptions were made of core material from three wells (7119/12-1, 7120/12-1 and 6507/12-1). Throughout this chapter, percentage abundances are quoted by volume.

LITHOLOGY AND MINERALOGY

The Jurassic sandstone specimens analysed vary from arkosic wacke to quartz arenite in composition (Table 1, Fig. 2; Dott's classification, 1964).

TABLE 1. Lithology and texture

Wells	*Principal sandstone types (Dott's Classification, 1964)*	*Grain size*	*Sorting*
7119/12-1, marine	Quartz arenite	Fine lower—medium lower	Poor—well
7119/12-1, deltaic—estuarine	Sublitharenite—quartz arenite	Very fine lower—medium upper	Poor—moderately well
7120/12-1	Quartz wacke—quartz arenite	Very fine lower—medium lower	Poor—well
7120/12-2	Subarkose—quartz arenite	Coarse silt—medium lower	Moderate well—well
6507/11-1	Arkosic arenite—subarkose	Very fine upper—coarse upper	Poor—well
6507/12-1	Arkosic arenite—subarkose	Very fine upper—coarse lower	Poor—well
7/12-4, marine	Arkosic arenite—subarkose	Very fine upper—medium lower	Poor—well
7/12-4, deltaic	Sublitharenite—arkosic arenite	Very fine upper—very coarse upper	Poor—moderately

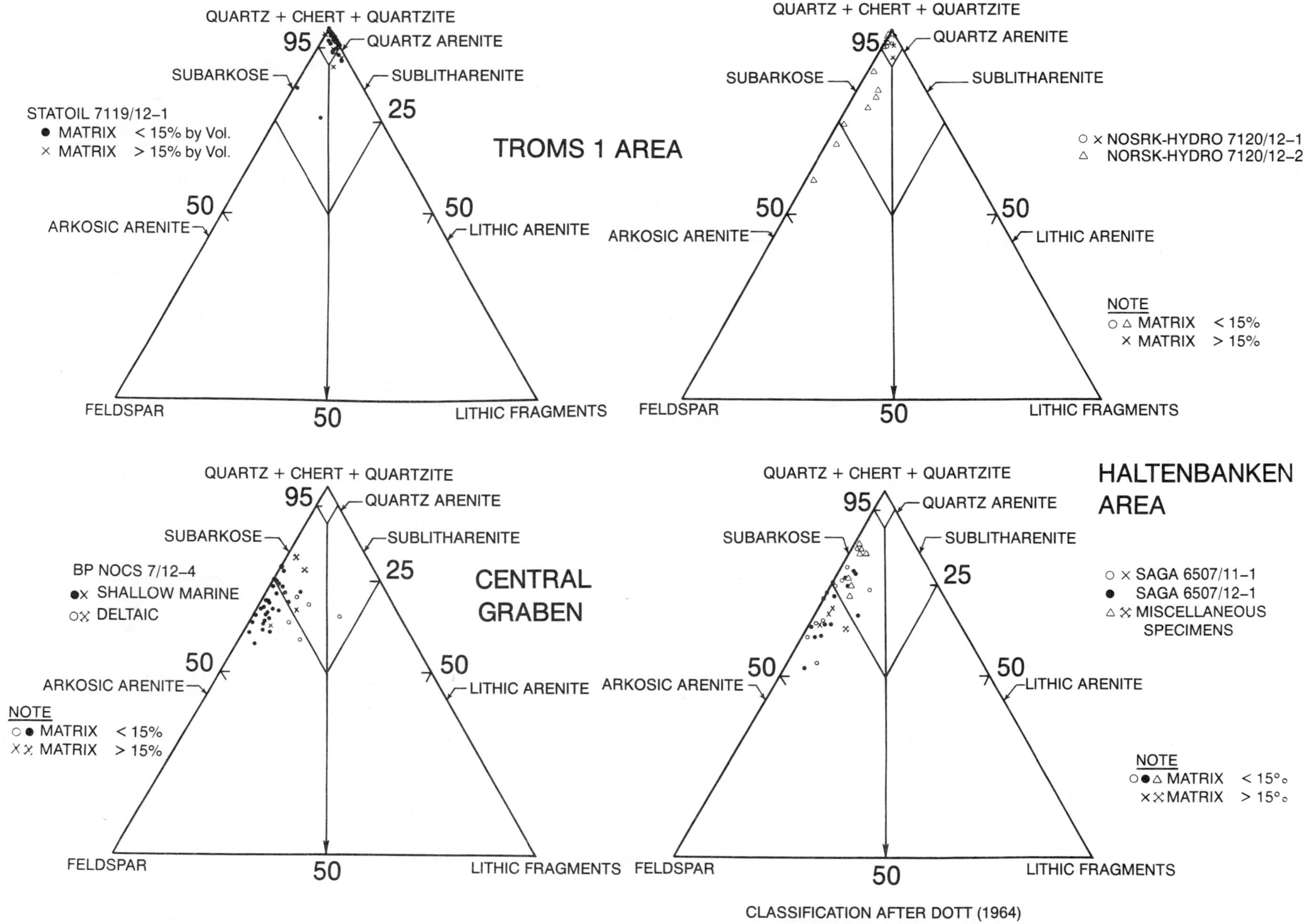

Fig. 2. Classification of original sandstone rock types observed in selected Jurassic core specimens from wells in the Norwegian and North Seas, Norway. Although most specimens represent arenite type sands (matrix <15 %), several wacke type sandstones (matrix >15 %) were noted and plotted on the 'arenite' diagrams for conservation of space. Classification modified after Dott, 1964. Specimens labelled X, ✕ represent wacke-type sandstones.

TABLE 2. Thin section modal analyses of detrital constituents[a]

	Quartz										Lithic fragments				Argillaceous					
	Monocrystalline		Polycrystalline[d]		Chert		Feldspar		Sedimentary		Igneous/metamorphic[c]		Origin?		(matrix-laminae)		Mica		Miscellaneous	
Troms 1																				
7119/12-1, marine	67.1	57.0–76.5	3.8	1.5–6.5	0.3	0.0–1.0	0.4	0.0–2.7	1.3	0.0–3.3	—	—	—	—	4.0	0.0–21.0	0.1	0.0–1.5	0.2	0.0–0.6
7119/12-1, deltaic—estuarine	59.0	53.0–81.0	7.0	2.0–17.5	0.5	0.0–2.0	2.4	0.0–9.0	2.3	0.0–9.0	—	—	0.1	0.0–0.5	7.5	0.0–23.0	1.7	0.5–3.0	2.5	0.5–5.0
7120/12-1	59.7	42.0–67.5	2.3	0.0–4.0	0.1	0.0–0.5	1.2	0.0–2.5	0.7	0.0–1.5	0.1	0.0–0.4	—	—	13.0	0.0–25.5	2.1	0.0–11.0	0.9	0.0–1.5
7120/12-2	45.7	35.0–53.0	4.5	2.5–7.6	0.60	0.0–1.9	6.4	0.5–19.5	0.9	0.0–6.0	0.1	0.0–0.5	1.2	0.0–3.5	2.4	0.0–11.5	6.1	0.0–15.0	2.7	0.0–5.5
Haltenbanken																				
6507/11-1	36.1	24.9–48.0	11.2	1.8–22.5	0.4	0.0–1.0	11.7	5.0–20.0	0.3	0.0–1.5	1.9	0.0–9.0	0.7	0.0–3.4	4.4	0.0–19.5	1.5	0.0–8.6	0.8	0.0–2.8
6507/12-1	25.0	22.1–33.5	13.0	3.3–21.8	<0.1	0.0–0.3	12.2	9.9–15.0	0.2	0.0–1.3	3.0	0.0–7.9	2.0	0.7–2.6	0.4	0.0–1.7	5.5	0.0–15.2	1.8	0.0–7.9
Central Graben																				
7/12-4, marine	42.4	25.5–56.0	6.8	2.0–11.0	0.5	0.0–1.5	16.9	8.5–23.0	1.8	0.0–10.5[b]	0.7	0.0–2.5	0.1	0.0–1.0	5.1	0.0–32.0	0.8	0.0–10.5	0.4	0.0–2.0
7/12-4, non-marine	44.7	37.0–50.0	7.3	1.5–17.5	0.2	0.0–0.5	12.3	3.5–18.5	7.3	0.0–17.0	0.4	0.0–1.5	—	—	12.6	0.5–43.5	2.6	0.0–5.0	0.8	0.0–1.0

[a] First column gives average of individual specimens.
Second column gives range among specimens. Percent by volume.
[b] Fossil fragments.
[c] Dominantly feldspathic/quartzose rock fragments.
[d] Some polycrystalline quartz of igneous/metamorphic origin.

TABLE 3. Thin section modal analyses of authigenic constituents[a]

	Quartz		Carbonate total[b]		Kaolinite		Pyrite		Feldspar		Illite		Chlorite	
Troms 1														
7119/12-1, marine	13.0	1.5–29.5	2.5	0.0–30.6	0.1	0.0–0.5	0.5	0.0–2.0	—	—	0.2	0.0–3.0	—	—
7119/12-1, deltaic—estuarine	9.2	0.0–25.5	2.4	0.0–15.0	5.2	0.0–17.5	—	—	—	—	—	—	—	—
7120/12-1	0.5	0.0–1.0	8.8	0.0–32.0	1.8	0.0–5.5	1.0	0.0–6.0	—	—	—	—	—	—
7120/12-2	8.4	1.0–12.2	4.8	0.0–30.0	3.4	0.0–9.2	0.2	0.0–1.0	tr[c]	0.0–tr	—	—	—	—
Haltenbanken														
6507/11-1	0.4	0.0–2.4	8.3	0.0–43.6	0.7	0.0–3.0	0.8	0.0–3.6	0.7	0.0–2.4	—	—	—	—
6507/12-1	0.7	0.0–2.3	16.7	0.0–49.3	5.5	3.3–9.3	0.1	0.0–0.7	0.4	0.0–3.0	—	—	—	—
Central Graben														
7/12-4, marine	4.5	0.0–13.0	7.5	0.0–46.0	0.3	0.0–3.0	0.5	0.0–3.0	0.8	0.0–2.0	—	—	—	—
7/12/4, deltaic	1.5	0.0–5.0	3.3	0.0–23.5	0.3	0.0–0.5	0.3	0.0–1.0	0.3	0.0–2.0	—	—	0.4	0.0–3.0

a First column gives average of individual specimens. Second column gives range among specimens. Percent by volume.
b Total carbonate as cement and mineral replacements.
c tr = trace amounts.

TABLE 4. Porosity and permeability from core measurements. Sources of data: Statoil, Norsk-Hydro A/S, Saga Petroleum Co., BP Petroleum Dev. of Norway A.S.

	Horizontal porosity (%)			*Horizontal permeability (mD)*		
	Average		*Range*	*Average*		*Range*
Wells	*Arithmetic*	*Geometric*		*Arithmetic*	*Geometric*	
7119/12-1	7.09	6.15	0.20—16.30	16.92	0.43	0.02—627.00
7120/12-1	12.88	10.69	3.50—20.60	11.64	0.54	0.0—134.38
7120/12-2	17.38	16.08	2.30—25.80	186.43	13.12	0.01—4294.70
6507/11-1	21.93	19.76	2.20—35.90	1433.43	45.82	0.03—18390.00
6507/12-1	16.24	13.21	0.90—33.10	444.82	3.10	0.04—6819.00
7/12-4	16.03	15.11	3.40—23.80	147.20	14.22	0.06—1968.00

Specimens from the Haltenbanken and Central Graben areas are considerably more feldspathic than those from several Jurassic intervals in the Troms 1 area (7119/12-1; 7120/12-1). The presence of coarse feldspathic—quartzose rock fragments in Haltenbanken cores suggests a nearby igneous/metamorphic source similar to pre-Jurassic crystalline bedrock outcropping along the Central Norwegian coastline (Holtedahl and Dons, 1960).

Mineralogically, detrital constituents comprise approximately 63—88% of the Jurassic sandstone specimens analysed (Table 2). Principal detrital components consist of quartz (mono and polycrystalline; core averages approximately 38—71%) with subordinate amounts of feldspar (0.4—16.9%), lithic fragments (0.8—7.7%), argillaceous material as matrix/laminae (0.4—13.0%), and mica (biotite—muscovite; 0.1—6.1%). Trace to minor concentrations of heavy minerals (garnet, leucoxene, rutile, brookite, tourmaline, zircon, etc.), organic detritus, and glauconite are present as well. Organic matter occurs in both structured (woody/plant) and amorphous forms. Vitrinite reflectance measurements (%R_0) of this material in a limited number of shale, coal, and sandstone specimens yield values ranging from approximately 0.4 to 0.92% R_0. On average, the majority of measurements in Norwegian Sea samples were in the 0.4—0.6% R_0 range, which suggests a relatively immature organic maturation stage.

In general, K-feldspars (microcline, orthoclase) constitute the dominant feldspar type observed. It appears likely that the dominance of K-feldspar over plagioclase can be attributed in part to provenance controls. However, a subtle relationship involving K-feldspar—plagioclase ratios with grain size exists in some intervals of 7/12-4 cores. In these, there is a tendency for plagioclase content to increase relative to K-feldspar with decreasing grain size. Aside from source area, the apparent increased concentration of plagioclase in finer grained sandstone intervals may be attributed to: (1) disintegration of larger plagioclase grains with erosion—transportation and subsequent accumulation in finer grained sediment fractions, and/or to (2) inability of corrosive formation waters to permeate through finer grained and commonly more argillaceous sands to facilitate dissolution of relatively more unstable plagioclase grains.

Authigenic constituents (Table 3) consist mainly of quartz, carbonate and kaolinite with trace to minor amounts of pyrite, chlorite, and illite. Texturally, specimens range from very fine to coarse grained and poorly to very well sorted (Table 1). Based on core analyses, porosities and permeabilities vary from 0.0 to 35.9% and 0.0 to 18 390 mD, respectively (Table 4).

SEDIMENTOLOGIC AND PROVENANCE POROSITY CONTROLS

The effects of provenance and sedimentologic controls on porosity and permeability development in Jurassic sediments are significant. Compositional aspects of mineral and lithic constituents are extremely important with burial, as in many instances they serve to define a unique pathway of diagenetic reactions (Hayes, 1979) which include both porosity reduction and enhancement mechanisms. Secondary porosity, developed by dissolution of detrital mineral constituents (feldspar) and lithic fragments, is dependent upon the composition of the sediment deposited and, ultimately, of the source area. Correspondingly, cementation and compactional effects are controlled in large part by the nature of precursor detrital constituents (feldspar, mica, shale). In compactional processes, ductile lithic fragments (shale clasts) are especially prone to plastic deformation with burial. Porosity reduction by this mechanism is greatest in relatively immature fluvial-deltaic channel sandstones (e.g., 7/12-4, 3528—3556.0 m).

Petrographic analysis indicates that the principal sedimentologic cause of porosity reduction is the presence of argillaceous material as primary interstitial matrix and bedding laminae. As defined by X-ray diffractometry, detrital clay material consists mainly of illite, chlorite, and kaolinite with lesser amounts of mixed layered smectite—illite. The greatest concentrations are in very fine (0.08m) to

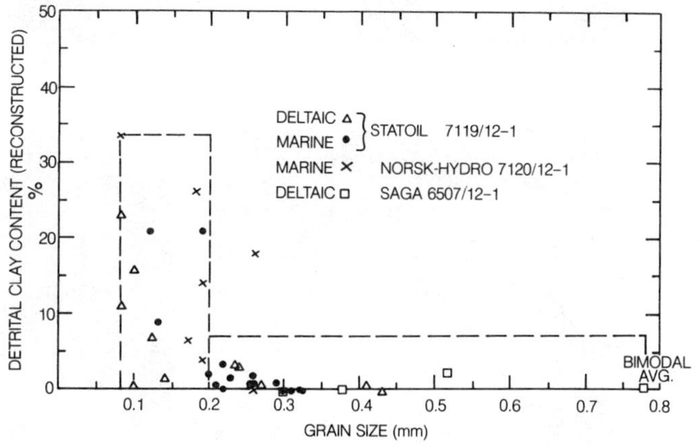

Fig. 3. Relationship of reconstructed detrital clay content (laminae, matrix) and average grain size in Jurassic sandstone specimens from the Troms and Haltenbanken areas, Norwegian Sea, Norway.

TABLE 5. Summary of diagenetic processes

Porosity reduction	*Porosity preservation*	*Porosity increase*
Cementation	*Grain coatings*	*Dissolution*
Quartz	Illite	Feldspar grains—cement
Carbonate	Mixed layered	Igneous/metamorphic rock fragments
Clay (i.e. kaolinite)	(illite/smectite)	Polycrystalline quartz
Feldspar	Chlorite	Carbonate cement/clasts
	Organic	Lithic grains
Compaction	Siderite	Argillaceous matrix
Pressure solution	Collophane	Mica
Ductile grains		Heavy minerals (garnet)
		Glauconite
		Organic detritus
		Chert
		Fractures (grain)

fine (0.200 mm) grained, laminated (overbank) and bioturbated (lower shoreface) sandstone specimens (Fig. 3). In coarser grained intervals to (0.200—0.800 mm), the detrital clay content is significantly less. For example, in 7/12-4 fine-grained lower shoreface sandstones (3517—3525 m) have 17.8% clay matrix and 2.8% intergranular porosity, whereas medium-grained middle shoreface sandstones (3462.1—3469.4 m) have 1.1% and 12.0%, respectively. These relationships (% clay vs. grain size) are consistent with the energy characteristics of the interpreted depositional environments, especially in bioturbated lower shoreface facies.

In addition, grain size, sorting and clay laminae can be closely associated with chemical compaction effects involving pressure solution. The significance of this will be discussed later.

DIAGENETIC POROSITY CONTROLS

Diagenetic processes (Table 5) have exerted a significant influence on porosity and permeability development. Individually, these processes have decreased, preserved, and enhanced porosity. The factors are commonly interrelated and the effects of one have been partially offset by the effects of another.

Porosity reduction

Porosity reduction in Jurassic sediments entails a combination of cementation and compaction effects. Cementation effects are more widespread, with primary cementing agents consisting of quartz, carbonate (calcite, dolomite/ankerite, siderite), and kaolinite (Fig. 4).

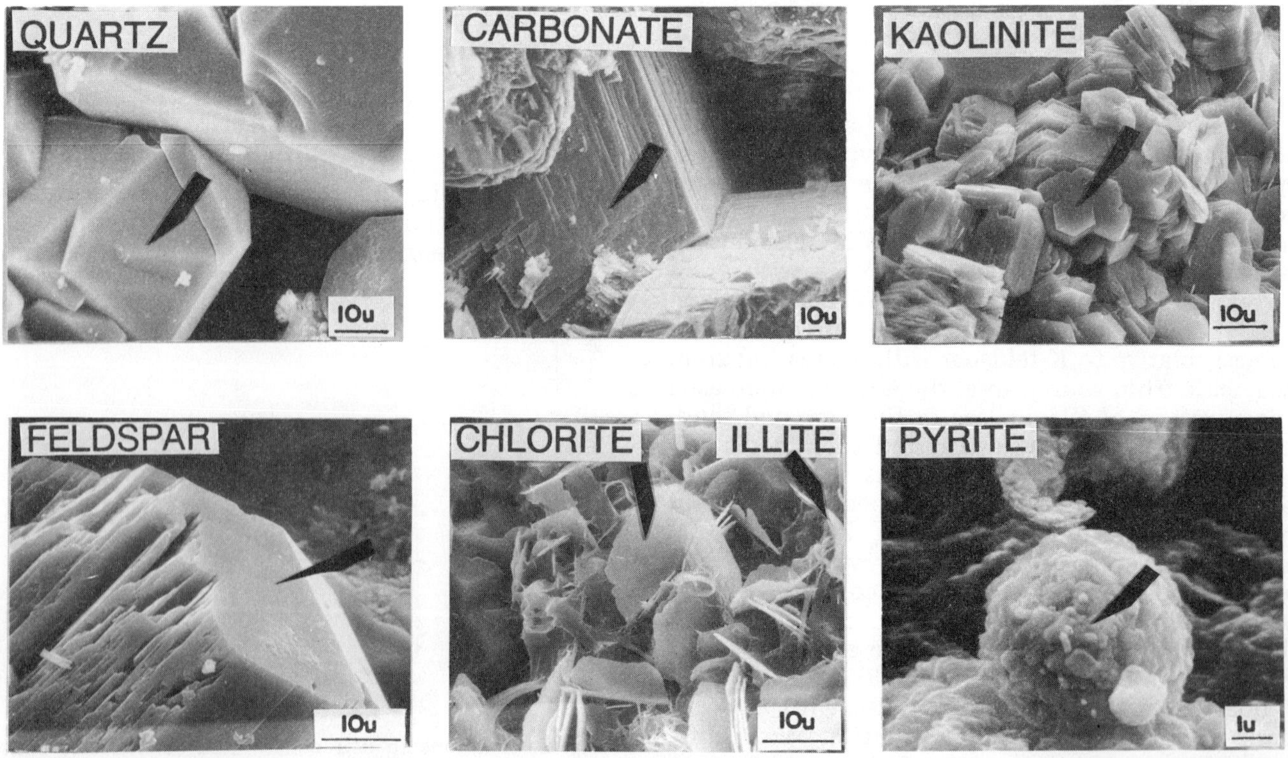

Fig. 4. Scanning electron micrographs depicting authigenic cements in selected Jurassic sandstone core specimens.

Quartz cementation and pressure solution.

Quartz cementation was noted in most specimens analysed. The effects appear most pronounced in the 7119/12-1 cores (0—29.5%) and least obvious in selected specimens from the 6507/11-1 and 6507/12-1 cores (0—2.4%). In the latter, authigenic quartz is present mainly as incipient growths on detrital quartz surfaces and as linings along fractures in mono- and polycrystalline quartz. In general, precipitation of quartz as overgrowths represents a relatively early diagenetic event which post-dates feldspar authigenesis *but* pre-dates major stages of carbonate cementation.

Petrographic evidence suggests that a major source of silica for quartz cementation in Jurassic sediments is pressure solution, on both grain and stylolitic scale. Where quartz cement is prevalent, a positive relationship exists with relatively intense pressure solution effects. This association is best demonstrated in cores from 7119/12-1 (2749.1—2760.4 m) in which illitic—micaceous clay and carbonaceous matter are accentuated as wisp-like lineations (some detrital coatings) and thicker laminae. Apparently, these clays act as catalysts (Heald, 1956), promoting pressure solution of detrital quartz at points of contact, by serving as diffusion conduits (Weyl, 1959) for enhanced silica migration and/or to increase pH on a microscale (Thomson, 1959), which subsequently facilitates quartz solution.

Stylolitic seams are best developed along carbonaceous—argillaceous bedding laminae (Lerbekmo and Platt, 1962). Where laminae are lacking, stylolitic solution is less pronounced. Correspondingly, stylolites may be well developed at boundaries of abrupt grain-size change.

Evidence for pressure solution is also common in several specimens from 7120/12-2. Its effects are accentuated along fine grained, quartzose—micaceous (mica, biotite) laminae (Plate 2). Cathodoluminescene microscopy supports this observation and suggests that quartz overgrowths are lacking on detrital surfaces along these laminae. In contrast, silica (quartz) cementation effects are well developed in nearby less-micaceous and coarser-grained quartzose intervals.

Pressure solution represents an important chemical compactional mechanism in many finer grained intervals (7119/12-1, 2708.5 m). Silica generated in these finer grained intervals has probably migrated into nearby coarser grained zones, where it has precipitated as syntaxial quartz overgrowths. These effects (stylolites) appear to be dependent upon sedimentary facies characteristics within the Troms 1 area (7119/12-1) sediments. Possibly, tectonic activity (rifting) within the basin constituted a 'triggering' mechanism for initiation of stylolitic solution within the sedimentary facies.

Porosity reduction by quartz cementation varies greatly in the Jurassic cores studied. Silica cementation effects are extensive in both marine and nonmarine intervals in 7119/12-1 but are less in the 7120/12-1 and 6507 wells. Authigenic quartz varies from 0 to 2.4% in discrete specimens from these wells (Table 3). Development of quartz overgrowths in the 6507 cores is most prevalent in the very fine to fine grained intervals. In part, this may be a reflection of grain size as silica growths on finer quartz surfaces tend to terminate (achieve euhedralism) at faster rates than those on coarser surfaces (Heald and Renton, 1966). It should be noted that a lack of well-defined dust rims compounds the problem of quartz overgrowth identification in many instances.

The reasons for the lack of appreciable quartz cementation in the 7120/12-1 and 6507 cores are unclear. In several fine-grained sand intervals, sporadic 'clots' of detrital clay have restricted quartz growth (Plate 3), but these do not appear to have been extensive enough to account for the paucity of well-defined quartz cementation effects. Possibly, formation waters attending diagenesis were saturated with respect to silica but not conducive to significant cementation. Moreover, inadequate flushing of sufficient pore volumes of water through the sands caused by early sealing of the system (i.e. geopressuring) may have reduced silica concentrations. There is little evidence that carbonate cementation (followed by dissolution) served to restrict the development of quartz. Although limited pressure solution effects were observed, evidence of extensive pressure solution is lacking. The absence of this potentially significant silica source may represent a major reason for the paucity of authigenic quartz.

Carbonate cementation

Carbonate (Plate 4) is present in varying amounts as cement, mineral replacements, and detrital clasts throughout the Jurassic interval studied. In discrete samples total authigenic carbonate varies from about 0.0 to 49.3%. Principal carbonate types include calcite (i.e. ferroan), ferroan dolomite/ankerite, and siderite. Although siderite, and in some instances ankerite and calcite (i.e. concretions), may represent early 'pre-quartz' cementation effects, most carbonate precipitation appears to have followed the development of authigenic feldspar and quartz.

Calcite and ferroan dolomite/ankerite may account for as much as about 43.6% (6507/11-1, 2510.55 m). In addition to its presence as cement, much of the authigenic carbonate has selectively replaced detrital mineral constituents (11—19%). Extensive replacement affects detrital feldspar, organic-argillaceous clasts/matrix, igneous/metamorphic rock fragments, polycrystalline quartz, and mica. Ferroan dolomite/ankerite and siderite in finer grained bioturbated intervals have selectively replaced organic-rich argillaceous (illitic) matrix. Most likely, these organic clays exhibit microenvironmental characteristics (i.e. pH, eH, Fe, Mg fixation) favourable for authigenesis of these carbonate types. Siderite (and pyrite-Fe oxides) is associated with detrital Fe-bearing biotite mica, as well.

Local occurrences of detrital carbonate are present as clasts and pelletal segregations (hydraulically) along bedding laminae. Authigenic carbonate cements have sometimes selectively nucleated on these detrital carbonate surfaces. Commonly, siderite clasts are segregated along bedding laminae in finer grained sediments and apparently represent 'rip up' clasts (replaced micaceous hash) which have been subjected to deformation attending compaction. Detrital siderite clasts in coarser

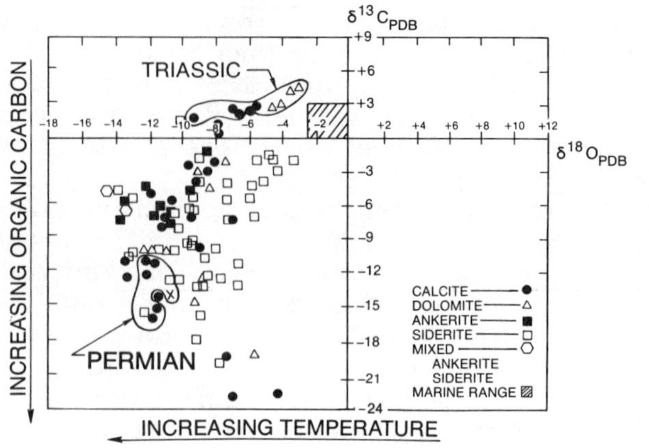

Fig. 5. Results of isotopic analyses ($\delta^{13}C$, $\delta^{18}O$) of carbonate cements.

grained intervals are more widely disseminated.

Carbon and oxygen stable isotopic analyses were made on the carbonate cements from 85 sandstone samples taken from the six wells in the study area. The Jurassic, Triassic, and Permian sections were sampled, with the majority of samples coming from the Jurassic interval. A discussion of the sample preparation techniques and data precision are given in Appendix 1. The analyses were made to determine the importance of organic diagenesis and temperature changes during carbonate cementation of the Jurassic sand units over a widely spaced area.

A covariation in the $\delta^{18}O$ and $\delta^{13}C$ composition of the carbonate cements analysed is apparent (Figs. 5 and 6). A depleted or more negative $\delta^{13}C$ composition is generally accompanied by a more negative $\delta^{18}O$ composition. This covariation indicates that the diagenetic environments of cementation in these sands was under significant organic influence during burial, evidenced by the large $\delta^{13}C$ shift from normal marine carbonate range of 0.0 to 3.0‰ and fresh water carbonate range of —1.0 to —5.0‰ (PDB Standard) to observed values as negative as —21‰.

The $\delta^{13}C$ composition of precipitated marine carbonate material is controlled by the carbon isotopic composition of the dissolved bicarbonate in the oceans. Thus, the $\delta^{13}C$ composition of a precipitated marine carbonate falls within a range of a +3.0 to 0.0‰ (relative to PDB). In order to alter this composition significantly during diagenesis solution-repricipitation), a new source of depleted carbon in the form of CO_2 or HCO_3^- must be introduced into the diagenetic system. This $\delta^{13}C$-depleted CO_2 can be generated *in situ*, or transported into the diagenetic system. The main mechanisms for generating a significant amount of $\delta^{13}C$-depleted CO_2 or HCO_3^- in a subsurface clastic system during burial have been discussed by Curtis (1978). These are:

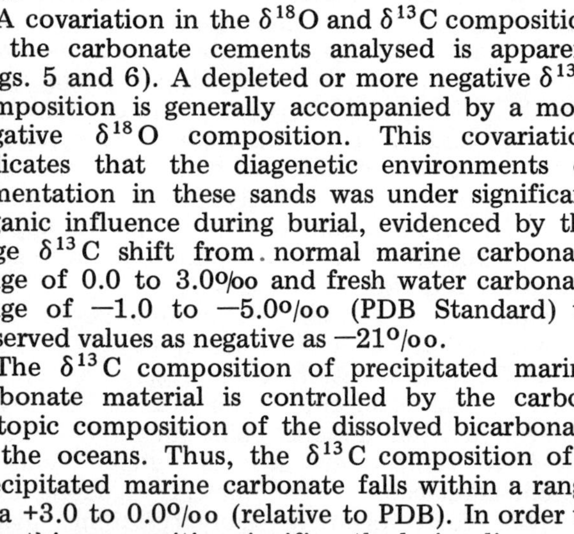

Fig. 6. Diagram showing covariation of $\delta^{18}O$ and $\delta^{13}C$ composition of carbonate cements in selected Jurassic sandstone core specimens from 7/12-4 and 7120/12-1.

(1) Microbial oxidation of organic material can generate CO_2 with $\delta^{13}C$ compositions of —25‰ (relative to PDB). This process is effective only in the upper tens of centimetres of the sediment column and is controlled by the rate of oxygen diffusion into the sediment from the water—sediment interface.

(2) Bacterial sulphate reduction (written simply $2CH_2O + SO_4^{2-} \longrightarrow H_2S + 2HCO_3^-$) allows the breakup of $\delta^{13}C$-depleted organic material and sulphate to form H_2S and $\delta^{13}C$-depleted CO_2. This CO_2 can be incorporated into any precipitated carbonate cements giving them a distinctly negative $\delta^{13}C$ composition. This process is only effective if SO_4^{2-} is present. Rapid utilization of *in-situ* SO_4^{2-} and the slow rate of SO_4^{2-} diffusion from the water—sediment interface during early burial limits this process to the upper tens of metres of the sediment column (Berner, 1980; Irwin *et al.*, 1977).

3. Thermal alteration of organic material during burial generates CO_2, H_2S, and CH_4 (Laplante, 1974). As much as 25‰ of the kerogen present in a sediment at the time of deposition can be converted to $\delta^{13}C$ depleted CO_2 during burial (Laplante, 1974) ranging from tens of metres to several kilometres. This CO_2 can have a $\delta^{13}C$ compositional range of —10.0 to —25.0‰ (PDB) (Irwin *et al.*, 1977).

A carbonate having 100% of its carbon derived from the thermal alteration of organic material during burial would have a $\delta^{13}C$ compositional range of —18 to —21‰ (PDB). This value range was determined by making the following assumptions: (1) $\delta^{13}C$ composition of derived CO_2 during thermal alteration of organic material was —25‰ (Irwin *et al.*, 1977), (2) precipitation took place at temperatures between 50 and 100°C, and (3) the $\delta^{13}C$ fractional relationship determined by Bottinga (1968) between carbon dioxide (gas) and calcite is valid for natural subsurface systems.

Some of the carbonate cements analysed have derived nearly 100% of their carbon from the alteration of associated organic material. We believe that this alteration was predominantly thermal in nature. A significant $\delta^{18}O$-depletion trend and the petrographic sequence of the carbonate cements argue against a significant input of $\delta^{13}C$-depleted CO_2 from the surface and near-surface processes of microbial oxidation and sulphate reduction.

The $\delta^{18}O$ composition of a carbonate is controlled by the $\delta^{18}O$ composition of the fluid from which it precipitated and the temperature of precipitation (Epstein *et al.*, 1953). Thus, the observed $\delta^{18}O$-depletion shift away from the surface, marine carbonate compositional range of 0 to —2.5‰ (PDB) (Fig. 5) could have been the result of precipitation from a $\delta^{18}O$-depleted water (meteoric water) and/or an increase in temperatures of cementation. The meteoric water potentially associated with the deposition of these fluvial/deltaic and shallow marine sediments would have had a depleted $\delta^{18}O$ composition relative to the Jurassic marine waters (—1.2°/oo, SMOW) (Irwin *et al.*, 1977). Although it is impossible to determine the exact $\delta^{18}O$ composition of the fluvial water, a value of approximately 6.0°/oo would not be unreasonable, based on analogy to the present day $\delta^{18}O$ composition of Mississippi River water of —4.9°/oo (Epstein and Mayeda, 1953). A calcite precipitated at surface conditions from such a water would have a $\delta^{18}O$ composition of approximately —5 to —7.5°/oo. Clearly the majority of the carbonate cements analysed have a more depleted $\delta^{18}O$ composition. Thus, we believe that the bulk of the $\delta^{18}O$ depletion observed in the carbonate cements is the result of increasing temperature of precipitation during progressive burial. Petrographic observations of early-to-late, relative, diagenetic placement of the carbonate cements support this conclusion.

The general $\delta^{13}C$-depletion trend observed (Fig. 5) reflects the dynamic interaction between the inherited marine and fresh-water carbon reservoir (i.e. carbonate skeletal material, carbonate lithoclast, and surface carbonate cements) and the $\delta^{13}C$-depleted carbon reservoir derived from the alteration of organic material. The corresponding $\delta^{18}O$-depletion trend in the carbonate cements analysed strongly suggests that increasing burial is followed by the introduction and incorporation of increasing amounts of $\delta^{13}C$-depleted carbon into the precipitated carbonate cements. The rate of $\delta^{18}O$) depletion (dominantly controlled by increases in precipitation temperature) and the incorporation of $\delta^{13}C$-depleted carbon in the carbonate cements varies from well to well, reflecting differences in the organic/inorganic diagenetic history of specific sand bodies (Fig. 6). These differences are due to variations in burial rates, amounts and types of organic matter, and hydrologic and sedimentologic characteristics of the sediment package.

Kaolinite cementation

Authigenic kaolinite (Plate 5), occurring as pore filling cement and mineral replacements, represents the most prevalent authigenic clay in the Jurassic specimens analysed, ranging upwards to 17.5% (Table 3). Development of kaolinite is most pronounced in fluvial-deltaic sediments (7119/12-1, 2818—2845 m), although appreciable amounts are found in marine sands as well. Possibly, the pore-water chemistry of these deltaic intervals was mildly acidic during this diagenetic interval, a condition favouring kaolinite authigenesis (Degens, 1965).

Petrographic analysis suggests that hydrolysis of detrital feldspar, muscovite, mica and detrital argillaceous material (kaolinite, illite) constituted a major source of cations (Al_2O_3, SiO_2) needed for kaolinite precipitation (Degens, 1965). Kaolinite commonly appears pseudomorphous after replaced detrital constituents (Plate 5B, feldspar). Extensive kaolinite cementation has caused appreciable micro-porosity and greatly diminished permeability.

Miscellaneous authigenic constituents

These mineral components (Fig. 4) include feldspar, pyrite, illite, and chlorite. As cements, these have had only minimal effects on porosity reduction. In some cases, illite and chlorite as coating material have served to restrict quartz cementation on a minor scale. These aspects will be discussed later.

Although not significant as a porosity-reducing mechanism, authigenic feldspar overgrowths are relatively common in Jurassic sediments in the Haltenbanken and Central Graben areas and vary from 0 to 3.0% (6507/12-1, 2532 m). Na-cobaltinitrite staining techniques and energy dispersive X-ray analyses suggest that the majority of these overgrowths are compositionally pure end-member (?) K-feldspar. In a few instances development of albitic overgrowths on detrital plagioclase was observed as well.

Authigenic K-feldspar (Fig. 4) overgrowths vary from anhedral to euhedral in form and may be turbid, apparently due to inclusions of organic matter (?) and clay. Overgrowth textures are not interlocking and thus have had little impact on porosity—permeability reduction. In contrast, incipient dissolution of feldspar overgrowths has resulted in minute quantities of secondary porosity (0—0.7%).

Petrographic evidence suggests that precipitation of authigenic feldspar was a very early diagenetic event in the Jurassic sediments, probably contemporaneous with *or* before quartz cementation. Apparently, the early formation waters possessed the favourably high alkalinities, low CO_2 levels and sufficiently high concentrations of K^+ (Degens, 1965) needed for precipitation of this cement.

Authigenic pyrite is present in most Jurassic sediments as a replacement of detrital organic matter, organic-argillaceous laminae and/or biotite mica comprising as much as 6% in some specimens.

Porosity preservation controls

Grain coatings

Detrital and authigenic grain coatings serve to inhibit the effects of quartz cementation by restricting the silica present in formation waters from nucleating on detrital quartz surfaces as syntaxial overgrowths. Coating types documented in Jurassic sandstone specimens include illite, mixed layered illite/smectite, chlorite, siderite, collophane, and organic matter (Table 5). The mechanisms responsible for development of these include sediment infiltration (Walker and Crone, 1974), bioturbation, and direct chemical precipitation/recrystallization. Distribution of coatings is sporadic in Jurassic specimens analysed. However, it has been documented that grain coatings constitute a significant mechanism for porosity preservation in several major reservoir sandstones (e.g. Tuscaloosa Ss., Thomson, 1982).

In some instances, bioturbated intervals in lower shoreface—lagoonal sediments are more porous than the encasing, less argillaceous, quartzose intervals due to grain coatings (Plate 6A, B). The development of these coatings appears to be related to bioturbation processes in which organisms rearranged detrital matrix clay material. Much of this coating material may represent organic (mucus) clay complexes resulting from the digestive processes of burrowing organisms. In other Jurassic marine intervals (7119/12-1, 2667.6 m), authigenic illite coatings (Plate 6C, D) have similarly impeded

TABLE 6. Porosity from thin-section analysis

Area/well	*Total porosity (%)*		*Ratio secondary porosity/total porosity*	
	Average	*Range*[a]	*Average*	*Range*
Troms 1				
7119/12-1, marine	6.7	0.0—20.5	0.31	0.00—1.00
7119/12-1 deltaic—estuarine	2.1	0.0—8.0	0.43	0.00—1.00
7120/12-1	7.9	0.0—25.5	0.03	0.00—0.17
7120/12-2	12.6	2.0—24.2	0.37	0.00—0.46
Haltenbanken				
6507/11-1	21.1	0.0—36.7	0.45	0.00—0.68
6507/12-1	13.5	4.0—21.4	0.56	0.39—0.74
Central Graben				
7/12-4, marine	10.9	0.0—23.5	0.39	0.00—1.00
7/12-4, deltaic	4.2	0.0—9.5	0.79	0.00—1.00

[a] Range for individual specimens.

development of normal continuous quartz growths. Although most authigenic illite coatings appear to have precipitated directly on detrital grain surfaces (quartz), some illitic growths developed by preferential nucleation on *detrital* clay laminated surfaces.

Authigenic chlorite coatings have effectively retarded silica cementation on a limited scale in some Jurassic sandstones. One source for this material may be the alteration of biotite mica.

Porosity enhancement controls

Porosity development in Jurassic sandstones represents a combination of 'primary' intergranular and 'secondary' type interstices.

Petrographic analysis indicates that secondary porosity accounts for approximately 3—80% (core averages) of total pore space (Table 6) and is primarily the result of dissolution processes involving detrital and authigenic mineral constituents (Plate 7). Identifiable secondary pores can often be attributed to dissolution of detrital feldspar grains and feldspar in igneous/metamorphic rock fragments. The dissolution of lithic grains, argillaceous pellets (glauconite) and matrix, carbonate clasts—cements, organic detritus, mica, heavy minerals, chert, and authigenic feldspar overgrowths contributes to porosity enhancement as well (Table 7).

Dissolution of feldspar

Feldspar dissolution (Plate 8) is an important mechanism for porosity development. In some samples (7/12-4, 3451 m) it accounts for as much as 10.5% pore space). However, incipient dissolution of feldspathic grains commonly resulted in microporosity development.

Feldspar dissolution processes affect both K-feldspar and plagioclase types. Although microcline is generally more resistant to dissolution than other feldspar types, it too may be extensively altered. Dissolution effects in igneous/metamorphic rock fragments containing both feldspar types are most pronounced in plagioclase. Moreover, dissolution of perthitic feldspars is accentuated commonly along plagioclase lamellae. The low amount of plagioclase in porous intervals of Jurassic sandstone may be explained *in part* by its higher susceptibility to dissolution. In some intervals in 7/12-4, plagioclase concentrations relative to K-feldspar *generally* increase with decreasing grain

TABLE 7. Interpretation of porosity types

Porosity types	*7119/12-1*		*7120/12-1*	*7120/12-2*	*6507/11-1*	*65-7/12-1*	*7/12-4*	
	Marine	*Estuarine*					*Marine*	*Deltaic*
'Primary' intergranular[a]	4.60	0.60	5.90	7.40	11.10	5.40	6.70	1.20
Secondary-dissolution								
Detrital feldspar								
Discrete grains	0.15	0.50	0.05	3.10	3.20	3.70	2.80	1.90
Feldspathic/quartzose frags.					0.10	0.30		
Polycrystalline quartz[b]				0.07	0.60	0.70		
Authigenic feldspar overgrowths							0.70	
Lithic grains (siltstone, shale, glauconite)				0.03	0.50	0.20		
Argillaceous matrix/lithic grains					0.30	0.10		
Mica (carbonate leaching?)				0.03	0.20	0.10		
Carbonate (cement/fragments)	1.20				1.50	0.40		
Origin(?)	0.70	0.40	0.20	1.30	3.10	1.90	0.70	1.10
Miscellaneous (heavies, organic, opaques, chert, etc.)				0.08		0.10		
Secondary-grain fractures						0.10		
Micro-porosity		0.60	1.70	0.60	0.50	0.50		

[a]Some pores interpreted as 'primary' intergranular may be secondary by dissolution of pore-filling carbonate cement.
[b]Pores in polycrystalline quartz most likely attributed to removal of feldspar

size (medium grained, 1:8 to 1:5; fine grained, 1:4.5 to 1:1). Possibly, the lower permeability of the finer grained intervals resulted in less 'flushing' by the corrosive formation waters which caused dissolution.

Coarser grained specimens from the Haltenbanken cores exhibit dissolution effects in feldspathic components of igneous/metamorphic rock fragments. A weakened detrital framework resulting in the collapse of interstices may have been triggered in part by faulting attending basin subsidence. In some instances, dissolution or alteration of micas in schistose/gneissic metamorphic fragments has resulted in the disaggregation of stable quartz constituents within the fragments.

Secondary pores due to feldspar dissolution range from 'microporous' interstices to large moldic-type voids. Although dissolution effects are present in both detrital and authigenic phases, most (about 92%) are restricted to detrital feldspar surfaces. Feldspar overgrowths are generally less affected by direct dissolution. As demonstrated by Pittman (1979), authigenic overgrowths consist of relatively pure K^+ or Na^+ feldspar end members, whereas detrital cores are less pure and consequently more susceptible to alteration.

In the Norwegian Sea Jurassic sediments, dissolution of detrital feldspar took place after the precipitation of quartz and carbonate cements. Some of the feldspathic material released probably provided a source of SiO_2 and Al_2O_3 for development of later (some contemporaneous) kaolinite cements.

Dissolution of lithic fragments and argillaceous material

Some secondary porosity (0—7.4%) in several specimens is the apparent result of direct dissolution of lithic fragments and argillaceous material. In several samples, siltstone—shale clasts, glauconite pellets, and organic detritus exhibit extensive corrosion features. Locally, 'leaching' of plant and woody material results in the formation of large interstices. Alteration of this organic material may have provided a potential source of organic carbon for carbonate cements. Correspondingly, voids in detrital argillaceous (illitic, kaolinitic, chloritic) matrix material and laminae suggest that they too have been subjected to incipient-extensive 'solution'. The presence of large isolated booklets of kaolinite in secondary 'matrix-leached' voids indicates that these pores are real and not artifacts of thin-section preparation. Furthermore, petrographic evidence does not support the possibility that these interstices are the result of chemical removal of replacement carbonate. However, this latter mechanism has been noted in other sands. Similar occurrences of matrix 'leaching' have been documented in the Morrow Sandstone (L. Pennsylvanian) of the US mid-continent region by Kasino and Davies (1979) and Z. El Shaieb (personal communication 1983).

Dissolution of other mineral constituents

These include garnet (pyrope), leucoxene, chert and mica, but they contribute only minor amounts of porosity.

Although classified as secondary pores (Table 7), most grain fractures probably reflect framework readjustment response to compaction.

Dissolution of carbonate

This has been observed in several Jurassic cores, but its full extent is somewhat problematic. Dissolution effects (Plate 9), involving removal of carbonate as detrital clasts, cement, and mineral replacements, are seen locally. Possibly, some pores classified as 'primary' intergranular may be secondary in origin, having resulted from complete removal of void filling carbonate. Dissolution effects are more pronounced in calcite cements and less so in dolomite and siderite cements.

Excellent examples of carbonate dissolution involving removal of ankerite cement are present in one cored interval (2718—2722.5 m) in 7119/12-1 (Plate 9 A, B). In this interval, ankerite cementation took place relatively early ('dominantly pre-quartz') as evidenced by the open packing of framework constituents (Fig. 14A) and paucity of quartz overgrowths. The significance of this early carbonate cement was to inhibit subsequent diagenetic processes, most notably quartz cementation. Petrographic evidence suggests that extensive corrosion of ankerite took place in the upper portions of the interval (2719.3 m) where sands are relatively porous (approximately 20.5%, 2718.75 m) and oil stained (Plate 9 B). Serrated quartz grain margins and remnant carbonate strongly suggest this porosity is secondary. Possibly, the continuing thermal alteration of organic matter associated with the sand bodies, triggered a transition from the previously documented constructive interaction between organic diagenesis and carbonate cementation, to a destructive relationship characterized by corrosive formation waters and secondary porosity formation. This change may have been triggered by the increasing input of carbonic (Schmidt and McDonald, 1979) and/or organic acids (Huang and Keller, 1970; I.D. Meshri, personal communication, 1983) giving rise to more acidic formation waters. These observations suggest that the potential may exist for enhanced porosity development in Jurassic sediments of the Troms Basin (7119/12-1) by carbonate dissolution.

Carbonate decementation effects are also prevalent in a few specimens from the upper cored intervals of the 7120/12-2 (1888.65 m). Incomplete dissolution of ferroan dolomite cement in this interval resulted in the development of an irregular pore network. However, specimens from other cored intervals (1888—2049.5 m) show few carbonate dissolution effects, except those associated with isolated detrital clasts.

In Jurassic specimens from Haltenbanken, dissolution of carbonate, although observed, is not considered to be a primary mechanism of porosity enhancement. However, it could be argued that the presence of small incipient quartz overgrowths (0.4—0.7 vol. %) in several porous intervals indicates that pervasive carbonate cementation (after quartz cementation had started) served to suppress continued quartz growth, while subsequent removal of this carbonate resulted in the rejuvenation of original intergranular pore textures. Results of artificial leaching experiments (Fig. 7) conducted as part of this study do not support this contention.

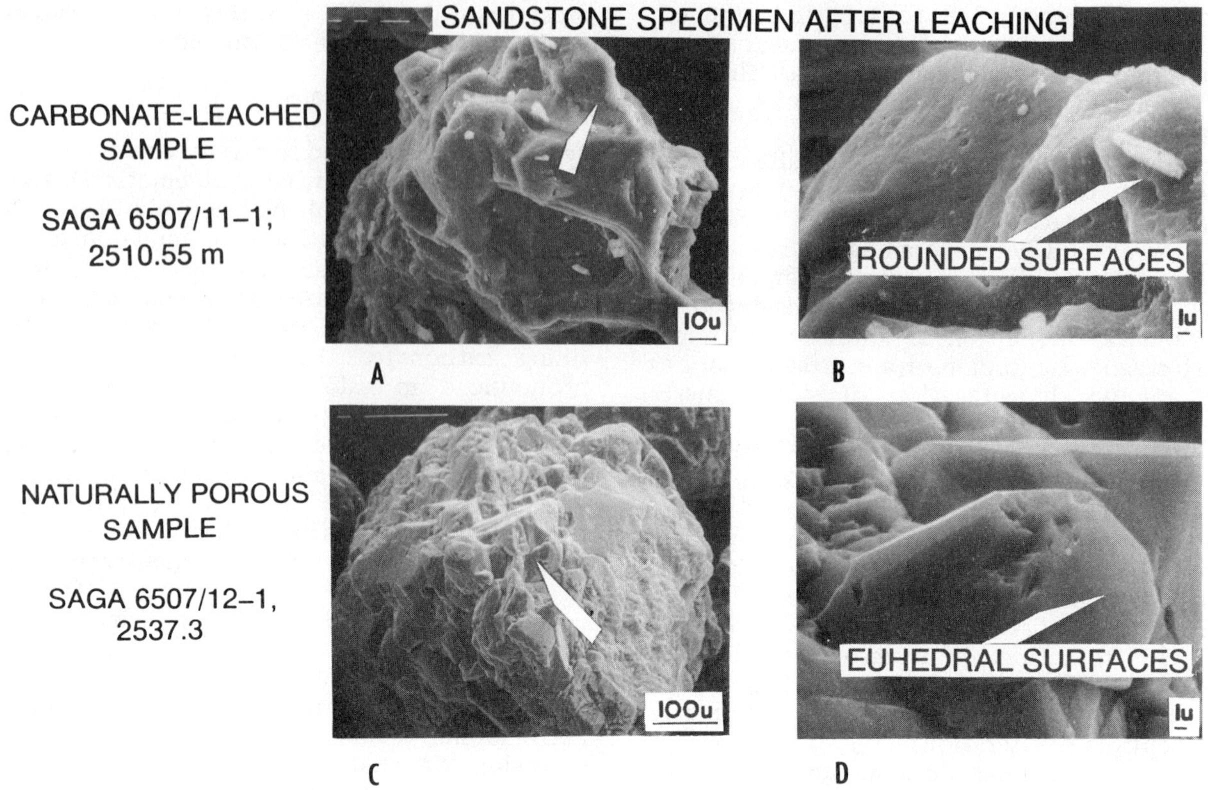

Fig. 7. Scanning electron micrographs comparing quartz overgrowth surface textures in experimentally carbonate-leached sandstones and naturally porous sandstones.

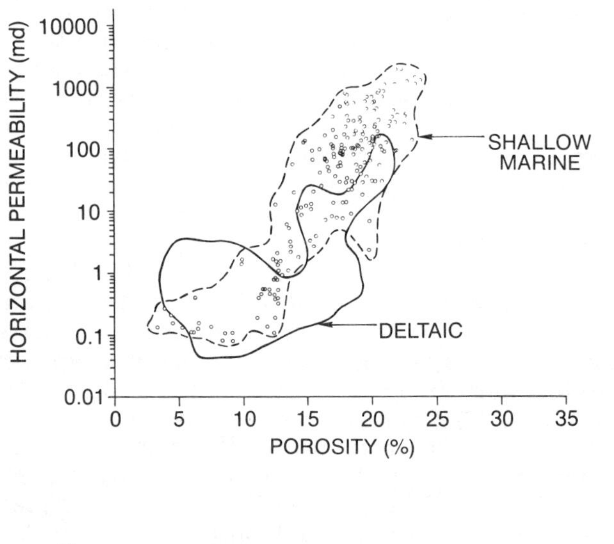

Fig. 8. Porosity—permeability and depositional environments for 7/12-4 cores. Data points shown in upper diagram are for shallow marine facies only.

Calcite and dolomite/ankerite were removed from extensively carbonate cemented sandstone specimens by gently etching the specimens in a buffered solution of sodium acetate and acetic acid (J.C. Hoffman, personal communication, 1983). Analyses of the leached specimens by scanning electron microscopy indicated that pre-carbonate authigenic quartz surfaces had been extensively modified by chemical replacement processes attending carbonate cementation. Commonly, authigenic quartz overgrowths exhibited rounded and serrated textures (Fig. 7B). In contrast, quartz overgrowths in carbonate-free, naturally porous specimens possessed well-defined crystal faces (some euhedral) although incipient in many cases (Fig. 7D). Moreover, fractured quartz surfaces in these naturally porous sands exhibit well-developed silica overgrowths along fracture walls. These observations suggest that porous sands in the Haltenbanken specimens studied have not experienced extensive carbonate cementation—dissolution effects during the course of their diagenetic history. Moreover, well-defined euhedral dolomite crystals showing few, if any, dissolution textures are present in several porous specimens in 6507/12-1.

RESERVOIR CHARACTERISTICS AND DEPOSITIONAL ENVIRONMENT IN 7119/12-1 AND 7/12-4 CORES

Sedimentologic and petrographic analysis of core material from 7/12-4 and 7119/12-1 suggest that, *at least locally*, variations in porosity and permeability may be related to depositional environment. Core analyses for these wells (Table 8, Figs. 8 and 9) indicate that the *potential* for porosity—

TABLE 8. Porosity and permeability of cores in 7119/12-1 and 7/12-4

	Porosity (%)			*Permeability (mD)*		
	Average (%)		*Range*	*Average (mD)*		*Range (mD)*
Well	*Arithmetic*	*Geometric*		*Arithmetic*	*Geometric*	
7/12-4, marine	16.78	16.03	3.4—23.8	180.97	28.06	0.083—1968.0
7/12-4, deltaic	12.88	11.81	4.0—21.5	5.89	0.83	0.055—131.0
7119/12-1, marine	7.91	6.87	0.2—16.3	28.52	1.04	0.038—627.0
7119/12-1, deltaic-estuarine	5.89	5.25	0.8—14.7	0.27	0.14	0.019—4.6

Source of Data: BP Petroleum Development of Norway A/S (7/12-4) and Statoil (7119/12-1).

permeability development is greatest in nearshore marine environments, as opposed to associated fluvial/deltaic—paralic marine sediments. This is largely due to the thicker sand sequences in the marine environment, but diagenetic processes have also contributed to significant porosity—permeability reduction of deltaic intervals.

In 7/12-4, deformation of shale clasts and other ductile lithic fragments (3535.8 m) and pressure solution have reduced the porosity in the poorly sorted, somewhat more immature distributary channel sandstones comprising the Middle Jurassic deltaic facies. Thin-section derived porosities average 4.2% for selected specimens from the deltaic facies (3528.25—3556 m) and 10.9% for specimens in the shallow marine facies (3447.5—3528.5 m).

In 7119/12-1 (Fig. 9), modal analyses of thin sections show porosity to average about 6.7% in nearshore marine facies (2658—2810 m) and about 2.1% in underlying lagoonal and deltaic—estuarine sediments (2810—3000 m). Authigenic kaolinite cement has reduced reservoir quality in the non-marine sands. The effects of kaolinite cementation are most pronounced in deltaic distributary channel sandstones (2828.5—2837 m) in which they may vary from about 14.5 to 17.5%. As such, these sands are commonly microporous. Locally, in marine intervals, enhancement of porosity has occurred by carbonate dissolution (2718—2722.5 m).

Although reservoir characteristics are poorly developed in Jurassic deltaic—paralic marine sediments in these two wells, many Jurassic deltaic sandstones (e.g. Brent Sandstone) exhibit excellent reservoir quality. Even in the present study fluvial-distributary channel sandstones in 6507/12-1 show good porosity potential. This appears to be due to the apparent lack of appreciable ductile lithic fragments (as opposed to the channels in 7/12-4), the higher energy environmental characteristics, less cementation, and the significant secondary porosity developed by dissolution of detrital constituents.

CONCLUSIONS

Based on sedimentological, petrographic, and geochemical analysis of selected Jurassic sandstone core specimens from wells in the Norwegian and North Sea, Norway, the following general conclusions can be made.

(1) Porosity and permeability development in Jurassic sandstones has been influenced significantly by provenance and by sedimentologic and diagenetic controls.

(2) The principal sedimentologic controls on porosity reduction are caused by the presence of argillaceous material as matrix and laminae.

(3) *Locally*, bioturbated marine intervals may be slightly porous due to the development of detrital clay coatings on quartz surfaces by burrowing organisms. These clay coatings apparently serve to restrict quartz cementation effects.

(4) Appreciable pore space in the Troms 1 cores (7119/12-1) has been eliminated by pressure solution effects on grain and stylolitic scale. The effects are mostly in finer grained, somewhat argillaceous intervals.

(5) Cementation by quartz, carbonate, and kaolinite is important for porosity reduction in several Jurassic sandstones studied. Trace—minor amounts of authigenic feldspar, pyrite, illite, chlorite, and mixed layered illite/smectite are present as well.

(6) Carbon-stable isotopic analysis of authigenic carbonate cements reveals a major input and incorporation of $\delta^{13}C$-depleted carbon during precipitation. This $\delta^{13}C$-depleted carbon was derived

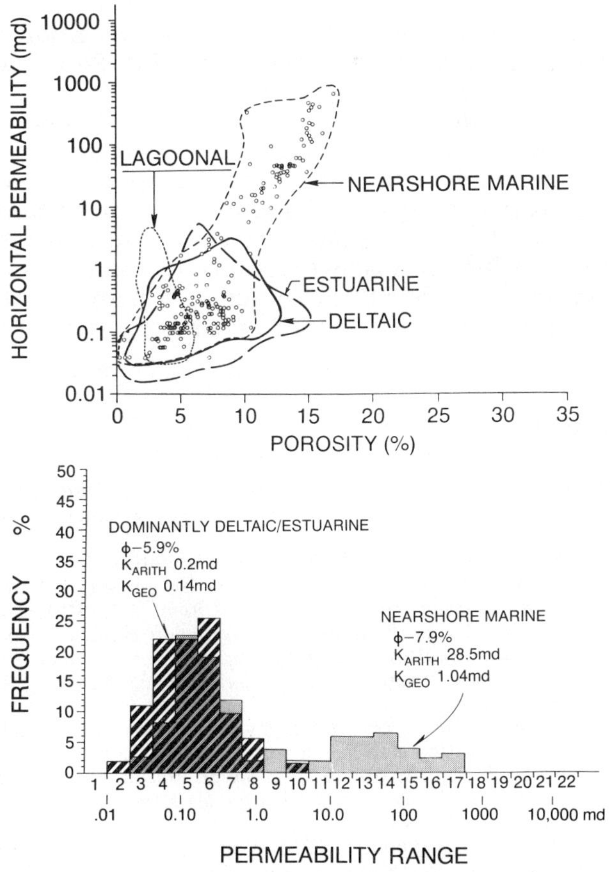

Fig. 9. Porosity—permeability and depositional environments for 7119/12-1 cores. Data points shown in upper diagram are for nearshore marine facies only.

from diagenetic alteration of organic material during burial. The δ^{18}O-depletion trend observed in carbonate cements is controlled to a large extent by increasing temperatures of precipitation (burial).

(7) Organic—inorganic diagenetic interactions are constructive during early burial of the sand bodies. These interactions, during continued burial, evolve toward destructive relationships, characterised by corrosive formation waters causing secondary porosity formation.

(8) Porosity development in the Jurassic sandstones studied is a combination of 'primary' intergranular and secondary 'dissolution' type interstices. Secondary porosity by dissolution of detrital and authigenic mineral constituents accounts for about 3.0—80.0% (core average) of total pore space.

(9) Readily identifiable secondary pores result mainly by dissolution of detrital feldspar as discrete grains and as constituents in igneous/metamorphic rock fragments. Additionally, dissolution effects involving lithic fragments, argillaceous matrix/clasts, carbonate clasts/cement, heavy minerals, organic detritus, and feldspar overgrowths contribute to porosity enhancement as well.

(10) At present, an accurate evaluation of the extent of carbonate dissolution in Jurassic sandstones is problematic.

(11) In two Jurassic cores (7119/12-1; 7/12-4), porosity and permeability is greatest in nearshore—shallow marine shelf facies, due largely to the better development of sand sequences in the marine facies but partly due to diagenetic effects.

ACKNOWLEDGEMENTS

The authors express their sincere appreciation to the management and staffs of BP Petroleum Development of Norway A/S, Norsk Hydro A/S Saga Petroleum Company, and Statoil for their aid in providing us with core specimens and geological data to conduct the present study. Appreciation is also extended to Dr C.J. Campbell, Gunnar Knag, and Ray Leonard of Amoco Norway Oil Company for their help in making this investigation possible and to Dr A.M. Spencer for his many helpful comments and suggestions to the manuscript.

REFERENCES

Berner, R.A. 1980. *Early Diagenesis — A Theoretical Approach.* Princeton University Press, Princeton, NJ, 241 pp.

Bottinga, Y. 1968. Calculation of fractionation factors for carbon and oxygen exchange in the system calcite—carbon dioxide—water. *J. Phys. Chem.* **72**, 800—808.

Curtis, C.D. 1978. Possible links between sandstone diagenesis and depth-related geochemical reactions occurring in enclosing mudstones. *J. Geol. Soc. Lond.* **135**, 107—117.

Degens, E.T. 1965. *Geochemistry of Sediments, A Brief Survey.* Prentice Hall, Englewood Cliffs, 342 pp.

Dott, R.H. Jr. 1964. Wacke, graywacke and matrix — what approach to immature sandstone classification. *J. Sed. Petrol.* **34** (3), 625—632.

Epstein, S. and Mayeda, T. 1953. Variations of O^{18} content of waters from natural sources. *Geochim. Cosmochim. Acta*, **4**, 213—224.

Epstein, S., Buchsbaum, H.A., Lowenstam, H.A. and Urey, H.C. 1953. Revised carbonate—water isotopic temperature scale. *Bull. Geol. Soc. Am.*, **64**, 1315.

Friedman, I. and O'Neil, J.R. 1977. *Compilation of Stable Isotope Fractionation Factors of Geochemical Interest.* United States Geological Survey Professional Paper 440-KK, 12 pp., 49 Figs.

Hayes, J.B. 1979. Sandstone diagenesis — the hole truth. In: P.A. Scholle and P.R. Schluger (eds.), *Aspects of Diagenesis*, SEPM Special Publ., No. 26, p. 127—139.

Heald, M.T. 1956. Cementation of Simpson and St. Peter sandstones in parts of Oklahoma, Arkansas and Missouri. *J. Geol.* **64**, 16—30.

Heald, M.T. and Renton, J.J. 1966. Experimental study of sandstone cementation. *J. Sed. Petrol.* **36** (4), 977—991.

Holtedahl, O. and Dons, J.A. 1960. Geological map of Norway (Bedrock). *Norges Geol. Unders.*, No. 208.

Huang, W.H. and Keller, W.D. 1970. Dissolution of rock-forming silicate minerals in organic acids: simulated first stage weathering of fresh mineral surfaces. *Am. Mineral.*, **55**, 2076—2094.

Irwin, H., Curtis, C., and Coleman, M. 1977. Isotopic evidence for source of diagenetic carbonates formed during burial of organic-rich sediments. *Nature, London*, **269**, 209—213.

Kasino, R.E. and Davies, D.K. 1979. Environments and diagenesis, Morrow sands, Cimmaron Co. (Oklahoma), and significance to regional exploration, production and well completion processes. In: N.J. Hyne (ed.), *Pennsylvanian Sandstones of the Mid Continent*, Tulsa Geological Society Spec. Publ. No. 1, pp. 169—194.

Land, L.S. 1973. Holocene meteoric dolomitization of Pleistocene limestone, North Jamaica. *Sedimentology*, **20**, 411—424.

Laplante, R.E. 1974. Hydrocarbon generation in Gulf Coast Tertiary sediments. *Am. Assoc. Petrol. Geol. Bull.*, **58** (7), 1281—1289.

Lerbekmo, J.F. and Platt, R.L. 1962. Promotion of pressure solution of silica in sandstones. *J. Sed. Petrol.* **32**, 514—519.

McCrea, J.M. 1950. On the isotope chemistry of carbonates and a paleotemperature scale. *J. Chem. Phys.* **18**, 849—857.

Pittman, E.D. 1979. Porosity, diagenesis and productive capability of sandstone reservoirs. In: P.A. Scholle and P.R. Schluger (eds.), *Aspects of Diagenesis*, SEPM Special Publ., No. 26, pp. 159—173.

Schmidt, V. and McDonald, D.A. 1979. Texture and recognition of secondary porosity in sandstones. In: P.A. Scholle and P. Schluger (eds.). *Aspects of Diagenesis*, SEPM Special Publ., No. 26, pp. 175—207.

Thomson, A. 1959. Pressure solution and porosity. In: H.A. Ireland (ed.), *Silica in Sediemnts*, SEPM Special. Publ. No. 7, pp. 92—111.

Thomson, A. 1982. Preservation of porosity in the deep Woodbine Tuscaloosa Trend, Louisiana. *J. Petrol. Technol.*, May, 1156—1162.

Walker, T.R. and Crone, A.J. 1974. Mechanically-infiltrated clay matrix in desert alluvium. Abstr., *Geol. Soc. Am. Annu. Meet.*, **6**, 998.

Weyl. P.K. 1959. Pressure solution and the force of crystallization — a phenomenological theory. *J. Geophys. Res.*, **64**, 2001—2005.

APPENDIX 1

STABLE ISOTOPIC SAMPLE PREPARATION AND PRECISION OF DATA

Samples were carefully selected for stable isotopic analysis by thin-section evaluation. This procedure allowed the selection of the most representative carbonate cemented sands samples. Care was taken to avoid or minimize possible detrital carbonate contamination.

Whole rock samples (1—2g) were ground to fine powder which was X-rayed to determine percentage and type of carbonate material present. Samples ranging from 15 to 60 mg in weight were used for the stable isotopic analysis. The standard method outlined by McCrea (1950) was used to generate

CO_2 gas from the reaction of phosphoric acid (H_3PO_4) with carbonate material at 25°C in a high vacuum reaction flask.

Fractional gas separation was used during the reaction of samples containing calcite, dolomite, ankerite, and siderite mixtures. The initial CO_2 gas generated during the first 20 min of reaction at 25°C was considered representative of the calcite fraction in the sample (Land, 1973). The CO_2 gas generated during the remaining reaction time (>72 h) was considered representative of the dolomite, ankerite, and siderite components. It is important to note that the reactability of carbonates with phosphoric acid is highly variable. As a result, the total CO_2 gas yield from the less soluble siderite was in the neighbourhood of 15—20%. The gas from this partial reaction is believed to be representative of the bulk siderite, evidenced by our ability to reproduce results.

Fractional gas separation works well when the carbonate mineral components are in roughly equal proportions. As the relative differences in weight percent of the calcite—(dolomite, ankerite, and siderite) pair increases, the precision of the analysis for the minor carbonate component decreases, being increasingly contaminated by the CO_2 gas generated from the major component. Stable isotopic analysis was not attempted on minor carbonate components whose volume was 10% or less of the major carbonate phase. No distinction could be made between coexisting dolomite, ankerite, and siderite components.

The CO_2 gas samples were analysed by a Finnigan MAT-250 triple collector, constant ratio mass spectrometer located at the Amoco Research Center, Tulsa, Oklahoma, USA. All analyses are reported relative to the PDB standard. The fractionation between dolomite and phosphoric acid as listed by Friedman and O'Neil (1977) was used for the ankerite and siderite reactions.

A total of 18 duplicate sample analyses were made during the course of this study. The maximum $\delta^{13}C$ variation observed was 0.2°/oo with the average being 0.08°/oo. The maximum $\delta^{18}O$ variation observed was 0.3°/oo and the average 0.12°/oo. Multiple analyses of an intra laboratory standard during each day's run showed no change in $\delta^{13}C$ and $\delta^{18}O$ values beyond the 0.1°/oo level over the course of this study.

7 Wandel Sea Basin – The North Greenland equivalent to Svalbard and the Barents Shelf

Eckart Håkansson and Lars Stemmerik
Institute of Historical Geology and Palaeontology, Copenhagen, Denmark

Recent investigations of the North Greenland Wandel Sea Basin confirm, to a large extent, previous assumptions of a parallel development with Svalbard and the Barents Shelf. Deposits from the Wandel Sea Basin have now been recorded in a coastal belt in North Greenland almost 500 km in length, but in many parts their distribution is limited and/or highly patchy. A composite sequence totalling well over 7 km has been recorded, which gives evidence for a complex depositional and structural history of the basin encompassing the Carboniferous to Tertiary interval. Depositional environments in the Carboniferous to Triassic, as well as in the Palaeocene, compare well with the situation recorded in Svalbard. The later Mesozoic history of the two areas deviates to varying degrees, and there is some evidence that the North Greenland development may be more closely paralleled in the Barents Shelf area than in on-shore Svalbard. Late Jurassic to early Cretaceous deposition testifies to the separation of smaller basin units; however, the presence of thick late-Cretaceous on-shore sequences in North Greenland constitutes the most conspicuous difference from Svalbard. During the late Cretaceous, several pull-apart basins were filled and deformed in a dextral strike-slip mobile belt, largely paralleling the subsequent suture between Greenland and Svalbard. The type of deposition in individual basins varies widely. Some contain fluvial-deltaic and marine sediments whereas a single basin is dominated by continental vulcanogenic sediments and extrusives. Severe post-Palaeocene thermal alteration centrally in the Wandel Sea Basin constitutes the last major event so far recorded in North Greenland.

INTRODUCTION

The most commonly favoured North Atlantic pre-drift reconstructions locate Svalbard in a position NE of Greenland in close association with the Wandel Sea Basin (Fig. 1). This position more or less explicitly implies a mutual, long-lasting geological frame for the two continent areas that are now separated by several hundred kilometres. Thus, at least from Carboniferous time, eastern North Greenland and Svalbard were part of an extensive continental platform which also included the Barents Shelf and, most likely, was connected to both the Sverdrup Basin and the Russian/Siberian basins. So far, assumptions of this nature have been founded on very scanty evidence from the Wandel Sea Basin. However, the recently completed regional field work in North Greenland has substantiated such assumptions and we can now include depositional similarities as independent factual evidence for these pre-drift reconstructions. Although differences abound, a Group by Group comparison reveals remarkable similarities through the entire Carboniferous—Palaeogene interval (Håkansson *et al.*, 1981b) (Fig. 2). The majority of contrasts are associated with differences in the present understanding of the structural histories. Thus, Håkansson and Pedersen (1982) interpreted much of the post-Caledonian geological development in North Greenland as being intimately related to a sequence of large-scale transcurrent movements. Although only some of these phases have been recognized or emphasized in accounts on Svalbard geology, they may indeed all have involved Svalbard. In eastern North Greenland, transcurrent movements peak in Middle to Late Devonian, Late Permian, Jurassic, and Late Cretaceous phases which show a general change in trends from E—W in the Devonian to predominantly NW—SE in the Cretaceous. The Tertiary separation of Greenland and Svalbard along the Spitsbergen Fracture Zone thus constitutes the ultimate expression of this change in strike-slip direction.

However, as a result of the immense differences that exist in the general character of the geological information of the regions compared, a word of

Petroleum Geology of the North European Margin ©

(Graham & Trotman, 1984) pp. 97—107.

caution is required prior to any detailed attempt to compare the North Greenland and Svalbard developments. While the on-shore geology of the Svalbard Archipelago has been investigated by innumerable geologists for well over a century, the information available from the Barents Shelf is based largely on geophysical data acquired over the last decade. In strong contrast to these levels of knowledge, the extreme remoteness of North Greenland has severely precluded acquisition of a comparable data base here. The very patchy history of exploration goes back to the turn of the century. However, the outline of a coherent regional geological understanding for the period under consideration has emerged only recently, in the wake of the geological mapping of eastern North Greenland, conducted by the Geological Survey of Greenland (GGU) from 1978 to 1980. Furthermore, pertinent geophysical information is essentially lacking, as regards both the wide shelf areas off North and Northeast Greenland and on-shore. The apparent differences between the two areas, therefore, may

Fig. 1. General pre-drift configuration of Greenland, Svalbard, and the Barents Shelf. The stippled area indicates the part of North Greenland where sediments of the Wandel Sea Basin have been recorded.

Fig. 2. Stratigraphy of the Wandel Sea Basin compared to Svalbard and the Barents shelf. (Wandel Sea Basin based on Håkansson *et al.*, 1981b and Håkansson and Pedersen, 1982; Svalbard largely based on Worsley and Møork, 1980 and Birkenmajer 1981; Barents shelf: age of major seismic boundaries based on Rønnevik *et al.*, 1982. Note — structural outliers to the west of the main Wandel Sea Basin outcrops not included).

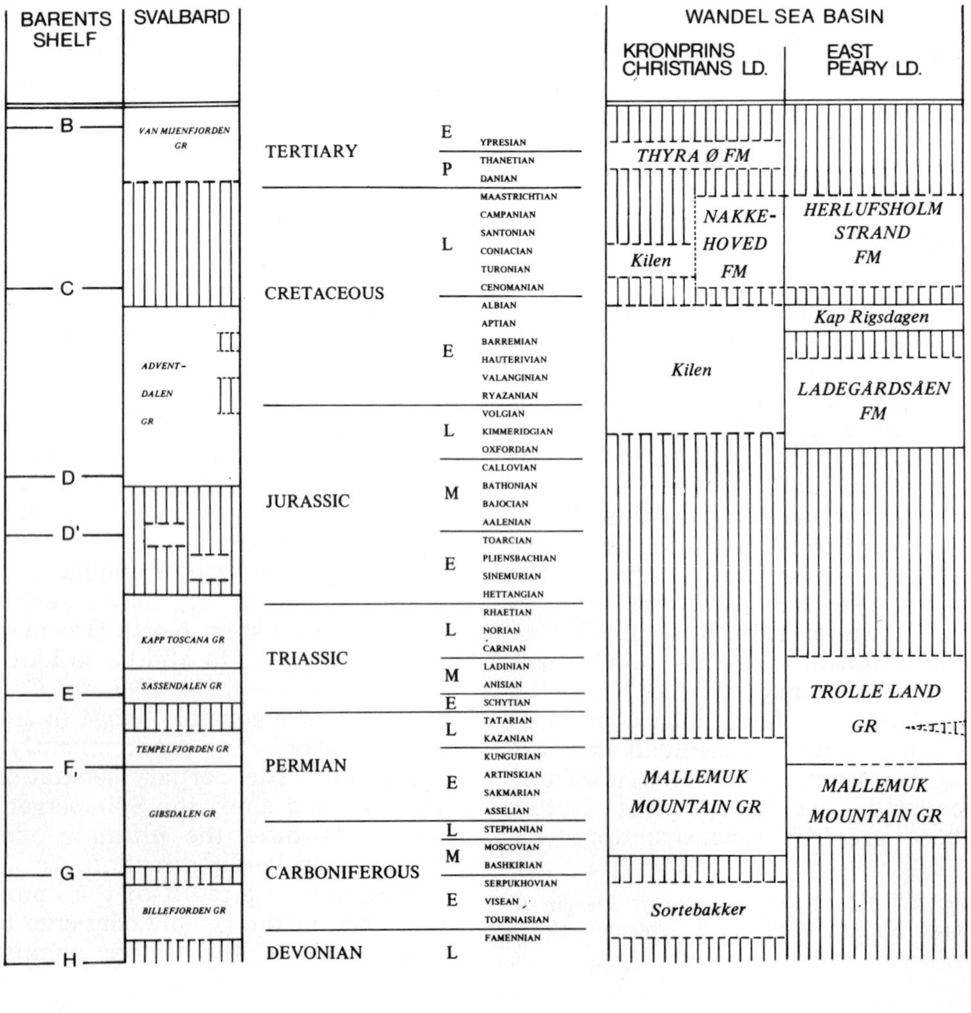

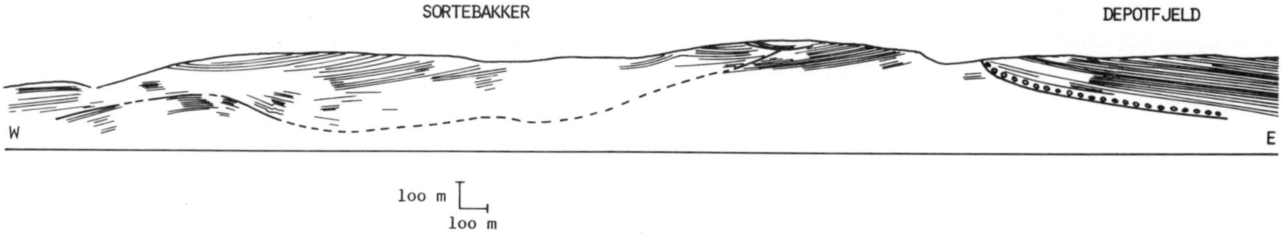

Fig. 3. Conspicuous beds in the coastal cliffs of the southern Holm Land traced from photographs. Internal unconformities in the continental early Carboniferous (Sortebakker) indicated by thick, in part dashed line; basal conglomerate of the marine late Carboniferous sequence (Depotfjeld) indicated by row of open dots.

become less significant when more balanced sets of data ultimately become available.

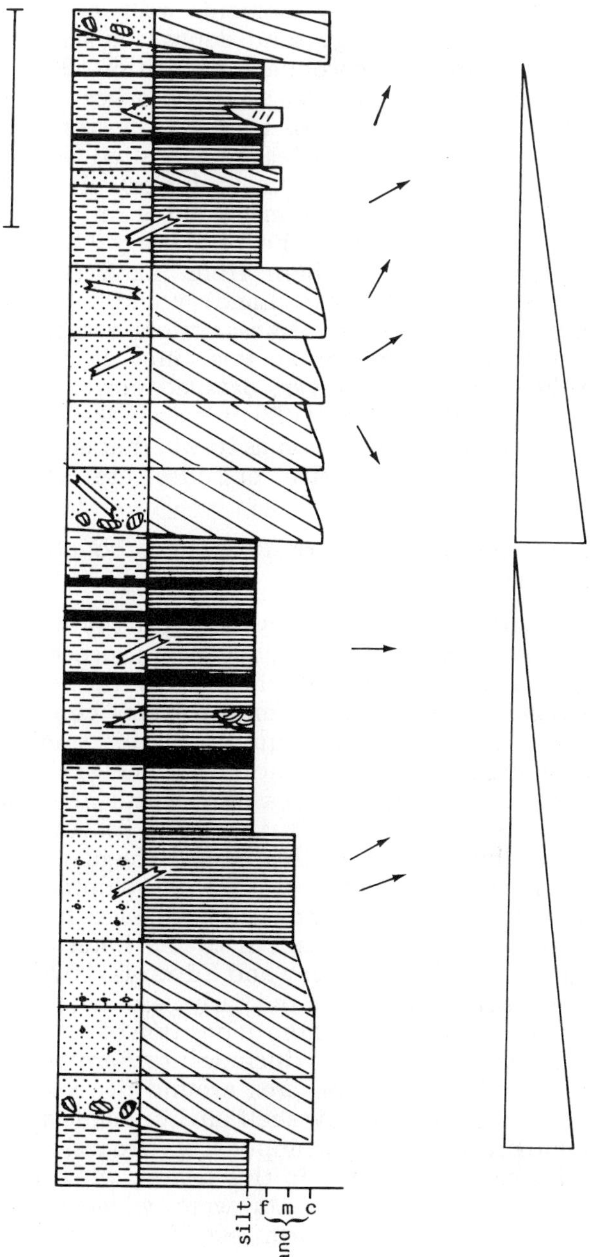

Fig. 4. Representative log comprising two complete cycles from the upper, sandy part of the continental early Carboniferous in southern Holm Land. Cross-hatched nodules are shale intraclasts; elongate bars are trunk fragments.

DEPOSITIONAL RECORD

Late Devonian—Early Carboniferous

In North Greenland, continental Early Carboniferous rocks are confined to the southern parts of Holm Land, where some 600 m of sandstone, shale and minor coal beds are preserved.

The sequence includes more than 70 cycles of rhythmically deposited sandstone and shale units, which may be traced laterally for several kilometres (Fig. 3). An internal low-angle unconformity subdivides the sequence into a lower part, more than 350 m thick characterized by relatively thin cycles dominated by shale, and an upper, sand-dominated part with thick cycles. Only the upper part has been noted to contain coal beds. The upper sandy part of the sequence includes more than 40 cycles. Each cycle (Fig. 4) is initiated by an erosional surface cutting into the sediments of the cycle below; above this surface the base of the sand contains abundant shale clasts. The sandstone units are 3—7 m thick, medium to coarse grained and show a variety of internal structures indicating an overall direction of transport from SW to NE (see Fig. 4). The contact with the overlying shale is sharp. The shales are composed of evenly laminated siltstone with abundant thin sandstone intercalations, which are often cross bedded (Fig. 4). In the higher part of the sequence the thicker shale units often contain 20—25 cm thick beds of coal. Both the sandstone and the shale contain abundant fragments of leaves and stems, which Nathorst (1911) referred to the early Carboniferous.

Each cycle is interpreted as representing deposits of a meandering river system during lateral accretion of the river channel, and related environments. The control of sedimentary fluvial deposits is determined by a number of factors, including climate, relief and tectonism (Miall, 1980). The large-scale change in the cyclicity and the relation of this to one, or perhaps two, internal unconformities indicates repeated tectonic pulses during deposition.

The development of the succession corresponds well to that described by Gjelberg (1981) and Gjelberg and Steel (1981) from Svalbard and Bjørnøya. However, the sudden change from humid to arid climate at the Namurian—Bashkirian boundary noted in these areas (Gjelberg and Steel, 1981) is not observed directly in North Greenland. Here, a similar change in climatic conditions took place during the time span represented by N—S faulting and erosion prior to the early Moscovian transgression. In contrast to the known on-shore development, sedimentation in the Barents Shelf area was perhaps initiated with marine, in part evaporitic, deposition in the late Devonian, which only gradually gave way to the continental deposits typical of the region (Rønnevik *et al.*, 1982).

Middle Carboniferous—Late Permian

Marine Middle Carboniferous—Permian sediments are widespread in eastern North Greenland (Fig. 5), where locally more than 1100 m of shelf carbonates, clastics and subordinate amounts of evaporites occur, which show an overall similarity to the sediments in Svalbard. The sediments are confined to the areas outside the major fault zones delimiting the stable Greenland Craton. During the late Palaeozoic these fault-zones probably separated the cratonic area from the active fault blocks along the Greenland margin (Fig. 5). Here, differential movements along NW—trending faults caused a varied sedimentary development and different timing of the initial transgression of individual fault blocks (Figs. 5 and 6).

Middle—Late Carboniferous cyclic shelf sedimentation

The marine Middle—Late Carboniferous sequence is developed as three major, fining-upwards cycles up to 400 m thick, each corresponding to a stepwise tectonic controlled transgression (Stemmerik and Håkansson, in preparation). Fusulinids indicate early Moscovian and Late Moscovian ages for the transgressions of Holm Land and southern Amdrup Land (Dunbar *et al.*, 1962), while the basal Carboniferous sediments in eastern Peary Land are of Stephanian age (Petryk, 1977).

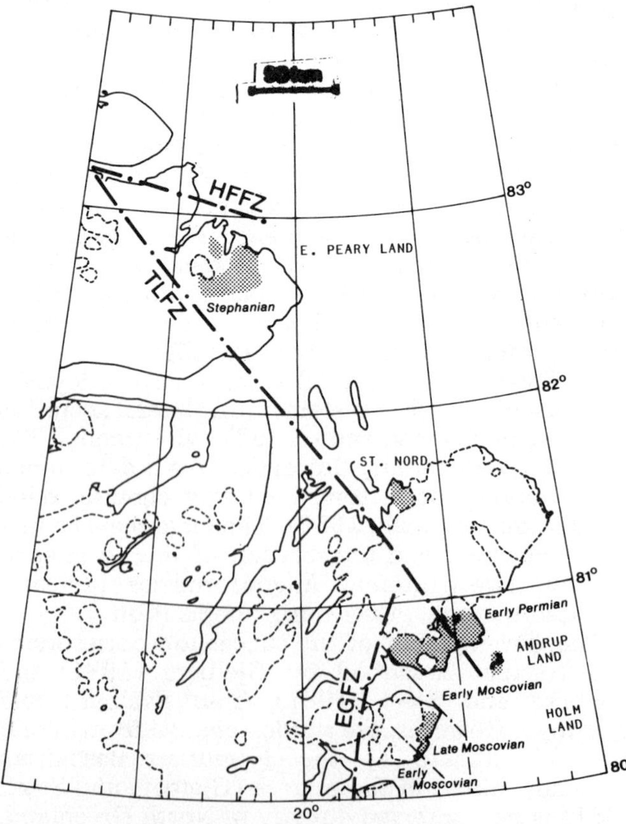

Fig. 5. Outcrop areas of marine late Carboniferous to early Permian strata in eastern North Greenland, with transgression time on individual fault blocks indicated in italics. HFFZ, Harder Fjord Fault Zone; TLFZ, Trolle Land Fault Zone; EGFZ, East Greenland Fault Zone.

Each transgressive megacycle is initiated by wedges of red bed conglomerates and sandstones up to 100 m thick. The conglomerates are dominated everywhere by crystalline clasts reaching approximately 50 cm in diameter. Internal structures are comparatively rare in these sequences and are recognized only in the more fine grained parts where large-scale, planar and trough cross bedding is visible. The conglomerates are succeeded by rhythmically alternating limestones and clastics, where individual beds typically have a large areal extension. Most widespread are alternating beds of medium—coarse grained, variegated sandstones and grey, purely biomicritic limestones (Fig. 7). Internal structures in the sandstones, together with the regular occurrence of bellerophontacean gastropods (E. Nielsen, unpublished data), indicate a continuously marine environment with alternating periods of high-energy sand deposition and more quiet episodes of carbonate deposition. Towards the top of a megacycle, sandstones become rare and thick sequences of bedded limestone with minor shales and evaporites are deposited (Fig. 6).

Stratigraphic and sedimentological similarities between the Hårbardbreen Member in Svalbard (Lauritzen, 1981) and the basal early Moscovian conglomerate on southern Holm Land (and perhaps southern Amdrup Land) could indicate a basin-wide Moscovian transgression rather than local tectonic control at this time. In the succeeding period, however, detailed correlation to Svalbard and the Barents Shelf is again less evident, and it is an open question as to whether any of the subsequent transgressive episodes are similarly related.

Late Carboniferous—early Permian carbonate deposits

The Carboniferous megacycles with rhythmically deposited clastics and carbonates are succeeded by thick accumulations of shallow to deeper shelf carbonates. There are some problems concerning the stratigraphic interpretation of this transition, but apparently sedimentation became exclusively calcareous throughout eastern North Greenland during late Carboniferous time (Fig. 6). According to fusulinid evidence, this North Greenland event seems to correspond to the development on Bjørnøya, where sedimentation changes from the clastics of Kapp Hanna Fm to the limestones of Kapp Duner Fm (Worsley and Edwards, 1976).

Subsequently, during the Early Permian, renewed transgression expanded the deposition of carbonates in North Greenland to include northern Amdrup Land and perhaps the area around St. Nord (see Figs. 5 and 6). Shallow shelf environments with an abundant fauna of brachiopods, bryozoans, crinoids and bivalves prevailed in this region, while deeper depositional environments were widespread towards both south and northwest. This Early Permian development corresponds well to the situation on both the Barents Shelf and Svalbard. In these areas, however, tectonic activity is witnessed by a slight angular unconformity beneath the sediments of the widespread Early Permian transgression (Cutbill and Challinor, 1965; Rønnevik *et al.*, 1982).

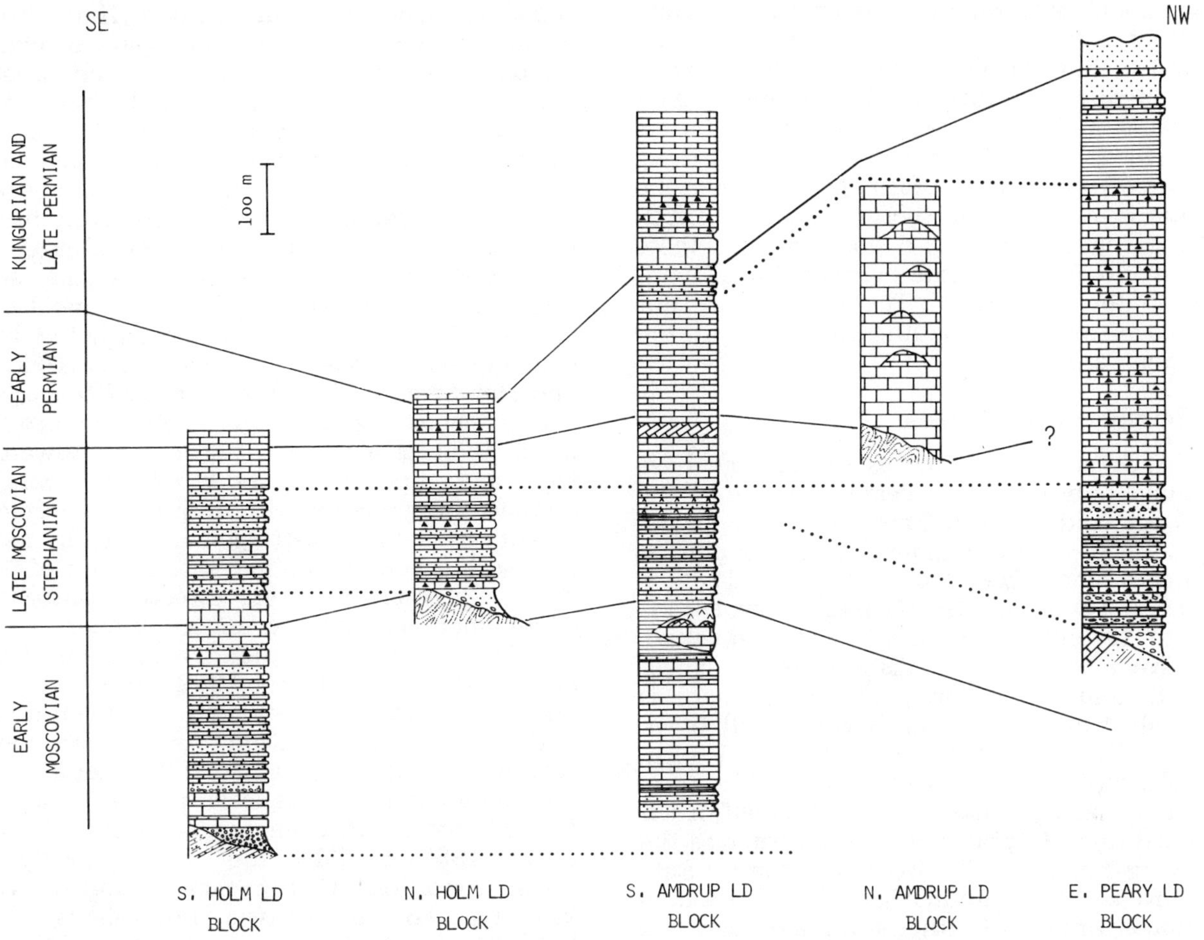

Fig. 6. Tentative correlation of the marine Permo-Carboniferous strata in eastern North Greenland. Solid lines indicate correlations based on biostratigraphical evidence; dotted lines indicate correlations based on major depositional events. Lithology based on Håkansson (1979) and Håkansson *et al.* (1981b).

Mid—Permian tectonism and renewed clastic sedimentation

The widespread mid-Permian tectonic episodes affecting central East Greenland (Surlyk *et al.*, 1984) as well as Bjørnøya, Svalbard and the Barents Shelf (Cutbill and Challinor, 1965; Worsley and Edwards, 1976; Rønnevik *et al.*, 1982) are recognized in North Greenland only as a renewed influx of clastic material. In Peary Land the clastics of Midnatfjeld Fm abruptly overlie the limestones. Continuous deep shelf environments are indicated by deposition of dark, calcareous shales of Early Permian age (Håkansson, 1979). The sea gradually shallowed and marine sandstones were deposited with a highly diverse Late Permian fauna, dominated by brachiopods and bryozoans.

In the southern half of the Wandel Sea Basin, Early Permian carbonates are the youngest sediments present through most areas, but in southern Amdrup Land in particular the exposures continue to include younger parts of the Permian. Here, the increased clastic influx is marked by a new sequence of interbedded sandstones and limestones (Fig. 6). Simultaneous shallowing of the sea is marked by highly fossiliferous, brachiopod-dominated faunas

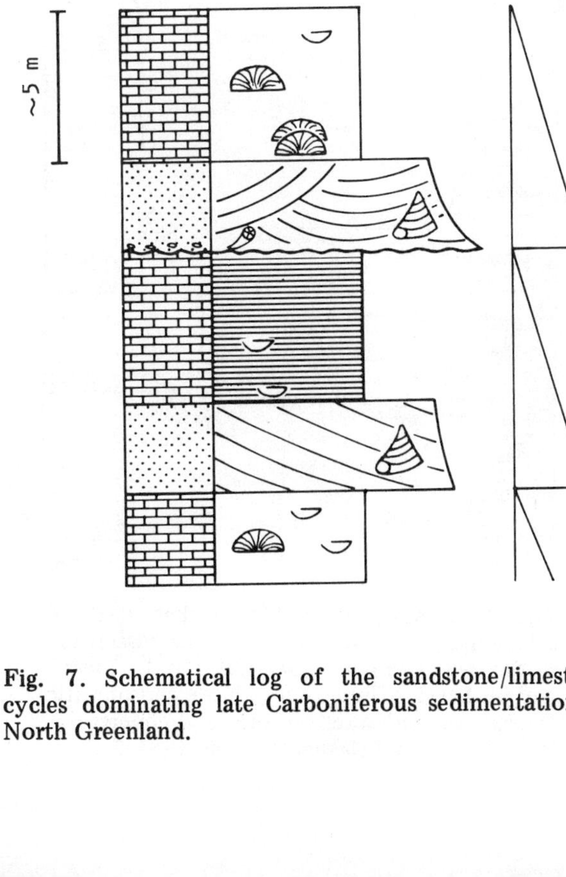

Fig. 7. Schematical log of the sandstone/limestone cycles dominating late Carboniferous sedimentation in North Greenland.

and occasional occurrences of stromatolites. Subsequently the deposition of carbonates increased again and shallow carbonate shelf conditions were established with rich, diverse faunas dominated by brachiopods and bryozoans, but also including corals, bivalves, etc. The brachipod fauna of these later parts of the Permian resembles that from the Kapp Starostin Fm of Spitsbergen (Dunbar, 1962) suggesting contemporary deposition of these sediments. So far, truly late Permian faunas comparable to those in eastern Peary Land have not been recognized in the southeastern half of the Wandel Sea Basin.

Latest Permian—Early Jurassic

In North Greenland, sediments of this age have been found only in eastern Peary Land where the Parish Bjerg and Dunken Fms comprise approximately 1 km of clastic sediments corresponding to the Svalbard Sassendalen Gp (Mørk *et al.*, 1982; = lower part of Torell Land Gp of Birkenmajer, 1981). Owing to tectonically controlled erosion, these strata have a very limited distribution in North Greenland, and the majority of the sequence is known only from a single, fairly small fault block (Håkansson, 1979; Håkansson and Pedersen, 1982).

In Svalbard, the base of the Sassendalen Gp rests conformably on the Late Permian Tempelfjorden Gp in most areas (Mørk *et al.*, 1982), whereas the Parish Bjerg Fm in North Greenland overlies Late Permian rocks of the Midnatfjeld Fm with a low-angle unconformity. Furthermore, although the Parish Bjerg Fm is poorly dated, it appears to be slightly older than the Sassendalen Gp, and possibly it encompasses the Permo—Triassic boundary (Håkansson, 1979). The age of the lower half of the overlying Dunken Fm is well established as Early to Middle Triassic (Anisian), whereas the

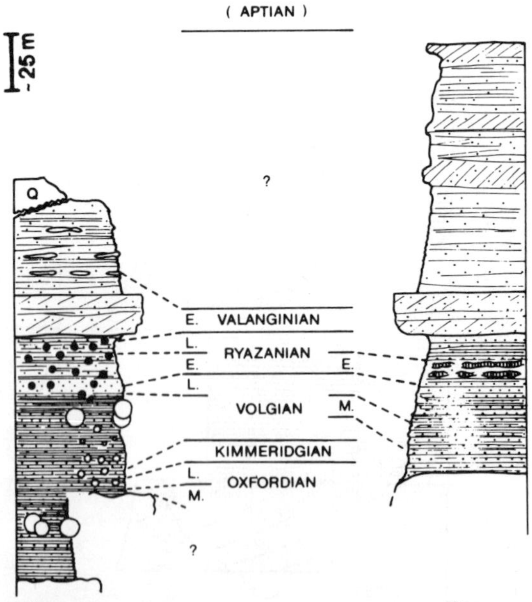

Fig. 8. Stratigraphy and somewhat idealized sedimentary sequences of the Ladegårdsåen Fm in the two main fault blocks (cf. Fig. 10, locs. 5 and 6). Levels of limestone (white), phosphoritic (black), and sideritic (vertical lines) concretions indicated. (From Håkansson *et al.* 1981a.)

remaining strata are still undated (Håkansson and Heinberg, 1977). Since their general character relates them closely to the lower, dated parts, they are considered to be of largely similar age.

Two major coarsening upwards sequences (corresponding to the two formations) can be distinguished in North Greenland, but their actual relation has yet to be clarified. At many horizons in both sequences trace fossil assemblages occur (including abundant *Rhizocorallium*) that suggest a fully marine provenance, and the general marine nature of the sequence is substantiated by the sporadic occurrence of marine faunas (Håkansson and Heinberg, 1977; Håkansson, 1979). Dispsersed carbonaceous detritus of probable terrigenous origin is quite abundant, particularly towards the top, and repeated episodes of shallowing and coastal progradation in response to tectonic activity is a likely setting. Likewise, in most of Svalbard, accumulation and distribution of Sassendalen Gp sediments were directly controlled by synsedimentary activity in a number of faults paralleling the Billefjorden Fault Zone (Harland *et al.*, 1974; Mørk *et al.*, 1982). During the Early and Middle Triassic deep water sedimentation, in part with anoxic bottom conditions, prevailed towards the east in Svalbard (Mørk *et al.*, 1982), but simultaneous large-scale progradation across the Barents Shelf from the southeast (Rønnevik *et al.*, 1982) heralded the shallow, partly freshwater conditions that became dominant in the later parts of the Triassic. Equivalents of these Late Triassic—Early Jurassic deposits of the Kapp Toscana Gp have not been encountered in North Greenland.

In palaeogeographical reconstructions of the early part of the Triassic there is a general agreement on the palaeopresence of a land mass to the west of Svalbard (e.g. Birkenmajer, 1981, fig. 14; Mørk *et al.*, 1982, fig. 10). The isolated occurrence of equivalent, fully marine sediments in Peary Land therefore allows speculations on a regional model with a fault controlled sedimentation pattern covering areas considerably larger than those actually documented in Svalbard by Mørk *et al.* (1982) and perhaps comprising a number of major sub-basins. In eastern North Greenland, independent evidence for activity in the dominant NW-trending faults of the Trolle Land Fault System have been demonstrated for the Jurassic—Tertiary period only (Håkansson and Pedersen, 1982). However, restored to their pre-drift position, the North Greenland Trolle Land Fault System and the Svalbard complex of Billefjorden plus associated faults become largely parallel. The two systems therefore may be closely related, and models involving a common origin and history appear quite attractive, also at this early period.

Middle Jurassic—Early Cretaceous

The sub-basin division, which was perhaps effective already in earlier periods, now became fully apparent and, in North Greenland, two areas of deposition are clearly distinguishable.

Centrally in the Wandel Sea Basin, in eastern Peary Land (Fig. 10, locs. 5 and 6), the Ladegårdsåen Fm comprises about 250 m of fairly mature clastic sediments containing marine faunas and microfloras

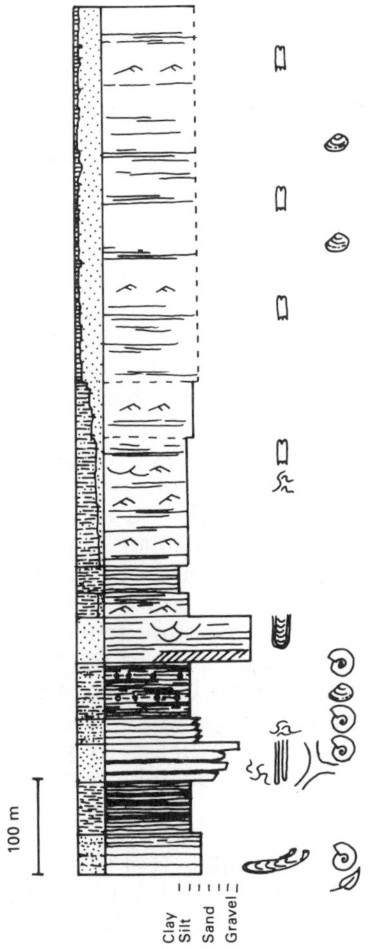

Fig. 9. Composite lithological log of the late Jurassic to early Cretaceous sequence at Kilen (Fig. 10, loc. 10). Main fossil and trace fossil levels indicated (except in the Albian). (From Håkansson *et al.*, 1981b.)

of Middle Oxfordian to Valanginian age in the lower half (Håkansson *et al.*, 1981a) (Fig. 8). Subsequent to a rapid Middle Oxfordian transgression, the sequence records a gradual change from near-shore marine to limnic conditions across the Jurassic—Cretaceous boundary. Coastal progradation was probably towards the south and southeast and, in the restricted areas now preserved, limnic conditions were attained during the Valanginian. Owing to complete lack of determinable organic remains, the age of the higher parts of the Ladegårdsåen Fm has not been ascertained. No later than the Aptian, however, the central part of the Wandel Sea Basin was again transgressed, and marine conditions prevailed here at least into the basal Albian (Håkansson *et al.*, 1981b; Birkelund and Håkansson, 1983). Very little remains of this mid-Cretaceous transgressive event, but the combination of open marine siltstones in one faultblock (Fig. 10, loc. 5) and near-coastal, lagoonal sand- and siltstone in the other (Fig. 10, loc. 7) indicate that it followed a distribution pattern deviating somewhat from the Oxfordian transgression (Håkansson *et al.*, 1981b).

Quite distinct from this development, some 900 m of mature clastic sediments, which are confined to the semi-nunatack Kilen (Fig. 10, loc. 10), seem to have been continuously marine throughout the entire Kimmeridgian—Albian period (Birkelund and Håkansson, 1983). The moderately to strongly folded sequence comprises a lower, mainly transgressive phase where most sediments accumulated in sheltered, coastal environments (Fig. 9); following a prominent unit of high-energy coastal sands, sediments of the last phase accumulated on a muddy shelf where increasing proportions of storm sand layers probably reflect a general shallowing (Håkansson *et al.*, 1981b).

It seems evident, therefore, that the sedimentary development follows different patterns in the two sub-basins in North Greenland, but since their macrofaunas as known presently have virtually nothing in common, correlation of possible common events has so far been unsuccessful. Compared with Svalbard, only the development in the eastern Peary Land Ladegårdsåen Fm would appear to follow roughly the same pattern as the Adventdalen Gp — marine (Janusfjellet Fm)/fluvio-deltaic (Helvetiafjellet Fm)/marine (Carolinefjellet Fm). However, a number of significant age differences in the scheme point to quite distinct, locally determined developmental histories in these two areas as well. Most conspicuous in this respect are the differences in transgression time (Middle Oxfordian versus Bathonian (Harland *et al.*, 1974) or Callovian (Birkenmajer, 1981)), and the differences in attaining limnic conditions (Valanginian versus Barremian (Harland *et al.*, 1974; Birkenmajer, 1981)). Furthermore, it should be pointed out that no parallel to the Volgian—Ryazanian (late Kimmerian) unconformities (Parker, 1967) has been detected in North Greenland.

Whether the widespread and, in part, related Volgian—Barremian intrusive and volcanic activity in Svalbard (Harland *et al.*, 1974) has any counterparts in the Wandel Sea Basin is less certain. Only in and around the westernmost structural outliers have magmatic rocks of any kind been found in association with sediments of this basin. Here, a number of dykes have yielded K/Ar ages ranging from 136 to 66 Ma (Dawes and Soper, 1979), and, therefore, they may be of comparable age. However, based on their limited distribution centered around the Kap Washington Group volcanics (cf. Higgins *et al.*, 1981, fig. 16) it seems more reasonable to

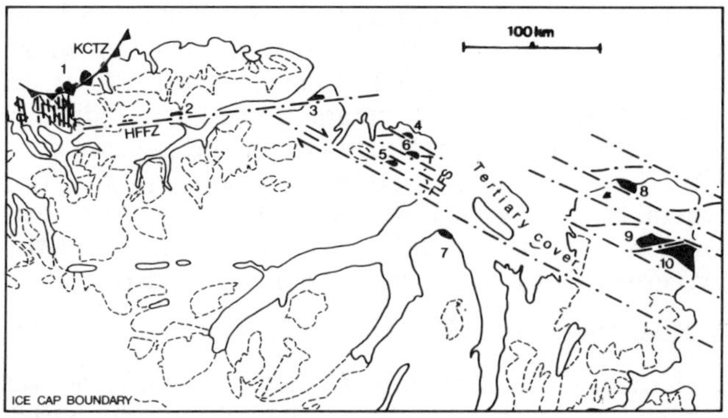

Fig. 10. Distribution of late Jurassic to late Cretaceous deposits in relation to the main faults of the late Cretaceous Wandel Hav Strike-Slip Mobile Belt in North Greenland. Numbers refer to Fig. 11; KCTZ, Kap Cannon Thrust Zone; HFFZ, Harder Fjord Fault Zone; TLFS, Trolle Land Fault System. (From Birkelund and Håkansson, in press).

consider these dykes as precursor elements of this important Late Cretaceous volcanic province (Håkansson and Pedersen, 1982). The possible relation of the North Greenland volcanic province to the late Mesozoic magmatic activity in Svalbard and the Franz Joseph Land Archipelago (cf. Birkenmajer, 1981) therefore must remain speculative until more precise dating and detailed geochemical information become available.

No equivalents of the early–Middle Jurassic remanié faunas from the Brentskardhaugen Bed (Birkenmajer, 1981; Mørk *et al.*, 1982) have been found in North Greenland, where the entire Late Triassic–Middle Jurassic interval constitutes a major hiatus. Håkansson and Pedersen (1982) suggested that at some time during this period large-scale sinistral transpression among other mechanisms caused differential block movements and extensive erosion in the central part of the Wandel Sea Basin. Contemporaneous activity in at least the Billefjorden Fault Zone (cf. Mørk *et al.*, 1982, fig. 18) may be taken to indicate that this transpressive phase was active also in Svalbard. Certainly, in the Barents Shelf area several structural elements have their origin in the later part of this period (Rønnevik *et al.*, 1982). Thus, the new Hammerfest and Bjørnøya basins could be structural analogues to the two North Greenland sub-basins. More important, however, is the likely presence of a major Jurassic wrench system along the western margin of the Barents Shelf (Rønnevik *et al.*, 1982). Considering its age and position in a pre-drift situation, it is of course tempting to suggest that this wrench system is related to the North Greenland transpressive phase. Rønnevik *et al.* (1982) stated that the structural pattern of the Barents Shelf during this period is best explained by assuming a dextral shear in the wrench zone. On the other hand, Håkansson and Pedersen (1982) found evidence for sinistral shear, at least in the Trolle Land Fault System in eastern Peary Land, so with the present understanding the two systems cannot be directly united.

Late Cretaceous—Early Tertiary

Late Cretaceous rocks have been found to be widespread and varied in North Greenland (Håkanssen *et al.*, 1981b), whereas equivalent rocks are completely absent in Svalbard and have been encountered only recently in the southern part of the Barents Shelf (Marty *et al.*, 1979). In contrast, Early Tertiary strata are of limited distribution and thickness in North Greenland as compared with the huge Palaeogene basin occupying much of central Spitsbergen. This period, therefore, exhibits some of the most striking differences between the two continental platforms under comparison.

Wandel Hav Strike-Slip Mobile Belt

In North Greenland late Cretaceous geology was controlled by a complex series of events which together characterize a right-lateral structural belt termed the Wandel Hav Strike-Slip Mobile Belt by Håkansson and Pedersen (1982). The dominant structural element in the on-shore part of this mobile belt is a series of at least five fault-bounded pull-apart basins in which thick, highly immature deposits were rapidly accumulated and subsequently deformed by thrusting or domal folding. The pull-apart basins have been recognized right across North Greenland (Fig. 10), containing quite varied infill ranging from extrusive volcanics and continental clastics to fully marine, fine-grained clastics (Fig. 11). The precision in dating of the individual sequences varies tremendously (Fig. 11) (Birkelund and Håkansson, 1983), rendering correlation difficult; the detailed series of events within the mobile belt, therefore, has yet to be determined. From the present data it can be stated that (1) formation of all pull-apart basins was initiated subsequent to the Albian; (2) the oldest strata so far dated from any of the basins are of Middle Turonian age, while the youngest strata correspond to the Cretaceous–Tertiary boundary; and (3) compressional deformation in all basins was apparently terminated prior to renewed deposition in the Late Palaeocene (Håkansson) and Pedersen, 1982; Birkelund and Håkansson, 1983). It appears, therefore, that tensional and compressional forces were most likely active simultaneously along the structural belt through most of the late Cretaceous and were terminated no later than the mid Palaeocene. Furthermore, a possible trend in basin age from oldest in the southeast to youngest in the northwest has been noted (Håkansson and Pedersen, 1982).

It is important to note that the structural elements of the Wandel Hav Strike-Slip Mobile Belt, to a large extent, are inherited from previous deformations, particularly in the form of widespread rejuvenation of the NW-trending fault systems (Fig. 10). Similarly, in the southern part of the Barents Shelf, Rønnevik *et al.* (1982) found that the Kimmerian wrench system was reactivated in the latest Cretaceous or early Tertiary since here, in for instance the Tromsø Basin, late-Cretaceous marine sediments appear to be overlain with angular unconformity by Palaeocene strata (Bjærke and Manum 1979; Rønnevik *et al.*, 1982). This similarity may be quoted in support of the proposal by Birkelund and Håkansson (1983) that the Late Cretaceous basins in North Greenland and on the Barents Shelf are directly related and were probably formed and deformed within a single megascopic, right-lateral shear zone expressed in the Wandel Hav Strike-Slip Mobile Belt. The trend of the major faults in the mobile belt appears to be largely parallel to the Spitsbergen Fracture Zone (Håkansson, 1979; Dawes and Peel, 1981; Håkansson and Pedersen, 1982), but lateral displacement may have been very modest in this late Cretaceous phase. Subsequently, in the Palaeocene, the megascopic shear was concentrated and ultimately released in the complex movements along the Spitsbergen Fracture Zone which eventually led to the separation of Greenland and Svalbard.

West Spitsbergen orogeny

In North Greenland, Tertiary deposits are essentially confined to the central part of the Wandel Sea Basin (Fig. 10), where virtually flat-lying, very poorly exposed sediments of the Thyra Ø Fm are indicative of fluvial plain/alluvial fan environments (Håkansson *et al.*, 1981b). The composite exposed thickness is probably well below 100 m, and the base of the formation is unknown; rare dinoflagellate cysts

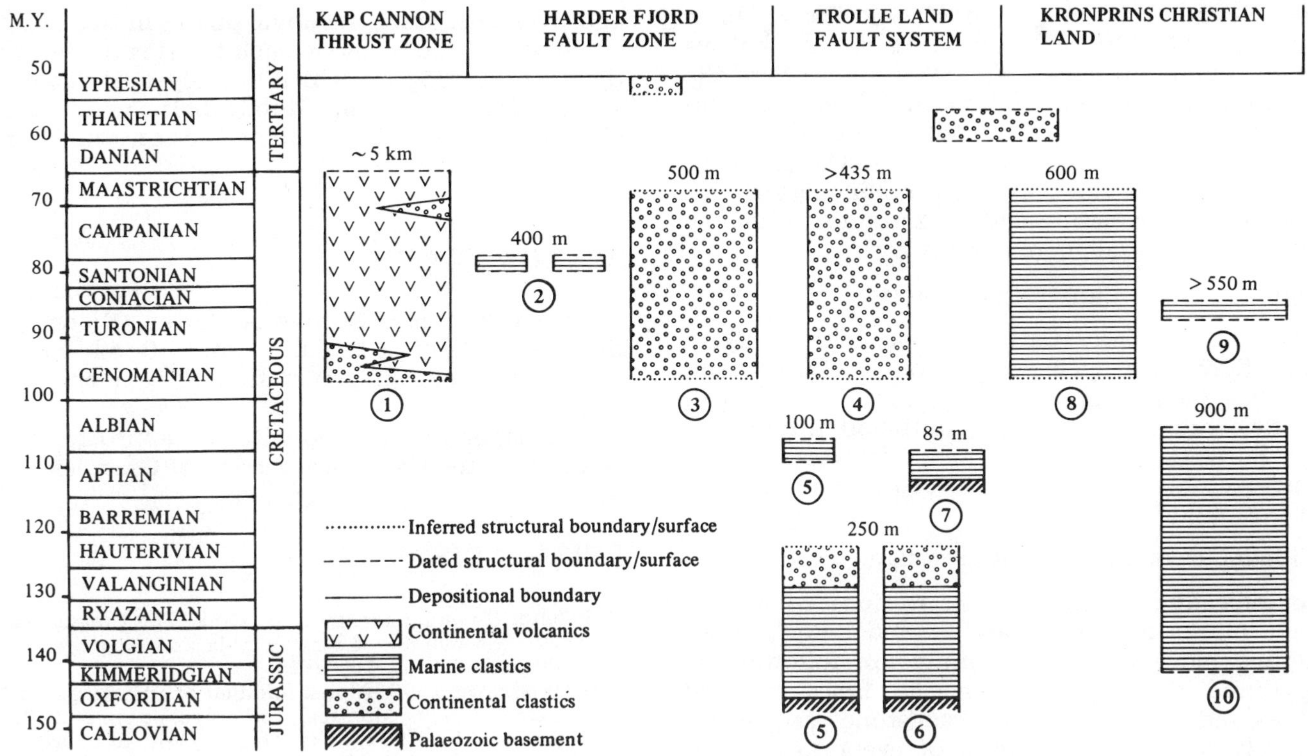

Fig. 11. Stratigraphy, thickness and depositional environment of late Jurassic to early Tertiary strata in North Greenland. Numbers refer to Fig. 10. (From Birkelund and Håkansson, 1983.)

indicate a late Palaeocene age for the sequence (Håkansson and Pedersen, 1982).

In Svalbard the partly comparable Van Mijenfjorden Gp rests unconformably on Permian and Triassic strata to the north and successively younger members of the Aptian—Albian Carolinefjellet Fm towards the south (Harland *et al.*, 1974; Birkenmajer, 1981). A total of 2500 m have been recorded; the age of the sequence is still a matter of some debate, but an early (?) Palaeocene to possibly early Eocene interval seems most probable (Harland *et al.*, 1974; Birkenmajer, 1981; Steel *et al.*, 1981). Along the western margin of the central basin in Spitsbergen strata of the entire group have been deformed in the West Spitsbergen orogeny. The main phases in this orogeny reflect the succession of transtensional, transpressional, and renewed transtensional forces resulting from the dextral translocation of the Greenland and Svalbard continental blocks along the Spitsbergen Fracture Zone (Harland and Horsfield, 1974; Kellogg, 1975; Birkenmajer, 1981; Steel *et al.*, 1981). The West Spitsbergen orogeny started to evolve during deposition of the Van Mijenfjorden Gp, so most likely transpression was initiated already in the Palaeocene (Steel, *et al.*, 1981) to culminate sometime in the mid-Tertiary (Birkenmajer, 1972; Harland *et al.*, 1974; Kellogg, 1975). Apparently, therefore, it entirely post-dates the Wandel Hav Strike-Slip Mobile Belt.

Given this background, it is perhaps somewhat surprising that clear evidence of post-Cretaceous compressional forces has not been encountered in North Greenland (Håkansson and Pedersen, 1982). The anticipated existence of mirror images on the Greenland Plate of the Tertiary wrench faulting in the West Spitsbergen orogeny, therefore, has not been directly substantiated. However, as speculated by Birkelund and Håkansson (1983), it is probable that they are present as an overprint on the structures of the Wandel Hav Strike-Slip Mobile Belt on the part of the broad Greenland Shelf bordering the Spitsbergen Fracture Zone.

One additional feature of the Tertiary geology in North Greenland may be accentuated here, although it is probably not directly related to the transcurrent movements separating the two continental blocks. Thus, an episode of substantial heat-flow increase, perhaps associated with some doming, has affected a mosaic of strata as young as the late Palaeocene Thyra Ø Fm in the central part of the Wandel Sea Basin (Håkansson and Pedersen, 1982; Håkansson *et al.*, in preparation). Very pronounced thermal gradients away from a focal temperature of at least 250° C points to a brief but very intense thermal event. The focus is situated at the north coast of Kronprins Christian Land (just west of loc. 8 in Fig. 10), very close to the present-day southward termination of the Nansen Ridge in the Eurasian Basin, and it is suggested that the thermal event is associated with the thermal activity in this ridge and, thus, may be very young (Håkansson and Pedersen, 1982; Håkansson *et al.*, in preparation).

SUMMARY

The main objective of this chapter is to compare the post-Devonian geology of the recently investigated eastern part of North Greenland with that of the classic Svalbard/Barents Shelf area. In this way, the possible relationship of the two regions may be assessed. Naturally, therefore, we have emphasized apparent differences or similarities rather than attempt a balanced presentation of the history of the North Greenland Wandel Sea Basin. As is already evident from the title, we believe that we have found sufficient similarity to sustain the traditionally assumed relationship of the continental platforms under consideration. On the other hand, where seemingly important deviations have been noted, we can only make preliminary speculations on their nature and background within the frame of an overall, common history.

Late Devonian—Early Carboniferous

Early in the Early Carboniferous, fairly uniform conditions with cyclic deposition of fluvial, commonly coal-bearing sediments became established. In Svalbard sedimentation continued throughout the Early Carboniferous, whereas in North Greenland tectonic activity has disturbed the later part of this period.

Middle Carboniferous—Late Permian

Tectonically controlled transgressions of several fault blocks during the Middle and Late Carboniferous more or less uniformly are followed by rhythmic deposition of marine limestone and variegated (mainly red) sandstone sequences. Locally significant intervals of evaporites occur. By the Late Carboniferous, deposition of pure shelf carbonates becomes dominant and prevails almost everywhere until interrupted by mid-Permian tectonic activity. Subsequently, in the Late Permian, clastic sediments once more became widespread, and local deviations in depositional environments are apparent.

Latest Permian—Early Jurassic

From this period onwards deposition is exclusively clastic. In spite of very sparse occurrences in North Greenland it appears that fault-controlled deposition dominates everywhere. Certain differences in age and local depositional environment could be taken to indicate a separation into distinct sub-basins as early as from the Permo-Triassic boundary.

Middle Jurassic—Early Cretaceous

Sub-basin separation seems well established, perhaps in association with the major early to Middle Jurassic wrench movements that affect most of the platform. Significant volcanic activity in the later part of the period is restricted to Svalbard (and Franz Joseph Land), and it remains uncertain whether it is related to the weak onset of vulcanicity in North Greenland.

Late Cretaceous—Early Tertiary

In North Greenland and the southern Barents Shelf area, a series of comparatively small pull-apart basins were formed in the dextral Wandel Hav Strike-Slip Mobile Belt through the late Cretaceous. The basins are characterized by rapid accumulation of thick sequences of highly immature deposits, but otherwise both depositional environment and age varies tremendously from basin to basin. All basins were compressed prior to the late Palaeocene. The trend of the Wandel Hav Strike-Slip Mobile Belt largely parallels both the preceding Jurassic wrench faults and the Spitsbergen Fracture Zone which subsequently, in the Tertiary, became instrumental in the actual separation of Greenland and Svalbard. Transpressional phases in this dextral translocation led to the wrench faulting of the West Spitsbergen orogeny which entirely post-dates the Wandel Hav Strike-Slip Mobile Belt. So far, these phases have not been identified on the North Greenland Plate.

ACKNOWLEDGEMENT

We wish to thank Pernille Andersen, Richard G. Bromley, Henrik Egelund, and Bodil Sikker-Hansen for help in preparation of text and illustrations.

The paper is published with the approval of the Director of the Geological Survey of Greenland.

REFERENCES

Birkelund, T. and Håkansson, E. 1983. The Cretaceous of North Greenland — a stratigraphic and biogeographical analysis. *Zitteliana*, **10**, 7—25.

Birkenmajer, K. 1972. Alpine fold belt of Spitsbergen. *24th Int. Geol. Congr.*, Montreal, 1972, Sec. 3, pp. 282—292.

Birkenmajer, K. 1981. The Geology of Svalbard, the western part of the Barents Sea, and the continental margin of Scandinavia. In: A.E.M. Nairn, *et al* (eds.), *The Ocean Basins and Margins*, Vol. 5, *The Arctic Ocean*, Chapter 6, pp. 265—329.

Bjærke, T. and Manum, S.B. 1979. Geology of the Barents sea shelf; Evidence from palynological studies of drift material. In: *Norwegian Sea Symposium* (Norsk Petroleumsforening), Article 17.

Cutbill, J.L. and Challinor, A. 1965. Revision of the stratigraphical scheme for the Carboniferous and Permian of Spitsbergen and Bjørnøya. *Geol. Mag.* **102**, 418—439.

Dawes, P.R. and Peel, J.S. 1981. The northern margin of Greenland from Baffin Bay to the Greenland Sea. In: A.E.M. Nairn, *et al.* (eds.), *The Ocean Basins and Margins*, Vol. 5, *The Arctic Ocean*, Chapter 5, pp. 201—264.

Dawes, P.R. and Soper, N.J. 1979. Structural and stratigraphic framework of the North Greenland fold belt in Johannes V. Jensen Land, Peary Land. *Grønlands Geol. Unders.*, Rapp. 93, 40 pp.

Dunbar, C.O. 1962. Faunas and correlation of the late Paleozoic rocks of Northeast Greenland. Part III. Brachiopoda, *Meddr Grønland* **167**(4), 1—14.

Dunbar, C.O., Troelsen, J., Ross, C., Ross, J.P. and Norford, B. 1962. Faunas and correlation of the late Paleozoic rocks of Northeast Greenland. Part I. General discussion and summary. *Meddr Grønland*, **167**(4), 1—16.

Gjelberg, J.G. 1981. Upper Devonian (Fammenian) to Middle Carboniferous succession of Bjørnøya. *Norsk Polarinst. Skr.*, **174**, 1—67.

Gjelberg, J.G. and Steel, R.J. 1981. An outline of Lower—Middle Carboniferous sedimentation on Svalbard: Effects of tectonic, climatic and sea level changes in rift basin sequences. In: J.W' Kerr and A.J. Fergusson (eds.), Geology of the North Atlantic Borderlands, *Can. Soc. Petrol. Geol., Mem.*, **7**, 543—561.

Harland, W.B. and Horsfield, W.T. 1974. West Spitsbergen orogeny. In: A.M. Spencer (ed.), *Mesozoic—Cenozoic Orogenic Belts; Data for Orogenic Studies*, Spec. Publ. Geol. Soc. Lond., No. 4, pp. 747—755.

Harland, W.B., Cutbill, J.L., Friend, P.F. Gobbett, D.J., Holliday, D.W., Maton, P.I., Parker, J.R. and Walls, R.H. 1974. The Billefjorden Fault Zone, Spitsbergen—the long history of a major tectonic lineament. *Nor Polarinst. Skr.*, No.161—172.

Håkansson, E. 1979. Carboniferous to Tertiary development of the Wandel Sea Basin, eastern North Greenland. *Grønlands Geol. Unders., Rapp.* **88**, 72—83.

Håkansson, E. and Heinberg, C. 1977. Reconnaissance work in the Triassic of the Wandel Sea Basin, Peary Land, eastern North Greenland. *Grønlands Geol. Unders., Rapp.* **85**, 11—15.

Håkansson, E. and Pedersen, S.A.S. 1982. Late Paleozoic to Tertiary tectonic evolution of the continental margin in

North Greenland. In: A.F. Embry, and H.R. Balkwill (eds.), Arctic Geology and Geophysics, *Can. Soc. Petrol. Geol., Mem.*, **8**, 331—348.

Håkansson, E., Birkelund, T., Piasecki, S. and Zakharov, V. 1981a. Jurassic—Cretaceous boundary strata of the extreme Arctic (Peary Land, North Greenland). *Bull. Geol. Soc. Denmark*, **30**, 11—42.

Håkansson, E., Heinberg, C. and Stemmerik, L. 1981b. The Wandel Sea Basin from Holm Land to Lockwood Ø, eastern North Greenland. *Grønlands Geol. Unders.*, Rapp. 106, 47—63.

Higgins, A.K., Friderichsen, J.D. and Soper, N.J. 1981. The North Greenland fold belt between central Johannes V. Jensen Land and eastern Nansen Land. *Grønlands Geol. Unders.*, Rapp. 106, 35—45.

Kellogg, H.E. 1975. Tertiary stratigraphy and tectonism in Svalbard and Continental Drift. *Am. Assoc. Petrol. Geol. Bull.*, **59**(3), 465—485.

Lauritzen, Ø. 1981. The Carboniferous and Permian stratigraphy of the Wahlenbergfjorden area, Nordaustlandet, Svalbard. *Norsk Polarinst. Skr.*, **176**, 23—44.

Marty, A., Rabate, J., Cazes, M., Jacquart, G. and Auxietre, J.L. 1979. CEMP Barents Sea seismic survey. In: *Norwegian Sea Symposium* (Norsk Petroleumsforening), Article 16.

Miall, A.D. 1980. Cyclicity and facies model concept in fluvial deposits. *Bull. Can. Petrol. Geol.*, **28**(1), 59—80.

Mørk, A., Knarud, R. and Worsley, D. 1982. Depositional and diagenetic environments of the Triassic and Lower Jurassic succession of Svalbard. In: A.F. Embry and H.R. Balkwill (eds.) Arctic Geology and Geophysics, *Can. Soc. Petrol. Geol. Mem.*, **8**, 371—398.

Nathorst, A.G. 1911. Contributions to the Carboniferous flora of Northeastern Greenland. *Meddr Grønland*, **43**(12), 1—10.

Parker, J.R. 1967. The Jurassic and Cretaceous sequence in Spitsbergen. *Geol. Mag.*, **104**(5), 487—505.

Petryk, A.A., 1977. Upper Carboniferous (Late Pennsylvanian) microfossils from the Wandel Sea Basin, Peary Land, eastern North Greenland. *Rapp. Grønlands Geol. Unders.*, **85**, 16—21.

Rønnevik, H., Beskow, B. and Jacobsen, H.P. 1982. Structural and stratigraphic evolution of the Barents Sea. In: A.F. Embry and H.R. Balkwill (eds.), Arctic Geology and Geophysics, *Can. Soc. Petrol. Geol., Mem.*, **8**, 431—440.

Steel, R.J., Dalland, A. Kalgraff, K. and Larsen, V. 1981. The Central Tertiary Basin of Spitsbergen: Sedimentary development of a sheared-margin basin. In: J.W. Kerr and A.J. Fergusson (eds.), Geology of the North Atlantic Borderlands, *Can. Soc. Petrol. Geol., Mem.* **7**, 647—664.

Surlyk, F., Piasecki, S., Rolle, F., Stemmerick, L., Thomsen, E. and Wrang, P. 1984. The Permian basin of East Greenland. This volume.

Worsley, D. and Edwards, M.B. 1976. The Upper Palaozoic succession of Bjørnøya. *Norsk Polarinst. Årbok* **1974**, 17—34.

Worsley, D. and Mørk, A. 1980. *Excursion Guide to Isfjorden, Svalbard.* Continental Shelf Institute, Trondheim, 50 pp.

8 Svalbard's post-Caledonian strata – an atlas of sedimentational patterns and palaeogeographic evolution

Ronald J. Steel and David Worsley*†
*Norsk Hydro Research Centre, Bergen, Norway
† Paleontologisk Museum, Oslo, Norway

The Carboniferous to Palaeogene sedimentational history of the Svalbard Archipelago is documented with the help of representative cross-sections, palaeogeographical maps and brief syntheses of major depositional phases.

The succession comprises eight lithostratigraphical groups, with a general composite thickness of more than 7 km, which accumulated between the late Devonian 'Svalbardian' deformation and the Tertiary orogeny. Varying tectonic regime is reflected in the succession by changing patterns of sediment thickness, facies configurations, provenance and transport directions. Climatic changes, resulting largely from northwards palaeocontinental movement, also had a great influence on stratigraphical variation and facies development.

Tectonic activity was concentrated along a series of NNW—SSE lineaments of Caledonian, or earlier, origin. Lineamental instability, with both normal and oblique-slip movements, was most marked in the earliest and latest phases of the depositional epoch. Facies of the Carboniferous and early Permian developed in a horst and graben setting. Early Carboniferous grabens were the sites of equatorial fluviatile deposition, with coal swamps developed on flood-plains and around lakes. Mid-Carboniferous to early Permian red bed, carbonate and sabkha sequences reflect arid terrestrial climates and ongoing stabilization of most lineaments. The transition to more stable platform environments during the Permian probably reflects the relocation of tensional tectonic regimes to intracratonic rift areas west of Svalbard. This transition was accompanied by a regional transgression in the mid-Permian and the establishment of temperate climates in the Svalbard area.

Thickness and facies variations in the Mesozoic clastic sequence demonstrate continued flexuring and down-warping along existing lineaments; active faulting was accompanied by intrusive and extrusive activity in the late Jurassic and early Cretaceous. Periodic northern uplift of the platform became more marked through the Mesozoic, perhaps reflecting mantle expansion related to the evolution of the present Arctic Basin.

Tertiary basins and the West Spitsbergen Orogenic Belt developed from transtensile and transpressive motion between Greenland and Svalbard. Deep basins developed subsequently off western Spitsbergen, with the change from a sheared to a rifted margin, as the Svalbard Platform finally separated from Greenland.

INTRODUCTION

The Svalbard Archipelago comprises all islands in the area 74—81°N, 10—35°E (Fig. 1). The largest islands of Spitsbergen, Nordaustlandet, Barentsøya and Edgeøya, together with innumerable smaller islands and skerries comprise a total land area of 62 000 km^2. Approximately 60% of this area is permanently covered by glaciers and inland ice, but coastal and mountain exposures display a varied and superbly exposed geological succession ranging in age from the Precambrian to the Palaeogene. In view of its extreme latitude the archipelago has a relatively mild climate, and western sea approaches are ice-free most of the year; however, snow cover permits effective geological field work only in the short summer season of July and August. The archipelago has a somewhat unusual status. Originally a political no-man's land, it has been under Norwegian sovereignty since 1925. However, all signatory nations to the treaty which granted that sovereignty have equal rights of exploration and economic exploitation of the area. This unique combination of geological, climatic and political factors has made the Svalbard Archipelago the subject of active international geological research for 150 years.

The Norwegian geologist B.M. Keilhau paid a short visit to the area in 1827, but geological expeditions through the remainder of the 19th century were very much dominated by Swedish expeditions under the leadership of S. Loven and, later, of A.E. Nordenskiold, A.G. Nathorst and G. De Geer. These expeditions established the strati-

Petroleum Geology of the North European Margin © Norwegian Petroleum Society (Graham & Trotman, 1984) pp. 109—135.

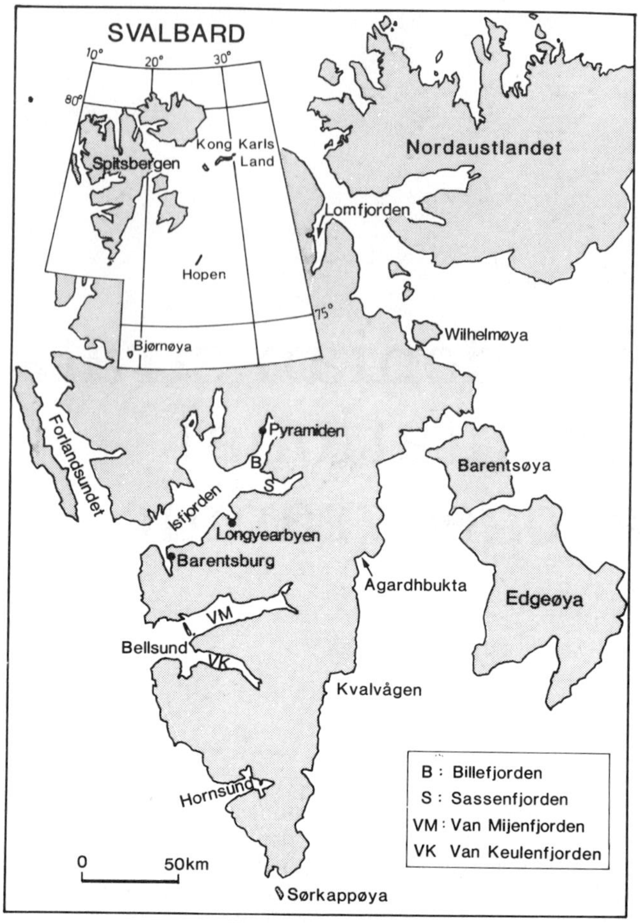

Fig. 1. A location map showing the major geographical features referred to in the text.

graphic framework of the area and resulted in many major contributions on the fossil floras and faunas (see Nathorst, 1910). Significant coal deposits of Carboniferous and Tertiary age were discovered at an early stage of investigation, and exploitation of these started around the turn of the century. This was accompanied by decreasing Swedish activity; from 1906 onwards Norwegian expeditions, led first by G. Isachsen and later by A. Hoel, played an increasingly important role in further work. The early Norwegian expeditions were privately sponsored, but increasing government support led finally to the establishment of 'Norges Svalbard og Ishavets undersøkelser' (later Norsk Polarinstitutt) in 1929. These expeditions contributed to increased knowledge both of coal-bearing sequences and of the palaeontology and biostratigraphy of the entire succession. Of especial note are works on the Festningen section on the southwestern coast of Isfjorden; continuous exposures from the mid-Permian to the basal Tertiary made this area a source of study as a biostratigraphical standard for the entire archipelago (Hoel and Orvin, 1937). Other important palaeontological work was carried out on the Devonian agnathan faunas of northern Spitsbergen; a Norwegian—British—Swedish expedition working in this part of the succession represented the final pre-war scientific effort in the summer of 1939 (Føyn and Heintz, 1943).

Frebold (1935) presented a synthesis of available knowledge on the geology of Spitsbergen, an important milestone after his own earlier presentation of 10 palaeogeographical maps for the Mesozoic succession (Frebold, 1931). A geological map of the archipelago (scale 1:1 000 000) compiled by Orvin (1940) was also accompanied by a general synthesis and Orvin (1947) presented a bibliography of geological and other scientific literature on Spitsbergen up to 1944.

The break in field investigations caused by the Second World War was followed by increasing multi-national activity, which has continued to the present day. Research activities have to some extent reflected interest in the archipelago's petroleum potential (seven boreholes have been drilled to date, most in the 1960s and early 1970s), although petroleum interest in land areas has decreased in recent years and moved to adjacent areas of the Barents and northern Norwegian shelves. Continuing geological studies have produced a gradually more coherent lithostratigraphical scheme for the entire geological column, with increased biostratigraphical refinements using microfaunas and palynofloras. Understanding of Svalbard's geology has also increased significantly in the last 20 years as a result of increased activity in the less accessible parts of the archipelago. Three 1:500 000 geological maps have now been published covering southern and northern Spitsbergen and Barentsøya/Edgeøya (Flood *et al.*, 1971; Winsnes and Worsley, 1981; Hjelle and Lauritzen, 1982) and the final sheet for northeastern areas is now in preparation.

Important studies which have led to a better conceptual understanding of Svalbard's geological history have discussed the major NNW—SSE trending lineaments on Svalbard (Fig. 2). Early studies centred on the Billefjorden Fault of central Spitsbergen (Harland *et al.*, 1974) but later work has shown the importance of other structures in the late Palaeozoic (Gjelberg and Steel, 1981, 1983) and Mesozoic (Mørk *et al.*, 1982). Several lineaments were rejuvenated in the Palaeogene transform or Neogene rift regions during the development of this segment of the Greenland and Spitsbergen continental margins. The implications of some of these Tertiary movements for the evolution of Spitsbergen have been discussed by Birkenmajer (1972), Kellogg (1975), and Steel *et al.* (1981). Geophysical investigations in the Norweigian-Greenland Sea (Talwani and Eldholm, 1977) and along the Spitsbergen Continental Margin to the west and north of the archipelago (Myhre *et al.*, 1982) have complemented geological work on land. Increasing geophysical activity on the Barents Shelf, both to the south and east of Spitsbergen in recent years, has also given important general information on the structure of this important region (Schlüter and Hinz, 1978; Marty *et al.*, 1979). However, recent reconstructions of regional geology (e.g. Rønnevik, 1981) are still almost as much hampered by the lack of relevant information from onshore Svalbard as was the first pioneering attempt at a synthesis of the Barents Shelf evolution by Frebold (1951). In spite of syntheses in recent years (e.g. Harland 1969, 1972; Birkenmajer 1981) there is still no published information on spatial and temporal variations in bio- and lithofacies or interpretations of the palaeogeographical and tectonic controls on these which permit extrapolations from land to adjacent shelf areas.

This chapter is intended as a first step towards

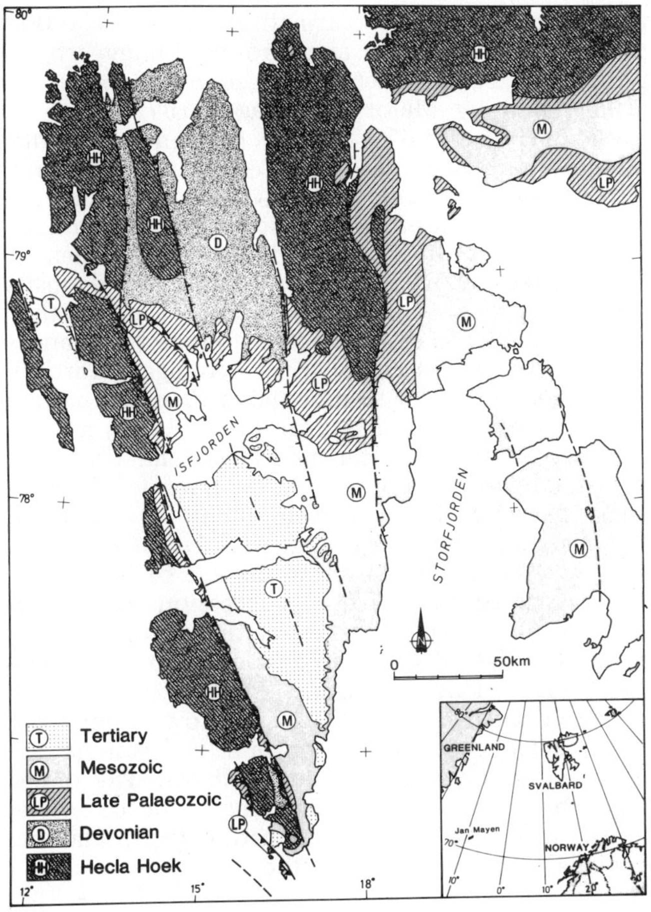

Fig. 2. A simplified geological map of Spitsbergen and adjacent areas.

a more detailed synthesis which will, hopefully, fill this gap in our knowledge. Using illustrative sections and reconstructions, we intend to summarize the results of detailed studies carried out since 1975 on the Carboniferous to Palaeogene strata of Svalbard by both staff and students at the Universities of Bergen and Oslo. This work has included sedimentological and palaeogeographical analyses of the post-Caledonian succession (see Appendix 1 for a list of theses from these Norwegian universities). With the notable exception of Devonian facies and basin analyses by Friend and Moody-Stuart (1972), this type of work had not been attempted previously; the efforts of the Norwegian universities have therefore heralded a new phase in the study of Svalbard's strata. Prior to the presentation of detailed results of this work we would also note that syntheses of single geological periods or lithostratigraphical groups are also being produced (e.g. Steel *et al.*, 1981; Mørk *et al.*, 1982; Hellem and Worsley, in press). Work is also in progress on a synthesis of the late Palaeozoic evolution of Bjørnøya; although we do not include Bjørnøya on our palaeogeographical maps, points of regional significance provided by this island are noted in the text.

STRATIGRAPHICAL SCHEME

The most recent total review of stratigraphical units from Svalbard was presented by Major *et al.* (1956). Increasing research activity and the general adoption of evolving international recommendations on stratigraphical nomenclature during the last 20 years has led to the development of a coherent lithostratigraphical scheme for most units of post-Caledonian strata on Svalbard. In view of the international engagement in research on the archipelago, conflicting stratigraphical proposals and use of synonomous units has perhaps been inevitable. A definitive revision of all lithostratigraphical terms now in use must await the recommendations of the newly established Norwegian Stratigraphical Committee; in the interim period we largely adhere to the scheme used by the Norwegian Polar Institute in their 1:500 000 Geological Map Series, with some modifications resulting from our own work. Figure 3 presents a schematic review of groups comprising the Carboniferous to Palaeogene sequence of Spitsbergen. Formational and member units assigned to these groups will be referred to in the text: the following short review should make it possible for the interested reader to ascertain the derivation of all stratigraphical units referred to herein. Precise datings of many lithostratigraphical units are still lacking and outstanding correlation problems will be discussed in this section.

The lithostratigraphical scheme for Carboniferous and Permian units assigned to the Billefjorden, Gipsdalen and Tempelfjorden Groups by Cutbill and Challinor (1965) has worn well, although some modifications are necessary. An important change concerns the junction between the Billefjorden and Gipsdalen Groups in the Billefjorden area. We propose that use of the upper Hultberg Member of the Svenbreen Fm should be modified to include only coal and shale sequences found in the lower parts of the member as it is presently defined. A new unit — the Anservika Member — is proposed for overlying red beds, which should now be used to define the base of the overlying Ebbadalen Fm

M.y.B.P. / LITHOLOGY	AGE	MIN./MAX. TH. (M)	GROUP	FACIES
	← TERTIARY DEFORMATION			
	Eocene / Palaeocene	1500+	VAN MIJENFJORDEN	DELTAIC/RESTRICTED MARINE CLASTICS
	Alb./Aptian / Callovian	550/1700+	ADVENTDALEN	MARINE CLASTICS, MAJOR PROGRADATION IN L. CRET.
	Bathonian / Rhaetian / Ladinian	70/510+	KAPP TOSCANA	DELTAIC/SHALLOW MARINE CLASTICS
	Lad. Griesb.	60/870	SASSENDALEN	FG MARINE CLASTICS
	«late P.» / Artinsk.	5/460	TEMPELFJORDEN	SILICIFIED MARINE CLASTICS/LSTS
	Artinskian / ?Bashkir.	270/1800	GIPSDALEN	LSTS/DMS/EVAPORITES, LOCAL CLASTICS ALONG FAULTED BLOCK MARGINS
	?Namur. / Famenn.	0/1250	BILLEFJORDEN	FLUVIAL/ALLUVIAL CLASTICS
	↖ SVALBARDIAN DEFORMATION			

Fig. 3. A schematic lithostratigraphical scheme showing the general development of the seven post-Devonian groups of central Spitsbergen.

of the Gipsdalen Gp. The biostratigraphical basis for the dating and correlation of formational units varies between the three major groups in this succession, reflecting to some degree the different depositional regimes which these represent. The Billefjorden Gp is dated on the basis of palynofloras and important works are those of Playford (1962/1963) for central Spitsbergen, Siedlecki and Turnau (1964) for Hornsund, and Kaiser (1970, 1971) on Bjørnøya. The Gipsdalen Gp has been dated mainly on the basis of fusulinids (Cutbill and Challinor, 1965; Sosipatrova, 1972) with some additional information from brachiopods (Gobbett, 1963). Our main modification to Cutbill and Challinor's datings is assignation of Gipsdalen Gp sequences in southern Spitsbergen and on Nordaustlandet to the entire depositional phase from the Bashkirian/Moscovian to the Artinskian and not just to the early Permian. Brachiopods (Gobbet, 1963), non-fusulinid forams (Sosipatrova, 1972), and conodonts (Szaniawski and Malkowski, 1979) have been used to date the Tempelfjorden Group; there remains much uncertainty as to how far sequences assigned to the group extend into the late Permian and opinions vary considerably; we indicate a ?Kazanian age for the youngest beds and hope that studies of palynomorphs, conodonts and foraminifera now in progress may elucidate this problem.

Buchan *et al.* (1965) introduced a lithostratigraphy for Triassic to early Jurassic units which are now assigned to the Sassendalen and Kapp Toscana Groups. Subsequent extensive studies by several groups have led to a wealth of new proposals for formational and lower rank units and the reader is referred to Mørk *et al.* (1982) for an interim synthesis of present status. These sequences have traditionally been dated on the basis of their molluscan macrofaunas (see reviews of Tozer and Parker, 1968; Korcinskaja, 1973), but subsequent work has proved the potential of palynostratigraphical studies, especially in the upper parts of the Kapp Toscana Gp (see e.g. Bjærke and Manum, 1977). There is, however, doubt on intercorrelation of palynofloras and macrofaunas at some levels, e.g. around the Triassic/Jurassic boundary, and little palynostratigraphical work has been done as yet on early to mid-Triassic sequences. However, extensive palaeontological and palynological studies are now in progress and a refined biostratigraphical zonation of the entire sequence may be expected in coming years.

Mid-Jurassic to early Cretaceous deposits of the Adventdalen Gp were first assigned to a formal lithostratigraphical scheme by Parker (1967), with subsequent modifications and refinements by Nagy (1970) and Major and Nagy (1972). The group has been dated on the basis of molluscan macrofaunas (Sokolov and Bodylevsky, 1931; Frebold, 1930, 1931; Nagy, 1970), augmented by recent palynological studies of the groups Callovian to Hauterivian black shales (Bjærke 1978, 1980). Detailed biostratigraphical studies are now needed in the remaining early Cretaceous sequences in order to test the model of diachronous, northwards younging, deltaic deposition presented herein.

Alternative lithostratigraphical schemes for the Tertiary sequence of central Spitsbergen now assigned to the Van Mijenfjorden Gp have been presented by Norsk Polarinstitutt in a 1:100 000 geological map of the area around Longyearbyen printed in 1964 (later described by Flood *et al.* (1971) and by Major and Nagy (1972)) and by Livsic (1965, 1967, 1974). Elements of both schemes have been used by Steel *et al.* (1981) in a new lithostratigraphical subdivision of the group. Tertiary deposits of northwestern Spitsbergen (the Forlandsundet Gp of Harland (1969) have been reinvestigated by Rye-Larsen (1982)). The deposits there are probably largely of Eocene age, and thus partly time-correlative with the youngest sequences of central Spitsbergen. Still younger Tertiary sequences which lie along Svalbard's rifted margin (Sundvor *et al.*, 1977; Myhre *et al.*, 1982) are the product of uplift, after mid-Oligocene times, of the northwestern corner of the Barents Platform and marginal basin development off western Spitsbergen. Results of Deep Sea Drilling Project, Leg 38 (White, 1976) show the age of the uppermost parts of the offshore succession to be Plio/Pleistocene. Despite some continued uncertainty about the age of the youngest on-shore sequences (see Ravn, 1922; Manum, 1962; Livsic, 1967, 1974; Vonderbank, 1970), we now know that both Palaeogene and Neogene successions are represented on and around Svalbard. The oldest deposits in the central basin may, indeed, extend down into the latest Cretaceous (S.B. Manum, personal communication, 1983).

TECTONIC FRAMEWORK

Tectonic setting has been an important controlling influence on sedimentation patterns of post-Caledonian Svalbard, as in most regions of thick sediment accumulation. Previous reviews of the development of Svalbard have emphasized tectonic 'episodes' or 'movements', particularly in late Devonian and early Tertiary times (e.g. Orvin, 1940; Harland, 1969; Harland *et al.*, 1974; Birkenmajer, 1981). Although recognizing the importance of these two tectonic 'highlights' we are impressed also by the considerable record of tectonic movement seen in the entire post-Caledonian succession of Svalbard. Variation in both local and regional tectonics is reflected by changing patterns of sediment thickness and by varying facies configurations, provenance and transport direction. Better understanding of these changing depositional patterns and of the reasons for major stratal breaks, documents a more complete tectonic record, and gives a better insight of the timing and character of the major deformational episodes themselves.

Svalbard's tectonic framework is dominated by a number of structural lineaments, mostly N—S to NW—SE oriented, across which there has been repeated differential movement or inversion since the Devonian. The most important of these are the Lomfjorden/Agardhbukta, Billefjorden, Inner Hornsund and Palaeo-Hornsund Fault Zones. Tectonic movement along these lineaments varied from slight flexuring to discrete faulting, with the latter varying from dip-slip to strike-slip dominance. Figure 4 outlines how these lineaments have affected sedimentation patterns in post-Devonian Svalbard.

The Carboniferous was characterized by rapid

subsidence along or between the major lineaments, with sequences up to 3-km thick accumulating in southern Spitsbergen.

Three marked phases of instability are distinguished:

(1) Tournaisian—Viséan subsidence along narrow (< 30-km wide), isolate zones on both Spitsbergen and Bjørnøya.

(2) Broad, asymmetric Namurian down-warping against the Palaeo-Hornsund Fault, resulting in a thick (< 2 km) clastic wedge which thinned rapidly eastwards.

(3) Bashkirian—Gzhelian return to localized subsidence in a series of narrow zones along the major lineaments. We speculatively suggest that the apparent *en echelon* arrangement of some of the troughs (Fig. 10) may indicate some oblique-slip movement across the region at this time.

A marked feature of Carboniferous tectonics was the tendency of the sense of dip-slip movement across the lineaments to reverse with time. This produced almost total inversion of the early Wedel Jarlsberg Basin in the Namurian, partial inversion of the primitive Billefjorden Basin in the Bashkirian, and Gzhelian to Sakmarian inversion of parts of the Bjørnøya Basin. It is tempting to relate these basinal inversions and the suggested early Carboniferous thrusting in inner Hornsund (Birkenmajer, 1981) to oblique-slip tectonics. Major oblique-slip movements have been proposed for Spitsbergen in the late Devonian (Harland, 1971; Harland *et al.*, 1974) and for the North Atlantic generally in the Carboniferous (Kent and Opdyke, 1979; van der Voo and Scotese, 1981; Ziegler, 1982; Russel and Smythe, 1983). Our own work on a local scale thus strengthens these larger-scale speculations. One consequence of early—mid. Carboniferous strike-slip movements along some of the Svalbard lineaments is that our early Carboniferous maps (Figs. 7 and 8) become more speculative. However, as yet we have no evidence of great movements within Svalbard in Carboniferous times and most of the large-scale translation may have been taken up off western Svalbard, along the Palaeo-Hornsund Fault Zone.

The early Permian to late Cretaceous is characterized by flexuring and minor instability across most lineaments when compared to the preceding and succeeding periods. In late Carboniferous times the Inner Hornsund and, to a lesser extent, the Billefjorden lineaments still divided 'high' areas to the west (Sørkapp-Hornsund High and Nordfjorden Block, respectively) from 'low' areas to the east. Eastwards influx of clastic debris over the Billefjorden lineament had ceased, however, and by the early Permian a relatively stable platform existed throughout most northern areas of Svalbard. In contrast, the Sørkapp—Hornsund High was a positive feature throughout the Permian (Hellem and Worsley, in press). Parts of the western edge of the southerly continuation of the Nordfjord Block were also active, at least into earliest Permian times (Kleinspehn *et al.*, 1983).

Platform conditions persisted through the Mesozoic, although marked thickness and facies variations reflect flexuring and downwarping along earlier lineaments, especially in the early Mesozoic. Differential (?submarine) movement has also been documented along the Lomfjorden-

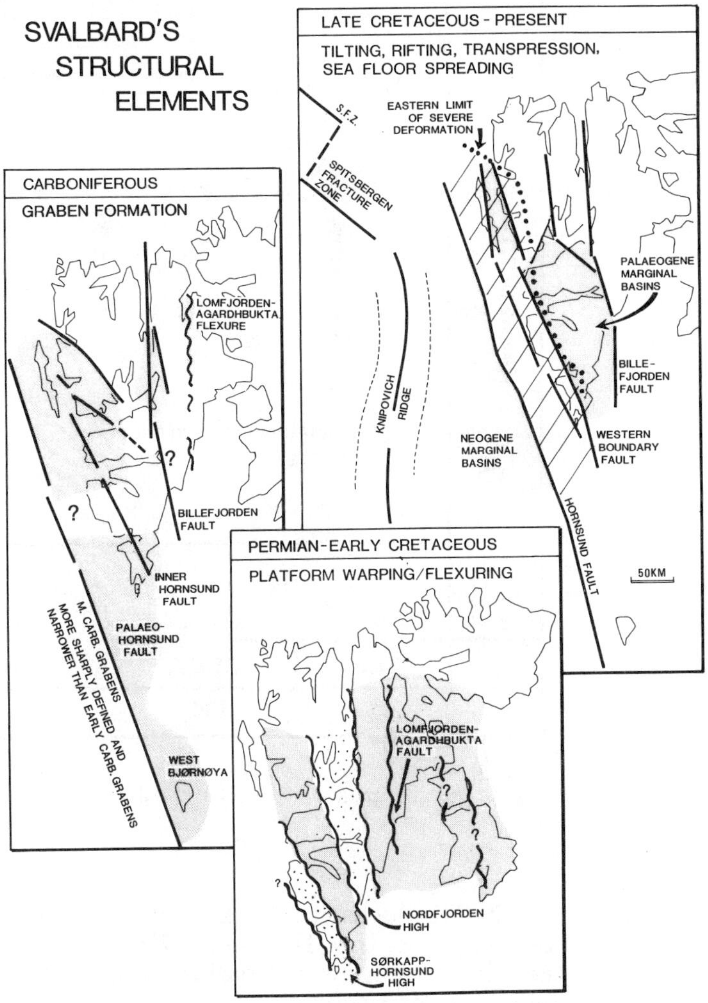

Fig. 4. Svalbard's structural elements during the three main phases of the sedimentary basin's development.

Agardhbukta and Billefjorden lineaments around the Jurassic—Cretaceous boundary. An additional tectonic constraint on deposition, increasingly important through the Mesozoic, was a broad northerly uplift of the platform area. This is evident from sediment transport patterns, and inferred basinal closure to the north from at least Jurassic times. This depositional phase culminated in marked late Cretaceous uplift, with consequent southerly tilting and erosion of the entire Spitsbergen region.

With a marked change in tectonic regime during Tertiary times, the main lineaments again defined the edges of major subsidence zones on Svalbard. They were initially the focus of a continental transform system, then the framework of a new continental margin (Steel *et al.*, 1981). Palaeogene oblique-slip movements along the western lineaments caused up to 3 km of subsidence against the Western Boundary Fault (partly located along the older Inner Hornsund Fault, Fig. 4), and a similar amount of subsidence in a narrow zone (Forlandsundet Graben) east of the Hornsund Fault. Compression across these lineaments also produced major overthrusts on to the flanking basinal areas at this time. The establishment of a sea-floor spreading axis in the Greenland Sea in the

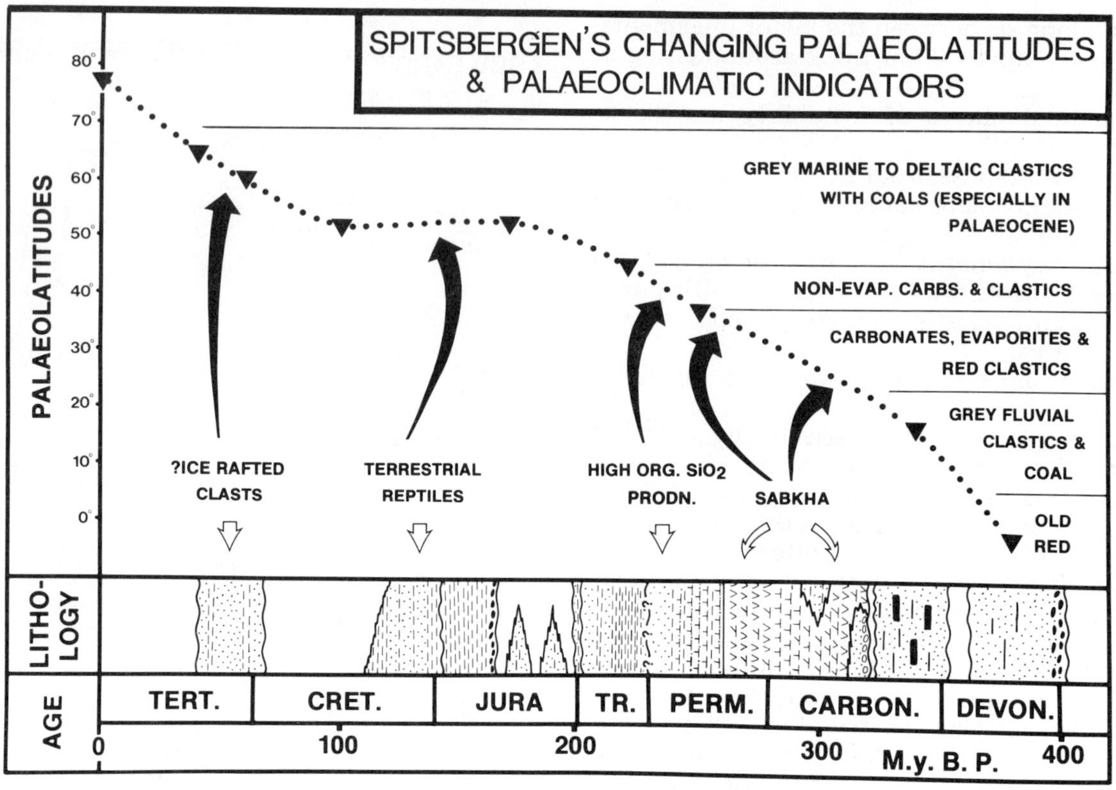

Fig. 5. Possible palaeolatitudes and palaeoclimatic indicators in relation to the major lithofacies observed.

Oligocene shifted the locus of subsidence westwards. The region west of the Hornsund Fault then became the site of Neogene rift basins on the new continental margin of Svalbard.

Changes in tectonic regime clearly reflect different phases in the palaeocontinental evolution of the present Arctic and Norwegian-Greenland Seas. Important information for facies interpretations is also given by relating palaeoclimatic indicators in Svalbard and adjacent basins' strata to possible palaeolatitudes suggested by the global reconstructions of Briden *et al.* (1974) and Smith and Briden (1977). Early Carboniferous coal sequences of the Billefjorden Gp formed in climates comparable to those of today's equatorial humid zone (Fig. 5). Rapid climatic shifts mark the base and top of the Gipsdalen Gp, whose sediments show all the indications of deposition in arid conditions. Although there appears to have been rapid northwards plate movement in the late Carboniferous and early Permian, it is doubtful whether this alone can explain the rapidity of the transition to and from arid conditions. It is probable that additional contributory factors were changing hydrographic regimes related to intracratonic rifting in this sector of the Laurasian Plate at the time.

More humid and temperate conditions persisted from the late Permian to the Cretaceous: terrestrial dinosaurs which inhabited the area in the early Cretaceous indicate continued mild climates at that time (Heintz, 1963; Edwards *et al.*, 1978). Renewed acceleration of movement northwards is suggested in the late Cretaceous, but Palaeocene coals witness a rich vegetation which could not have developed in extreme climatic conditions. However, Dalland (1977) has explained the occurrence of erratic clasts in marine deposits as representing ice-rafting, reflecting freezing of coastal embayments and/or inland lakes in winter months. A major, although elementary point has often been overlooked in interpretations of geophysical data from nearby shelf areas — the climatic shifts noted above did not occur at the same times as in southern Norwegian and North Sea shelf regions; it is therefore unrealistic to expect Triassic red beds in the subsurface of the Barents Shelf, to name only one possible example.

LATE PALAEOZOIC GRABEN DEVELOPMENT

Billefjorden Group

This thick succession of non-marine, coal-bearing strata, lying unconformably on Old Red Sandstone or Hecla Hoek sequences, was early referred to as the 'Kulm' (Nathorst, 1910) or 'Culm' (Orvin, 1940) sandstones. The succession is characterized by plant-bearing quartz arenites, black shales and coals. Early interest in the group was stimulated by the search for coal, now mined only by the Soviet community at Pyramiden. The succession is best known from the type area around Billefjorden, largely through the work of numerous Cambridge University expeditions (summarized by Harland *et al.*, 1976). More recently, the sedimentology of the Billefjorden Gp basins of Svalbard was briefly summarized by Gjelberg and Steel (1981) and a synthesis of Bjørnøya data has been presented by Gjelberg (1982). The tectonic setting of late Devonian to early Carboniferous sedimentation on Svalbard was one of graben development. The main basins were comparable in size to the earlier Old Red Sandstone graben, but lay mainly on the

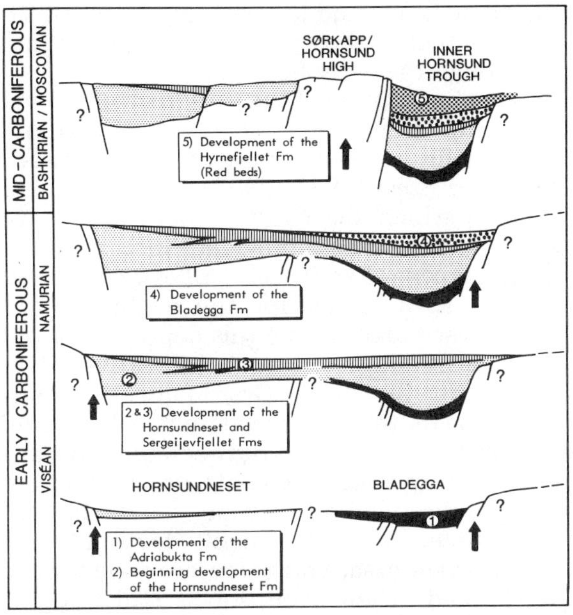

Fig. 6. Early to mid-Carboniferous basinal development in southern Spitsbergen (data by courtesy of J. Gjelberg).

flanks of the latter, which was now partly inverted. This configuration resulted in great thickness variations (0—2000 m) in strata assigned to the Billefjorden Gp (Fig. 6), with major depocentres lying along and east of the Palaeo-Hornsund Fault Zone, off western Spitsbergen and Bjørnøya, and in the Billefjorden area. The palaeogeography was dominated by large, humid alluvial fans which built out from the graben edges into swamps, lakes and river floodplains (Gjelberg and Steel, 1981).

Tournaisian to Visean sequences (extending back into the Fammennian on Bjørnøya, and possibly also elsewhere) are preserved only poorly and exposed on Spitsbergen, but appear to have formed in narrow, elongate, restricted basins (Fig. 7). The Adriabukta Fm (Hornsund and Van Keulenfjorden areas) is about 300 m thick and consists of black shales and a basal conglomerate sequence representing sublacustrine fans which built out from the east. Sedimentation in the Billefjorden area (Hørbyebreen Fm) was dominated by alluvial fans which prograded west and northwestwards into coaly swamps and flood-basins, and partly eastwards into a black shale facies. The Rødevika Fm of Bjørnøya is analogous, representing alternations of fluvial and flood basin deposits (Gjelberg, 1981). Coal seams from these levels were mined until 1925 on Bjørnøya, and are still mined today at Pyramiden.

Namurian times on Svalbard saw the following important changes (Fig. 8):

(1) Basinal areas had significantly increased in size, particularly along the eastern flank of the Palaeo-Hornsund Fault.

(2) Drainage and dominant sediment transport direction was now from westerly regions rather than from the east.

(3) The same spectrum of sedimentary environments was present, but the Namurian braided fans were much larger than earlier systems, probably in

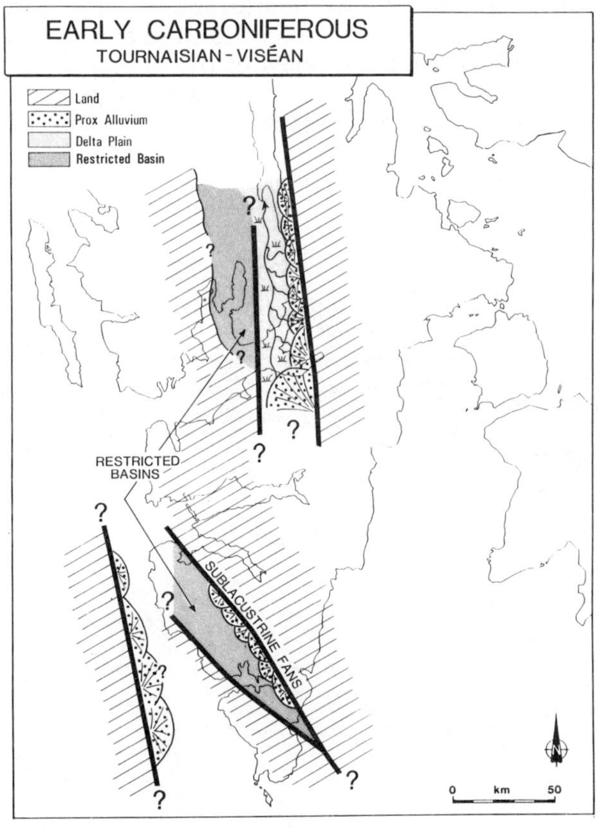

Fig. 7. Tournaisian/Viséan palaeogeography. Map is speculative due to limited data base and to uncertainty as to the relative position of the two basinal areas, due to possible later strike-slip faulting.

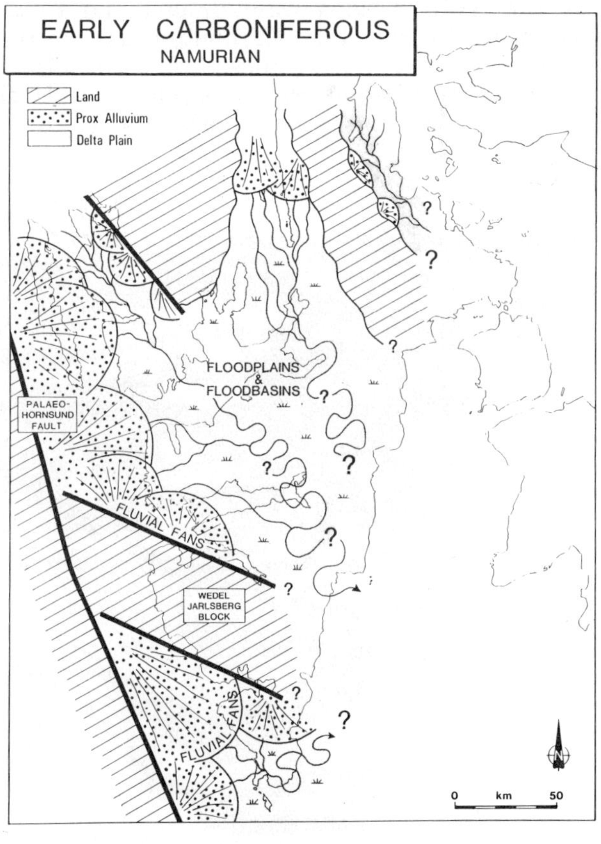

Fig. 8. Namurian palaeogeography. The position of the northeastern area relative to the western area is uncertain, due to possible later strike-slip faulting.

response to increased tectonic activity along western lineaments.

The result of these changes was that thick quartz arenitic successions, largely of braided stream origin, were deposited across Svalbard at this time; these include the Hornsundneset Fm (>1500 m) in areas south of the Wedel Jarlsberg Block and the Orustdalen Fm (>1000 m) in western Spitsbergen north of this block (Fig. 8). Subsidence in the primitive Billefjorden Trough was less impressive, with the deposition of the <100 m sequence of the Sporehøgda Mbr. The analogous Nordkapp Fm on Bjørnøya appears to have developed in the Viséan (Gjelberg, 1981).

Decreased subsidence rates in the late Namurian are suggested by a widespread retreat of the braided fans and their replacement in many areas by fine-grained marsh and flood-plain environments. This produced the youngest Carboniferous coal-bearing sequences of the Sergeijevfjellet Fm in the south, the Vegard Fm in the north, and uppermost part of the Sporehøgda Mbr in the Billefjorden area.

The Gipsdalen Group

Bashkirian times heralded the transition to semi-arid terrestrial climates and a general rise in sea-level. Rapid subsidence and faulting along the main lineaments continued to be important. Although sometimes difficult to define precisely, the junction between the Billefjorden and Gipsdalen Gps in basinal areas is consequently marked by the appearance of red alluvial fan sequences, with their laterally equivalent fan delta deposits, evaporites and carbonates. During trough infilling in northern areas the fault margins were draped and finally overstepped in the late Carboniferous, so that carbonate platform sequences then developed directly upon a basement of Devonian or Hecla Hoek units in the previously positive areas of the Nordfjorden Block and the eastern platform. These areas subsequently show a well-developed regressive trend through the early Permian.

Some of the aspects of the early tectonic development of the group, and facies inter-relationships with changing sea-level, have been discussed by Gjelberg and Steel (1981, 1983). The initial setting was one in which the early Carboniferous basinal system had become subdivided into a series of narrower grabens (Fig. 9). The Billefjorden Trough, often used as a model for the evolution of the entire Svalbard area during this depositional phase, may in reality show an unrepresentative development of the Gipsdalen Gp. Major depositional activity appears to have been situated in a series of *en echelon* arranged basins between the Billefjorden and Palaeo-Hornsund lineaments, possibly reflecting an oblique-slip tectonic regime. These basins are less accessible, and more highly tectonized, than the Billefjorden area, and therefore have tended to be overlooked. Their evolution is still not understood in detail, but a striking feature is a trend of southwestwards directed migration of lineamental instability throughout the depositional phase. By early Permian times tectonic activity had all but ceased along the margins of the Billefjorden and St. Jonsfjorden troughs, although changes in carbonate facies still betray a subtle lineamental influence in these areas. In contrast, the Sørkapp-Hornsund High and the adjacent Inner Hornsund Basin were tectonically active throughout the depositional phase. Bjørnøya shows a complex tectonic development, with alternating periods of stability and faulting; partial basinal inversion occurred in the late Carboniferous and the entire Bjørnøya area was extensively faulted and uplifted in the early Permian. In view of the varied tectonic settings sketched above, we will discuss separately three major depositional systems, *viz.* that of north-

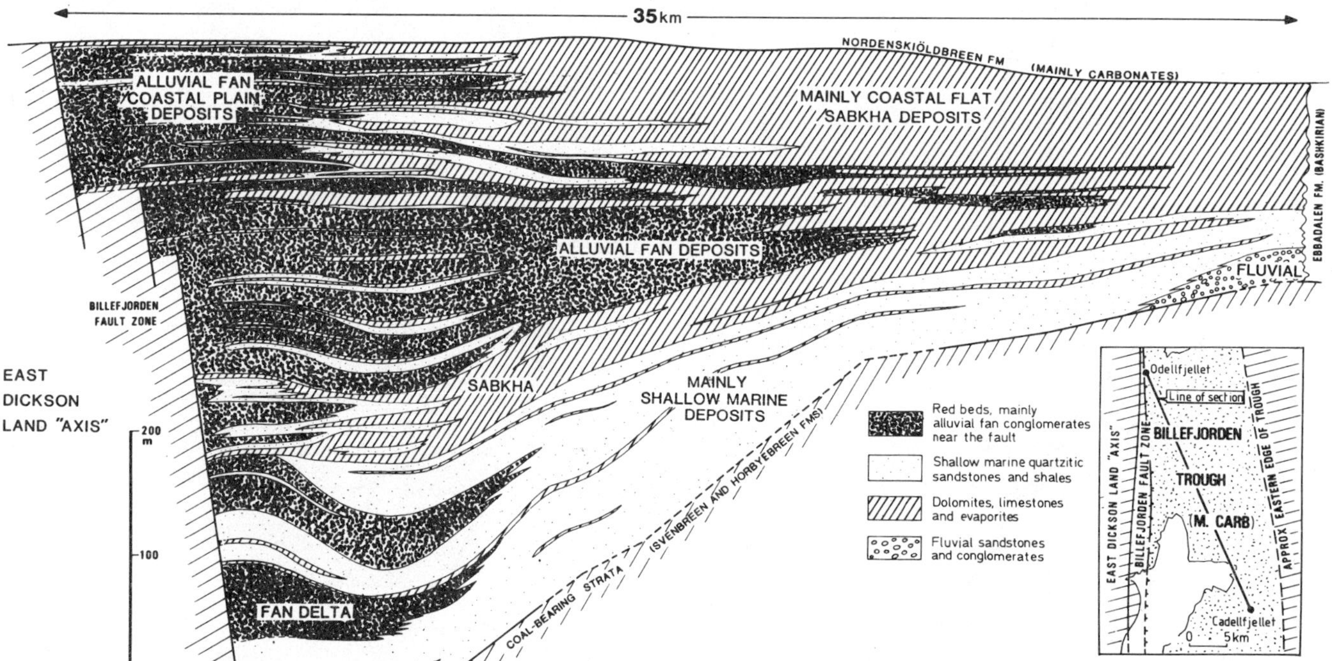

Fig. 9. The Bashkirian development of the Ebbadalen Fm in the Billefjorden Trough.

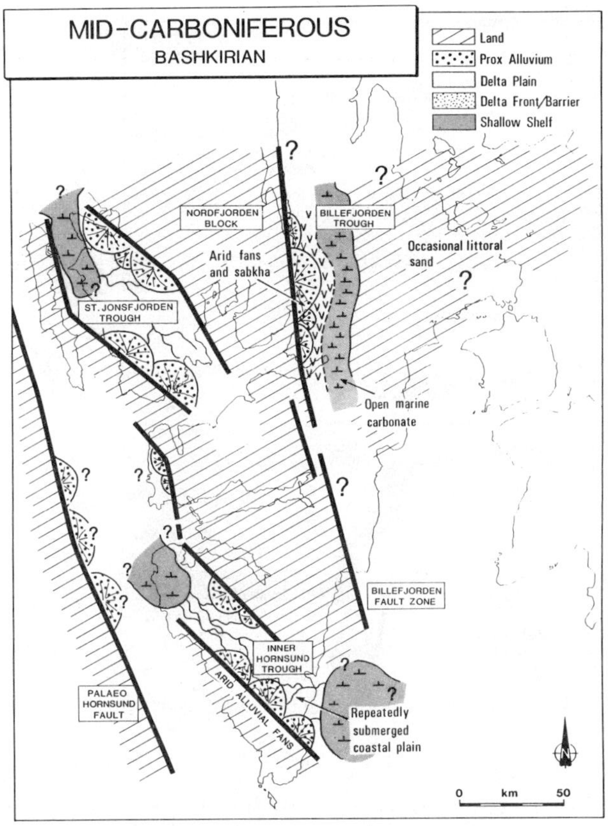

Fig. 10. Bashkirian palaeogeography.

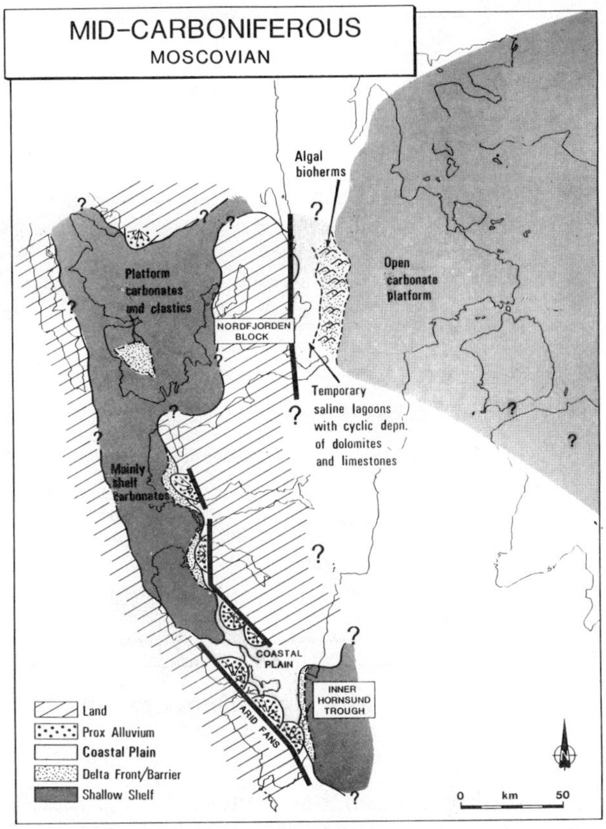

Fig. 11. Moscovian palaeogeography. By latest Carboniferous times much of the Nordfjorden Block and the Inner Hornsund Trough was submergent, though the margins of the latter were still tectonically active.

ern, central and eastern areas, as distinguished from the Inner Hornsund Basin of southern Spitsbergen, and from the Bjørnøya area.

Mid-Carboniferous sedimentation in central and eastern areas of Spirtsbergen was centred on the Billefjorden Trough (Figs. 9 and 10). The varied development of the (<800-m thick) Ebbadalen Fm represents alluvial fans and fan deltas which built out eastwards from the Billefjorden lineament into restricted marine and sabkha environments (Gjelberg and Steel, 1981). Rhythmic intercalations of these, often organized into thin upwards coarsening sequences (Johannessen and Steel, 1981), reveal an intricate interplay of local lineamental activity and regional transgression; the trough developed as an embayment opening to normal marine environments to the north. A general fining-upwards succession reflects ongoing transgression, finally leading to development of the carbonate-dominated Nordenskioldbreen Fm. Platform areas to the east of the trough were submerged in the course of the Moscovian (Fig. 11).

In the St. Jonsfjorden Trough, the mid-Carboniferous clastic sequences of the Brøggertinden, Petrelskardet and Tårnkanten Fms suggest the development of alluvial fans out from fault scarps along both eastern and western basinal margins; sequences appear to become finer northwards through floodplain deposits to open marine environments. The Moscovian saw the establishment of marine carbonate sedimentation also in this area. The late Carboniferous development was characterized by stabilization and the progressive eastwards transgression of the adjacent Nordfjorden Block. This structure was finally transgressed by the close of the Carboniferous, but until then it formed an effective barrier between marine depositional sites to the west and east. Consequently Billefjorden Trough areas adjacent to the block's margins show restricted lagoonal developments behind barrier and shoal systems formed by bioclastic banks and algal bioherms (Fig. 12). There appears to have been minor uplift of the entire area around the Carboniferous/Permian transition, resulting in the development of laterally extensive discontinuity surfaces and intraformational conglomerates. Renewed transgression in the early Permian led to the establishment of open marine environments on both sides of the earlier block. Although there are no significant thickness variations in Lower Permian units at the block's margins, shoals and bioherms still tended to develop along these sites (Fig. 13). A major regressive sequence from the Sakmarian to the Artinskian led to the eastwards progradation of sabkha sequences (Fig. 14) of the (approximately 200 m) Gipshuken Fm (Lauritzen, 1981). The margins of the Nordfjorden Block represent the approximate easternmost extent of sabkhas, although dolomitization is prevalent throughout eastern platform areas and these were evidently the site of restricted marine circulation.

Tertiary tectonism and the present distribution of outcrops along the western margins of the Inner Hornsund Basin make detailed interpretation of this area's evolution difficult. Present data suggest a mid-Carboniferous development of the Sørkapp-Hornsund High (Fig. 6) from which alluvial fans of the Hyrnefjellet Fm built out eastwards. Ongoing transgression produced the fan delta systems and interbedded carbonates of the Treskelodden and

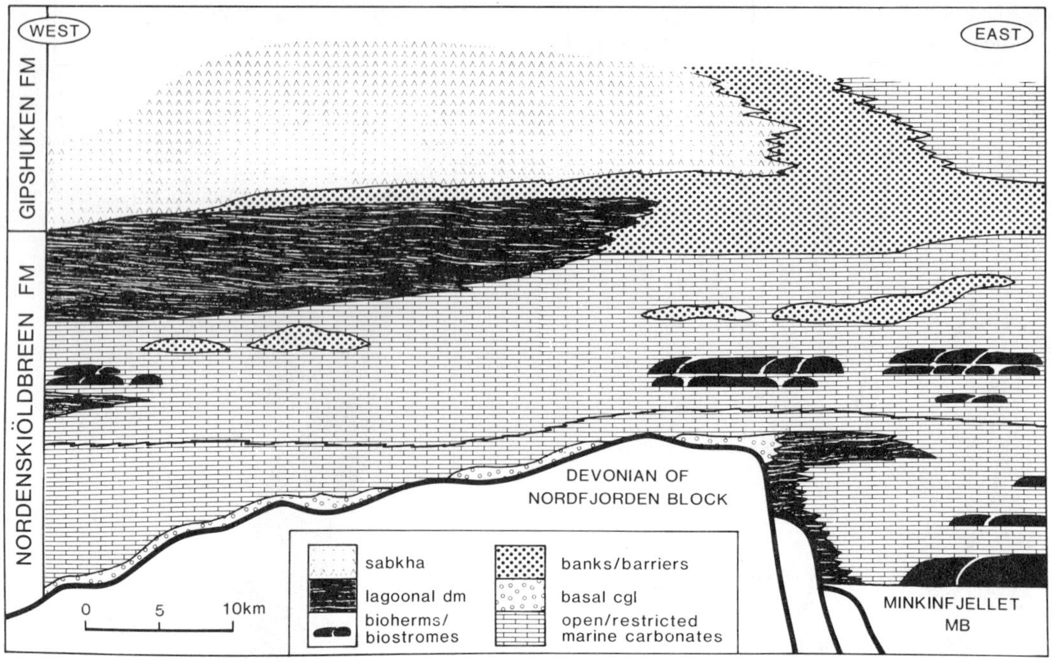

Fig. 12. Late Carboniferous to early Permian facies developments over the eastern and western margins of the Nordfjorden Block.

Reinodden Fms, which may represent progradations from western and eastern margins, respectively. Although datings are uncertain, active fan deltas probably persisted into the early Permian (Kleinspehn *et al.*, 1983). Marginal areas were then uplifted and exposed, but the transition from clastic domination into sabkhas and then open marine carbonates in exposures towards the centre of the basin (Nysæther, 1977) may suggest a transgression which is not detected clearly in the dolomites and sabkhas of the Gipsuken Fm to the north.

Local representatives of the Gipsdalen Gp on Bjørnøya suggest a complex depositional development of this southern sector of the Svalbard Platform in the late Carboniferous and early Permian. Red beds dominate the ?Bashkirian development of the Landnørdingsvika Fm and conglomeratic fans built out eastwards from fault scarps probably forming a southern continuation of the Palaeo-Hornsund Fault Zone (Gjelberg and Steel, 1981, 1983). Lineamental stabilization and rising sea-level led to the transition

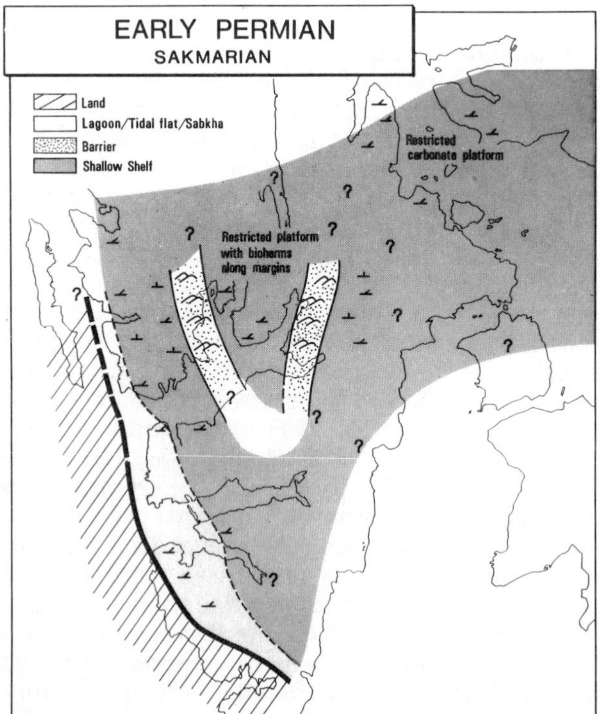

Fig. 13. Sakmarian palaeogeography.

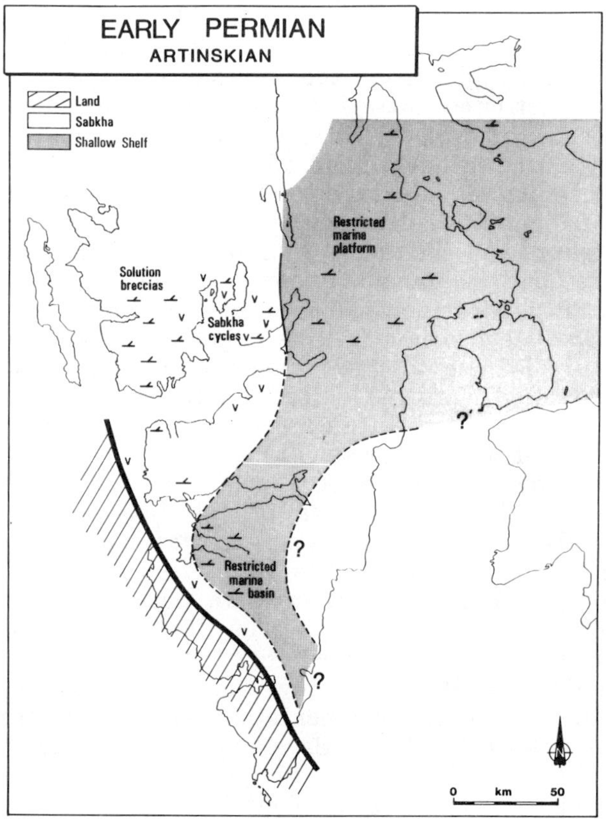

Fig. 14. Artinskian palaeogeography.

to platform carbonate sequences in the middle of the Moscovian Kapp Kåre Fm. So far, depositional patterns had closely paralleled the northern Spitsbergen model, but basinal inversion in the late Moscovian and Gzhelian produced *westerly* prograding alluvial sequences of the clastic-dominated Kapp Hanna Fm from new fault scarps established in the eastern parts of the area. Temporary stabilization around the Carboniferous–Permian boundary, resulting in the carbonate sequences of the Kapp Duner Fm, was followed by extensive faulting, uplift and erosion in the early Permian. The Artinskian transgression noted in the Inner Hornsund Basin may be represented on Bjørnøya by the thin easterly onlapping wedge of the mixed clastic and carbonate Hambergfjellet Fm; however, the horst which had developed in the east of the area in the late Carboniferous does not appear to have been submerged at that time.

LATE PERMIAN TRANSITIONAL REGIMES

The sharp contact between the carbonate-dominated Gipsdalen Gp and the cherty sequences of the Tempelfjorden Gp represents a major facies change which can be traced throughout Svalbard. This change, dated to the late Artinskian–early Kungurian, reflects a large-scale transgression which was either responsible for, or accompanied, a transition from warm arid to temperate humid climatic conditions. Resultant changes in lithofacies from the Gipsdalen to Tempelfjorden Gps include: the cessation of evaporitic mineral formation, even in marginal marine depositional areas; the shift from carbonate to clastic-dominated sediments throughout the area; the replacement of chlorozoan by foramol related biotic associations in the sediments.

The high frequency of cherts and of silica cements in the Tempelfjorden Gps shales, siltstones, sandstones and limestones is striking. Siliceous sponges appear to have been the primary and dominant source of the silica and sponge spicules form an important biogenic component in all lithofacies.

The tectonic setting of the group is also transitional between that of under- and overlying units. Unlike older groups, there is no evidence of major syn-sedimentational faulting in central Spitsbergen or on Bjørnøya. However, preferential downwarping did occur along the lineaments bounding the Nordfjorden Block, and thickest sequences (<490 m) fill the troughs on either side of this structure. Eastern areas show a stable platform development, thinning gradually to 150 m on Nordaustlandet (Fig. 15). An even thinner development (120 m) is seen on Bjørnøya, where basal deposits of the group rest on a peneplaned surface of all older units.

In contrast, the tectonic setting of southwest Spitsbergen was more akin to that of Carboniferous times; the Sørkapp-Hornsund High was clearly a positive structure through the late Permian and sequences thin dramatically against the eastern margins of this structure. A small exposure area around the southwestern tip of Spitsbergen displays a 425-m thick sequence which was probably deposited off the western margins of the high; however, the original position of this sequence is uncertain because of crustal shortening in the Tertiary deformation.

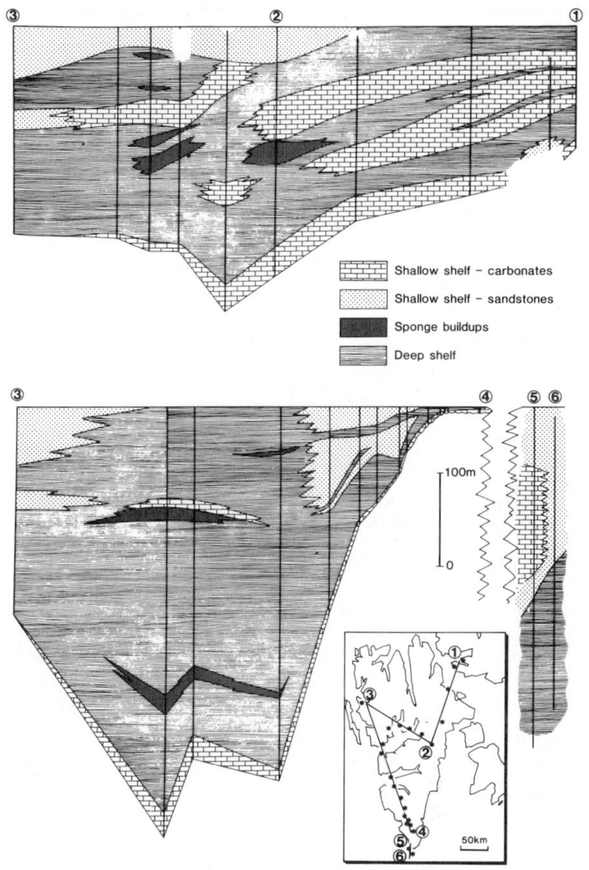

Fig. 15. Cross-sections showing the development of the Tempelfjorden Gp throughout Spitsbergen.

Tempelfjorden Group

The Tempelfjorden Gp comprises three laterally equivalent formations developed in geographically separated areas: the Kapp Starostin Fm of all exposures east and northeast of the Sørkapp-Hornsund High; the Tokrossøya Fm in the area southwest of the high; the Miseryfjellet Fm of Bjørnøya.

The depositional history of the group is best reviewed by discussing the main platform and trough development of central and eastern areas of Spitsbergen separately from the special developments observed along the margins of the Sørkapp-Hornsund High and on Bjørnøya.

Basal beds of the Kapp Starostin Fm commonly consist of bioclastic limestones with a sharp and often erosive contact to the underlying Gipshuken Fm. The limestones are interpreted as shoreface deposits formed by the transgression of barrier sequences over the restricted marine platform and sabkha environments represented by Gipshuken Fm strata (Fig. 16). This basal development grades upwards into the spiculitic shales/siltstones and cherts which form the most abundant lithofacies of the formation, especially in trough areas.

Abundant trace fossils, such as *Zoophycos*, clearly indicate low energy, but oxygenated, bottom environments well below normal wave base. Various interbedded units include bioclastic or sandy shoals and massive chert bodies interpreted

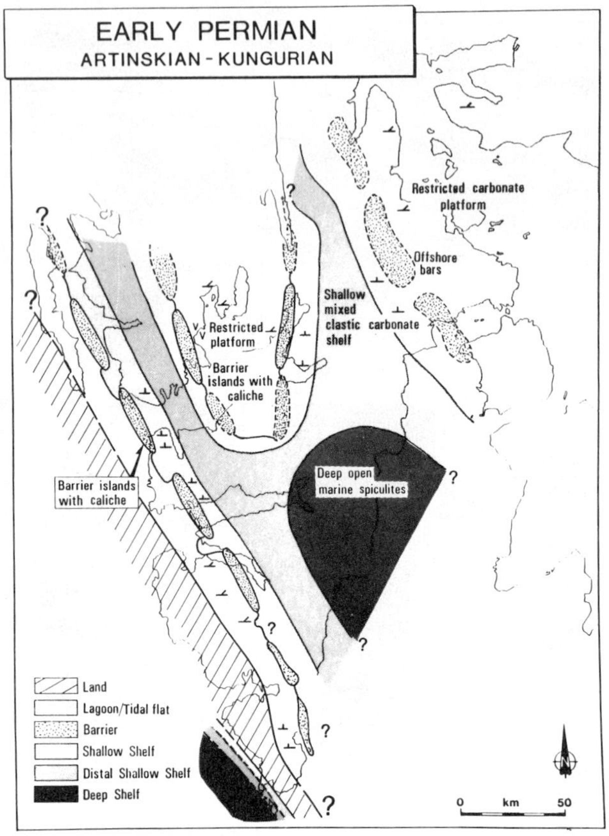

Fig. 16. Artinskian—Kungurian palaeogeography.

as organic (sponge) buildups. Shoals are most common on eastern platform areas (Figs. 15 and 17), whereas organic buildups tend to be concentrated along platform margins. Thin developments of these facies in trough areas may have developed at times of regional low sea-level stand. Sandstones are found only in the uppermost part of the sequence in northwestern exposure areas — these have a high glauconite content (<30%) and are heavily bioturbated; both features suggest relatively low rates of deposition in intermediate water depths.

Southwestern exposure areas of the Kapp Starostin Fm show a major coarsening-upwards sequence wedging onto the margins of the Sørkapp-Hornsund High (Fig. 15). Thinnest sequences immediately adjacent to the high show either fine-grained silty limestones or complex highly condensed sequences with both intra- and extraformational clasts. Both facies types rest on eroded or karstic surfaces of the Gipsdalen Gp and no sequences of the Kapp Starostin Fm have yet been found overlying Lower Carboniferous or earlier units on the high itself. The Tokrossøya Fm to the southwest of the high also shows a large-scale coarsening-upwards sequence, passing from spiculitic siltstones into sandstones interpreted as beach sands (Siedlecka, 1970) or peritidal sand waves (Hellem and Worsley, in press). The total development along both margins of the Sørkapp-Hornsund High suggests that the transgression which initiated the group's deposition may have submerged this structure, but subsequent uplift produced the coarsening-upwards sequences observed on both sides of the high. The sandstones of northwestern exposures may reflect similar lineamental movements rather than a general northern tilting of the entire depositional platform: note that youngest units in northeastern areas show a deep shelf spiculitic facies (Fig. 18).

The Miseryfjellet Fm of Bjørnøya consists of sandy limestones and well-sorted sandstones deposited in shallow well-agitated environments. The newly submerged structure which had de-

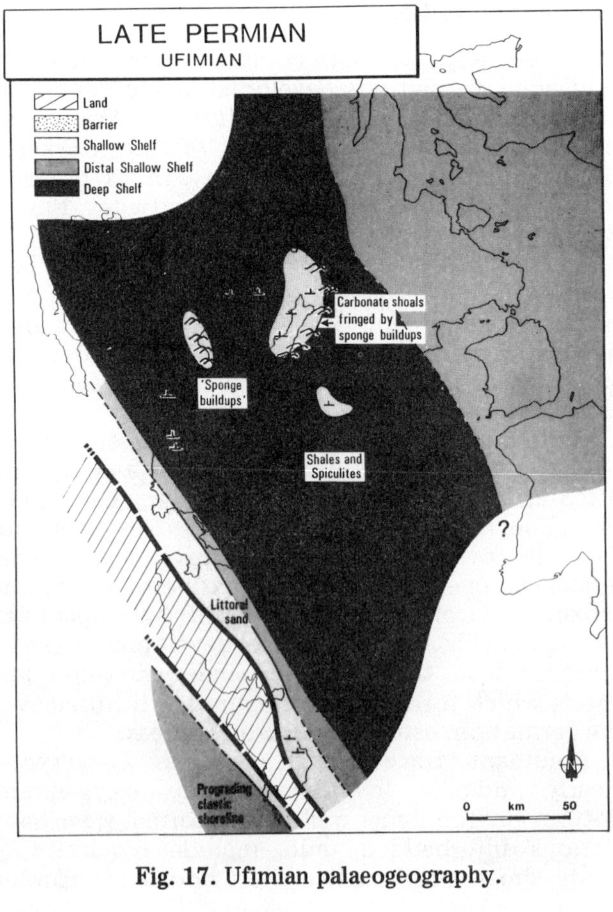

Fig. 17. Ufimian palaeogeography.

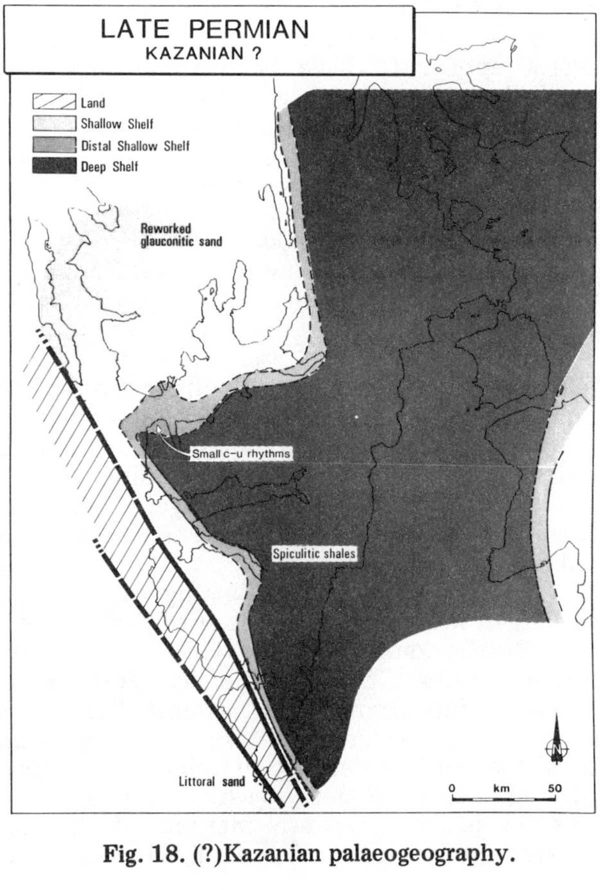

Fig. 18. (?)Kazanian palaeogeography.

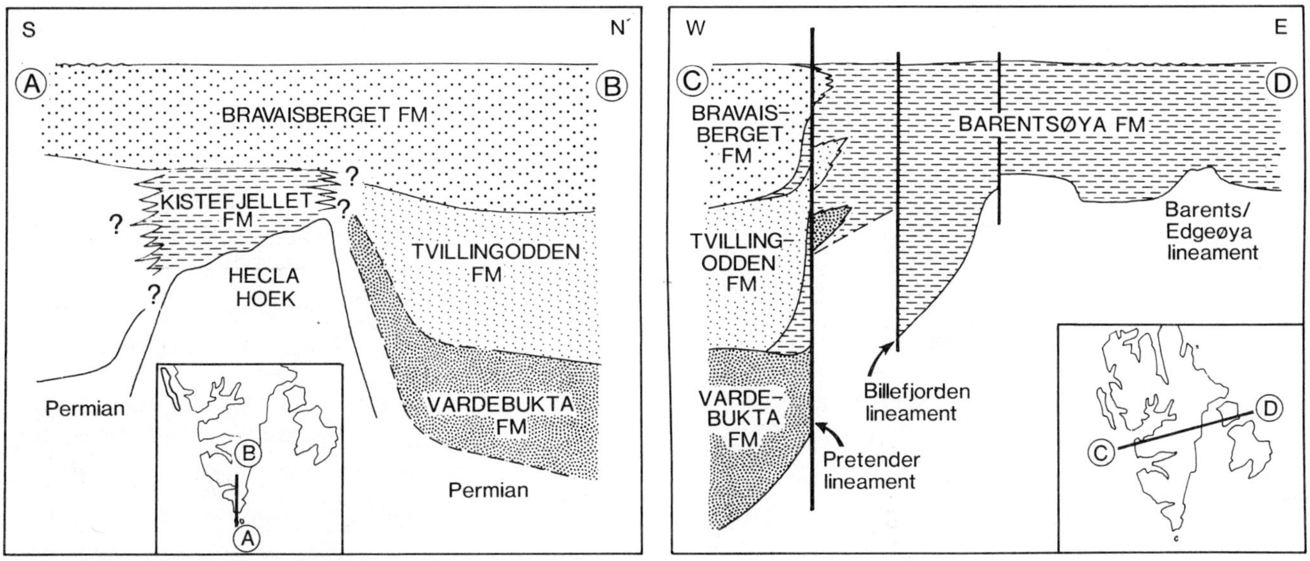

Fig. 19. Schematic cross-sections showing the development of the Sassendalen Gp.

veloped after the cessation of faulting activity in this area was evidently still a positive feature.

Although datings are uncertain and conflicting, the uppermost beds of the group throughout Svalbard are probably not younger than the Kazanian; the duration, cause and nature of the hiatus between Permian and Triassic deposits is, as in many other areas, poorly understood.

THE MESOZOIC DEPOSITIONAL PLATFORM

The Mesozoic development of Svalbard is characterized by genererally stable platform environments throughout the archipelago. There is little evidence of active faulting, except around the Jurassic—Cretaceous boundary — a period also marked by intrusive and volcanic activity. Major breaks in deposition are characteristic of the early Jurassic of western areas and of the late Cretaceous of the entire region. The clastic sequences of this depositional phase are dominated by marine shales; several major sandstone sequences occur, however, reflecting coastal and deltaic progradations from both the Palaeo-Hornsund Fault Zone and from northern and eastern uplifted areas.

Sassendalen Group

The group consists of marine shales, with subordinate siltstones and sandstones, ranging in age from Griesbachian to Ladinian. Appreciable thickness variations are closely related to downwarping over the sites of earlier active lineaments and a maximum thickness of 700 m is seen in the St. Jonsfjorden trough between the Sørkapp-Hornsund High to the southwest and the Nordfjorden Block to the east (Fig. 19). Western areas of Spitsbergen display a clear three-fold division, passing eastwards into the laterally correlative but finer grained Barentsøya Fm and southwestwards into an attenuated sequence on the Sørkapp-Hornsund High.

Basal shales and siltstones with Griesbachian faunas disconformably overlie siliceous siltstones and sandstones of the Tempelfjorden Gp, but an erosive contact is only rarely demonstrable. Basal conglomerates are only developed on the Sørkapp-Hornsund High; these overlie various Hecla Hoek and early Carboniferous units and mark the Dienerian submergence of this structure.

The three coarsening-upward sequences of the western exposure belt suggest a regionally transgressive regime interrupted by repeated coastal progradations caused by ongoing uplift, perhaps along the Palaeo-Hornsund Lineament. Sandstones in the (late Griesbachian) upper parts of the Vardebukta Fm around Isfjorden are interpreted as barrier systems with tidal inlets and are associated with lagoonal back-barrier units (Fig. 20). Less prominent sandstone units in southern localities suggest more open marine environments, while the Sørkapp-Hornsund High was either the site of very shallow shoals or was intermittently emergent throughout the Griesbachian. Transgression in the early Dienerian submerged the high completely and led to the development of finely-laminated dark shales in the lower parts of the Tvillingoden Fm in basinal areas. This formation's marked coarsening-upward sequences in southern areas terminate in delta-front sandstones. A major transgression in the Lower Anisian is marked by the black phosphatic shales of the lower parts of the Bravaisberget Fm. These again coarsen up through delta-front deposits to tidally influenced channels developed on a delta plain located on the Sørkapp-Hornsund High (Fig. 21).

Central and eastern areas of Svalbard show a more homogeneous shale development now assigned to the Barentsøya Fm. This formation has its thickest development (420 m) in the Billefjorden Trough, thinning gradually to about 200 m in eastern Edgeøya. Its three component members are general time equivalents to the formations of western areas; the two lowest members show varying contents of interbedded thin silts in the shales. The uppermost Botneheia Mbr shows a weak coarsening-upward trend from soft, organic-rich, shales to somewhat more silty beds. This uppermost member is rich in organic carbon (mean 5—7%) composed mainly of marine kerogens, and has an interesting source rock potential. Although there is no suggestion of large-scale sediment

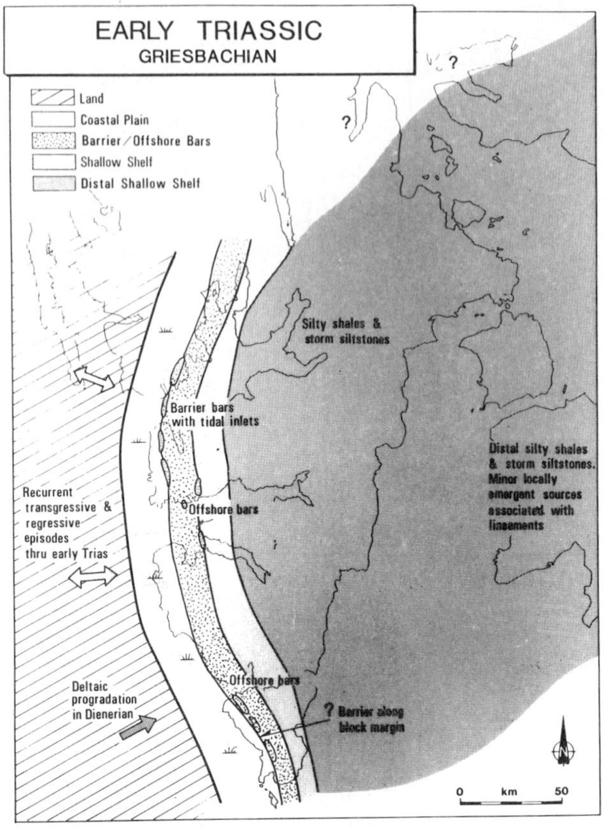

Fig. 20. Griesbachian palaeogeography.

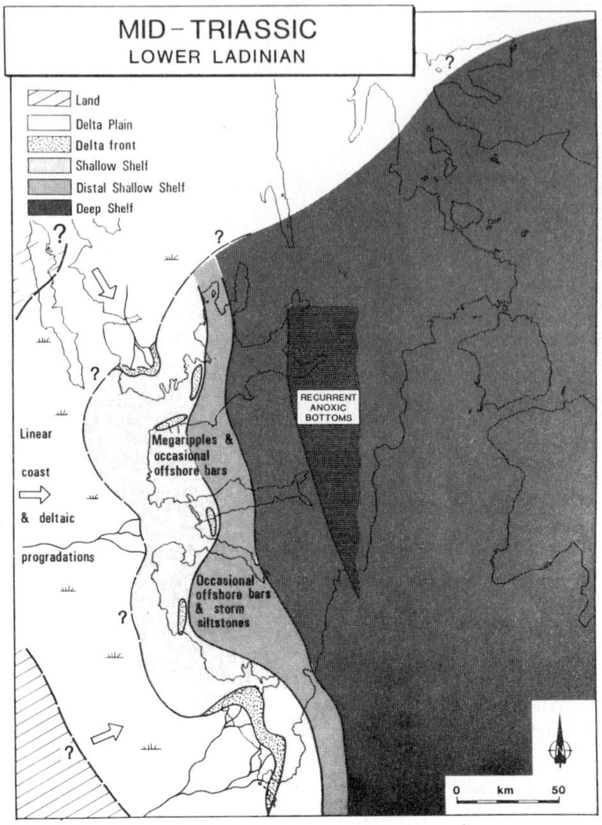

Fig. 21. Lower Ladinian palaeogeography.

transport from the east or northeast we note that organic geochemical studies show no simple W—E reduction in land-derived organic matter, as might be expected if the only source area was to the west.

Kapp Toscana Group

The junction between the Sassendalen and Kapp Toscana Gps may reflect a mid-Ladinian hiatus in many localities; present evidence suggests a period with low sedimentation rates which produced marked horizons with condensed faunal assemblages.

The deposits of the Kapp Toscana Gp represent two broadly regressive sequences; sedimentational patterns reflect an important change in depositional and tectonic regimes around the Norian/Rhaetian transition.

The Ladinian to Norian sequence of the combined (<550-m thick) Tschermakfjellet and De Geerdalen Fms forms an eastwards-thickening wedge, with significant local increases over the margins of the Sørkapp-Hornsund High and the Nordfjorden Block (Fig. 22). We note that although the De Geerdalen Fm's strata have earlier been

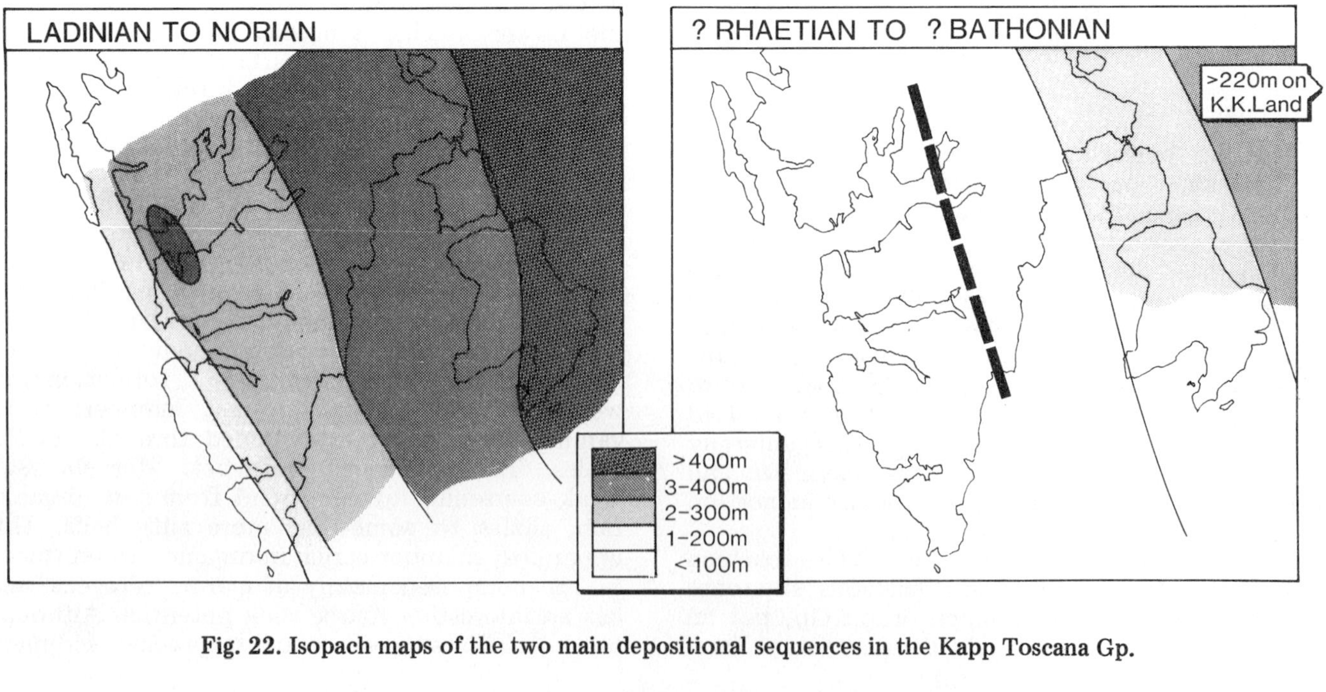

Fig. 22. Isopach maps of the two main depositional sequences in the Kapp Toscana Gp.

referred to as 'continental' or 'fresh-water' in origin, delta plain environments have only been detected in restricted parts of some northeastern and southwestern exposures. Deltaic influx clearly influenced late Triassic depositional regimes in many other areas, but most exposures demonstrate wave and storm reworking of deltaically introduced sediments in marine environments (Fig. 23).

The late Triassic sequences of western areas appear to represent a natural continuation of the coastal progradations seen in the underlying Sassendalen Gp. Fluvial dominated channels low in the sequence around Hornsund show that delta plains now extended further east and north than earlier. Channel sequences have not yet been found in northwestern exposures; repeated, small (5—15 m), coarsening-upward rhythms there suggest deposition in marine environments adjacent to major delta lobes. Present data suggest decreasing deltaic influence and more extensive marine reworking upwards in the west as eastern deltaic systems become dominant; maximal basin subsidence also shifted eastwards to Barentsøya, Edgeøya and Hopen during this depositional phase. Sequences in these eastern exposure areas show a progradation from prodelta shales into delta front deposits with wave-reworked mouth bar and barrier sands. Delta plain environments, with occasional thin coals in interdistributary areas, advanced westwards and southwestwards through the Carnian and Norian and at their maximum extent were developed over large areas of Barentsøya and Edgeøya (Lock *et al.*, 1978). Synsedimentary deformation, interpreted as growth faults in delta-front areas by Edwards (1976a), is now known from several localities on these islands. Eastern and central areas of Spitsbergen also show coarsening-upward sequences, but sandstone units there represent offshore bars reworked by wave and storm processes on a marine platform. Some of these sand bodies had temporarily emergent tops and the rhythmicity developed clearly reflects deltaic processes to the east; however, the entire development in these areas shows the overriding imprint of marine distribution and deposition of sediments introduced onto the platform. A major question, not yet resolved, is whether these sequences formed in a marine embayment closed to the north. On balance, present evidence suggests an open seaway between western and eastern source areas; this contrasts with the situation which developed in the Cretaceous and Tertiary, with uplifted northern land areas.

The uppermost part of the Kapp Toscana Gp in western Spitsbergen is represented by thin sequences which represent intermittent deposition between the Norian/Rhaetian and the mid-Jurassic. Thicker developments in eastern areas led Worsley (1973) to define the Wilhelmøya Fm, comprising a coarsening-upward unit developed east of the Billefjorden lineament (Fig. 24; see also Pcelina, 1980).

This interval has now been recognized in all exposure areas west of the Billefjorden lineament; thin (5—45 m) developments of sandstone and shale in central and northwestern areas represent widespread transgression in the Norian/Rhaetian and again in the Toarcian/Bajocian, with an intervening long period of non-deposition (although

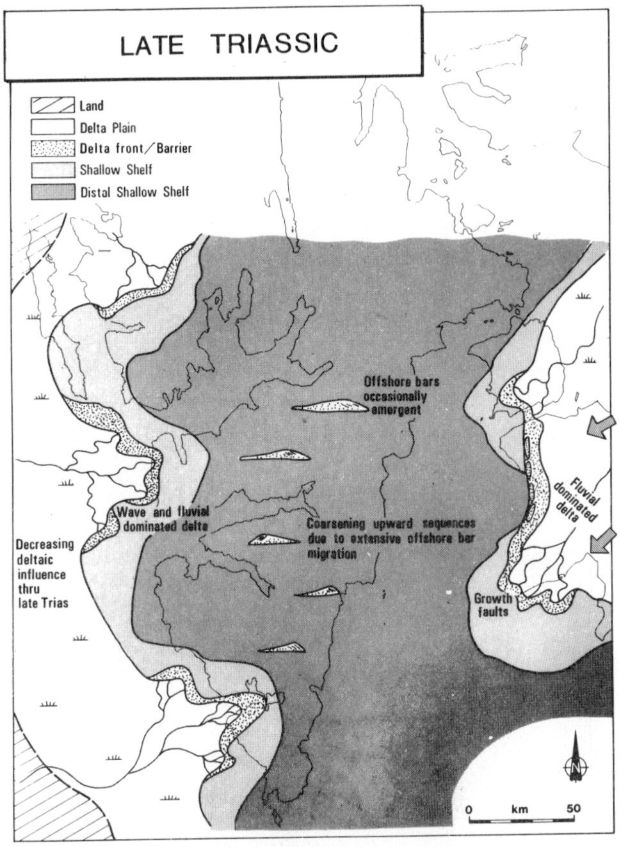

Fig. 23. Late Triassic palaeogeography, composited for the Carnian—Norian.

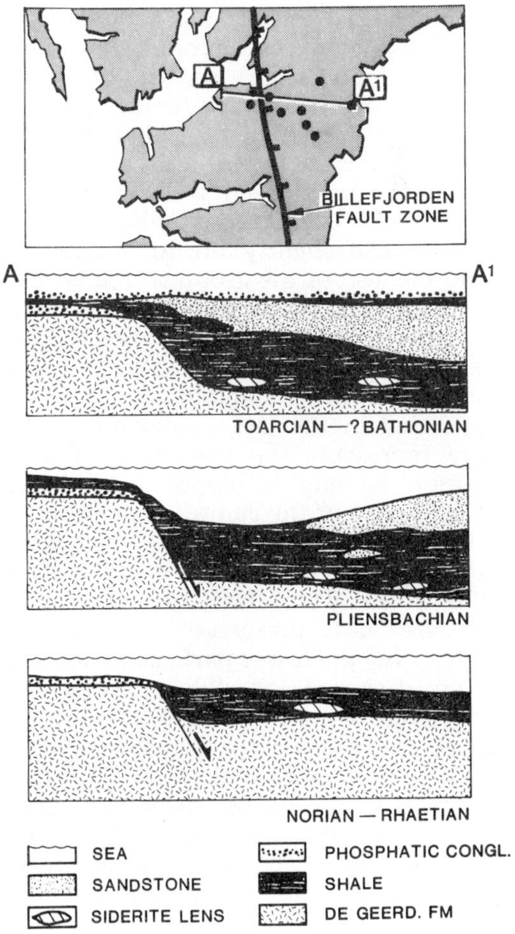

Fig. 24. Schematic cross-section showing the development of the Wilhelmøya Fm over the Billefjorden Lineament.

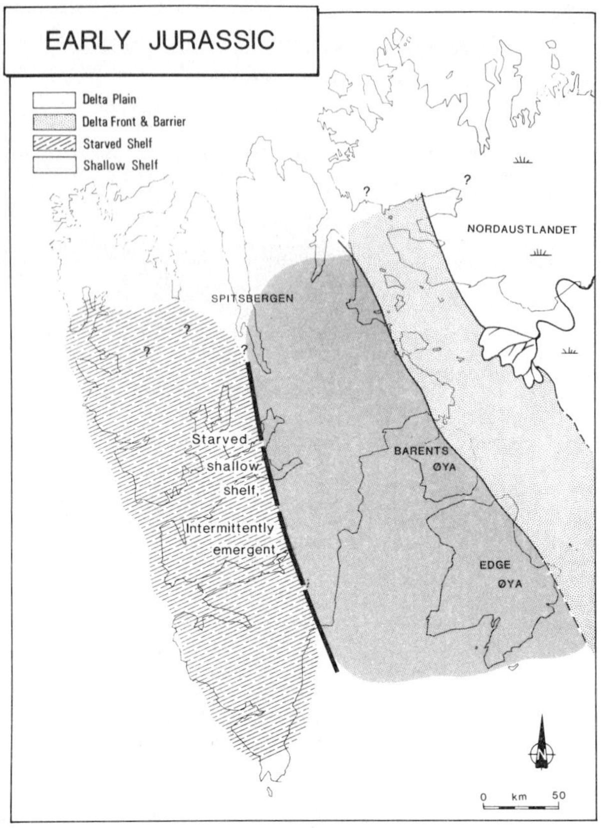

Fig. 25. Early Jurassic palaeogeography, composited Hettangian—Pliensbachian.

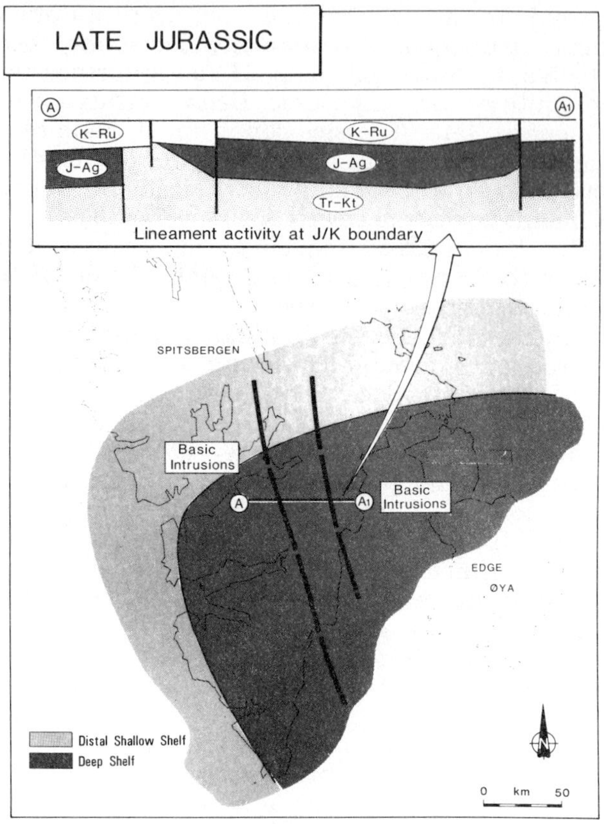

Fig. 26. Late Jurassic palaeogeography, composited for Callovian—Kimmeridgian. Abbreviations: Jurassic (J), Cretaceous (K), Triassic (Tr), Kapp Toscana (Kt), Agardhfjellet (Ag), Rurikfjellet (Ru).

highly condensed Pliensbachian sequences are also seen locally). Exposures on or near to the earlier Sørkapp-Hornsund High consist of well-sorted quartz sandstones with trace fossils and structures suggesting high energy depositional environments; frequent lag horizons and erosive surfaces also indicate highly condensed sedimentation in this area. In contrast to all older units, this sequence tends to thicken slightly on to the earlier horst structure. The top of the group is clearly marked in all areas by the phosphatic and quartzitic conglomerates of the Brentskardhaugen Bed. Phosphatic nodules contain remanié faunas of Toarcian/?Bajocian age, but the conglomerate itself was probably deposited in the late Bajocian or Bathonian. It clearly represents the beginning of a regional transgression leading to deposition of the overlying black shales of the Janusfjellet Fm.

Equivalent sequences east of the Billefjorden lineament thicken to 120 m on Wilhelmøya, and to >220 m on Kong Karls Land. Such young deposits are not preserved on Edgeøya and Barentsøya, but the lower parts of a similar unit on Hopen are 135-m thick. When traced eastwards the succession thickens, contains fewer breaks and takes on a more proximal character. The section on Wilhelmøya coarsens upwards from marine shales to mouth-bar or coastal sands while Kong Karls Land shows a development from marine shales through barrier sands into back-barrier lagoons (Worsley and Heintz, 1977; Løfaldli and Nagy, 1980). Thus, while western areas of Spitsbergen developed as a positive block only intermittently transgressed at times of high sea-level, eastern areas were the site of more continuous sedimentation with renewed deltaic progradation from the northeast (Fig. 25). There is a marked mineralogical contrast between the immature lithic wackestones of the De Geerdalen Fm and the mature quartz sandstones of the Wilhelmøya Fm. Regional and stratigraphical evidence suggests that this reflects lower rates of deposition and consequent more extensive reworking of the younger unit's sands, rather than palaeoclimatic variation or changing source areas. Although the Brentskardhaugen Bed is also seen in areas east of the Billefjorden lineament, more continuous deposition eastwards led to this unit becoming less well marked and it has not been found on Kong Karls Land.

Adventdalen Group

The group comprises three superimposed, but diachronous, formational units: open marine shales (<800 m) of the Janusfjellet Fm (Callovian to Hauterivian); deltaic sandstones (<150 m) of the Helvetiafjellet Fm (Barremian to ?Aptian); mixed marine sequences (<850 m) of the Carolinefjellet Fm (Aptian to Albian).

Barremian alluvial influx was a response to regional uplift of Spitsbergen, particularly in the north; thickening of alluvium along the west of the basin and local sediment dispersal from the same direction suggest, however, that uplands along the line of the incipient Hornsund Fault Zone were also source areas. Terrigenous input was evidently great, but it was outpaced both by basinal subsidence and by a eustatically rising sea-level, so that the alluvial-deltaic system and its distal equivalents progressively onlapped older strata northwards.

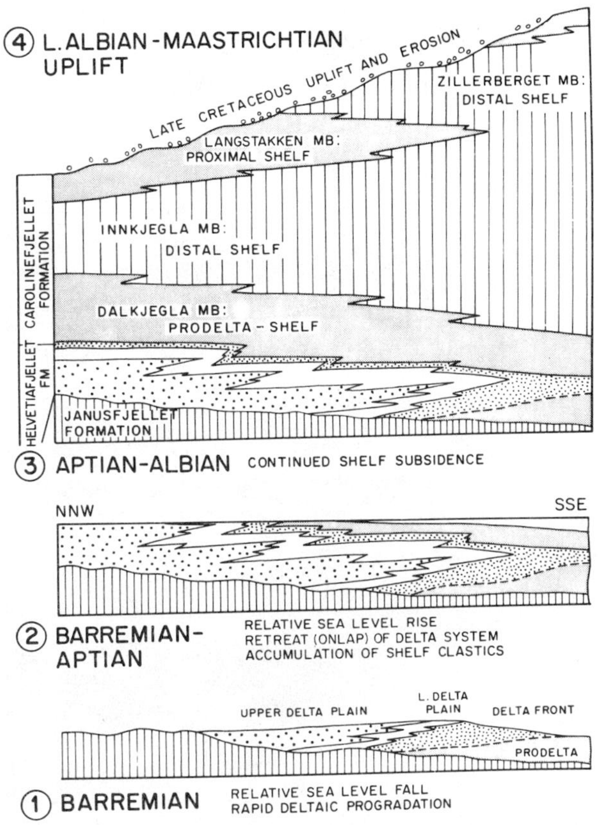

Fig. 27. Schematic cross-sections showing the early Cretaceous development of Spitsbergen on a north to south traverse.

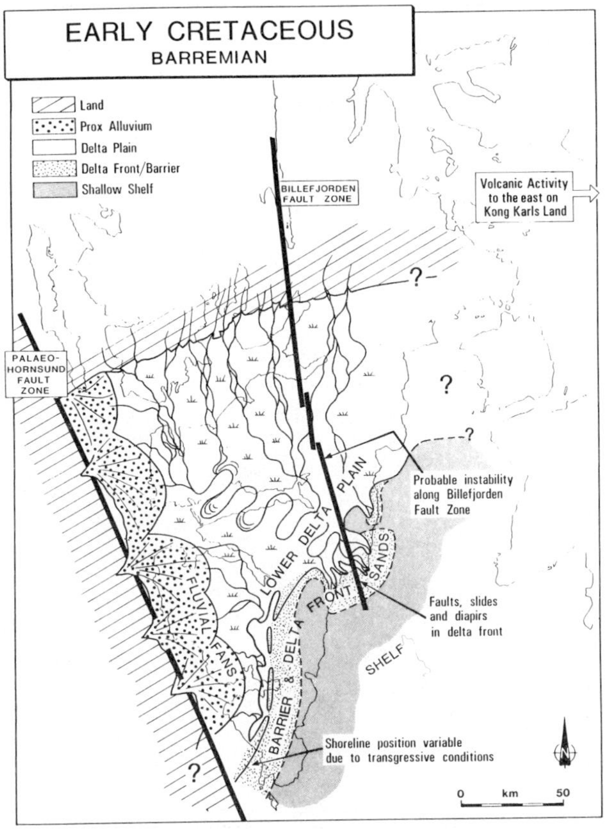

Fig. 28. Barremian palaeogeography.

Apart from these large-scale trends, local tectonic unrest is otherwise indicated by: thickness changes in the Janusfjellet Fm's shales over the Billefjorden and Agardhbukta Fault Zones; synsedimentary faulting and sliding in the Barremian delta front sequences adjacent to the Billefjorden lineament at Kvalvågen; and Barremian to ?Aptian volcanic activity in eastern Svalbard (notably on Kong Karls Land), as well as dolerite sill intrusions in the late Jurassic and early Cretaceous throughout the region.

The Janusfjellet Fm comprises variable thicknesses of dark shales and mudstones; thickest developments (<800 m) are seen along the western basin margins. Thinner sequences in central and eastern areas show marked thickness variations over the Agardhbukta and Billefjorden lineaments. The dark shales of the Agardhfjellet Mbr (Callovian to Volgian) suggest continuing transgression and the development of deep and often anoxic shelf environments. Organic carbon contents achieve maxima of 12% in Callovian paper shales (Dypvik, in press) but otherwise vary between 1% and 4%. Increasing interbeds of bioturbated sandstone to the west and north, especially in the lower parts of the member, prompt the reconstruction of Fig. 26. Differential subsidence along the Billefjorden and Agardhbukta lineaments in the late Jurassic culminated in marked movements in the earliest Cretaceous (Parker, 1966). A contemporaneous minor break in deposition throughout central and eastern Spitsbergen was followed by renewed fine clastic sedimentation; the bioturbated mudstones of the Rurikfjellet Mbr (Berriasian to Hauterivian) are enriched in volcanogenic debris and have a higher silt and lower organic carbon content than the underlying shales. The Rurikfjellet Mbr of central areas shows a weak upward coarsening sequence, often not detectable in the field, and prodelta shales may be sharply and directly overlain by deltaic sand (Fig. 27). A different situation is seen in the south, where deposition appears to have been continuous through the late Jurassic to early Cretaceous and gradual transitions from prodelta through delta-front to delta plain sequences are well developed (Edwards, 1976b). This geographic variation is further discussed below.

The non-marine (albeit coastal) nature of much of the Helvatiafjellet Fm has long been known, not least because it is coal-bearing. Analyses of the unit show that it contains fluvial, bay, mouth-bar, wave-dominated shoreline and shelf elements (Fig. 28) and that a deltaic rather than purely alluvial setting is likely (Steel, 1977; Steel *et al.*, 1978). Spectacular examples of rotational block faults, slides, and mudflows are seen in the distributary channel and delta front deposits of the unit on coastal cliffs north of Kvalvågen.

Evidence that the Helvetiafjellet Fm sandstone belt is diachronous, younging northwards includes: the lack of significant lateral facies changes southeastwards towards the palaeocoast, over a distance of more than 200 km; alluvium is no finer distally (Kvalvågen—Agardhbukta) than it is some 150 km to the northwest in exposures near to source areas; marine sandstones are more common in northwestern than southeastern exposures.

These anomalies can be best accounted for by postulating a diachronous sandbody complex, where alluvial facies always pass southeastwards, along time lines, into marine facies within less than 100 km. A gradual northwestwards retreat of both alluvial and coastal tracts, onlapping the Janusfellet

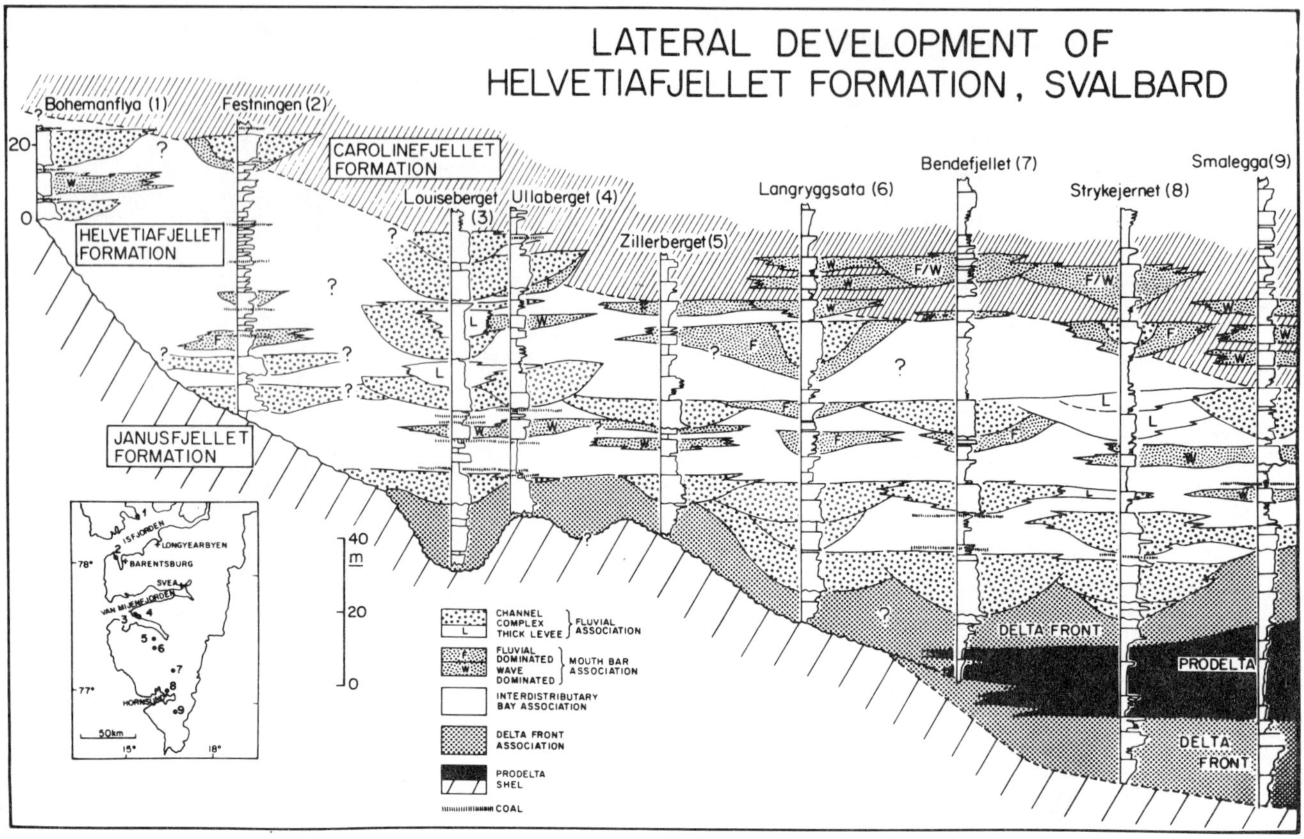

Fig. 29. Lateral development of the Helvetiafjellet Fm, suggesting the northwards-younging diachronous model presented herein.

Fm surface possibly through both Barremian and Aptian time, caused the present great lateral extent of the sandbody. Independent evidence of an important eustatic sea-level rise in early Cretaceous times (Vail *et al.*, 1977) is consistent with the proposed transgressive nature of the succession on Spitsbergen, though we believe that most of the relative sea-level rise may be accounted for by subsidence.

We have already noted the important contrasts in early Cretaceous development in the region north of Van Keulenfjorden as compared with that in the south. In the north, fluvial sediments commonly overlie the Janusfjellet Fm shelf deposits abruptly, whereas in the south there is a transition through thick marine sandstones (delta front/barrier) and lagoonal shales (Fig. 29). This contrast may reflect a general downcutting of the fluvial systems on the delta plain as well as the erosive and diachronous nature of the sandstone in the north. Alternatively, it may simply be due to a northwestwards change in the coastal configuration, from a wave-dominated regime in the south to a more strongly fluvial-dominated deltaic regime in the north.

Up to 850 m of alternating sandstones, siltstones and shales of the Carolinefjellet Fm overlie (and are probably partly laterally equivalent to) the coalbearing deltaic sequence. The overall environmental interpretation is of shelf sedimentation, although three sandier levels in the succession suggest lower delta front facies. The latter are commonly organized into upward coarsening sequences (<15-m thick) representing delta lobe progradation (Steel, 1977), although abandonment (carbonate-rich) and transgressive (upward fining) sequences have also been identified (H. Ramberg-Moe, personal communication, 1980). Overall wave-domination and the importance of storm waves and storm-induced bottom currents on the shelf is emphasized by the abundance of wave-generated ripple lamination and of hummocky cross-stratification in the Carolinefjellet Fm.

In the speculative palaeogeographic reconstruction of Fig. 30, the coastline, compared with its Barremian position, has receded northwards, and there is evidence for the importance of a northeasterly provenance in addition to some sediment input from the northwest.

Late Cretaceous Uplift

Spitsbergen was uplifted and subject to erosion in late Cretaceous times, preferentially in the north because erosion cuts deeper into Carolinefjellet Fm strata there (Fig. 27). Lower Palaeocene strata thus overlie Albian strata in southeastern areas and Aptian strata in the northwest of the basin (Nagy, 1970). This uplift of the northwestern edge of the Barents Shelf was probably part of a doming related to development of the Arctic Basin to the north, and was also an immediate precursor to large-scale transtensional tectonics along the western margins of Svalbard.

THE TERTIARY RIFT BASINS

The Tertiary basins of Svalbard developed during considerable tectonic upheaval, in contrast to the relative quiet of Mesozoic conditions. Detailed

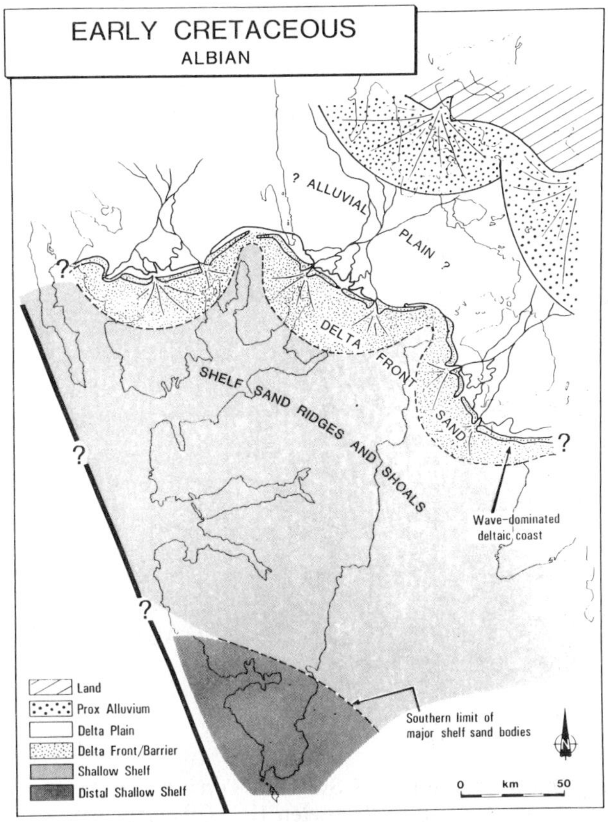

Fig. 30. Albian palaeogeography.

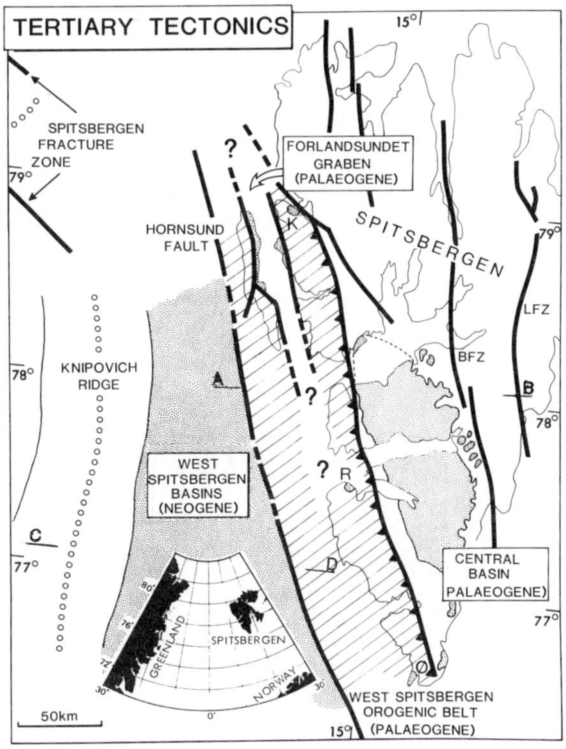

Fig. 31. Tertiary tectonic framework and sedimentary basins. Small Tertiary exposures in Kongsjorden (K), Reinardodden (R) and Øyrlandet (Ø) are indicated. The Billefjorden (BFZ) and Lomfjorden–Agardhbukta (LFZ) Lineaments are also shown. The rift basins west of the Hornsund Fault are mainly of Neogene age, but probably originated in late Palaeogene times.

interpretations of Tertiary sedimentation are critical to an understanding of the development of the oblique-slip mobile belt in West Spitsbergen and of the later continental margin, discussed in many papers (Harland, 1969; Birkenmajer, 1972; Lowell, 1972; Kellogg, 1975; Steel *et al.*, 1981). Figure 31 shows the three main basins of Tertiary sedimentation in relation to the West Spitsbergen Orogenic Belt. The Central Basin (Van Mijenfjorden Gp and probably also the Forlandsundet Graben (Forlandsundet Gp) are of Palaeogene age. The former probably originated somewhat earlier than the latter, but both are a consequence of oblique-slip movement along the orogenic belt. The larger basinal area west of the fold belt is probably mainly Neogene in age and developed after the change from sheared to rifted margins around 36 Ma (Talwani and Eldholm, 1977).

Van Mijenfjorden Group

The Van Mijenfjorden Gp is subdivided schematically as shown in Fig. 32. The succession is generally accepted as being of Palaeocene–Eocene age (Harland, 1969), though Livsic (1965) has suggested that it may, in part, be Oligocene. The latter is now less likely in view of a late Palaeocene dating for the upper regressive sequence (Manum and Throndsen, 1978).

The Central Basin succession thickens from northeast (1.5 km) to southwest (2.5 km), reflecting mainly a Palaeocene basin asymmetry towards the sheared margin (Fig. 32). The original thickness of the Palaeogene pile was <1.5 km greater, as judged by vitrinite reflectance data (Manum and Throndsen, 1978).

The early Palaeocene development of the Central Basin is shown in Fig. 34. Sedimentation took place in (and possibly, in part, prior to) a tensional tectonic setting on the flank of the shear zone between Greenland and the Barents Platform. That this early setting was tensional (superimposed on the regional 'shear' regime) is suggested by: the marked thickness asymmetry of the basinfill towards the west; the overall transgressive nature of the succession from coaly sequences up through wave-dominated deltaic sequences to shelf and prodelta sequences; the scarcity of sediment supply from the shear zone into the basin.

The last observation reflects a lack of compression (uplift) along the sheared margin, while the former two probably indicate that rift-floor subsidence outpaced terrigenous influx. The transgressive succession is too thick to be explained merely by eustatic sea-level rise, though this factor may also have been important.

Figure 33 illustrates the main features of early Palaeocene deltaic sedimentation. A basal estuarine system and three overlying regressive deltaic units (Todalen Mbr) are of fluvial/tide-dominated character and have been discussed by Steel *et al.* (1981) and Nøttvedt (1981). It is from those sequences that coal is presently taken by Norwegian and Soviet mines at Svea, Longyearbyen and Barentsburg. The main reason for abundant coal at these lowest levels in the succession is evident from Fig. 33. The early deltas in the north were pro-

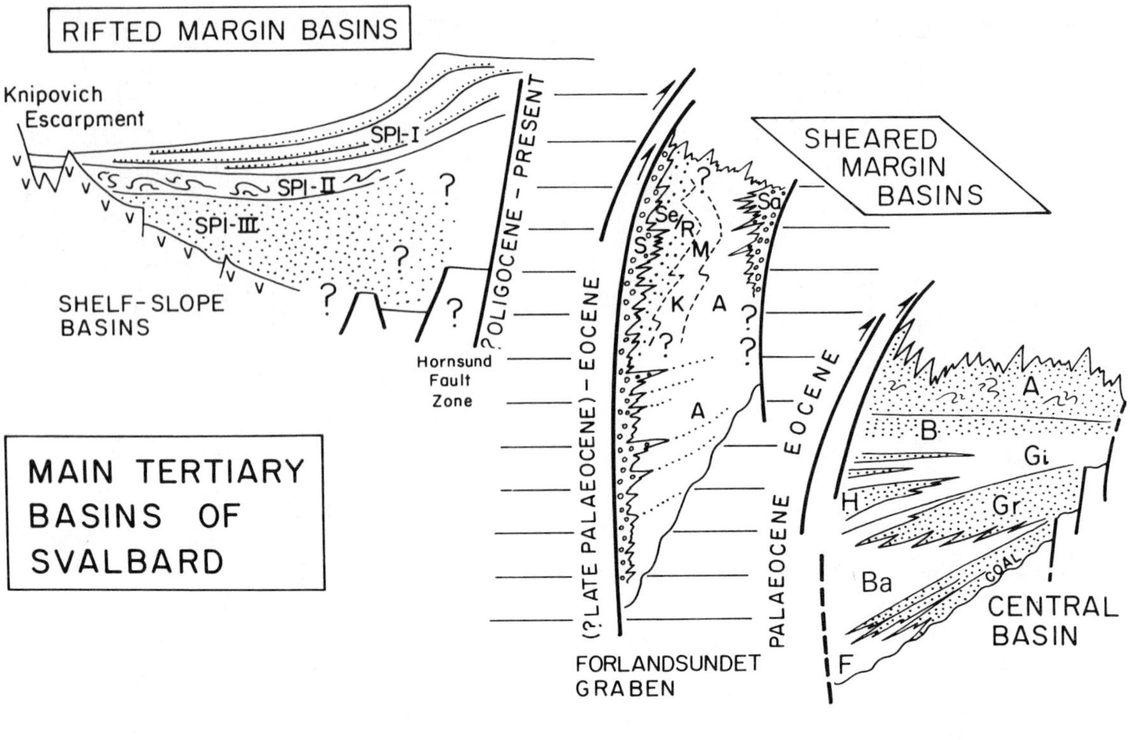

Fig. 32. Schematic cross-sections (not to scale) showing the development of Svalbard Tertiary basins. Central Basin: the Firkanten (F), Basilika (Ba), Grumantbyen (Gr), Hollenderdalen (H), Gilsonryggen (Gi), Battfjellet (B) and Aspelintoppen (A) Fms.

Forlandsundet Graben: the Aberdeenflya (A), Marchaise Lagune (M), Krokodillen (K), Reinhardpynten (R), Sesshøgda (Se), Selvågen (S) and Sarstangen/Sarsbukta (Sa) Fms. Marginal rift basin: sequences SPI—III to SPI—I represent depositional sequences suggested by seismic studies of Schlüter and Hinz (1978).

tected in a semi-enclosed embayment by the high-standing edge of a southern block. Later drowning of this barrier changed the coastal configuration and at least three more sand sequences, now wave-dominated (Endalen—Kalthoffbergen Mbrs), developed. The more open coastal regime was not conducive to coal development as there were no longer extensive delta plains. The early Palaeocene development is concluded by the prodelta-shelf siltstones and sandstones of the Basilika Fm, marking high sea-level stand and the culmination of transgressive (transtensional) conditions in the basin.

Late Palaeocene sedimentation in the Central Basin is less well understood as yet, but was dominated by a major regressive pulse of sedimentation (Kellogg, 1975; Steel *et al.*, 1981). This segment of the basin infill is also markedly asymmetric, thickening from northeast (<200 m) to southwest (>500 m), away from source areas and towards the shear zone. The sediments reflect shallow marine environments in the northwest (Grumantbyen Fm) and give way southwestwards to offshore/shelf siltstones and shales of the Basilika Fm (Fig. 32). During this phase of basin development rift subsidence no longer outpaced sediment influx. However, sediment influx is judged to have been low, as recorded by unusually intense bioturbation throughout this succession. This implies decreased subsidence rates in the basin, probably heralding the latest Palaeocene—Eocene transpressive conditions.

Eocene sedimentation in the Central Basin was dominated by a transpressive tectonic regime along the western margin. Signs of this type of setting are:

(1) the regressive nature of the entire latest Palaeocene—Eocene succession, and an upwards coarsening succession more than 1.5 km thick in places;

(2) influx of sediment from the *west* (already in late Palaeocene times) implying western uplift and a major shift in drainage patterns into the basin;

(3) unusual abundance and intensity of syn-sedimentary deformation in the uppermost 1 km of strata, probably earthquake-induced.

The lower 200—400 m of the succession consist of black shales and occasional siltstones of the Gilsonryggen Fm with a wedge-shaped deltaic sequence at its base (Dalland, 1977; Steel *et al.*, 1981). Intercalations of turbidites, slump sheets and occasional conglomerates formed on a fringe of easterly directed submarine fans (Fig. 35) occur increasingly upwards in the succession. Upwards coarsening is most marked by the incoming of barrier and wave-dominated deltaic sand sheets of the Battfjellet Fm, representing a prograding coastline. The marine sequence is then capped by the very thick succession of coastal and deltaic plain deposits of the Aspelintoppen Fm which show unusually intense signs of liquefaction and soft sediment deformation.

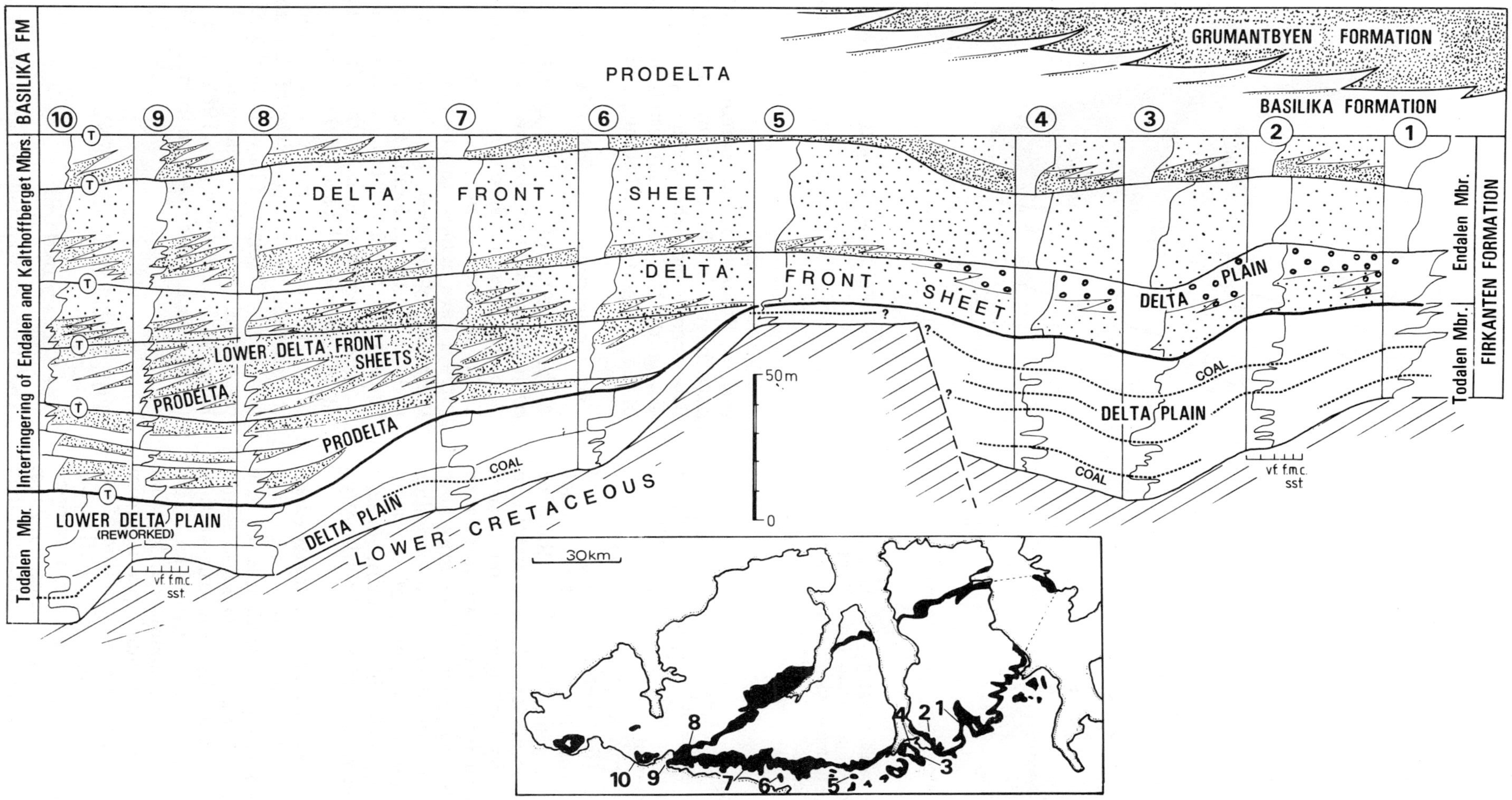

Fig. 33. Cross-section showing the tectonic control on coal development in the lower part of Firkanten Fm, in contrast to the later, more open, wave-dominated conditions (from Steel *et al.*, 1981).

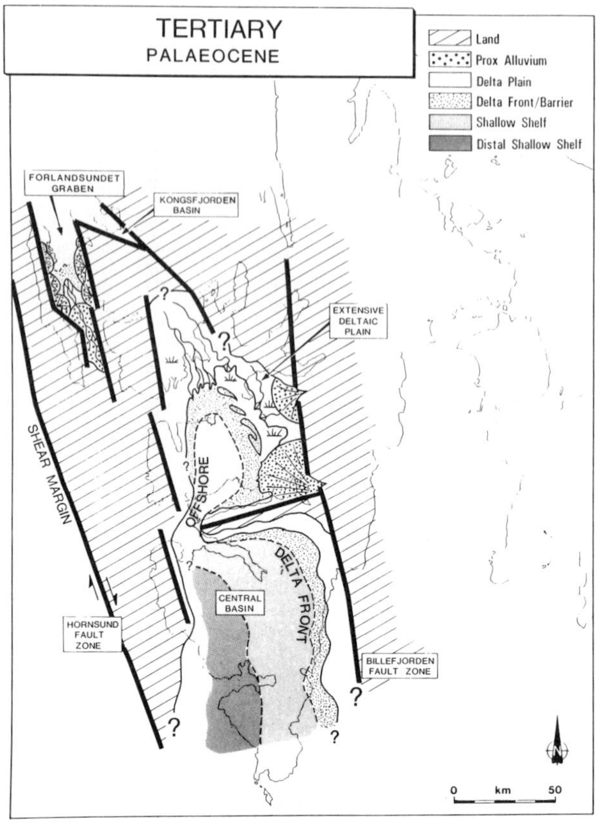

Fig. 34. Palaeocene palaeogeography. Forlandsundet Graben probably developed from latest Palaeocene times. The age of the Kongsfjorden Basin is uncertain.

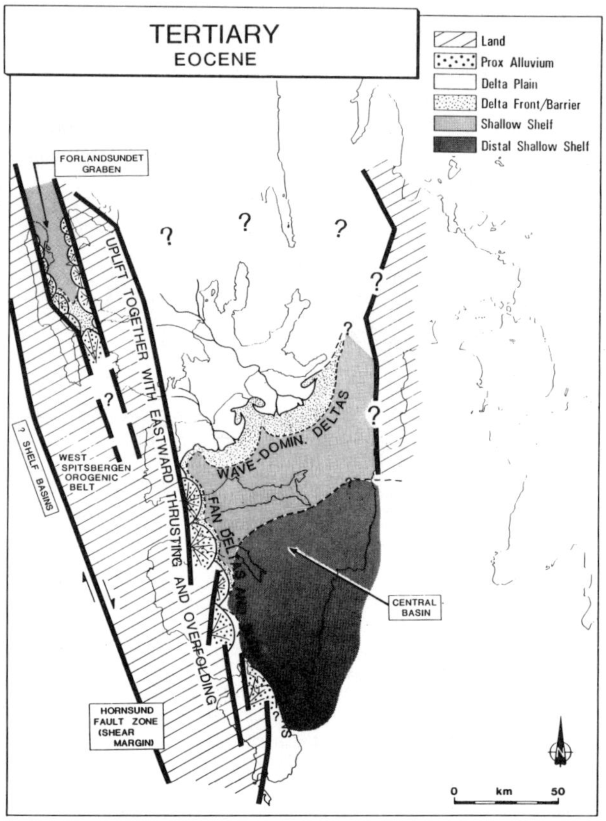

Fig. 35. Eocene palaeogeography.

Forlandsundet Group

Forlandsundet Graben is an elongate (<50 km), but narrow (<20 km) trough which is entrenched in the West Spitsbergen Fold Belt (Fig. 31). It displays a sequence with a stratigraphic thickness of more than 5 km (Rye-Larsen, 1982). Marine geophysical results from southern Forlandsundet suggest a thickness of some 3 km there (Sundvor *et al.*, 1977). Along both sides of the trough there are alluvial fan deposits (Selvågen, Sarsbukta and Sarstangen Fms). These interfinger basinwards with fan delta deposits (Sesshøgda and Reinhardpynten Fms) and other nearshore deposits (Krokodillen and Marchaise Lagune Fms). Much of the axial region of the basin, especially in the north, consists of submarine fan deposits of the Aberdeenflya Fm (Fig. 32).

Sedimentation patterns, trough asymmetry (deep to west), present structural attitude and the apparently great stratigraphic thickness of the succession suggest that the graben migrated southwards during infilling, and that the true vertical thickness at any point is less than 3 km. This evidence from the sediments themselves, together with evidence of NE-trending faults splaying from the western master fault and of deformation of the Tertiary deposits by NW-trending folds (Atkinson, 1962), strongly suggest that the graben developed and was infilled in an oblique-slip tectonic regime. We therefore suggest that Forlandsundet Graben formed somewhat earlier than has been previously assumed (Harland, 1969; Kellogg, 1975) and was probably a feature of the sheared margin rather than of the later rifted margin. Infilling is likely to have begun in the north of the graben in latest Palaeocene times, though development was mainly though the Eocene, as suggested by tentative dating of the youngest strata (Manum, 1962; Livsic, 1965).

Tertiary deposits west of Spitsbergen

A thick wedge of Tertiary deposits (Fig. 32) has been shown to underlie the continental shelf and slope off western Spitsbergen (Malod and Mascle, 1975; Sundvor *et al.*, 1977), while more recent surveys have defined this as a complex basinal area lying largely between the Knipovich Escarpment and the Hornsund Fault (Fig. 31). (Schlüter and Hinz, 1978; Sundvor *et al.*, 1979). Although there is general agreement as to the Plio-Pleistocene age of the uppermost part of this more than 6—7 km thick succession, there continues some debate as to whether the lower parts of the sequence are also largely Neogene (Talwani and Eldholm, 1977) or extend into the Palaeogene (Schlüter and Hinz, 1978). Steel *et al.* (1981) and Myhre *et al.* (1982) discuss this question in more detail.

THE TERTIARY OROGENY

The Tertiary fold belt

The Tertiary fold belt extends along the western coast of Spitsbergen for some 300 km, usually in a belt less than 50 km wide (Fig. 31). The first clear suggestions that this orogenic belt related to a

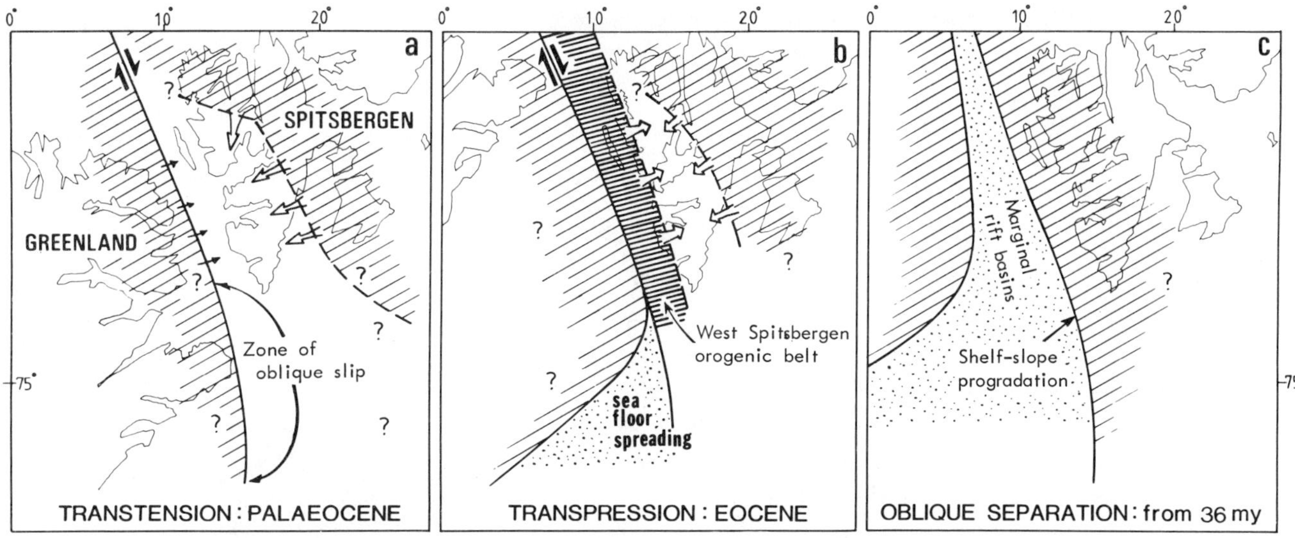

Fig. 36. **The Tertiary evolution of the Norwegian-Greenland Sea (from Steel *et al.*, 1981).**

major continental transform fault and to the opening of the Norwegian-Greenland Sea were made by Harland (1969), while Lowell (1972) and Kellogg (1975) elaborated upon details of the oblique-slip motion and associated tension and compression. The fold belt occurs along and adjacent to the steep western limb of the broad asymmetric syncline which presently contains Tertiary outcrops. Folds are generally strongly asymmetric and associated with major overthrusts in the southern and northern regions, all symptoms of eastwards directed compression. Signs of uplift and compression die out rapidly eastwards, so that deformation in central and eastern Spitsbergen is gentle, and often restricted to minor folds, faults or local thrusts associated with pre-existing major lineaments.

Sedimentation and tectonics: sequence of events

The following sequence of events summarizes the development of the oblique-slip mobile belt in western Spitsbergen, and its eventual transformation to a passive, rifted margin (Fig. 36).

Early—mid Palaeocene

Regional tension (probably transtension by mid-Palaeocene) and the initial development of a major continental transform fault zone; Central Basin subsidence leading to deltaic sedimentation in an overall transgressive setting; some of the major normal faults (e.g. those preserving Carboniferous strata along the west coast) may have originated at this time.

Late Palaeocene—Eocene

Growing compression (transpression) along the embryonic fold belt, uplift in the west and drainage reversal into the Central Basin, leading to the development of a megasequence of regressive character. Forlandsundet Graben probably originated at this time as an elongate, (southerly migrating) transpressional trough within the mobile belt.

Late Eocene—(?) early Oligocene

Climax of transpression (maximum impingement of Greenland and Svalbard) causing extensive overthrusting in the west; probable continued sedimentation in Central Basin, and minor folding.

Early Oligocene

Major change in tectonic regime; growing tension along western Spitsbergen with thinning of continental crust and location of spreading axis in Greenland Sea; sea-floor spreading and new oceanic crust formed west of Hornsund Fault; initial development of rifted margin and possible deposition of lower parts of sequence SPI—III of Schlüter and Hintz (1978) in grabens west of Hornsund Fault.

Mid Oligocene—Miocene

Increasing uplift of northwestern part of Barents block (Spitsbergen) and terrigenous influx (SPI—III to II, Fig. 32) to the new marginal basin.

Early Pliocene—Present

Eastwards shift of spreading axis in Greenland Sea to just beyond the Svalbard continental slope, 5—6 Ma (Eldholm and Sundvor, 1979), causing damming and overspill of terrigenous input from uplifted Svalbard (SPI—I, Fig. 32).

DISCUSSION

The main aim of this brief review has been to outline the status of present knowledge of the evolving depositional regimes reflected by the post-Caledonian strata of Svalbard. Our maps, figures and explanatory texts make a summary redundant; it is more relevant at this stage to note the limitations and perspectives of our work to date.

In spite of considerable research activity on Svalbard in recent years, much work remains to be done to test our models. The lack of precise biostratigraphical control on many units is a serious limitation to a full palaeogeographical application of many of the sedimentological analyses carried out to date. Reconstructions of the late Palaeozoic and Mesozoic evolution of western exposure areas are also still at an early

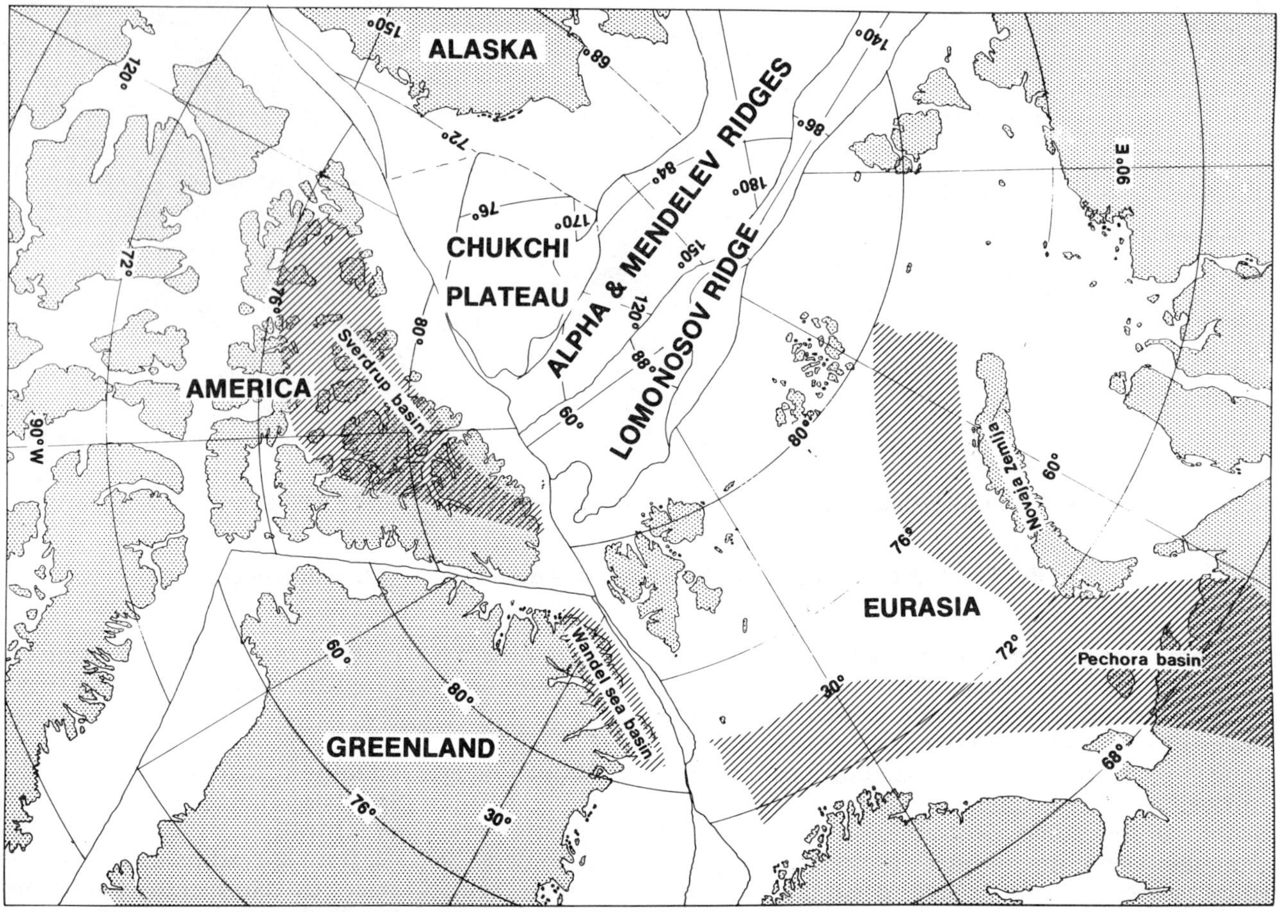

Fig. 37. Probable geographic relationships of the Sverdrup, Wandel Sea and Pechora Basins in relation to the Svalbard Platform during the late Palaeozoic and early Mesozoic.

stage; further work will demand extensive interdisciplinary co-operation in the fields of biostratigraphy, sedimentology and structural geology to establish the primary relations of the clearly parautochthonous sequences there. We hope, at least, that our interpretations will encourage a dynamic approach to continued investigations, not least to enable more meaningful extrapolations on to the Barents Shelf and adjacent northern Norwegian Shelf areas.

The Svalbard Archipelago's critical position near the northwestern margins of the Eurasian Plate is clear. Increased understanding of the dynamic stratigraphy of Svalbard's post-Caledonian strata and relations to the Pechora, Wandel Sea and Sverdrup Basins (Fig. 37) are essential for meaningful reconstructions of the evolution of both the Norwegian-Greenland Sea and the present Arctic Ocean. Comparisons of the sedimentational patterns displayed by these different basins will significantly augment present knowledge of the evolution of Arctic land, shelf and oceanic areas.

ACKNOWLEDGEMENTS

This synthesis has been made possible by the enthusiasic and inspiring participation of 30 research students in the co-operative research programme which we led from the Universities of Bergen and Oslo between 1975 and 1982. Individual thesis titles produced in the course of this programme are listed in Appendix I. We also gratefully acknowledge the active participation in project work by our University colleagues and co-operation in our studies with geologists from Statoil, the Continental Shelf Institute, Trondheim and the Norwegian Polar Institute, Oslo. Financial and logistical support to our programmes has been given by Statoil, SNSK A/S, Norwegian Petroleum Directorate, NTNF (Norwegian Council for Scientific and Industrial Research), NAVF (Norwegian Research Council for Science and the Humanities), Store Norske Spitsbergen Kullkompani A/S and Norwegian Polar Institute. The final preparation of the atlas was made possible by support from Norsk Hydro and Statoil and, not least, by the expert draughtmanship of Masaoki Adachi.

REFERENCES

Atkinson, D.J. 1962. Tectonic control of sedimentation and the interpretation of sediment alternation in the Tertiary of Prince Charles Foreland, Spitsbergen. *Bull. Geol. Soc. Am.*, 73, 343–364.

Birkenmajer, K. 1972. Tertiary history of Spitsbergen and continental drift. *Acta Geol. Polon.*, 22, 193–218.

Birkenmajer, K. 1981. The Geology of Svalbard, the western part of the Barents Sea and the continental margin of Scandinavia. In: A.E.M. Nairn, M. Churkin Jr., and F.G. Stehli (eds.), *The Ocean Basins and Margins, Vol. 5 The Arctic Ocean* Plenum Press, New York, pp.265–329.

Bjærke, T. 1978. Mesozoic palynology of Svalbard III. Dinoflagellates from the Rurikfjellet Mbr, Janusfjellet Fm (Lower Cretaceous) of Spitsbergen. *Palinol., num. extraord.*, 1, 69–93.

Bjærke, T. 1980. Mesozoic palynology of Svalbard V. Dinoflagellates from the Agardhfjellet Mbr (Middle and Upper Jurassic) in Spitsbergen. *Norsk Polarinst. Skr.*, **172**, 145—167.

Bjærke, T. and Manum, S.B. 1977. Mesozoic palynology of Svalbard I. *Norsk Polarinst. Skr.*, **165**, 40pp.

Briden, J.C., Drewrey, G.E. and Smith A.G. 1974. Phanerozoic equal-area world maps. *J. Geol.*, **82**, 555—574.

Buchan, S.H., Challinor, A., Harland, W.B. and Parker, J.R. 1965. The Triassic stratigraphy of Svalbard. *Norsk Polarinst. Skr.*, **135**, 94pp.

Cutbill, J.L. and Challinor, A. 1965. Revision of the stratigraphical scheme for the Carboniferous and Permian of Spitsbergen and Bjørnøya. *Geol. Mag.*, **102**, 418—439.

Dalland, A. 1977. Erratic clasts in the Lower Tertiary deposits of Svalbard — evidence of transport by winter ice. *Norsk Polarinst. Årbok*, **1976**, 151—166.

Edwards, M.B. 1976a. Growth faults in Upper Triassic deltaic sediments, Svalbard. *Bull. Am. Assoc. Petrol. Geol.*, **60**, 341—355.

Edwards, M.B. 1976b. Depositional environments in Lower Cretaceous regressive sediments, Kikutodden, Sørkapp Land , Svalbard. *Norsk Polarinst. Årbok*, **1974**, 35—50.

Edwards, M.B., Edwards, R. and Colbert, E. 1978. Carnosaurian footprints in the Lower Cretaceous of eastern Spitsbergen. *J. Paleontol.*, **52**, 940—941.

Dypvik, H. Jurassic and Cretaceous black shales of the Janusfjellet Fm, Svalbard, Norway. *Sediment. Geol.*, (in press).

Flood, B., Nagy, J. and Winsnes, T.S. 1971. Geological map of Svalbard 1:500 000, Sheet 1G, Spitsbergen, Southern part. *Norsk Polarinst. Skr.* **154A**.

Føyn, S. and Heintz, A.1943. The Downtonian and Devonian vertebrates of Spitsbergen, VIII. The English—Norwegian—Swedish Expedition 1939, geological results. *Skr. Svalbard og Ishavet*, **85**, 52 pp.

Frebold, H. 1930. Verbreitung und Ausbildung des Mesozoikums in Spitsbergen. *Skr. Svalbard og Ishavet*, **31**, 126pp.

Frebold, H. 1931. Fazielle Verhältnisse des Mesozoikums im Eisfjordgebiet Spitsbergens. *Skr. Svalbard og Ishavet*, **37**, 94pp.

Frebold, H. 1935. Geologie von Spitsbergen, der Bäreninsel, des König Karl- und Franz-Joseph-Landes. *Geologie der Erde*, Berlin, 195pp.

Frebold, H. 1951. Geologie des Barentsschelfes. *Abh. Dtsch. Akad. Wiss. Berl. Kl. Math. Naturwiss.*, **5**, 51pp.

Friend, P.F. and Moody-Stuart, M. 1972. Sedimentation of the Wood Bay Fm (Devonian) of Spitsbergen. Regional analysis of a late orogenic basin. *Norsk Polarinst. Skr.*, **157**, 77pp.

Gjelberg, J. 1981. Upper Devonian (Famennian) to Middle Carboniferous succession of Bjørnøya. *Norsk Polarinst. Skr.*, **174**, 67pp.

Gjelberg, J. and Steel, R.J. 1981. An outline of lower-middle Carboniferous sedimentation on Svalbard: effects of climatic, tectonic and sea-level changes in rift basin sequences. *Mem. Can. Soc. Petrol. Geol*, **7**, 543—561.

Gjelberg, J. and Steel, R.J. 1983. Middle Carboniferous marine transgression, Bjørnøya, Svalbard: facies sequences from an interplay of sea level changes and tectonics. *Geol. J.*, **18**, 1—19.

Gobbett, D.J. 1963. Carboniferous and Permian brachiopods of Svalbard. *Norsk Polarinst. Skr.*, **127**, 201pp.

Harland, W.B. 1969. Contribution of Spitsbergen to understanding of tectonic evolution of North Atlantic Region. In: M. Kay (ed.), North Atlantic Geology and Continental Drift, *Mem. Am. Assoc. Petrol. Geol.*, **12**, 817—851.

Harland, W.B. 1972. Mesozoic geology of Svalbard, In: M.G. Pitcher (ed.), Arctic Geology, *Mem. Assoc. Petrol. Geol.*, **19**.

Harland, W.B., Cutbill, J.L., Friend, P.F., Gobbett, D.J., Holliday, D.W., Maton, P.I., Parker, J.R. and Wallis, R.H. 1974. The Billefjorden Fault Zone, Spitsbergen. *Norsk Polarinist. Skr.*, **164**, 89 pp.

Harland, W.B., Pickton, C.A.G. Wright N.J.R., Croxton, C.A., Smith, D.G., Cutbill J.L. and Henderson, W.S., 1976. Some coal-bearing strata of Svalbard. *Norsk Polarinst. Skr.*, **164**, 89pp.

Heintz, N. 1963. Dinosaur foot-prints and polar wondering. *Norsk Polarinst. Årbok*, **1962**, 35—43.

Hellem, T.A. and Worsley, D. Permian depositional environments of Svalbard. *Polar Res.* (in press).

Hjelle, A. and Lauritzen, Ø. 1982. Geological map of Svalbard 1:500 000. Sheet 3G, Spitsbergen northern part. *Norsk Polarinst. Skr.*, **154C**, 15pp.

Hoel, A. and Orvin, A.K. 1937. Das Festungsprofil auf Spitsbergen. Karbon-Kreide. I. Vermessungsresultate. *Skr. Svalbard og Ishavet*, **18**, 59pp.

Johannessen, E. and Steel, R.J., 1981. Alluvial fan and fan delta facies sequences in a Bashkirian rift basin, Billefjorden, Spitsbergen. *Abstr. Int. Assoc. Sed., 2nd Eur. Mtg. Bologna*, pp. 87—88.

Kaiser, H. 1970. Die Oberdevon-Flora der Bäreninsel. 3. Mikroflora des hoeheren Oberdevons und des Unterkarbons. *Paleontogr. Abt. B*, **129**, 71—124.

Kaiser, H. 1971. Die Oberdevon-Flora der Bäreninsel. 4. Mikroflora der Misery-Serie und der Floezleeren Sandstein-Serie. *Paleontogr. Abt.B*, **135**, 127—164.

Kellogg, H.E. 1975. Tertiary stratigraphy and tectonism in Svalbard and continental drift. *Bull. Am. Assoc. Petrol. Geol.*, **59**, 465—485.

Kent, D.V. and Opdyke, N.D. 1979. The early Carboniferous Palaeomagnetic field of North America and its bearing on tectonics of the Northern Appalachians. *Earth Planet. Sci. Lett.*, **4**, 365—372.

Kleinspehn, K.L., Steel, R.J., Johannessen, E., and Netland, A. 1983. Conglomeratic fan delta sequences (late Carboniferous—early Permian), western Spitsbergen. *Abstr. Int. Assoc. Sed., 4th Eur.Mtg.*, Split, pp.90—91.

Korcinskaja, M.V. 1973. Biostratigraphy of Triassic deposits of Svalbard. In: A. Logan and L.V. Hills (eds.), Permian and Triassic Systems and their Mutual Boundary, *Mem. Can. Soc. Petrol. Geol.*, **2**, 261—268.

Lauritzen, Ø. 1981. Investigation of Carboniferous and Permian sediments in Svalbard. *Norsk Polarinst. Skr.*, **176**, 44pp.

Livsic, Ju. Ja. 1965. Tektonika central'noj casti Zapadnogo Spicbergena. In: U.N. Sokolov (ed.), *Materialy på Geologii Spicbergena*, NIIGA, Leningrad.

Livsic, Ju. Ja. 1974. Treticnye otlozenija zapadjon casti archipelago Spicbergen. In: V.N. Sokov (ed.), *Materialy på Geologii Spicbergena*, NIIGA, Leningrad, pp. 185—204.

Livsic. Ju. Ja. Palaeogene deposits and the platform structure of Svalbard. *Norsk Polarinst. Skr.*, **138**, 58pp.

Lock, B.E., Pickton, C.A.G., Smith, D.G., Batten, D.J. and Harland, W.B. 1978. The geology of Edgeøya and Barentsøya, Svalbard. *Norsk Polarinst. Skr.*, **168**, 64pp.

Lowell, J.D. 1972. Spitsbergen Tertiary orogenic belt and the Spitsbergen Fracture Zone. *Bull. Geol. Soc. Am.*, **83**, 3091—3102.

Løfaldi, M. and Nagy, J. 1980. Foraminiferal stratigraphy of Jurassic deposits on Kongsøya, Svalbard. *Norsk Polarinst. Skr.*, **172**, 63—95.

Major, H., Harland, W.B. and Strand, T. 1956. *Lexique stratigraphique* 1 (fasc. 1d Svalbard), 21—97.

Major, H. Nagy, J. 1972. Geology of the Adventdalen map area. *Norsk Polarinst. Skr.*, **138**, 58pp.

Malod, J. and Mascle, J. 1975. Structures geologiques de la marge continentale a l'ouest du Spitsberg. *Mar. Geophys. Res.*, **2**, 215—229.

Manum, S.B. 1962. Studies in the Tertiary flora of Spitsbergen, with notes on Tertiary floras of Ellesmere Island, Greenland and Iceland. *Norsk Polarinst. Skr.*, **125**, 127pp.

Manum, S.B. and Throndsen, T., 1978. Rank of coal and dispersed organic matter and its geological bearing in the Spitsbergen Tertiary. *Norsk Polarinst. Årbok*, **1977**, 159—177.

Marty, A., Rabate, J., Cazes, M., Jacguart, G. and Auxietre, J.L. 1979. CEPM Barents Sea Seismic Survey. *Norwegian Sea Symposium* (Norwegian Petroleum Society) NSS/16, 17pp.

Myhre, A.M., Eldholm, O. and Sundvor, E. 1982. The margin between Senja and Spitsbergen fracture zones: implications from plate tectonics. *Tectonophysics*, **89** 33—50.

Nagy, J. 1970. Ammonite faunas and stratigraphy of Lower Cretaceous (Albian) rocks in southern Spitsbergen. *Norsk Polarinst.Skr.*, **152**, 58pp.

Mørk, A., Knarud, R. and Worsley, D. 1982. Depositional and diagenetic environments of the Triassic and Lower Jurassic succession of Svalbard. In: M. Embry and H.R. Balkwill (eds.), Arctic Geology and Geophysics, *Can. Soc. Petrol. Geol. Mem.*, **8**, 371—398.

Nathorst, A.G. 1910. Beiträge zur Geologie der Bären-Insel, Spitsbergen und des König-Karl-Landes. *Uppsala*

Univ. Geol. Inst. Bull., **10**, 261-416.
Nøttvedt, A. 1981. Characteristics and evolution of the Askeladden delta system (Palaeocene) on Spitsbergen. *Int. Assoc. Sed., 2nd Eur. Mtg.*, Bologna. Abstract.
Nysæther, E. 1977. Investigations on the Carboniferous and Permian stratigraphy of the Torell Land area, Spitsbergen. *Norsk Polarinst. Årbok*, **1976**, 22—41.
Orvin, A.K. 1940. Outline of the Geological History of Spitsbergen. *Skr. Svalbard og Ishavet*, **78**, 57pp.
Orvin, A.K. 1947. Bibliography of literature about the geology, physical geography, useful minerals and mining of Svalbard. *Skr. Svalbard og Ishavet*, **89**, 121pp.
Parker, J.R. 1966. Folding, faulting and dolerite intrusions in the Mesozoic rocks of the fault zone of central Spitsbergen. *Norsk Polarinst. Årbok*, **1964**, 47—55.
Parker, J.R. 1967. The Jurassic and Cretaceous sequence in Spitsbergen. *Geol. Mag.*, **103**, 487—505.
Pcelina, T.M. 1980. New data on the Triassic/Jurassic boundary in the archipelago of Svalbard. In: D.V. Semevskij (ed.), *Geology of the Sedimentary Platform of the Archipelago of Svalbard*, NIIGA, Leningrad (In Russian).
Playford, G. 1962/1963. Lower Carboniferous Microfloras of Spitsbergen, Parts 1 and 2. *Palaeontology*, **5**, 550—678.
Ravn, J.P.J. 1922. On the Mollusca of the Tertiary of Spitsbergen. *Res. norske Spitsbergen eskp.*, **1**(2), 28pp.
Rønnevik, H.C. 1981. Geology of the Barents Sea. In: L.V. Illing and G.D. Hobson (eds.)., *Petroleum Geology of the Continental Shelf of North-West Europe*, Institute of Petroleum, London, pp. 395—406.
Russell, M.J. and Smythe, D.K. 1983. Origin of the Oslo Graben in relation to the Hercynian—Alleghenian Orogeny and lithospheric rifting in the N. Atlantic *Tectonophysics*, **94**, 457—72.
Rye-Larson, M. 1982. Forlandsundet Graben (Palaeogene) — an oblique-slip basin on Svalbards western margin. *Abstr. Int. Assoc. Sed., 3rd Eur. Reg. Meet.*, Copenhagen, pp. 31—34.
Schlüter, H.U. and Hinz, K. 1978. The geological structure of the western Barents Sea. *Mar. Geol.*, **26**, 199—230.
Semevskij, D.G. (ed.) 1980. *Geology of the Sedimentary Platform of the Archipelago of Svalbard.* NIIGA, Leningrad (in Russian).
Siedlecki, S. and Turnau, E. 1964. Palynological investigations of the Culm in the area south-west of Hornsund, Vestspitsbergen. *Stud. Geol. Polon.*, **11**, 125—138.
Siedlecka, A. 1970. Investigations of Permian cherts and associated rocks in Southern Spitsbergen. *Norsk Polarinst. Skr.*, **147**, 70pp.
Smith, A.G. and Briden, J.C. 1977. *Mesozoic and Cenozoic Paleocontinental Maps.* Cambridge University Press, Cambridge, 63pp.
Smith, D.G., Harland, W.B., Hughes, N.F. and Pickton, C.A.G. 1976. The geology of Kong Karls Land, Svalbard. *Geol. Mag.*, **113**, 193—232.
Sokolov, D. and Bodylevsky, W. 1931. Jura- und Kreidefaunaen von Spitsbergen. *Skr. Svalbard og Ishavet*, **35**, 151 pp.
Sosipatrova, G.P. 1972. Upper Palaeozoic foraminifera of Spitsbergen. In: V.N. Sokolov and N.D. Vasilevskaja (eds.). *Stratigraphy of Spitsbergen*, NIIGA, Leningrad, pp. 126—163 (in Russian).
Steel, R.J. 1977. Observations on some Cretaceous and Tertiary sandstone bodies in Nordenskiöld Land, Svalbard. *Norsk Polarinst.Årbok*, **1976**, 43—68.
Steel. R.J., Gjelberg, J. and Haarr, G. 1978. Helvetiafjellet Fm (Barremian) at Festningen, Spitsbergen — a field guide. *Norsk Polarinst. Årbok*, **1977**, 111—128.
Steel. R.J., Dalland, A., Kalgraff, K. and Larsen, V. 1981. The central Tertiary basin of Spitsbergen — sedimentary development of a sheared-margin basin. In: J. Wm. Kerr and A.J. Fergusson (eds.), *Mem. Can. Soc. Petrol. Geol.*, **7**, 647—664.
Sundvor, E. and Eldholm, O. 1979. The western and northern margin off Svalbard. *Tectonophysics*, **59**, 239—250.
Sundvor, E., Eldholm, O., Giskenhaug, A and Myhre, A. 1977. Marine geophysical survey of the western and northern continental margin of Svalbard. *Univ. Bergen. Seism. Obs. Sci. Rep.*, 4, 35pp.
Sundvor, E., Myhre, A. and Eldholm, O. 1979. The Svalbard continental margin. *Norwegian Sea Symposium* (Norwegian Petroleum Society) NSS/6, 25 pp.
Szaniawski, H. and Malkowski, K. 1979. Condonts from the Kapp Starostin Fm (Permian), Spitsbergen. *Acta Palaeontol. Polon.*, **24**, 231—264.
Talwani, M. and Eldholm, O. 1977. Evolution of the Norwegian-Greenland Sea. *Bull. Geol. Soc. Am.*, **88**, 969—999.
Tozer, E.T. and Parker, J.R. 1968. Notes on the Triassic biostratigraphy of Svalbard. *Geol. Mag.*, **105**, 526—42.
Vail, P.R., Mitchum, R.M. and Thompson, S. 1977. Seismic stratigraphy and global changes of sea level, Part 4. Global cycles of relative change of sea level. *Mem. Am. Assoc. Petrol. Geol.*, **26**, 83—97.
van der Voo, R. and Scotese, C. 1981. Palaeomagnetic evidence for a large (200 km) sinistral offset along the Great Glen fault during Carboniferous time. *Geology*, **9**, 583—589.
Vonderbank, K. 1970. Geologie and Fauna der Tertiären Ablagerungen Zentral-Spitzbergens. *Norsk Polarinst. Skr.* **151**, 119pp.
White, S.M. (ed.). 1976. Leg. 38. *Initial Report Deep Sea Drilling Project.* U.S. Govt. Printing Office, Washington, D.C., 1256 pp.
Winsnes, T.S. and Worsley, D. 1981. Geological map Svalbard 1:500 000. Sheet 2G, Edgeøya. *Norsk Polarinst. Skr.*, **154B**.
Worsley, D. 1973. The Wilhelmøya Fm — a new lithostratigraphical unit from the Mesozoic of eastern Svalbard. *Norsk Polarinst. Årbok*, **1971**, 7—16.
Worsley, D. and Edwards, M.B. 1976. The Upper Palaeozoic succession of Bjørnøya. *Norsk Polarinst. Årbok*, **1974**, 17—34.
Worsley, D. and Heintz, N. 1977. The stratigraphical significance of a marine vertebrate fauna of Rhaetian age, Kong Karls Land. *Norsk Polarinst. Årbok*, **1976**, 69—81.
Ziegler, P.A. 1982. *Geological Atlas of Western and Central Europe.* Elsevier, Amsterdam, 130pp.

APPENDIX I

Theses on Svalbard

Aakvik, R. 1981. Fasies analyse av undre karbonske kullførende sedimenter, Billefjorden, Spitsbergen. Cand. real., Univ. i Bergen, 219pp.
Adgestein, T. 1980. En stratigrafisk, sedimentologisk og diagenetisk undersøkelse av karbon-perm sedimenter (Kapp Hanna og Kapp Duner formasjonene) på Bjørnøya, Svalbard. Cand. real., Univ. i Oslo. 144pp.
Backstrøm, S. 1980. Paleontologisk og sedimentologisk undersøkelse av Brentskardhaugen laget (Jura) på Spitsbergen. Cand.real., Univ. i Oslo. 138pp.
Bjærke, T. 1974. A palynological investigation in the Upper De Geerdalen Fm—Lower Wilhelmøya Fm (Rhaetic) of Hopen, Svalbard, (including a note on the correlation with beds on Kong Karls Land). Cand.real., Univ. i Oslo, 134pp.
Eien, M. 1979. Sammensetning og termisk omdannelse av organisk materiale i Spitsbergens underkritt. Cand. real., Univ. i Oslo, 134pp.
Finnerud, E. 1982. Sammensetning, sedimentasjon og termisk omdanning av klastisk organisk materiale i jurassiske og krittassiske svartskifre, Svalbard. Cand. scient., Univ. i Oslo, 165pp.
Forsberg, A.W. 1982. En sedimentologisk og organisk geokjemisk undersøkelse av Barentsøya Formasjonen (Trias), Svalbard. Cand. scient., Univ. i Oslo, 197pp.
Dons, C.E. 1983. Fasies og paleostrømanalyse av Nordenskiøldbreen formasjonen (overkarbon-underperm), Sentrale Spitsbergen. Cand. scient., Univ. i Oslo, 344pp.
Gjelberg, J. 1978. The Upper Devonian (Famennian) to Middle Carboniferous (?Moscovian) strata on

Bjørnøya (Svalbard) — A study in alluvial and coastal marine sedimentation. Cand. real., Univ. i Bergen, 265pp.

Haarr, G. 1979. A regional analysis of the Helvetiafjellet Fm (Barremian), Svalbard. Cand. real., Univ. i Bergen, 155pp.

Hansen, O.K. 1982. Den sedimentologiske utvikling og dens innvirkning på kulldannelse i nedre Firkanten Formasjonen (Palaeocene) på nordsiden av Adventdalen, Spitsbergen. Cand. real, Univ. i Bergen. 155pp.

Hellem, T. 1981. En sedimentologisk og diagenetisk undersøkelse av utvalgte profiler fra Tempelfjorden gruppen (Perm) i Isfjorden området, Spitsbergen. Cand. real., Univ. i Oslo, 214pp.

Johannessen, E. 1980. Fasies analyse av Ebbadalen Formasjonen, Mellom Karbon, Billefjorden Trau, Spitsbergen. Cand. real., Univ. i Bergen, 314pp.

Johnsen, S. 1980. A sedimentological, palaeontological and palaeoecological investigation of Lower Triassic outcrops on the west coast of Spitsbergen. Cand. real., Univ. i Oslo, 206pp.

Kalgraff, K. 1978. Aspects of sedimentation in Firkanten Fm, Svalbard. Cand. real., Univ. i Bergen, 178pp.

Kirkemo, K. 1979. En sedimentologisk undersøkelse av Kapp Kåre formasjonen (moskov), Bjørnøya. Cand. real., Univ. i Oslo, 234pp.

Knag, G. 1980. Gipshuken- og Kapp Starostin Formasjonen, mellom til øvre perm, langs vestkysten av Svalbard. Cand. real., Univ. i Bergen, 210pp.

Knarud, R. 1980. En sedimentologisk og diagenetisk undersøkelse av Kapp Toscana formasjonens sedimenter på Svalbard. Cand. real., Univ. i Oslo, 208pp.

Lønøy, A. 1981. Fasies analyse av undre permiske karbonater i Tyrrellfjellet Ledd av Nordenskjoldbreen Formasjonen, Billefjordområdet, Spitsbergen. Cand. real, Univ. i Bergen, 218pp.

Moe, H.R. 1980. Sedimentologiske og diagenetiske undersøkelser i krittlagrekken på Svalbard. Cand. real., Univ. Oslo, 140pp.

Netland, A. 1982. Fasies analyse av Drevbreen beds og Nordenskiøldbreen Formasjonen, øvre karbon til undre perm, Bellsund området, Svalbard. Cand. real., Univ. i Bergen, 204pp.

Nøttvedt, A. 1982. Characteristics and evolution of the Askeladden deltaic sequence (Palaeocene) on Spitsbergen — with comparisons to the Ravenscar Group deltaic sequences (Bajocian) of northeast England. Dr. Scient. thesis, Univ. i Bergen, 146pp.

Rye Larsen, M. 1982. Forlandsundet Graben (Paleogen) — Svalbard's vestmargin — sedimentasjon og tektonisk utvikling av et basseng ved en transform plategrense. Cand. real., Univ. i Bergen, 380pp.

Skaug, M. 1982. Bentiske fossile assosiasjoner og faciesvariasjoner i Nordenskiøldbreen Formasjonen (overkarbon-underperm), sentrale Spitsbergen. Cand. scient., Univ. i Oslo, 220pp.

Sørensen, L. 1981. Et regionalt studie av mineralogi og diagenese i Helvetiafjellet Formasjonen (Barrem), Svalbard. Cand. real., Univ. i Bergen, 125pp.

Sundsbø, G. 1982. Facies analyses of Late Carboniferous and Early Permian carbonates in the Billefjorden Area, Spitsbergen. Cand. real., Univ. i Bergen, 161pp.

Tangen, O. 1981. A sedimentological and environmental interpretation of the upper part of the Nordenskiøldbreen Fm and the Gipshuken Fm on Western Spitsbergen. Cand. real., Univ. i Bergen, 256pp.

Throndsen, T. 1977. Sammensetning og temisk omdannelse av organisk materiale i Svalbards tertiær. Cand. real., Univ. i Oslo, 266pp.

Tønseth, D. 1981. The sedimentary history of Firkanten Fm (Palaeocene) in the Adventdalen Area (Spitsbergen). Cand. real., Univ. i Bergen, 181pp.

Ytreland, G. 1980. Sedimentation along the western margin of the Central Tertiary Basin (Firkanten Fm), Spitsbergen. Cand. real., Univ. i Bergen, 183pp.

9 Outline of Arctic post-Silurian continental displacements

W.B. Harland, B.A. Gaskell, A.P. Heafford, E.K. Lind and P.J. Perkins
Department of Earth Sciences, University of Cambridge, UK

The positions and relative motions of Arctic land masses from Devonian to present times are presented in palinspastic maps at selected intervals. These are accompanied by summaries of the post-Silurian tectonic evolution of the Arctic.

This chapter is already in abstract form, having been taken from detailed reports written for the Cambridge Arctic Shelf Programme. In these reports, account was taken of available stratigraphic, structural and geophysical data in order to evaluate as many palinspastic options as possible. Those which best fit the geological data were incorporated into an epoch-by-epoch series of palinspastic reconstructions, a selection of which is given.

INTRODUCTION

In this paper we offer a sample from a sequence of palinspastic maps reconstructing Arctic palaeogeological relationships indicated by the present-day coastlines (Fig. 1). Our brief was to restrict this consideration to post-Caledonian history and to focus on Svalbard. As it happens, Svalbard provides critical clues for the reconstruction of the geological history of what is now the Arctic and our maps are centred thereon. However, the reconstructions take into account available stratigraphic evidence from all the areas within the Arctic Circle. This has been possible through our participation in the Cambridge Arctic Shelf Programme (CASP), initiated in 1975 as a comprehensive study and made possible by subscriptions from industry, especially oil companies (e.g. Meyerhoff, 1978). Their identity is not revealed because their subscriptions were for a commensurate service. The Programme is non-profit-making and we retain the right to publish our work; this is part of it.

Not all the reasoning behind the maps can be given and we offer only a brief commentary on the events implied by the changing relationships shown on them. However, it may be useful to indicate steps that led to the whole series of maps. For some years CASP geologists, working mostly from the literature and occasionally from first-hand knowledge, summarized available information about each area in the Arctic. These areal studies were then synthesized into regional studies on a stratigraphic basis so as to show the sequence of rock formation and the history of deformation. About half the effort was directed to the Soviet sector.

The next phase was to treat the whole Arctic, abstracting data for each geologic period from the regional reports with an attempt to depict, for example, the facies distribution for each epoch. These were first plotted on present-day maps which, although obviously misleading for ancient times, nevertheless provided a common objective starting point. All available published reconstructions were assembled for the Arctic region for each period and tested against aspects of facies such as sedimentary provenance, and the points in favour and against each were noted.

Because most published palaeocontinental reconstructions were based on limited parts of the Arctic, we then reconsidered them in sequence for each of three sectors: North American, North Atlantic and Soviet, and attempted to evaluate available options while at the same time noting how they could be broken into their component parts and combined in a simple classification (see Fig. 2). For the present stage, we selected those reconstructions that seemed to pass the various stratigraphic and tectonic tests applied. Then maps were drawn, one for each epoch from Vendian time onwards, at first in outline as sampled in this paper, and then plotted on a large scale with the already available compilation of stratigraphic information. The programme to date represents more than 25 geologist-years work and is by no means finished. We delayed using computer map-

Petroleum Geology of the North European Margin © Norwegian Petroleum Society (Graham & Trotman, 1984) pp. 137—148.

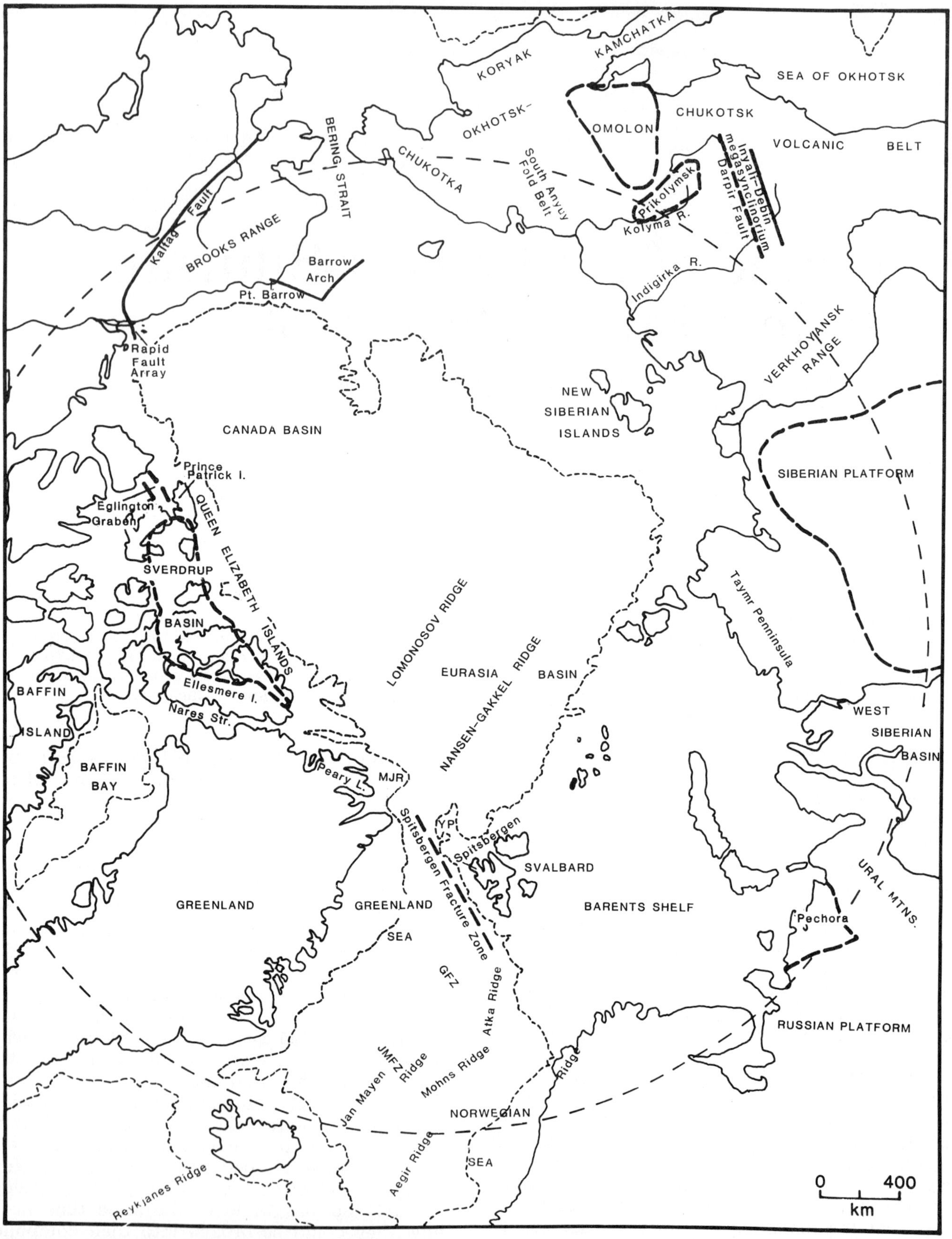

Fig. 1. Location map of places referred to in the text. MJR, Morris Jessup Rise; YP, Yermak Plateau; GFZ, Greenland Fracture Zone; JMFZ, Jan Mayen Fracture Zone.

ping methods systematically because we were dealing with many hundreds of different options and it would have been premature to take time to enter each into computer files. However, we have arrived at the point where many options have been eliminated and the remaining few need to be tested on computer maps against global reconstructions and re-evaluated against recent literature, which continues to challenge our thinking.

We believe our study is unusual in applying systematically so wide a range of constraints (sedimentary, igneous, structural, radiometric and palaeomagnetic, both pole positions and ocean striping) for a region as large as the whole Arctic.

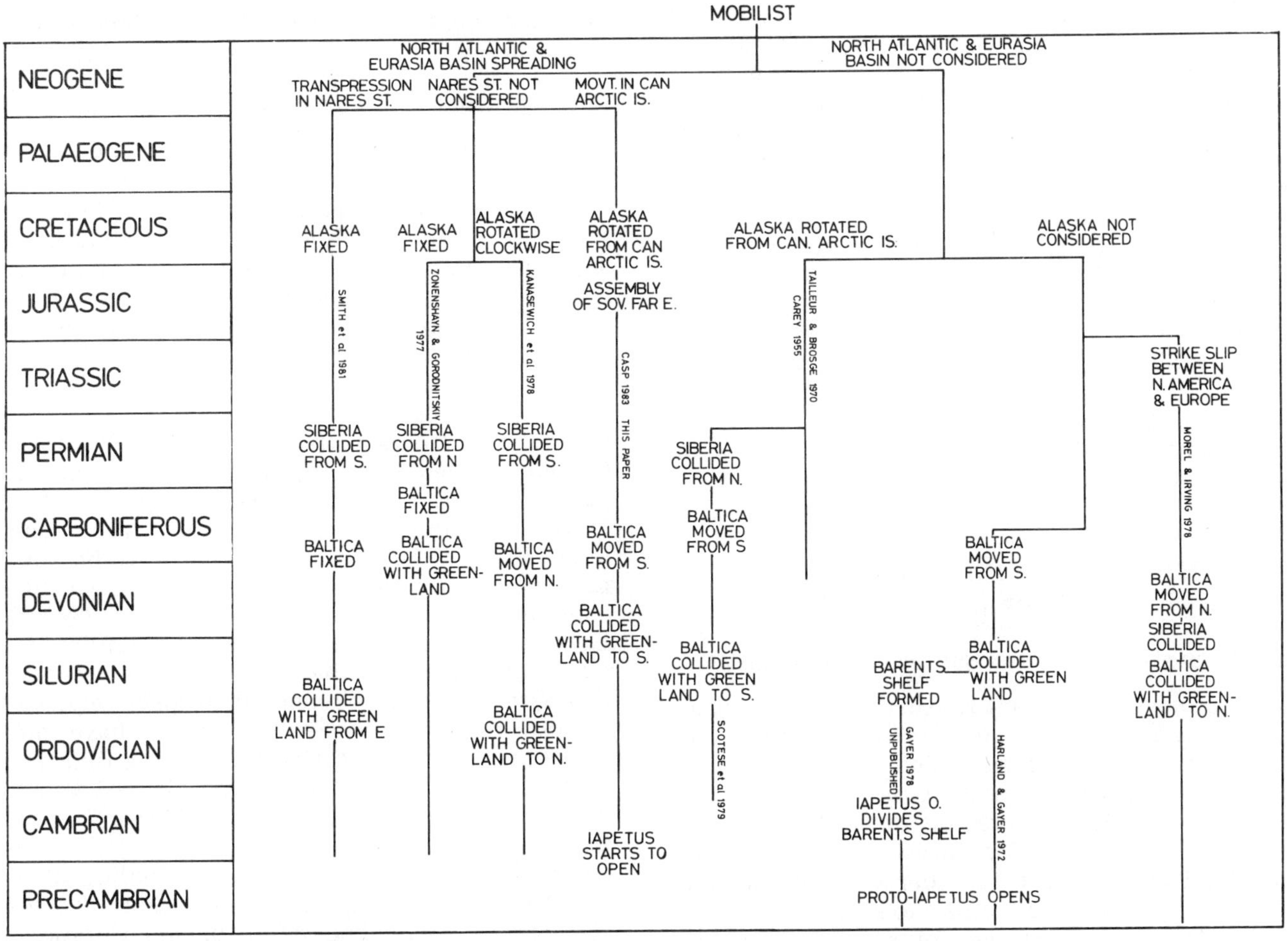

Fig. 2. Diagrammatic scheme to show the relationship between a number of options for palinspastic reconstructions that have been published and assessed against stratigraphic and tectonic evidence by CASP. Only key references which indicate a publication representative of a particular class of palinspastic hypotheses are listed.

It is clearly based mostly on secondary literature sources and in this respect must be superficial. Nevertheless, from an early position when no options were ruled out we have developed confidence that our preferred sequence of reconstructions is substantially correct. It will, however, be subjected to a further phase of rigorous testing before we offer the complete sequence of maps. The principal events considered here are the differential motions between land masses (identified by present coastlines) needed to explain the evolution from one map to the next. As suggested by the Editor, we selected five out of 33 such maps.

Finally we repeat that this outline is restricted to post-Caledonian events according to the plan of this symposium.

DEFINITION AND CLASSIFICATION OF PALINSPASTIC OPTIONS

Any comprehensive tectonic synthesis of such a large and complex region is a composite of many elements which may be combined differently and so yield many options. Most published attempts at the geological evolution of the Arctic have focussed on one sector of it or on a particular problem. Comprehensive hypotheses generally have components well founded in the author's experience, while in order to complete the picture the remainder is speculation beyond his sure knowledge. In attempting to treat the matter systematically we identified and defined the various component hypotheses and classified them according to the particular relationships they purported to explain. The classification is given in Fig. 2, in outline only, to show the approach.

The classification focussed consideration on the main hypothetical events and so identified the controversies surrounding them. This directed attention to the critical evidence that might be available to distinguish the 30 or more published models and thence to see what variations on them we might also experiment with.

The classification scheme clearly shows six major events in the evolution of the Arctic: (1) the opening of the Iapetus Ocean; (2) the closure of the Iapetus Ocean causing collision of Baltica with Greenland; (3) the relative movements between Eurasia and North America (manifested as major strike-slip movements between the different elements of Svalbard); (4) the collision of Siberia with the Russian Platform to form the Urals; (5) possible movements of Arctic Alaska with respect to North America; (6) the opening of the North Atlantic and Eurasia Basin and associated events in the Labrador Sea and Nares Strait. Other major events not emphasized in this condensed scheme include the accretion of a series of microplates to form the Soviet Far East, and the

Ellesmerian orogeny.

As an example, one controversy concerns the relative movements of Baltica (we use this term here to include the Barents Shelf) and Greenland. According to various authors, Baltica collided with Greenland either from the south followed by sinistral displacement, from the north followed by dextral movement, or from the east (using present directions). Further, there is some difference of opinion about the age of strike-slip movement. Morel and Irving (1978) showed the major movements to be of Devonian age while Scotese *et al.* (1979) indicated a mid-Carboniferous age. These component options are illustrated in Fig. 3. Each strand in the scheme represents a variety of components or competing options, each with its own literature.

EARLY/MIDDLE DEVONIAN TO MID-CARBONIFEROUS KINEMATICS

Relative motions between Baltica and Laurentia

Svalbard plays a dominant role in the determination of the relative positions of Northern Europe and North America. We have relied much on stratigraphic and palaeomagnetic data from this area for the Middle Devonian reconstruction in Fig. 4.

Emplacement ages for thrust nappes in East Greenland and Norway together with radiometric dates for metamorphic events show that the climax of the Caledonian orogeny was reached in late Silurian to early Devonian times. This event marked the suturing of the Laurentian (N. American) and Baltic plates in a configuration often referred to as the Old Red continent. Most workers have considered this configuration to have been stable. We follow the minority, who argued that this was not the case, in suggesting that major strike-slip motions took place within the continent. The argument has been strengthened by recent palaeomagnetic results and the constraints imposed by a range of data are discussed in more detail in a parallel paper together with the earlier history (Harland *et al.*, in press). The conclusion of that study, however, is as follows.

The Caledonian orogeny was dominantly a compressive event, although a transpressive regime may have developed in the later stages (Harland, 1971). This transpression was expressed as major sinistral strike-slip movements which displaced Baltica 1500—2000 km northwards relative to North America in Devonian and Carboniferous times. Evidence of sinistral displacement between eastern Svalbard and Greenland (i.e. between Baltica and Laurentia) during the Ordovician to Permian interval (possibly Silurian to Pennsylvanian) is based partly on the similarity of pre-Silurian successions in Eastern Svalbard and Central East Greenland and this requires between 1500 and 2000 km of strike-slip motion as argued by Harland in 1965. From this evidence it is, of course, impossible to date the motion precisely.

Motion may have been continuous throughout this time, though the major part was taken up in two phases: (1) a late Devonian (Svalbardian) phase in which about 500 km of sinistral displacement occurred; and (2) a mid-Carboniferous phase in which the displacement may have been 1000 km or more.

Late Devonian phase (Svalbardian)

Evidence for the pre-Carboniferous phase is seen in Svalbard where a minimum of 200 km of sinistral displacement was claimed along the Billefjorden Fault Zone alone, while a figure of 1000 km was suggested as possible (Harland *et al.*, 1974). Five-hundred kilometres is suggested as a reasonable figure for motion along this zone in late Devonian time and the constraints on that figure are tightening.

The minimal value of 200 km depends on the nature of the present Old Red Sandstone Graben in northern Spitsbergen. The Caledonian Ny Friesland Orogen immediately to the east could not have been in that position during early and Middle Devonian sedimentation and must have been at least 200-km farther south. But if only 200-km farther south another conflict with the evidence emerges. It would then interpose the Hornsund Geosyncline between the notably similar Hecla Hoek and Greenlandian geosynclines of eastern Svalbard and East Greenland, respectively, so the eastern province of Svalbard must have moved more than 200-km sinistrally along the Billefjorden Fault Zone with respect to the Central Province, with the Hornsund Geosyncline (Harland *et al.*, in press).

The late Devonian Svalbardian folding of Vogt (1928) is evident in the deformation of the Old Red Sandstone Graben of Spitsbergen. Compression structures are manifest and these were interpreted as the product of normal convergence (e.g. McWhae, 1952). However, they were later interpreted as the effect of sinistral transpression (Harland *et al.*, 1974).

Steel (1976, 1978) considered that sediments in the molasse type Old Red Sandstone basins of southern Norway had been folded and thrust in a late Devonian Svalbardian event.

In Svalbard the Billefjorden Fault Zone ceased significant strike-slip motion by about Tournaisian time when the Eastern Province (of Harland and Wright, 1979) became united with the Central Province.

Early Carboniferous (Mississippian) phase

The second phase of motion is supported by: (1) palaeomagnetic evidence from the North American Craton, the maritime provinces of North America, northern Scotland and Europe (Kent and Opdyke, 1978, 1979; van der Voo *et al.*, 1979; van der Voo and Scotese, 1981) which seems to imply that sinistral strike-slip in the range 1000—2000 km took place between North America and Europe in Carboniferous times; and (2) by recent palaeomagnetic work on Lower Namurian rocks from eastern Svalbard which suggests that, at that time, eastern Svalbard was located well to the south of its Permian position with respect to Greenland and North America (D. Watts, personal communication, and 1983). Stratigraphical evidence implies that the three provinces of Svalbard were united by Bashkirian times in a position to the north-east of Greenland (Harland and Wright 1979; Harland *et al.*, in press), so major strike-slip motion probably took place in mid-Carboniferous times (late Mississippian). It is from that position that, as is

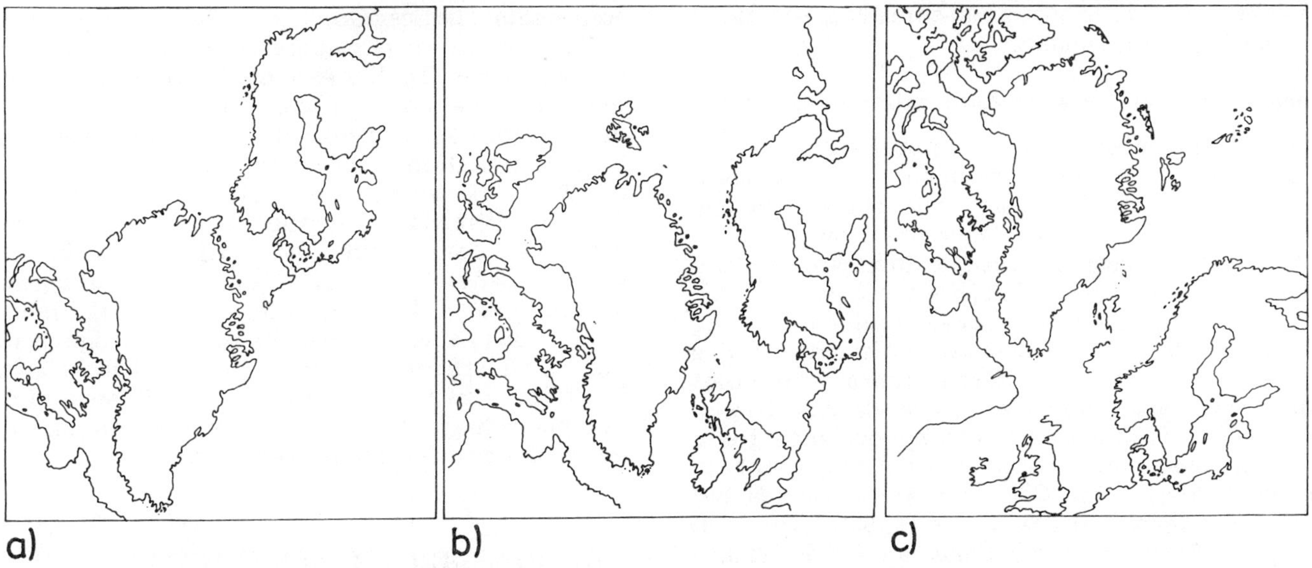

Fig. 3. Reconstructions of the relative positions of Baltica and Greenland in late Silurian to early Devonian time, after: (a) Morel and Irving, 1978; (b) Smith *et al.*, 1981; and (c) Harland, 1981.

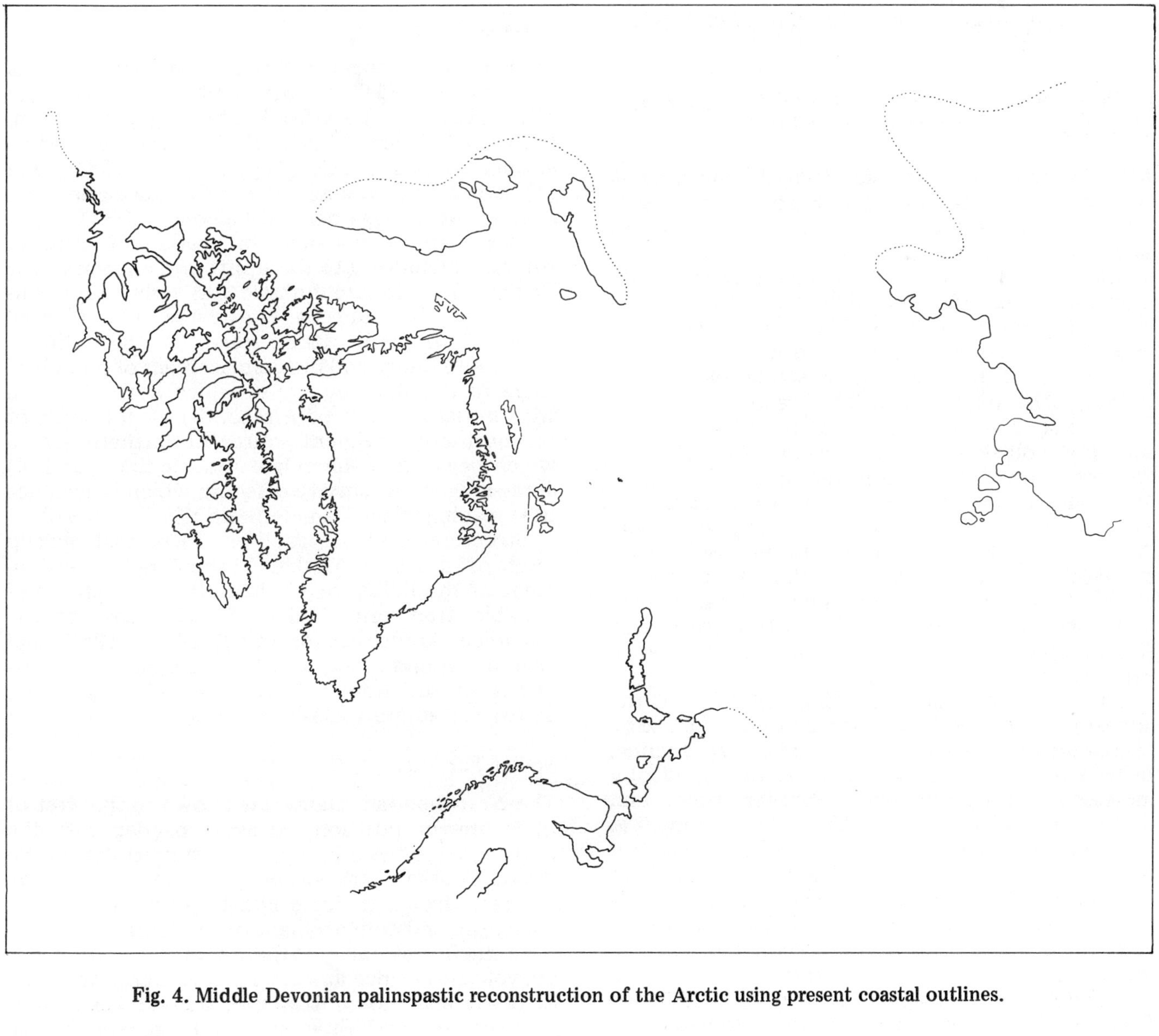

Fig. 4. Middle Devonian palinspastic reconstruction of the Arctic using present coastal outlines.

well known, Cenozoic ocean-spreading separated Svalbard from Greenland.

Relative motions between Laurentia and other units

The Arctic Alaska Plate comprising northern Alaska and Chukotka collided with northern Greenland and the Canadian Arctic Islands, causing the late Devonian Ellesmerian orogeny. The collision may have included a dominant sinistral transcurrent component because subsequent to this the Arctic Alaska Plate moved southwards along the margin of the Canadian Arctic Islands, coming to rest in early Carboniferous times with Point Barrow roughly opposite Prince Patrick Island. This model is supported: (1) by evidence for a simultaneous mid-Palaeozoic orogenic event in northern Alaska, the Canadian Arctic Islands and northern Greenland; (2) by facies distributions in northern Alaska and the Canadian Arctic Islands (Sweeney, 1982; confirmed by our own work); and (3) by the presence of major sinistral strike-slip faults in northern Ellesmere Island (Trettin, 1983) and northern Greenland (Håkansson and Pedersen, 1982).

The exact position of the Siberian Plate throughout late Palaeozoic times is not well constrained and we have relied, in these reconstructions, on the palaeomagnetic evidence of Khramov *et al.* (1981).

LATE CARBONIFEROUS (PENNSYLVANIAN) TO EARLY TRIASSIC KINEMATICS

After the mid-Carboniferous strike-slip movements between North America and Europe had ceased, these continents remained fixed with respect to each other until late Cretaceous time (Fig. 5). The uplifted areas in eastern Greenland gradually subsided, and parts of eastern Greenland were invaded by the late Permian sea. Vertical fault movements in Svalbard also gradually ceased and by Permian time Svalbard had become a stable block (e.g. Wright, 1976: unpublished CASP report).

The major tectonic event of this time interval was the collision of the Russian and Siberian Platforms, which resulted in the uplift of the Ural Mountains (Hamilton, 1970) (Fig. 6). Because the suture zone for this collision is not exposed in the Polar or Subpolar Urals, but lies to the east beneath the West Siberian Lowlands, much is still unknown about the nature of this collision, but palaeomagnetic evidence (Khramov *et al.*, 1981) suggests, and stratigraphic evidence confirms, that it was a Permian event.

The first sign of an impending collision was the appearance, in late Carboniferous time, of a clastic source area to the east of the Uralian Geosyncline. Influence from this sediment source gradually increased throughout early Permian time, and during Artinskian time, the Urals Geosyncline was uplifted and eroded. In the Pechora Basin just west of the Ural Mountains some 5 km of late Permian coal-bearing fluvial, deltaic, and lagoonal sediments were deposited (Yaroslavtsev, 1965), indicating very rapid subsidence in the Pechora Basin, and erosion of uplifted areas to the east.

During Triassic time sedimentation in the Pechora Basin continued to be dominated by continental clastics (Gorskiy, 1973), but the rate of sedimentation was considerably less than during late Permian time. To the east of the Ural Mountains, the West Siberian Basin began to form on the site of the former orogenic belt. Downfaulting and graben formation took place, and across much of the West Siberian Lowlands, Siberian Platform, and Taymyr Peninsula, significant volumes of trap volcanics were extruded, and tuffaceous sediments were deposited. This indicates a period of extension in the area. Palaeomagnetic evidence (Khramov *et al.*, 1981) suggests that the northern edge of the Siberian Platform may have moved slightly away from the Russian Platform during Triassic time, but palaeomagnetism is hardly precise enough for establishing such small movements.

MID-TRIASSIC TO MID-CRETACEOUS KINEMATICS

The main Mesozoic events were the rotation of Alaska and Chukotsk from a position adjacent to the Canadian Arctic Islands (Fig. 6), in late Jurassic—earliest Cretaceous times, and the assembly of microplates to form the Soviet Far East.

Canada Basin

Evidence in favour of rotation of the Arctic Alaska plate includes: (1) the controversial data of Newman *et al.* (1977); (2) sea-floor spreading anomalies in the Canada Basin (Vogt *et al.*, 1982); (3) a possible extinct spreading axis (Taylor *et al.*, 1981); and (4) estimates of the age of the Canada Basin from seismic data (Baggeroer and Falconer, 1982).

Stratigraphic evidence in favour of Alaskan rotation includes: (1) the uplift and erosion of the Barrow Arch in late Jurassic and early Cretaceous times, which could be considered as a thermal effect of the early rifting stage, (2) the uplift of the Brooks Range and the emplacement of ophiolites there in late Jurassic times, which correlates well with initial rotation of Alaska, (3) the influence of the northerly sediment source in northern Alaska which began to wane in late Jurassic times, and (4) Jurassic igneous activity offshore which is evidence of spreading in the Canada Basin (Tailleur, 1973).

Structural evidence such as (1) normal dip-slip faults off northern Alaska, (2) the Jurassic development of the Eglington Graben, (3) the existence of possible transform faults off the coast of the Canadian Arctic Islands (Vogt *et al.*, 1982) and Alaska (Grantz *et al.*, 1979, 1981), and (4) movements of the Rapid Fault Array (Young *et al.*, 1976) also support Alaskan rotation.

Soviet Far East

The New Siberian Islands are shown to the west of their present position, to avoid overlap with the rotating Arctic Alaska Plate. We suggest that as the Arctic Alaska Plate swung past them, they were dragged from their original position, causing volcanism and emplacement of ophiolites.

In the Soviet Far East there is evidence for island arc volcanism since at least Triassic times. Arkhipov *et al.* (1981) noted the approach of island arcs towards the East Siberian margin in late Triassic

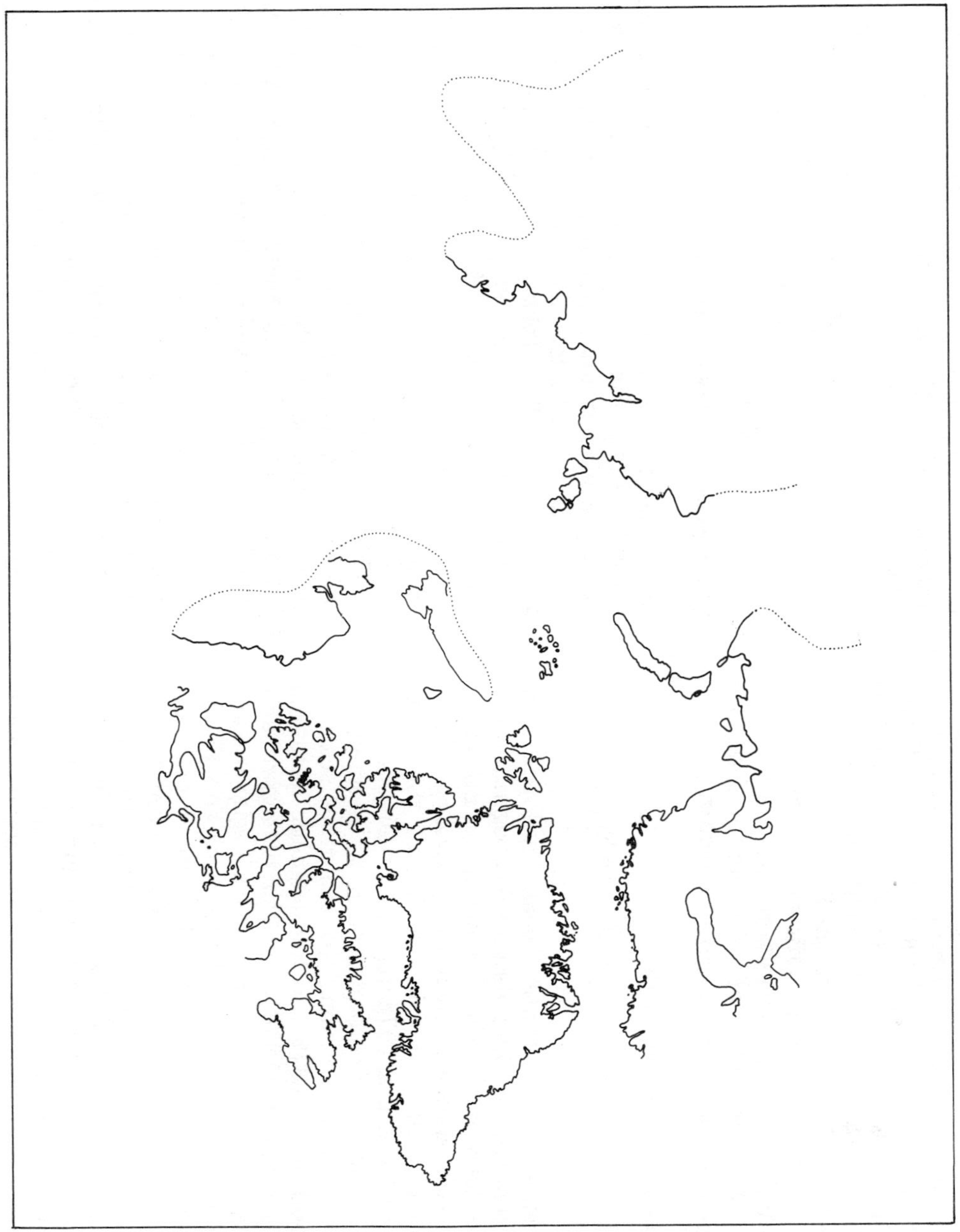

Fig. 5. Late Carboniferous (Pennsylvanian) palinspastic reconstruction of the Arctic using present coastal outlines.

Fig. 6. Late Permian palinspastic reconstruction of the Arctic using present coastal outlines.

and Jurassic times, the time of development of the Inyali-Debin Megasynclinorium which is interpreted as a trailing trough (Parfenov *et al.*, 1981). The occurrence of ophiolites is also evidence for subduction along the East Siberian margin. Omolon collided with Prikolymsk (Fig. 7) in late Jurassic times and the rotating Alaska Plate then collided with Omolon in Neocomian (early Cretaceous) times causing folding, uplift and emplacement of ophiolites in the South Anyuy Foldbelt (Fujita and Newberry, 1982).

Atlantic Arctic sector

Svalbard was affected by minor fault movements which caused the difference in thicknesses of Upper Jurassic rocks across the Billefjorden Fault Zone. Late Jurassic to early Cretaceous magmatic activity is evident along the northern margin of the Barents Shelf. These could have been related to initial rifting events which affected the East Greenland and Norwegian margin or else Alaskan rotation. The magmatic activity could have been a manifestation of the high mantle heat flux in the general region of later fission of the Eurasia Basin.

MID-CRETACEOUS TO HOLOCENE KINEMATICS

The Cenozoic evolution of the Arctic was dominated by the opening of the Labrador Sea, Baffin Bay, the Greenland-Norwegian Sea and the Eurasia Basin. These motions are associated with the Atlantic opening in which story Svalbard was again in a key position for the Arctic. In addition, there were complex motions related to the North Pacific sector in both Alaska and the Western Cordillera with its many accreted and allochthonous terranes; and also on the other side in the Soviet Far East. In both cases the Cenozoic Pacific events were a direct continuation of Mesozoic tectonic developments.

From the perspective of this symposium we give prominence to the Atlantic—Arctic sector.

Atlantic-Arctic sector

Labrador Sea to the Nares Strait (the Eurekan orogeny)

Divergence in the Atlantic Ocean developed from south to north and the Labrador Sea began to open prior to the separation of Greenland and Europe.

Magnetic anomaly lineations provide evidence for the earliest Cenozoic opening of the North Atlantic (Srivastava, 1978, fig. 4). Crustal generation in the Labrador Sea began prior to anomaly 32 (about 73 Ma according to the time scale of Hailwood *et al.*, 1979 and the magnetic lineations given by Srivastava, 1978). However, rifting probably pre-dates this event and may have begun in early Cretaceous times (Pierce, 1982; Rice and Shade, 1983). Sea floor-spreading in the Labrador Sea and Baffin Bay continued until late Eocene times.

Spreading between Greenland and North America was most extensive in the south and Greenland rotated anti-clockwise while moving north. The result was transpression in Svalbard and sinistral transpression in the Nares Strait area and the Queen Elizabeth Islands to the north. This impact was recorded in the Eurekan orogeny (e.g. Harland, 1965, 1969; Kerr, 1981). The northwest-trending Cornwall and Princess Margaret Arches were developed on the site of more ancient structures in the Sverdrup Basin at this time (Balkwill, 1978).

There has been some debate as to whether the Nares Strait marks a zone of substantial or insignificant strike-slip faulting (the Wegener Fault). There is also the embarrassing gap commonly depicted between Ellesmere Island and Greenland which derives from Bullard *et al.* (1965) who assumed, for the purpose of map-making, a purely geometrical displacement between Greenland and a wholly undistorted North America, i.e. two rigid plates without internal deformation.

Both problems can be resolved by transpressive tectonics:

(1) A combination of sinistral displacement along Nares Strait and transpression in the Canadian Arctic Archipelago has been invoked to compensate for the opening of Baffin Bay and the Labrador Sea. This is supported by a recent reappraisal of the evidence which suggests that displacement along the Nares Strait is limited to 90 km or less (Rice and Shade, 1983) and that crustal shortening by thrust faulting occurred in Ellesmere Island in Palaeogene time (Håkansson and Pedersen, 1982; Håkansson, personal communication 1983.). Moreover there is new evidence that at least many thrusts reveal oblique, i.e. transpressive, motion (Irving, personal communication, 1983).

(2) Ellesmere Island must have been attached to Greenland throughout Cenozoic time. Although some northward movement of Greenland relative to Ellesmere Island is possible (Kristofferson and Talwani, 1977), the geologic data indicate that it could not have been as much as suggested by Sclater *et al.* (1977). In any case, the Eurekan structures throughout the Queen Elizabeth Islands, if restored to their Cretaceous undeformed plan, would account for significant shortening and possibly enough to fill the gap.

The Greenland-Norwegian Sea

Spreading of this extension of the North Atlantic began later than that of the Labrador Sea, although exactly when has not been agreed. Spreading began sometime prior to magnetic anomaly 24 (Pitman and Talwani, 1972; Talwani and Eldholm, 1977). Although this has traditionally been dated at 63 Ma using the time-scale of Heirtzler *et al.* (1968), recent workers now use the Hailwood *et al.* (1979) scheme and date this event at 53 Ma (cf. Soper *et al.*, 1982).

This coincided with spreading on the Nansen—Gakkel Ridge (Pitman and Talwani, 1972) as well as the eruption of tholeiitic basalt on to East Greenland (Brooks and Nielson, 1982). After the initial separation Greenland moved northward and anti-clockwise with respect to the North American Plate, and dextral transpression occurred along the Spitsbergen Fracture Zone with concomitant Eocene deformation in Svalbard — the West Spitsbergen orogeny (Harland, 1965, 1969; Pitman and Talwani, 1972; Lowell, 1972; Harland and Horsfield, 1974; Harland, 1975). This motion continued till late Eocene when spreading ceased

Fig. 7. Early Cretaceous palinspastic reconstruction of the Arctic using present coastal outlines.

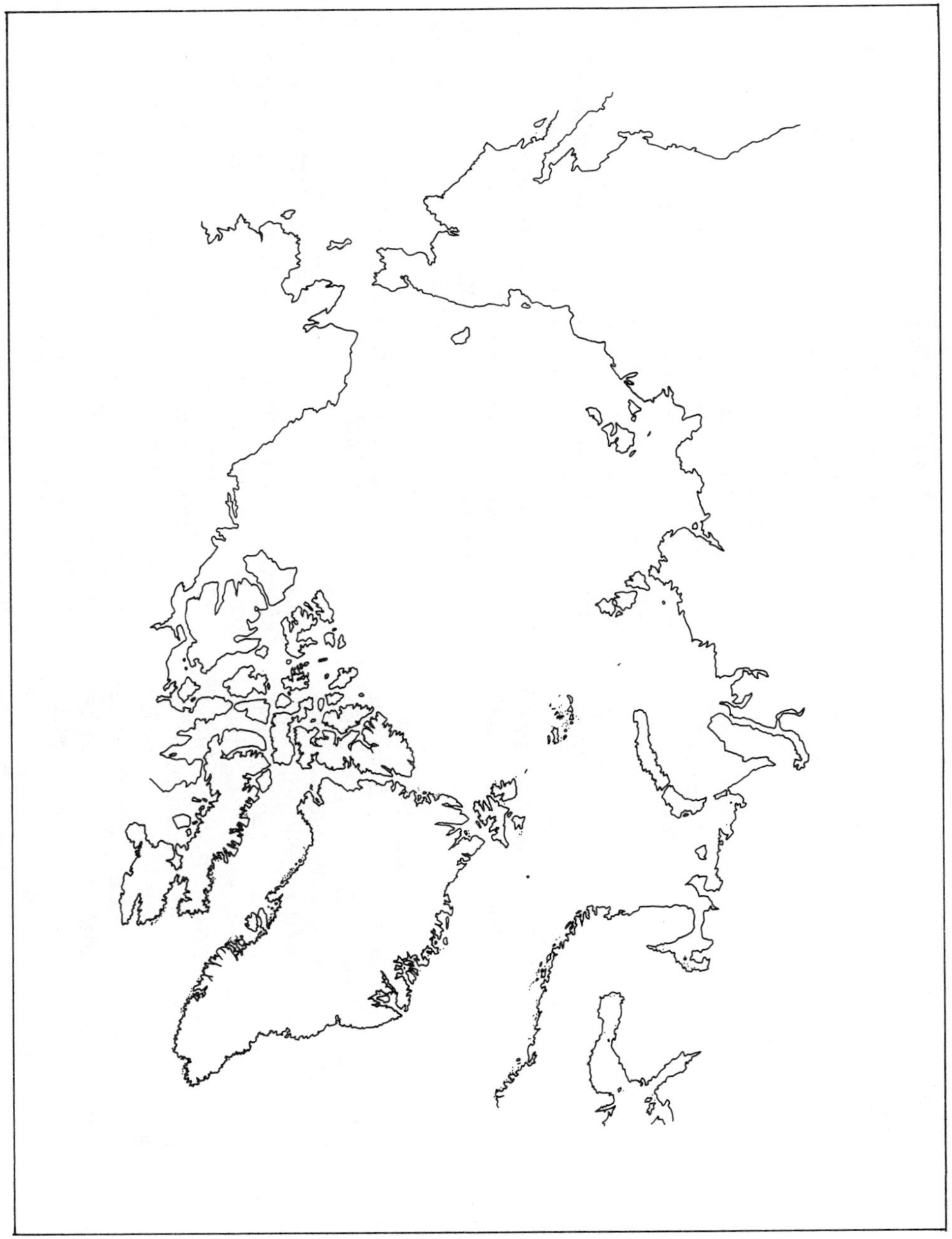

Fig. 8. Mid-Palaeogene reconstruction of the Arctic using present coastal outlines.

in the Labrador Sea (Fig. 8).

The cessation of Greenland's separation from America was reflected in a change of direction of ocean spreading between Greenland and Europe and the formation of the Greenland Sea. Prior to this time only the Norwegian Sea began opening (Pitman and Talwani, 1972). This two-stage development of the Greenland Norwegian Sea is recorded in the configuration of spreading centres and transforms as well as the distribution of magnetic lineations. These indicate that the shift in spreading occurred between magnetic anomalies 13 and 19 i.e. about 35—42 Ma (the magnetic anomaly ages given in Pitman and Talwani, 1972 and Talwani and Eldholm, 1977, have been adjusted to fit the Hailwood *et al.*, 1979, magnetic time scale).

The Eurasia Basin

Heezen and Ewing (1961) first noted the Eurasia Basin as a locus of spreading extending from the Atlantic Ocean. The Gakkel Ridge (or Nansen—Gakkel Ridge) is not a direct continuation of the Mid-Atlantic Ridge (as suggested by Wilson, 1963) but terminates the dextral transform fault system of the Spitsbergen Fracture Zone and related fault zones in the Greenland-Norwegian Sea. Initial opening of the Eurasia Basin commenced prior to magnetic anomaly 24 (Pitman and Talwani, 1972; Mair and Forsyth, 1982; Vogt *et al.*, 1982). Prior to substantial accretion along the Nansen—Gakkel Ridge, the areas now containing the Yermak Plateau, the Morris Jessup Rise and Peary Land of northern Greenland were approximately adjacent (Feden *et al.*, 1979). The Lomonosov Ridge was also attached to the Barents Shelf prior to Gakkel spreading (Wilson, 1963; Sweeney, 1982; Mair and Forsyth, 1982).

Spreading in the Eurasia Basin and the Greenland-Norwegian Sea continues today with activity along the Nansen—Gakkel, Reykjanes, Kolbeinsey, Mohns and Atka Ridges.

The Pacific—Arctic sector

In mid-Cretaceous times, a continental margin volcanic belt (Okhotsk—Chukotsk Volcanic Belt) began to develop on the Pacific margin of the Soviet Far East as oceanic material was subducted under the augmented Eurasian continent. This volcanic activity terminated in Senonian and Palaeogene times with the collision of the Sea of Okhotsk Plate and accretion of oceanic thrust slices (Koryak terranes) to the Pacific margin. The boundaries of each of these elements is marked by a belt of ophiolites. Volcanism was then further developed on the Pacific margins of Koryak and Kamchatka in essentially its modern configuration. The effects of the opening of the North Atlantic involved plate motions and tectonism elsewhere. These events are very much less well understood. However they may include:

(1) Oroclinal deformation of the Bering Strait and related fault movements in Chukotsk and Alaska. As much as 800 km of compression could be involved (Patton and Tailleur, 1977). The best known of these faults is the Kaltag Fault where up to 135 km of movement may have occurred since mid-Cretaceous times.

(2) Dextral movements of 'transform' faults in the Kolyma-Indigirka region such as the Darpir Fault (Kosygin and Parfenov, 1981).

(3) The possible extension of Eurasia Basin sea-floor spreading into the Soviet Far East.

(4) Further compressive deformation in the Verkhoyansk Foldbelt.

(5) Possible further extension of the West Siberian Lowlands.

Taken together, these movements eliminate the need for a large pre-Cenozoic ocean basin between North America and Eurasia in the Bering Strait as shown, for example, by Smith *et al.* (1981).

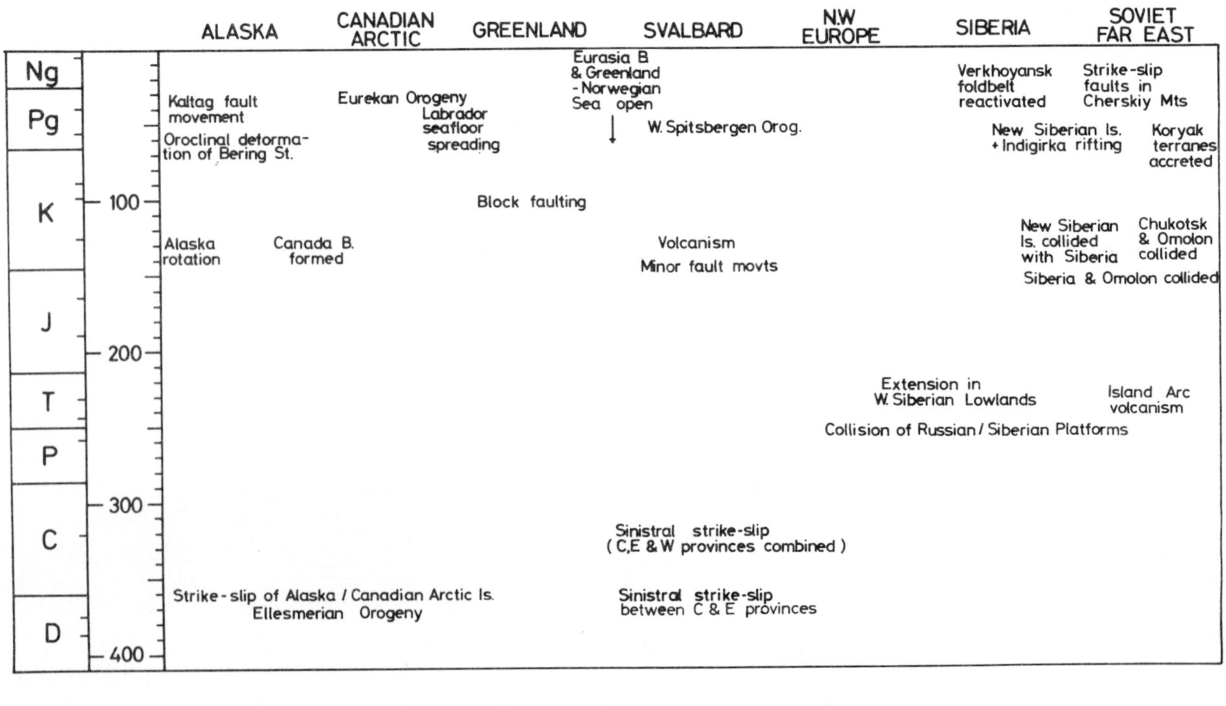

Fig. 9. Summary diagram of principal kinematic events referred to in the text.

CONCLUSION

The tectonic evolution of any large area is difficult to envisage when, as in the Arctic, it contains a number of lithospheric plates dividing or combining and moving relative to one another in the process. Hence the timing of the post-Caledonian Arctic events discussed in the text has been summarized in Fig. 9

From the preceding kinematic history of the Arctic it can be seen that Svalbard occupies a position of great importance. It is a major key to the problem of the relative movements of Greenland and Baltica, and the opening of the Eurasia Basin and Greenland-Norwegian Sea. However, it may also be able to provide important constraints on the rotation of Arctic Alaska away from the Canadian Arctic Islands and the earlier strike-slip movement of Alaska along the Canadian Arctic Islands. According to our reconstruction (Figs. 4 and 5), only the narrow Lomonosov Ridge separated Svalbard from Chukotsk, a part of the Arctic Alaska Plate. The strike-slip movement between Alaska and the Canadian Arctic Islands appears to have occurred at about the same time as the initial strike-slip movements in Svalbard. We thus conclude that a pan-Arctic perspective is needed to fully understand the geological history of any single region of the Arctic.

REFERENCES

Arkhipov, Yu.V., Klimash, V.P., Popov, L.N. Trushelev, A.M. and Shashkina, J.A. 1981. Structure of the Blyn'ya Block in the central part of the Adycha-Taryn Fault (Upper course of the Adycha River). *Geotectonics*, 15(4), 342—351.

Baggeroer, A.B. and Falconer, R. 1982. Array refraction profiles and crustal models of the Canada Basin. *J. Geophys. Res.*, **87**, 5461—5476.

Balkwill, H.R. 1978. Evolution of Sverdrup Basin, Arctic Canada. *Am. Assoc. Petrol. Geol. Bull.*, **62**, 1004—1028.

Brooks, L.K. and Nielsen, I.F. 1982. The Greenland continental margin: A transition between oceanic and continental magmatism. *J. Geol. Soc. Lond.*, **139**, 265—275.

Bullard, E.C., Everett, J.E. and Smith, A.G. 1965. The fit of the continents around the Atlantic. *Phil. Trans. R. Soc. Lond.*, A, **258**, 45—51.

Carey, S.W. 1955. The orocline concept in geotectonics. *R. Soc. Tasmania Pap. Proc.*, **89**, 255—288.

Churkin, M. Jr. and Trexler, J.H. Jr. 1980. Circum-Arctic accretion isolating part of a Pacific plate to form the nucleus of the Arctic Basin. *Earth Planet. Sci. Lett.*, **48**, 356—362.

Feden, R.H., Vogt, P.R. and Fleming, H.S. 1979. Evidence for the Yermak hot spot, Arctic Ocean. *Earth Planet. Sci. Lett.*, 4, 18—38.

Fujita, K. and Newberry, J.T. 1982. Tectonic evolution of northeastern Siberia and adjacent regions. In: G.L. Johnson and J.F. Sweeney (eds.). Structure of the Arctic, *Tectonophysics*, **89**, 337—357.

Grantz, A., Eittreim, S. and Dinter, D.A. 1979. Geology and tectonic development of the continental margin north of Alaska. *Tectonophysics*, **59**, 263—291.

Grantz, A., Eittreim, S. and Whitney, O.T. 1981. Geology and physiography of the continental margin north of Alaska. In: A.E.M. Nairn, M. Churkin, Jr. and F.G. Stehli (eds.). *The Ocean Basins and Margins*, Vol. 5 *The Arctic Ocean.* Plenum Press, New York, pp. 439—492.

Gorskiy, V.P. 1973. The Triassic system. In: L.O. Kiparisova, G.D. Radchenko and V.P. Gorskiy (eds.). *The Stratigraphy of the USSR.* Nedra, Moscow, pp. 145—159.

Hailwood, E.A., Bock, W., Costa, L., Dupeuble, P.A., Muller, C. and Schnitker, D. 1979. Chronology and biostratigraphy of Northeast Atlantic sediments, DSDP Leg 48. In: Initial Reports Deep Sea Drilling Project, Vol. 48, U.S. Govt. Printing Office, Washington, D.C. pp. 1119—1141.

Håkansson, E. and Pedersen, S.A.S. 1982. Late Palaeozoic to Tertiary tectonic evolution of the continental margin in North Greenland. *Can. Soc. Petrol. Geol. Mem.*, **8**, 331—348.

Hamilton, W. 1970. The Uralides and the moti… of the Russian and Siberian Platforms. *Geol. Soc. … Bull.*, **81**, 2553—2576.

Harland, W.B. 1965. The tectonic evolution of the Arctic—North Atlantic Region. *Phil. Trans. R. Soc. Lond.*, B, **258**, *59—75*.

Harland, W.B. 1969. Contribution of Spitsbergen to understanding of tectonic evolution of North Atlantic region. *Am. Assoc. Petrol. Geol. Mem.*, **12**, 817—851.

Harland, W.B. 1971. Tectonic transpression in Caledonian Spitsbergen. *Geol. Mag.*, **108**, 27—42.

Harland, W.B. 1975. Essay Review: Palaeogene correlation in and around Svalbard. *Geol. Mag.*, **112**, 421—429.

Harland, W.B. and Gayer, R.A. 1972. The Arctic Caledonides and earlier oceans. *Geol. Mag.*, **109**, 289—364.

Harland, W.B. and Horsfield, W.T. 1974. West Spitsbergen Orogen. In: A.M. Spencer (ed.). *Mesozoic-Cenozoic orogenic belts, Data for orogenic studies.* Geol. Soc. Lond. Spec. Pub. 4, pp. 747—755.

Harland, W.B. and Wright, N.J.R. 1979. Alternative hypothesis for the pre-Carboniferous evolution of Svalbard. *Norsk Polarinst. Skr.*, **167**, 90—117.

Harland, W.B., Cutbill, J.L., Friend, P.F. Gobbett, D.J., Holliday, D.W., Maton, P.I., Parker, J.R and Wallis, R.H., 1974. The Billefjorden Fault Zone, Spitsbergen. *Norsk Polarinst. Skr.*, No. 161, 72pp.

Harland, W.B., Perkins, P.J. and Gaskell, B.A. Outline of Arctic pre-Permian continental displacements. *Geol. Mag.* (in press).

Heezen, B.C. and Ewing, M. 1961. The mid-oceanic ridge and its extension through the Arctic Basin. In: G.O. Raasch (ed.). *Geology of the Arctic*, Univ. Toronto Press, Toronto, pp.622—642.

Heirtzler, J.R., Dickson, G.O., Herron, E.C., Pitman, W.C. and Le Pichon, X. 1968. Marine magnetic anomalies geomagnetic field reversals, and motions of the ocean floor and continents. *J. Geophys. Res.* **73**, 2119—2136.

Kanasewich, E.R., Havskov, J. and Evans, M.E. 1978. Plate tectonics in the Phanerozoic. *Can. J. Earth Sci.*, **15**(6), 919—955.

Kent, D.V. and Opdyke, N.D. 1978. Palaeomagnetism of the Devonian Catskill Beds: Evidence for motion of the coastal New England—Canadian Maritime region relative to cratonic North America. *J. Geophys. Res.*, 83(B9), 4441—4450.

Kent, D.V. and Opdyke, N.D. 1979. The Early Carboniferous palaeomagnetic field of North America and its bearing on tectonics of the northern Appalachians. *Earth Plan. Sci. Lett.*, 44, 365—372.

Kerr, J.W. 1981. Evolution of the Canadian Arctic Islands: a transition between the Atlantic and Arctic Oceans. In: A.E.M. Nairn, M. Churkin Jr. and F.G. Stehli (eds.). *The Ocean Basins and Margins*, Vol. 5, *The Arctic Ocean*, Plenum Press, New York, pp.105—199.

Khramov, A.N., Petrova, G.N. and Perchersky, D.M. 1981. Paleomagnetism of the Soviet Union. In: M.W. Elhinny and D.A. Valencio (eds.). Palaeoreconstruction of the Continents, *Geodynamics*, Ser. 2, pp.177—194.

Kosygin, Y.A. and Parfenov, L.M. 1981. Tectonics of the Soviet Far East. In: A.E.M. Nairn, M. Churkin Jr. and F.G. Stehli (eds.). *The Ocean Basins and Margins*, Vol. 5, *The Arctic Ocean*, Plenum Press, New York, pp.377—412.

Kristoffersen, Y. and Talwani, M. 1977. Extinct triple junction south of Greenland and the Tertiary motion of Greenland relative to North America. *Geol. Soc. Am. Bull.*, **88**, 1037—1049.

Lowell, J.D. 1972. Spitsbergen Tertiary orogenic belt and Spitsbergen fracture zone. *Bull. Geol. Soc. Am.*, **83**(10), 3091—3101.

Mair, J.A. and Forsyth, D.A. 1982. Crustal structures of the Canada Basin near Alaska, the Lomonosov Ridge and adjoining basins near the North Pole. *Tectonophysics*, **89**, 239—254.

McWhae, J.R.H. 1952. The major fault zone of Central Vestspitsbergen. *Q. J. Geol. Soc. (London)*, **108**, 209—232.

Meyerhoff, A.A. 1978. Cambridge Arctic Shelf Programme (CASP). *J. Petrol. Geol.*, 1(1), 112.

Morel, P. and Irving, E. 1978. Tentative paleocontinental maps for the Early Phanerozoic and Proterozoic. *J. Geol.*, **86**(5), 535—561.

Newman, G.W., Mull, C.G. and Watkins, N.D. 1977. Northern Alaskan paleomagnetism, plate rotation, and tectonics in the relationship of plate tectonics to Alaskan geology and resources. *Proc. Alaska Geol. Soc. Symp.*, C1—C7.

Parfenov, L.M., Natal'in, B.A., Voynova, F.P. and Popeko, L.I. 1981. Tectonic evolution of active continental margins along the northwest margin of the Pacific Ocean. *Geotectonics*, 15(1), 54—67.

Patton, W.W. and Tailleur, I.L. 1977. Evidence in the Bering Strait region for differential movement between North America and Eurasia. *Geol. Soc. Am. Bull.*, **88**, 1298—1304.

Peirce, J.W. 1982. The evolution of the Nares Strait lineament and its relation to the Eurekan orogeny. In: P.R. Dawes and J.W. Kerr (eds.), Nares Strait and the drift of Greenland: a conflict in Plate Tectonics. *Medd. om Grønland, Geoscience*, 8, 237—254.

Pitman, W.C. III and Talwani D. 1972. Sea-floor spreading in the North Atlantic. *Geol. Soc. Am. Bull.*, **83**, 619—646.

Rice, P.D. and Shade, B.D. 1983. Reflection seismic interpretation and sea-floor spreading history of Baffin Bay. *Can. Soc. Petrol. Geol. Mem.*, 8, 245—265.

Sclater, J.C., Hellinger, S. and Tapscott, C. 1977. The paleobathymetry of the Atlantic Ocean from the Jurassic to the Present. *J. Geol.*, **85**, 509—552.

Scotese, C.R., Bambach, R.K., Barton, C., van der Voo, R. and Ziegler, A.M. 1979. Paleozoic base maps. *J. Geol.*, **87**(3), 217—278.

Smith, A.G., Hurley, A.M. and Briden, J.C. 1981. *Phanerozoic Palaeocontinental World Maps.* Cambridge University Press, Cambridge, 102 pp.

Soper, N.J., Dawes, P.R. and Higgins, A.K. 1982. Cretaceous—Tertiary magmatic and tectonic events in North Greenland and the history of adjacent ocean basins. In: P.R. Dawes and J.W. Kerr (eds.), Nares Strait and the Drift of Greenland: a conflict in Plate Tectonics. *Medd. om Grønland, Geoscience*, 8, 205—220.

Srivastava, S.P. 1978. Evolution of the Labrador Sea and its bearing on the early evolution of the North Atlantic. *Geophys. J. R. Astron. Soc.*, 52(2), 313—357.

Steel, R.J. 1976. Devonian basins of western Norway — sedimentary response to tectonism and to varying tectonic context. *Tectonophysics*, **36**, 207—224.

Steel, R.J. 1978. Late orogenic Devonian basin formation in the western Norwegian Caledonides. *Geol. Surv. Can. Pap.* 78-13, pp. 57—62.

Sweeney, J.F. 1982. Mid-Palaeozoic travels of Arctic—Alaska. *Nature, Lond.*, **298**, 647—649.

Sweeney, J.F., Weber, J.R. and Blasco, S.M. 1982. Continental ridges in the Arctic Ocean: LOREX constraints. *Tectonophysics*, **89**, 217—237.

Tailleur, I.L. 1973. Probable rift origin of the Canada Basin, Arctic Ocean. In: M.G. Pitcher (ed.), Arctic Geology, AAPG *Mem.*, **19**, 526—535.

Tailleur, I.L. and Brosgé, W.P. 1970. Tectonic history of northern Alaska. In: W.L. Adkinson and M.M. Brosgé (eds), *Proceedings of the Geological Seminar on the North Slope of Alaska,* Palo Alto, California, 1970. Los Angeles, AAPG Pacific Section, E1—E19.

Talwani, M. and Eldholm, O. 1977. Evolution of the Norwegian-Greenland Sea. *Geol. Soc. Am. Bull.*, **88**, 969—999.

Taylor, P.T., Kovacs, L.C., Vogt, P.R. and Johnson, G.L. 1981. Detailed aeromagnetic investigation of the Arctic Basin. *J. Geophys. Res.*, **86**(B7), 6323—6333.

Trettin, H.P. 1983. Innuitian Orogen and Arctic Platform. In: A.R. Palmer (ed.), *Perspectives in Regional Geological Synthesis. Planning for the Geology of North America,* Geol. Soc. Am., Boulder, Colorado.

van der Voo, R., French, A.N. and French, R.B. 1979. A palaeomagnetic pole position from the folded Upper Devonian Catskill red beds, and its tectonic implications. *Geology*, **7**(7), 345—348.

van der Voo, R. and Scotese, C. 1981. Palaeomagnetic evidence for a large (2000 km) sinistral offset along the Great Glen fault during Carboniferous time. *Geology,* **9**(12), 583—589.

Vogt, T. 1928. Den Norske Fjellkjedes revolusjons — historie. *Norsk Geol. Tidsskr.*, Bd. 10, 97—115.

Vogt, P.R., Taylor, P.T., Kovacs, L.C. and Johnson, G.L. 1982. The Canada Basin: aeromagnetic constraints on structure and evolution. In: G.L. Johnson and J.F. Sweeney (eds.). Structure of the Arctic, *Tectonophysics,* **89**, 295—336.

Watts, D.R. 1983. Lower Carboniferous palaeomagnetic results from the Billefjorden region of Spitsbergen. Abstr., European Geophysical Union Meeting, Spring 1983.

Wilson, J.T. 1963. Continental drift. *Sci. Am.*, **208**, 86—100.

Yaroslavtsev, G.M. (ed.) 1965. Geologiya mestorozhdeniy uglya i goryuchikh slautsev SSSR. Tom 3. Pechorskiy ugol-nyy basseyn i drugiye mestorozhdeniya uglya komi ASSR i NenetskogoNational' Nogo Okgruga [Geology of coal and comb. shale USSR. Tom 3, Pechora, etc.] Moscow, Izdatel'stva *Nedra,* 491 pp.

Young, F.G., Myhr, D.W. and Yorath, C.J. 1976. Geology of the Beaufort-Mackenzie Basin. *Geol. Surv. Can. Pap.* 76-11.

Zoneshayn, L.P. and Gorodnitsky, A.M. 1977. Palaeozoic and Mesozoic reconstructions of the continents and oceans. *Geotectonics*, II, 83—94; 159—172.

10 The Lower Tertiary Balder Formation: An organogenic and tuffaceous deposit in the North Sea region

O.A. Malm, O. Bruun Christensen,* H. Furnes,† R. Løvlie,† H. Ruselåtten‡ and K. Lorange Østby**
*Statoil, Geologisk Laboratorium, Stavanger, Norway
† Universitet i Bergen, Bergen, Norway
‡ Norges Tekniske Høyskole (NTH), Trondheim, Norway

The Lower Tertiary tuffaceous Balder Formation is described on the basis of a 16-m long core from well 30/2-1 in the north Viking Graben. The core comes from a 47-m thick main tuff-zone within the 76-m thick formation. It contains 168 tuff-beds which comprise 44% of the sediment volume. The beds are 0—28-cm thick, graded (FU) and represent first-generation ash deposits. They result from explosive eruptions of magma with a composition transitional between within plate and mid-ocean ridge basalt.

Comparisons of grain-size, bed thickness and abundance with other wells in the northern North Sea, indicate that the source-volcanism for this area was located to the north and that the ashes were blown generally southwards. Comparison with the onshore Danish ash-deposits support the supposition that the latter were sourced by a separate volcanism in the Skagerak. This volcanism and other possible volcanism along the tectonic Tornquist Zone are suggested as a major source for the Balder ashes in the southern part of the North Sea. Multiple volcanic centres situated north to northwest of the North Sea and in the Skagerak can best explain the regional distribution of the Balder tuffs.

The Balder tuffs are interbedded with siliceous shales composed of microcystalline quartz and altered volcanic ash. They contain abundant pyrite pseudomorphs after siliceous microfossils. A rich, well-preserved flora of diatoms and silicoflagellates is found in carbonate-concretions. The flora is identical to that of the Mo-Clay diatomite containing the onshore Danish tuff-deposits. Certain silicoflagellate species make correlation with the upper part of the Mo-Clay possible. This correlation is confirmed by the dinoflagellate flora, which can also be recognized in the Danish Ølst Formation and in the main phase of pyroclastic deposits in the North Sea.

It is proposed that the siliceous shales represent recrystallized equivalents to diatomite. This greatly extends the known distribution of such deposits which were earlier interpreted as localized Danish deposits or suggested to be confined to the shelf areas along the Norwegian coastline. In their undeformed state both tuffs and interbedded diatomite could have excellent reservoir properties. It is suggested that the Balder Formation may have reservoir-potential in some areas of the shelf.

The volcanic ash of the tuff-beds and the siliceous shales are altered into swelling clay minerals like chlorite, chlorite-smectite and smectite. The alterations result in an exceptionally high secondary porosity (40%) for the tuffs. Unfortunately it is accompanied by very poor permeabilities (1 mD). The clay minerals formed during alteration are responsible for the swelling problems in the Lower Tertiary of the North Sea.

INTRODUCTION

The Lower Tertiary tuffaceous Balder Fm (Deegan and Scull, 1977) represents an excellent seismic and stratigraphic marker over most of the North Sea and northwards along the Norwegian Shelf in the Haltenbanken and Trænabanken areas. In spite of its importance in exploration, the composition and genesis of the formation is poorly known. This is due to a general lack of coring, as the formation is not considered of reservoir potential. However, last year the formation was cored in the Norwegian well 30/2-1, situated in the Viking Graben in the northern part of the North Sea (Fig. 1). The core, which is 16-m long and from the central part of the Balder Fm, is the subject of the present study.

The aim of the study was to increase our knowledge of the structure, composition, stratigraphy, and origin of the Balder Fm. In a broader sense, we wished to contribute to the understanding of the North Sea Basin and its environment during these times of early rifting between NW Europe and Greenland.

Petroleum Geology of the North European Margin © Norwegian Petroleum Society (Graham & Trotman, 1984) pp. 149—170.

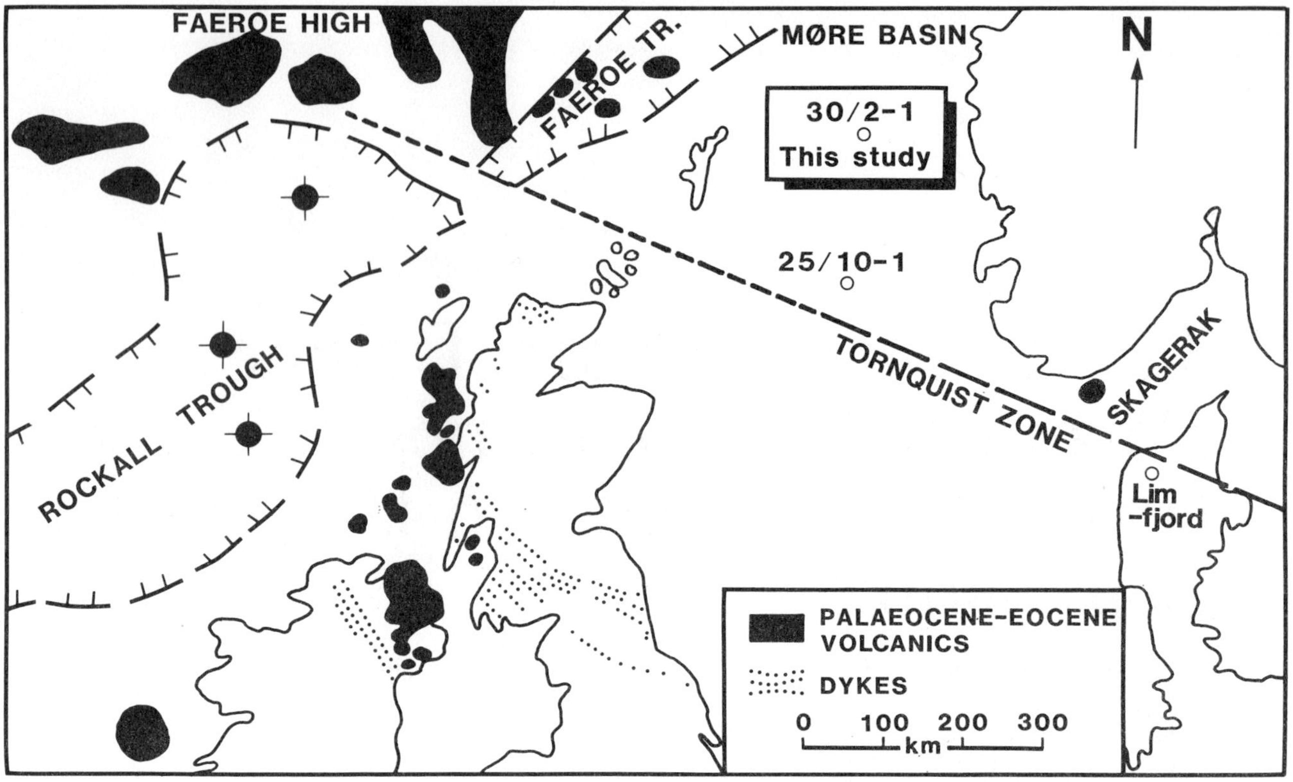

Fig. 1. Location map.

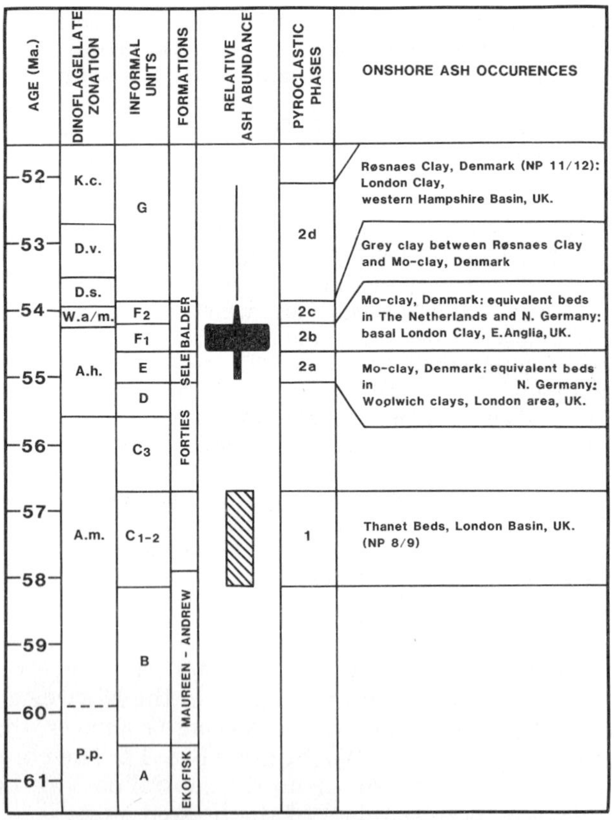

Fig. 2. Stratigraphy of the North Sea Palaeogene ashes and their on-shore equivalents. Reproduced after Knox and Morton (1983). Dinoflagellate cyst zonation: A.m., *Alisocysta margarita;* A.h., *Apectodinium hyperacanthum;* W.a., *Wetzeliella astra;* W.m., *Wetzeliella meckelfeldensis;* D.s., *Dracodinium simile;* D.v., *Dracodinium varielongitudum;* K.c., *Kisselovia coleothrypta.*

PALAEOGENE PYROCLASTIC STRATIGRAPHY

Discussions of the distribution, stratigraphy and timing of Palaeogene pyroclastic deposition in the North Sea are given by Jaqué and Thouvenin (1975) and Knox and Morton (1983). Jaqué and Thouvenin (1975) distinguished four phases of pyroclastic deposition: one main phase corresponding to the major zone of regional distribution (i.e. the Balder Fm) and three subordinate phases corresponding to zones with less abundant pyroclastic material and more local distribution.

Based on a study of cuttings and cores from more than 200 North Sea boreholes, Knox and Morton (1983) reassess the distribution of the lower Palaeogene tuffs reported by Jaqué and Thouvenin (1975). This is based on recent advances in dinoflagellate cyst correlation of both onshore and offshore sequences. The dinoflagellate zonation used is that of Costa *et al.* (1978), which is combined with the informal Palaeocene lithostratigraphic zonation of Knox *et al.* (1981). Knox and Morton (1983) divide the North Sea pyroclastic sedimentation into two main phases (Fig. 2).

Their first main Phase 1, which corresponds to Jaqué and Thouvenin's (1975) lowermost phase, is restricted to Knox *et al.*'s units C_1 and C_2 of the Montrose Gp. It is represented by sandy tuffs and tuffaceous sands in the Moray Firth Basin and southeastwards into adjacent parts of the northern Central Graben, by a well defined sandy tuff unit (C_1). Abundant graded ash layers are also identified in the mudstones of unit C_1 and C_2 from the central and northern North Sea. The ashes of Phase 1 belong to the upper part of the *Alisocysta margarita* dinocyst Zone (Fig. 2).

The second main Phase 2 of Knox and Morton expands the upper three phases of Jaqué and Thouvenin (1975). This phase is subdivided into four subphases (Fig. 2). The lowermost three subphases (2a—2c) are all present in a core from Norwegian well 25/10-1. According to descriptions of this core by Ræstad (1976) and Ræstad and Prestvik (1982) it contains a continuous series of 430 ash layers over an interval of 100 feet. They divide the sequence into three units on the basis of thickness and abundance of the ash-beds. In a similar way Knox and Morton use the same core to define their three subphases 2a, 2b and 2c.

In well 25/10-1, subphase 2a is represented by 20 ash layers between 1 and 2.5-cm thick in 27 m of shale, which Knox and Morton assign to the Sele Fm. In contrast subunits 2b and 2c are assigned to the Balder Fm (Fig. 2).

Subphase 2b constitutes the main tuff zone of the North Sea responsible for the widespread seismic marker defined in downhole logs by their low gamma-ray response and high sonic velocity This corresponds to Ræstad's (1976) unit II of well 25/10-1 which is 52-feet thick and contains 333 recorded ash layers comprising about 30% of the sediment volume. These main-phase ashes are recognized throughout the North Sea, except in the inner Moray Firth Basin and parts of the Viking Graben where sands occupy the relevant interval.

In well 25/10-1 subphase 2c (Unit III of Ræstad 1976) contains about 12 ash beds, 1-cm thick, in a 17-m thick sequence of laminated shale.

According to Knox and Morton (1983) subphases 2a and 2b both belong to the *Apectodinium hyperacanthum* dinoflagellate Zone (Fig. 2). Subphase 2c contains only a restricted facies-controlled dinocyst assemblage, but based on lithostratigraphical and log correlation with the southern North Sea and on-shore eastern England, the phase appears to range through the *Wetzeliella astra* Zone and into the *Dracodinicum simile* Zone, according to Knox and Morton (Fig. 2). They correlate the different subphases of main Phase 2 with different parts of the on-shore Danish ash-series (Fig. 2). Subphase 2d has been recognized only in cores from well 30/1-1 by Knox and Morton. Here it is represented by several graded, highly altered ash layers in distinctive red and variegated shales found above the Balder Fm over much of the North Sea area, and also in the present well.

On-shore equivalents are thought to be the greenish tuffaceous layers in the Røsnaes clays of Denmark. Subphase 2d appears to range from the upper part of the *Dracodinium simile* Zone and into the *Kisselovia coleothrypta* Zone (Fig. 2).

Based on an indirect correlation with the stratigraphical time-scales of Lowrie *et al.* (1982) and Hailwood *et al.* (1979), Knox and Morton (1983) assign an absolute time-scale to their Palaeogene pyroclastic stratigraphy (Fig. 2). This results in an age of 57—58 Ma for their main Phase 1 and an age of 54—55 Ma for their subphases 2a+2b+2c.

NORWEGIAN WELL 30/2-1

Based on log-patterns, the Balder Fm in well 30/2-1 can be divided into three distinct zones (Fig. 3). A

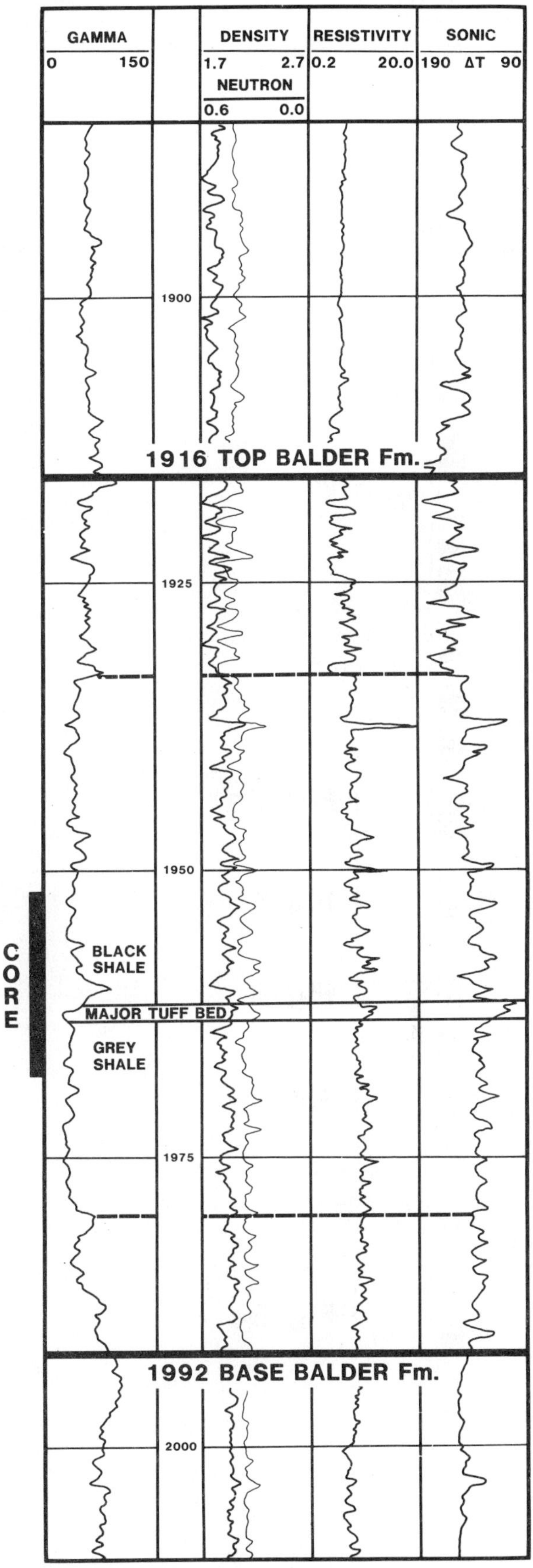

Fig. 3. The Balder Fm in Norwegian well 30/2-1 (Fig. 1). Dashed lines indicate boundaries between a lower, a central (and major) and an upper tuff-zone. A major lithological break is indicated in the core described in Fig. 4.

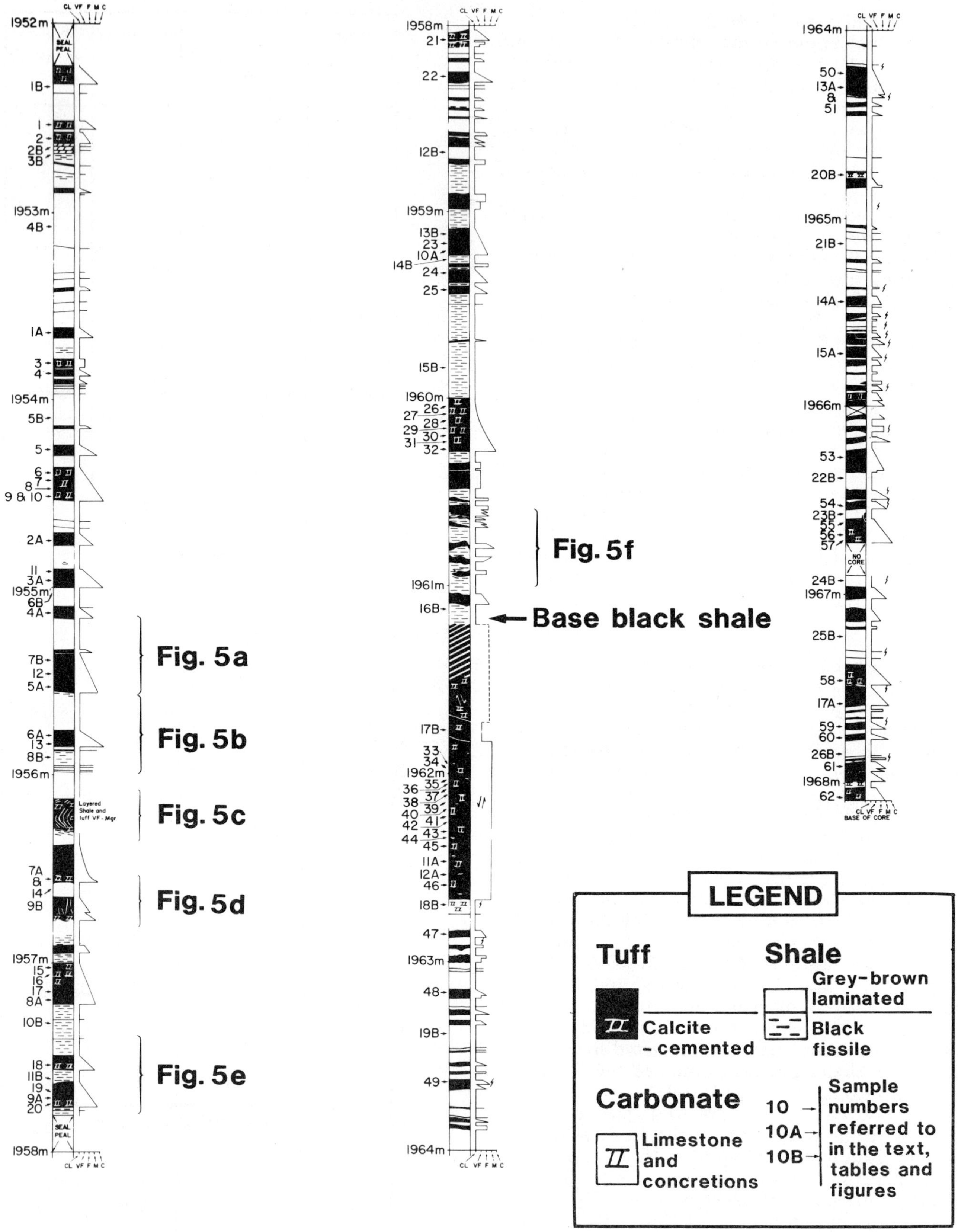

Fig. 4. Simplified lithologic description of the 16-m long core from the main tuff-zone of the Balder Fm in well 30/2-1 (see also Fig. 3).

Samples 1—62: Orientated 1-inch horizontal plugs from tuff-beds analysed for palaeomagnetism and bulk rock trace element chemistry (Fig. 11 and Table 4).

Samples 1A—17A: One-inch horizontal plugs from tuff-beds analysed for porosity—permeability (Table 3), bulk rock and clay-fraction mineralogy (Fig. 10 and Table 6) and bulk chemistry (Table 4).

Samples 1B—26B: Core-pieces of shale (and three tuff-beds) analysed for palynology (Fig. 6) and micropalaeontology (Fig. 8). Samples also analysed for bulk-rock and clay-fraction mineralogy (Fig. 10 and Table 6).

With few exceptions all samples were examined in thin sections. Selected samples and thin sections were studied with scanning electron microscopy and microprobe (Fig. 9, Table 2).

lower zone with fairly high gamma readings due to a micaceous shale, a central zone with low gamma and sonic and high resistivity readings, and an upper thinner zone with higher gamma and sonic and lower resistivity readings. The logs reflect different types of shales and different contents of tuff in the three zones. The central zone has a higher tuff/shale ratio than the upper zone. The central zone is referred to as the *main* tuff-zone of the Balder Fm. In terms of Knox and Morton's (1983) pyroclastic stratigraphy (Fig. 2), the uppermost zone corresponds to subphase 2c. The studied core comes from the central part of the main tuff-zone (subphase 2b), (Fig. 3).

CORE DESCRIPTION

The core has a length of 16.1 m. It is well preserved and disintegrated parts only amount to about 0.5 m. Two seal peals cover a core length of 0.4 m. A simplified core-log is given in Fig. 4. Four lithologies are distinguished: tuff-beds, grey-brown shales, black shales and carbonate beds. The latter comprise less than 20 cm, so the core consists predominantly of tuff-beds interbedded with grey-brown and black shales. To the right of the lithological columns are plotted grain-sizes determined on the core with the help of a rig-microscope.

The tuff-beds

The tuff-beds comprise about 685 cm or 44% of the core (Table 1). Altogether 168 individual beds have been distinguished, 49 of which are less than 1-cm thick.

With few exceptions each of the 117 beds with a thickness between 1 and 28 cm show a strong normal grading (fining upwards) with respect to grain-size (Fig. 4 and Fig. 5a, b, e). They commonly range from medium to coarse (< 1 mm) at the base to clay-size at the top. The basal contacts with underlying shales are sharp and well-defined (Fig. 5a, b, e). The top contacts are also easily distinguished due to a difference in colour and lamination between tuffs and shales. The tuff-beds have an overall massive appearance while the shales have a distinct lamination or fissility (Fig. 5a, b, e). The tuffs are dark greenish-grey, except where calcite-cementation gives them a light-grey colour. Some beds are completely cemented while others are cemented only in their coarse basal parts. Many beds are uncemented.

The core contains one exceptionally thick interval of tuff between 1962.65 and 1961.25 m (Figs. 3 and 4). This consists of two major beds with a thickness of 85 cm and 51 cm, and one minor bed (10 cm) in between. The lowermost and thickest bed has a massive appearance with a uniform grain-size (Fig. 4). The bed is heavily calcite-cemented. Scattered ellipsoid-shaped carbonate granules (a few millimetres in diameter) occur throughout the bed. In thin-section some of them appear to represent nodular concretions enclosing and partly replacing volcanic shards. However, some also have smoothly defined grain-boundaries which indicate that they represent detrital clasts. This bed also contain a few larger (0.5 cm) angular clasts of fine-grained tuff/shale. The textural data thus suggest that this bed can represent a massflow/high density current deposit.

The other bed comprising the top of this thick tuff interval is similar to that just described, but it is strongly fractured. Two generations of fractures seem to be present: early calcite-cemented fractures and later less well-cemented ones. Late, brittle and high-angle fractures are also observed in the lower bed described above.

Shales

The shales are divided into two types based on colour and lamination (Fig. 4) and content of organic carbon and mineral composition (see below): (1) light grey-brown shales with well-defined lamination and (2) black fissile shales without lamination. The two types have a distinct distribution through the core with respect to the major tuff interval between 1962.65 and 1961.25 m (Fig. 3 and 4).

The shales below this interval belong to type (1). Immediately above the major tuff interval (at 1961 m) the shales are black (type 2) and stay dominantly so for an interval of about 4 m up to 1957 m. From there to the top of the core (at 1952 m) the shales are dominantly of type (1), light grey-brown. In addition to colour and lamination, these shale types are also distinguished with respect to visible content of pyrite.

In the core, type (1) light laminated shales contain abundant pyrite (Fig. 5a and b) while type (2) black shales apparently are free from this mineral (Fig. 5e). On X-ray graphs many pyrite concretions display a dark core (light organic material?) and a light rim (heavy pyrite). Possibly they represent pyrite crystallization on the fringes of burrows. Much of the pyrite, however, represents replacements of microfossils as evidenced by pyritized diatoms. The lack of pyrite in the black shales is only apparent as X-ray diffraction analyses (Fig. 10) indicate pyrite contents in the range of 20%. This is present as microscopic crystals (Fig. 8b) not observed on the core.

The laminations in the type (1) shales are defined by colour ranging from light grey to dark brown (Fig. 5a and b). The colour is linked to the content of organic carbon, the darker laminae being richer in carbon than the lighter.

TABLE 1. Review of number of tuff-beds and their thickness measured on the core described in Fig. 4

Bed thickness (cm)	*Number of beds*	*Average bed thickness (cm)*
<1	49	0.5
1—4.5	81	2.2
5.0—8.5	20	5.9
10—13	11	11.1
18—22	4	20.3
28	1	28.0
51	1	
85	1	

Total number of beds	168
Total bed thickness	685 cm
Net core length	1550 cm
Volume % tuff	44%

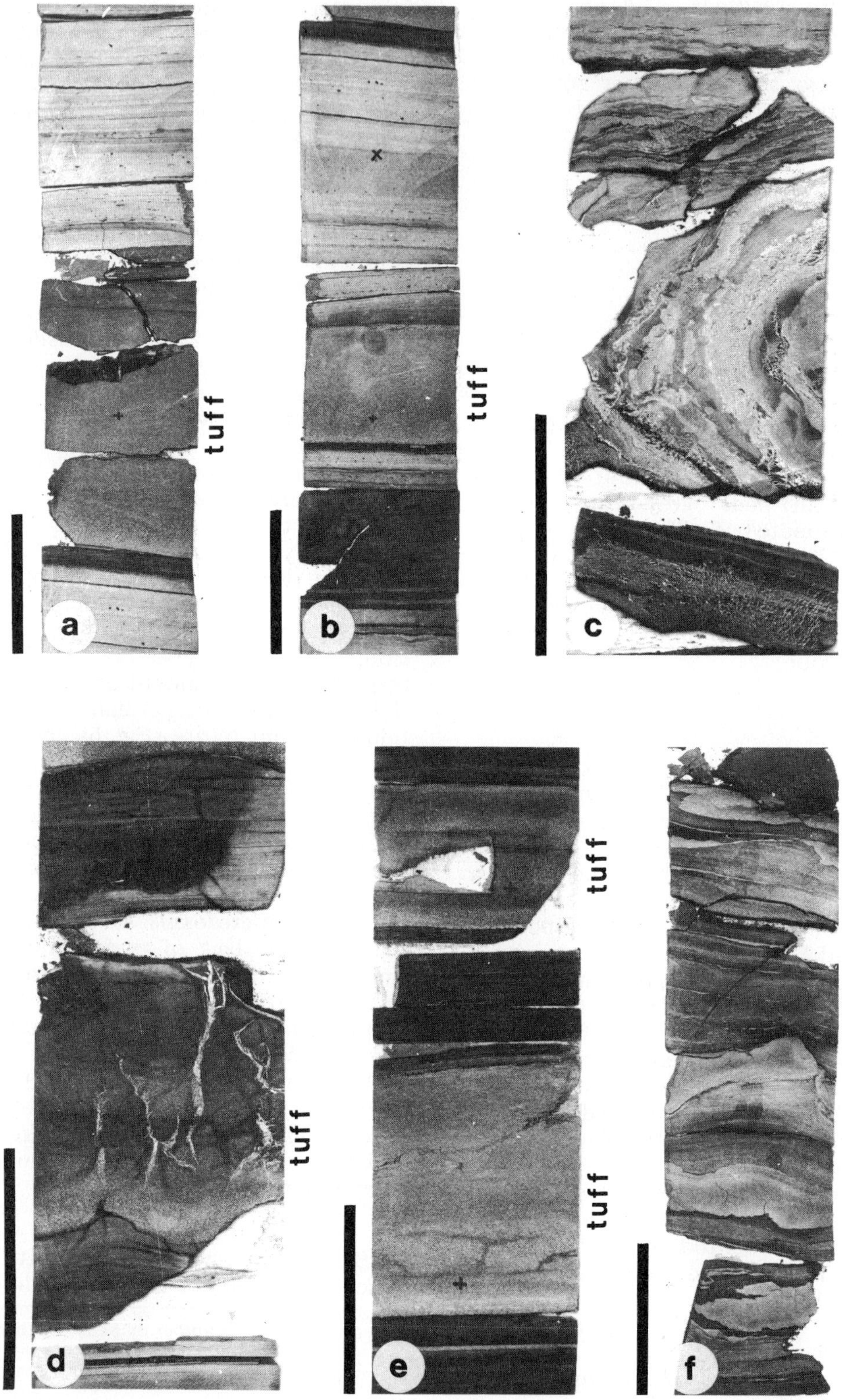

Fig. 5. Core-photographs (for location see Fig. 4). All scalebars = 10-cm. (a) A major tuff-bed approximately 20-cm thick overlain by finely laminated grey shale with abundant pyrite (black spots). Calcite-cementation is indicated in the coarse grained basal parts of the tuff-bed by lighter shades. **(b)** A tuff-bed less than 10-cm thick overlain by finely laminated grey shale (black spots, pyrite). Minor tuff-beds and black shale below. (c) Overturned slump-fold with later calcite-cementation. Undeformed shale below and above. (d) A 10-cm thick strongly graded tuff-bed with irregular calcite cemented fractures. (e) Graded and undeformed tuff-beds interbedded with black fissile shale. (f) Interbedded tuff-beds and black shale which have undergone soft sediment deformation. Note that the tuff-beds are boudinaged and partly folded between the softer shale.

Carbonate beds

The carbonate beds comprise less than 20 cm of the core (Fig. 4). One of these beds seems to represent a sedimentary limestone. It occurs directly beneath the major tuff-interval at about 1962.7 m. The bed, which is very fine-grained, contains a few scattered calcareous fossils and fossil fragments which have not been identified. The bed also displays an anomalously dense bioturbation on X-ray graphs. The bioturbation consists of 1-mm thick, very irregular worm-like burrows. Other carbonate beds occur at 1952.65 m (2B) and at 1964.75 m (20B), and probably represent concretions. They contain well-preserved siliceous microfossils which are used in the biostratigraphic zonation.

Deformation

Minor parts of the core are heavily fractured as evidenced by calcite-cemented fractures (Fig. 5d) and less cemented fractures, which often display slickenside textures. In addition there is good evidence of early soft and possibly synsedimentary deformation. For example, a 20-cm thick interval (1956.10—1956.30 cm) of interbedded tuff and shale defines one large overturned slump fold (Fig. 5c) with post-deformational calcite-cementation. The beds above and below the fold are flat-lying and undeformed. Immediately above the major tuff-interval there is a 1-m thick interval (1961.20—1960.20 m) of interbedded tuff and black shale which displays strong soft sediment deformation. The tuff-beds are here boudinaged and folded in between the softer black shales (Fig. 5f).

BIOSTRATIGRAPHY

In most late Palaeocene—early Eocene deposits of the North Sea there is a paucity of calcareous microfossils (foraminifera, ostracods, nannoplankton). Therefore, the biostratigraphy over such intervals has to be based on palynomorphs (dinoflagellates, pollen, spores) and fossils with siliceous skeletons (diatoms, silicoflagellates, radiolarians).

Palynology

A dinoflagellate zonation for Upper Palaeocene to Oligocene in North West Europe, based on the evolution of the Wetzeliellaceae, was established by Costa and Downie (1976), and later refined around the Palaeocene—Eocene boundary by insertion of the *Wetzeliella astra* Zone (Costa *et al.*, 1978). For the North Sea area the zonation has been further refined by subdivision of the *Apectodinium hyperacanthum* Zone based on an acme occurrence of *Deflandrea oebisfeldensis* in its upper part (Knox and Harland, 1979; Knox *et al.*, 1981) (Fig. 7).

The early Tertiary phases of pyroclastic deposition in the North Sea have been related to palynofloras and dinoflagellate zonations by Jaqué and Thouvenin (1975), Knox *et al.* (1981), and Knox and Morton (1983).

This study

The palynological analysis of the core shows a relatively sparse dinoflagellate cyst flora. The highest diversity is recorded from a grey shale in the lowermost sample (26B) of the core (Fig. 4). Generally, two species are dominant: *Glaphyrocysta ordinata* (Williams and Downie, 1966) Stover and Evitt, 1978 (Fig. 6a—c) and *Deflandrea oebisfeldensis* Alberti, 1959 (Fig. 6d—k). Other common or abundant dinoflagellate cysts represent species from the genera *Hystrichosphaeridium*, *Spiniferites* and in some samples from *Ceratiopsis.* The terrestrial microflora is richly represented by abundant *Taxodiaceaepollenites* and *Caryapollenites;* bisaccate pollen are also abundant.

Only one of the three tuff samples analysed yielded palynomorphs. This sample (7B) is from the uppermost part of a major tuff bed (Fig. 4), and is dominated by *Ceratiopsis wardenensis* (Williams and Downie, 1966) Bujak *et al.*, 1980 (Fig. 6l—n). This species was rare or absent in the other samples analysed.

In the black-shale sequence above the main tuff interval (Fig. 4) there is an upward decrease in dinoflagellate cyst diversity. The ratio of marine to non-marine palynomorphs is very low, but shows the same upward decreasing tendency. Gymnosperm and angiosperm pollen may totally dominate the flora. Generally, in the black shales the amount of amorphous material is high. In this black-shale sequence there is also an increase in cortex content. The relative increase in terrestrially derived kerogen implies an increased continental influence for this interval.

The dinoflagellate cyst assemblage in the lower part of the black shale sequence is dominated by *Ceratiopsis* sp. 1 (Fig. 6o) and *G. ordinata. Hystrichosphaeridium* spp. are common. In the lowermost sample *D. oebisfeldensis* and *Paleostomocystis* sp. are also common. Below this black shale sequence *Ceratiopsis* sp. 1 becomes rare or is absent. Above it is generally common.

Discussion

The most striking feature of the dinoflagellate cyst assemblages is the dominance of *D. oebisfeldensis* and *G. ordinata* together with the absence of species from the genera *Apectodinium* and *Wetzeliella.* Another striking feature is the great amount of *Taxodiaceapollenites* and *Caryapollenites* compared to marine organisms.

The main phase of pyroclastic deposition in the North Sea was subdivided by Jaqué and Thouvenin (1975) into three according to their palynological zones nt2a upper, middle and lower. This corresponds to Knox and Morton's (1983) subphases 2a and 2b (Fig. 2). The deposition starts in the upper part of palynozone nt2a lower which is dominated by *Apectodinium hyperacanthum.* This phase probably corresponds to subphase 2a of Knox and Morton which belongs to the *Apectodinium hyperacanthum* Zone (Fig. 2). This flora was not recorded from our core.

Most of Jaqué and Thouvenin's main phase occurs during the most regressive phase of the Lower Tertiary, in their palynozone nt2a middle. This zone is dominated by pollen of taxodiacea. Such a dominance is found also in our core, especially in the black-shale sequence.

The final volcaniclastic deposition is recognized by Jaqué and Thouvenin in the basal part of the

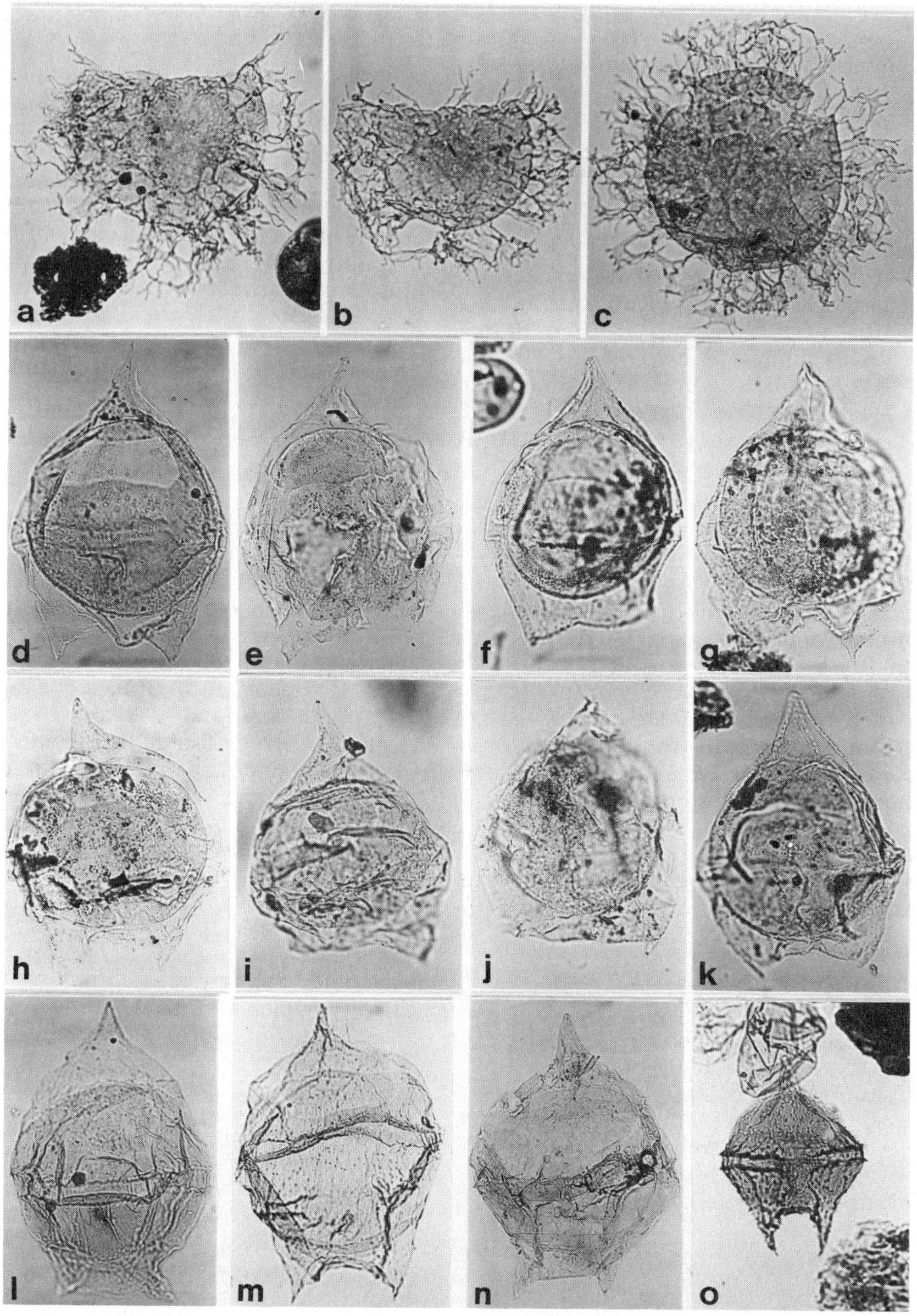

Fig. 6. Dinoflagellate cysts. All magnifications × 500. **(a–c)** *Glaphyrocysta ordinata* (a, b: sample 1B; c: sample 6B). **(d–k)** *Deflandrea oebisfeldensis* (d, f, k: sample 7B; e: sample 6B; g, h: sample 1B; i, j: sample 26B). **(l–n)** *Ceratiopsis wardenensis* (l, m: sample 7B; n: sample 8B). **(o)** *Ceratiopsis* sp. 1 (sample 7B).

overlying transgressive zone nt2a upper. This zone is characterized by *G. ordinata* and *D. oebisfeldensis*, the same two dinoflagellate cyst species that dominate our core samples throughout. The nt2a upper zone may be attributable to the uppermost part of the *A. hyperacanthum* Zone (Fig. 2), which is characterized in the North Sea by the *D. oebisfeldensis* acme. Knox and Morton's subphase 2b belongs to the *D. oebisfeldensis* acme. Knox and Morton state that their subphase 2b constitutes the "main tuff zone" of Jaqué and Thouvenin. However, the lower part represented by the nt2a lower zone, may as well be correlated with Knox and Morton's subphase 2a.

Dominance of *G. ordinata* and *D. oebisfeldensis* is also recorded from the Mo-Clay and the Ølst Formation in Denmark (Heilmann-Clausen, 1982). Above ash layer −19b *G. ordinata* dominates until the upper part of the formations where *D. oebisfeldensis* becomes dominant. The *D. oebisfeldensis* acme in Denmark correlates with the *D. oebisfeldensis* acme of the *A. hyperacanthum* Zone (Heilmann-Clausen, 1982). Although the two dominant species are represented about equally throughout the core, the section examined may be correlated with the upper part of the Mo-Clay and Ølst Formation in Denmark. This correlation is supported by the low to moderate frequency of *Apectodinium* spp. above ash layer +130 at Silstrup, where *D. oebisfeldensis* is abundant (Heilmann-Clausen, 1982).

Conclusion

According to the dominant floral features, the core interval investigated may be assigned to the *D. oebisfeldensis* acme in the *A. hyperacanthum* Zone (Fig. 7). It belongs to the main tuff sequence of the North Sea represented by subphase 2b of Knox and Morton (1983) (Fig. 2). It also equates with the *G. ordinata* and *D. oebisfeldensis* zone (nt2a upper) of Jaqué and Thouvenin (1975), and probably also with parts of the nt2a middle zone. Onshore it can be correlated with the upper part of the Mo-Clay and the Ølst Formation in Denmark.

Micropalaeontology

The routine micropalaeontology investigations of well 30/2-1 were carried out by Paleoservices Ltd., Watford. In their report the core interval with tuffs (1952—1968 m) is dated as late Palaeocene—early Eocene. The microfossils reported from the cored section are rather poorly preserved diatoms and a few other fossils commonly known from many other tuff and clay deposits in the North Sea area. Further biostratigraphical investigations were desired.

This study

To start with this study was based on fossils picked from the residues of disintegrated and washed core samples. From these a few agglutinated foraminifera (Fig. 8g) were picked from two samples (12B and 24B, Fig. 4). More or less pyritized diatoms and radiolaria occur in many samples from the sequences of grey shales. Only plant-remains were picked from the black-shale sequence in the central core interval (between samples 12B and 18B), and in the lower part of the upper grey shale sequences. Vertebrate-remains were only picked from the upper grey shale sequence.

Pyritized diatoms

Diatoms are well-known from the Lower Tertiary all over the North Sea area. So far, no well-preserved diatom material has been recognized in the offshore area of the North Sea Basin. Taxonomically, some of the most well-preserved individuals of pyritized diatoms are recorded into an uncertain species rank, but mostly the pyritized diatoms are reported in a general rank with open nomenclature. Some of the pyritized diatoms in the core samples from well 30/2-1 are relatively well-preserved. The commonly known *Coscinodiscus* sp. 1, *Coscinodiscus* sp. 2, *'Triceratium'*, and some other taxa are illustrated in Fig. 8.

Selected samples and correlation

With further micropalaeontological study, a flora of well-preserved diatoms and silicoflagellates were

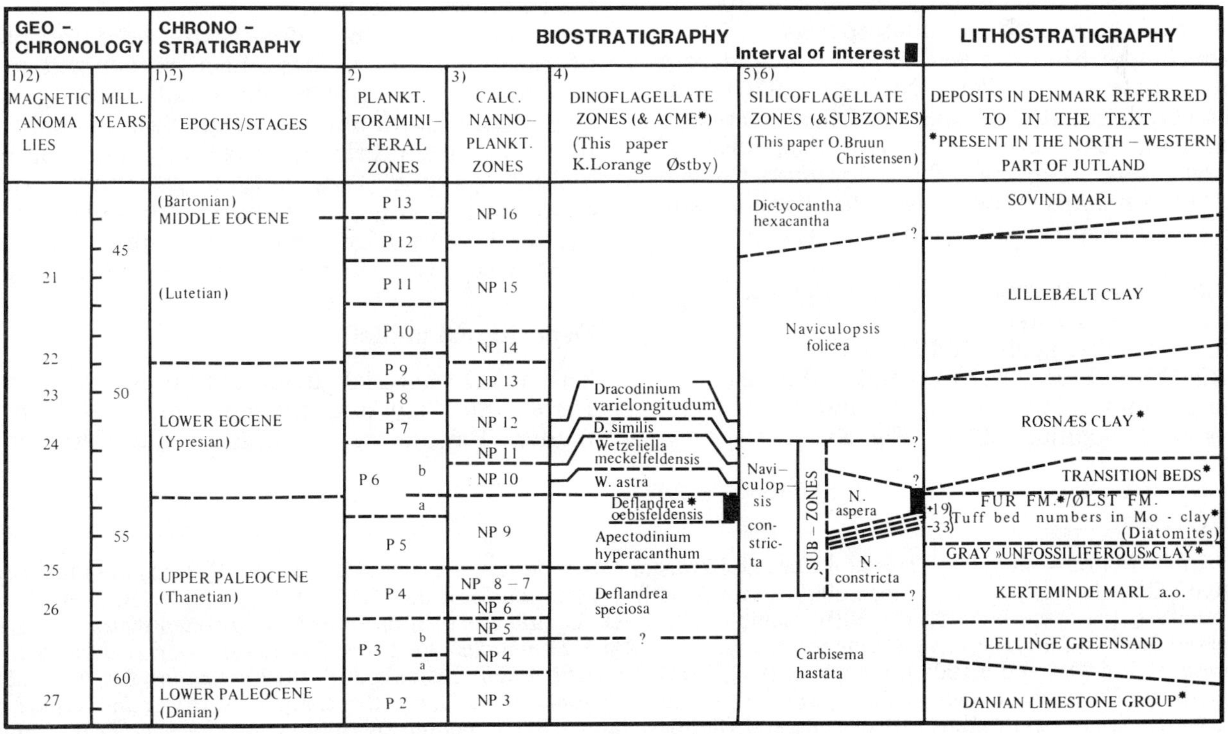

Fig. 7. A stratigraphical outline constructed from the following sources: (1) Hardenbol and Berggren (1978), (2) Berggren and Gradstein (1981), (3) Martini (1971), (4) Heilmann-Claussen (1982), (5) Perch-Nielsen (1976) and (6) Bukry (1981a, b). The lithostratigraphy is a generalized outline in which the most relevant units in NW-Jutland are incorporated. Detailed information on these may also be found in Hansen (1979), Pedersen (1981), and from papers in preparation (e.g. by C. Heilmann-Claussen, O. Bjørslev Nielsen and F.L. Gersnes, and by O. Brunn Christensen).

traced to a few thin horizons in the core-samples. Therefore, a correlation of the core and the Danish Palaeocene 'Clay with volcanic ash' (Dinesen *et al.*, 1977) rich in diatoms, was most relevant for the study. In North-West Jutland the clay is a diatomite informally called Mo-Clay, most recently Fur Fm. An Upper Palaeocene and Lower Eocene stratigraphy for the area, constructed from various papers, is outlined in Fig. 7. Tuff beds also have been recognized recently in the Røsnæs Clay as well as in the lower part of Lillebælt Clay by Heilmann-Clausen *et al.* (in preparation).

Diatoms and silicoflagellates

In some selected rock pieces, split from two samples (2B and 20B) with calcareous bands or concretions, a rich fossil assemblage of diatoms and silicoflagellates was located and retrieved. Some of the fossils are illustrated in Fig. 8. The flora is closely related to that of the Mo-Clay diatomite in Denmark, where they are best preserved in calcareous concretions (cement-stones) occurring at certain levels in the diatomite. The Mo-Clay diatom flora has not been studied and described in great detail, but Benda (1972) has recorded and illustrated some of the most common species.

The Mo-Clay belongs to the lower part of the silicoflagellate zone of *Naviculopsis constricta*, an Upper Palaeocene and Lower Eocene zone known from localities around the Atlantic Oceans (Bukry, 1981b). Perch-Nielsen (1976) subdivided the zone in the Mo-Clay into five subzones (Fig. 7). The *Naviculopsis aspera* Subzone is the uppermost of these, found above the subzones of *Corbisema naviculoidea*, *Dictocha elongata*, *N. danica*, and lowermost the subzone of *N. constricta*. The *N. aspera* Subzone contains only *N. aspera* and *N. constricta* of the mentioned index species.

In the Balder Fm from well 30/2-1 the *N. aspera* Subzone assemblage (Perch-Nielsen, 1976) was found to be dominated by species of *Corbisema* and *Naviculopsis*, and not so common species of *Dictyocha*, but no species of the common Eocene genera *Distephanus*. These are all characteristic features for the subzone assemblage:

C. apiculata (Lemmermann, 1901) — Fig. 8t.
C. hastata (Lemmermann, 1961).
D. deflandrei (Frenguelli, 1940).
D. fibula (Ehrenberg, 1838) s. ampl. — Fig. 8s.
N. aspera (Schulz, 1928) — Fig. 8v and y.
N. constricta (Schulz, 1928) — Fig. 8x.

Only one specimen of *N. constricta* was found in the core samples. The *N. aspera* Subzone assemblage was established above tuff bed +19 in the Danish Mo-Clay (Fig. 7).

Palaeoecology

The relatively large variation in the degree of preservation of the diatoms picked indicates both environmental and diagenetic innovations for at least some parts of the deposits. A cover of microcrystalline pyrite on many of the fossils (Fig. 8b) and collections of well-preserved quartz crystals (Fig. 9i) picked mostly from the samples with a few completely pyritized diatoms, indicate strong diagenesis in the deposits. A common presence of diagenetically developed quartz and pyrite in the shales was confirmed by the petrographical analysis (see 'Shale Petrology').

The palaeontological and ecological variations recognized in the samples throughout the core section are not affected by major biostratigraphical breaks. The large amounts of microcystalline pyrite and organogenetic quartz in the deposits indicate a high production rate of pelagic life mostly in plankton. A few elements of the flora and the fauna have been fossilized by pseudomorphosis. However, in some collections of relatively well-preserved more or less pyritized diatoms a cover of microcrystalline pyrite is most probably a result of syn-depositional processes. In some levels calcareous concretions similar to those known from the Danish Mo-Clay, were developed more or less syn-depositionally. During post-depositional diagenesis, the siliceous microfossils were affected by dissolution on a large scale. Only in a few levels, fossils remain preserved.

No calcareous microfossils or pseudomorphosis of calcareous fossils have been recognized in the samples. A few benthonic fossils, among those agglutinated foraminifera (Fig. 8g), indicate that the bottom conditions, at least in some intervals, include oxygen. Stagnant, oxygen-poor bottom conditions are most common for areas rich in pyrite, favoured by large organic production and limited supply with bottom water rich in oxygen.

Depositional model

The fossil material from the upper part of the Danish Mo-Clay and the core material from the Balder Fm, is very similar. Stratigraphically,

Fig. 8 (*right*). Diatoms, silicoflagellates and Foraminifera. (a) *Coscinodiscus* sp. 1, c. 37×, sample 12B. (b) *Coscinodiscus sp.* 1, covered with microcrystalline pyrite, c.2000×, sample 12B. (c) *Coscinodiscus muelleri*, c.470×, sample 2B. (d—e) *Coscinodiscus muelleri*, the same individual, c.500×, sample 2B. (f—i) Assemblage with more or less pyritized diatoms and a few agglutinated foraminifera (g, c.24×). *Coscinodiscus* sp. 2 (f, c.24×). *Hemiaulus?* sp. 1 (h, c.24× and c.37×). *Trinacria* aff. *wittii* (*'Triceratium'*, i, c.24× and c.100×). Sample 12B. (j) *Trinacria wittii*, c.250×, sample 20B. (k) *Hemiaulus* cf. *affinis*, c.500×, sample 20B. (l) *Hemiaulus elegans*, c.320×, sample 2B. (m) *Brightwellia?* sp. 1, c.750×, sample 2B. (n) *Actinotychus?* sp. 1, c.900×, sample 2B. (o—q) Residual with diatoms and silicoflagellates, c.320×, sample 2B. o: *Hemiaulus elegans* (?), p: *Corbisema flexuosa*, q: *Stepanopyrix turris.* (r) *Corbisema archangelskiana*, c.250×, sample 2B. (s) *Dictyocha fibula*, c.1000×, sample 2B. (t) *Corbisema apiculata*, c.250×, sample 2B. (u) *Corbisema flexuosa*, 320×, sample 2B. (v) *Naviculopsis aspera*, c.500×, sample 2B. (x) *Naviculopsis constricta*, c.320×, sample 2B. (y) *Naviculopsis aspera*, apical view of basal ring, c.1600×, sample 2B. SEM-micrographs (a—c, f—i, m—q, s, v, y). Microscope photographs in transmitted light (d—e, j—l, r, t, u, x).

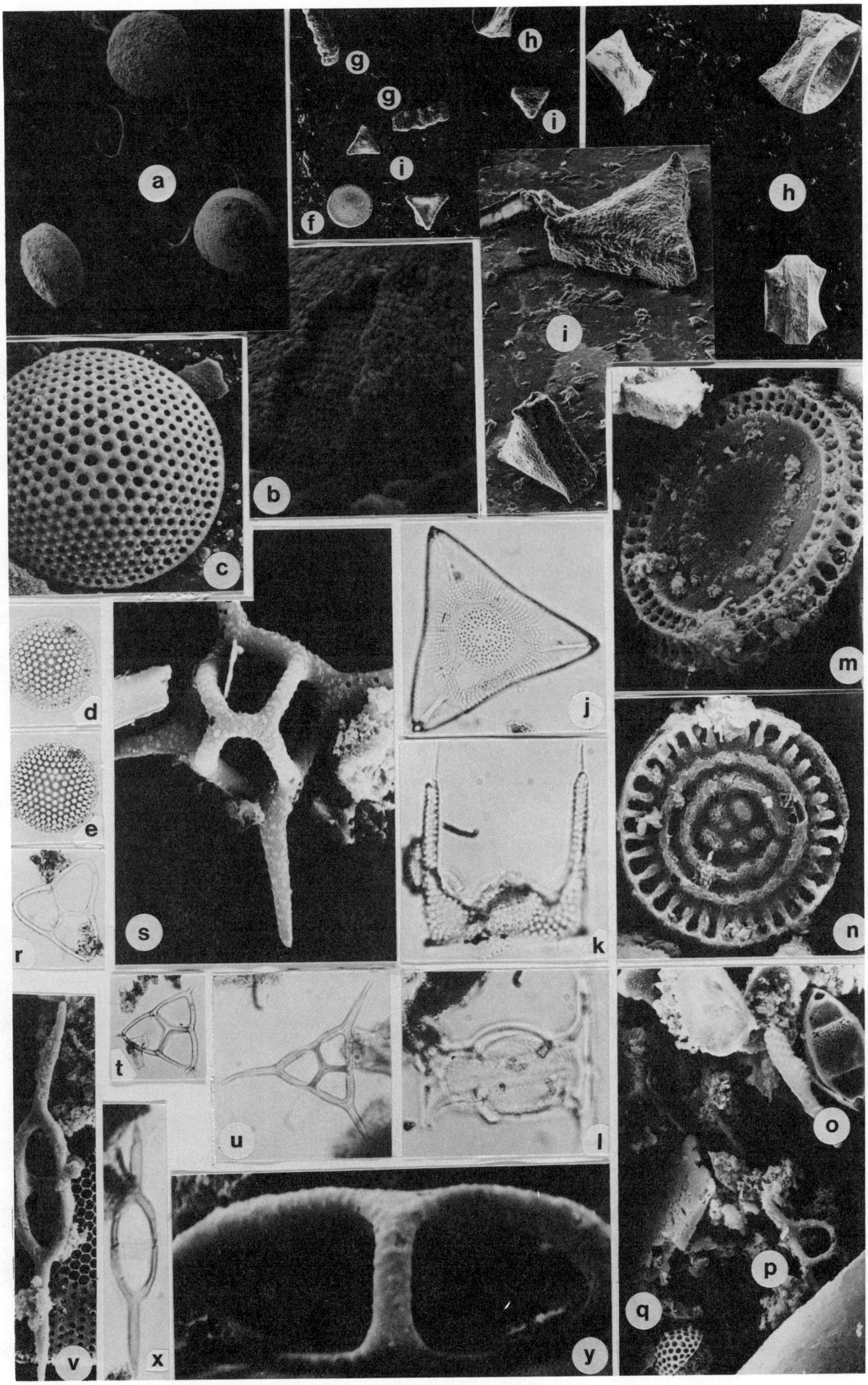
a
g
g
h
i
i
f
h
i
b
c
m
j
d
e
s
r
k
n
t
u
l
o
v
x
y
p
q

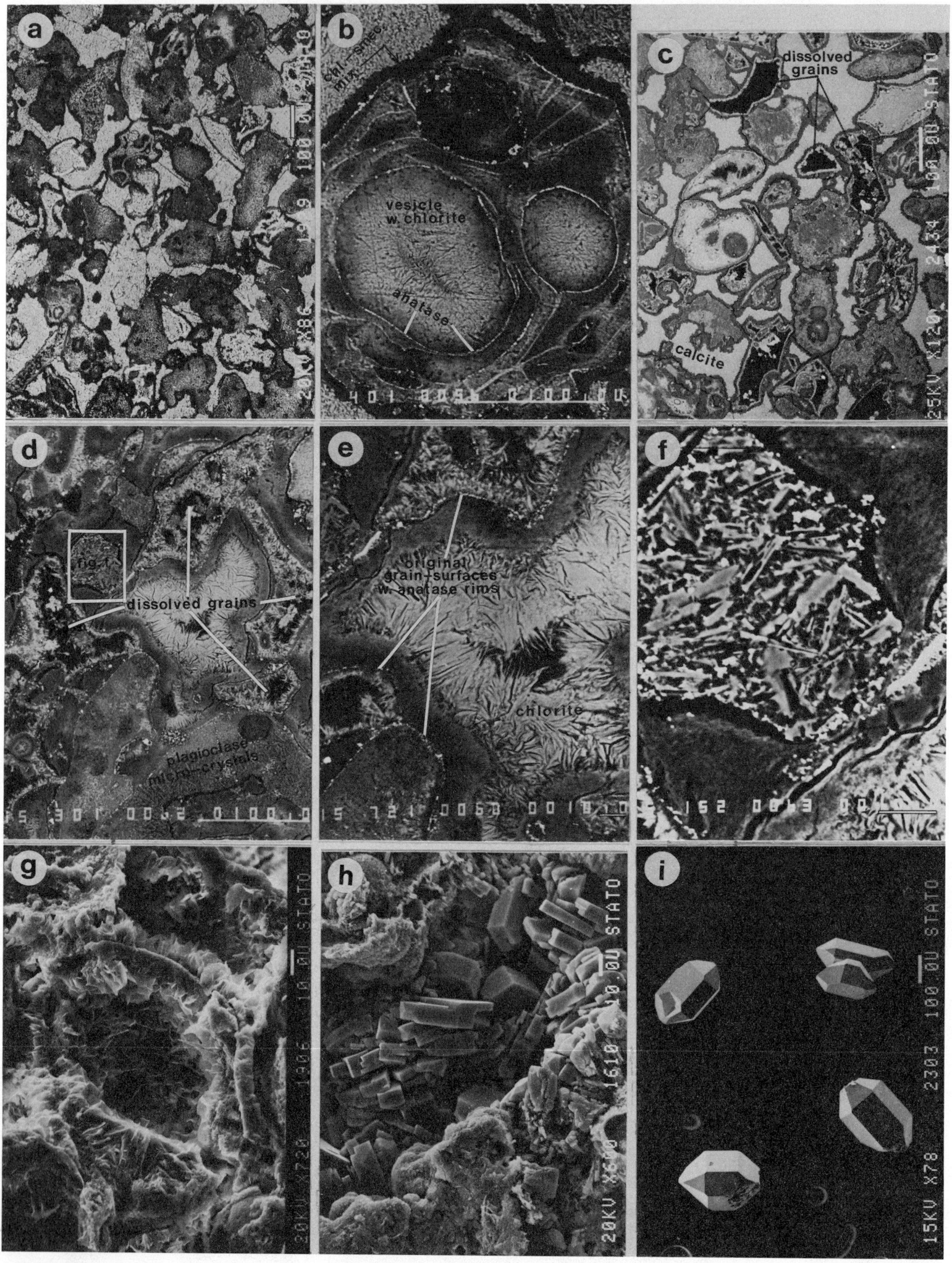
a
b
chl.-smec.
mix.
vesicle
w. chlorite
anatase
c
dissolved
grains
calcite
d
fig f
dissolved grains
plagioclase
micro-crystals
e
original
grain-surfaces
w. anatase rims
chlorite
f
g
h
i

palaeoecologically, and possibly depositionally, too, the two areas are closely connected. We interpret the original deposits in well 30/2-1 as a sediment very rich in diatoms, most probably a diatomite with tuffs.

A model for the deposition of Palaeocene diatomite in the North Sea has been suggested by Bonde (1973, 1974, 1979, 1982). Regional variations in the thickness of some Danish tuff beds, probably those from nearby sources in Skagerak, indicate wind directions from the north during deposition. Bonde stresses that prevailing northerly winds affect the current system of the sea causing upwelling situations supporting rich planktonic life. The palaeoecological studies of the core from well 30/2-1 verify some basic conditions for working of the model: prevailing northerly winds, rich planktonic life, restricted bottom conditions and sedimentation conditions not unlike those for the diatomites in the Danish area.

PETROLOGICAL ANALYSES

Detailed textural, mineralogical and chemical analyses were undertaken on both the tuffs and the shales. The analyses include light and scanning electron microscopy, X-ray diffraction of whole-rock samples and clay-fractions, and whole-rock chemical analyses of major and trace elements. The stratigraphic position of the analysed samples and the type of analyses undertaken on them are shown in Fig. 4. Details on sample preparations and analytical techniques are available upon request to the authors.

Tuff petrology

Deposition

It is suggested above that the two exceptionally thick tuff-beds comprising a major tuff-interval in the central part of the core (Figs. 3 and 4) may represent high density/mass flow deposits. One might feel tempted to interpret the strong normal grading of the other tuff-beds (1—28-cm thick) as a result of similar depositional processes. Our arguments against such an interpretation are these: the graded beds are dominantly massive and do not display any other textures of a typical Bouma sequence. The beds contain no detrital components except those of primary volcanic ash. The regional distribution of the Balder tuffs at a distinct stratigraphic level is incompatible with mass-flow/high density current deposition. This would be expected to result in more localized deposition centres. We therefore conclude that the tuff-beds represent first generation air-fall deposits.

Texture and mineralogy

In thin-sections the tuff-particles appear dominantly green, but dark brown and light yellow grains are also observed. The grains commonly contain spherical vugs or vesicles formed through expansion of vapour and gas in the magma during eruption and pressure release (Fig. 9a—c). The grains have irregular shapes and arcuate grain boundaries often curved inwards (Fig. 9a, c). These represent the walls of broken vesicles. The grains are comparable with those produced during Surtseyan type eruptions. Such eruptions of basaltic composition attain their explosive nature due to a quenching of the magma with water under shallow water conditions. The particles are predominantly vitric (glassy). However, many grains contain microscopic quench crystals of plagioclase formed during the rapid chilling of the magma (Fig. 9d and f). Plagioclase also occurs as larger porphyric crystals. The quench crystals have a composition corresponding to labradorite (Table 2) which is compatible with a basaltic magma composition. Microprobe analyses of the vitric particles have failed to demonstrate fresh glass.

X-ray diffraction analyses (Fig. 1), Table 6) show that the tuff-beds are composed mainly of swelling chlorite, chlorite-smectite mixed layered clays and chamositic clays. In addition primary plagioclase is very abundant (25%). Minerals present in subordinate amounts are: pyrite, anatase, siderite and zeolites.

The textural relations between these minerals were identified by combining electron microscope images (Fig. 9a—h) and microprobe analysis (Table 2).

Figure 9b show a typical ash-grain from a calcite-cemented tuff-bed. The original surface of the grain and the vesicle walls are defined by thin rims of anatase which represent titanium expelled from the glass during alteration and crystallization of secondary clay-minerals. The vesicles are filled with well-crystalline chlorite that has grown from the vesicle

Fig. 9. *(Left)*. **Micrographs from scanning electron microscopy of polished thin-sections (a—f) and core-chips (g—h) from tuffs. (a) Scalebar = 100μm. Tuff-grains (dark grey) with intergranular calcite-cement (white). Note irregular arcuate grain-boundaries of tuff-particles and their content of spherical vesicles/vugs. (b) Scale-bar = 100μm. Detail of ash-grain from sample shown in perspective in (c). The grain illustrates the alterations of the tuff that took place before calcite-cementation. This includes (i) growth of chlorite within vesicles, (ii) crystallization of anatase (white) as rims on vesicle walls and grain surfaces, (iii) deposition of mixed layered chlorite—smectite clays as grain-coatings (dark, grey), (iv) internal dissolution and transformation of the glass into an amorphous or microcrystalline phase, possibly chamosite. (c) Scale-bar = 100μm. Calcite-cemented (white) tuff grains (grey-black) which display primary textures as described in (a) and alteration features described in (b). Note that the interior of some grains is completely dissolved (black). (d) Scale-bar = 100μm. Alterations observed in tuff-beds** ***without*** **calcite-cementation. Intergranular porespace is filled with chlorite. Most ash-grains show strong internal dissolution (dark shades) and growth of chlorite whiskers from grain-walls and inwards (see also g). (e) Scale-bar = 10μm. Detail of (d). Chlorite growth on and at the expense of grain coatings. Original grains defined by anatase rims (white) are dissolved and replaced by chlorite-whiskers growing inwards from grain-walls. (f) White scale-bar = 10μm. Detail of (d) showing a typical grain with abundant lathlike plagioclase quench microcrystals. (g) White scale-bar = 10μm. Depth-view of a dissolved grain with chloritic clay-minerals that grows inwards and outwards from an early formed grain-coating. (h) White scale-bar = 10μm. Typical view of a zeolite-aggregate formed in porespace, due to dissolution of tuff-grains. (i) White scalebar = 100μm. Bipyramidal quartz micro-crystals of biogenic origin from shale sample 5B (Fig. 4).**

TABLE 2. Chemical compositions of plagioclase (Fig. 9f), secondary chlorites and microcrystalline amorphous phases shown in Fig. 9b, d and e[a].

	Plagioclase	*Analcite*	*Chlorite*			*Amorphous or microcrystalline material*	
			Within vesicles	*Within grains*	*Between grains*	*Within grains*	*Rim outside grains*
	(n=*1*)	(n=*1*)	(n=*3*)	(n=*3*)	(n=*2*)	(n=*8*)	(n=*4*)
SiO_2	51.9	57.4	35.9	34.4	36.7	37.1	40.0
TiO_2	—	0.3	0.7	0.1	—	0.4	0.4
Al_2O_3	27.1	21.9	16.9	18.5	16.9	17.1	19.6
FeO	2.4	1.6	32.3	33.0	31.4	30.3	25.1
MnO	—	—	0.3	0.1	—	0.1	—
MgO	—	—	12.0	13.0	14.3	14.0	12.5
CaO	11.5	4.4	1.5	0.7	0.7	0.7	2.3
Na_2O	5.1	5.7	—	—	—	—	—
K_2O	0.5	0.6	0.4	0.2	0.2	0.3	0.1
TOTAL	98.5	91.8	100	100	100	100	100

[a] Determined by microprobe analyses. All analyses except those of plagioclase and analcite are recalculated to 100 and a water free basis.

walls. The outer surface of the grain is coated by a dark rim of less dense material. The chemical analyses show that it has higher SiO_2 and Al_2O_3 and lower FeO and MgO than the chlorite within the grains (Table 2). We suggest, therefore, that this outer rim represents the mixed layered chlorite—smectite clays identified by the X-ray diffraction analyses.

The original grains between the rims of anatase are partly dissolved and partly transformed into an amorphous phase. The chemical composition of this is similar to that of the chlorite (Table 2). It could represent a micro-crystalline chlorite or chamositic phase (Fig. 10).

In many grains this amorphous-looking phase is replaced by coarser crystalline chlorite which grows inwards from the grain wall and outwards from the surface of the vesicles. This growth is accompanied by a net dissolution and in its more prolonged state, the major part of the original grains are dissolved. (Fig. 9c).

These observations from calcite-cemented tuff-beds show that alteration of the volcanic ash took place before cementation. It should be noted that during this alteration chlorite crystallized only in the vesicles and the interior parts of the grains and not on the grain-surfaces. The formation of chlorite requires a certain activity/concentration of magnesium-ions, so that brucite layers can form. In the semi-closed interior of the grains the magnesium concentration would be higher than in the porespace, especially if the porewaters are under exchange with the basinal waters. We suggest, therefore, that the alteration before calcite cementation took place while the tuff-beds remained open to exchange of porewaters. By comparing beds or parts of beds with and without calcite-cementation a distinct difference in the alteration is observed. Where no calcite cement is present the alterations described above are followed by a later stage of chlorite crystallization in the porespace between the grains (Fig. 9d and e). This is accompanied by a further dissolution of the interior of the grains (Fig. 9d). The dissolution creates abundant macropores isolated from each other by a network of chlorite—smectite grain coatings and chlorite filling the original porespace between the grains (Fig. 9d,g). This texture explains the exceptionally *high porosity* (40%) and *low permeability* (< 1 mD) measured on uncemented tuff-beds (Table 3). As argued above, dissolution and simultaneous chlorite-crystallization is best achieved in stagnant porewater. We suggest, therefore, that this late stage of alteration took place during burial after the tuff-beds were effectively sealed off from contact with the basinal waters.

Chemistry

Volcanism is related to tectonically active areas, e.g. rift-zones or subduction zones. The chemistry of volcanic rocks shows fairly systematic variation with respect to tectonic setting (e.g. Pearce and Cann, 1973). In theory it is possible to analyse a volcanic rock and suggest its tectonic setting. With

TABLE 3. Conventional porosity—permeability data on tuff-beds without calcite-cementation (for sample location see Fig. 4) from 1-inch horizontal core plugs

Depth (m)	*Plug No.*	*Hor. Perm. KA (mD)*	*Hel. por.* (%)	*G. dens.* (g/cm^3)
1954.60	1A	NMP	38.2	2.76
1954.69	2A	0.080	38.2	2.81
1954.91	3A	0.198	37.9	2.77
1955.14	4A	0.83	37.3	2.81
1955.45	5A	0.082	34.5	2.81
1956.49	6A	NMP	41.4	2.71
1957.45	8A	0.23	38.4	2.78
1957.90	9A	0.157	35.8	2.75
1959.24	10A	NMP	36.9	2.74
1961.61	—	0.31	28.5	2.67
1962.39	11A	NMP	39.0	2.74
1962.53	12A	0.113	40.1	2.75
1964.35	13A	NMP	44.3	2.78
1965.48	14A	NMP	40.2	2.75
1965.78	15A	0.125	37.3	2.77
1966.63	—	3.7	31.6	2.76
1967.68	17A	NMP	38.0	2.81
1968.03	—	0.05	32.1	2.78

NMP — no measurement possible.

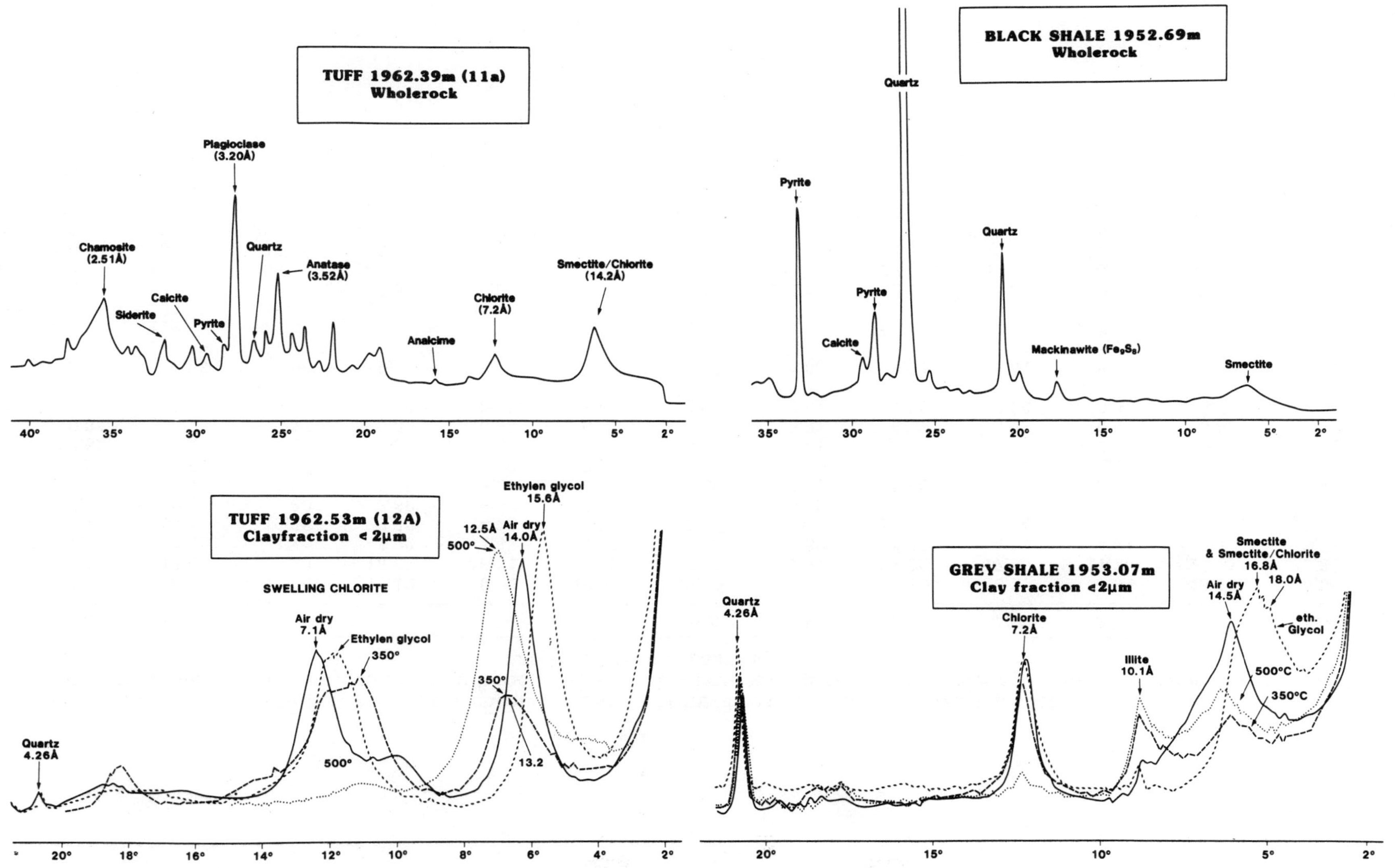

Fig. 10. X-Ray diffractograms of tuff and shale whole-rock samples and clay fractions less than 2μm.

TABLE 4. Whole-rock chemical analyses of tuff-beds without calcite (sample location; Fig. 4). Determinations by X-ray spectrometry on fused beads of 1—9 mixtures of ignited rockpowder and lithium tetraborate (SiO_2—P_2O_5) and beads of pressed rockpowder (Cf—Nb)

	1A	*2A*	*3A*	*4A*	*5A*	*6A*	*9A*	*10A*	*11A*	*12A*	*13A*	*14A*	*15A*	*17A*
SiO_2	41.76	39.92	42.64	41.19	40.96	46.23	42.89	45.20	44.70	43.27	41.94	43.17	40.58	41.96
Al_2O_3	16.59	16.61	17.28	16.67	16.31	17.79	17.19	17.37	17.22	17.06	17.10	17.41	15.35	17.08
TiO_2	4.14	4.52	4.44	4.22	4.47	4.53	4.62	4.80	4.33	4.31	4.81	3.72	3.58	3.83
Fe_2O_3	22.17	24.65	19.76	23.62	24.13	16.39	20.34	17.36	16.40	19.94	21.18	21.31	17.25	22.74
MgO	7.10	9.09	7.80	8.69	8.80	6.33	7.66	6.86	6.32	7.78	7.86	8.23	7.21	8.83
CaO	3.87	2.42	4.32	2.49	2.36	3.99	3.69	3.98	5.46	3.37	3.27	2.52	10.44	2.34
Na_2O	1.85	1.56	2.22	1.57	1.70	2.87	2.23	2.59	2.76	2.57	2.25	2.30	2.39	1.88
K_2O	0.65	0.31	0.46	0.51	0.39	0.51	0.42	0.61	0.59	0.53	0.42	0.51	0.50	0.59
MnO	0.32	0.29	0.27	0.30	0.33	0.30	0.29	0.31	0.27	0.27	0.31	0.33	0.39	0.30
P_2O_5	1.56	0.63	0.80	0.72	0.55	1.07	0.66	0.92	1.95	0.90	0.85	0.51	2.30	0.46
L.o.i.[a]	10.42	9.56	9.00	9.77	9.89	9.46	9.53	9.53	8.94	9.01	9.43	9.61	10.46	9.59
Cr (ppm)	143	119	126	118	115	131	163	133	162	182	198	146	172	147
Ni (ppm)	26	37	32	29	31	41	41	27	77	79	35	37	52	60
Rb (ppm)	3.7	n.d	0.7	2.0	0.5	2.1	2.1	2.1	2.9	3.3	2.6	1.1	2.9	2.9
Sr (ppm)	268	154	209	163	160	271	220	308	362	227	223	192	261	169
Y (ppm)	86	50	60	56	53	67	61	77	60	53	57	45	103	39
Zr (ppm)	280	269	294	248	255	311	275	352	307	268	274	216	220	217
Nb (ppm)	30	30	31	28	26	33	33	54	32	30	28	21	28	24

[a] Loss on ignition.

the assumption that the ash is representative of the erupting magma, analyses of such material can be used to suggest the tectonic setting of its source.

Despite strong alteration and recrystallization, the chemistry of the bulk ash might be more or less conserved in its primary state, depending on what degree ions in solution can be transported out of, or into the tuffs, during alteration. Ideally the alteration can take place in a stagnant porewater and this would have little effect on the bulk chemistry of the ash. Against this background chemical analyses of bulk ash samples were carried out to evaluate if this could indicate the primary composition of their source.

Fourteen full major element analyses were done of 13 individual tuff beds (Table 4). The samples were picked due to their lack of or low content of calcite. Characteristically the 'loss on ignition' values are high, showing that the samples are hydrated and altered. The analyses indicate a uniform composition of the tuff beds. Compared with fresh basaltic rocks, as for example the Danish tertiary tuffs (Table 5), they are lower in SiO_2 and CaO and generally higher in Al_2O_3, TiO_2 and FeO^t.

Concerning the classification of these basaltic tuffs from major element chemistry, the conclusions should be confined to suggesting that they represent rather TiO_2- and FeO^t-rich basalts. In comparison, the Danish tuffs have been classified as high Ti—Fe tholeiites (Pedersen *et al.*, 1975). The much lower CaO content in our tuffs (Table 5) must be attributed to depletion of calcium during alteration. However, X-ray diffraction analyses undertaken by us on the basaltic Danish tuffs show that their plagioclase is different from the plagioclase in our tuffs. This suggests that the two tuff-series were sourced from magmas with different contents of CaO.

The samples analysed for major elements were also analysed for the trace elements Cr, Ni, Rb, Sr, Y, Zr, and Nb (Table 4). Similar analyses were also performed on about 30 other samples from individual tuff beds (Fig. 4). Analyses of the latter are not listed, but they are used in the chemical discrimination diagrams in Fig. 11a—d. Some of the samples are from calcite-cemented tuff-beds. The dilution effect of calcite on the element concentrations for these samples were corrected for through modal analyses and substraction of calcite.

Also the trace elements may become depleted or enriched during alteration of the glass. However, the incompatible elements like Zr, Nb and Y show a sympathetic variation and their ratios remain constant (Fig. 11a), indicating that they are not affected by alteration. They can be used, therefore, in discrimination diagrams such as those in Fig. 11b—d, which indicate that the tuffs have a continental tholeiitic composition with a transitional character between within-plate basalt (WPB) and mid-ocean ridge basalt (MORB) (Fig. 11d). A similar result is obtained by cross-plotting the incompatible element Y with the compatible element Cr (Fig. 11c).

Therefore, we conclude that the magmatism producing the Balder tuffs straddled the continental/oceanic environment.

TABLE 5. Average chemical composition of tuff-beds calculated from the individual analyses shown in Table 4 (sample 15A excluded), compared with the average composition of vitric ash-grains from Danish ash-layers (the latter do not include the contributions of plagioclase phenocrysts).

	Samples 1A to 17A (n=*13*)[a]	*Danish ash-layers* (n=*6*)[b]
SiO_2	42.8	49.6
Al_2O_3	17.1	12.7
TiO_2	4.4	3.5
Fe_2O_3	20.8	15.0
MgO	7.8	5.1
CaO	3.4	10.1
Na_2O	2.2	2.8
K_2O	0.5	0.6
MnO	0.3	0.2
P_2O_5	0.9	0.4
	100.0	100.0
Loss on ignition	9.5	

[a]Whole-rock X-ray spectrometry.
[b]Microprobe on ash-grains by Pedersen *et al.* (1975).

Palaeomagnetic properties

Palaeomagnetic and rock magnetic properties of 62 up—down oriented 1-inch horizontal plugs from 34 different tuff-beds (Fig. 4) were investigated. After thermal cleaning the samples show normal polarity inclinations of 62 ± 12° which is compatible with the suggested inclinations for the Palaeocene/Eocene geomagnetic field of Europe (Løvlie and Kvingedal, 1975). The core contains one minor interval (1967—1968 m) with negative inclinations and reversed polarity. Correlation with the suggested reversal stratigraphy (Ness *et al.*, 1980) has not been attempted. Presentation of measurements and results will be given elsewhere (Løvlie and Malm, in preparation).

Shale petrology

Thin-section studies and X-ray diffraction analyses (Fig. 10, Table 6) reveal that the shales contain abundant aggregates (50—80% volume) of microcrystalline quartz and a matrix of clay-minerals and volcanic dust.

The quartz-aggregates which are composed of idiomorphic microcrystals of diagenetic origin (Fig. 9i) are believed to have a biogenic origin. Thus the dominant microfossils preserved in calcite concretions are siliceous microfossils like diatoms and silicoflagellates (Fig. 8). Pseudomorphic pyrite replacements of such fossils are common in the shales. The original microfossils are similar to those composing the so-called Mo-Clay, a diatomite containing the Lower Tertiary ash-series of Denmark. We propose that the siliceous shales are recrystallized deposits very rich in siliceous microfossils and probably represent equivalents to the Danish Mo-Clay diatomite (see 'Micropalaeontology').

X-ray diffraction analyses show that the clay-minerals of the shales are swelling chlorite, chlorite—

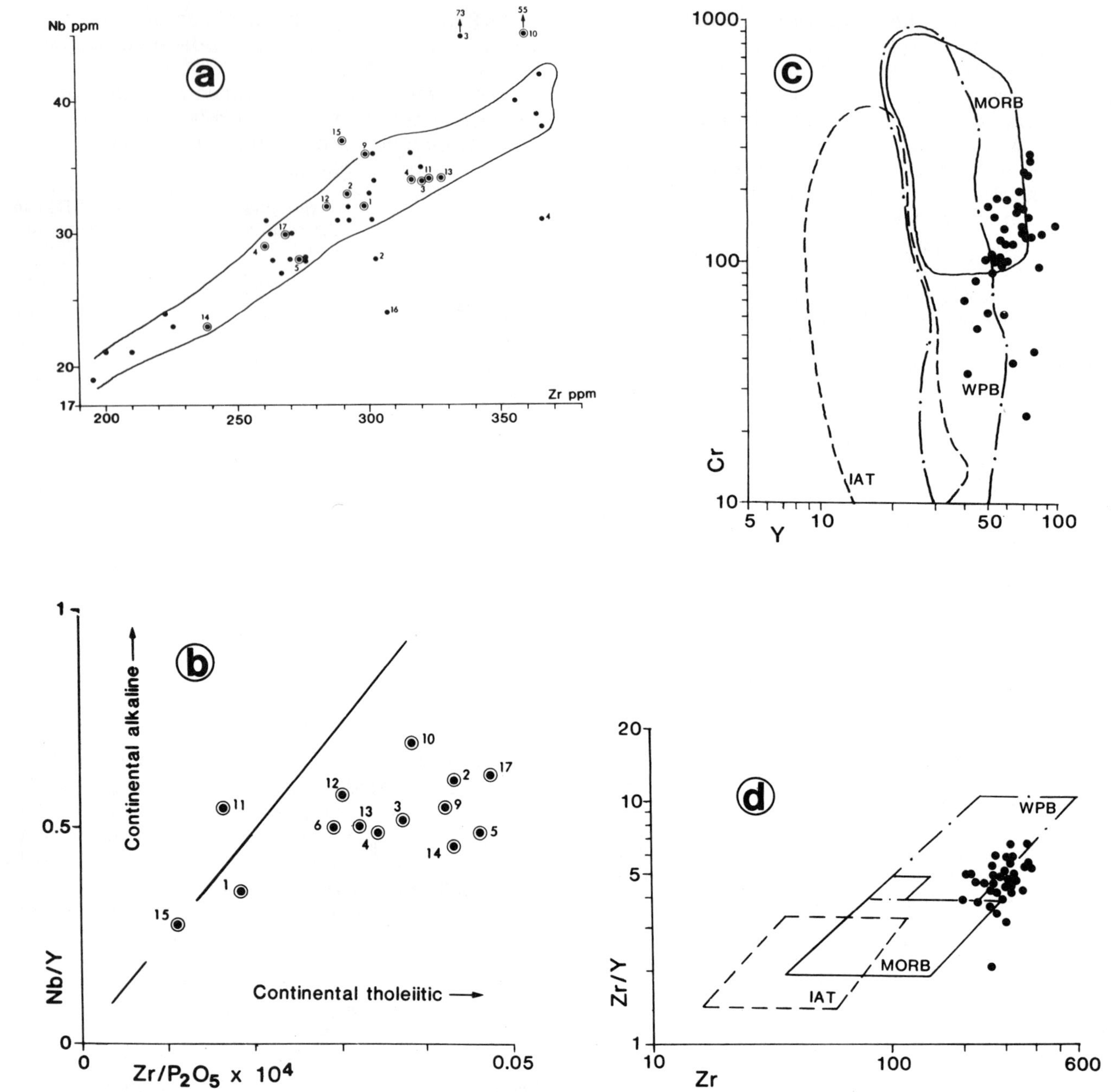

Fig. 11. Discrimination diagrams used to classify basaltic rocks. All points plotted represent whole-rock analyses of tuffs (Fig. 4) partly listed in Table 4. (a) Shows a covariance between concentrations of the two incompatible elements Nb and Zr. This indicates that they are unaffected by alteration. Circled samples represent analyses presented in Table 4. (b) Discrimination between continental alkaline and tholeiitic basalts. Plotted samples are from Table 4. (c and d) Discrimination between mid oceanic ridge basalt (MORB) and within plate basalt (WPB).

smectite mixed layers and smectite (Fig. 10, Table 6). The chlorite and chlorite—smectite mixed layered clay minerals are similar to those found in the tuffs. Therefore, we suggest that the clay-minerals of the shales represent alteration products of volcanic dust. The contents of smectite in the shales show a positive correlation with the contents of organic matter (Fig. 12). We suggest that this reflects diagenetic alteration of volcanic dust during burial. During degradation of organic matter the the CO_2 produced will lower the pH and favour the stability of smectite to chlorite. The amount of CO_2 produced and the degree of smectite formation will thereby show positive correlation with the amount of organic carbon in the shales. Although the tuffs do not contain pure smectite, it is found that their mixed-layered chlorite—smectite clays have higher smectite/chlorite ratios where the tuffs are interbedded with black shales.

The abrupt change from grey shales to black shales in the central part of the core (Fig. 4) is related to an increase in terrestrially-derived kerogen (see 'Palynology'). This implies increased continental input to the basin and possibly an elevation of the source area. However, the low content of detrital minerals like mica in the black shales indicates that the area lay distant to any major depositional centres also during this period.

TABLE 6. Semi-quantitative mineral-compositions of average shale and tuff (without calcite) determined by whole-rock X-ray diffractometry. Both rocktypes might contain amorphous/poorly crystalline alteration products of volcanic ash which are not accounted for by these analyses

	Average shale composition (sample 1B—26B) (n=20)	*Average tuff composition (sample 1A—17A) (n=14)*
Quartz	65.6	0.9
Plagioclase	3.7	26.0
Chlorite + Smectite	8.1	28.1
Chl./smec.mix.lay.	6.0	8.2[a]
Illite	4.2	—
Chamosite	—	24.4
Pyrite	10.9	4.6
Anatase	—	4.1
Siderite	tr.	1.3
Calcite	0.7	1.1

[a] No pure smectite, only chlorite/smectite mixed layers.

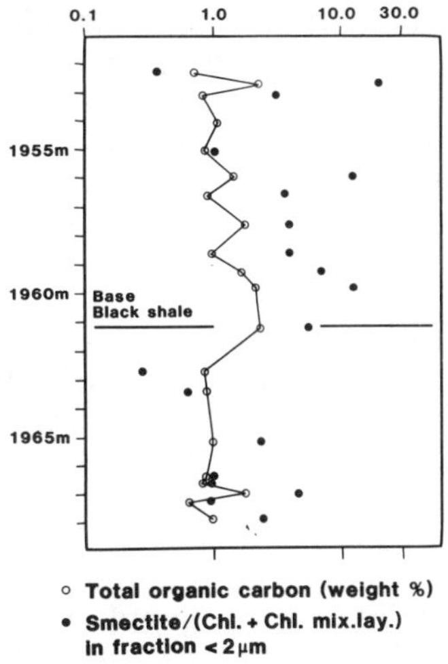

Fig. 12. Correlation between content of organic carbon and smectite/chlorite-ratios in the shales.

SOURCE LOCATION OF THE NORTH SEA AND DANISH TUFF-DEPOSITS

Considerations on the location of the source areas of the Balder tuffs should take into account the following:

(1) Distribution of known volcanic centres of Lower Tertiary age within or in the vicinity of the North Sea.

(2) The age, composition and character of these centres.

(3) Data on the Balder tuffs themselves, including stratigraphy, composition, texture, regional distribution and variation of bed abundance, thickness and grain-size.

(4) Studies on recent volcanism and volcaniclastic deposits, including their eruptive style, composition, grain-size and dispersal area of volcanic ashes.

Concerning (1), evidence of volcanic centres of Palaeocene--Eocene age is widespread in the N.W. British Isles (Hebridean Province), in the Faeroe-Shetland region (Thulean Province) and in the Rockall-Faeroe Troughs (Fig. 1). These provinces, and especially the Thulean Province, contain abundant basaltic rocks of the right composition and age to match the Balder tuffs. Thus, earlier workers on the Balder tuffs in general concluded that their source or sources are to be found in these volcanic provinces. Such a conclusion has not been verified by tephrological studies, as detailed core-descriptions of bed abundance, thickness and grain-size have been available only from a limited area (e.g. well 25/10-1: Ræstad 1976; Ræstad and Prestvik, 1982).

The present study adds a new well with a different location and this makes some considerations concerning (3) possible. Data on the number of beds, bed-thickness and grain-size from the present well, well 25/10-1 and the onshore Danish deposits (Fig. 1) are therefore listed and compared in Table 7.

The data show that the *main* tuff zone (defined by bed-abundance and log-characteristics) of the wells 30/2-1 and 25/10-1 are distinctly different, with more numerous and thicker beds with a coarser grain size in the former. This points to a distinct decrease in bed abundance, thickness and grain-size from north to south over relatively short distances (about 200 km) in the northern North Sea. This again indicates a source for the tuffs located north of the North Sea possibly in the Møre Basin area (Fig. 1). Radiometric dating by Bugge *et al.* (1980) have confirmed magmatic activity of Lower Tertiary age in this area.

TABLE 7. Comparison of total number of tuff-beds, bed thicknesses and grainsize of the *main* tuff zones in well 30/2-1 (Northern Viking Graben), well 25/10-1 (Balder area) and Limfjorden (Denmark on-shore) (Fig. 1)

	Northern Viking Graben	*Balder area*	*Denmark on-shore*
Total number of beds	?	430	186
Fraction of *main* tuff-zone	1/3 (complete)	Complete	Complete
Number of beds in *main* tuff-zone	168 (500)	333	122
5—10 cm	20 (60)	21	10
10—20 cm	15 (45)	None	5
Max. bed thickness	28 cm	10 cm	20 cm
Max. grain size	Coarse	Medium	Medium
Reference	This study	Ræstad (1976), Ræstad and Prestvik (1982)	Bøggild (1918)

The present data do not exclude that more westerly located volcanism in the Faeroe-Shetland region (Thulean Province) also has contributed to the Balder tuffs. Thus the tuff-beds might become more abundant, thicker and coarser also towards that direction from well 25/10-1. However, this still has to be confirmed by well data.

The Danish onshore deposits fall into the N—S-trend of decreasing bed abundance (Table 7), but their grain-size is comparable with that of well 25/10-1 and they also contain thicker beds. Therefore, they do not simply fit the trend defined by wells 30/2-1 and 25/10-1. However, Pedersen *et al.* (1975) have suggested that the Danish tuffs have their source in a volcanic centre in the Skagerak (Åm, 1973) (Fig. 1), as some of the tuff-beds display a N—S-thinning of bed thickness (Bøggild, 1918; Andersen, 1937; Norin, 1940). We therefore conclude that the Balder tuffs in the northern North Sea and the time-equivalent Danish tuffs were sourced from different areas. In both cases the ashes were transported by winds blowing dominantly southwards.

The proposed volcanic centre in Skagerak lies close to or within the Fennoscandian border zone. This major tectonic zone is shown by Pegrum (1984) to extend from the Baltic, across southern Sweden, northern Denmark and into the central part of the North Sea, and speculatively all the way to the Wyville—Thomson Ridge (Fig. 1). Larsen *et al.* (1983) suggest that the extensive Palaeozoic volcanism in the Norwegian—Danish sectors is linked to tectonic movements along this zone proposed by Pegrum (1984). Pegrum and Ljones (1983) have shown that in the central North Sea the extended Tornquist Zone was tectonically active at least into the early Tertiary. We suggest, therefore, that the Skagerak volcanism is linked to this fundamental fracture zone where it is cross-cut by other tectonic lineaments related to the Bamble Trough of southern Norway. We also suggest that the Skagerak volcanism and other possible volcanism along the Tornquist Zone, represent a major source for the Balder tuffs in the southern part of the North Sea.

Such a model of several source areas for the Balder tuffs can best explain their large regional distribution. Analogy with recent volcanism shows that basaltic ash similar to that of the Balder Fm has a very limited dispersal area compared with the total area of the North Sea.

A concept of the dispersal area, grain-shape and grain-size variation of different types of recent volcanic ash deposits has been developed by Walker and Croasdale (1972), Walker (1973, 1980) and Self and Sparks (1978). Basaltic eruptions of Strombolian and Surtseyan types are shown to have dispersal areas (defined by the 0.01 T_{max} isopach) in the range of 1000 km^2, while Plinian types of dacitic to rhyolitic composition in extreme cases have dispersal areas in the range of 10 000 km^2.

The Balder tuffs have a basaltic composition and their grain-morphology is comparable to Surtseyan type ashes as described by Walker and Croasdale (1972). Like Surtseyan type deposits, the Balder tuffs also have a finer grain-size (normally < 1 mm) over most of their dispersal area, compared with Plinian types, which are coarser and show a strong decrease in grain-size away from their eruptive centres. An analogy with Surtseyan type deposits therefore results in an expected dispersal area of less than 1000 km^2 for the Balder ashes. However, Williams (1983) has shown that basaltic eruptions can have dispersal areas of about 1200 km^2 within the range of Plinian types, as defined by earlier workers. Plinian eruptions seldom exceed 10 000 km^2. Even with such exceptionally large dispersal areas, multiple volcanic source areas are necessary to account for the wide regional distribution of the Balder tuffs. According to Jaqué and Thouvenin (1975) they cover an area in the range of 400 000 km^2 in the North Sea.

SUMMARY AND CONCLUSIONS

The Balder Fm of well 30/2-1 is 76-m thick and consists of three distinct tuff-zones: a lower and an upper zone which are relatively thin and contain less-abundant tuff-beds than the central zone which is 47-m thick and constitutes the main tuff-zone of the formation.

The main tuff-zone consists of interbedded tuffs and shales in a ratio of approximately 1:1. From the available core it is calculated to contain about 500 tuff-beds, ranging from a few millimetres up to 28-cm in thickness. The beds, which show strong normal grading from coarse grain-size to clay-size, represent first-generation airfall volcanic ashes. They are composed of highly vesicular ash-particles which resulted from explosive volcanic eruptions probably of Surtseyan type. They have a basaltic composition comparable to the high Fe—Ti tholeiites. They show a composition transitional between 'within plate basalt' (WPB) and 'mid ocean ridge basalt' (MORB). Compatible with a magmatism that straddles the continental/oceanic environment they can be linked to an active rifting phase between N.W. Europe and Greenland.

The tuff-beds of well 30/2-1 are more abundant. thicker and coarser than those further south in the northern North Sea. This suggests that the volcanic sources for this area were located north of the North Sea and that the volcanic ashes were transported by winds blowing dominantly southwards.

In comparison, the Danish tuffs were probably sourced by volcanism in the Skagerak, south of Norway. This and other possible volcanic centres were located along a fundamental tectonic zone, the Tornquist Zone, which cross-cuts the central part of the North Sea. With prevailing winds from the north the areas south of the Tornquist Zone could receive volcaniclastic debris from two different source areas simultaneously. Multiple volcanic centres situated north to northwest of the North Sea and in the Skagerak can best explain the large regional distribution of the Balder tuffs.

Deposition of the Balder tuffs took place during a period of low clastic input to the North Sea Basin. Thus the shales interbedded with the tuff-beds have a very low content of detrital minerals. They are dominantly composed of microcrystalline quartz which results from recrystallization of siliceous microfossils. Pyrite pseudomorphs after such organisms are widespread in the shales, but a rich, well-preserved flora is found in some carbonate concretions. The species of diatoms and silicoflagellates are identical to those of the so-called Mo-Clay, a diatomite containing the on-shore Danish tuff deposits. More specifically, certain silicoflagellates make correlation possible with the upper part of the Mo-Clay and the main Danish tuff-zone. This greatly extends the known distribution of diatomite which was earlier interpreted as a localized Danish deposit, or suggested to be confined to the shallow shelf areas along the Norwegian coast and Denmark. The correlation with the Mo-Clay is confirmed by the dinoflagellate flora which can also be recognized in the Danish Ølst Formation and in the main phase of pyroclastic deposits in the North Sea. In its undeformed state, a diatomite such as the Danish Mo-Clay and its North Sea equivalents, would have a high porosity and permeability, comparable to that of undeformed chalk.

The poro-perm of the tuffs could be similar to or even better than the best reservoir-sands in the North Sea. One should not, therefore, reject the possibility that the regionally distributed Balder Fm can have reservoir-potential in some areas of the shelf.

In the more deeply buried areas, such as in the Viking Graben, this reservoir potential is largely destroyed. This is especially true for the diatomites. The tuffs are strongly altered but this has resulted in an exceptionally high secondary porosity. Unfortunately, it is accompanied by very poor permeabilities.

The alteration of volcanic ash in the tuff-beds and shales represents transformations into swelling chlorite, chlorite—smectite mixed layered clays and smectite. These clay-minerals are responsible for the swelling problems often encountered within the Lower Tertiary sequence in the North Sea. Detailed investigations of the properties of these clays, in line with the data presented here, may assist in finding an effective mud-treatment to prevent these problems.

ACKNOWLEDGEMENTS

We thank the following members of the staff at Statoil's Laboratory for their assistance: L.E. Aasberg, K.A. Gibbons, A.E. Gilje, I. Hervik, J. Johannesen, I. Nordvall and B. Tan. Special thanks go to Anne Beth Fløtre who typed the manuscript, and Arild Kaada who made the illustrations. We also thank Statoil and the licence partners in block 30/2, Tenneco Oil Co. and Union Oil for permission to publish this study.

REFERENCES

Andersen, S.A. 1937. De vulkanske Askelag i Vejgennemskaeringen ved Ølst og deres udbredelse i Danmark. *Danm. Geol. Unders.*, Ser. 2, **59**, 1—52.

Benda, L. 1972. The diatoms of the Moler formation of Denmark (Lower Eocene). A preliminary report. In: Simonsen (ed.), *First Symp. Recent and Fossil Marine Diatoms*, Vol. 39. Beih. Nova Hedwegia, pp. 251—266.

Berggren, W.A. and Gradstein, F.M. 1981. Agglutinated benthonic foraminiferal assemblages in the Palaeogene of the Central North Sea: their biostratigraphic and depositional environmental significance. In: Illing, L.V. and Hobson, G.D. (eds.). *Petroleum Geology of the Continental Shelf of North-West Europe*, Institute of Petroleum, London, pp. 283—285.

Bonde, N. 1973. Fiskefossiler, diatomiter og vulkanske askelag. *Dansk Geol. Foren., Årsskr.*, **1972**, 136—143.

Bonde, N. 1974. Palaeoenvironment as indicated by the 'Mo-Clay Formation' (Lowermost Eocene of Denmark). *Tertiary Res.* (Tertiary Times), **2** (1), 29—36.

Bonde N. 1979. Palaeoenvironment in the 'North Sea' as indicated by the fish bearing Mo-Clay deposit (Paleocene/Eocene), Denmark. *Meded. Werkgr. Tert. Kwart. Geol.*, **16** (1), 3—16.

Bonde, N. 1982. Teleostei (bony fish) from the Palaeocene of the Norwegian North Sea Drilling. *Norsk Geol. Tidskr.* **62**, 59—65.

Bugge, T., Prestvik, T. and Rokoengen, K. 1980. Lower Tertiary volcanic rocks off Kristiansund, mid Norway. *Mar. Geol.* **35**, 277—286.

Bukry, D. 1981a. Cenozoic Coccoliths from the Deep Sea Drilling Project. *SEPM Spec. Pub.*, No. 32, pp. 335—353.

Bukry, D. 1981b. Synthesis of Silicoflagellate Stratigraphy for Maestrichtian to Quaternary Marine Sediment *SEPM Spec. Pub.*, No. 32, pp. 433—444.

Bøggild, O.B. 1918. Den vulkanske Aske i Moleret samt en oversigt over Danmarks eldre Tertiærbjærgarter. *Danm. Geol. Unders.* Række 2, **33**, 159pp.

Christensen, O. Bruun. Studies of the formerly exposures in the transition beds (Mo-Clay/Røsnæs Clay) in Knudeklint, Fur. (in preparation).

Costa, L. and Downie, C. 1976. The distribution of the dinoflagellate *Wetzeliella* in the Palaeogene of northwestern Europe. *Palaeontology* **19**(4), 591—614, pl. 92.

Costa, L., Denison, C. and Downie, C. 1978. The Paleocene/Eocene boundary in the Anglo-Paris Basin. *J. Geol. Soc. Lond.* **135**, 261—264.

Deegan, C.E. and Scull, B.H. (compilers) 1977. A proposed standard lithostratigraphic nomenclature for the Central and Northern North Sea. Rep. Inst. Geol. Sci., No. 77/25; *Bull Norw. Petrol. Direct.*, No. 1.

Dinesen, A., Michelsen, O. and Lieberkind, K. 1977. A survey of the Paleocene and Eocene deposits of Jylland and Fyn. *Geol. Surv. Denm.*, B, **1**, 1—15, Copenhagen.

Hailwood, E.A., Bock, W., Costa, L., Dupeuble, P.A., Miller, C. and Schnitker, C. 1979. Chronology and biostratigraphy of northeast Atlantic sediments, DSDP Leg 48. In: L. Montadert *et al.* (eds.), *Initial Reports Deep Sea Drilling Project, Vol. 48*, U.S. Govt. Printing Office, Washington D.C., pp. 1119—1141.

Hansen, J.M. 1979. Age of the Mo-Clay Formation. *Bull. Geol. Soc. Denmark*, **27**, 89—96.

Hardenbol, J. and Berggren, W.A. 1978. A New Paleogene numerical timescale. *AAPG Stud. Geol.* **6**, 213—234.

Heilmann-Clausen, C. 1982. The Paleocene—Eocene boundary in Denmark. *Newsl. Stratigr.*, **11**(2), 55—63.

Heilmann-Clausen, C., Nielsen, O.B. and Gersner, F.L. Late Paleocene and Eocene lithostratigraphy of Denmark (in preparation).

Jaqué, M. and Thouvenin, J. 1975. Lower Tertiary tuffs and volcanic activity in the North Sea. In: A.W. Woodland (ed.), *Petroleum and the Continental Shelf of Northwest Europe*, Applied Science Publishers, London, pp. 455—465.

Knox, R.W. O'B and Harland, R. 1979. Stratigraphical relationships of the early Palaeogene ash-series of NW Europe. *J. Geol. Soc. Lond.*, **136**, 463—470.

Knox, R.W. O'B and Morton, A.C. 1983. Stratigraphical distribution of early Palaeogene pyroclastic deposits in the North Sea basin. *Proceedings of the Yorkshire Geol. Soc.*, Vol. 44, Part 3, No. 25, pp. 355—363.

Knox, R.W. O'B, Morton, A.C. and Harland, R. 1981. Stratigraphical relationships of Palaeocene sands in the UK sector of the central North Sea. In: L.V. Illing, and G.D. Hobson (eds.), *Petroleum Geology of the Continental Shelf of North-west Europe.* Heyden, London, pp. 267—281.

Larsen, B., Sundvoll, B., Neumann, E.-R and Kirkemo, K. 1983. Upper Palaeozoic volcanics in the North Sea (*MS*).

Lowrie, W., Alvarez, W., Napoleone, G., Perch-Nielsen, K., Premoli-Silva, I. and Tourmarkine, M. 1982. Paleogene magnetic stratigraphy in Umbrian pelagic carbonate rocks: the Contessa sections, Gubbio. *Bull. Geol. Soc. Am.*, **93**, 414—432.

Løvlie, R. and Kvingedal, M. 1975. A palaeomagnetic discordance between a lava sequence and an associated interbasaltic horizon from the Faeroe Islands. *Geophys.*

J.R. Astron. Soc., **40**, 45—54.

Martini, E. 1971. Standard Tertiary and Quaternary calcareous nannoplankton zonation. *Proc. II. Planktonic Conf.*, **2**, 739—785.

Ness, G., Levi, S. and Couch, R. 1980. Marine magnetic anomaly timescales for the Cenozoic and Late Cretaceous: A precise critique and synthesis. *Rev. Geophys. Space Phys.*, **18**, 753—770.

Norin, R. 1940. Problems concerning the volcanic ash layers of the Lower Tertiary of Denmark. *Geol. Fören. Stockh. Förh.*, **62**, 31—44.

Pearce, J.A. and Cann, J.R. 1973. Tectonic setting of basic volcanic rocks determined using trace element analyses. *Earth Planet. Sci. Lett.*, **12**, 290—300.

Pedersen, A.K., Engell, J. and Rønsbo, J.G. 1975. Early Tertiary volcanism in the Skagerrak: New chemical evidence from ash-layers in the Mo-Clay of northern Denmark. *Lithos*, **8**, 255—268.

Pedersen, G.K. 1981. Anoxic events during sedimentation of a Palaeogene diatomite in Denmark. *Sedimentology*, **28**, 487—504.

Pegrum, R.M. 1984. The extension of the Tornquist zone in the Norwegian North Sea. *Norsk. Geol. Tidskr.* (in press).

Pegrum, R.M. and Ljones, T.E. 1983. The 15/9 Gamma Gasfield: a new trap type for the North Sea with regional implications. *Am. Assoc. Petrol. Geol.* (in press).

Perch-Nielsen, K. 1976. New silicoflagellates and a silicoflagellate zonation in the European Palaeocene and Eocene diatomites. *Bull. Geol. Soc. Denmark*, **25**, 27—40.

Ræstad, N. 1976. Characteristics of the lower Tertiary volcanic ashes in the northern North Sea. Thesis (unpublished). Norwegian Institute of Technology, Univ. of Trondheim, 118 pp. + Appendixes.

Ræstad, N. and Prestvik, T. 1982. Vulkanske askelag fra undre Tertiær i nordlige Nordsjø. Abstr. Nordisk Geologisk Vintermøte, Reykjavik 1982, p.7.

Sélf, S. and Sparks, R.S.J. 1978. Characteristics of widespread pyroclastic deposits formed by interaction of silicic magma and water. *Bull. Volcanol.*, **41**(3), 1—17.

Walker, G.P.L. 1973. Explosive volcanic eruptions — a new classification scheme. *Geol. Rundsch.*, **62**, 431—446.

Walker, G.P.L. 1980. The Taupo pumice: product of the most powerful known (ultraplinian) eruption? *J. Volcanol. Geotherm. Res.* **8**, 69—94.

Walker, G.P.L., 1981. Plinian eruptions and their products. *Bull. Volcanol.* **44**, 223—240.

Walker, G.P.L. and Croasdale, R. 1972. Characteristics of some basaltic pyroclastics. *Bull. Volcanol.* **35**, 303—317.

Williams, S.N. 1983. Plinian airfall deposits of basaltic composition. *Geology*, **11**, 211—214.

Ziegler, P.A. 1982. *Geological Atlas of Western and Central Europe.* Shell Interationale Petroleum Maatschappij B.V. Elsevier Scientific Publishing Company, Amsterdam, 130 pp. with enclosures.

Åm, K. 1973. Geophysical indications of Permian and Tertiary igneous activity in the Skagerrak. *Norges Geol. Unders.*, **287**, 1—25.

11 Marginal models

Christopher Beaumont, Ross Boutilier* and Charlotte E. Keen†*
*Department of Oceanography, Dalhousie University, Canada
† Atlantic Geoscience Centre, Geological Survey of Canada

During rifting of a continent the lithosphere does not fracture along a single discrete boundary. Instead, it is believed that extension by stretching and necking occurs prior to the eventual break-up and formation of oceanic crust and lithosphere at an oceanic ridge. In this behaviour the lithosphere is analogous to that of a hot metal rod when drawn. The thermal and rheological aspects of the analogy further show that extension may cause initial tectonic uplift or subsidence of the margin according to which of the competing effects of crustal thinning, or the replacement of cold lithosphere by hot lithosphere and asthenosphere, dominates.

During the post-rift phase the lithosphere cools, contracts and subsides with a time constant that is proportional to the square root of the initial lithospheric thickness. The rheological properties of the lithosphere, which are also temperature dependent, determine its flexural isostatic response to loading by the sediments that accumulate on the margin, and by the ocean.

In this chapter we present a summary of results from a thermomechanical model of the evolution of rifted continental margins. It is shown that the physical consequences of stretching can accurately reproduce the first-order properties of the 185-Ma old Nova Scotian Margin. The results predicted by archetypal stretching profiles further demonstrate that the margins also develop characteristics that are diagnostic of necking style. Examples from the east-coast margin of North America show that linear, concave and convex necking profiles occur on the same margin. The reasons for such variability are, as yet, not well understood.

INTRODUCTION

This chapter provides a summary and explores the geological consequences of our recent work on the rift and post-rift stages of the evolution of rifted Atlantic-type continental margins, or as they are sometimes called, passive margins. These margins are the subject of intense research interest because they may contain significant hydrocarbon resources.

Of the possible mechanisms responsible for their formation and subsequent evolution, the proposal that there is significant horizontal extension during rifting which thins both the crust and subcrustal lithosphere can explain most of the first-order properties of the margins and can be tested by modelling. These properties include: the thinning of the crust landward of the ocean—continent boundary (Keen and Hyndman, 1979; Montadert *et al.*, 1979; Sheridan *et al.*, 1979), subsidence (Sleep, 1971; Watts and Ryan, 1976, Keen, 1979; Watts and Steckler, 1979; Royden and Keen, 1980), and listric faulting of the upper crust (Montadert *et al.*, 1979).

The thermal and isostatic consequences of the lithospheric stretching model were first investigated by McKenzie (1978). In this chapter three kinematic models of rifting are reviewed: the uniform extension model (McKenzie model); the uniform extension and melt segregation model (Beaumont *et al.*, 1982; Foucher *et al.*, 1982); and the depth-dependent extension model (Royden and Keen, 1980; Beaumont *et al.*, 1982). The thermal and subsidence differences among these models are important only to the pre- and syn-rift sediments and to those post-rift sediments that were deposited within approximately 30 Ma of the end of rifting. Reactivation of the stretching process can, however, significantly modify the thermal history of the sediments from that predicted by a single extensional event.

The rifting processes control the post-rifting history, which is traced by a thermo-mechanical model in which subsidence and temperature are the result of conductive cooling. The isostatic response to loading by sediment and water is found by assuming that these loads are supported by an elastic layer. The thermal and rheological

Petroleum Geology of the North European Margin © Norwegian Petroleum Society (Graham & Trotman, 1984) pp. 171—186.

aspects of the model are coupled in that the space available for sediments and water primarily depends on the thermal history, but the rheological properties of the lithosphere, which determine the isostatic amplification in the mechanical model, are also determined by the temperature distribution.

The two-dimensional models predict the form of cross-sections of continental margins for comparison with observations. Here results are compared with a cross-section of the approximately 185-Ma old Nova Scotian Margin and with other margin archetypes. The amount of crustal extension in the Nova Scotian example is estimated from the seismic measurements of the present crustal thickness, assuming that prior to rifting the crust had a uniform thickness equal to the present thickness at the landward end of the section. The sediment budget was calculated by decompacting the observed stratigraphic section.

The archetypal models illustrate the effect of variations in the amount of stretching across a magin on the sedimentary and thermal history. They also show that the gravity anomaly may be used as a fingerprint for the stretching profile.

It is shown that the first-order predictions of margin cross-sections may be accurately matched with observations when a suitable choice of model parameters is made. That these parameters (for example, pre-rift crustal and lithospheric thickness and corresponding densities, and the properties of the margin sediments) are not accurately and independently known means that observations must be used to constrain parameter values. The sensitivity of model predictions to variations in model parameters is also important when assessing the significance of the results. Properties, such as sediment thermal conductivity and radiogenic heat production, which have only a minor influence on subsidence, may be very important with regard to the thermal maturation of the sediments. Consequently, from an academic viewpoint it could be claimed that we have a model for the formation, subsidence and stratigraphy of rifted continental margins. However, from the perspective of those engaged in hydrocarbon exploration the model will be of real use only if it can be shown to provide accurate temperature histories as well as subsidence histories.

RIFTING MODELS

The three models we use to describe the rifting processes are: (1) uniform extension (McKenzie, 1978); (2) uniform extension and melt segregation (Beaumont *et al.*, 1982); and (3) depth-dependent extension (Royden and Keen, 1980; Sclater *et al.*, 1980). In all three models, extension and thinning of the lithosphere are the dominant processes during rifting (Fig. 1). Uniform extension involves stretching of both the crust and underlying lithosphere by a factor β, which varies with position across the margin. This model predicts immediate initial tectonic subsidence during extension followed by thermal subsidence as the lithosphere cools. Both phases of tectonic subsidence are, of course, isostatically amplified by the sediment and water load.

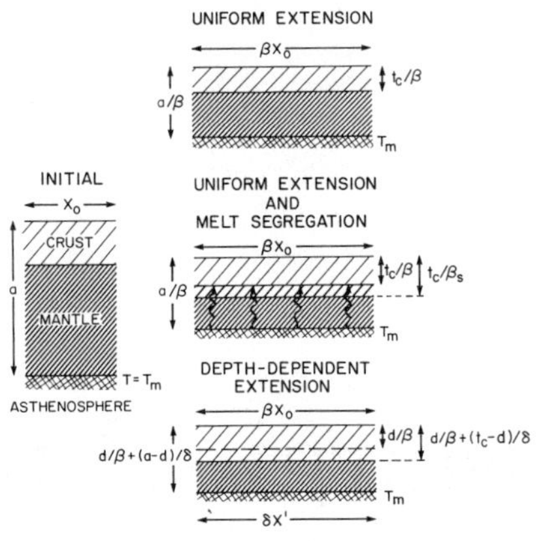

Fig. 1. One-dimensional illustrations of the three rifting processes discussed. A segment of the lithosphere of initial length x_o and depth a is extended by β in the uniform extension model. The other two models include the possibility that segregated melt from the asthenosphere migrates to the extended crust, and that the amount of extension may change with depth in the lithosphere. The temperature of the asthenosphere is T_m, t_c is the initial crustal thickness, d is the depth at which extension changes from β to δ, and t_c/β_s is the final thickness of the crust when segregated melt is included (see Fig. 3).

The model involving uniform extension and melt segregation is similar to uniform extension, except that it allows realistic description of the generation of oceanic crust and assessment of the relative importance of basaltic intrusion into the margin. As before, crust and lithosphere are extended by β but, in addition, fractionated basaltic magma migrates from the asthenosphere to the base of the crust. This thickens the crust from the t_c/β of the uniform extension model to t_c/β_s, where $\beta_s = t_c/((t_c - t_{oc}/\beta) + t_{oc})$ (Figs. 2 and 3), a change that effectively reduces the values of β as determined from seismic measurements of crustal thickness. The melt segregation results in small changes in density which are included in calculations of the initial subsidence. The thermal subsidence predicted by this model is almost identical to that predicted by uniform extension. A maximum amount of melt is generated when oceanic crust is formed, i.e. when $\beta = \infty$. Proportionately smaller amounts segregate as β decreases. The main difference from the uniform extension model is that the uniform extension and melt segregation model provides a smooth transition from unextended continental crust, through extended continental crust with an added basaltic fraction, to oceanic crust formed solely from the basaltic fraction (Fig. 3).

The depth-dependent extension model differs from the uniform extension model in that it assumes different amounts of extension in the upper and lower lithosphere, separated at a depth, d (Fig. 2). In this paper $d=t_c$, the original thickness of continental crust. The crust is thinned by an amount β, while the lithosphere beneath the crust is thinned by δ. This model predicts either initial uplift or initial subsidence during rifting, depending on the choice of β and δ, whereas for normal continental crust the other two rifting models predict only subsidence.

Details of the formulae controlling the thermal and subsidence behaviour of these rift models are given in a uniform manner by Beaumont *et al.* (1982) or in the original papers. It must be emphasized that these are kinematic models which assume instantaneous extension and that the dynamical evolution could be substantially different (Keen, in preparation.). For example, the form of subsidence in the uniform extension and melt segregation model could be altered by the trapping of melt during extension and its subsequent sudden lateral flow from beneath the crust and out into nascent oceanic crust during final break-up of the margin.

The two-layer stretching model is also an idealization of the true process of extension which may be visualized through the analogy of the stretching and necking of a metal plate. A uniform temperature plate necks equally from above and below, but one in which temperature increases with depth can be expected to fail in a brittle manner near the surface. At greater depths and increasing temperature there is first a change to ductile creep, which itself shows less and less resistance to flow as temperature increases. The depth-dependent extension model crudely represents this increasing ease of flow with increasing depth in a kinematic manner by making δ, the stretching in subcrustal lithosphere, greater than that, β, in the surface layer. The transition between the two regions of stretching is a region of high shear and can be equated with a decollement zone at the bottom of a region of brittle listric faulting in the crust, δ itself certainly varies with depth, but it would be of little use to add more parameters, whose values are not independently known, to the kinematic model. A more useful approach to the details of rifting is through a dynamic model in which the temperature dependence of creep is included (Keen, in preparation.).

It is also important to remember that the total amounts of extension in the upper, $\beta(x)$, and lower $\delta(x)$, layers must be equal when summed across the entire extensional zone. This means that areas where $\beta(x)$ is larger than $\delta(x)$ must be compensated by other areas where the converse is true. Features diagnostic of the areal distribution of β and δ are: $\delta(x) \gg \beta(x)$, significant uplift without evidence of extension, followed by an increase in heat flux which then decays along with the uplift with a 50—60-Ma time constant; $\beta(x) \gg \delta(x)$, rapid subsidence concomitant with the extension followed by little thermal subsidence. If the depth-dependent extension model is correct, we should not be surprised to see areas of uplift and subsidence adjacent to each other in an extensional environment. Moreover, the thermal anomalies in the apparently unextended but uplifted regions may be just as significant as those in the regions of obvious crustal extension.

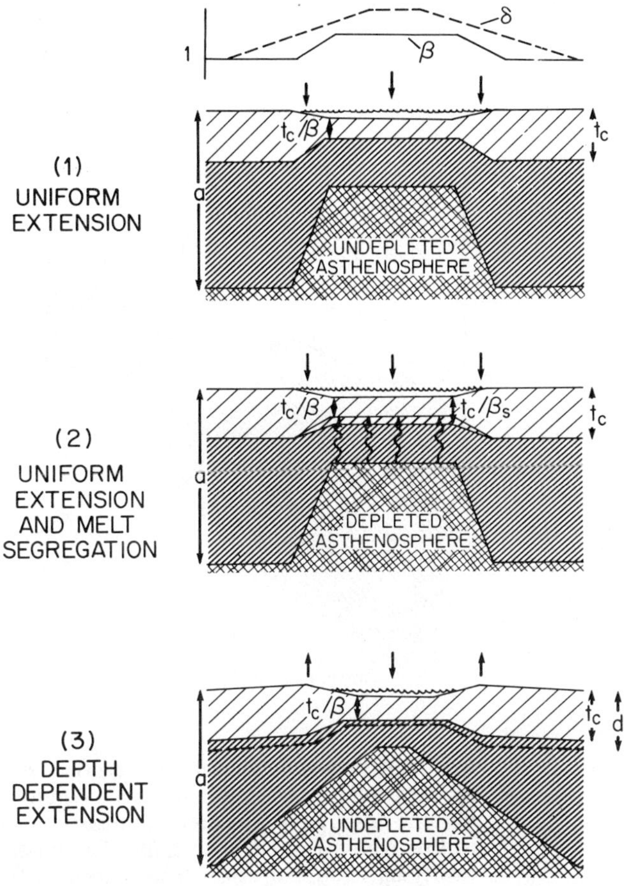

Fig. 2. Two-dimensional cross sections showing lithospheric extension during rifting for the three rifting models. In the uniform extension model both crust and lithosphere are extended and thinned by β, with β varying with position as shown in the upper panel. Initial subsidence occurs for all regions where $\beta > 1$. Extended lithosphere is replaced by the passive upwelling of undepleted asthenosphere. Similar extension by β occurs in the uniform extension and melt segregation model. However, the extended crust is thickened by basaltic melt that segregates from the incoming asthenosphere. This melt is shown underplating the crust, but can equally well be considered to intrude it. In the depth-dependent extension model an upper layer $z \leqslant d$ is extended by β whereas the layer beneath $z=d$ is extended by δ as shown in the upper panel. d may be greater or less than the initial crustal thickness, t_c. Such a model can give initial uplift, for regions were β is small, as well as subsidence where β and δ are large. No melt segregation is postulated.

Before using the results of this and other papers to predict the initial subsidence on rifting, readers should explore the sensitivity of the subsidence to assumptions concerning densities, coefficients of thermal expansion, crustal and lithospheric thickness, and temperature distribution in the lithosphere. Small variations in densities, for example, can have significant effects, as can pre-existing conditions. A second phase of extension that affects an already extended region can result in either uplift or subsidence depending on the relative amounts of extension during each of the phases and the elapsed time between the phases. Most authors (e.g. Parsons and Sclater, 1977, McKenzie, 1978; Royden *et al.*, 1980; Beaumont *et al.*, 1982) have chosen to use similar values of the important parameters so that the results may be compared

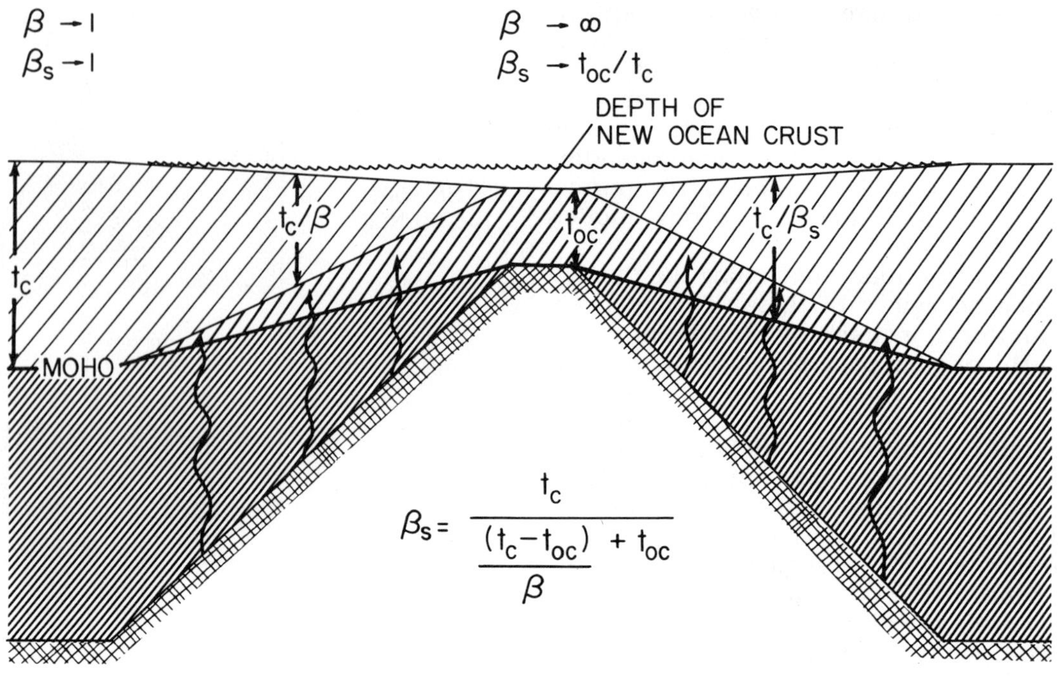

Fig. 3. A two-dimensional cross section showing how the uniform extension and melt segregation model produces oceanic crust and a smooth transition from unextended continental crust, β=1, to oceanic crust, β=∞. The medium-density shading illustrates the basaltic melt that thickens the crust in this region from t_c/β to t_c/β_s. The basaltic melt can equally well be considered to intrude the overlying extended crust. The amount of melt that segregates is proportional to the amount of incoming undepleted asthenosphere.

(but see Dewey, 1982, for a greater range of possibilities); therefore, the sensitivity aspect has not received enough emphasis.

The reason that the initial tectonic subsidence is so sensitive to parameter values is that it pits two processes against each other; uplift that results from the replacement of relatively cooler subcrustal lithosphere by hot, less dense lithosphere and asthenosphere; and subsidence that results from extension of the crust. Uniform extension of average continental lithosphere results in subsidence, but the converse is true for average oceanic lithosphere (McKenzie, 1978).

THE POST–RIFT EVOLUTION

The post-external phase of subsidence is largely controlled by the thermal evolution of the lithosphere. Extension results in advective mass movements which, if rapid, carry heat with them and perturb the thermal balance of the lithosphere (McKenzie, 1978). The thermal gradient is raised above the ambient value in proportion to the amount of extension, and the isotherm considered to define the base of the lithosphere (1300–1350°C) migrates upward towards the surface (Figs. 1–3). Thermal equilibrium is restored by the conductive cooling and thickening of the lithosphere. Thermal contraction which accompanies the cooling results in additional tectonic subsidence of the surface and isostatic amplification by the loading sediment and water.

The first-order thermal aspects of this evolution have been discussed by McKenzie (1978), Royden *et al.* (1980), Keen (1979), Royden and Keen (1980) and Watts and Steckler (1979). The results show that the subsidence rate should be proportional to $(\text{time})^{-1/2}$ for approximately one lithospheric thermal time constant and then trend to an exponentially decaying form. All these authors assumed a local model of isostatic equilibrium.

We have used a coupled thermo-mechanical model to trace the evolution of a rifted margin as it cools, contracts and is loaded by sediment and water. The thermal model is based on interpolations between one-dimensional finite difference solutions for the conductive cooling and thickening of a self-contracting subsiding lithosphere as a function of changing β (and δ) with position across the margin. The finite-element mechanical model calculates the isostatic response to loading.

The one-dimensional, time-dependent equation of heat diffusion in the absence of radioactive heat generation is:

$$\frac{\partial}{\partial z}\left(K(t)\frac{\partial T}{\partial z}\right) = \rho c\left(\frac{\partial T}{\partial t} + v_z \frac{\partial T}{\partial z}\right)$$

where T = temperature, K = thermal conductivity, ρ = density, and c = specific heat: $\kappa = K/\rho c$ = thermal diffusivity. The last term is the heat advected by isostatic subsidence of the lithosphere at velocity v_z. The initial temperature distribution is that predicted by the rifting models. The upper boundary, at the sea-floor, is maintained at T=0°C, while the bottom boundary remains at $T = T_m \simeq 1350°C$ (to limit the maximum lithospheric thickness to 'a', the value before extension. The model has two regions, sediments and lithosphere, each of which can have different thermal properties. Sediments are deposited at 0°C and are then considered part of the model to be conductively heated or cooled. The effects of radioactive heat generation in the sediments and crust can also be included. A finite difference discretization is used to solve the diffusion equation with a spatial step size of 0.5 km and thermal timestep of 0.1Ma.

The thermal contraction tectonic subsidence, $w(\Delta t)$, for the interval Δt, is equal to the self-contraction of the lithosphere.

$$w(\Delta t) = a_v \left[\int T(z,t)dz - \int T(z,t+\Delta t)dz\right].$$

where a_v is the volume coefficient of thermal expansion and the integral is taken over the entire lithosphere plus sedimentary cover. There is also a small additional subsidence related to the contraction of the finite difference grid (Beaumont *et al.*, 1982).

The thermal evolution is accompanied by mechanical effects which, though secondary, have important consequences for margin evolution, as can be judged by comparison with local isostatic models. The lithosphere, the Earth's thermal boundary layer, behaves as a mechanically strong competent region that can undergo flexure during isostatic adjustment to surface loads. The importance of flexural effects is most simply considered a function of the flexural rigidity, $D = Ed_L^3/12(1-\sigma^2)$, where E is the average Young's modulus of the lithosphere, d_L is its mechanical thickness, and σ its Poisson's ratio. The flexural parameter, a $\propto |D^{1/4}|$(Wallcott, 1980) is proportional to the distance to which a load can drag down (downflex) the lithosphere against the buoyancy forces exerted by the relatively much more fluid asthenosphere or mantle. a varies from $\leqslant$10 km for oceanic ridges. Where the mechanically strong part of the lithosphere is very thin, to $\geqslant$100 km for shield regions of continents, where the lithosphere is thermally old.

Rifted continental margins cannot be characterized by one value of the flexural rigidity. On the continental side, where there is little or no extension, the lithosphere remains strong giving a a large value, whereas in a highly stretched region thinning of the lithosphere and fracture of the upper crust both contribute to making the lithosphere mechanically weak. Thinned regions cool as they age, in much the same way as oceanic lithosphere (Parsons and Sclater, 1977), and the mechanical strength returns.

No simple analytical mathematics exist to predict the flexure and vertical deformation of a lithosphere that has properties which vary in time and space. We have used a two-dimensional finite-element model to determine the isostatic response to loading by water and sediments (Beaumont *et al.*, 1982) but a finite-difference approach (Bodine, 1981) is equally valid.

The finite-element grid (Fig. 4) represents the rheological lithosphere, that part of the lithosphere which remains elastic and undergoes flexure. What region is this? It is the region below that which has failed by faulting during extension and above that which is sufficiently hot to undergo sufficient viscous creep on the timescales of margin evolution, 10—100 Ma. Because creep is a thermally activated process, with activation energy Q (Weertman, 1970; Ashby and Verrall, 1977, for example), the viscosity of rocks, η, is an exponential function of temperature and changes very rapidly with depth in the lithosphere.

In Fig. 5 the temperature and viscosity (normalized by the basal value, η_b) variations as a function of depth in the lithosphere are shown in

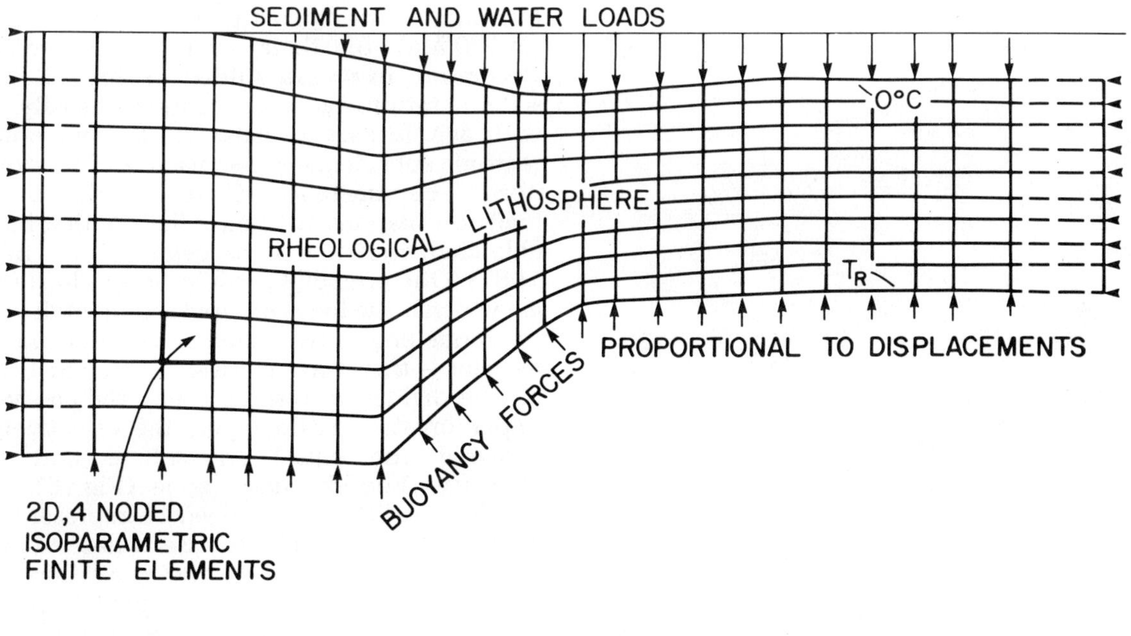

Fig. 4. Simplified version of the finite-element mesh used in the mechanical model. The mesh of 9×46 elastic elements is uniform in size with depth, z, at any location, but has a gradation in the horizontal, x, direction. The smallest elements in the extended zone have $\Delta x \simeq$ 12 km, whereas those near the boundaries have $\Delta x \simeq$ 500 km. The elements will have the same material properties. Poisson's ratio, ν = 0.25; Young's modulus, E=2×10^{11} Pa.

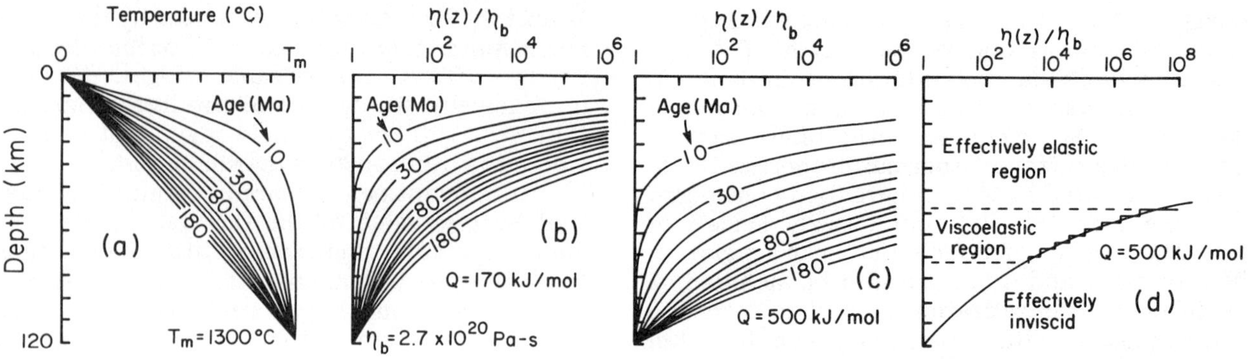

Fig. 5. (a) The time evolution of temperature in a cooling plate model of the oceanic lithosphere. Parameter values are from Parsons and Sclater (1977). (b) Corresponding evolution of viscosity normalized by the viscosity of the underlying asthenosphere, η_b, as a function of thermal age for Q=170 kJ/mol. (c) As for (b) but with Q=500 kJ/mol. (d) Model representation of the viscosity distribution for Q=500 kJ/mol and 100 Ma age illustrating the effective behaviour. The elastic shear modulus, $\mu=2.7\times10^{10}$ Pa.

relation to the thermal age of the oceanic lithosphere, on the assumption that creep is a linear function of deviatoric stress. The two values of Q, 170 and 500 kJ/mol, bound the possible range for mineralogies considered reasonable for the lithosphere (Kirby, 1980). A region where $\eta \geqslant 10^6 \eta_b$ will not creep significantly on the timescales of margin evolution unless plastic failure occurs; therefore, such regions can be considered to be elastic and undergo flexure when loaded. Regions where $\eta \leqslant 10^3 \eta_b$ will creep so rapidly that they appear fluid on the timescales of margin evolution

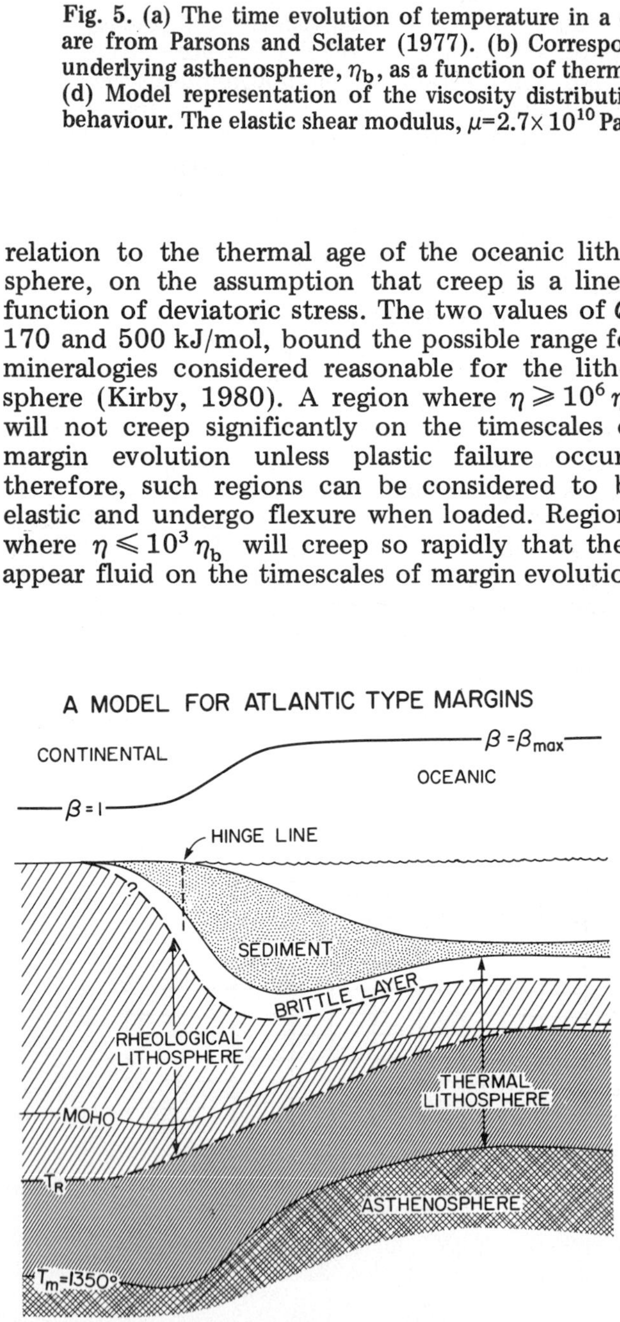

Fig. 6. The mechanical model showing: the distribution of $\beta(x)$; the brittle crustal layer, faulted during extension; the rheological lithosphere, that region which responds elastically to sediment and water loads; the thermal lithosphere, that region from the base of the sediments to the melting isotherm, T_m; and the crust and asthenosphere. Note that the region below the relaxation isotherm, T_R, is assumed to act as an inviscid fluid which buoys up to the rheological lithosphere.

and can be modelled as such. The intervening region (Fig. 5) is transitional and will show significant creep which can be modelled as a viscoelastic process (Courtney and Beaumont, in press). This extrinsic zoning of the lithosphere is primarily a function of temperature and is not an intrinsic property; the whole lithosphere could equally well and equivalently be modelled as a temperature-dependent viscoelastic material (Courtney, 1982).

It is not always necessary to include explicitly the viscoelastic transition in models. Consider the cases of young stretched lithosphere beneath a continental margin and young oceanic lithosphere. The viscoelastic transition layer in these examples is sufficiently thin to be ignored. Furthermore, cooling is proceeding sufficiently rapidly that it effectively freezes the region in which creep would occur in a thermally stable environment, so that the extrinsic effects of creep do not appear in the geological record of oceanic lithosphere or rifted margin evolution. This is true for a significant part ($\geqslant$100 Ma) of their evolution (Courtney and Beaumont, in press). Added to this is the fact that sedimentation rates on margins usually decrease with age, in part, because the thermal contraction tectonic subsidence is also waning. Therefore, under normal circumstances there is little potential for loads on margins to elicit a flexure that has observable characteristics of viscoelastic flow (Beaumont, 1981). One possibility would be an old starved margin which is the locus of rapid sediment deposition.

Modelling of the rheological properties of rifted margins is therefore relatively easy. The lithosphere can be divided into three zones, the passive faulted upper crustal brittle layer, the effectively elastic flexured rheological lithosphere and the low viscosity fluid-like basal zone (Fig. 6). Only the rheological lithosphere need be represented in the finite-element model. Its base is determined by an isotherm which we term the relaxation isotherm, T_R, corresponding to the negligibly thick viscoelastic transition zone. Its value cannot be specified *a priori* because it is a function of the mineralogy and activation energy, Q, which are not precisely known.

It is now evident why a coupled thermo-mechanical model is used. The thermal evolution determines the tectonic subsidence and the position of the relaxation isotherm. The mechanical model then determines the flexural response to water and sediment loads which fill and isostatically amplify tectonic subsidence. Feedback from sediment blanketing and erosion modify the cooling history, thermal contraction, and tectonic subsidence. There are additional secondary interactions that have only small effects on the model configuration but must be included to ensure mass and heat conservation, themselves related through thermal expansion, and therefore, accurate gravity anomaly calculations.

The variation of sediment properties in the model are discussed in detail by Beaumont *et al.* (1982). Each lithology undergoes compaction by dewatering according to $\phi(z) = \phi_0 \exp(-hd_z)$, where ϕ is the porosity with surface value ϕ_0, d_z is the depth and h is an empirical compaction constant. ϕ_0 and h were determined for each of the major lithologies using sonic and density logs from explanatory wells.

The computational methods used in the coupled thermo-mechanical model may be summarized as follows. The required input variables are: (1) the variation in $\beta(x)$ and $\delta(x)$; (2) the value of T_R; (3) the values of the thermal, mechanical, and sedimentological parameters; and (4) the sediment budget. Calculations are then performed by stepping the models through time in model timesteps, usually chosen to be 10—20 Ma. The procedure is as follows:

(1) At t=0, extension occurs resulting in the initial subsidence as predicted by local isostatic adjustment. This depression is then filled with a combination of sediment and water to simulate loading of the continental margin just after formation. Isostatic adjustment to this load is calculated using the mechanical model described above.

(2) A time interval Δt elapses during which the lithosphere cools and subsides as predicted by the thermal model: the position of the isotherm T_R also changes and a further load of water and sediment accumulates. Isostatic adjustment is again calculated by the mechanical model, adjusted for the new depth of T_R.

(3) This process is continued over the number of time steps necessary to simulate the evolution of the margin. After each time step the depths to the sea floor and to the basement, the position of the Moho Discontinuity, the temperature distribution, the temperature distribution, and the gravity anomaly across the margin are computed.

THE NOVA SCOTIAN MARGIN

We use the example of the approximately 185Ma-old rifted continental margin off Nova Scotia to illustrate the application of models to specific transects. Figures, 7, 8, and 9a illustrate the location of the transect (line 108); what is known of the basic stratigraphy from wells and seismic reflection profiles; the inferred lithologies and chronostratigraphy (when the effects of salt diapirism have been removed); and the locations of lines from which other geophysical data have been obtained. All these data and their primary sources are discussed in greater detail by Beaumont

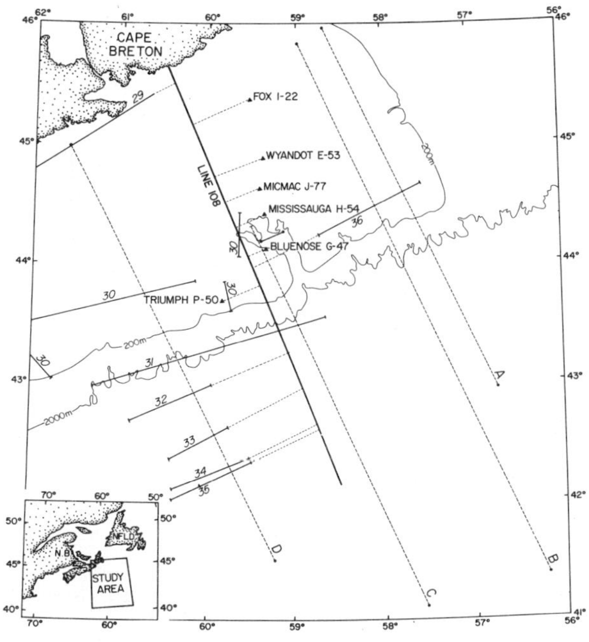

Fig. 7. Location map of the sources of data used for the Nova Scotian study. The heavy solid line is the line of the cross section (shown in Fig. 8 and 9); it is also the location of seismic reflection line 108. The triangles are deep exploratory wells. The dashed lines are the tracks along which free air gravity anomalies have been measured. Solid lines show the positions of deep crustal seismic refraction lines. The numbers associated with these refer to the publications in which the results are described: 29—30, Dainty *et al.* (1966); 31—34, Keen *et al.* (1975); 35—36, Keen and Cordsen (1981). The well data and the refraction data have been projected on to line 108, as shown by the fine dashed lines.

et al. (1982). It is assumed that the observations can be projected along strike on to the transect.

The observations, in particular those of Fig. 9a, were compared with predictions in which the rifting models; T_R; and $\beta(x)$ and $\delta(x)$ were varied across the margin to achieve the best fit. The seismically determined crustal thickness was used to estimate $\beta(x)$ on the assumption that the crust had a uniform thickness of 35 km, like that adjacent to Nova Scotia, before rifting. Small variations from these estimates were allowed in order to achieve the desired stratigraphy (Fig. 9a).

T_R is most strongly constrained by the gravity anomalies and an acceptable value was found to be 250°C±100°C in models that did not include a near surface faulted crustal region. Had such a layer approximately 20 km thick been included, T_R would have increased to about 650°C. The importance of the flexure implied by these values of T_R is that the maximum basin depth is decreased by approximately 1 km by comparison with a locally compensated model. An equivalent amount of sediment occupies downflexed regions landward of the hinge line (Wyandot well location) and beneath the rise.

The sediment budget takes the form of the decompacted section of Fig. 9a. That is, for each timestep sediment, equal to the amount shown

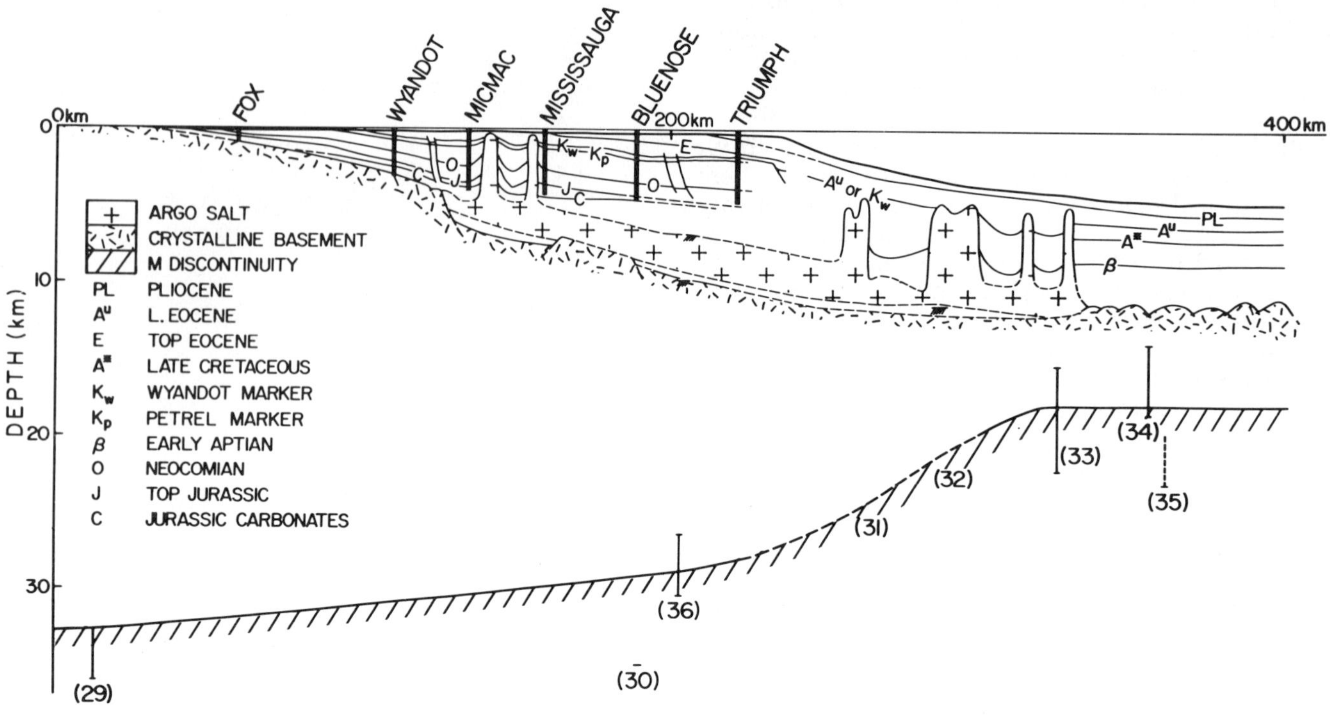

Fig. 8. Composite cross section of the Scotian Basin and crust along the line shown in Fig. 7. The depths to the Moho at the ends of the seismic refraction lines (Fig. 7) are shown as the vertical bars which are thought to represent the variability of crustal structure along strike. The refraction lines are labelled in the same manner as those in Fig. 7. Regions where the positions of interfaces remain somewhat uncertain are shown as dashed lines. The position of the basement—sediment interface beneath the basin is specified on the basis of wells that intersect it (Fox and Wyandot), seismic reflection, and seismic refraction (shown by observations. This basement is thought to be that of syn- and post-rift sediments: older sediments may extend to greater depth. Stratigraphy of the basin is based on evidence from deep exploratory wells (shown as vertical bars and seismic reflection data (Jansa and Wade, 1975; J.A. Wade, personal communication, 1981). Lithologies are regional averages representing the dominant component. Where unshaded, the sediments are mainly clastics.

beween the chonostratigraphic horizons, is added in a decompacted form to the model. The weight of this sediment causes compaction of itself and underlying sediments, and isostatic adjustment. Sediment that remains above sea level is eroded, resulting in sedimentary layers that are ultimately too thin. An acceptable model is one that can accommodate the sediments, without erosion, while maintaining realistic palaeobathymetries for the shelf and slope.

Any combination of uniform extension, or uniform extension and melt segregation, with instantaneous rifting that gave the desired final stratigraphy (Fig. 9a) predicted excessive water depths during the early stages of evolution. The depth-dependent extension model addresses this problem. For example, uplift of 420 m is predicted landward of the Micmac well in virtue of the large values of $\delta(x)$ beneath this region. The net initial uplift remains even under the downflexing of seaward loads, thereby preventing early sedimentation or the excessive water depths of the other models. Thermal contraction eventually allows sedimentation at the Fox and Wyandot wells approximately 25 Ma after rifting. Other than this property and a more realistic early shallow-water pattern of sedimentation, there is little to differentiate this model from others which have the same asymptotic properties yet differing rifting histories. It is not so necessary to invoke depth-dependent extension if it is believed that stretching occurred over an interval of time such that one part of the margin is stretched before another instead of infinite rifting rates. Jarvis and McKenzie (1980) have already shown that finite rates of extension delay part of the initial subsidence. An acceptable model section which closely reproduces Fig. 9a is shown in Fig. 9b. This result can be achieved with any of the rifting mechanisms, but the depth-dependent one gives the best palaeobathymetry. Three steps in the evolution of the gravity anomaly, basin structure, and Moho and isotherm positions for the depth-dependent version of this model are shown in Fig. 10. These correspond to time of rifting, the end of the Jurassic, and the present.

The sediments act as a low thermal conductivity blanket over the crust and lithosphere. Their insulating effect is included in the thermo-mechanical model and can be seen in Fig. 10 causing a bowing-up of the isotherms beneath the basin and into the sediment. Sediment thermal conductivities vary with lithology and compaction. The sediments comprise a matrix of grains and interstitial fluids such that their combined conductivity in the model is a parallel sum of the separate components (Beaumont *et al.*, 1982).

Subsidence is not very sensitive to variations in the thermal properties of the sediments. The main effect of blanketing is to reduce the subsidence of the basin by up to about 1 km in comparison with models with infinitely conductive sediments, because there is less thermal contraction in the

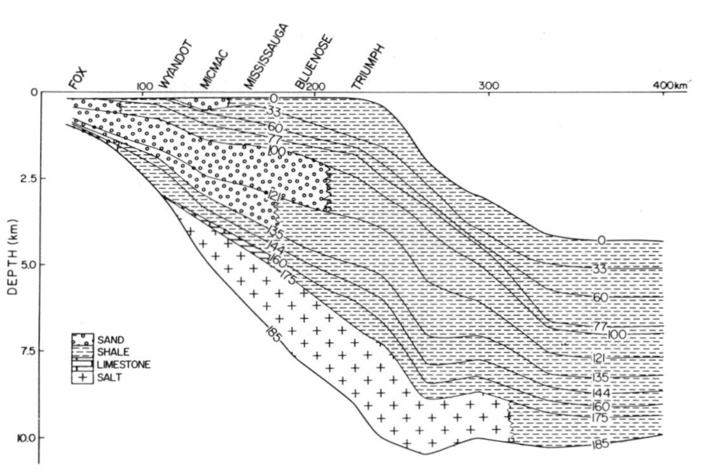

Fig. 9a. Generalized and interpreted chrono- and lithostratigraphy of the Scotian Basin along the section of Fig. 7. This interpretation of the section shown in Fig. 8 had a chronostratigraphy that was also used in the theoretical models (e.g. Fig. 9b). The section has been restored to remove the effects of diapirism and the lithology simplified to include the four components shown. In the cross section only the dominant sediment type in an area for a given time interval is shown. However, the calculations allowed for mixtures of two components and such mixtures were used. This is the section that was decompacted and used as the sediment budget for the theoretical models.

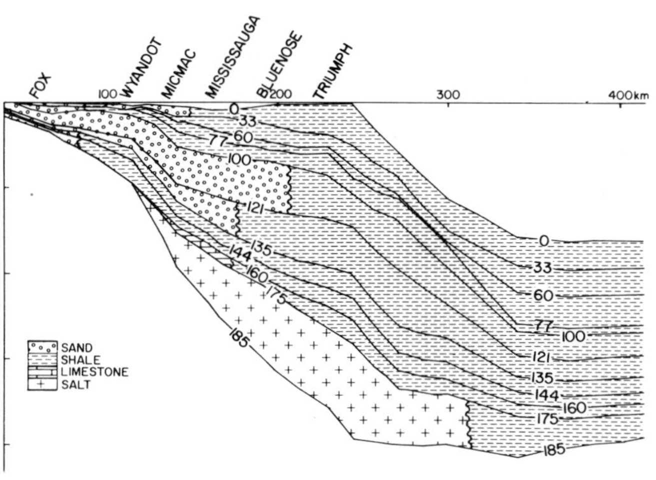

Fig. 9b. Stratigraphy of a T_R = 250°C model basin 185 Ma after rifting for comparison with Fig. 9a. Good agreement exists between the observed and computed strata, suggesting an underlying validity to the model.

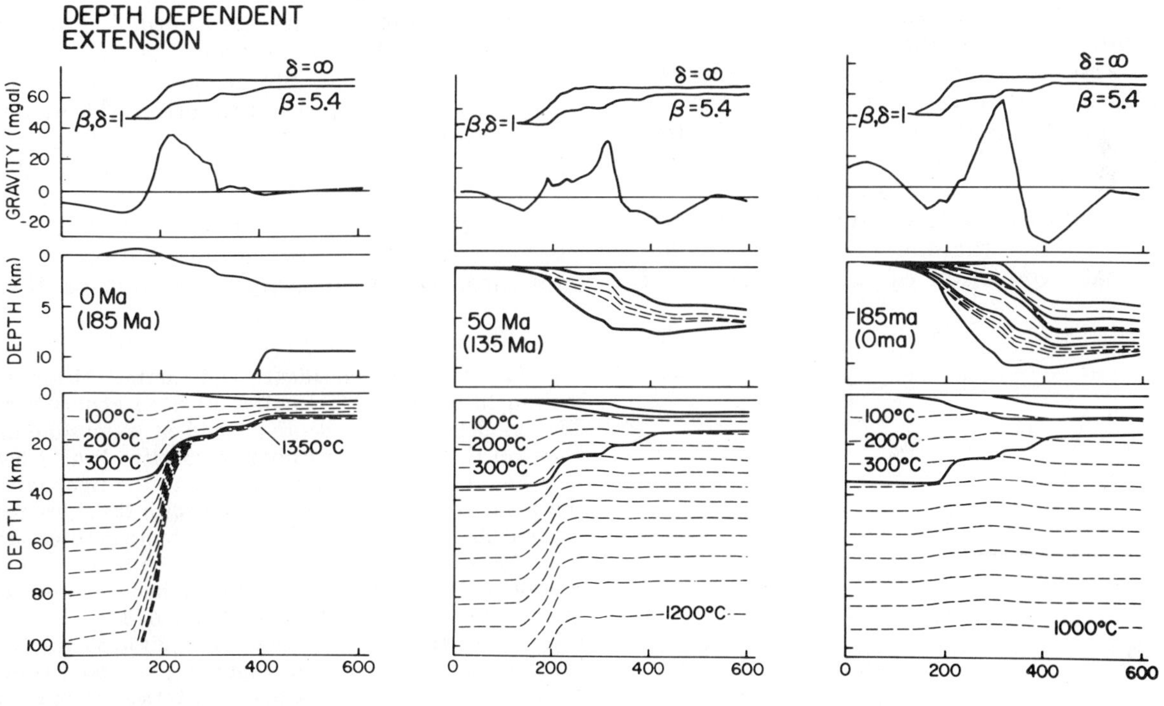

Fig. 10. Three stages in the evolution of a depth-dependent extension model corresponding to 0, 50, and 185 Ma after rifting. Each of the three panels shows: the β and δ distribution; the free air gravity anomaly; the form of the basin; and the temperature distribution. The short wavelength features in the gravity field are caused by palaeobathymetric undulations. The lines in the basin are chronostratigraphic horizons. Those that are continuous correspond to the ends of the Jurassic, Lower and Upper Cretaceous, and the sediment—water interface. Note the syn-rift uplift in the region where $\delta \gg \beta$ and the close approach of the 1350°C isotherm to the base of the crust.

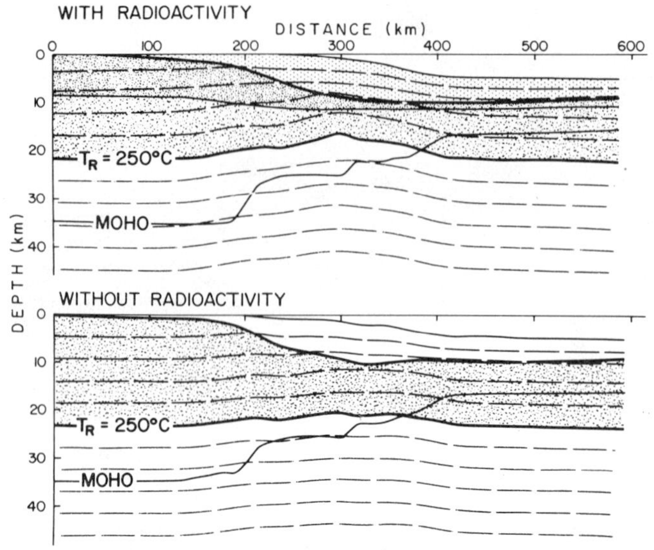

Fig. 11. Comparison of depth-dependent extension models, T_R=250°C (d=t_c, and β and δ as shown in Fig. 10) 185 Ma after rifting with and without radioactivity. Temperatures are contoured at 50°C intervals. Light stipple shows sediments with radiogenic heat production based on measured values. Medium stipple shows the extended crustal layer of uniform heat production which had a pre-rift thickness of 7.5 km. Dense and medium stipple taken together show the rheological lithosphere. Note the thinning of the rheological lithosphere beneath the basin.

insulated, hotter model. This effect is comparable in size to that due to flexure. Only pre-, syn-, or early post-rift sediments are sensitive to the thermal differences inherited from the rifting models. This is because the model differences, which are confined mainly to the upper lithosphere, dissipate rapidly in response to the much shorter thermal time constant of the upper lithosphere than the 63 Ma time constant of the whole lithosphere.

Thermal histories of the sediments however, are sensitive to the thermal conductivities and radiogenic heat production of the sediments and crust. The final forms of a model which differ only in radiogenic heat production (Fig. 11) demonstrate that temperatures at the bottom of the basin can be raised by about 30°C by this effect. Radioactive heat-production values that are appropriate for the crust and sediments of the Nova Scotian Margin were used (Beaumont *et al.*, 1982; Keen and Lewis, 1982). The corresponding effect on the temperature histories predicted for sediments at a typical well, Triumph, (Fig. 12) shows that the isotherm depths for the model with radiogenic heat production (long dashed lines) diverged from the non-radiogenic model (short dashed lines) about 50 Ma after rifting, once significant amounts of sediment had accumulated.

Accurate values of both thermal conductivity and radiogenic heat production are needed if the model is to hindcast temperatures accurately enough to be used in kerogen maturation predictions. The only direct test of model temperatures is that they match corrected bottom hole values from wells. The relationship of these corrected temperatures to true formation temperatures is not known and more emphasis should be placed on accurate temperature measurements within the sediment column.

In our recent work we have constrained model thermal properties by using measured thermal conductivities, radiogenic heat production, and bottom hole temperatures. With the exception of the value for salt, the results imply that our values (Beaumont *et al.*, 1982) for sediment thermal conductivity should be reduced by 30–40%, thereby increasing the geothermal gradient by comparison with that shown in Fig. 12.

We have also used simple geochemical reactions, thought to be primarily a function of temperature, as an indirect test of past temperatures. The reactions (Mackenzie and McKenzie, 1983; Beaumont *et al.*, in preparation; MacKenzie *et al.*, in preparation); isomerization of hopanes at C-22; isomerization of steranes at C-20; and aromatization of C-ring monoaromatic steroid hydrocarbons to their tri-

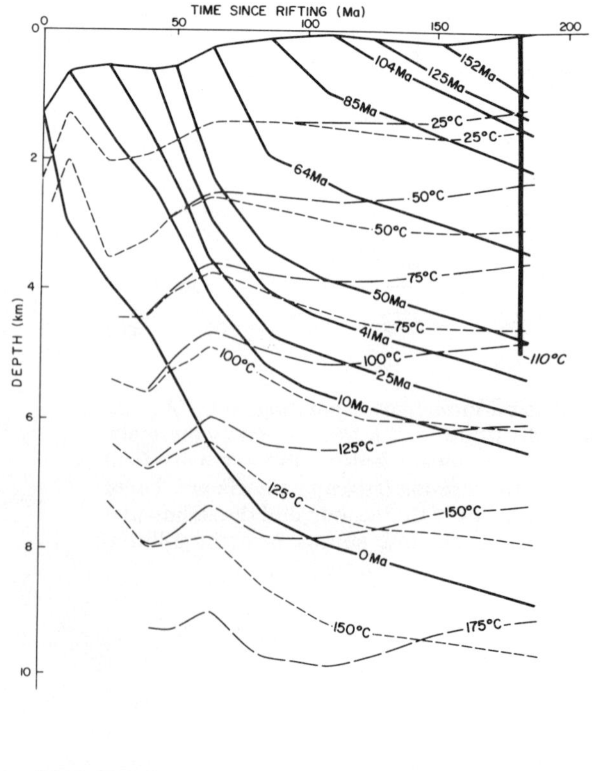

Fig. 12. Predicted subsidence and thermal histories for a section through the Triumph P-50 deep exploratory well. The models are the depth-dependent extension models, T_R=250°C, with and without radioactive heat generation (see Fig. 11). The column furthest to the right represents the predicted present configuration with the short and long dashed lines indicating predicted temperatures without and with radioactivity, respectively. The bold vertical lines represents the Triumph well which had an observed but uncorrected bottom hole temperature of 110°C. Successive columns to the left represent the equivalent situation for progressively earlier times. Therefore, the bold solid lines trace the predicted subsidence histories of each of the stratigraphic horizons and are marked by their time of deposition measured from the time of rifting. Similarly, the dashed lines trace the temperature changes at depth with time. The 0°C isotherm, shown as a solid line, is also the water–sediment interface and its depth illustrates the changes in model palaeobathymetry for this location. The relationship between the subsidence curves and the isotherms describes the temperature histories of the sediments.

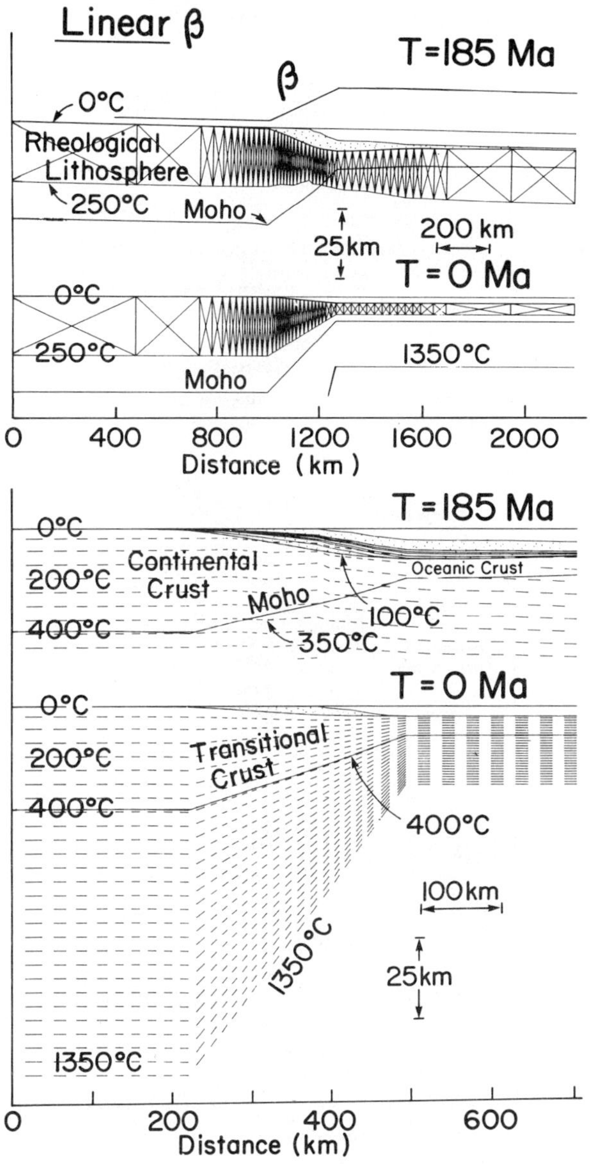

Fig. 13. Crustal and lithospheric structure of the linear β model at 0 and 185 Ma after rifting. Upper panel shows the β distribution. The two panels following illustrate the change in the rheological lithosphere (cross lined), Moho position and basin shape (stippled). The lower panel shows the change in the thermal regime in the central region of the model: isotherms (dashed lines) are contoured at 50°C intervals. Note the change in temperature at the bottom of the basin and at the Moho.

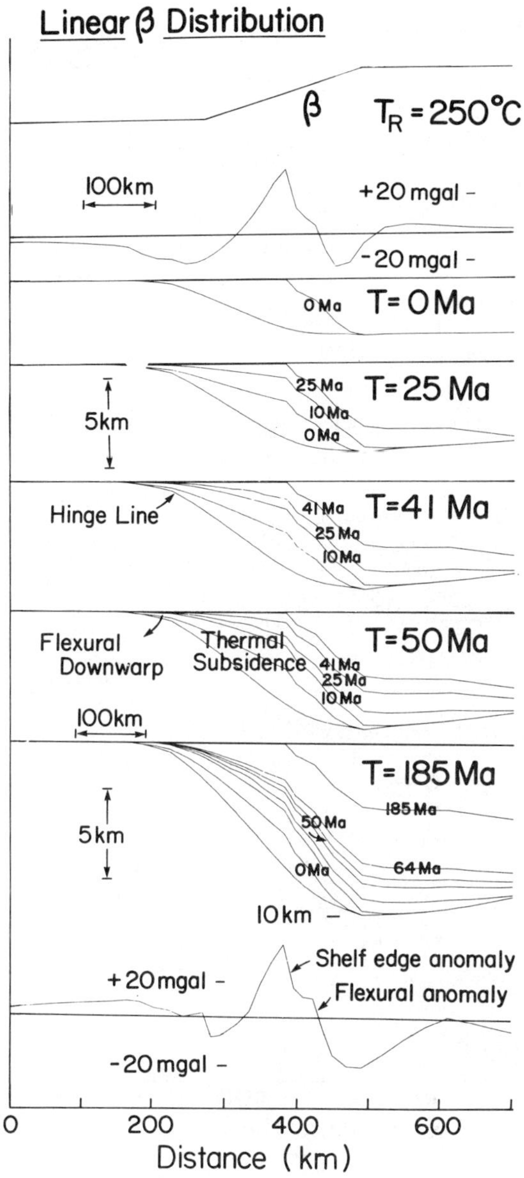

Fig. 14. Free air gravity anomalies and sedimentary basin evolution for the linear β model. The upper panel shows the β distribution. The central panels, showing the basin evolution at selected times after rifting, are bracketed by the gravity anomalies at 0 and 185 Ma after rifting. Note the weakly developed hinge line and flexural downwarp in the unstretched region.

aromatic form, are measured on extracts from sediments and predicted using theoretical reaction kinetics in which the thermo-mechanical model temperatures are used. The method is very sensitive and can differentiate between the effects of temperature and time. Potentially, it provides an excellent method for constraining temperature histories in the 65—100°C range. Such constraints may be used to calibrate the thermo-mechanical model to ensure accurate predictions in the 100—130°C range where the main phase of kerogen to oil breakdown occurs.

ARCHETYPAL MODELS

The archetypal models (Figs. 13—18) are included to illustrate how the form of $\beta(x)$ influences the character of a rifted continental margin. All three uniform extension models have exactly the same input parameters except for $\beta(x)$, which takes a linear, convex and concave form in the approximately 300-km wide transition region. The model properties are initial crustal and lithospheric thickness and densities, 35 and 125 km and, 2820 and 3330 kg/m^3, respectively; base temperature of the lithosphere, 1350°C; T_R=250°C. No faulted crustal region or radiogenic heat production are included and the sedimentation is designed to keep the basin filled to a bathymetric profile that corresponds to

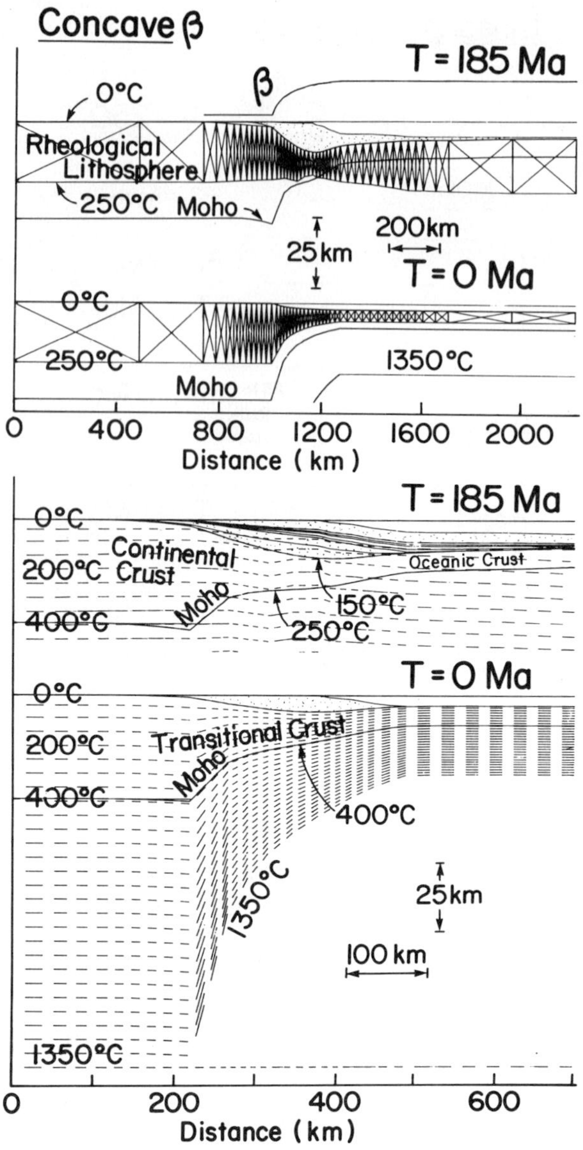

Fig. 15. Crustal and lithospheric structure of the concave β model at 0 and 185 Ma after rifting. Upper panel shows the β distribution. The next two panels illustrate the change in the rheological lithosphere (cross lines), Moho position, and basin shape (stippled). The lower panel shows the change in thermal regime in the central region of the model: isotherms (dashed lines) are contoured at 50°C intervals. Note the change in temperature at the bottom of the basin and at the Moho.

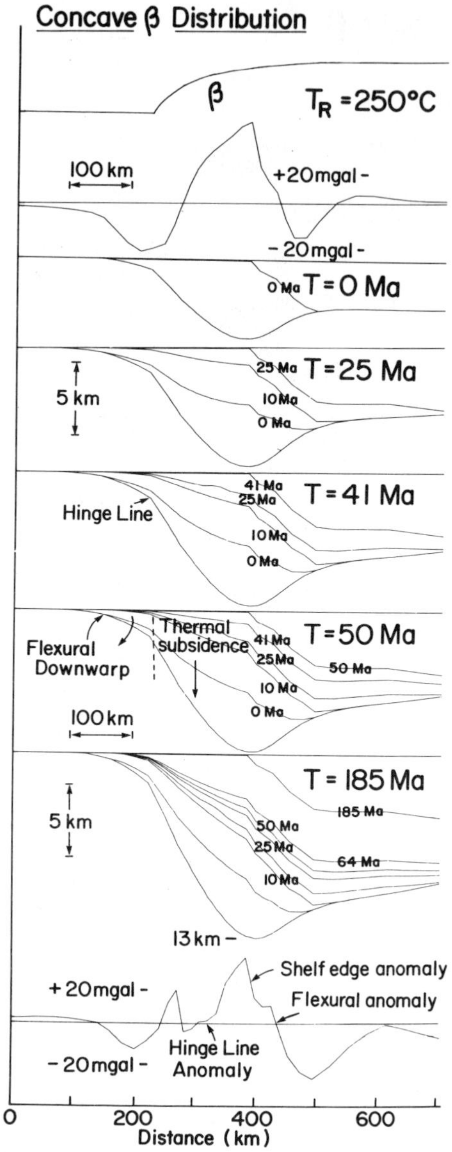

Fig. 16. Free air gravity anomalies and sedimentary basin evolution for the concave β model. The upper panel shows the β distribution. The central panels, showing the basin evolution at selected times after rifting, are bracketed by the gravity anomalies at 0 and 185 Ma after rifting. Note the strongly developed hinge line, flexural downwarp, and associated gravity anomaly in the unstretched region; and the spoon shape of the basin.

the Nova Scotian cross-section. The position of the shelf break is held fixed. Normal compaction occurs in the sediments and the same relationships are as used in Beaumont *et al.* (1982).

The models are designed to investigate the extent to which the necking process during rifting leaves its imprint during the post-rift evolution of the margin. The analogy of lithospheric necking during extension with that of a drawn metal rod suggests various necking profiles, the most simple of which, linear β, has uniformly increasing attenuation across the margin to a point where break-up and infilling by oceanic crust occurs. The second model is termed concave β because a plot of $\beta(x)$ is concave when viewed from below, giving the Moho and lithosphere the corresponding shape. Extension that progresses incrementally so that successive amounts of extension are proportional to the thickness of the lithosphere results in a profile of this form. The converse produces a convex β distribution.

Although we do not know which of these processes is preferred, or even if there is a preferred mode of extension, it is probable that dynamical models of rifting will predict attenuation that

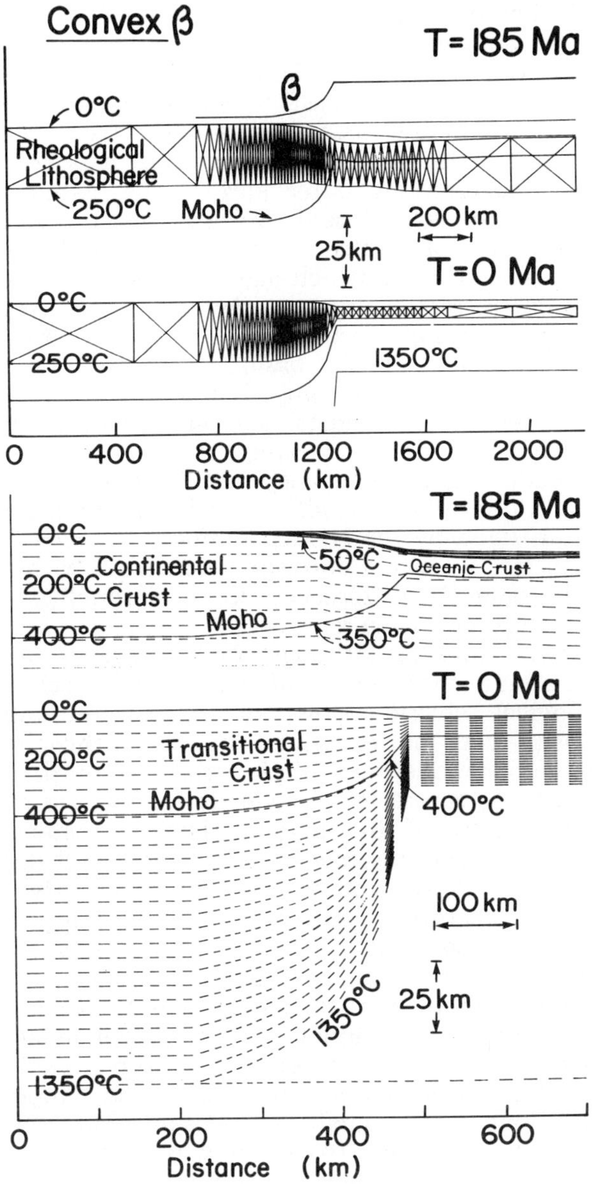

Fig. 17. Crustal and lithospheric structure of the convex β model at 0 and 185 Ma after rifting. Upper panel shows the β distribution. The next two panels illustrate the change in the rheological lithosphere (cross lined), Moho position, and basin shape (stippled). The lower panel shows the change in thermal regime in the central region of the model: isotherms (dashed lines) are contoured at 50°C intervals. Note the change in temperature at the bottom of the basin and at the Moho.

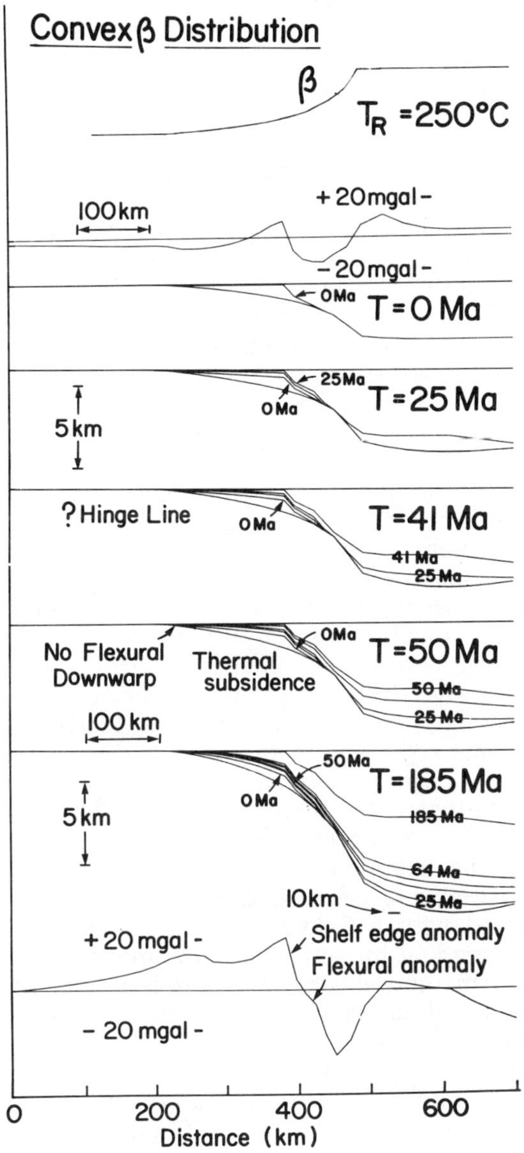

Fig. 18. Free air gravity anomalies and sedimentary basin evolution for the convex β model. The upper panel shows the β distribution. The central panels, showing the basin evolution at selected times after rifting, are bracketed by the gravity anomalies at 0 and 185 Ma after rifting. Note that there is little evidence of flexure despite the same value of T_R as the other models. The basement shape below the shelf mirrors the β distribution.

conforms with at least one of these profiles. Can we detect which profile is appropriate for a given margin transect? The results (Figs. 13–18) suggest that this is possible. All things being equal, the character of the sediment-thickness distribution, the position of the Moho, and the gravity anomalies are good fingerprints of $\beta(x)$ and these fingerprints, although modified, are preserved during margin evolution. The β distribution also has a significant effect on the maximum temperature within the sediment column, suggesting that thermal maturation of kerogens may not occur on sections like convex β.

The position of the Moho preserves the β character on the assumption that the crust was uniformly thick before stretching. Therefore, the character could be determined by refraction and reflection measurements designed to locate the Moho; not always an easy task, particularly on heavily sedimented margins. The depth to basement, even if in error by 1–3 km, can be used to retrieve $\beta(x)$; spoon shapes and great sediment thicknesses for concave β; no over deepened part for linear β; and an almost partitioned sediment distribution for convex $\bar{\beta}$. Of course these will be modified if the shelf edge position is different, but some of the character will be retained.

The gravity anomalies are made up of three parts (Beaumont *et al.*, 1982) from: basin shape including distributions of sediment and water, Moho position, and a thermal contribution, the last resulting from density perturbations from thermal expansion. It is mainly the relative contribution of flexural effects from sediment loading that gives these models differing gravity anomalies which are illustrated for model ages of 0 and 185 Ma. Intermediate ages show a smooth transition in behaviour between the end points.

The 'hinge line', dividing regions of flexural subsidence from extensional and thermal subsidence, is well developed in the concave β model because the sediment spoon is deep and sufficiently close to unextended regions to cause significant flexure. Flexure gives the gravity anomaly the characteristic central high, flanked by two negatives. Increasing flexure as the margin ages produces a separate hinge line anomaly for concave β. On margins where the shelf is narrower than our example this anomaly merges with the main flexural positive. When T_R is larger or smaller, the amplitudes and wavelengths of the characteristic anomalies increase or decrease, respectively, reflecting the differing degree of flexural support of the sediment load. There is much less sediment in the convex β model; therefore the gravity anomalies are much smaller and less dominated by flexural effects. The characteristic low over the shelf is absent and a hinge line cannot be detected. The smooth increase in sediment thickness beneath the shelf could be easily confused with a flexural downwarp, but it is due to the gradual increase in β. The linear β model has an intermediate character. There is a hinge line and the flexural character of the gravity anomaly is partly developed.

Sedimentation landward of the hinge line exhibits on-lap during the early stages of model development, followed by off-lap. On-lap is due to the increasing strength of the stretched lithosphere. Off-lap may not be anticipated from a cursory examination of the model, but is explained by the seaward migration of the centre of mass of the sediment load with increasing time.

Are these models realistic? A preliminary investigation of published margin cross-sections indicates that examples of all three types exist, together with more complex ones, in which stretching does not increase smoothly across the margin.

The Nova Scotian cross-section (Beaumont *et al.*, 1982; this chapter) is an example which conforms closely to the linear β model. The style and 'fingerprints' of gravity, basin shape, hinge line, and Moho depth can all be matched by linear β.

USGS Line 5 south of Cape Cod (Grow *et al.*, 1979, figs. 2 and 10; and Klitgord and Behrendt, 1979, fig. 8) is an example of a slightly more heavily sedimented convex β model than the example given here. More sediment increases flexure and depresses the positive gravity signature over the shelf to near zero: otherwise the style of the margin is very similar to that of our model. The inferred Moho position (Grow *et al.*, 1979, fig. 10) may be more convex, by comparison with Fig. 17. USGS Line 9 off Long Island and the IPOD/USGS Line off North Carolina (Grow *et al.*, 1979, figs. 2 and 11) are additional examples of transects that have convex β distributions but a more prograded shelf than our example.

The concave β model is best exemplified by USGS Line 2 off New Jersey (Grow *et al*, 1979, figs. 2 and 12; and LASE Study Group, in press). The sediment distribution within the basin is not the same as that of our model because a carbonate bank occupied the outer part of the margin for much of its development. Nevertheless, the cumulative effect of sediment loading has developed a spoon shaped basin which may be as deep as 16 km. Seismic refraction measurements (LASE Study Group, in press) clearly show the attenuated crust beneath much of the basin as required by the convex β model. The hinge line is not as strongly developed as we predict, but our model assumes that initially no topography existed in the unstretched region. This may not be correct. That flexure does play a significant role for this section is confirmed by the gravity anomaly (Grow *et al.*, 1979, fig. 2) which corresponds in form with our model once the differences due to sharpness of the shelf edge have been removed.

These results, although preliminary, give us confidence that many of the geophysical and geological characteristics of rifted continental margins can be explained by the lithospheric stretching model and that variations in β, and perhaps δ, across the margin determine the style of that part of the margin. The next steps in developing this approach are to include β distributions that are not monotonic; to classify the margins according to style of stretching and sedimentation; and to consider margins, such as the Norwegian Margin, which have been subject to more than one stretching event. Once the classification is complete, margin style must be correlated with other tectonic features, such as inherited lithospheric structure and thermal conditions, and proximity to hot spots and fracture zones, in an attempt to explain why there is such a large variation in style of stretching.

CONCLUSIONS

In summary, our conclusions are that:

(1) Extension models are successful in explaining the first-order processes that occurred during rifting of the Nova Scotian Continental Margin and of other margins where the necking profile is somewhat different.

(2) The thermo-mechanical model correctly predicts the large-scale development of simple continental margins from rifting to the present.

(3) The development of the margin exhibits a fading memory of the details of the rifting process. The uniform-extension rifting model can, therefore, explain most of the features of old margins. The depth-dependent extension model is, however, more realistic in its approach to the kinematics of lithospheric stretching. Of the proposed rifting models, the uniform extension and melt segregation model provides the best explanation of the processes by which the transition from continental to oceanic crust at continental margins is created.

(4) The concept of a rheological lithosphere, bounded by an isotherm and the brittle—ductile transition, which responds elastically to loading, is

useful in explaining departures from local isostatic equilibrium and the form of the free air gravity anomalies across the margin. The use of an elastic lithosphere in the thermo-mechanical model is justified because recognizable effects of thermally activated creep are suppressed in a cooling lithosphere.

(5) The thermo-mechanical model predicts that thermal blanketing by sediments and departures from local isostatic equilibrium significantly reduce the maximum potential depth of rifted margin basins with respect to that predicted by more simple theory.

(6) Landward of the basin hinge, subsidence is dominated by flexural downwarping, whereas seaward of the hinge line thermal effects predominate.

(7) The models are potentially capable of accurately tracing the subsidence history of sediments within the basin, and, insofar as conductive models of heat transport within the basin are correct, the models can also hindcast thermal histories for the sediments. Accurate measurements of sediment thermal conductivities and formation temperatures are, however, necessary. Thermally activated geochemical reactions are potentially very useful in constraining sediment-temperature histories.

(8) Crustal and sedimentary radioactive heat production almost certainly plays a significant role in determining the temperature distribution within the sedimentary basin and, therefore by implication, the degree of thermal metamorphism of organic material within the sediments.

(9) The form of the stretching function, $\beta(x)$, is strongly reflected in the gravity anomalies and sedimentary basin shapes and stratigraphies. Convex β gives predictions which closely correspond to platform regions along the rifted margin of eastern North America. Concave β corresponds to the deepest sedimentary basins, such as the Baltimore Canyon Trough, while linear β is intermediate and corresponds to the areas like the Scotian Basin part of the Nova Scotian Margin.

ACKNOWLEDGEMENTS

We wish to thank Donna Nelson for her help in compiling much of the data for the cross-section of the Nova Scotian Margin and Kathleen Sawler for typing the manuscript. Financial support for this study was provided by the Natural Sciences and Engineering Research Council, Canada, through operating and strategic grants, and a Steacie Fellowship to C. Beaumont.

REFERENCES

Ashby, M.F. and Verrall, R.A. 1977 Micromechanisms of flow and fracture and their relevance to the rheology of the upper mantle. *Phil. Trans. R. Soc. Lond. Ser. A*, **288**, 59—95.

Beaumont, C. Boutilier, R., Mackenzie, A.S., and Rullkotter, J. Isomerization and aromatization of hydrocarbons and the palaeothermometry and burial history of the Alberta foreland basin. (in preparation).

Beaumont, C., Keen, C.E. and Boutilier, R. 1982. On the evolution of rifted continental margins: comparison of models and observations for the Nova Scotian margin. *Geophys. J. R. Astron. Soc.*, **70**, 667—715.

Bodine, J.H. 1981. The Thermo-mechanical Properties of Oceanic Lithosphere. Ph.D. thesis, Columbia Univ., New York, NY, USA.

Courtney, R.C. 1982. On the rheology of the oceanic and continental lithosphere. Unpubl. M.Sc. Thesis, Dalhousie Univ., Halifax, N.S., Canada.

Courtney, R.C. and Beaumont, C. Thermally activated creep and flexure of the oceanic lithosphere, *Nature, Lond.* (in press).

Dainty, A.M., Keen, C.E., Keen, M.J. and Blanchard, J.E. 1966. Review of geophysical evidence on crust and upper mantle structure on the eastern seaboard of Canada. In: The Earth Beneath the Continents. *Am. Geophys. Union. Mon.*, **10**, 349—369.

Dewey, J.F., 1982. Plate tectonics and the evolution of the British Isles. *J. Geol. Soc. Lond.*, **139**, 371—412.

Foucher, J-P., Le Pichon, X. and Sibuet, J.-C. 1982. The ocean—continent transition in the uniform lithospheric stretching model: role of partial melting in the mantle. *Phil. Trans. R. Soc. Lond., Ser. A*, **305**, 27—40.

Grow, J.A. Mattick, R.E. and Schlee, J.S. 1979. Multichannel seismic depth sections and interval velocities over outer continental shelf and upper continental slope between Cape Hatteras and Cape Cod. In: Watkins, J.S., et al. (eds.). Geological and Geophysical Investigations of Continental Margins, *Am. Assoc. Petrol. Geol., Mem.*, **29**, 65—83.

Jansa, L.F. and Wade, J.A. 1975. Geology of the continental margin off Nova Scotia and Newfoundland. In: *Offshore Geology of Eastern Canada, 2, Regional Geology.* Geol. Surv. Can. Paper 74-30, pp.51—105.

Jarvis, G.T. and McKenzie, D.P. 1980. Sedimentary basin formation with finite extension rates. *Earth Planet. Sci. Lett.*, **48**, 42—52.

Keen, C.E., 1979. Thermal history and subsidence of rifted continental margins — evidence from wells on the Nova Scotian and Labrador shelves. *Can. J. Earth. Sci.*, **16**, 505—222.

Keen, C.E. The dynamics of rifting: extension and viscous flow in the lithosphere. (in preparation).

Keen, C.E. and Cordsen, A. 1981. Crustal structure and seismic stratigraphy of the rifted continental margin off eastern Canada: Ocean bottom seismic refraction results off Nova Scotia. *Can. J. Earth Sci.*, **18**, 1523—1538.

Keen, C.E. and Lewis, T. 1982. Radiogenic heat production from the continental margin of eastern North America: Implications for petroleum generation. *Am. Assoc. Petrol. Geol. Bull.*, **66**, 1402—1407.

Keen, C.E. and Hyndman, R.C. 1979. Geophysical review of the continental margins of eastern and western Canada. *Can. J. Earth Sci.*, **16**, 712—747.

Keen, C.E., Keen, M.J., Barrett, D.L., and Heffler, D.E. 1975. Some aspects of the ocean—continent transition at the continental margin of eastern North America. In: W.J.M. van der Linden and J.A. Wade (eds.), *Offshore Geology of Eastern Canada*, Geol Surv. Can. Paper. 74-30, pp.189—197.

Kirby, S.H. 1980. Tectonic stresses in the lithosphere: constraints provided by the experimental deformation of rocks. *J. Geophys. Res.*, **85**, 6353—6363.

Klitgord, K.D. and Behrendt, J.C. 1979. Basin structure of the U.S. Atlantic margin. In: J.S. Watkins *et al.*, (eds.). Geological and Geophysical Investigations of Continental Margins. *Am. Assoc. Petrol. Geol., Mem.* **29**, 85—112.

LASE Study Group. Continuity of the deep crustal layer across a rifted continental margin. *Science*, (in press).

Mackenzie, A.S. and McKenzie, D.P. Isomerisation and aromatisation of hydrocarbons in sedimentary basins formed by extension. *Geol. Mag.* (in press).

Mackenzie, A.S., Beaumont, C., Boutilier, R. and Rullkotter, J. Aromatization and isomerization of hydrocarbons and the thermal evolution of the Nova Scotian continental margin. (in preparation).

McKenzie, D.P., 1978. Some remarks on the development of sedimentary basins. *Earth Planet. Sci. Lett.*, **40**, 25—32.

Montadert, L., de Charpal, O., Roberts, D., Guennoc, P. and Sibuet, J-C. 1979. *Northeast Atlantic Passive Continental Margins: Rifting and Subsidence Processes*, In: Maurice Ewing Symposium Series, No. 3, Am, Geophys. Union, pp.154—186.

Parsons, B. and Sclater, J.G. 1977. An analysis of the variation of ocean floor bathymetry and heat flow with age, *J. Geophys. Res.* **82**, 803—827.

Royden, L. and Keen, C.E. 1980. Rifting processes and thermal evolution of the continental margin of eastern Canada determined from subsidence curves. *Earth Planet. Sci. Lett.*, 51, 343—361.

Royden, L., Sclater, J.G. and Von Herzen, R.P. 1980. Continental margin subsidence and heat flow, important parameters in formation of petroleum hydrocarbons. *Am. Assoc. Petrol. Geol. Bull.*, 64, 173—187.

Sclater, J.G., Royden, L., Horvath, F., Burchfiel, B.C., Semkin, S. and Stegena, L. 1980. Subsidence and thermal evolution of the intra-Carpathian basins. *Earth Planet. Sci. Lett.*, 51, 139—162.

Sheridan, R.E., Grow, J.A., Berherdt, J.C. and Bayer, K.C., 1979. Subsidence and thermal evolution of the intra-Corpathian basins. *Earth Planet. Sci. Lett.*, 51, 139—162.

Sleep, N.H. 1971. Thermal effects of the formation of Atlantic continental margins by continental break-up. *Geophys. J. R. Astron. Soc.*, 24, 325—350.

Walcott, R.I. 1970. Isostatic response to loading of the crust in Canada. *Can. J. Earth Sci.*, 7, 2—13.

Watts, A.B. and Steckler, M.S. 1979. *Subsidence and Eustasy at the Continental Margin of Eastern North America.* In: Maurice Ewing Symposium Series, No. 3. Am. Geophys. Union, pp.218—234.

Watts, A.B. and Ryan, W.B.F. 1976. Flexure of the lithosphere and continental margin basins. *Tectonophysics.* 36, 25—44.

Weertman, J. 1970. The creep strength of the Earth's mantle. *Rev. Geophys. Space Phys.*, 8, 145—168.

12 West Spitsbergen Fold Belt and Cretaceous opening of the Northeast Atlantic

J. Hanisch Federal Institute for Geosciences and Natural Resources, Hannover, Federal Republic of Germany

The evolution of the West Spitsbergen Orogenic belt is reviewed from a literature study. It is subdivided into three independent phases: (1) a folding phase which should have taken place during the Cretaceous, between the Aptian/Albian and the Palaeocene, (2) a phase of overthrusting during the late Palaeocene and Eocene, and (3) a tensional folding phase during the Oligocene.

The Cretaceous compressional movements are thought to reflect a plate tectonic movement of Greenland, the pole of rotation of which is inferred to have been situated at about 74°N, 21°E (in the Bjφrnφy Basin). The corresponding extensional regime south of the pole of rotation is suggested to have resulted in a continuous, northward-narrowing rift system from the Rockall Trough through the Faeroe, Mφre, Vφring, Traen, Harstad and Tromsφ Basins up to the Bjφrnφy Basin. Fault-controlled subsidence of this rift system probably started in Middle Aptian time. Sea-floor spreading succeeded in its southern part. This oceanic crust is thought to diminish northward in width and to disappear in the Mφre Basin. North of the Mφre Basin the crustal extension probably was compensated by stretching of continental crust.

This rift/spreading system apparently intersected a series of older basins such as the West Shetland Basin, the North Sea Graben, any older structures beneath the Norwegian Atlantic Shelf and, to the north, the east—west striking Hammerfest Basin. The observed obliqueness of structural trends between the Northeast Greenland Shelf and the Norwegian Shelf and the unusual depths of the Mφre, Vφring and Tromsφ Basins can be explained by this Cretaceous rift. The Rona Ridge, Margareta's Spur, the Nordfjord and Frφya Highs and the Nordland Ridge, which all were deeply eroded during the Cretaceous, are supposed to be the uplifted eastern shoulders of the rift system.

INTRODUCTION

In plate tectonic reconstructions of the Northeast Atlantic Ocean the West Spitsbergen orogeny plays an important role. Its relationship to the Tertiary opening of the Norwegian-Greenland Seas has been repeatedly stressed. Harland (1965) proposed a transpressional origin of the fold belt signifying that the compressional deformations occurred during the strike-slip movement between Greenland and Svalbard. According to Talwani and Eldholm (1977) this motion took place between late Palaeocene and mid-Oligocene times.

This interpretation has found general acceptance, although Harland (1969, p. 839) did not exclude an earlier age for part of the compression. He stated that with the opening of the Northeast Atlantic in late Cretaceous time "Greenland would have floated north before space was opened for it ahead, and the marginal compression in the Canadian Arctic Islands and in Spitsbergen would have resulted".

REVISION OF THE WEST SPITSBERGEN FOLDBELT

The foldbelt of western Svalbard displays a number of unusual features (for discussion see Lowell, 1972 and Birkenmajer, 1972, 1981). Several subdivisions of its genesis have been proposed. Birkenmajer recognized three stages of deformation: (1) compression from the southwest; (2) rifting (Forlandsundet Graben); and (3) wrench faulting with horizontal stress from the west. The compressional stage was tentatively dated as late Palaeocene/early Eocene by correlation with "the youngest phase of the Laramide orogeny". Lowell (1972), however, assumed a post-Oligocene—Miocene(?) age for this compression.

Kellogg (1975) stated that from late Palaeocene

Petroleum Geology of the North European Margin © Norwegian Petroleum Society (Graham & Trotman, 1984) pp. 187—198.

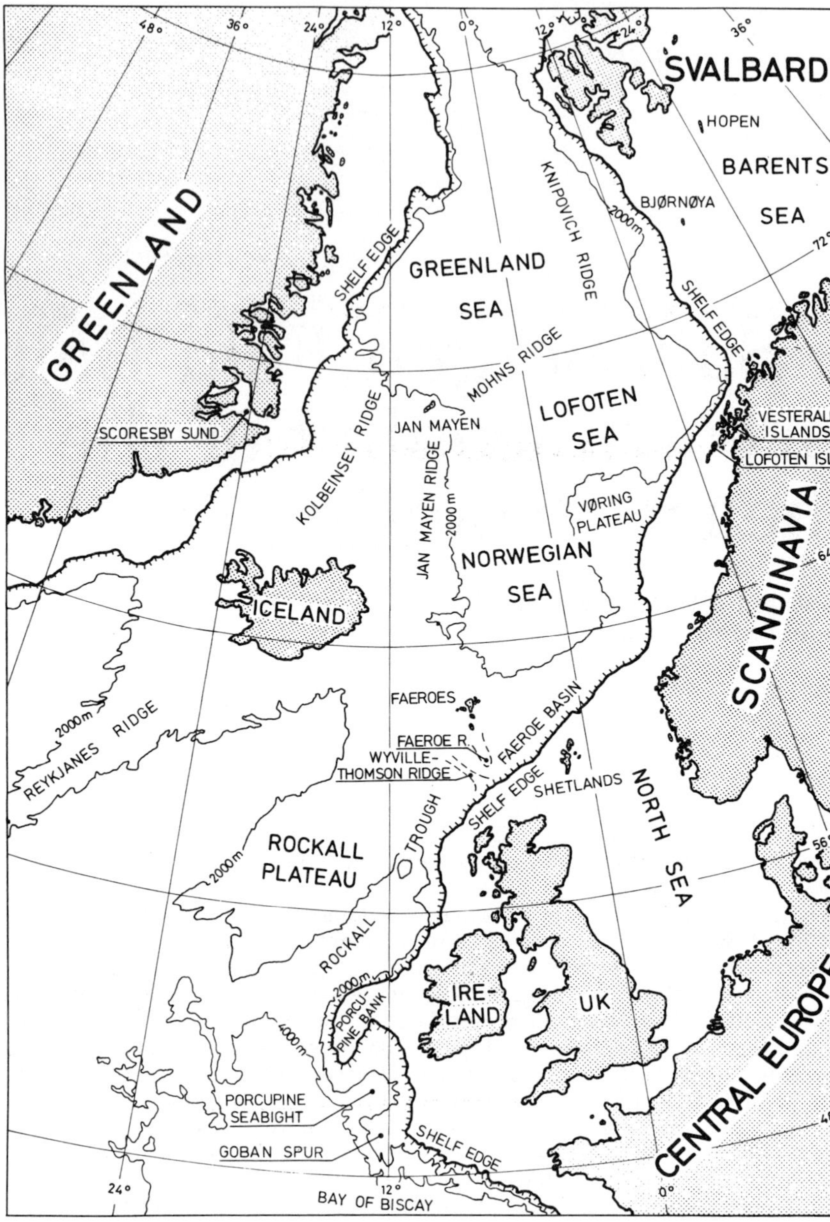

Fig. 1. Map of the Northeast Atlantic (based on Tectonic Map of Europe and Neighbouring Areas, Moscow, 1979).

to Eocene, there was little tectonism in Svalbard "because transform faulting was associated with perpendicular tensional forces (transtension)". The "transpressional collision of the northward-moving Greenland block against Svalbard during the late Eocene—early Oligocene resulted in upthrusting and compressional deformation" (p. 484). Steel *et al.* (1981) concluded from detailed investigations on the sedimentary cycles within the Tertiary sequences that transtension occurred in the Palaeocene, transpression in the Eocene and oblique separation from Oligocene time on.

In a detailed evaluation of ocean-type magnetic anomalies, Talwani and Eldholm (1977) showed that Greenland moved first to the northwest relative to Europe (from late Palaeocene to Oligocene) and changed to WNW in the Oligocene. These authors adopted Harland's (1965) transpression model and concluded that the final rifting stage on West Spitsbergen took place with this change of the direction of spreading.

This brief summary shows that there is great controversy about the development and the exact age of the West Spitsbergen orogeny. "All these events need to be better dated . . ." (Birkenmajer, 1981). General accord exists only in the Tertiary age of the orogenic deformations. Many of the structural models and speculations are by no means in accordance with the very detailed descriptions by Orvin (1934, 1940) already acknowledged by Lowell (1972). From his extensive field studies the following conclusions can be drawn: the structural deformations of western Spitsbergen can be subdivided into three independent steps. The first was a phase of folding and upthrusting, the second a phase of overthrusting (north of Isfjorden), and the third a phase of extensional faulting. The folding and upthrusting (south of Isfjorden) originated from horizontal pressure from the WSW (Orvin, 1940; Birkenmajer, 1972), whereas the overthrusts north of Isfjorden had a northeastern throw. On Brøggerhalvøya (Fig. 2) "it is clearly seen that the

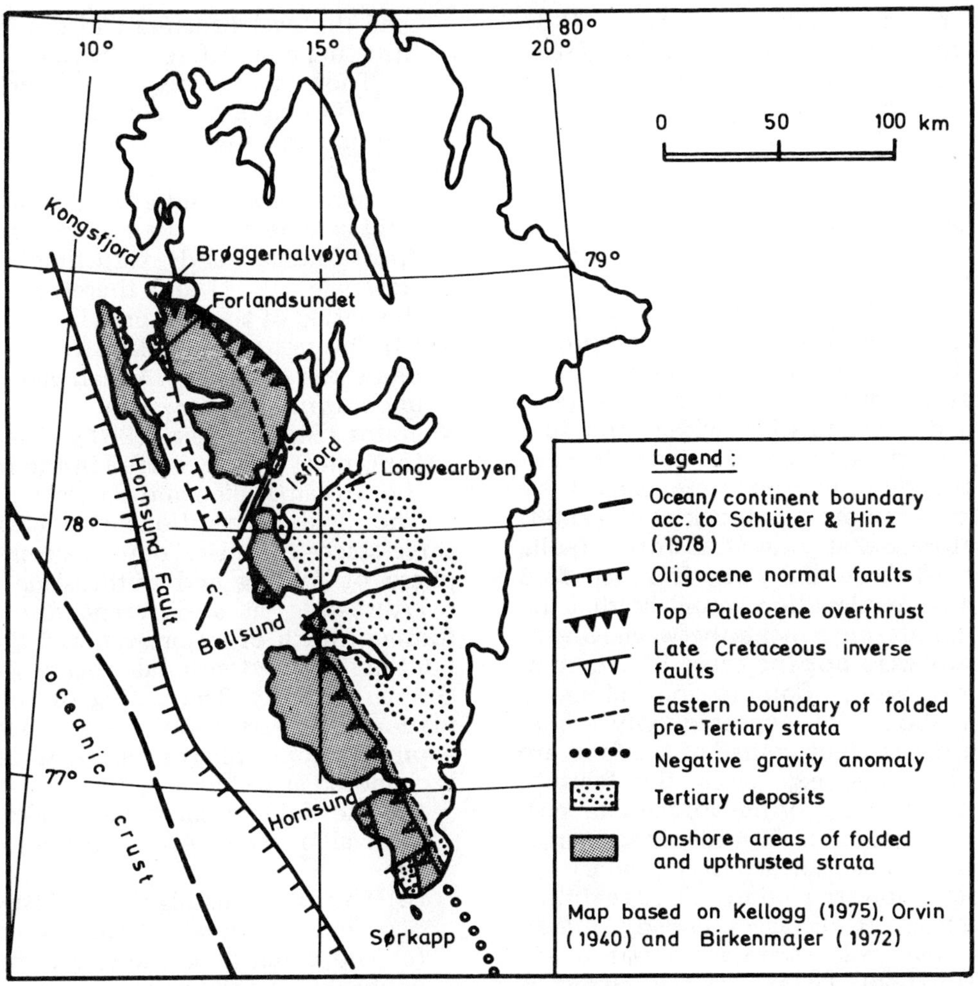

Fig. 2. Revised structural map of western Spitsbergen.

thrust planes have intersected already folded strata". The thrust plane has "cut through the fold obliquely to the folding axis" or "has cut the original fold" (Orvin, 1934, p. 77 ff). Furthermore he stated that the "pressure at Kings Bay must have been considerably stronger than further south, because it has not only caused the overfolding but also the great thrusts".

These descriptions signify that north of Isfjorden the structural deformations were governed by a different tectonic regime. South of Isfjorden the folds and upthrusts of the Precambrian to early Cretaceous rocks expose a more simple style of deformation: no large-scale overthrusts with throws of several kilometres are observed; the thrusts have the same NNW—SSE strike as the fold axes, signifying that both deformations are certainly of the same age and are caused by the same compressional forces.

The proof for a Tertiary age of these deformations, however, is still pending because south of Isfjorden Tertiary deposits are nowhere involved in these deformations. This may have at least three different reasons: (1) no Tertiary layers were present during a Tertiary orogenic event, (2) possible Tertiary layers involved in the structural deformations have been removed by subsequent erosion, or (3) the structural deformations took place in pre-Tertiary time.

Birkenmajer (1972), as proof of a Tertiary age of the compression south of Isfjorden, refers to a small outcrop near Longyearbyen (Fig. 2). The corresponding cross section is given in Birkenmajer (1966, fig. 5E) and shows a basal Tertiary conglomerate lying unconformably on early Cretaceous shales. The deformations of the beds shown there, however, are clearly caused by drag effects of a fault dipping steeply westward and by no means by compressional tectonic forces. Furthermore, this site lies about 40 km outside of the fold belt and therefore is not valid for the dating of the fold belt south of Isfjorden.

The Tertiary main basin of Spitsbergen is a wide intracratonic basin (Fig. 2). The Tertiary sediments lie unconformably and nearly horizontally on basement rocks or Lower Cretaceous deposits (Harland, 1969; Birkenmajer, 1972; Lowell, 1972; Kellogg, 1975). The unconformity marks an angle of up to 10° between the Tertiary and the Cretaceous layers (Birkenmajer, 1972). Kellogg (1975) pointed out that the strong folding of the west coast belt dies out abruptly eastward: sharply folded Mesozoic rocks lie adjacent to "more gentle folds in Tertiary beds". The deformation of the Tertiary layers "mostly took the form of folding over basement faults" (Kellogg, 1975, p. 482). The Tertiary main basin was not involved in the crustal compression.

At the southern end of the island (Fig. 2) a small outlier of Tertiary beds is preserved within the area of the fold belt. The poor outcrop conditions there do not allow examination of the base of the Tertiary, but the sediments "appear to lie disconformably on Cretaceous rocks" (Kellogg, 1975, p. 478). They are flat-lying and exhibit no folding (W.B. Harland, personal communication, 1983). "It is likely that they were originally contiguous with those of the main Tertiary basin" (Kellogg, 1975, p. 478). If so, this would be a strong indication for a pre-Tertiary age of the compressional movements south of Isfjorden. In any case, south of Isfjorden, the proof for a Tertiary age of the compressional event is still pending.

Craddock *et al.* (1983) have recently queried the Tertiary age of the compression south of Isfjorden. They stressed that, at least in the area around Bellsund, indications of SSE—NNW trending transcurrent movements are missing within the fold belt. Such strike-slip faults should certainly have developed if the deformations were caused by transpressional movements. Craddock *et al.* (1983) propose, therefore, that the transform faults developed only off-shore and away from the fold belt. An alternative explanation would be that the Tertiary transcurrent movements between Greenland and Svalbard were not the cause of the compressional deformation. Compression along a transform fault should be expected only where deviations from the straight course of the fault are present (as is the case, e.g., along the Levante Fracture Zone to the north of the Jordan Graben).

The overthrusts north of Isfjorden had northeasterly throws. The southeastern boundary of these thrust sheets (along the Isfjord, Fig. 2) exhibits the same direction. It displaces *already folded* Precambrian to Mesozoic strata by about 5 km (Harland and Horsfield, 1974). Brøggerhalvøya is the only zone where Tertiary deposits are involved in the compressional deformations. According to the cross sections presented by Orvin (1934) and Challinor (1967), *all* these deformations with Tertiary layers involved can be demonstrated to be the results of *drag effects of overthrusts.* Pre-overthrust folding of Tertiary beds is not observed anywhere. The age of the overthrust movement north of Isfjorden can be determined from the following datings: the youngest rocks involved in the deformations by the drag effects have been dated as Palaeocene by Vonderbank (1970) on the basis of foraminiferal studies. According to Atkinson (1963) and Livshits (1965), beds of Eocene to Miocene age were not affected by the deformations (cf. Harland and Horsfield, 1974). Therefore, the *overthrusting* north of Isfjorden will certainly have occurred during the *late Palaeocene.* As the overthrusting there is younger than the folding, the latter must have taken place in Palaeocene or earlier times.

The final extensional faults which cut through all fold axes and thrust planes (Orvin, 1940) "are probably all of about the same age and all of them younger than the folding and thrusting" (Orvin, 1934, p. 76). This faulting "must have taken place in a period when the horizontal pressure had ceased to act and had been replaced by a tensional stress. This fault system is certainly of the same age as the great faults along the west coast of Spitsbergen" (Orvin, 1934, p. 79).

The discussion can be summarized as follows:

(1) There is no evidence for a simultaneity of the folding and overthrusting in the area north of Isfjorden. The folding episode was earlier than the overthrusting and the folding and overthrusting had different throws.

(2) There is no evidence that Tertiary beds south of Isfjorden have been involved in the folding and upthrusting. The few Tertiary outliers all show no folding but lie disconformably upon strongly folded and eroded strata. The Tertiary infill sediments of the Forlandsundet Graben are "post-folding and thrusting" (Harland and Horsfield, 1974) and therefore not relevant for the dating of the compressional events.

(3) The youngest rocks affected by the folding phase are of Aptian/Albian age and are covered unconformably by basal Palaeocene conglomerates (Birkenmajer, 1981). The folding, therefore, must have occurred in the interval between the Aptian/Albian and the Palaeocene.

The field observations give clear evidence for three stages of crustal deformation: The first was a phase of folding and upthrusting, the second an overthrusting in a different direction from the folding (north of Isfjorden); and the third a downfaulting in a tensional regime (Forlandsundet Graben, Fig. 2). The dating of the crustal deformations in western Spitsbergen, independently of Alpine tectonic phases and plate tectonic models, is proposed as follows:

(a) The folding must have taken place in the interval between Aptian/Albian and Palaeocene times.

(b) The overthrusting north of Isfjorden probably occurred during late Palaeocene to Eocene time.

(c) The extensional faulting occurred later, probably during the Oligocene. The total absence of late Cretaceous deposits on Spitsbergen (despite the highest global sea level in earth history) has been generally explained by a strong epirogenetic upheaval. Following the proposed new dating of the folding, the non-deposition or erosion during the late Cretaceous could well have been the result of the Cretaceous compression.

OFF-SHORE DATA

Compared with the intense fieldwork done on Spitsbergen, the data available from the off-shore area west of the Svalbard Archipelago are rather scarce. Sundvor and Eldholm (1976) demonstrated that the shelf area between southwest Spitsbergen and Bear Island is structurally subdivided by a major fault running SSE—NNW with downthrow to the west. It was called the 'Hornsund Fault' and was tentatively suggested to mark the ocean—continent boundary (Sundvor *et al.*, 1978). The fault extends to 79°N (Fig. 2) and runs parallel to the Forlandsund Graben.

An important conclusion from the geophysical work done in the area west of Spitsbergen, which has been summarized by Vogt *et al.* (1978), is that the structural deformations of Spitsbergen continue off-shore to the west. The same folds as on-shore occur between the coastline and the Hornsund Fault (their fig. 8). According to Schlüter and Hinz (1978) this fault does not coincide with the ocean—continent boundary. The boundary is expected to lie between the easternmost oceanic crust distinguishable on the seismic lines and the Hornsund Fault. Similarly the interpretive section presented by Vogt *et al.* (1978, fig. 8) also shows that the Hornsund Fault does not represent the ocean—continent boundary, but rather the boundary of a pre-spreading rift.

The orogenic belt of West Spitsbergen extends to the southern end of the island (Fig. 2). An offshore continuation to the SSE, therefore, can be expected. Such a continuation has not yet been observed on reflection seismic lines, perhaps because of strong sea-floor multiples which obscure the seismic information (Hinz and Weber, 1975; Schlüter and Hinz, 1978). The only indication for an extension of the orogenic belt to the SSE is a negative gravity anomaly (Rønnevik and Motland, 1979; Vogt *et al.*, 1981). This anomaly runs from the southern on-shore end of the orogenic front to the northern edge of the Bear Island Basin (about 74.2°N 21°E) in a gentle arc (Fig. 2). Therefore, the gravity anomaly is interpreted as the structural contact between highly consolidated folded basement rocks to the west and flat-lying Mesozoic(?) deposits to the east. This relationship is observed in the on-shore geology at the southern end of Spitsbergen. Hence, the West Spitsbergen orogenic belt probably continues down to the northwestern border of the Bjørnøy Basin. At this site, Rønnevik (1982) has observed compressional features on recent reflection seismic lines.

PLATE TECTONIC IMPLICATIONS OF THE CRETACEOUS EPISODE ON SPITSBERGEN

The Cretaceous compressional phase in western Spitsbergen claimed from the above revision of this fold belt should have consequences on the opening history of the Northeast Atlantic. It must be determined as to whether this plate convergence was either a "somewhat accidential collision" of the American and Eurasian plates (Harland and Horsfield, 1974), or whether it was the result of a more important movement of the two plates. In the second case, a corresponding divergence of the two plates must be expected elsewhere. Such a Cretaceous plate divergence between the American and Eurasian plates resulting in rifting and stretching of continental crust or even in ocean-floor spreading has been repeatedly assumed. It was suggested to have occurred either between Greenland and the Rockall Plateau (Harland, 1969) or between Europe and the Rockall Plateau (Scrutton, 1972; Roberts, 1974; Bott, 1975; Kristoffersen, 1977; Burke, 1978), or even at both sites (Laughton, 1975). Recent investigation (Bott, 1978; Roberts *et al.*, 1981) shows that the pre-Tertiary opening of the NE-Atlantic took place in the Rockall Trough and a pre-Tertiary spreading between the Rockall Plateau and Greenland can be excluded (Talwani and Eldholm, 1977).

The Cretaceous opening of the Rockall Trough and its northeastern continuation

The Rockall Trough is situated between the British Isles and the submarine Rockall Plateau. The latter is regarded as a subsided continental fragment (Scrutton, 1972; Roberts, 1974; Bott, 1975). The opening of the Rockall Trough occurred very probably during Cretaceous time according to Roberts (1974), Bott (1978) and Roberts *et al.* (1981) and not during the Permian as suggested by Russell (1976), Russell and Smythe (1978) and Smythe (1983). Carboniferous to Jurassic basins, grabens or halfgrabens in this area (Ziegler, 1982a) cannot be completely excluded (although there is no evidence) but the "major rifting episode" leading to the Rockall Trough probably took place during the Aptian (Roberts *et al.*, 1981, p. 126). Smythe is reexamining his earlier view (D.K. Smythe, personal communication). The assumed oceanic crust in the centre of the Rockall Trough is thought to be 160-km (Bott, 1978) or 120-km (Roberts *et al.*, 1981) wide. Price and Rattey (in press) suggest a width of oceanic crust in the Rockall Trough of about 190-km.

The northeastward continuation of the Rockall Trough and its oceanic crust into the Faeroe Basin now seems quite probable. Not only based on geometric constraints (Roberts, 1974; Bott, 1975; Kristoffersen, 1977; Hanisch, 1983, in press) but also from refraction seismic data (Bott, 1984), the Faeroe Basin is thought to be underlain by oceanic crust. The present highs separating it from the Rockall Trough, the Wyville—Thomson and the Faeroe Ridges, are suggested to consist of huge Palaeogene volcanic buildups (Bott, 1984) and a connection of the two basins may well have existed during Cretaceous time.

Further to the north, at the transition from the Faeroe to the Møre Basin, the presence of the buried Brendan Seamount points to an oceanic basement (Smythe *et al.*, 1983). From seismic lines across the West Shetland Basin and the Faeroe Basin (e.g. Ziegler, 1982a, fig. 13) it is evident that the main fault-controlled subsidence of the Faeroe Basin occurred from early Cretaceous time on (cf. Ridd, 1981).

The Cretaceous age of the Møre Basin is now generally accepted (De'Ath and Schuyleman, 1981; Hinz *et al.*, 1982; Hanisch, 1983; Price and Rattey, in press; Gabrielsen *et al.*, 1984; Hamar and Hjelle, 1984). The late Cimmerian unconformity is downfaulted by up to 3.5 seconds TWT at the so-called Frøya High which forms the eastern boundary of the Møre Basin (Hinz *et al.*, 1982, fig. 9).

The western edges of the Faeroe and Møre Basins are not known. They seem to be seismically masked by widespread Palaeogene flood basalts (Fig. 3). Only on line BGR 79-02 (Hinz *et al.*, 1982, fig. 10), in the western part of the Møre Basin, are some indications for eastward-dipping listric faults present which point to a western bordering of the basin. The central high (called Vigra High by Hamar and Hjelle, 1984) presumably represents a mid-Cretaceous volcanic centre or even a volcanic ridge from where the supposed pre-Cenomanian volcanic beds in the Møre Basin could have been derived (cf. Price and Rattey, in press).

To the north the Møre Basin seems to merge into the Vøring Basin. The situation at the eastern boundary of this basin is quite similar to that of the Møre Basin. The boundary is formed by the Nordland Ridge (Fig. 3) which forms a NE-striking elongated high where most of the pre-Cretaceous Mesozoic sediments have been eroded (Price and Rattey, in press; Gowers *et al.*, 1984; Larsen and Skarpnes, 1984). The northwestern flank of the Nordland Ridge consists of a pronounced fault

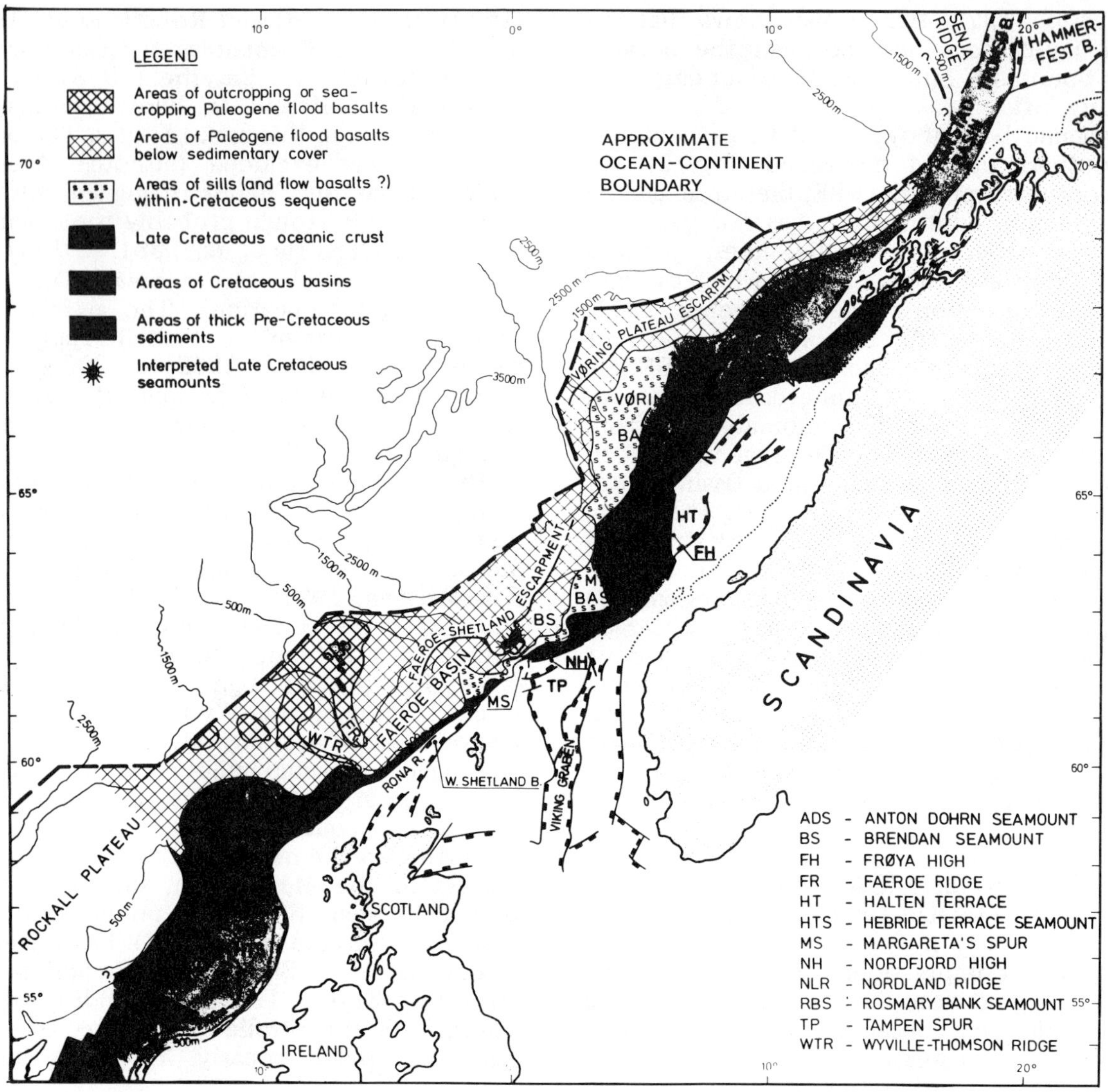

Fig. 3. Structural sketch map of the British—Norwegian Continental Margin (based on GEBCO, Ottawa 1978, references cited in text).

zone which was clearly active in post-Jurassic time.

The Vϕring Basin itself seems to be divided into several sub-basins (Gabrielsen *et al.*, 1984). Its northwestern edge is not known. Some indications for a block-faulted boundary zone come from line BGR 79-05 (Hinz *et al.*, 1982, fig. 7). On this line the boundary of the Vϕring Basin can be interpreted to lie at about 65.35°N, 3.5°E and not at the Vϕring Plateau Escarpment as suggested by previous workers. This escarpment does not form the ocean—continent boundary (Talwani and Eldholm, 1977; Talwani *et al.*, 1981; Mutter *et al.*, 1982), but rather the edge of Palaeogene flood basalts (Smythe, 1983).

The Traen Basin possibly forms the northeastwards continuation of the Vϕring Basin (Jϕrgensen and Navrestad, 1981). However, this has recently been questioned (Price and Rattey, in press; Gowers and Lunde, 1984). An E—W-striking transform fault is thought to separate the two basins. The Cretaceous age of the Traen Basin, however, is beyond doubt (Gabrielsen *et al.*, 1984).

North of 68°N and west of the Lofoten and Vesteralen Islands the available data are scarce. Eldholm *et al.* (1979) have indicated a narrow sedimentary basin there, the boundary of which lies just off-shore the Lofoten Islands (Fig. 3). From on-shore studies it is known that NW-dipping downthrow faults were active in post-Aptian and pre-Tertiary time on the Lofoten and Vesteralen Islands (Dalland, 1979; Bartley, 1981).

Another sedimentary basin, the Harstad Basin, is present north of the Vesteralen Islands (Fig. 3). Its eastern boundary constitutes a major downthrow fault which has been identified by shallow seismic studies (Gunleiksrud and Rokoengen, 1980, fig. 4). According to Bergsager (1981, fig. 2) this fault runs into the "faulted transition zone" between the Hammerfest and Tromsϕ Basins and continues northward up to 71.4°N, 21°E. The western edge of the Harstad Basin is unknown.

The western boundary of the Tromsϕ Basin is formed by a major fault zone which separates it from the Senja Ridge (Fig. 3). The development of the Tromsϕ Basin post-dates the Middle Jurassic and had a fast late Cretaceous subsidence (Gabrielsen *et al.*, 1984). The observed tectonic contrast to the Hammerfest Basin gives rise to the assumption that this basin was intersected by the N—S-trending Tromsϕ Basin, resulting in the "faulted transition zone" (Øvrebϕ and Talleraas, 1977).

The northern fault trend of the Tromsϕ Basin

continues to the Bjørnøy Basin and ends at about 74°N (Rønnevik, 1982). In the northwestern part of the Bjørnøy Basin the tensional regime seems to be replaced by a compressional one. The compressional features at this site have a NNW trend (Rønnevik, 1982). This compression has been explained by Rønnevik and Jacobsen (1984) by a southward movement of the Svalbard Block.

Summing up, a chain of sedimentary basins is present along the British—Norwegian Continental Margin, all of which have developed or had their major subsidence in Cretaceous time. It is therefore suggested that the basins are part of *one continuous Cretaceous rift system* which stretched from the Rockall Trough through the Faeroe, Møre, Vøring, Traen, Harstad and Tromsø Basins up to the Bjørnøy Basin. A general northward narrowing of the rift can be inferred. In its southern part a narrow stripe of oceanic crust (Bott, 1978: 160-km wide; Roberts *et al.*, 1981: 120-km wide; Price and Rattey, in press: 190-km wide) was formed during the late Cretaceous. The oceanic crust is believed to decrease in width towards the north and to disappear somewhere in the Møre Basin (Hanisch, in press; Price and Rattey, in press). A plate tectonic movement of Greenland together with the Rockall Plateau has been inferred to be the cause of this rifting/spreading system (Hanisch, 1983, in press; Price and Rattey, in press). The first author proposes a pole of rotation of this movement at 74°N, 21°E, while the second authors assume a rotation pole to the northwest of the Vøring Plateau.

North of the pole of rotation a compressional regime must be expected. The revision of the West Spitsbergen orogeny does not exclude a Cretaceous folding phase which may have taken place between the Aptian/Albian and the Palaeocene. As the West Spitsbergen fold belt apparently continues down to the northwestern edge of the Bjørnøy Basin, the proposed pole of rotation at about 74°N, 21°E seems to fulfil perfectly the requirements of plate tectonics: extension on one side of the pole of rotation and compression on the other.

SYNTHESIS: EXTENSION AND COMPRESSION IN THE NORTHEAST ATLANTIC DURING THE CRETACEOUS

The supposed structural configuration between Europe and Greenland by the end of the Cretaceous is sketched in Fig. 4. A continuous Cretaceous rift system is inferred to have developed from the Rockall Trough to the Bjørnøy Basin. The oceanic crust in the centre of the southern Rockall Trough (according to Roberts *et al.*, 1981) is thought to diminish continuously in width toward the northeast and to disappear at about 65.5°N (cf. Hanisch, in press).

The southeastern boundary of the proposed rift system is quite well documented. There is a pronounced rifted continental margin along the Porcupine Bank and the eastern border of the Rockall Trough (Roberts *et al.*, 1981). The Rona Ridge, situated between the Faeroe Basin and the West Shetland Basin, the Margareta's Spur and the Nordfjord High (Fig. 3) were deeply eroded during Middle and late Cretaceous time (Fyfe *et al.*, 1981; Ridd, 1981; Ziegler, 1982a, fig. 13). These structural highs are therefore interpreted to represent the uplifted shoulders of the Cretaceous rift. Similar erosion took place at the Frøya High (cf. Hinz *et al.*, 1982; Price and Rattey, in press; Gabrielsen *et al.*, 1984) and on the Nordland Ridge (Jørgensen and Navrestad, 1981). Bukovics *et al.* (1984) have proposed the emplacement of an asthenolith during the Cretaceous which resulted in the uplift of the Nordland Ridge at the eastern boundary of the Vøring Basin.

The situation between the Vøring and the Harstad basins is not yet completely understood. The mentioned post-Aptian but pre-Tertiary faults on the Lofoten and Vesteralen Islands, however, also point to a Middle to late Cretaceous rifting. Finally, the rifting in the transition zone between the Hammerfest and the Tromsø Basins probably occurred during the Cretaceous.

The suggested Cretaceous rift system has apparently cross-cut a series of older basins: the northward disappearance of the West Shetland Basin, the northern end of the North Sea Graben and, to the north, the intersection of the Hammerfest Basin by the younger Tromsø Basin can be convincingly explained by the Cretaceous rift. Vogt *et al.* (1981, p. 30) wondered at the marked obliqueness between the structures in northeast Greenland (and its shelf) and the structural trends of the Norwegian Shelf and Margin. Price and Rattey (in press) demonstrate that in the Haltenbanken area, younger faults have cross-cut older trends.

The western boundary of the proposed rift system is not nearly so clear. Only in its southern and northern parts are such boundary fault zones present (Fig. 4). In between there are only weak indications for a western limitation of the Faeroe Basin (flexuring of the Palaeogene flood basalts, cf. Ridd, 1981; Hanisch, in press) and of the Møre and Vøring Basins as mentioned above. Most of the area, however, is occupied by abundant Palaeogene flood basalts and igneous intrusions of unknown age (Hinz *et al.*, 1982; Price and Rattey, in press; Smythe *et al.*, 1983; Hamar and Hjelle, 1984). These are seismically masking older features.

The missing link of the rift between the Vøring/Traen Basin and the Tromsø Basin may be expected partly below the basalts around the northernmost Vøring Plateau Escarpment and partly below the eastern corner of the Northeast Greenland Shelf (Fig. 4). Between 68°N and 69.5°N these basalts have subsided to a considerably greater depth than at the Vøring Plateau (Talwani and Eldholm, 1977; Eldholm *et al.*, 1973; Hinz *et al.*, 1982). This could be a consequence of the following configuration: the late Palaeocene line of opening is thought to have intersected the Cretaceous rift basin between about 69°N and 71°N (Fig. 4). It would therefore have run into already thinned and subsided crust (in contrast to the thicker crust beneath the Vøring Plateau, according to Hinz, 1972). The original Cretaceous subsidence would have been superimposed by the later subsidence at the Tertiary continental margin. This could have led to the present much greater depth of the basalts in this area than beneath the Vøring Plateau. A certain confirmation of this hypothesis can be deduced from the gravity map by Talwani and Eldholm (1972). A marked gravity low area runs from the

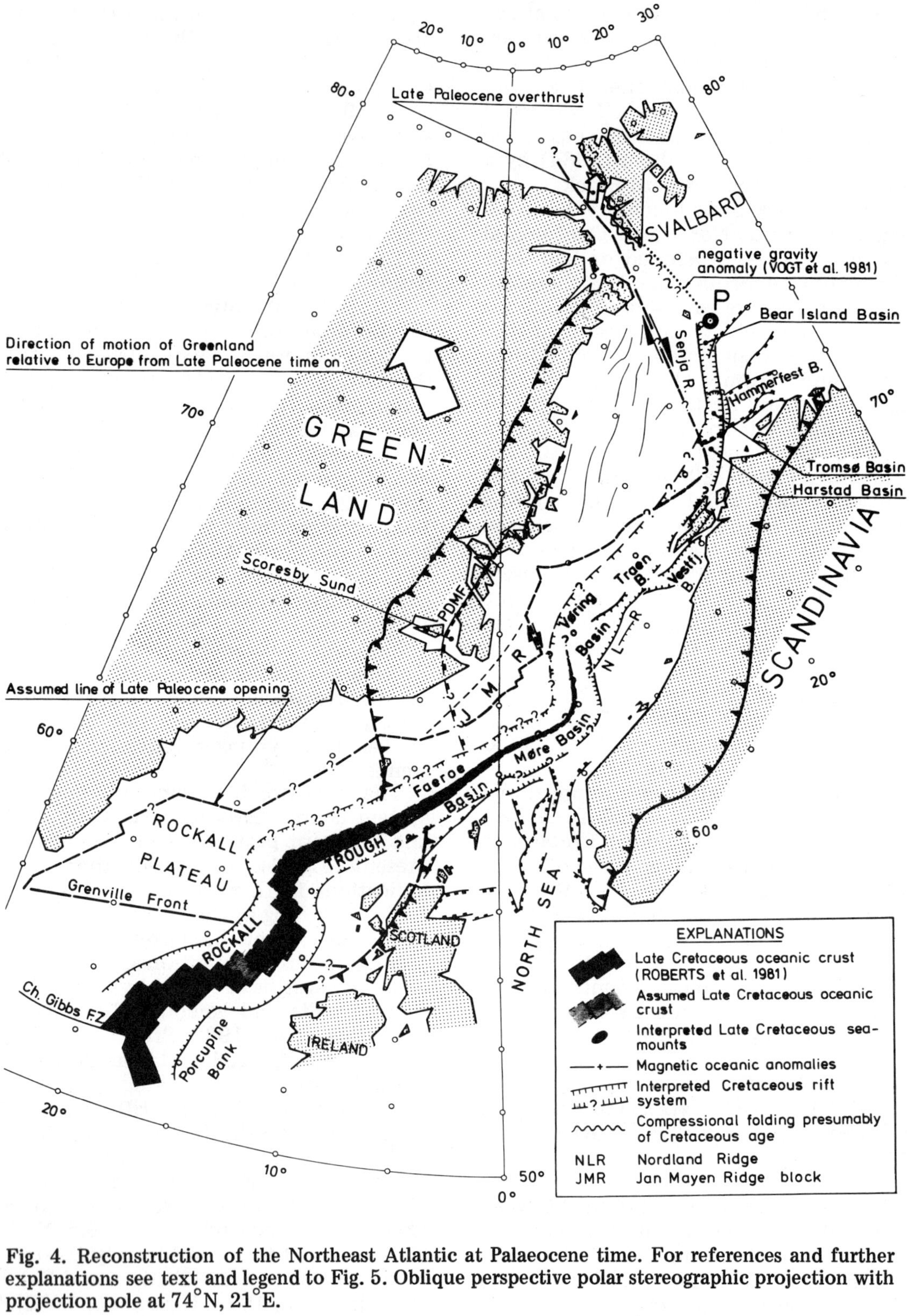

Fig. 4. Reconstruction of the Northeast Atlantic at Palaeocene time. For references and further explanations see text and legend to Fig. 5. Oblique perspective polar stereographic projection with projection pole at 74°N, 21°E.

northern Vøring Basin to the NNE and cross-cuts the Vøring Plateau Escarpment at 68°N, 7.6°E. This gravity low may well be the expression of the supposed Cretaceous rift basin.

EARLY CRETACEOUS RESTORATION

Figure 5 shows the reconstruction of the NE Atlantic region at early Cretaceous time as it results from the closure of the inferred Cretaceous rifting/ spreading system. Greenland with the Rockall Plateau has been rotated anti-clockwise by 5° around a pole at 74°N, 21°E. This pole of rotation is compatible with the pole calculated by Pitman and Talwani (1972) for the time before 81 Ma. This rotation pole was deduced from the spreading history of the North Atlantic north of the Azores and lies at 70.4°N, 15.1°E (north of Vesterålen Island). The pole proposed here (74°N, 21°E) lies in the same trend and is within the margins of error if the great distance to the central Atlantic is considered.

A schematic model of this plate tectonic rotation is sketched in Fig. 6. It is based on the model of formation of continental margins proposed by

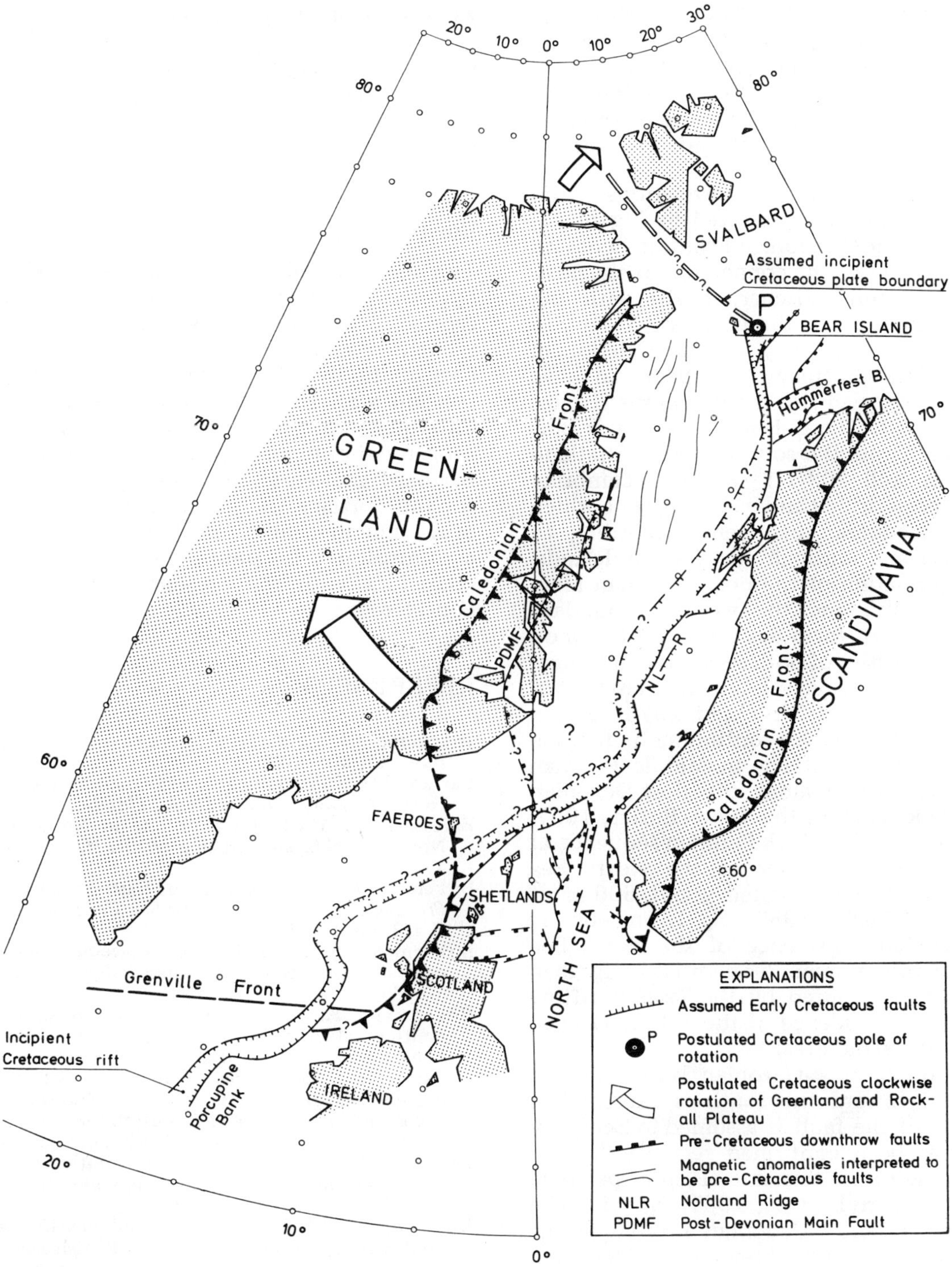

Fig. 5. Reconstruction of the Northeast Atlantic at early Cretaceous time. References quoted in text, see legend to Fig. 4. Same geographical projection as used in Fig. 4.

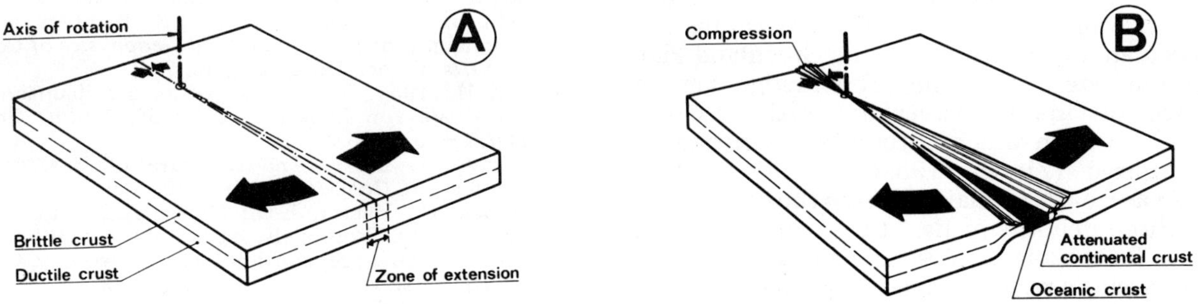

Fig. 6. Schematic model for the proposed plate tectonic movement between Greenland and Europe during the Cretaceous.

LePichon and Sibuet (1981). Details of this plate tectonic rotation are discussed in Hanisch (in press). The resulting position of the Rockall Plateau is in accordance with that indicated by LePichon *et al.* (1977). The amount of extension south of the pole of rotation is in good agreement with the amount of compression north of it.

Two important structural correlations from Europe to Greenland are suggested in Fig 5: (1) the connection of the Caledonian Fronts in Scotland and Greenland, and (2), the connection of the post-Devonian Main Fault (Vischer, 1943) with the western boundary fault of the northern North Sea Graben. The great similarity of the Caledonian fold belts of East Greenland and Scotland have been repeatedly stressed (for discussion see Higgins and Phillips, 1979). The Caledonian Front is thus thought to run across the southern West Shetland Basin (between well 202/1-1 where Precambrian basement has been drilled and well 205/26-1 which encountered Caledonian basement, according to Ziegler, 1982a) and through the Faeroe Islands.

The new correlation of the post-Devonian Main Fault on Greenland and the western boundary fault of the northern North Sea Graben is based on the extreme similarities of both areas in their faunal, lithological structural developments (Kempter, 1961; Surlyk *et al.*, 1973; Birkelund, 1976; Larsen, 1980; Hanisch, 1983). Such a continuous halfgraben probably came into existence in Middle Permian time. The main rifting along the post-Devonian Main Fault took place at this time (Bütler, 1957; Surlyk *et al.*, 1981). The thick Permian sequence at the northwesternmost corner of the North Sea Graben (well 210/4-1 perforated about 600 m of Permian beds, cf. Ziegler, 1982a, fig. 13 and encl. 1) points to a similar subsidence of the North Sea Graben during the Permian. The initial halfgraben was apparently overprinted by the Triassic rifting which led to the subsidence of the Viking Graben obliquely to older structures.

The course of the post-Devonian Main Fault on East Greenland toward the south is sketched in Fig. 5. A bend of the fault is assumed in Scoresby Sund. This has been confirmed recently by the seismic investigation in this area (Risum, personal communication; Surlyk *et al.*, 1981, fig. 11). The parallel courses of the Caledonian Front and the post-Devonian Main Fault observed on Greenland (Vischer, 1943; Surlyk, 1977) thus seems to have continued from East Greenland to Northwest Europe in pre-Cretaceous times.

The structural configuration proposed here for the early Cretaceous differs from that suggested by Ziegler (1982b, fig. 6) in several aspects. First of all, no rifting is shown between the Rockall Plateau and SE Greenland. Rockall Plateau is thought to have been an integrated part of the Greenland Plate up to the late Palaeocene. The dredge samples recovered on the SE Greenland Shelf (Johnson *et al.*, 1975) from which Carboniferous rifting has been inferred (Ziegler, 1982a, p. 26) must be regarded as ice-rafted material from the corresponding on-shore deposits in East Greenland. The complex late Palaeozoic to Mesozoic rifting systems in the NE Atlantic region estimated by Ziegler (1978, 1982a; 1982b) bear great resemblance to the Cenozoic configuration. This would signify that the late Mesozoic to Tertiary opening of the NE Atlantic had structural precursors during about 200 Ma. This concept is replaced by the view that the late Palaeozoic and early Mesozoic rift systems had a widely different pattern from the present ones and were intersected by completely new and independent crustal cracks during the Cretaceous and Lower Tertiary.

Final remarks

The plate tectonic model proposed here is a working hypothesis. In many aspects it is speculative, especially for those areas for which little or no reliable data are available. The most controversial part of the hypothesis deals with the unusual fold belt on western Spitsbergen. The attempt to call into question the exclusively Tertiary age of the orogeny should be regarded as a challenge to the Svalbard geologists to come up with the evidence for the Tertiary transpression (if possible). As long as no unequivocal proof of the exclusively Tertiary compressions is published, alternative models consistent with the present data should also be considered.

REFERENCES

Atkinson, D.J. 1963. Tertiary rocks of Spitsbergen. *AAPG Bull.*, **47**, 302—323.

Bartley, J.M. 1981. Mesozoic high-angle faults, east Hinnøy, North Norway. *Norsk Geol. Tidsskr.*, **3—4**, 291—296.

Bergsager, E. 1981. First drilling in Norwegian Sea off Norway yields encouraging results. *Oil Gas J.*, June, 8, 115—124.

Birkelund, T. 1976. Geology of East Greenland: A review *Proceedings, Offshore North Sea Conference, Stavanger*, G 4/3, pp. 1—23.

Birkenmajer, K. 1966. Lower Cretaceous tidal deposits of central Vestspitsbergen. *Norsk Polarinst. Årbok*, **1964**, 73—85.

Birkenmajer, K. 1972. Tertiary history of Spitsbergen and continental drift. *Acta Geol. Polon.*, **22**, 193—218.

Birkenmajer, K. 1981. The geology of Svalbard, the western part of the Barents Sea and the continental margin of Scandinavia. In: A.E.M. Nairn, M. Churkin, Jr. and F.G. Stehli (eds.). *The Ocean Basins and Margins Vol. 5, The Arctic Ocean*, Plenum, New York, pp. 265—329.

Bott, M.H.P. 1975. Structure and evolution of the Atlantic floor between northern Scotland and Iceland. *Norges Geol. Unders.*, **316**, 195—203.

Bott, M.H.P. 1978. The origin and development of the continental margins between the British Isles and southeastern Greenland. In: D.R. Bowes and B.E. Leake (eds.). Crustal Evolution in Northwestern Britain and Adjacent Regions, Spec. Iss., *Geol. J.*, **10**, pp. 377—392.

Bott, M.H.P. 1984. Deep structure and origin of the Faeroe-Shetland Channel. This volume, pp. 341—348.

Bukovics, C., Cartier, E.G.; Shaw, N.D. and Ziegler, P.A. 1984. Structure and development of the mid Norway continental margin. This volume, pp. 407—424.

Burke, K. 1978. Evolution of continental rift systems in the light of plate tectonics. In: I.B. Ramberg and E.-R Neumann (eds.). *Tectonics and Geophysics of Continental Rifts.* Reidel, Dordrecht, pp. 1—9.

Bütler, H., 1957. Beobachtungen an der Hauptbruchzone der Küste von Zentral-Ostgrönland. *Meddr. Grønland*, **160**, 1—79.

Challinor, A. 1967. The structure of Brøggerhalvøya, Spitsbergen. *Geol. Mag.*, **104**, 322—336.

Craddock, C., Hauser, E.G.; Maher, A.D. and Zhu Guo-Qiang, 1983. Tectonic Evolution of the West Spitsbergen fold belt. Oral presentation Int. Mn. Geodesy, Geophys., XVIII Gener. Assembly, Hamburg 1983.

Dalland, A., 1979. The sedimentary sequence of Andøy, Northern Norway — depositional and structural history. *Proc. Norw. Sea Symp.*, Tromsø, **26**, pp. 1—31.

De'Ath, N.G. and Schuyleman, S.F. 1981. The geology of

the Magnus oilfield. In: L.V. Illing and G.D. Hobson (eds.). *Petroleum Geology of the Continental Shelf of North-West Europe, Proc., 2. Conf. Petrol. Geol. Continental Shelf NW Europe, London,* Heyden, London, pp. 342—351.

Eldholm, O., Sundvor, E. and Myhre, A. 1979. Continental margin off Lofoten—Vesteralen, Northern Norway. *Mar. Geophys. Res.,* **4**, 3—35.

Fyfe, J.A., Abbots, I. and Crosby, A. 1981. The subcrop of the mid-Mesozoic unconformity in the UK area. In: L.V. Illing and G.D. Hobson (eds.). *Petroleum Geology of the Continental Shelf of North-West Europe, Proc., 2. Conf. Petrol. Geol. Continental Shelf NW Europe, London,* Heyden, London, pp. 236—244.

Gabrielsen, R.H., Faerseth, R., Hamar, G.P. and Rønnevik, H. 1984. Nomenclature of the main structural features on the Norwegian Continental Shelf north of the 62nd parallel. This volume.

Gowers, M.B. and Lunde, G. 1984. The geological history of Traenabanken. This volume, pp. 237—252.

Gunleiksrud, T. and Rokoengen, K. 1980. Regional geological mapping of the Norwegian Continental Shelf with examples of engineering applications. In: D.A. Aardus (ed.). *Offshore Site Investigation,* Graham & Trotman, London, pp. 23—35.

Hanisch, J. 1983. The structural evolution of the NE Atlantic region. *Geol. Jb.* Reihe B, **52**, 1—37.

Hanisch, J. The Cretaceous opening of the Northeast Atlantic. *Tectonophysics,* (in press).

Hamar, G.P. and Hjelle, K. 1984. Tectonic framework of the Møre Basin and the northern North Sea. This volume.

Harland, W.B. 1965. The tectonic evolution of the Arctic—North Atlantic region. *Phil. Trans. R. Soc. Lond., Ser. A,* **258**, 59—75.

Harland, W.B. 1969. Contribution of Spitsbergen to Understanding of Tectonic Evolution of North Atlantic Region. *Mem. Am. Assoc. Petrol. Geol. No.* 12, pp. 817—851.

Harland, W.B. and Horsfield, W.T. 1974. West Spitsbergen orogen. In: A.M. Spencer (ed.). Mesozoic—Cenozoic Orogenic Belts; Data for Orogenic Studies, Spec. Publ. *Geol. Soc. Lond.,* **4**, 747—755.

Higgins, A.K. and Phillips, W.E.A. 1979. East Greenland Caledonides — an extension of the British Caledonides. In: A.L. Harris, C.H. Holland and B.E. Leake (eds.). *The Caledonides of the British Isles — reviewed.* Pitman, Bath, pp. 19—32.

Hinz, K. 1972. Der Krustenaufbau des Norwegischen Kontinentalrandes (Vøring Plateau) und der Norwegischen Tiefsee zwischen 66° und 69°N nach seismischen Untersuchungen. *Meteor Forschungs.,* Reihe C, **10**, 1—16.

Hinz, K. and Schlüter, H.U. 1978. The geological structure of the western Barents Sea. *Mar. Geol.,* **26**, 199—230.

Hinz, K. and Weber, J. 1975. Zum geologischen Aufbau des norwegischen Kontinentalrandes und der Barents See nach reflexionsseismischen Messungen. *Compendium 75/76, Erg. Bd. Ztschr. Erdöl, Kohle, Erdgas, Petrochem.* 3—29.

Hinz, K., Dostmann, H.J. and Hanisch, J. 1982. Structural framework of the Norwegian Sea. *Proc. Northern Seas Symp.,* Stavanger, 1982, E/4, pp. 1—22.

Johnson, G.L., McMillan, N.H. and Egloff, J. 1975. East Greenland continental margin. In: C.J. Yorath, E.R. Parker and D.J. Glass (eds.). Canada Continental Margins and Offshore Petroleum Exploration. *Can. Soc. Petrol. Geol. Mem.,* **4**, 205—224.

Jørgensen, F. and Navrestad, T. 1981. The geology of the Norwegian Shelf between 62°N and the Lofoten Island. In: L.V. Illing and G.D. Hobson (eds.), *Petroleum Geology of the Continental Shelf of North-West Europe Proc., 2. Conf. Petrol., Geol. Continent. Shelf NW Europe.* Heyden, London, pp. 407—413.

Kellogg, H.E., 1975. Tertiary stratigraphy and tectonism in Svalbard and continental drift. *AAPG Bull.,* **59**, 465—485.

Kempter, E.H.K. 1961. Die jungpaläozoischen Sedimente von Süd Scoresby Land. *Meddr. Grønland,* **164**(1), 1—123.

Kristoffersen, Y. 1977. Late Cretaceous sea floor spreading and the early opening of the North Atlantic. In: *Mesozoic North North Sea Symp.,* Norskpetroleumsforening, paper 5, Oslo, pp. 1—25.

Larsen, H.C., 1980. Geological perspectives of the East Greenland continental margin. *Geol. Soc. Denmark, Bull.,* **29**, 77—101.

Larsen, R.M. and Skarpnes, O. 1984. Regional interpretation and hydrocarbon potential of the Traenbanken area. This volume, pp. 217—236.

Laughton, A.S., 1975. Tectonic evolution of the Northeast Atlantic Ocean: a review. *Norges Geol. Unders.,* **316**, 169—193.

LePichon, X. and Sibuet, J.-C. 1981. Passive margins: a model of formation. *J. Geophys. Res.,* **86**, 3708—3720.

Le Pichon, X., Sibuet, J.-C. and Francheteau, J. 1977. The fit of the continents around the North Atlantic Ocean. *Tectonophysics,* **38**, 169—209.

Livshits, Y.Y. 1965. (Palaeogene deposits of Nordenskiold Land, Vestspitsbergen). In: *Materiali po geologii Shpitsbergena,* V.N. Solokov (ed.), Leningrad Inst. Geol. of the Arctic, pp. 185—208 (in Russian).

Lowell, J.D. 1972. Spitsbergen Tertiary orogenic belt and the Spitsbergen fracture zone. *Geol. Soc. Am. Bull.,* **83**, 3091—3102.

Mutter, J.C., Talwani, M. and Stoffa, P.L. 1982. Origin of seaward-dipping reflectors in ocean crust off the Norwegian margin by "Subaerial sea-floor spreading". *Geology,* **10**, 353—357.

Orvin, A.K. 1934. Geology of the Kings Bay region, Spitsbergen. *Skr. Svalbard Ishavet,* **57**, 1—195.

Orvin, A.K. 1940. Outline of the geological history of Spitsbergen. *Skr. Svalbard Ishavet,* **78**, 1—57.

Pitman, W.C. and Talwani, M. 1972. Sea-floor spreading in the North Atlantic. *Bull. Geol. Soc. Am.,* **83**, 619—646.

Price, I. and Rattey, R.P. Cretaceous tectonics of mid-Norway: implications for the Rockall Faeroes-Shetland rift. *J. Geol. Soc. London.* (in press).

Ridd, M.F. 1981. Petroleum geology west of the Shetlands. In: L.V. Illing and G.D. Hobson (eds.), *Petroleum geology of the continental shelf of North-West Europe, Proc. 2. Conf. Petrol. Geol. Continent. Shelf NW Europe.* Heyden, London, pp. 414—425.

Roberts, D.G., 1974. Structural development of the British Isles, continental margin and Rockall Plateau. In: C.A. Burk and C.L. Drake (eds.), *The Geology of Continental Margins,* Springer, New York, pp. 343—359.

Roberts, D.G., Masson, D.G. and Miles, P.R. 1981. Age and structure of the southern Rockall Trough: new evidence. *Earth Planet. Sci. Lett.,* **52**, 115—128.

Rønnevik, H., 1982. Structural and stratigraphic evolution of the Barents Sea. *Proc. Northern Seas Symp., Stavanger, 1982,* E/3, 1—30.

Rønnevik, H. and Jacobsen, H.-P. 1984. Structural highs and basins in the western Barents Sea. This volume, pp. 19—32.

Rønnevik, H. and Motland, K. 1979. Geology of the Barents Sea. *Proc. Norweg. Sea Symp.* Tromsø 1979, NSS/15, pp. 1—34.

Russell, M.J. 1976. A possible Lower Permian age for the onset of ocean floor spreading in the northern North Atlantic. *Scott. J. Geol.,* **12**, 315—323.

Russell, M.J. and Smythe, D.K. 1978. Evidence for an early Permian oceanic rift in the North Atlantic. In: E.-R. Neumann and I.B. Ramberg (eds.), *Petrology and Geochemistry of Continental Rifts.* Reidel, Dortrecht, pp. 173—179.

Schlüter, H.U. and Hinz, K. 1978. The continental margin of West Spitsbergen. *Polarforschung,* **48**, 151—169.

Scrutton, R.A. 1972. The structure of Rockall Plateau microcontinent. *Geophys. J.R. Astron. Soc.,* **27**, 259—275.

Smythe, D.K. 1983. Faeroe-Shetland Escarpment and continental margin north of the Faeroes. In: M.H.P. Bott *et al.* (eds.). *Structure and development of the Greenland-Scotland Ridge, new methods and concepts. Proc. NATO Advanced Res. Inst., Bressanone, May 1981.* Plenum, New York, pp. 109—120.

Smythe, D.K., Chalmers, J.A., Skuce, A.G., Dobinson, A. and Mould, A. 1983. Early opening history of the North Atlantic: 1. Structure and origin of the Faeroe-Shetland Escarpment. *Geophys. J.R. Astron. Soc.,* **72**, 373—398.

Steel. R.J., Dalland, A., Kalgraff, K. and Larsen, V. 1981. The Central Tertiary Basin of Spitsbergen: Sedimentary development of a sheared-margin basin. In: J.W. Kerr and A.J. Fergusson (eds.). Geology of the North Atlantic borderlands. *Can. Soc. Petrol. Geol., Mem.,* **7**, 647—664.

Sundvor, E. and Eldholm, O. 1976. *Marine Geophysical Survey on the Continental Margin from Bear Island to Hornsund, Spitsbergen.* Univ. Bergen. Seism. Obs., Sci. Rep., **3**, pp. 1—29.

Sundvor, E., Sellevoll, M., Gidskehaug, A. and Eldholm, O., 1978. Seismic investigation on the western and northern margin off Svalbard. *Polarforschung*, **48**, 41—43.

Surlyk, F., 1977. Mesozoic faulting in East Greenland, In: R.T.C. Frost and A.J. Dikkers (eds.), Fault Tectonics in NW Europe. *Geol. Mijnb.*, **56**, 311—327.

Surlyk, F., Callomon, J.H., Bromley, R.G. and Birkelund, T., 1973. The stratigraphy of the Jurassic—Lower Cretaceous sediments of Jameson Land and Scoresby Land, East Greenland. *Grønl. Geol. Unders., Bull.*, **123**, 1—56.

Surlyk, F., Clemmensen, L.B. and Larsen, H.C., 1981. Post-Palaeozoic evolution of the East Greenland continental margin. In: J.W. Kerr, and A.J. Fergusson (eds.), Geology of the North Atlantic Borderlands. *Can. Soc. Petrol. Geol., Mem.*, **7**, 611—646.

Talwani, M. and Eldholm, O. 1972. The continental margin off Norway. A geophysical study. *Geol. Soc. Am. Bull.*, **83**, 3575—3608.

Talwani, M. and Eldholm, O., 1977. Evolution of the Norwegian-Greenland Sea. *Geol. Soc. Am. Bull.*, **88**, 969—999.

Talwani, M., Mutter, J. and Eldholm, O., 1981. The initiation of opening of the Norwegian Sea. In: Geology of Continental Margins. *Suppl., Oceanol. Acta*, **4**, 23—30.

Vischer, A. 1943. Die postdevonische Tektonik von Ostgrönland zwischen 74°N and 75°N. Br., Kuhn Ø, Wollaston Forland, Clavering Ø und angrenzende Gebiete. *Meddr. Grønland*, **133**, 1—195.

Vogt, P.R., Feden, R.H., Eldholm, O. and Sundvor, E., 1978. The oceanic crust West and North of the Svalbard archipelago: synthesis and review of new results. *Polarforschung*, **48**, 1—19.

Vogt, P.R., Bernero, C., Kovacs, L. and Taylor, P., 1981. Structure and plate tectonic evolution of the marine Arctic as revealed by aeromagnetics. In: Geology of Continental Margins. *Suppl., Ocean. Acta*, **4**, 25—40.

Vonderbank, K., 1970. Geologie und Fauna der tertiären Ablagerungen ZentralSpitzbergens. *Norsk Polarinst. Skr.*, **153**, 1—119.

Ziegler, P.A. 1978. North-Western Europe: Tectonics and basin development. *Geol. Mijnb.*, **57**, 589—626.

Ziegler, P.A. 1982a. Geological Atlas of Western and Central Europe. (Shell Internationale Petroleum Maatschappij B.V.).

Ziegler, P.A. 1982b. Faulting and graben formation in western and central Europe. *Phil. Trans. R. Soc. Lond. Ser. A.* **305**, 113—143.

Øvrebø, O. and Talleraas, E., 1977. The structural geology of the Troms area (Barents Sea). *Geojournal*, **1**, 47—54.

13 Tertiary structural development of the western Barents Shelf: Troms to Svalbard

A.M. Spencer, P.C. Home* and L.T. Berglund*†
* BP Petroleum Development Ltd., Stavanger, Norway
† Norsk Hydro A/S, Bergen, Norway

Recent drilling in western Troms (7117/9-1) has yielded preliminary ages of the major, Tertiary wedge close to the continental margin at 71°N. Within the overall Tertiary wedge four main units are recognized. A widespread late Palaeocene to early Eocene unit (I) rests unconformably on older, faulted strata. Overlying strata of early Eocene to Oligocene ages (II) are developed thickly in the Tromsø Basin. A major westerly thickening and dipping prograding wedge (unit III) of ?Miocene to Pliocene age is developed west of the Senja Ridge, beneath the present continental slope. Plio-Pleistocene strata (unit IV) blanket the Barents Shelf widely and form a second westward-thickening wedge beneath the continental slope.

This major wedge of Tertiary strata underlies all of the margin west of the Barents Shelf from Troms (71°N) to Svalbard (79°N). To the west, ocean-floor spreading took place during Tertiary times, commencing prior to anomaly 24 (53 Ma, early Eocene) south of the Senja Fracture Zone. North from there the spreading history is known only tentatively. Transform movements between Greenland and Europe certainly took place and gave way to spreading in, probably, two stages: (1) between the Senja Fracture Zone and the Hornsund Fault Zone (72—74°N) at anomaly 21 (48 Ma, mid Eocene); (2) opposite the Hornsund Fault Zone (74—79°N) at anomaly 13 (37 Ma, early Oligocene).

The sequence in Troms agrees well in its ages and geometry with this tentative spreading history. Unit I mostly dates from just prior to the beginning of ocean spreading south of the Senja Fracture Zone (72°N). Unit II largely accumulated whilst spreading extended to 74°, and the main Tertiary wedge (Units III and IV) was formed after transform movements ceased and spreading extended along the whole margin to 79°N.

INTRODUCTION

The western margin of the Barents Shelf extends 1000 km, from 70°N to 79°N (Fig. 1). In the south, at the latitude of Tromsø (70°N), the continental shelf is less than 50 km wide from the outcropping Caledonian orogenic rocks to the most eastern oceanic crust. At 71°N, the Troms I licensed area has been explored for hydrocarbons since 1980. Nine deep wells have been drilled there so far and the region is covered with a dense grid of reflection seismic data. Further north to Svalbard (72—77°N), only outline, widely spaced reflection and refraction seismic data are available, complemented by knowledge of the sea-floor geology from dredging and from the outcrops on Bjørnøya (74°N). Western Svalbard, from 77° to 79°N, is a region affected by Tertiary orogenic movements in a zone lying west of the Central Tertiary sedimentary basin. The aim of this chapter is to relate the Tertiary history of this margin to the story of the Tertiary opening of the ocean to the west by, in particular, using the results obtained by a seismic stratigraphy study of the Tertiary in the west of Troms I and the subsequent drilling of the first well there (7117/9-1) in 1982.

In this study we use the geologic time scale of Harland *et al.* (1982), from which we have taken the calibration of the ages of the oceanic magnetic anomalies.

Petroleum Geology of the North European Margin © Norwegian Petroleum Society (Graham & Trotman, 1984) pp. 199—209.

Ocean-floor geology

Talwani and Eldholm (1977) have mapped ocean-floor magnetic lineations as old as anomaly 24 in the Norwegian-Greenland Sea, where today spreading is taking place at the Mohns Ridge (Fig. 1). This sector of the ocean basin terminates where the mid-ocean ridge changes to a northerly trend (Knipovich Ridge). A prominent fracture zone west of the ridge (Greenland Fracture Zone) is thought to be mirrored east of the ridge by its equivalent, the Senja Fracture Zone. But there

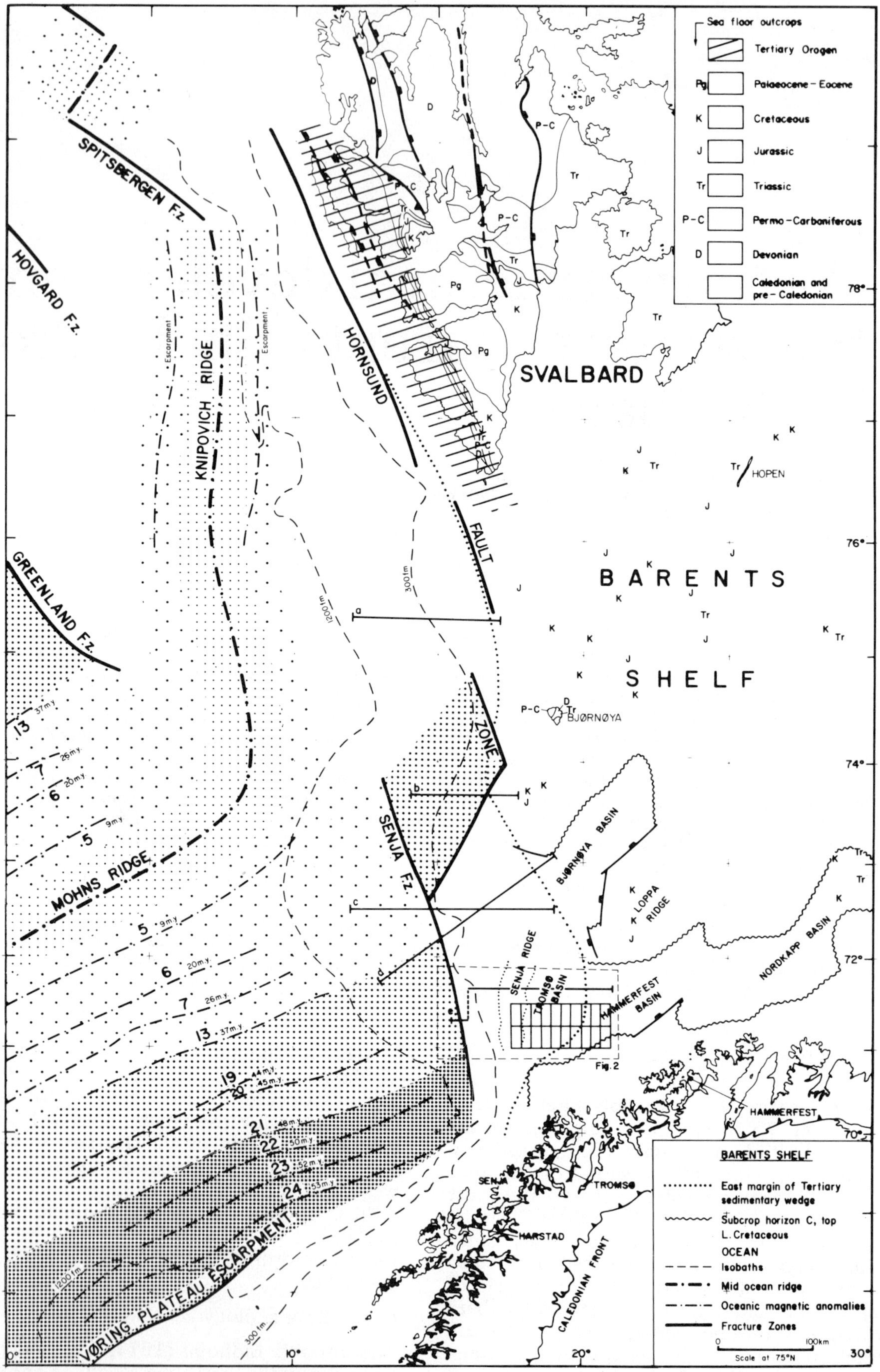

Fig. 1. Structural map of the western Barents Shelf. Data sources: ocean-floor features and bathymetry, Grønlie and Talwani (1978); Hornsund Fault Zone, Myhre *et al.* (1982); Barents Shelf sea-floor outcrops, Faleide and Gudlaugsson (1981, fig. 4.19); Barents Shelf Basin, Rønnevik *et al.* (1982, fig. 13).

the thick sediment pile on the continental slope obscures any topographic expression: the Senja Fracture Zone is revealed only by a gravity high (Talwani and Eldholm, 1972, p. 3592; 1977, p. 973). To the north of those two fracture zones, ocean-floor magnetic lineations have not been observed. In this area, Sundvor and Eldholm (1976) recognized the Hornsund Fault, a major fault or steep flexure trending-NNW and lying close to the shelf edge between Bjørnøya and Svalbard. They suggested that the Hornsund Fault marks the west boundary of the Svalbard-Barents Shelf. Myhre, *et al.* (1982) have proposed a detailed history of the shearing and spreading during Tertiary times of this northern sector between the Greenland—Senja and Spitsbergen fracture zones.

Margin of the Barents Shelf

From Troms to Svalbard, seismic refraction measurements have indicated the presence of a thick low-velocity sedimentary wedge (Eldholm and Ewing, 1971; Renard and Malod 1974; Sundvor, 1974; Houtz and Windisch, 1977; Eldholm and Talwani, 1977; Hinz and Schlüter, 1978). Eldholm and Talwani (1977) characterized this wedge as having refraction velocities of 1.8—2.4 km/s and suggested that it consisted of Tertiary strata deposited as a prograding wedge whilst the main part of the Barents Shelf to the east was undergoing erosion. A more detailed analysis of the velocity structure of the Tertiary wedge was given by Faleide and Gudlaugsson (1981, pp. 53—56).

Barents Shelf

The earlier seismic refraction and gravity studies had suggested the widespread occurrence of Mesozoic and Palaeozoic sedimentary rocks in several separate sedimentary basins. Bottom sampling (e.g. Edwards, 1975; Faleide and Gudlaugsson, 1981, p. 109) have confirmed that Mesozoic rocks crop out widely (Fig. 1). Regional reflection seismic data, especially that available to the Norwegian Petroleum Directorate, have allowed the mapping of these Mesozoic—late Palaeozoic basins (Rønnevik *et al.*, 1975; Øvrebø and Talleraas, 1976; Hinz and Schlüter, 1978; Marty *et al.*, 1979; Rønnevik, 1979, 1981; Faleide and Gudlaugsson, 1981; Rønnevik *et al.*, 1982). In the Troms area, Mesozoic strata are developed thickly in the Hammerfest, Tromsø and Bjørnøya Basins. Salt diapirs intrude the Mesozoic strata and in the Tromsø Basin penetrate Tertiary strata (Fig. 2). West of there, the Tertiary strata are arched in a N—S ridge, a feature named the Senja Ridge by Rønnevik *et al.* (1975).

TERTIARY WEDGE IN WEST TROMS 1 (71°N)

Tertiary seismic stratigraphy

The Tertiary sediments of the Senja Ridge and surrounding area have been divided into five seismic sequences of depositionally related strata, using the approach described by Mitchum *et al.* (1977). These have been grouped into three super-sequences (termed I, II and III from the oldest upwards), which are separated by major regional discontinuities (Fig. 3). Each of these discontinuities marks an important episode in the evolution of the Tertiary wedge. The Tertiary sediments in turn are overlain by a further depositional sequence (IV) of Plio-Pleistocene glacio-marine sediments, which form a thin sheet over much of the Barents Shelf.

The basal Tertiary unconformity is itself only mildly faulted but truncates a suite of pre-Tertiary fault blocks developed from the Senja Ridge to the edge of the shelf. The unconformity marks the change from pre-Tertiary shelf sedimentation to the accumulation of late Palaeocene bathyal mudstones and is a major hiatus.

The lowermost Tertiary unit (I) is relatively uniform in thickness, but does show some thicken-

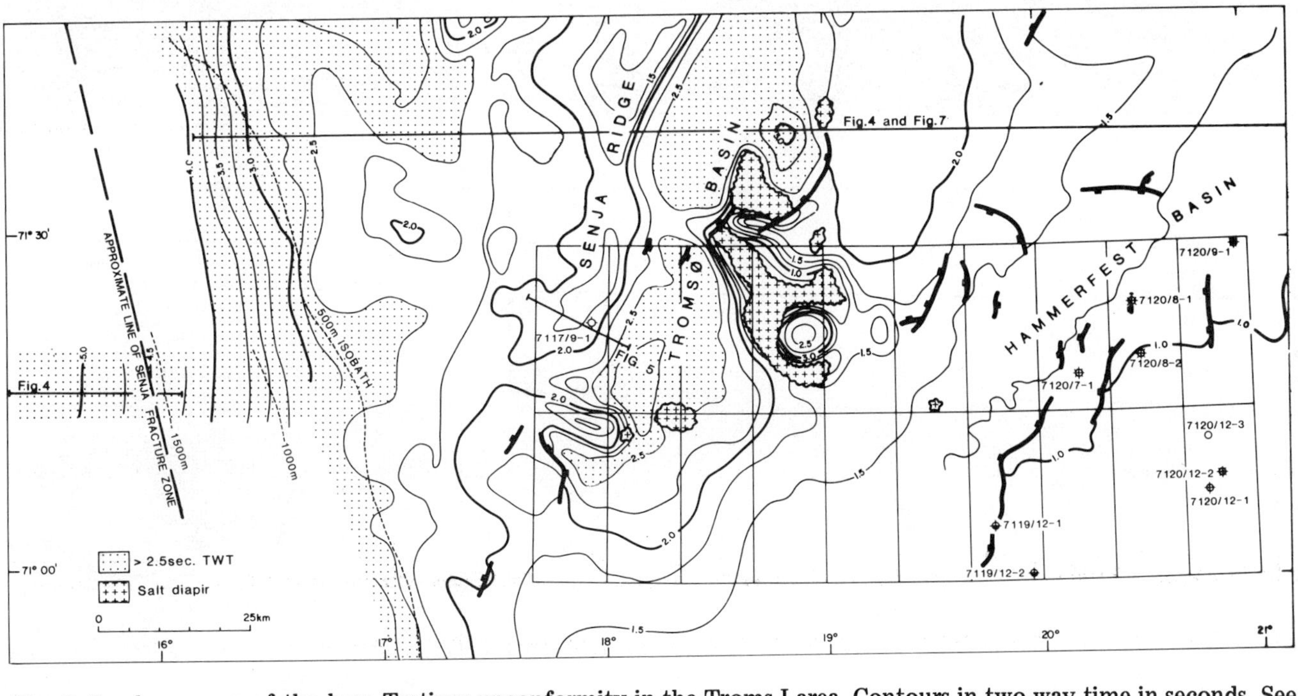

Fig. 2. Isochron map of the base Tertiary unconformity in the Troms I area. Contours in two-way time in seconds. See Fig. 1 for location.

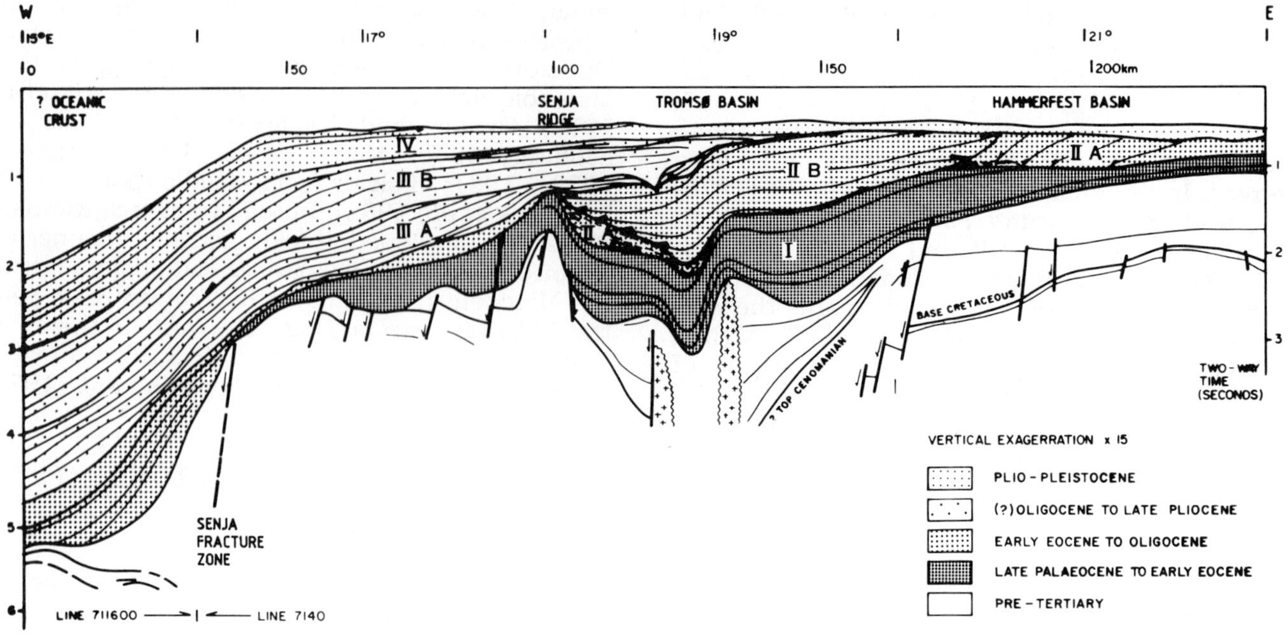

Fig. 3. Geoseismic profile across the western Barents Shelf in the Troms I area. Based on seismic lines 7140 and 711600; see Fig. 2 for location.

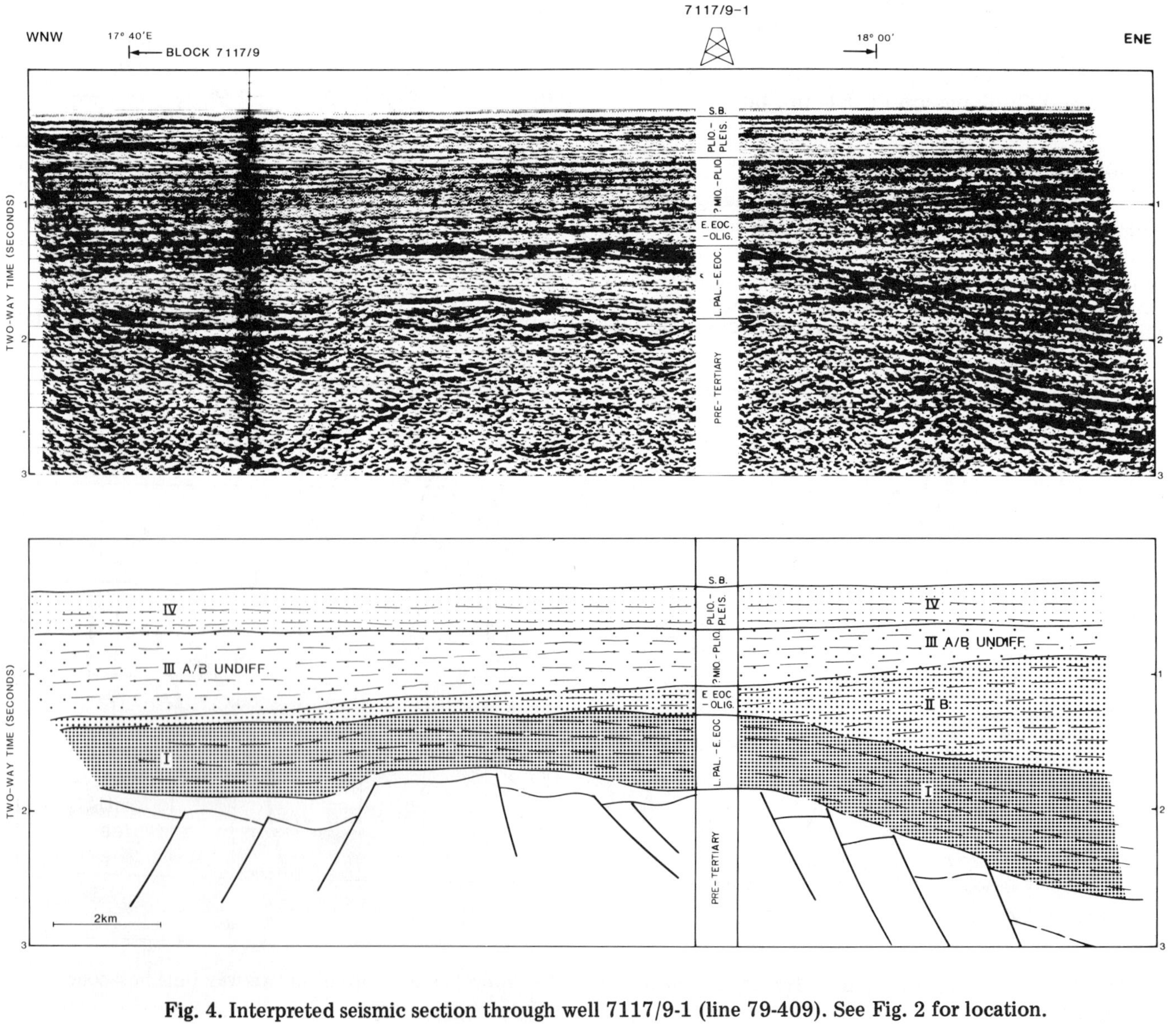

Fig. 4. Interpreted seismic section through well 7117/9-1 (line 79-409). See Fig. 2 for location.

ing in the Tromsø Basin relative to the Senja Ridge (Fig. 5a). This trend appears to be largely of depositional origin modified by minor erosion at the end of Unit I times. Internally, Unit I is characterized by parallel low-amplitude reflections which can be followed laterally over long distances. The lowermost reflections onlap the flanks of the Senja Ridge, which was overstepped during the early stages of Unit I deposition. These features are particularly well illustrated on seismic line 79—409 through well 7117/9-1, as is the angular discordance between Units I and II (Fig. 4).

Unit II shows a marked degree of thickening in the Tromso Basin (Fig. 5b). However, the lowermost of the two constituent sequences, Sequence IIA, is restricted to the Hammerfest Basin and the eastern flank of the Senja Ridge (Fig. 3). In the former area it has a progradational form, composed of westerly dipping clinoforms. It downlaps the upper surface of Unit I in a westerly direction and pinches out in the vicinity of the transition zone between the Hammerfest and Tromsø Basins. A possible stratigraphic equivalent of this prograding sequence is located on the eastern flank of the Senja Ridge. Here a thin lens of sediment, elongated in a N—S direction, infills a former depression on the present ridge flank. It has an irregular easterly directed prograding form and was probably derived from the erosion of Unit I over the crest of the ridge.

Sequence IIA is separated from IIB by a minor discontinuity, which in the Hammerfest Basin is marked by the onlap of the latter against the upper surface of IIA. Sequence IIB is largely restricted to the Tromsø Basin where it has a progradational profile, and continues the overall pattern of westerly progradation initiated at the beginning of Unit II.

Clinoforms and toplap can only rarely be distinguished, however, because of the seismically transparent nature of the interval. The progradation and thickening of this interval into the Tromsø Basin is combined with the pronounced downlap of the internal reflectors against the upper boundary of Units I and IIA in a westerly direction, which is especially marked on the eastern flank of the Senja Ridge (Fig. 4). Sequence IIB is relatively thin over the crestal part of the Senja Ridge. It is interpreted as a westerly progradational wedge with its maximum thickness in the Tromsø Basin and with rapid depositional thinning occurring above or just east of the eastern flank of the present-day Senja Ridge. The upper boundary dips towards the west, but shows only minor truncation of the internal reflections, although this is more pronounced towards the west. This dip is partly due to the depositional slope of the prograding system and partly due to post-depositional tilting.

The youngest of the Tertiary depositional units (Unit III) forms a westerly thickening and dipping prograding wedge of shelf and slope clastic sediments, which updip onlap the tilted upper surface of Unit II in an easterly direction (Figs. 3 and 4). Sequences IIIA and IIIB are most thickly developed on the continental slope and pinchout updip within the Tromsø Basin (Figs. 5c and 5d).

The boundary at the top of Unit II is, judging from the seismic record, a major unconformity. The overlying strata onlap the surface eastwards and thicken greatly in the west of the Troms I area. The overall geometry of the sequences above and below this surface and their internal reflection patterns show that this unconformity is a major hiatus in the area.

Dating in well 7117/9-1

Well 7117/9-1 penetrated representatives of all three Tertiary super-sequences (Fig. 4). However, the biostratigraphy of this well proved to be difficult to resolve owing to the poor and sporadic nature of the microfauna, discrepancies in the previously recognized ranges of the dinoflagellate cysts, and reworking of palynomorphs. The latter phenomenon is, we think, particularly marked in Units II and III, in which *in-situ* forms are sparse.

Unit I ranges in age from late Palaeocene (Thanetian) to within the early Eocene and consists exclusively of bathyal mudstones. Unit II is less confidently dated. Although thin, it appears to represent an interval of continuous deposition from early Eocene to intra-Oligocene, suggesting the sequence was deposited slowly and under starved conditions. We interpret Unit III as ranging in age from possibly Miocene to late Pliocene. In its lower half many Eocene forms are present, but in view of the major unconformity shown by our seismic stratigraphic study we interpret these as reworked. Unit IV is of Plio-Pleistocene age.

Ages of the unit boundaries

On a regional scale, Units I and II cannot differ significantly in age from their representatives dated in the well. This places the Unit I/II boundary in the early Eocene, whereas Unit III, due to its onlapping character and wedge-like form, may extend into the Oligocene downdip to the west. This would then place the Unit II/III boundary within the Oligocene. This suggestion is supported by the absence of Unit III in the interior of the Tromsø Basin and in the Hammerfest Basin, which implies that the boundary coincided with a major relative fall in sea level. This fall could be the intra-Oligocene event of Vail *et al.* (1977) at about 30 Ma.

Tertiary subsidence history

The early Palaeocene (Danian) marks a major hiatus in the area, associated with the truncation and submarine planation of the faulted pre-Tertiary strata. This was followed in late Palaeocene times by regional subsidence which just pre-dated the initiation of sea-floor spreading in the Norwegian—Greenland Sea (pre-anomaly 24 times). Subsequent sedimentation (Unit I) was characterized by the accumulation of parallel-bedded bathyal mudstones over the whole of the Troms I licence area, extending westwards towards the present shelf edge. The seismic character of Unit I suggests that the pre-Tertiary substrate was fairly flat-lying, although the Senja Ridge was probably a low-relief positive feature during the early stages of deposition. This phenomenon is exaggerated by a degree of syn-sedimentary differential subsidence in the Tromsø Basin (Fig. 3).

This pattern of sedimentation stopped in early Eocene times. This coincided with a regeneration of some of the pre-late Palaeocene faults on the Senja Ridge and the shelf to the west and resulted in the major positive expression of the ridge (Fig. 6).

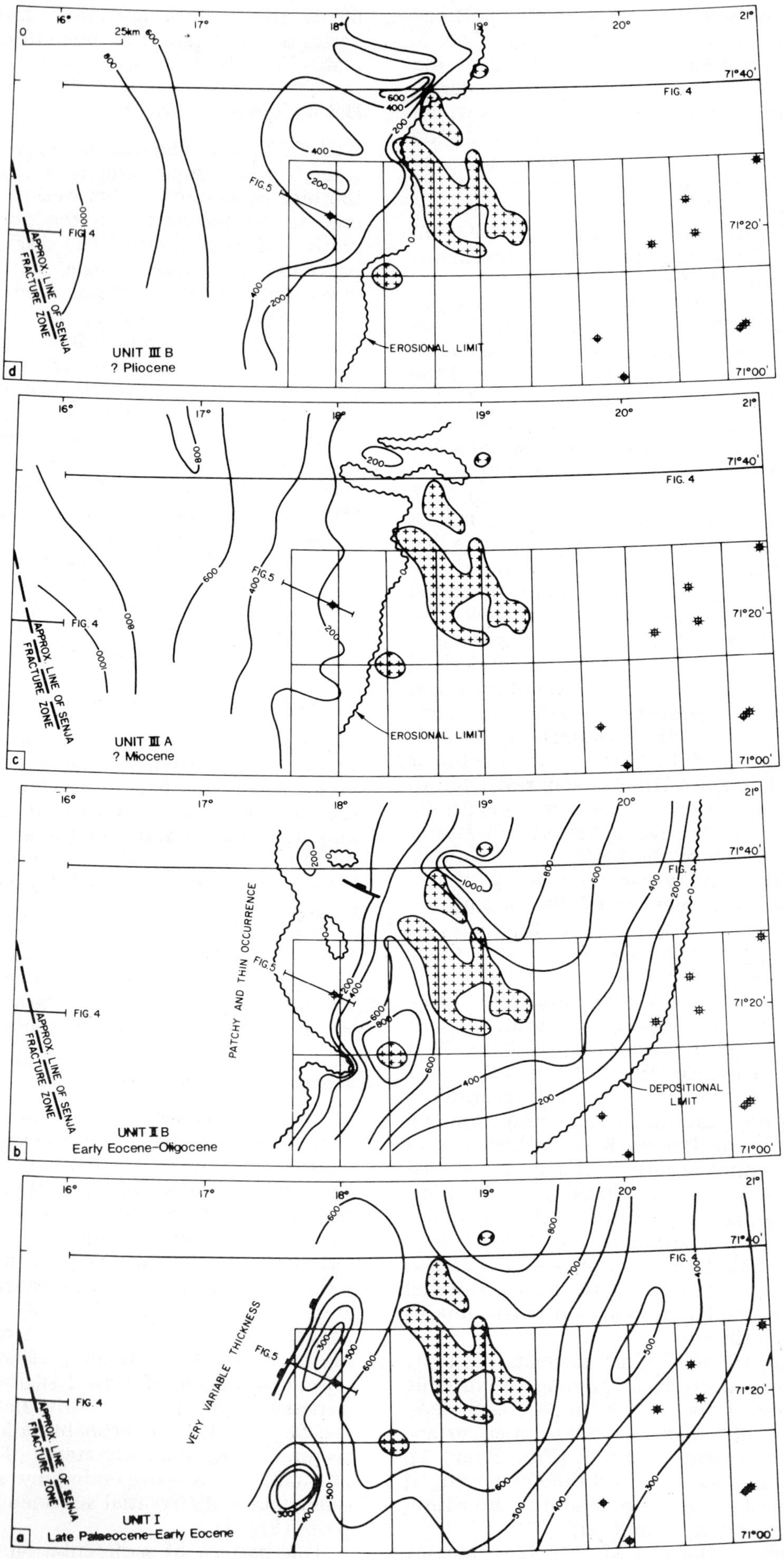

Fig. 5. Sketch isopach maps of the Tertiary depositional units. (a) Unit I. (b) Unit IIB. (c) Unit IIIA. (d) Unit IIIB.

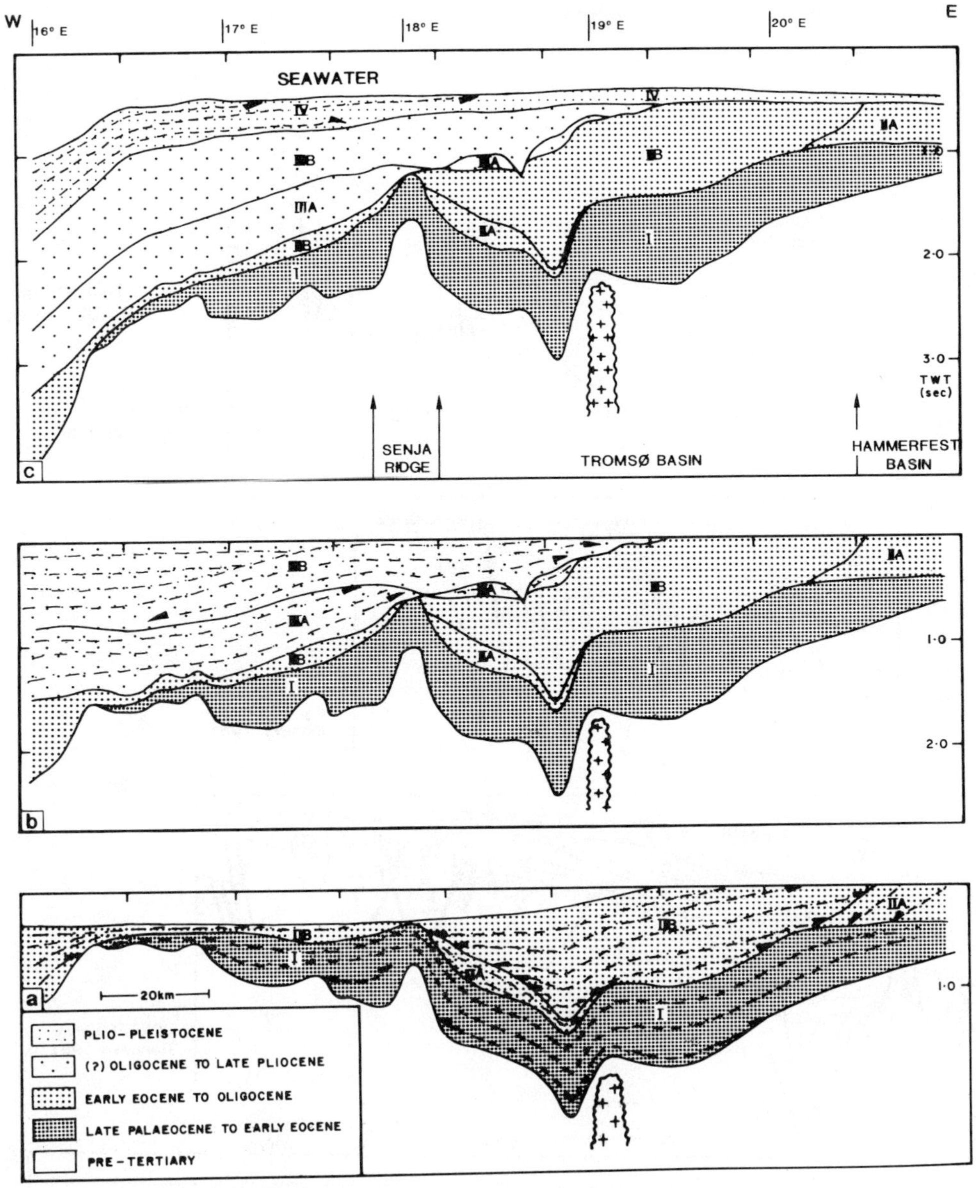

Fig. 6. Backstripped profiles of line 7140. (a) Post-Unit II, intra-Oligocene. Before the onset of tilting. Reconstructed to illustrate the depositional slope of the system. (b) Post-Unit III, late Pliocene. (c) Present. No decompaction correction or allowance for variations in water depth have been made.

Subsequent, early Eocene to intra-Oligocene, sedimentation was restricted largely to the Tromsø Basin and was characterized by the westerly progradation of shelf sediments. Sedimentation was accompanied by passive subsidence of the basin, in which the thickness of Sequence IIB exceeds 1 second of two-way time. The Senja Ridge was overstepped during the early part of the interval, following a period of minor erosion. Major halokinesis in the Tromsø Basin accompanied the later stages of deposition, resulting in the diapiric emplacement of probable late Palaeozoic salt into Mesozoic and Tertiary strata.

A second major change in the pattern of Tertiary sedimentation occurred during Oligocene times, resulting from the regional westerly tilting of the shelf edge along a hinge line in the centre of the Tromsø Basin. This produced the present deceptive symmetry of the Senja Ridge and the inversion of the edge of the shelf (Fig. 6b). Later Oligocene to late Pliocene (Unit III) sedimentation was restricted largely to the outer area of the continental shelf and slope, and resulted in the development of thick seaward thickening clastic wedges during continuing phases of tilting. However, the initial tilting coincided with a major relative fall in sea level.

The final phase of sedimentation during Quaternary times appears to have continued the pattern initiated in Oligocene times, although sediments of this age extend across the whole of the Barents Shelf.

TERTIARY WEDGE FROM TROMS TO SVALBARD (71–77°N)

The extension to the north of the history derived from the detailed studies in Troms can now be attempted. Figure 7 shows serial sections of the margin selected as, perhaps, representative from amongst those published by Houtz and Windisch (1977), Hinz and Schlüter (1978), Rønnevik (1981), Faleide and Gudlaugsson (1981), Rønnevik *et al.* (1982) and Myhre *et al.* (1982). The general character of the wedge is clear: it progrades westwards and everywhere is most thickly developed on the edge of the oceanic crust. The wedge is thicker

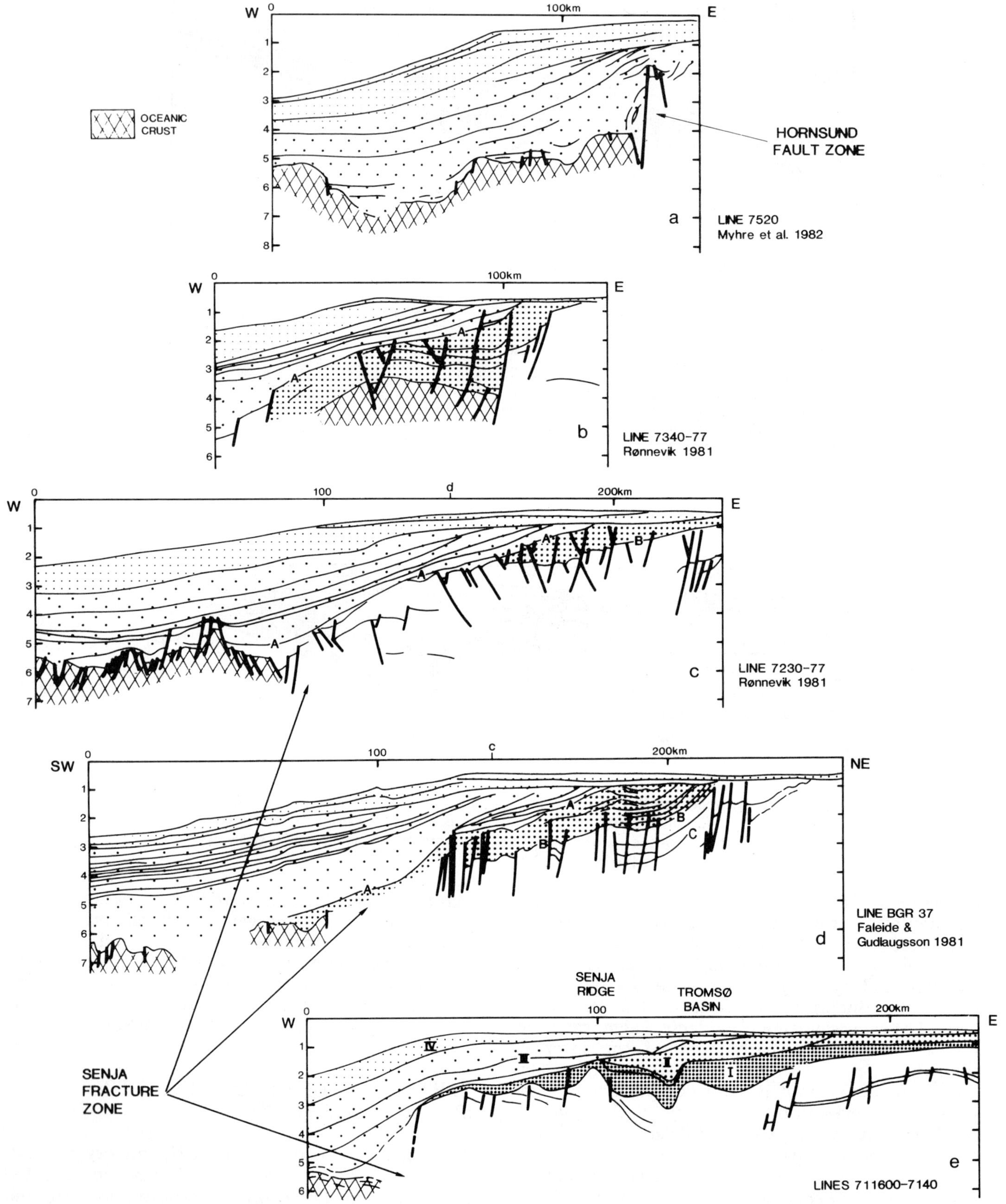

Fig. 7. Profiles of the continental margin from 71° to 75°N. See Fig. 1 for locations. Note that lines c and d cross and their intersection is shown.

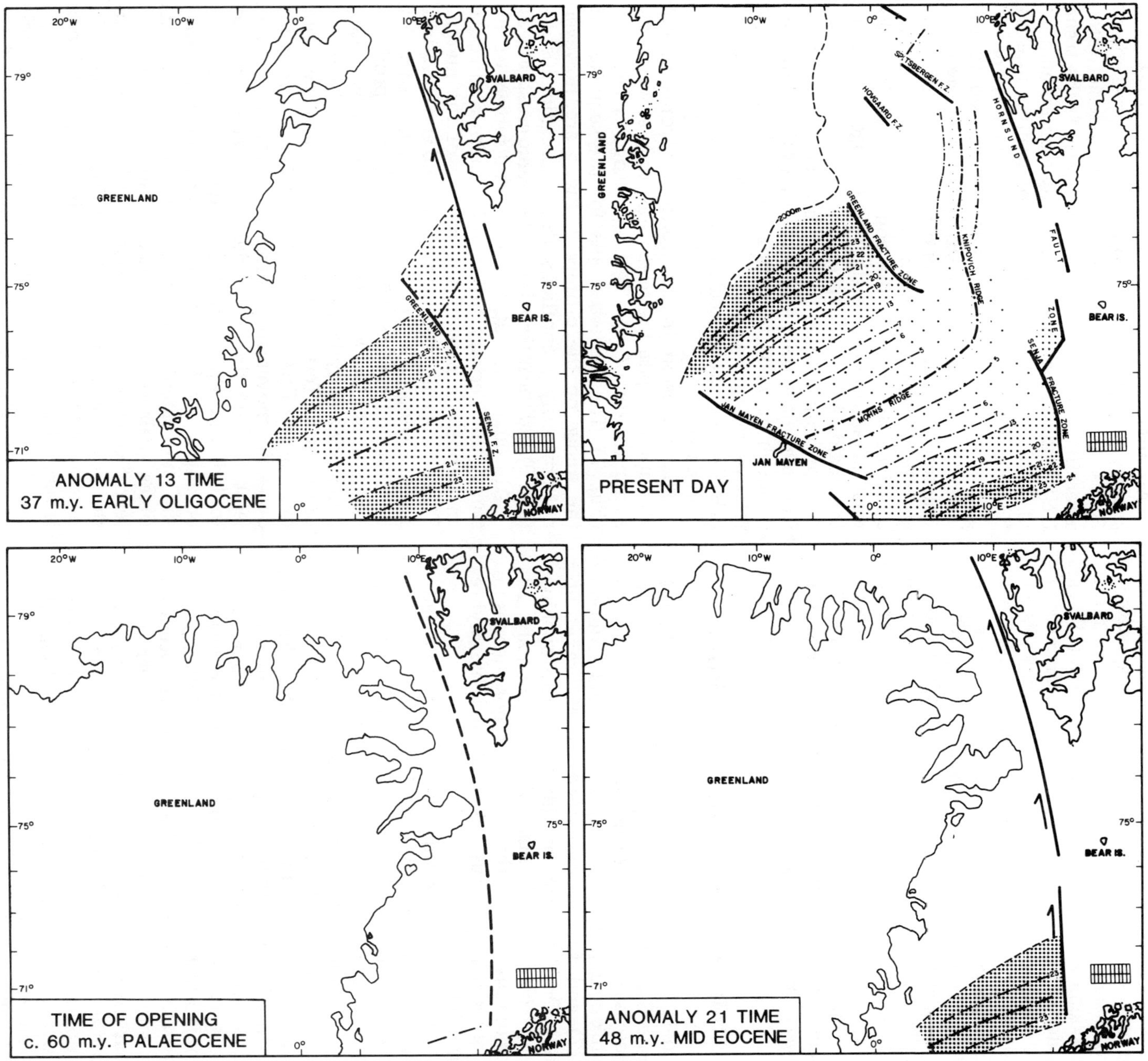

Fig. 8. Stages in the evolution of the Norwegian-Greenland Sea. In part modified from Talwani and Eldholm (1977, fig. 17) Myhre *et al.* (1982, fig. 7).

north from Troms, attaining 5 seconds TWT opposite the Hornsund Fault Zone (Myhre *et al.*, 1982, fig. 6). Lacking the original seismic sections, we are unable to tie the seismic lines shown on Fig. 7 together in the normal way. We can thus only suggest, on character, how the sequences seen on Figs. 7a—d relate to those on Fig. 7e; the diagram is ornamented to illustrate our suggestions.

To attempt more detailed inferences we must first examine the previous work. Rønnevik (1981), Faleide and Gudlaugsson (1981) and Rønnevik *et al.* (1982) had suggested ages for the Tertiary seismic sequence boundaries. For example, Rønnevik (1981) proposed that reflector A was either mid Oligocene or younger; that B was base Tertiary or mid Oligocene; and that C was mid Cretaceous or Base Tertiary (Fig. 7c). Following the results from the first drilling in East Troms, Rønnevik *et al.* (1982) gave slightly modified suggestions. Faleide and Gudlaugsson's (1981) suggestions were: A, Eocene; B, base Upper Palaeocene; C, near top Cretaceous (Fig. 7d).

TERTIARY EVOLUTION OF THE MARGIN

The detailed history derived by using well 7117/9-1 to date the seismic stratigraphy Units I—IV in west Troms is in good agreement with the story of ocean opening proposed by Talwani and Eldholm (1977) and further developed by Myhre *et al.* (1982). On Fig. 8 we have illustrated this opening history, choosing anomaly 21 and anomaly 13 times to illustrate the times of the major changes in the spreading patterns, as suggested by Myhre *et al.* (1982).

(1) Spreading was initiated prior to anomaly 24 time (53 Ma, early Eocene) in the region south of the Greenland—Senja Fracture Zone (about 70—72°N). To the north, movements were taken up by right-lateral strike-slip movements between the Greenland and North European plates. All of the west margin of the Barents Shelf underwent these strike-slip movements.

(2) Later this pattern changed and the spreading ridge extended to the north of the Greenland—Senja Fracture Zone. Myhre *et al.* (1982, p. 46) tentatively suggested that this change occurred at anomaly 21 time (48 Ma, mid Eocene) and that spreading then extended as far north as the southern limit of the Hornsund Fault Zone (approximately 74°N). Further north strike-slip movements continued, probably along the Hornsund Fault Zone.

(3) The last major change in the opening pattern took place at anomaly 13 time (37 Ma, early Oligocene) when strike-slip movement ceased and was replaced (to the north of 74°N) by spreading.

The timing of the changeover from (1) to (2) is, as Myhre *et al.* (1982) indicated, tentative, but, as we can show, is supported by our analysis of the history of the Tertiary wedge in West Troms. Our Unit I (late Palaeocene—early Eocene) records the widespread, general subsidence of the margin accompanying the earliest phase (1) of spreading south of the Greenland—Senja Fracture Zones. In the second phase (2) of the opening history, spreading extended much further north and this is reflected in west Troms by a change in deposition: Unit II (early Eocene—Oligocene) is largely restricted to the Tromsø Basin. The final phase (3) of the opening history, when spreading occurred west of the whole Barents Shelf margin, equates to the major change to the westward thickening and dipping Unit III (?Miocene—Pliocene).

An interesting observation at this point is that in the west Troms I area some of the pre-Tertiary faults appear to have been selectively re-activated after the deposition of Unit I. These faults trend NNE and are associated with only a minor amount of displacement. This renewal of faulting in early Eocene times may have developed in response to the episode of crustal stretching related to the initiation of sea-floor spreading between the Senja Fracture Zone and Hornsund Fault Zone. This may also have resulted in isostatic readjustment and uplift of the edge of the continental shelf to the west of and including the Senja Ridge, leading to the major positive expression of the latter. The change from bathyal to shelf sedimentation at this time may have been a direct result of the tectonism, although it may in part be related to a major relative fall in sea level identified by Vail *et al.* (1977) on a global scale at about this time ('mid' Ypresian).

The extension to the north of Troms of our attempt to relate the ocean-spreading story to the accumulation of the Tertiary wedge is less clear (Fig. 7). The suggested ages for the sequences in the wedge on the profiles in Fig. 7 are, however, not obviously in conflict with the spreading history (Fig. 8) and the inferred ages of oceanic crust (Fig. 1).

Although it is outside the central subject of this study, the Tertiary history of western Svalbard is an important part of the history of the margin. There, orogenic movements (Harland and Horsfield, 1974; Kellogg, 1975) affected a zone at least 50-km wide stretching from 76° to 79°N and produced major uplift (>4 km) relative to the approximately 2-km thick early Palaeocene to Eocene sequence of the Central Tertiary Basin. The Palaeocene strata provide the first evidence of uplift to the west and orogenic movements certainly continued into at least Eocene times, for the Palaeogene strata are folded and thrusted. This agrees well with the oceanic story, for opposite western Svalbard, strike-slip movements occurred from pre-anomaly 24 times (Palaeocene) through to anomaly 13 times (early Oligocene). The contrast between the strike-slip/orogenic nature of the margin here and the passive extensional nature of the margin at Troms is very clear.

It is interesting to note that Bjørnøya lies only about 50 km from the southern section of the Hornsund Fault. The strata on the island are Devonian to Triassic in age, but they have not been affected by Tertiary orogenesis: a fact which is not in conflict with the ocean-opening story recounted here.

A further corollary of our seismic stratigraphic analysis of west Troms I concerns the origin of the Senja Ridge in Tertiary times. The earliest suggestion (Rønnevik *et al.*, 1975, p. 127) was that it was analogous to the West Spitsbergen orogen and could be related to the early shear of the strike-slip movements between the Greenland and the North European plates. This suggestion is not correct. The tectonic structure of the Senja Ridge at Tertiary levels is quite unlike that of the West Spitsbergen orogen. Even its gentle anticlinal symmetry at Tertiary levels is misleading, caused as it is by early (Unit II, early Eocene to Oligocene) subsidence of the east flank, followed by later (Unit III, ?Oligocene to Pliocene) subsidence of the west flank. Having this extended composite history, it long post-dates the cessation of transform movements on the adjacent Senja Fracture Zone.

ACKNOWLEDGEMENTS

The seismic stratigraphy results described here were found in a study by John D. Roseway (BP). The authors wish to thank Statoil, Norsk Hydro, Total, Saga and BP — the companies in the block 7117/9 licence group — for kindly releasing the information which forms the basis of this chapter.

REFERENCES

Edwards, M.B. 1975. Gravel fractions on the Spitsbergen Bank, NW Barents Shelf. *Nor. Geol. Unders.*, **316**, 205—517.

Eldholm, O. and Ewing, J. 1971. Marine geophysical survey in the southwestern Barents Sea. *J. Geophys. Res.*, **76**, 3832—3841.

Eldholm, O. and Talwani, M. 1977. Sediment distribution and structural framework of the Barents Sea. *Bull. Geol. Soc. Am.*, **88**, 1015—1029.

Faleide, J.I. and Gudlaugsson, S.T. 1981. Geology of the Western Barents Sea — A Regional Study based on Marine Geophysical Data. Cand. Real. Thesis in Applied Geophysics, Dept. of Geology, University of Oslo, 160 pp.

Grønlie, G. and Talwani, M. 1978. *Geophysical Atlas Norwegian Greenland Sea.* Vema Research Series, Vol. IV. Lamont-Doherty Geological Observatory.

Harland, W.B. and Horsfield, W.T. 1974. The West Spitsbergen orogen. In: A.M. Spencer (ed.), *Mesozoic—Cenozoic Orogenic Belts.* Geological Society, London, pp. 747—755.

Harland, W.B., Cox, A., Llewellyn, P.G., Pickton, C.A.G., Smith, A.G. and Walters, R. 1982. *A Geological Time Scale.* Chart. Cambridge University Press, Cambridge.

Hinz, K. and Schlüter, H.U. 1978. The geological structure of the western Barents Sea. *Mar. Geol.*, **26**, 199—230.

Houtz, R. and Windisch, C. 1977. Barents Sea continental margin sonobuoy data. *Bull. Geol. Soc. Am.*, **88**, 1030—1036.

Kellogg, H.E. 1975. Tertiary stratigraphy and tectonism in Svalbard and continental drift. *Bull. Am. Ass. Petrol. Geol.*, **59**, 465—485.

Marty, A., Rabate, J., Cazes, Jacquart, G. and Auxietre, J.L. 1979. CEPM Barents Sea seismic survey. In: *Norwegian Sea Symposium* (Norsk Petroleumsforening), Article 16.

Mitchum, R.M., Jr., Vail, P.R. and Thompson, S. 1977. Seismic stratigraphy and global changes of sea level, Part 2: The depositional sequence as a basic unit for stratigraphic analysis. *Mem. Am. Ass. Petrol. Geol.*, **26**, 53—62.

Myhre, A.M., Eldholm, O. and Sundvor, E. 1982. The margin between Senja and Spitsbergen fracture zones: implications from plate tectonics. *Tectonophysics*, **89**, 33—50.

Øvrebø, O. and Talleraas, E. 1976. The structural geology of the Troms area. In: *Offshore North Sea* (Norsk Petroleumsforening), Article G-IV/6.

Renard, V. and Malod, J., 1974. Structure of the Barents Sea from seismic refraction. *Earth Planet. Sci. Lett.*, **24**, 33—47.

Rønnevik, H.C. 1979. Geology of the Barents Sea. In: *Norwegian Sea Symposium* (Norsk Petroleumsforening), Article 15.

Rønnevik, H.C. 1981. Geology of the Barents Sea. In: L.V. Illing and G.D. Hobson (eds.). *Petroleum Geology of the Continental Shelf of North-West Europe.* Heyden, London, pp. 395—406.

Rønnevik, H.C., Bergsager, E., Moe, A., Øvrebø, O. and Navrestad, T. 1975. The geology of the Norwegian Continental Shelf. In: A.W. Woodland (ed.), *Petroleum and the Continental Shelf of North-West Europe,* Applied Science Publishers, London, pp. 117—129.

Rønnevik, H.C., Beskow, B. and Jacobsen, H.P. 1982. Structural and stratigraphic evolution of the Barents Sea. *Offshore Northern Sea Symposium,* Stavanger, August 1982.

Sundvor, E. 1974. Seismic refraction and reflection measurements in the southern Barents Sea. *Mar. Geol.*, **16**, 255—273.

Sundvor, E. and Eldholm, O. 1976. Marine geophysical survey on the continental margin from Bear Island to Hornsund, Spitsbergen *Sci. Rep. Seismol. Obs. Bergen,* **3**.

Talwani, M. and Eldholm, O. 1972. Continental margin off Norway: a geophysical study. *Bull. Geol. Soc. Am.*, **83**, 3575—3606.

Talwani, M. and Eldholm, O. 1977. Evolution of the Norwegian-Greenland Sea. *Bull. Geol. Soc. Am.*, **88**, 969—999.

Vail, P.R., Mitchum, R.M. Jr. and Thompson, S. 1977. Seismic stratigraphy and global changes of sea level, Part 4; Global cycles of relative changes of sea level. *Mem. Am. Assoc. Petrol. Geol.*, **26**, 83—97.

14 The continental margin/crustal structure off the Lofoten Islands, Northern Norway

G. Drivenes, M.A. Sellevoll,* V. Renard,† F. Avedik† and J. Pajchel‡*
*Seismological Laboratory, University of Bergen, Norway
† Centre Océanologique de Bretagne, Plouzane, France
‡ Polish Academy of Science, Warsaw, Poland

INTRODUCTION

In order to study the crustal transition zone in the Vøring Plateau area and off the Lofoten Islands, Northern Norway, a series of long seismic refraction profiles were shot as part of a co-operative research programme between the Centre Océanologique de Bretagne at Brest, France and the Seismological Observatory at the University of Bergen, Norway. The experiment was a combined land—sea operation during which 19 ocean-bottom seismograph stations (OBS) and four land stations were utilized. The main energy source was a prototype air-gun with a volume of 45 l and service pressure of 250 bars. The marine part of the programme was carried out by the French research vessel *Jean Charcot*.

Figure 1 shows the seismic profiles and the OBS and land stations during various stages of the project. The shooting interval was 2 minutes at a ship's speed of 5 knots, giving a distance between shots of about 300 m along the profile lines. The instrumentation and methods used at sea are described by Avedik *et al.* (1978). The land stations employed three component stations of the Mars 66 type.

Several unknown parameters, including the maximum range obtainable with the air-gun, rendered the design of the experiments difficult. The profile lines were mainly placed along or perpendicular to the main structural trend of the margin.

DATA

All field data have been converted to digital form and merged with navigational data. T—X seismic sections have been computer-plotted for all profiles. Figure 2 is an example of a deep-water profile in an L-shaped pattern (OBS station 3), and Fig. 3 is an example of a record-section obtained at OBS station 10E, located in shallow water.

A study of the OBS-record sections shows that the signal-generated noise level is much higher in deep water than in shallow water. An explanation of this may be that the energy loss in water multiple reflections is higher in shallow water than in deep water, simply because of the larger number of bottom reflections as a function of time. This may explain why the recording range in deep water is limited to about 80 km, while on an OBS shelf station and a land station, P_n phases have been recorded at distances of 180 km and 220 km, respectively.

A major problem concerning the interpretation of the OBS and land-station data has been the lack of information about the sediments along the profile lines (Fig. 1). Profiles 10 and 11 were, however, located along a multichannel seismic reflection line, shot in 1974 and described by Eldholm *et al.* (1979). Eldholm *et al.* (1979) also published 11 sonobuoy profiles distributed along profiles 10 and 11. These earlier results concerning the sediments have made it possible to interpret the OBS and land-station (11A) data with respect to the deeper crustal structure. Profile lines 10 and 11, together with part of profile 5, constitute a seismic transverse across the continental margin off Lofoten. This chapter deals with the interpretation and modelling of the data obtained at OBS stations 11B, 11C, 10C, 10E, 5C and land station 11A (Hamnøy).

THE UPPER CRUSTAL STRUCTURE

Figure 4 gives the result of combining information based on mechanical seismic data and sonobuoy measurements, as well as OBS data. The dominant P-velocities in the sediments along the deep oceanic

Petroleum Geology of the North European Margin © Norwegian Petroleum Society (Graham & Trotman, 1984) pp. 211—216.

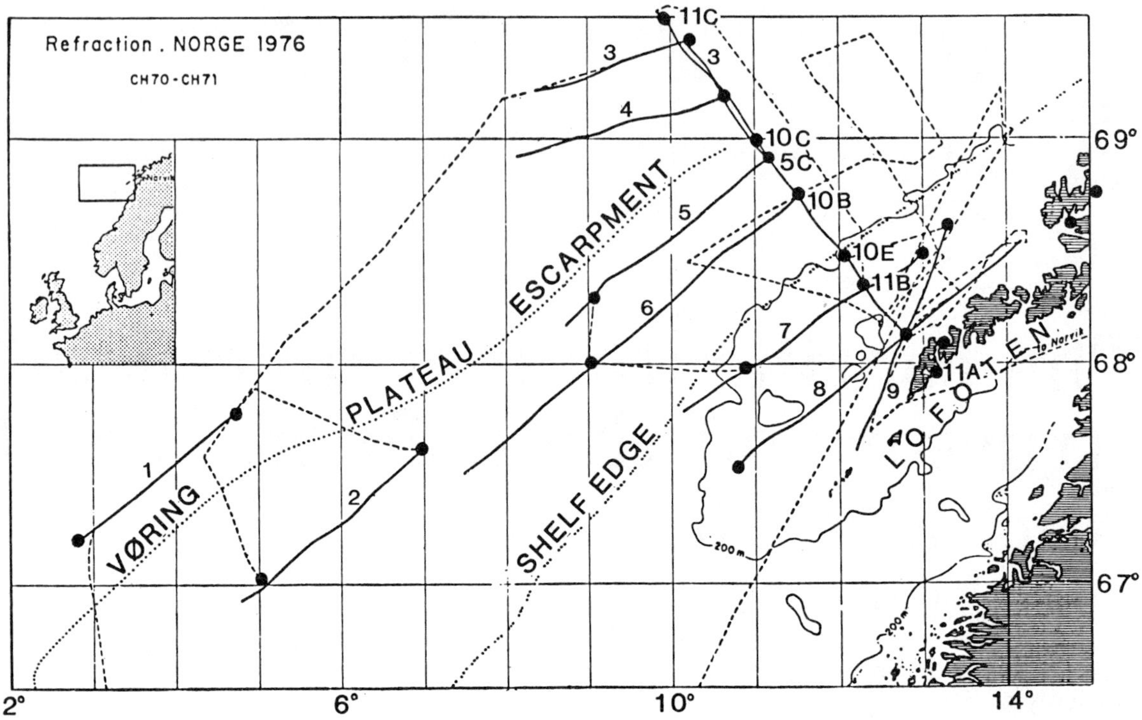

Fig. 1. Location map of seismic profiles.

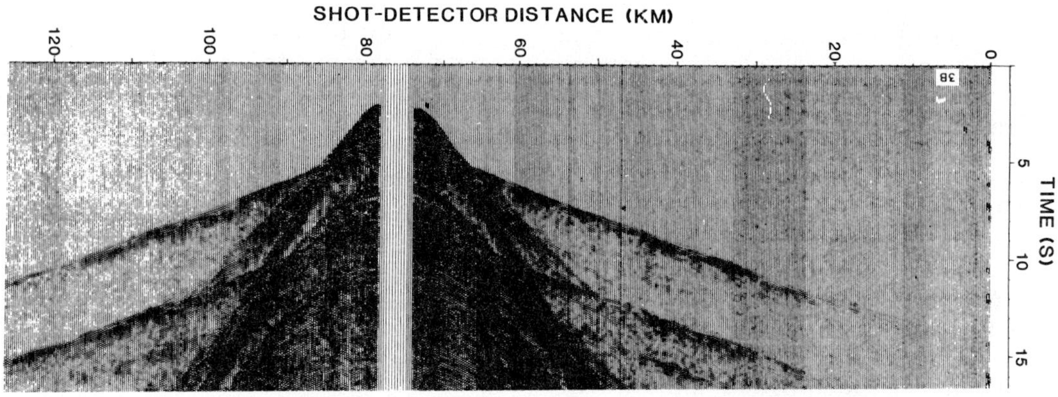

Fig. 2. T—X section of Lshaped (profile 3), deep water profile.

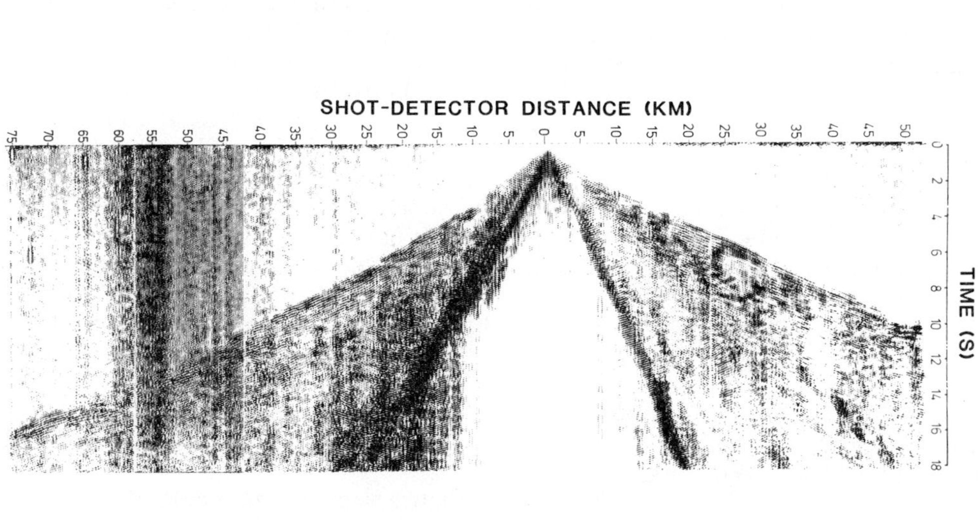

Fig. 3. T—X section for the shallow-water station 10E.

part of the profile are 1.9 km/s and 2.6 km/s. The 5.2 km/s P-velocity is interpreted as originating in the uppermost part of a basaltic oceanic crust. The OBS data indicate a lateral change in velocity from 5.2 to 3.9 km/s on the landward side of the sub-basement reflectors and the Vøring Plateau Escarpment (VPE). It is likely that the 3.9-km/s layer beneath the continental slope represents a sedimentary sequence which is also present beneath the shelf. The available data from the shelf area are, however, not sufficient to elaborate the structure of the sediments and the corresponding velocities along this part of the profile. The velocity of 3.55 km/s (Fig. 4) is the average P-velocity obtained above the crystalline basement.

As a first step in calculating a preliminary model for the crust along profiles 10 and 11, all primary and secondary onsets recorded on the OBS stations and the land station were read and plotted into a time-distance diagram. A phase-correlation was carried out, and the parallelity method employed to find travel-time branches which may have originated from common discontinuities. By

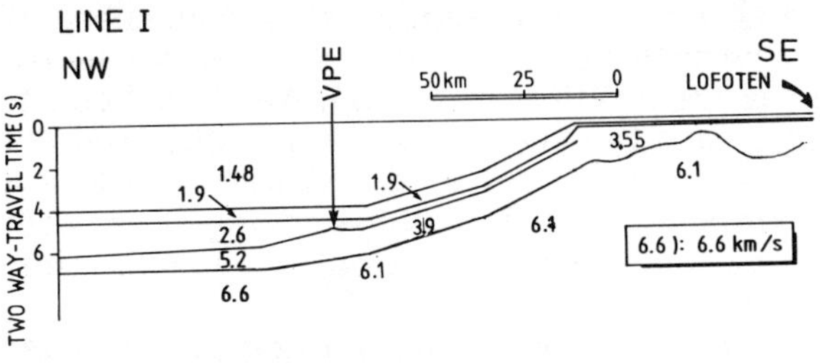

Fig. 4. Model of uppermost crust.

application of this approach, information was obtained about the number and approximate location of crustal discontinuities.

Figure 5 shows the identification of the travel-time branches and the corresponding preliminary crustal model. The thickness of the sedimentary layer increases rapidly westwards from Lofoten (3.5 km). Within the central part of the shelf area a sedimentary layer about 1 km thick is observed, increasing in thickness to 3.5 km further to the west beneath the outer shelf and along the con-

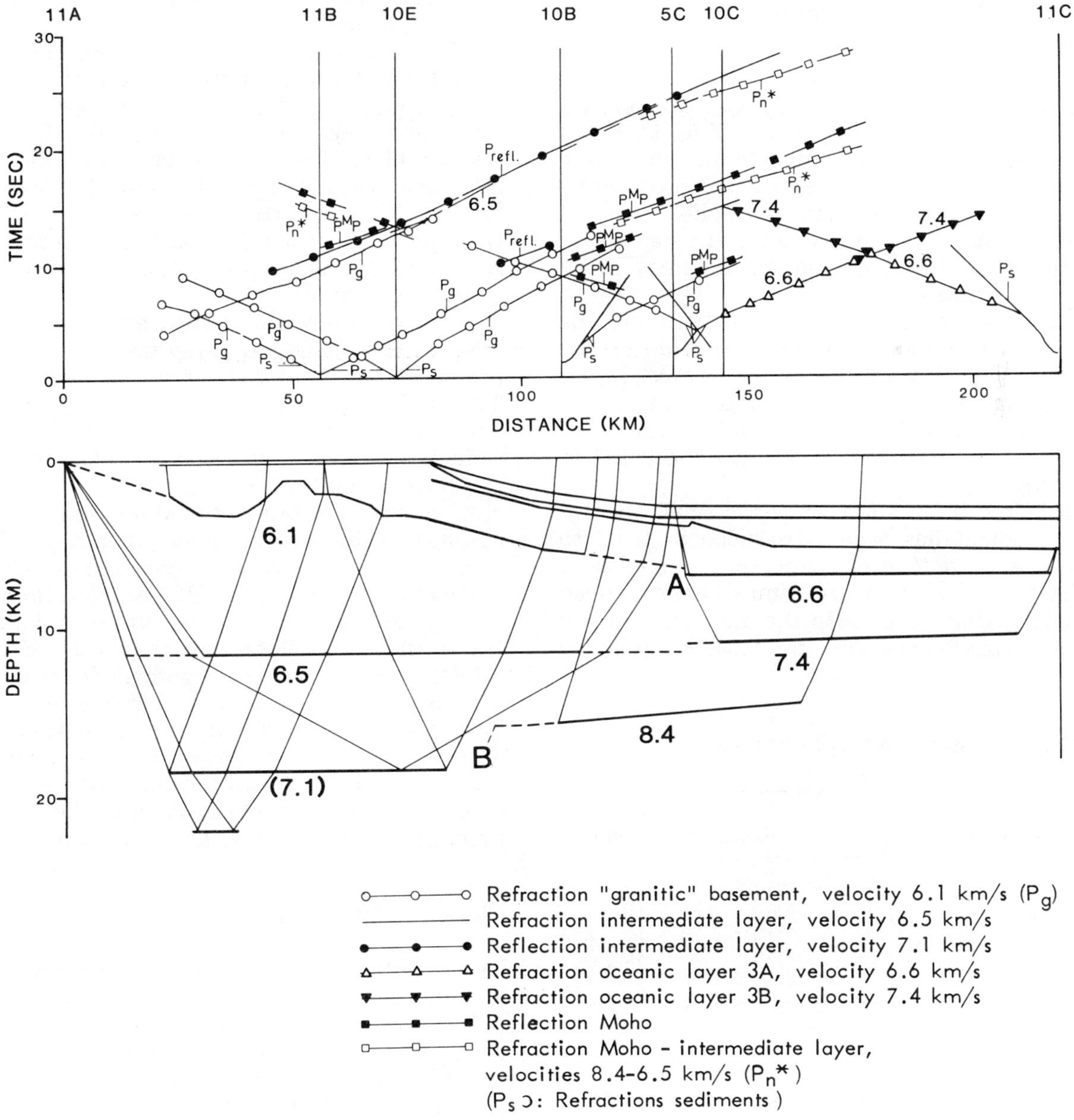

Fig. 5. Identification of the travel-time branches and the corresponding crustal model (preliminary).

tinental slope. The sedimentary sequences thin again near the escarpment (in the area where the sub-basement reflectors occur). An almost constant sedimentary thickness of 2.5 km is observed along the outermost 60 km of the profile. The P_g-phase (6.1 km/s) is related to the uppermost crystalline crust within 21—130 km from the land station at Hamnøy (11A).

A P-velocity of 6.6 km/s is observed at the depth of 7 km on the oceanic side of the escarpment. This velocity is assumed to be associated with oceanic layer 3A. Another P-phase (6.5 km/s) seems to have originated at a discontinuity on the landward side of VPE at a depth of 11.5 km. Below this interface two additional discontinuities are observed: one at a depth of 18.5 km (P_{refl}) and the other at a depth of 22 km (P^MP), defined as the Moho. Based on dynamic considerations, a P-phase velocity of 7.1 km/s is estimated for the top of the layer at 18.5 km depth. A 7.4 km/s P-velocity is observed on the oceanic side of the escarpment at a depth of 10—11 km, and the associated layer is assumed to be oceanic layer 3B.

Modelling of the crustal structure

The preliminary model (Fig. 5) is based primarily on data obtained at stations 11A, 11B, 11C, 10E and 5C. The recordings from stations 10B and 10C have been used with the assumption that the P_g-phase velocity is equal beneath the slope and shelf. By application of a ray-tracing programme (Pajchel, 1983) the preliminary model (Fig. 5) has been iteratively modified until the calculated travel times fit the observed travel times at all OBS and land stations.

The seismic waves propagating between the continental and oceanic part of the profile can be explained as head waves or diving waves. It has been impossible to model continuous interfaces from the continental to the oceanic region. For this reason, the refracted waves are modelled as diving waves.

From below the 6.1-km/s layer, a small positive velocity gradient has been introduced in order to explain the wave transmission as diving waves. The difference in the travel times between head waves and diving waves with the gradient applied, is without significance for the crustal modelling (Table 1).

TABLE 1. P-wave velocities applied in crustal modelling

Velocity for head-waves (km/s)	*Velocity for diving-waves (km/s)*
6.1	6.1
6.5	$6.47 + 8.55 \times 10^{-3}$ Z[a]
6.6	$6.60 + 5.55 \times 10^{-3}$ Z
7.1	$7.10 + 5.25 \times 10^{-3}$ Z
7.4	$7.40 + 5.25 \times 10^{-3}$ Z
8.4	$8.40 + 6.65 \times 10^{-3}$ Z

[a] Z = depth (km)

Two regions, A and B on Fig. 5, were particularly problematic during modelling:

(A) The region where the seismic velocities change from 'continental' to 'oceanic' velocities.

(B) The region with a strong change in depth to the Moho.

Modelling in region A can be carried out under different assumptions:

(1) The lateral velocity-change takes place at a first-order vertical discontinuity.

(2) The velocity change is associated with a lateral velocity gradient.

(3) The velocity change is related to an atypical crustal 'block' between the two adjacent crustal regions.

Assumption (3) can be ruled out on the basis of the data recorded at stations 10B and 10C. Modelling by introducng lateral gradients (assumption 2) shows that it is possible to obtain a good fit between the observed and the calculated travel times. An unique solution cannot, however, be obtained for either the width of the transition zone or the value of the velocity gradient. The fit between the model and experimental data is also good for the simple-transition model (assumption 1). A first-order vertical discontinuity, therefore, has been chosen as a model for the transition between the oceanic and the continental crust. Differences in travel time for assumptions 1 and 2 are below the resolution of the survey.

The escarpment is situated at a distance of 137 km from Hamnøy, and a first-order lateral discontinuity located at this distance shows a good fit with the OBS recordings, as well as the recordings at land station 11A. The 'granitic' layer (6.1 km/s) passes into the oceanic layer 3A and the intermediate layer (6.5 km/s) into the oceanic layer 3B. Such a transition is probable since the depth to the respective layers is almost equal on opposite sides of the escarpment.

The geometrical configuration of region B (Fig. 5) is modelled in a way which explains the P_n onsets at 11A and 11B as well as the location of 10E within a shadow-zone for the P_n phase. The depth of the Moho is determined by fitting P^MP reflections from stations 10B, 10C and 10E. The P_n-phase velocity is calculated to be 8.4 km/s by combining the depth to the Moho and the apparent P_n-phase velocities observed at stations 11A and 11B. A strong change in the depth to the Moho takes place at a distance of 92—94 km from Hamnøy.

The good fit between the observed data and the travel times calculated on the basis of the final model (Fig. 6) is a strong argument for the reliability of our results. The modelled travel-time branches have been plotted directly on the seismic sections. Two of the plots are shown in Figs. 7 and 8. The results obtained explain the most significant records at all stations, and none of the modelled travel-time branches contradict one another (Drivenes, 1983).

FINAL REMARKS

The results presented are a summary of some aspects of a continuing study of the crustal structure of the continental margin off Lofoten. Even if very good agreement has been obtained between the observed and the travel times calculated from the model, the work so far has been hampered by lack of information on the sedimentary sequences on the margin. In order to obtain more information about the sediments, 2500 km of multichannel

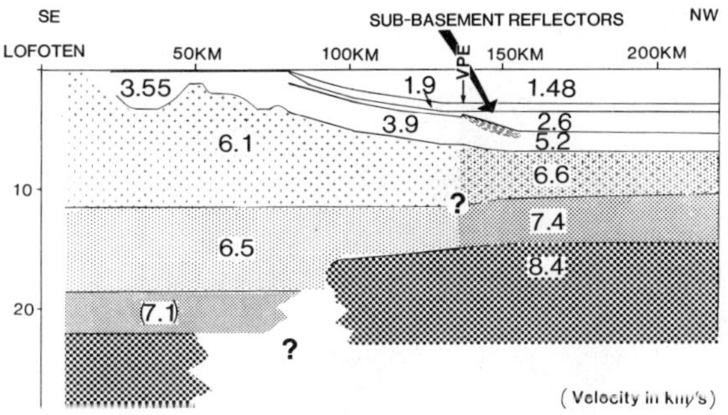

Fig. 6. Crustal-structure section across the continental margin off Lofoten, Northern Norway.

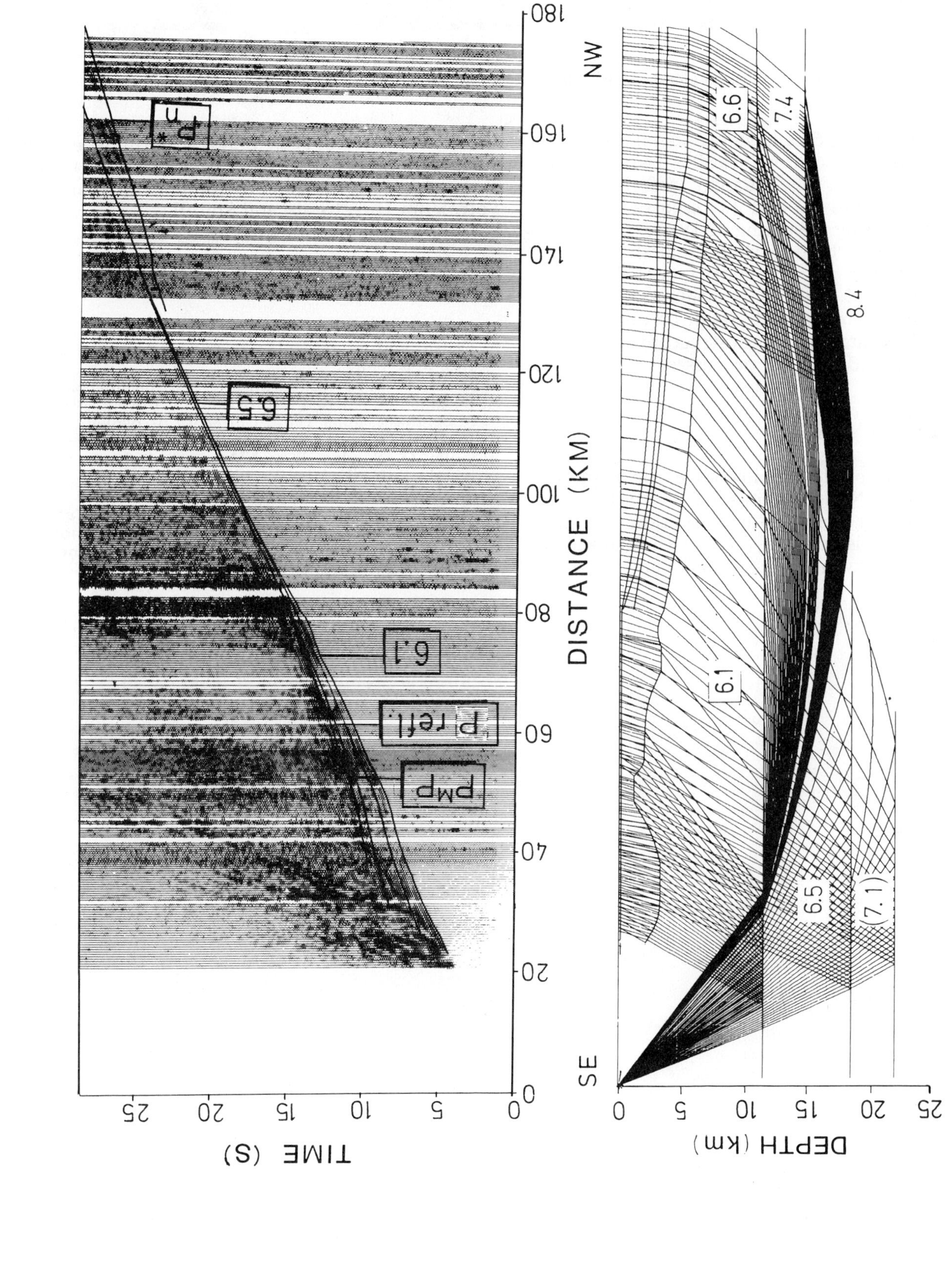

Fig. 7. Recordings obtained at land station 11A and the result of kinematic modelling with ray pattern.

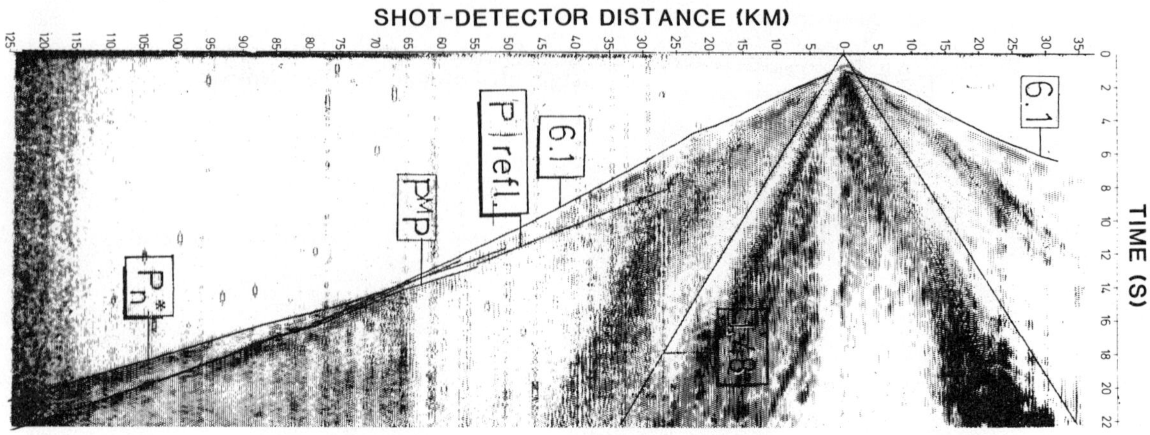

Fig. 8. Recordings obtained at OBS station 11B and the results of kinematic modelling.

seismic reflection profiles have been shot recently on the margin off Lofoten (1982). Two further OBS profiles across the margin off Lofoten, analogous to profiles 10 and 11 discussed herein, were recorded during the spring of 1983.

Finally it should be underlined that the crustal model presented is a preliminary model from an on-going project where all data available from the continental margin off Lofoten will be considered in order to elaborate the crustal structure as well as the underlying geodynamical processes.

ACKNOWLEDGEMENTS

The authors gratefully acknowledge the support from the Centre National pour l'Exploitation des Océans (CNEXO) for the construction of the Ocean Bottom Seismograph (OBS). They are also grateful for help given by the officers and crew of the research vessel *Jean Charcot* during the cruise. The authors also wish to express their thanks for support given to the project by the Norwegian Petroleum Directorate and Statoil.

REFERENCES

Avedik, F., Renard, V. Buisine, D. and Cornic, J.Y. 1978. Ocean bottom refraction seismograph. *Mar. Geophys. Res.* 3, 357—377.

Drivenes, G. 1983. En seismisk jordskorpeundersøkelse på kontinentalmarginen vest for Lofoten. Hovedfagsoppgave. M.Sc.-thesis, June 1983, University of Bergen, 122 pp., 61 Figs.

Eldholm, O., Sundvor E. and Mvhre, A. 1979. Continental margin off Lofoten-Vesterålen, Northern Norway. *Mar. Geophys. Res.* 4, 3—35.

Pajchel, J. 1983. *Modelling Kinematics of Reflected, Refracted and Diffracted Waves in Inhomogenous Media with Curve—linear Boundaries.* Seismological Observatory, University of Bergen.

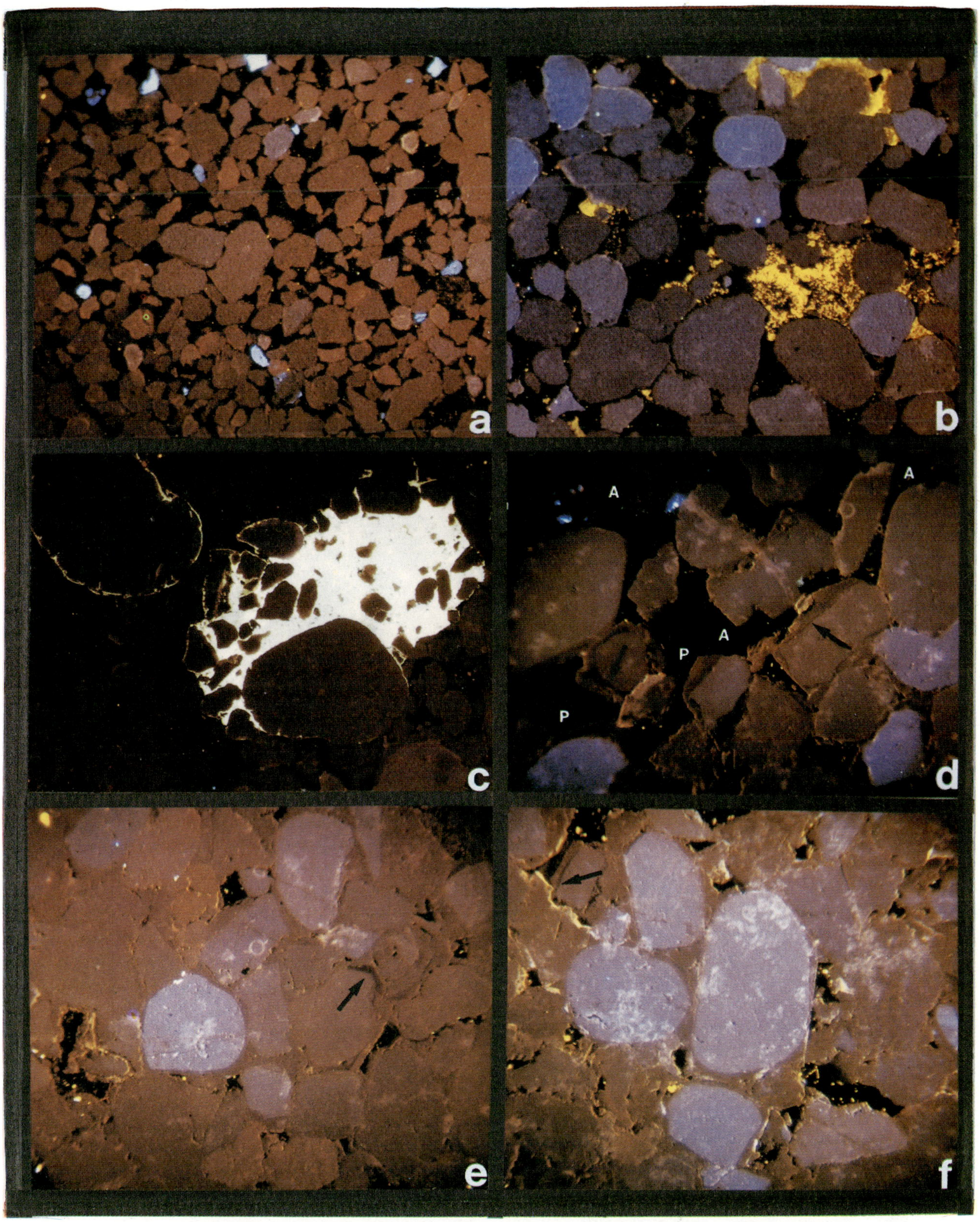

Plate 1 (see p. 69). Luminescence picture of quartz arenites from the near shore/inner shelf deposits. **(a)** Fine-grained sandstone with sutured contacts and minor quartz overgrowth suggesting grain to grain pressure solution. Blue colour: feldspar; brown/reddish, quartz grains; black, porosity (1389 m, 7119/12-2). **(b)** Medium grained sandstone with both floating grains and sutured contacts textures. Minor brown quartz overgrowths occur. Bright yellow is calcite cement; note rotten appearance and patchy development and oversized pores (black) in non-cemented areas (black), suggesting dissolution of calcite cement. All grains, quartz (2120.2 m, 7120/8-1). (c) Phosphate as patchy cement and coatings (white) in a medium grained sandstone. Compare with Fig. 7a (2178.8 m, 7120/8-1). (d) Patchy ankerite cemented, medium grained sandstone. Quartz overgrowth in non-cemented areas (P) and no overgrowth in ankerite cemented areas (A) showing that the quartz overgrowth postdate the carbonate cement (2728 m, 7119/12-1). (e, f) Tight cemented, medium grained quartz arenite. All grains quartz. Differences in colour suggest different origin of quartz grains. Arrows show a first generation of quartz cement. Black, porosity. Note also roundness of grains (7119/12-1; e, 2751 m; f, 2749 m).

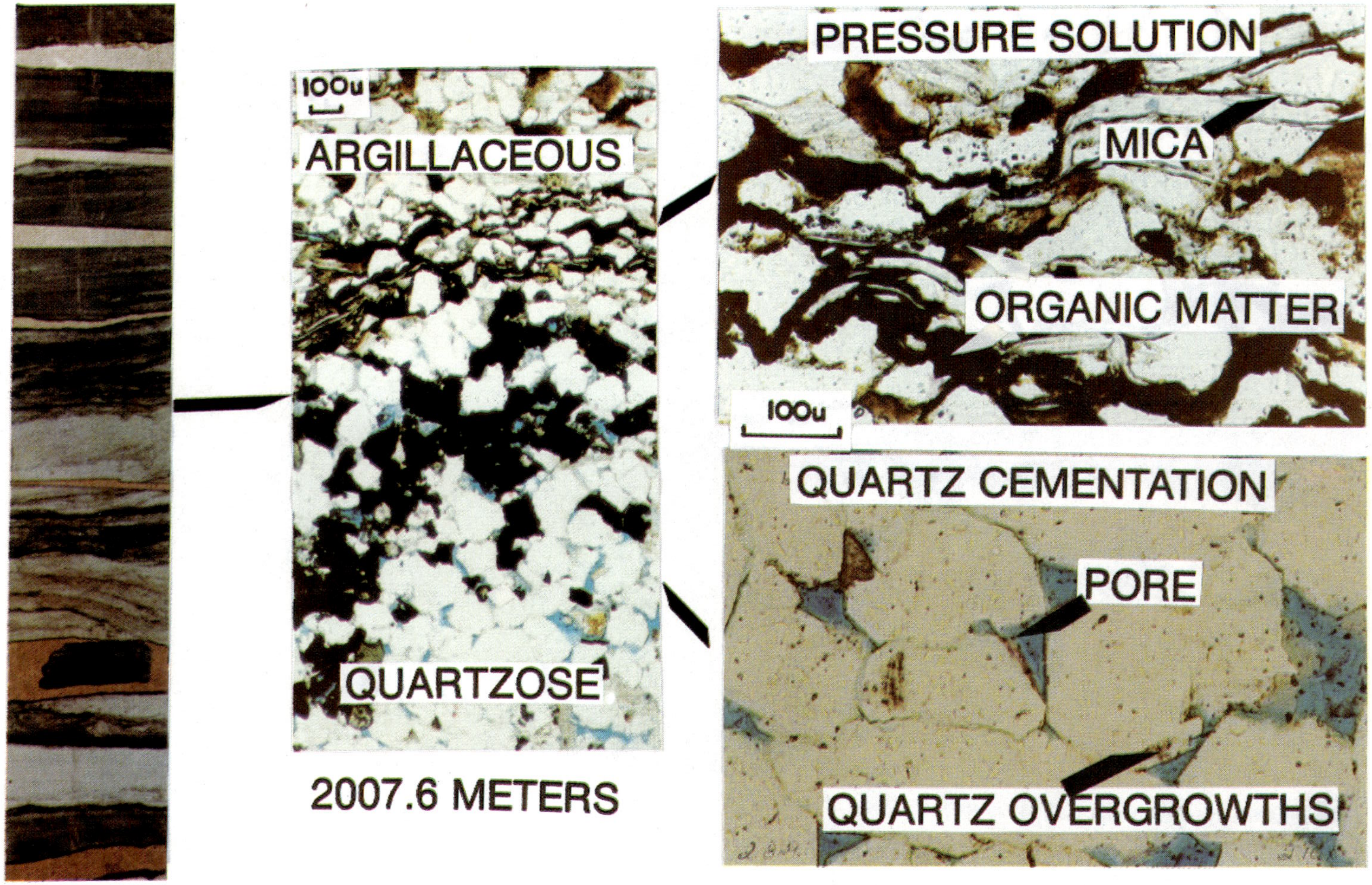

Plate 2 **(see p. 87)**. Core photographs and photomicrographs showing porosity reduction by pressure solution and quartz cementation in well 7120/12-2 (2007.6 m).

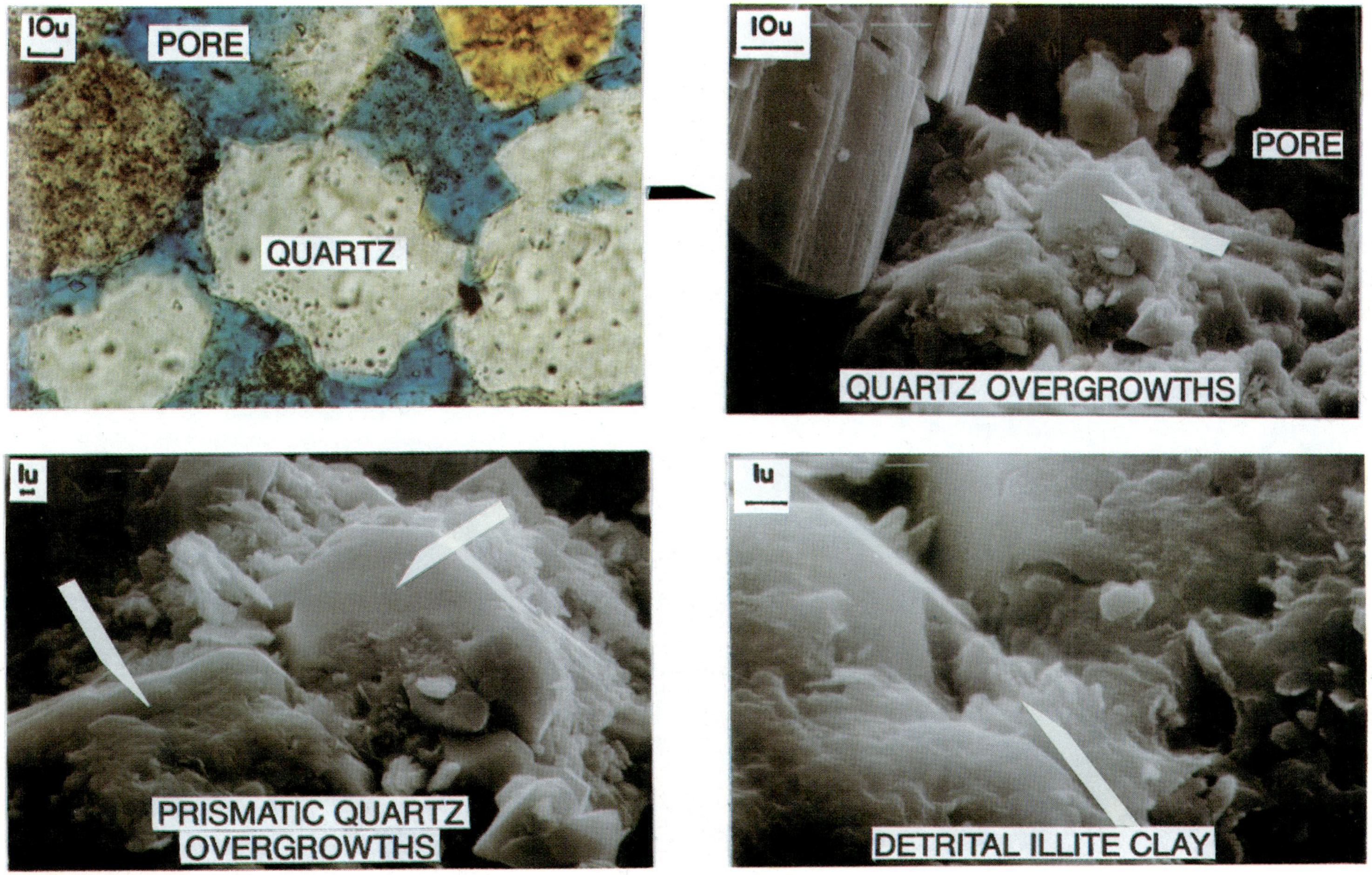

Plate 3 (see p. 87). Relationship of disseminated detrital clay and irregular quartz overgrowth development (well 6507/11-1, 2536.1 m).

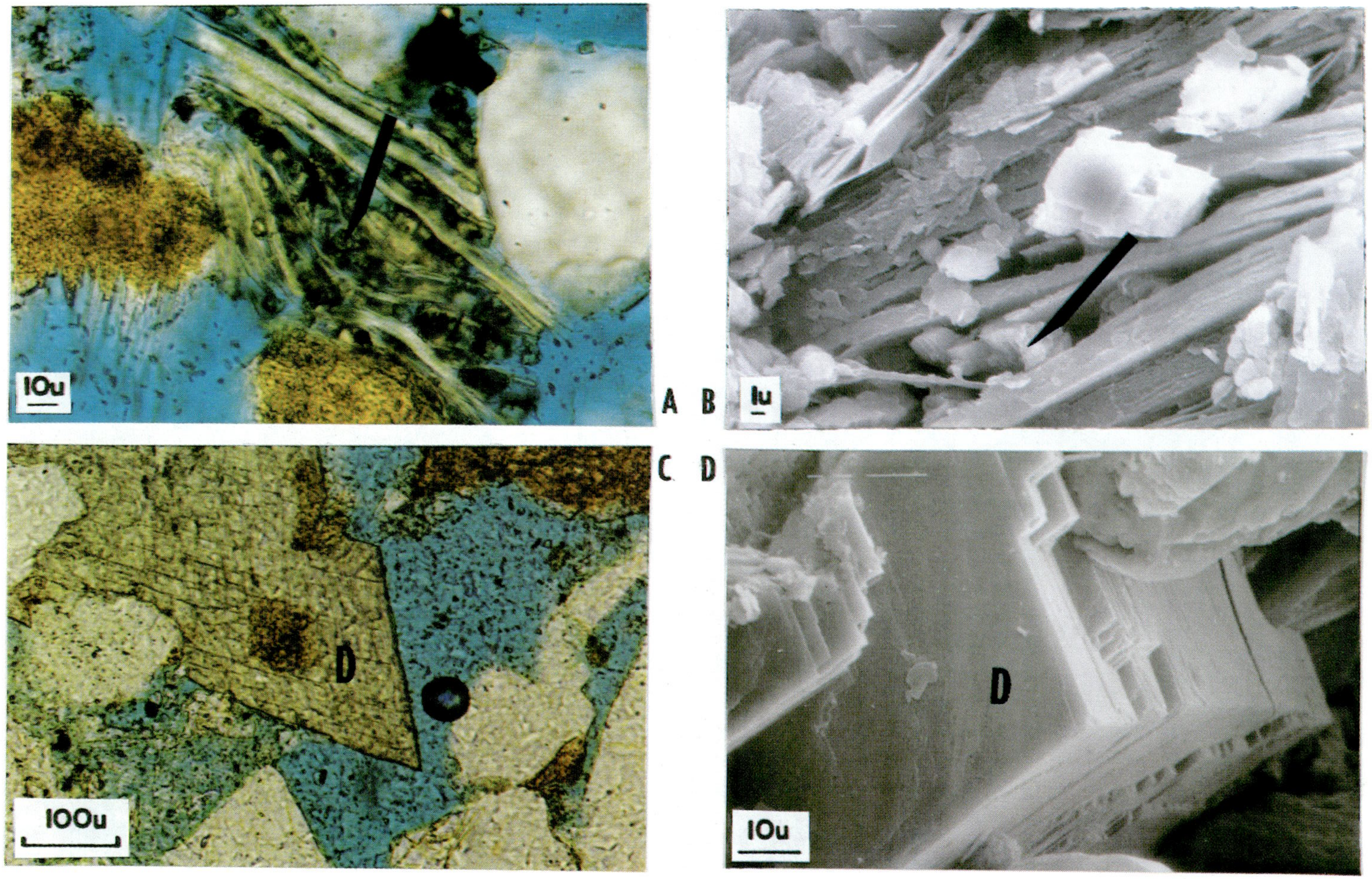

Plate 4 (see p. 87). Photomicrographs and scanning electron micrographs showing development of carbonate cement: (A and B) siderite (arrows) associated with biotite mica (well 6507/12-1, 2404.9 m); (C and D) Fe-dolomite cement (well 6507/12-1, 2533 m).

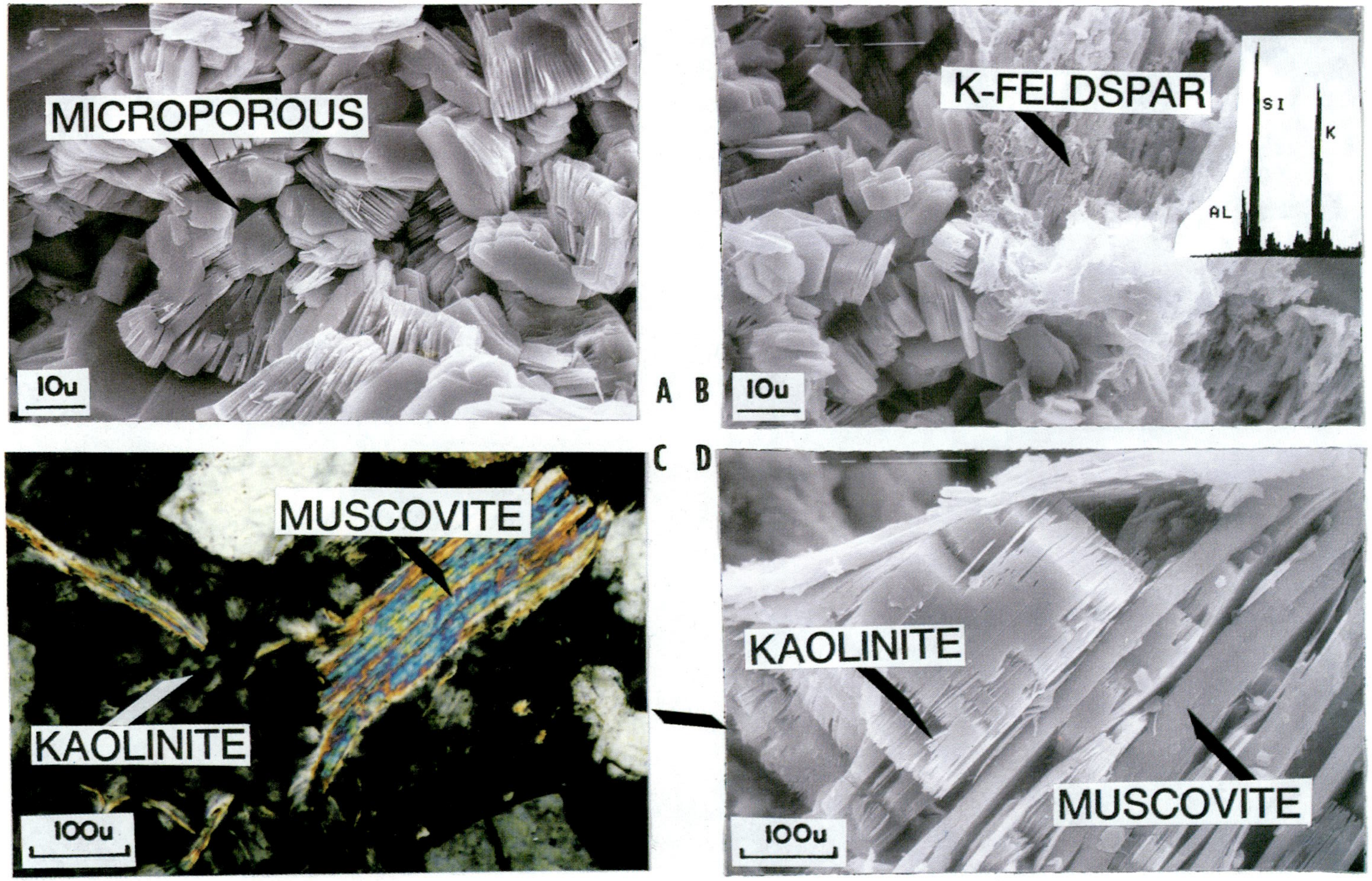

Plate 5 (see p. 89). Photomicrographs and scanning electron micrographs showing development of authigenic kaolinite: (A) microporous kaolinite cement; (B) kaolinite which apparently developed by pseudomorphous replacement of K-feldspar; (C and D) development of kaolinite by alteration of muscovite mica. A, B, D, scanning electron micrographs. C, photomicrograph, crossed-polarized light.

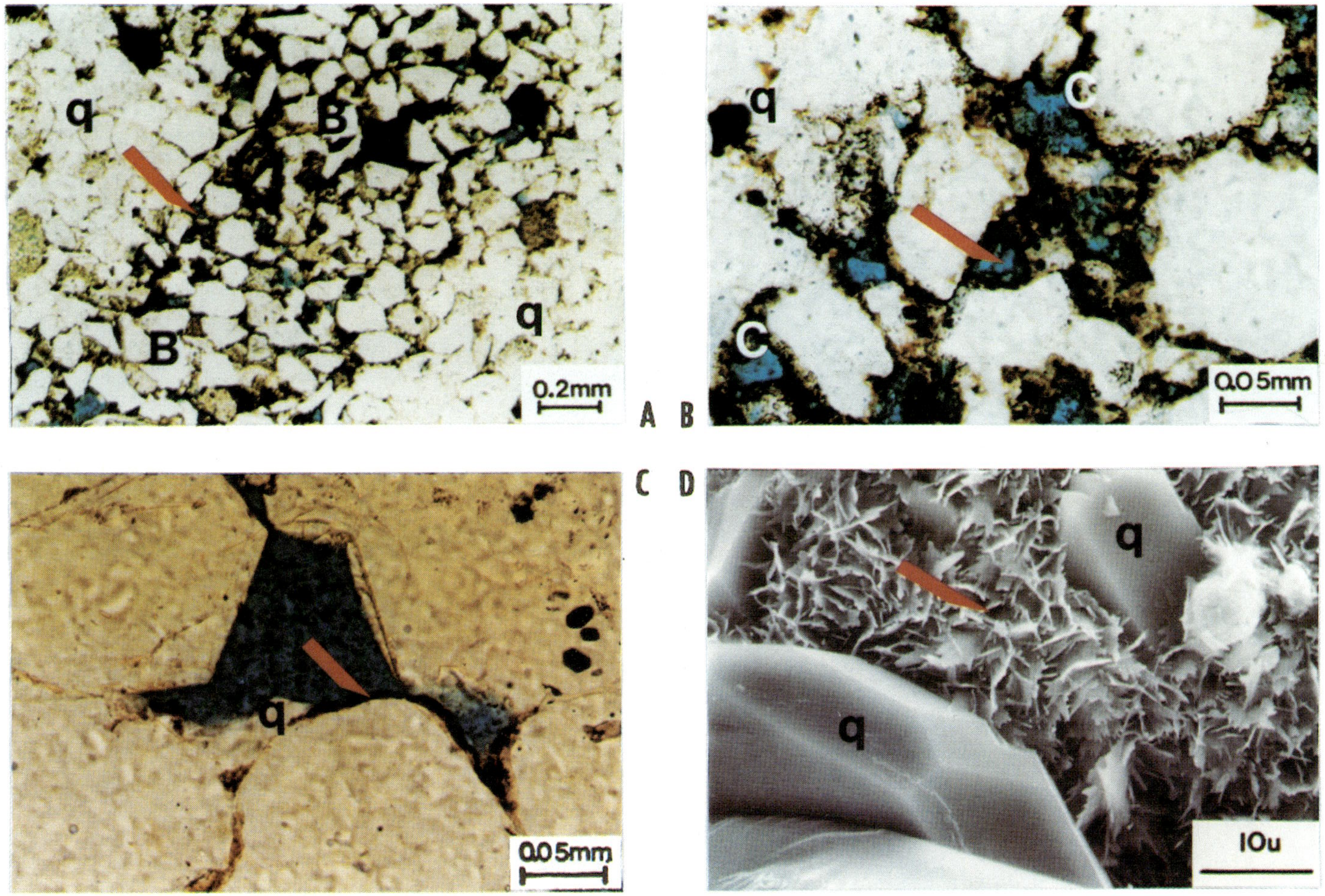

Plate 6 (see p. 89). Photomicrographs and scanning electron micrographs showing development of clay coatings in sandstone specimens from well 7119/12-1. (A and B) organic-clay coating created as a result of burrowing processes. Note pervasive quartz overgrowth development (q) outside of burrow. However, within burrow lining, rearrangement of detrital matrix clay to coatings (c) has served to inhibit nucleation of silica as quartz overgrowths (2821.3 m). (C and D) development of authigenic illite coatings (arrows) in sandstone (2667.6 m). Note prismatic quartz overgrowths (q) where coatings are lacking. Plane polarized light.

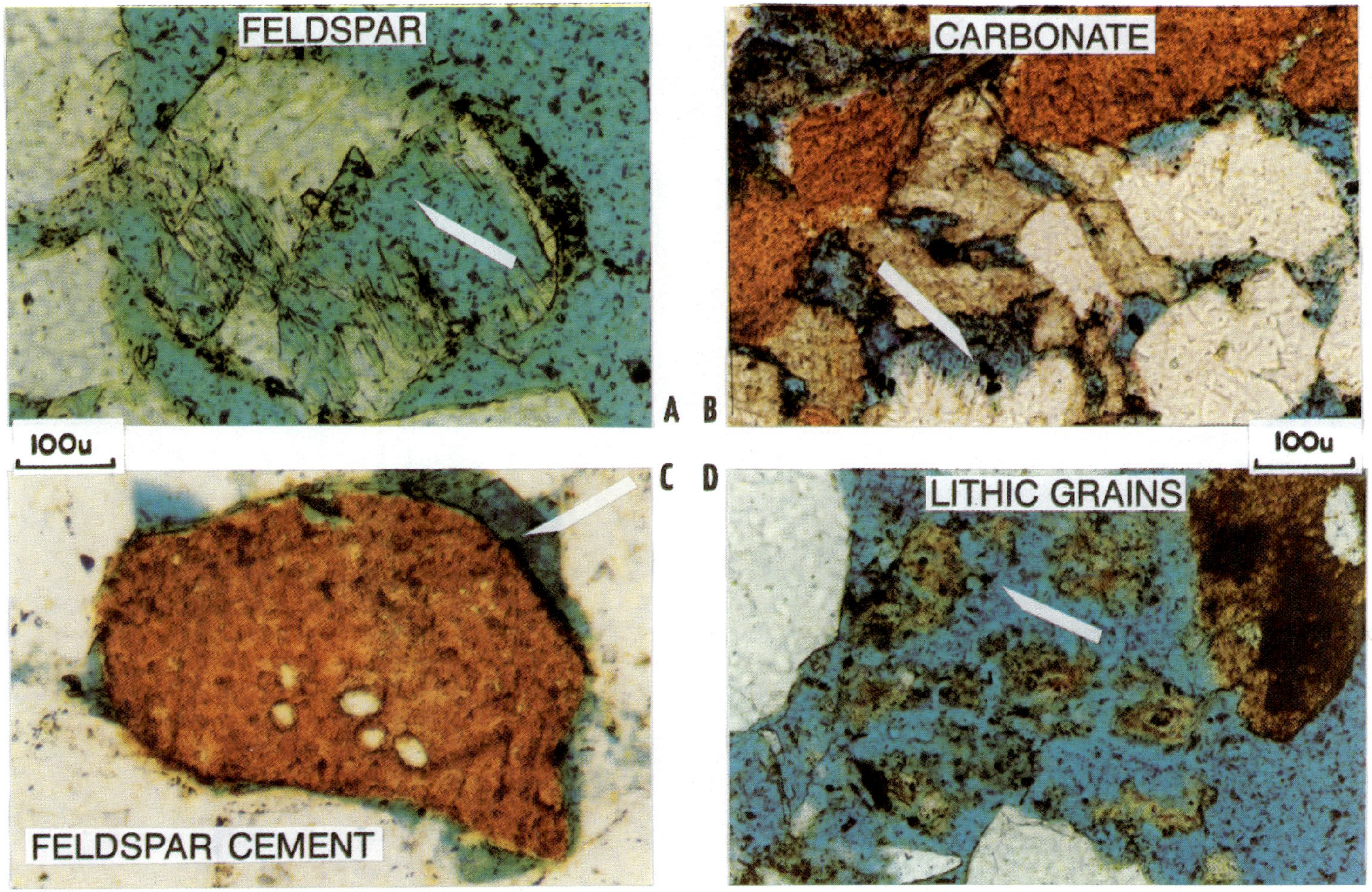

Plate 7 (see p. 90). Photomicrographs illustrating porosity enhancement by dissolution of detrital and authigenic mineral constituents. Plane polarized light.

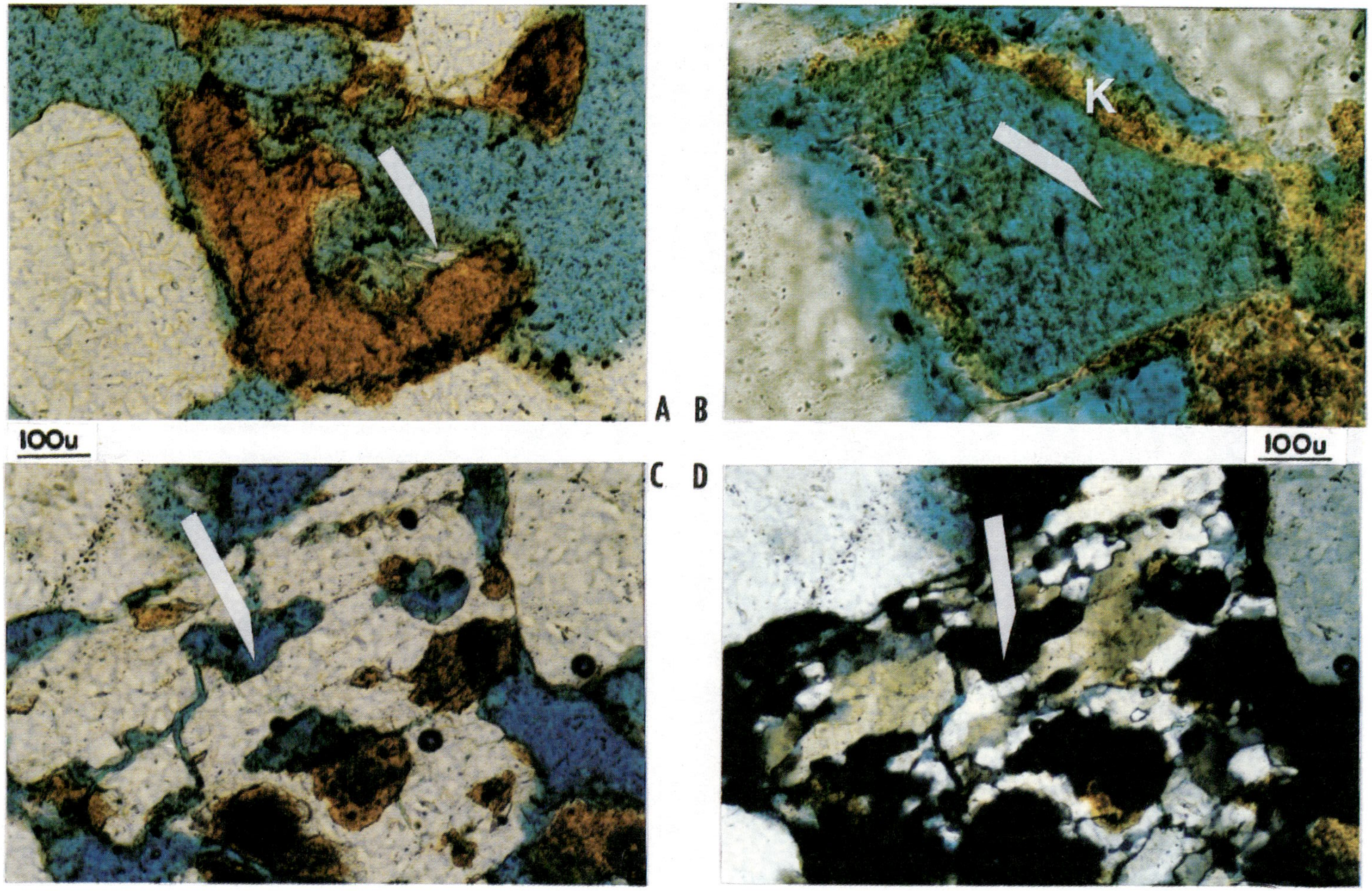

Plate 8 (see p. 90). Photomicrographs depicting secondary porosity development by dissolution of feldspar. (A) selective dissolution of plagioclase (arrows) in plagioclase/K-feldspar rock fragment. (B) selective dissolution of detrital feldspar core (arrows). Dissolution effects in K-feldspar overgrowth (K) are much less pronounced. (C and D) dissolution of feldspathic constituents (arrows) in feldspathic/quartzose metamorphic rock fragment. A and C, plane polarized light. D, crossed polarized light.

Plate 9 (see p. 91). Photomicrographs illustrating dissolution of carbonate cement in selected Jurassic sandstone core specimens. (A and B) dissolution of ankerite cement (A) in 7119/12-1 (2718–2722.5 m). Note remnant ankerite and serrated quartz grain margins (arrows). (C and D) dissolution of calcite (c) cement in 6507/11-1 (2510.55). Note calcite (c) and dolomite (D) associated with muscovite mica (m). Apparently, this calcite has been subjected to extensive dissolution effects resulting in good porosity development. Note remnant dolomite (D) associated with mica (m). A, crossed polarized light. B–D, polarized light.

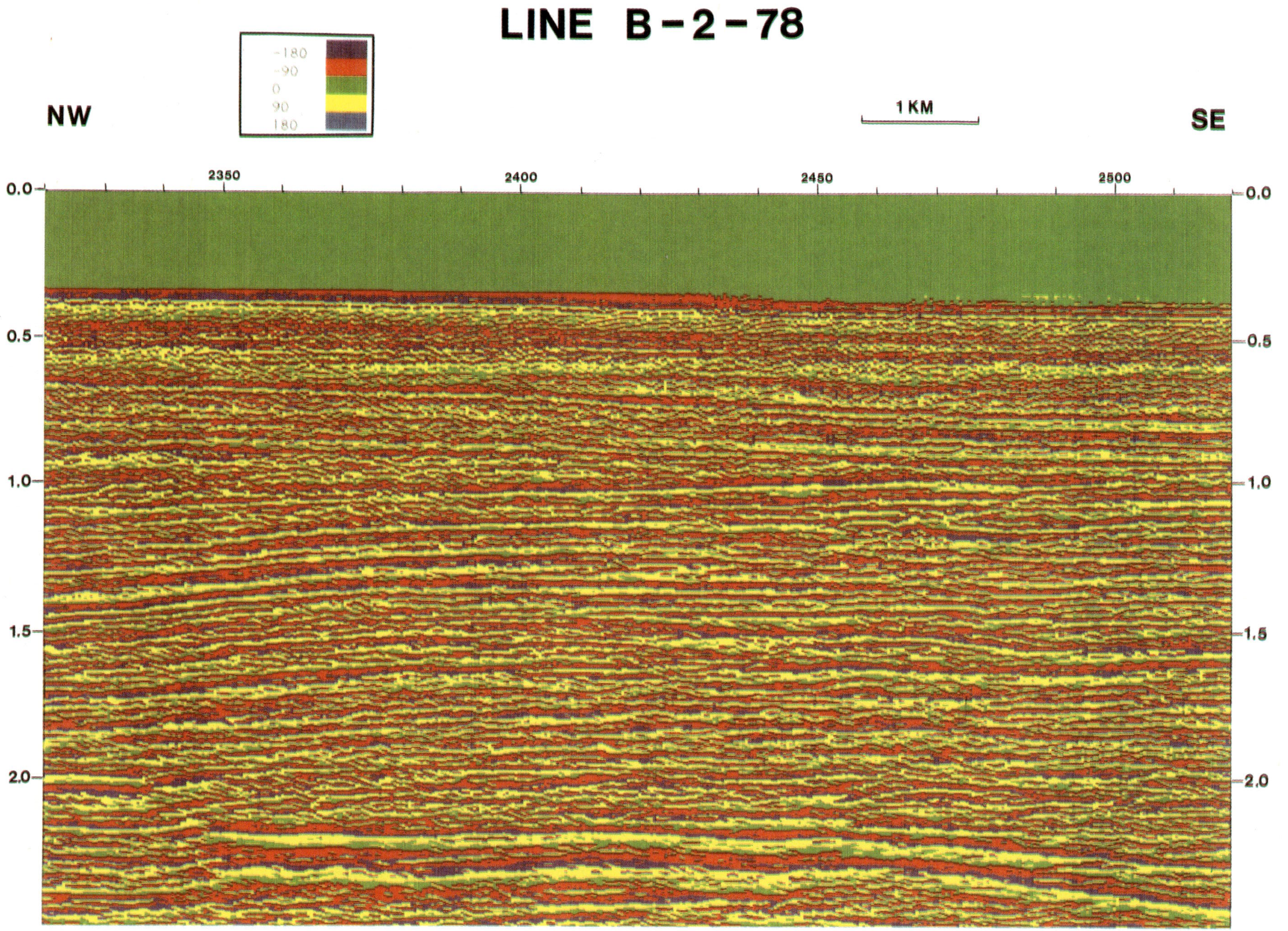

Plate 10 (see p. 300). Instantaneous phase display of line B-2-78. 'Flat spot' is at approximately 1.35 seconds.

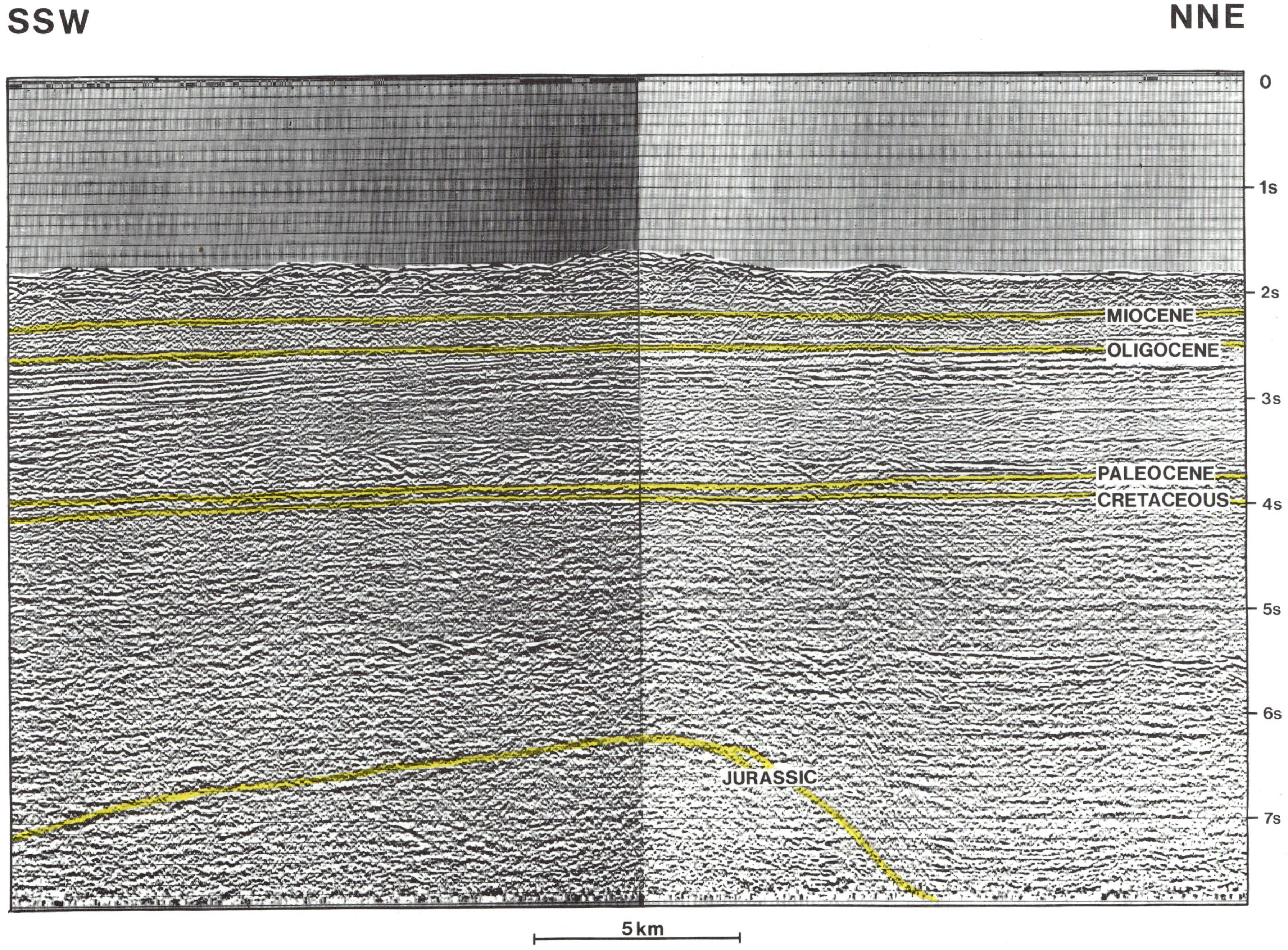

Plate 11 (see p. 352). The Vigra High (seismic section B-3-74 sh.p. 1620—1760).

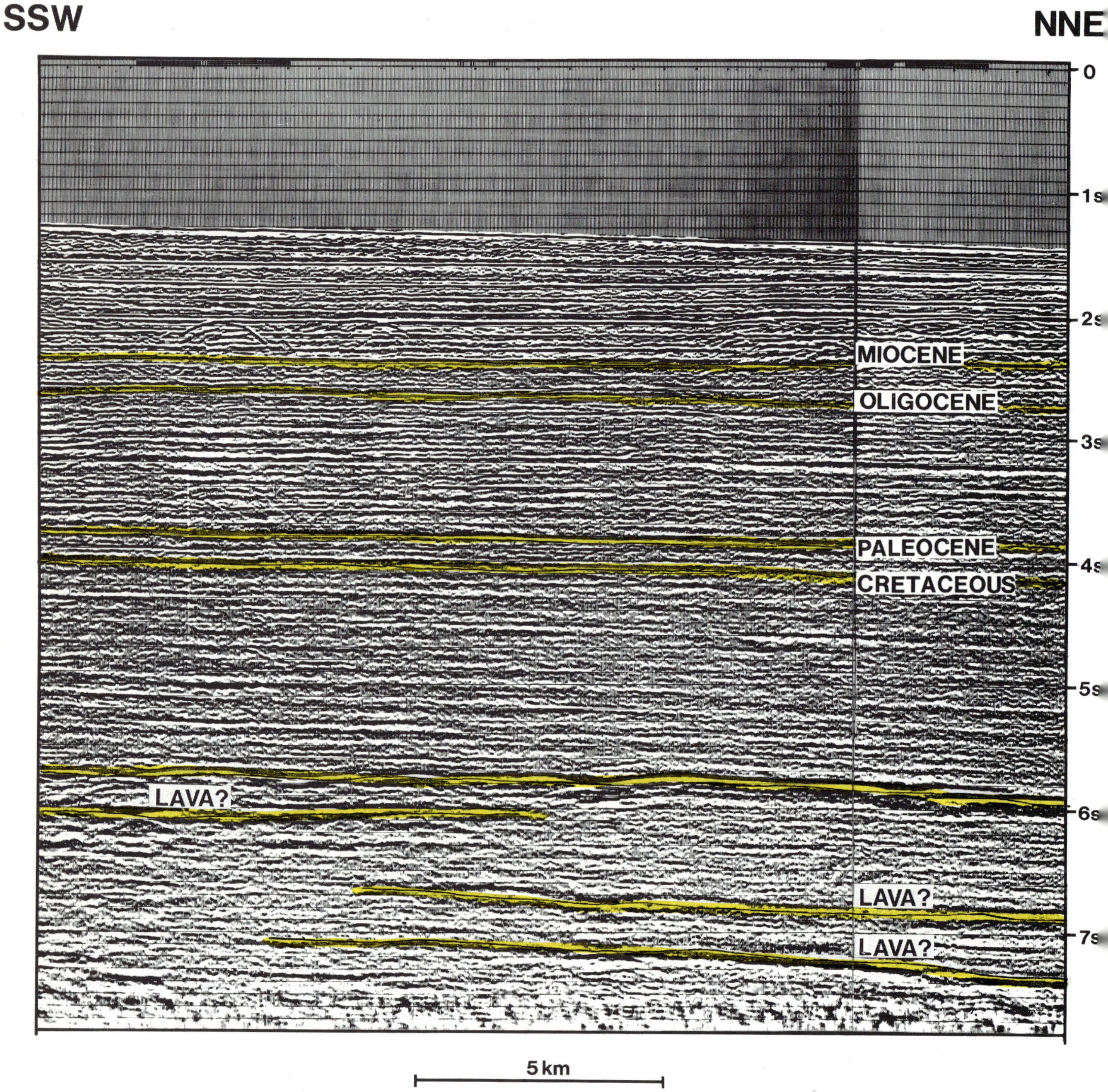

Plate 12 (see p. 353). The Møre Basin. Early Cretaceous lava flows (B-3-74, sh.p. 1980–2080).

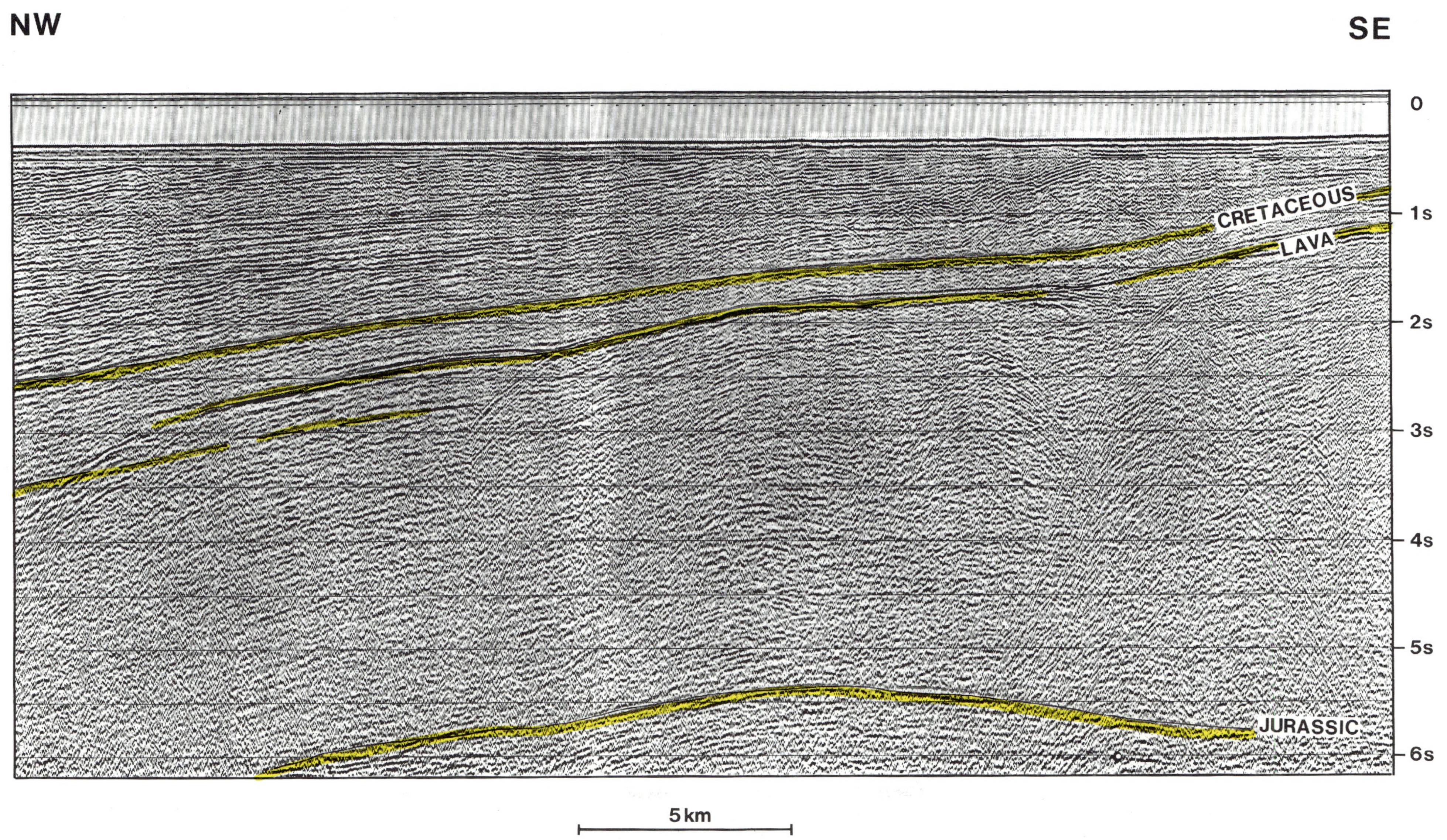

Plate 13 (see p. 353). The Frøya High. Early Cretaceous lava (ST 516-432, sh.p. 50—700).

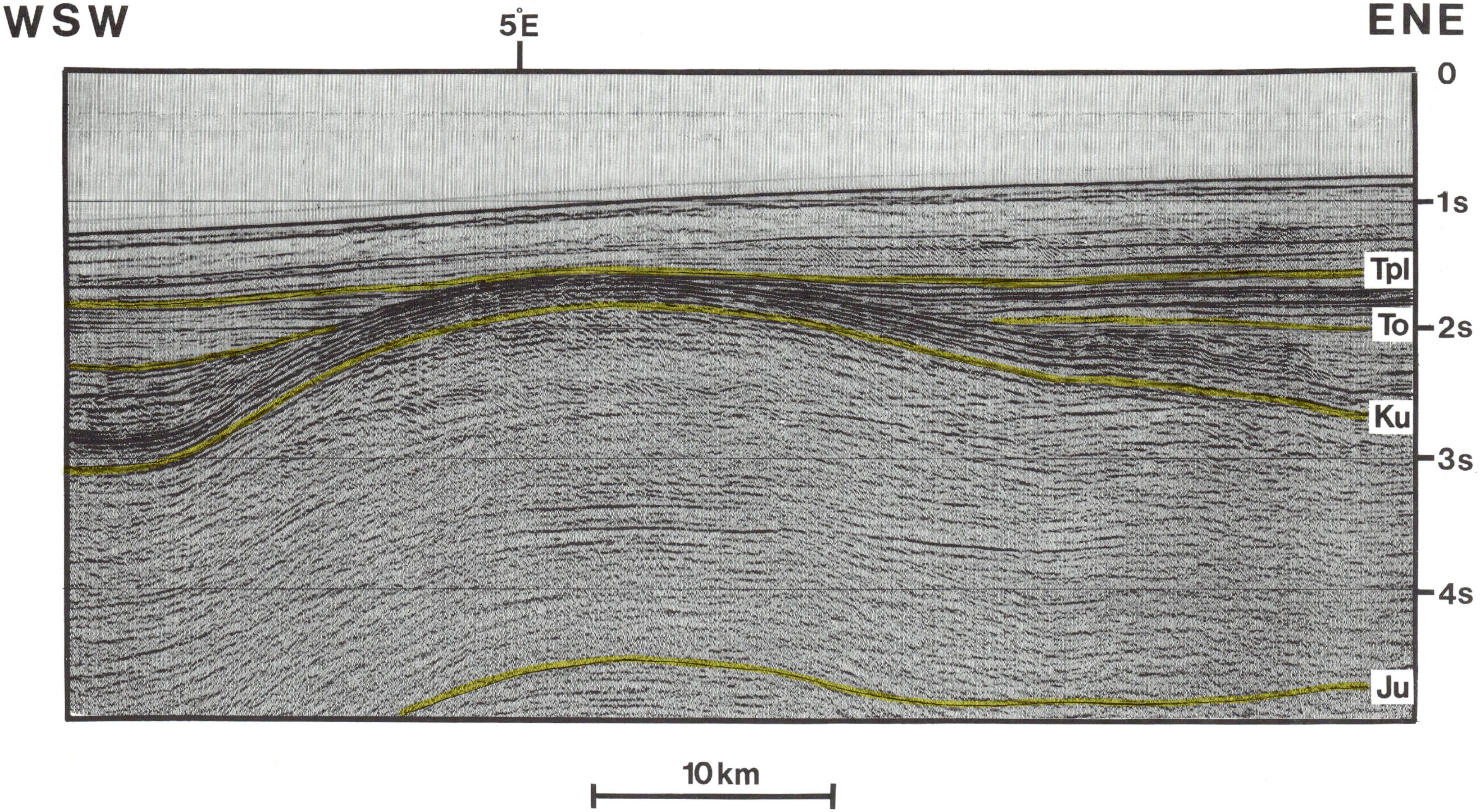

Plate 14 (see p. 354). The Molde High (UiB 225, sh.p. 4660—5680).

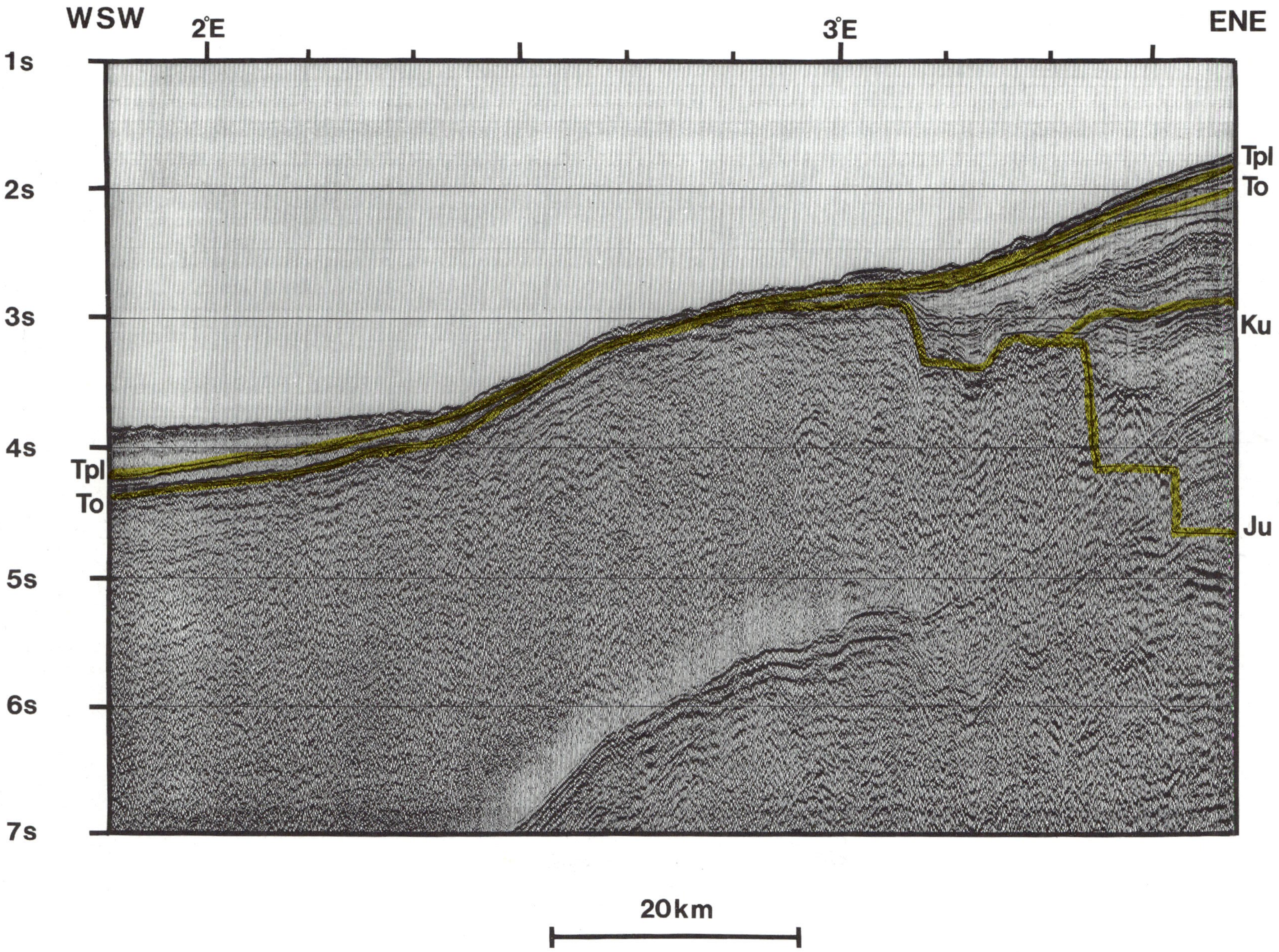

Plate 15 (see p. 354). The Møre Platform (UiB 225, sh.p. 6550–8236).

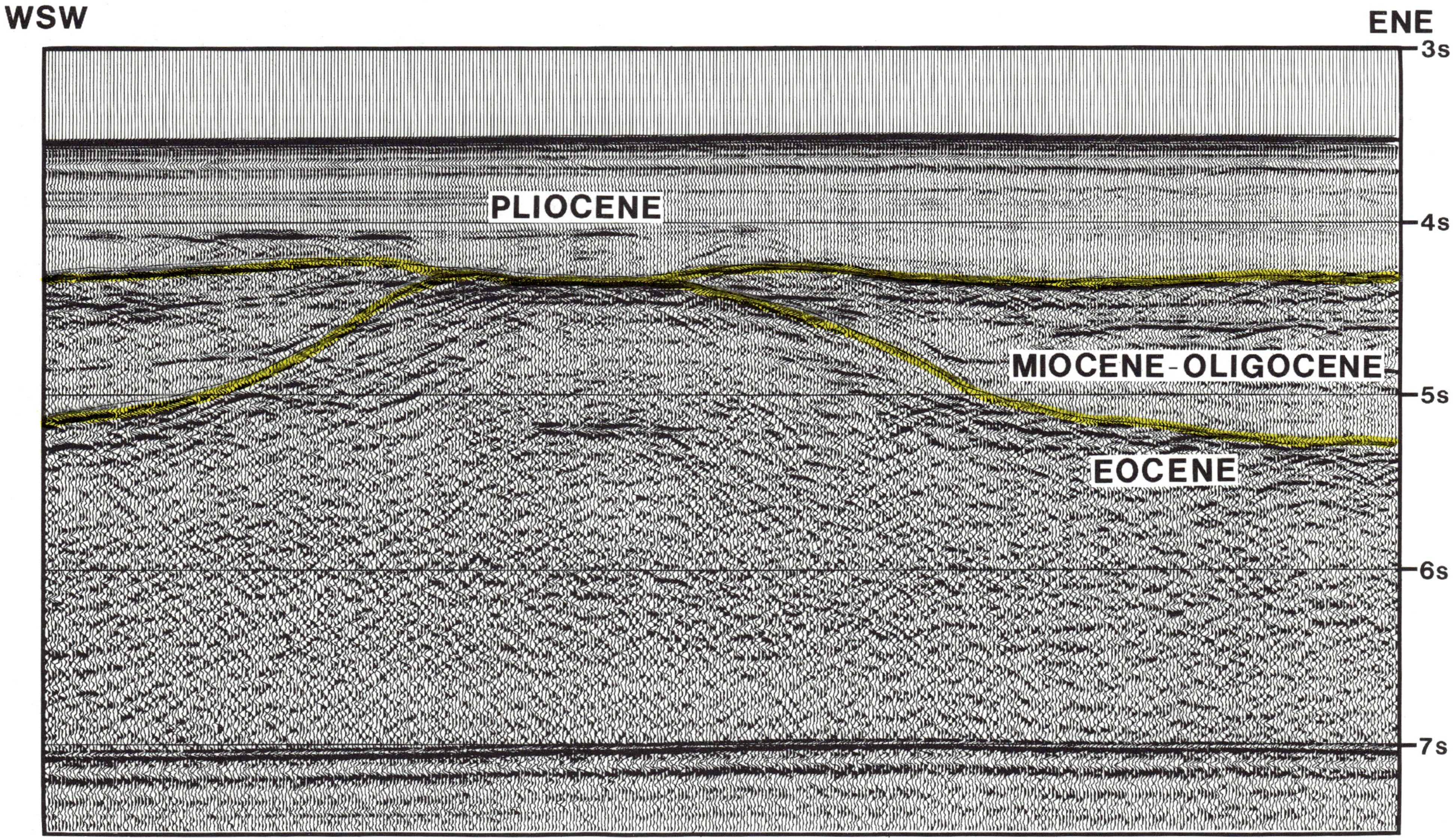

Plate 16 (see p. 356). Outer Ridge 'D' (UiB 225, sh.p. 10 049—10 445).

15 Regional interpretation and hydrocarbon potential of the Trænabanken area

Rolf M. Larsen and Oddvar Skarpnes
Statoil, Stavanger, Norway

The Trænabanken area covers 5000 km^2 and is located on the Norwegian Continental Shelf 100-km offshore mid-Norway and 100-km north of Haltenbanken, where drilling started in 1980. Recently five blocks have been awarded in the Trænabanken area. Three major tectonic elements are observed inside the area: the Helgeland and Træna Basins subdivided by the NE-trending Nordland Ridge. Pre-Mesozoic tectonic history in the area is largely masked by late Jurassic—early Cretaceous events. During this timespan the ridge was exposed subaerially and heavily eroded, removing the whole Jurassic and parts of the Triassic section from the crest developing a peneplain. The basins filled with Cretaceous sediments, the Træna Basin subsiding faster and over a longer period than the Helgeland Basin. The ridge and basins developed in response to rifting. The Upper Cretaceous and Tertiary sequences are thin, broken by major hiatuses due to sea-level variations. During the Oligocene a compressional event deformed parts of the area. The Tertiary and Cretaceous sequences are expected to be dominantly shaly, with possibly interbedded sandstones in the Lower Cretaceous as a result of the erosion of the Nordland Ridge. The Upper Jurassic consists of organic-rich oil-prone shales, while thick sandstones are expected in Middle and Lower Jurassic, the latter interbedded with coal layers with excellent gas sourcing potential. The Triassic is expected to be mainly shales interbedded with sandstones and evaporites. Palaeozoic sediments are dominated by sandstones. Lower and Middle Jurassic sandstones represent the main objectives, while Palaeozoic and possible Triassic sandstones are the main target where Jurassic sediments are eroded. Numerous minor structures are defined in the basins and transition zones to the Nordland Ridge, but the greatest potential is expected along the ridge axis. Highest exploration risk is connected to generation/migration of hydrocarbons: hydrocarbons are generated in the Vøring and Træna Basins, a great distance from the larger structures, while generation of hydrocarbons in the Helgeland Basin, closer to most of the structures, is more questionable.

INTRODUCTION

The Trænabanken area is located on the mid-Norwegian Continental Shelf, approximately 100 km off the coast of Helgeland (Fig. 1). The Haltenbanken area, where drilling started in 1980, lies 100 km further south.

Twelve blocks on Trænabanken, between 9° and 10°20′, 66° and 66°45′, were opened for seismic acquisition in 1981. These blocks cover an area of approximately 5000 km^2, each block being about 415 km^2. In 1982 all 12 blocks were announced in the seventh round of Norwegian licensing, and five blocks were awarded later the same year. Early in April 1983 the first well was spudded by Statoil in block 6610/7.

The water-depth increases from 250 m centrally to more than 400 m in the northeastern corner. Between the Trænabanken and Haltenbanken areas the water-depth varies between 300 and 450 m.

This chapter is based mainly on the evaluation done by Statoil in connection with the 7th round of licensing.

Petroleum Geology of the North European Margin © Norwegian Petroleum Society (Graham & Trotman, 1984) pp. 217—236.

STRUCTURAL SETTING

The Trænabanken area is situated marginally to the deep Vøring and Træna Basins (Fig. 2). These developed during Cretaceous times, and are believed by some to be the northward extension of the Viking and Rockall Troughs. Combined, these troughs/basins can fit a failed rift, triple junction model (Whiteman *et al.*, 1975).

The Nordland Ridge is the dominating tectonic element within Trænabanken (Fig. 3). It trends NE from Haltenbanken, but shifts to a NE-direction inside Trænabanken. The Kristiansund-Bodø Fault

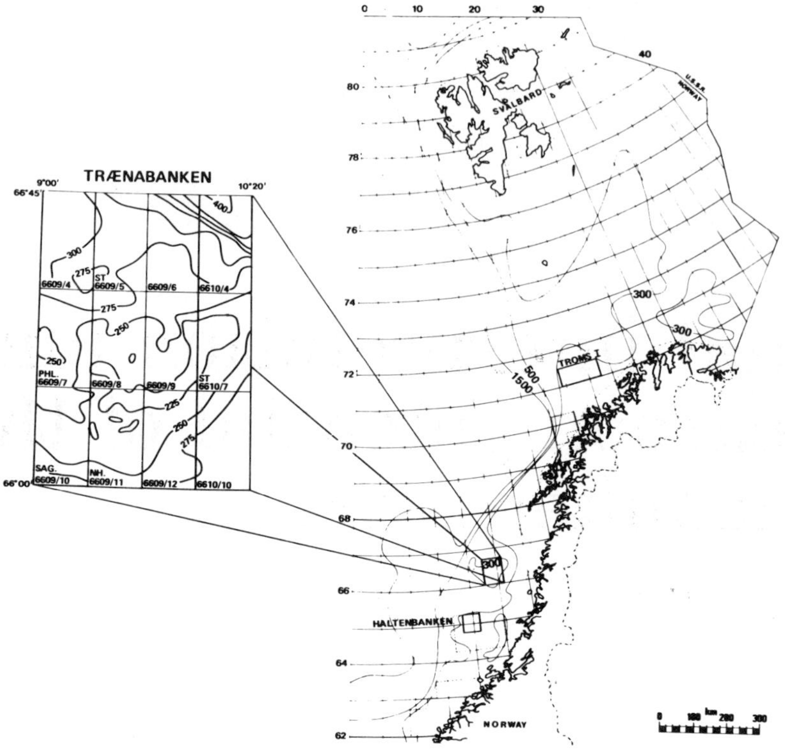

Fig. 1. Location of Trænabanken on the northern Norwegian Continental Shelf.

Complex separates the ridge from the Træna Basin in the northwestern corner of the area. The Rødøy and Grønnøy Highs are defined by the highest areas along the ridge-axis. The former was already a positive area during Palaeozoic, while the latter formed during Oligocene times.

East of the Nordland Ridge, the shallow Helgeland Basin also developed during the Cretaceous. It constitutes the southeastern corner of the area, with the Cretaceous depocentre just within the opened acreage. The Helgeland Basin gradually shallows towards the south. To the north the basin is believed to butt against the Nordland Ridge as the ridge bends in a still more easterly direction northeast of the area. Both the Nordland Ridge and the Helgeland Basin are part of the Trøndelag Platform.

The Vestfjord Basin, to the north of the Nordland Ridge along the Norwegian coast (Fig. 2), is mainly pre-Cretaceous (Jørgensen and Navrestad, 1981).

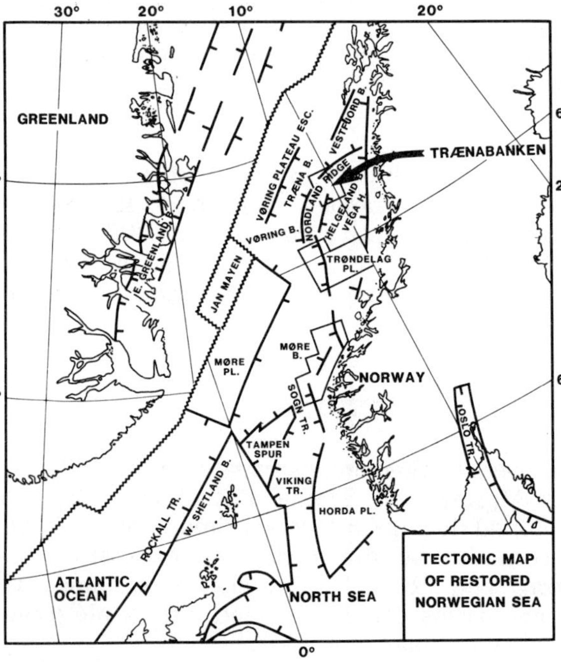

Fig. 2. Tectonic map of restored Norwegian Sea. TR, trough; B, basin; PL, plateau; ESC, escarpment; H, high.

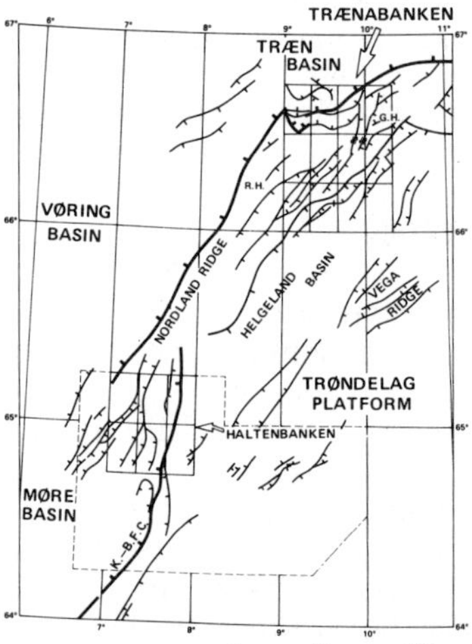

Fig. 3. Main structural elements in the Haltenbanken—Trænabanken area. K.-B.F.C., Kristiandsund-Bodø Fault Complex; G.H., Grønnøy High; R.H., Rødøy High.

EARLIER WORK

Several papers have been published which treat the early Tertiary opening of the Norwegian Sea (e.g. Talwani and Eldholm, 1972). Still, uncertainties exist as to the landward extension of the Tertiary oceanic crust (Rønnevik *et al.*, 1983), and the nature of seaward dipping reflectors west of the Faeroe-Shetland, and Vøring Plateau Escarpments (Talwani *et al.*, 1981). These problems are, however, related mostly to a region west of the area of major interest.

Knowledge about pre-Tertiary geology in the area has been achieved mainly through various papers by the Norwegian Petroleum Directorate (Jørgensen and Navrestad, 1979; Rønnevik *et al.*, 1979; Jørgensen and Navrestad, 1981). In these publications the major tectonic elements have been defined and described. Major tectonic activity took place from late Jurassic until early or probably mid-Cretaceous times.

Following the start of drilling on Haltenbanken, more detailed information on stratigraphy and hydrocarbon potential has been made available (Hollander, 1982; Bleie *et al.*, 1982; Rønnevik *et al.*, 1983). Main reservoir rocks are of early and Middle Jurassic age (Hollander, 1982), and these are also expected within the Trænabanken area (Bleie

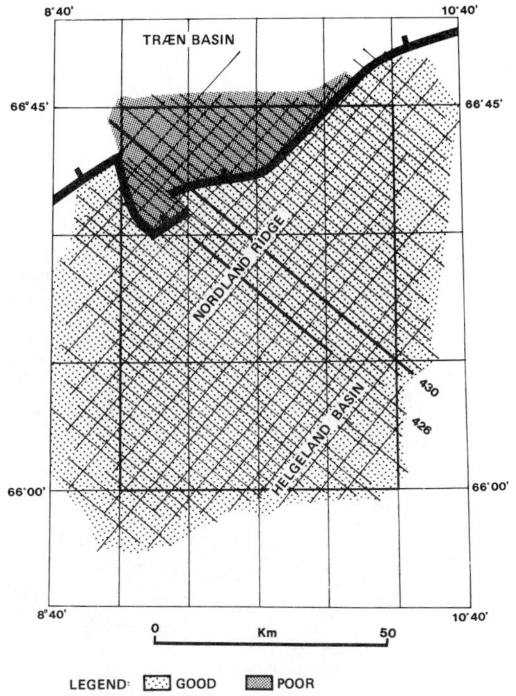

Fig. 4. Seismic coverage and quality on Trænabanken. The lines total about 4500 km. Heavy lines, 430 and 426, are shown in Figs. 6 and 7, respectively.

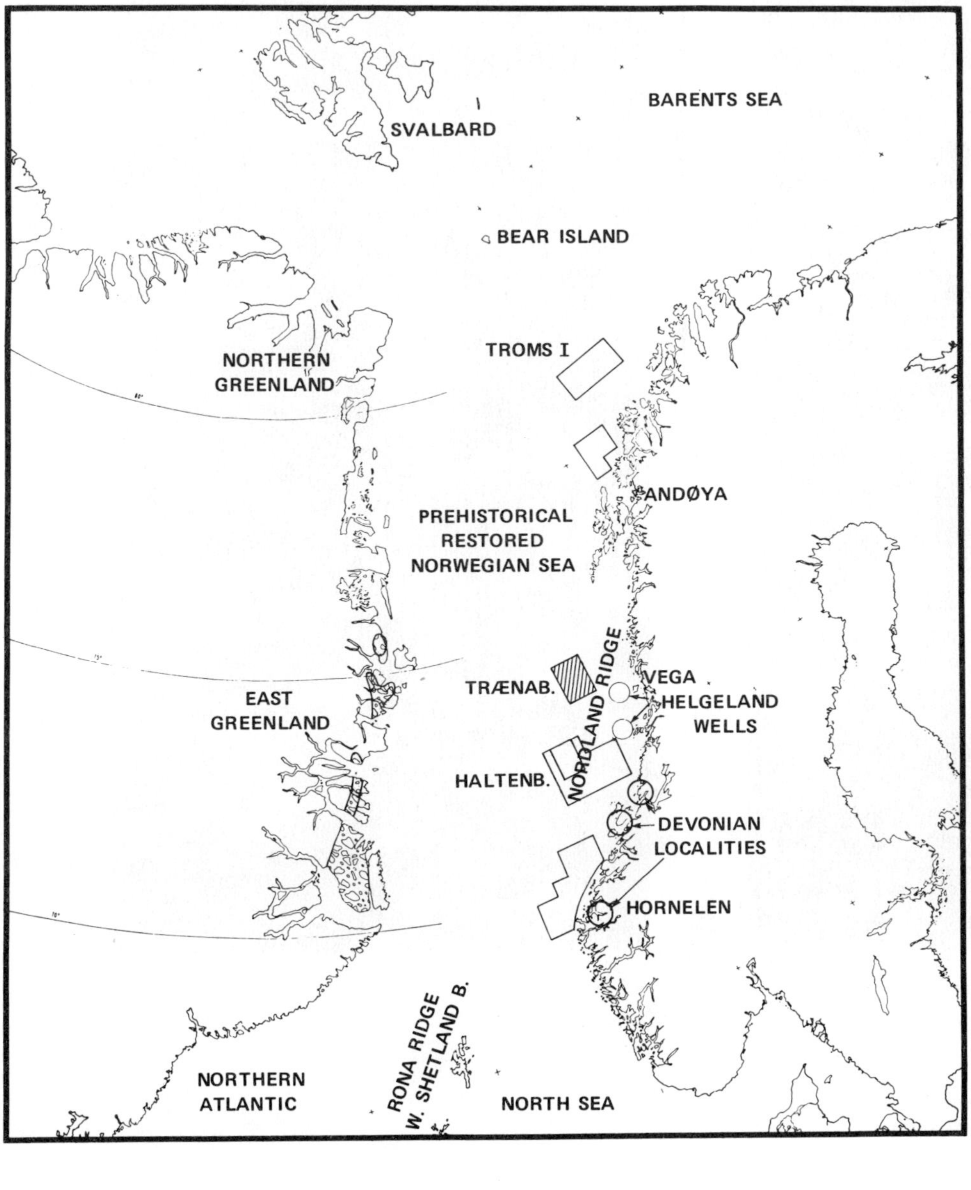

Fig. 5. Basis for seismic tie and stratigraphic control. Seismic tie and post-Lower Triassic stratigraphic information from the Haltenbanken wells and the Helgeland coast. Pre-Middle Triassic stratigraphy from the other areas.

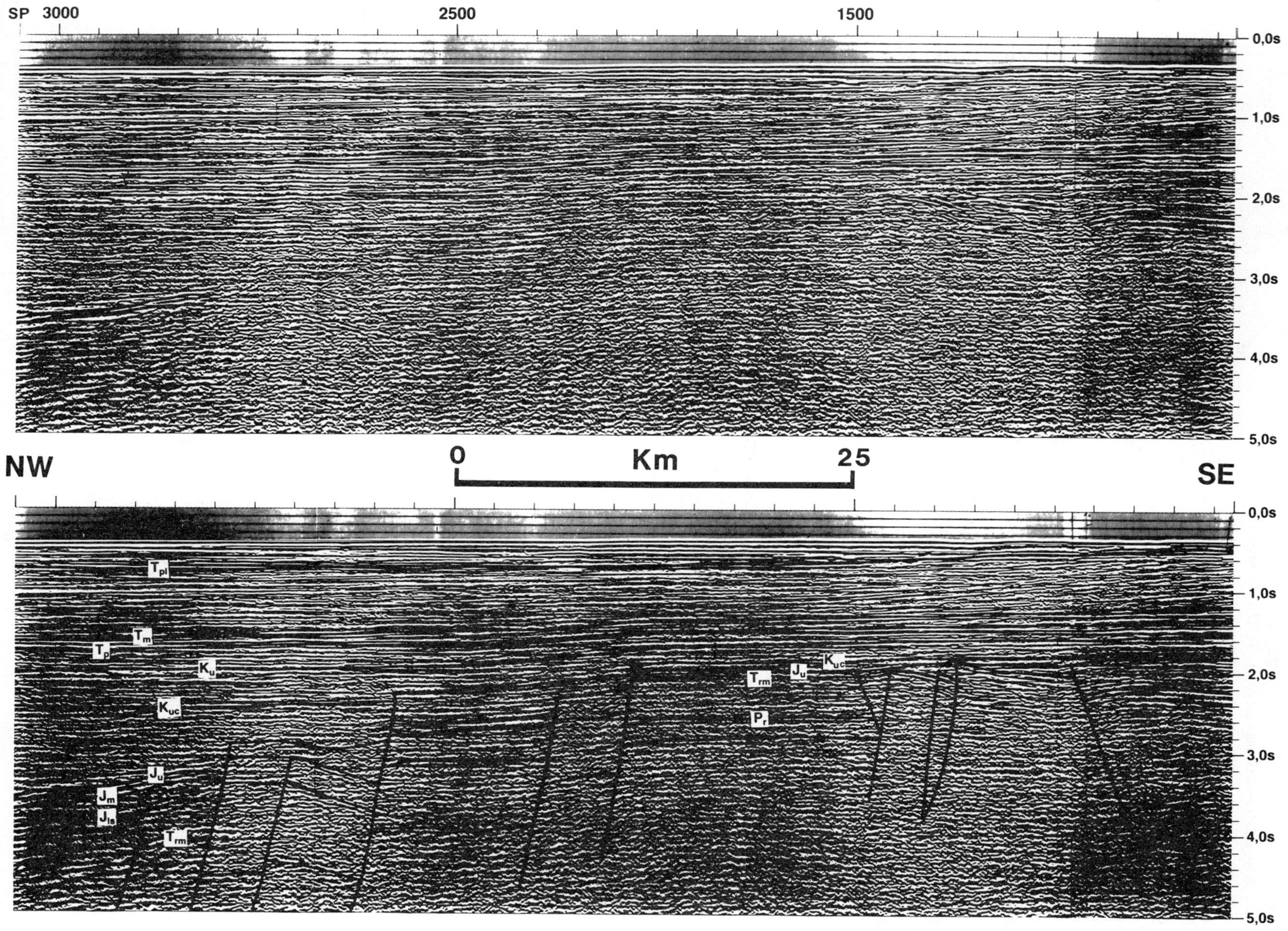

Fig. 6. Migrated seismic line NH 8102-430, orientation shown on Fig. 4. Upper, uninterpreted; lower, interpreted.

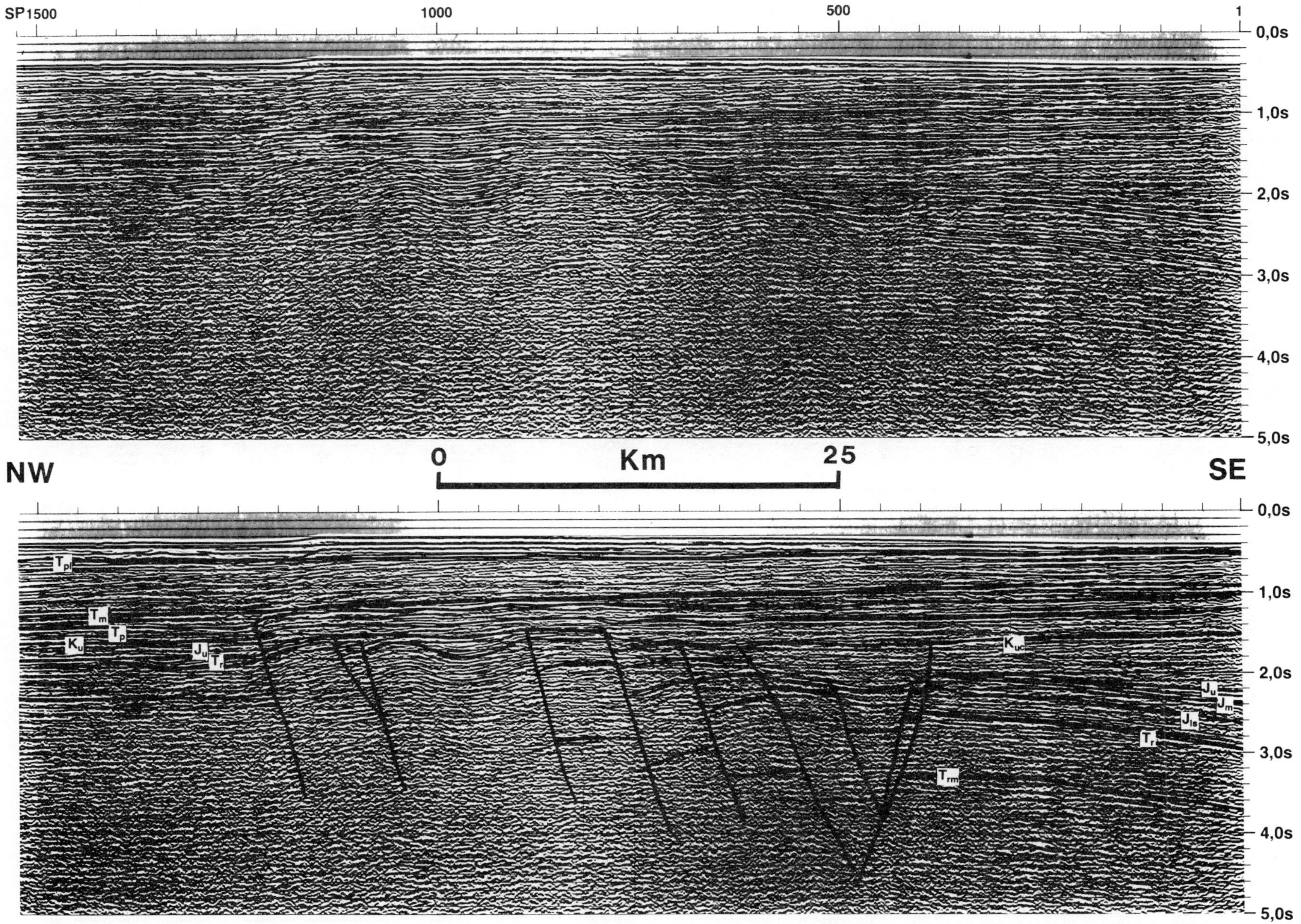
SP1500
1000
500
1
0,0s
1,0s
2,0s
3,0s
4,0s
5,0s
NW
SE
0
Km
25
T_{pl}
T_m
T_p
K_u
J_u
T_r
K_{uc}
J_u
J_m
J_{ls}
T_r
T_{rm}

Fig. 6b.

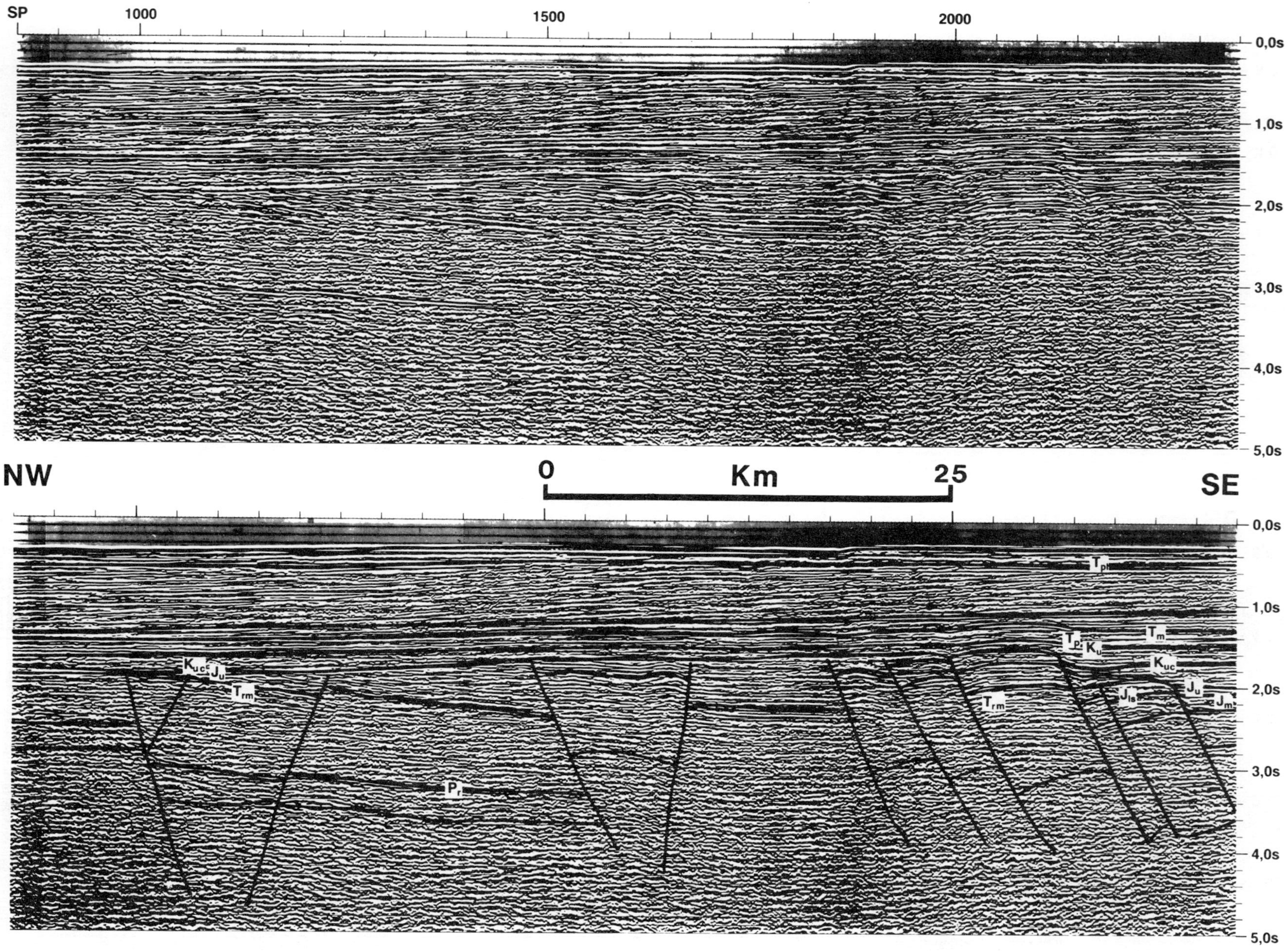

Fig. 7. Migrated seismic line NH 8102-426, s.p. 850—2350. Orientation shown on Fig. 4. Upper, uninterpreted; lower, interpreted.

et al., 1982). However, due to early Cretaceous relative uplift and subaerial exposure, with subsequent erosion of the Nordland Ridge, the Trænabanken area will be partly dependent on older reservoir rocks (Bleie *et al.*, 1982). Dominantly gas-prone source rocks are found in the Lower Jurassic, and thin oil-prone source rocks are found in the Upper Jurassic. Due to late Tertiary burial, the source rocks are immature to moderately mature where drilling has taken place so far on Haltenbanken (Rønnevik *et al.*, 1983).

As yet, the drilling has not disclosed the age and nature of the crust underlying the Møre, Vøring, and Træna Basins. Rønnevik *et al.* (1983) state that the Møre Basin has been rapidly subsiding since late Jurassic, thereby implying a crust not younger than Jurassic. Price and Rattey (in press) argue for a possible mid-Cretaceous oceanic crust extending through a transtensional regime, northward from the Rockall Trough. True oceanic crust is not believed to exist north of 65°30′, but the age of the basin-fill north of this point will still be dominantly younger than mid-Cretaceous. According to Price and Rattey (in press), dextral movements along ENE-trending faults affected the Trænabanken area at these times.

Rønnevik *et al.* (1983) and Bleie *et al.* (1982, figs. 17 and 19) point to a compressional event and reverse faulting of mid-Tertiary age affecting the Nordland Ridge. This is coincident with the start of spreading between Greenland and Svalbard, and a southward shift of the Jan Mayen Fracture Zone.

DATA

The study of the Trænabanken area is based on interpretation of 4500-km seismic data within the area opened for licensing (Fig. 4). The bulk of the data was acquired and processed in 1981 (Træna Group Shoot). This survey is supplemented by some of the earlier data collected by the Norwegian Petroleum Directorate during the years 1972–1979. The latter surveys have been used to tie the Trænabanken to the Haltenbanken area. The seismic data-quality is good except in the Kristiansund–Bodø Fault Complex Zone (Fig. 4).

All wells drilled prior to 1983 inside the Haltenbanken area have been available for stratigraphic and geochemical information. Surrounding areas have been studied to obtain information on pre mid-Triassic stratigraphy (Fig. 5). During the summer of 1982 shallow cores were collected off the Helgeland coast where Mesozoic rocks outcrop. Valuable information from parts of the stratigraphic sequence was obtained this way.

STRUCTURAL HISTORY

The present structural configuration of the Trænabanken area is the result of a long and complex structural evolution, with evidence of both extensional and compressional tectonic events.

The deepest reflector partly mapped is the top of the pre-Triassic sequence (Figs. 6 and 8). This

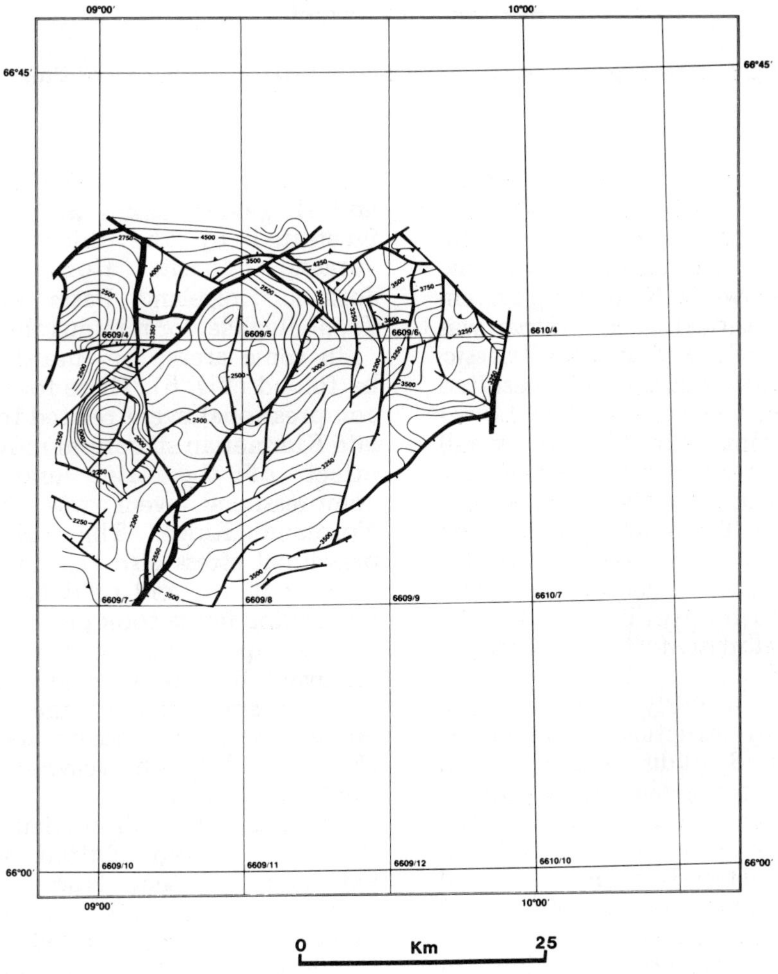

Fig. 8. Pre-Triassic, structural map in two-way seismic travel time.

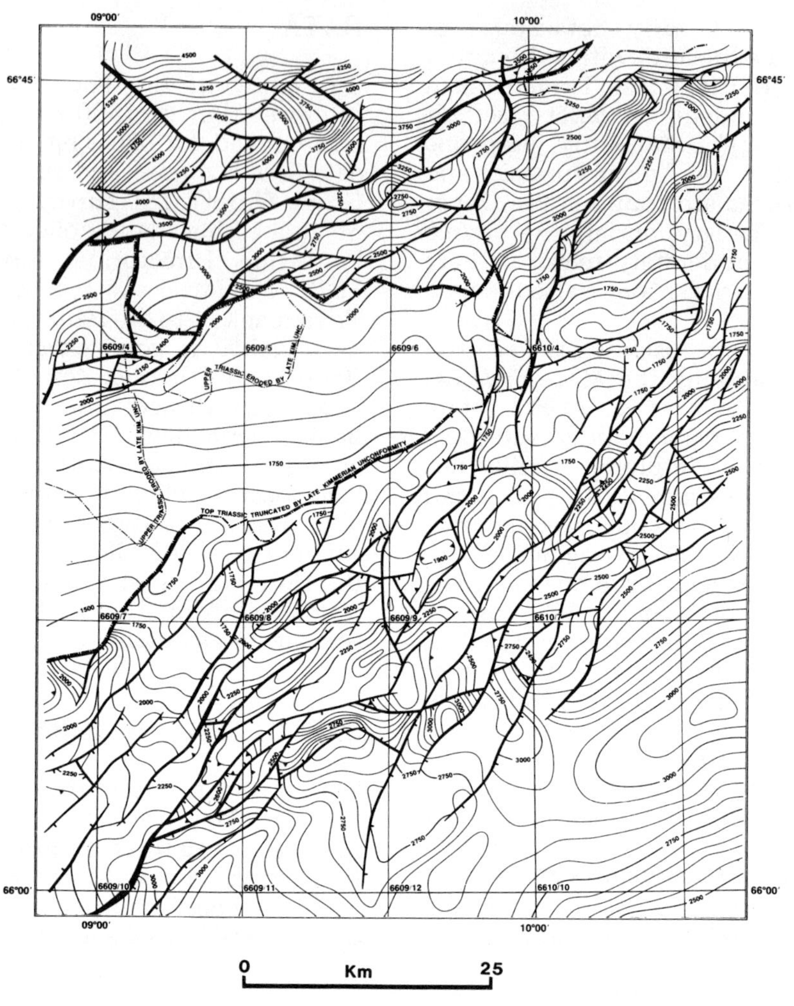

Fig. 9. Near top Triassic, structural map in two-way seismic travel time.

reflector represents a marked angular unconformity (Fig. 6). The tectonic activity leading to this pre-Triassic unconformity probably followed the same pattern as the first post-unconformity tectonic activity. This faulting followed a N-trending system (Fig. 8). The E—W extensional stress system was active throughout the early and middle Triassic. The syn-depositional extensional tectonics resulted in considerable thickness variations of the Lower and Middle Triassic section (Fig. 6). The overall structuring of the area throughout this period was the development of a horst to the west, with a basin to the east (Fig. 14). The horst represents the initial Rødøy High. The Triassic basin east of the high incorporated both the Helgeland Basin and what later developed into the Nordland Ridge. To the west an early development of the Vøring—Træna Basins is observed.

The late Triassic was a fairly quiet tectonic period, with only minor structural movements. Earlier-formed relief levelled out during late Triassic.

During early Jurassic times the stress pattern changed, and NW—SE extensional forces resulted in a new NE-trending fault system (Figs. 9 and 10). This parallels a well-defined trend onshore (Aanestad *et al.*, 1981). Syn-depositional faulting is illustrated by thickness variations across faults.

The early and middle Jurassic fault activity in the NW—SE extensional stress system represents the initial structuring of the Trænabanken area as seen today. The early development of the Træna and Helgeland Basins and the Nordland Ridge is illustrated by the thickness variations of the Jurassic deposits. However, the early—middle Jurassic Helgeland Basin included most of the Trøndelag Platform minus the Nordland Ridge.

In some areas, the Jurassic sequence appears to be folded (Fig. 6). These folds are local effects of a compressional event related to an early stage of the late Kimmerian structural movements. The regional effect of the late Jurassic—early Cretaceous compressional event was the updoming of the Nordland Ridge. The SE—NW oriented compressional stress also rejuvenated some of the old Triassic faults, and right lateral movements along N-trending faults took place.

The updoming of the Nordland Ridge was followed by extensive erosion, and most of the Jurassic and parts of the Triassic section were eroded over the ridge, developing a peneplain (Figs. 6 and 7). The heaviest erosion took place on the Rødøy High.

After the late Kimmerian compressional event, the stress pattern shifted back to the NW—SE extensional system. Normal extensional faulting along the same NE-trend continued along the flanks of the Nordland Ridge throughout the early Cretaceous. In the same period, the region as a whole subsided towards the NW. The downwarping to the NW is illustrated by the gradual onlap to the

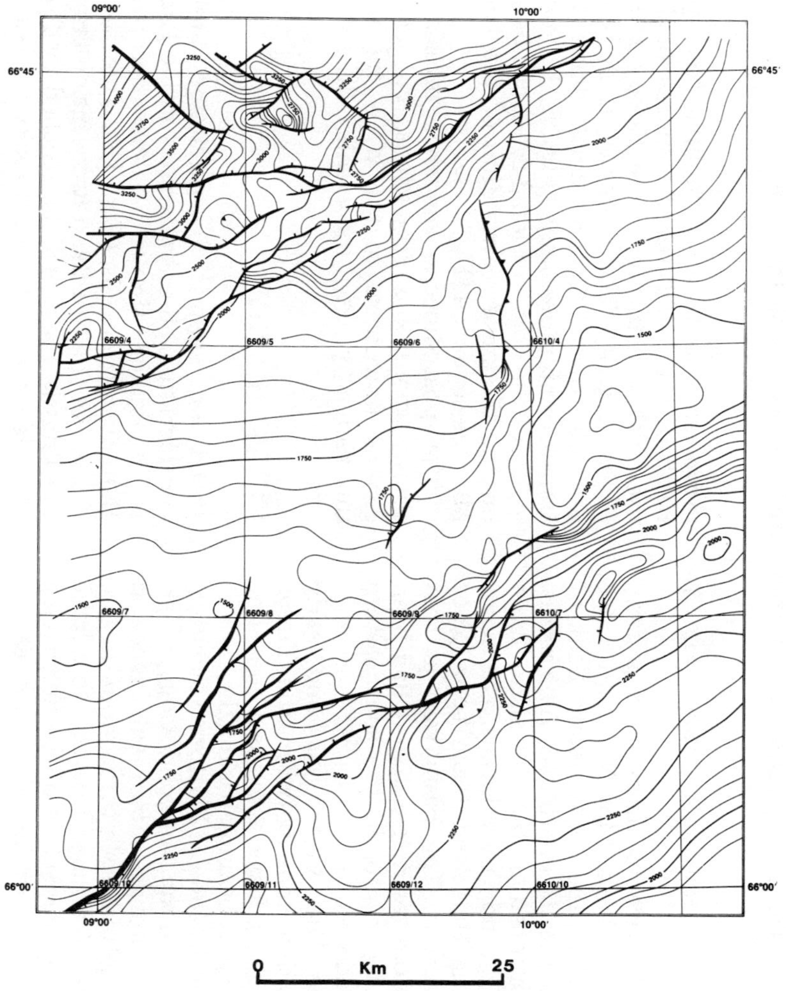

Fig. 10. Late Kimmerian unconformity, structural map in two-way seismic travel time.

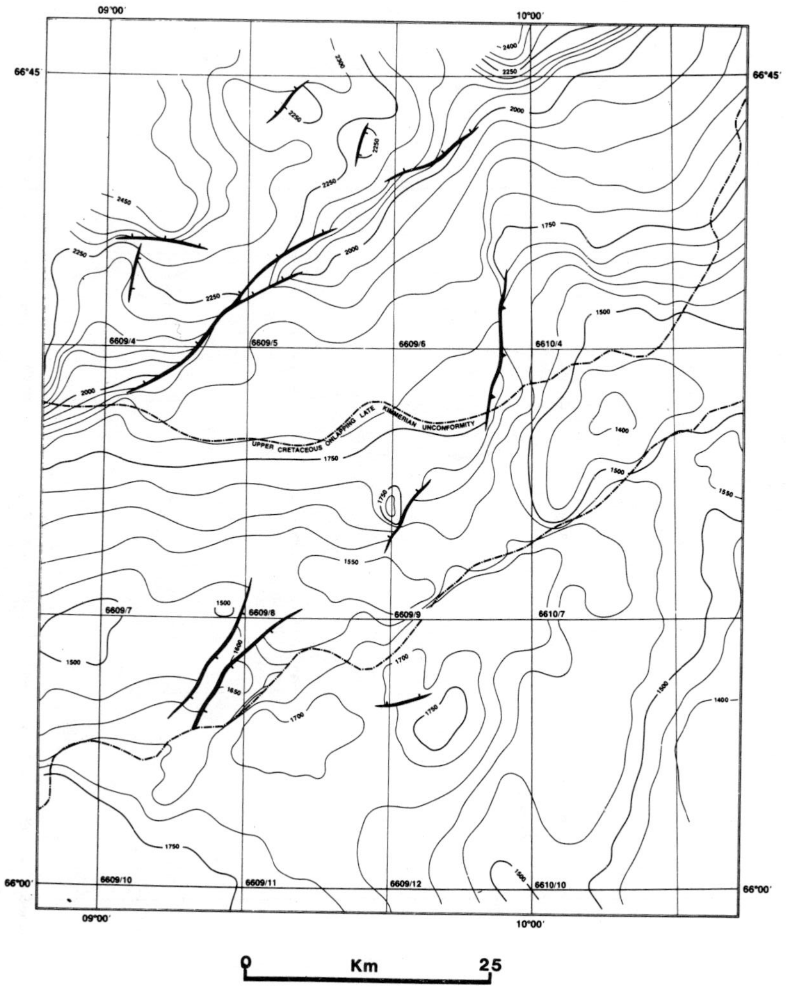

Fig. 11. Top Cenomanian, structural map in two-way seismic travel time.

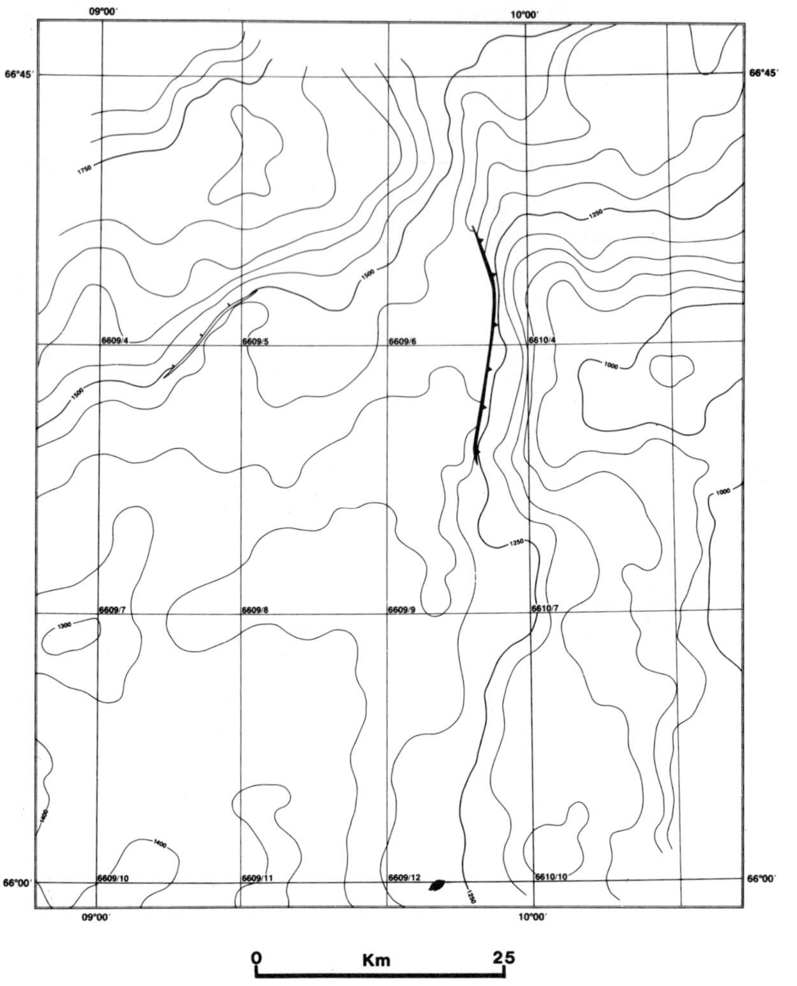

Fig. 12. Top Palaeocene, structural map in two-way seismic travel time.

SE of the Lower Cretaceous on the ridge (Figs. 7 and 11).

Through early Cretaceous times fault activity along the southeastern flank of the Nordland Ridge gradually ceased. Along the northwestern flank of the ridge the fault activity continued. As a result, subsidence rates dropped in the Helgeland Basin relative to the Vøring and Træna Basins. The major development of the Helgeland Basin took place during early Cretaceous. The basin was, at that time, areally far more restricted than it had been earlier, with the depocentre in the southeastern corner of the opened area.

Throughout the late Cretaceous and early Tertiary the transition zone between the Træna Basin and the Nordland Ridge acted as a hinge-line between an actively subsiding basin (the Træna Basin) and the stable Trøndelag Platform (including the Helgeland Basin and the Nordland Ridge) (Figs. 7, 12 and 14). Although minor subsidence is observed in the Helgeland Basin throughout the late Cretaceous and early Tertiary, this can to a large extent be ascribed to compaction of the underlying Lower Cretaceous and Upper Jurassic sequences.

The last tectonic event affecting the structuring of the Trænabanken area was a compressional event, probably of Oligocene age (35 Ma?). Compressional forces acted in a NNW—SSE direction, and as a result the N-trending Triassic faults were rejuvenated. Depending on the faults' orientation relative to the compressional forces, left lateral movements or reversed faulting took place. The result of the mid-Tertiary compressional event was the final structuring of the Rødøy and Grønnøy Highs (Fig. 3). The Grønnøy High is a classical example of a compressional dome related to transcurrent movements (Figs. 6, 12 and 14). The effect of the compressional event can be seen easily on the thickness distribution of the Palaeocene—Miocene sequence (Fig. 14).

The post-Miocene sequence reveals numerous transgressions and regressions, but except for a general downwarp to the WNW the sequence is unaffected by structural events (Fig. 13).

STRATIGRAPHY

Based on seismic interpretation and regional considerations, the stratigraphy inside Trænabanken is considered similar to that observed in the Haltenbanken wells (Fig. 15). Due to the heavy erosion of the Nordland Ridge inside Trænabanken, pre-Triassic rocks will be reached by future wells. Knowledge of Palaeozoic rocks has been obtained from more distant areas (Figs. 5 and 16). In the basins more complete Cretaceous and Middle—Upper Jurassic sequences are expected than observed inside the Haltenbanken area. Additional information has been revealed by the shallow drilling off the Helgeland coast (IKU, 1983) (Fig. 5). The expected stratigraphy inside the Trænabanken

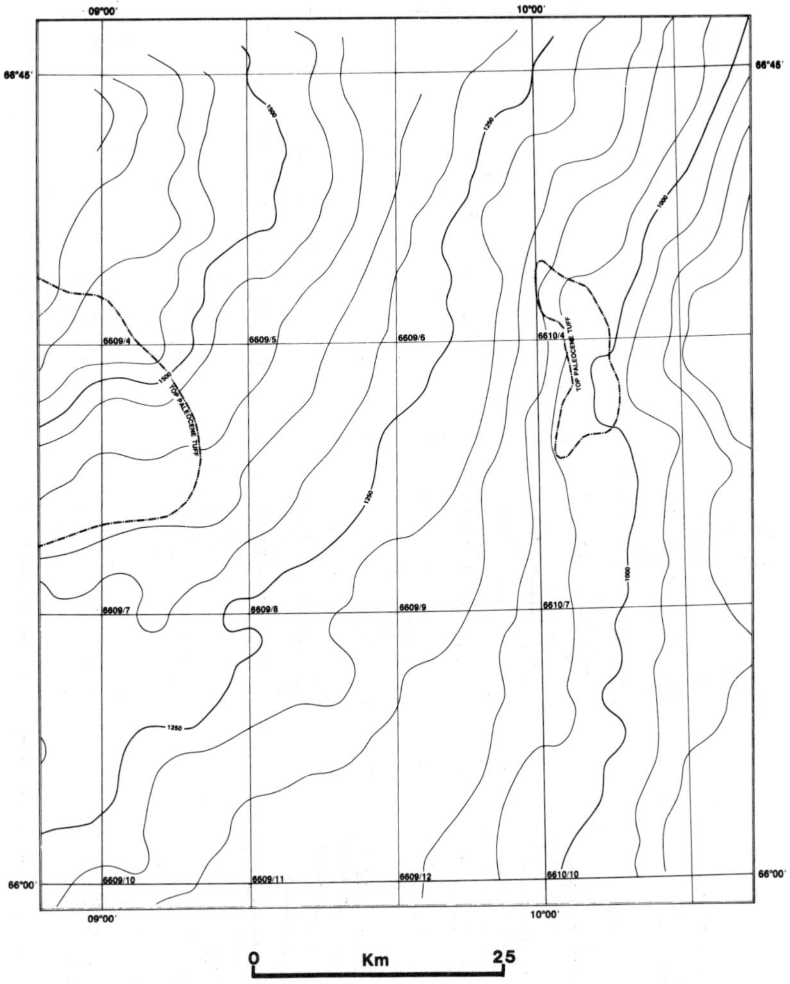

Fig. 13. Top Miocene, structural map in two-way seismic travel time.

area is shown in Fig. 17, and continuous reference is made to Figs. 15—17.

Basement

Magnetic and gravimetric interpretation indicate that basement lies relatively deep. To the north the gravimetric effect of the dense Lofot Massif is observed. In general, the basement is considered to be similar to the Caledonian and Precambrian rocks observed on the Helgeland coast.

Palaeozoic

Following the Caledonian orogeny, clastic sedimentation has dominated the North-Atlantic geologic province. The orogeny formed the structural framework for later sedimentation. On Trænabanken the Palaeozoic sequence will be dominated by sandstones with conglomerates and minor shale- and ashlayers. Carbonates deposited in marginal marine environments may be found in the Upper Permian. Expected unconformities, ashlayers and thickness-variations are due to synsedimentary tectonic activity. Seismic interpretation indicates an erosional unconformity at or close to the top of the Palaeozoic sequence on the Nordland Ridge, probably removing at least parts of the Permian sequence.

Mesozoic

The boundary between Palaeozoic and Mesozoic rocks varies from being gradual to defined by hiatal gaps in the surrounding areas. InsideTrænabanken the boundary will be unconformable, at least on the Palaeozoic highs where the unconformity is observed.

Triassic

Marginal marine sandstone and siltstone of early Triassic age were cored off the Helgeland coast in 1982 (IKU, 1983). From East Greenland, Lower Triassic is represented by marine shale and sandstone overlain by continental sandstone. A similar development is expected on Trænabanken where seismic interpretation locally indicates thicknesses in excess of 1000 m.

Seismic interpretation within Trænabanken does not indicate salt flowage, and thin salt layers or immobile carbonate facies are expected interbedded with sandstones and shales in the Middle Triassic. The total thickness of Middle Triassic varies, but it is generally several hundred metres. On the Rødøy High, the erosion has removed Upper and Middle Triassic sediments.

Marginal marine to continental shale and sandstone are expected in the lower part of the Upper Triassic. A very typical seismic character from the Triassic—Lower Jurassic coal sequence (Fig. 6) can be followed from Haltenbanken to Trænabanken, demonstrating its presence also in the latter area.

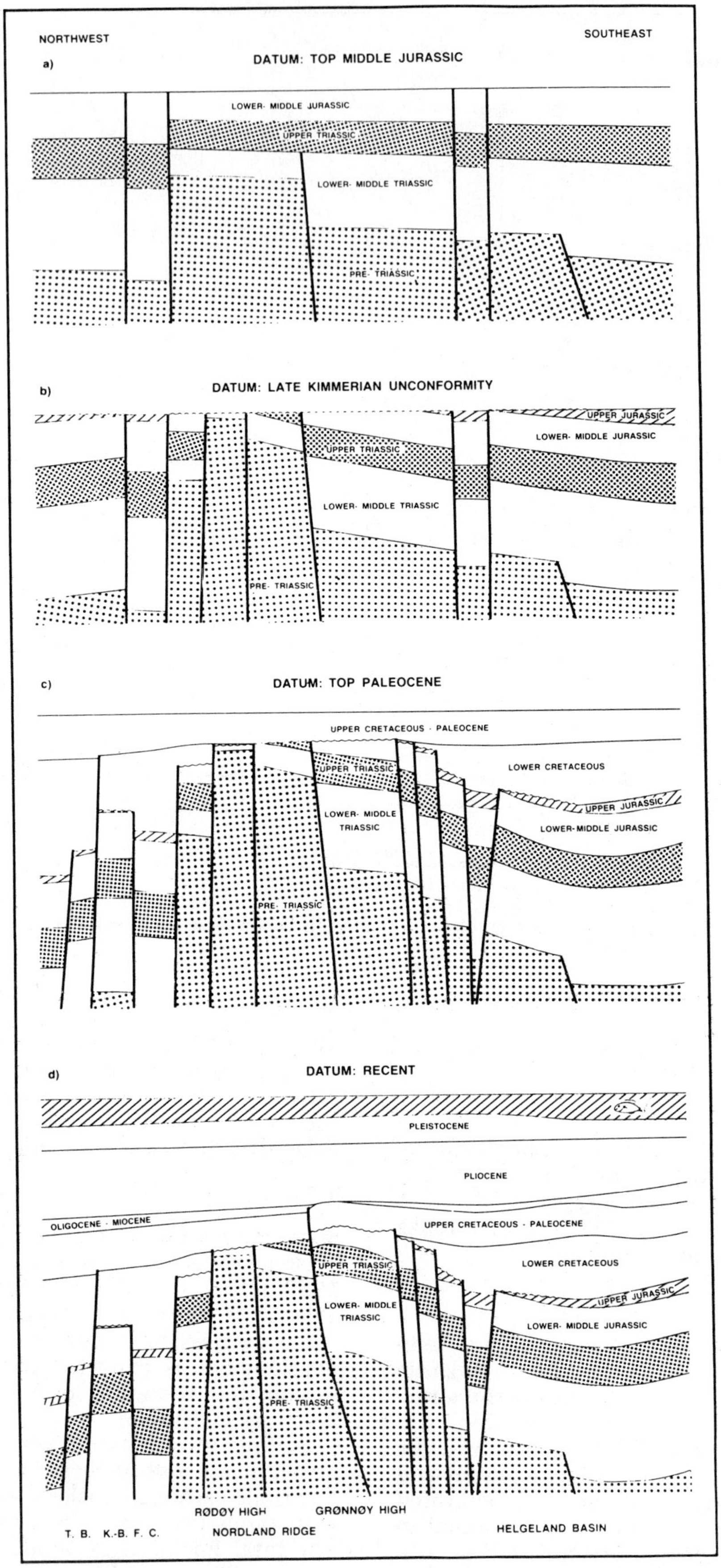

Fig. 14. Restored schematic cross-sections from Trænabanken. Orientation roughly as seismic lines NH 8102-426 and -430 shown on Fig. 4.

Jurassic

The main part of the coal sequence is of early Jurassic age. It is overlain by cleaner, marginal sandstone, and the whole sandstone—coal sequence is typically 500-m thick, but with variations as discussed earlier.

A shale sequence of relatively constant thickness is expected to overlie the sandstone—coal sequence. This grades up into a sandstone of Middle Jurassic age, deposited in a marginal marine environment. The Middle Jurassic thickness, where the sequence apparently is not eroded, will be greater than that observed in the Haltenbanken wells, indicating generally more or less continuous Middle—Late Jurassic deposition. The shallow wells off the Helgeland coast suggest that the additional thickness will be dominated by shales. Locally, the seismic data demonstrates onlap of Upper Jurassic on to Middle Jurassic, and syn-sedimentary faulting is also observed to have occurred during this epoch. Shale will be the dominating lithology in the Upper Jurassic sequence. Thickness variations are due to both varying subsidence, locally fault controlled, and erosion.

Cretaceous

A continuous deposition is expected over the Jurassic/Cretaceous boundary in the Træna Basin and the deeper parts of the Helgeland Basin. As on Haltenbanken, shale will be the dominating lithology. Due to erosion of the Nordland Ridge, sandstone may be found towards the Træna Basin. As the basins developed during the Cretaceous cooling stage, huge variations in thickness are observed. The Lower Cretaceous sequence onlaps out against the Nordland Ridge (Figs. 6 and 7). The Upper Cretaceous sequence is generally thin on the ridge, and onlaps out against the Rødøy High southwest of the area studied.

Tertiary

Except for areas subject to mid-Tertiary compressional tectonics, marine shales are expected to dominate inside the Trænabanken area. Thin sandstones may have been deposited around the compressional highs. As on Haltenbanken the main Tertiary subsidence occurred post-Miocene. The seismic character of a Pliocene prograding unit (Figs. 6 and 7) may suggest some sand development.

As the whole Trøndelag Platform acted as a stable, high area during most of the Upper Cretaceous and Tertiary, sea-level fluctuations have resulted in frequent hiatuses in the condensed sequence.

HYDROCARBON POTENTIAL

Reservoir rocks

Reservoir rocks are expected at different stratigraphic levels inside the Trænabanken area (Fig. 17).

Where present, Middle and Lower Jurassic sandstones represent the main objectives. As discussed earlier, these sandstones are truncated on the Nordland Ridge, and absent over parts of the ridge (Figs. 6 and 7). The Middle Jurassic sandstones, where not truncated, are expected to be of at least equal thickness to those observed in the

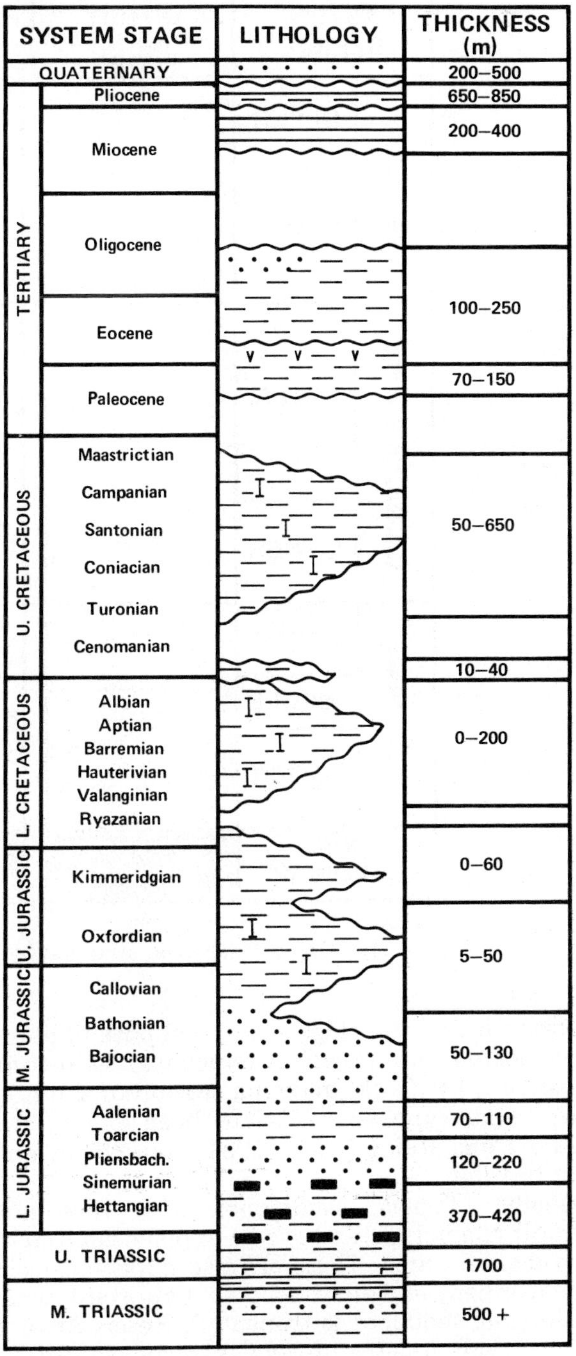

Fig. 15. Haltenbanken, generalized stratigraphy. Based on wells drilled prior to 1983.

Haltenbanken wells (50—130 m). By seismic interpretation the gross Lower Jurassic sandstone—coal sequence thins from 600 m in the Helgeland Basin to 400—500 m in the transition zone to the ridge, before thinning more by truncation. The thickness on the northwestern side of the ridge is more uncertain, but indications are that the sequence is even thicker than on the southeastern side. The Jurassic sandstones will exhibit good reservoir characteristics.

The area where the Jurassic sequence is completely stripped off is dependent on the presence of older reservoir rocks. The dating of the sediments in this region is uncertain, but sandstones are expected to dominate in the Permo-Devonian sequence. Palaeozoic sandstones represent the reservoir rocks on the Rona Ridge in the West Shetland Basin (Ridd, 1981).

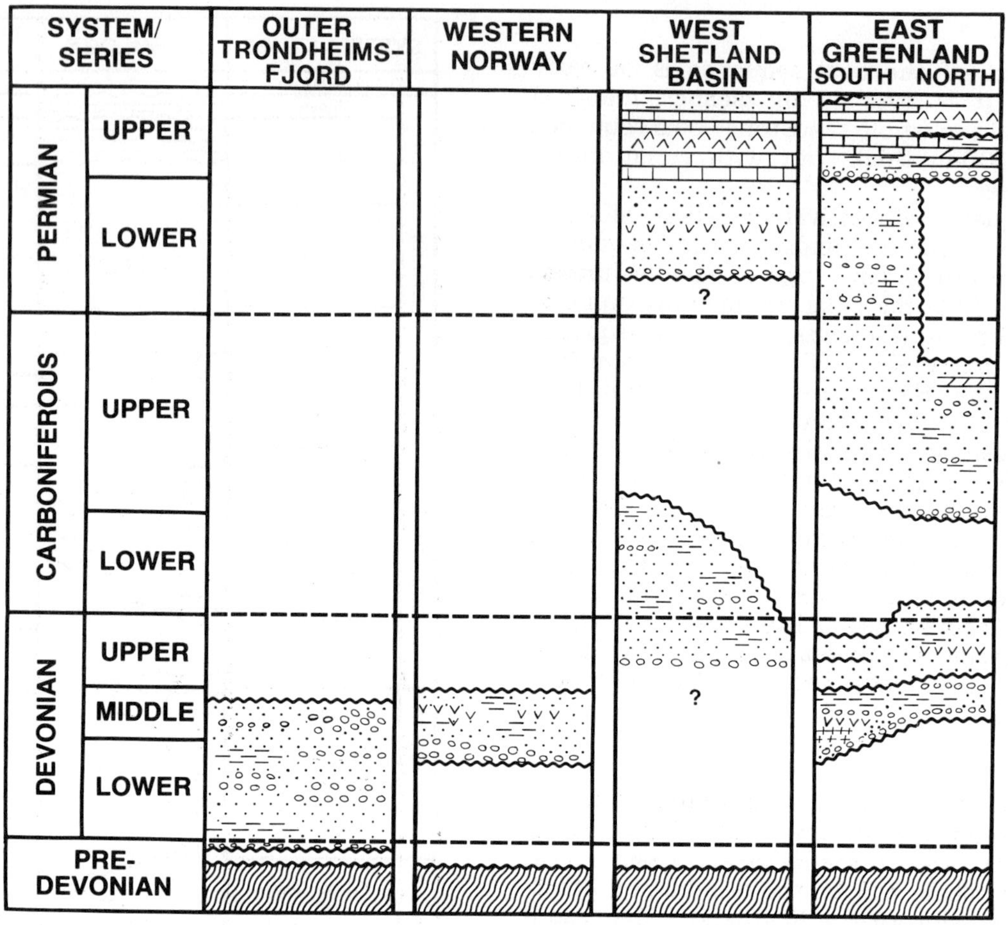

Fig. 16. Upper Palaeozoic stratigraphy from different areas. Locations shown on Fig. 5.

Northwest of the ridge additional potential may be expected in Lower Cretaceous sandstones deposited downflank following erosion of the ridge itself. Similar type sediments have been encountered off the Rona Ridge (Ridd, 1981), but so far only water-bearing.

Finally, Triassic sandstones with reasonable reservoir characteristics have been penetrated in the Haltenbanken area. These are also expected inside Trænabanken, and may become important where the Jurassic sequence is truncated. Reservoir rocks younger than possible Lower Cretaceous sandstones are not believed to be of major importance.

In conclusion, reservoir rocks will generally not represent a major risk inside the Trænabanken area. However, the central western area is dependent on the presence of pre-Jurassic reservoir rocks.

Traps

The complex tectonic development of the Trænabanken area has resulted in the formation of numerous traps. Structural, stratigraphic and combination type traps are observed and, based on type and genesis, they can be grouped as follows (Figs. 6, 7 and 18):

(1) Type I — Structural traps on the faultblocks down from the Nordland Ridge into the flanking basins. These formed during the late Jurassic rifting phase. They may be limited both by synthetic or antithetic faults. Due to a general tilt of the faultblocks towards the basin, most type I traps are dependent on closure against faults. Type I traps are generally small.

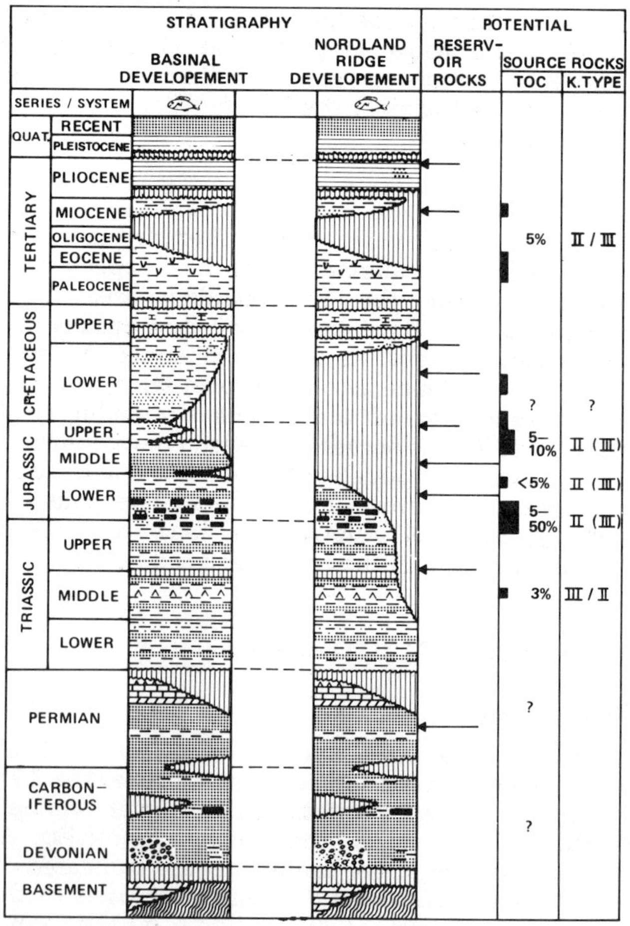

Fig. 17. Trænabanken, generalized stratigraphy, and potential reservoir and source rocks.

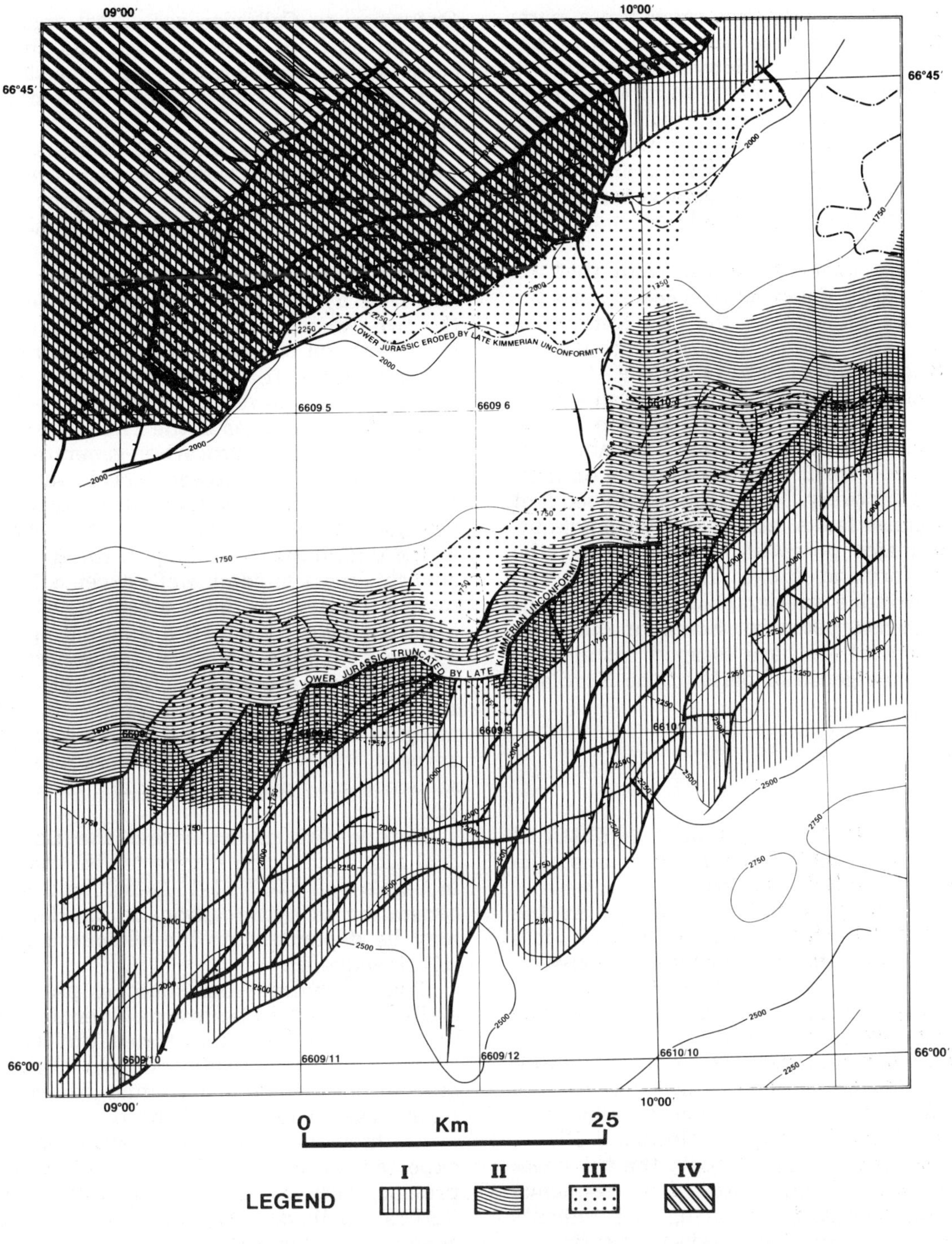

Fig. 18. Location of various trap types on Trænabanken. See also Figs. 6 and 7.

(2) Type II — Structural and combination traps along the axis of the Nordland Ridge. Structurally, type II traps were formed by late Jurassic rifting, creating the ridge, later Cretaceous and Tertiary tilting towards northwest locating the axis on the Helgeland Basin side of the ridge, and finally by mid-Tertiary compression deforming parts of the axis in an articlinal manner; some of these traps are partly closed by erosional truncation, making them combination traps. The erosion took place during early Cretaceous. Some type II traps are dependent on closure against faults. Type II traps generally have a great potential.

(3) Type III — Stratigraphic unconformity traps on the Nordland Ridge. Type III traps formed by early Cretaceous erosion, and by later structural tilting away from the truncation line. Also some of these traps are partly dependent on fault seal, and they are of limited size.

(4) Type IV — Stratigraphic facies change traps in the Kristiansund-Bodϕ Fault Complex zone. Type IV traps may have formed during early Cretaceous erosion of Jurassic and Triassic sandstones on the Nordland Ridge, and subsequent redeposition out towards the Træna Basin. These traps are always dependent on closure against the ridge itself and are, in fact, combination traps. Type 4 traps are high risk, both because of the uncertainty concerning the presence of reservoir rock, and concerning closure.

Caprocks generally do not represent any risk, as shales are expected to be the dominant lithology in the upper part of Lower Jurassic, Upper Jurassic, Cretaceous and Tertiary. However, after the erosional event during early Cretaceous times, parts of the Nordland Ridge were not onlapped before late Cretaceous times. The Rødøy High, subject to mid-Tertiary tectonic movements southwest of the opened area, may not have been adequately covered before the post-Miocene subsidence. Hydrocarbon generation commenced in the Vøring and Træna Basins about 100-Ma ago. Consequently, appreciable amounts of generated hydrocarbons may have escaped entrapment.

Seal may be a problem for the traps that are completely or partly dependent on seal against the various pinchout lines below the late Kimmerian unconformity. This is due to possible continued migration along the unconformity surface.

The following conclusions can be based on the mapped traps: numerous traps of different origin and configuration are present within Trænabanken. Most traps are too small to be economical on their own in case of discovery. Many traps are dependent on closure against faults, and many on pinchout-closure. The major potential is found along the axis of the Nordland Ridge. Lower Cretaceous fault scarp-sandstone stratigraphic traps represent additional high-risk potential. Some of the early generated hydrocarbons have escaped entrapment due to late submergence of parts of the Nordland Ridge.

Source rocks

Potential

Knowledge of source-rock potential is obtained mainly from the Haltenbanken area and the shallow wells on the Helgeland coast.

Although some potential is expected in the Upper Cretaceous-Tertiary sequence (Fig. 17), it is too shallowly buried in the Trænabanken area to be of interest as a hydrocarbon generator.

So far, few geochemical data are available from the Lower Cretaceous sequence. The indication is that Lower Cretaceous shales have a poor to moderate gas sourcing potential. In the Trænabanken area the Lower Cretaceous sequence may be complete in the basins flanking the Nordland Ridge. Even in the shallow Helgeland Basin the thickness locally may reach 1000 m. Parts of the sequence may have been deposited in reducing environments as observed in the North Sea (Hesjedal and Hamar, 1983) and on Tromsøflaket (Bleie *et al.*, 1982). At optimum maturity appreciable amounts of both oil and gas may be generated from the Lower Cretaceous shaly sequence flanking the Nordland Ridge.

Both Kimmeridge Clay and Heather Formation equivalents are expected in the Upper Jurassic sequence. The former, of Kimmeridgian to Volgian (—Ryazanian?) age, has a total organic content from 5 to more than 10%. Kerogen type II is slightly dominating over type III. Samples from two shallow boreholes off the Helgeland coast and from East Greenland support the impression that this interval in the Trænabanken area will be capable of generating large amounts of oil and gas. The total Upper Jurassic thickness is expected to exceed 200 m in the basins flanking the Nordland Ridge.

Large oil and gas sourcing potential is also expected in the Heather Formation equivalent of Oxfordian to Callovian age. Total organic content in samples from Haltenbanken and the Helgeland coast varies considerably, but exceeds 5% in the best intervals. Kerogen type II dominates over type III.

Thin shale layers and coal stringers interbedded with sandstones of Middle Jurassic age will locally demonstrate high content of organic matter, but the potential is low due to the small volume of source rock.

Large variations in total organic content are observed in the Lower Jurassic shales separating the two Jurassic reservoir rocks. The best intervals have a total organic content above 5%, with kerogen type II dominating. This sequence is expected to have a moderate potential for generating oil and gas in the Trænabanken area.

The highest hydrocarbon-generating potential is expected in the Lower Jurassic sandstone—coal sequence and interbedded shale layers. The total thickness of this sequence is 500—600 m, and coal is the dominating lithology in the lower part. Net coal thickness in some wells exceeds 100 m. Total organic content in the coalbeds is, of course, very high. Kerogen type III is dominating, giving this sequence a huge potential for generation of gas with minor amounts of heavy oil. Interbedded shales and siltstones are partly inertinitic and meagre, but some contain appreciable amounts of kerogen types II and III. Consequently the coal sequence may also be important for generation of oil.

Isotope data indicate that the Lower Jurassic coal sequence is the source for the gas condensate in Saga's well 6507/11-1. As will be discussed later, generation of hydrocarbons in the Helgeland Basin may be dependent on the oil sourcing capabilities of this sequence.

Triassic hydrocarbon-sourcing potential is generally expected to be low on Trænabanken. A thin sequence has been penetrated on Haltenbanken, showing a total organic content above 3%. This may also have some potential for generation of oil and gas on Trænabanken.

As discussed earlier, Permian or older rocks are expected at shallow depths on Trænabanken. If present, Permian carbonates and Lower Carboniferous coalbeds may have some hydrocarbon-generating potential.

Maturity

From the previous discussion it is apparent that potential source rocks will be present at different levels in the Trænabanken area. To evaluate the level of maturation in this area, the maturation profiles in the Haltenbanken wells have been analysed. Maturation varies with temperature history, burial history and, to some extent, kerogen type.

The temperature history is mainly dependent on magnitude and timing of tectonic events, and on heat transfer by convection. To analyse the former, stretching factors have been estimated for the late Jurassic/early Cretaceous and early Tertiary events. This results in a variable temperature gradient with time. Using this and the present-day relatively high temperature gradient for the area of about 3.5°C/100 m, the wells can be reasonably modelled by use of different kinetic models for kerogen maturation (Wapples, 1980; Hood *et al.*, 1975; Gretener, 1981). A complicating factor in using this approach is that the present-day high temperature gradient may in part be due to transfer of heat by convection from the basin up into the Kristiansund-Bodø Fault Complex zone and the Trøndelag Platform (Fig. 3).

The low maturity in the Haltenbanken wells compared with wells in the Viking Graben area is mainly due to late Tertiary burial. A burial history for a point within the Lower Jurassic coal sequence (180-Ma old) for typical wells east in the fault complex zone and on the Trøndelag Platform is shown in Fig. 19 (well A). Both on this, and on the curve typical for wells nearer to the basin (Fig. 19, well B), the large burial during the latest 10 Ma is apparent. The burial curves are not decompacted and not corrected for palaeo-waterdepth.

The timing of the late Tertiary burial, set at 10 Ma in Fig. 19, is uncertain. But by applying the burial histories as shown, and a *constant* temperature gradient of 3.5°C/100 m, three wells were modelled very closely by use of Wapples' method (Wapples, 1980). Wapples' equation reads:

$$\text{Maturity index} = \sum_{n_{\min}}^{n_{\max}} \Delta tn \cdot r^n$$

where: n denotes different temperature intervals (20°C—30°C) > $n = -8$; n increases by 1 for each 10°C increase in temperature); Δtn denotes the timespan the source rock stayed in interval n and; r is a factor used to describe the increase in rate of maturation. It is a kerogen-type dependent constant thought to lie in the general range 1.6—2.5. Calibration by Wapples (1980) gave a most likely value for r of 2. This means the rate of maturation increases by a factor of 2 for every 10°C rise in temperature.

The observed and modelled maturity values for the burial curves shown in Fig. 19 are Ro = 0.4 and Ro = 0.6, respectively. This indicates immaturity and moderate maturity, respectively.

The use of a constant palaeogeothermal gradient is definitely not the correct solution. However, due to uncertainties in establishing the heat-flow effect of the various tectonic events, an unknown convection effect, and acceptable modelling with a constant gradient, the latter was used in parallel with a varying geothermal gradient.

The burial history from the two points in the Trænabanken area is also shown in Fig. 19; one in the Træna Basin and one in the Helgeland Basin. The point chosen in the Helgeland Basin lies in the deeper parts of that basin. Burial curves are shown both on intra Lower Jurassic (180 Ma) and top Jurassic (130 Ma) level. Both points chosen are in principle following the same burial trend as seen in well B on Haltenbanken. The effect of the Cretaceous cooling is, however, far more pronounced in the Træna Basin, and a thicker Jurassic and Cretaceous sequence also resulted in a more rapid burial in the Helgeland Basin than in well B. The age of the Cretaceous basin fill is somewhat uncertain. In the Træna Basin a constant subsidence throughout Cretaceous times is used. In the Helge-

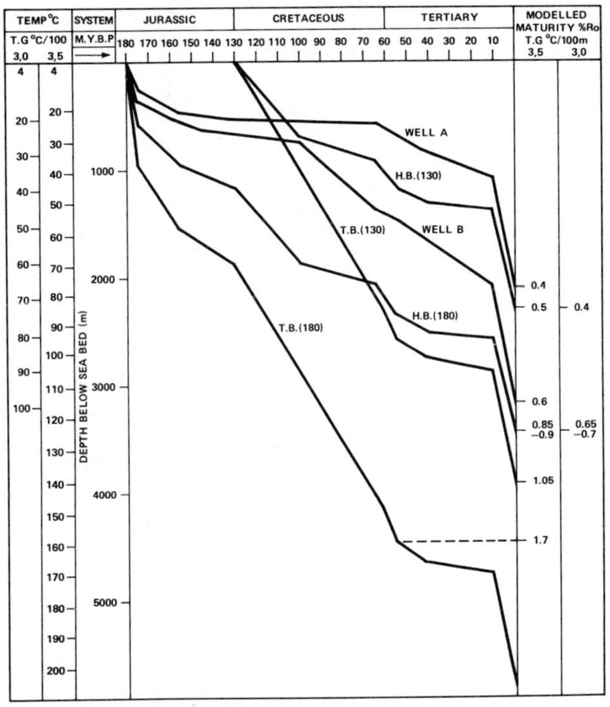

Fig. 19. Burial history and modelled temperature and source-rock maturity from Haltenbanken wells, Helgeland and Træna Basins. Well A typical of eastern Haltenbanken area, well B of western area. H.B.(130), 130-Ma old bed in the Helgeland Basin; H.B.(180), 180-Ma; T.B.(130), 130-Ma old bed in the Træna Basin; T.B.(180), 180-Ma.

land Basin the preferred interpretation with rapid early Cretaceous burial is shown.

The maturity has been modelled using both variable and constant palaeogeothermal gradients. For the constant gradient case, 3.5°C/100 m was used in the Træna Basin, and both 3.5°C/100 m and 3.0°C/100 m in the Helgeland Basin. Probably the heat flux in the Helgeland Basin has been lower than indicated from Haltenbanken. Judging from that, 3.0°C/100 m is probably the more realistic value. The results from the modelling (Fig. 19) indicate that the Lower Jurassic coal sequence today generates dry gas in the Træna Basin, while in the Helgeland Basin it is in the oil-generating zone, close to peak generation. An alternative, later, Cretaceous burial does not change this conclusion. The Upper Jurassic is today in the main oil-generating zone in the Træna Basin, while in the Helgeland Basin early oil generation has just begun.

Modelling with variable geothermal gradient indicates slightly greater maturity, and the present-day maturity maps for top Jurassic and intra Lower Jurassic (Figs. 20 and 21) are the results of the variable gradient modelling.

The following conclusions can be drawn from the maturity analysis:

(1) Large quantities of oil and gas have been generated from the Jurassic sequence in the Træna and Vøring Basins north and west of the Nordland Ridge. Generation started more than 100 million years ago.

(2) In the Helgeland Basin the Upper Jurassic shales have just entered the early oil-generating zone. It is doubtful whether any generated hydrocarbons have been expelled.

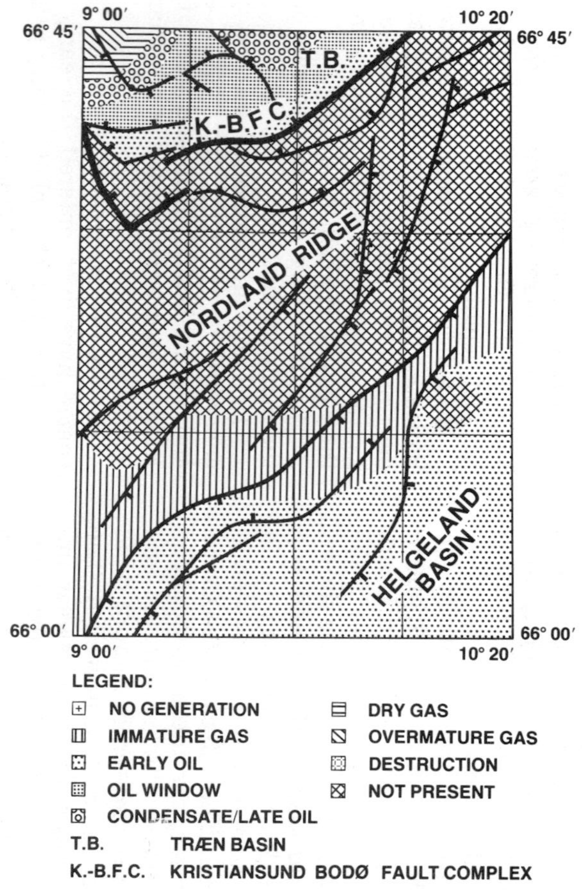

Fig. 20. Present-day maturity map, Upper Jurassic.

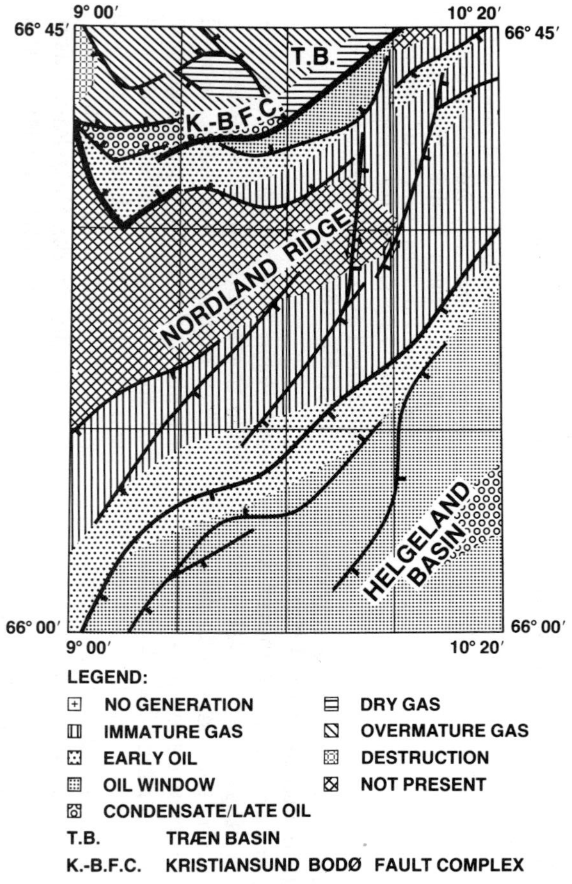

Fig. 21. Present-day maturity map, Lower Jurassic coal sequence.

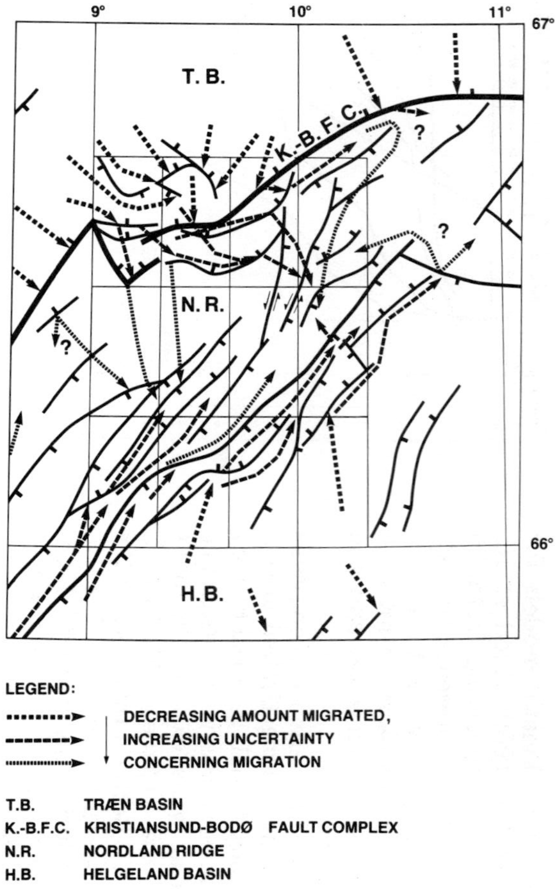

Fig. 22. Migration map from Trænabanken.

(3) Deeper Jurassic source rocks in the Helgeland Basin are more mature, and the Lower Jurassic coal sequence is in the oil window. However, maximum generation has only gone on for a few million years.

Secondary migration and entrapment for hydrocarbons generated in the Helgeland Basin is consequently dependent on the presence of appreciable quantities of kerogen type II in the Lower Jurassic coal sequence and the generation of quantities large enough to allow expulsion from the source rock (primary migration).

Secondary migration

Generation and expulsion of hydrocarbons in the Helgeland Basin is questionable, whereas those processes represent no problems in the Vøring and Træna Basins. The hydrocarbon potential of the area is then mainly dependent on whether hydrocarbons have been made available for secondary migration in the Helgeland Basin, and to what degree hydrocarbons generated in the basins have been able to migrate on to the ridge. The largest mapped structures are found along the axis of the Nordland Ridge.

The fact that most fault blocks dip towards the basins is a positive factor for secondary migration on to the Nordland Ridge. Even though most faults probably will have some sealing capacity, they will probably not withstand the pressure from a large hydrocarbon column. Over most faults, sandstone lies against sandstone.

Secondary migration from the Vøring and Træna Basins (and the Kristiansund-Bodø Fault Complex zone) is in addition partly dependent on migration paths along the unconformity surface, and/or migration in pre-Jurassic permeable rocks where the Jurassic sequence is stripped off. These questions are related to sealing against pinchout lines and presence of pre-Jurassic reservoir-rocks, respectively, as discussed earlier.

Finally, due to late submergence of parts of the ridge, some of the early generated hydrocarbons will have escaped.

A secondary migration map has been constructed based on the present-day top Lower Jurassic sandstone map (Fig. 22). The deepest structures towards the Helgeland Basin are located favourably for entrapment of any hydrocarbons generated in that basin. Additional amounts will probably migrate in a northeastward direction and may fill higher-lying structures. Migration from the southwest will be limited, as the Nordland Ridge in that direction probably will trap most hydrocarbons. Depending on the depth of possible pre-Jurassic reservoir rocks below the late Kimmerian unconformity surface of the Rødøy High, that high is favourably located. Depending on sealing capacity of faults and pinchout lines, on migration paths along the unconformity surface, and again on pre-Jurassic reservoir rocks, the Grønnøy High is also favourably positioned. The complex horst block in the fault complex zone towards the Træna Basin has probably the least risk with regards to secondary migration.

In conclusion, the major structures in the Trænabanken area, along the axis of the Nordland Ridge, have a reasonable chance of being reached by hydrocarbons generated in the flanking basins. Both oil and gas are generated in the Vøring and Træna Basins, and gas may have forced out earlier trapped oil. The relatively complicated migration path is more easily accessible for gas than oil.

SUMMARY AND CONCLUSIONS

The Trænabanken area has undergone a complex structural history. The present-day configuration is mainly the result of compressional and extensional forces that acted during late Jurassic—early Cretaceous times. During Cretaceous times the Nordland Ridge was heavily eroded, while the flanking basins subsided. The Rødøy High was also subject to pre-Triassic erosion, while both the Rødøy and Grønnøy Highs experienced mid Tertiary compressional forces. Main reservoir rocks are Lower and Middle Jurassic sandstones. Areas where these are eroded are dependent on older reservoir rocks. Lower Cretaceous sandstones may represent additional potential. Large quantities of oil and gas are generated from the Jurassic sequence in the Vøring and Træna Basins. Due to lower maturation, generation from the Helgeland Basin is dependent on oil-generating potential in the Lower Jurassic coal sequence. Numerous traps are defined throughout the area, but the major potential is found along the axis of the Nordland Ridge. The migration path into those traps is complicated. Some hydrocarbons may have escaped entrapment due to late burial of parts of the ridge. Due to tectonic, stratigraphic and hydrocarbon generation—migration uncertainties, the Trænabanken area must be considered a relatively high-risk area in terms of petroleum exploration. However, the hydrocarbon potential of the Trænabanken area is considered to be large.

ACKNOWLEDGEMENTS

We wish to thank Statoil for permission to publish this paper.

Special thanks to S.A. Tonning who participated in the 7th round evaluation, to T. Knutsen for typing, to I. Giltner for suggesting linguistic improvements, and to Statoil's draughting department.

REFERENCES

Aanstad, K.M., Gabrielsen, R.H., Hagevang, T., Ramberg, I.B. and Torvanger, O. 1981. Correlation of offshore and onshore structural features between 62°N and 68°N, Norway. In: *Norwegian Symposium on Exploration* (Norsk Petroleumsforening), Article 11, 24 pp.

Bleie, J., Oppebøen, K.A. and Nysæther, E. 1982. The hydrocarbon potential of the Northern Norwegian Shelf in the light of recent drilling. In: *The Geological Framework and Hydrocarbon Potential of Basins in Northern Seas. Offshore Northern Seas Conference.* (Norsk Petroleumsforening), Article, E/5, 39 pp.

Gretener, P.E. 1981. *Geothermics: Using Temperature in Hydrocarbon Exploration.* AAPG Education Course Note Ser. No. 7, 156pp.

Hesjedal, A. and Hamar, G.P. 1983. Lower Cretaceous stratigraphy and tectonics of the south-southeastern Norwegian offshore. *Geol. Mijnb.* **62**, 135—144.

Hollander, N.B. 1982. Evaluation of the hydrocarbon potential offshore mid-Norway. *Oil Gas J.* April 12, 168—172.

Hood, A., Gutjaha, C.C.M. and Heacock, R.L. 1975. Organic metamorphism and the generation of petroleum. *Bull. Am. Assoc. Petrol. Geol.* 59, 986—996.

IKU, 1983. *Shallow Drilling of Mesozoic Rocks off Helgeland, 1982.* Report prepared for oil companies by IKU, 332pp.

Jørgensen, F.H. and Navrestad, T. 1979. Main structural elements and sedimentary succession on the shelf outside Nordland (Norway). In: *Proceedings of Norwegian Sea Symposium*, Tromsø, NPF, NSS-79, paper 11, 20pp.

Jørgensen, F.H. and Navrestad, T. 1981. The geology of the Norwegian Shelf between 62°N and the Lofoten Islands. In: L.V. Illing and G.D. Hobson (eds.), *Petroleum Geology of the Continental Shelf of North-West Europe.* Heyden, London, pp. 407—413.

Price, I. and Rattey, R.P. Cretaceous tectonics off mid-Norway: Implications for the Rockall—Faeroes—Shetland rift. *J. Geol. Soc. Lond.* (in press).

Ridd, M.F. 1981. Petroleum geology west of the Shetlands. In: L.V. Illing and G.D. Hobson (eds.), *Petroleum Geology of the Continental Shelf of North-West Europe.* Heyden, London, pp. 414—425.

Rønnevik, H.Chr., Eggen, S. and Vollset, J. 1983. *Exploration of the Norwegian Shelf.* NPD-contribution No. 5, 39pp.

Rønnevik, H.Chr., Jørgensen, F.H. and Motland, K. 1979. The geology of the northern part of the Vøring Plateau. In: *Norwegian Sea Symposium* (Norsk Petroleumsforening), Article 11, 20 pp.

Strønen, L.K., 1981. Main Geophysical Study of the Norwegian Continental Margin between 62° — 69°. Unpublished Cand.real. Thesis, Seismological Observatory, University of Bergen, 146pp.

Talwani, M. and Eldholm, O. 1972. Continental margin off Norway: A geophysical study. *Geol. Soc. Am. Bull.*, 83, 3575—3606.

Talwani, M., Mutter, J. and Eldholm, O. 1981. The initiation of opening of the Norwegian Sea. *Oceanol. Acta*, 23—30.

Wapples, D.W. 1980. Time and temperature in petroleum exploration: application of Lopatin's method to petroleum exploration. *Bull. Am. Assoc. Petrol. Geol.*, 64, 916—926.

Whiteman, A.J., Rees, G., Naylor, D. and Pegrum, R.M. 1975. North Sea troughs and plate tectonics. *Tectonophysics*, 26, 39—54.

16 The geological history of Trænabanken

Martin B. Gowers and Geir Lunde
Norsk Hydro, Sandvika, Norway

After the Caledonian orogeny the folded and metamorphosed Cambro-Silurian sediments of the Nordland Shelf were subjected to erosion, and local Devonian intra-montane clastic basins formed around the eroding highs. In mid-Devonian times sinistral wrench faulting established an important tectonic trend which controlled the later development of the Trænabanken area. Major normal faulting at the start of Permian times created a deep basin in the east of Trænabanken. This basin was filled progressively with Permian clastics and marine carbonates and was subsequently covered with a thick continental Trias sequence. Extensive coal formation in earliest Jurassic times marked the transition from continental to marine depositional environments. Apart from some synsedimentary faulting in mid-Jurassic times, the Permian to end Middle Jurassic sequence is relatively undisturbed by tectonic movements.

A fully marine late Jurassic environment witnessed extensive local faulting along a zone cutting diagonally northeast across Trænabanken. However, there was no major relief in the area at this time, although there are indications that the west and northwest of the Trænabanken area may have undergone some erosion.

In early Cretaceous times the old Devonian trend was reactivated to form a distinct flexure across Trænabanken (the hinge line) which separated a strong erosional area in the northwest from the thick Lower Cretaceous deposits of the Helgeland Basin. Lower Cretaceous deposition northwest of the Nordland Ridge is difficult to define, but is thought to be significantly different from that of the Helgeland Basin.

During late Cretaceous and Tertiary times tectonic conditions in Trænabanken were generally stable, except in the Oligocene period, when a distinct wrench movement, connected with changes in the sea-floor spreading axis between Norway and Greenland, resulted in folding, reverse faulting and reverse rejuvenation of normal faults along a NNW trend in Trænabanken.

INTRODUCTION

In this chapter we attempt to summarize the geology of an area of 12 blocks between 66° and 66°45′N, and 9° and 10°20′E on the Norwegian Continental Shelf (Fig. 1). Herein, these blocks are collectively referred to as Trænabanken. Interpretation of the available data has led to an understanding of the geological evolution of the area and it is hoped that the presentation of the main ideas of the interpretation will advance the regional knowledge discussed in earlier publications. The area is, at present, undrilled and under active exploration. Hence we expect that refinement of the ideas presented will be possible in the near future.

Early regional geophysical work (Åm, 1970; Talwani and Eldholm, 1972; Sellevoll, 1975) gave a useful broad regional understanding, but relatively little data concerning the Trænabanken area. Rønnevik *et al.* (1975), Jørgensen and Navrestad (1979), and Rønnevik *et al.* (1979) presented interpretations covering Trænabanken based on the various regional seismic surveys made by the Norwegian Petroleum Directorate since 1972. Detailed interpretation was limited due to the quality of the seismic data and lack of sufficient data.

The main data base for this study is an extensive seismic programme shot in 1981. This survey gave a grid of approximately 2 × 4 km of good quality seismic data which enabled detailed interpretation of Trænabanken for the first time. Figure 1 shows this survey and the older regional lines in Trænabanken. In addition to the seismic data, we have analysed the gravity and magnetic data acquired in 1981 and earlier. Sea-bed sampling data (IKU, 1982) and an extensive review of the literature covering West Shetlands to Svalbard, and East Greenland to Norway were used to refine inter-

Petroleum Geology of the North European Margin © Norwegian Petroleum Society (Graham & Trotman, 1984) pp. 237—251.

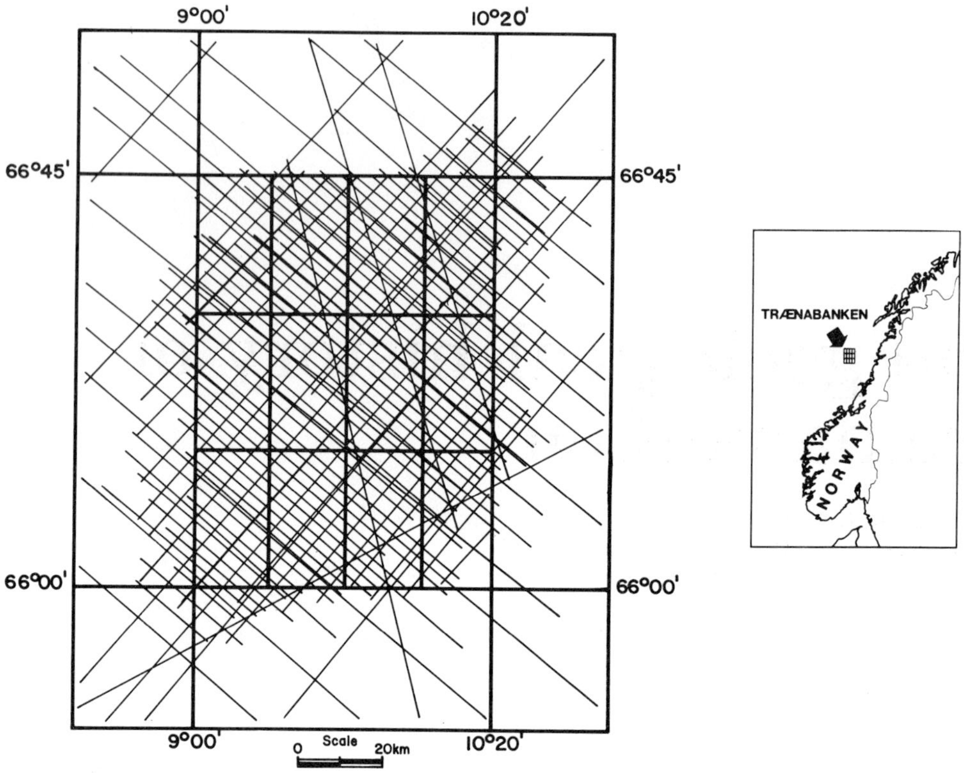

Fig. 1. Seismic data-base used for this study.

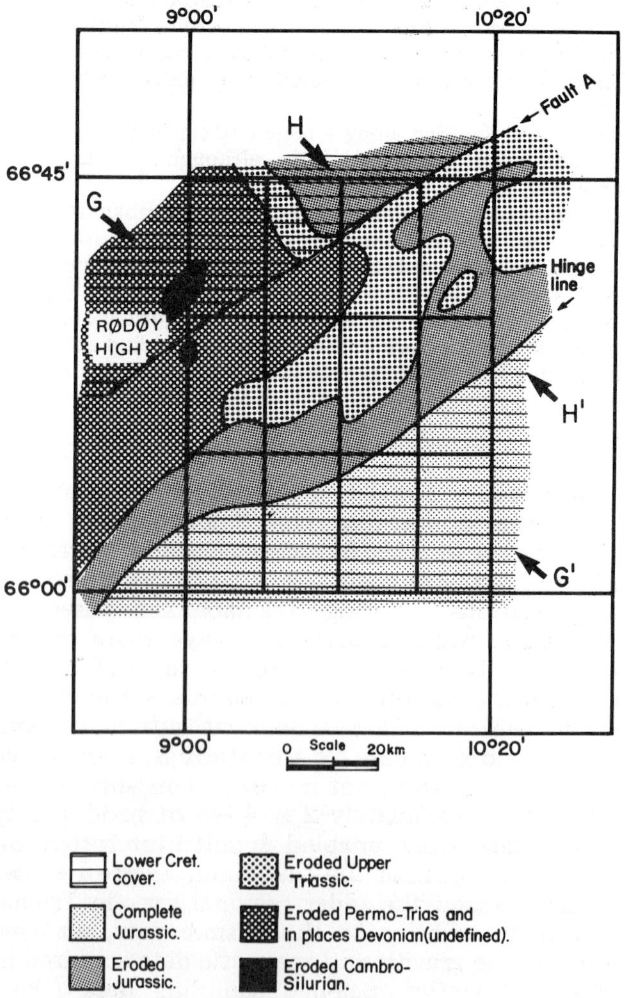

Fig. 2. Geology of the pre-Cretaceous subcrop and the distribution of Lower Cretaceous sediments. (See Fig. 3 for cross sections G—G′ and H—H′.)

pretations of the seismic lines, as was information from exploration wells in Haltenbanken and from the considerable seismic data-base there.

STRUCTURE AND STRATIGRAPHY

The distribution of the Lower Cretaceous sediments conveniently divides Trænabanken into three structural provinces, separated by 'Fault A' and the 'hinge line' (Figs. 2 and 3). Southeast of the hinge line (SE area) a thick Lower Cretaceous sequence conformably overlies a thick parallel bedded and relatively undisturbed Jurassic, Triassic and Permian sequence. Between Fault A and the hinge line (central area), the Upper Cretaceous rests unconformably on eroded older strata. The subcropping sequence is Jurassic near the hinge line and becomes progressively older to the west, culminating in an area of eroded Cambro-Silurian sediments, the Rødøy High (Fig. 2).

The third structural province lies northwest of Fault A (NW area). Due to poor seismic data quality, interpretation in this area is less certain than in the rest of Trænabanken. A Lower Cretaceous sequence is thought to rest on eroded Cambro-Silurian rocks of the Rødøy High and on progressively younger rocks to the northeast.

The area is dominated by the Rødøy High and the hinge line (Fig. 4). Fault A has a large vertical throw in the northeast, where it forms the southeastern border of the Træna basin. The vertical throw becomes insignificant to the southwest over the Rødøy High. The terms Nordland Ridge and Helgeland Basin were introduced in previous publications. We interpret both features to have been tectonically active only in early Cretaceous times

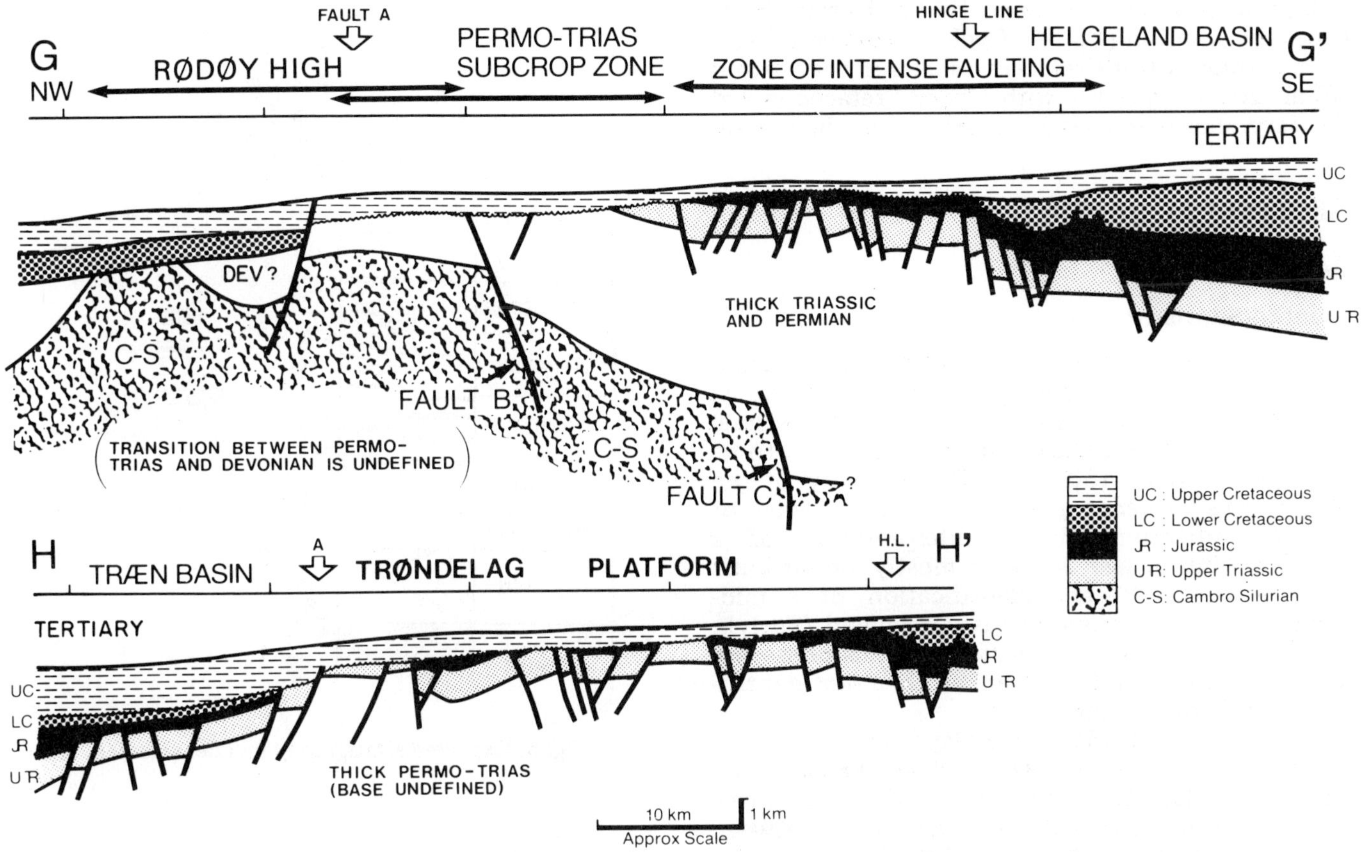

Fig. 3. Schematic geological cross sections over Trænabanken (see Fig. 2 for index map).

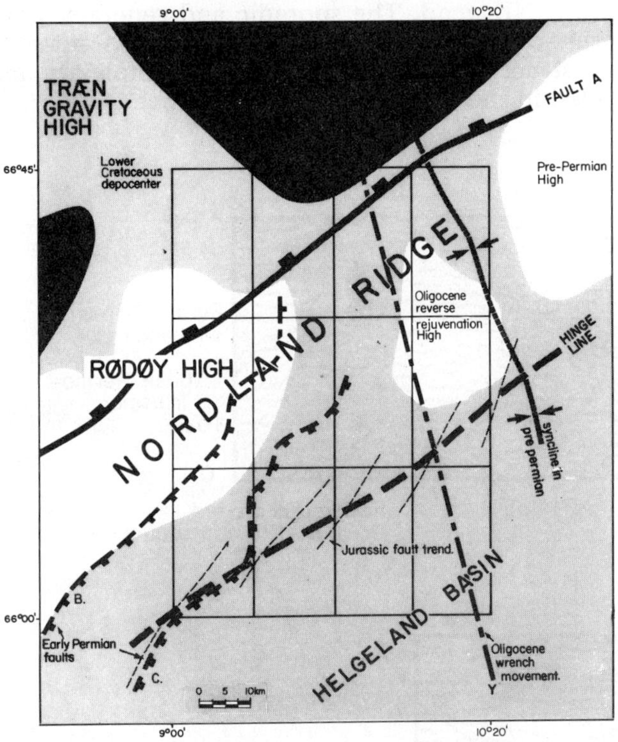

Fig. 4. Structural elements in Trænabanken.

and we use the term Nordland Ridge to refer to the area of early Cretaceous erosion between Fault A and the hinge line (central area).

Extensive faulting of the Jurassic and Upper Triassic sequence is concentrated along the hinge line, and to a lesser extent in the northeast of the central area. Fault trends are dominantly NNE, roughly parallel to the earlier Permian faults, and ENE parallel to the hinge line and Fault A. The Oligocene reverse rejuvenation high is thought to be caused by a sinistral movement along a deep wrench fault Y—Y′ which is subparallel to the deep syncline in the pre-Permian surface.

Figure 5 summarizes the geological sequence expected in Trænabanken. Since no wells have been drilled in the area as yet, the lithologies shown are derived from regional studies. Three major tectonic events (a, b and c) divide the sequence into four major stratigraphic cycles (1—4).

Cycle 1 encompasses the Cambro-Silurian marine sediments which were folded and metamorphosed during the Caledonian orogeny (a). Cycle 2 sediments of Devonian to Carboniferous age onlap and infill the Cambro-Silurian topography. In Trænabanken we expect coarse clastics of Devonian age in the NW area and little or no Carboniferous.

Tectonic event b in the early Permian significantly changed the structure of the area and initiated stratigraphic cycle 3. A thick Permo-Triassic sequence progressively onlaps the faulted pre-Permian surface from the east. Cycle 3 sedimentation continued until end Jurassic times without significant tectonic disturbance. The whole sequence is dominantly parallel bedded and over 10 000 m thick along the axis of the syncline in the pre-Permian surface (Fig. 4). Except for a possible marine environment in late Permian times, the Permo-Trias sediments of cycle 3 are mainly continental. Jurassic sediments are mostly marine and marginal marine.

Tectonic event c in early Cretaceous times terminated cycle 3 by establishing the three structural provinces outlined by the Lower Cretaceous distribution (Fig. 2). Cycle 4 sediments begin with marine Lower Cretaceous shales in the NW and SE areas and continue with Upper Cretaceous and Tertiary shales and sands covering the whole area.

GEOLOGICAL HISTORY

Pre-Permian

Cambro-Silurian

The oldest sequence we can identify shows discontinuous, folded and strongly dipping seismic events beneath a strong erosion surface (Fig. 6a). Seismic stacking velocities indicate interval velocities to be consistently over 4500 m/s. The majority of the data in this sequence gives little or no evidence of internal reflections, hence interpretation of a Cambro-Silurian age is based mostly on regional considerations. Recent identification of a mid-Permian tectonic phase in east Greenland (Surlyk, *et al.*, 1984) may be important. If this phase has also been active in Trænabanken then our identifications will be too early.

According to Harland and Gayer (1972), Gayer (1973), and Siedleka (1975), in late Precambrian through to late Ordovician times clastic and carbonate marine sediments were deposited between the Baltic Craton to the east and the Iapetus Ocean to the west. The sequence was probably deformed in two phases: late Cambrian—early Ordovician, and in the main Caledonian orogeny of the late Silurian. This latter phase resulted in extensive thrusting from the axial zone west of Trænabanken towards the cratonic area to the east. Therefore, a thick deformed and metamorphosed sequence of Cambro-Silurian sediments may be expected in Trænabanken.

The few structural features we can define from the seismic in the oldest sequence of Trænabanken have a NE-trend. The sporadic and generally poor nature of the seismic reflections are consistent with the structural and metamorphic complexity expected in the Cambro-Silurian sequence.

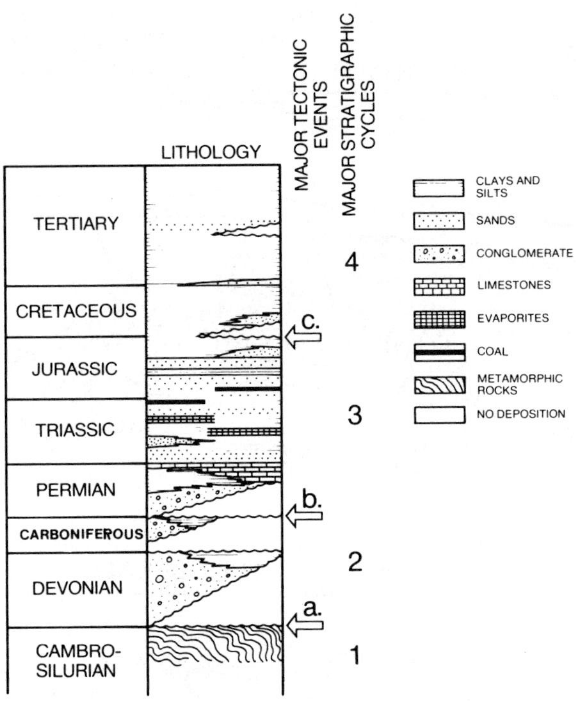

Fig. 5. Expected stratigraphy in Trænabanken.

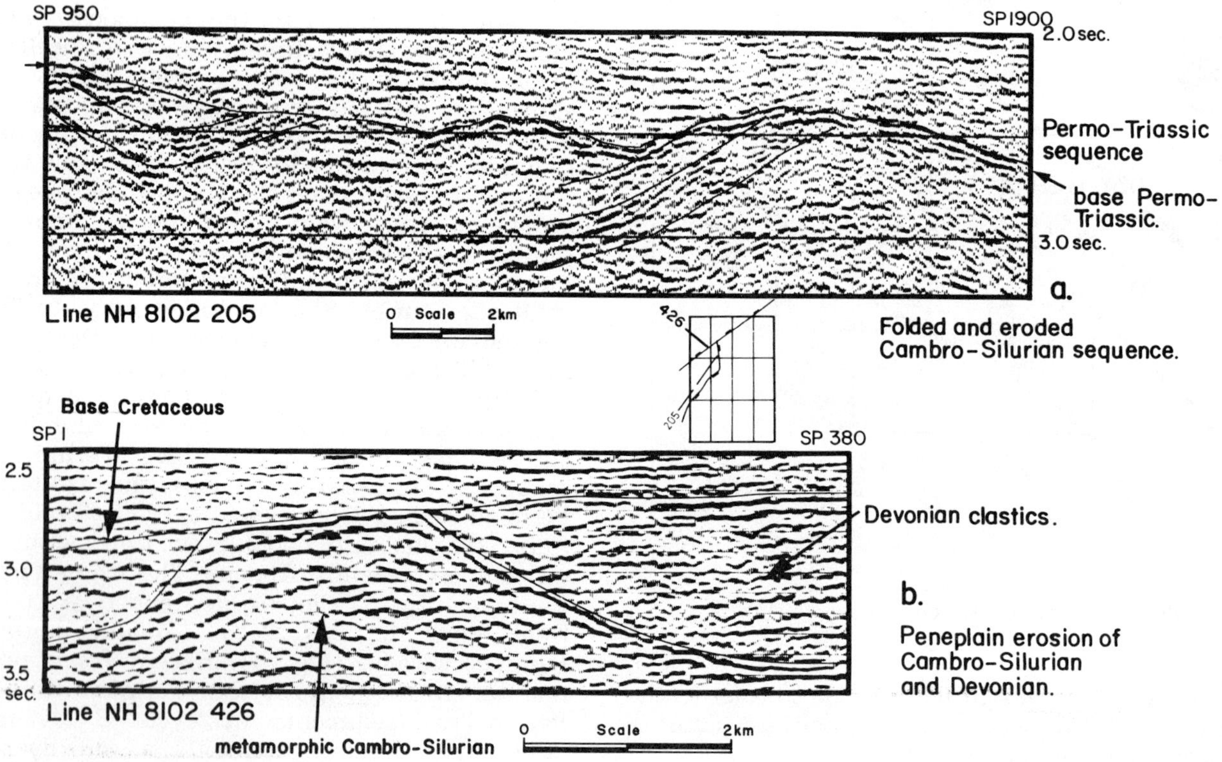

Fig. 6. Seismic sections illustrating some aspects of the Cambro-Silurian sequence.

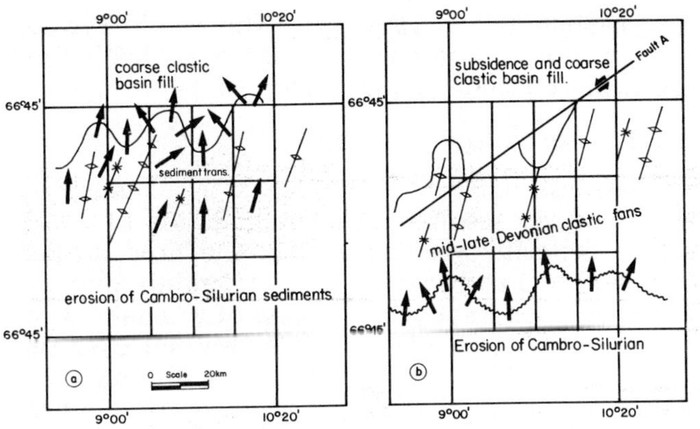

Fig. 7. Devonian geology. (a) Early Devonian northward tilting followed by basin infill with coarse clastic sediments, progressively onlapping further south. (b) Late Devonian sinistral wrench movement along Fault A. Continuing progressive southward onlap to eroded Cambro-Silurian sediments.

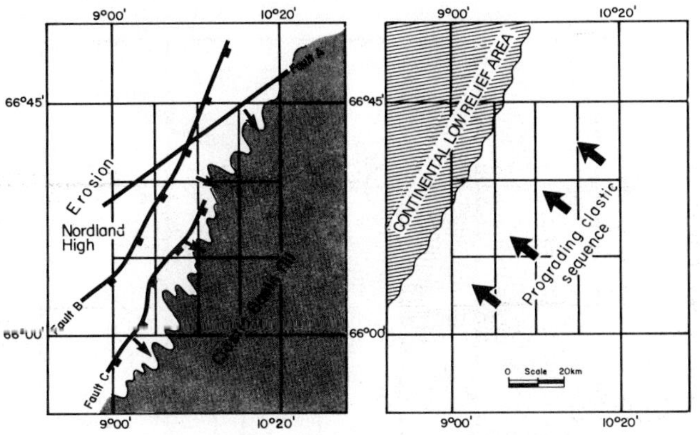

Fig. 8. (a) Early Permian faulting establishes a basin in the east. Subsequent clastic basin fill is followed by marine carbonates in late Permian. (b) Early Triassic uplift of sediment source area to the east initiates a new clastic influx. Continental/fluvial conditions then dominated the early Triassic.

Devonian

The uplifted mountainous Cambro-Silurian rocks were subjected to intense erosion in Devonian times. An intermontane basin was established north of Trænabanken. Coarse-grained fluvial sediments were deposited contemporaneously with continued uplift and erosion of the Cambro-Silurian sediment source area to the south, imparting a northerly basinwards tilt to the Trænabanken area (Fig. 7a). A seismic sequence with no apparent internal reflections onlaps the Cambro-Silurian topography from the north. Figure 6b shows a seismic section across a NNW-trending Caledonian ridge, showing the featureless Devonian clastic fill, both subsequently eroded to a peneplain. A similar geological situation can be found in the Devonian rocks of the Trondheimsfjord district 300 km to the south (Nilsen, 1973).

Sinistral transcurrent faulting along Fault A accelerated the subsidence of the intermontane basin in the north. This faulting is recognized in Trænabanken by an offset of the Cambro-Silurian surface along Fault A (Fig. 7b). These movements are probably related to the major sinistral transcurrent movements between Norway and Greenland in late Devonian times (Harland, 1973). Similar wrench controlled Devonian clastic basins have been described from western Norway (Steel, 1976).

The featureless sequence identified as Devonian in Trænabanken shows some internal reflection character further north, corresponding with a more distal depositional environment. These reflections indicate some open folding but are not consistent enough to enable us to properly define the fold geometry. It is probable that the Devonian movements continued on a diminishing scale into Carboniferous times (see Nilsen, 1973).

Carboniferous

In Carboniferous times we expect Trænabanken to have been an area of positive relief, experiencing a continuation of the erosion begun in Devonian times. Continental sediments may have been deposited in the Devonian clastic basin to the north, to be eroded in later times. Dalland (1979) interpreted early Carboniferous deep weathering followed by deposition of a thick Upper Palaeozoic sedimentary succession (later removed) in Andøy, about 400 km northeast of Trænabanken.

We found no evidence on the seismic data to suggest that Carboniferous sediments are present in Trænabanken. Any thick Carboniferous sequence which may have been deposited would have been continental and has since been eroded. A long-term sea-level rise in the late Carboniferous (documented on Svalbard by Gjelberg and Steel, 1979) may have resulted in some thin, uppermost Carboniferous sediments conformably underlying the Lower Permian in the east of Trænabanken.

Permian to Jurassic

Early Permian faulting

Late Carboniferous to early Permian faulting found in East Greenland (Larsen, 1979) was probably connected to the Variscan Fault System further to the south (Ziegler, 1978). In Trænabanken we interpret faulting of the eroded Cambro-Silurian surface (Faults B and C, Fig. 8a) to be of late Carboniferous to early Permian age. The faults, which cannot be followed in the north due to the thick post-Permian cover, formed a deep basin in the east of Trænabanken in which Lower Permian synsedimentary continental sedimentation derived from a source area to the west can be expected. This situation is mirrored in East Greenland, where the Permian basin of Jameson Land was bounded to the east by a mountainous region (Birkelund and Perch-Nielsen, 1976).

Late Permian transgression/regression cycle

Further analogies with East Greenland (Birkelund and Perch-Nielsen, 1976) allow us to interpret a late Permian transgression and the deposition of a marine carbonate/shale sequence, which continued the progressive infill of the Permian basin. Various strong seismic events in the Permian sequence may be due to limestone beds.

Regional data indicate that the sea receded northwards in early Triassic times and the progressive basin infill continued with continental Triassic-

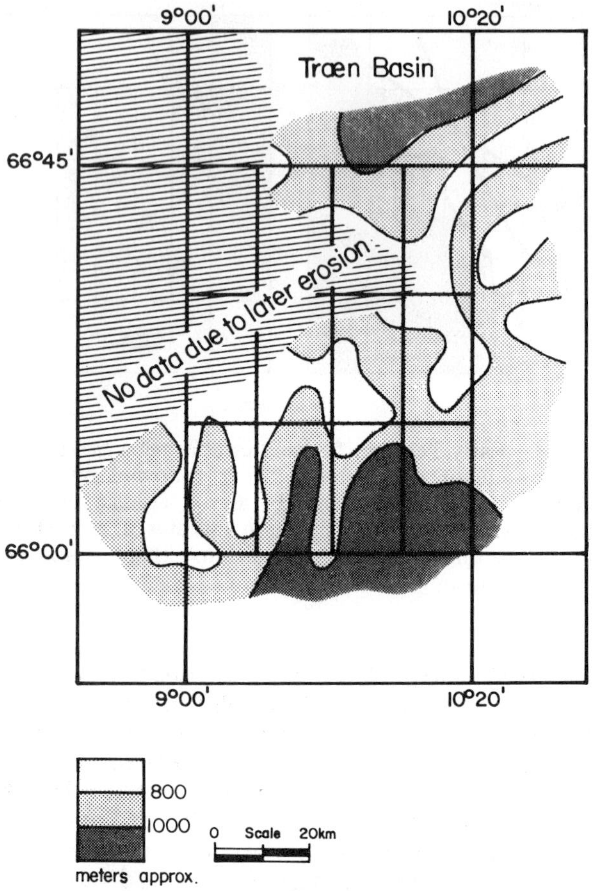

Fig. 9. Isopach of uppermost Triassic sequence illustrating the subsidence in the Helgeland and Træn basins. Continental conditions and a flat depositional plain characterize the Rhaetian.

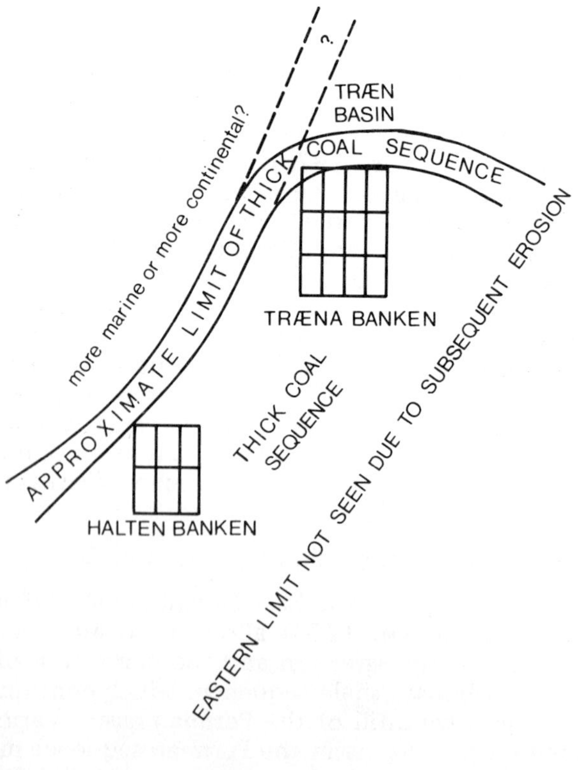

Fig. 10. Palaeogeography of the Lower Jurassic coal sequence. A flat fluvio-deltaic plain with no significant topography may be limited to the northwest by a facies change.

time sedimentation (Fig. 8b). The whole Permo-Triassic sequence is parallel bedded, indicating no significant tectonic disturbance.

Mid Triassic continental plain

An increase in clastic sediment input in early Triassic times resulted in a progradation of continental deposits westwards over the now infilled Permian basin and over the eroded Cambro-Silurian/Devonian topography in the west of Trænabanken. A flat continental plain was established over the region by mid-Triassic times. Within the assumed Middle and Lower Triassic sections, we find indications of thick sequences prograding from the east, lacustrine deposits and evaporite horizons, all of which can be expected from regional studies. Some limited marine incursions can also be expected from regional data.

Late Triassic

Continental low relief conditions persisted in Trænabanken until end Triassic times. The uppermost Triassic can be more confidently correlated on the seismic data than older sequences, and interpretation is thus more reliable for this and later intervals. In addition, information from wells in the Haltenbanken area lends greater confidence to environment and facies interpretations.

Subsidence in Rhaetian times (Fig. 9) did not reflect the underlying Permian basin but shows a slight relative subsidence to the south and to the north. Apart from this, no major structures were active during this interval.

Coal sequence

A regional relative rise in sea-level towards the end of Triassic times resulted in a shift from continental to marginal marine/deltaic environments and the deposition of an extensive coal sequence. This coal sequence, which extends into the Lower Jurassic, has a distinctive seismic character which can be followed from the wells in Haltenbanken to Trænabanken and eastwards to its subcrop line near the Norwegian coast. The western limit (Fig. 10) is seen as a gradual deterioration of reflection strength and character. This is interpreted as a facies change, either to more continental or to more marine sediments. To the north of Trænabanken, the facies change is indistinct and our data quality and density poor. It is possible that the coal sequence is also found in the Træna Basin.

Early Jurassic transgression/regression cycle

There is no evidence of faulting in late Triassic—earliest Jurassic times in Trænabanken, which contrasts with reported Rhaetian faulting in east Greenland (Surlyk, 1977). The increase in marine influence during the early Jurassic times culminated with the deposition of fully marine late Lower Jurassic shales. At this time, there was no significant relief in the region, such that the regional regressive phase at the end of the early Jurassic resulted in the regional deposition of shallow marine Middle Jurassic sandstones (Bajocian).

Mid-Jurassic faulting

The first Mesozoic faulting seen in Trænabanken occurred early in mid Jurassic times. The faults are

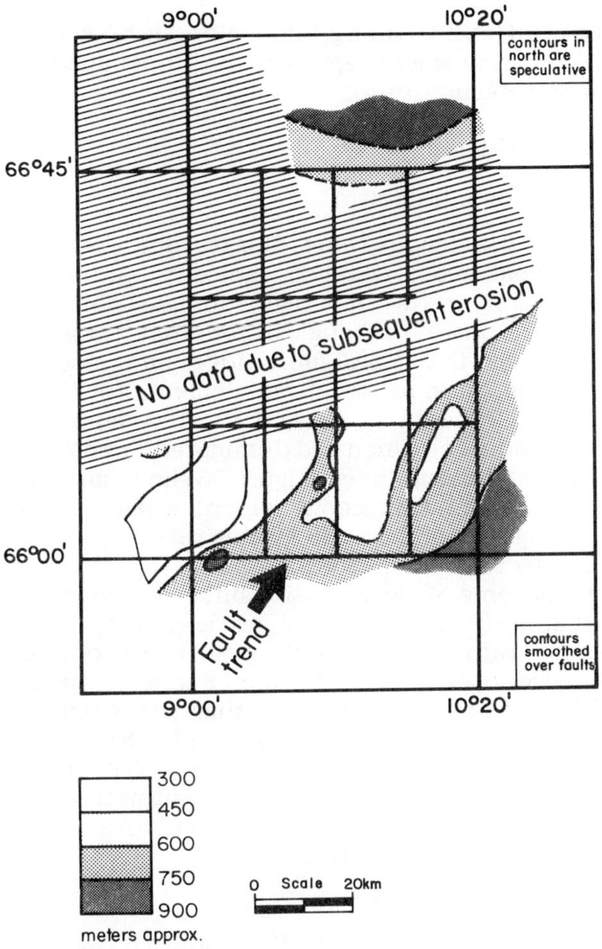

Fig. 11. Isopach of the Middle and Lower Jurassic sequence. A shallow marine environment with a flat depositional surface prevailed during the mid Jurassic, with some synsedimentary faulting.

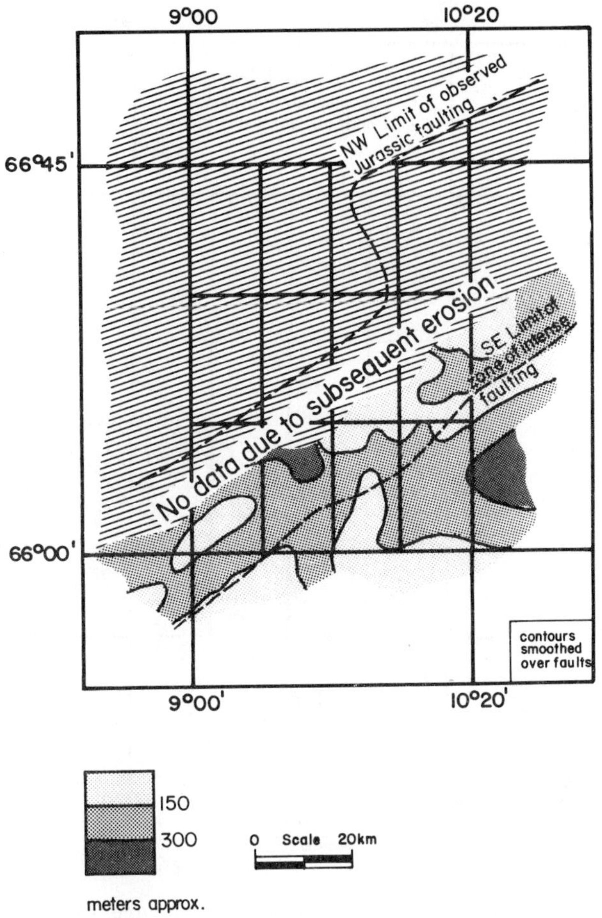

Fig. 12. Isopach of the Upper Jurassic sequence illustrating the absence of influence of major structural elements.

mostly small and synsedimentary, such that the deposition surface remained flat. Fault trends are NNE, possibly inherited from the underlying Permian trend. Direct rejuvenation along the Permian Fault C (Fig. 8a) produced the only major fault of this tectonic phase. Other faults appear to die out in the Middle Triassic section, possibly due to accommodation within the salt. The initiation of faulting, in late early to early mid-Jurassic times, can be correlated with similar movements in east Greenland (Surlyk, 1977) and on a more regional basis with the main mid-Cimmerian movements (Ziegler, 1981). Mid-Jurassic faulting was not accompanied by any significant relative subsidence, illustrated by the isopach of Middle and Lower Jurassic sediments (Fig. 11). Some influence of the Permian Faults B and C (Fig. 8a) and a speculative subsidence to the north can be seen.

Synsedimentary faulting continued during mid Jurassic times until a new tectonic phase occurred in Bathonian to early Callovian times. These end mid-Jurassic movements initiated many new faults in Trænabanken in addition to reactivating the earlier Jurassic growth faults. The faulting of the shallow marine Bajocian/Bathonian sandstones resulted in contemporaneous erosion of some local faulted highs.

Faulting continued into Callovian times accompanied by a regional transgression also found in Andøy (Dalland, 1979). Some of the latest faults, with throws of over 400 m, show no signs of erosion of the Bathonian sands, indicating that faulting was active after the establishment of deeper water conditions in the area. The absence of Callovian strata on some of the fault blocks in Haltenbanken suggests that deep water conditions did not exist in the Callovian times. The late Middle Jurassic faulting in Trænabanken may, therefore, have continued until Oxfordian times, or perhaps consisted of more than one phase, similar to the faulting in east Greenland (Surlyk, 1977).

Bathonian/Callovian faulting is concentrated along the hinge line (Fig. 4) in an ENE direction and faulting intensity decreases away from this line. The faulting is confined mostly to the Jurassic and Upper Triassic, deeper levels being unaffected. In the northeast of Trænabanken, where Upper Triassic strata have not been completely removed by erosion, faulting is found from the hinge line as far to the northwest as Fault A. However, in the southwest the observed faulting dies out rapidly away from the hinge line towards the northwest (Fig. 12). This may be due to the greater stability of the Rødøy High area, or merely to the subsequent erosion of the evidence.

Late Jurassic

Although the local structure of the top of the Middle Jurassic during late Jurassic times was relatively intense, there is no indication of any major relative subsidence in Trænabanken. The Upper Jurassic isopach (Fig. 12) suggests a continuation

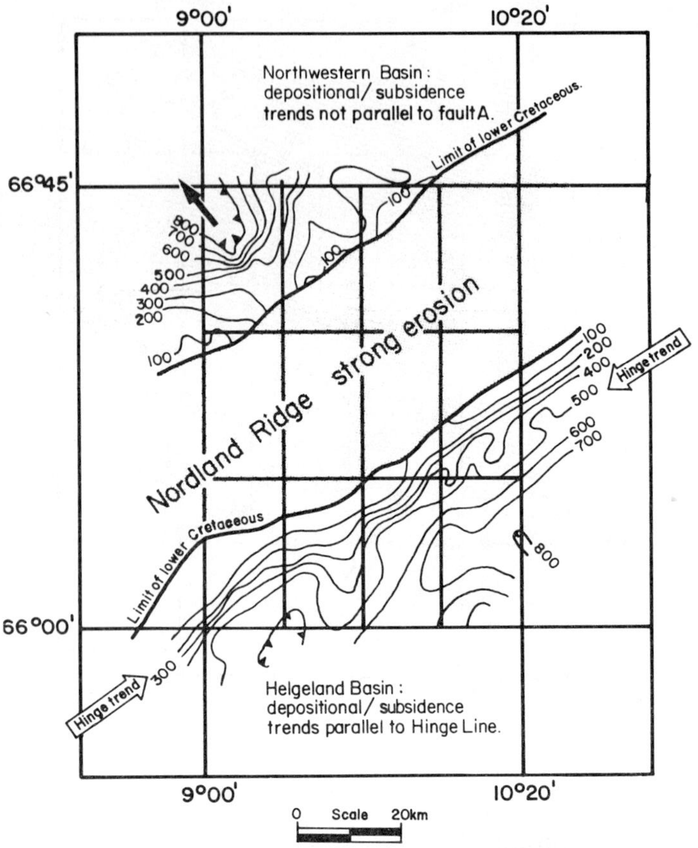

Fig. 13. Isopach of Lower Cretaceous sequence. Contours are in milliseconds, two-way seismic time. Note the asymmetry over the Nordland Ridge.

of the relatively flat topography of mid-Jurassic times. The faulting along the hinge line in Bathonian/Callovian times, therefore, was not a response to vertical movements about that line, but more likely due to some sinistral wrench movement. This would also explain the *en-echelon* nature of the faulting (NNE-trending faults along a ENE trend). Fully marine shales represent the Upper Jurassic in Trænabanken.

Post Jurassic

New tectonic framework

In early Cretaceous times, the tectonic framework of Trænabanken changed radically, probably in connection with the regional late Cimmerian tectonic phase. Strong differential subsidence established the Nordland Ridge and associated Helgeland Basin as structural elements for the first time and brought to an end the flat depositional plain conditions which had dominated Trænabanken since mid Triassic times. Figure 13 shows an isopach of the Lower Cretaceous which, when compared with Figs 11 and 12, illustrates this major change in the tectonic framework.

Only some of the Jurassic faults were reactivated during early Cretaceous times and it seems that flexure rather than faulting was the dominant response over the hinge line. Some gentle local folding at this time suggests that some strike-slip movement is involved. (Surlyk (1978) describes similar folding in East Greenland.) The isopach (Fig. 13) also illustrates the different nature of the subsidence to the northwest of Fault A compared with that of the Helgeland Basin.

Complex early Cretaceous history

Flexure over the hinge line resulted in Lower Cretaceous isopach contours strongly parallel to the hinge line. A similar pattern is not seen northwest of Fault A, since no such flexure has taken place. Subsidence was partly fault controlled (vertical movement on Fault A), but was more strongly influenced by basin subsidence in the Træna Basin to the north. Detailed study of the seismic data shows the early Cretaceous history to

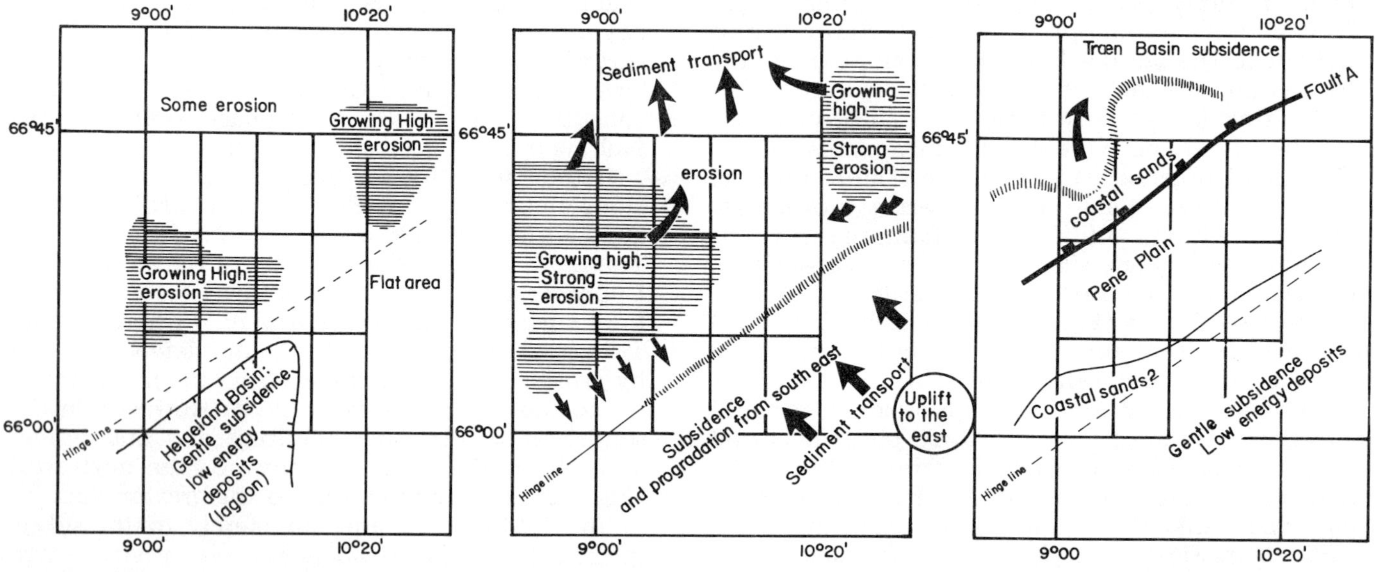

a RYAZANIAN - VALANGINIAN

Hinge line activated. Synsedimentary deposition in south. No strong relief. No high energy deposits. No strong movements.

b VALANGINIAN - BARREMIAN

Strong uplift on Nordland Ridge highs. Subsidence south of active hingeline. Basin fill during subsidence. Sediments prograding from active uplift area to the east or southeast. Sediment transport over Trænabanken area and toward north. Some Jurassic faults reactivated along hinge line.

c APTIAN - CENOMANIAN

Some movement along fault A. Previous highs still active. Peneplain surface on Nordland Ridge.

Fig. 14. Tentative Lower Cretaceous history.

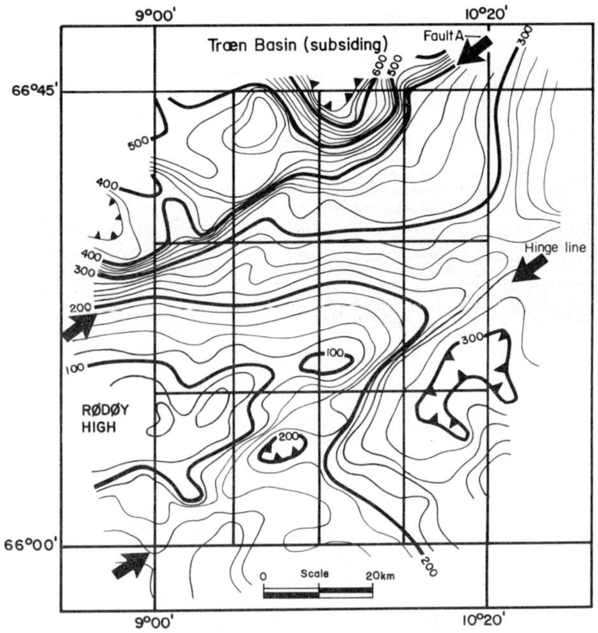

Fig. 15. Isopach of Upper Cretaceous sequence. Contours are in milliseconds, two-way seismic time. Note the stable Rødøy High, the subsidence in the Træn Basin, the small movement on Fault A and the general northward tilting. (Faults are omitted.)

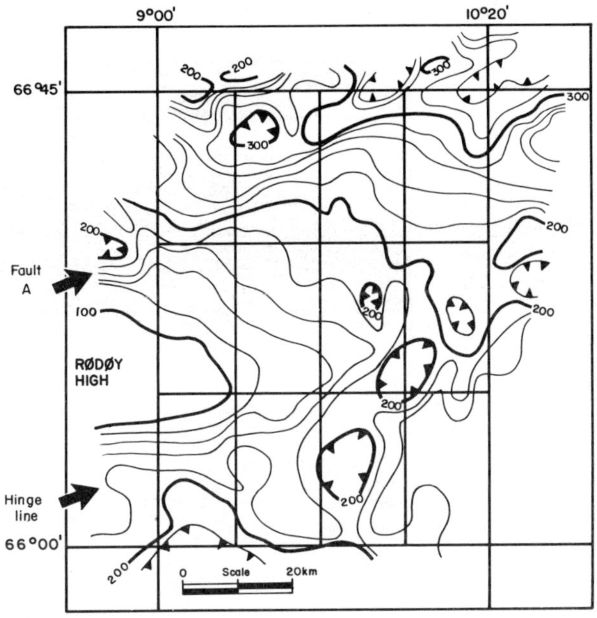

Fig. 16. Isopach of the Palaeocene interval. Contours are in milliseconds, two-way seismic time. Note the stable Rødøy High and the tilting to the North.

be rather complex. For example, in the southwest of Trænabanken it can be shown that basin subsidence did not occur at the start of Cretaceous times, but after the deposition of at least 200 m of sediment conformable to the underlying Upper Jurassic. This and other details are weak on the seismic and interpretation in the northwest of Trænabanken is uncertain. A possible synthesis is presented in Fig. 14.

We have interpreted three tectonic phases in early Cretaceous times. Dating of the phases is speculative and based entirely on regional data. The first phase, from Ryazanian to Valanginian (Fig. 14a), is characterized by gentle activation of hinge line with some related subsidence to the south and possibly erosion to the northwest. The main tectonic phase occurred in the Valanginian to Barremian interval (Fig. 14b). A major movement (flexure) along the hinge line resulted in strong subsidence in the Helgeland Basin and strong uplift and erosion to the northwest. Uplift and erosion was greatest over the Rødøy High and the high in the northeast. Sediment transport in the Helgeland Basin was from the east or southeast and was probably more than adequate to fill the basin as it subsided.

A regional Aptian—Albian transgression marked the end of the intense erosion and subsidence and the start of the third phase. The Nordland Ridge was a peneplain with only minor subsidence in the flanking basins. Fault A, which had been inactive since Devonian times, experienced a small vertical movement and late Lower Cretaceous marine sediments were deposited on the Cambro-Silurian to Jurassic peneplained erosion surface north of Fault A (Fig. 14c). The Albo-Aptian conditions probably continued into late Cenomanian times, but we include any such Cenomanian deposits in the 'Lower Cretaceous' section for convenience. It seems likely that marine shales were also deposited over the Nordland Ridge towards the end of this phase.

Late Cretaceous

Regional data, particularly in east Greenland (Birkelund and Perch-Nielsen, 1976), suggest that the Cenomanian, if deposited, was subsequently eroded from the Nordland Ridge in Turonian times. It is likely that the pre-Turonian erosion surface was not submerged until the onset of the widespread Santonian transgression. No Turonian or Coniacian sediments are therefore expected in Trænabanken and Upper Cretaceous sedimentation continued with marine shales of Santonian and Campanian age. Maestrichtian sediments are generally absent in the region and are not expected in Trænabanken. The late Cretaceous ended with a period of erosion/non-deposition.

A small vertical movement along Fault A occurred in late Cretaceous times, as can be seen in the Upper Cretaceous isopach (Fig. 15). Subsidence in the Træn Basin, the stable nature of the Rødøy High, and a northerly tilt between the two may also be inferred from this isopach. The apparent subsidence along the hinge line is misleading and entirely accounted for by a compaction of the Lower Cretaceous section to 80% of depositional thickness during late Cretaceous times. (A few of the early Cretaceous faults show some small movement during late Cretaceous times.)

Palaeocene transgression

The northerly tilt from the Rødøy High continued during the Palaeocene transgression. An isopach of the Palaeocene thickness (Fig. 16), shows that the Rødøy High remained the dominant structural feature with some relict influence of Fault A also evident. A transgressive sand followed by marine clays can be expected in this interval.

Oligocene wrench tectonics

The start of sea-floor spreading in late Palaeocene times between Norway and Greenland (Talwani

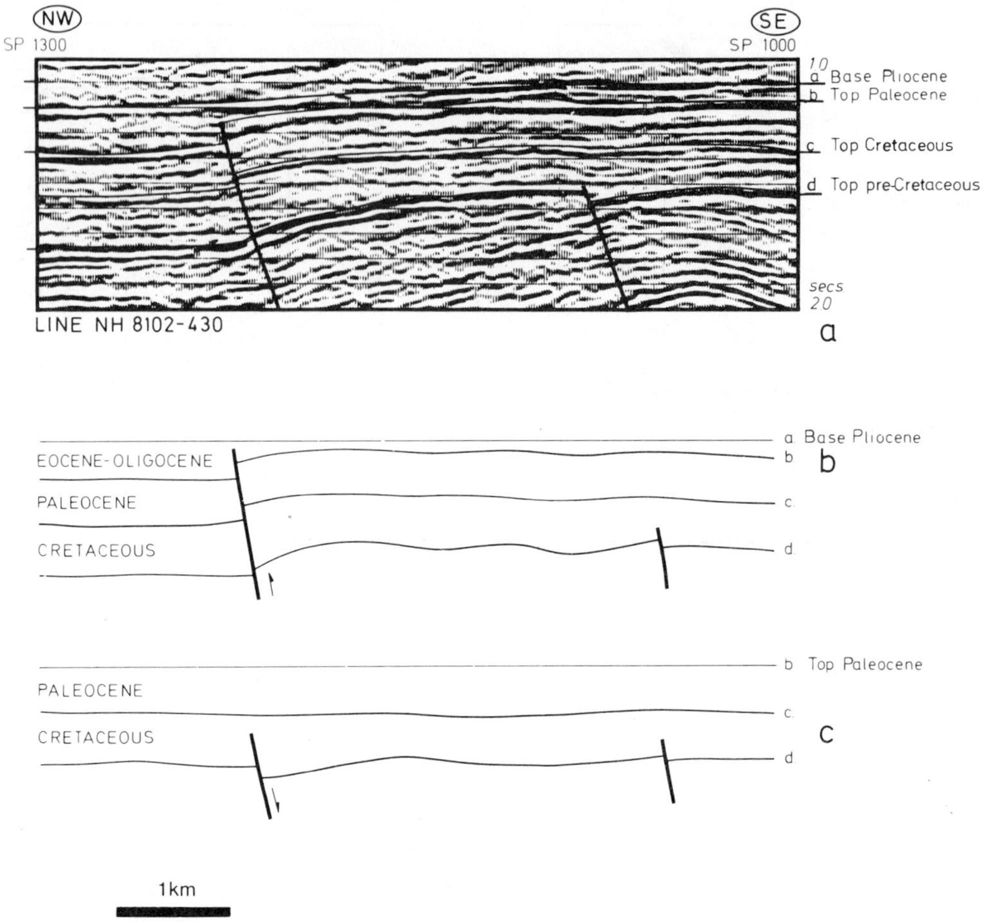

Fig. 17. Reverse rejuvenation. The seismic line interpreted (a) can be palinspastically reconstructed with an end Palaeocene datum (c) to show the normal fault which becomes reversed by the start of Pliocene times (b). (See Fig. 18 for the index.)

et al., 1981) did not significantly affect the Træna-banken area and the Eocene to Oligocene interval is presumed to have followed the Palaeocene with quiet conditions and little relief. However, Oligocene to Miocene times witnessed compressive stresses, which resulted in reverse faulting, mostly by reverse rejuvenation of previous normal faults. An example from the seismic data (Fig. 17a) when palinspastically reconstructed shows that the original Cretaceous normal fault, dominant during Palaeocene times (Fig. 17c), became reversed in the Eocene—Miocene interval (Fig. 17b).

The compressional effects were concentrated along a distinct NNW-trend (Y—Y′) and can be followed both north and south of the Nordland Ridge as flexures in the basinal area (Fig. 18). Individual faults along the Y—Y′ trend show an *en-echelon* configuration; they trend N—S. Thus we can interpret a sinistral wrench movement along the Y—Y′ trend as the cause of the compressive movements. The movements resulted in a structural high, illustrated in the isopach (Fig. 19). Larsen and Skarpnes (1984) propose the name Grønnøy High for this feature. If we correlate this wrench movement with the sea-floor spreading history (Talwani and Eldholm, 1977), we might date the wrench movement as early Oligocene to coincide with the change in the spreading axis at that time.

Post Oligocene

In late Oligocene times a global fall in sea level resulted in some erosion of the Eocene—Oligocene sequence and it is probable that Miocene and Oligocene sediments are absent in Trænabanken. The erosion surface is the base of an extensive prograding sequence which we date as Pliocene. The isopach of the sediments between the base Pliocene and the top Palaeocene (Fig. 19) clearly shows the Oligocene reverse rejuvenation high and the wrench trend Y—Y′. Note also that the Rødøy High dominates the map as it has done since the end of early Cretaceous times, and that subsidence to the east is evident.

Subsidence and tilting to the WNW followed the Miocene erosion, and Pliocene clastic sediments from the uplifted Baltic Shield prograded towards the ocean. This represents a new tectonic regime in Trænabanken, since the Rødøy High ceased to be a positive structural element and, for the first time since the Silurian, the west of Trænabanken became structurally more basinal than the east.

An irregular thickness of Pleistocene deposits covered the late Pliocene coastal shelf and is largely responsible for the sea-floor irregularities seen today.

TECTONICS

The main structural elements of the region (Fig. 20) are taken from Jørgensen and Navrestad (1979) and Aanstad *et al.* (1981), except for the Træn Gravity High, which we introduce in this chapter. Larsen and Skarpnes (1984) have proposed the

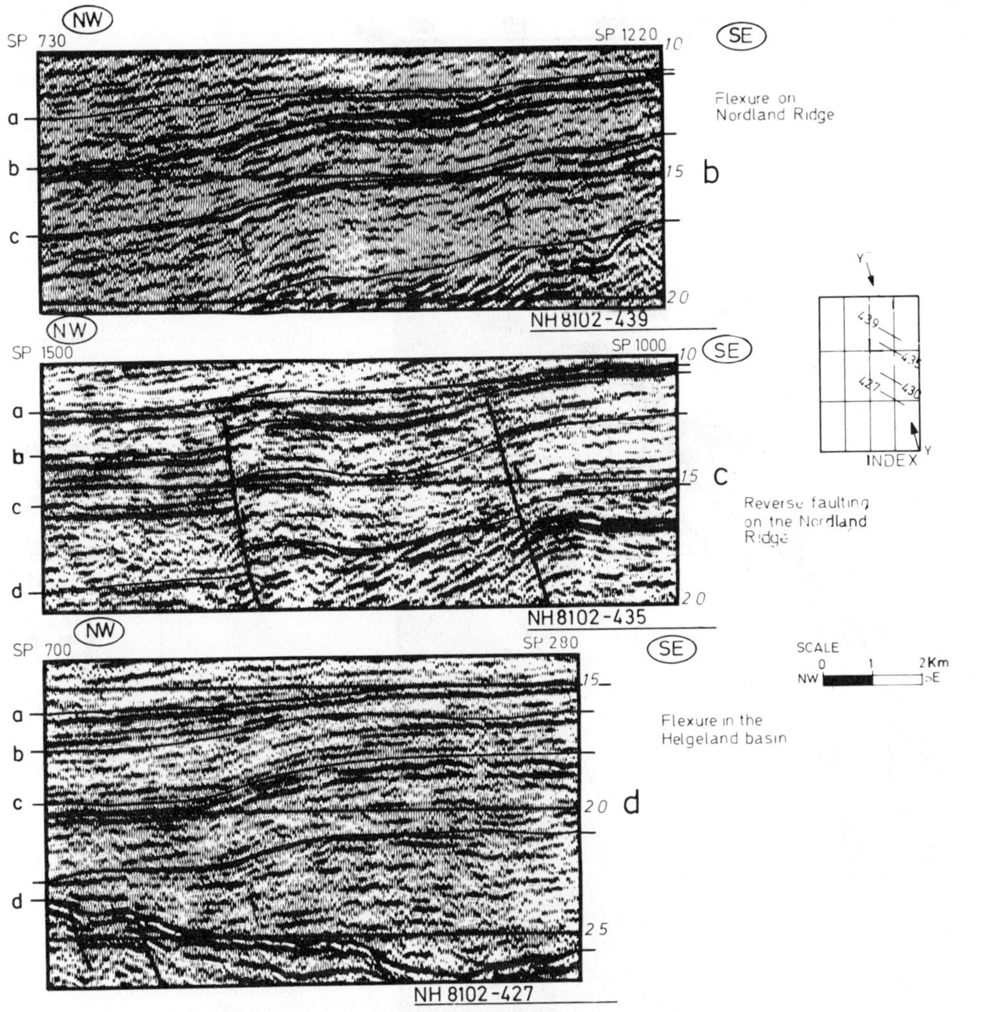

Fig. 18. Reverse rejuvenation on the Nordland Ridge (b, c) can be followed as a flexure along the trend Y—Y′ into the Helgeland Basin (d).

Rødøy High, which Gabrielsen *et al.* (1984) adopt as standard nomenclature.

Rødøy High

The term 'Nordland Ridge' as used by Jørgensen and Navrestad (1979) defines a linear area trending NNE from Haltenbanken. In Trænabanken this ridge was active only in early Cretaceous times and belongs to a different ENE tectonic trend. Therefore we prefer not to include it in Fig. 20. Instead we define the Rødøy High which had a more extensive influence on the geology of Trænabanken. The Rødøy High is a faulted structural high of eroded Palaeozoic rocks to the west of Trænabanken. The feature dates from early Permian times and was a dominant stable high from late Cretaceous to end Miocene times.

Træn Gravity High

A large positive linear gravity anomaly occurs to the northwest of Trænabanken. Talwani and Eldholm (1972) related this high to intrabasement density contrasts, while Rønnevik *et al.* (1979) suggested that it is due to thickness variations in the crust. The gravity high has a magnitude of about 60 mGal on the free-air gravity map of Talwani and Eldholm (1972) and an associated

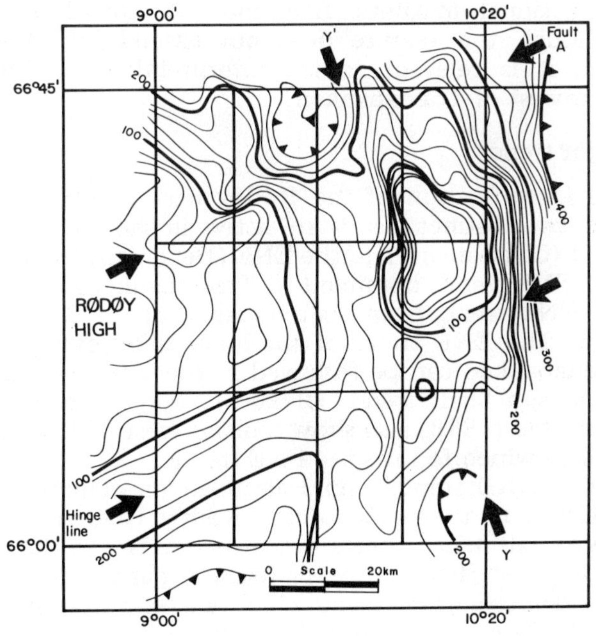

Fig. 19. Isopach of the Eocene to Miocene interval. Contours are in milliseconds, two-way seismic time.

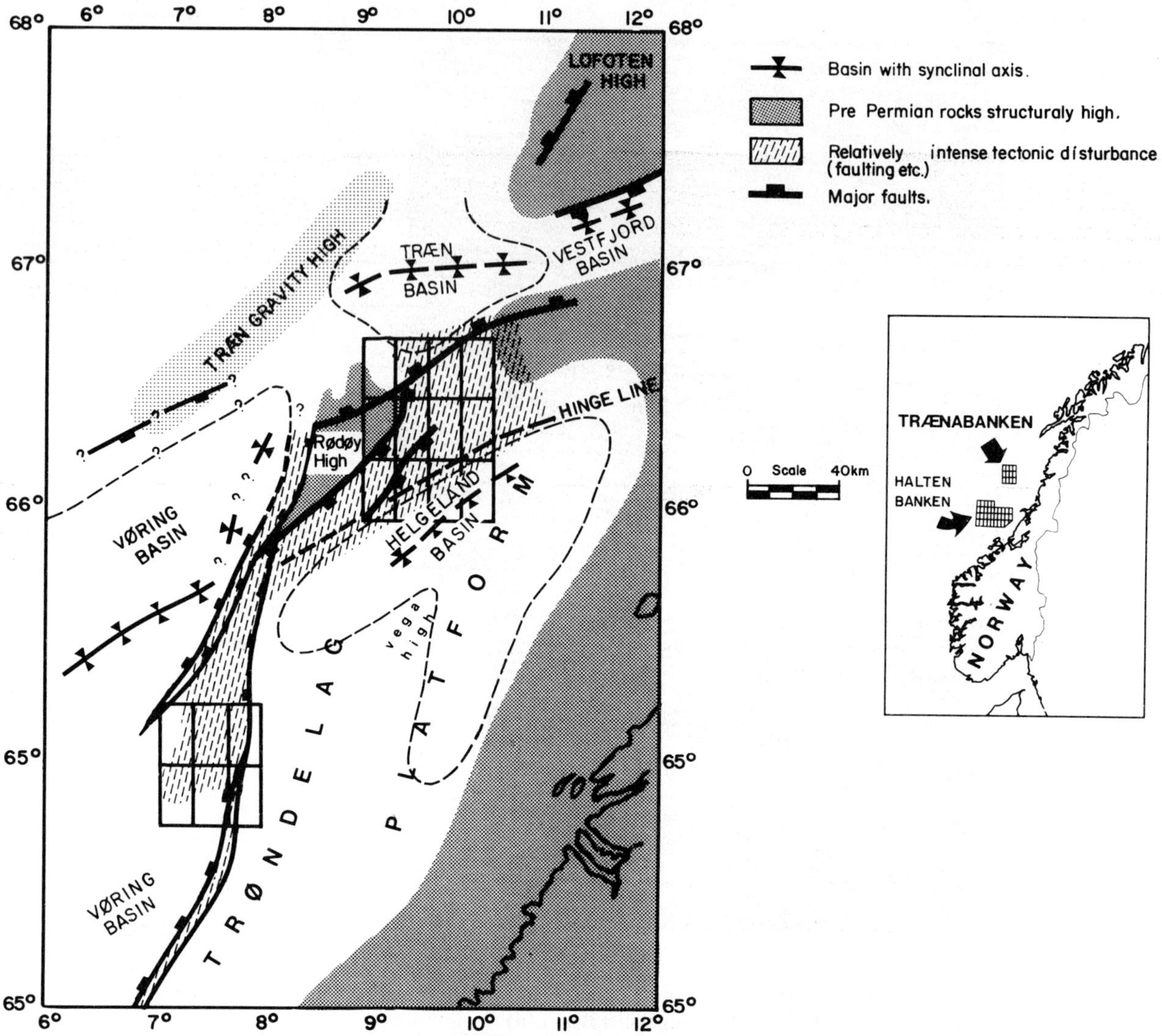

Fig. 20. Major structural elements of the Nordland Shelf.

magnetic anomaly of the order of 100 gammas on the residual magnetic field map of Åm (1970). Although the feature does not extend into the area, it is relevant to the interpretation of the northwest of Trænabanken.

Major trends

The identification of the Rødøy High emphasizes that two distinct trends are active in the area, the NNE (Caledonian) and the ENE (Devonian) trends. The ENE trend, dominant in Trænabanken, meets the NNE trend from Haltenbanken at the Rødøy High. The ENE trend of the hinge line flexure in Trænabanken can be followed in the Vøring Basin as a synclinal trend (flexure). Jørgensen and Navrestad (1979) also show faulting west of Trænabanken which follows the ENE trend.

The NNE trend from Haltenbanken cannot be followed to the north of the Rødøy High in the Træna Basin. It appears north of the Træna Basin as the Røst High and other features parallel to the Lofoten structural trend (Eldholm and Sundvor, 1980). Permian and Jurassic faulting are dominated by the NNE trend, which is probably inherited from the Caledonides. The Træn Basin has a distinct approximately E—W trend and can be followed into the Vestfjord Basin, which trends ENE. We conclude that the ENE trend in northwest Trænabanken is a major regional trend. It marks the southern limit of the NNE-trending high gravity zones (Lofoten, Røst and Træna Gravity Highs). It is interesting to note that the continuation of this trend to the WSW, using the pre-continental drift map of Talwani and Eldholm (1977), aligns with the early Tertiary dyke swarm and sedimentary basin boundary in southeast Greenland (Larsen, 1978, 1979, Henderson, 1976).

Permo-Triassic basins

The interpretation of a NNE-trending Permian basin in the east of Trænabanken, bounded to the west by a mountainous area, fits well with the Permian geology of East Greenland. Birkelund and Perch-Nielsen (1976) describe a NNE-trending Permian basin in East Greenland, bounded to the east by a mountainous area. It seems likely that the Caledonian Fold Belt between Greenland and Norway persisted as a high area in Permian times, whilst faulted basins developed on its flanks. A complex rift situation, with several such Permian basins, seems reasonable (Ziegler (1978) shows a similar idea in his fig. 7).

The Permian basin conditions continued into Triassic times and in East Greenland the regression in the late early Triassic times was accompanied by an increase in sediment supply from both west and east (Birkelund and Perch-Nielsen, 1976). We assume a similar situation in Trænabanken as the Triassic continental sediments progressively onlapped the Caledonian high to the west.

Jurassic basins

The regional context of the Jurassic is difficult to establish due to our very limited information. The main transgressive and regressive cycles can be correlated with East Greenland, but the existence of a relatively flat Jurassic deposition surface in the Trænabanken and Haltenbanken areas cannot so easily be compared. The difference in Middle Jurassic faunas between East Greenland and Northwest Europe allows Birkelund and Perch-Nielsen (1976) to postulate a barrier between them. We link this idea with indications that the Jurassic thins to the west from Trænabanken and Haltenbanken. We can speculate that the Caledonian High (or highs) between east Greenland and the Trøndelag Platform, which were active in the Lower Triassic, continued to influence the Jurassic palaeogeography. The facies change illustrated in Fig. 10 would then represent the landward (western) limit of the coal sequence.

The interpretation of wrench movements in the end Middle Jurassic faulting in Trænabanken can be correlated with wrench movements postulated in the northern North Sea at this time (Hay, 1978). Surlyk (1977) also notes *en-echelon* faulting in East Greenland at this time but attributes it to normal movement along older wrench systems. This cannot be the case in Trænabanken, due to the thick unfaulted Trias section underlying the fault zone. Active wrenching along an older trend at the end of mid Jurassic times is indicated.

Early Cretaceous basins

Subsidence and faulting in the major late Cimmerian tectonic phase is often assumed to be at the Jurassic—Cretaceous boundary. In East Greenland Surlyk (1977) dated this phase as Volgian to Valanginian. In the Vøring Basin, west of Haltenbanken, the earliest Lower Cretaceous rests conformably on the strong top Jurassic reflector, indicating that this basin did not subside until after the start of Cretaceous times (Fig. 21). A similar conclusion can be drawn from evidence in the Helgeland Basin in Trænabanken, as previously discussed. Ziegler (1978) relates the late Cimmerian tectonic phase to a sharp eustatic drop in sea-level, which would then indicate an end Ryazanian date for the main subsidence in Trænabanken.

The early Cretaceous movements marked the end of major tectonic events in Trænabanken. We have found no evidence in the Upper Cretaceous or early Tertiary sections to suggest that Trænabanken was influenced by pre-rift doming, or rifting leading to sea-floor spreading between Norway and Greenland.

Oligocene wrench movement

The reserve rejuvenation and wrench movement of the Oligocene in Trænabanken can be viewed in the context of the sea-floor spreading between Norway and Greenland. Rønnevik *et al.* (1979) describe inversion movements in mid Oligocene times to the northwest of Trænabanken and these are likely to have been caused by the same wrench activity detected in Trænabanken. Our present data are insufficient to draw conclusions concerning the cause of the wrenching, but we can make some observations. The Oligocene shear trend, Y—Y′ in Fig. 5, is subparallel to the Jan Mayen Fracture Zone and sub-perpendicular to the Mohns Ridge

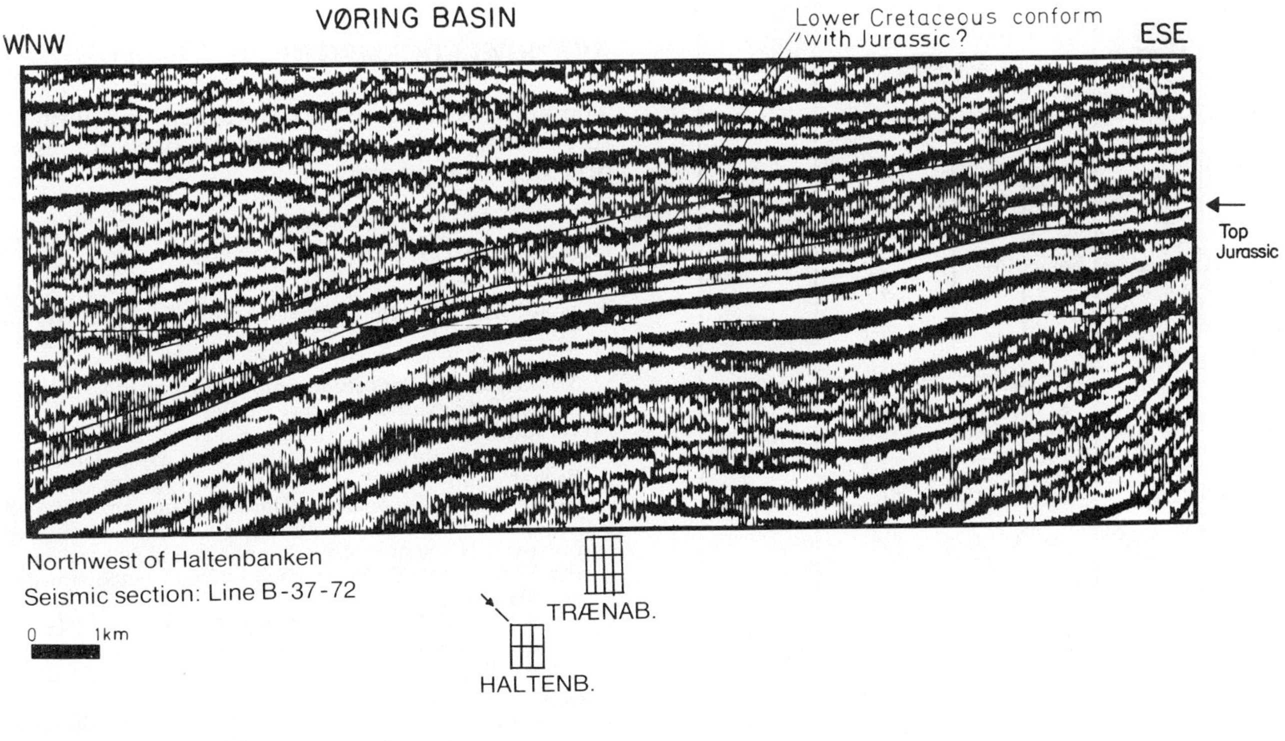

Fig. 21. Seismic line B-37-72 from the Vøring Basin showing indications that basin subsidence did not start until after the start of the Cretaceous.

and its magnetic spreading anomalies. It is also parallel, in the pre-drift position (Talwani and Eldholm, 1977), to the major northwesterly cross-faulting of the Jurassic Basins of East Greenland (Surlyk, 1977). It is likely that the wrench movements are related to the change in the spreading configuration in the Norwegian Sea as described by Talwani and Eldholm (1977).

DISCUSSION

Interpretation uncertainties

Confidence in dating of the Jurassic is high, especially south of the hinge line, but dating of pre-Jurassic reflectors becomes more speculative with increasing age. However, a consistent match with the regional geology is found with the interpretation at all levels. The assumption of the absence of Carboniferous is probably the weakest point as regards identification of the pre-Permian section. The interpretation of the seismic northwest of Fault A is open to doubt, due to the poor data quality and the lack of adequate regional control.

Northwest of Fault A

The seismic interpretation here suggests that there was no deep Cretaceous basin similar to the Vøring Basin to the south. The Træn Basin to the north, which can be linked to the NW area, has an E—W axis orientation, and a different seismic response from the Vøring Basin and we have interpreted a structural high between them, joining the Rødøy High with the Træn Gravity High. The Jurassic, Permian and much of the Triassic is expected to be absent over this structural high.

Further support for such an interpretation comes from the gravity data. Figure 22 shows a simplified 2D Bouguer Gravity Map after the removal of a regional gradient illustrating the dominant Rødøy High and its relation to the features northwest of Fault A. The Træn Gravity High could be explained more easily if the Cretaceous in the area rested on Palaeozoic rocks. A more extensive regional seismic data base is required before the problems in this area can be solved.

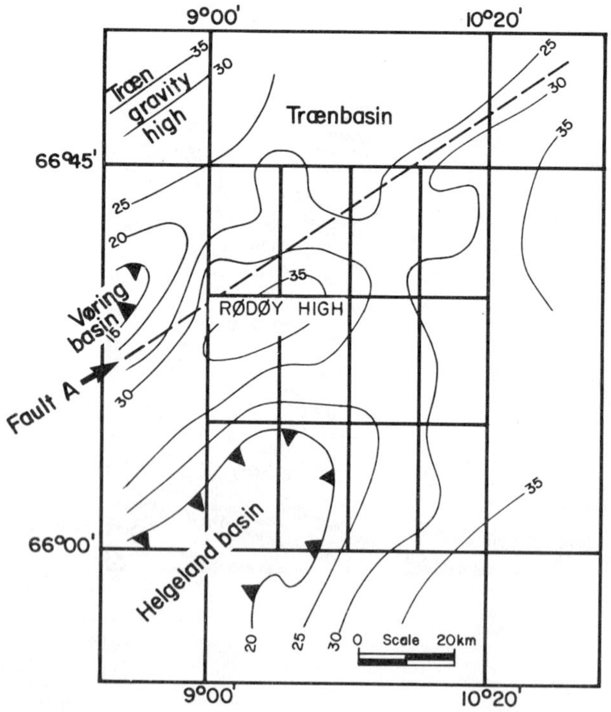

Fig. 22. 2-D Bouguer Gravity after removal of a regional gradient, showing the main structural features in Trænabanken.

Pre-spreading doming and rifting

In view of the great interest in sea-floor spreading and its initiation, it is natural to review our findings for evidence of influence of pre-spreading tectonics in Trænabanken. Using our interpretation and similar reports from East Greenland, we have concluded that the area between Trænabanken and Greenland was relatively high in Triassic and possibly in mid Jurassic times. This does not, of course, preclude some rifting within this high area. However, these events seem to be too far removed from the eventual Tertiary spreading to be reasonably connected. They may, instead, result from a Jurassic dome/rift situation postulated by Surlyk (1977) in the southern part of East Greenland.

If the Jurassic high was related to the old Caledonian orogenic zone, we might speculate that the Shetland Platform and Møre Platform (Hamar *et al.*, 1980), the Jan Mayen Ridge, the Træna Gravity High and the basement highs in the Lofoten Shelf area formed a linear feature at the time. Jurassic sediments would then be thin or absent in this area and we could explain the linear gravity highs as remnants of the old Caledonian suture line.

After Jurassic times there is little to suggest doming or rifting in the Trænabanken data. The eastward tilting in late Cretaceous and Palaeocene times is not found to the west of the area in the interpretations shown by Rønnevik *et al.* (1979) and is not evidence of doming.

ACKNOWLEDGEMENTS

The authors thank Norsk Hydro AS for permission to publish this work, and Michael Råberg for drafting the diagrams.

REFERENCES

Aanstad, K., Gabrielsen, R., Hagevang, T., Ramberg, I. and Torvanger. O. 1981. Correlation of offshore and onshore structural features between 62°N, Norway. In: *Norwegian Symposium on Exploration,* (NSE-81) (Norsk Petroleumsforening) MSE/11.

Birkelund, T. and Perch-Nielsen, K. 1976. Late Paleozoic—Mesozoic evolution of Central East Greenland. In: A. Escher, and W.S. Watt (eds.). *Geology of Greenland.* Geol. Surv. Greenland, Copenhagen, pp. 304—339.

Dalland, A. 1979. The sedimentary sequence of Andøy, northern Norway — depositional and structural history. In: *Norwegian Sea Symposium.* Norsk Petroleumsforening, Oslo NSS/26.

Eldholm, O. and Sundvor, E. 1980. The continental margins of the Norwegian—Greenland Sea, recent results and outstanding problems. *Phil. Trans. R. Soc. Lond., Ser. A,* **294**, 77—86.

Gabrielsen, R.H., Færseth, R., Hamar, G. and Rønnevik, H. 1984. Nomenclature of the main structural features on the Norwegian Continental Shelf north of the 62nd parallel. This volume, pp. 41—60.

Gayer, R.A. 1973. Caledonian geology of Arctic Norway. In: M.G. Pitcher (ed.). Arctic Geology, *AAPG Mem.*, **19**, 453—468.
Gjelberg, J. and Steel, R. 1979. Middle Carboniferous sedimentation in relation to tectonic, climate and sea level changes on Bjørnøya and Spitsbergen. In: *Norwegian Sea Symposium.* Norsk Petroleumsforening, Oslo, NSS/27.
Hamar, G.P., Jakobson, K.H., Ormansen, D.E. and Skarpnes, O. 1980. Tectonic development of the North Sea north of the central highs. In: *The Sedimentation of the North Sea Reservoir rocks.* Norsk Petroleumsforening, Oslo, section III, pp. I—II.
Harland, W.B. 1973. Tectonic evolution of the Barents Shelf and related plates. In: M.G. Pitcher (ed.), Arctic Geology. *AAPG Mem.* **19**, 599—608.
Harland, W.B. and Gayer, R.A. 1972. The Arctic Caledonides and earlier oceans. *Geol. Mag.*, **109**, 289—314.
Hay, J.T.C. 1978. Structural development in the Northern North Sea. *J. Petrol. Geol.*, **1**, 65—77.
Henderson, G. 1976. Petroleum geology. In: A. Escher and W.S. Watt (eds.). *Geology of Greenland,* Geol. Surv. Greenland, Copenhagen, pp. 488—505.
IKU, 1982. *Examination of the Outcropping Pre-Quarternary Rocks of the Shelf of Helgeland, mid Norway.* Institutt for kontinentalundersøkelser, Trondheim, proprietary report.
Jørgensen, F. and Navrestad, T. 1979. Main structural elements and sedimentary succession on the shelf outside Nordland (Norway). In: *Norwegian Sea Symposium.* Norsk Petroleumsforening, Oslo, NSS/11.
Larsen, H.C. 1978. Offshore continuation of East Greenland dyke swarm and North Atlantic Ocean formation. *Nature, (London)*, **274**, 220—223.
Larsen, H.C. 1979. Geological perspectives of the East Greenland Continental Margin. *Norwegian Sea Symposium.* Norsk Petroleumsforening, Oslo, NSS/7, preprint.
Larsen, R.M. and Skarpnes, O. 1984. Regional interpretation and hydrocarbon potential of the Trænabanken area. This volume, pp. 217—236.
Nilsen, T.H. 1973. Devonian (Old Red Sandstone) sedimentation and tectonics of Norway. In: M.G. Pitcher (ed.), Arctic Geology, *AAPG Mem.*, **19**, 471—481.
Rønnevik, H. Bergsager, E.I., Moe, A., Øvrebø, O., Navrestad, T. and Stangenes, J. 1975. The geology of the Norwegian Continental Shelf. In: A.W. Woodland (ed.), *Petroleum and the Continental Shelf of Northwest Europe,* Vol. 1, Applied Science Publishers, London, pp. 117—129.
Rønnevik, H., Jørgensen, F. and Motland, K. 1979. The geology of the Vøring Plateau. In: *Norwegian Sea Symposium.* Norsk Petroleumsforening, Oslo, NSS/12.
Sellevoll, M.A. 1975. Seismic refraction measurements and continuous seismic profiling on the continental margin off Norway between 62° and 69°N. *Norges Geol. Unders.*, **316**, 219—236.
Siedlecka, A. 1975. Late Precambrian stratigraphy and structure of the northeastern margin of the Fennoscandinavian Shield (East Finnmark-Timan Region). *Norges Geol. Unders.*, **316**, 313—348.
Steel, R.J. 1976. Devonian basins of western Norway — sedimentary response to tectonism and to varying tectonic context. In: M.H.P. Bott *et al.* (eds.), Sedimentary basins of Continental Margins and Cratons, *Tectonophysics,* **36**(1—3), 207—224.
Surlyk, F. 1977. Mesozoic faulting in East Greenland. In: R.T.C. Frost and A.J. Dikkers (eds.), Fault Tectonics in N.W. Europe, *Geol. Mijnbouw,* **56**, 311—327.
Surlyk, F. 1978. Jurassic basin evolution of East Greenland. *Nature (London)*, **274**, 130—133.
Surlyk, F., Piasecki, S., Rolle, F., Stemmerick, L., Thomsen, E. and Wrang, P. 1984. The Permian basin of East Greenland. This volume, pp. 303—316.
Talwani, M. and Eldholm, O. 1972. The continental margin off Norway: a geophysical study. *Geol. Soc. Am. Bull.*, **83**, 3575—3608.
Talwani, M. and Eldholm, O. 1977. Evolution of the Norwegian-Greenland Sea. *Geol. Soc. Am. Bull.*, **88**, 969—999.
Talwani, M., Mutter, J. and Eldholm, O. 1981. The initiation of opening of the Norwegian Sea. In: Proc. 26th Int. Geol. Cong. Symp. Geology of Continental Margins, *Oceanol. Acta,* 23—30.
Ziegler, P.A. 1981. Evolution of sedimentary basins in North-West Europe. In: L.V. Illing and G.D. Hobson (eds.). *Petroleum Geology of the Continental Shelf of North-West Europe.* Heyden, London, pp. 3—39.
Åm, K. 1970. Aeromagnetic investigations on the continental shelf of Norway, Stadt-Lofoten (62°—69°N). *Norges Geol. Unders.*, **266**, 49—61.

17 Structures and basins of the margin from 62° to 69°N and their development

F. Bøen, S. Eggen and J. Vollset
Norwegian Petroleum Directorate, Stavanger, Norway

An updated synthesis of the development of the main sedimentary basins between 62°–69°N is presented with emphasis on the structural elements and the tectonic history and geological evolution. Young tectonic events make mapping of old Palaeozoic basins difficult. Only in the Vestfjord Basin can sedimentary deposits as old as Carboniferous be indicated. West of the Vestfjord Basin, younger Mesozoic and Tertiary tectonics have overprinted the old basin trends. The main structural elements in this area are results of Kimmerian movements.

Parts of the Nordland Ridge were probably representing positive features as early as Palaeozoic times acting as source areas for coarse clastics deposited both into the Vestfjord Basin and the Vøring Basin. Periodically the ridge was submerged, and in post Cretaceous time, the whole area was one epeirogenically subsiding basin. The main subsidence both in the Møre-, Vøring- and Træn basins took place in Upper Cretaceous times.

Kimmerian structural elements can be mapped west of the Nordland Ridge, but are difficult to map in the Møre Basin, due to deposition of very thick post Jurassic sediments. In Tertiary times volcanic activity related to the opening of the Norwegian Sea masks both deeper structural trends on the outer part of the shelf as well as the border between oceanic and continental crust. The area north of the Jan Mayen Fracture Zone has been affected by Tertiary compressional movements resulting in domal uplifts in the Vøring Basin and over the Nordland Ridge.

INTRODUCTION

The intention in this chapter is to give a review of the history of structure and basin development offshore mid Norway from late Palaeozoic through Cenozoic time.

A number of papers have been published which deal with the geology of this area (Rønnevik *et al.*, 1975; Rønnevik and Navrestad, 1977; Rønnevik *et al.*, 1979; Jørgensen and Navrestad, 1981). Geological problems of special interest, such as the nature of the crust underlying the Vøring Basin and the transition between oceanic and continental crust, have been discussed in great detail (Talwani and Eldholm, 1972; Talwani *et al.*, 1981; Hinz, 1972; Hinz *et al.*, 1982). These studies will not be reviewed thoroughly here.

The bathymetric map of the Norwegian Sea shows two main topographic elements between 62° and 69°N:

(1) The shelf area, shallower than about 500 m.
(2) The Vøring Plateau, between 1000 and 1500 m.

Petroleum Geology of the North European Margin © Norwegian Petroleum Society (Graham & Trotman, 1984) pp. 253–270.

The main geological subdivisions of the area are (Fig. 1):

(1) The Møre Basin.
(2) The Vøring Basin (between the Nordland Ridge and the Vøring Plateau Escarpment (VPE).
(3) The platform areas.
(4) The areas west of the escarpments.

The area between 62° and 69°N has been covered by regional and semi-regional seismic grids by the Norwegian Petroleum Directorate in the years 1969–1981 (totalling approximately 25 000 km). In addition, data acquired by scientific institutions (Bundesanstalt für Geowissenschaften und Rohstoffe and Lamont Doherty Geological Observatory) have been made available for this study. In the Møre area data from a special survey has also been used. The regional data coverage in the area totals approximately 35 000 km. In addition, large quantities of detailed seismic data have been acquired within the opened areas (the Haltenbanken and the Trænabanken).

GEOLOGICAL EVOLUTION

Pre-Jurassic

According to Jørgensen and Navrestad (1979, 1981)

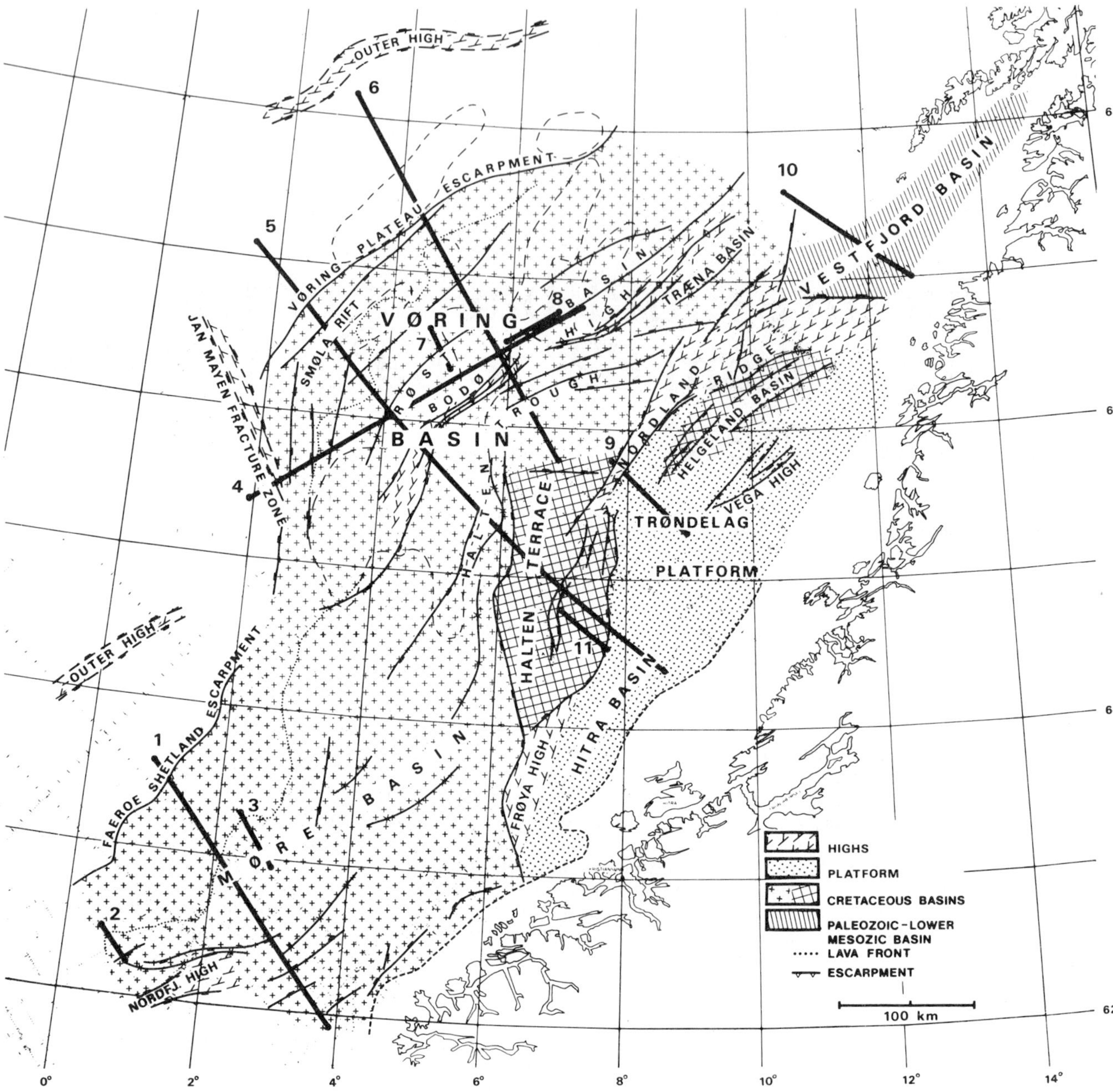

Fig. 1. The mid-Norwegian Shelf. The main structural elements and location of geoseismic profiles and seismic sections are shown.

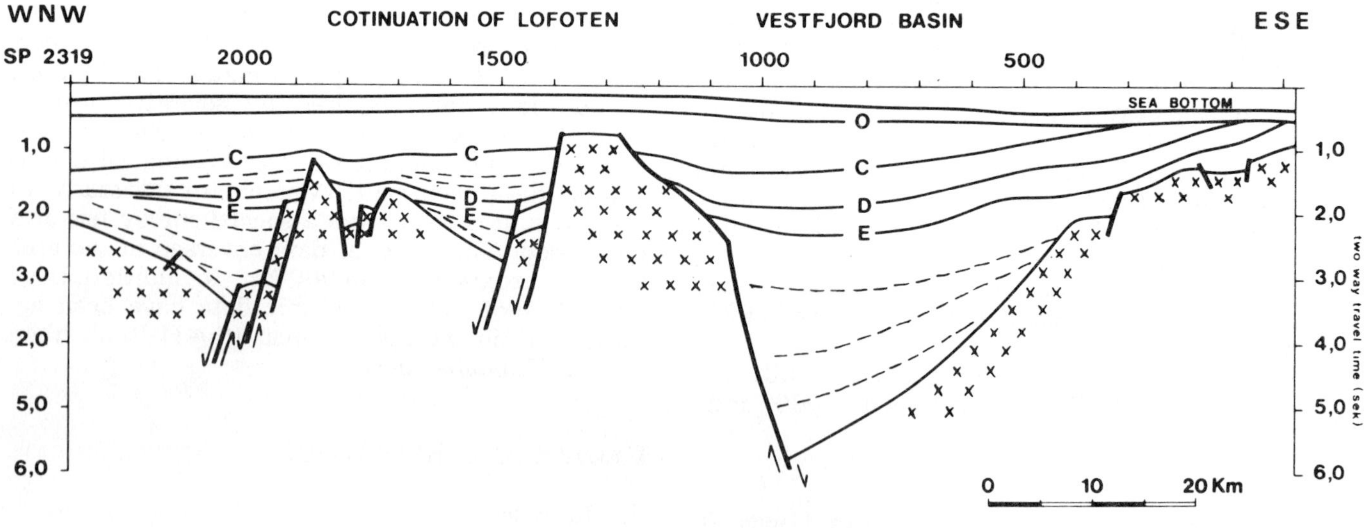

Fig. 2. Geoseismic profile 10, Vestfjord Basin.

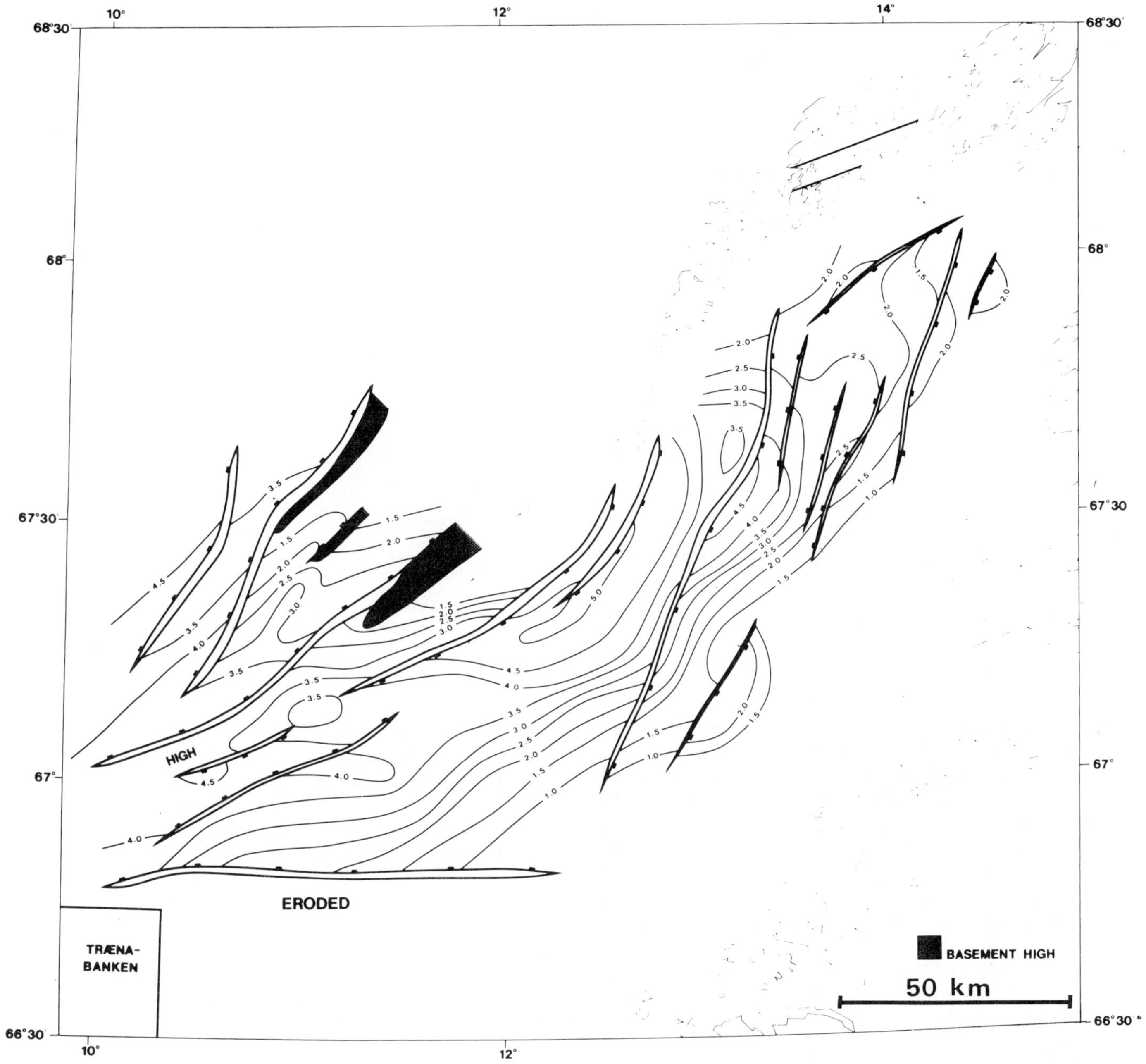

Fig. 3. Isochron map, Palaeozoic reflector in the Vestfjord Basin.

the Vestfjord Basin was an area of major pre-Jurassic sedimentation and subsidence. A geoseismic section across the basin (Fig. 2) shows several rotated basement blocks separated by narrow half-grabens. These basement blocks are the subsea continuation of the Lofoten Islands.

Based on charts of global eustatic sea-level variations (Vail *et al.*, 1977) tentative ages were proposed for the different reflectors in the area. For the deepest interpreted reflector a Middle Carboniferous age was proposed, assuming post-Caledonian peneplaining during Devonian to early(?) Carboniferous.

Kimmerian and younger tectonic movements seem to have affected the Vestfjord Basin only to a minor degree, which makes mapping of the old basin trends possible. Figure 3 is an isochron map of the deep reflector (Middle Carboniferous?) in the Vestfjord Basin. Based on this map the basin can be subdivided into several parallel-oriented half-grabens separated by NNE—SSW, down to the east, normal faults. The basin is bounded by ENE—WSW-oriented fault systems to the south and north. The thicknesses of the Jurassic to mid-Carboniferous sediments exceeds 3 seconds two-way time (TWT) in the deepest parts of the half-grabens.

Mapping of the old unconformity from the Vestfjord Basin towards the younger Træn Basin is obscured by a combination of younger tectonic movements, depth of burial and present regional data coverage. Southwards the old unconformity can be mapped to the above-mentioned ENE—WSW fault system that marks the transition to the eroded Nordland Ridge. South of this fault system Palaeozoic unconformities cannot be mapped continuously. An old unconformity can be followed in the most heavily eroded parts of the Nordland Ridge, e.g. the SW part of the Trænabanken area. The exact age of this unconformity, however, is uncertain, but a Permian age is suggested.

Two-dimensional gravity modelling over the

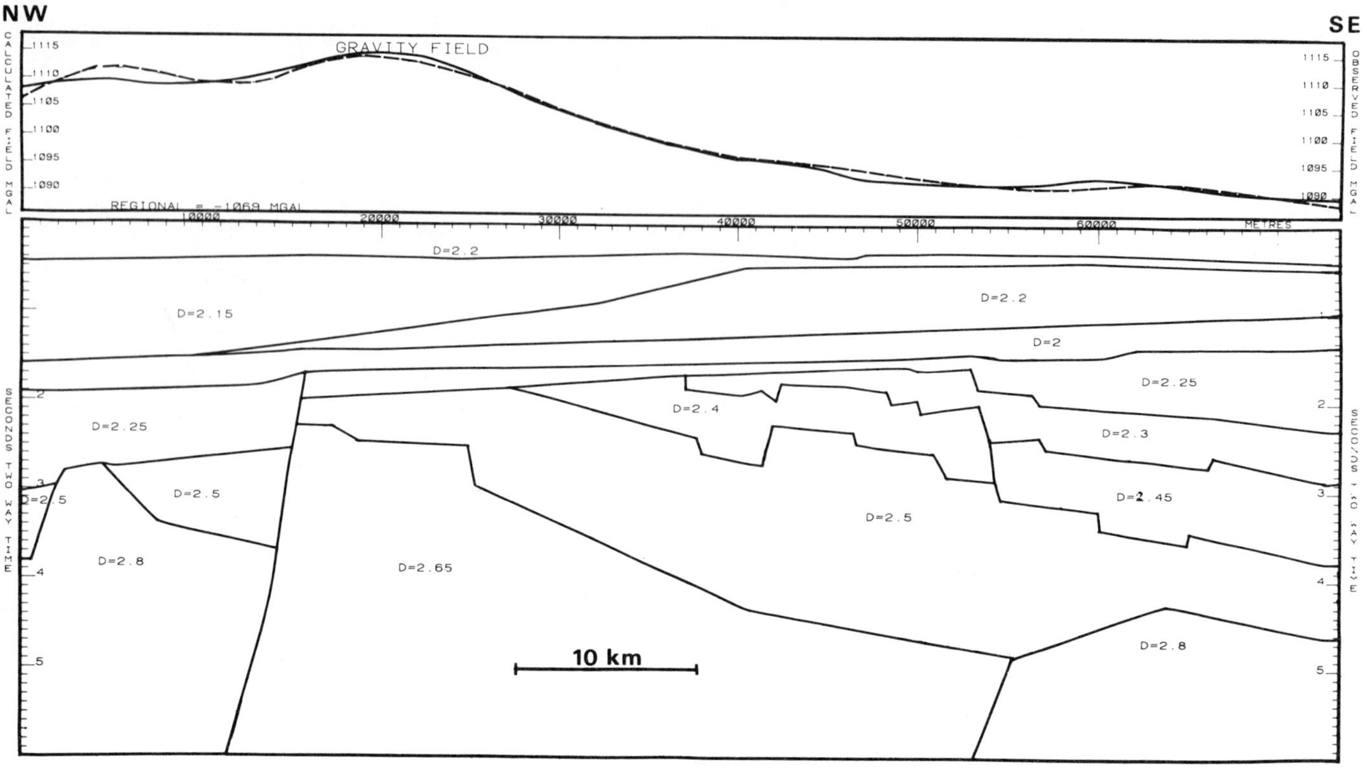

Fig. 4. Two-dimensional gravity modelling along a seismic line crossing the Nordland Ridge in the Trænabanken area.

Trænabanken area (Fig. 4) shows that one possible interpretation of the gravity field is that part of the Nordland Ridge is an inverted old sedimentary basin, and presence of high-gravity material can be deduced both on the outer side of the ridge and inside towards the Helgeland Basin. One of the trends inferred from the gravity data has a N—S direction. This direction may correspond to Palaeozoic—Lower Mesozoic basin trends, as shown in the Vestfjord Basin.

On the east flank of the Møre Basin remnants of an old asymmetric basin can be seen on a seismic line (Rønnevik and Navrestad, 1977). Unlike the Vestfjord Basin this area has been subjected to uplift and deep erosion during early Mesozoic time.

Subsidence in these old basins is believed to be mainly fault related. This fault-generated relief was then subsequently filled in during Upper Palaeozoic and Lower Mesozoic time.

Jurassic

Upper Triassic and Jurassic sediments can be confidently mapped only on the inner part of the mid-Norwegian Shelf, where Cretaceous subsidence was low to moderate. This area comprises the Vestfjord

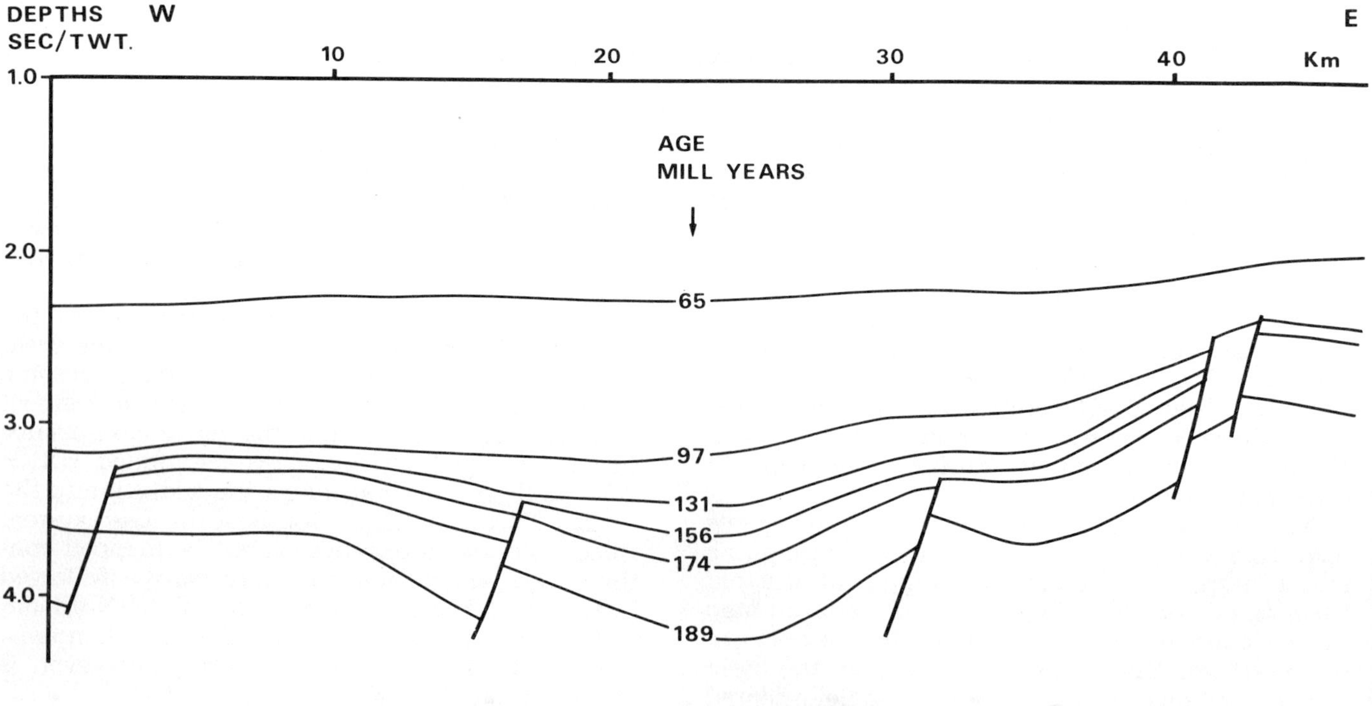

Fig. 5. Geoseismic profile 11, Halten Terrace.

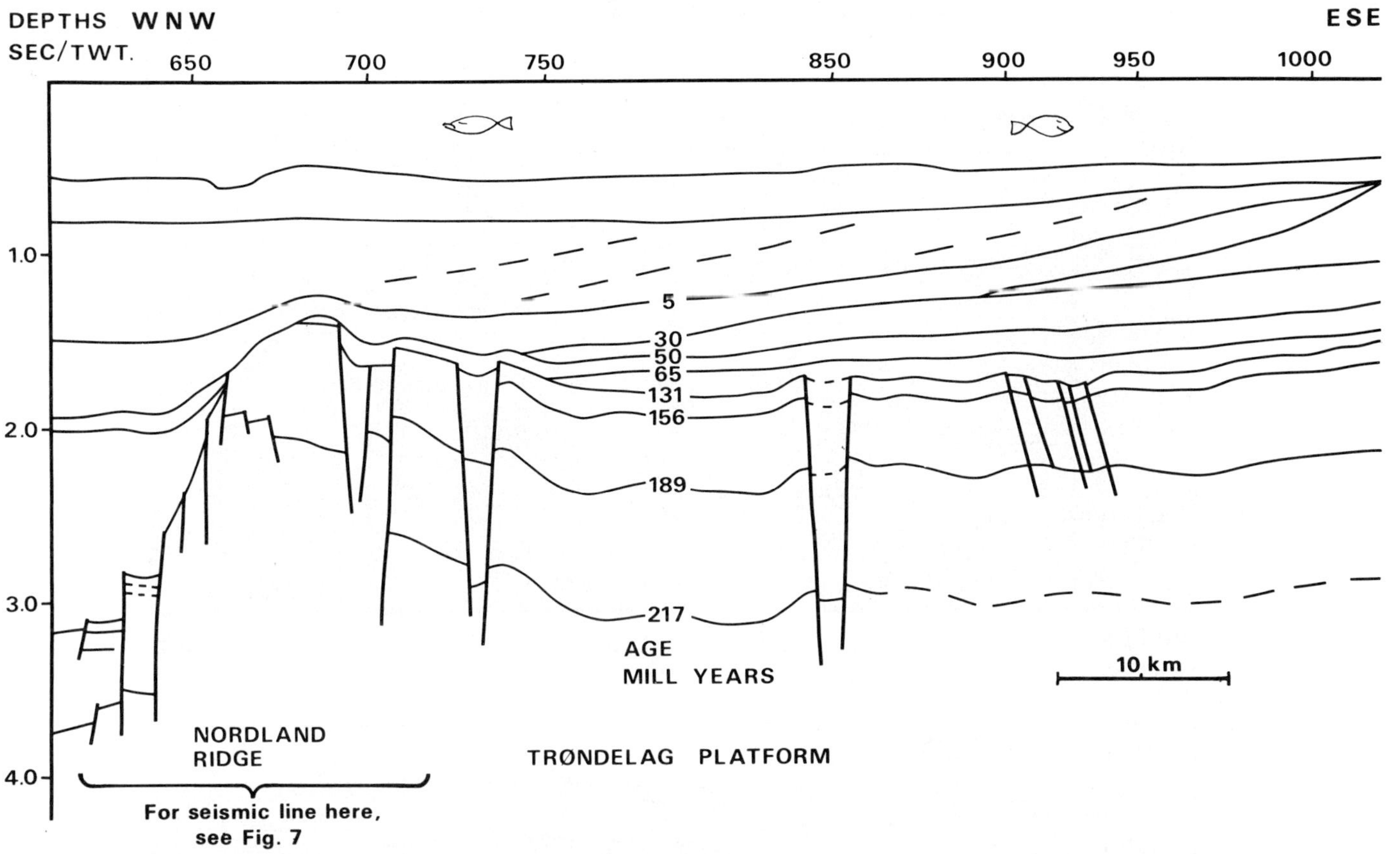

Fig. 6. Geoseismic profile 9, B10-75, crossing the Trøndelag Platform and the Nordland Ridge.

Basin, the Trøndelag Platform, the Halten Terrace and parts of the Nordland Ridge (see Fig. 1).

Wells drilled on the Haltenbanken have penetrated a clastic Jurassic sequence of approximately 700—1000 m, which can be mapped over the entire Trøndelag Platform, and with less certainty into the Vestfjord Basin.

On the Trøndelag Platform rather uniform thicknesses of Jurassic sediments were deposited. No major depocentres can be identified. A small thinning of the sequence can, however, be observed when approaching the Nordland Ridge. Slight thickening in the Vega Ridge and the Helgeland Basin (Upper Jurassic) indicates that the Nordland Ridge became active in least in the Upper Jurassic, probably following initial rifting further west. The Vega Ridge was later inverted in Cretaceous time, and is presently mapped as a ridge at base Cretaceous level.

Growth faulting on the Trøndelag Platform in Lower and middle Jurassic indicates limited crustal activity in the area.

On the downfaulted Halten Terrace to the south, more active growth faulting has taken place in the same period. Significant thickness variations can be mapped in this sequence, as shown on the geoseismic profile (Fig. 5).

The Nordland Ridge (Fig. 6) was exposed to erosion during the Cretaceous, and subsequently Jurassic sediments were removed in some areas, as in the southwestern part of the Trænabanken area. Similar erosion on the western part of the Halten Terrace can be observed (Fig. 12).

West of the Nordland Ridge the Jurassic is downfaulted from approximately 1.5 seconds TWT on the ridge to more than 6 seconds in the trough (Figs. 6 and 10). The data quality in this transition zone has previously been poor. Reprocessing of old seismic lines has, however, shown major improvements as indicated on Fig. 7. The existence of Jurassic sediments in rotated fault blocks west of the Nordland Ridge is quite probable.

On the east flank of the Møre Basin the pre-Cretaceous sediments seem to be generally thin, and are located in the structural lows (Rønnevik and Navrestad, 1977).

Under the very thick Cretaceous sedimentary cover in the Vøring and Møre basins, reflectors of possible Jurassic age can be identified. The oldest reflector that can be mapped continuously in these basins is interpreted to be base Cretaceous. Rønnevik *et al.* (1979) proposed a mid-Cretaceous age for this reflector. However, the drilling results so far on the Haltenbanken, as well as the seismic character of the reflector, favour a base Cretaceous age for this unconformity. Below the base Cretaceous reflector, sub-parallel reflectors can be identified (Figs. 8 and 9), indicating the presence of Jurassic and/or older sediments. However, the depth of the reflectors and the regional seismic grid make them difficult to map at present.

The base Cretaceous isochron map of the Vøring and Møre Basins (Fig. 10) shows that the area is severely affected by tectonic activity. The relief on this map is a result of the combined effects of Kimmerian tectonic movements, strong Cretaceous subsidence and later Tertiary movement.

Evidence of Jurassic rifting can be observed in both the Vøring and Møre Basins (Figs. 11 and 12). Rifting resulting in major Jurassic depocentres cannot be verified presently, but may be inferred from the variations in thickness of Upper Jurassic

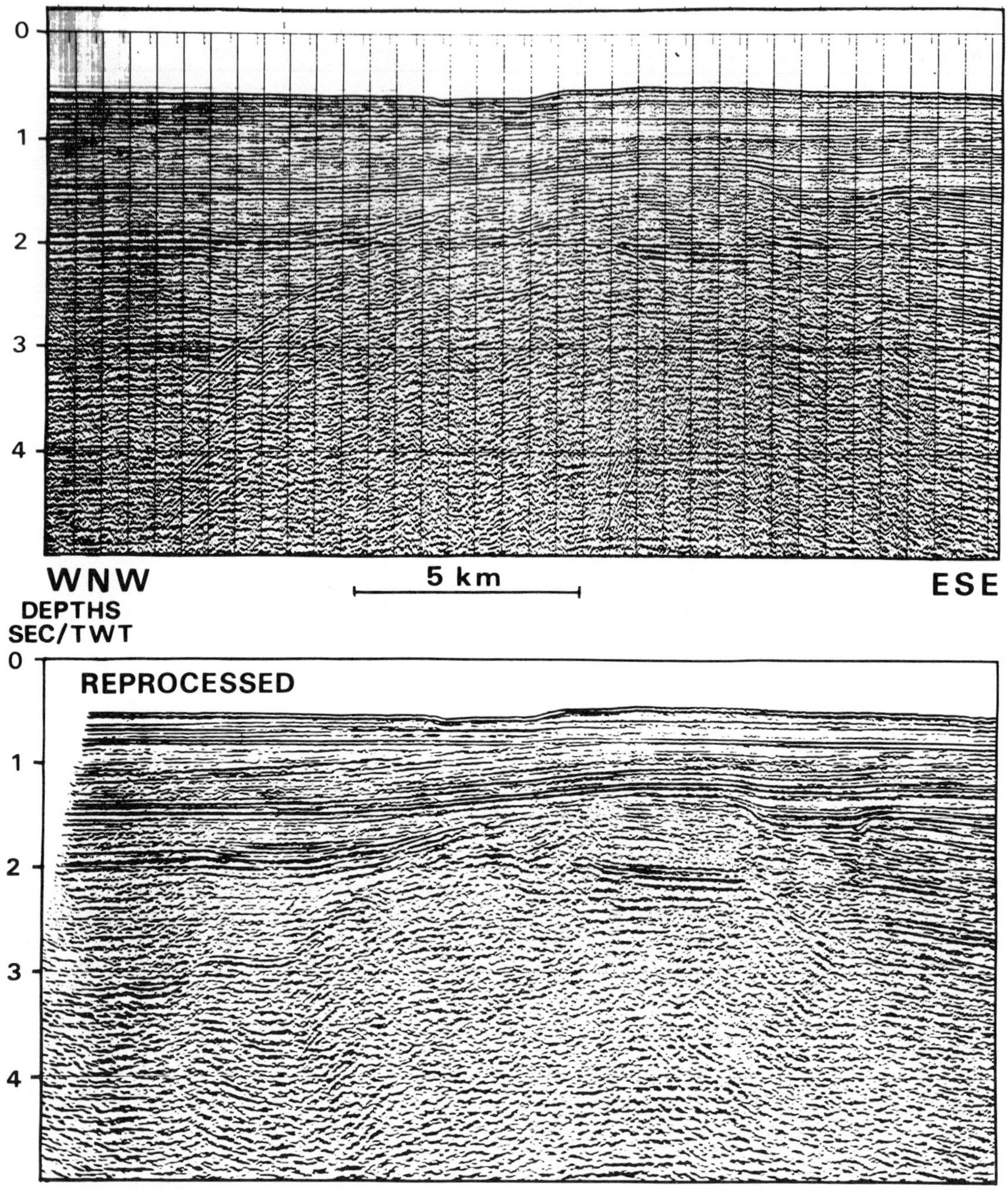

Fig. 7. Original and reprocessed version of B10-75, showing the transition zone between the Nordland Ridge and the Halten Trough.

sediments on the Halten Terrace (Fig. 5).

Cretaceous

At base Cretaceous level several lows and highs can be mapped (Fig. 1):

(1) The Smøla Rift immediately east of the Vøring Plateau Escarpment is defined by a complex set of NE—SW-oriented faults in the north, and by a more N—S direction in the south. Evidence of possible late Jurassic to early Cretaceous rifting can be seen in this rift (Figs. 12 and 13). The rift has been inverted in early Tertiary time.

(2) The Bodø High is a poorly defined, narrow, NE—SW-oriented, deeply buried high, bounded to the east and the west by large normal faults (Figs. 12 and 13). Indications of rotated fault blocks can be observed in parts of the high.

(3) The Røst Basin (earlier defined as Røst Syncline by Rønnevik *et al.*, in press) between the Smøla Rift to the west and the Bodø High to the east, is subdivided into a northern narrow syncline and a southern broader basin by a WNW—ESE fault (66°30′N, 6°00′E). This fault shows compressional features and may represent dextral movements stopping in base Tertiary (Figs. 14 and 9). The Røst Basin is thus a result of both Cretaceous subsidence and Tertiary lateral movements. This may explain the difference in configuration between the narrow, deep northern segment and the broader southwestern segment.

(4) Another complex rift zone can be mapped between the Bodø High and the Nordland Ridge. This rift, which Rønnevik *et al.* (in press) named the Halten Trough, shows an overall NE—SW trend north of 66°N. Southwards it continues into the Møre Basin, here obtaining a N—S

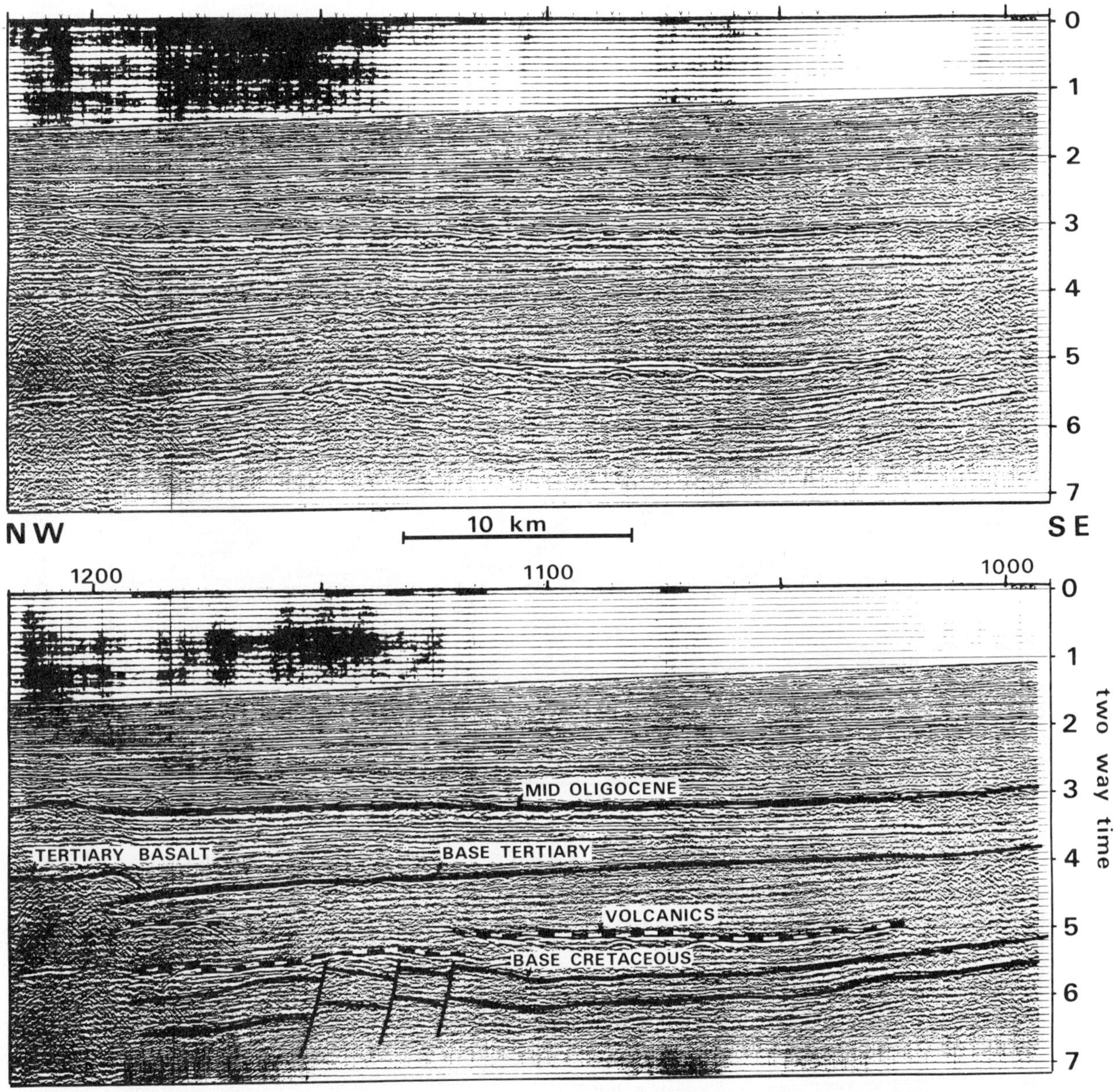

Fig. 8. Seismic profile 3, OS-E (Mϕre Basin), showing sub-parallel reflectors below base Cretaceous, and volcanics in the Cretaceous sequence.

orientation. In the northern part of the Mϕre Basin this rift splits into two parts, separated by a deeply buried high (Fig. 10).

(5) The Træna Basin (Rϕnnevik and Navrestad, 1977) is the northern continuation of the Halten Trough (Fig. 1). This NE—SW-oriented basin shows, at present, a better definition of the base Cretaceous unconformity than in the Halten Trough.

Due to Tertiary inversion and doming in the Vϕring Basin, the base Cretaceous isochron map gives a poor delineation of the Cretaceous depocentres.

The isopach map (Fig. 15) shows that the Cretaceous deposits vary in thickness from zero to more than 4 seconds TWT. The thinnest deposits are found on the Trϕndelag Platform and over the Nordland Ridge where, in parts, the Cretaceous is eroded/non-deposited. Another area with thin or no Cretaceous deposits is the high west of the Vϕring Plateau Escarpment. However, mapping the base Cretaceous in this area is highly tentative, due to the Tertiary lavas.

The greatest thickness is found in the Mϕre Basin, where thicknesses exceed 4 seconds TWT in the central parts of the rifts. Some of the seismic surveys are not recorded deep enough in these areas to map the exact thicknesses.

The Vϕring basin was another main depocentre during the Cretaceous. Sediment thicknesses show distinct variations within this megabasin. The main depocentre is the Halten Trough. Sediment thicknesses greater than 3.5 seconds TWT are recorded here, in an area just south of the Træna Basin.

A prominent ENE-WSW fault trend is mapped crossing the northern part of the Halten Trough (Fig. 1). This fault trend separates the Halten Trough from the Træna Basin, and possibly from the Halten Terrace. A fault zone of the same direction also defines the boundary between the Vestfjord Basin and the Nordland Ridge. There is no definite evidence of lateral movements along these faults, but doming above some of them might be an indication of this.

Another main depocentre is situated just east of the Vϕring Plateau Escarpment (Fig. 15). It is difficult to map its areal extent, because Tertiary

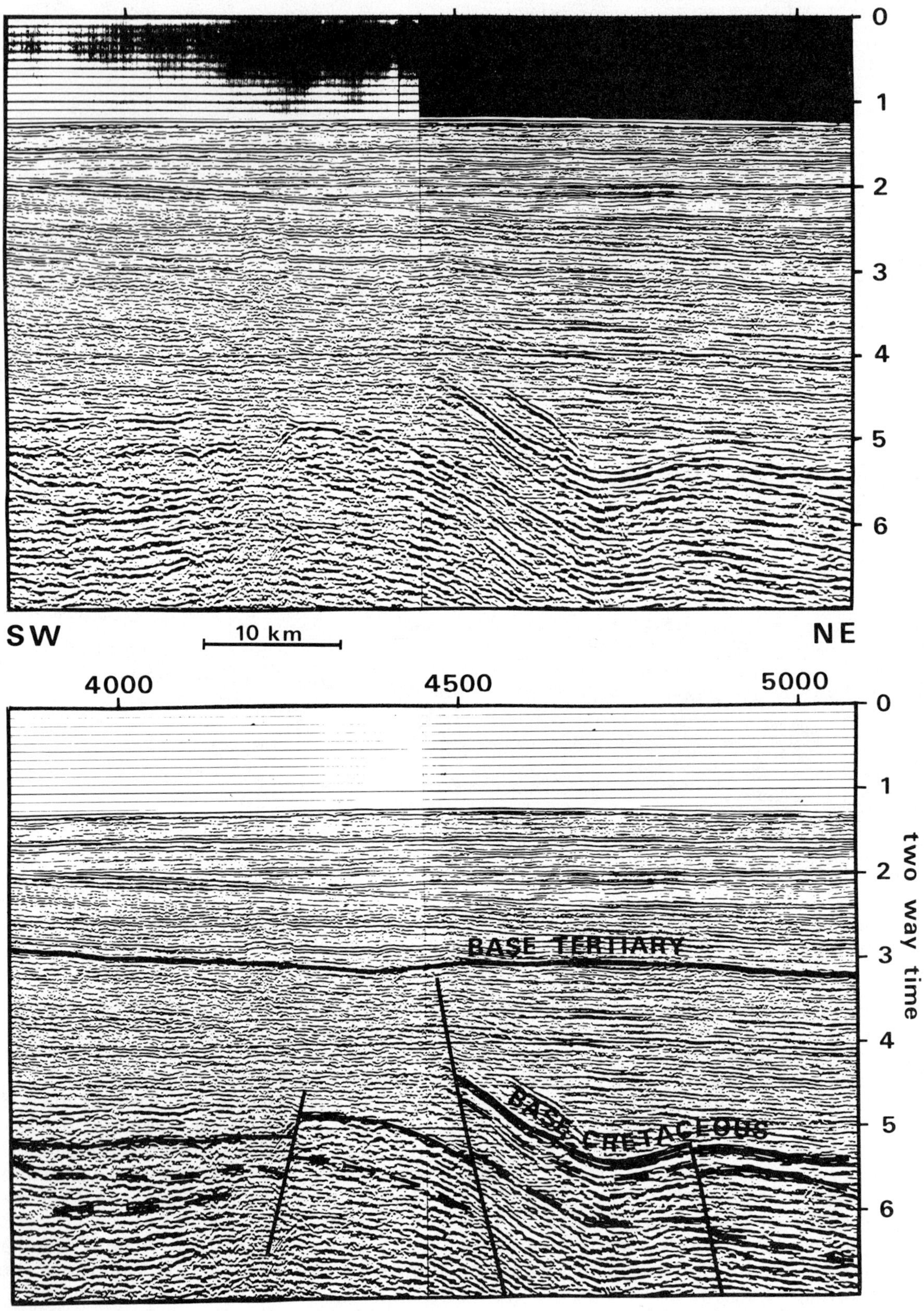

Fig. 9. Seismic profile 8 (Vøring Basin) showing inverted fault and sub-parallel reflectors below base Cretaceous level.

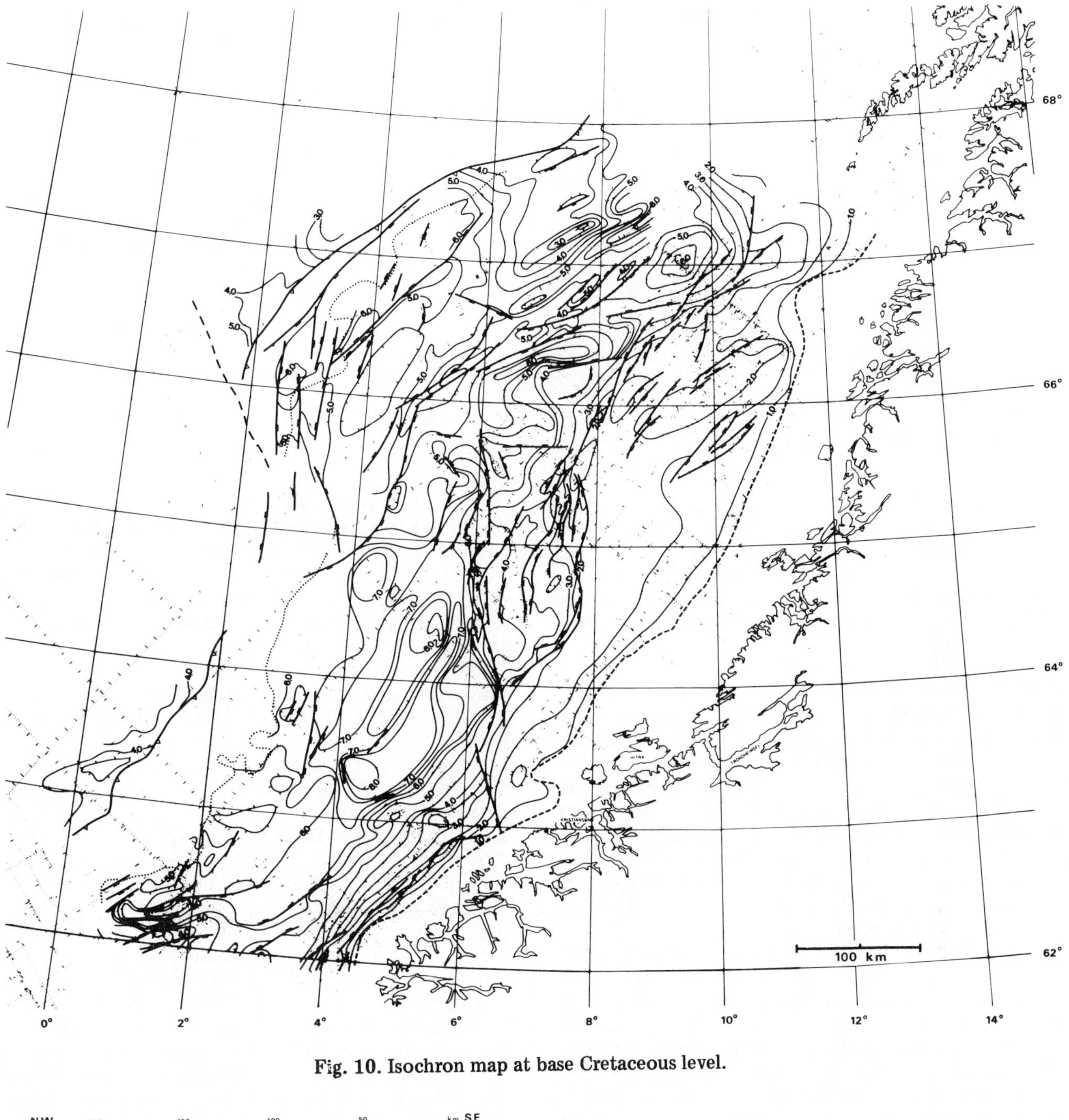

Fig. 10. Isochron map at base Cretaceous level.

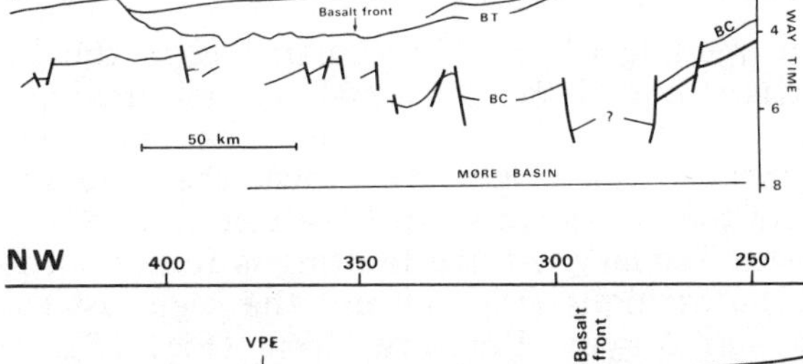

Fig. 11. Geoseismic profile 1 (Møre Basin). MO, mid Oligocene; BT, base Tertiary; BC, base Cretaceous; FSE, Faeroe-Shetland Escarpment.

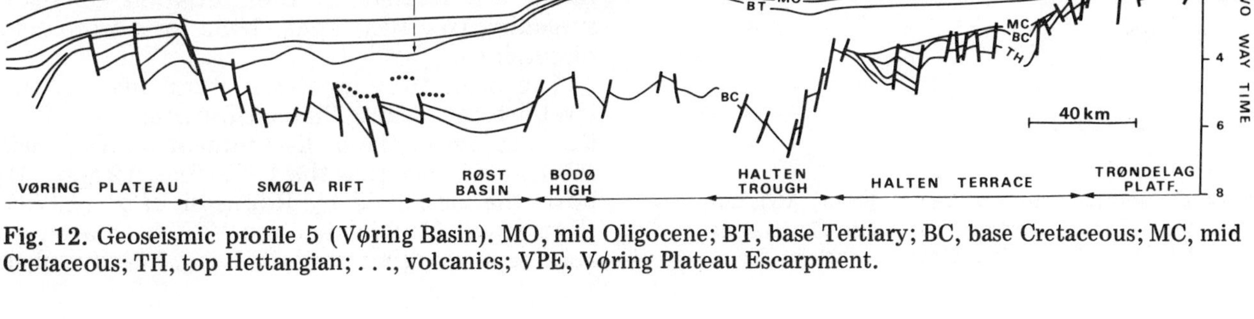

Fig. 12. Geoseismic profile 5 (Vøring Basin). MO, mid Oligocene; BT, base Tertiary; BC, base Cretaceous; MC, mid Cretaceous; TH, top Hettangian; . . ., volcanics; VPE, Vøring Plateau Escarpment.

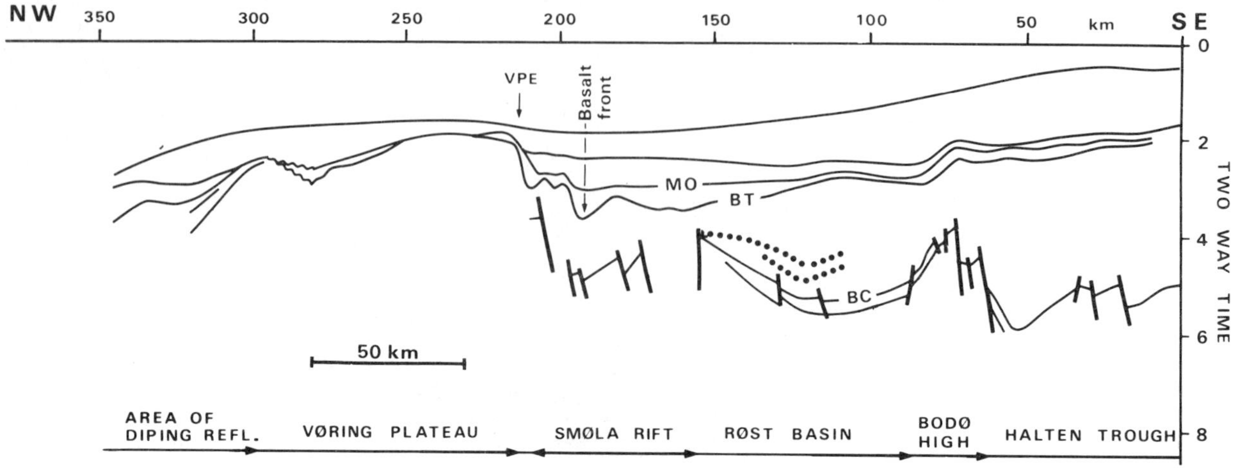

Fig. 13. Geoseimic profile 6 (Vϕring Basin). MO, mid Oligocene; BT, base Tertiary; . . ., volcanics; VPE, Vϕring Plateau Escarpment.

lavas mask underlying reflectors.

An area with thinner Cretaceous sediments is mapped between the northern segment of the Rϕst Basin and the Vϕring Plateau Escarpment.

On the Trϕndelag Platform the Helgeland Basin is defined as a Cretaceous depocentre (Rϕnnevik and Navrestad, 1977).

Jurassic—Cretaceous tectonics

During the Jurassic and early Cretaceous the area off mid Norway was subjected to strong tensional forces. These forces resulted in crustal thinning and rifting. The most dramatic crustal thinning is believed to have taken place in the Mϕre Basin and the Vϕring Basin between the Nordland Ridge and the Vϕring Plateau Escarpment (Hinz, 1972). This is the area where rifts of presumed Jurassic age can be observed, and which suffered the largest subsidence during Cretaceous time.

Basin formation in this area is thus similar both in age and mechanism to the better known Viking Trough in the northern North Sea. Rifting and crustal stretching in Jurassic and early Cretaceous was followed by upwelling of deep mantle material under an area of thinned crust. Cooling of this material resulted in subsidence in Upper Cretaceous time. The difference in sedimentary thicknesses between the Vϕring and Mϕre Basins is most probably a result of difference in crustal thinning between the two basins.

Thinning by a factor of more than 3 can be estimated from Hinz's cross-section for the outer part of the Vϕring Basin. This is nearly twice the

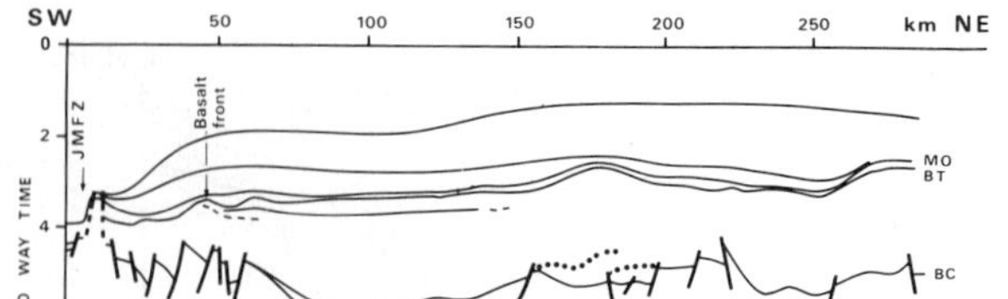

Fig. 14. Geoseismic profile 4 (Vϕring Basin). MO, mid Oligocene; BT, base Tertiary; BC, base Cretaceous; . . ., volcanics; JMFZ, Jan Mayen Fracture Zone.

crustal thinning observed and estimated for the North Sea (Sclater and Christie, 1981). No refraction profiles are available over the inner part of the Vϕring Basin, the Halten Trough, where the largest subsidence can be observed. If subsidence is related to crustal thinning, the thinnest crust should be expected along this trough.

Earlier authors have proposed Cretaceous sea-floor spreading in this area (Bott, 1975, 1978; Roberts, 1974; Roberts *et al.*, 1981; Price and Rattey, in press). Price and Rattey (in press) extended the Cretaceous sea-floor spreading from the Rockall Trough northwards through the Mϕre Basin to the southeastern part of the Vϕring Basin. The zone of sea-floor spreading was confined to the axial portion where pre-Cretaceous strata cannot be demonstrated. A mid-Cretaceous (Cenomanian) age was proposed. This was based on the assumed age of the sills which, according to them, do not penetrate sediments younger than Cenomanian. Our data in the area (Figs. 8, 16 and 17) indicate, however, that volcanics can be found throughout the whole Cretaceous sequence. Based on this, the volcanics are interpreted as intrusions (sills and dykes), most probably of Lower Tertiary age, related to the opening of the Norwegian Sea. Therefore, we do not see any positive evidence favouring sea-floor spreading in the Mϕre Basin.

Tertiary

The opening of the Norwegian Sea in the Lower Tertiary ended the tensional tectonic regime that was active in Mesozoic time. The base Tertiary isochron map (Fig. 18) shows that the growth faults that were active in Mesozoic time die out in Lower Tertiary. Active faulting is found bordering the outer highs (Fig. 1) and the high just east of the Jan Mayen Fracture Zone (Fig. 18). These faults are related to the initiation of sea-floor spreading (see Hinz *et al.*, 1982 for a more detailed discussion).

The most prominent structural features at this level are the Vϕring Plateau Escarpment (VPE) and the Faeroe-Shetland Escarpment (FSE). Seismic interpretation across the VPE (Figs. 12 and 13) supports the view held by Rϕnnevik *et al.* (1979) and Hinz *et al.* (1982) that the VPE is an active fault zone separating an easterly lying, heavily subsiding

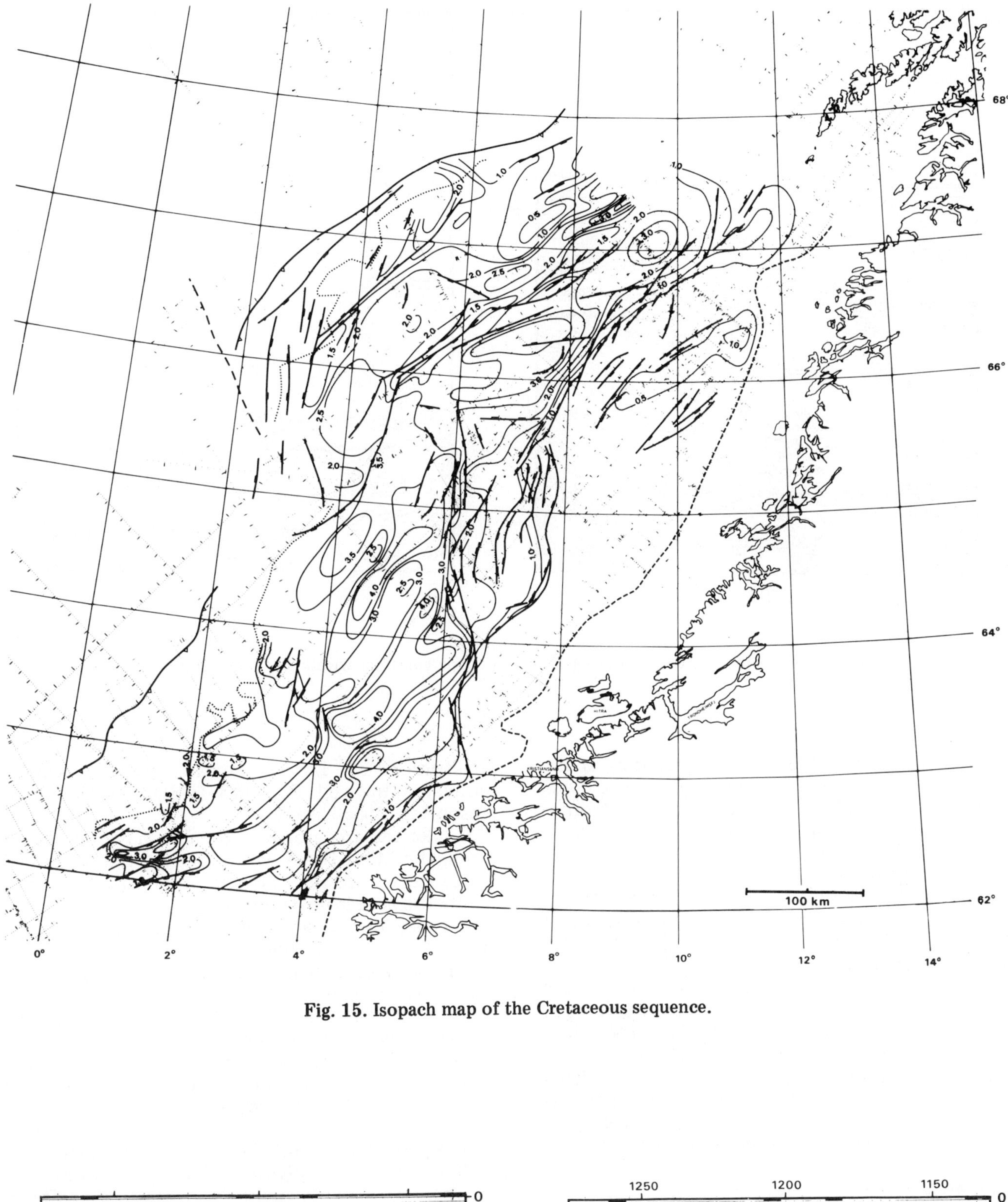

Fig. 15. Isopach map of the Cretaceous sequence.

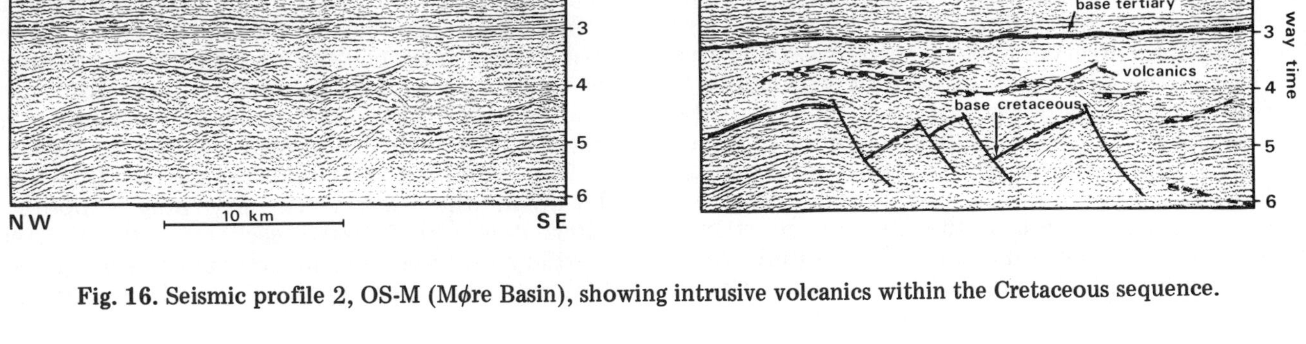

Fig. 16. Seismic profile 2, OS-M (Møre Basin), showing intrusive volcanics within the Cretaceous sequence.

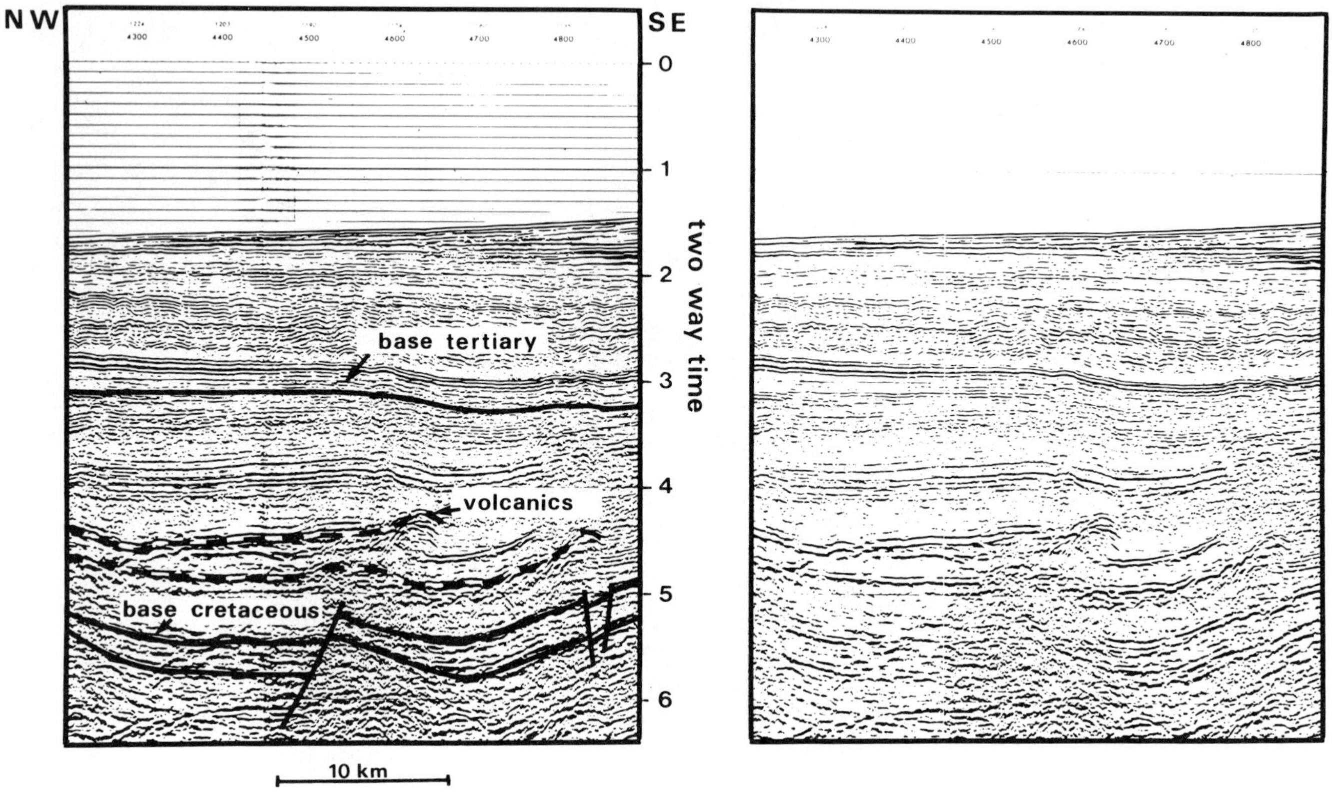

Fig. 17. Seismic profile 7, B10-81 (Vøring Basin), showing volcanics within the Cretaceous sequence.

area from a westerly lying area undergoing more slow subsidence. Its present shape may be a combined result of faulting and Tertiary lava-fronts.

A comparable origin for the FSE is unlikely, since deeper reflectors, as illustrated in Fig. 11, seem to cross this escarpment without any sign of faulting. The FSE is thus interpreted as a lava-front. Other, less prominent, lava fronts can be mapped both west and east of the FSE, all within the Lower Tertiary sequence.

Radiometric datings of these lavas in DSDP boreholes on the outer Vøring Plateau give ages of 44—46 Ma (Kharin *et al.*, 1976). Age determination based on microfossils in the sediments above the lavas at site 338 gives ages of 49—53 Ma. This discrepancy in age may be due to uncertainties in the radiometric method (Talwani *et al.*, 1981). A late Palaeocene age for the lavas is thus the most probable.

The areas north and south of the landward continuation of the Jan Mayen Fracture Zone show striking differences in tectonics and sedimentation.

The Lower Tertiary isopach map (Fig. 19) shows that the central parts of the Vøring Basin and the areas northwest of the VPE received small amounts of sediments in this period. Major depocentres are found in the central parts of the Møre Basin, the area just east of the VPE, and north of the Vøring Basin.

Thin sedimentary cover in the central part of the Vøring Basin may be a result of regional uplift of this area during Lower Tertiary or a combination of low subsidence and low rate of sedimentation. The local maxima east of the VPE are most probably a result of erosion and sediment supply from a westerly lying high area (Rønnevik *et al.*, 1979). Penetration of Lower Tertiary Eocene clastic sediments in well 338/343 (DSDP) is the main reason for this assumption.

Unlike the area north of the Jan Mayen Fracture Zone, the Møre Basin was under continuous subsidence during Lower Tertiary. Apart from the lava fronts the area is unstructured.

Tertiary domes are found in the entire Vøring Basin area. The Nordland Ridge and the area west of VPE also show signs of doming (Fig. 20). On the base Tertiary map (Fig. 18) the largest domes can be shown to have an overall NNE—SSW trend. This is best illustrated by the central zone of domes crossing the Vøring Basin diagonally. Seismic interpretation shows that doming is restricted to two separate episodes; it occurs at base Tertiary and mid-Oligocene levels (Fig. 20). In some areas, doming may have continued into the Miocene.

Doming over the Nordland Ridge is a result of sinistral movements along N—S-oriented faults in Middle Oligocene to Miocene time.

Other evidence of a compressional regime is found in inverted Mesozoic faults and basin trends (Figs. 12 and 13). The doming in Lower Tertiary is thus believed to be a result of compressive tectonics in the area north of the Jan Mayen Fracture Zone. Hinz *et al.* (1982) propose doming as a result of clay diapirism for the northernmost dome in the Vøring Basin (Fig. 20). Such a mechanism may explain the highly complex geology in this area.

The complex tectonic regime in the Vøring Basin in the Lower Tertiary may support the alternative put forward earlier for the thin early Tertiary sediments in the central parts of the basin. This area may have been subjected to uplift or no

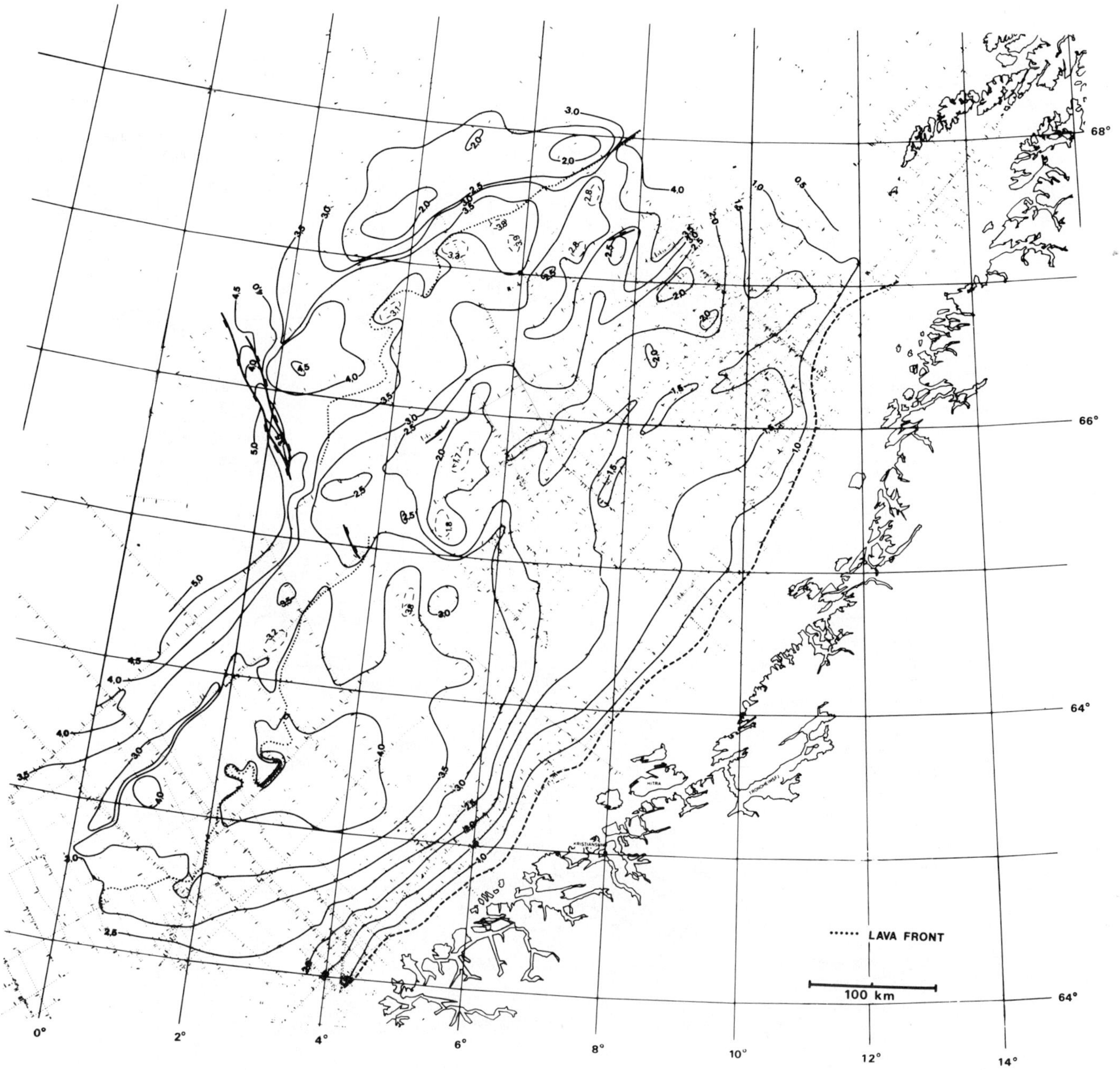

Fig. 18. Isochron map at base Tertiary level.

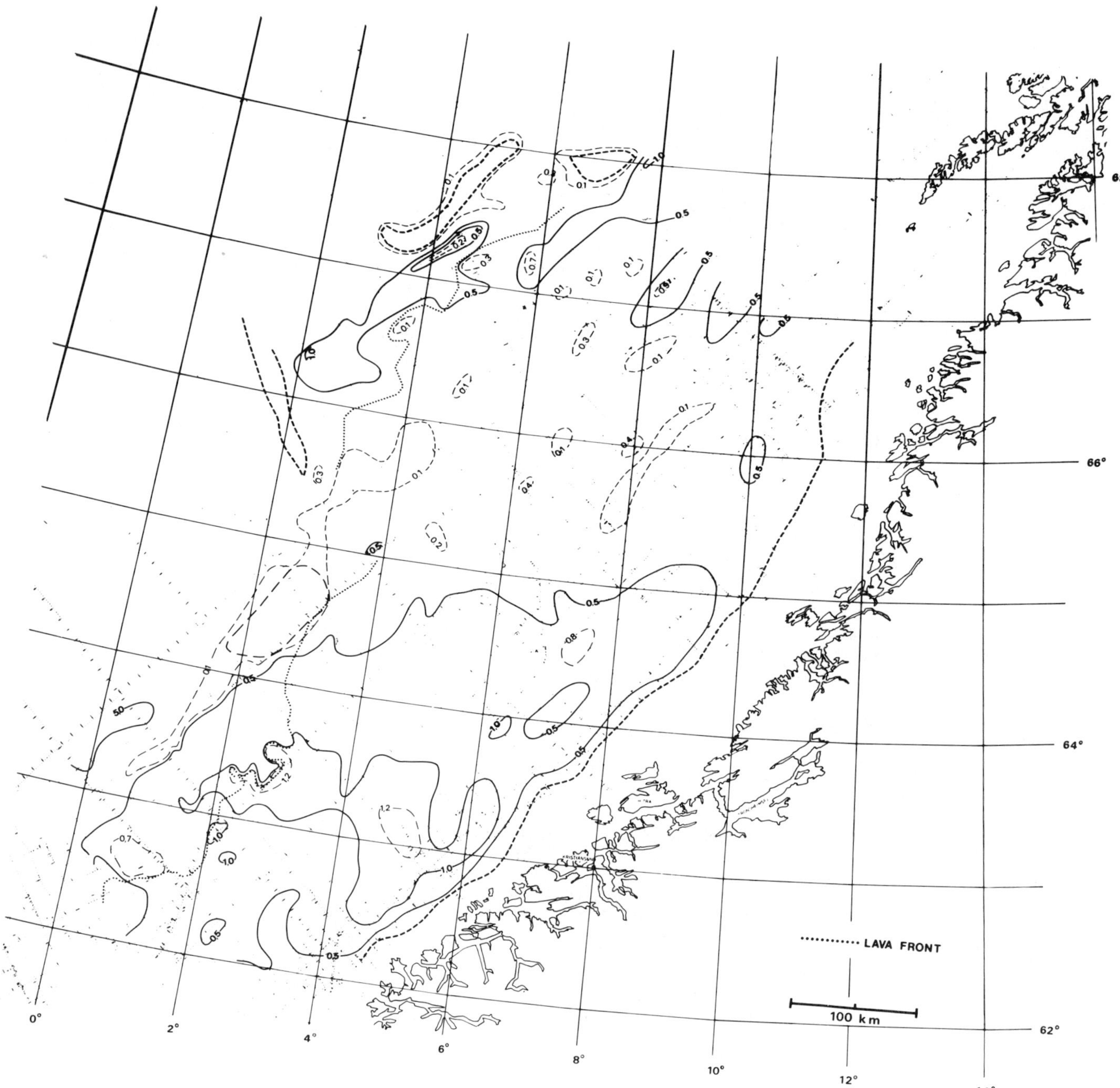

Fig. 19. Isopach map of the Lower Tertiary sequence.

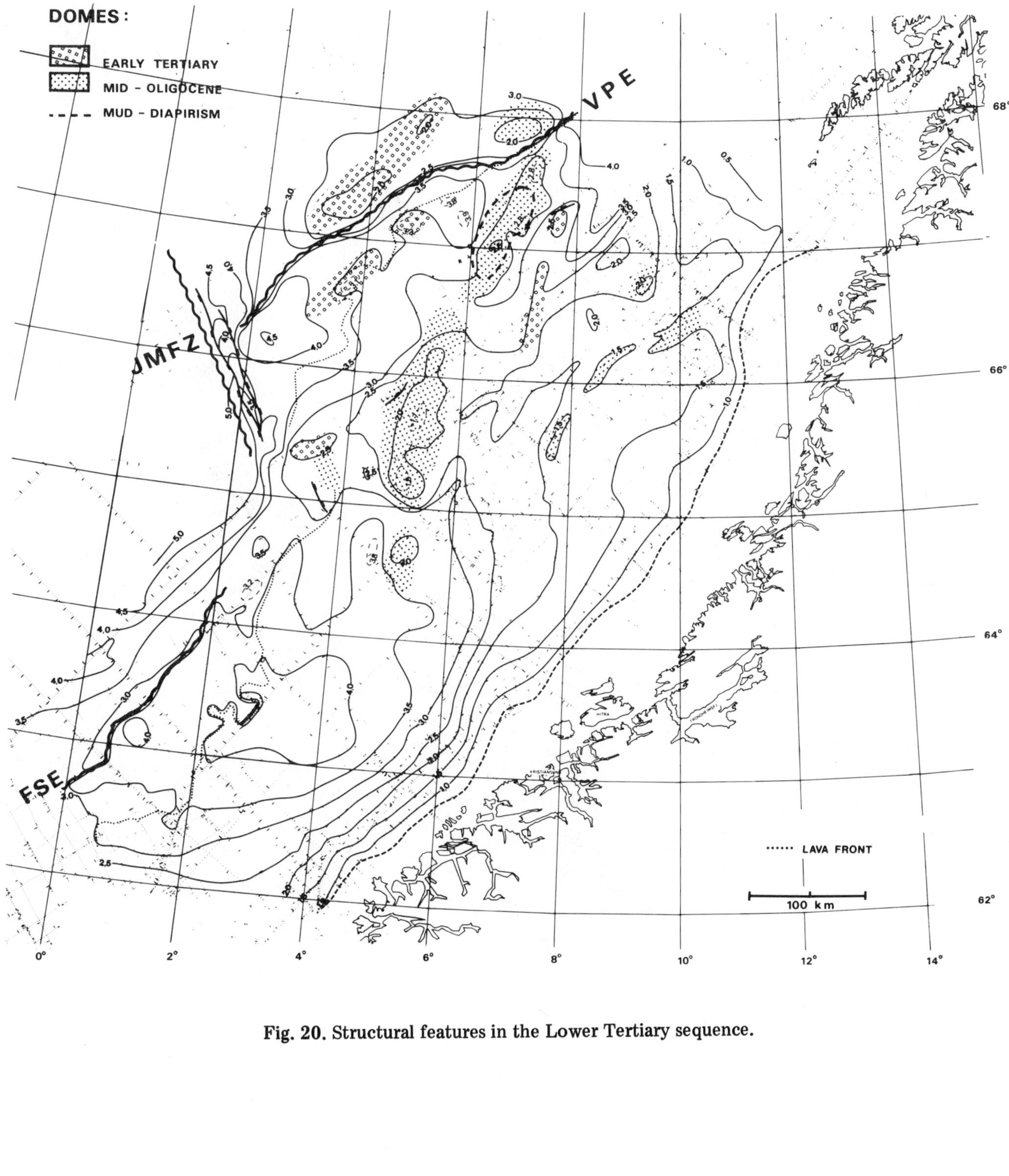

Fig. 20. Structural features in the Lower Tertiary sequence.

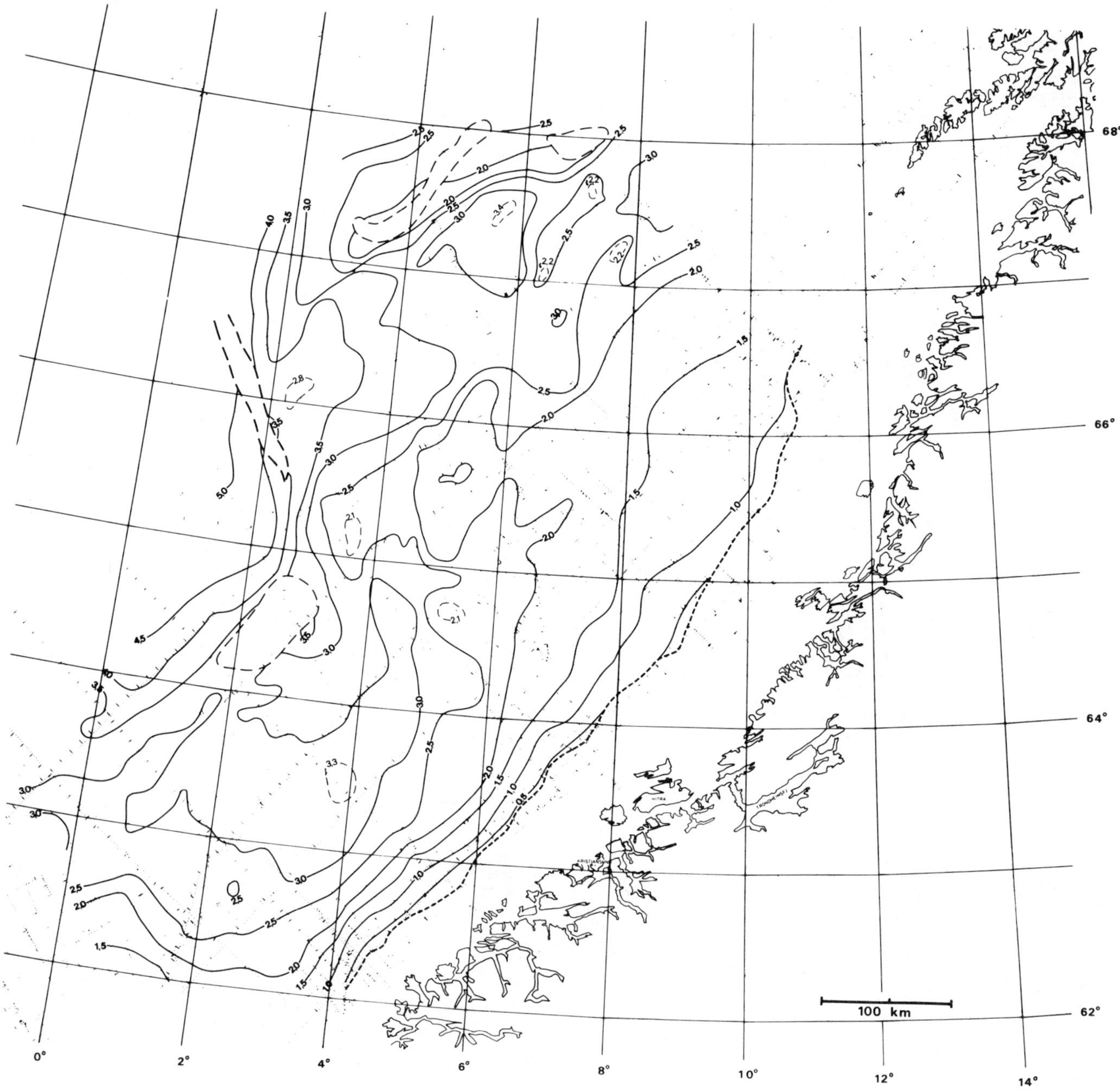

Fig. 21. Isochron map at mid-Oligocene level.

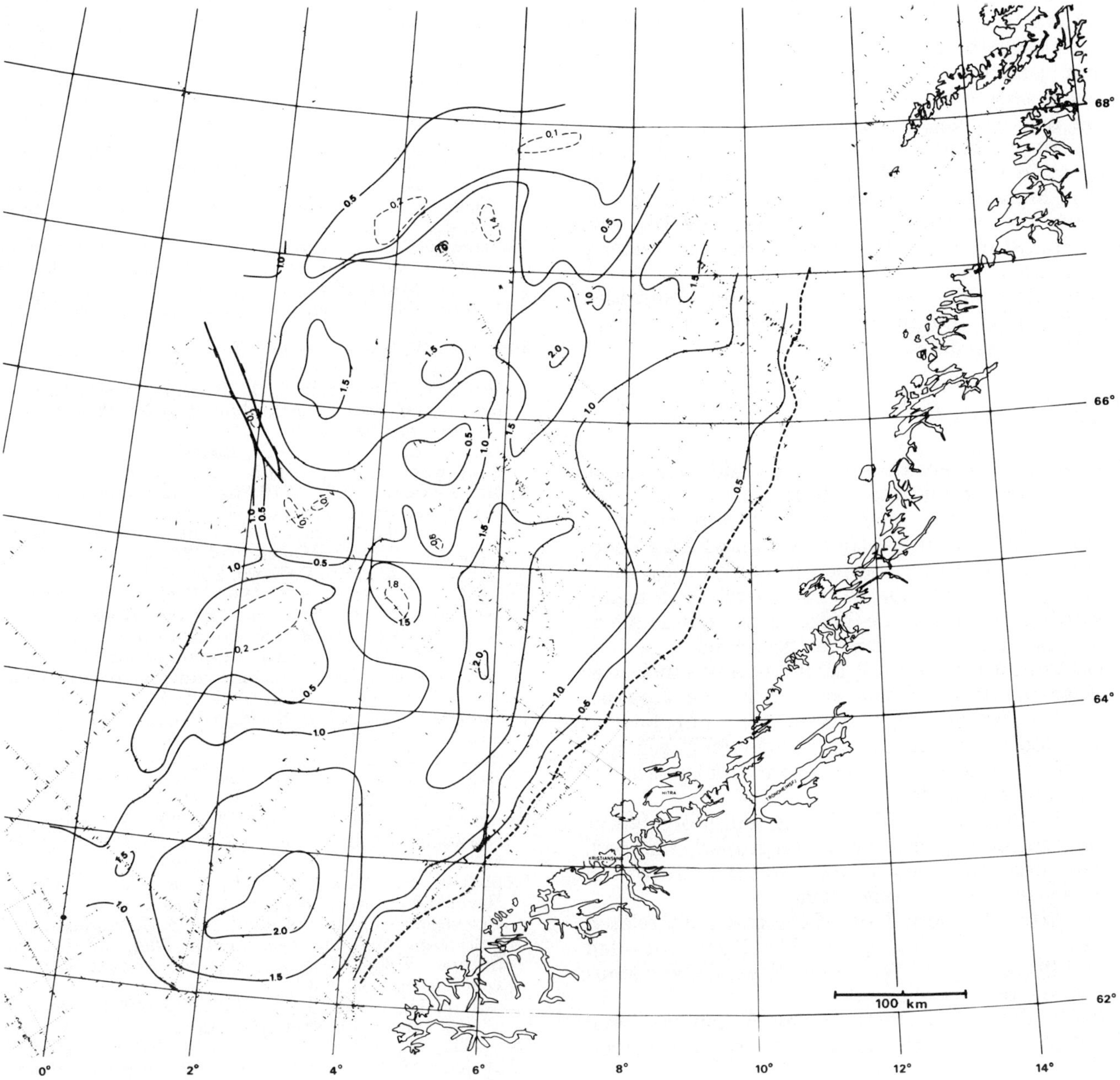

Fig. 22. Isopach map of the Upper Cenozoic sequence.

subsidence due to compressional forces. This compressional regime died out in Upper Cenozoic time.

Structural features at the mid-Oligocene level visible on the isochron map for this horizon (Fig. 21) are the VPE and the mid-Oligocene domes in the Vøring Basin. The Møre Basin is unstructured.

The isopach map over Upper Cenozoic sediments (Fig. 22) shows that sediments fill in the relief defined at the mid-Oligocene level. In the Vøring Basin the main depocentres are found on both sides of the central domal zone. Sediments younger than mid-Oligocene to Lower Miocene show onlap termination on these domes. The Oligocene and Miocene sediments penetrated on the outer Vøring Basin are of pelagic origin (Caston, 1976). Shelf-edge progradation throughout Upper Tertiary resulted in an increased input of terrestrial material. In the Møre Basin, deposition in Upper Cenozoic time shows the same pattern as in Lower Tertiary, with undisturbed sedimentation from an easterly lying source area.

The entire area between 62° and 69°N subsided during Upper Cenozoic time with only minor tectonic disturbances.

CONCLUSIONS

The geological evolution of the mid Norwegian Shelf can be summarized as follows:

(1) In Palaeozoic time sedimentation was fault controlled. In the Vestfjord Basin, where Palaeozoic basins can be mapped properly, they show an asymmetric shape with an overall N—S trend. Further south in the Møre Basin remnants of deeply eroded old basins are preserved.

(2) Following the infill of these basins in Lower Mesozoic time, rifting started in the Jurassic.

(3) The Trøndelag Platform was not heavily affected by this rifting, but underwent even subsidence resulting in uniform thicknesses of Jurassic sediments.

(4) Further west in the Vøring and Møre Basins the Jurassic sequence cannot be properly mapped. The base Cretaceous unconformity shows an area affected by strong rifting.

(5) After the period of rifting and fault related subsidence, the Vøring and Møre Basins subsided rapidly in Upper Cretaceous. Several Cretaceous depocentres are mapped.

(6) In the Lower Tertiary time a compressional regime affected the area north of Jan Mayen Fracture Zone.

(7) Doming took place at base Tertiary and in Middle Oligocene time.

(8) This compressional regime affected sedimentation in this area, resulting in thin sediments in the central Vøring Basin.

(9) This regime died out in Middle Oligocene—Miocene time, resulting in even subsidence and shelf-edge progradation in Upper Cenozoic time. The Møre Basin was not affected by the Tertiary tectonics, but was subsiding evenly throughout Cenozoic time.

REFERENCES

Bott, M.H.P. 1975. Structure and evolution of the North Scottish Shelf, the Faeroe Block, and the intervening region. In: A.W. Woodland (ed.), *Petroleum and the Continental Shelf of Northwest Europe*, Vol. 1 *Geology*, Applied Science Publishers, London, pp. 105—115.

Bott, M.H.P. 1978. The origin and development of the continental margins between the British Isles and south-eastern Greenland. In: D.R. Bowes and B.E. Leake (eds.), Crustal Evolution in Northwestern Britain and Adjacent Regions. *Spec. Iss., Geol. J.*, **10**, 377—392.

Caston, V.N.D. 1976. Tertiary sediments of the Vøring Plateau, Norwegian Sea, recovered by Leg 38 of the Deep Sea Drilling Project. In: *Initial Reports of Deep Sea Drilling Project*, Vol. 38, M. Talwani *et al.* (eds.), U.S. Govt. Printing Office, Washington D.C., pp. 761—782.

Hinz, K. 1972. The seismic crustal structure of the Norwegian Continental Margin in the Vøring Plateau, in the Norwegian deep sea, and on the eastern flanks of the Jan Mayen Ridge between 66° and 68°. *24th Int. Geol. Congr.*, Montreal, Sec. 8, pp. 28—37.

Hinz, H.K., Dostman, H. and Hanisch, J. 1982. Structural framework of the Norwegian Sea. In: *The Geologic Framework and Hydrocarbon Potential of Basins In Northern Seas.* Norsk Petroleumsforening, Stavanger E/4, pp. 1—22.

Jørgensen, F. and Navrestad, T. 1979. Main structural elements and sedimentary succession on the shelf outside Nordland (Norway). In: *Norwegian Sea Symposium*, Norsk Petroleumsforening, Tromsø NSS/11, pp. 1—20.

Jørgensen, F. and Navrestad, T. 1981. The geology of the Norwegian Shelf between 62° and the Lofoten Islands. In: L.V. Illing and G.D. Hobson (eds.), *Petroleum Geology of the Continental Shelf of Northwest Europe*, London, pp. 407—413.

Kharin, S.N., Udintsev, G.B., Bogatikov, O.A., Dimitriev, J.L., Raschka, H., Kruezer, H., Mohr, M., Harre, W. and Eckhards, F.J. 1976. K/Ar age of the basalts of Norwegian-Greenland Sea Glomar Challenger, Leg 38 DSDP. In: M. Talwani *et al.* (eds.). *Initial Reports the Deep Sea Drilling Project*, Vol. 38, U.S. Govt. Printing Office, Washington, D.C., pp. 755—760.

Price, I. and Rattey, R.P. 1984. Cretaceous tectonics of mid-Norway: Implications for the Rockall Faeroes-Shetland Rift. *J. Geol. Soc. London* (in press).

Roberts, D.G. 1974. Structural development of the British Isles, the continental margin and the Rockall Plateau. In: C.A. Burke and C.L. Drake (eds.), *The Geology of Continental Margins*, Springer-Verlag, New York, pp. 343—359.

Roberts, D.G., Masson, D.G. and Miles, P.R. 1981. Age and structure of the southern Rockall Trough: new evidence. *Earth Planet. Sci. Lett.*, **52**, 115—128.

Rønnevik, H. and Navrestad, T. 1977. Geology of the Norwegian Shelf between 62°N and 69°N. *Geojournal*, **1**(1), 33—46.

Rønnevik, H., Bergsager, E.I., Moe, A., Øvrebø, O., Navrestad, T. and Stangenes, J. 1975. The geology of the Norwegian Continental Shelf. In: A.W. Woodland (ed.), *Petroleum and the Continental Shelf of Northwest Europe*, Vol. 1, *Geology*, Applied Science Publishers, London, pp. 117—130.

Rønnevik, H., Eggen, S. and Vollset, J. 1984. *Exploration of the Norwegian Shelf* NPD Contrib. No. 5 (in press).

Rønnevik, H., Jørgensen, F. and Motland, K. 1979. The geology of the northern part of the Vøring Plateau. In: *Norwegian Sea Symposium*, Norsk Petroleumsforening, Tromsø, NSS/12, pp. 1—21.

Sclater, J.G. and Christie, P.A.F. 1980. An extensional origin for the Buchan and Witchground Graben in the North Sea. *Nature (London)* **283**, 729—732.

Talwani, M. and Eldholm, O. 1972. The continental margin off Norway. A geophysical study. *Geol. Soc. Am. Bull.*, **83**, 3575—3608.

Talwani, M., Mutter, J. and Eldholm, O. 1981. The initiation of opening of the Norwegian Sea. *Oceanol. Acta*, Spec. Issue, 23—30.

Vail, P.R., Mitchum, R.M. Todd, R.G., Widmier, J.M., Thompson, S., Sangree, J.B., Bubb, J.N. and Hatlelid, W.G. 1977. Seismic stratigraphy and global changes of sea level. In: Seismic Stratigraphy — Applications to Hydrocarbon Exploration. *Am. Assoc. Petrol. Geol. Mem.*, **26**, 49—212.

18 Bedrock geology on the mid-Norwegian continental shelf

T. Bugge, R. Knarud and A. Mørk
Continental Shelf Institute (IKU), Trondheim, Norway

Based on shallow seismic surveys, sea-bed sampling and shallow drilling, the bedrock geology from Møre (62°N) to Lofoten (67°N) has been mapped by the Continental Shelf Institute (IKU) during the last ten years. The sediments range in age from Lower Triassic to Pliocene and have been divided into 11 seismic units; all with a slight dip to the west and northwest. The strike is mainly parallel to the present coastline. These Mesozoic and Tertiary deposits are covered by Quaternary tills and glaciomarine and marine clays of variable thickness (up to 500 m).

Off the southern part of Møre, only Pliocene sediments subcrop beneath the Pleistocene, while gradually older sediments are exposed onlapping the basement northwards towards the Trænabanken area.

North of 66°30′N and east of the Trænabanken area the structural pattern is markedly different due to the presence of the Nordland Ridge. The ridge has a complex internal structure with locally tilted fault blocks. Samples from these local structures are dated Triassic to Lower Tertiary, becoming successively younger to the east. Our data suggest that the Nordland Ridge in this area was active from the Upper Jurassic to Cretaceous, with signs of reactivation up to the Oligocene.

INTRODUCTION

Since 1973 the Continental Shelf Institute (IKU) has carried out a regional mapping programme of the shallow bedrock geology on the Norwegian Continental Shelf from the northern North Sea to Tromsøflaket (60—72°N). The programme can be divided into three steps: (1) shallow seismic profiling; (2) bedrock sampling; and (3) shallow drilling. The shallow seismic surveys were done by a high resolution sparker, with a typical penetration of 0.3—0.5 seconds, supplemented with side-scan sonar and pinger or deep-towed boomer. For the sampling, gravity corer, piston corer, vibrocorer, grab and an underwater electric diamond core drill (2.5-cm core diameter) were used. Shallow drilling (6-cm core diameter) was carried out for the first time in 1982, using the drill-ship MV *Pholas*.

The area examined offshore Møre—Lofoten comprises approximately 120 000 km^2 and is covered by about 19 000 km of seismic profiling (Fig. 1a). More than 90 successful bedrock samples had been obtained prior to 1982, and 14 cores with penetrations ranging from 5 to 30 m were obtained from the shallow drilling programme in 1982.

Bedrock mapping based on data obtained by the methods described above is dependent on a relatively thin Quaternary overburden. On the mid Norwegian Continental Shelf there is a general trend of thin Quaternary overburden on the innermost part of the shelf, with increasing thickness towards the shelf break. Accordingly there are areas of almost no overburden on the eastern shelf and thicknesses of up to 300—500 m along the shelf break. In the areas with minimum Quaternary cover the bedrock can be reached with a 0.5-m^3 grab. In areas with soft and thin Quaternary sediments a 3-m gravity corer was required for penetration, while firmer overburden was penetrated only with the vibrocorer and the electric diamond core drill. With the drill-ship it was possible to penetrate thicker overburden and thus select drill sites over a wider area, giving access to deeper and less weathered samples.

The mapping programme was funded by NTNF (The Royal Norwegian Council for Scientific and Industrial Research) for the first eight years. Various oil companies joined the programme with financial aid in 1981 and this enabled bedrock sampling on the Vega (65—66°N) and Trænabanken (66—67°N) map sheets, in addition to making shallow drilling possible in the Vega map area.

Petroleum Geology of the North European Margin © Norwegian Petroleum Society (Graham & Trotman, 1984) pp. 271—283.

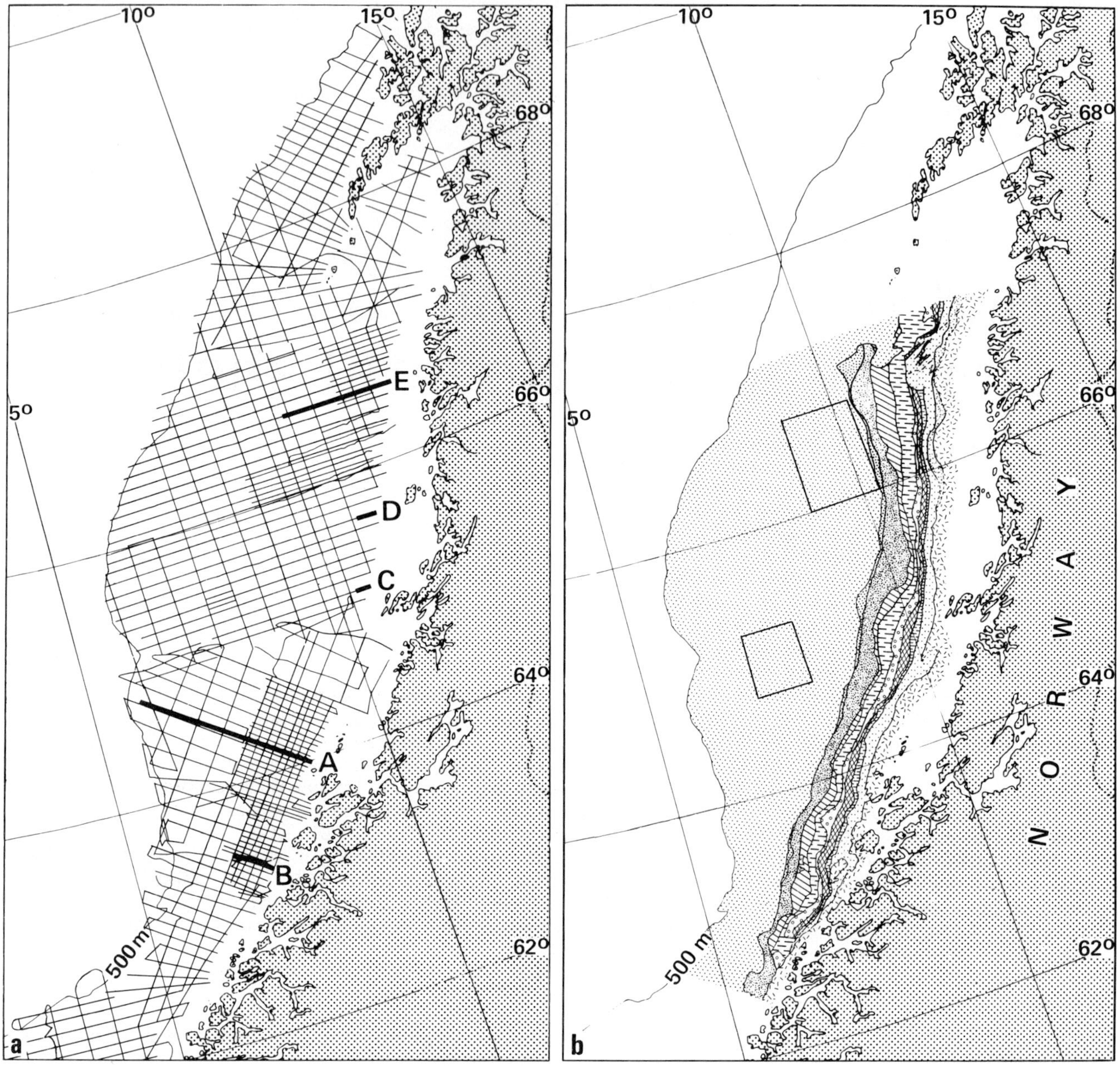

Fig. 1. (a) IKU's shallow seismic profile grid. Locations of interpreted profiles presented herein are shown as heavy lines: A, Fig. 2; B, Fig. 5; C and D, Fig. 7; E, Fig. 8. (b) Bedrock map compiled from IKU's regional geological mapping programme off mid Norway. Legend is given in Figs. 4 and 6. Frames show the Trænabanken and Haltenbanken licence areas.

REGIONAL GEOLOGY

Seismic signature of the individual bedrock units

Eleven different units were distinguished from their seismic signatures (Figs. 1b, 2 and 3). On the seismic records the metamorphic basement (*Unit I*) usually has a very rough sea-floor topography and the rock gives no distinct internal seismic reflectors (Fig. 3a). Based on the topography, Unit I is also easily recognized from the bathymetric maps. Although the basement consists of various lithologies throughout the area, the seismic appearance does not vary significantly.

The Triassic (*Unit II*) is usually characterized by acoustically dense layers, relatively steeply dipping (Fig. 3b). In the southern part of the outcropping area of Unit II, the seismic signature changes to one of more openly spaced and gently dipping layers. Penetration is rarely more than 100—130 ms.

The Jurassic (*Unit III*) is seismically one of the most characteristic zones in the mid Norway Shelf area (Fig. 3c). It has a distinct, more or less parallel, thinly layered appearance. In parts of the area the unit can be further divided into six sub-units.

The Upper Jurassic—Lower Cretaceous (*Unit IV*) is distinguished by a lack of internal structures and reflectors (Fig. 3d). It is very often located in depressions on the sea floor, which would imply a low resistance to erosion. The appearance of the unit is widespread and can be followed from south of 63° to 66°30′N.

The Cretaceous (*Unit V*) has a layered lower part, which is usually heavily folded and faulted (Fig. 3d). The upper part is more homogeneous and has no distinct reflectors. The boundary between Units V and VI (base Palaeocene) is often

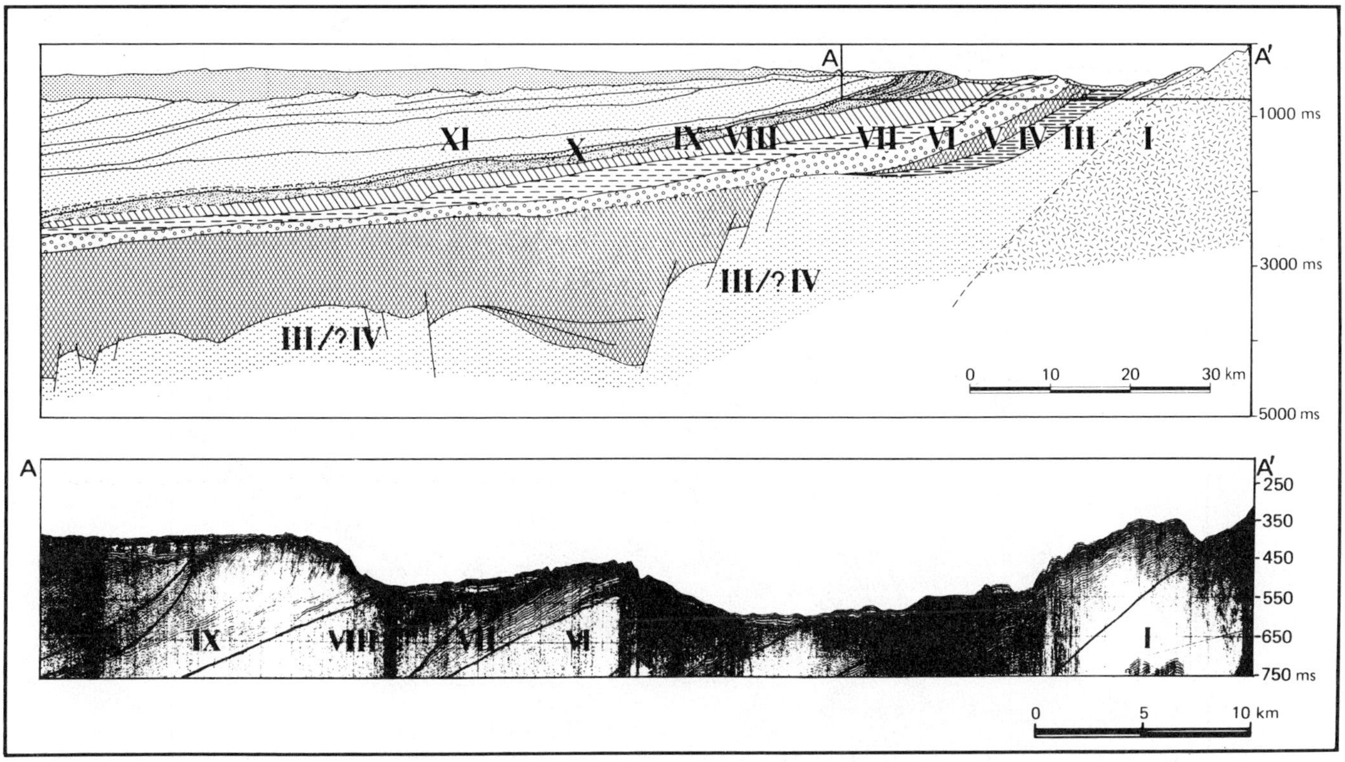

Fig. 2. Interpreted profile compiled from the Norwegian Petroleum Directorate deep seismic line B-33-72 and the IKU-lines B73-153 and B74-153. Location of the profile is shown as line A in Fig. 1a. Inset A—A′ is a sparker record of the eastern part of the section. See Fig. 4 for legend.

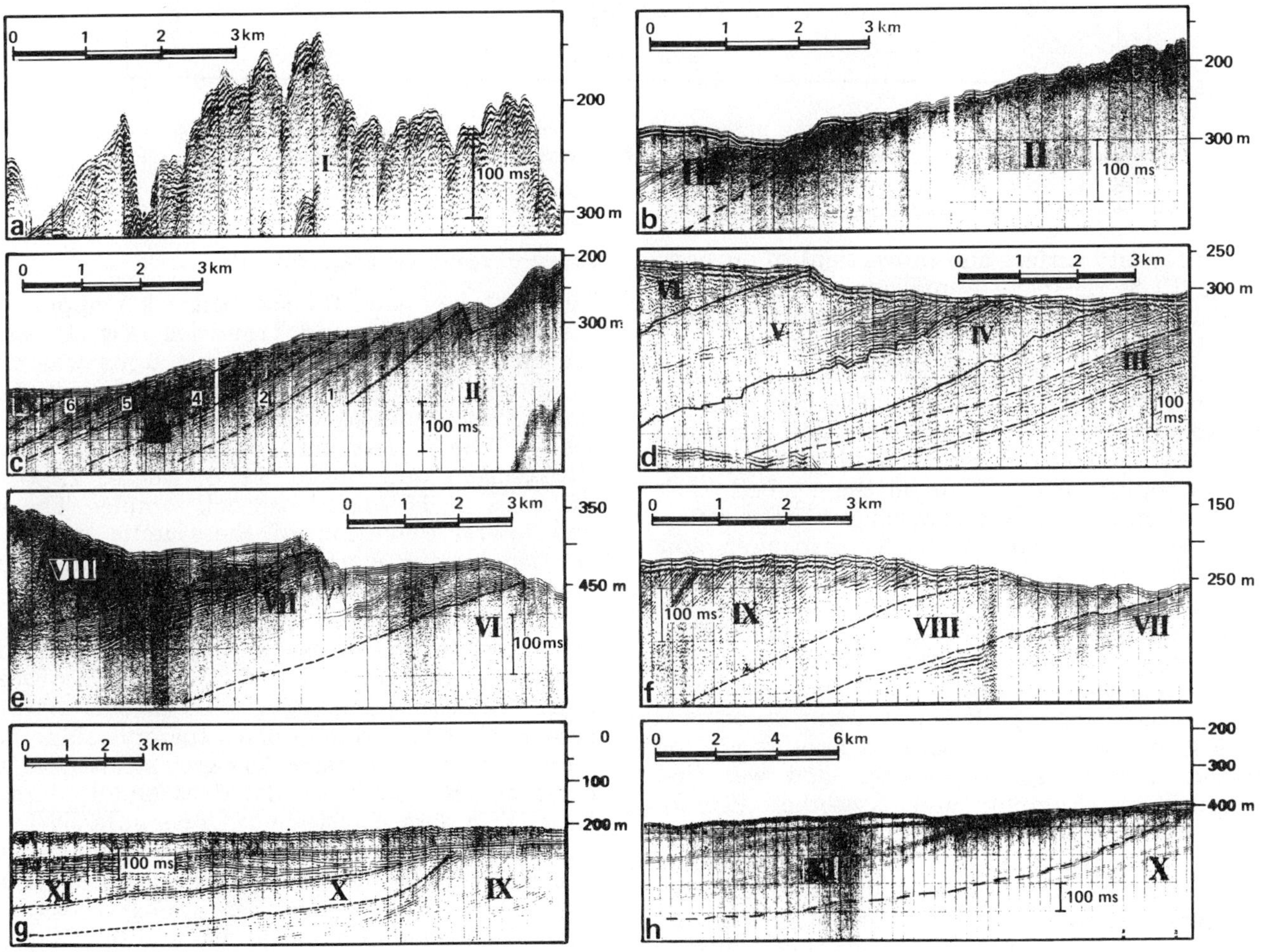

Fig. 3. Sparker records showing the seismic characteristics of the individual units. All examples are from the Vega area. Roman numbers refer to the eleven main units, and the numbers 1—6 in Fig. 3c refer to the subdivision of Unit III in the Vega area.

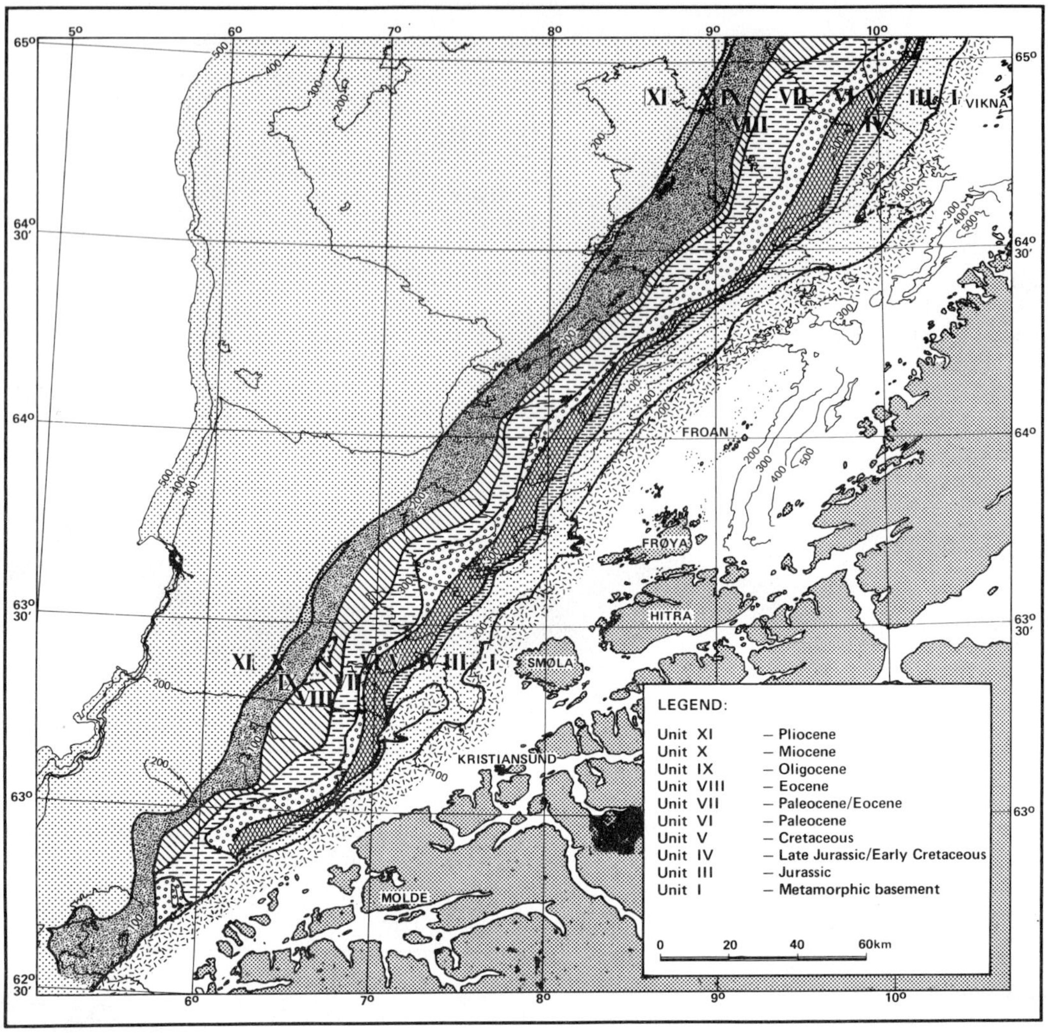

Fig. 4. Bedrock map showing the units subcropping below the Quaternary cover, off Møre—Trøndelag. The v-raster west of Smøla in Unit VIII shows the occurrence of volcanic peaks (see Fig. 5).

a disconformity surface and shows signs of erosion.

Unit VI is relatively homogeneous and has no especially characteristic reflection pattern (Fig. 3e). In the Palaeocene/Eocene (*Unit VII*) the lower part often has a weak layering, while the middle and upper parts usually show strong layering (Fig. 3e). The sediments seem to have high physical resistance towards erosion and this unit is often found as a positive topographical form on the sea-bed. *Unit VIII* (Eocene) has only a few, unsystematic reflectors (Fig. 3f), except in the Trænabanken area where it shows a dense pattern of extensively folded layers.

The Oligocene (*Unit IX*) has a delta-like appearance in cross-section and is the most characteristic seismic unit on the mid Norwegian Continental Shelf (Figs. 3f and 3g). It can be easily recognized in most of the shallow seismic lines from the northern North Sea to Lofoten and was first described by Eldholm and Nysæther (1969). Above this unit there is a thin outcropping wedge of Miocene sediments (*Unit X*). The outcropping part of the unit is sporadic in the southern part, but more extensive in the northern area (Fig. 3g). The Pliocene (*Unit XI*) is a prograding sequence with very marked, and slightly dipping, reflectors (Fig. 3h). Some of them are continuous and can be traced over long distances.

Møre—Trøndelag area (62—65°N)

The first key area for the bedrock mapping was the shelf off Møre and Trøndelag (Fig. 4), where the first subdivision into seismic units was made by Bugge *et al.* (1976) and the first bedrock sample was obtained as early as 1973 (Bugge *et al.*, 1975). The various units were tentatively dated (Rokoengen and Bugge, 1976; Bugge, 1980) on the basis of 15 *in-situ* bedrock samples. The ages and lateral extension of these units were later verified by the work within the Vega map sheet area. From the seismic profile interpretation the two areas are linked together and the result is presented in the bedrock map (Fig. 1b) which covers the shelf from 62 to 67°N (Møre—Lofoten).

The oldest sedimentary unit in this southern area is Unit III, which onlaps the crystalline basement. This unit contains dark grey sandstones with coals which have been dated as Middle Jurassic. Unit IV, a dark claystone of Upper Jurassic/Lower Cretaceous age, is very characteristic in this area, and is heavily eroded in the (>500-m deep) Suladjupet depression. More competent against erosion are the Palaeocene/Eocene Unit VII and the Oligocene Unit IX; both can be observed as ridges on the present-day sea floor. The Oligocene unit has delta-like bottom- and foreset structures and contains

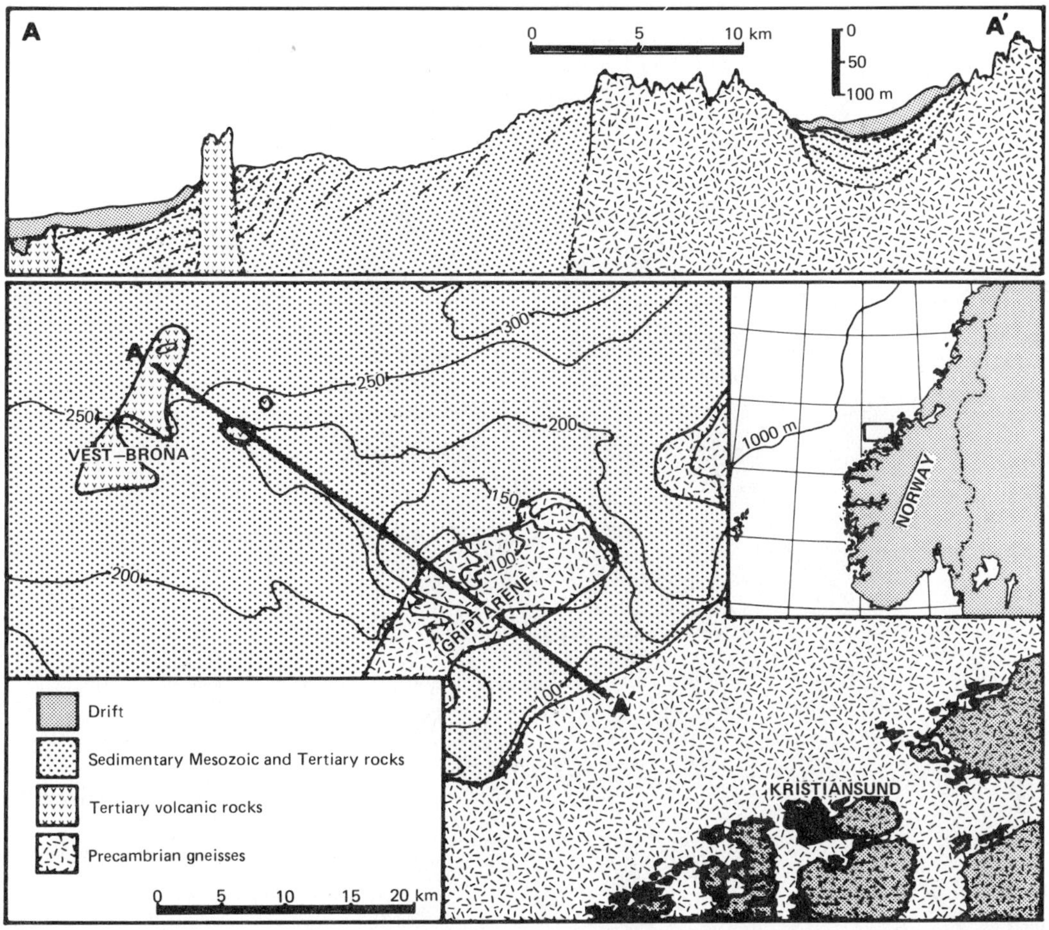

Fig. 5. Simplified map showing the occurrence of volcanic peaks off Kristiansund. Location of profile line A—A′ is indicated as line B in Fig. 1a. (After Bugge *et al.*, 1980.)

evidence of the Tertiary upheaval and increased sediment supply. The Palaeocene/Eocene Unit VII consists of carbonate cemented sandstone beds interlayered with tuffs, while the Oligocene Unit IX seems to be dominated by sandstone beds.

The bedrock map (Fig. 4) shows the regular, parallel outcrop pattern in the middle and northern part of the Møre—Trøndelag area. In the southern part (between 62° and 63°N), the Tertiary sediments cover most of the Mesozoic units, while south of 62°N the subcrops are covered by Upper Pliocene sediments (Rokoengen, 1980; Rokoengen and Rønningsland, 1983).

About 60-km northwest of Kristiansund, a number of volcanic peaks occur within the sedimentary sequence (Fig. 5). At least five of these penetrate the sedimentary cover and due to higher erosional resistance they protrude up to 80-m above the surrounding sea floor. Other peaks which do not penetrate the sediment cover are apparent from the seismic records. Some of these are covered by both Tertiary and Quaternary sediments. Samples of ultrabasic ankaratrite obtained from one peak were dated to 55.7 ± 0.9 Ma (Bugge *et al.*, 1980), which is roughly the same age as the opening of the Norwegian-Greenland Sea in this area (LaBreque *et al.*, 1977; Talwani and Eldholm, 1977). Structurally the volcanic peaks are situated on the Møre High (Hamar and Hjelle, 1984).

The Vega map sheet (65—66°N)

All of the ten bedrock units from the Møre—Trøndelag area can also be mapped further to the north through the Vega map-sheet area (Fig. 6). From about 65°30′N and northwards a Triassic unit also outcrops on the sea floor, lying unconformably on the crystalline basement. In 1981, 37 bedrock samples were obtained from this area. All of the units were sampled, except for the two outermost units of supposed Miocene and Pliocene age (due to the thickness of the Quaternary overburden). In the northwestern part of this map area the Quaternary overburden was also too thick to allow sampling of any of the units younger than the Jurassic. Core holes drilled in the Vega area by the drill-ship *Pholas*, however, penetrated and provided good samples from less-accessible units. The coreholes were located in all the units ranging in age from Triassic to Palaeocene (Units II—VII).

Based on the seismic data and the shallow drilling, the Jurassic Unit III was further divided into six sub-units (III_1—III_6) (Fig. 7). In the southern part of the map sheet Sub-unit III_4 is the oldest outcropping Jurassic unit. Northward, progressively older Jurassic sediments outcrop, and from about 65°30′N Sub-unit III_1, presumably of Pliensbachian—Hettangian age, also onlaps a wedge of Triassic sediments.

The Trænabanken area (66—67°N)

The same structural trend as observed within the Vega map-sheet area can be followed up to about 66°30′N into the Trænabanken area. The uniform geology coincides with the presence of the deeper Helgeland Basin below, as earlier concluded by Jørgensen and Navrestad (1979). The basin is

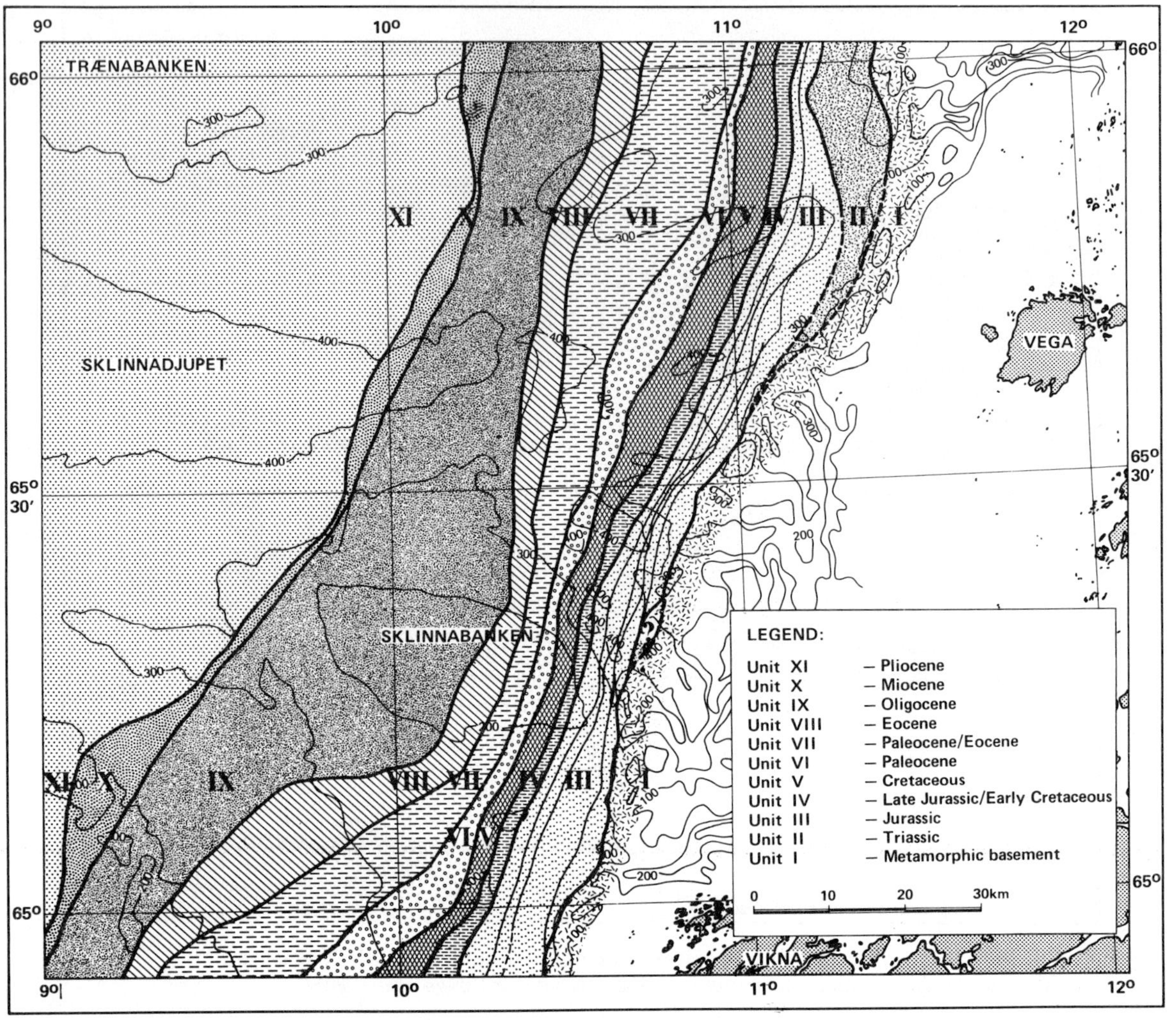

Fig. 6. Bedrock map showing the units subcropping below the Quarternary cover off Helgeland; the Vega map sheet.

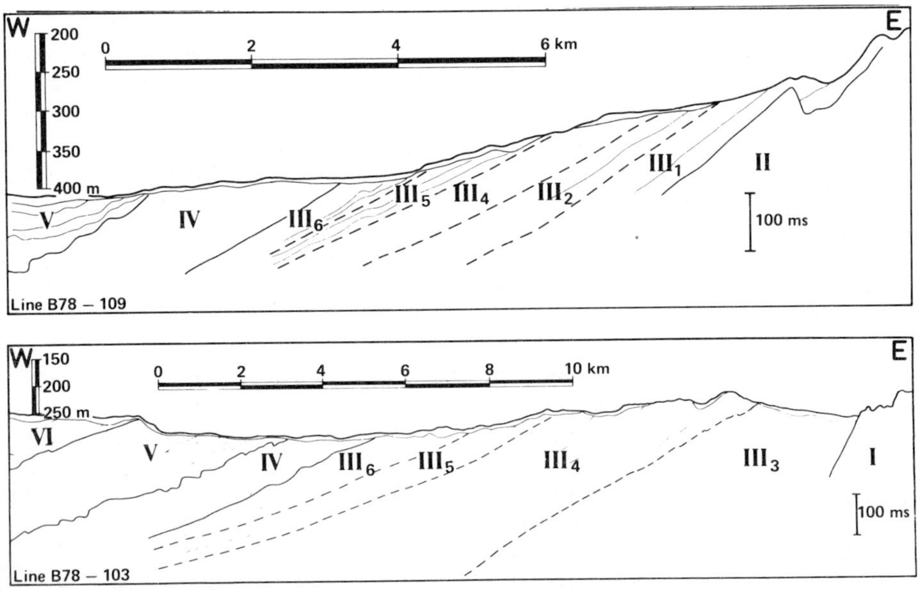

Fig. 7. Sketches of seismic lines B78-109 and B78-103, with tentative sub-division of Unit III in the Vega area. Locations are given in Fig. 1a, profiles C and D. Note the direct onlap of Sub-unit III_3 on the basement on the southern line (line 103).

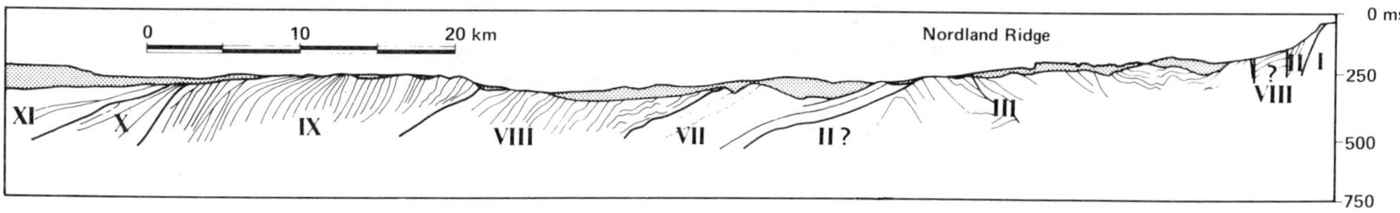

Fig. 8. Interpreted sparker record showing the development of the Nordland Ridge. Location is shown in Fig. 1a.

restricted to the west by the Nordland Ridge. At about 66°30′N, however, the Nordland Ridge bends towards the ENE and influences progressively younger stratigraphic units. To the east of about 11°30′E the ridge outcrops on the sea floor. The seismic data show that the Nordland Ridge was formed mainly during the late Jurassic and the Cretaceous. A slight flexing of the Palaeocene to Oligocene sediments (Units VII—IX) indicates that the ridge may have been tectonically active up to the Oligocene in the eastern part. On the southern flank of the ridge the Jurassic and Cretaceous sediments were apparently deposited from the north and northeast, and local onlap structures can be observed in the Jurassic unit. Due to the thick Quaternary overburden in the southern area only the Triassic and Jurassic units could be sampled south of the Nordland Ridge. The interpretation of the age of the younger units is therefore based on results from the Vega map sheet and from the datings in the central and northern part of the Trænabanken map area. Reliable samples from the northern side of Trænabanken and in the Trændjupet depression have proved that these sedimentary sections are most likely of Eocene (Unit VIII) and Oligocene (Unit IX) age. The first confirmed Miocene and Pliocene datings from outcrops on the shelf north of 62°N were also obtained in this area. The earlier suggested ages of Miocene for Unit X and Pliocene for Unit XI were thus confirmed.

The seismic data show intense local block faulting on the Nordland Ridge (Fig. 8). On the ridge the sediments became progressively younger toward the east; however, a zone of Triassic sediments occurs between the younger Mesozoic sequence and the basement. This is in agreement with the interpretation of the ridge further to the southwest, as proposed by Jørgensen and Navrestad (1979). Bedrock samples ranging in age from Triassic to Lower Tertiary also support this interpretation. This implies that the ridge formed a topographic high in the Jurassic and Cretaceous, with some movement in the Lower and Middle Tertiary. Local basins containing sediments of these periods were downfaulted and the sediments preserved. The basins are local and a line spacing of 5 km is too wide to correlate these basins from one line to another. However, the faulting seems to be roughly parallel with the present coastline, which is also parallel with the strike of the bedrock units away from the ridge.

The Nordland Ridge was strongly eroded in late Cretaceous/early Tertiary time. Most of the younger Mesozoic sediments on the ridge, apart from the sediments in the downfaulted basins, were therefore removed. On the western side of the exposed part of the ridge the base of the Tertiary is marked by a strong angular unconformity (Fig. 8). North of the Nordland Ridge no Mesozoic samples were obtained. Units II—VI are mapped, therefore, only from the seismic data and tentatively assigned unit numbers from their seismic signatures. They seem to continue north of the Nordland Ridge in a similar pattern as to the south of the ridge, i.e., parallel to the coastline and with dips slightly to the northwest. Further sampling, preferably by shallow drilling, is needed to confirm this interpretation.

STRATIGRAPHICAL AND SEDIMENTOLOGICAL DEVELOPMENT

The Quaternary deposits vary significantly in thickness and extent, and a typical example is given in the isopach map from the Vega map area (Fig. 9). Consequently not all the areas and stratigraphic units are equally represented or documented. In our material the most complete section is found in the area seaward of Vega (65—66°N) and therefore geological modelling is based on the information we have from this area. Comments on lateral changes in development to the north and south are made only where additional information has been available.

Metamorphic basement (Unit I)

This unit reflects the same rock types that occur on the islands and westernmost land areas (cf.: Map sheets of Vega and Mo i Rana from the Geological Survey of Norway). Different types of mica schists are the dominant rock types; however, both metamorphosed plutonic rocks and metasediments also occur. Some of the seismic lines in this area of crystalline basement also indicate the presence of local basins of Mesozoic sediments. Although micaschist and laminated gneissose rocks may give distinct seismic reflectors, we consider the seismic signature of the Mesozoic sediments so typical, that a more comprehensive seismic investigation would reveal possible local basins of Mesozoic sediments within the crystalline area. Such basins are noticed by Oftedahl (1975) in Beitstadfjorden and Frohavet near Trondheim.

Triassic (Unit II)

Triassic sediments onlap the outcropping crystalline basement from about 65°30′N and northwards into Vestfjorden. Structurally this unit can be divided into two main zones; an inner zone which onlaps the basement, and an outer zone which is fond outcropping on the exposed part of the Nordland Ridge. The innermost part of the inner zone is generally more fine-grained (Fig. 10), while the outermost and also southernmost part contains

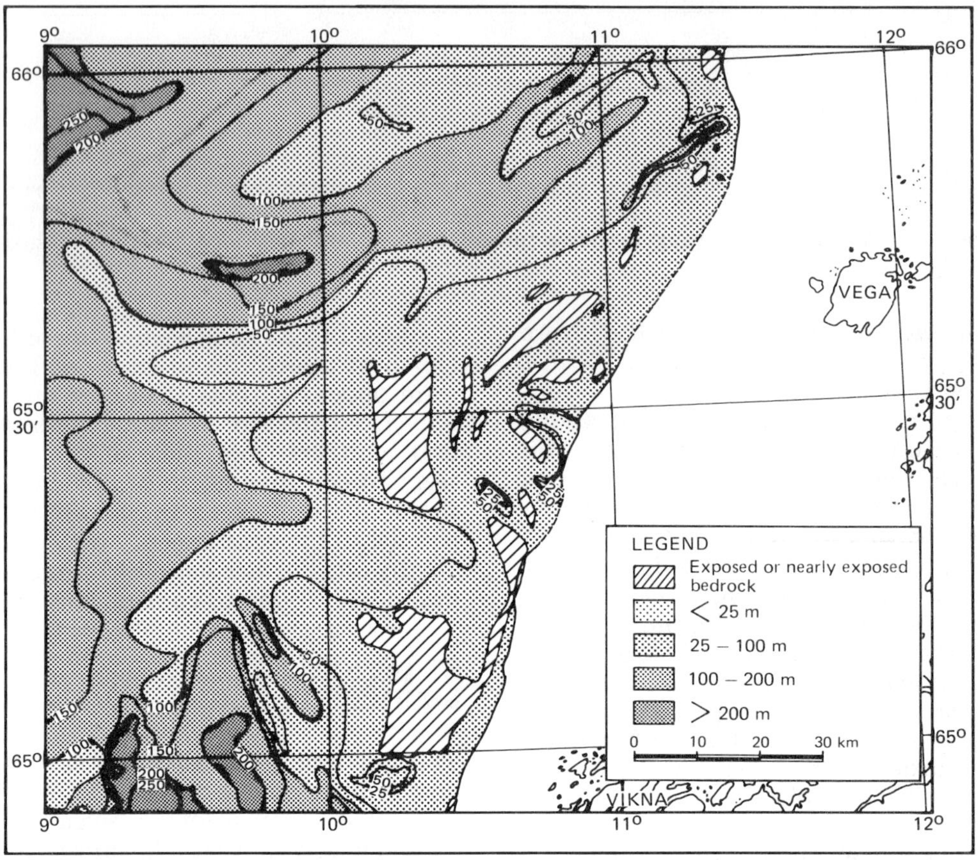

Fig. 9. Map showing the Quaternary overburden in the Vega map area, 65—66°N.

coarse-grained sandstones and conglomerates, often with a reddish staining. These coarse-grained sediments may represent either a lateral or more proximal equivalent to the Lower Triassic section or it may represent remnants of continental deposits of Middle to Upper Triassic age. None of these samples has provided a conclusive age determination and the latter interpretation (above) is therefore based mainly on structural evidence and on the similarity in lithological development to that observed from East Greenland (Clemmentsen, 1980) and Haltenbanken (Hollander, 1982).

All of the samples within Unit II are characterized by a very immature mineralogy, reflected both in the claystones and in the sandstones. The claystones are laminated and have no sign of bioturbation. The colour alternates between pale red and grey. Mica/illite and Fe-chlorite dominate, but there are also significant amounts of quartz and plagioclase. Calcite occurs as micritic mud in variable amounts, locally up to as much as 40%. The K-feldspar content (2—4%) shows only a small variation relative to plagioclase. Other characteristic minerals in these claystones are siderite and pyrite, in addition to finely dispersed haematite. Both siderite and pyrite are probably authigenic, while the haematite was most likely formed in the prediagenetic stage in a continental/lacustrine depositional environment, and is allochthonous.

The Lower Triassic sandstones contain, in general, 20—40% rock fragments and mud clasts while mica (often rich in biotite) contributes from 5 to 10% and feldspar (mostly plagioclase) from 10 to 40%. The quartz, which shows a slight undulose extinction, varies from 25 to 35%. Occurrence of garnets (up to 1%) also demonstrates the mineralogical immaturity of these sediments (Fig. 11).

The effective porosity in the sandstones examined is low (<1%) due to strong diagenetic alteration, caused by both chemical reactions and mechanical breakdown. Frequent transformations of feldspar to clay minerals (both illite and chlorite), dissolution of clastic mica and feldspar and calcite replacement of weathered and sericitized feldspar grains are observed in the sandstones. These processes are combined with lithostatic compaction processes, which in turn have resulted in a widespread grain readjustment, with plastic squeezing of ductile grains and brittle fracturing of altered feldspar grains and aggregates. The primary grain textures and mineral composition in these sandstones are therefore strongly modified and transformed.

Based on the structural pattern and the gradation of the material, the main source area for these Triassic sediments was most likely to the east (Caledonian basement). A further westward sediment transport into the location for the present drilling activity in the Trænabanken area would probably result in a slightly higher mineralogical maturation, but the general immature character will probably also be dominating within equivalent sections in this area. Burial to 2000—3000 m has, therefore, most likely resulted in an even stronger physical breakdown and compaction of these sediments. Although the sedimentary facies would indicate the possibility of sedimentary traps in the Trænabanken area, the porosity and permeability are likely to be low, resulting in a low reservoir potential for these Lower Triassic sandstones.

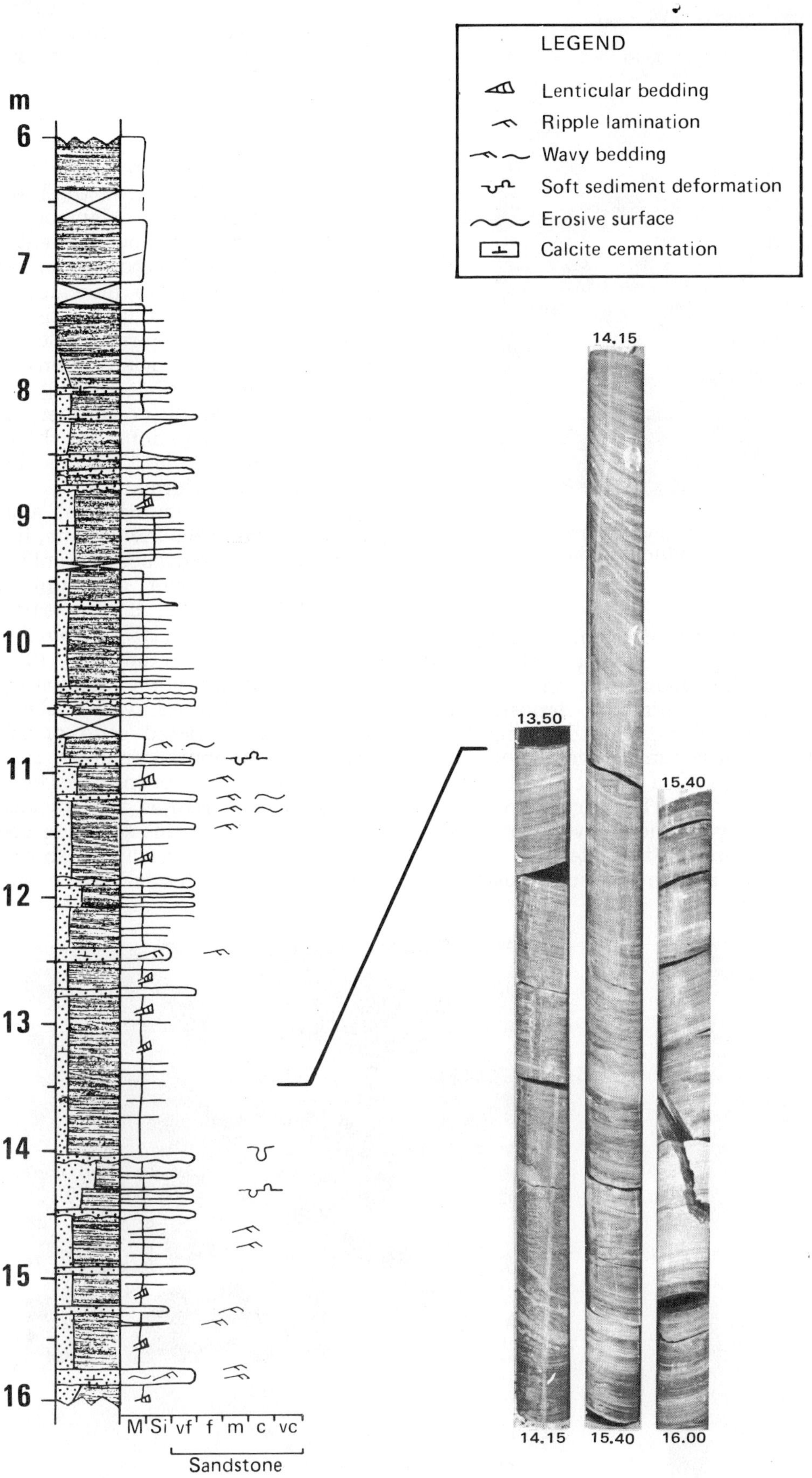

Fig. 10. Interpreted lithological log and photographs from the Lower Triassic core in Unit II drilled in the northern part of the Vega map area.

Fig. 11. Thin-section photograph showing the immature nature of the sandstone in the Lower Triassic, Unit II.

The great variation in grain sizes and rock types, both laterally and stratigraphically, as well as the reddened nature of the sediments, indicates a probable dominance of continental depositional environments. In some of the claystones, marine microflora and fauna suggest marine incursions. In the drilled core (Fig. 10) such a marine to estuarine facies segment is noticed. The very immature mineralogy, together with the above-mentioned characteristics, may therefore indicate high relief, rapid sedimentation and interaction between marine and continental facies segments. This facies mixture is most likely controlled by regional faulting activity along the basin margins, as noticed in time-equivalent rocks on East Greenland (Clemmentsen, 1980).

Jurassic (Unit III)

The Jurassic unit is heterogeneous, with generally coarse-grained sediments in the lower part, grading up to an organic-rich claystone sub-unit. This general transgressive development is periodically interrupted by local tectonism and eustatic sea-level fluctuations, which are indicated by the marked lithofacies changes throughout the unit.

The two oldest Jurassic Sub-units III_1 and III_2 (Fig. 7) (?Hettangian—Pliensbachian) contain kaolinitic claystones alternating with both mud- and grain-supported conglomerates. Microfossils indicate marine incursions, but there is a dominance of continental facies within these sediments. Although our information is fragmentary, we suggest an interpretatation of alluvial fan facies prograding into a shallow marine, low energy basin. The abundant occurrence of kaolinite also suggests that the source area was subjected to intense chemical weathering; a property which is also reported by Dalland (1979) from Andøya.

The Sub-units III_3 and III_4 (Toarcian to early Callovian) are sandstone dominated. The porosity in these sandstones varies significantly, from almost unconsolidated sand of 35% porosity to completely dolomite cemented sandstones of less than 1% effective porosity. The sandstones are K-feldspar-rich arenites and wackestones, and the original feldspar grains seems to accept K-feldspar overgrowths more readily than the other clastic grains (Fig. 12). This feldspar overgrowth is found

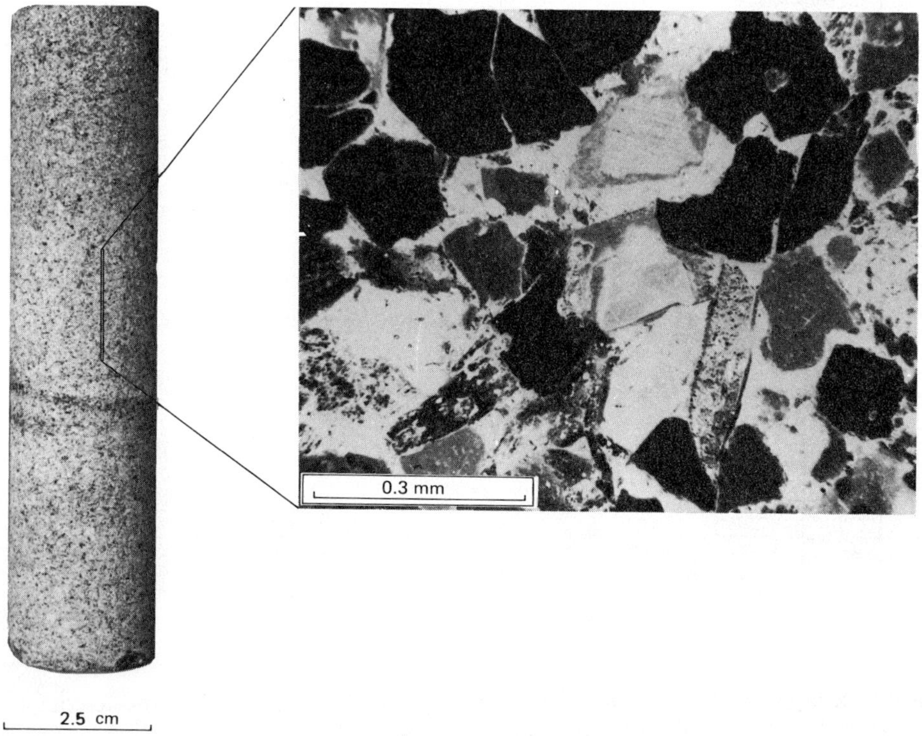

Fig. 12. Sandstone sampled by the diamond core drill in the Jurassic Unit III from the Trænabanken area. The thin-section photograph from the middle part of the core shows K-feldspar overgrowth on clastic grains.

over a large lateral extent and the occurrence is also common in the Lower and Middle Jurassic at Haltenbanken (Karlsson, 1984). Interbedded clay- and siltstones of these Jurassic sub-units show some of the same mineralogical characteristics as in the underlying sub-units, with high kaolinite content. The feldspar overgrowth, therefore, can be closely related to extensive weathering processes onshore. The formation of kaolinite (in a humid climate) gives K^+ ions in excess, resulting in a rise of K^+ activity in the water phase. This water was then poured out into the deltaic, sandy sediments basinwards and enabled precipitation of K-feldspar in this system, preferentially on K-feldspar grains. The porosity variation in this sandy unit is, however, not only controlled by the processes related to the siliciclastic phases. Early percolation of dolomite supersaturated porewater may have resulted locally in a strong dolomite cementation. A combination of grain readjustment, feldspar overgrowth, partial precipitation of dolomite and finally a late and local carbonate dissolution is then responsible for the large variation in porosity.

Local occurrence of coal in some of the samples examined also supports our interpretation of delta-influenced sedimentation on this part of the shelf (the Vega map sheet). Based on the seismic interpretation further south, and also verified with a few samples, it seems that coal-bearing strata are even more frequent in the equivalent unit basinwards and in the Møre—Trøndelag region. It should be remembered that onshore along Beitstadfjorden, ice-transported blocks of Middle Jurassic coals, clay and ironstones have been recorded earlier (Kjerulf, 1870; Carstens, 1929; Horn, 1931) and also have been studied recently by Vigran (1970) and Oftedal (1972, 1975). This material has yielded remnants (leaf cuticles) related to the Middle Jurassic flora of Yorkshire (Vigran, 1970) and spores and pollen of Middle Jurassic affinity.

The two uppermost sub-units (III_5 and III_6) are dated late early Callovian to Oxfordian and differ significantly from the sub-units beneath. Both are clay dominated and have a marked increase in total organic carbon towards the top of the uppermost sub-unit. The lower part is completely dominated by a terrestrial assemblage of clay minerals and contains a mixture of kaolinite, chlorite and illite. Mineralogically this rock is similar to the claystones examined in the sub-units beneath. In the uppermost part of III_6 there is a marked input of smectite, which may represent a product of diagenetically altered pyroclastic debris. The palaeontological record suggests open marine conditions; however, some proximity to land is indicated by the presence of terrestrial kerogen and by the palynomorph assemblage, especially in the lower part of these sub-units. Bioturbation shows that normal oxic conditions prevailed during deposition of the lower part of the section, while the unbioturbated, organic rich, upper part was probably deposited under anoxic conditions.

Upper Jurassic—Lower Cretaceous (Unit IV)

By its seismic signature this unit can be easily followed laterally through the entire map area. Lithologically the unit consists of a black, organic-rich claystone with occasional thin strings or interlaminae of calcite. The claystone represents the termination of a general Upper Triassic to Jurassic transgression in the area. Although this unit has only minor lithological variations over large areas, there are signs of a new regime starting in the Upper Jurassic/Lower Cretaceous, with volcanic/tuffaceous material which can be related to the late Kimmerian tectonic phase, and with the initiation and elevation of the Nordland Ridge.

The mineralogical composition, therefore, is influenced both by a clastic input (indicated by the presence of illitic and kaolinitic material) and by pyroclastic components — seen by the presence of abundant smectite and cristobalite. The change in terrestrial clay components through the Triassic and Jurassic from illite—chlorite to the dominance of kaolinite also indicates a marked shift in climate and topographical properties within this period. A change from an arid to a humid climate is indicated, in addition to a gradual transition towards a much more gently inclined and denuded source area.

Organic carbon values of up to 10%, absence of bioturbation and the frequent occurrence of pyrite indicate deposition in anoxic marine conditions, with restricted circulation in the lower water column. The kerogen composition is a mixture of type II and III, and has a very rich petroleum potential.

Cretaceous (Unit V)

The Cretaceous unit seems complex with marked lithological facies variations both stratigraphically and laterally. The Helgeland Basin, with its prolongation southward, was subjected to tectonic movements during the early Cretaceous, and a structural high (the Vega High) within the basin seems to have controlled the sedimentation in the middle and inner part of this shelf area. The Cretaceous unit offshore Vega is represented by several sedimentary sequences, which were separated by periods of non-deposition or erosion, together with local reworking.

Superimposed on the black shale in Unit IV, at a marked lithological break, there is a limestone series of late Ryazanian to Valanginian age which defines a lower part of Unit V. This carbonate series with intraformational conglomerates represents a large time-interval with very little or no siliciclastic sediment input. The carbonate interclasts demonstrate an early consolidation. At a later stage the beds were broken, probably by marine activity and redeposited. These clasts, and also the presence of boring, may therefore indicate formations of hardgrounds.

The succeeding sequence, dated Hauterivian to Barremian, is claystone dominated. This claystone is calcareous and has a clay mineral composition rich in kaolinite, but also with significant amounts of illite, chlorite and irregular mixed-layer clay minerals of smectite/illite/chlorite. This complexity is probably due to local redeposition, a process which is supported by the palaeontological study, where well preserved specimens of Jurassic and Lower Cretaceous ages are found together with the Hauterivian—Barremian assemblages.

A reddish stained silty claystone section overlies the grey claystone. The fossil recovery from this series is poor and gives contradictory datings,

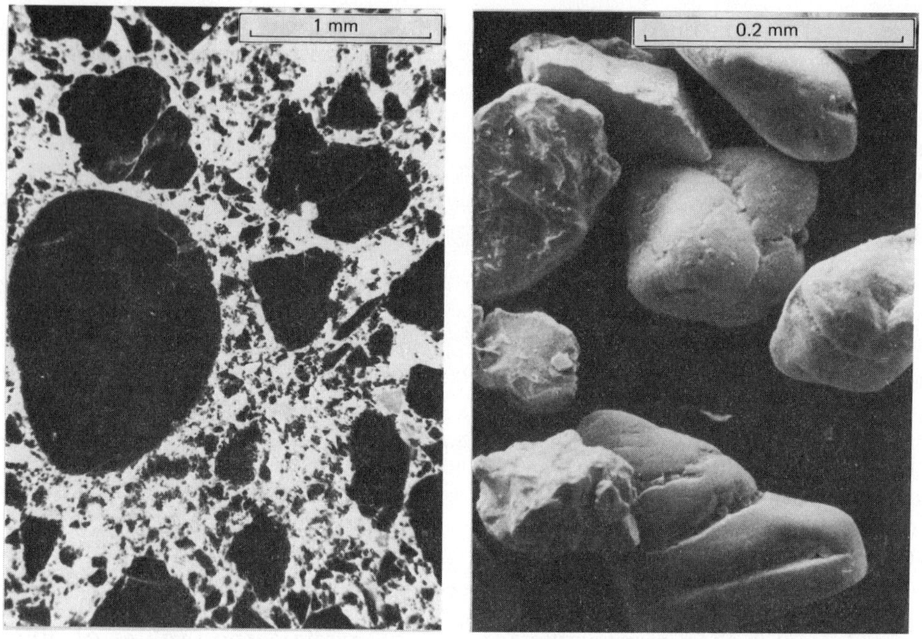

Fig. 13. Thin-section photograph (left) and scanning electron microscope (SEM) photograph (right) of sandgrains from the Oligocene Unit IX. The thin-section photograph shows poor sorting of the sand, with both well rounded and angular quartz grains which occur in a glauconite and goethitic matrix. The SEM photograph shows preserved infillings of foraminifera tests consisting of glauconite.

showing flora and fauna elements of different ages within the same sample. The X-ray diffractometry analyses demonstrate almost identical claystone mineralogy as further down in the unit. The colour of the sediment may therefore be explained by upheaval and subaerial exposure of the marine Hauterivian—Barremian sediment, with subsequent oxidation and redeposition in ventilated marine conditions.

Due to incomplete sample recovery, we cannot conclude or document a complete stratigraphical sequence through this unit. The succession up towards the youngest sub-unit of Middle Cenomanian—late Coniacian age, therefore, may contain sedimentary sections which are not yet sampled or lack fossils for age documentation.

The uppermost sub-unit differs significantly in mineralogy from the Lower Cretaceous section, mainly due to a considerable content of smectite, semiquantitatively estimated to approximately 70% in some of the samples examined. The clayey terrestrial input to the basin still contains a mixture of illite, chlorite and kaolinite, but this influx was diluted by micritic calcite mud and also by tuffaceous material, which gave rise to the smectitic clay mineralogy. An assemblage rich in planktonic and benthic foraminifera, together with a rich occurrence of *Inoceramus* prisms and a diverse dinocyst flora, indicates deposition in a shallow marine environment with good water circulation.

Tertiary (Units VI—XI)

The seismic signatures for the various Tertiary units are relatively distinct, but the sample recovery is sparse over most of the mapped area. Sedimentological and biostratigraphic data, therefore, are not as complete as for the Mesozoic units.

Unit VI of Palaeocene age contains a very characteristic sandstone/sand lithology as exposed on the sea floor off Vega. The pebbly sand contains material from a variety of sources, as shown by the heterogeneity in mineralogy and the significant variation in roundness. Quartz (often yellow stained), and K-feldspar, together with goethitized fragments, dominate in the coarse-grained fraction, whereas glauconite, quartz and goethite are most frequent in the fine-grained sand and in the silt fraction (Fig. 13, left). The glauconite is an unexpandable, illitic variety and has partly been formed both in tests of foraminifera (Fig. 13, right) and as glauconitized mica. The goethite content is up to 20%. This phase probably represents a weathering product and may be directly correlatable to the redeposited phase of glauconite. Both of these phases have survived the physical stress occurring in normal bedload transport, and the glauconite is considered to be allochthonous. The poor sorting of this material and the restricted lateral extent of the lithofacies (seismically interpreted) may indicate rapid sedimentation in a high energy, fluvially influenced deltaic environment. Lateral facies segments have not been sampled, and the lithofacies is thought to wedge out seaward into the basin.

Although few in numbers the samples recorded from Unit VII (Palaeocene—Eocene) define a distinct lithological change from the coarse-grained unit beneath. Above a dark grey claystone which occurs in the basal part of the unit, a zone of tuffaceous claystones dominate and alternate with terrestrial claystone laminae and beds. Smectite is the dominant mineral, but the occurrence of clinoptilolite (zeolite) is also observed. The distinct, parallel, reflectors seen in the seismic sections within parts of this unit are thus interpreted as representing a formation with tuff and tuffaceous layers.

At the present stage of study the four outermost units (VIII—XI) of Eocene to Pliocene age, give only incomplete lithological information. Some local outcrops in the northwestern map area (Fig.

1b), however, are promising and it is hoped that further surveys will give additional information from these units. We restrict our comments to samples from Units IX and XI.

The Oligocene (Unit IX), as exposed in the northern part of the map area, is sandy with poorly consolidated sediments. As in Unit VI this unit has both glauconite and goethite intercalated with quartzitic and feldspathic sand grains. From the seismic records the unit is seen to have a delta-like structure with bottom- and foresets prograding westwards, along the entire shelf area examined. The over-all prograding development is thought to be a response to a more regional geological event, probably related to the opening of the Greenland Sea in 'post-anomaly-13' time (Talwani and Eldholm, 1977), resulting in an oblique upheaval of the inner shelf area along a hinge-line near the shelf break. The development of this delta-like prograding structure, however, may also be related to the marked fall in sea level during late Oligocene (Vail and Hardenbol, 1979).

The Pliocene (Unit XI) has a considerable thickness (Fig. 2). The sediments are poorly consolidated, and most of the sampled material has a coarse-grained composition (sandy to gravelly). In a few samples (grab material) this coarse-grained sediment is mud supported or has a very high content of clayey material. The textural properties, the seismic signature, the large thickness of the unit and the very immature mineralogy (lithic), may indicate a partially glacial influenced sedimentation.

The Pliocene sequence was glacially eroded in the Pleistocene, and the upper boundary is usually seen as a slight angular disconformity. Some Pliocene sediments are also found on top of the older sedimentary units in the eastern area, as flat-lying erosional remnants restricted to depressions. These are usually covered by Pleistocene, glacial sediments.

REFERENCES

Bugge, T. 1980. *Øvre lags geologi på kontinentalsokkelen utenfor Møre og Trøndelag.* (Shallow geology on the continental shelf off Møre and Trøndelag, Norway). *IKU Publ. No.* **104**, 44pp.

Bugge, T., Lien, R. and Rokoengen, K. 1976. *Kartlegging av de prekvartære lag utenfor Møre/Trøndelag (63°—65° N).* (Geological mapping of preQuaternay units offshore Møre/Trøndelag, mid Norway). *IKU Publ. No.* **69**, 48pp.

Bugge, T., Løfaldi, M., Maisey, G.H., Rokoengen, K., Skaar, F.E. and Thusu, B. 1975. Geological investigations of a Lower Tertiary/Quarternary core, offshore Trøndelag, Norway. *Nor. Geol. Unders.*, **316**, 253—269.

Bugge, T., Prestvik, T. and Rokoengen, K. 1980. Lower Tertiary volcanic rocks off Kristiansund — mid Norway. *Mar. Geol.*, **35**, 277—286.

Carstens, C.W. 1929. Brunkullfund på Tun, Verran. *Kgl. Nor. Vidensk. Selsk. Forh.*, **2(9)**, 32—33.

Clemmentsen, L.B. 1980. Triassic rift sedimentation and palaeogeography of central East Greenland. *Bull. Grønland Geol. Unders.* **136**, 72pp.

Dalland, A. 1979. The sedimentary sequence of Andøya, Northern Norway — Depositional and structural history. In: *Norwegian Sea Symposium.* Norsk Petroleumsforening, Article 26.

Eldholm, O. and Nysæther, E. 1969. *Seismiske undersøkelser på den norske kontinentalsokkel 1968. Tekn. rapp. 2c, Seism. Obs. Univ. Bergen,* 17 pp.

Hamar, G.P. and Hjelle, K. 1984. Tectonic framework of the Møre Basin and the northern North Sea. This volume.

Hollander, N.B. 1982. Evaluation of the hydrocarbon potential offshore mid Norway. *Oil Gas J.*, **12**, 168—172.

Horn, G. 1931. Über Kohlen-Gerölle in Norwegen. *Norsk Geol. Tidsskr.*, **12**, 341—361.

Jørgensen, F. and Navrestad, T. 1979. Main structural elements and sedimentary succession on the shelf outside Nordland (Norway). In: *Norwegian Sea Symposium,* (Norsk Petroleumsforening), Article 11.

Karlsson, W. 1984. Sedimentology and diagenesis of Jurassic sediments of mid Norway. This volume.

Kjerulf, T. 1870. Undersögelse av nogle Kulslags og Torv. *Vidensk. Selsk. Forh. Christiana* **1870**, 404—413.

LaBreque, J.L., Kent, D.V. and Cande, S.C. 1977. Revised magnetic polarity time scale for late Cretaceous and Cenozoic time. *Geology*, **5**, 330—335.

Oftedahl, C. 1972. A sideritic ironstone of Jurassic age in Beitstadfjorden, Trøndelag. *Norsk Geol. Tidsskr.*, **52**, 123—134.

Oftedahl, C. 1975. Middle Jurassic Graben Tectonics in mid Norway. In: *Jurassic Northern North Sea Symposium.* (Norsk Petroleumsforening), Article 21.

Rokoengen, K. 1980. De øvre lags geologi på kontinentalsokkelen utenfor Møre og Romsdal. (Shallow geology on the continental shelf off Møre and Romsdal). *IKU Publ., 105,* 49 pp.

Rokoengen, K. and Bugge, T. 1976. Grunnforholdene på den norske kontinentalsokkel sør for 72° N. IKU *Publ.*, **81**, 45s.

Rokoengen, K. and Rønningsland, T.M. 1983. Shallow bedrock geology and Quaternary thickness in the Norwegian sector of the North Sea between 60°30′N and 62° N. *Norsk Geol. Tidsskr.* **63**, 83—102.

Talwani, M. and Eldholm, O. 1977. Evolution of the Norwegian Greenland sea. *Geol. Soc. Am. Bull.*, **88**, 969—999.

Vail, P.R. and Hardenbol, J. 1979. Sea-level changes during the Tertiary. *Oceanus,* **22**, 71—79.

Vigran, J.O. 1970. Fragments of a Middle Jurassic flora from Northern Trøndelag, Norway. *Norsk Geol. Tidsskr.*, **50**, 193—214.

19 The Tyrihans* Discovery – preliminary results from well 6407/1-2

S.M. Aasheim and V. Larsen
Statoil, Bergen, Norway

Block 6407/1 is located in the Haltenbanken area northeast of the Møre Basin, south of the Nordland Ridge and west of the Haltenbanken Platform. Tectonically, the area is dominated by NE—SW faults downthrown to the west, occasionally off-set by NW—SE fault trends. The 6407/1-2 discovery is on an anticlinal structure within a southeasterly tilted fault block. The structure is partly connected in the north by a horst, bounded by faults to the west, north and east. The reservoir sandstone is the thickest one encountered so far in the Haltenbanken area, with more than 250 m of shallow marine sandstone of Middle Jurassic age, and having an average porosity of 15% and permeabilities ranging from 10 to 400 mD. Reserves proven by the well are marginal. However, further drilling in the area is expected to increase the reserves.

INTRODUCTION

The blocks awarded on Haltenbanken (officially called Møre Trøndelag II), are located between 130 and 150 km west of the mainland. The water depth varies from 240 to 300 m (Fig. 1). The 6407/1-block was awarded on 23 April 1982, under licence 073, to a group made up of the following companies: Statoil (operator) 50%, Amoco 20%, Conoco 20% and Norsk Hydro 10%. The work obligation includes a seismic coverage equal to a 1 × 1 km grid and two wells drilled into Triassic rocks or to 5000 m. The block has been covered with a 1 × 1 km seismic grid in the most interesting western part and elsewhere with a grid of maximum 2 × 2 km (1975, 1977 and 1981 data). There is also a seismic tie between the Statoil well 6407/1-2 and the Saga well 6407/2-1, and a survey of approximately 290 km was planned for 1983.

The first well on Haltenbanken was drilled by Saga during the summer of 1980. Seven wells have been drilled in this area, and two more are planned for 1983. Although the wells have improved our geological understanding of the area, the results have not been very positive with regard to hydrocarbons.

Well 6407/1-1 was spudded on 19 October 1982 by the drilling rig 'Dyvi Delta', but had to be plugged at approximately 900 m due to operational problems. The well was respudded as 6407/1-2 a few days later, and a total depth of 4560 m was reached in April 1983. This was the first well to be drilled through the rough winter season north of 62° N. The well has now been tested. A discussion on the hydrocarbons encountered will follow after a description of the block in a regional context, as well as some preliminary geological results from the well.

REGIONAL TECTONICS

Regionally, block 6407/1 is located within the down-faulted edge of the Trøndelag Platform on the boundary between the Møre Basin and the Vøring Basin, and represents a southwesterly elongation of the Nordland High. The area was heavily affected by fault tectonics during the evolution of the basins, platforms and highs in Mesozoic and Cenozoic times (Figs. 2 and 3).

Three pronounced fault trends are recognized in the area: N, NE and NW (Fig. 3). The evolution of the faults through time is not fully understood, but the N and NE trends are both locally and regionally known to produce a 'dog leg' fault pattern. The NE faults are parallel to the Caledonian trend, the Møre-Trøndelag Fracture Zone, the Kristiansund-Bodø Fracture Zone and other NE-trending tectonic elements. It has been speculated (Price and Rattey,

*The name Tyrihans is not yet officially approved.

Petroleum Geology of the North European Margin © Norwegian Petroleum Society (Graham & Trotman, 1984) pp. 285—291.

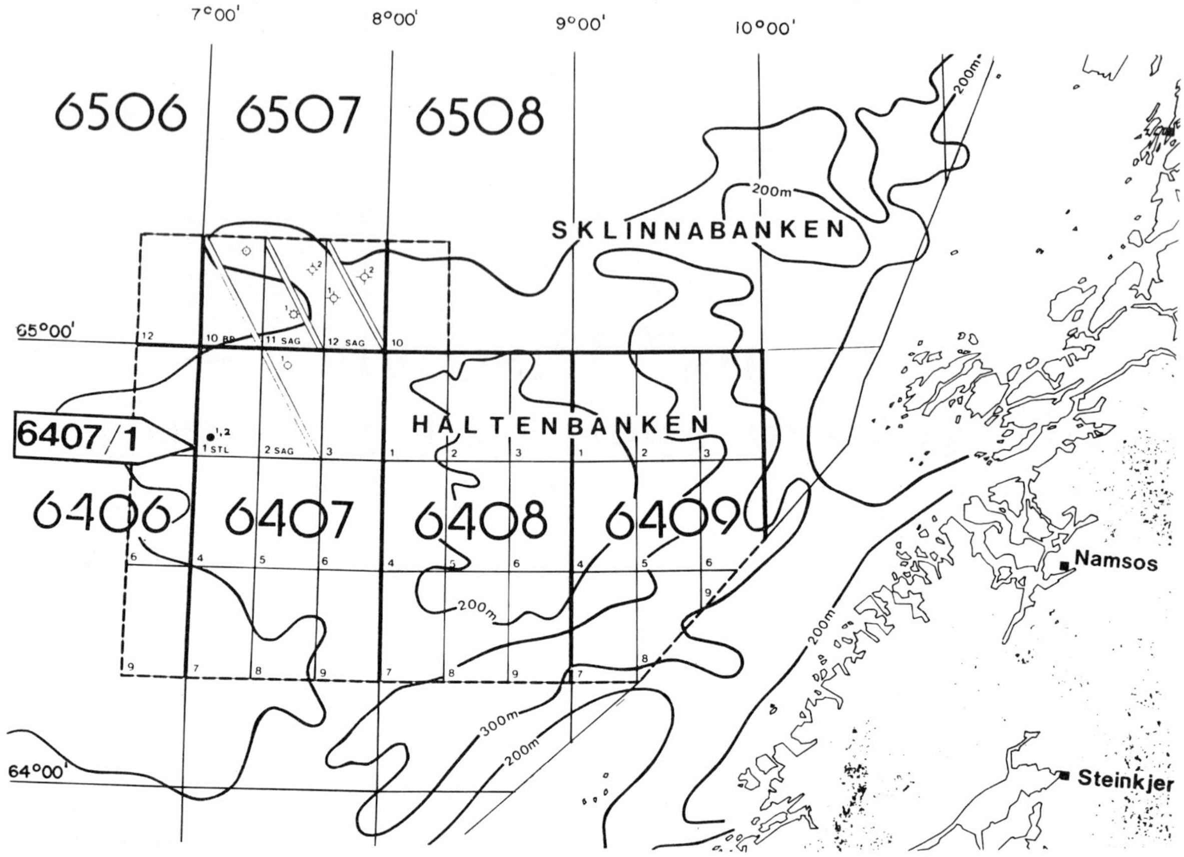

Fig. 1. Map showing location and water depths.

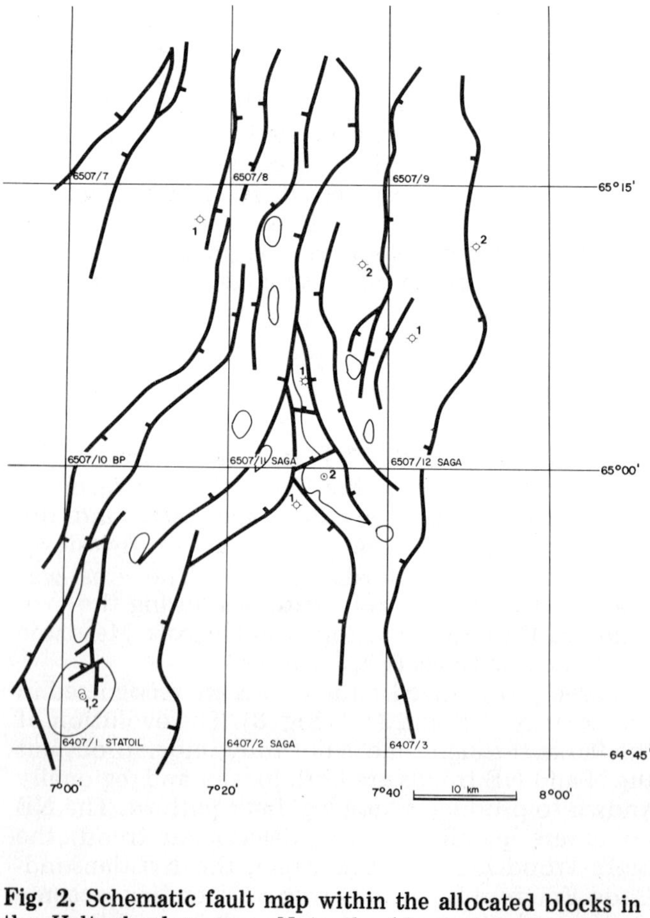

Fig. 2. Schematic fault map within the allocated blocks in the Haltenbanken area. Note the 'dog-leg' fault pattern.

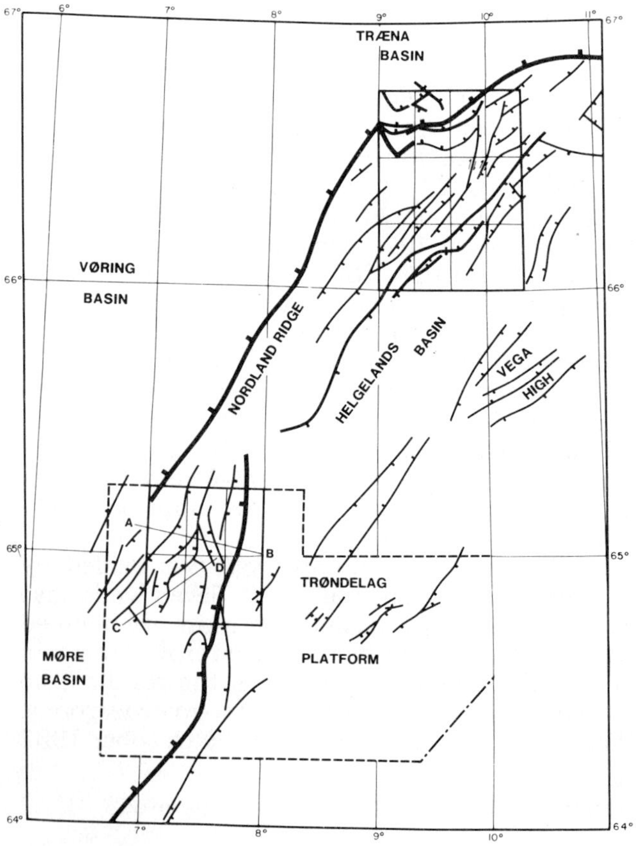

Fig. 3. Main structural elements within the Haltenbanken—Trænabanken area.

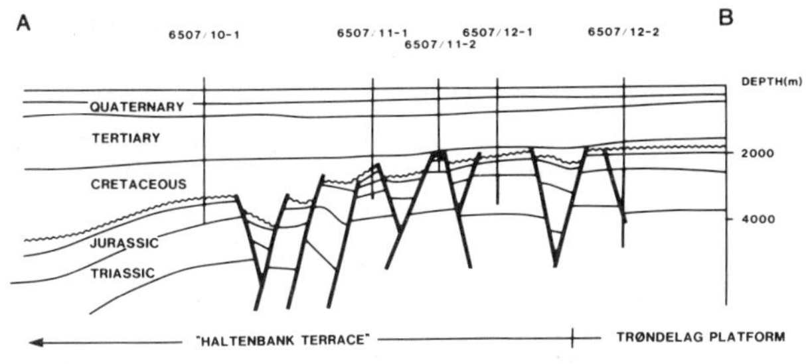

Fig. 4. East—West structural cross-section from the Trøndelag Platform across the 'Haltenbanken terrace'. Marked as A—B on Fig. 3.

in press) that this trend was re-activated during the mid Cretaceous in a dextral fault regime associated with the opening of the North Atlantic.

The interplay between the N- and NE-trending faults is, without doubt, very complex. Both trends appear to have been reactivated several times during the Mesozoic. It is likely that the two trends played an active role in the Jurassic and Cretaceous palaeogeography.

The third fault zone (NW—SE) is believed to be the most recent one and to be of a sinistral type, parallel to the Jan Mayen Fracture Zone.

REGIONAL STRATIGRAPHY

The stratigraphic control of the Upper Palaeozoic is very poor. Total sediment thickness in the area is suggested to be about 7000 m, which allows for a 3000-m pre-Middle Jurassic sediment package.

A N—S tectonic regime developed after the deposition of Lower Permian clastic sediments. Faulting initiated a transgression which deposited what are inferred to be evaporites and salt in the Møre Basin. The Permian evaporites were restricted to the west of the Nordland High, upon which Triassic clastics were deposited. Characteristic of the Triassic sequence are two 300+ m thick salt sections separated by more than 500 m claystones/shales and siltstones. The salt sections of both Triassic and Permian age are thought to have been mobilized during Jurassic—Cretaceous times.

The Upper Triassic salt section is overlain by 850 m of red claystones and siltstones interbedded with some stringers of continental sandstone. In the upper 100 m of this sequence, there is a change of colour from red to grey (grey beds). Well 6407/1-2 stopped above these grey beds within the Triassic part of an overlying coal unit (Fig. 5), a sequence of coal, claystones and sandstones. A more humid climate in a fluvio-deltaic environment is thought to be responsible for the change from Triassic red beds to the continental post-Rhaetian coal units. The Lower Jurassic coal unit is 400—500-m thick (Fig. 6). The associated sandstones are normally fine to medium grained, and the unit sometimes contains calcareous horizons. Dating gives a Rhaetian—Pliensbachian age, thus implying a prolonged continental environment in this area compared with the North Sea. A 200-m thick Lower Jurassic sand sequence overlying the coal unit is often referred to as the Cook equivalent. However, we think that these fluvio-deltaic fine to coarse sandstones, with marine incurions towards the top,

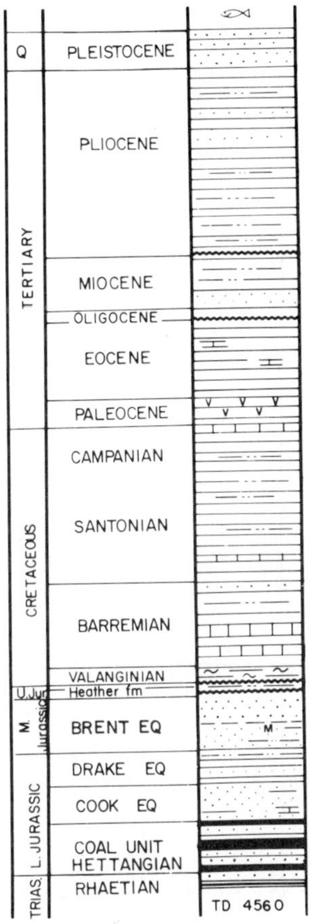

Fig. 5. Generalized post-Triassic stratigraphy on Haltenbanken.

rather represent a prolonged deposition of the Statfjord and Johansen-like sediments.

On top of this sandstone lies a 150-m thick Lower Jurassic marine claystone unit with sand stringers. The sand content in this unit increases towards the top, a fact which possibly can be related to the regressive nature of the overlying Middle Jurassic sand unit. The shale sequence appears to decrease in thickness towards the east.

Well 6407/1-2 found a 250-m thick Middle Jurassic sand sequence. This is more than twice the thickness previously encountered for this sequence in the area (Fig. 6). Thickening of individual units, as well as preservation of younger sandstones, accounts for a relative subsidence in this area compared to the areas drilled previously.

The penecontemporaneous subsidence and the relatively immature texture of the sandstone could suggest local drainage patterns and source areas, which again would be important for the facies evolution through the sequence.

Regionally, the Middle Jurassic sand sequence is part of a major regressive episode found in all the wells on Haltenbanken. Preliminary datings give a Bajocian—Bathonian age which is thus comparable to the North Sea Brent Group.

The top of the sequence is defined by an apparently regional unconformity which is transgressed by Oxfordian marine shales defining the Heather Fm. These are overlain by a dark brown to black organic shale of Portlandian age, which is comparable to the 'hot shale' of the North Sea, thus being an excellent source rock for the area

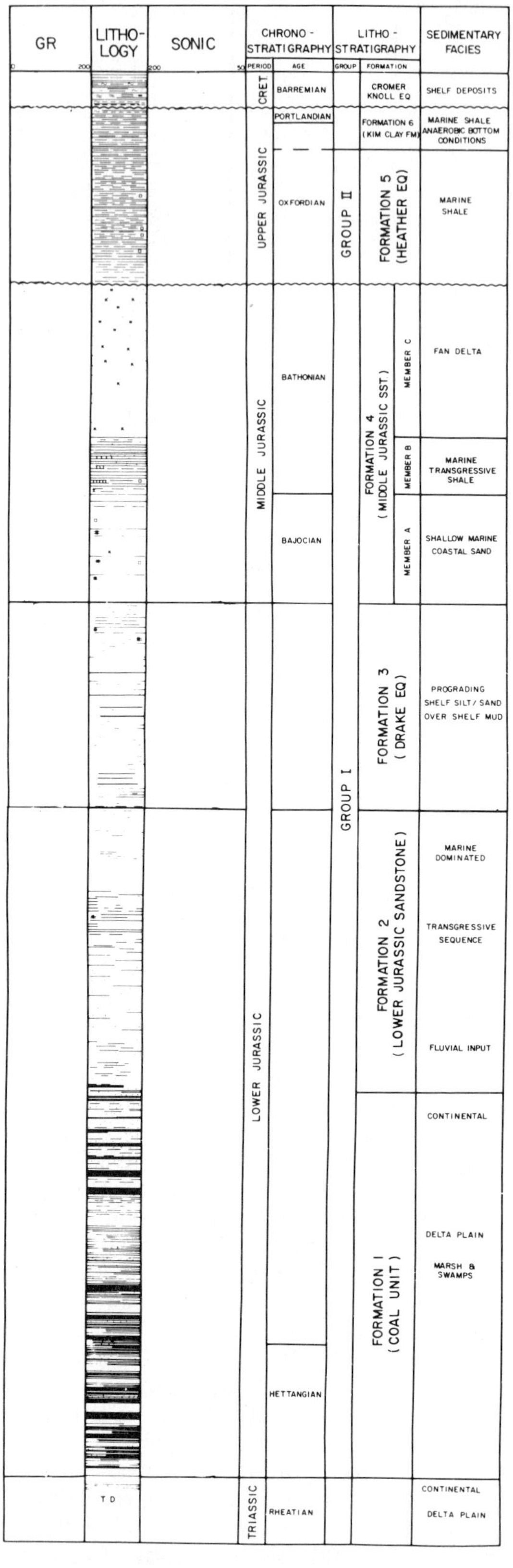

Fig. 6. The Jurassic—Triassic sequence in well 6407/1-2.

where it is sufficiently buried for maturation.

The Lower Cretaceous is generally thin on the Trøndelag Platform. Long periods of non-deposition seem to have occurred between Portlandian and Barremian times. However, thicker sequences may be expected in down-faulted areas. The sediments consist of claystones, marl and limestones with occasional sand stringers, and were deposited in a sublittoral environment. A regional westerly tilt during mid and late Cretaceous times resulted in an overall onlap situation, with transition from more than 1700 m of clay west of the Trøndelag Platform to non-deposition in the east. The tilting was possibly related to Cretaceous faulting and gave rise to several unconformities in the Cretaceous.

In well 6407/1-2 an approximately 45-m early Palaeocene claystone overlies the Maastrichtian sequence, which again is overlain by about 70 m of tuffaceous sediments correlated with the North Sea Balder Fm.

A second group of Tertiary sediments (early Eocene—early Pliocene), some 770-m thick, is dominated by claystones with some silt and limestone streaks. Some interfingering fine-grained sandstones are occasionally observed. An unconformity is present some 300 m below the top. This group is found in all wells on Haltenbanken and can be correlated with the Hordaland Group in the northern North Sea. Another unconformity defines the top of the group.

The Hordaland Group equivalent is overlain by light grey claystones and poorly sorted sandstones deposited during the Pliocene. These rocks are equivalent to the Nordland Group in the North Sea. The Pliocene is some 900-m thick and the Quaternary about 200 m.

SEISMIC INTERPRETATION

Seven horizons have been interpreted in block 6407/1 (Fig. 7): top Palaeocene; top Cretaceous; a Lower Cretaceous reflector; late Kimmerian unconformity; Middle Jurassic unconformity; top Dunlin equivalent; near top Hettangian.

Well 6507/11-1 has been used as the tie well for the seismic interpretation. The interpretation was made on data collected in 1977 and 1981 and only migrated sections were used. The well 6407/1-2 confirmed the seismic interpretation which is now generally accepted.

STRUCTURAL INTERPRETATION

Block 6407/1 is situated in a complex zone of faults in an area of structural element intersections (Fig. 3). The tectonic movements, which culminated at the Jurassic/Cretaceous transition, resulted in a domination of rotated fault blocks in the southern part of block 6407/1. In the eastward tilted fault block, an anticlinal structure is superimposed and forms a horst structure northwards (Fig. 8). Well 6407/1-2 was drilled on this anticlinal structure and the next well in the block will be on the northern horst-like structure.

The thickness variations of individual units indicate a synsedimentary fault activity. Truncation of the crests of rotated fault blocks has led to periods of erosion in the late and main Kimmerian periods.

A graben-like trough can be traced from block 6406/3, diagonally through blocks 6407/1—6507/10 and into 6507/11 (Fig. 2). The Kimmeridge 'hot shale' is thought to have reached a sufficient degree of maturation to generate hydrocarbons within this structural low and seems to be in contact with the reservoir sandstone across the eastern fault.

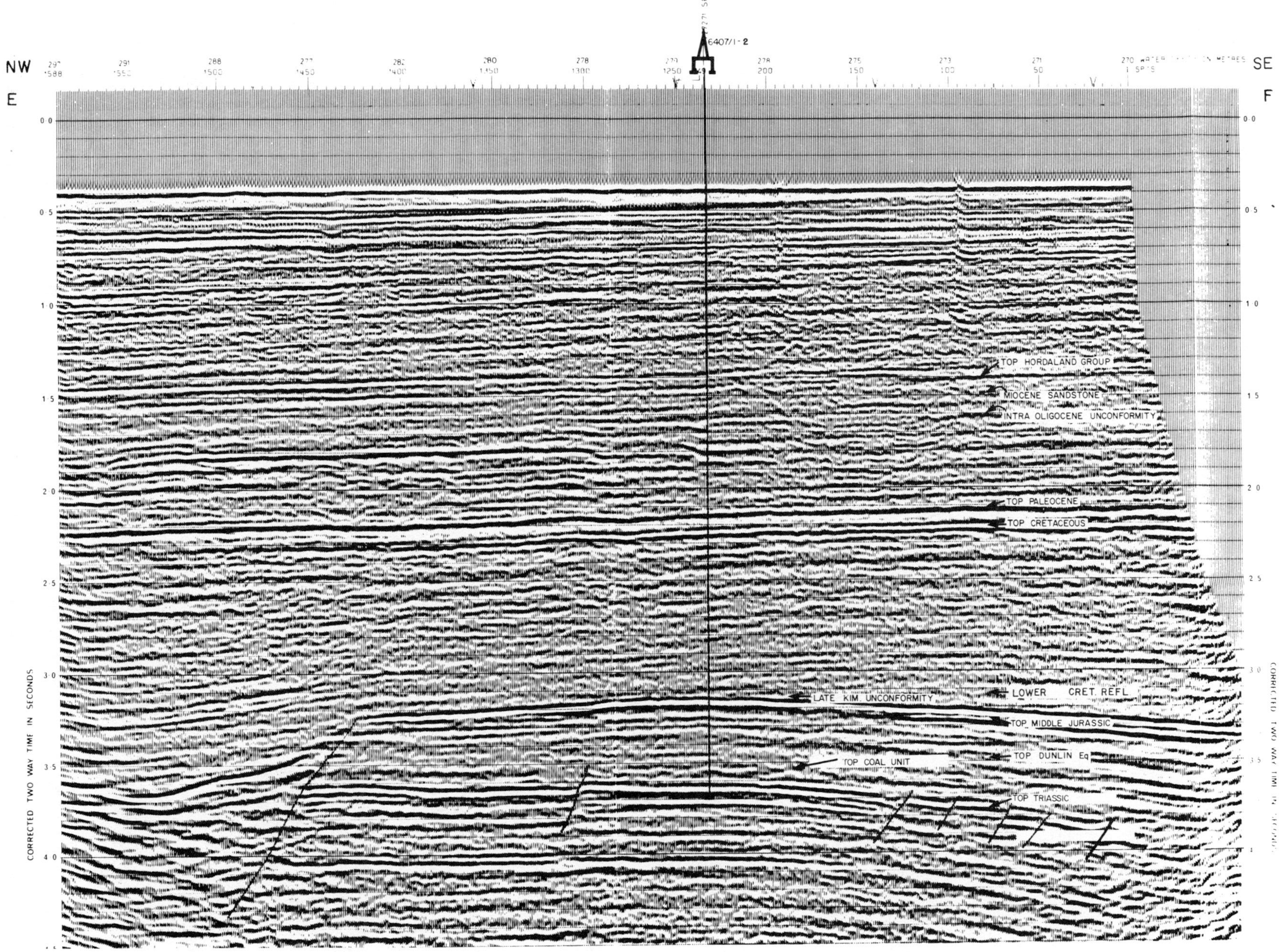

Fig. 7. A NW—SE seismic line through well 6407/1-2. See Fig. 8 for location.

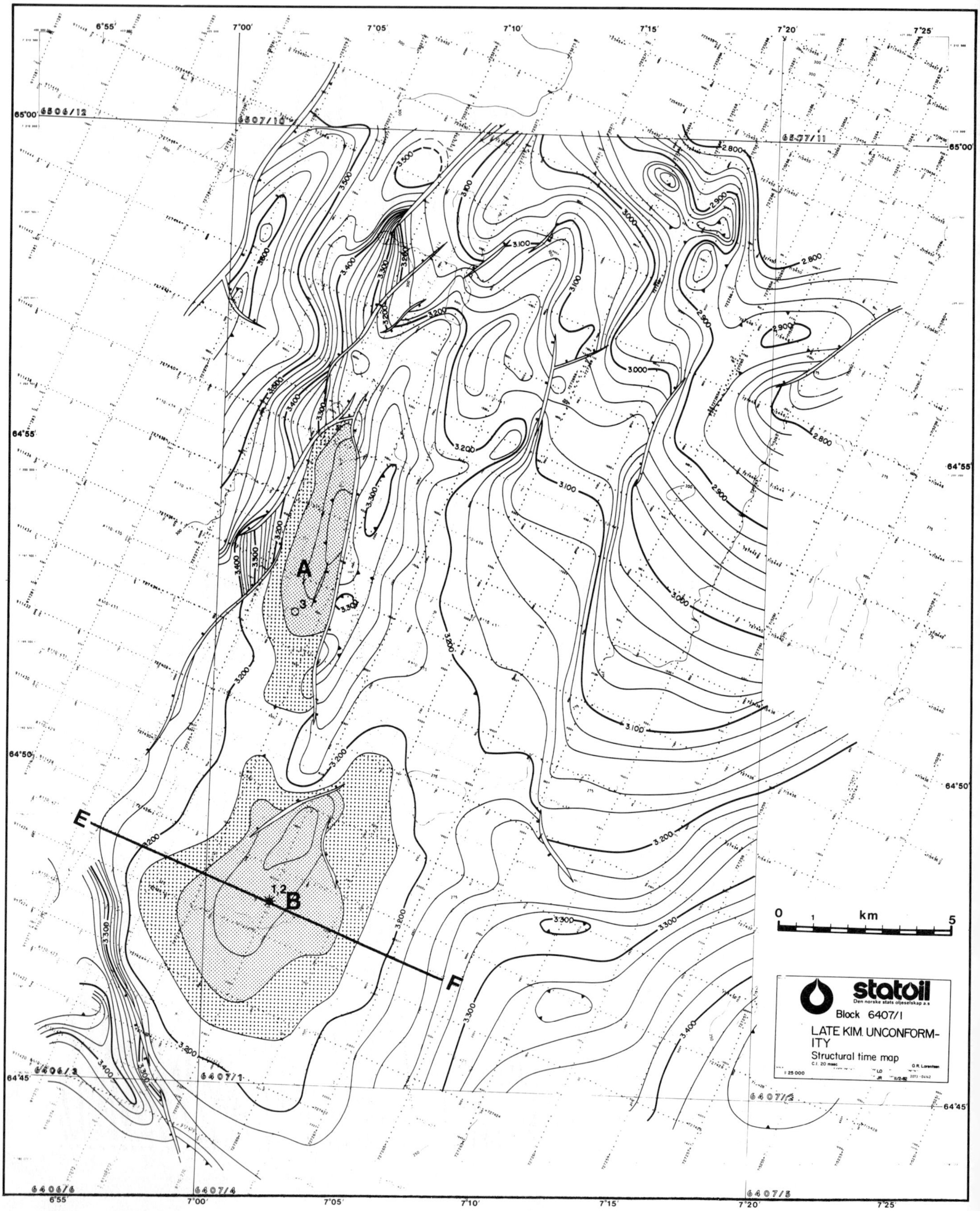

Fig. 8. A two-way time structural map of the top Jurassic (late Kimmerian unconformity) block 6407/1.

RESERVOIR GEOLOGY

The 250-m thick Middle Jurassic sandstone, which is the reservoir rock in the well, has been subdivided into three units.

The basal unit consists of subrounded—subangular, fine—medium, occasionally coarse sandstone interbedded with scattered stringers of limestone and claystone. The unit has a low net/gross ratio of 15% and the average porosity is 14%. The unit is not yet finally evaluated, but is probably of Bajocian age and represents a regressive shallow marine sequence, possibly tidally influenced.

An apparently regional transgression defines the base of the middle unit, which is a 43-m thick sequence of claystones interbedded with some calcareous sand layers and streaks of dolomitic limestones. Preliminary pressure data indicate that this sequence may act as a seal.

The upper reservoir unit consists of angular—subangular, fine—medium sandstone with frequent layers of coarse-pebbly sandstone. The unit gives a strikingly uniform log pattern reflecting a very high net/gross ratio. The average porosity is 15%.

The cored section of the Middle Jurassic sandstone contains subarkosic arenites with 5—10% mica and K-feldspar as the dominant feldspar mineral. Quartz overgrowth and clay minerals are the major diagenetic porosity reducing agents. Secondary porosity is observed, resulting from dissolution of the feldspar.

Investigation of the cored section (member C, Fig. 6) revealed sedimentary structures indicating deposition in standing shallow waters. Massive reworked sandstone, occasionally arranged into less than 1-m coarsening up sets and with very few wave-influenced structures characterizes the lower part of the cored section, suggesting a rather constant supply of clastics into the subsiding basin. The upper half of the cored section shows frequent layers of pebbly sandstones, arranged both in a fining up (erosive base) and coarsening up pattern; the sandstones are also frequently crossbedded.

As a preliminary depositional model for the upper reservoir unit, we suggest a regressive fan delta building out into fairly shallow waters, with decreasing wave influence and increasing fluvial processes towards the top. The immature texture and the poor rounding of quartz grains, together with dipmeter indications as well as regional data, could support a westerly source for the Middle Jurassic sandstones.

HYDROCARBONS

A production test was carried out in May 1983, and only preliminary results are available. The well tested approximately 3000 bbls/day of a 45° API gravity oil and 13 mcf/day of gas through a ¾-inch choke. Detailed analysis has begun to determine more precisely whether the hydrocarbon system is a light oil or gas/condensate discovery.

The geophysical maps are presently being revised and when the next well has been drilled on the northern structure, we will know more about the size of the discovery and the possibility of communication between the two structural highs. However, the discovery may be classified today as a marginal field.

CONCLUSION

Well 6407/1-2 has so far proved hydrocarbons and as the name 'Tyrihans' implies, we expect that further drilling in this area will give us more discoveries linked together, as 'Tyrihans' linked people with his golden goose when he won the princess and half the kingdom. We hope that that 'Tyrihans' discovery turns out to be such a princely find, if not for half the kingdom, perhaps at least for the coast from 62° to 66° north.

ACKNOWLEDGEMENTS

The authors wish to thank Conoco, Amoco, Norsk Hydro and Statoil for permission to use the data presented. We also thank Statoil's exploration staff, drafting office and secretaries for their support and help.

REFERENCES

Gabrielsen, R.H., Fœrseth, R., Hamar, G. and Rønnevik, H. 1984. Nomenclature of the main structural features on the Norwegian Continental Shelf north of the 62nd parallel. This volume, pp. 41–60.

Hamar, G.P. and Hjelle, K. 1984. Tectonic framework of the Møre Basin and the Northern North Sea. This volume, pp. 349–358.

Price, I. and Rattey, R.P. Cretaceous tectonics of mid-Norway: implications for the Rockall Faeroes—Shetland rift. *J. Geol. Soc. Lond.* (in press).

Rønnevik, H. Chr., Eggen, S. and Vollset, J. 1983. *Exploration of the Norwegian Shelf.* In: NPD-Contrib. No. 5.

20 Anomalous seismic event in Lower Tertiary strata, Trænabanken

D.A. Herron, A. Øvretveit and J. Almaas
Norwegian Gulf Exploration Company, Bergen, Norway

Interpretation of regional and detailed seismic data from the Trænabanken lease area has identified a 'flat spot' in Lower Tertiary strata on the northwestern flank of the Helgeland Basin. This anomalous seismic event is located on the crest of a broad, NNE-trending anticline in block 6610/10 and is evident on several vintages of seismic data. The results of special seismic processing, including inversion and analysis of seismic attributes, indicate that the 'flat spot' might represent a gas—liquid contact. The occurrence of this feature suggests that the Lower Tertiary section in the Helgeland Basin contains potential reservoirs and that hydrocarbons have been generated in the Trænabanken area.

INTRODUCTION

The seismic anomaly was first recognized during interpretation of regional seismic data from the Helgeland Basin (Figs. 1 and 2). The anomaly appears as a pronounced flattening and increase in amplitude of the base Tertiary reflection on the crest of a gentle, NNE-trending anticline in Trænabanken blocks 6610/7 and 10. This 'flat spot' is clearly evident on both super-long air-gun (SLAG) and conventional air-gun data (Figs. 3 and 4). Of particular significance on these data is the character of the reflection package overlying the 'flat spot' (Fig. 4), which is very similar to that of the Palaeogene interval in the Frigg, Balder and Cod fields in the Norwegian North Sea (Heritier *et al.*, 1979; Skjold, 1980; Kessler *et al.*, 1980). Since the gas—liquid contact in the Frigg reservoir is marked by a seismic 'flat spot', it is possible that the anomaly in Trænabanken might be a direct indicator of hydrocarbons. In order to evaluate this hypothesis, detailed study of the 'flat spot' proceeded along two lines of investigation: interpretation and mapping of seismic horizons within the Tertiary section in the area of the anomaly, and special processing of selected seismic lines over the anomaly.

Petroleum Geology of the North European Margin © Norwegian Petroleum Society (Graham & Trotman, 1984) pp. 293—301.

SEISMIC INTERPRETATION

Seismic interpretation in the area of the 'flat spot' was conducted to determine the extent of the anomaly and to provide stratigraphic characterization of Tertiary sediments at the level of, and above, the anomaly. Regional geologic studies indicate that the Tertiary section in the Helgeland Basin is a clastic sequence which was deposited during subsidence and seaward tilting of the Norwegian Continental Shelf. Four seismic horizons within the Tertiary section (Fig. 5) were interpreted and mapped over an area of approximately 400 km^2, centred on the anomaly, in blocks 6609/9 and 12 and 6610/7 and 10. Two of these horizons, the base Tertiary and mid-Tertiary (Chattian) unconformity, were identified by regional correlation from well control in the Haltenbanken area to the south. The remaining two horizons were identified as the boundaries between distinct reflection packages within the Palaeogene section; specific ages were not assigned to these horizons.

Base Tertiary (Fig. 6)

The base Tertiary horizon is a high-amplitude trough which extends continuously across the study area. The NNE-trending anticline on which the 'flat spot' is located is clearly evident in Fig. 6, bounded to the east by a shallow syncline and to the west by regional westerly dip. The anticline is doubly plunging, as suggested by convergence of contours at the top and bottom of the figure. The

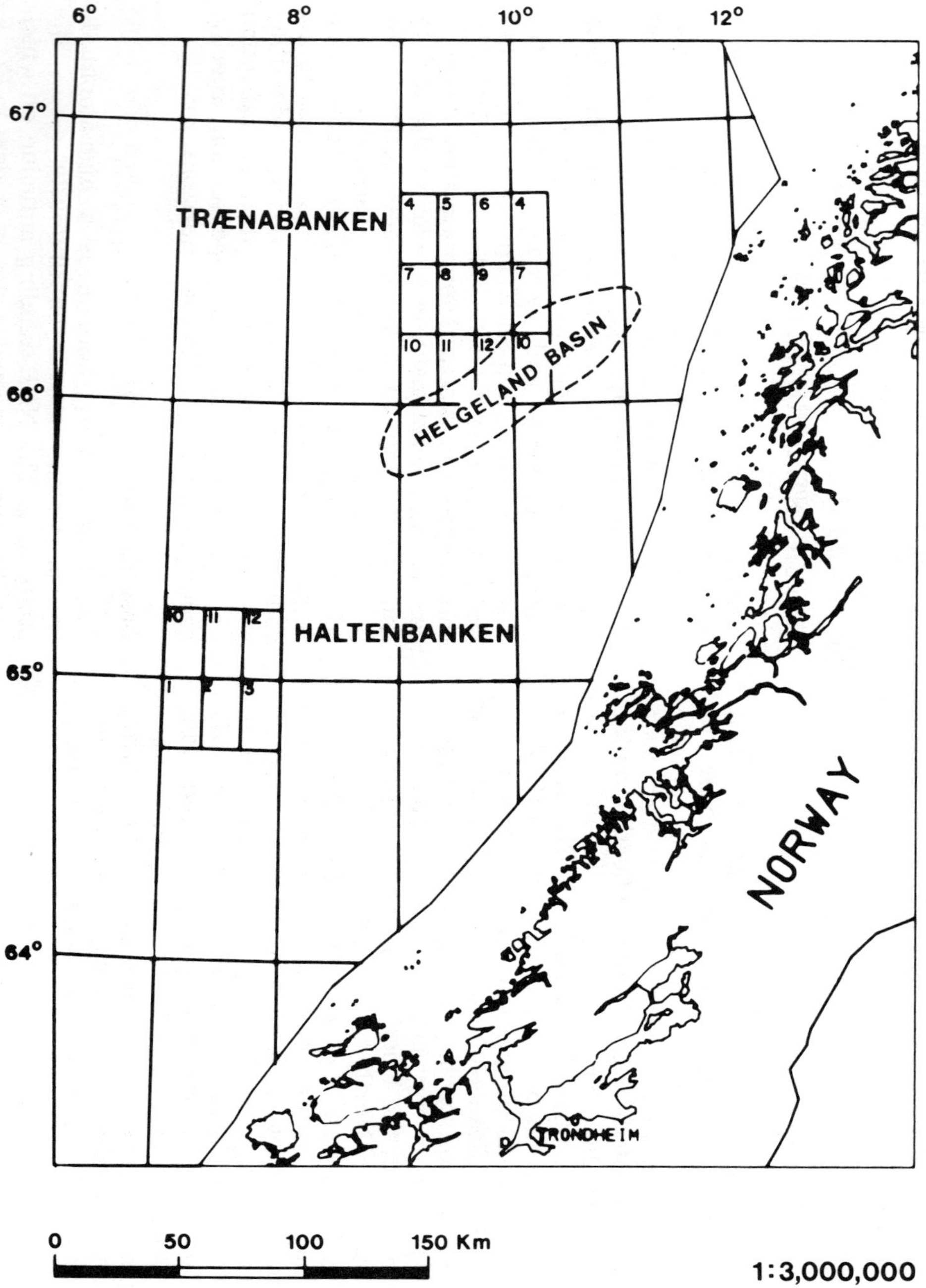

Fig. 1. Regional setting.

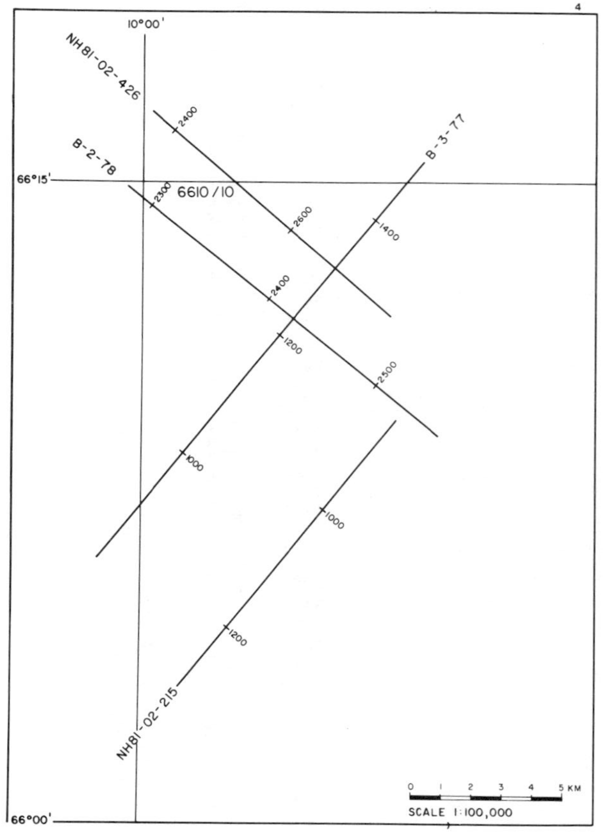

Fig. 2. Trænabanken block 6610/10 index map.

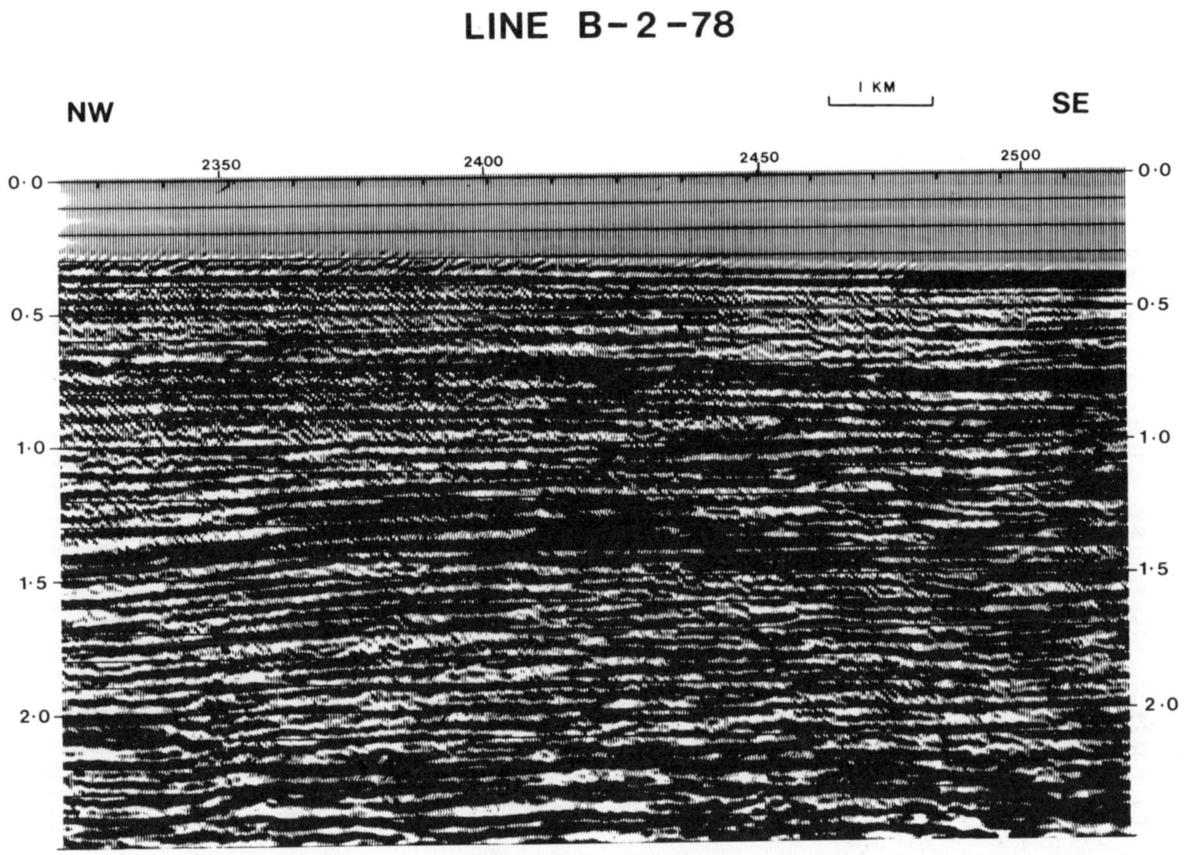

Fig. 3. Original processing of line B-2-78. The 'flat spot' is at approximately 1.35 seconds. Polarity of this and all succeeding displays is S.E.G. standard.

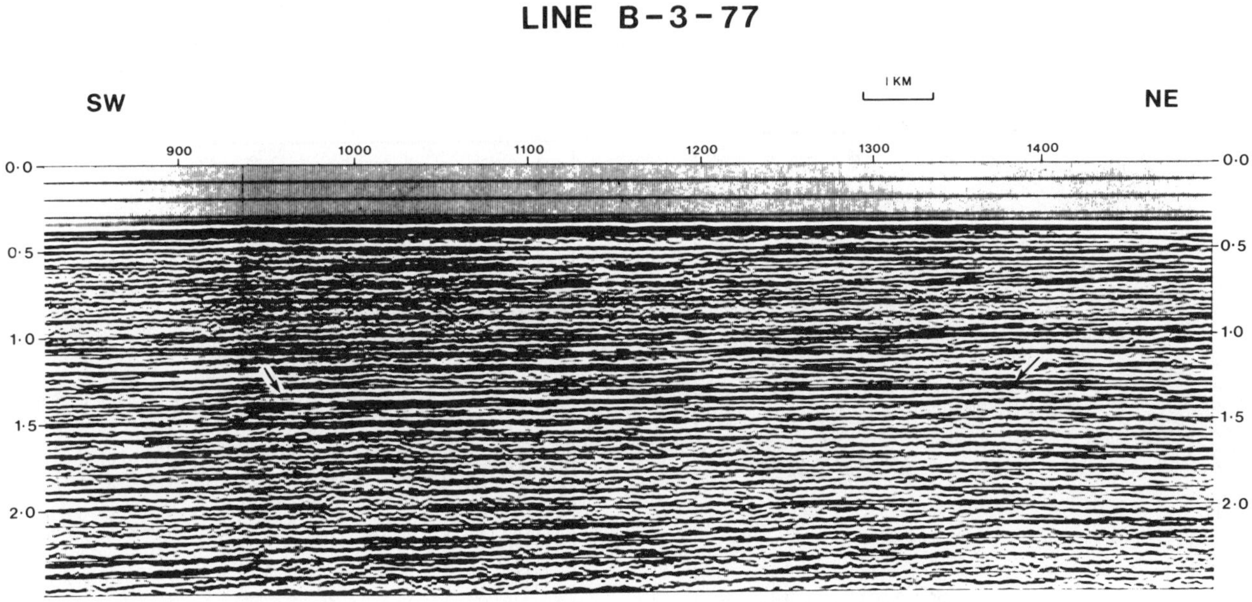

Fig. 4. 1981 Reprocessing of line B-3-77. The 'flat spot' is at approximately 1.35 seconds.

'flat spot' is approximately coincident with the closed area of the anticline (at 1.355—1.360 seconds in Fig. 6) and covers an area of approximately 40—45 km^2. Referring to Fig. 4, note that the anomaly is not absolutely flat along the strike of the anticline but plunges gently to the southwest.

Horizon B (Fig. 7)

Horizon B, which probably represents strata of Palaeocene(?) age, is a relatively low-amplitude, high-frequency trough which often marks the top of a relatively reflection-free interval above the base Tertiary event. As shown in Fig. 8, this interval occasionally contains prograding clinoform reflections. The dominant NNE structural trend at the base Tertiary level has been replaced at this stratigraphically higher level by a prominent NW trend. The major elements of this trend are two elongate positive features, the larger of which (to the north) is coincident with the crest of the previously described anticline at the base Tertiary

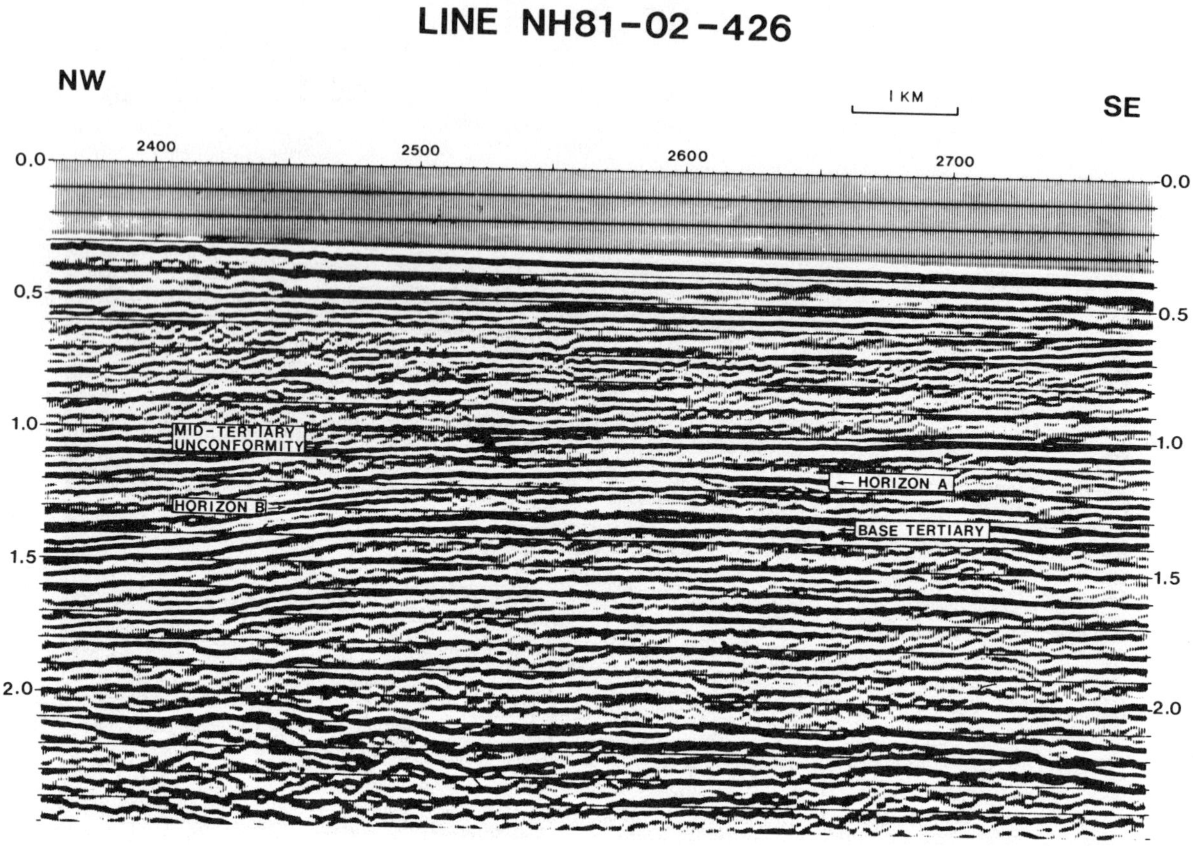

Fig. 5. Original processing of line NH81-02-426 showing interpreted Tertiary seismic horizons.

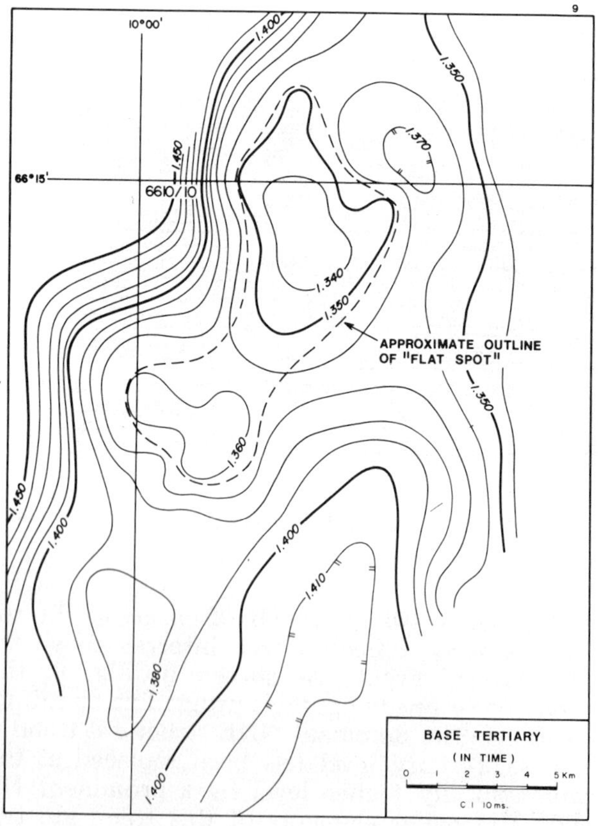

Fig. 6. Two-way time contours on base Tertiary.

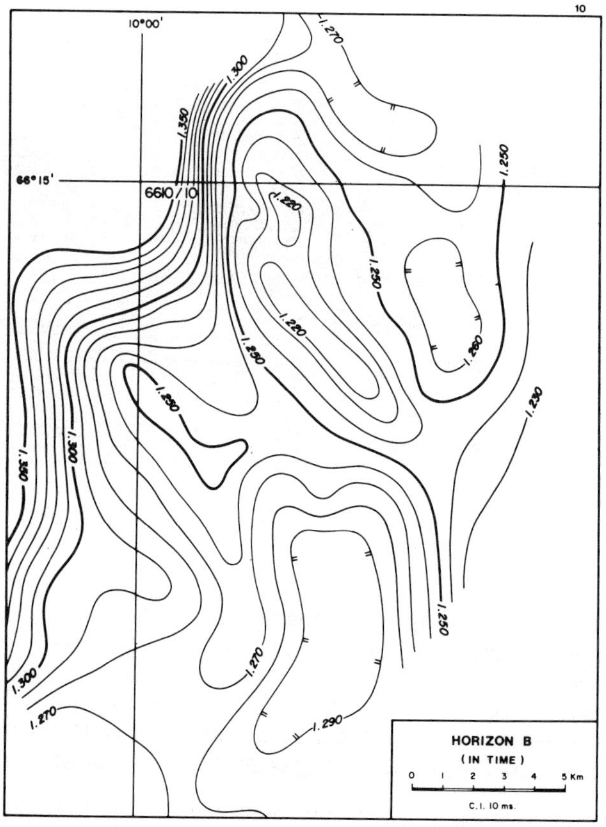

Fig. 7. Two-way time contours on horizon B.

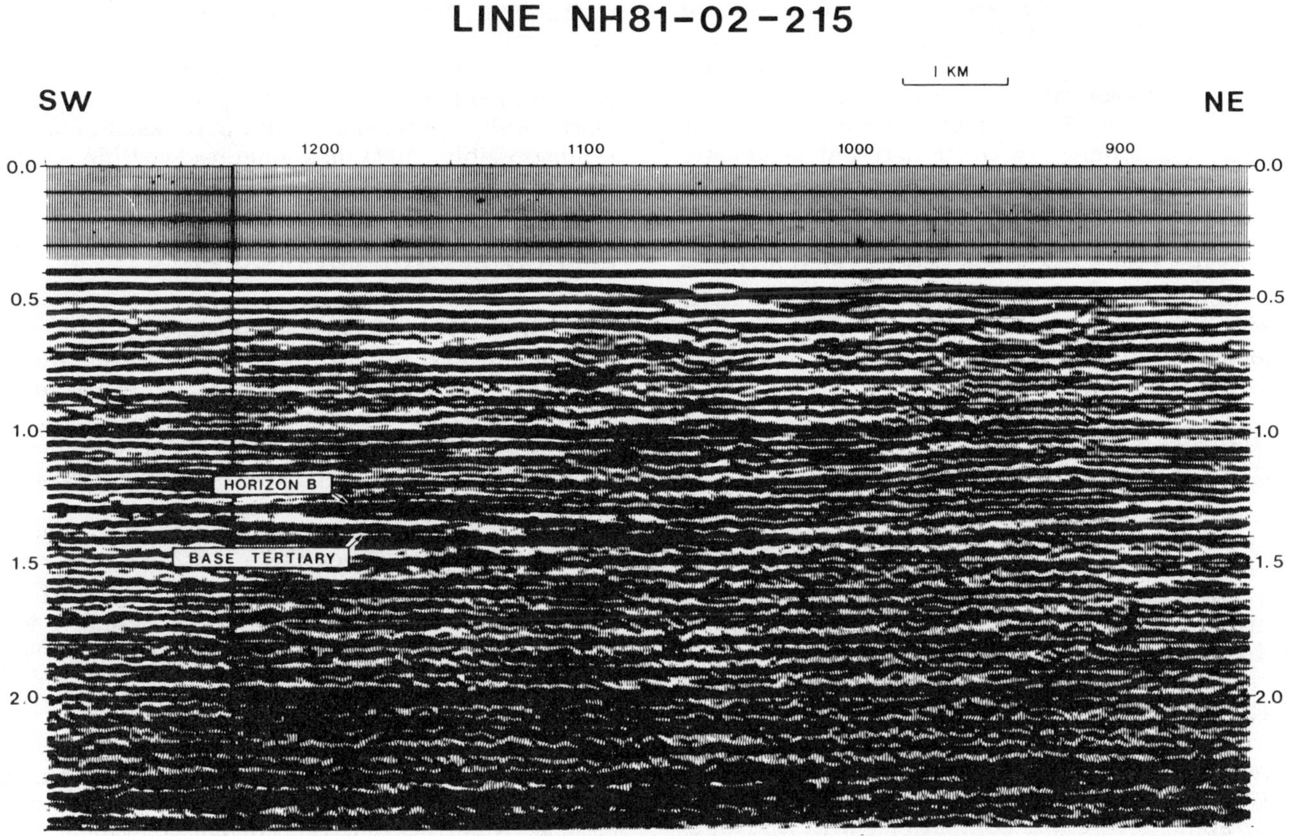

Fig. 8. Original processing of line NH81-02-215 showing clinoform reflections in the base Tertiary—horizon B interval.

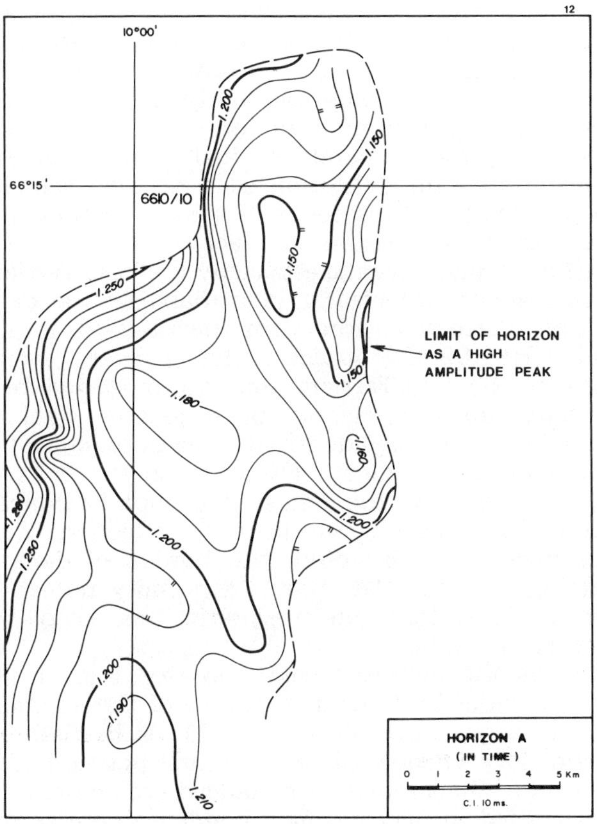

Fig. 9. Two-way time contours on horizon A.

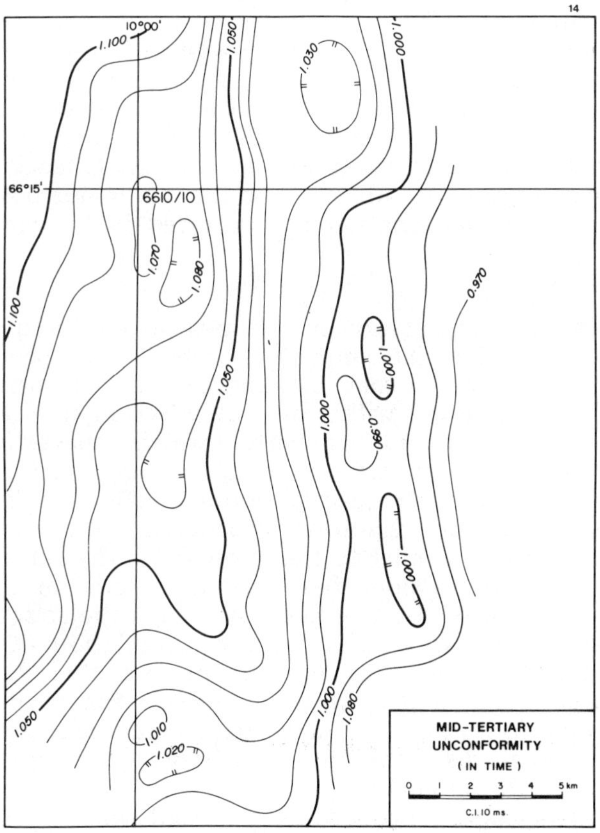

Fig. 10. Two-way time contours on mid-Tertiary unconformity.

level. Despite their NW trend, these features are bounded by the same shallow syncline and regional dip as is the underlying anticline. The pronounced angularity between the strike of these features and the regional trend of the Helgeland Basin, and their apparent discordance with underlying structure, strongly suggest that the features were formed primarily by depositional rather than structural processes. By analogy to the Frigg, Balder and Cod fields of the North Sea, these features may be lobes of a submarine fan; however, the lack of a tight seismic grid and the absence of well control in the Helgeland Basin preclude formulation of a detailed depositional model for the area.

Horizon A (Fig. 9)

Horizon A, which possibly represents strata of Eocene(?) age, changes character across the study area and was mapped only where it appears as a high-amplitude peak (see Fig. 5). As in Fig. 7, the most prominent structures in Fig. 9 are elongate features which appear to radiate from the south-east to the north and west, forming a lobate pattern across the study area. These features are usually located on the flanks of similar underlying features (see Fig. 4) and were also probably formed by depositional processes. Note that the shifts in locations of these features at successively younger stratigraphic levels are compatible with the submarine fan model described by Skjold (1980) for the Balder field.

Mid-Tertiary unconformity (Fig. 10)

The mid-Tertiary unconformity, which developed during a sea-level lowstand in mid-Oligocene time, truncates structural highs in the Helgeland area, including the anticline on which the flat spot is located (see Fig. 5). Figure 10 shows that the unconformity dips very gently westward, reflecting post-Oligocene tilting and subsidence; underlying structures in Palaeogene strata are not evident at this level, indicating that these structures most likely formed in latest Eocene—earliest Oligocene time. Figure 10 indicates also that no major post-Oligocene deformation has occurred in the area.

Discussion

Seismic interpretation in the area of the flat spot indicates that the anomaly occurs in Palaeocene sediments which may be submarine fan deposits. Elongate features within the Lower Tertiary section in the area form a lobate pattern which is suggestive of a submarine fan complex. In cross-section, these features frequently contain reflection-free sequences and clinoform prograding reflections, and are very similar to the 'mounds' in the Palaeocene of the Balder field and the 'pods' in the Palaeocene of the Cod field. Although the seismic data alone are insufficient for formulation of a detailed depositional model, the available data suggest that a deep-water submarine fan model, similar to that for the Balder field, could be reasonably applied for the Lower Tertiary section in the area of the anomaly.

The coincidence of the flat spot with a well-defined structural closure, its flatness and discordance with overlying reflections, and its geometric similarity to the flat spot at the gas—liquid contact in the Frigg field all suggest that the anomaly may be a reflection from a gas—liquid contact. If this is so, then it follows that Palaeocene and possibly Eocene sediments in the Helgeland Basin are, at least locally, potential reservoir rocks, and that gas (and possibly oil?) has migrated within or into the basin.

SPECIAL SEISMIC PROCESSING

The objectives of special seismic processing in detailed study of the 'flat spot' were twofold:

(1) To determine conclusively whether or not the anomaly is a primary reflection.

(2) To determine as precisely as possible the physical nature of the interface which gives rise to the anomaly.

These objectives were approached through processing of seismic lines B-3-77, B-2-78 and NH81-02-426 (see Fig. 2 for locations). Processing concentrated on velocity analysis, inversion, and study of seismic attributes such as amplitude, instantaneous phase and instantaneous frequency.

Anomaly as a primary reflection

A very important step in special processing of the flat spot was velocity analysis, in which it was critical to ensure that the correct stacking velocity for the anomaly was determined. Initially, it was hoped that the anomaly could be identified as a primary event on the basis of its stacking velocity; however, inspection of CDP gathers indicated that seismic data from the area, even the most recently acquired data, are pervaded by strong sea-floor and interbed multiples and generally have low signal/noise, particularly in the far traces. Preliminary velocity analysis indicated that in these data primary reflections cannot always be distinguished from multiples, especially interbed multiples, even after having improved signal/noise by techniques such as common offset summation or extended receiver array simulation. Therefore, final velocity analysis concentrated on definition of a regional velocity trend for the Cretaceous—Tertiary section and accurate determination of stacking velocities for the few primary events, especially the base Tertiary event, which are clearly identifiable in the data.

NMO-corrected gathers for lines B-2-78, B-3-77 and NH81-02-426 indicate that primary events such as the mid-Tertiary unconformity and base Tertiary horizon clearly have greater stacking velocities than easily identifiable sea-floor multiples. These displays indicate also that on the basis of stacking velocity, the flat spot cannot be conclusively distinguished as a primary reflection. With this uncertainty in mind, the burden of demonstrating that the flat spot is a primary reflection necessarily shifts to an interpretive analysis of the reprocessed data.

As on the unreprocessed data (e.g. Fig. 3), the most obvious and striking feature of the anomaly on the reprocessed data (Fig. 11) is its flatness, particularly in view of its structural position. The flatness of the anomaly is enhanced even more on a reprocessed display having a condensed horizontal scale (Fig. 12). There is positive evidence on these displays that the flat spot is primary, because there are no horizontal events above it of which it could

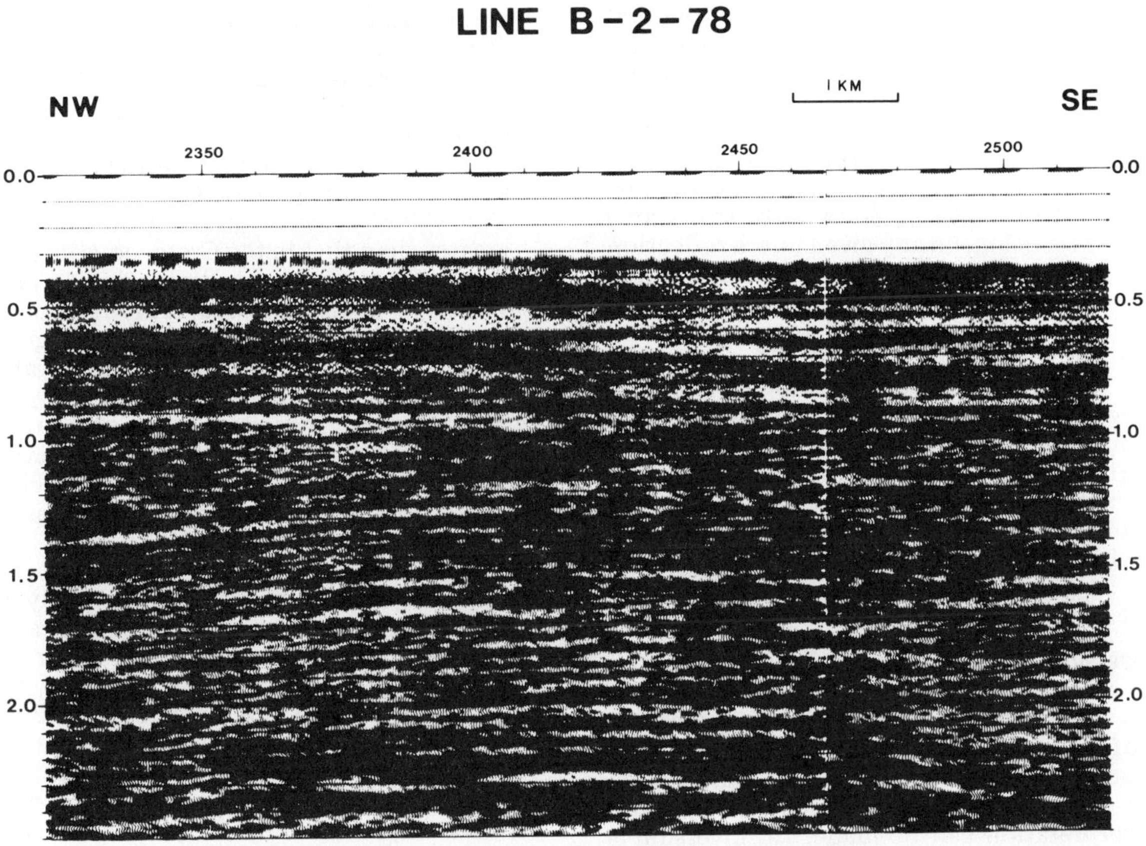

Fig. 11. Gulf relative amplitude reprocessing of line B-2-78. The 'flat spot' is at approximately 1.35 seconds.

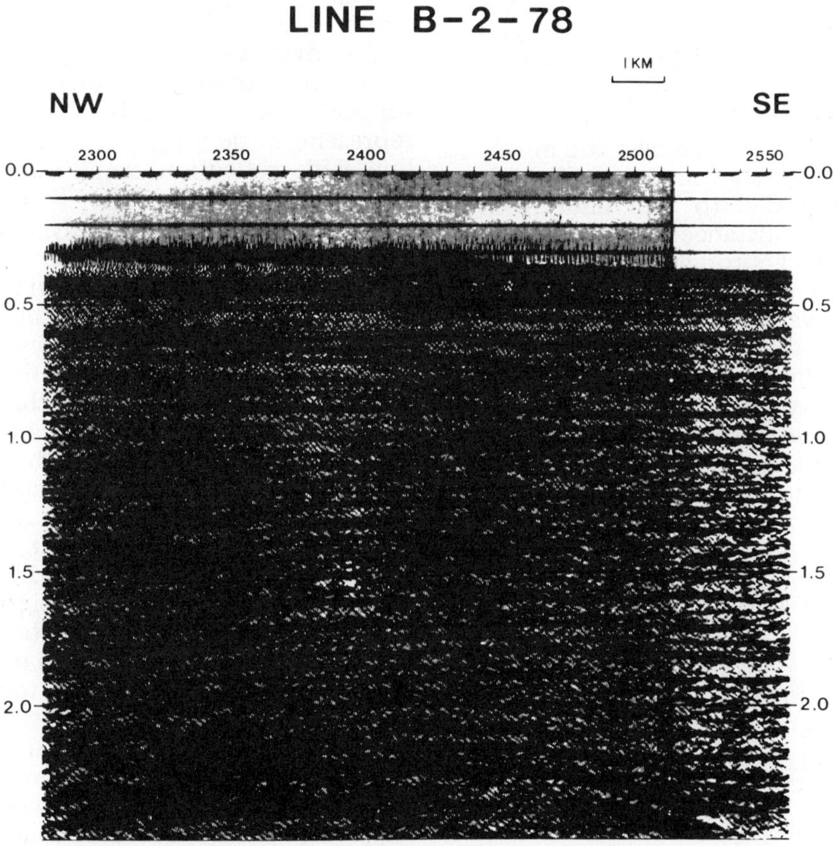

Fig. 12. Gulf relative amplitude reprocessing of line B-2-78 displayed with condensed horizontal scale.

be a multiple. Referring to Figs. 3 and 11, note that the sea floor dips SE above the anomaly and that reflections from overlying Palaeogene strata dip NW. These dipping interfaces indicate that the flat spot cannot be a simple sea-floor, peg-leg or interbed multiple. In addition to its flatness, the areal consistency of correlation of the anomaly and the low structural relief of Tertiary strata in the study area argue against its being, on any given line, a reflection from out of the plane of that section. The possibility that the anomaly represents a geologic interface which was, at its time of formation, non-horizontal and was subsequently tilted to its present horizontal position requires invocation of a mechanism for which there is no supporting evidence and which would have operated in a very fortuitous manner. The consistency of correlation of the flat spot, its structural position, and its geometric relation to overlying reflections all argue convincingly that it is a primary reflection.

Physical nature of the anomaly

The processing techniques used to study the physical nature of the interface which the flat spot represents were seismic inversion and calculation/display of seismic attributes. Seismic inversion involves generation of pseudo-velocity logs from reflection seismic data, from which direct lithological interpretation can be made. An inversion is produced by adding high-frequency velocity information extracted from a stacked seismic trace to low-frequency velocity information derived during stacking velocity analysis. The composite pseudo-velocity trace thus contains gross interval velocity data, indicative of major lithologic packages, on which detailed velocity data, indicative of finer lithologic units, have been superimposed. The inversion process is very sensitive to data quality and the fidelity of processing which precedes it, and so great care must be used in the interpretation of lithology from inversions.

Inversions for part of line B-2-78 over the northwestern edge of the flat spot are shown in Fig. 13. These inversions were scaled to well data from Haltenbanken, and the distance between that control and Trænabanken strongly suggests that caution be exercised in interpretation of details and absolute velocity values on these displays. Three intervals within the Cretaceous—Tertiary section can be distinguished in Fig. 13 on the basis of their velocity character (these characterizations generally hold for the entire area):

(1) The Cretaceous interval, in which velocity gradually increases with depth. There is no major velocity discontinuity at the Cretaceous—Tertiary boundary.

(2) The mid-Tertiary unconformity to base Tertiary interval, in which velocity is relatively constant.

(3) The Recent to mid-Tertiary unconformity interval, in which velocity increases with depth. The velocity/depth gradient in this interval is greater than that in the Cretaceous interval. There is usually no velocity discontinuity at the mid-Tertiary unconformity, although occasionally the underlying section exhibits slightly lower velocity than that immediately above the unconformity.

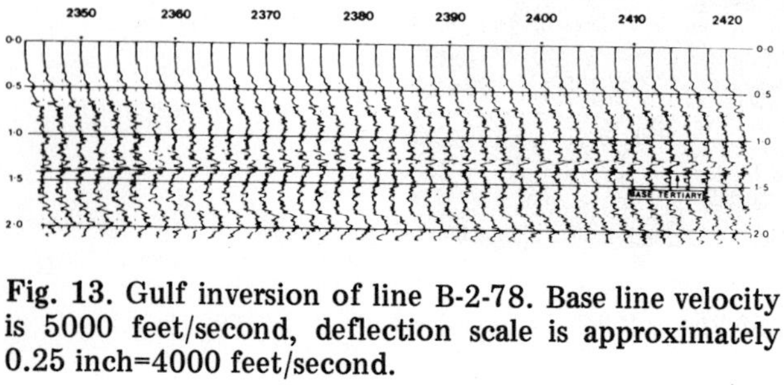

Fig. 13. Gulf inversion of line B-2-78. Base line velocity is 5000 feet/second, deflection scale is approximately 0.25 inch=4000 feet/second.

The anomaly is evident on Fig. 13 to the right of shot point 2380, at which point the base Tertiary horizon flattens. Note that the velocity of the intervals just above and below the base Tertiary horizon decreases from left to right across the anomaly. This effect is particularly noticeable on the two high-velocity peaks immediately above the base Tertiary horizon; the upper peak remains relatively constant, while the lower decreases markedly over the anomaly. Figure 13 clearly indicates that there is a Palaeocene interval within the study area which exhibits a decrease in velocity coincident with the occurrence of the flat spot, and this same observation has been made also on inversions for lines B-3-77 and NH81-02-426. In the context of the previous discussion on the flatness and structural position of the anomaly, an explanation for the observed decrease in velocity is that it may be a manifestation of gas saturation in a sand reservoir, and so the flat spot may be a reflection from the gas—liquid contact in the reservoir. The occurrence of such a feature as a direct indicator of hydrocarbons is by no means new to North Sea explorationists, as evidenced by the well-known seismic flat spot in the Frigg field.

Displays of seismic attributes of line B-2-78 support the hypothesis that the flat spot might represent a gas—liquid contact. A version of this line with reflection amplitudes squared is shown in Fig. 14 (see Fig. 11); this display emphasizes the flatness of the anomaly and indicates that it is of higher amplitude than surrounding reflections, particularly between shot points 2410 and 2430. The higher amplitude might be indicative of increased reflectivity caused by gas saturation, but accurate evaluation of the significance of amplitude changes in the vicinity of the anomaly requires more detailed stratigraphic information for the study area (i.e. well data) than is now available. Displays of instantaneous phase and instantaneous frequency for line B-2-78 can also be used to study details of the flat spot, but, as is the case for the amplitude data in Fig. 14, these require coordination with good stratigraphic control for accurate quantitative interpretation. The instantaneous phase plot (see Plate 10) outlines the flat spot very well, in particular its terminations at shot points 2380 and 2470, and the changes in phase across the anomaly possibly represent differences in reflectivity of the gas — and water — saturated portions of the inferred reservoir. Instantaneous frequency plots show some loss of high frequencies below the anomaly, as is often observed in association with gas accumulations.

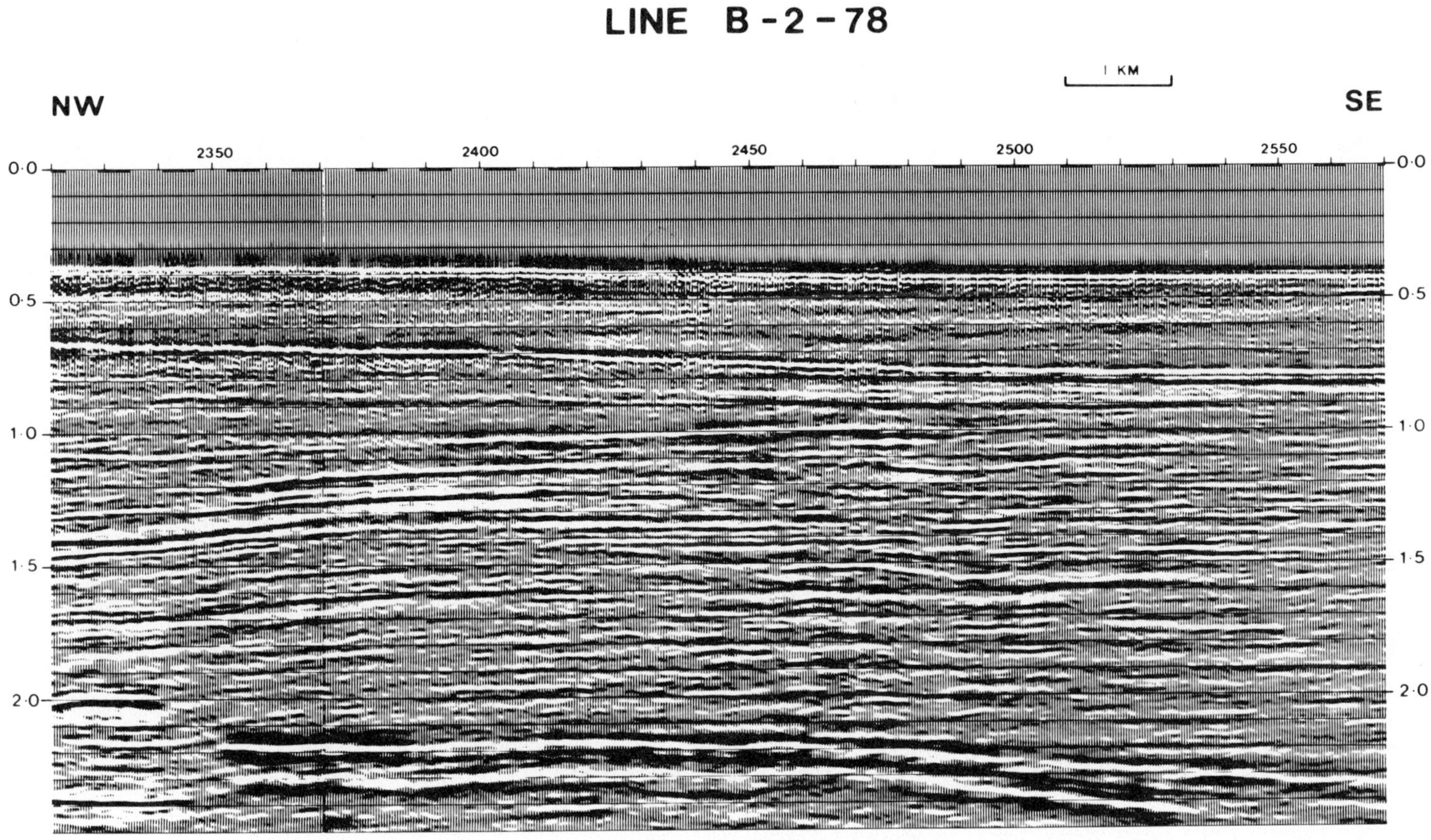

Fig. 14. Gulf reprocessing of line B-2-78 with reflection amplitudes squared.

Discussion

Special seismic processing, especially inversion, has provided the most important information for determining the physical nature of the interface which the flat spot represents. Inversions clearly show that the anomaly coincides with a structurally closed area in which Palaeocene (and possibly Eocene) strata exhibit a noticeable decrease in interval velocity, one possible cause of which is gas trapped in the section. Even in view of the limitations and uncertainties involved in analysing inversions, this interpretation is sound, and, therefore, is central to understanding the nature of the flat spot. It is a well-known geophysical phenomenon and a proven concept in North Sea exploration that gas—liquid contacts in clastic reservoirs can give rise to seismic flat spots (note, however, that these features are not quantitative indicators of gas saturation). All available data indicate that the anomaly in Trænabanken is a flat spot in this sense, and as such is a direct indicator of hydrocarbons.

CONCLUSIONS

The results of special seismic processing and detailed interpretation in the vicinity of a seismic flat spot in block 6610/10 indicate that this anomaly is a reflection from the gas—liquid contact in a Lower Tertiary reservoir. The flat spot is therefore a direct hydrocarbon indicator, with the implications that the Lower Tertiary section in the Helgeland Basin has reservoir potential and that hydrocarbons have migrated within or into the basin.

ACKNOWLEDGEMENTS

The writers express their thanks to Gulf Oil Exploration and Production Company for permission to publish this study, and especially to P. Cramer, R. Sanislo and B. Hastings for their contributions to the data processing portion of the work.

REFERENCES

Heritier, F.E., Lossel, P. and Wathne, E. 1979. Frigg field — large submarine-fan trap in Lower Eocene rocks of North Sea Viking Graben. *A.A.P.G. Bull.*, **63**(11), 1999—2020.

Kessler, L.G., Zang, R.D., Englehorn, J.A. and Eger, J.D. 1980. Stratigraphy and sedimentology of a Paleocene submarine fan complex, Cod field, Norwegian North Sea. In: *The Sedimentation of North Sea Reservoir Rocks,* Article VIII. Norsk Petroleumsforening, Oslo.

Skjold, L. 1980. Paleocene sands of the Balder field. In: *The Sedimentation of North Sea Reservoir Rocks,* Article VII, Norsk Petroleumsforening, Oslo.

21 The Permian base of East Greenland

F. Surlyk, S. Piasecki, F. Rolle, L. Stemmerik, E. Thomsen and P. Wrang
Grønlands Geologiske Undersøgelse, København, Denmark

A major transgression took place in East Greenland in late Permian times. The resulting marine basin can be traced over 400 km, from Jameson Land in the south to Wollaston Forland in the north. More than 300 m of limestone, evaporite, shale and sandstone were deposited in mainly shallow marine and hypersaline lagoonal environments. The lateral facies changes are rapid and very complex and seem to some extent to have been controlled by older tectonic structures formed during an important mid-Permian phase of block faulting. A widespread black shale facies is a good source rock for oil and the Permian and Lower Triassic sequence must be characterized as attractive from the point of view of exploration.

INTRODUCTION

The onshore part of the continental margin of East Greenland exposes an extremely thick, relatively complete late Precambrian–Cenozoic sequence. The extensive literature focuses mainly on the Caledonian mountain belt (Haller, 1970, 1971; Henriksen and Higgins, 1976), the intramontane Devonian 'Old Red' type basins (summary in Friend *et al.*, 1983), the Mesozoic rift basins (Birkelund and Perch-Nielsen, 1976; Surlyk *et al.*, 1981), and the Tertiary plateau basalts (Brooks and Nielsen, 1982a, b; Brooks *et al.*, 1976; Noe-Nygaard, 1976). The late Palaeozoic systems have in marked contrast received only cursory treatment. The Carboniferous–Lower Permian continental sediments are virtually unknown, while the Upper Permian of northern East Greenland was studied by Maync (1942). Maync's demonstration of the complex Upper Permian facies pattern was remarkably 'modern', even if it was based on reconnaissance-type fieldwork.

In the last few years we have studied the Upper Permian carbonates, evaporites and terrigenous clastic rocks (e.g. Stemmerik, 1979, 1980, 1982; Stemmerik and Sørensen, 1980; Surlyk, 1983). Also the underlying continental Lower Permian rocks have received considerable attention.

The background for this work is two-fold. First, the Upper Permian sequence contains the main potential onshore petroleum source rock and is thus essential for any evaluation of the onshore hydrocarbon potential. Second, the late Palaeozoic basins are of considerable scientific interest as they were formed during the first extensional rifting stages heralding the formation of an incipient North Atlantic Ocean.

During our work in 1982, normal geological fieldwork was supplemented by a helicopter-portable shallow core drill to obtain fresh cores for source rock and sedimentological analyses. The majority of the holes were drilled down to about 20-m depth, but one hole was drilled to approximately 60 m (Surlyk, 1983). The fieldwork was finished in the summer of 1983 and the laboratory work and final reports are planned to be completed in 1984–1985. A large number of samples have, however, already been subject to analyses of organic geochemistry, palynofacies and stratigraphy, and vitrinite reflectance.

The present chapter is therefore to be considered a preliminary progress report only. The main emphasis is placed on Permian tectonics and sedimentary facies, and on the occurrence and distribution of potential source rocks, reservoirs, and seals.

EARLY PERMIAN CONTINENTAL SEDIMENTATION

The Carboniferous period in East Greenland was characterized by continental sedimentation in rifted

Petroleum Geology of the North European Margin © Norwegian Petroleum Society (Graham & Trotman, 1984) pp. 303–315.

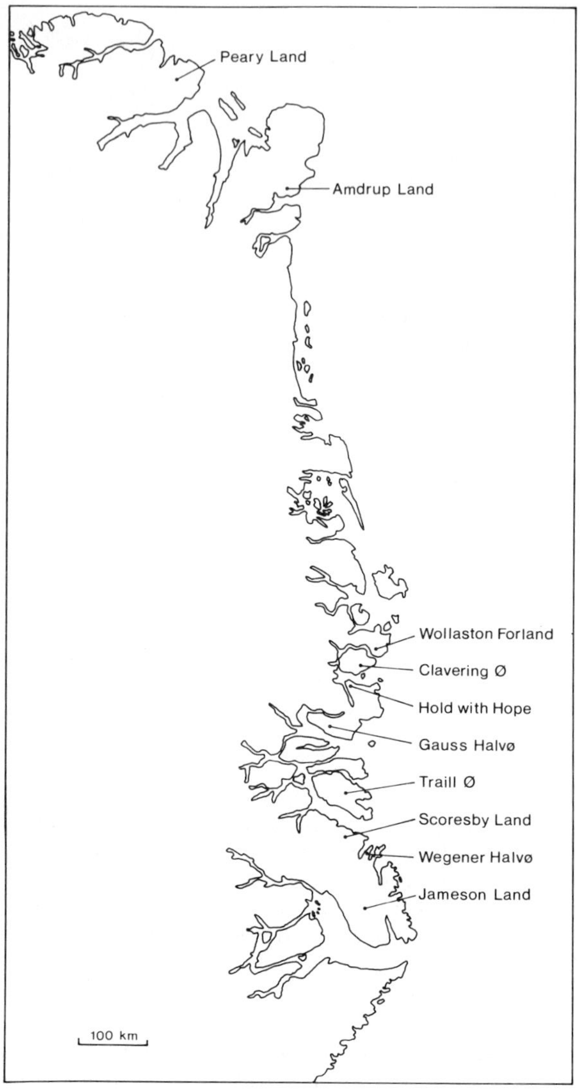

Fig. 1. Outcrops of Upper Permian strata.

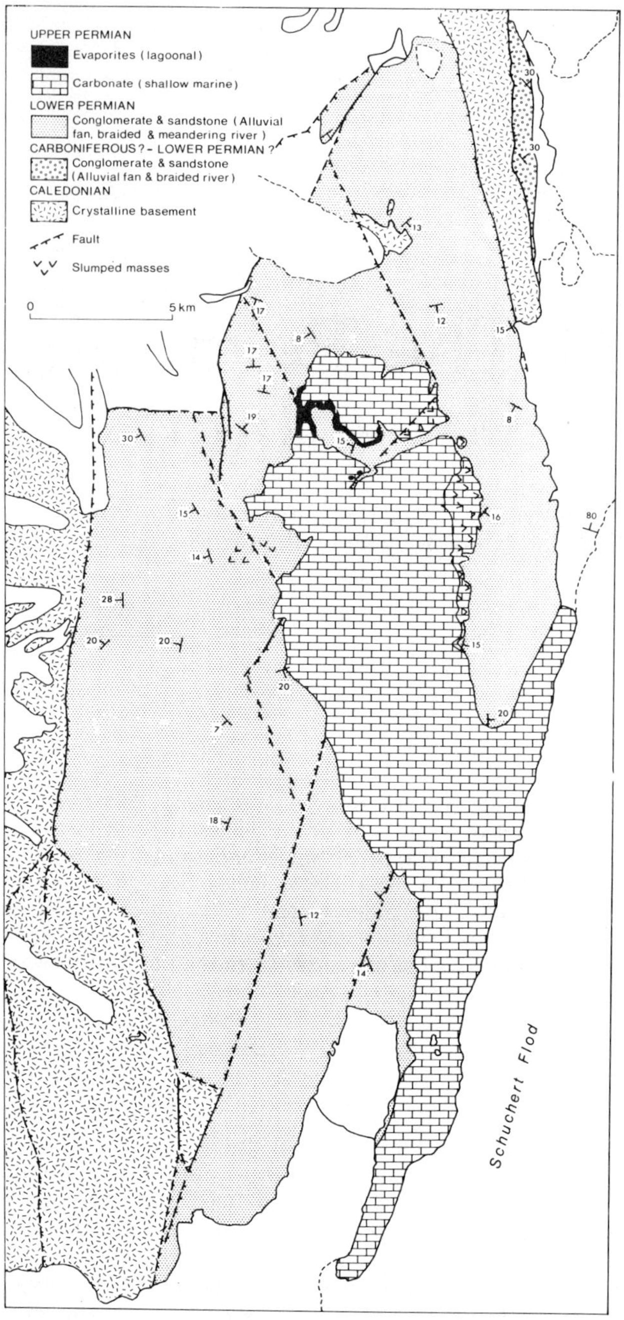

Fig. 2. Geologic map of Karstryggen area, western margin of the Jameson Land Basin. The main pre-Upper Permian fault block is tilted towards the west and is limited to the west by the main N-trending fault along the left margin of the map, and to the east by the N-trending fault seen in the upper right-hand corner. The NNW and NNE faults which cut the Lower Permian sequence are only of minor, local importance.

basins. Sediments of this age are known only from the northern part of the East Greenland Basin from Traill Ø to Clavering Ø (Fig. 1). Lower Permian deposits, on the other hand, are known only from the southern part of the basin in Jameson Land and Scoresby Land (Fig. 1) (Kempter, 1961; Collinson, 1972). They represent a seemingly unbroken continuation of the Carboniferous tectonic—sedimentologic regime.

The early Permian basin was limited to the west by a set of major N-trending fault-zones. The main fault zone limits the Caledonian highlands to the west from the subsiding Jameson Land Basin to the east (Figs. 2 and 3). The sediments are dominated by conglomerate, mainly red pebbly sandstone, and sandstone of alluvial fan origin along the western margin. Transport directions are towards the east away from the faults. A gradual transition to well-sorted grey fluviatile cyclic sandstone—shale facies takes place in the Schuchert Dal area. These facies types are uniformly developed through great thicknesses of strata, and the sandstone unit can be traced laterally over long distances (Fig. 3).

The majority of the sandstone units were deposited in extensive meander-belts. They are characterized by basal elongate current parallel scours, and complex longitudinal cross-bedding formed by migration of point-bars (Figs. 4 and 5). The top of each sandstone unit was deposited in levée environments and shows a great variety of sedimentary structures indicative of very shallow water and rapid fluctuations in current velocity (Figs. 5A and 6).

Most of the sandstone units are well sorted, fine-grained, and do not show any fining upwards tendency. They are abruptly overlain by black, laminated shale of flood-plain origin with lenticular bodies of crevasse sandstone. The shale units commonly contain well preserved fossil fish.

Fig. 3. (A) Thick sequences of Lower Permian fluviatile sandstone—shale couplets. Mountain in the foreground is about 800-m high. In the background the Caledonian Northern Scoresby Land Mountains.

(B) Close-up of cyclic Lower Permian sandstone—shale sequences of the same type as shown in Fig. 3A. Lower sandstone unit is shown in Figs. 4 and 5. North of Trias Elv, western Jameson Land.

Fig. 5. (A) Sand-filled elongate scour at the base of meander-belt sandstone body. Center of picture shows laminated shale of flood-plain origin. Hammer head on top of underlying meander-belt sandstone body. Palaeocurrent towards north (out of the picture).

(B) Stream bed exposes top of sandstone body deformed by liquefaction, overlain by flood-plain shales with crevasse splay sandstones (at hammer), followed meanderbelt sandstone body (shown on Fig. 4 between 67 m and 77 m).

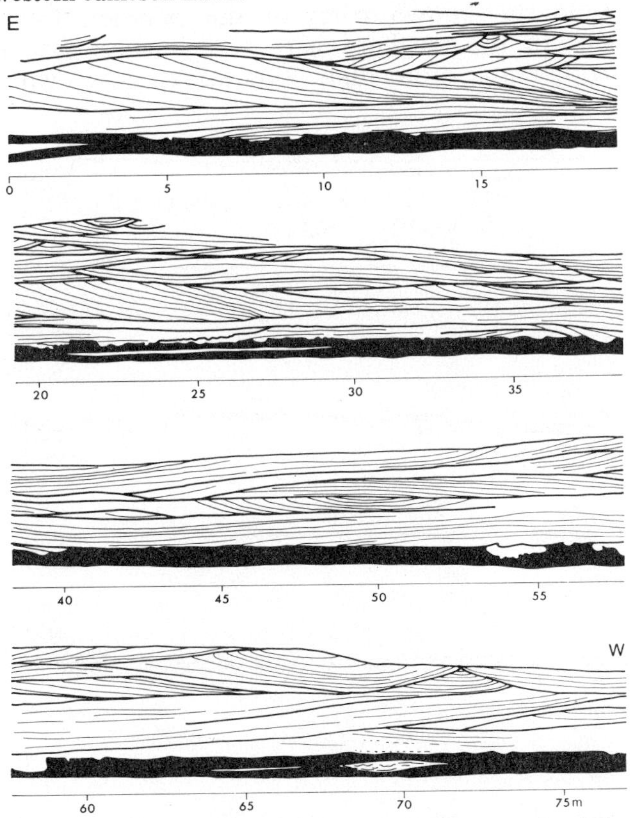

Fig. 4. Section of meander-belt sandstone unit (see Fig. 3A, B) showing longitudinal cross-bedding of point-bar origin, with sand-filled scours at the base, overlying flood-plain shales with thin crevasse splay sandstones. Palaeocurrent towards the observer.

Intercalated with the meander-belt sandstones occur extremely poorly sorted, red, arkosic upwards-fining sandstone—shale units. They show large-scale trough cross-bedding and characteristically have a basal pebble or cobble lag. These units reflect deposition under higher energy in less water logged areas and are genetically more closely associated with the alluvial fan sequences along the basin margin.

All palaeocurrent data from this big river system which flowed through the central part of the Jameson Land Basin are from south to north. This northwards dipping palaeoslope is remarkable, since all later palaeoslopes are dipping towards the south (Surlyk, 1978; Surlyk *et al.*, 1981).

The total thickness of the Lower Permian continental sequence is not precisely known due to limited stratigraphic control, and later tectonic disturbances. It is assumed to be of the order of 2—3 km. Below the Lower Permian probably occurs an equally thick sequence of Carboniferous continental sediments, but their existence in the Jameson Land Basin has not yet been proved.

An outlier of continental red beds, the Røde Ø conglomerate, occurs SW of Jameson Land (Collinson, 1972). It is limited to the west by a major fault zone and was deposited on a system of alluvial fans which towards the east pass into a

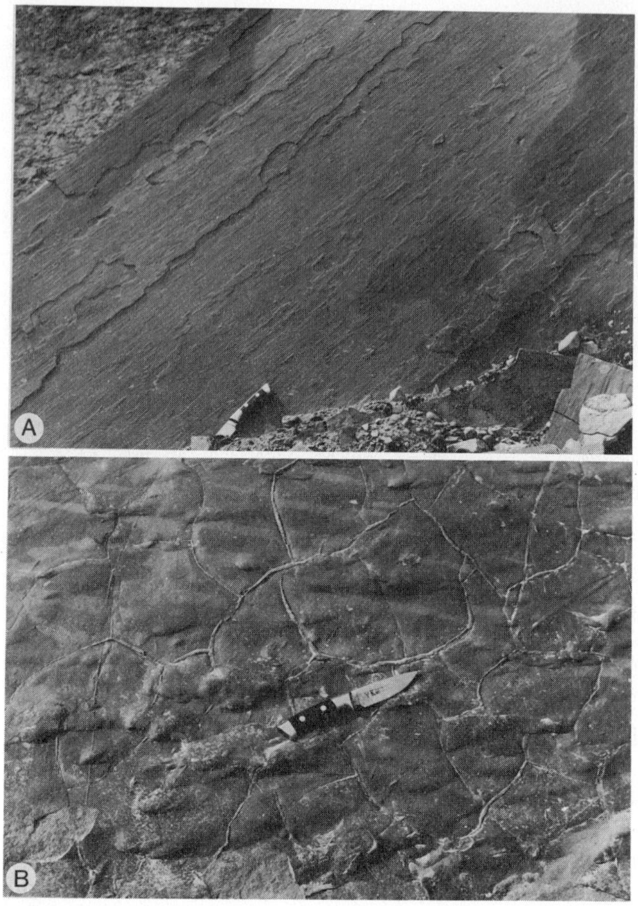

Fig. 6. (A) Parting lineation in horizontally laminated fine-grained sandstone from Lower Permian meander-belt sequence. North of Trias Elv, western Jameson Land.

(B) Sole of Lower Permian fluviatile sandstone bed showing flute casts with superimposed mud cracks. North of Trias Elv, western Jameson Land.

flood-plain. The facies types, colour, and general setting is highly reminiscent of the occurrences from the western part of Jameson Land described above and a general early Permian age is assumed by analogy.

MID PERMIAN BLOCK FAULTING

Deposition of the thick Lower Permian alluvial fan and fluviatile sequences was followed by a very important episode of block faulting (Figs. 7, 10) (Vischer, 1943; Kempter, 1961). This phase has been somewhat overlooked in the literature, but it must be classed as the most important post-Devonian tectonic event in East Greenland together with the phase of block faulting at the Jurassic—Cretaceous boundary, and the onset of final early Tertiary plate separation.

The mid Permian faults in the Karstryggen area and western Jameson Land show trends towards N, NNE and NNW (Fig. 2). The downthrow along the main faults is towards the east, and the strata are tilted about 20° towards the west.

The mid Permian block faulting can be dated only with reasonable precision in southern Scoresby Land and western Jameson Land, where faulted and tilted Lower Permian sediments are peneplained and overlain with marked angular unconformity by Upper Permian sediments (Figs. 8—11) (Kempter, 1961). Elsewhere in East Greenland Upper Permian sediments rest on Carboniferous or older rocks. The erosional peneplain surface is remarkably even, as can be seen by the structural contour map of the base Upper Permian (Fig. 9). The quite enormous amount of eroded material was deposited mainly outside the known parts of the Jameson Land Basin. In the northern part of Schuchert Dal a conglomerate, more than 50-m thick, occurs with

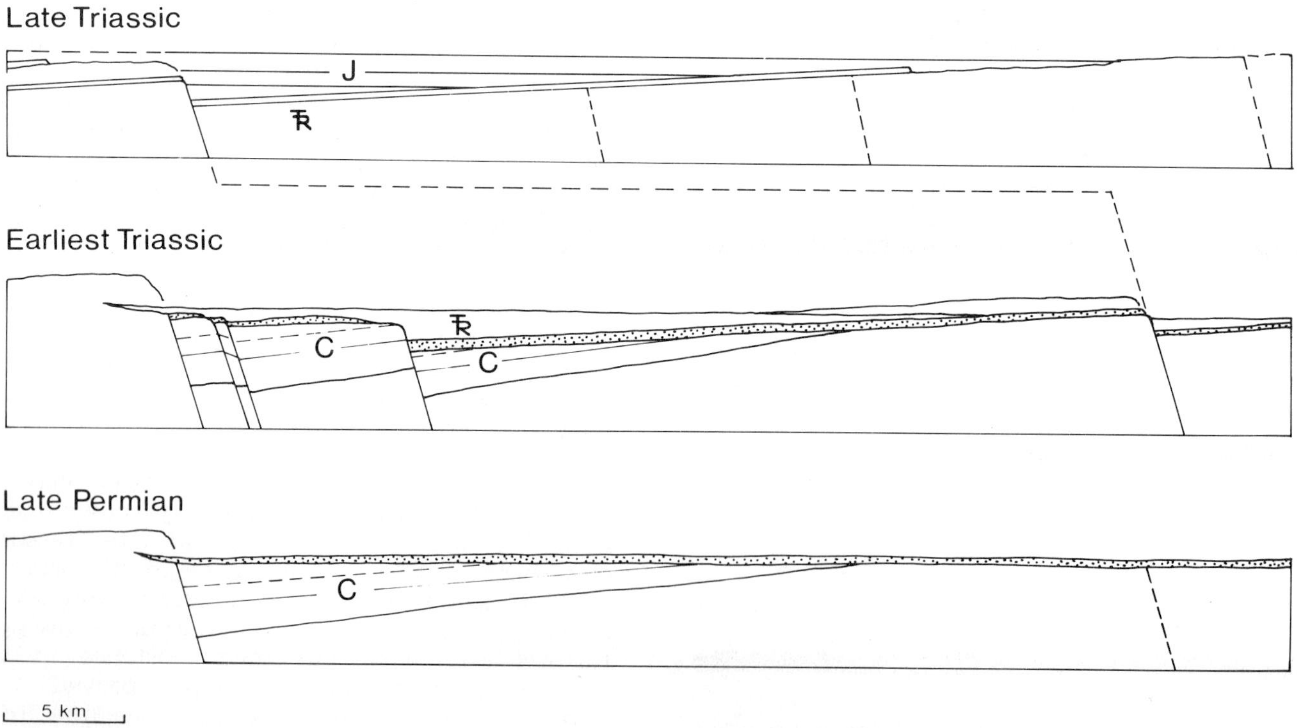

Fig. 7. Palinspastic E—W cross sections of the sedimentary basin of the Hold with Hope—Clavering Ø region in northern East Greenland. The stippled areas indicate the transgressive Upper Permian sequence. C, Carboniferous; T, Triassic; J, Jurassic. Note the existence of pre-late Permian, and earliest Triassic tectonic phases. (Modified from Vischer, 1943.)

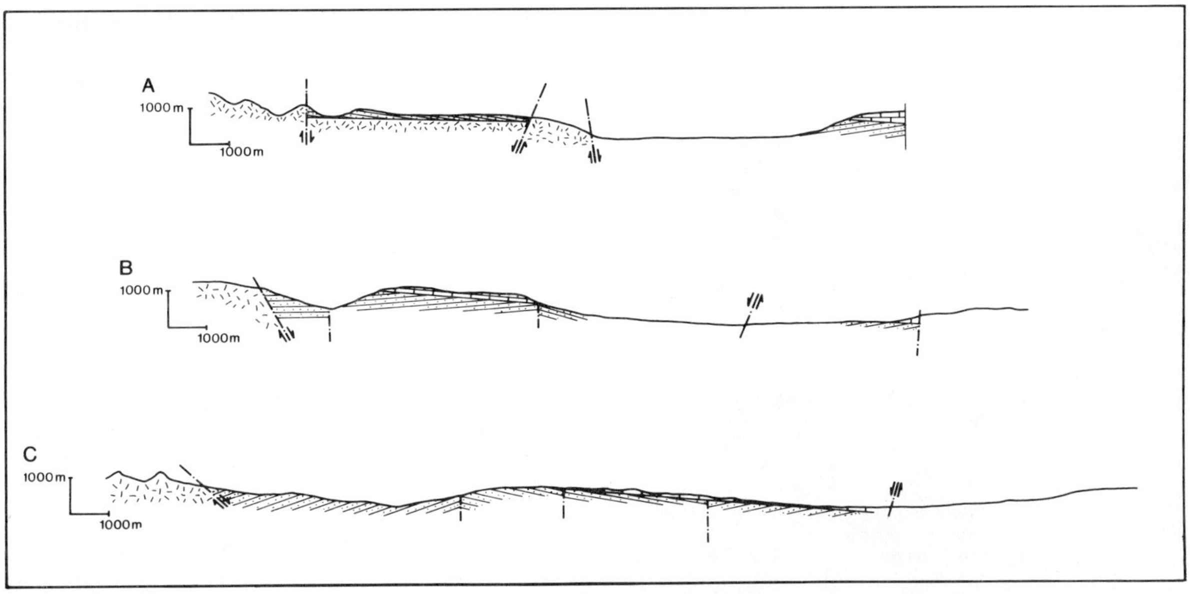

Fig. 8. Cross-section from west (left) to east (right) of Karstryggen (see Fig. 2).

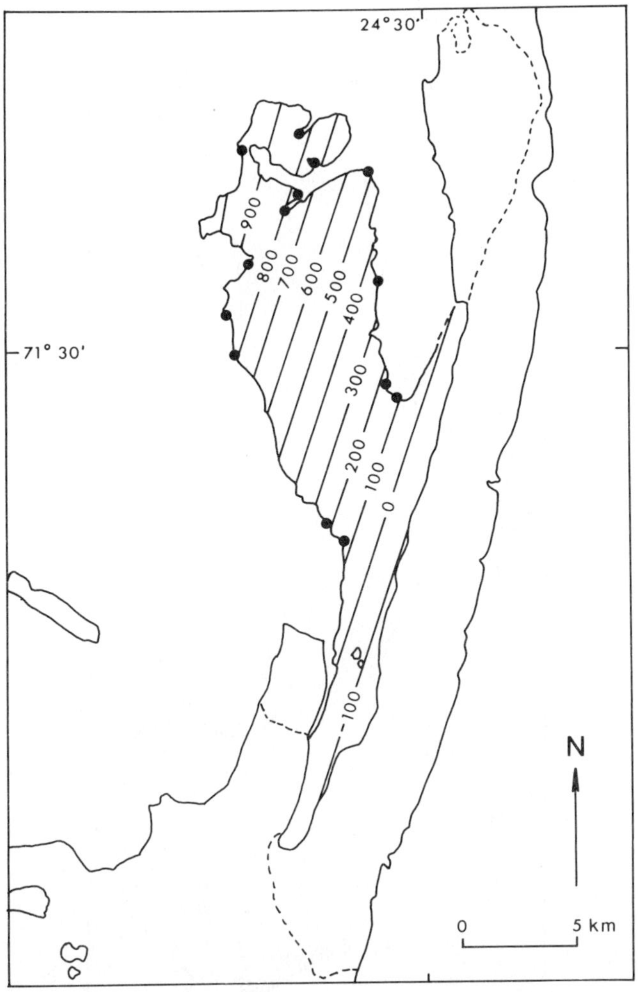

Fig. 9. Structure contour map of the flat, peneplained base of the Upper Permian. The plain was tilted towards ESE during a later tectonic phase. (Modified from Kempter, 1961.)

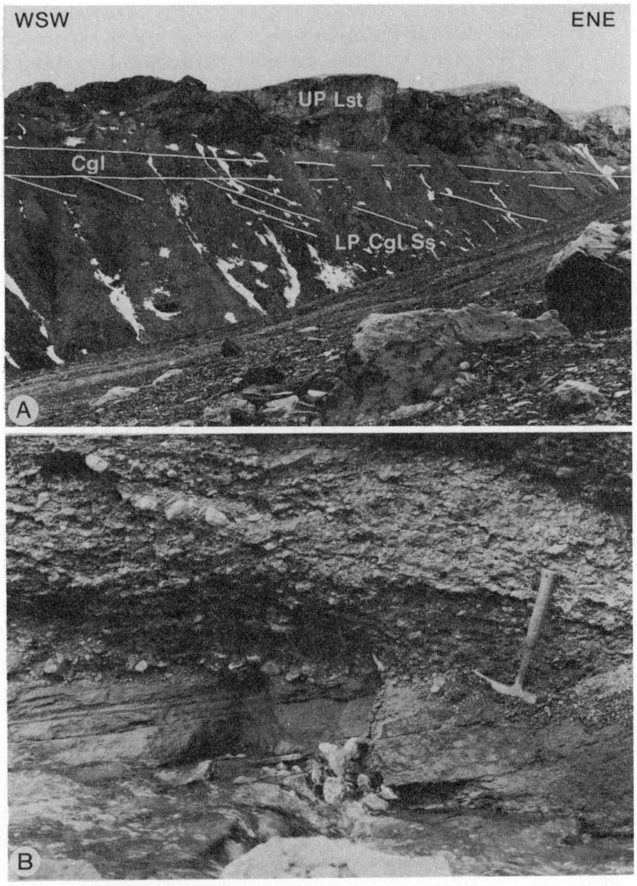

Fig. 10. (A) Strongly tilted Lower Permian red-beds (LP Cgl Ss) overlain by uniform conglomerate, followed by Upper Permian limestone (Lst), Revdal.

(B) Detail of angular, base Upper Permian unconformity, Huledal, Karstryggen.

Fig. 11. (A) Tilted Lower Permian continental red-beds overlain by Upper Permian gypsum and carbonate (upper right corner). Note the extremely flat and even nature of the pre-Upper Permian peneplain. Arkose Dal, west of Karstryggen.

(B) Tilted Lower Permian red-beds (LP Cgl Ss) overlain by Upper Permian gypsum erosively truncated by major conglomerate. Location back landward side of central ridge in A.

metre-sized blocks of Lower Permian sandstone together with abundant clasts from the Caledonian basement. The conglomerate is unconformably overlain by the Upper Permian basal conglomerate, suggesting that it represents erosional material from the formation of the mid Permian peneplain.

LATE PERMIAN FAN DELTA AND BRAID PLAIN SEDIMENTATION

The peneplained surface of the faulted and tilted Lower Permian red beds are overlain with a sharp contact by an extremely widespread red conglomerate of relatively uniform thickness varying between 20 and 50 m (Fig. 10). Greater variations in thickness are met where the conglomerate fills important erosional valleys, and in the Hold with Hope area where the thickness decreases from 160 m to 8 m over 3 km distance. The dominant facies are irregular and indistinctly bedded conglomerates and pebbly sandstones. The clasts are mainly Caledonian granites and gneisses of pebble to cobble size (Fig. 21B). Along the basin margin clast size may reach 2—3 m. Palaeocurrents are extremely varied and deposition seems to have taken place by sheet floods in an extensive inland braid plain. The highest few metres of the conglomerate are well bedded and contain scattered marine fossils. This suggests that the depositional environment during the initial stages of the Late Permian transgression changed from a braid plain to a system of coarse clastic short-headed fan deltas.

THE LATE PERMIAN TRANSGRESSION

Sedimentation of terrigenous clastic sediments was strongly reduced due to rapid transgression across the low relief area, and a variety of carbonates and evaporites were deposited on the reworked top of the conglomerates. A palaeogeographic map shows the extent of the late Permian sea (Fig. 12).

The tectonic activity still continued, for a mild unconformity can be observed between the conglomerate and the overlying sediments at several localities. The conglomerate is furthermore intruded by a number of lamprophyric dykes, which are truncated at their top, and erosional products from the dykes are found in the immediately overlying carbonates and evaporites (Stemmerik and Sørensen, 1980). Intrusive activity continued during the late Permian, but the tectonic activity seems otherwise to fade out in concert with the ongoing late Permian transgression (Fig. 12). Renewed tectonism occurred at the end of the Upper Permian and basin sedimentation once more became dominated by terrigenous clastics.

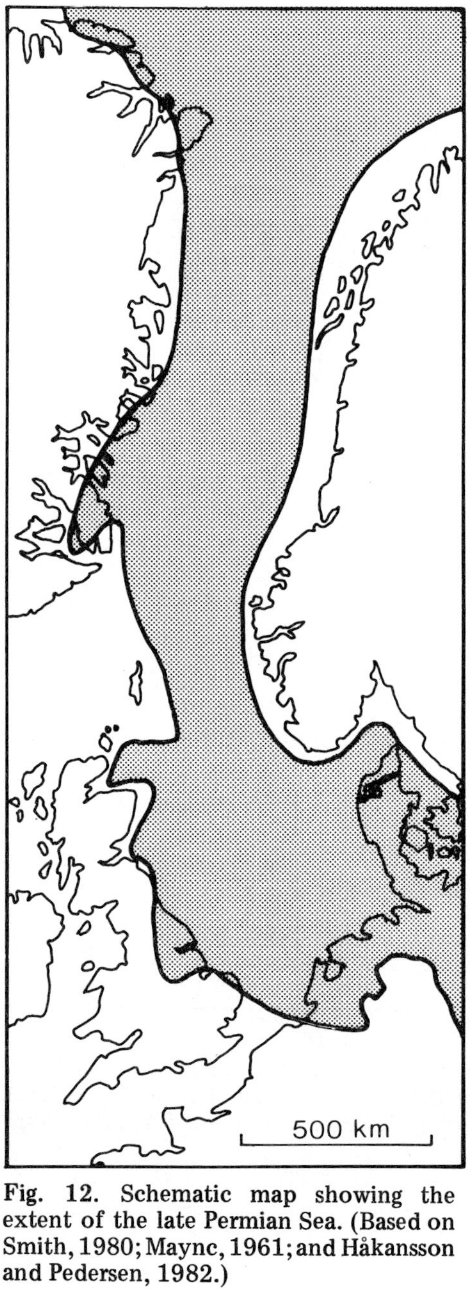

Fig. 12. Schematic map showing the extent of the late Permian Sea. (Based on Smith, 1980; Maync, 1961; and Håkansson and Pedersen, 1982.)

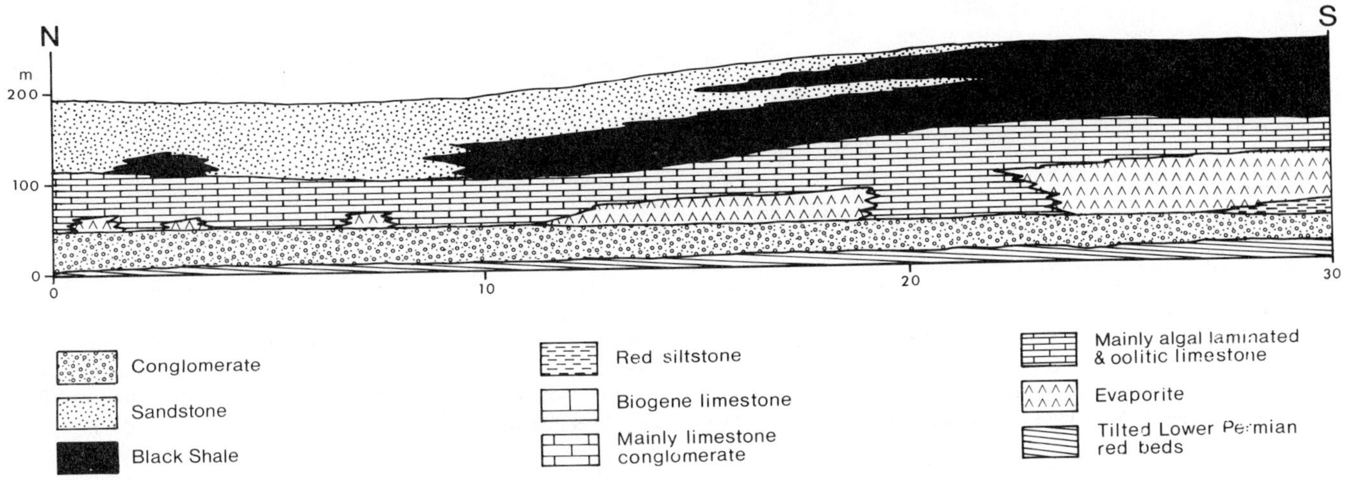

Fig. 13. Section through the Upper Permian sequence parallel to the basin margin East of Schuchert Flod (Fig. 2). To the south the lower part is dominated by sabkha to lagoonal evaporites, which to the north pass into algal laminated limestone. Carbonate deposition was eventually interrupted by terrigenous shale and sandstone reflecting progradation from the north and continued transgression. Horizontal scale in kilometres. (Modified from Stemmerik, 1982.)

The initial transgressive stages were characterized by sedimentation of intertidal to shallow subtidal carbonates and subordinate evaporites. A number of characteristic subenvironments or facies belts soon became established. They seem to be strongly controlled by the tectonic structures, but also to some extent by the proximity to the coast-lines. Main areas of carbonate deposition probably occurred over structural highs such as horsts and crests of tilted blocks, while the adjacent lows received evaporites followed by carbonates and terrigenous mudstones.

It is essential to understand the structural control of facies in order to make valid facies extrapolations to the subsurface.

Shallow-water hypersaline carbonate and evaporites

The low, peneplained relief of the land areas, and the hot, arid climate resulted in a strongly reduced input of terrigenous clastics to the basin during the initial phases of the late Permian transgression (Maync, 1942, 1961; Stemmerik, 1982). Sedimentation took place under intertidal to shallow subtidal conditions, and the main facies includes algal boundstones, lagoonal micrites, and intraclast floatstones and rudstones deposited in channels. The very sparse fauna, which includes brachiopods and bryozoans, is almost restricted to the northern part of the basin (Maync, 1942). The southern and western areas experienced increased salinity as seen by deposition of evaporites and by intensive very early diagenetic growth of celestite.

A phase of increased subsidence associated with terrigenous clastic influx terminated carbonate sedimentation over large areas, probably due to the last movements of the mid Permian tectonic episode. This facies shift is especially characteristic in the central parts of the basin, where black shale overlies the basal carbonates (Fig. 13).

The carbonate facies received detailed study by P.A. Scholle and J.M. Hurst in the 1983 field season and it is thus considered premature to go into any detailed discussion here. Carbonate deposition seems to have continued uninterrupted over structural highs. The facies development varied considerably, however, between the different areas due to differences in salinity and structural configuration.

The general environment of carbonate deposition can be characterized as shallow marine, meta- to hypersaline, platforms. Occasional high turbulence can be inferred from the presence of intraclast floatstone and rudstone formed by reworking of early cemented, stromatolitic limestones. Relatively high salinity is suggested by the presence of spar-filled cavities after gypsum crystals, and partial dolomitization of the micrites. This is supported by the virtual absence of fossils, except for rare and isolated occurrences of monospecific gastropod faunas associated with larger cryptalgal laminated structures.

The carbonate platform area towards the SW of the Karstryggen area (Figs. 2, 15) was situated close to the coast-line. Here the initial extensive intertidal flats were characterized by deposition of cryptalgal laminates, with occasional evaporites

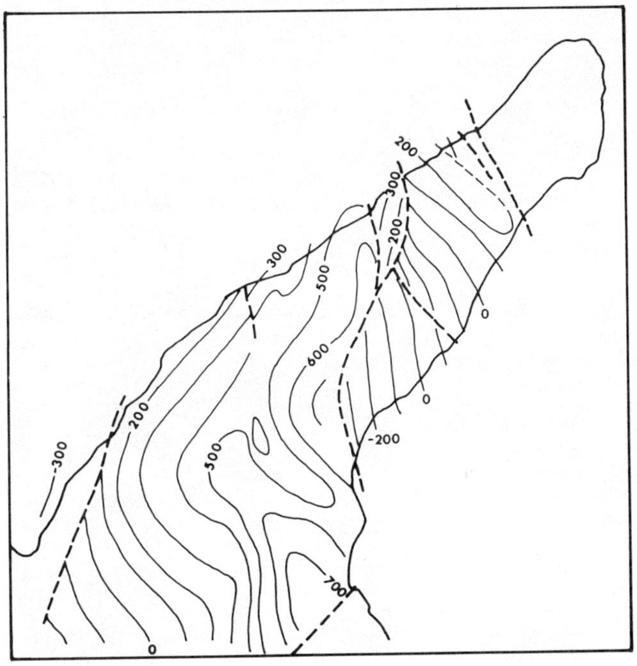

Fig. 14. Structure contour map of the base of Permian of Wegener Halvø; contours 100-m equidistant. (Modified from Grasmück and Trümpy, 1969.)

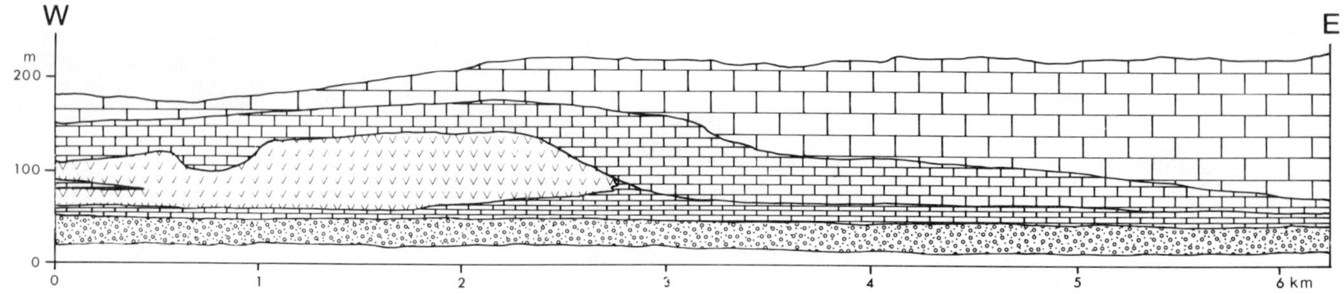

Fig. 15. Section through the Upper Permian deposits in the SW part of the basin at the Karstryggen area. The western end of the section starts at the main evaporite outcrop shown in Fig. 2. The very uniform basal terrigenous conglomerate is overlain by lagoonal evaporites to the west. To the east a major carbonate build-up forms a barrier to the lagoon. The whole sequence is truncated to the west by a major conglomerate-filled channel. (Modified from Stemmerik, 1982.)

and important early diagenetic formation of celestite. This uniform facies was gradually replaced by SSW-trending facies belts which seem to be controlled by the architecture of the underlying fault-blocks.

The western part of the platform area in Karstryggen, corresponding to the down-tilted part of the main underlying fault block, was isolated from the open sea by the formation of a barrier complex. The barrier was composed of stromatolitic limestones and associated intraclast conglomerates. Hypersaline conditions rapidly developed in the lagoon and large amounts of gypsum/anhydrite were deposited (Figs. 11A and 15). The evaporitic conditions persisted for a considerable length of time, interrupted only by short, less saline events marked by deposition of floatstone beds derived

Fig. 16. Finely laminated anhydrite with recrystallized gypsum nodules. From the lower, evaporitic part of the Upper Permian sequence. Trias Elv, SW-Jameson Land.

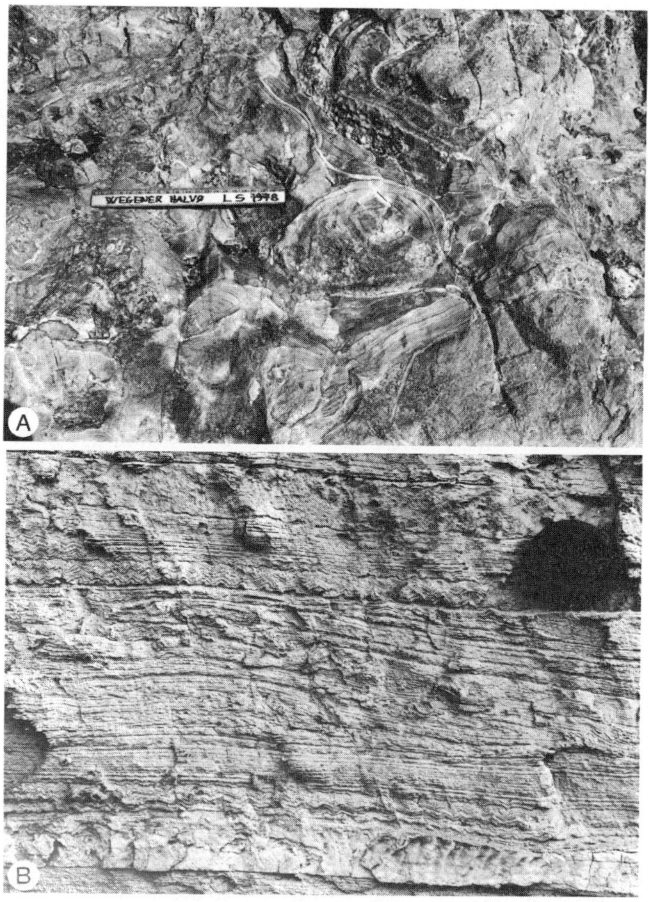

Fig. 17. (A) Stromatolitic limestone from the lower part of the Upper Permian sequence in Jameson Dal, Southern Wegener Halvø.

(B) Laminated anhydrite with very early diagenetic deformation structures and internal unconformities. From the lower, evaporitic part of the Upper Permian sequence, Trias Elv, SW-Jameson Land.

from the barrier overlain by thin stromatolitic limestone.

A similar environmental setting with a carbonate barrier enclosing an evaporitic lagoon was developed to the north on Gauss Halvø (Maync, 1942). While the initial basin-wide phase of carbonate deposition resulted in deposits of only 5—10-m thick, the barrier—lagoon complex corresponds to thicknesses of more than 100 m.

Sabkha-oolite bar transition

The facies on the eastern side of Schuchert Dal are remarkably different from those on the western

Fig. 18. Upper Permian limestone mounds.

(A) Bedded, biomicrites interbedded with thin shales. Fauna is dominated by brachiopods and bryozoans. Southern part of Wegener Halvø.

(B) Transition from non-bedded biogenic limestone (lower left corner) to bedded limestones mainly composed of crinoid fragments in a micritic matrix. Central part of Revdal.

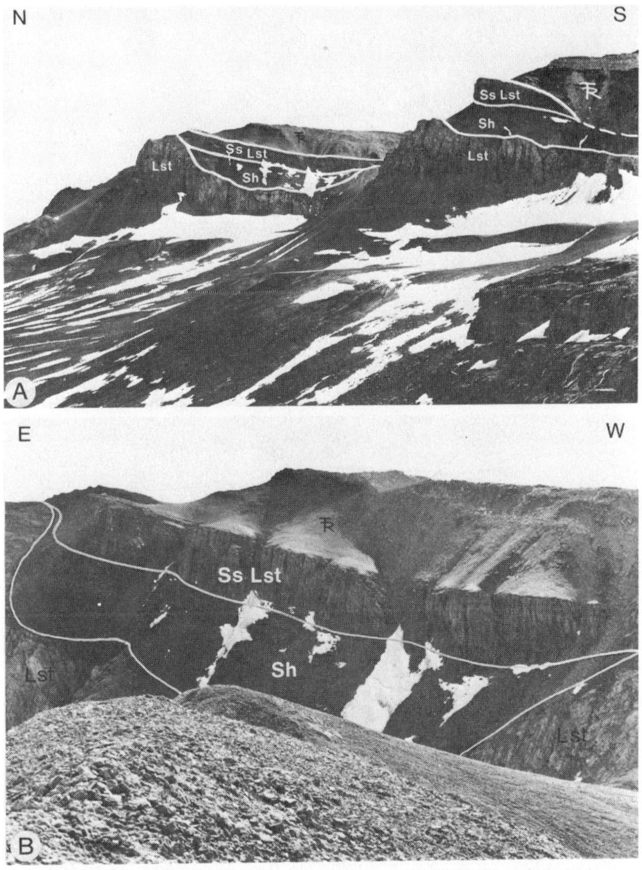

Fig. 19. (A) Panoramic view showing the complex Upper Permian facies transitions between carbonate mounds (Lst), black shale (Sh) and sandy fossiliferous limestones (SsLst). The sequence is overlain unconformably by Triassic sandy shales of the Wordie Creek Fm. In the lower right-hand corner red conglomerates and sandstones of presumed Early Permian age.

(B) Detail of the left part of A showing how the black shales were deposited in low areas between the carbonate mounds. Calamites Dal, Northern Wegener Halvø.

side in the Karstryggen area described above (compare Figs. 13 and 15). Deposition comprised shallow subtidal to supratidal carbonates and evaporites (Figs. 16 and 17). Sequences of finely laminated gypsum and carbonate, 15—20-m thick, were deposited in lagoons. The extensive occurrence of diagenetic gypsum is characteristic of supratidal sabkha-type environments. Sabkha cycles in the present case seem to include only cryptalgal laminates with or without gypsum nodules, and nodular chicken-wire gypsum with sparse remains of dolomitic material.

Erosive channels interpreted as having a tidal origin are filled with intraclast floatstones. The clasts include algal limestone, gypsum and rarer oncolites.

In this area the progressive transgression also resulted in the development of less saline and more normal marine conditions. This is reflected by the development of a normal marine oolite bar limiting the lagoon and tidal flat complexes. Gypsum became almost restricted to the tidal flats where it formed during early diagenesis. The faunas include brachiopods and trepostome bryozoans. Upwards the carbonates gradually give way to deeper-water black shale.

Organic build-ups and associated sediments

Improved water circulation related to the transgressive deepening of the basin resulted in decreasing salinity until normal marine conditions were established throughout the basin. The platform areas were colonized by rich faunas of brachiopods, bryozoans, crinoids and bivalves, and in the northern part of the basin and on Wegener Halvø were characterized by the development of carbonate build-ups (Figs. 18—21).

The majority of build-ups developed on Wegener Halvø and were localized above small topographic highs. They range in size from several metres to many tens of metres, their size and extent being related to their age durations. There are several types of build-ups characterized by distinct facies. A characteristic type consists of a core of early cemented lime mud with pockets of smooth athyrid brachiopods and trepostome bryozoans flanked by brachiopod shell beds. Small crinoidal grainstone shoals developed on local highs within the complex.

The build-ups both accreted and prograded quickly, often forming a mosaic of interconnected highs and lows. In between individual build-ups debris beds were deposited as distinct fringing aprons.

Marine sandstone and shale

Carbonate deposition eventually came to an end and black shale was deposited almost throughout

Fig. 20. (A) Upper Permian carbonate mound (Lst) overlain by black shale (Sh) followed by sandy fossiliferous limestone (SsLst). The exposed part of the mound is about 60-cm thick. Northern Wegener Halvϕ.

(B) Upper Permian carbonates. At the base, hypersaline, non-fossiliferous limestone (1), is overlain by non-bedded, biogenic limestone mounds (2), followed by bedded fossiliferous limestone (3). Tent for scale (encircled). South slope of Paradignebjerg, central Wegener Halvϕ.

Fig. 21. (A) Erosive conglomerate in the top of the Upper Permian sequence indicated to the left in the section in Fig. 15. Clasts of gypsum (G), limestone (Lst), and red sandstones in a matrix of sandy micrite. Western Revdal.

(B) Conglomerate at the base of the Upper Permian carbonate—evaporite sequence. Clasts are mainly red and gneissic in poorly sorted sandy matrix. Bedding indistinct and poorly developed imbrication. Ruler 1-m long. Revdal.

the basin except for Karstryggen (Fig. 2) and Wollaston Forland (Fig. 1). The onset of shale and sandstone deposition was highly diachronous, starting early in northern Scoresby Land and southern Clavering Ø (Maync, 1942). Prograding sandstone deposition dominates along the basin margin, especially in northern Scoresby Land (Fig. 3). In the Schuchert Dal area the sequence contains storm deposits of graded sandstone and oolitic limestone, indicating contemporary deposition of terrigenous and carbonate facies. The black shale unit reaches thicknesses of up to about 100 m. The shales are silty and show irregular alternation of laminated and bioturbated horizons. They are rich in the presumed pseudoplanktonic bivalve *Posidonia permica*, and diverse fish faunas have been described from a number of localities. The content of acritarchs suggests normal marine salinities over much of the basin. Towards the SW they decrease strongly in density, perhaps as a response to abnormal salinities.

Upwards the black shale gradually gives way to a 300-m thick sequence of turbiditic sandstone, followed by interbedded bioturbated siltstone and sandstone alternating with thick bioturbated siltstone units. A terminal Permian tectonic phase resulted in up to 30 m of conglomerate with metre-sized slabs of Permian limestone and well rounded quartzite pebbles. The conglomerate is overlain by shallow marine, coarse, often pebbly, sandstone. The whole sequence above the black shale reflects a major regressive phase at the end of the Permian. It can be recognized throughout the basin.

EARLY TRIASSIC TRANSGRESSION

After the marked regression and accompanying weak block faulting and westwards tilting at the end of the Permian period, the sea rapidly transgressed the area at the beginning of the Triassic Period. The progressive onlap of marine Triassic on Permian and older strata is beautifully illustrated in Wegener Halvϕ by the recognition of the increasing eastwards extent of successive ammonite zones (Grasmück and Trümpy, 1969). The initial transgression was followed by rapid progradation, and a wide, well-aerated shelf sea was developed dominated by shale and fine sandstone deposition. Along the margins thick bodies of conglomerates and coarse pebbly sandstones of alluvial fan and fan delta origin were rapidly deposited. Where fan delta progradation took place into relatively deep-water over structural troughs, the sequences are initiated by turbidite sequences rapidly coarsened upwards into coarse, thick, channelized sandstone bodies (Fig. 22).

The areal extent of these sandstone units is not very well known; however, one major sandstone body can be followed over tens of kilometres along the length of Schuchert Dal (Fig. 22).

Progradation and basin filling continued and the brief early Triassic marine interlude represents the only important phase of marine deposition during the Triassic. The remainder of this period was characterized by deposition in a variety of continental environments.

Fig. 22. Upwards coarsening fan delta sequence starting with thin-bedded sandstone turbidites in black shale passing upwards into beds of structureless and finally horizontally laminated pebbly sandstone. Note the channelled base of the sequence. Thickness about 20-m. Trias Elv, western side of Schuchert Dal.

PETROLEUM GEOLOGY

Hydrocarbon source rocks

Lower Permian

Possible source rocks in the continental Lower Permian sequence only occur as up to 1-m thick shales of flood-plain origin. It is, however, possible that the shales reach greater thicknesses in the non-exposed central parts of the basin. The organic matter is mainly of terrestrial origin except for a few horizons rich in liptinite. West of Schuchert Dal the sediments are weakly mature to mature, and east of the valley highly mature (Fig. 23). This jump in coalification is probably due to the mid Permian westwards tilting and peneplaination of the sequence, resulting in west to east exposure of progressively older Lower Permian strata.

Upper Permian.

Possible source rocks include the major black shale sequence which occurs throughout the basin in the higher parts of the Upper Permian sequence. The content of organic matter in the shales varies, but is generally showing an upwards increase with 2–5% TOC in the upper part of the shale unit (Figs. 24 and 25). This part is rich in liptinite associated with varying amounts of vitrinite and inertinite. Optical and chemical studies indicate that a part of the shale sequence is a good source rock for oil.

The Upper Permian sediments on both sides of Schuchert Dal are immature to weakly mature. Thus there is no jump in coalification across the valley in contrast to the conditions for the underlying Lower Permian sequence. This supports the interpretation given above for this difference, as

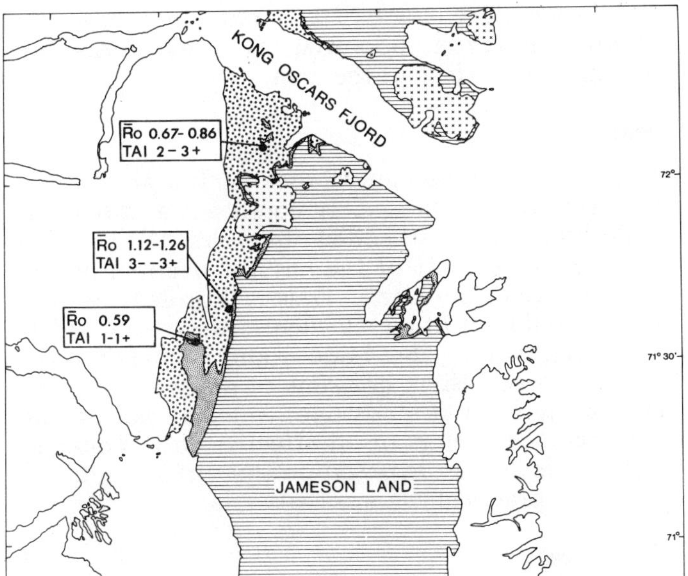

Fig. 23. Mean random reflectance (R̄o) and TAI index from three Lower Permian localities along the western basin margin. The rocks are premature to very mature.

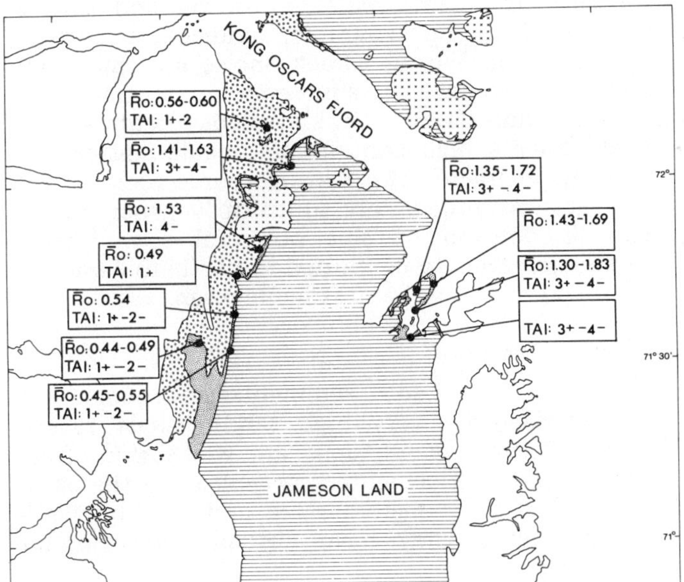

Fig. 24. Mean random reflectance (R̄o) and TAI index from 11 Upper Permian localities. The rocks are immature along the western basin margin except for two localities close to a major igneous centre. The Wegener Halvø samples are all post mature.

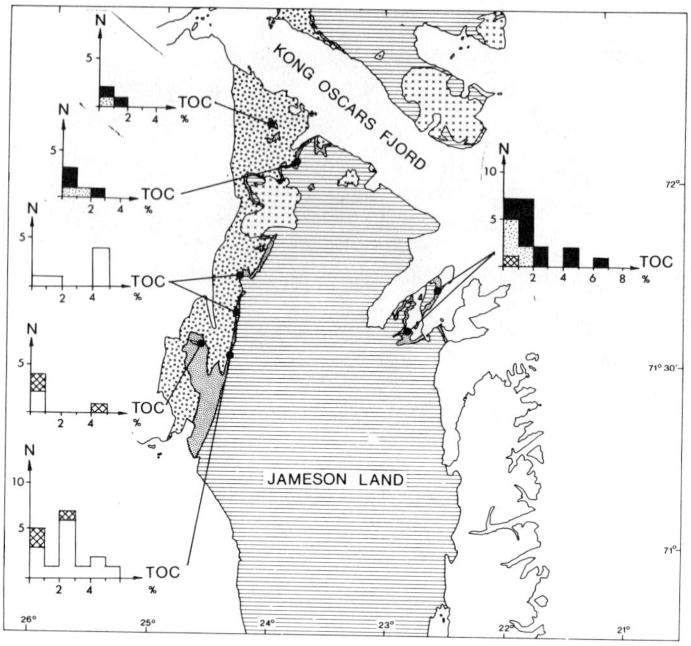

Fig. 25. Histograms from eight Upper Permian localities showing total organic content (TOC) plotted against number of samples. Black, T_{max} = 460° (post-mature); stippled, T_{max} between 435° and 460° (mature); white, T_{max} below 435° (immature); and cross-hatched, T_{max} not determined. The results were obtained by Rock-Eval pyrolysis.

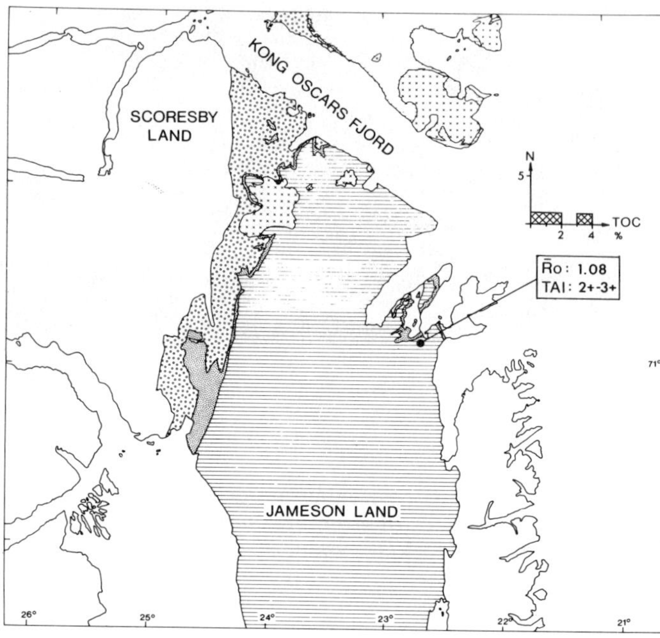

Fig. 26. Mean random reflectance (R̄o) from one Middle Triassic (Anisian?) locality south of Wegener Halvø. The sequence is mature.

the tectonic phase of tilting and erosion of the Lower Permian succession took place before the onset of Upper Permian sedimentation. The sequence reaches post-mature conditions in local areas around major Tertiary intrusions. On Wegener Halvø the Upper Permian sediments are likewise postmature.

Triassic

The Lower Triassic shales of the Wordie Creek Fm resemble the Upper Permian shales. They show low contents of organic matter, however and this is dominated by unfavourable, oxidized, terrestrial material. The Wordie Creek shales are thus considered not to have any source rock potential.

The Middle Triassic Gråklint Beds (Clemmensen, 1980) have a high content of organic matter and have the character of a mature source rock for oil in localities immediately south of Wegener Halvø. The distribution of this potential source rock seems, however, to be limited to northern Jameson Land and Scoresby Land (Clemmensen, 1980).

Potential reservoirs

The Jameson Land sedimentary sequence is characterized by very high sand : shale ratios. Lower Permian sandstones are well-sorted, fine-grained and well cemented, except along the basin margins where conglomerates and pebbly sandstones occur. They have virtually no reservoir potential.

The thick Upper Permian limestone units have not yet been systematically investigated for porosity and permeability. Their lateral extent over often tens of kilometres, combined with their thickness of more than 100 m, suggest that they are major targets if sufficiently good porosity—permeability values are found. The limestones seem to occur mainly over structural highs rather than just along basin margins. Subsurface occurrences of major limestone units are, therefore, expected to be fairly widespread in central and probably also southern Jameson Land.

The basal Triassic sequence contains a large number of fairly extensive coarse-grained sandstone bodies of fan delta origin. They are relatively mature in composition and thus not as susceptible to diagenetic changes as the contemporaneous and younger, extremely immature, arkosic alluvial fan sequences. They have, however, not yet been studied for porosity and permeability.

In contrast, the Jurassic sandstones have excellent reservoir characteristics, but they have no onshore reservoir potential as they form the topographic surface of the area.

Seals

The main possible source rock in the area is the Permian shale sequence. It could act as a seal for underlying and interbedded reservoirs. The Permian sequence over most of the basin is overlain by the thick shales of the Lower Triassic Wordie Creek Fm which could act as an excellent seal to any underlying reservoirs. The Middle and Upper Triassic include several thick horizons of dense limestones, evaporites and red mudstones which all possess the necessary characters of an effective seal. Younger, Jurassic shales are exposed on the surface and thus are of no practical value as seals.

Structures

The exposed part of the Jameson Land sedimentary sequence is flat lying and undisturbed except in the northern part and along the basin margins. Thus there is no direct surface evidence for major structures. Important phases of folding and faulting in the Devonian and faulting and tilting in the mid Permian have, however, been demonstrated.

Younger sediments will probably show draping and compaction effects over buried structural highs, as can be seen over the build-ups in the Wegener Halvø area. Furthermore, carbonate build-ups of Permian age are expected to be located over such highs. There is good sedimentological evidence for Triassic block faulting along the basin margins (Clemmensen, 1980), but it is not known if the more basinal areas were also subjected to faulting. An important E-trending geophysical anomaly can be traced across central Jameson Land. It seems to have exerted a strong influence, at least during Mesozoic times. From the Middle Jurassic the anomaly marks the position of a major hinge line. The most obvious potential play is thus considered to include Upper Permian shale as source rock, Upper Permian 'fore reef' and 'reef' limestone and Lower Triassic fan delta conglomerate and sandstone as potential reservoirs, combined stratigraphic—structural traps, and Lower Triassic shale of the Wordie Creek Fm as seal.

ACKNOWLEDGEMENTS

P.H. Larsen is thanked for able assistance in the field and for supplying information on fault tectonics. J.M. Hurst contributed with information on carbonate build-ups on Wegener Halvø. L. Stemmerik acknowledges support from the Danish Natural Science Research Council. B. Sikker Hansen, J. Lautrup and N. Turner are thanked for their help in preparing the manuscript. The contribution is published with permission of the Director of the Geological Survey of Greenland.

REFERENCES

Birkelund, T. and Perch-Nielsen, K. 1976. Late Palaeozoic—Mesozoic evolution of central East Greenland. In: A. Escher and W.S. Watt, (eds.). *Geology of Greenland.* (Geological Survey), pp. 305—339.

Brooks, C.K. and Nielsen, T.F.D. 1982a. The E. Greenland continental margin: a transition between oceanic and continental magmatism. *J. Geol. Soc. Lond.* **139**, 265—275.

Brooks, C.K. and Nielsen, T.F.D. 1982b. The Phanerozoic development of the Kangerdlugssuaq area, East Greenland. *Meddr Grønland, Geosci.*, **9**, 1—30.

Brooks, C.K., Nielsen, T.F.D. and Petersen, T.S. 1976. The Blosseville coast basalts of East Greenland: their occurrence, composition and temporal variations. *Contrib. Mineral. Petrol.*, **58**, 279—292.

Clemmensen, L.B. 1980. Triassic rift sedimentation and palaeogeography of central East Greenland. *Bull. Grønlands Geol. Unders.*, **136**, 1—72.

Collinson, J.D. 1972. The Rødde Ø conglomerate of inner Scoresby Sund and the Carboniferous(?) and Permian rocks west of the Schuchert Flod. *Meddr. Grønland,* **192**, 1—48.

Friend, P.F. *et al.* 1983. Devonian sediments of East Greenland VI. Review of results. *Meddr Grønland,* **206**, 1—96.

Grasmück, K. and Trümpy, R. 1969. Triassic stratigraphy and general geology of the country around Fleming Fjord (East Greenland). *Meddr Grønland,* **168**, 5—71.

Haller, J. 1970. Tectonic map of East Greenland (1:500 000). An account of tectonism, plutonism, and volcanism in East Greenland. *Meddr. Grønland,* **171**, 1—286.

Haller, J. 1971. *Geology of the East Greenland Caledonides.* Interscience, London.

Henriksen, N. and Higgins, A.K. 1976. East Greenland Caledonian fold belt. In: A. Escher and W.S. Watt (eds.), *Geology of Greenland,* pp.183—246.

Håkansson, E. and Petersen, S.A.S. 1982. Late Paleozoic to Tertiary tectonic evolution of the continental margin in North Greenland. *Can. Soc. Petrol. Geol., Mem.,* **8**, 331—348.

Kempter, E. 1961. Die jungpalaozoischen Sedimente von Süd-Scoresby Land (Ostgrönland), 71 1/2° N), mit besonderer Berücksichtigung der kontinentalen Sedimente. *Meddr Grønland,* **164**, 1—123.

Maync, W. 1942. Stratigraphie und Fazies verhältnisse der ober-permischen Ablagerungen Ostgrönlands (olim Oberkarbon-Unterperm) zwischen Wollaston Forland und dem Kejser Franz Josephs Fjord. *Meddr Grønland,* **115**, 1—128.

Maync, W. 1961. The Permian of Greenland. In: G.O. Raasch (ed), *Geology of the Arctic,* Vol. 1, Toronto University Press, pp. 214—223.

Noe-Nygaard, A. 1976. Tertiary igneous rocks between Shannon and Scoresby Sund, East Greenland. In: A. Escher and W.S. Watt (eds.), *Geology of Greenland,* p. 386—404.

Smith, D.B. 1980. The evolution of the English Zechstein basin. *Contrib. Sedimentol,* **9**, 7—34.

Stemmerik, L. 1979. Observations on the Upper Permian of the Wegener Halvø, East Greenland. *Rapp Grønlands Geol. Unders.*, **95**, 90—92.

Stemmerik, L. 1980. Observations on Upper Permian sediments in southern Scoresby Land, East Greenland. *Rapp. Grønlands Geol. Unders.*, 105—107.

Stemmerik, L. 1982. Stratigrafi, sedimentologi og basinudvikling i Øvre Perm aflejringsbassinet syd for Kong Oscars Fjord, Østgrønland. Unpublished thesis, Univ. Copenhagen, 200pp.

Stemmerik, L. and Sørensen, M. 1980. Upper Permian dykes in southern Scoresby Land, East Greenland. *Rapp. Grønlands Geol. Unders.*, **100**, 108.

Surlyk, F. 1978. Jurassic basin evolution of East Greenland. *Nature (London)* **174**, 130—133.

Surlyk, F. 1983. Source rock sampling, stratigraphical and sedimentological studies in the Late Palaeozoic of the Jameson Land basin, East Greenland. *Rapp. Grønlands Geol. Unders.*

Surlyk, F., Clemmensen, L.B. and Larsen, H.C. 1981. Post-Paleozoic evolution of the East Greenland continental margin. *Bull. Can. Petrol. Assoc. Mem.*, **7**, 611—645.

Vischer, A. 1943. Die postdevonische Tektonik von ostgrönland zwischen 74° und 75°N. Br., Kuhn Ø, Wollaston Forland, Clavering Ø und angrenzende Gebiete. *Meddr Grønland,* **133**, 1—195.

22 The Triassic offshore Norway north of 62°N

Vigdis Wiik Jacobsen and Paul van Veen*
Norwegian Petroleum Directorate, Stavanger, Norway

The Triassic of the Norwegian Shelf north of 62°N has been studied in wells in the Haltenbanken and Troms I areas. The Triassic of Troms I shows a 1900-m thick, almost complete succession of open to marginal marine to occasionally terrestrial sediments. Climatic conditions during deposition were predominantly temperate. On Haltenbanken 2500-m of Upper and possible Middle Triassic is present, consisting of alluvial to shallow marine deposits. Two halite successions are conspicuous, of latest Ladinian to earliest Karnian and late Karnian age. Climatic conditions were arid to semi-arid during deposition.

A comparison and integration of the data from the wells with the Triassic on Svalbard and East Greenland allow several trends to be correlated. An interpretation of these trends (lithological/transgressive—regressive, biostratigraphical) in terms of relative subsidence and tectonic episodes, eustatic sea-level changes and the occurrence of climatic variation is discussed. This suggests that the Troms area was part of the Svalbard Basin during the Triassic and that this area was responding to a tectonic regime also recorded on East Greenland. The Haltenbanken area was a separate, small fault-bounded basin, reminiscent in part of its development to the North Sea during the late Triassic.

The palaeogeographical development of this area is described, accompanied by a series of sketch maps. The source rock, reservoir and cap rock potential are discussed briefly.

INTRODUCTION

The exploration of the Norwegian Shelf north of 62°N, which started with the spudding of 7120/12-1 in 1980, has provided a considerable amount of new stratigraphic information from the Troms and Haltenbanken areas (Fig. 1). The Triassic, rarely a primary prospect, has been penetrated in a number of wells. Earlier interest in the North Sea Triassic sedimentation (Clemmensen *et al.*, 1980) has led to a natural extension north of 62°N. The lithology and biostratigraphy of the Triassic of Troms I and Haltenbanken have been examined in detail. A detailed sedimentological study has been hampered by the near lack of cores, however.

The main aims of this chapter are: (1) to present a chronostratigraphical and limited lithostratigraphical framework for the Triassic in the Troms and Haltenbanken areas; (2) to correlate the well successions with the Triassic on Svalbard and in East Greenland; (3) to detect major common trends and factors determining the sedimentation pattern in each individual area; (4) to integrate the available information into a series of palaeogeographical reconstructions; and (5) to provide information concerning the distribution of source, reservoir and cap rocks in the wells.

*Presently with Statoil, Stavanger, Norway.

Petroleum Geology of the North European Margin © Norwegian Petroleum Society (Graham & Trotman, 1984) pp. 317—327.

LOCATION AND GEOLOGICAL SETTING

In the Troms I area Norsk Hydro's wells 7120/12-1 and 2 together supply information from the whole Triassic sequence. These wells are situated in the Hammerfest Basin immediately north of the Troms-Finnmark fault (Fig. 2a). The Triassic found in the Hammerfest Basin can be followed on seismic in most of the Barents Sea (Rønnevik *et al.*, 1982).

In the Haltenbanken area Saga's 6507/12-1 and 2 penetrate into thick late and possible Middle Triassic deposits. The deepest well, 6507/12-2, is situated on the inside of the platform boundary fault of the Kristiansund-Bodø Fault Complex (Fig. 2b), while 6507/12-1 tested a roll-over structure in a graben west of this fault.

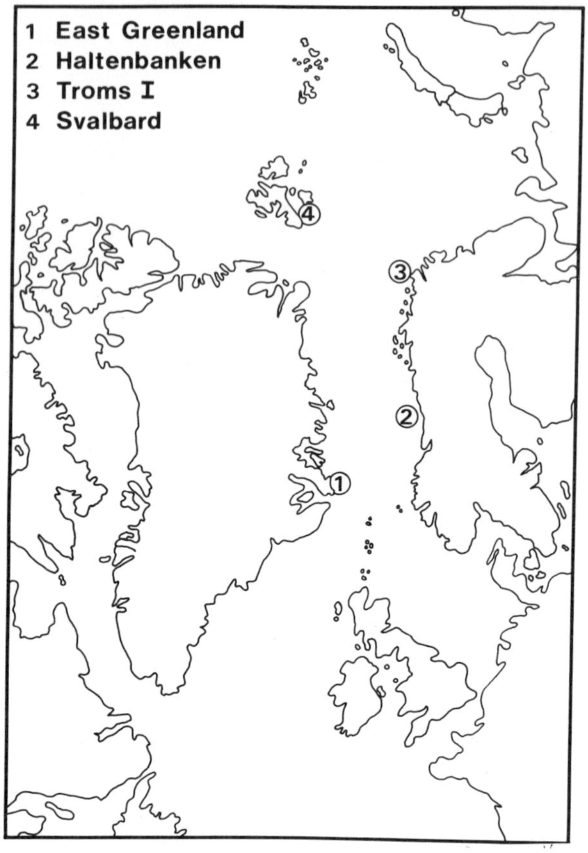

Fig. 1. Palinspastic reconstruction with location of the areas discussed. (After Ziegler, 1982, fig. 8.)

CHRONOSTRATIGRAPHY AND CHRONOLOGY

Age determinations in the Triassic in the Troms I wells are based entirely on a detailed palynological analysis of side wall cores. This allows a subdivision of the Triassic up to substage level. In the Haltenbanken wells the age determinations are based predominantly on cuttings and, therefore, the ages have to be considered as more tentative.

The age determinations have been approximated to the nearest substage or stage, applying the chronostratigraphy shown in Fig. 3. It is here adopted to consider the Lower Griesbachian as being of Permian age (cf. Visscher and Brugman, 1981). The Ladinian–Anisian boundary is calibrated to the Alpine Triassic (Van der Eem, 1983). Furthermore, an early Norian age for the Flatsalen Member of the Wilhelmøya Fm at Hopen (Korchinskaja, 1980) has been reservedly accepted, corroborated by results from the 7120/12 wells. The status of the Rhaetian has been left open, pending a future decision (see, Visscher and Brugman, 1981).

According to Odin and Letolle (in Odin, 1982) the Triassic period lasted from 204 to 245 Ma, a duration of approximately 41 Ma. As for chronology, a modified version of Odin's (1982) proposals has been applied. The duration of the early Triassic has been extended from 6 to 10 Ma, on the assumption that the Scythian (early Triassic) should be regarded as a series rather than a stage. Consequently we have reduced the Norian–Rhaetian interval from 16 to 12 Ma.

LITHOLOGICAL DESCRIPTIONS AND CORRELATIONS

A short description of the lithological successions of Svalbard and East Greenland is given, as both areas will be referred to frequently. For the Troms and Haltenbanken areas the lithological descriptions are accompanied by age determination of the various intervals, correlation to other areas, as well as brief references to the depositional environment. Most of the following information is summarized in Figs. 4 and 5.

Svalbard

Mørk *et al.* (1982) recently presented a revised lithostratigraphical nomenclature for the Triassic of Svalbard and described its depositional and diagenetic history. The succession of clastic sediments assigned to the Triassic varies in thickness from 250 m over local highs (e.g. Sørkapp-Hornsund high) to 1200 m, mainly due to differential downwarping over a series of NNW-trending fault lineaments (Mørk *et al.*, 1982, fig 3.).

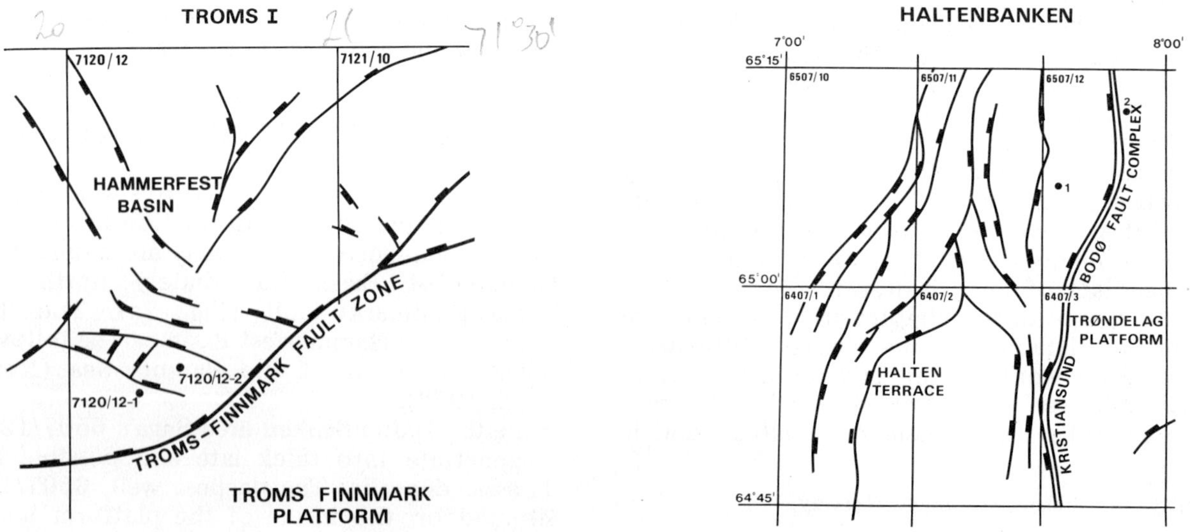

Fig. 2. Location maps of Haltenbanken and Troms I (block 7120/12).

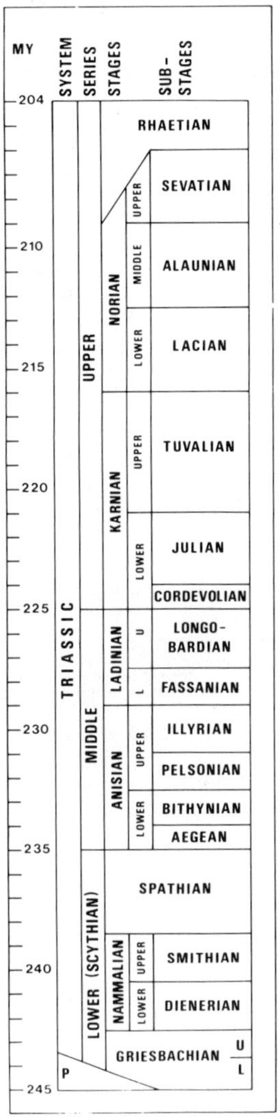

Fig. 3. Triassic chronostratigraphy and chronology.

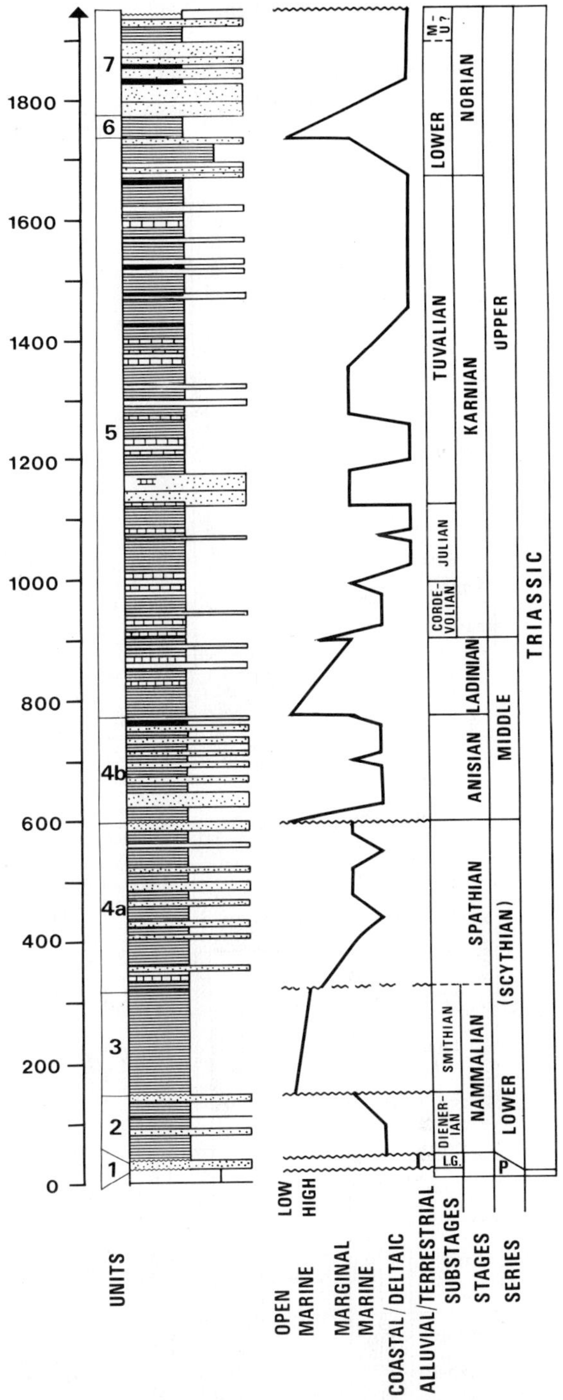

Fig. 4. Composite Triassic succession from the Troms 7120/12 wells with a tentative environmental interpretation. LG, Lower Griesbachian (see chronostratigraphy). Conventional lithology. Thickness of thinner beds exaggerated.

In general, one is able to recognize (Figs. 5 and 6): (1) a western succession developed in the 'central basin'. This succession shows thick, open to marginal marine, Lower to Middle Triassic (Sassendalen Gp) followed by an often relatively thin marginal marine Middle to Upper Triassic (Kapp Toscana Gp); (2) more easterly situated successions, with a relatively thin, shale-dominated Lower to Middle Triassic (Barentsøya Fm), followed by an often relatively thick marginal marine (mainly deltaic) Middle to Upper Triassic (Kapp Toscana Gp).

According to Mørk *et al.* (1982) this depositional pattern reflects intermittent coastal progradation in a shallow marine basin from the west during early and middle Triassic times and deltaic progradation predominantly from the northeast and less from the west during middle and late Triassic times.

In all these successions the thin, mainly open marine, Wilhelmøya Fm (Norian–Toarcian) terminates the Triassic. Rhaetian to early Jurassic erosion has resulted in a considerable hiatus of this time interval, its magnitude decreasing from west to east.

East Greenland

The lithostratigraphy and depositional history of the fault-bounded East Greenland Basin has recently been described in detail by Clemmensen (1980a, b). For the correlation of the various lithostratigraphical units, most of the age determinations cited in Clemmensen (1980b) have been retained. A small number of samples was kindly supplied by Dr. L. Clemmensen of which only one, from the Ødepas Member of the Wordie Creek Fm, proved productive. The palynological assemblage suggested a late Nammalian (Smithian) age relative to the Troms material.

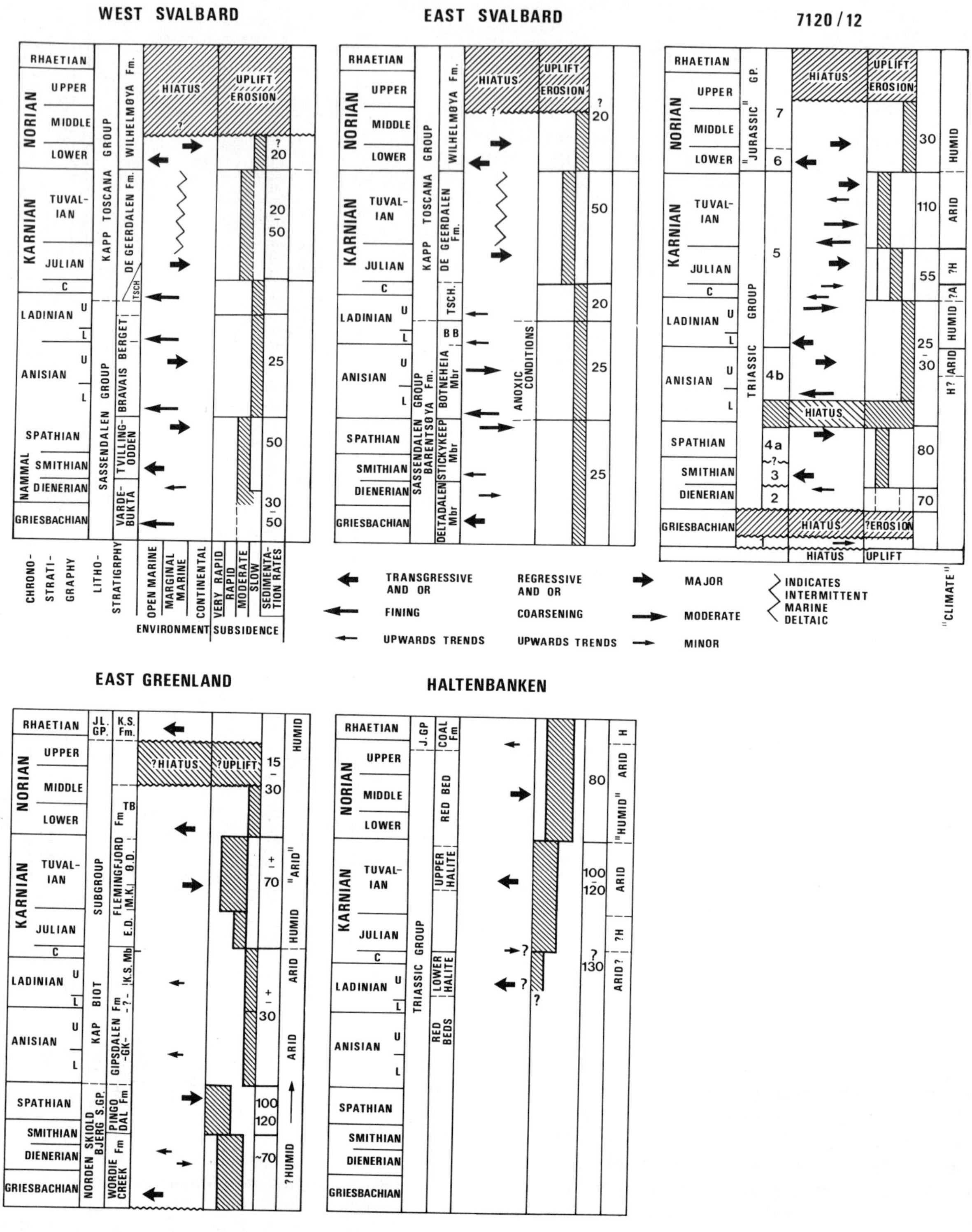

Fig. 5. Major trends and events recorded in the Triassic north of 62°N. TSCH, Tschermakfjellet Formation (Svalbard); BB, Blanknuten Beds (East Svalbard); KS, Kap Seaforth Member (East Greenland); ED, Edderfugledal Member; MK, Malmros Klint Member; ØD, Ørsted Dal Member; TB, Upper part of ØD, Tait Bjerg Beds; JL, Jameson Land Group; KS, Kap Stewart Formation.

The Lower Triassic is represented by the marine Wordie Creek and continental Pingo Dal Fms. The evaporitic Gipsdalen Fm is taken to represent the Middle Triassic, with the topmost Kap Seaforth Member probably extending into the Karnian (Clemmensen, 1980b). The clastic continental Fleming Fjord Fm is mainly of Karnian age. Our present interpretation suggests a Norian age for the topmost part of the Ørsted Dal Member, the marginal marine Tait Bjerg Beds.

According to Clemmensen (1980b) a hiatus separates the Tait Bjerg Beds from the overlying fluvial to deltaic Kap Stewart Fm. In more south-westerly sections on Jameson Land, however, the Kap Stewart Fm overlies red beds, assigned to the Ørsted Dal Member by Pedersen and Lund (1980), apparently without a hiatus.

Troms I

Only the 7120/12-1 and 12-2 (Norsk Hydro) wells have penetrated the Triassic sufficiently deep to give an impression of the succession. 7120/12-1 terminated in Upper Anisian sandstones; a 1250-m

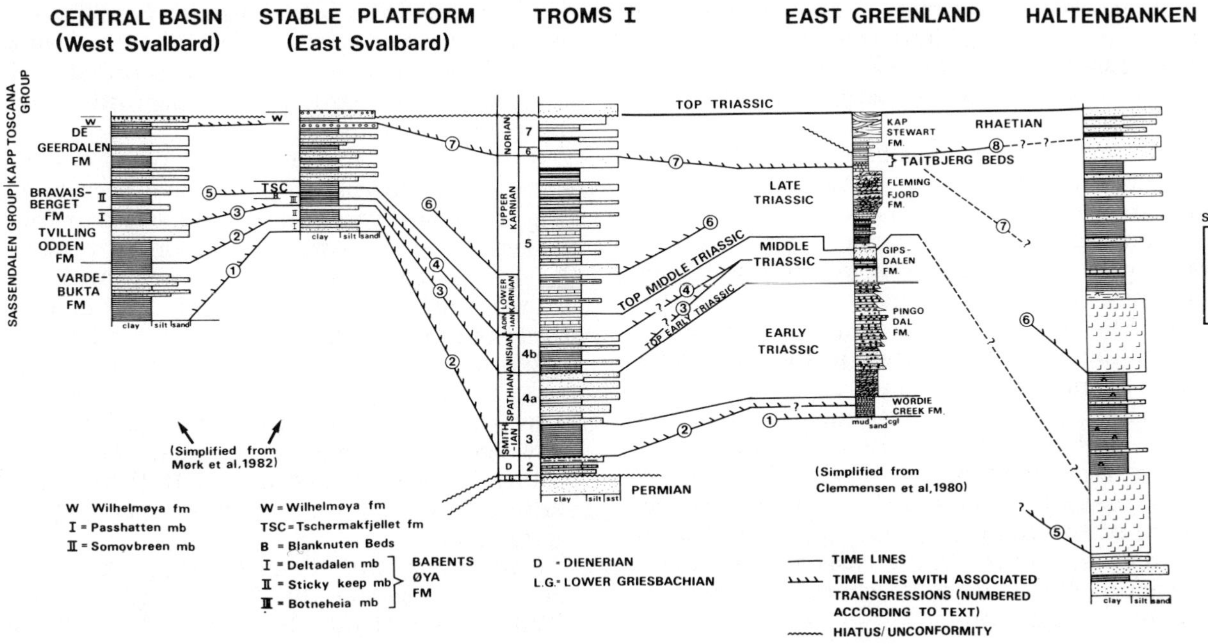

Fig. 6. Correlational scheme, major transgressions indicated.

thick Middle and Upper Triassic is apparently completely preserved. The deepest well in the Troms area, 7120/12-2, shows approximately 1500-m Triassic before entering the Permian. Some 450 m of Tuvalian (Upper Karnian) deposits are faulted out.

The composite section presented in Fig. 4 shows a more than 1900-m thick, almost complete Triassic. Since a formal lithostratigraphy has not yet been approved, the succession has been subdivided into seven more or less distinct lithological units. The major lithological trends have been summarized in Fig. 5.

Unit 1, of earliest Griesbachian (latest Permian) age, is a thin interval consisting of an intercalation of sand- and siltstones, probably of fluvial origin. The unit is bounded by unconformities, the lower unconformity marked by a considerable change in maturation of the kerogen. A comparable horizon has not yet been described from Svalbard, but it is similar in age to the much disputed basal 'Triassic' interval from Cape Stosch on northeastern Greenland (see, e.g., Balme, 1979).

Lower Triassic

Unit 2, of early Nammalian (Dienerian) age, consists of grey shales with occasional interbeds of white to grey, very fine sandstones. The lower and upper boundaries are apparently unconformable. The appearance of marine microplankton in the upper part indicates a marginal marine environment. Unit 2 compares in age as well as in its slightly transgressive trend with the upper part of the Vardebukta Fm present in the 'Central Graben' in western Svalbard.

Unit 3, of tentative late Nammalian (Smithian) age, consists of black to greyish, very homogeneous shales. The abundance of marine microplankton and the type of kerogen suggest a low energy, open marine environment of deposition. The kerogen indicates a slight upwards coarsening within this unit. The rather abrupt lithological change from Unit 3 to Unit 4 may suggest the presence of a minor hiatus at that level.

Unit 4, of Spathian to Anisian age, consists of an intercalation of sandstones, silt- and claystones. The unit is here subdivided into two parts, a lower Spathian part (4a) and an upper, Anisian part (4b). Biostratigraphy suggests a hiatus of latest Spathian to earliest Anisian age between the two parts, indicating a disconformable contact. The lower part (4a) shows an overall coarsening upwards trend. Glauconite is fairly widespread which, together with the presence of microplankton, suggests a shallow marginal marine environment of deposition.

Unit 3 and the lower part of Unit 4 conform in age and coarsening upwards trend with the Tvillingodden Fm, present in the 'Central Basin' on western Svalbard. Møork *et al.* (1982) suggest a deep to shallow shelf environment of deposition for the Svalbard succession, with a succession of shelf muds, storm generated siltstones, shoals and barriers (Møork *et al.*, 1982, fig. 8).

Middle Triassic

A thin, fine-grained (marine) interval occurs at the base of the upper part of Unit 4. This upper (Anisian) part (4b) differs from the lower part by the near absence of glauconite and marine microplankton. More and thicker sandstones are present and coal stringers are recorded towards the top. A marginal marine to deltaic depositional environment is suggested. The coarsening upwards development resembles the (lower part of the) Bravaisberget Fm described from western Svalbard, for which Møork *et al.* (1982, fig. 8) suggest a similar setting. Phosphatic nodules, present at the base of the Bravaisberget Fm and characteristic for the eastern Svalbard basinal Middle Triassic have not been recorded, however.

Unit 5 of Fassanian (early Ladinian) to earliest Norian age consists predominantly of shales and siltstones, with minor interbeds of sandstone, limestone and some coals. An important lithological change marks the transition from Unit 4 to Unit 5. Open marine shales at the base of Unit 5 fairly abruptly overly the deltaic sandstones. For biostratigraphical reasons we have chosen to approximate this (Anisian—Ladinian) transition with the base of the Blanknuten Beds — the topmost part of the Botneheia Member of the Barentsøya Fm — on eastern Svalbard (Fig. 6). A correlation in the succession on western Svalbard is uncertain, but approximates with a level within the Bravaisberget Fm.

The lowermost (Ladinian) part of Unit 5 consists of dark to grey shales with some minor limestones and siltstones. Some coarsening upwards cycles are apparent. The succession is generally regressive with an open marine environment at the base and a (restricted?) marginal marine environment at the top. This part is correlated mainly with the Blanknuten Beds, present on East Svalbard.

A lateral equivalent of the marine Tschermakfjellet Fm, present in parts of the Svalbard Triassic at the base of the Kapp Toscana Gp, has not been recognized in the wells. On the basis of the present data it cannot yet be decided whether a hiatus of ?latest Ladinian age is present in the wells.

Upper Triassic—Karnian

The remainder of Unit 5, of which the bulk is of Karnian age, consists of varicoloured shales and siltstones, minor sandstones and some limestones. In the Tuvalian some of the finer grained lithologies show a red or green colouration. Coals appear towards the top of the succession. The biostratigraphical information and the facies suggest a marginal marine to terrestrial (lacustrine, alluvial) environment of deposition.

The Julian—Tuvalian transition is marked by the occurrence of an interval of thick, probably transgressive, sandstones with good reservoir properties, which contained gas in 7120/12-2.

The De Geerdalen Fm on Svalbard is the contemporaneous equivalent of the Karnian part of Unit 5. The latter formation is, however, mainly of deltaic origin (Mørk *et al.*, 1982), and as such in contrast to the slightly more continental nature of Unit 5 in the Troms area. Individual trends in the upper part of Unit 5 conform with the trends observed in the Iversenfjellet Member of the De Geerdalen Fm at Hopen (e.g. Smith *et al.*, 1975; Bjærke and Manum, 1977).

Upper Triassic—Norian

The transition from Unit 5 to Unit 6 is marked by a rapid change from marginal marine sand- and siltstones to open marine shales. The following thin Unit 6 consists of black to grey shales. Marine microplankton are ubiquitous which, together with the type of kerogen, suggests an open marine depositional environment. Unit 6 is the lithological equivalent of part of the Flatsalen Member at Hopen (Smith *et al.*, 1975; Bjærke and Manum, 1977), and its lateral correlatives (Mørk *et al.*, 1982), the basal part of the Wilhelmøya Fm. The latter authors stressed the importance of this interval on Svalbard, which is now emphasized by its presence in the wells. The Flatsalen Member at Hopen is of earliest Norian age (Korchinskaja, 1980).

Unit 7, of Norian to ?Rhaetian age, is dominated by fairly thick to thick, well sorted sandstones, intercalated with silt- and claystones and some coals. A marginal marine to fluvial environment of deposition is inferred. A hiatus of mainly Rhaetian to ?Hettangian age is inferred above Unit 7. A hiatus of variable extent is also recorded in the successions on Svalbard, with a magnitude decreasing from west to east (Mørk *et al.*, 1982).

Summarizing the present data of the 7120/12 wells, the following features have been recognized (Fig. 4). The succession shows a series of four more or less distinct upwards coarsening cycles (Units 3 + 4; Unit 4b; Unit 5; Units 6 + 7). Unconformities are recorded at the base of Unit 1 (coinciding with a marked change in maturity of the kerogen) and at the base and top of Unit 2. Hiatuses are present between the Upper Permian and the base of Unit 1, between Unit 1 and Unit 2 (late Griesbachian), between Units 2 and 3 (Smithian—Dienerian transition), between Units 4a and 4b (latest Spathian—earliest Anisian) and within Unit 7 (?Norian—Hettangian).

Correlation with Svalbard suggests (Fig. 6): (1) the Lower to Middle Triassic (Anisian) in the 7120/12 area (Units 2—4) is similar in lithological trends and characteristics to the equivalent successions in the 'Central Graben' on western Svalbard; (2) the Middle to Upper Triassic (Ladinian to Karnian) in the 7120/12 area (Unit 5) is similar in lithological trends to the successions on eastern Svalbard; and (3) the uppermost Triassic (Norian) has a similar development as recorded over all of Svalbard (Units 6, 7).

It can be concluded, therefore, that from the Dienerian onwards the 7120/12 area formed an integral part of the Svalbard Basin.

Haltenbanken

As in the Troms area, the main information from the Triassic in the Haltenbanken area is derived from the first two wells, 6507/12-1 and 12-2 (Figs. 5 and 6). The other wells only penetrate the top of the Triassic (Rhaetian—Norian). 6507/12-1 terminated in claystones and redbeds of presumed late Triassic age, after drilling through more than a 900-m thick Rhaetian—Norian succession. The deepest well, 6507/12-2, terminated in redbeds of unknown age and showed a 2500-m thick, mainly late Triassic succession. A formal lithostratigraphy has not yet been proposed, apart from the top (Rhaetian) part.

The lowermost interval in 6507/12-2 consists of an intercalation of red to brownish, calcareous silt- and claystones and white to light grey to light brown very fine sandstones. The age of this interval is not known, but a (?Middle) Triassic age is inferred. The lowermost clastic interval is overlain by a thick evaporitic unit. Two thick halites are recorded in this unit. The halite layers are interbedded with thin, often dolomitic shales. A restricted marine depositional environment is suggested on the basis of the presence of scarce microplankton. The age of the lower part of the lower salt is uncertain, but the upper part appears to be of latest Ladinian to Cordevolian (early Karnian) age. The age of the upper part of the upper salt is Tuvalian (late Karnian). The two halites are separated by an interval consisting of grey to black to varicoloured calcareous to dolomite shales and claystones and some sandstones. Anhydrite is frequent throughout the sequence. A sabkha depositional environment is suggested. A single age determination from the lower part of this interval indicates a tentative Julian (early Karnian) age. The upper salt is overlain by an interval of similar lithology, as recorded between the halites. A latest Tuvalian age is inferred.

The lowermost part of the succession overlying the evaporitic unit is dominated by often calcareous to sideritic claystones with a few limestones, probably of lacustrine origin. Evaporitic intervals are notably absent. Limited data suggest an early Norian age. The appearance of sandstones defines a gross coarsening upwards trend. The clay- and siltstones, which still dominate the succession, become brick red, red brown, green to grey. A fluvial depositional environment is suggested. The age is presumably Middle to Late Norian. Typical redbed facies disappears towards the Norian—Rhaetian transition. The succession becomes dominated by fine-grained, well sorted thick sandstones and coals appear. The depositional environment is mainly fluvial. The appearance of coal marks the base of the coal unit (Fig. 6).

Correlation

By its internal organization and age, the evaporitic unit shows a striking similarity to, e.g. the Oddesund Fm in the Norwegian-Danish Basin (Bertelsen, 1980). The succession of the Keuper Halite Member of the Dudgeon Saliferous Fm and the Keuper Anhydritic Member of the Triton Anhydritic Fm in the North Sea (Brennand, 1975) or the Lower and Upper Gipskeuper in Germany and Poland (see Bertelsen, 1980, p. 36) are also comparable. The available data suggest, therefore, that the Triassic in the Haltenbanken area is genetically related to the North Sea and the Norwegian-Danish Basin rather than to East Greenland.

Only the upper part of the succession at Haltenbanken (Upper Norian—Rhaetian) can be compared lithologically and partly biostratigraphically with the transition from the Fleming Fjord Fm (Ødepas Member) to the Kap Stewart Fm at Jameson Land (Pedersen and Lund, 1980). Coals are also present in the latter formation.

MAJOR TRENDS

Subsidence

The authors are aware of the problems related to the interpretation of sediment thicknesses and sedimentation rates in terms of subsidence. Yet it is thought that a simple indication of relative rates of subsidence provides a considerable insight into the tectonic circumstances during Triassic sedimentation. Therefore, sedimentation rates have been calculated considering only maximum thicknesses cited in relevant publications, without accounting for compaction. In cases of continental to marginal marine deposition these values are thought to represent net subsidence. The values are translated as follows: slow (< 40 m/Ma), moderate (40—70 m/Ma), rapid (70—110 m/Ma) and very rapid subsidence (>110 m/Ma).

An illustration of the rates of subsidence is presented in Fig. 5. The following pattern emerges:

(1) The early Triassic is characterized by moderate to rapid to ?very rapid subsidence and infill in, respectively, western Svalbard, 7120/12 and on East Greenland. The eastern basinal part of Svalbard is characterized by low sedimentation rates.

(2) The Middle Triassic is characterized by slow subsidence.

(3) The Karnian is characterized by moderate to rapid to very rapid subsidence. An increase in magnitude is seen when going from West to East Svalbard and further to 7120/12, or from East Greenland to Haltenbanken.

(4) Subsidence decreases fairly abruptly at the Karnian—Norian transition in all areas except Haltenbanken.

(5) There is a period of uplift and presumably erosion some time during the (?late) Norian on Greenland and during the (?late) Norian to early Jurassic in the Svalbard Basin.

Succession of transgressions and regressions

Integration of data from the wells with results published from Svalbard and Greenland establishes the following pattern of transgressions and regressions in the Triassic north of 62°N (see Fig. 6). The trends are further summarized in Fig. 5.

(1) An early Griesbachian transgression affects Svalbard as well as East Greenland. A regressive trend is recorded in the Dienerian on Svalbard and presumably also on Greenland. In the latter area the coarse grained Svinhufvuds Bjerge Member interferes in the marine Wordie Creek Fm (Clemmensen, 1980b).

(2) A Smithian transgression is recorded on Svalbard and in the 7120/12 area. On Greenland this transgression is presumably represented by the (?Smithian) Ødepas Member of the Wordie Creek Fm, which overlies the Svinhufvuds Bjerge Member. A strong regressive and/or coarsening upwards trend characterizes the Spathian of Svalbard, the 7120/12 area and East Greenland.

(3) A latest Spathian to earliest Anisian transgression is recorded on Svalbard and the 7120/12 area. The event may be represented on Greenland by the Gråklint Beds of the Gipsdalen Fm (Clemmensen, 1980b). On western Svalbard and in the 7120/12 area a subsequent regressive trend is apparent in the Anisian.

(4) An early Ladinian transgression is recorded in the 7120/12 area. Despite the fact that this event has not yet been identified with certainty on Svalbard, it should be regarded as being of major importance from a regional point of view. It coincides with the transgression recorded at the Middle/Upper Muschelkalk transition (cf. Ziegler, 1982; Ager, 1981). An alternative opinion suggests that the Gråklint Beds on East Greenland represent this event rather than (3). A regressive trend during the Langobardian (late Ladinian) is recorded in the wells and, less apparently, on Svalbard.

(5) An (?intra) late Ladinian transgression is recorded on Svalbard at the base of the Tschermakfjellet Fm. Clear recognition of this event in the wells is not possible at present. The introduction of the first salt in the Haltenbanken area may be related to this event.

A strong regressive trend characterizes the Karnian of Svalbard, the 7120/12 wells and East Greenland.

(6) There appear to be at least two Tuvalian transgressions in the 7120/12 area (Fig. 4). Deposition of the second salt in the Haltenbanken area may be related to these transgressions.

(7) A very important regional Norian transgression is recorded on Svalbard as well as in the 7120/12 area. It is here suggested that the marginal marine Tait Bjerg Beds at the top of the Fleming Fjord Fm on East Greenland are a phenomenon of the same event. This idea is at variance with previous authors who considered these beds to be of late Norian or early Rhaetian age (see Clemmensen, 1980b). The event is not evident in the Haltenbanken Triassic.

A subsequent regressive trend is recorded in all areas. There is evidence for non-deposition or erosion in the Svalbard Basin and on East Greenland (Clemmensen, 1980b).

(8) There exist indications for an early Rhaetian transgression on East Greenland (Clemmensen, 1976). The introduction of coals on Haltenbanken may be related to the same event. It is further supported by the presence of scarce ostracods at that level in the wells.

Palaeoclimatology

Climate has been an important factor during sedimentation in the Triassic. Several recent papers cover the subject (e.g., Visscher and Van der Zwan, 1981; Parrish and Curtis, 1982; Parrish *et al.*, 1982).

During the early Triassic, Greenland was located at approximately 30°N, and Svalbard at 45°N. According to Parrish *et al.* (1982, fig. 2), during this time a strong SE—NW gradient in rainfall existed over the northern North Atlantic. East Greenland and West Norway apparently received low and Svalbard moderately high precipitation, resulting in (semi) arid to humid conditions, respectively. Palynological data indicate that a similar contrast still existed during the Karnian. Following the procedure outlined in Visscher and Van der Zwan (1981), the palynological assemblages from the Karnian in Haltenbanken can be classified as 'arid', whereas the assemblages from the Troms area in general reflect more temperate conditions.

Superimposed on this generally temperate pattern in the Troms area, we find distinct palynologically defined 'arid' events. These events are sometimes only slightly expressed, e.g. in the uppermost Anisian (Illyrian) and in the Lower Karnian, with a humid Ladinian in between, but very distinctly in the 'Middle' Tuvalian. The latter event is apparently followed by a gradual increase in humidity towards the Norian. Yet, the Karnian—Norian transition may well represent a very distinct climatic change, since it coincides more or less with the disappearance of evaporitic facies in the Haltenbanken area. In the latter area a further change from 'arid' to humid conditions is recorded at the Rhaetian—Norian transition.

A similar pattern during the latest Anisian—Ladinian—early Karnian has been deduced by Van der Eem (1983) from compositional variations in arid palynological assemblages from the Alpine Triassic. Furthermore, these events appear to be recorded in Poland, interpreting palynological data from Orlowska-Zwolinska (1979). In Poland additional 'arid' intervals seem to be represented in the lowermost Anisian (Röt) and the uppermost Ladinian to lowermost Karnian (Lower Gipskeuper). These two latter events have not yet been recorded from the Troms area. The widespread occurrence of these variations shows that they are not caused by local variations in environment.

The existence of a pattern of increasing aridity during the Karnian (at least in the Troms area) stands quite in contrast to the explanation offered by Clemmensen (1980a) for the vertical succession of climatically sensitive rocks in the Triassic of East Greenland. This vertical succession, with evaporite rocks in the Middle Triassic and coals in the Rhaetian, suggests a simple, gradual climatic shift from semi-arid to semi-humid conditions thoughout the Middle to late Triassic, due to a northwards continental drift.

The present authors concur with the existence of northwards continental drift of Eurasia during the Triassic. It is inferred, however, that important climatic variations have affected and counterbalanced a general shift towards more temperate conditions in this area, especially during the Karnian. Alternatively, it can be supposed that a main phase of continental drift of this area took place during the Norian and Rhaetian. The occurrence of climatic variations has been indicated in Fig. 5.

Previously, it has been suggested (e.g., Bertelsen, 1980) that the deposition of salts in the North Sea Basin was related to climatic variations. The present data from the Troms area, outside the arid area, lend credibility to this suggestion. Confirmation of the hypothesis of climatic variations must await further research, however.

Palaeogeographical and tectonic development (Fig. 7)

In this section the available data are integrated into a description of the palaeogeographical evolution of the Triassic north of 62°N. Two terms are informally introduced for ease of description. The Svalbard Basin includes the 7120/12 area. The use of the term northern North Atlantic Rift Basin is thought to be justified by its tectonic development (see also Ziegler, 1982, fig. 8).

Early Triassic

The latest Permian obviously witnessed a period of uplift, probably accompanied by a thermal event, resulting in the formation of local highs and subsequent erosion (East Greenland, see Surlyk *et al.*, 1982, fig. 2; Svalbard, Sørkapp-Hornsund High, Mørk *et al.*, 1982, fig. 3; Troms I, 7120/12).

Subsidence started in early Triassic times, when the Griesbachian sea transgressed the lower areas and extended into the northern North Atlantic Rift Basin. But during the Dienerian, subsidence progressed so far that the local highs were onlapped. A slight regression was probably the result of initial uplift of the basin margins.

Figure 7A shows the situation in Smithian times. A strong transgression again established marine conditions in large parts of the Svalbard Basin. Shallow marine conditions prevailed in the East Greenland Basin. The presence of a carbonate shelf, favoured by climatic conditions, is indicated in the intermediate areas. Poorly dated Triassic

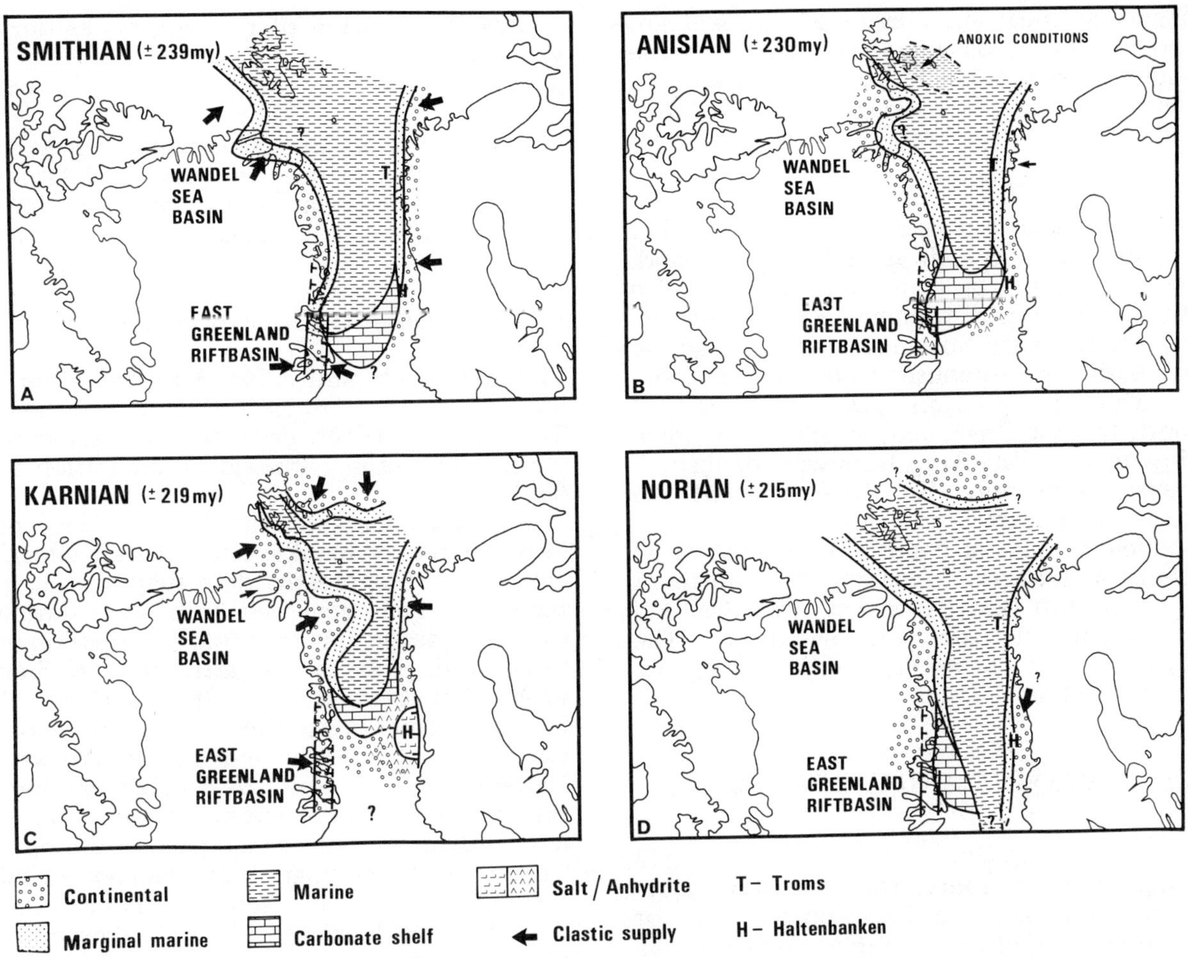

Fig. 7. Sketch maps showing the palaeogeographical development during the Triassic.

deposits of marginal marine to marine origin have been reported from the Wandel Sea Basin of North Greenland (Håkansson, 1979).

Like the North Sea areas (Ziegler, 1982), the northern North Atlantic Rift Basin and the Svalbard Basin were apparently affected by a phase of rifting during late early Triassic times. The increasing clastic input, caused by erosion of rapidly uplifted source areas, enhanced by gradually more arid conditions, produced the progradation of continental deposits in the East Greenland Basin. The clastic input emphasized a regressive trend at the southern and western margins of the Svalbard Basin during the Spathian.

Middle Triassic

From the change in magnitude of the subsidence, it is concluded that rifting ceased during Middle Triassic times. Major relief probably had disappeared. The Svalbard Basin and the northern North Atlantic Rift Basin were subsiding only slowly or were even stable. The climate was generally more arid with some humid interludes.

Figure 7B shows the approximate palaeogeography during Anisian times. A latest Spathian to earliest Anisian transgression had again established open marine conditions in the Svalbard Basin and probably penetrated towards the East Greenland Basin (Gråklint Beds, Gipsdalen Fm). In the latter area low precipitation and high evaporation resulted in the repeated deposition of evaporites along the margins of the sea and in the fault-bounded basin. In the Svalbard Basin a large amount of organic material was deposited and preserved in the deeper parts due to anoxic conditions. Deltaic deposition and paralic conditions prevailed along the margins of the Svalbard Basin, e.g. in the Troms I and Svalbard areas.

Late Triassic—Karnian (Fig. 7C)

By the end of Middle Triassic times the northern North Atlantic Rift Basin and the Svalbard Basin started to subside again. In the Svalbard Basin this subsidence is more or less balanced by the input of clastics from the uplifted Greenland and Fennoscandian Shields, resulting in a gross regressive trend. A N—S-trending trough was continuously filled by deltaic progradation from the northeast and southwest (Mørk *et al.*, 1982, fig. 17; Rønnevik *et al.*, 1982, fig. 11).

During this time the Haltenbanken area probably developed as a (small) fault bounded basin with a restricted connection to the open sea. High evaporation resulted in the deposition of evaporites. It is tentatively suggested that the entrance to the open sea was towards the northwest and that anhydrites and dolomites were deposited in front of a 'barrier' (cf. Brunstrom and Walmsley, 1969) and on surrounding coastal sabkhas (cf. Taylor and Colter, 1975, Z2 model). In the basin centre, hypersaline brines were formed and pure halites were repeatedly deposited. Evaporitic conditions apparently also existed during the earliest Karnian in East Greenland.

During Tuvalian times the increasing aridity resulted in the deposition of redbed facies at the eastern and southeastern margin of the Svalbard Basin, behind the deltas. Erosion, enhanced by the arid conditions, resulted in the easterly progradation of a clastic wedge in East Greenland (Malmros Klint and Ørsted Dal Members, Fleming Fjord Fm).

Late Triassic—Norian (Fig. 7D)

Subsidence changed in magnitude and decreased in the Svalbard Basin as well as in East Greenland. Climatic conditions became more humid. The deposition of evaporites terminated in the Haltenbanken area. The early Norian seas transgressed the Svalbard Basin and established open marine conditions. The transgression probably extended southwards to East Greenland, where the marginal or shallow marine Tait Bjerg Beds were deposited. This transgression may also have affected the northern North Sea area.

Subsequently, alluvial deposits prograded in all areas. Uplift at the western margin of the northern North Atlantic Rift Basin, as well as in the Svalbard Basin, interrupted the sedimentation. The Haltenbanken area, however, continuously subsided until Rhaetian times.

SOURCE, RESERVOIR AND CAP ROCKS

Source rocks

An evaluation of the source rock potential of the Triassic in the Troms area, based on the study of kerogen, suggests that considerable amounts of oil- and gas-prone deposits are present in the area. Because of the marginal position of the Troms area during most of the Triassic the distribution of marine sapropelic kerogen ('type II') is more restricted than in the basinal parts of Svalbard (Forsberg and Bjorøy, 1982). Yet this type of kerogen has been recorded from Units 3 and 6 in the Troms wells. Important amounts of terrestrial sapropelic kerogen ('types II—III') have been recorded from various intervals, especially the Middle Triassic. In the Haltenbanken area significant source rocks are not present, except for the Rhaetian coals.

Reservoir rocks

The Triassic of Troms I is dominated by clay and siltstones, but moderately- to well-sorted, fine to medium grained, relatively clean sandstones with good porosity have been recorded. Such intervals occur, e.g. within Unit 6, the (?barrier) sandstones of Tuvalian age, as mentioned earlier, which delivered gas in 7120/12-2, as well as near the top of the Unit 4 (Anisian). The thicknesses of the individual sandstones rarely exceed 20 m, but where separated with only thin interbeds of shale, the possibility of finding a hydrocarbon column of prospective interest should exist.

On the basis of the present knowledge, the reservoir potential in the Haltenbanken area is restricted to the Norian and Rhaetian. In this interval well sorted, fine to medium grained sandstones with good porosity have been recorded, some of which have oil shows.

Since good reservoir rocks are to be expected in the marginal marine environment, the palaeogeographical maps give an indication of the spatial and vertical distribution of reservoir rocks during the Triassic.

Cap rocks

It is clear that the widespread occurrence of transgressive horizons, resulting in laterally continuous marine claystones, provides important cap rocks in the Troms area.

DISCUSSION AND CONCLUSIONS

The present account is based on still rather limited data from the Troms and Haltenbanken wells. Therefore, it is too early to draw elaborate conclusions concerning Triassic sedimentation north of 62°N.

As far as the similarity in subsidence patterns between the Svalbard Basin and East Greenland is concerned, the present data suggest that a single tectonic regime governed both areas during most of the Triassic. A four-fold subdivision can be recognized. This can be attributed to changes in interaction between the Greenland and Fenno-Scandian Shields. In the absence of factual data concerning the Lower and Middle Triassic in the Haltenbanken, it cannot be decided whether or not this tectonic regime extended to this area. The largely different subsidence pattern during the late Triassic suggests that Haltenbanken acted as a fault-bounded basin, separate from the North Atlantic Rift Basin, with a near continuous very rapid subsidence.

At present any interpretation of transgressive—regressive and/or coarsening upwards cycles in terms of eustatic sea-level changes is thought to be premature. The considerable subsidence during the early Triassic and the Karnian, with its related tectonic activity and clastic input, biasses the cycles recorded. On the other hand, the absence of major subsidence during most of the Middle Triassic and the Norian indicates that eustatic sea-level changes are responsible for the cycles recorded. In this respect, the early Anisian, early Lanidian and early Norian transgressions are outstanding.

Predicted palaeoclimatological patterns and lithological data from the wells, as well as from Svalbard and East Greenland, seem to match each other. Nevertheless, rather than applying a static model, it is here suggested that climatic variations and a northwards continental drift had a considerable impact on the distribution and type of sediments in this area during the Triassic.

ACKNOWLEDGEMENTS

The authors wish to thank Saga Petroleum A/S and their affiliates in licence 059 (6507/12), and Norsk Hydro A/S and their affiliates in licence 061 (7120/12) for their approval to publish this chapter; The Norwegian Petroleum Directorate for allowing preparation and offering facilities; Torhild Grytten for repeatedly typing the text; and Astrid Byberg Knutsen and Eila Røheim for skilfully drawing the figures.

REFERENCES

Ager, D.V. 1981. Major marine cycles in the Mesozoic. *J. Geol. Soc. Lond*, **138**; 159—166.

Balme, B.E. 1979. Palynology of the Permian—Triassic boundary beds at Kap Stosch, East Greenland. *Medd. Grønl. Bd*, **200**, 6.

Bertelsen, F. 1980. Lithostratigraphy and depositional history of the Danish Triassic. *Danm. Geol. Unders., Serie B*, no. 4, 59pp.

Bjærke, T. and Manum, S.B. 1977. Mesozoic palynology of Svalbard-I. The Rhaetian of Hopen, with a preliminary report on the Rhaetian and Jurassic of Kong Karls Land. *Norsk Pol. Inst. Skr.*, no. 165, 67pp.

Brennand, T.P. 1975. The Triassic of the North Sea. In: A.W. Woodland (ed.). *Petroleum and the Continental Shelf of North-west Europe*, Vol. 1, *Geology* Applied Science Publishers, London, pp. 295—311.

Brunstrom, G. and Walmsley, P.J. 1969. Permian evaporites in the North Sea Basin. *Am. Assoc. Petrol. Geol. Bull.*, **53**(4), 870—883.

Clemmensen, L.B. 1976. Tidally influenced deltaic sequences from the Kap Stewart Formation (Rhaetian—Liassic), Scoresby Land, East Greenland. *Bull. Geol. Soc. Denmark*, **25**, 1—13.

Clemmensen, L.B. 1980a. Triassic rift sedimentation and paleogeography of central East Greenland. *Bull. Grønlands Geol. Unders.*, 136, 72pp.

Clemmensen, L.B. 1980b. Triassic lithostratigraphy of East Greenland between Scoresby Sund and Kejser Franz Josephs Fjord. *Bull. Grønlands Geol. Unders.*, **139**, 56pp.

Clemmensen, L.B., Jacobsen, V.W. and Steel, R. 1980. Some aspects of Triassic sedimentation and basin development, East Greenland, North Sea and West Scotland. *The sedimentation of North Sea Reservoir Rocks.* Norsk Petroleumsforening Article 17.

Forsberg, A. and Bjorøy, M. 1982. A sedimentological and organic geochemical study of the Botneheia Formation, Svalbard with special emphasis on the weathering effect on the organic matter in shales. In: M. Bjorøy (ed.). *Advances in Organic Geochemistry 1982.* Heyden, London.

Håkansson, E. 1979. Carboniferous to Tertiary development of the Wandel Sea Basin. *Rapp. Grønlands Geol. Unders.*, **88**, 73—83.

Korchinskaja, M.V. 1980. Early Norian fauna of the archipelago of Svalbard. In: *Geology of the Sedimentary Platform Cover of the Archipelago of Svalbard.* NIIGA, Leningrad, pp. 30—43. (In Russian.)

Mørk, A., Worsley, D. and Knarud, R. 1982. Depositional environments of the Triassic and Lower Jurassic succession of Svalbard. *Can. Soc. Petrol. Geol. Mem.*, **8**, 371—398.

Odin, G.S. 1982. *Numerical dating in stratigraphy*, Vols. 1 and 2. John Wiley, New York.

Orlowska-Zwolinska T. 1979. Miospory. In: L. Malinowska (ed.), *Budowa Geologiczna Polski*, Tom III, *Mesozoic-Trias*, Wyd. Geol., Warszawa. (In Polish.)

Parrish, J.T. and Curtis, R.K. 1982. Atmospheric circulation, upwelling and organic-rich rocks in the Mesozoic and Cenozoic aras. *Palaeogeogr., Palaeoclim., Palaeoecol.*, **40**, 31—66.

Parrish, J.T., Ziegler, A.M. and Scotese, C.R. 1982. Rainfall patterns and the distribution of coals and evaporites in the Mesozoic and Cenozoic. *Palaeogeogr., Palaeoclimatol.*, *40*, 67—101.

Pedersen, K.R. and Lund, J.L. 1980. Palynology of the plant-bearing Rhaetian to Hettangian Kap Stewart Formation, Scoresby Sund, East Greenland. *Rev. Paleobot. Palynol.*, **31**, 1—69.

Rønnevik, H. Chr., Beskow, B. and Jacobsen, H.P. 1982. *Structural and Stratigraphic Evolution of the Barents Sea.* OD-publ. 82—08.

Smith, D.G. 1975. The stratigraphy of Wilhelmøya and Hellwaldfjellet, Svalbard. *Geol. Mag.*, **112**, 481—491.

Smith, D.G., Harland, W.B. and Hughes, N.F. 1975. The geology of Hopen, Svalbard. *Geol. Mag.*, **112**, 1—23.

Surlyk, F. and Clemmensen, L.B. 1983. Rift propagation and eustacy as controlling factors during Jurassic inshore and shelf sedimentation in Northern East Greenland. *Sed. Geol.*, **34**, 119—143.

Surlyk, F., Clemmensen, L.B. and Larsen, H.C. 1981. Post-Palaeozoic evolution of the East Greenland margin. *Mem. Can. Soc. Petrol. Geol.* **7**, 611—645.

Taylor, J.C.M. and Colter, V.S. 1975. Zechstein of the English sector of the southern north Sea Basin. In: A.W. Woodland (ed). *Petroleum and the Continental Shelf of North west Europe*, Vol. 1, *Geology* Applied Science Publishers, London, pp. 249—263.

Van der Eem, J.G.L.A., 1983. Palynological investigations in the Ladinian and Lower Karnian of the western Dolomites, Italy. *Rev. Palaeobot. Palynol.* **39(3/4)** 189—300.

Visscher, H. and Brugman, W.A. 1981. Ranges of selected palynomorphs in the Alpine Triassic. *Rev. Palaeobot. Palynol.*, **34**, 115—128.

Visscher, H. and Van der Zwan, C.J. 1981. Palynology and the Circum-Mediterranean Triassic: phytogeographical and palaeoclimatological implications. *Geol. Rundschau*, **70**(2), 625—636.

Ziegler, P.A. 1982. *Geological Atlas of Western and Central Europe.* Shell International Petroleum.

23 Geology of the East Greenland Shelf

H.C. Larsen
Grønlands Geologiske Undersøgelse, København, Denmark

A four-year geophysical data acquisition programme on the East Greenland shelf was completed in 1982. This includes 63 000 km of aeromagnetic data, 7800 km of multichannel reflection seismic data, 50 sonobuoy refraction seismic profiles and 7800 km of marine gravity data.

The extreme southern part of the East Greenland shelf consists only of Precambrian bedrock and some Tertiary volcanics. The remainder of the southern shelf south of lat. 70°N is a post-rift feature consisting of clastic sediments derived from the nearby rising continent and deposited on the subsiding, post-rift oceanic crust. In areas of little subsidence, the sedimentary shelf consists of a fairly thin and strongly prograding sequence deposited on flat basement. Up to 5 or 6 km of sediment have been deposited in areas of large subsidence. The Palaeogene is characterized by sediment aggradation whereas the shelf progrades in the Neogene. The crustal flexure between the rising continental crust and the subsiding oceanic crust show variations in amplitude and position in relation to the coast. Where most intense, it is developed as a huge escarpment with 5 to 8 km relief just outside the coast. North of lat. 70°N a different shelf style is present comprising a number of pre-rift basins of Mesozoic and older age. In the region between latitudes 70°N to 74.5°N these pre-rift basins subsided together with and in some cases more than the adjacent oceanic crust. They received a cover of post-rift clastic sediments of 1 to 5 km in thickness. Seismic data are lacking from the northernmost shelf, but aeromagnetic data indicate the presence of a number of fault-bounded, elongate coast-parallel deep basins of Mesozoic and older age. The cover of post-rift (Tertiary) sediments appear absent to thin (inner shelf) or of intermediate thickness (outer shelf).

INTRODUCTION

Geologically, the East Greenland shelf remained virtually unexplored for many years while the onshore East Greenland Mesozoic basins were intensely studied and served as a geological model for the interpretation of the offshore NW-European margin. This development was a natural consequence of the almost year-around presence of polar pack-ice on the East Greenland shelf. The first sparse geophysical information from the East Greenland shelf was published by Vogt (1970), Johnson *et al.* (1975a, b), B. Larsen (1975), Larsen (1975, 1978), Henderson (1976) and Featherstone *et al.* (1977).

Results of the first multichannel reflection seismic surveying on the shelf were reported by Hinz and Schlüter (1978, 1980). On the basis of this information and initial results from a new aeromagnetic survey (Larsen and Thorning, 1979, 1980), a regional geological model for the East Greenland shelf was proposed by Larsen (1980). This model involved early Tertiary oceanic crust to be present below parts of the outer shelf off central East Greenland and subsided continental crust to be present below the shelf slope and rise off SE-Greenland. The actual position of the shelf edge was found to be controlled mainly by post-rift sedimentation rather than deep crustal features (Larsen, 1980, figs. 4, 8 and 9). The predominance of Tertiary sediments on the southern half of the shelf compared to the prevalence of Mesozoic and older basins on the northern shelf was also pointed out (Larsen, 1980, p. 95).

The present study follows a four year marine geophysical acquisition programme conducted by the Geological Survey of Greenland. This includes a regional coverage of high quality multichannel reflection seismic, refraction seismic and marine gravity data (Larsen, 1983 fig. 1). These data provide, in the area south of latitude 72°N, good control of the sediment thickness, the general stratigraphy and the structure of the sedimentary shelf. The data also provide information on the

Petroleum Geology of the North European Margin © Norwegian Petroleum Society (Graham & Trotman, 1984) pp. 329—339.

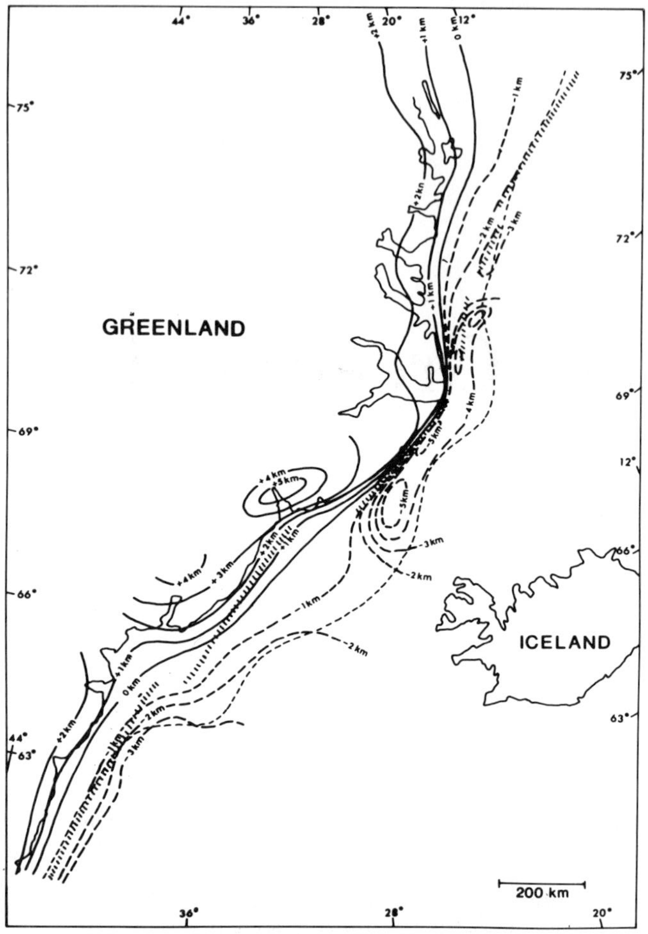

Fig. 1. Post-rift (post-Palaeocene) vertical movement of the crust along the East Greenland margin.

structure and origin of the basement below the sedimentary shelf. Seismic data are sparse north of latitude 72°N and were never recorded north of latitude 74°N. Hence, the interpretation of the stratigraphy and the basin structures on the northern shelf still remains based on information deduced from aeromagnetic data.

The first integrated examination of all the geophysical data now available provides an image of a rather non-typical passive rifted margin. The main characteristic features are:

(1) the presence of layered early Tertiary oceanic crust below most of the southern half of the sedimentary shelf, implying non-presence of pre-rift sediments in that area;

(2) the limited or absent cover of post-rift sediments within the Mesozoic and older basins on the northern shelf. These observations imply for East Greenland a pattern, different from most other passive margins, of strong vertical post-rift movements of the crust (Fig. 1).

Along the southern margin, the differential vertical movements concentrated near to the ocean-to-continent transition (OCT), leading to the formation of a crustal flexure or escarpment with down-warped oceanic crust and uplifted continental crust. Along the northern-most margin, differential vertical movements spread over a much broader zone reaching from far inland (uplift) over the broad continental shelf to the shelf slope (subsidence). The inner half of the northern shelf is considered to have remained almost neutral with regard to post-rift vertical movements. Only along a fairly small part of the margin (middle part), pre-rift crust and pre-rift sediments subsided either together with or more than the adjacent oceanic crust, thus coming close to the supposed typical passive margin setting.

From a plate tectonic viewpoint, the finding of a wide zone of strongly layered pre-anomaly 24 oceanic crust off SE-Greenland (Larsen, in preparation) is important and suggests the existence of an anomaly 25 to anomaly 24 high-rate spreading phase in the NE-Atlantic.

PHYSIOGRAPHY

The detailed morphology of the East Greenland shelf is badly known especially in the northern part where even information on the regional bathymetry is sparse. Studies of the shelf morphology have been published by Sommerhoff (1973), B. Larsen (1975, 1980, 1983) and Johnson *et al.* (1975b). Larsen (1980) presented a small scale compilation of regional shelf and slope bathymetry. The regional bathymetric maps by Perry *et al.* (1980) and Grønlie and Talwani (1978) cover part of the East Greenland shelf.

In general, the East Greenland shelf shows rather large water depths typically in the range of 250 m to 350 m. A number of banks, often situated at the outer shelf, show water depths around 200 m and a number of mainly transverse shelf channels originating from the major fiords show water depths between 400 m and 600 m on the sedimentary shelf, but can be as much as 1000 m deep where basement is exposed. Some smaller banks situated close to the coast are reported to show water depths of only 20 m to 100 m.

The morphology of the sea bed is irregular where basement is exposed. Precambrian basement is characterized by "hilly" topography whereas the areas of Tertiary lava exposure show a cuesta like morphology (B. Larsen, 1983, fig. 5). The sedimentary shelf is normally fairly flat with only slight changes in water depth. The monotony is only disturbed by the broad and mainly transverse channels and by the intense iceberg scouring of the shallower banks (B. Larsen, 1983, fig. 8).

The lithology of the sea bed on the sedimentary shelf is poorly known. Some dredge samples were described by Sommerhoff (1973) and Johnson *et al.* (1975a, b). Clastic sediments and volcanics ranging from the Carboniferous to the Oligocene in age were reported, suggesting a major component of glacial drift material. The sea bed of the sedimentary shelf off East Greenland appears very hard and initiates a sequence of strong water bottom multiples during seismic work.

SETTING

The East Greenland shelf is bounded landwards by a variable geological scenario changing with latitude (Fig. 2). A summary of the onshore geology was given by Larsen (1980) and Surlyk *et al.* (1981).

In short, the southern half of the East Greenland onshore area is made up by Precambrian basement rocks of old Proterozoic and Archaean age. The northern half comprises a coast-parallel mountain

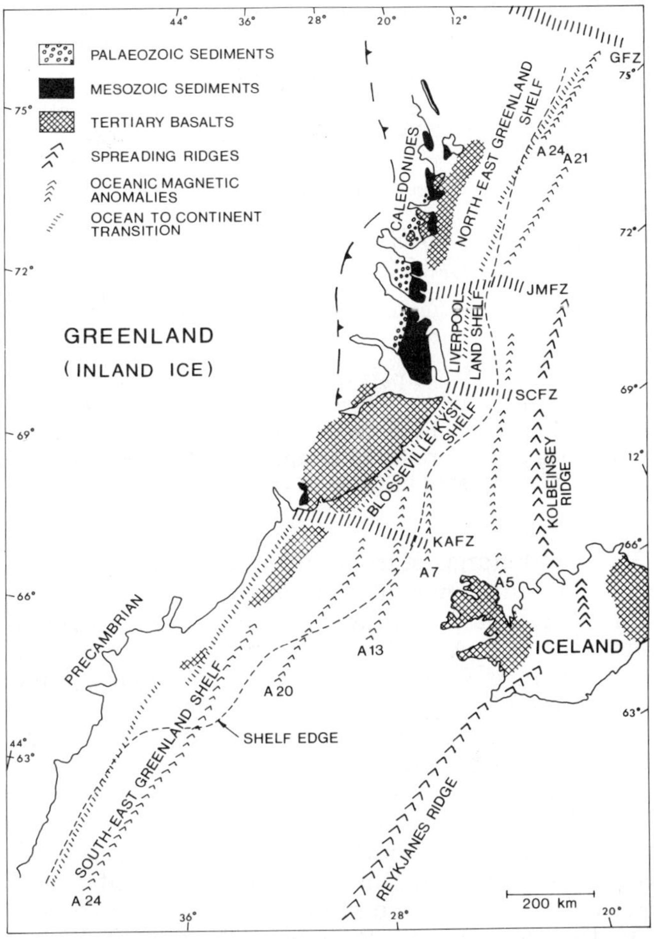

Fig. 2. Main structural provinces off East Greenland with simplified onshore geology: (KAFZ) Kangerdlugssauq Fracture Zone; (SCFZ) Scoresby Sund Fracture Zone; (JMFZ) Jan Mayen Fracture Zone; (GFZ) Greenland Fracture Zone.

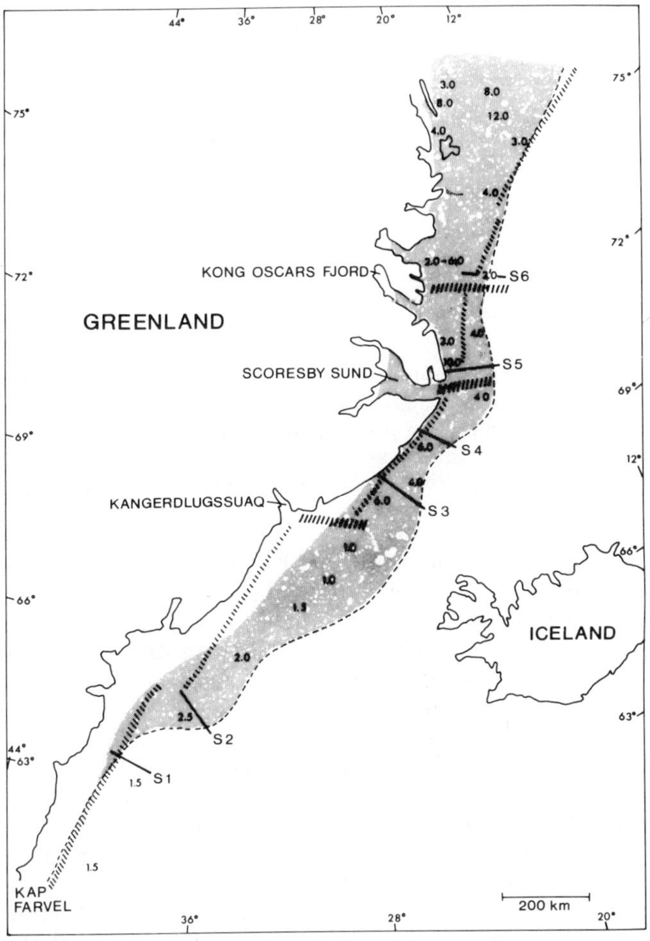

Fig. 3. Sediments on shelf (shaded) and depth to acoustic basement (south of JMFZ) and magnetic basement (north of JMFZ). Position of seismic profiles S1 through S6 shown. Seismic profiles shown in Figs. 4—7, 9 and 11.

belt of Caledonian age which was later followed by Palaeozoic and Mesozoic basin formation along the same trend. At the conjunction between these two geologically different regions, vast volumes of early Tertiary volcanics have overflooded the margin and subaerially erupted lavas make up the coast for several hundred kilometres (see Fig. 2). Important intrusive and tectonic activity of the same age also took place in even wider regions. A well known example of the latter is the formation of the so-called East Greenland dyke swarm and coastal flexure (Wager and Deer, 1938). Finally, most of the East Greenland land area was uplifted from 1 to 3 km during the Cenozoic.

The seawards boundary of the East Greenland continental margin is marked by the onset of Tertiary oceanic crust formed along the Reykjanes Ridge, the Icelandic spreading centre, the Kolbeinsey Ridge, the Mohns Ridge and the Knipovitch Ridge all listed from south to north. The type of transition from continental to oceanic crust is believed mainly to be of the passive rifted type except for the very northern part of the margin (north of Mohns Ridge) where both sheared-type passive margin and rifted type passive margin along the Greenland Fracture zone (GFZ) and the Knipovitch Ridge may occur.

There are, however, very few data published in support of any specific location and development of the continent to ocean transition (see Larsen, 1980 for review). Also the so-called marginal escarpments suggested to form the ocean-to-continent transition on the NW-European margin (Talwani and Eldholm, 1977) have not been located off East Greenland except for the Greenland Escarpment (Eldholm and Windisch, 1974; Talwani and Eldholm, 1977). Both Vogt *et al.* (1980) and Larsen (1980) have suggested that oceanic crust may be present below substantial parts of the East Greenland shelf in the Denmark Strait area and northwards to the mouth of the Scoresby Sund fiord.

STRUCTURE AND STRATIGRAPHY

The main structural elements of the East Greenland margin are shown in Figs. 2 and 3. Three transform zones are believed to separate four different structural provinces. Important structural changes take place across these transform zones while each of the four tectonic provinces appears fairly uniform. The tectonic diversity includes changes in crustal type, amount of basin subsidence, contrasting basement tectonics and different timing of basin formation.

The three transform zones are situated as follows:
(1) in the northern Denmark Strait just outside the mouth of the Kangerdlugssuaq fiord;
(2) at or just south of the mouth of the Scoresby Sund fiord;

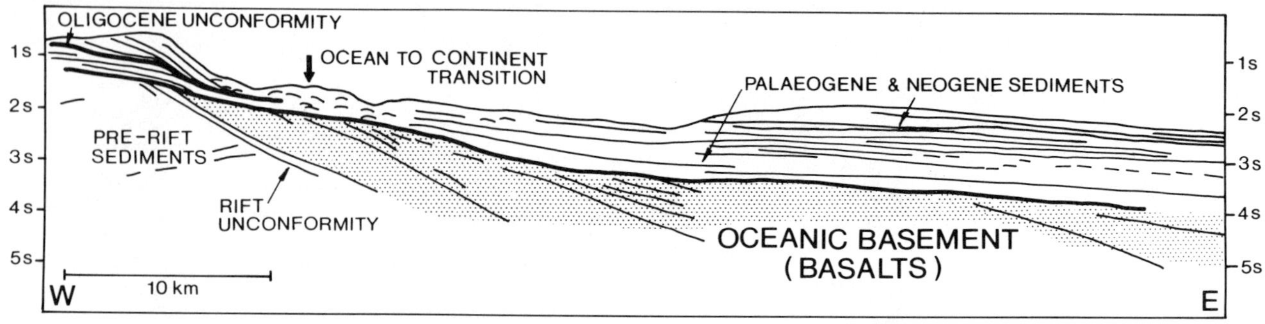

Fig. 4. Interpreted multi-channel seismic profile GGU81-13 across the southern part of the south-east Greenland shelf and slope. Location indicated in Fig. 3 (S1). The oceanic basement is strongly layered and shows exclusively seaward dip. The presence of landwards tilted pre-rift sediments are indicated in the western end of the profile. The nature and thickness of the pre-rift sediments are not known. A strong unconformity of assumed mid-Oligocene age is present within the shelf sediments. The oceanic magnetic anomaly 24 is situated just east of the profile.

(3) outside the mouth of Kong Oscars fiord.

The recent equivalent to (1) is to be found in the transform zone of southern Iceland and it is named the Kangerdlugssuaq Fracture Zone (KAFZ). The next fracture zone (2) may have its recent equivalent in the Spar Fracture zone north of Iceland (Vogt *et al.*, 1980) and is named the Scoresby Sund Fracture Zone (SCFZ). The third important fracture zone (3) is a direct landwards continuation of the wellknown Jan Mayen Fracture Zone (JMFZ).

The four areas separated by the three fracture zones have been named the following: the south-eastern shelf (SEAS), the Blosseville Kyst shelf (BLKS), the Liverpool Land shelf (LILS) and the North-eastern shelf (NEAS).

South-eastern Shelf (SEAS)

A major structural element in the Cenozoic development of this area is the formation of a coast-parallel monoclinal flexure separating uplifted continental crust from subsided oceanic crust. The ocean-to-continent transition (OCT) and the associated flexure follow the shelf edge in the very southern part of the SEAS and come very close to the coast in the north. Thus, most of the very southern shelf is made up by exposed continental basement (Precambrian) with some Tertiary volcanics and the northern part of the SEAS is floored by subsided oceanic crust covered with Tertiary sediments. The southernmost (basement) shelf is quite narrow (75 km), but widens northwards to about 175 km.

Sea-floor spreading along this part of the East Greenland margin started under subaerial conditions at or just after magnetic anomaly 25 about 80—90 km inland of the oldest well-developed magnetic anomaly (24 A and B, Larsen, 1980; Larsen, in preparation). Following the start of spreading and the incipient formation of the crustal monoclinal flexure, a fairly shallow sea overflooded the recently formed oceanic crust from the south. This marine embayment was open to the south and may have extended as far north as to the transverse escarpment within the southern part of the Denmark Strait (see below). During the very early stages of this marine basin, an eastern basin limit was formed by the subaerially spreading Reykjanes Ridge. However, an open sea-way was established to the east with the Eocene Rockall basin following subsidence of the Reykjanes Ridge below sea level during the Eocene (Larsen, in preparation).

A total of up to 700 m of presumed shelf carbonates, cherts, clastic sediments (mainly western and northern part) and volcanic-clastic sediments (mainly eastern part) were laid down in this south-east Greenland basin during the latest Palaeocene to early Oligocene.

The palaeo-setting of the basin was changed at mid-Oligocene time by a strong sea level drop (Vail *et al.*, 1977) and by accelerated uplift of the Greenland block to the west. The approximately 300—400 m drop in sea level caused a 50—60 km seawards displacement of the depot centre axis and led to the formation of a pronounced unconformity along the western part of the basin (Figs. 4, 5). However, in response to both repeated sea level

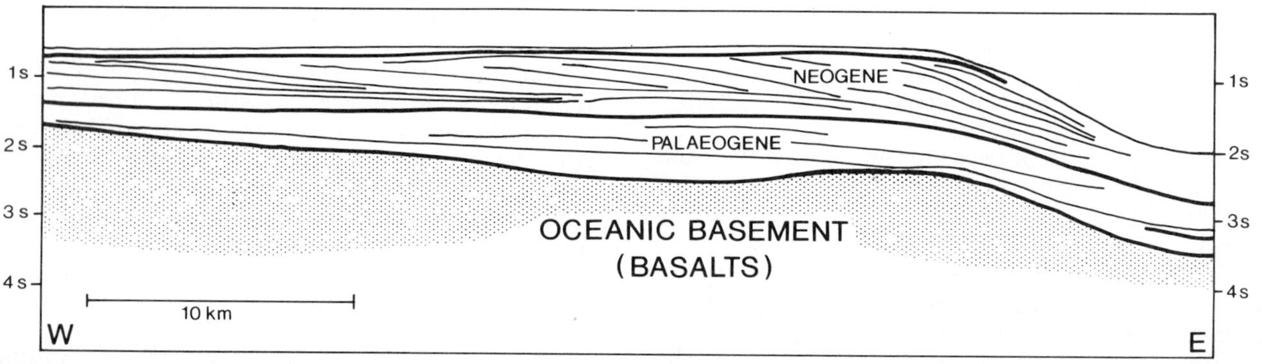

Fig. 5. Interpreted multi-channel seismic profile GGU81-05 across the middle part of the south-east Greenland shelf. Location indicated in Fig. 3 (S2). Compared to profile in S1 in Fig. 4, the shelf has prograded further onto the oceanic basement. Note the presence of the strong mid-Oligocene unconformity in the middle of the sedimentary section.

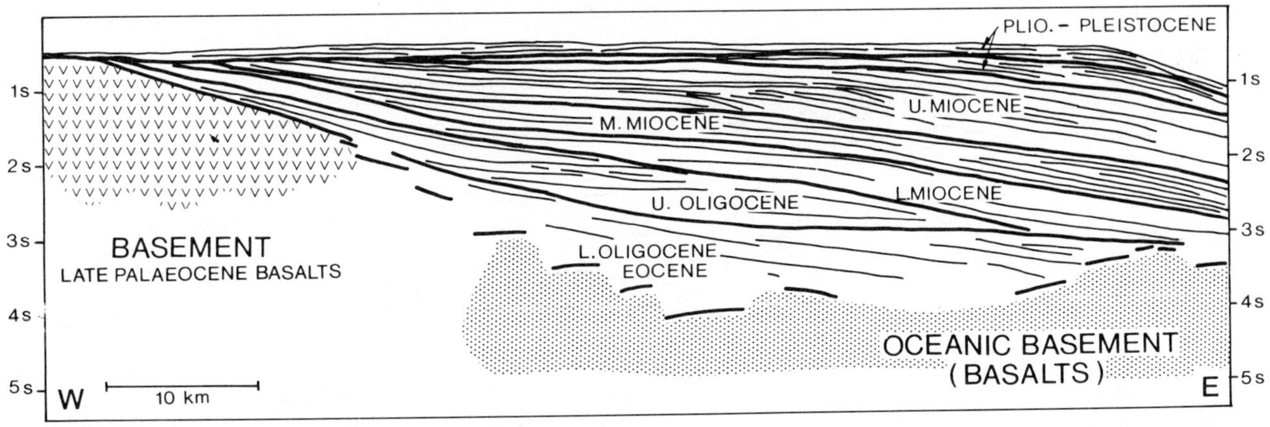

Fig. 6. Interpreted multi-channel seismic profile GGU82-11 across the southern part of the Blosseville Kyst shelf. Location indicated in Fig. 3 (S3). The volcanic basement within the western end of the profile continues into the late Palaeocene coastal basalts. The basement at the eastern end of the profile is oceanic basement of approximately anomaly 13 age and continues eastwards into a smooth, unbroken basement surface. The apparent basement topography in the middle of the profile is partly an artifact (see text). Note the eastward down-flexuring of the volcanic basement. Ocean-to-continent transition below flexure zone. Part of the upper Miocene sequence may be of only Plio-Pleistocene age.

rise and increased crustal flexure along the initial line of rifting (OCT) a new pattern of basin filling developed during the Miocene. Deposition of a strongly progradational wedge of clastic sediments in front of the crustal flexure zone was initiated along with abyssal current deposition/erosion within the more distal and deeper lying parts of the basin (Fig. 4).

The outbuilding clastic wedge making up most of the present day sedimentary shelf has prograded only 5 to 10 km within the very southern area and between 50 and 70 km along the middle and northern part of the south-east Greenland shelf. The little prograded areas show very steep (5—10°) sigmoidal clinoforms (Fig. 4) and are probably the result of reduced clastic input and more intense contour current activity compared with the much more prograded areas of comparable thickness (Fig. 5) further north (1—3° dipping clinoforms).

Evidence for the existence of pre-rift sediments on this part of the East Greenland margin is virtually absent except for one small exposure at the coast (Kap Gustav Holm, Larsen, 1980) and some sub-volcanic reflections (see Fig. 4).

The basement for this post-rift sedimentary development is in most cases very flat, seismically well defined and affected very little by faulting. Its oceanic character is inferred either from the presence of linear, oceanic magnetic anomalies or from the presence of seawards dipping, divergent sub-basement reflectors (Fig. 4). The latter phenomenon has been discussed by various authors (Mutter *et al.*, 1982; Hinz, 1981; Smythe, 1983; Larsen, in preparation), and is used in this contribution as an oceanic crust indicator.

The basement is cut by a transverse escarpment of about 500 m relief along the southern edge of the Denmark Strait. North of this escarpment, i.e. within the Denmark Strait area, the oceanic basement has only subsided to very shallow depth (1 km), and hence, the coast-parallel flexure zone associated with the OCT in this area is only weakly developed. The pre-mid-Oligocene marine sequence is absent in large parts of the Denmark Strait area which apparently formed an emergent or very shallow water transverse ridge until the early Miocene.

Blosseville Kyst Shelf (BLKS)

The tectonic province called the Blosseville Kyst shelf (BLKS) in this contribution is bounded to the south by the KAFZ and the uplifted basement below the Denmark Strait. The BLKS is bounded northwards by the SCFZ. It thus stretches for about only 300 km along the coast in contrast to the much larger southern tectonic province (SEAS). The continental shelf in the BLKS is quite wide to the south (around 200 km) but narrows northwards to about 100 km.

The continental shelf in this region has the same general setting as the south-east Greenland shelf. It is made up of post-rift sediments of mainly clastic origin and deposited in front of a pronounced crustal flexure. The BLKS is different, however, from the SEAS, by showing much larger subsidence and greater sediment thickness and by having a seismically less well defined basement. The basement displays a fairly strong topography and shows no obvious sign of seawards dipping sub-basement reflectors (Fig. 6). Also, the landwards bounding crustal flexure zone is much more intensely developed in the BLKS compared with the flexure zone on the southern shelf, and it locally develops into a major crustal escarpment (northern part of the BLKS, Fig. 7). The flexure/escarpment zone is named the Blosseville Kyst escarpment (BKES, Fig. 8).

The basement flooring the sedimentary shelf is evidently of volcanic origin because of its layered nature and its continuation into proven oceanic volcanic basement (seawards) and into the onshore continental volcanics (landwards). The rather 'hefty' topography displayed by this volcanic basement appears very little related to faulting, but can be explained by, in part synvolcanic, clastic sedimentation on a highly irregular volcanic basement surface.

The volcanic basement in the BLKS is believed to be mainly of oceanic origin seawards of the coast-parallel flexure and escarpment zone, the latter forming the approximate ocean-to-continent transition. Such an interpretation is supported by the studies of magnetic lineations, gravity modelling, local seawards continuation of basement into un-

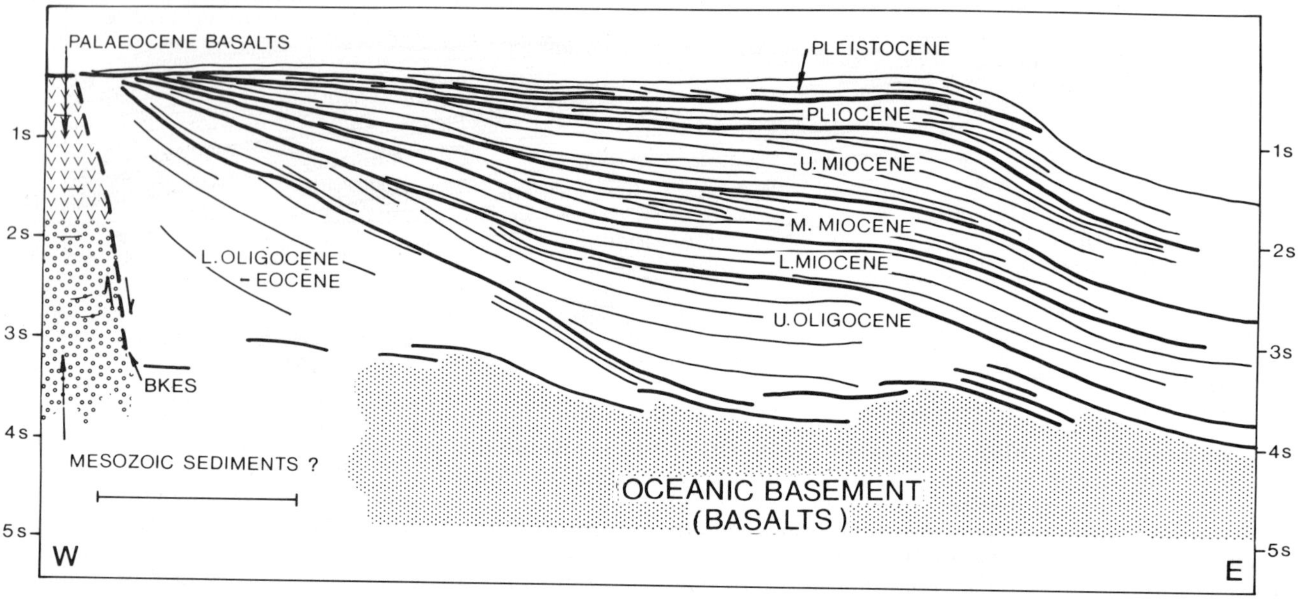

Fig. 7. Interpreted multi-channel seismic profile GGU82-28 across the northern part of the Blosseville Kyst shelf. Location indicated in Fig. 3 (S4). The western end of the profile comes almost to the coast and shows outcrop of late Palaeocene lavas at sea bed. The eastward down-flexuring of the basement also seen in Fig. 6 is here developed as a major seawards facing escarpment (Blosseville Kyst Escarpment, BKES). Possible Mesozoic sediments below the Late Palaeocene lavas may be in contact with the mainly Eocene wedge of fan sediments below the inner shelf.

questionable oceanic basement, refraction seismics and the non-faulted character of the strong subsidence of the area in post-rift time.

The volcanic basement believed to be present below the BLKS was formed between anomaly 23 and 6. Magnetic lineations are not coast-parallel (Larsen, 1980, Fig. 6; Vogt *et al.*, 1980) and transform terminations of anomalies against the coast-parallel flexure/escarpment are consequently a necessity (see Fig. 8). The lack of post-anomaly 24 volcanic extrusives in the nearby onshore region (young intrusives and faulting are present) means that the coast-parallel flexure/escarpment was initiated just after initial rifting and prevented the younger offshore (oceanic) volcanism in over-flooding the continental areas. This taken together with the general stratigraphical development offshore (see Figs. 6 and 7) suggests initial high

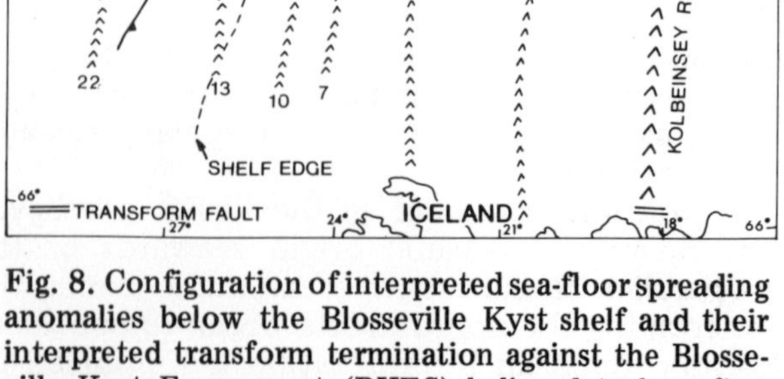

Fig. 8. Configuration of interpreted sea-floor spreading anomalies below the Blosseville Kyst shelf and their interpreted transform termination against the Blosseville Kyst Escarpment (BKES) believed to be a first order crustal feature (approximate ocean-to-continent transition).

rate subsidence in the Palaeogene followed by more moderate subsidence in the Neogene on the inner shelf.

The basement of the outer shelf was mainly formed during the Neogene and subsided as much as 3 km during the latest Neogene in response to thermal contraction and loading.

Sedimentation within the BLKS area started in the early Eocene with deposition of fluviatile and lacustrine sediments into local depressions within the volcanic surface. The fast formation of the flexure/escarpment just outside the present day coast, however, caused predominant fan-sedimentation along this zone from about mid-Eocene time. The sediments were derived from the rising, volcanic land area to the west and deposited into the subsiding area to the east of the escarpment and below the present day inner shelf. Fan-sedimentation was mainly of submarine nature.

The depression and the sea into which these fan systems were deposited was initially an inter-continental sea with very restricted or absent marine sea-ways to the north Atlantic/Arctic ocean. The sea was bounded to the west by the rising Greenland landmass, to the south by the uplifted Denmark Strait ridge, and to the east by the subaerially spreading Kolbeinsey Ridge.

A possible, but narrow sea-way to the north, into the Greenland Sea, may have been present from the very beginning but is, however, evidently present from not later than the mid to late-Oligocene time when a dramatic 300-400 m sea level drop changed the whole pattern of sedimentation and gave rise to the formation of a marked unconformity (see Fig. 7). This marked sea level drop is believed to be associated with the well known regional eustatic sea level drop reported by Vail *et al.* (1977) and many others. Following the Oligocene sea level drop, the depo-centre moved seawards and mainly vertical shelf aggradation took place during the lower and middle Miocene. During this period, a balance between subsidence, sea level rise and sedimentary input was maintained.

In the late Miocene and the Plio-Pleistocene, increasing fluctuations in sea level stand most likely in combination with increased input of sediments (renewed uplift of Greenland block?) led to episodes of strong shelf progradation alternating with more vertical aggradation (Fig. 7). The upper Miocene and Plio-Pleistocene sequence locally achieves thicknesses of about 2.5 km and the shelf edge advanced seawards by between 10 and 30 km during this period.

Liverpool Land Shelf (LILS)

The Liverpool land shelf (LILS) is separated from the BLKS in the south by the Scoresby Sund Fracture Zone (SCFZ) and extends northward to the Jan Mayen Fracture Zone (see Figs. 2 and 10). The LILS thus stretches for about 350 km along the East Greenland coast and the shelf varies in width only between 100 and 125 km. The outer shelf is evidently floored by oceanic crust of about anomaly 6 age. Landwards, this part of the shelf faces Precambrian and Caledonian basement, Mesozoic basins and locally, Tertiary plutonic centres.

Structurally, the LILS can be divided into two coast-parallel units, one comprising the inner shelf and the other comprising the outer shelf. The inner shelf area is bounded both landwards and seawards by approximate coast-parallel escarpments and thus provides a graben-like structure. The landwards escarpment is situated just outside the coast along Liverpool Land (Precambrian to Caledonian basement) and thinned continental crust is downfaulted 4—6 km on its seawards side.

The down-faulted crust shows evidence of faulting and landwards rotation of fault blocks. The escarpment named the Liverpool Land Escarpment (LIES, see Fig. 10) was active in the Palaeogene, but may well have taken up the position of older Mesozoic and Palaeozoic fault-systems, in which case the crystalline basement is downfaulted in excess of the 4—6 km mentioned above. Aeromagnetic depth-mapping and the presence of reflectors below the rift-unconformity indicate very deep basement. Pre-rift sediments may thus be present within the deeply buried and rotated fault-blocks.

The inner shelf basin is bounded seawards by another less well developed and lower amplitude, coast-parallel escarpment separating the deeply buried basement of the inner shelf basin from the shallower basement on the outer shelf (Fig. 9). This escarpment is not believed to be associated with any major vertical crustal movement but is merely caused by the landwards termination of the volcanic basement of the outer shelf. The escarpment is not thought as a well-defined edge of the outer shelf volcanic basement, but is rather of transitional character with interdigitating volcanics and sediments. Notable linear, deeply sourced magnetic and gravity anomalies are however present parallel to, and slightly seawards of the escarpment suggesting important crustal changes, possibly the ocean-to-continent transition, across the escarpment zone.

The outer shelf basin is floored by a flat and in most cases non-faulted, well defined acoustic basement, interpreted as being volcanic and oceanic in origin. The interpretation of the basement is based on its evidently layered nature, its simple continuation into younger oceanic basement of anomaly 5 to 6 age, the presence of weak but very linear magnetic anomalies of inferred anomaly 6 to 6c age, refraction seismics and gravity. The local presence of seawards dipping sub-basement reflectors further support this interpretation.

The inner shelf area of the Liverpool Land shelf (LILS) was most likely continuous with the Jan Mayen Ridge before separation of this crustal segment from the Greenland block (Talwani and Eldholm, 1977; Hinz and Schlüter, 1980; Vogt *et al.*, 1980; and Larsen, 1980). Thinning of the crust in association with formation of landwards rotated fault-blocks was followed by a strong seawards subsidence along the coastal escarpment (see Fig. 9). Dating of this tectonic episode is impeded by the lack of late Palaeocene to early

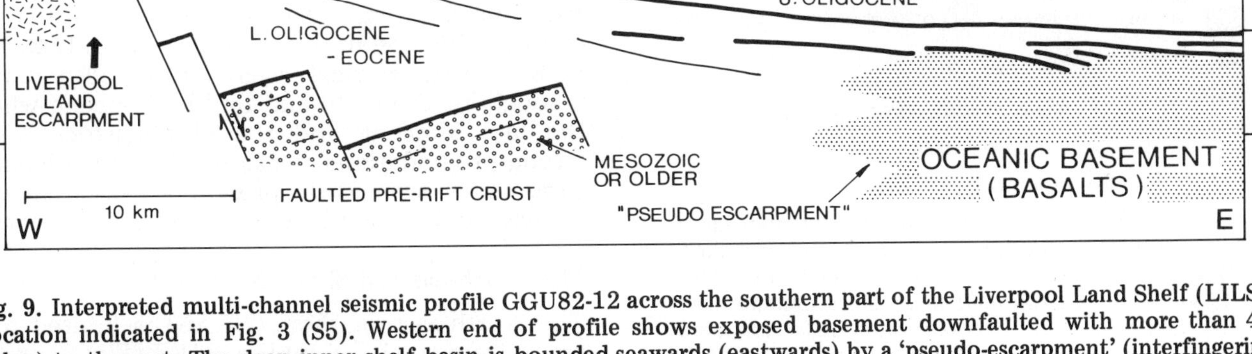

Fig. 9. Interpreted multi-channel seismic profile GGU82-12 across the southern part of the Liverpool Land Shelf (LILS). Location indicated in Fig. 3 (S5). Western end of profile shows exposed basement downfaulted with more than 4 s (6 km) to the east. The deep inner shelf basin is bounded seawards (eastwards) by a 'pseudo-escarpment' (interfingering volcanics and Palaeogene sediments).

Eocene volcanics providing good age control of the very similar development in the northern part of the Blosseville Kyst Shelf (BLKS). By analogy, however, and as a fair amount of non-disturbed Eocene to early Oligocene sediments are present, the main tectonic episode of block-faulting and escarpment formation is suggested to be of early to mid Eocene age. This tectonic episode therefore partly correlates in time with the main tectonic episode in the BLKS area and the rifting east of the Jan Mayen Ridge, i.e. within the Norway basin (Talwani and Eldholm, 1977). This implies that a similar type of block-faulting might be present below the Jan Mayen Ridge area if this area is a continental fragment. In contrast, however, to Hinz and Schlüter (1980), we do not find evidence for any younger strong tectonism in the area in connection with the later separation of the Jan Mayen Ridge area from the Greenland margin.

The crustal thinning and associated strong subsidence rapidly changed the setting from an emergent or shallow sea area of fairly low relief into a deep tectonic depression separated from the present day land area by a large topographical escarpment. The floor of this tectonic depression also showed fairly strong topography due to rotation of fault-blocks.

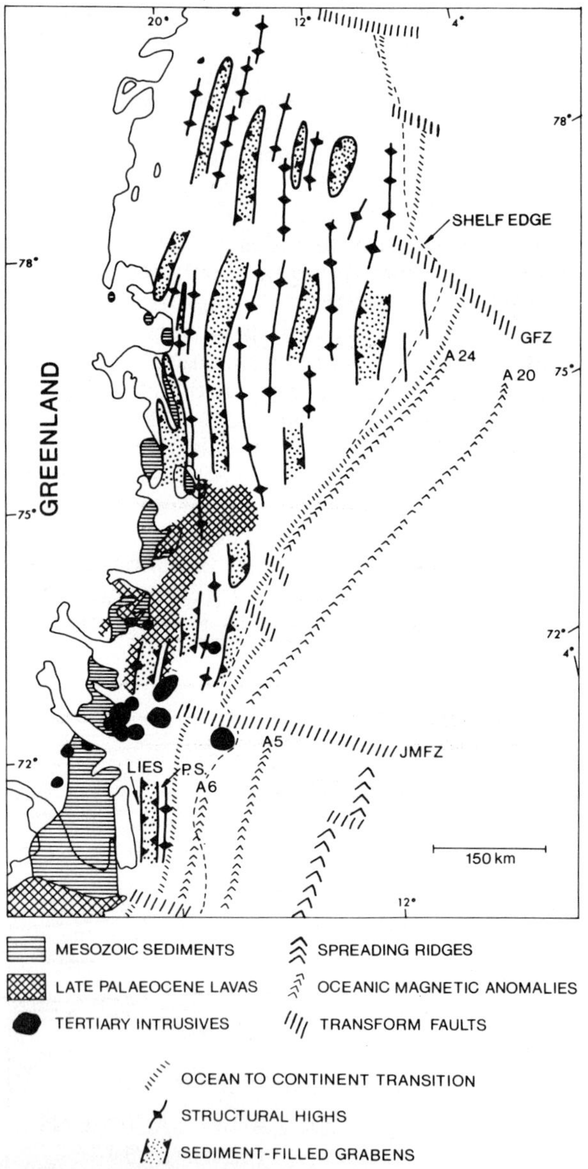

Fig. 10. Regional tectonic map of the Liverpool Land shelf area (LILS) and the North-east Greenland shelf (NEAS). Sediment filled grabens assumed to contain 6-12 km of Palaeozoic and Mesozoic sediments (see text). Sediment cover over elongated highs estimated to be of approximately 2-4 km in thickness. (LIES) Liverpool Land escarpment, (P.S.) Pseudo escarpment.

Some continental to shallow water clastic deposits of early Eocene age may locally be present between the tilted blocks. However, as a consequence of the rapid subsidence and the associated escarpment formation, sedimentation soon became dominated by fairly deep water fan sedimentation in front of the coastal escarpment (Fig. 9). The more distal parts of this sequence were deposited on the Jan Mayen Ridge as deep water, complex fan sediments and turbidites (Talwani *et al.*, 1976a, b). Unimpeded passage of sediments from the Greenland margin into the Jan Mayen Ridge area thus apparently existed throughout the Eocene and early Oligocene.

This scenario changed during the mid-Oligocene. This was mainly because of incipient formation of a volcanic ridge between the Greenland margin and the Jan Mayen Ridge and because of a strong sea level drop during the later part of the Oligocene. The strong sea level drop, like in other parts of the East Greenland margin, gave rise to the formation of a strong unconformity.

The volcanic ridge coincides with the position of the ocean-to-continent transition inferred in this paper. The ridge acted as a positive structure from well before to somewhat after the formation of the marked unconformity. The western flank of the volcanic ridge constitutes the 'pseudo-escarpment' discussed above and forms the seawards boundary of the inner shelf basin within the LILS area (Figs. 9, 10). The volcanism associated with the ridge formation apparently commenced well before the start of spreading between anomaly 7 to 6c (this paper) and was most likely submarine during the earlier stages. However, the spreading ridge later formed (Kolbeinsey Ridge) was mainly emergent during the early and middle Miocene and first became permanently submarine around anomaly 5 to 5a (late-middle to early-upper Miocene).

Sedimentation within the LILS area from the late Oligocene to present follows the same trend as seen in the BLKS area to the south. Repeated sea level rise and continued subsidence provided the formation of a shallow to intermediate deep marine embayment open to the north. The marine embayment extended as far south as the Denmark Strait area, i.e. through the BLKS area and was bounded to the west by the huge coastal escarpment and to the east by the active Kolbeinsey Ridge until anomaly 5a to 5 time. Since that time open marine conditions have prevailed. Like in the BLKS area, vertical aggradation predominated throughout the early and middle Miocene whereas the late Miocene and the Plio-Pleistocene show strong progradational features. Some erosional features can be seen on the present day slope and could be caused by incipient contour current activity.

The North-eastern Shelf (NEAS)

The North-eastern shelf (NEAS) is limited to the south from the LILS by the JMFZ (Figs. 2 and 10). To the north, the NEAS is believed to terminate against the Greenland Fracture Zone (GFZ). The NEAS thus stretches for about 700 km along the coast and varies in width from 125 km (southern part) to about 300 km (northern part).

Information on the deep structure of this northern shelf segment mainly has to be derived from aeromagnetic data as only limited seismic information is available and only from the very southern part. However, the aeromagnetic data coverage is good (Larsen and Thorning, 1980; Larsen, 1980), and provides along with onshore Mesozoic basin exposures and fault studies (Surlyk, 1977; Surlyk *et al.*, 1981) fairly firm structural/stratigraphic conclusions.

Compared especially with the seaward part of the LILS, the NEAS is structurally different by showing a much stronger basement topography caused mainly by N—S to NE—SW trending faults and by showing much less or no subsidence (see Fig. 1) in post-rift time (post-Palaeocene). As a consequence of this, the strong coastal flexuring or escarpment-formation present within the BLKS and the LILS area to the south is not seen in the NEAS. The general post-rift subsidence pattern is that of a gentle, low amplitude (1—1.5 km) seawards down-flexuring along the coast and a somewhat stronger (2—3 km) seawards down-flexuring below the outer shelf. Tertiary deposits on the inner shelf may therefore locally be absent and are not likely to exceed about 1 km in thickness. The Tertiary deposits on the outer shelf may achieve about 3 km in thickness (Fig. 11).

The NEAS can be divided into a southern area affected by Tertiary volcanism (and faulting?) and a northern area without any sign of major Tertiary events (Fig. 10).

The Tertiary volcanism in the southern area does comprise both emplacement of major plutons and extrusion of lavas as it is seen onshore (see Noe-Nygaard, 1976). The plutonic activity seems to have been concentrated close to the JMFZ whereas a partly erratic occurrence of lavas at sea bed level has a much wider distribution (inner shelf). The lavas also seem present at greater depth below younger deposits at the outer, southern part of the NEAS.

The aeromagnetic depth-mapping of the basement topography in this region is somewhat impeded by the presence of the Tertiary volcanics at different levels within the sedimentary section. Ridge crests at about 2 km depth and basins of more than 6 km thickness are indicated.

The northern part of the NEAS (74°N—78°N) is dominated by a number of very long, coast-parallel graben-ridge systems (Fig. 10). Correlation of these structures can be made with the onshore basins (Larsen, 1980; Surlyk *et al.*, 1981). Since little movement between the northern shelf and the Barents Sea area took place within the period late Carboniferous to late Cretaceous (Harland *et al.*, 1984) comparison can also be made with the south-western part of the Barents Sea, especially with the Bjørnøya basin/Loppa Ridge system (Rønnevik and Jacobsen, 1984).

The basement topography derived from the aeromagnetic depth-mapping shows large variations from more than 10 km depth within the graben areas to about only 3 km depth over some of the ridges. The strong 'magnetic basement' topography is likely to exceed the true basement topography because of the general very weak magnetization of East Greenland Caledonian metamorphic rocks (Larsen, 1980).

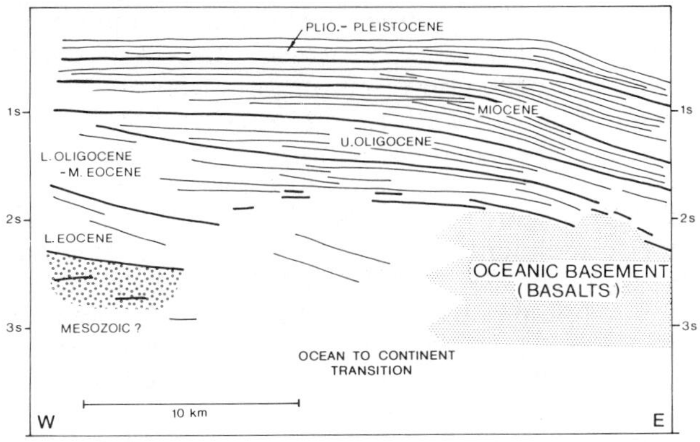

Fig. 11. Interpreted multi-channel seismic profile GGU82-16 from the very southern part of the North-east Greenland shelf (NEAS) just north of the Jan Mayen Fracture zone. Location indicated in Fig. 3 (S6).

Altogether, it appears more than likely that as much as 8—10 km of late Palaeozoic and Mesozoic sediments are present within fault-bounded graben-like depressions on the north-eastern part of the Greenland shelf. The sequence may include salt layers of Devonian to Triassic age.

PETROLEUM POTENTIAL

Compared with Larsen (1980) and Hinz and Schlüter (1980), the structural and stratigraphical information now available provides improved possibilities for discussing the petroleum potential of the various structural provinces along the East Greenland margin.

South-east Greenland shelf (SEAS)

Source rocks

The limited thickness of the post-rift (Tertiary) sedimentary sequence is not likely to provide mature source rocks of post-rift age. The presence of deep syn- to pre-rift source rocks is very unlikely in most of the SEAS. This is because of the continuous presence of early Tertiary oceanic crust below this basin. Coal-bearing horizons could theoretically be present within this subaerially formed oceanic crust, but is not considered important. Presence of pre-rift sediments below the early Tertiary volcanic rocks are, however, indicated in a very limited 10—15 km wide zone below the shelf edge and close to the ocean-to-continent transition along the southernmost part of the SEAS (60° to 63°N). The nature and the source rock potential of these possible pre-rift sediments are not known.

Reservoirs/Structures

Virtually no structuring of the post-rift sediments is present. Stratigraphical trapping within this sequence may be a possibility, but has not been explored. Discarding the actual lithology, the geometry of the unconformity between the seawards dipping sub-basement reflectors and the overlying Eocene marine sediments is favourable to petroleum accumulation (see Fig. 4), The sub-basement reflectors interpreted mainly as oceanic crust basalts, however, have a negligible source potential and reservoir potential.

The few indications of sub-basaltic, pre-rift sediments being present in a narrow zone just landwards of the ocean-to-continent transition are associated with an angular rift-unconformity. The pre-rift sediments are apparently tilted landwards (block-faulting?) and are covered by slightly seawards dipping basalts of limited thickness and 1—2 km of post-rift sediments. The structure is in general well suited for trapping of hydrocarbons (Fig. 4). The basalts may, however, not yield effective ceiling nor is the source and reservoir potential of the underlying rocks known.

In general, the East Greenland margin shows no petroleum potential within and south of the Denmark Strait with the exception of a narrow zone (10—15 km) just landwards of the ocean-to-continent transition where hydrocarbons may have been generated and trapped below rift-basalts in sediments of unknown nature and thickness.

The Blosseville Kyst shelf (BLKS)

Source rocks

Likewise the SEAS area, the larger part of the BLKS area is floored by oceanic crust of early Tertiary age, reducing very much the spectrum of possible source rocks. The post-rift sediments are, however, thick enough to provide maturity within the Palaeogene and locally may also be the lowermost Neogene sediments. During this period, sedimentation was dominated by fan and turbidite infill from the west and to a lesser degree volcano clastic infill from the east into a fairly deep marine environment with restricted water exchange (to the north). This palaeo-setting may have provided at least temporary euxinic conditions within the basin and shales with a source rock potential may have formed. Parallel developments at higher levels within the Neogene sequence are a possibility, but are not likely to have achieved sufficient maturity for hydrocarbon generation.

Although early Tertiary oceanic crust floors most of the basin, some restricted possibilities exist for older Mesozoic and Palaeozoic hydrocarbon sources. Attenuated, pre-rift crust may thus be present in a narrow belt just seawards of the coastal flexure/escarpment. If so, large thicknesses of Palaeozoic and Mesozoic sediments with a late Palaeocene lava cover can be expected at depth. From the onshore exposures (Surlyk *et al.* 1981; Surlyk *et al.*, 1984), Permian, Triassic and Jurassic organic-rich shales can be expected.

Another possibility is that the same rocks are present immediately below the coastal basalts, and hence, as a consequence of the offshore post-rift subsidence, the Eocene to Oligocene fan sediments are now in contact with these older rocks across the escarpment zone and horizontal migration of hydrocarbons from the onshore area into the offshore basin may have taken place (Fig. 7).

Larsen (1980) mentioned the existence of methane rich warm springs in the coastal area. Recently found indications of migrated hydrocarbons from one coastal locality (unpublished data, Geological Survey of Greenland) further strengthen the probability for hydrocarbon generating source rocks below the onshore basalts.

Structures/reservoirs

Apart from the strong coastal flexure/escarpment, the BLKS area is little structured. The apparent fairly strong basement topography is partly an artifact due to contemporaneous clastic and volcanic infill and, when real, was mainly established prior to sedimentation. Stratigraphical and combined stratigraphical/structural trapping is considered the most likely option in the BLKS area. The Eocene to early Oligocene marine fan system deposited in front of the coastal flexure is considered a large potential reservoir. Other potential but smaller scale stratigraphical reservoirs are present within the post-Eocene section.

In conclusion, the BLKS area is considered to have an intermediate petroleum potential, the primary objective being the early Tertiary submarine fan sediments in front of the coastal escarpment.

The Liverpool Land shelf (LILS)

Source rocks

The early Tertiary development of this area is similar to the BLKS area and the same possibilities for the formation of Eocene and lower Oligocene source rocks of sufficient maturity exist. However, as more continental crust (rotated fault-blocks) is believed to be present in this area, improved possibilities for deeper lying Palaeozoic and Mesozoic source rocks exist. On the other hand, horizontal migration from the land area into the offshore area is hindered because of the Liverpool Land basement horst. Possible pre-rift source rocks are buried by 2.5 to 5.0 sec. of younger sediments. No mature source rocks are expected to be present below the outer shelf area because of the very recent subsidence in this area and the young oceanic basement present (see Fig. 9).

Structures/reservoirs

Large structural and structural/stratigraphic traps are likely to be present within the inner part of the LILS area. The most conspicuous potential traps are within the tilted fault-blocks at about 3 s depth (Fig. 9) and within smaller fault-blocks and Eocene to lower Oligocene fan sediments close to the coastal escarpment (1 to 2 s depth). There are no obvious structural trapping possibilities on the outer shelf area. Stratigraphic trapping possibilities are present both on the inner and outer shelf area.

In conclusion, the outer part of the LILS area has a poor to absent hydrocarbon potential because of the expected lack of maturity and because of limited trapping possibilities. The inner part of the LILS area does have an intermediate to high hydrocarbon potential because good source rocks known from nearby onshore areas (Surlyk *et al.*, 1984) are likely to be present at depth and because of the presence of several and large trapping possibilities.

The North-eastern shelf (NEAS)

The stratigraphy and tectonics of this very large and inaccessible area is not known to the same detail as the southern and middle part of the East

Greenland shelf (see Figs. 10 and 11). The area is, however, viewed as having a high hydrocarbon potential, especially the northern and wide part of the shelf (74°N — 78°N). This is because of the expected presence of very thick sequences of late Palaeozoic and Mesozoic sediments likely to provide mature source rocks of Permian and Triassic age. Middle and Upper Jurassic shales can only be expected mature in areas with thick Tertiary or unexpected thick Cretaceous sequences. Good potential reservoir rocks should be present within both the Triassic and Jurassic sediments. Secondary reservoir possibilities may exist within Devonian, Permian and Cretaceous rocks. Details of trapping possibilities are not known. The strong basement topography and the large horizontal extension of the coast-parallel graben-ridge systems provide a potential for large structural traps. Structures caused by salt tectonics may also be present.

ACKNOWLEDGEMENTS

B. Sikker Hansen and N. Turner are thanked for their help in preparing the manuscript. This contribution is published with permission of the Director of the Geological Survey of Greenland.

REFERENCES

Eldholm, O. and Windisch, C.C. 1974. The sediment distribution in the Norwegian-Greenland Sea. *Geol. Soc. Am. Bull.*, **86**, 1661—1676.

Featherstone, P.S., Bott, M.H.P. and Peacock, J.H. 1977. Structure of the continental margin of southeastern Greenland. *Geophys. J.R. Astron. Soc.*, **48**, 15—27.

Grønlie, G. and Talwani, M. 1978. *Geophysical Atlas of the Norwegian-Greenland Sea.* VEMA Res. Ser. IV, Lamont—Doherty Geol. Observatory, Palisades, NY, 26 pp.

Harland, W.B. *et al.* 1984. Outline of Arctic post-Silurian continental displacements. This volume, pp. 137—148.

Henderson, G. 1976. Petroleum geology. In: A. Escher and W.S. Watt (eds.). *Geology of Greenland,* Geol. Surv. Greenland, Copenhagen. pp. 488—505.

Hinz, K. 1981. An hypothesis on terrestrial catastrophes wedges of very thick oceanward dipping layers beneath passive continental margins — their origin and paleoenvironmental significance. *Geol. Jb.*, **E22**, 3—28.

Hinz, K. and Schlüter, H-U. 1978. The North Atlantic — results of geophysical investigations by the Federal Institute of Geosciences and Natural Resources on North Atlantic continental margins. *Erdöl-Erdgas-Z.* **94**, 271—280.

Hinz, K. and Schlüter, H-U. 1980. Continental Margin off East Greenland. *Proc. Tenth World Petroleum Congress, Vol. 2. Exploration Supply and Demand.* Bucharest. Heyden, London, pp. 405—418.

Johnson, G.L., McMillan, N.J. and Egloff, J. 1975a. East Greenland Continental Margin. In: C.J. Yorath, E.R. Parker, and D.J. Glass (eds). *Canada's Continental Margins and Offshore Petroleum Exploration.* Can. Soc. Petrol. Geol. Mem. 4, pp. 205—224.

Johnson, G.L., Sommerhoff, G., and Egloff, J., 1975b. Structure and morphology of the West Reykjanes basin. *Mar. Geol.*, **18**, 175—196.

Larsen, B. 1975. Marine geophysical survey of the East Greenland shelf south of Angmagssalik. *Rapp. Grønlands geol. Under.* **75**, 87—88.

Larsen, B. 1980. A marine geophysical survey of the continental shelf of East Greenland 60—71 N — Project Dana 79. *Rapp. Grønlands geol. Unders.* **100**, 94—98.

Larsen, B. 1983. Geology of the Greenland-Iceland Ridge in the Denmark Strait. In: M.H.P. Bott *et al.* (eds.), *Structure and Development of the Greenland—Scotland Ridge. Plenum, New York.*

Larsen, H.C. 1975. Aeromagnetic investigations in East Greenland. *Rapp. Grønlands geol. Unders.* **75**, 88—91.

Larsen, H.C. 1978. Offshore continuation of East Greenland dyke swarm and North Atlantic Ocean formation. *Nature, London* **274**, 220—223.

Larsen, H.C. 1980. Geological perspectives of the East Greenland continental margin. *Bull. geol. Soc. Denmark,* **29**, 77—101.

Larsen, H.C. 1981. A high-pressure granulite facies complex in north-west Payers Land, East Greenland fold belt. *Bull. geol. Soc. Denmark,* **29**, 161—174.

Larsen, H.C. 1983. Marine geophysical investigations offshore East Greenland. *Rapp. Grønlands geol. Unders.* **115**, 93—100.

Larsen, H.C. (in preparation). Seawards dipping basement reflectors off E. Greenland and their bearing on early Tertiary rifting.

Larsen, H.C. and Thorning, L. 1979. Project EASTMAR: planning of an aeromagnetic survey off East Greenland. *Rapp. Grønlands geol. Under.* **95**, 93—96.

Larsen, H.C. and Thorning, L. 1980. Project EASTMAR: acquisition of high sensitivity aeromagnetic data off East Greenland. *Rapp. Grønlands geol. Unders.* **100**, 91—94.

Mutter, J.C., Talwani, M. and Stoffa, P.L. 1982. Origin of seaward-dipping reflectors in oceanic crust off the Norwegian margin by 'subaerial sea-floor spreading'. *Geology,* **10**, 353—357.

Noe-Nygaard, A. 1976. Tertiary igneous rocks between Shannon and Scoresby Sund, East Greenland. In: A. Escher and W.S. Watt, (eds.), *Geology of Greenland.* Geological Survey of Greenland, Copenhagen, pp. 387—402.

Perry, R.K. *et al.* 1980. Bathymetry of the Norwegian—Greenland and Western Barents Seas. U.S. Naval Res. Lab. Rep., Washington, D.C.

Rønnevik, H. and Jacobsen, H.P. 1984. Structures and basins in the western Barents Sea. (This volume.)

Smythe, D.K. 1983. Faeroe—Shetland Escarpment and Continental Margin North of the Faeroes. In: M.P.H. Bott *et al.* (eds.), *Structure and Development of the Greenland—Scotland Ridge. Plenum, New York.*

Sommerhoff, G. 1973. Formenschatz und morphologische Gliederung des südostgrönlandischen Schelfgebietes und Kontinental-abhanges. *Meteor Forschungsergeb.* **C15**, 1—54.

Surlyk, F. 1977. Mesozoic faulting in East Greenland. In: R.T.C. Frost and A.J. Dikkers (eds.) *Fault Tectonics in NW Europe. Geol. Mijnb.* **56**, 311—327.

Surlyk, F., Clemmensen, L.B. and Larsen, H.C. 1981. Post-Paleozoic Evolution of the East Greenland Continental Margin. In: J.W. Kerr and A.J. Ferguson (eds.), *Geology of the North Atlantic Borderlands.* Can. Soc. Petrol. Geol. Mem. 7.

Surlyk, F. 1984. The Permian basin of East Greenland. This volume, pp. 303—316.

Talwani, M. and Udintsev, G. 1976a. *Tectonic Synthesis.* Initial Rep. 38 Deep-Sea Drilling Project, pp. 1213—1242.

Talwani, M. *et al.* 1976b. *Sites 346, 347 and 349.* Initial Rep. 38 Deep-Sea Drilling Project, pp. 521—594.

Talwani, M. and Eldholm, O. 1977. Evolution of the Norwegian—Greenland Sea. *Bull. Geol. Soc. Amer.* **88**, 969—999.

Vail, P.R. *et al.* 1977. Seismic stratigraphy and global changes of sea level. In: C.E. Payton (eds.), *Seismic Stratigraphy — Applications to Hydrocarbon Exploration.* Amer. Assoc. Petrol. Geol. Mem. 16, pp. 49—212.

Vogt, P.R. 1970. Magnetized basement outcrops on the Southeast Greenland continental shelf. *Nature (London)* **226**, 743—744.

Vogt, P.R. *et al.* 1980. Morphology and magnetic anomalies north of Iceland. *J. Geophys.*, **47**, 67—80.

Wager, L.R. and Deer, W.A. 1938. A dyke swarm and coastal flexure in East Greenland. *Geol. Mag.* **75**, 39—46.

24 Deep structure and origin of the Faeroe–Shetland Channel

M.H.P. Bott
Department of Geological Sciences, University of Durham, UK

Evidence from explosion seismology indicates that the Faeroe-Shetland Channel is probably underlain by pre-Tertiary oceanic crust, forming an extension of the Rockall Trough and separating the continental crustal regions of the West Shetland Shelf and the Faeroe Block. The morphology of the Channel has been subsequently greatly modified by the extensive early Tertiary volcanism associated with the Faeroe Islands, the Faeroe Ridge and the Wyville-Thomson Ridge. Mesozoic sediments may be sandwiched between the volcanics and the crust beneath.

Results from the North Atlantic Seismic Project indicate the following crustal model at 61°N: water layer 1.1 km (1.5 km/s), sediment layer 1.7 km (2.24 km/s assumed), early Tertiary volcanics and trapped sediments about 6 km (4.5 km/s), main crustal layer 6–8 km (6.6 km/s). The Moho (7.8–8.2 km/s) is thus at an estimated depth of 15–17 km, contrasting with a depth of about 26–27 km beneath the West Shetland Shelf. There is weak evidence of seismic anisotropy in the uppermost mantle beneath the Channel, consistent with oceanic origin. Arrivals from shots in the Channel recorded at land stations in the Shetland–North Scotland region display abundant scattering, indicating a major intervening boundary interpreted as the junction between the Channel and the shelf.

The arrivals at land stations from shots in the Channel at around 60°20′N, after correction for the sediments, show Moho delays about 1 second greater than those at 61°N. This might be interpreted in terms of a crustal thickening of over 10 km along the axis of the channel, but a more realistic interpretation is in terms of a substantial thickening of the low velocity volcanic pile (with trapped sediments), with downsagging of the underlying crust beneath the load.

INTRODUCTION

The Faeroe-Shetland Channel is separated from the Rockall Trough by the transverse Wyville-Thomson Ridge. It is much narrower than Rockall Trough, and it widens northeastwards as it deepens from just over 1000 m to over 1700 m at its junction with the Norwegian Sea. The Cainozoic sediments which form the seabed thicken substantially towards the northeast, but are thinnest (or absent locally) over a buried extension of the Faeroe Ridge. The underlying acoustic basement is formed of strongly magnetic, early Tertiary volcanic rocks.

The origin of the Faeroe-Shetland Channel is controversial. The sea-depth is neither typically oceanic nor typically continental, but in the vicinity of the Icelandic transverse ridge depths are generally anomalous. According to a gravity interpretation by Bott and Watts (1971), the underlying crust of about 60°30′N is about 7 km thinner than beneath the adjacent Shetland Shelf. In the absence of any direct evidence prior to the North Atlantic Seismic Project, both oceanic and continental origins have been suggested.

The oceanic hypothesis is based on the assumption that the Rockall Trough and the Faeroe-Shetland Channel formed as a continuous narrow seaway during an early stage of opening of the northeastern North Atlantic (Botts and Watts, 1971; Roberts, 1975; Bott, 1975). The Rockall Trough is probably oceanic for reasons summarized by Roberts *et al.* (1983), although the magnetic anomalies are not diagnostic as most of it forms a magnetic quiet zone. The continuity of the eastern margin of the Rockall Trough and the Faeroe-Shetland Channel suggests a common origin. Furthermore, there is insufficient space in the Faeroes or East Greenland Shelf regions to accommodate a northward extension of the Rockall Trough displaced westwards by a transform fault, as the region appears to be entirely occupied by continental crust or Tertiary ocean floor. An

Petroleum Geology of the North European Margin © Norwegian Petroleum Society (Graham & Trotman, 1984) pp. 341–347.

obvious difficulty of the oceanic hypothesis is that the channel is narrower than the trough. The Wyville-Thomson Ridge and the Faeroe Ridge are also enigmatic. These difficulties, however, are avoided if the morphology of the region has been grossly modified by the early Tertiary volcanism which appears to have formed the Wyville-Thomson Ridge and to have swamped the channel, particularly on its western side, depressing the earlier oceanic crust beneath the weight of volcanics.

The age of the postulated oceanic crust is also controversial. Two recent views are as follows. Roberts *et al.* (1981) summarized evidence for 120 km of post-Albian pre-Maestrichtian oceanic crust beneath the Rockall Trough, probably formed during the Cretaceous normal polarity period. Alternatively, Russell and Smythe (1978) inferred a much older (early Permian) origin.

The alternative continental hypothesis for the origin of the Faeroe-Shetland Channel arose from the discovery of the Faeroe-Shetland Escarpment by Talwani and Eldholm (1972). They interpreted the escarpment as the fundamental boundary between oceanic crust to the west and continental crust to the east. This implies that the Faeroe Islands are underlain by oceanic crust and that the Faeroe-Shetland Channel is a subsided continental region. A continental origin has been supported by Ridd (1981, 1983). The original basis for this hypothesis has been thrown in doubt by more recent investigations of the Faeroe-Shetland Escarpment based on a much larger amount of data (Robinson, 1980; Smythe, 1983). According to Smythe (1983) the escarpment is the sinuous eastern feather edge of a series of flood basalts which are probably correlateable with the Eocene middle and upper basalt series of the Faeroe Islands. These investigations are of further interest because they show that an underlying sequence of inferred Palaeocene volcanics, which may correlate with the lower lavas of the Faeroe Islands, forms the subsedimentary acoustic basement of the southwestern part of the Faeroe-Shetland Channel. The lavas in the vicinity of the Faeroe-Shetland Escarpment overlie a substantial thickness of Mesozoic sediments which are likely to be present at depth beneath the local acoustic basement of the Channel.

This chapter presents new seismic evidence on the structure and thickness of the crust beneath the Faeroe-Shetland Channel. This was obtained during the North Atlantic Seismic Project (NASP) of 1972. It was interpreted initially by Smith (1974), but a new appraisal follows the recognition that the Wyville-Thomson Ridge can be best explained in terms of depression of the crust through a thick early Tertiary volcanic pile (Roberts *et al.*, 1983). A fuller account of the seismic interpretation is given by Bott and Smith (in press).

CRUST BENEATH ADJACENT BORDERLANDS

The shelf region north of Scotland and west of the Shetlands and Orkneys is underlain by relatively thin continental crust. Precambrian basement rocks occur to the west of the Caledonian front. The crustal structure was investigated along line D and parts of lines B and C (Fig. 1) of the North Atlantic Seismic Project of 1972 (Smith and Bott, 1975). The Moho beneath the Caledonian foreland occurs at an estimated depth of 26 ± 2 km. A 6.5 km/s refractor was detected at between 2 and 16-km depth and was interpreted in terms of Lewisian granulites of Scourian type. P_n was determined to be 7.99 ± 0.02 km/s beneath the shelf.

The nature of the crust beneath the Faeroe Islands has been controversial. Pálmason (1965) obtained a basement velocity of 6.4 km/s beneath 2.5–4.5 km of lavas along a short refraction line, and this suggested that the Faeroe crust resembles that of Iceland, which is of anomalous oceanic type. Talwani and Eldholm (1972) also suggested that oceanic crust occurs west of the Faeroe-Shetland Escarpment and thus underlies the Faeroe Islands. Bott and Watts (1971) alternatively suggested that continental crust underlies the Faeroe region on the basis of a continental reconstruction of the north eastern North Atlantic. More recent seismic refraction results support this interpretation. Casten (1973) obtained an unreversed basement velocity of 5.9 km/s on the northern shelf. Interpretation of the NASP data from the Faeroe block yielded basement velocities of 5.9–6.2 km/s, which gave first arrivals out to ranges of at least 80 km (Bott *et al.*, 1974). This is interpreted as the upper continental Pg phase. A more detailed analysis of the NASP data by Nielsen (1976) showed that the basement velocity is 5.90 ± 0.04 km/s, with the possibility of an increase to 6.2 km/s beneath the northwestern shelf. The crust is estimated to be between 27 and 38 km thick (Bott *et al.*, 1974), which is consistent with continental crust but not diagnostic because similar thicknesses of Icelandic type oceanic crust underlie the adjacent Iceland-Faeroe Ridge. The well defined 6.7-km/s layer which underlies the Iceland-Faeroe Ridge at 6–8-km depth has not been detected beneath the Faeroes.

The evidence for a continental origin of the Faeroe region is further supported by the following:

(1) P^* and P_n arrivals from the Iceland-Faeroe Ridge are converted to slower travelling phases such as P_g at the boundary between the Ridge and the Faeroes block, indicating a major intervening crustal boundary (Bott *et al.*, 1976).

(2) A conspicuous negative gravity anomaly on the northwestern shelf is best explained by a granite batholith indicative of continental basement.

(3) There is lack of evidence of lithospheric subsidence of nearly 2 km since the Palaeocene, such as would be expected if the underlying crust is anomalously thick oceanic crust formed during the Palaeocene (Bott, 1983).

(4) Strontium isotope evidence indicates that some of the Faeroe-lavas have been contaminated by continental crustal material. Taken together, this evidence gives compelling support for a continental origin.

The basement velocity not far to the east of the Faeroe Islands appears to decrease abruptly to about 5.3 km/s (Bott *et al.*, 1974; Nielsen, 1976). This may mark the eastern edge of the ancient Faeroe-Shetland Channel which has subsequently been covered by early Tertiary lavas.

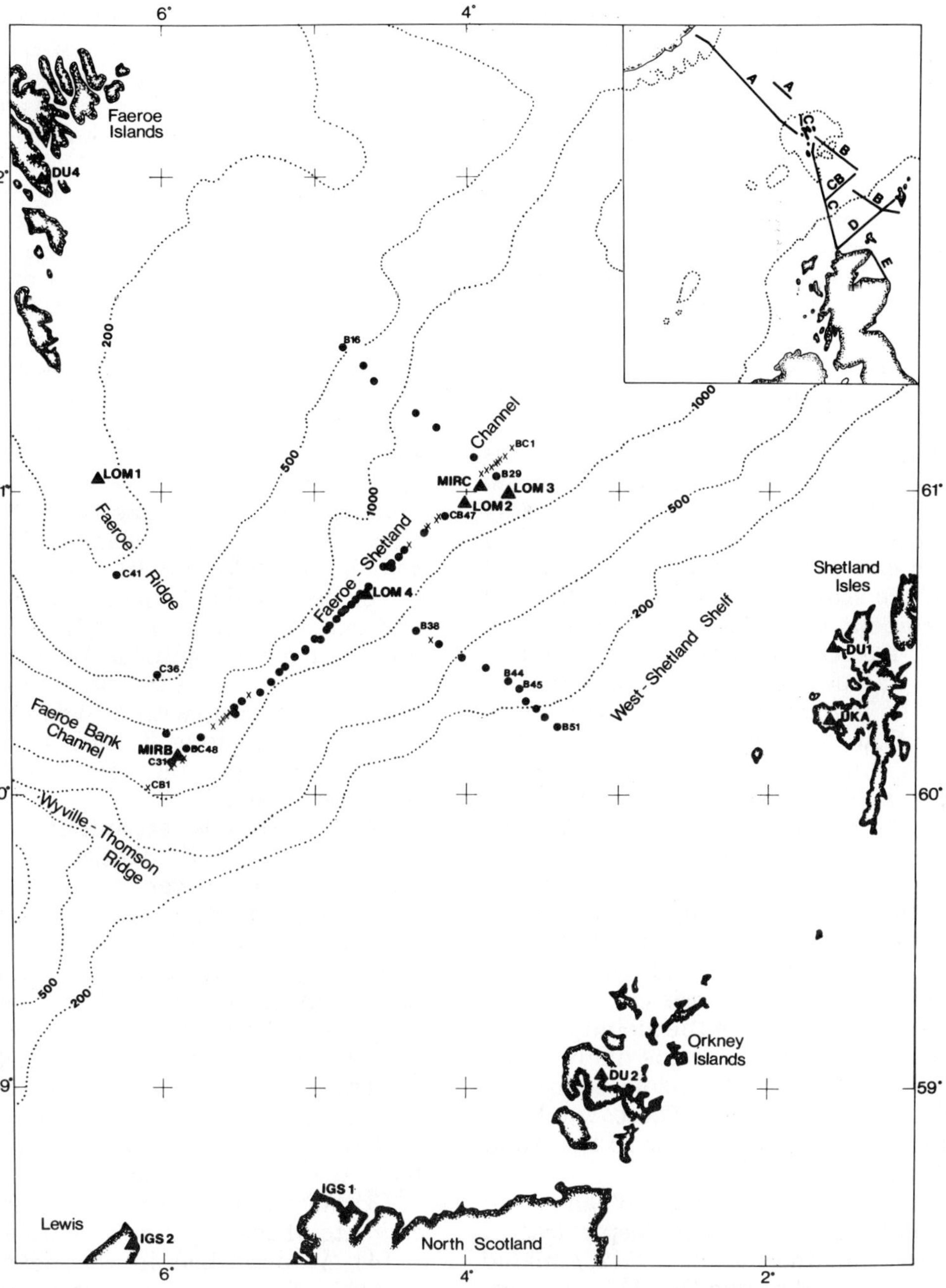

Fig. 1. Map of the Faeroe-Shetland region, showing shot and station locations used in determining the crustal structure of the Faeroe-Shetland Channel. Solid circles denote large shots (300 lb or more), crosses denote smaller shots, and solid triangles denote receiving stations on land and at sea. Inset: The overall shot lines of the North Atlantic Seismic Project. (After Bott and Smith, in press.)

CRUSTAL STRUCTURE AT ABOUT 61°N

The crustal structure beneath the southwestern end of the Faeroe-Shetland Channel was investigated as part of NASP in 1972. This was a sea-to-land explosion seismology experiment carried out between North Scotland and Iceland. As part of the project, a line of shots of up to 600 lb each was fired in both directions along the axis of the Faeroe-Shetland Channel between latitudes 60° and 61° (line CB towards the NE and BC towards the SW). Line C was shot across the Channel between Cape Wrath and the Faeroe Islands, and the two segments of line B were shot between the Faeroe Islands and the Shetland Isles. The shots were observed at a series of land seismic stations set up on the Faeroe Islands, Shetland Isles, Orkney Islands and North Scotland. They were also observed at a number of sites at sea by the NERC chartered ship *MV Miranda* and the Russian research ship *Mikhail Lomonosov*. The shots and stations relevant to the determination of the crustal structure of the Faeroe-Shetland Channel are shown in Fig. 1.

The uppermost layer is formed of Cainozoic sediments which overlie the acoustic basement. The sediment thickness and basement velocity were determined along a series of unreversed refraction segments observed at *Miranda* and *Lomonosov* during NASP. More precise delineation of the sediment thickness has subsequently been obtained by seismic reflection profiling. Reflection sections approximately along the NASP lines across and along the floor of the channel have kindly been provided by Britoil, to assist in the interpretation of the NASP data. The average sediment velocity of 2.24 km/s obtained in the reflection surveys has been used to compute depth to basement. The refraction data tend to yield greater depth estimates than the reflection data, probably mainly because of increased velocity with depth at the top of the basement, but otherwise the agreement is quite good (see Bott and Smith, in press, for details). The basement velocity estimated from the refraction data ranges from 3.90 to 4.56 km/s.

The sediment profile along line CB (Fig. 2) was determined using refracted arrivals of MIRB for the southernmost 50 km and a reflection profile kindly supplied by Britoil over the rest of the line. Near the northeastern end of the line and over adjacent parts of line B, the sediments beneath the floor of the channel are about 1.5 to 1.8-km thick. Further southwestwards, line CB is underlain by a buried extension of the Faeroe Ridge, which appears to be fairly rugged. The buried ridge is apparent from both the refraction data and from the nearby Britoil reflection line, along which it appears to crop out locally at the seabed. A local pocket of thick sediment lies between the buried ridge and the Wyville-Thomson Ridge, at the junction with the Faeroe Bank Channel.

As pointed out above, the acoustic basement is formed of volcanic rocks (probably of Palaeocene age). The magnetic anomalies and the seismic velocity of 4.0—4.5 km/s are consistent with this interpretation. From evidence in the vicinity of the Faeroe-Shetland Escarpment, it seems likely that a substantial thickness of late Mesozoic sediments may underlie the volcanic rocks. These would be expected to form a low velocity zone so that their presence would be hidden.

Underlying this composite layer of volcanics and Mesozoic sediments, a conspicuous deep refractor was observed at the marine station LOM2, situated near the northeastern end of line CB. This was best observed at the floating hydrophone station, where the apparent velocity for shots at ranges of between 20 and 60 km was observed to be 6.56 ± 0.04 km/s and the seabed-to-seabed intercept time was determined to be 3.36 ± 0.05 s. The same arrivals were observed at the nearby bottom seismometer station, but the apparent velocity of 6 km/s was less well determined. Assuming horizontal layering, a true refractor velocity of 6.56 km/s and a uniform overlying layer velocity of 4.5 km/s, the estimated depth to the refractor is 8.8 km. The corresponding thickness of the 4.5-km/s layer is 6.0 km.

P_n arrivals which travel just below the Moho were observed for most of the shots on the floor of the Faeroe-Shetland Channel at some of the land stations. P_n arrivals were also observed for the CB/BC shots beyond about 60-km range at LOM2. Because of the irregular distribution of receiving stations in relation to the shots, time-term analysis was used to interpret these arrivals. Four important results come from the study of the P_n arrivals:

(1) There is indication of P_n velocity anisotropy beneath the Faeroe-Shetland Channel. The shots *along* the axis of the channel yield an estimated P_n velocity of 7.76 ± 0.39, whereas the B segments *across* the channel yield estimates of 8.54 ± 0.26 and 8.33 ± 0.72 km/s. Although these results do not quite differ significantly at the 95% confidence level, there is nevertheless a good chance that the difference is real, with a higher velocity across the channel than along it. This result would be consistent with oceanic crust beneath the Faeroe-Shetland Channel.

(2) P_n arrivals at stations in the Shetlands, Orkneys and North Scotland display conspicuous scattering which can be explained by a major intervening structural boundary (see Bott and Smith, in press, for details). This scattering may be partly caused by the edge of the supposed Palaeocene lavas, but is probably partly caused by a major deep boundary which may represent the contact between oceanic and continental crust. Along the eastern segment of shot line B, the records abruptly change character between shots B44 and B45, suggesting that this position may approximately mark the continent—ocean contact.

(3) As described below, the P_n arrivals yield an estimate of the depth to the Moho at 61°N.

(4) As described in the next section, there is evidence for a major lateral change in structure along the axis of the channel associated with the buried extension of the Faeroe Ridge.

The time-term analysis yielded an average time term of 3.06 s at shots CB46, CB47 and station LOM2 at the northeastern end of the line for a

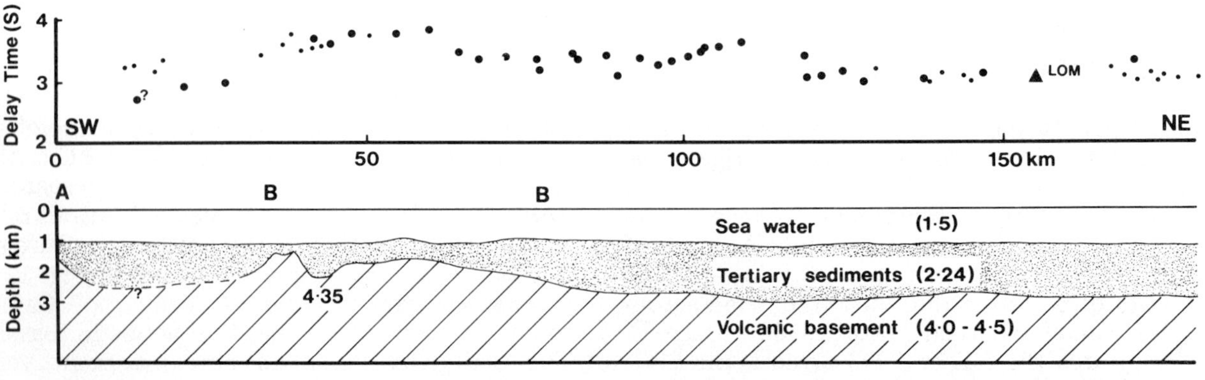

Fig. 2. Delay times (time terms) for the shot points of line CB along the axis of the Faeroe-Shetland Channel (Fig. 1), shown in relation to the thickness of Cainozoic sediments determined by refraction and reflection seismics. A denotes the northern flank of the Wyville-Thomson Ridge, and B—B denotes the buried extension of the Faeroe Ridge. Velocities are shown in km/s.

P_n velocity of 8.0 km/s. Assuming a 6.6 km/s velocity for the main crustal layer, the depth to the Moho is estimated to be 15.7 km and the thickness of the crustal layer to be 6.9 km. Slightly different values of the time term and the estimated depth are obtained if smaller or larger values of P_n are adopted, but the differences are not great over the range of realistic values of P_n. Using all the above data, the following model of crustal structure beneath the Faeroe-Shetland Channel at 61°N is obtained:

	Velocity (km/s)
Water layer	1.5
Layer A (sediments)	2.24
Layer B (volcanics, etc)	4.5
Layer C (crust)	6.6
Upper mantle	7.8—8.2

	Thickness (km)	*Depth to base (km)*
Water layer	1.1	1.1
Layer A (sediments)	1.7	2.8
Layer B (volcanics, etc)	6.0	8.8
Layer C (crust)	5.7	14.5 (for 7.8 km/s)
	6.9	15.7 (for 8.0 km/s)
	7.9	16.7 (for 8.2 km/s)

The crustal thickness is in excellent agreement with the slightly thicker than average crust at the edge of an ocean, such as the 140-Ma old oceanic crust 300 km southwest of Bermuda which was determined to be 7-km thick by Purdy (1983). However, extremely thinned continental crust cannot be conclusively ruled out on this evidence, although the likely anisotropy of P_n and the inferred boundary between the channel and the Shetland Shelf both support an oceanic origin.

LATERAL VARIATION OF CRUSTAL STRUCTURE

The main time-term analysis yielded P_n delay times (time terms) for most of the shots on the floor of the Faeroe-Shetland Channel. Each delay time is a function of the underlying velocity—depth distribution down to the Moho, with thicker layers or lower velocities causing increased delay. A feature of particular interest is the large variation of delay time along the line CB, with the largest delays occurring over the buried extension of the Faeroe Ridge (Fig. 2). The average delay time at stations over the buried ridge is 3.8 s, assuming P_n to be 8.0 km/s, which is about 0.7 s larger than that at the northeastern end of the line where the delay time is about 3.1 s. The variation in delay time along the line is fairly insensitive to the assumed value of P_n, thus a variation in P_n of ± 0.4 km/s only changes the relative delay times along the line by about ± 0.05 s, which is hardly significant (Bott and Smith, in press). There is a striking correlation between the delay times and the broad features of the buried topography of the acoustic basement. The largest delay times occur where the acoustic basement is shallowest, contrary to expectation. If the sediment delay is allowed for, then the variations in delay time are accentuated (Fig. 3). This evidence indicates that there must be a substantial low-velocity root beneath the buried ridge.

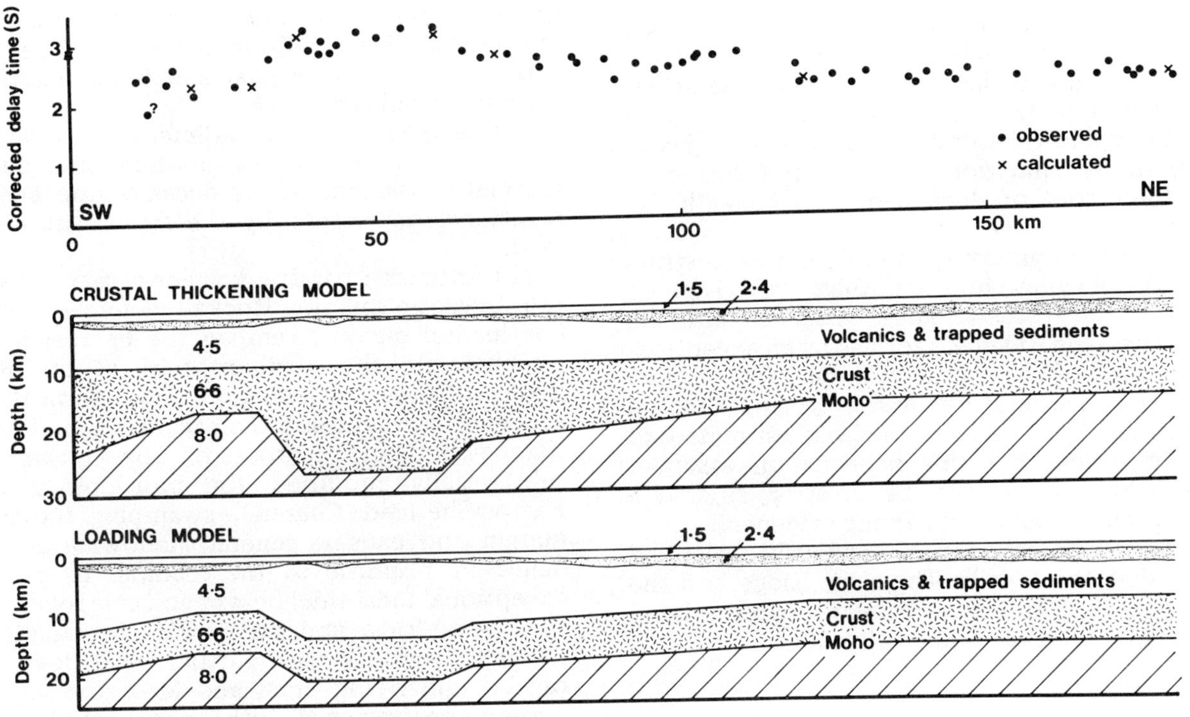

Fig. 3. Interpretation of the delay times along line CB, after correction for lateral variations of sediment and water layers, in terms of crustal models: (above) crustal thickening model, (below) crustal loading model. Velocities are shown in km/s.

Using the model of crustal structure at 61°N presented in the previous section, the variation in delay time along the line can be explained by a substantial lateral variation in thickness of either (1) the deep crustal layer, or (2) the overlying lava and sediment layer, with downsagging of the crust beneath the load, or (3) a combination of these. These two hypotheses are shown in Fig. 3. The main crustal layer would need to thicken by about 11 km beneath the buried ridge to explain the observations. Alternatively, they could be explained by a 5-km deepening of the base of the lava/sediment layer, with downsagging of the underlying uniform crustal layer.

The hypothesis that the lava layer is thickened, rather than the underlying crustal layer, is preferred on the following grounds:

(1) The good correlation between the buried ridge and the low-velocity root is readily explained in terms of downsagging beneath the excess volcanic load, but a relationship between the ridge and thickening of the crustal layer would be much less easy to understand.

(2) It would be consistent with a similar interpretation of the origin of the Wyville-Thomson Ridge proposed by Roberts *et al.* (1983).

(3) This hypothesis is easier to reconcile with the lack of a major negative free-air gravity anomaly associated with the buried ridge. This hypothesis, therefore, is the simplest explanation consistent with the available evidence.

WYVILLE-THOMSON RIDGE

The transverse Wyville-Thomson Ridge separates the Rockall Trough from the Faeroe-Shetland Channel. The ridge is formed of Palaeocene volcanic rocks which may locally be covered by thin sediments. The main evidence on the underlying deep structure comes from the gravity anomalies. There is a positive free-air anomaly of 30—60 mgal over the ridge. This is not as large as the calculated effect of the bathymetry, indicating that the ridge is underlain by a low-density root. This was originally interpreted by Himsworth (1973) in terms of a 10-km thick root of thickened crust beneath the Ridge.

A new gravity interpretation of the deep structure of the Wyville-Thomson Ridge was given by Roberts *et al.* (1983). They assumed that the Wyville-Thomson Ridge was formed by an exceptionally thick load of Palaeocene volcanic rocks erupted on top of older uniform crust, causing flexural downsagging. The volcanic pile was estimated to be 12-km thick, but this depends on assumed density values which may be in error, so that a better estimate must await seismic evidence.

The seismic interpretation previously described suggests that the Wyville-Thomson Ridge and the buried Faeroe Ridge have similar origins related to loading by the Palaeocene volcanics.

CONCLUSIONS AND DISCUSSION

Recordings on land and at sea of shots fired during the North Atlantic Seismic Project have enabled the crustal structure beneath the Faeroe-Shetland Channel between 60° and 61°N to be investigated. The resulting model of crustal structure at 61°N is as follows:

Water layer (1.5 km/s) 1.1 km
Cainozoic sediments (2.24 km/s) 1.7 km, depth to base 2.8 km
Lavas and trapped sediments (4.5 km/s) 6 km, depth to base 8.8km
Main crustal layer (6.6 km/s) 6—8 km, depth to Moho 15—17 km
Uppermost mantle (7.8—8.2 km/s)

There is evidence of anisotropy of P_n beneath the Faeroe-Shetland Channel, with higher velocity across the channel than along its axis. Land stations to the east of the channel exhibit considerable scattering of arrivals from the channel, indicating a possible major intervening boundary. This evidence, taken with the thickness of the main crustal layer, supports an origin as pre-Tertiary oceanic crust, although the evidence cannot be regarded as absolutely conclusive.

Major lateral variations in the Moho delay times for the shots along the axis of the channel between 60° and 61°N indicate that there is a substantial low-velocity root beneath a buried extension of the Faeroe Ridge. This is most realistically interpreted in terms of downsagging of uniform pre-existing oceanic crust under an excessive load of early Tertiary lavas, rather than by an underlying root of thickened crust. This is consistent with a previous gravity interpretation of the deep structure of the Wyville-Thomson Ridge as a volcanic load structure (Roberts *et al.*, 1983).

It is suggested that the Faeroe-Shetland Channel has evolved in the following four major stages:

(1) During an initial stage of formation of the northeastern North Atlantic, the Rockall Trough and Faeroe-Shetland Channel formed as a continuous seaway by sea-floor spreading. The age of formation is controversial and is not clarified by the new seismic evidence.

(2) Late(?) Mesozoic sediments were deposited on to the oceanic crust, contemporaneous with thermal subsidence of the oceanic lithosphere and additional depression by the significant sediment load.

(3) Extensive basaltic volcanism associated with the break-up of the Rockall—Faeroe—Greenland continental region swamped the Faeroe—Shetland Channel and the northern part of the Rockall Trough. These volcanic rocks now form the local acoustic basement and their feather edge forms the Faeroe-Shetland Escarpment. The volcanic rocks substantially modified the morphology of the Faeroe-Shetland Channel, swamping the western margin and causing general shallowing and subsidence in response to the loading. In particular, exceptional local thickness of volcanic rocks formed the Faeroe Ridge and the Wyville-Thomson Ridge as loading structures with further local downsagging of the underlying crust and Mesozoic sediments. Some re-heating of the lithosphere may have taken place at this stage.

(4) Further sedimentation and thermal subsidence occurred during Cainozoic times.

This evolutionary history explains the shallow

depth and other anomalous features of the region in consistent terms. The Wyville-Thomson Ridge may have acted as a land bridge between the microcontinent and the North Scottish region during Palaeocene times as suggested by Roberts *et al.* (1983); it is also possible that the Faeroe Ridge formed a short-lived land bridge.

REFERENCES

Bott, M.H.P. 1975. Structure and evolution of the Atlantic floor between northern Scotland and Ireland. *Norges Geol. Unders.*, **316**, 195—199.

Bott, M.H.P. 1983. The crust beneath the Iceland-Faeroe Ridge. In: M.H.P. Bott, S. Saxov, M. Talwani and J. Thiede (eds.). *Structure and Development of the Greenland-Scotland Ridge — new methods and concepts*, Plenum Press, New York pp. 63—75.

Bott, M.H.P., Nielsen, P.H. and Sunderland, J. 1976. Converted *P*-waves originating at the continental margin between the Iceland-Faeroe Ridge and the Faeroe Block. *Geophys. J.R. Astron. Soc.*, **44**, 229—238.

Bott, M.H.P. and Smith, P.J. Crustal structure of the Faeroe-Shetland Channel. *Geophys. J.R. Astron. Soc.* (in press).

Bott, M.H.P., Sunderland, J., Smith, P.J., Casten, U. and Saxov, S. 1974. Evidence for continental crust beneath the Faeroe Islands. *Nature, Lond.*, **248**, 202—204.

Bott, M.H.P. and Watts, A.B. 1971. Deep structure of the continental margin adjacent to the British Isles. In: F.M. Delany (ed.). *The Geology of the East Atlantic Continental Margin*, Vol. 2, Inst. Geol. Sci. Rep. No. 70/14, pp. 89—109.

Casten, U. 1973. The crust beneath the Faeroe Islands. *Nature (Phys. Sci.), Lond.*, **241**, 83—84.

Himsworth, E.M. 1973. Marine Geophysical Studies between Northwest Scotland and the Faeroe Plateau. Unpublished Ph.D. thesis, University of Durham, 125pp.

Nielsen, P.H. 1976. Seismic refraction measurements around the Faeroe Islands. *Fróoskaparrit*, **24**, 9—45.

Pálmason, G. 1965. Seismic refraction measurements of the basalt lavas of the Faeroe Islands. *Tectonophysics*, **2**, 475—482.

Purdy, G.M. 1983. The seismic structure of 140 Myr old crust in the western central Atlantic Ocean. *Geophys. J. R. Astron. Soc.*, **72**, 115—137.

Ridd, M.F. 1981. Petroleum geology west of the Shetlands. In: L.V. Illing, and G.D. Hobson (eds.), *Petroleum Geology of the Continental Shelf of North-west Europe.* Heyden, London, pp. 414—425.

Ridd, M.F. 1983. Aspects of the Tertiary geology of the Faeroe-Shetland Channel. In: M.H.P. Bott, S. Saxov, M. Talwani and J. Thiede (eds.), *Structure and Development of the Greenland-Scotland Ridge — New Methods and Concepts.* Plenum Press, New York, pp. 91—108.

Roberts, D.G. 1975. Marine geology of the Rockall Plateau and Trough. *Phil. Trans. R. Soc. Lond., Ser A*, **278**, 447—509.

Roberts, D.G., Bott, M.H.P. and Uruski, C. 1983. Structure and origin of the Wyville-Thomson Ridge. In: M.H.P. Bott, S. Saxov, M. Talwani, and J. Thiede (eds.), *Structure and Developments of the Greenland-Scotland Ridge — New Methods and Concepts.* Plenum Press, New York, pp. 133—158.

Roberts, D.G., Masson, D.G. and Miles, P.R. 1981. Age and structure of the southern Rockall Trough: new evidence. *Earth Planet. Sci. Lett. (Neth.)*, **52**, 115—128.

Robinson, J. 1980. A Geophysical Study of the Continental Margin West of Norway. Unpublished Ph.D. thesis, University of Durham, 100pp.

Russell, M.J. and Smythe, D.K. 1978. Evidence for an early Permian oceanic rift in the northern North Atlantic. In: E.R. Neumann and I.B. Ramberg (eds.), *Petrology and Geochemistry of Continental Rifts*, Reidel, Dordrecht, pp. 173—179.

Smith, P.J. 1974. A Seismic Refraction Study of Crustal Structure between the Faeroe Isles and Scotland. Unpublished Ph.D. thesis, University of Durham, 182pp.

Smith, P.J. and Bott, M.H.P. 1975. Structure of the crust beneath the Caledonian foreland and Caledonian belt of the North Scottish shelf region. *Geophys. J. R. Astron. Soc.*, **40**, 187—205.

Smythe, D.K. 1983. Faeroe-Shetland escarpment and continental margin north of the Faeroes. In: M.H.P. Bott, S. Saxov, M. Talwani and J. Thiede (eds.), *Structure and Development of the Greenland-Scotland Ridge — New Methods and Concepts.* Plenum Press, New York, pp. 109—119.

Talwani, M. and Eldholm, O. 1972. Continental margin off Norway: a geophysical study. *Bull. Geol. Soc. Am.*, **83**, 3575—3606.

25 Tectonic framework of the Møre Basin and the northern North Sea

G.P. Hamar and K. Hjelle
Statoil, Stavanger, Norway

The uppermost 10 km of the sedimentary sequence is illustrated by a base Cretaceous depth map and Cretaceous and Tertiary isopach maps. Substantial areas have been covered by Cretaceous lava flows, both west and east of the Mφre Basin. The Tertiary volcanism is mapped over and west of the Mφre Platform and at the southern tip of the Frφya High. Strike-slip lateral movements are very common both in the northern North Sea and in the Mφre Basin. The northern Mφre Platform formed a shallow shelf area until late Miocene, when it submerged to 2.5-km depth. The faults measured are oriented mainly in NE—SW and NNW—SSE directions in all three areas studied: the northern North Sea, the Mφre Basin and the Trφndelag Platform, the last having a N—S-oriented fault zone which is absent from the two other areas. Five major highs along the eastern margin of the Mφre Basin are described briefly. Two geologic cross sections are controlled by gravimetric modelling. The Mφre Basin is underlain by a thinned continental crust and the basin is extended to twice its original width. Both the Triassic and the Jurassic source rocks are overmature and the mid-Cretaceous sediments are generating oil in the Mφre Basin. The estimated ultimate recoverable reserves are found to be of substantial quantity in the Mφre Basin.

INTRODUCTION

The Mφre Basin is situated on the eastern side of the pre-Tertiary Norwegian Sea, which is described by many authors as a late Mesozoic depression filled with huge amounts of clastic sediments (Fig. 1). Rapid subsidence of the crust has been going on for the last 150 Ma. Early Cretaceous spreading and formation of oceanic crust in the Mφre Basin were also suggested by Bott (1973), Roberts (1974), Woodhall and Knox (1979), Hanisch (1983) and lately by Price and Rattey (in press). Other authors concluded that the Mφre Basin is a result of early crustal stretching, but the rifting did not open up for formation of oceanic crust in this area (Hinz and Moe, 1971; Hinz, 1972; Talwani and Eldholm, 1972; Hinz *et al.*, 1982). According to Whiteman *et al.* (1975), the North Sea, Faeroe-Shetland Channel and Vφring Basin rifts form a triple junction in the Mφre Basin. The stretching—rifting occurred during late Palaeozoic or early Mesozoic times (Jφrgensen and Navrestad, 1981).

The Mφre Basin is bordered by major fault complexes both on the westsouthwestern and on

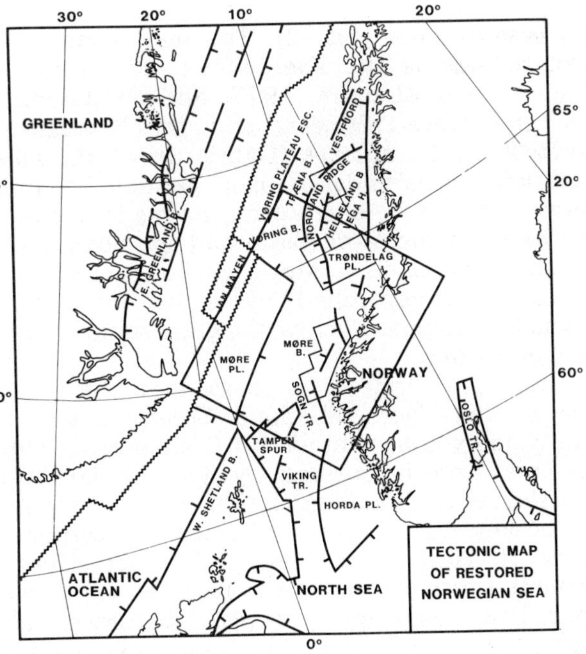

Fig. 1. Norwegian Sea before Tertiary times.

Petroleum Geology of the North European Margin
pp. 349—358.

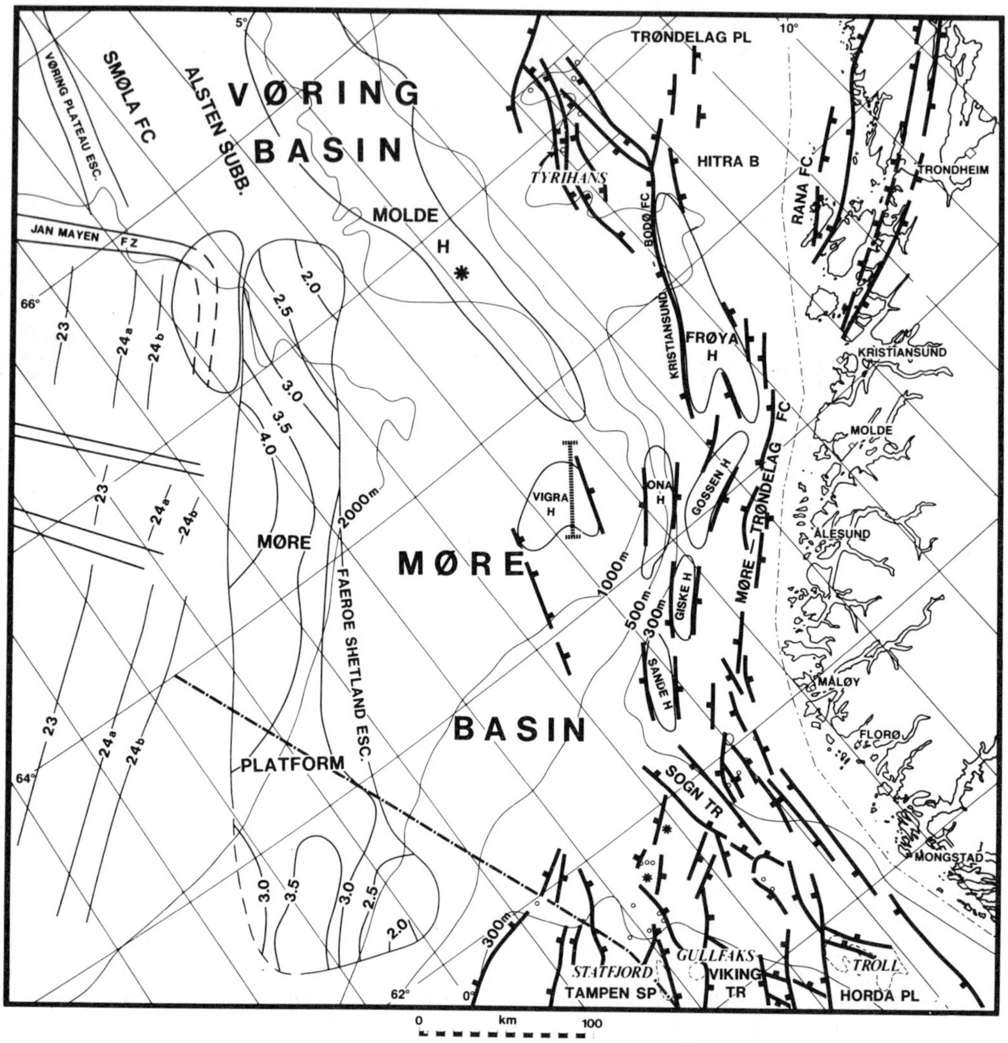

Fig. 2. Tectonic map of the Møre Basin. ESC, escarpment; FZ, fracture zone; FC, fault complex; SBB, sub-basin; B, basin; H, high; PL, platform; TR, trough; SP, spur. Asterisks denote 'compressional domes'; dashed lines, cross sections. Møre Platform contour lines in seconds (TWT). Magnetic anomalies are numbered 23—24, double lines denote transform faults.

the eastern sides (Fig. 2). The fault zone on the western side is the Faeroe-Shetland Escarpment (Talwani and Eldholm, 1977) and on the eastern side the Møre-Trøndelag and the Kristiansund-Bodø Fault Complexes (Gabrielsen *et al.*, 1984). Well-defined boundaries can neither be drawn between the Møre and Vøring Basins in the north, nor between the Møre Basin and the Sogn Trough to the south. The Møre Basin—Tampen Spur boundary is controlled by major faults which gradually disappear below the good seismic quality level eastwards.

The Møre Platform is located on the seaward side of the Faeroe-Shetland Escarpment and is underlain by continental crust (Hamar *et al.*, 1980). The platform is covered by Tertiary lava flows and sediments (Strønen, 1981) where a sequence of oceanward dipping reflectors is terminated (zone II, Hinz *et al.*, 1982). Another structural element south of the Møre Basin is the Tampen Spur. This is a triangular-shaped structural elevation consisting of a number of high ranges separated by depressions formed during the Kimmerian tectonic movements.

Most of the Møre Basin is situated in deep water. The sea floor off the Møre coast dips very steeply from a narrow shelf down to below 2000 m. Mass flows have been very common features on this very abrupt continental slope in recent times.

STRATIGRAPHY AND STRUCTURE

The hypothetical pre-Quaternary sedimentary sequence is illustrated by a simplified stratigraphic column (Fig. 3). The dominating constituents are the fine clastics from early Mesozoic to Holocene, occasionally interbedded by coarser sandstones and conglomerates. Coals and evaporites are present locally. The sedimentation is interrupted twice by volcanic activity and both submarine and subaerial lava flows are interbedded with Cretaceous and Tertiary beds. Very little is known about the pre-Mesozoic stratigraphy in the Møre Basin. However, information from East Greenland and the northern North Sea indicates the presence of non-marine coarse clastic sediments underlain by meta-sediments of early Palaeozoic age in this area. The

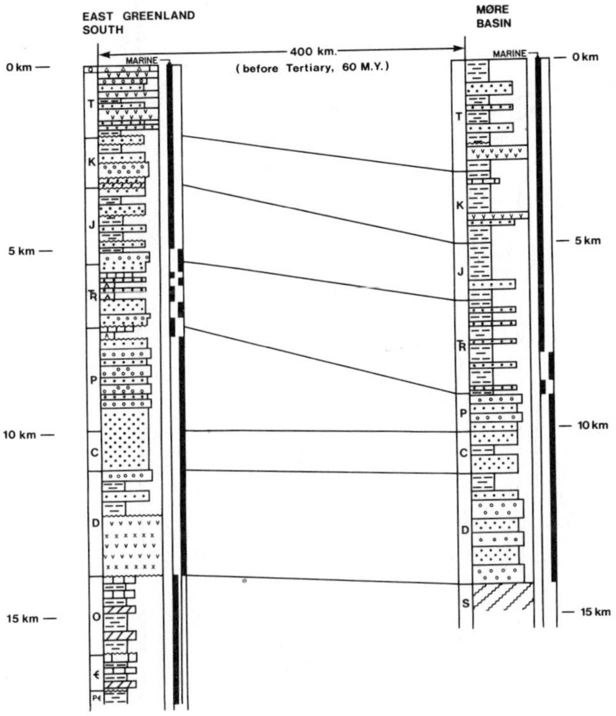

Fig. 3. Lithostratigraphy (Møre Basin and East Greenland). T, Tertiary; K, Cretaceous; J, Jurassic; TR, Triassic, P, Permian; C, Carboniferous; D, Devonian; S, Silurian; O, Ordovician, Є — Cambrian, PЄ Precambrian. V, ashes and extrusives; x, intrusives. Environment: black line, marine; white, continental.

late Phanerozoic stratigraphy from East Greenland and from the Trøndelag Platform shows close similarity to that of the Møre Basin. Cretaceous lava flows are absent in East Greenland, while the events associated with volcanism in the Møre Basin are observed only as hiatuses or unconformities over the Trøndelag Platform. Jurassic coals and Triassic evaporites are typical constituents both in East Greenland and on the Trøndelag Platform. Evaporites are also identified by gravity modelling at the eastern margin of the Møre Basin.

Regional geophysical mapping was carried out both over the shelf area and over the deep-sea region, using dense seismic coverage over the former and very sparse data over the latter. The deepest mappable event is identified as base Cretaceous (Fig. 4). The Møre Basin appears as a NE—SW-

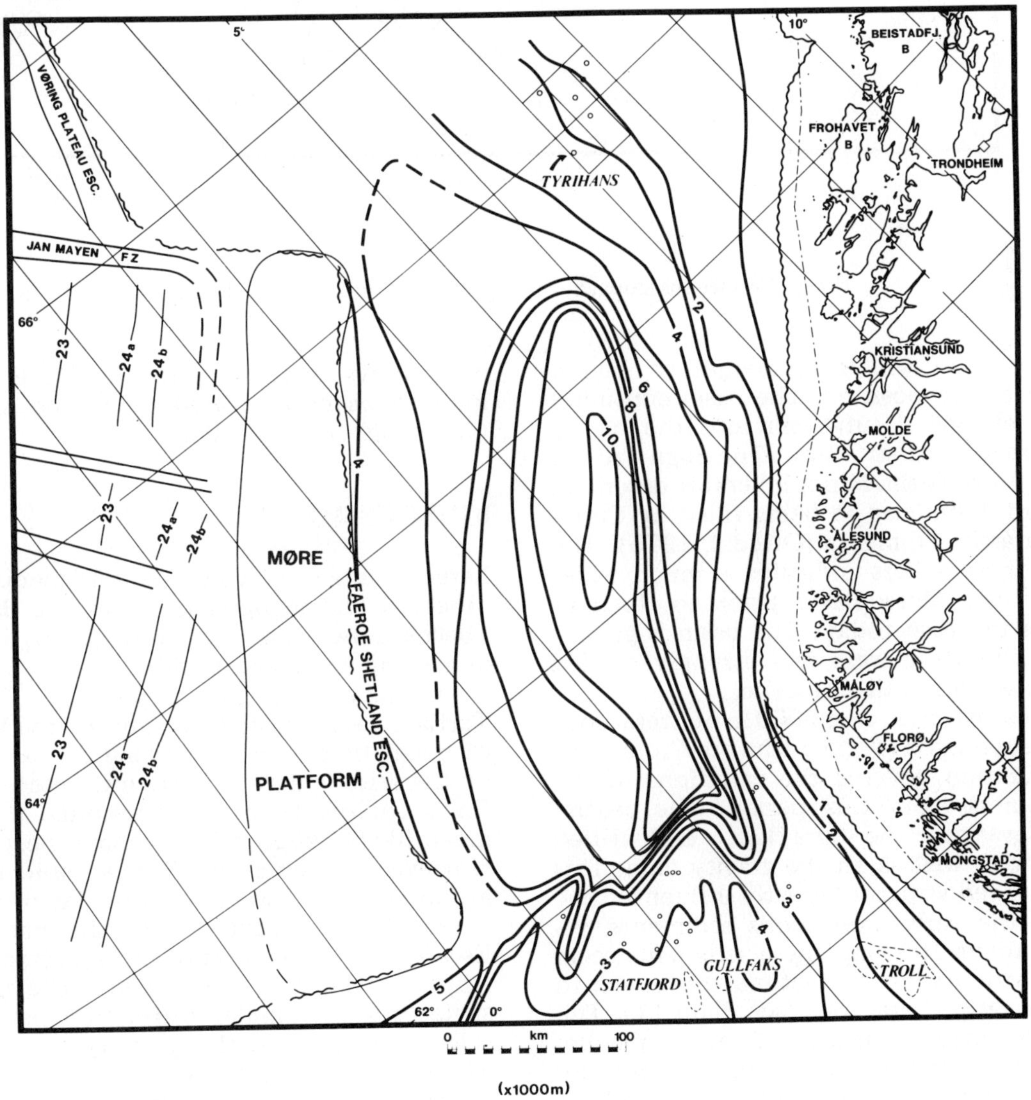

Fig. 4. Base Cretaceous isobath. For legend see Fig. 2. C.I. = 1000 m.

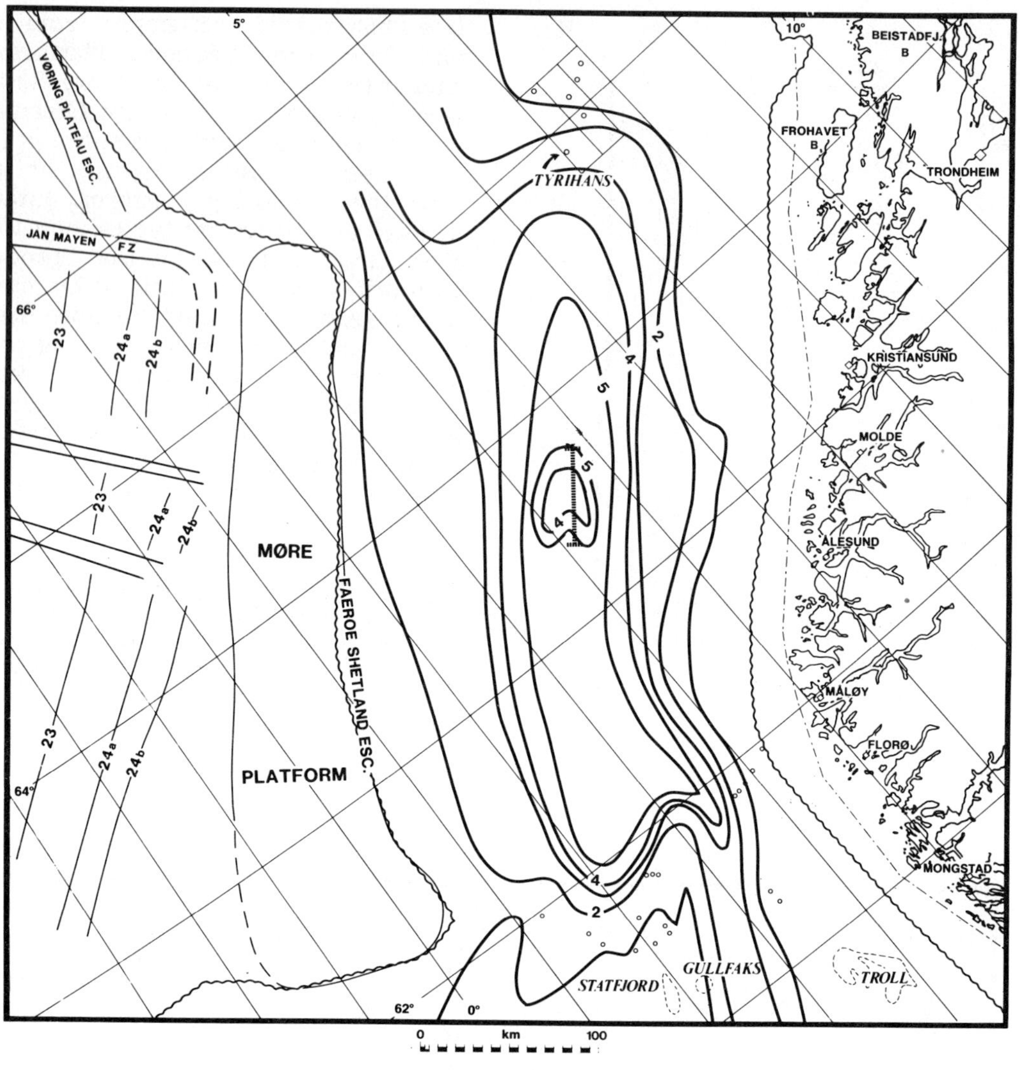

Fig. 5. Cretaceous isopach. For legend see Fig. 2. C.I. = 1000 m.

oriented depression which submerges to 10-km depth in the centre position. The basin continues as a narrow elongated sub-basin into the highly elevated Tampen Spur structure. The Sogn Trough is part of the Møre Basin. Minor inliers are described from southern Trøndelag as being filled by Cretaceous and Jurassic sediments (Oftedal, 1975). The presence of these inliers indicates a much wider extension of the Møre Basin prior to Tertiary times than is the case today. The basin is greatly diminished and its margin was strongly eroded from late Cretaceous until Holocene.

In the mid-basinal area, over 5000 m of Cretaceous sediments accumulated (Fig. 5). The pre-Cretaceous basin configuration governs the variations of the deposit thickness from this period. The central Møre Basin elevation — the Vigra High — is identified on the seismic as well as on the gravity data. This horseshoe-outlined structure greatly influenced the thickness of the early Cretaceous sediments, reducing it by about 1000 m. These sediments seem to onlap the high, dating the tectonic movement to be older than early Cretaceous (see Plate 11). The Cretaceous sequence consists of even, parallel events. Offlaps in the Tertiary sequence indicate prograding sediment build-out from the Trøndelag Platform. The seismic section crosses Quaternary mass flows originated off the Møre coast. These mass flows have created submarine hills, in some places higher than 200 m.

IGNEOUS ACTIVITY

Large numbers of volcanic and igneous bodies are observed and mapped from seismic data (Fig. 6). These can be divided into four groups on the criteria of age and areal distribution.

On the seaward side of the Møre Platform only Tertiary lavas exist. These are subdivided by means of magnetic anomalies. Strong diffraction patterns on the seismic data indicate high viscosity type lava flows (pillow lavas). It is assumed that these are most likely erupted in a submarine environment. However, subaerial conditions could have existed around the northwestern Møre Platform.

On the landward side of the Faeroe-Shetland Escarpment lava flows of early Cretaceous age are mapped (Hamar *et al.*, 1980). The irregular lava surfaces show similar diffraction patterns on the seismic sections to the Tertiary lavas to the west. These lavas are interbedded in early Cretaceous sediments at the southeastern part of the mapped area, but covered by Palaeocene sediments and

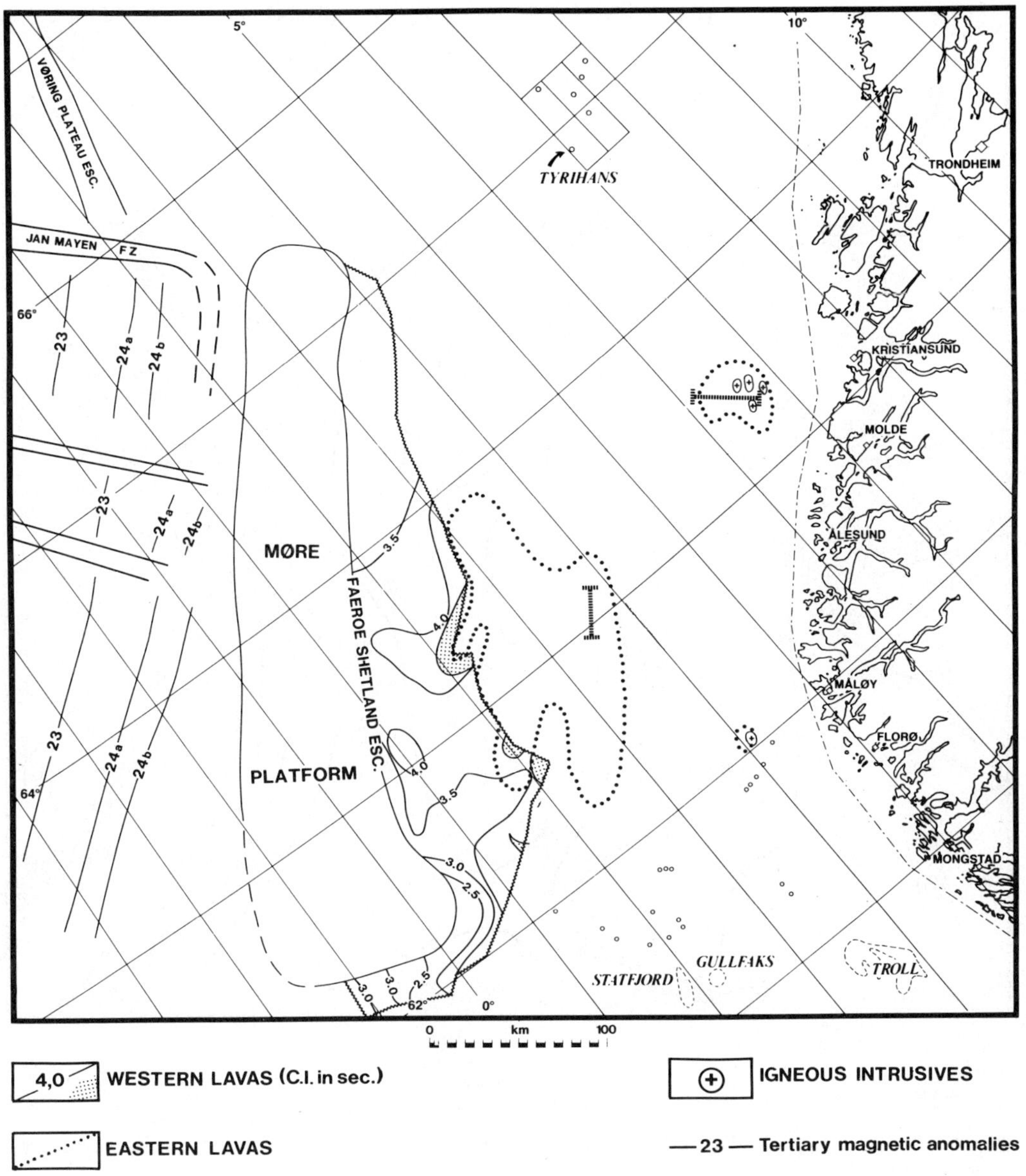

Fig. 6. Distribution of volcanic and igneous rocks.

ashes further north. This has often misled scientists to date these lavas as Tertiary (Rønnevik *et al.*, 1983; Hinz *et al.*, 1982). The uppermost surface of the lava flows range in 'depth' from 2.5 to 4.0 seconds (TWT) below the sea level.

High amplitude reflectors occur in the centre of the Møre Basin, at as many as four different stratigraphic levels (see Plate 12). These are interpreted as Tertiary sills intruded in early Cretaceous sediments (Hinz *et al.*, 1982). Price and Rattey (in press), however, suggest that they represent pre-Cenomanian sills or lavas related to mid-Cretaceous ocean-floor spreading. These volcanics could probably have erupted during the same time as the lava flows on the landward side of the Faeroe-Shetland Escarpment, mentioned above, but the latter belong to an eastern igneous province. If these are lavas, they must be of low viscosity types which form smooth upper surfaces. Such lava flows have been reported both from continental and submarine environments. The seismic section shows four levels of high-amplitude reflectors, some of which terminate suddenly.

The fourth igneous activity is recognized along the eastern margin of the Møre Basin (Fig. 6), identified as vertically intruding igneous bodies and subhorizontal high-amplitude reflectors. At the southeast of the Frøya High, four igneous bodies are found. These are covered by Quaternary sediments, and sampled rocks have been dated as Tertiary (Bugge *et al.*, 1980). This Tertiary igneous activity is correlated to tectonic activity along the Jan Mayen Fracture Zone further northwest. A few kilometres west of this igneous centre high amplitude reflectors are seen over a small area. These are covered by late Cretaceous sediments suggesting an early Cretaceous age (see Plate 13). The authors intend to favour Mr S. Bergseth's (personal communication) interpretation which correlates these lavas to the mid-Møre Basin volcanic event. Their seismic character is very similar, suggesting that low-viscosity type lavas probably erupted in a submarine environment. Another deeply buried strong reflector is observed on some of the seismic lines. This marks either the top of early Palaeozoic meta-sediments or an inter late Palaeozoic or Mesozoic sediment surface. The presence of an elevated basement block is confirmed by increased

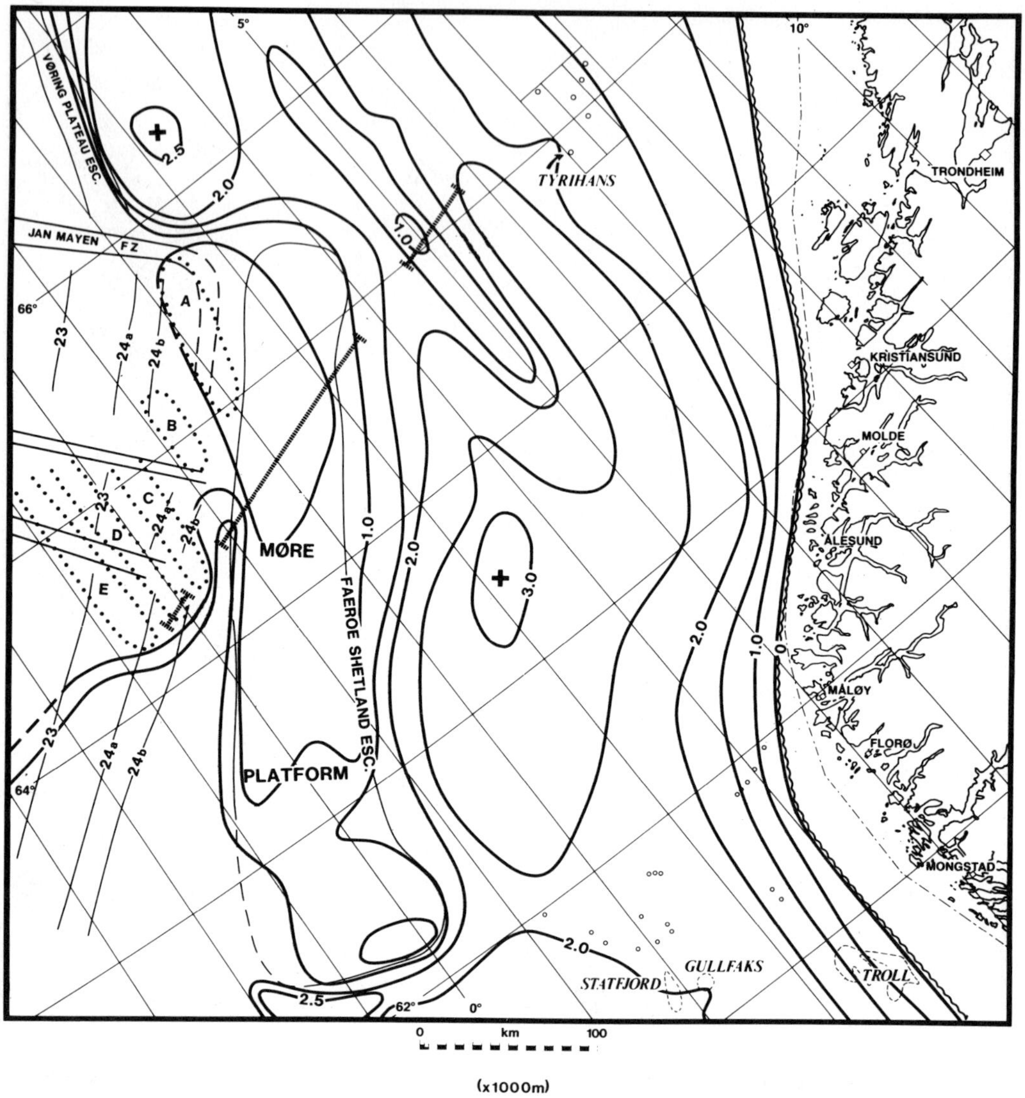

Fig. 7. Tertiary isopach. Dotted lines, outer ridges A, B, C, D and E; +, thick; dashed lines, cross sections.

gravity values in this area. Some of the intra-Cretaceous seismic events continue below the predicted lava surface. The distribution and thickness variation of this basin has not yet been clearly identified and therefore has not been mapped.

West of Målφy a columnar structure is found, covered by Cretaceous and Tertiary sediments (Fig. 6). This isolated igneous body is surrounded by high-amplitude reflectors which could represent an early Cretaceous eruptive cycle in southern mid-Norway.

TERTIARY SEQUENCES

The Tertiary sediments are distributed all over the studied area, but generally thin towards the northwest (Fig. 7). The maximum thickness is measured west of the base Cretaceous basin axis. The Alsten Sub-basin depression, at the southwestern Vφring Basin with about 2.5-km sediment thickness can easily be recognized. Thickness variations are very common over structural highs in several places. A cross-section through the strongly folded Molde High (structure D of Hinz *et al.*, 1982) is shown on Plate 14. The structural movement was initiated in late Oligocene and continued throughout Miocene. The thick Pliocene cover over the Molde High is still slightly folded, with decreasing amplitude toward the sea bottom. The present authors favour the deep-seated crustal movement explanation of this structure, proposed by Hinz *et al.* (1982), but neither evaporitic nor igneous origin can be ruled out. The Molde High is a N—S-oriented structure, its axis running parallel to 5°E (Fig. 7). The elevation amplitude changes along its axis; the cross section is narrow and high centrally, and broad and low towards south and north.

The Tertiary sedimentary and volcanic sequence is thinner both in the south and north on the Mφre Platform (see Plate 15); in the north about 200 m of early Tertiary lavas are buried below very thin Pliocene clastics. Nearly the whole Neogene is missing, which is explained by submergence of this northern Mφre Platform from above sea level to 2.5-km water depth during the last 2—3 Ma.

Further east, on the landward side of Faeroe-Shetland Escarpment, mid-Cretaceous lavas are covered by Eocene ashes and sediments (Fig. 7) caused by extensive erosion before the late Tertiary

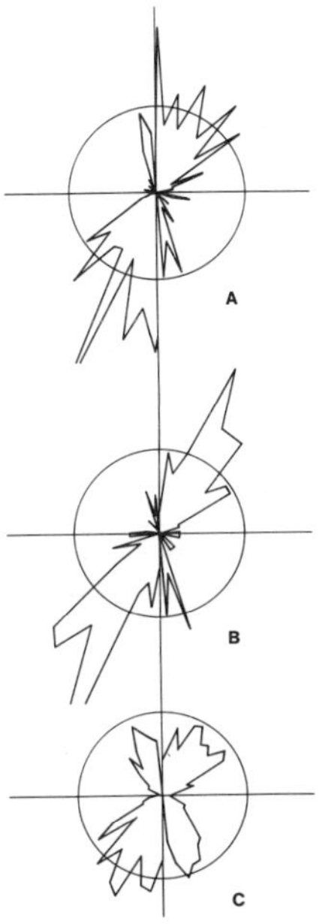

Fig. 8. Rose diagrams. (A) Trøndelag Platform (south), (B) Eastern Møre Basin and (C) northern North Sea. The directions are gathered in 30 classes, each covering 6°; the left half shows vector units, the right half shows the total number of faults, both in percentage. The circle indicates 10% cumulative length or frequency of faults.

submergence. West of the Møre Platform several ridges are observed, oriented in a N—S direction. Five of these can be correlated for several kilometres as marked with dotted lines on the map. The 'C' and 'D' ridges are discussed by Strønen (1981) and named 'Outer High'. On the map of Strønen (1981) the 'Outer High' runs subparallel to the Faeroe-Shetland Escarpment but 80—100 km to the west and trends NE. This direction is disproved by our recent data crossing the 'domal' structures. Oceanic crust of Eocene age (Hinz *et al.*, 1982) underlies these five ridges which are most probably cut by numerous transform faults parallel to the Jan Mayen Fracture Zone. The direction of both the Jan Mayen Fracture Zone and of the other transform faults diverge about 30° from that of the 'Outer Ridge' axis, making the tectonic relationship uncertain. The Molde High east of the Møre Platform is lined up in the same direction as these five ridges. The tectonic movement for all these structures is dated to late Oligocene, which is penecontemporaneous with a shift of the spreading axis in the Norwegian Sea (Talwani and Eldholm, 1977). It is possible that this major tectonic event influenced both the size and direction of the structures. The variable elevation of the surface along the axis of the 'Outer Ridges' suggests they were probably generated by strike-slip lateral movements or perhaps by perpendicularly acting compressional stress. The change of direction of the Jan Mayen Fracture Zone, proposed by Strønen (1981), could hardly explain the presence of all these ridge-like structures. A cross-section through the 'ridge' marked D is

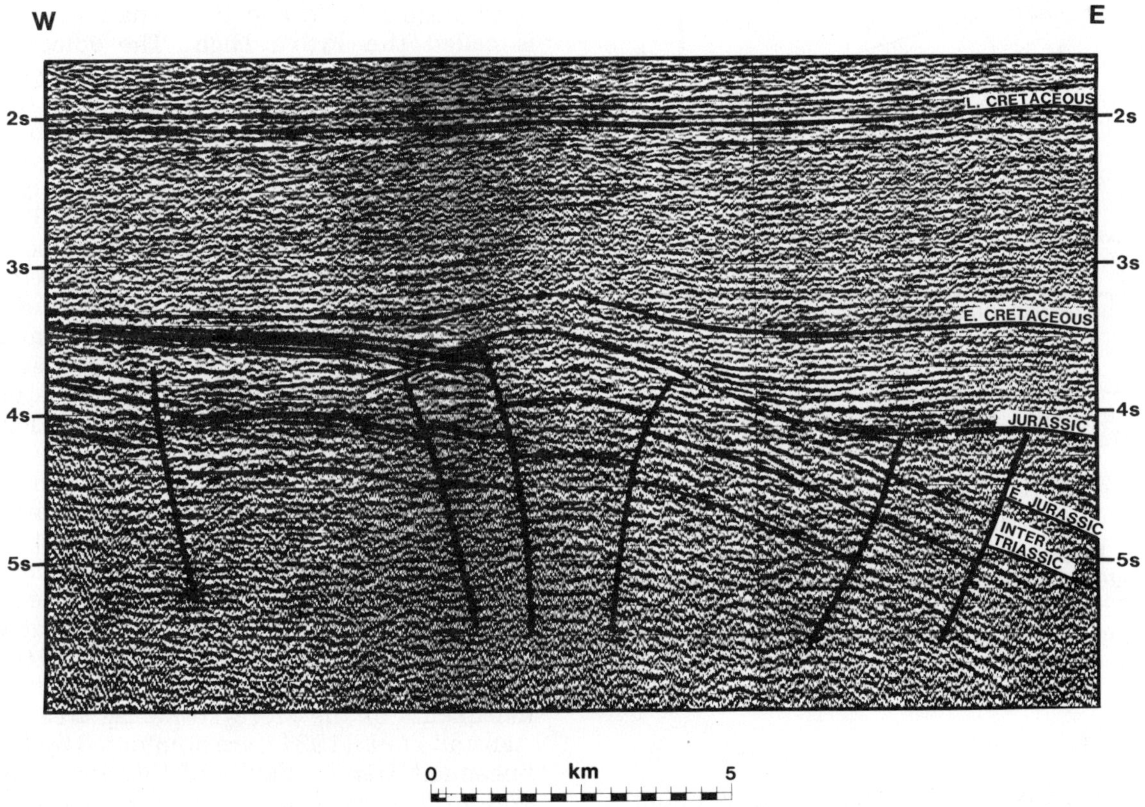

Fig. 9. Strike-slip faulting on Tampen Spur. 'Compressional dome' is assumed to be underlain by a strike-slip fault, Tampen Spur, (Line TLGS 78-1, shp. 4600—3930). (Reproduced with kind permission of Saga Petroleum Company.)

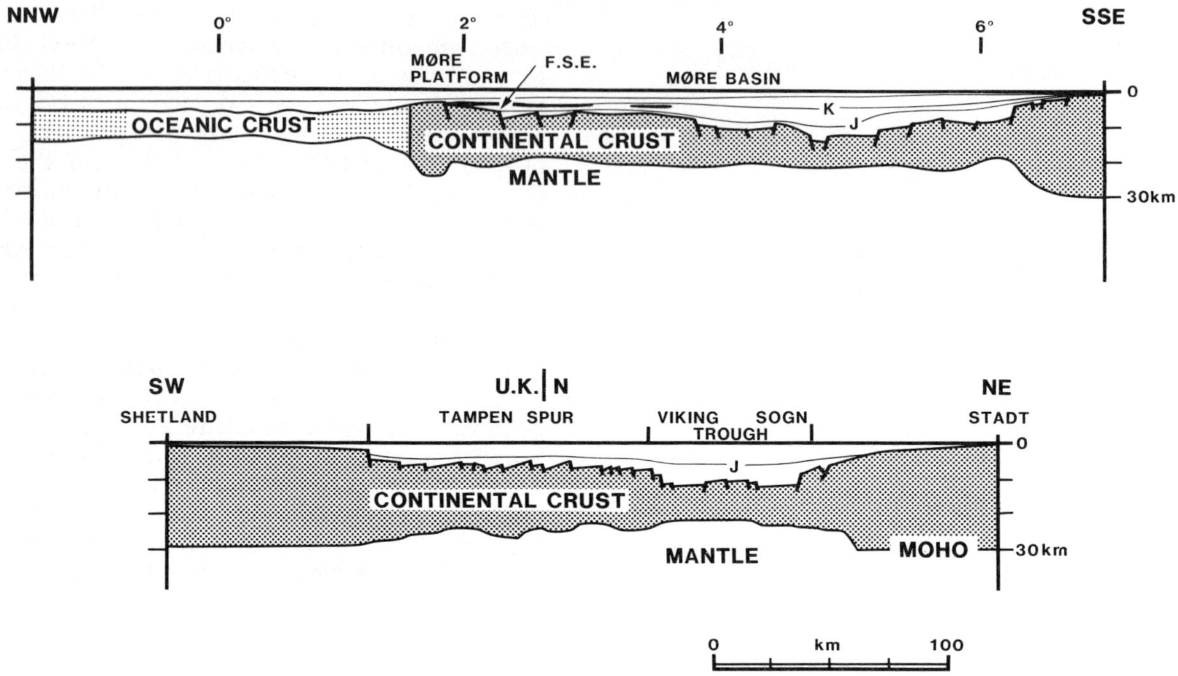

Fig. 10. Geologic model of Møre and East Shetland Basins. F.S.E. Faeroe-Shetland Escarpment; K, Cretaceous; J, Jurassic.

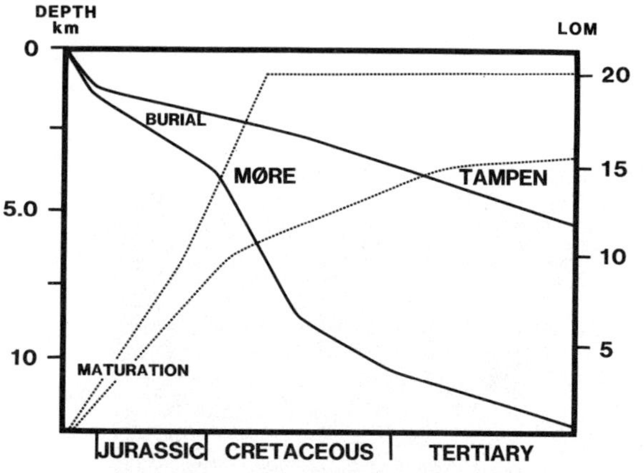

Fig. 11. Subsidence and maturation of late Jurassic sediments (Møre and East Shetland Basins). LOM, level of maturation.

shown on Plate 16. The Oligo-Miocene sediments onlap the ridge, which consists of igneous rocks of Eocene age and these are covered by Pliocene fine clastics. The other four 'ridges' show similar stratigraphy except that the base Pliocene reflector continues without break above the 'C' structure.

FAULTING

The eastern margin of the Møre Basin is bounded by major NE—SW and NNW—SSE-oriented normal faults, of which a selected set is marked on the tectonic map (Fig. 2). These have formed numerous tilted faultblocks as well as horst type structures. Five major highs are identified and named. A giant structure situated in the junction of three major fault complexes, the Møre-Trøndelag, the Kristiansund-Bodø and the Rana Fault Complexes, is called the Frøya High. The volcanic activity found in the southern part of this high indicates that the fault complexes extend very deeply into and most probably through the crust. The basement rock found at shallow depth in the central part of the Frøya High (6306/3) is successively onlapped by older sediment from all directions. Two parallel ridges originate from the bifurcated southern tip of the Frøya High. These are intersected by transverse depressions and form a landward row (the Gossen and the Giske Highs) and a seaward range (the Ona and the Sande Highs). All these highs are cut into several minor structures by a number of normal faults. The highs lie in the depth range from about 1500 to 6000 m.

All the mapped faults at the northern North Sea, the Møre coast and the Trøndelag Platform were digitized and, with the help of a computer program, displayed as Rose diagrams (Fig. 8). The measured directions (right half of diagrams) and length of the faults (left half of diagrams) show great variations for these three areas. (Circles indicate 10% cumulative length or frequency of faults.) The onshore lineaments of the Møre-Trøndelag area as given by Aanstad *et al.* (1981), are dominated by ENE—WSW lineament trends. Faults of this direction are very seldom observed from offshore areas. The N—S fault direction, found extensively on the Trøndelag Platform, is almost absent further south. The main direction of the synthetic faults is concentrated in the 30°—60°E sector and those about 60—70°

form an antithetic counterpart. The majority of the offshore fault movements have been going on since pre-Cretaceous times; onshore data, on the other hand, indicate syn- or post-Permian faultings (Aanstad *et al.*, 1981). More precise dating of the movements cannot be given at present.

The N—S-oriented axial direction of the Molde High and the origin of this structure, related to fault movements along the same direction, have been discussed earlier. Except for the correlation between the Molde High and the 'A to E Ridges', no other structures of similar type have been found within the Møre Basin. The absence of these structures here is simply explained by the very thick sedimentary cover and the presence of volcanics in the western part of the Møre Basin. The N—S-oriented fault systems of the Trøndelag Platform probably belong to the same Molde High — 'Outer Ridge' tectonic regime, even if none of them have moved in Tertiary. The only known late Tertiary movement outside the area studied is found over the Rødøy High in the Trænabank area (Larsen and Skarpnes, 1984). Basin inversions, as well as strike-slip lateral movements, are reported from the southern and central North Sea (Ziegler, 1982; Hamar *et al.*, in press; Pegrum and Ljones, in press). All these movements are dated mid and late Cretaceous and are certainly much older than those found in the northern Møre Basin. Strike-slip lateral faults have recently been observed also in the northern North Sea. The site of these faults is marked by an asterisk in Fig. 2. A section illustrates deep-seated strike-slip faulting cut into the early Cretaceous sequence, slightly folding the overlying Upper Cretaceous (Fig. 9).

'Compressional' dome-type structures have also been observed in several places on the downfaulted side of major faults in the northern North Sea. These have never been adequately explained and have usually been completely disregarded as valid information during interpretation of seismic data. They are probably initiated by the same strike-slip lateral movement as found on the seismic section presented (Fig. 9). Strike-slip lateral faulting is generally very common all over the Norwegian Continental Shelf, even if not many of these could be described from the Møre Basin.

GRAVITY

Geological models derived from interpretation of seismic sections, have been tested and controlled by using gravimetric data. A 'prism-oriented computer calculation system' is applied to gravimetric field variations to determine the depth to the Moho, the thickness of the crust and of the sediments above. Such computations were carried out along two seismic lines, LDGO line 160A crossing the Møre Basin and the line ST 805—203 in the northern North Sea (Fig. 2). Both sections show considerable thinning of the crust close to the landward side of the basins (Fig. 10). The crust reaches 20 km in thickness below the outer part of the Møre Platform. Similar crustal sections are found over the northern North Sea, where the thickness increases from 12 km below the Sogn Trough to 28 km under the Shetland Islands. Simple mass calculation shows a 100% extension in the Møre Basin. On the other hand, the extension is only 30% below the northern North Sea. The maximum sediment thickness is measured as 13 km in both basins. The very sudden rises of the Moho (5 km close to 2°E and 10 km just east of 6°E) on the Møre Basin sections are noticeable. They are interpreted as major crustal weakness zones where lavas were conveyed to the surface during the early Cretaceous at both sides of the Møre Basin.

SUBSIDENCE

The subsidence history of the Møre Basin was studied and compared with the northern North Sea (Fig. 11). Both basins have the same rate of burial during the Tertiary. The pre-Jurassic sediments have undergone slow and continuous subsidence from the Jurassic to the Tertiary in the Tampen Spur area. The same beds in the Møre Basin have subsided 9 km during the same period of time. According to the calculated maturation curves, the pre-Jurassic sediments in the Møre Basin reached the overmature level during early Cretaceous and in the Tampen Spur area at the beginning of Tertiary. The level of maturation (LOM) has also been calculated for late Jurassic and mid-Cretaceous sediments. The late Jurassic sedimentary sequence has just reached the early oil maturation zone, but the mid-Cretaceous level is immature on the Tampen Spur. In the Møre Basin the Jurassic was overmature 100-Ma ago (Albian). The Cretaceous has been mature for oil generation since late Tertiary times.

CONCLUSIONS

The Møre Basin, a major Cretaceous depression, is surrounded by highs of Precambrian intrusives and/or early Palaeozoic metasediments. The maximum sediment thickness exceeds 13 km in the deep basinal areas, decreasing gradually toward the margins. The highs centrally and along the rim of the basin, and possible stratigraphic traps in the Cretaceous and Tertiary sequence, could be the principal trap types. Early Cretaceous sediments are the primary, and Jurassic shales in certain places could be secondary, source rocks. The Møre Basin is characterized by large-scale rifting on a passive continental margin. It probably has a considerably higher hydrocarbon yield factor than indicated earlier by Bleie *et al.* (1982). We hope that in the future the area can be listed among the promising hydrocarbon-producing basins.

ACKNOWLEDGEMENTS

The authors would like to thank colleagues at Statoil for many valuable discussions and comments and, especially, to thank Mr Johannes Tryti and Øyvind Sylta for their contribution to this work. Our acknowledgements also to Dr O. Eldholm (Oslo), Professor M. Sellevoll and Dr E. Sundvor (Bergen) for their kind help in obtaining some of the geophysical data and for their valuable comments. The contribution is published with approval of Management at Statoil.

REFERENCES

Aanstad, K.M., Gabrielsen, R.H., Hagevang, T., Ramberg, I.B. and Torvanger, O. 1981. Correlation of offshore and onshore structural features between 62°N and 68°N, Norway. In: *Norwegian Symposium on Exploration*, Bergen (Norsk Petroleumsforening), Article 11, pp. 1—24.

Bleie, J., Nysæther, E. and Oppebøen, K.A. 1982. The hydrocarbon potential of the Northern Norwegian Shelf in the light of recent drilling. In: *The Geologic Framework and Hydrocarbon Potential of Basins in Northern Seas*, Stavanger (Norsk Petroleumsforening), E/5, pp. 1—39.

Bott, M.H.P. 1973. The evolution of the Atlantic north of the Faeroe Islands. In: *Implications of Continental Drift to the Earth Sciences, Vol. 1*, D.H. Tarling and S.K. Runcorn (eds.). NATO Advanced Study Institute, Univ. Newcastle upon Tyne, pp. 175—189.

Bugge, T., Prestvik, T. and Rokoengen, K. 1980. Lower Tertiary volcanic Rocks off Kristiansund, Mid-Norway. *Mar. Geol.* **35**, 277—286.

Gabrielsen, R.H., Færseth, R., Hamar, G. and Rønnevik, H. 1984. Nomenclature of the main structural features on the Norwegian Continental Shelf north of the 62nd parallel. This volume.

Hamar, G.P., Jacobsen, K., Ormaasen, D. and Skarpnes, O. 1980. Tectonic development of the Northern North Sea north of Central Highs. In: *The Sedimentation of the North Sea Reservoir Rocks.* Norsk Petroleumsforening, Article 3, 11 pp.

Hamar, G.P., Fjæran, T. and Hesjedal, A. Jurassic stratigraphy and tectonics of the south-southeastern Norwegian offshore. *Geol. Mijnb.* (in press).

Hanisch, J. (1983). The structural evolution of the NE Atlantic region — a new model. *Geol. Jahrb.* Reihe B, 52.

Hinz, K. 1972. The seismic crustal structure of the Norwegian continental margin on the Vøring Plateau, in the Norwegian deep sea and on the eastern flank of the Jan Mayen Ridge between 66 and 68°N. *Int. Geol. Congr.*, Montreal, Sect. 8, pp. 28—37.

Hinz, K. and Moe, A. (1971). Crustal structure in the Norwegian Sea. *Nat. Phys. Sci.*, **232**, 187—190.

Hinz, H.K., Dostman, H. and Hanisch, J. 1982. Structural framework of the Norwegian Sea. In: *The Geologic Framework and Hydrocarbon Potential of Basins in Northern Seas* Stavanger (Norsk Petroleumsforening) E/4, pp. 1—22.

Jørgensen, F. and Navrestad, T. 1982. The geology of the Norwegian Shelf between 62°N and the Lofoten Islands. In: L.V. Illing and G.D. Hobson (eds.), *Petroleum Geology of the Continental Shelf of Northwest Europe.* Institute of Petroleum, London, pp. 407—413.

Larsen, R.M. and Skarpnes, O. 1984. Regional interpretation and hydrocarbon potential of the Trænabanken area. This volume, pp. 217—236.

Oftedahl, Chr. 1975. Middle Jurassic Graben tectonics in mid-Norway. *Proceedings, Jurassic Northern North Sea Symposium, Stavanger*, Norwegian Petroleum Society. JNNSS 21/1-13.

Pegrum, R.M. 1984. The extension of the Tørnquist Zone in the Norwegian North Sea. *Norsk Geol. Tidskr.* (in press).

Price, I. and Rattey, R.P. 1984. Cretaceous tectonics of mid-Norway: implications for the Rockall-Faeroes-Shetland Rift *J. Geol. Soc. London* (in press).

Roberts, D.G. 1974. Structural development of the British Isles, the Continental Margin, and the Rockall Plateau. In: C.A. Burke and C.L. Drake (eds.), *The Geology of Continental Margins*, Springer Verlag, Berlin, pp. 343—359.

Rønnevik, H.Chr., Eggen, S. and Vollset, J. 1983. Exploration of the Norwegian Shelf. In: J. Brooks (ed.), *Petroleum Geochemistry and Exploration of Europe.* Blackwell, Oxford, pp. 71—94.

Strønen, L.K. 1981. *Marine Geophysical study of the Norwegian Continental Margin between 62°—69°N.* Cand. Real. Thesis, Seismological Observatory, University of Bergen, 146pp.

Talwani, M. and Eldholm, O. 1972. Continental margin off Norway: A geophysical study. *Geol. Soc. Am. Bull.*, **83**, 3575—3606.

Talwani, M. and Eldholm, O. 1977. Evolution of the Norwegian-Greenland Sea. *Geol. Soc. Am. Bull.*, **88**, 969—999.

Whiteman, A.J., Rees, G., Naylor, D. and Pegrum, R.M. 1975. North Sea troughs and plate tectonics. *Norges Geol. Unders.*, **316**, 137—161.

Woodhall, D. and Knox, R.W.O'B. 1979. Mesozoic volcanism in the northern North Sea and adjacent areas. *Bull. Geol. Surv. G.B.*, **70**, 34—56.

Ziegler, P. 1982. *Geological Atlas of Western and Central Europe.* Elsevier Scientific, Amsterdam.

26 Structural development of the southwestern margin of the Russian–Fennoscandian Platform

R.M. Pegrum
Statoil, Stavanger, Norway

The Russian-Fennoscandian Platform as a structural entity extends from Scandinavia eastwards to the Urals. The platform comprises Precambrian crystalline basement rocks, overlain in Eastern Europe and the USSR by a sedimentary cover ranging in age from early Palaeozoic to Cenozoic which is virtually undeformed and unmetamorphosed. Along its northwestern margin the Precambrian basement is overthrust by highly deformed and metamorphosed Lower Palaeozoic and structurally reworked Precambrian basement rocks of the Scandinavian Caledonides. Similar rocks extend westwards beneath the North Sea. The Precambrian basement platform is cut by a number of ancient, often rejuvenated, NW-trending tectonic zones. Two of these zones, the Varanger-Timian Zone and the Tornquist Zone, respectively, form the northeastern and southwestern margins of the platform. Both have complex histories of dip-slip and strike-slip faulting. The Tornquist Zone, the principal subject of this chapter, has been traced from the Black Sea coast of Roumania to the line of the Oslo Graben. Recently published studies demonstrate that it extends for at least a further 500 km into the central North Sea and possibly even to the margin of the continental shelf of northwestern Europe.

Intermittent structural activity has occurred along the Tornquist Zone from at least the late Precambrian through to the Cenozoic. Regional evidence suggests that during the Cadomian, Caledonian and Variscan orogenic cycles the Tornquist Zone marked a fundamental crustal-plate boundary and was the site of major transcurrent displacements. The Variscan deformation was of considerable importance in establishing the tectonic framework which exerted important controls on the subsequent late Permian to Cenozoic geologic history of the North Sea. A late Variscan right-lateral strike-slip displacement of 300—350 km along the Tornquist Zone is suggested. During the Triassic and Jurassic, regional extensional deformation established a complex pattern of rift basins in northwestern Europe and the North Sea. Transtension along the Tornquist Zone led to extensional faulting and graben formation. In the Cretaceous and early Tertiary, Alpine regional compression resulted in transpression along the Tornquist Zone, causing compressional folding, reverse faulting and structural inversion. Weak compressional effects observed in the central North Sea, 2000 km from the Carpathian fold front, are associated with rejuvenation of the deeply buried Tornquist Zone. Further evidence of instability along the Tornquist Zone is provided by volcanic extrusives of Palaeozoic, Mesozoic and Tertiary ages.

The Tornquist Zone, a fundamental zone of weakness established prior to the late Precambrian, can be shown to have exerted a structural influence from the Palaeozoic through the Mesozoic and into the Cenozoic. In block 15/9, in the central Norwegian North Sea, it is believed to be directly associated with the formation of a commercial gas-bearing structural trap in Palaeocene sandstones. The presence of similarly ancient NW- and complementary NE-trending tectonic zones in the pre-Caledonian basement platform west of the Norwegian coast can be anticipated to have had similar structural influences as the result of repeated rejuvenations.

THE PRECAMBRIAN PLATFORM

As a structural entity the Russian-Fennoscandian Platform is limited to the northwest by the Scandinavian Caledonides, to the east by the Urals, to the north by the Varanger-Timian Tectonic Zone and to the southwest by the Tornquist Zone (Fig. 1). The Tornquist Zone can be traced from the North Sea to the Black Sea. The southern margin of the platform from the Black Sea to the southern Urals has been modified by Alpine tectonism, but runs more or less parallel to the front of the Caucasus.

The platform is an ancient and complex feature, consolidated and stabilized by the end of the Precambrian. It was characterized throughout the Palaeozoic, Mesozoic and Cenozoic by its structural stability. The Precambrian basement comprises

Petroleum Geology of the North European Margin © Norwegian Petroleum Society (Graham & Trotman, 1984) pp. 359—369.

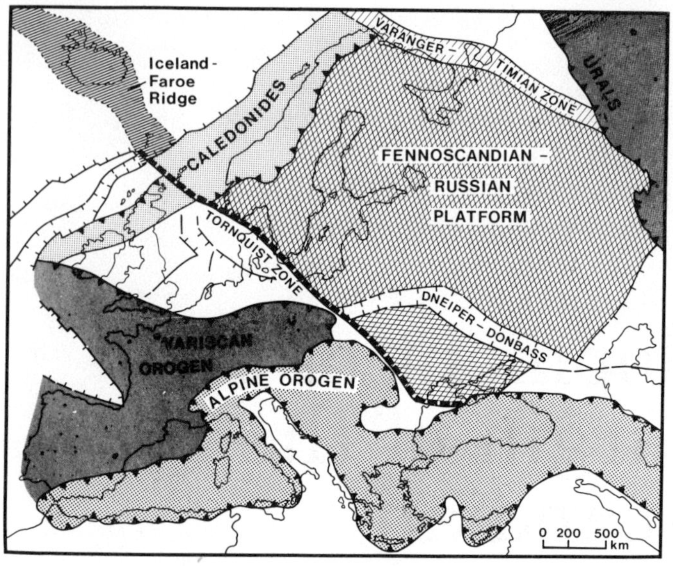

Fig. 1. Major tectonic units in Europe and the extent of the Fennoscandian-Russian Platform.

ovoid Archaean (2500 Ma and older) provinces welded together by a network of early Proterozoic (1800 Ma and older) tectonic belts (Watson, 1976). The southwestern corner of the platform, the Svecconorwegian Province, is younger, being dominated by a 1000-Ma tectonothermal event comparable to the Grenvillian phase of the North American Craton. The northeastern corner of the platform is also young, having been consolidated by late Proterozoic (Baikalian) orogenesis.

The crystalline basement platform is cut by a number of ancient, frequently rejuvenated, tectonic lineaments with a more or less consistent NW-trend. Based on stratigraphic, structural, magnetic, gravimetric and seismicity criteria, Strömberg (1976) has recognized and described a pattern of linear 'tectonic zones' in the Fennoscandian Platform. Some of the lineaments can be traced over distances of 1000 km or more. The northwesterly trends are intersected by, and often left-laterally offset by, a subordinate NE-trending set of lineaments. The lineament system is certainly Precambrian in origin and may be Archaean or older. Strömberg (1976) suggests that it may belong to a primordial crustal shear pattern. Important NW-trending lineaments are also recognized beneath the Palaeozoic and younger cover of the Russian Platform, and are considered by Watson (1976) to be at least 1000 Ma in age.

Two prominent, often rejuvenated, northwesterly lineaments are those bounding the platform to the northeast and to the southwest. The northeastern margin (Varanger-Timian Tectonic Zone of Strömberg, 1976) includes the Trollfjord-Komagelv Fault of northernmost Scandinavia and a probable parallel zone lying north of the present coastline. The latter has been interpreted from aeromagnetic data. The Trollfjord-Komagelv Fault Zone shows evidence of both left-lateral strike-slip faults with offsets measured in tens of kilometres as well as reverse dip-slip faults. The zone can be traced into the northern USSR through the Timan uplift southeastwards as far as the Urals. The Timan Zone exposes narrow outcrops of folded and metamorphosed late Precambrian rocks associated with conspicuous gravity and magnetic anomalies. Isotopic evidence suggests an age of at least 1200 Ma (Watson, 1976), although it has suffered repeated rejuvenations. The main phase of left-lateral transcurrent movement on the Trollfjord-Komagelv Fault System is believed to be late Precambrian (640—530 Ma, Gabrielsen, in press).

The southwestern margin of the Fennoscandian-Russian Platform is commonly referred to in the literature as the Tornquist Line or Zone. The Tornquist Zone has been considered as terminating at the line of the Oslo Graben (Ziegler, 1981) or at the projected line of the Caledonain Front (Watson, 1976), but evidence suggesting that it continues at least to the line of the Viking Graben (Fig. 2), and possibly to the edge of the continental shelf northwest of Scotland, has recently been reviewed by Pegrum (in press). The continuation of the Tornquist Zone beneath or through the NE-trending Caledonides (Fig. 1) has fundamental structural implications. Strömberg (1976) noted the coincidence of the Iceland-Faroe volcanic ridge with the projected Tornquist Zone and reviewed the evidence that several of the other northwesterly lineaments of the Precambrian basement have recognisable stratigraphical and structural affects on the Caledonian orogen.

Knowledge of the Precambrian basement southwest of the Tornquist Zone is fragmentary. Only a few deep boreholes in the North Sea and northern Europe have reached pre-Palaeozoic rocks (Frost *et al.*, 1981), but additional information is provided by limited outcrops in the British Isles and within the Variscan Massifs across central Europe (Zwart and Dornsiepen, 1978; Strömberg, 1981). Many of these outcrops carry evidence of a Cadomian orogenic event which occurred towards the end of of the Precambrian, between 700 and 550 Ma. Zwart and Dornsiepen (1978) relate the Cadomian event to the opening of the Iapetus Ocean between the North American and European crustal plates. Evidence of Cadomian deformation within the Precambrian platform north of the Tornquist Zone is generally lacking, the rocks having been consolidated by Svecconorwegian (1000 Ma) or earlier orogenesis. The evidence suggests that the Tornquist Zone was already an important tectonic boundary in the late Precambrian, forming the northeastern limit to the unstable block/basin framework produced by the Cadomian orogenic cycle (Watson, 1976). The detailed nature of the boundary is unknown as the basement fault system is now deeply buried beneath younger rocks, but it is not unreasonable to assume a complex zone of strike-slip and dip-slip faults similar to those exposed in the Trollfjord-Komagelv Fault Zone at the northern margin of the platform, which was tectonically active at approximately the same time.

EARLY PALAEOZOIC/CALEDONIAN HISTORY

The Caledonian orogenic cycle was intermittently active from the late Ordovician to the late Silurian, closing the Iapetus Ocean and suturing the North American-Greenland and European Continental Plates along the Scottish and Scandinavian Foldbelts. The early Palaeozoic rocks of the northwestern

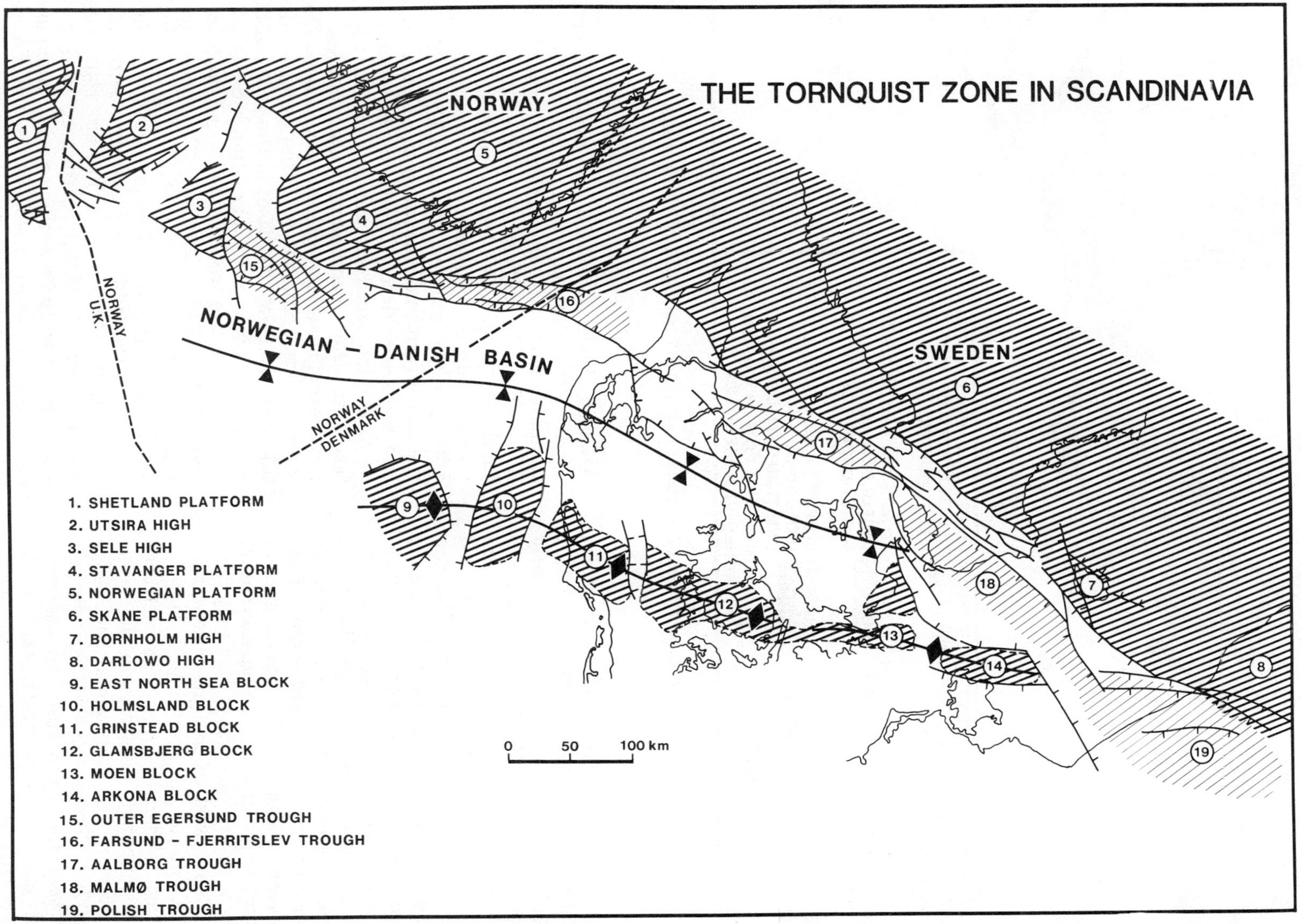

Fig. 2. Structural features associated with the Tornquist Zone between the Baltic coast of Poland and the central North Sea. (after Pegrum, in press.)

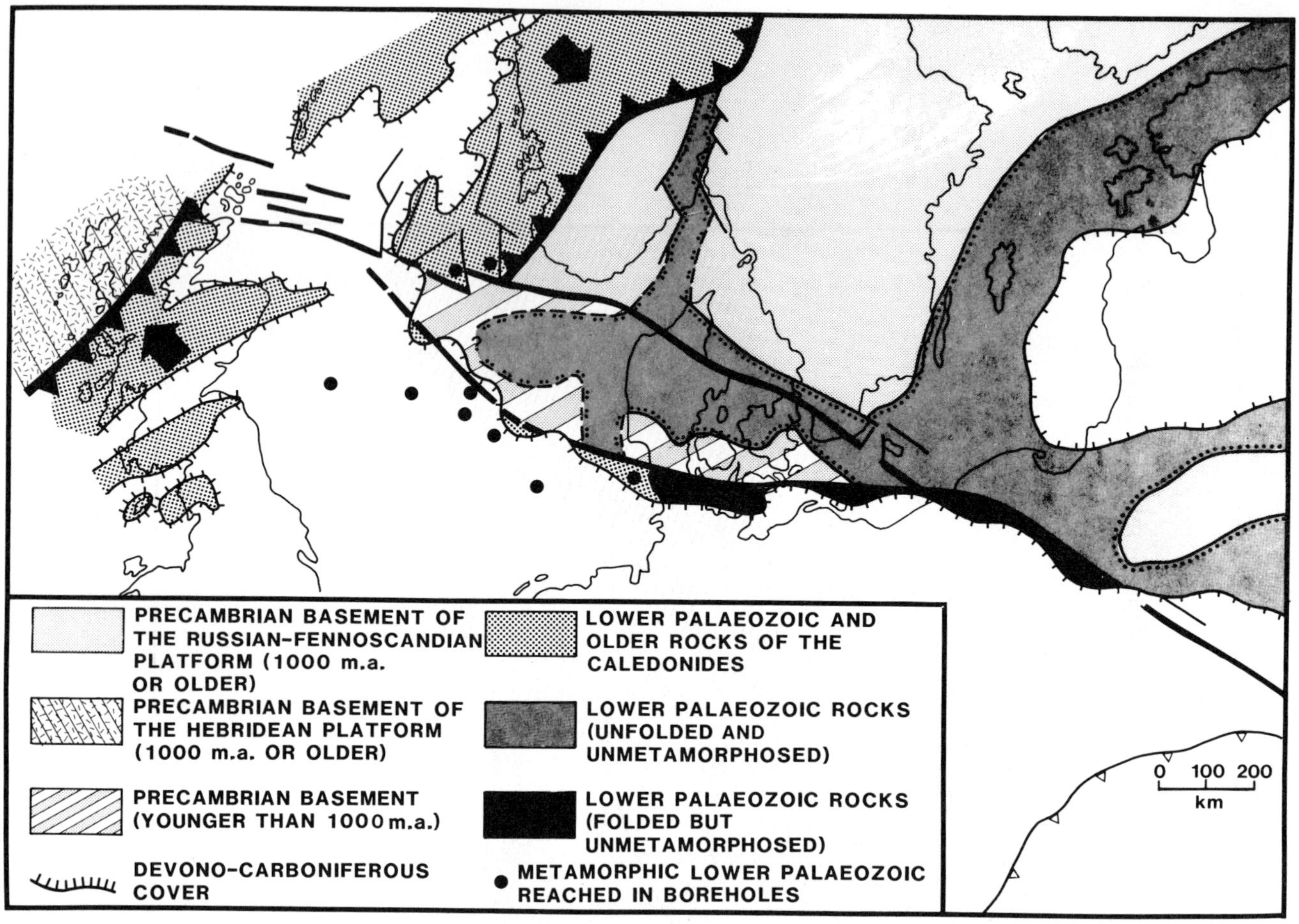

Fig. 3. Distribution of Precambrian and Lower Palaeozoic rocks in the neighbourhood of the Tornquist Zone. After Frost *et al.*, 1981; Ziegler, 1981.)

margin of the Russian-Fennoscandian Platform were highly deformed and metamorphosed during the Caledonian cycle. In strong contrast, the early Palaeozoic rocks of the southern Baltic and extending far to the east across the Russian Platform are virtually undeformed and unmetamorphosed (Fig. 3). Along the length of the Scandinavian Caledonides the metamorphosed early Palaeozoic and Precambrian sediments, together with crystalline basement rocks, are overthrust eastwards and southeastwards on to the basement platform and its cover of autochthonous sediments in a series of detached nappes. The northwestward extent of the basement platform beneath the allochthonous Caledonides and the location of the Iapetus suture are not known with any accuracy, but may lie a considerable distance west of the present coastline of Norway. Indications of rejuvenated tectonic zones in the Precambrian basement can be traced through the Caledonian orogen and in some cases west to the present continental shelf edge (Strömberg, 1976). These observations suggest the possibility that the Fennoscandian basement platform itself extends northwestwards to the continental shelf edge and that the Iapetus suture lies beneath the continental slope and, in part, may have controlled the line of Mesozoic—Tertiary rifting which led to the opening of the Norwegian Sea.

In Scotland, the Caledonides are overthrust northwestwards on to the Hebridean basement platform. The latter is clearly a fragment of the North American Craton, separated from it by the Mesozoic and Tertiary opening of the North Atlantic (Watson, 1976). The opposing vergences of the Scandinavian and Scottish Caledonides coincide with the projection of the Tornquist Zone (Fig. 3), indicating its fundamental structural significance. The Iapetus suture in the Scottish Caledonides is considered to run beneath the Solway Firth and the Southern Uplands, marking the line of collision between the North American and European Plates. Clearly this line cannot easily be projected northeastwards beneath the North Sea into the Scandinavian Caledonides. If the continental shelf edge off the west coast of Norway more or less coincides with the Iapetus suture in this area, a left-lateral offset in the order of 500 km along the Tornquist Zone could be invoked to match it with the projected Solway Firth—Southern Uplands line. This offset could be the result of an active transform fault system during the closing of Iapetus or of post-Caledonian transcurrent faulting.

Beneath the southern Baltic and across much of the Russian Platform the Precambrian basement has a cover of unmetamorphosed and virtually undeformed early Palaeozoic rocks, which are time equivalent to those involved in the Caledonian foldbelts of Scandinavia and Scotland. The cover is characterized by epicontinental stable platform deposits. Cambrian sandstones, siltstones and shales of shallow neritic facies overlie the eroded and weathered basement. The Cambrian ranges from less than 60-m to more than 700-m thick. Shallow neritic sedimentation continued through the Ordovician, depositing a further sequence of sandstones, shales and limestones. Minor breaks in sedimentation can be correlated with pulses of early Caledonian tectonism (Pozaryski, 1977). The Ordovician is generally 100-m thick, or less. The Silurian sequence is more or less complete in the southern Baltic and northern Poland, but locally it has been deeply truncated or entirely removed by subsequent erosion. The Silurian increases in thickness from about 300 m in northeast Poland, close to the USSR frontier, to more than 2250 m close to the Tornquist Zone. The thinner sequence is typified by shelfal limestones, marls and clays and the thicker sequences are predominantly basinal graptolitic shales and mudstones. Late Caledonian movements caused a progressive narrowing of the area of sedimentation and a trough developed along the Tornquist Zone in which sedimentation was continuous from the Silurian into the earliest Devonian (Sokolowski, 1970).

Northeast of the Tornquist Zone, the early Palaeozoic sediments are flat lying and undisturbed. Within the Tornquist Zone, in the Pomerania area of northwest Poland, the early Palaeozoic sediments are unmetamorphosed but tectonically disturbed, dipping at angles of up to 80°. Beneath Rügen Island, off the Baltic coast of the German Democratic Republic, steeply-dipping sequences of basinal Ordovician sediments have been encountered in boreholes. The sediments are well compacted, quartz veined, sometimes brecciated and have abundant slickensides. However, metamorphism is only incipient and the shales have a well preserved graptolite fauna. The Ordovician shales beneath Rügen may be more than 3000-m thick and contrast strongly with the thin (100 m) shelf sequences of the southern Baltic. Silurian sediments have not been proved beneath Rügen Island, but immediately to the south flat-lying Devonian and Carboniferous sediments have been drilled.

The presence of structurally deformed Ordovician and Silurian sediments close to the Tornquist Zone in the Rügen-Pomorze area led to speculation about the existence of a continuous zone of Caledonian folding paralling the margin of the Russian (East European) Platform (Bush *et al.*, 1974; Glushko *et al.*, 1975). Even more speculatively, these folded early Palaeozoic rocks have been considered to be part of the external zone of a North German-Polish Caledonian foldbelt converging with the Scandinavian-Scottish system beneath the North Sea (Krebs, 1978; Ziegler, 1978, 1981). Frost *et al.* (1981) support the concept of a NW—SE striking Caledonian foldbelt across northern and central Europe, but note that the evidence suggests that the rocks are less metamorphosed and possibly less deformed than in the main branch of the Scottish-Norwegian Caledonides. They suggest alternatively that the metamorphic grade could increase towards the southeast "into an internides zone as yet unseen" (Frost *et al.*, 1981, p. 54). Deep drilling in Poland, however, has shown that the Rügen-Pomorze folded Lower Palaeozoic basin is only of limited extent. Further to the southeast, boreholes penetrating beneath the Permian have found more or less flat lying Silurian sediments and in the Holy Cross Mountains, close to the Carpathian fold front, Palaeozoic sediments ranging in age from Cambrian to Carboniferous are considered to have been folded primarily by Variscan rather than Caledonian orogenesis. The Rügen-Pomorze Fold Belt is possibly one of several folded 'aulacogens' lying transverse to the trend of the Tornquist Zone (Pozaryski and Brochwicz-Lewinski, 1978). The folded aulacogens are separated by more stable blocks in which Caledonian folding is weak or absent. These relationships suggest that the folding may be the result of transpression associated with Caledonian strike-slip faulting along the Tornquist Zone (Pegrum, in press) rather than folding associated with the external zone of a buried Caledonian fold system crossing north and central Europe.

Brochwicz-Lewinski *et al.* (1981) suggested that the contact between the thick, folded Lower Palaeozoic basinal sequence of the Rügen-Pomorze area and the thin platform sequences found to the northeast is a tectonic one, the juxtaposition resulting from major strike-slip faulting along the Tornquist Zone. They suggest an early Devonian age for the faulting and postulate a left-lateral displacement in the order of 1500 km based on their correlation of the Vättern Tectonic Zone of Skåne with a similar zone in the Dobrogea Massif of the Roumanian Black Sea coast. Figures 4A and 4B show the effects of removal of 1500 km of left-lateral offset along the Tornquist Zone. The sense of movement is the same as that discussed above regarding the possible offset of the Iapetus suture, although the amount of displacement proposed by Brochwicz-Lewinski *et al.* (1981) is much greater. Lack of quantification of the amount of shortening

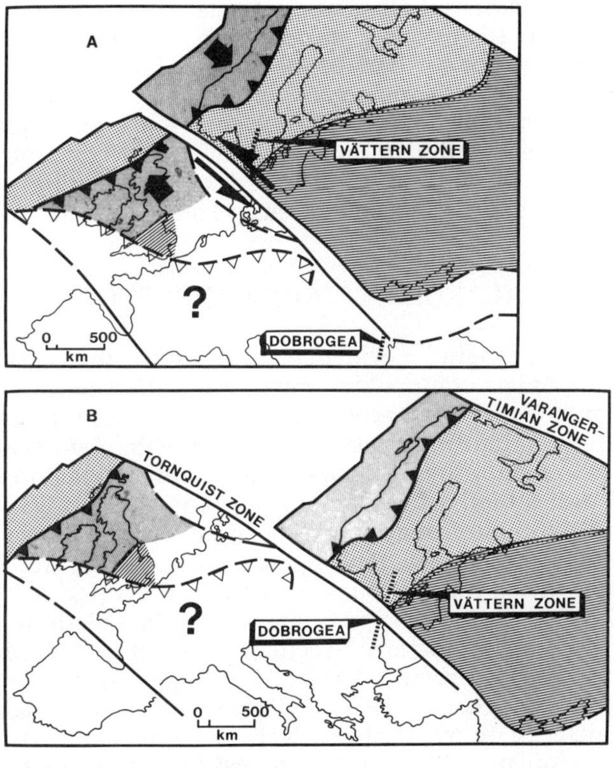

Fig. 4. (A) Possible juxtaposition of Precambrian and Lower Palaeozoic features across the Tornquist Zone at the close of the Caledonian orogenic cycle. (B) Relationship of Precambrian and Lower Palaeozoic features across the Tornquist Zone after the removal of 1500 km of left-lateral strike-slip displacement. Fine stipple; Caledonides; coarse stipple, Precambrian basement; line shading, undeformed Lower Palaeozoic cover.

involved within the Caledonian orogen itself and the modifications introduced by the Variscan orogenic cycle render accurate reconstruction impossible.

Basement rocks yielding Caledonian isotopic ages have been reached in a number of boreholes in the North Sea area (Frost *et al.*, 1981). North of a line running approximately from the Highland Boundary Fault in Scotland to Stavanger in Norway the Caledonian basement comprises high-grade metamorphic rocks and deformed plutonic rocks, as would be predicted from a straightforward projection from the outcrop areas in Norway and Scotland. To the southeast of this line the Caledonian basement, with few exceptions, consists of low-grade metamorphics and undeformed plutonic rocks. Seismic and limited well data suggest that thick sequences of more or less undeformed Palaeozoic rocks are present beneath the Skagerrak, the Kattegat and beneath the Danish Trough (Ziegler, 1981). Their northwestward limit beneath the Norwegian-Danish trough is not known. Across the Ringkøbing-Fyn High Precambrian basement rocks lie immediately beneath the Permo-Mesozoic cover. The distribution of early Palaeozoic and basement rocks in the North Sea area is summarized in Fig. 3. Both the Tornquist Zone and a zone of faulting extending along the southern margin of the Ringkøbing-Fyn High and beneath the Central Graben may have been active during the Caledonian orogenic cycle. The correspondence between the projected Tornquist Zone with the change in vergence of the Caledonides and the apparent truncation of the Iapetus suture have been discussed earlier. Left-lateral offsets in the order of 500 km to possibly more than 1500 km have been postulated. Part or all of the offset may have been contemporaneous with the Caledonian compressional deformation, and it is envisaged that the Tornquist Zone was an active transform system during both the Cadomian opening of Iapetus and its subsequent early Palaeozoic closing.

LATE PALAEOZOIC/VARISCAN HISTORY

Palaeomagnetic evidence suggests that the Caledonian compressive cycle was followed in the Devonian and early Carboniferous by major wrench faulting between the North American Craton and the Russian-Fennoscandian Platform. A left-lateral offset amounting to between 1500 and 2500 km may have been involved (Ziegler, 1981). Strike-slip faulting accompanied by rapid subsidence has been invoked for the formation of Devonian molasse type basins within the Caledonian Fold Belt in eastern Greenland (Haller, 1971), in Svalbard (Harland, 1973, 1978) and in western Norway (Steel, 1976). During the Devonian, much of northwestern Europe outside of the Caledonian foldbelt was effected by extensional or transtensional tectonics. Adjacent to the Caledonian orogen thick sequences of post-orogenic fluviatile and lacustrine red bed conglomerates, sandstones, siltstones and shales (Old Red Sandstone facies) were deposited, passing southwards into marine shales and limestones.

Devonian sediments are absent from outcrop over much of the emergent part of the Fennoscandian Platform but occupy wide areas of the sea floor of the southern Baltic, dipping southeastwards under a younger cover. In northern Poland, the basement was elevated and largely emergent, only thin sequences in a typical Old Red Sandstone facies being locally preserved. Along the platform margin, in the Tornquist Zone, a more or less complete Devonian sequence was deposited in a shallow marine basin. Sands and claystones dominate the Lower Devonian; shales, dolomites and limestones the Middle Devonian and marls, clays and dolomites the Upper Devonian. Reefal limestones occur towards the southeast and continue via the Podlasie Graben into the Dneiper-Donbass Graben system of the southwestern USSR (Sokolowski, 1970). Marine Devonian sediments may be extensively developed beneath northern Germany and the North Sea south of the Central Highs. In these areas, if present, they are very deeply buried beneath the Carboniferous and younger cover and have not been reached by drilling. Devonian sediments may also be widespread north of the Central Highs and have been proved in a number of deep wells. Marine Devonian sediments have been encountered as far to the northwest as the Argyll Oilfield on the western flank of the Central Graben.

Evidence of tectonic activity along the Tornquist Zone during the Devonian has not been documented in detail, but in northern Poland it more or less coincides with the transition from Old Red Sandstone continental facies to the marine facies, indicating that it may have exerted an important syndepositional control. Alternatively, post-Devonian strike-slip faulting may have juxtaposed the contrasting facies. Northwest from the Polish mainland no boreholes have penetrated Devonian sediments along the Tornquist Zone. Dipping reflectors truncated by Permian horizons occur beneath the Sele High (Fig. 2) and possibly represent bedded Devonian or Carboniferous sediments, but this remains unconfirmed.

Lower Carboniferous sediments in Poland show a general pattern of clastic sediments in the southwest, on the flank of the Sudeten (Bohemian) Massif, passing northeastward into a trough of argillaceous deep water to neritic clastics and, further to the northeast, a carbonate platform. Lower Carboniferous sediments do not extend across the Tornquist Zone on to the basement platform, although their absence may be due to subsequent erosion. In the Upper Carboniferous the Polish Basin was progressively filled with neritic sands and clays passing marginally into paralic, lacustrine and swamp facies comprising sandstones, siltstones, shales and coals. The last facies (coal measures) become dominant in the Westphalian and early Stephanian. The depositional basin filled with Upper Carboniferous clastics and coal measures extended westward from Poland, beneath northern Germany and the North Sea into eastern England, forming the Variscan Foredeep (Fig. 5). The late Carboniferous clastics in part represent a 'molasse' facies of the Variscan Fold Belt which began to develop in the south in the Devonian and progressively migrated northwards during the Carboniferous. The northward limit of the Variscan Foredeep coincides approximately with the southern flank of the Central Highs. The

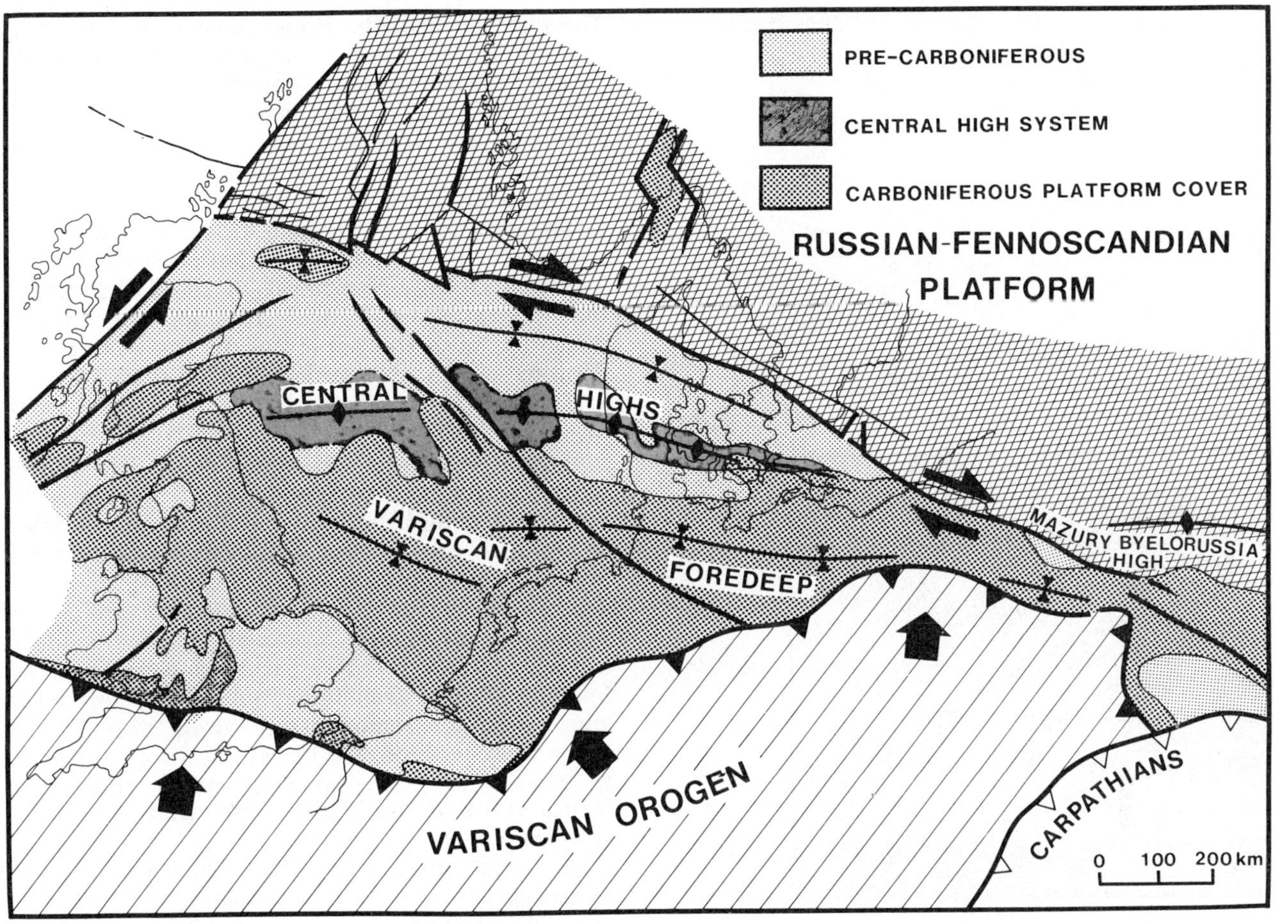

Fig. 5. The relationship between the Tornquist Zone and the major Variscan structural units in Europe.

latter can be considered as an E—W-trending crustal upwarp complementary to the great downwarp of the foredeep, both accentuated by the northerly directed Variscan compression (Fig. 5).

The northern external zone of the Variscan foldbelt has well developed folds and thrusts exhibiting a northerly vergence. The autochthonous Devonian and Carboniferous sediments along the southern margin of the Variscan Foredeep are similarly deformed by the northerly directed compression. Towards the end of the Variscan orogenic cycle, deformation migrated northward on to the northern foreland, initiating extensive wrench faulting and volcanism which persisted though the early Permian. The widespread strike-slip and oblique-slip faulting in northwestern Europe has been ascribed to regional transpression associated with a major E—W right-lateral shear zone extending from the southern Urals to the northern Appalachians (Arthaud and Matte, 1977). Pre-existing NW-trending zones of weakness were reactivated with right-lateral strike-slip offsets. Northeasterly trending fault zones were reactivated with left-lateral offsets and have been considered as forming a conjugate pair with the NW-trends (Ziegler, 1975).

The Tornquist Zone was undoubtedly reactivated by Variscan compression and the subsequent regional transpression. Ziegler (1981) recognizes right-lateral wrench movements along the 'Tornquist-Teisseyre Lineament' but terminates this feature at the line of the 'Oslo-Bamble-Horn rift'. Compilations of deep-seated fault trends in the Skagerrak, however, do not support the supposed continuity of the Oslo-Bamble Rift System with the Horns Graben and show clearly that the Tornquist Zone extends for several hundred kilometres towards the northwest (Pegrum, in press). The apparent alignment of the Oslo-Bamble Rift and the Horns Graben may be a fortuitous juxtaposition following significant right-lateral displacements along the Tornquist Zone. Based on a speculative correlation of the easterly truncated Ringkøbing-Fyn High and Variscan Foredeep with the Mazury-Byelorussia High and Podlasie-Dneiper-Donbass Depression, Pegrum (in press) suggests a late Carboniferous—early Permian right-lateral displacement in the order of 300—350 km along the Tornquist Zone. The effects of restoring this displacement are illustrated in Fig. 6. Interesting alignments arising from this reconstruction include the east facing Minch Fault System with the east facing Shetland Boundary Fault System, the Great Glen Fault with the Horda Fault System and the trilete convergence of the Midland Valley Graben, Central Graben and Oslo Graben. The occurrence of chemically similar late Carboniferous—early Permian volcanics in all three grabens has been noted previously.

The fracture pattern reactivated and established by the Variscan compressive cycle and the subsequent late Carboniferous—early Permian wrench faulting had a fundamental role on late Permian and Mesozoic deposition and structural deformation.

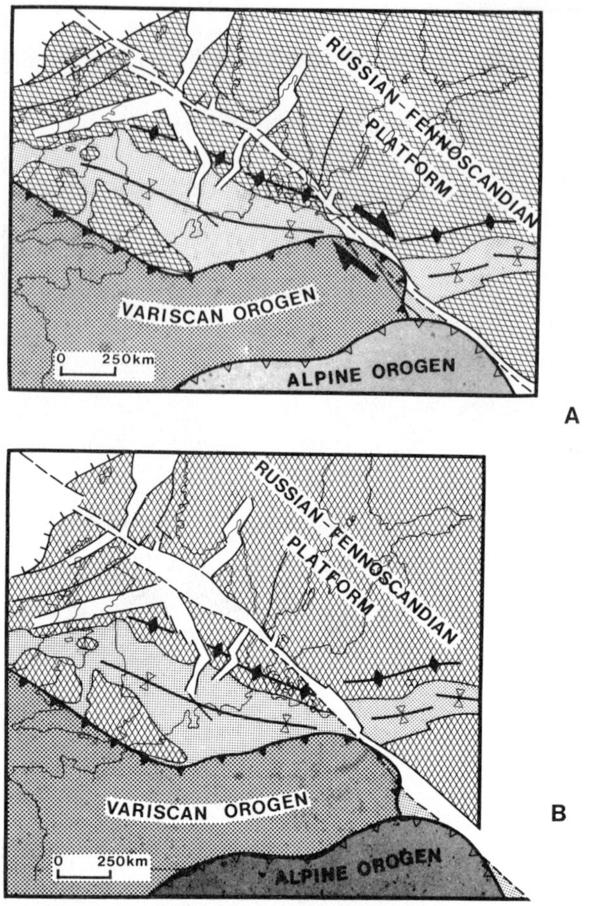

Fig. 6. (A) Present day juxtaposition of major structural features associated with the Tornquist Zone. (B) Juxtaposition of major structural features across the Tornquist Zone after the removal of 350 km of late- or post-Variscan right-lateral strike-slip movement.

LATE PERMIAN TO TERTIARY HISTORY

Beginning in the late Permian and extending through much of the Mesozoic, northwestern Europe and the North Sea were subjected to regional extensional stresses which resulted in widespread rifting and differential subsidence. Initial rifting in the late Permian and Triassic reactivated pre-existing zones of weakness, establishing a framework of elevated blocks and grabens extending from the proto-Atlantic in the northwest to the Tethyan Ocean in the southeast. The depositional and deformational framework of the area during the Mesozoic has been well summarized by Ziegler (1981). Triassic rifting segmented the Central Highs, especially the Ringkøbing-Fyn High which was cut into a number of discrete blocks separated by grabens and half grabens (Fig. 2). The Tornquist Zone was also cut by Triassic rifting, establishing the South Viking Graben and Egersund Trough. Rifting extends far to the north across the Horda Platform and along the western Norwegian coastline. Many, but not all, of the Triassic rifts were reactivated by renewed extensional deformation during the Jurassic.

The Tornquist Zone played an important role during the late Permian and Mesozoic. Figure 7 illustrates the gross thickness of the Upper Permian to Quaternary sedimentary cover in northwestern Europe and the North Sea (based on Ziegler, 1981). The alignment of the Polish Trough and Norwegian-Danish Trough along the Tornquist Zone at the margin of the Russian-Fennoscandian Platform is evident. Figure 8 emphasizes this relationship in showing the location of the principal Jurassic sedimentary troughs and fundamental fault zones. In unstable Europe, southwest of the Tornquist Zone, extension was widespread and affected an area stretching from the Atlantic Seaboard to the Tethyan Ocean. Northeast of the Tornquist Zone the basement platform retained its structural integrity, with the exception of the Viking Graben System which uncoupled the Shetland Platform, and the Horda Fault System which partially uncoupled the Horda Platform from the Fennoscandian Platform. In both cases the Mesozoic rifting probably utilized zones of weakness in the basement established during Variscan, Caledonian or earlier orogenesis. Evidence of Jurassic left-lateral displacements in the order of several kilometres (Pegrum, in press) suggests that stresses between the stable basement platform and the unstable stress to the southwest were resolved, at least locally, by transtensional oblique-slip or strike-slip faulting. Transtension along the Tornquist Zone was accompanied by Middle Jurassic volcanism in the Witch Ground Graben; at the southern termination of the Viking Graben; in the Egersund sub-basin and in Skåne.

Regional extension and pull-apart rifting waned in the early Cretaceous and was replaced by a more general downwarp. The onset of Alpine compression, resulting from convergence of the African—Arabian Plates with the European Plates, began in the eastern Tethys in the late Jurassic and early Cretaceous and spread progressively westward from the Carpathians to the western Alps. The northward directed Alpine compression rejuvenated many of the fundamental fault zones. The switch from Jurassic transtension to Cretaceous transpression resulted in widespread tectonic inversion. Tectonic inversion occurred during the intra-Senonian Sub-Hercynian phase (Ziegler, 1981), but reached its climax during the Laramide phase at the end of the Cretaceous. Pulses of tectonic inversion driven by Alpine compression persisted until the Middle Miocene. Figure 9 shows the location of the principal areas of positive inversion. Tectonic inversion is particularly well expressed along the Tornquist Zone and can be traced from the Carpathian front in Poland to the centre of the North Sea, a distance of 2000 km. Pegrum and Ljones (in press) have documented the effects of mid-Tertiary transpression at the southern margin of the Horda Platform, close to the southern termination of the Viking Graben, which has led to the deformation of Palaeocene submarine fan sands. The effects of inversion become progressively weaker from the Holy Cross Mountains in southeastern Poland where inversion is complete, exposing the Palaeozoic floor of the trough at the surface, to the central North Sea where the Cretaceous cover is unwarped but intact.

SUMMARY

The southwestern margin of the Russian-Fennoscandian Platform of Precambrian crystalline

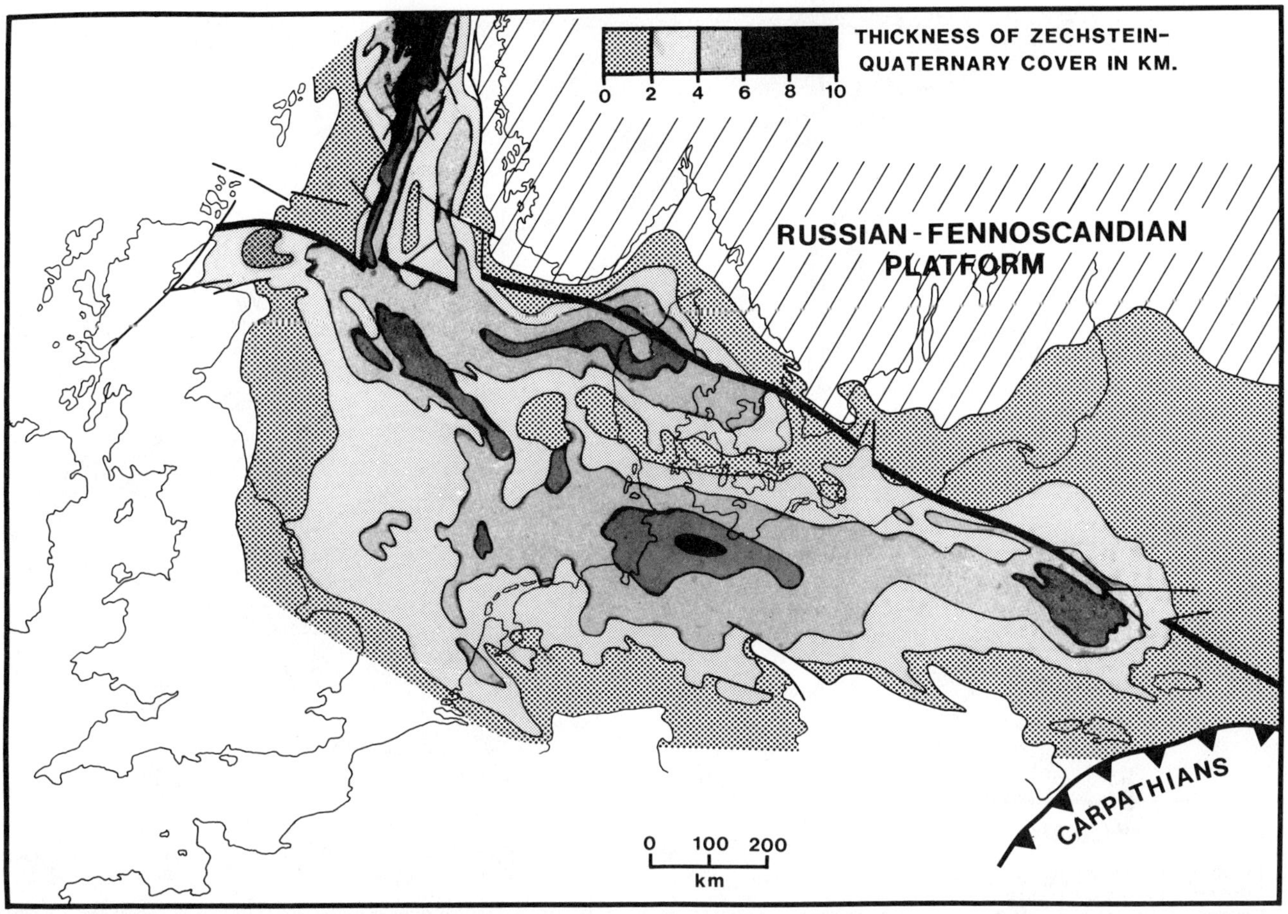

Fig. 7. Relationship between the Tornquist Zone and the distribution of Upper Permian to Quaternary sediments in northern Europe. (Modified after Ziegler, 1981.)

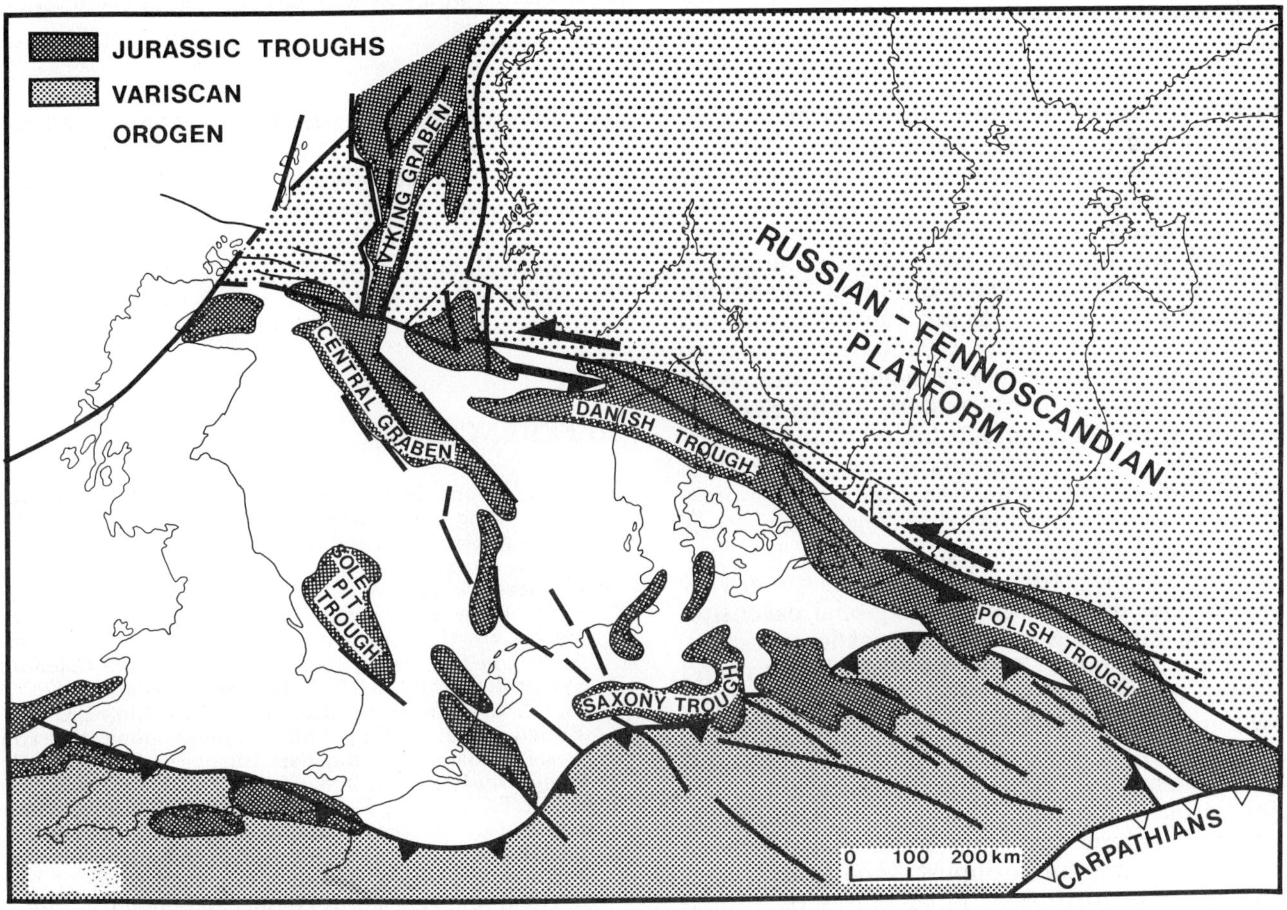

Fig. 8. Relationship between Jurassic sedimentary troughs and the Tornquist Zone in northern Europe and the North Sea. Left-lateral transtension along the Tornquist Zone associated with the opening of the proto-Atlantic and the Tethyan Ocean is indicated.

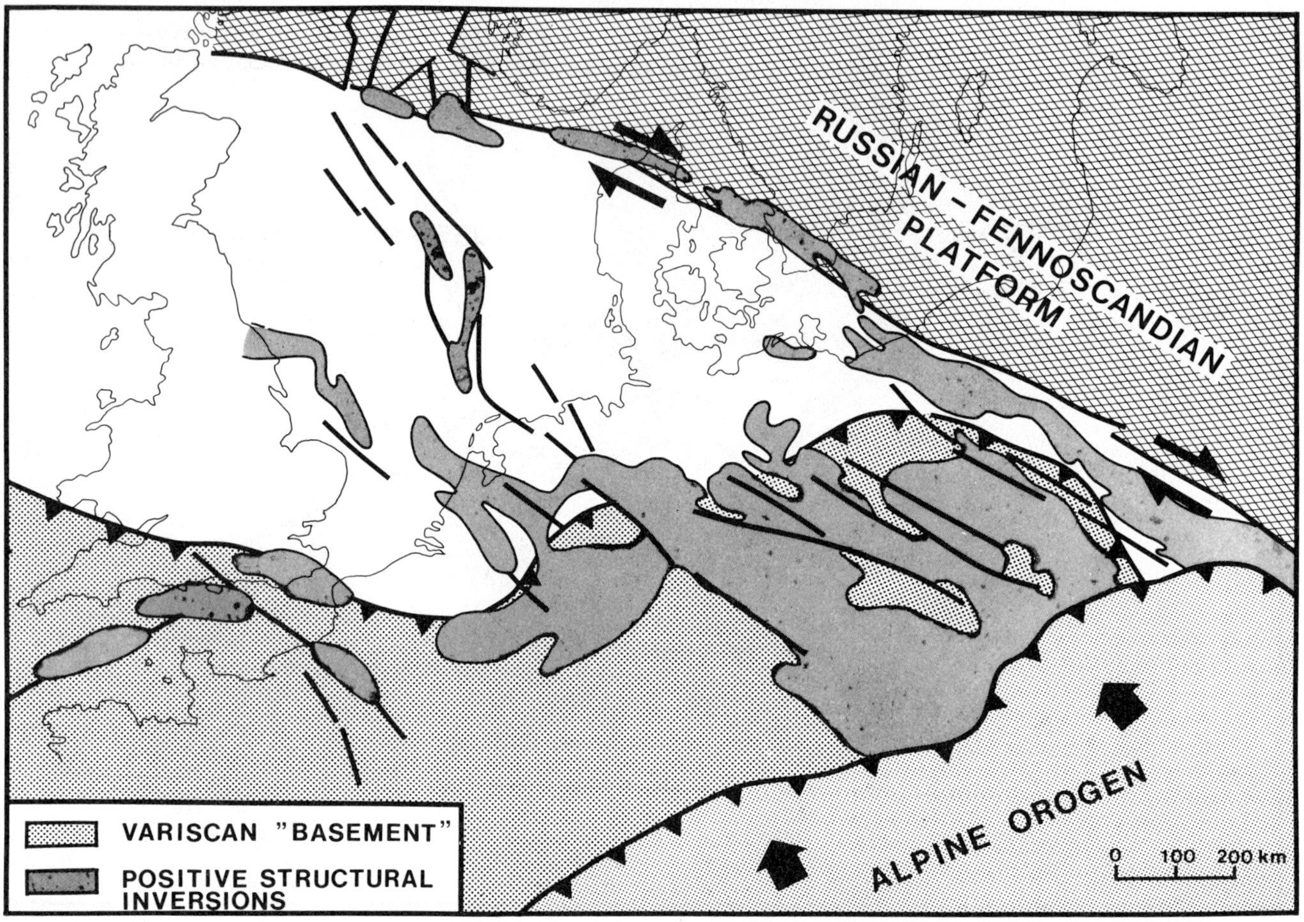

Fig. 9. Relationship between positively inverted Mesozoic troughs and the Tornquist Zone in northern Europe and the North Sea. Right-lateral transpression along the Tornquist Zone associated with northerly directed Alpine compression is indicated.

basement rocks correspond to the Tornquist Zone. By analogy with sub-parallel tectonic zones transecting the basement platform it is not unreasonable to assume an age of at least 1000 Ma for the Tornquist Zone. The Tornquist Zone may have performed an important role in the Cadomian opening of the Iapetus Ocean and in its Caledonian closing. The Tornquist Zone can be projected northwestwards across the Caledonian orogen to separate the southeasterly verging Scandinavian Caledonides from the northwesterly facing Scottish Caledonides. Pre-Variscan left-lateral displacements in the order of 1500 km have been suggested. During Variscan compression and early post-Variscan wrench faulting the Tornquist Zone was the locus of right-lateral strike-slip faulting. An offset in the order of 350 km is tenatively proposed. During the Mesozoic regional extension in unstable Europe was expressed, at least locally, along the Tornquist Zone by transtension and minor left-lateral oblique-slip or strike-slip faulting. A switch from left-lateral transtension to right-lateral transpression occurred during the Cretaceous with the closing of Tethys and the initiation of the Alpine orogenic cycle. This switch resulted in widespread tectonic inversion which can be traced for 2000 km along the Tornquist Zone from Poland to the central North Sea. Final pulses of Alpine transpression occurred in the mid Tertiary.

The southwestern margin of the Russian-Fennoscandian basement platform can thus be shown to be an ancient, and often rejuvenated, zone of tectonic weakness. The western margin of the basement platform lying beneath the continental shelf or slope of western Norway and the north-eastern margin of the platform, lying beneath the southern Barents Sea, are of equal antiquity and can be expected to have exerted a similar structural influence as a result of frequent rejuvenations.

REFERENCES

Arthaud, F. and Matte, Ph. 1977, Late Palaeozoic strike-slip faulting in southern Europe and North Africa: result of a right-lateral shear zone between the Appalachians and the Urals. *Bull. Geol. Soc. Am.*, **88**, 1305—1320.

Brochwicz-Lewinski, W.. Pozaryski W. and Tomczyk, H. 1981. Large scale strike-slip movements along the S.W. margin of the East European Platform in the early Palaeozoic. Abstr. of paper read at *Uppsala Caledonide Symposium*, August 1981, Uppsala, Sweden.

Bush, V.A., Garetskiy, R.G. and Kirjuchin, L.G. 1974. Buried zone of Caledonian folding along the south-western margin of the East European Platform. *Dolk. Akad. Nauk. S.S.R.*, **208**, 65—69.

Frost, R.T.C., Fitch, F.J. and Miller, J.A. 1981. The age and nature of the crystalline basement of the North Sea Basin. In: L.V. Illing and G.D. Hobson (eds.). *Petroleum Geology of the Continental Shelf of North-west Europe*, Institute of Petroleum, London, pp. 43—57.

Gabrielsen, R.H. Long lived fault zones and their influence on tectonic development of the southwestern Barents Sea. *Geol. Soc. Lond.* (in press).

Glushko, V.V., Dikenstein G.Ch. and Schmidt, K. 1975.

The Caledonides of Rügen Island. *Dok. Akad. Nauk. S.S.R.*, **214**, 89—90.

Haller, J. 1971. *Geology of the East Greenland Caledonides.* Interscience, London.

Harland, W.B. 1973. Tectonic evolution of the Barents Shelf and related plates. *Am. Assoc. Petrol. Geol., Mem,* **9**, 599—608.

Harland, W.B. 1978. The Caledonides of Svalbard, Caledonian-Appalachian orogen of the North Atlantic Region. *Geol. Surv. Canada*, Paper 78-13, pp. 3—11.

Krebs, W. 1978. Die Kaledoniden im nördlichen Mitteleuropa, *Z. Dtsch Geol. Ges.*, **129**(2), 403—422.

Pegrum, R.M. 1984. The extension of the Tornquist Zone in the Norwegian North Sea. *Norsk Geol. Tiddskr.* (in press).

Pegrum, R.M. and Ljones, T.E. The 15/9 Gamma Gasfield, Offshore Norway. A new trap type for the North Sea with regional implications. *Bull. Am. Assoc. Petrol. Geol.* (in press).

Pozaryski, W. (ed.). 1977. *The Geology of Poland*, Vol. 4, *Tectonics*, Warsaw.

Pozaryski, W. and Brochwicz-Lewinski, W. 1978. On the Polish Trough, *Geol. Mijn.*, **57**, 545—557.

Sokolowski, S. (ed.). 1970. *The Geology of Poland*, Vol. 1, Part 1, *Precambrian and Palaeozoic*, Warsaw.

Steel, R.J. 1976. Devonian basins of western Norway — sedimentary response to tectonism and varying tectonic context. *Tectonophysics*, **36**, 207—224.

Strömberg, A.G.B. 1976. A pattern of tectonic zones in the western part of the East European Platform. *Geol. Fören. Stockholm Förh.*, **98**, 227—243.

Strömberg, A.G.B. 1981. The European Caledonides and the Tornquist Lineament. *Geol. Fören. Stockholm Förh.* **103**, 167—171.

Watson, J. 1976. Eo-Europa: the evolution of a craton. In: D.V. Ager and M. Brooks (eds.), *Europe from Crust to Core*, Wiley, London, pp. 59—78.

Ziegler, P.A. 1978. North Sea Rift and basin development. In: I.B. Ramberg and E.R. Neumann (eds.), *Tectonics and Geophysics of Continental Rifts,* Riedel, The Netherlands.

Ziegler, P.A. 1981. Evolution of sedimentary basins in North West Europe. In: L.V. Illing and G.D. Hobson (eds.), *Petroleum Geology of the Continental Shelf of Northwest Europe.* Institute of Petroleum, London, pp.3—39.

Ziegler, W.H. 1975. Outline of the geological history of the North Sea. In: A.W. Woodland (ed.), *Petroleum and the Continental Shelf of Northwest Europe*, Vol. 1, *Geology*, pp. 165—190.

Zwart, H.J. and Dornsiepen, U.F. 1978. The tectonic framework of central and western Europe. *Geol. Mijn.*, **57**(4), 627—654.

27 Mesozoic source rocks on Svalbard

Atle Mørk and Malvin Bjorøy
Continental Shelf Institute (IKU), Trondheim, Norway

A review of the source rock potential of the Mesozoic formation on Svalbard is based on organic geochemical analyses of approximately 1000 samples collected during expeditions from 1976 to 1979. Regional and stratigraphic trends have been established for richness, quality and maturity of the potential hydrocarbon source rocks.

The Mesozoic sedimentary succession on Svalbard has a composite thickness up to 3 km and includes several possible source rocks. The maturities of the Mesozoic rocks on Svalbard vary considerably. Overmature rocks are found in the area of the Tertiary deformation zone at the western part of Spitsbergen and locally along the southwest coast of Edgeøya due to igneous activity. In the north and east of the study area the Mesozoic rocks are found to be immature, while they are found to be mature at the exposed area bounding the Tertiary trough in the middle and eastern part of Spitsbergen. The Mesozoic rocks are thus believed to have an optimum maturity for hydrocarbon production below the central Tertiary trough on Spitsbergen.

Organic carbon-rich shales and siltstones are generally abundant in the Mesozoic of Svalbard. Especially, sediments of the Lower to Middle Triassic Sassendalen Gp are very rich in organic carbon, in particular the Botneheia and Sticky Keep Members. The latter members are rated to be good to excellent quality petroleum source rocks due to the hydrogen-rich nature of their kerogens (type II). Regional variation in quality and in thickness of these petroleum source rock intervals is, however, shown to have a trend of increasing kerogen quality towards the north and northeast of the study area, reaching best quality on Barentsøya and Edgeøya.

Although shaly sequences are abundant throughout the remainder of the Mesozoic succession, the dominance of terrestrial and reworked kerogen (type III and IV) means that most other units have a low petroleum potential, except for the Janusfjellet Fm (Upper Jurassic—Lower Cretaceous) which has a high organic content. The Jurassic part of this formation has a fair amount of type II kerogen, giving the formation a good source rock potential for oil and gas. The dominance of type III kerogen in the Cretaceous part of the formation restricts this unit's possibilities for the production of gas.

INTRODUCTION

A renewed interest in the geology of the Svalbard Archipelago has arisen after commencement of exploratory drilling for hydrocarbons on the shelf off northern Norway. The few wells drilled on the Svalbard islands in the late 1960s and 1970s have not shown any commercial discoveries of hydrocarbons; however, the Svalbard Archipelago is situated at the northwestern corner of the Barents Shelf, and has good exposure of late Palaeozoic and Mesozoic rocks, and information on these is of considerable interest for the exploration area further south.

Investigations of the source rock potential of the late Palaeozoic and the Mesozoic succession have thus been conducted at the Continental Shelf Institute (IKU) to gather information for use in exploration work. This chapter summarizes the results of our studies of the Mesozoic sequence; a more thorough and detailed presentation will be given by Bjorøy *et al.* (in preparation).

After reconnaissance studies of samples of all the Mesozoic formations, our effort has been concentrated on the Lower and Middle Triassic units which proved to have the best source rock potential for hydrocarbons. The younger Mesozoic part of the succession has been partially covered in an earlier paper (Bjorøy and Vigran, 1980); a summary of this, including results of more analytical work, is discussed herein. The geographically important island, Bjørnøya, mid-way between Spitsbergen and Norway, mostly comprises Palaeozoic rocks and some Triassic clastic sediments, and is described in more detail by Bjorøy *et al.* (1983b).

Petroleum Geology of the North European Margin © Norwegian Petroleum Society (Graham & Trotman, 1984) pp. 371—382.

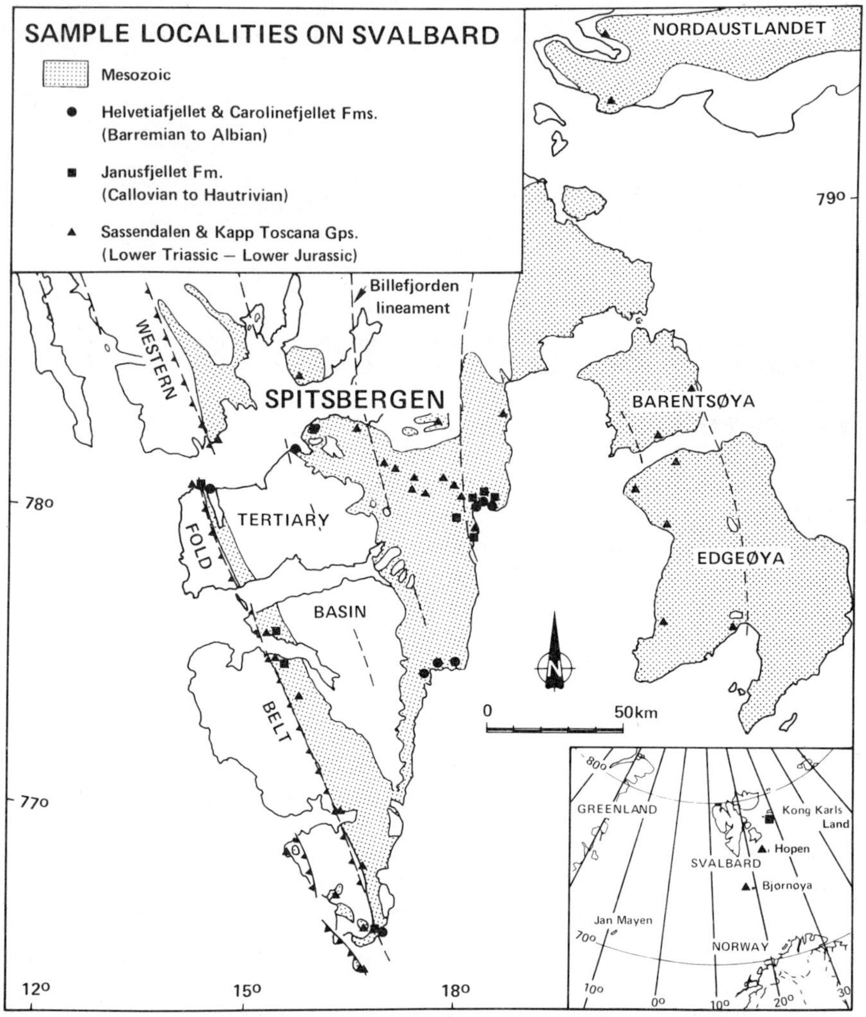

Fig. 1. Map of outcropping Mesozoic beds and sampling localities.

Geological framework

The Mesozoic succession of Svalbard has a composite thickness of 3000 m and readers are referred to the review by Steel and Worsley (1984) for information on sedimentological patterns and palaeogeographic evolution.

The Mesozoic rocks occur in two different settings (Fig. 1):

(1) Rocks of the western Tertiary fold belt of Spitsbergen.

(2) Rocks bordering and underlying the Tertiary basin of central Spitsbergen; those beds being more flat-lying towards eastern Spitsbergen and on the other islands east and south of Spitsbergen.

The sequence is subdivided into three groups:

(1) The Sassendalen Gp of Lower and Middle Triassic age consists of marginal to deeper marine clastic sediments which are organic-rich in part.

(2) The overlying Kapp Toscana Gp contains continental, deltaic and mostly shallow marine clastic sediments, which have some reservoir potential.

(3) On top is the Adventdalen Gp, with a lower, thick dark shale unit with hydrocarbon potential, occurring below a fluvial deltaic clean sandstone unit overlain by a thick sequence with shallow, marine silty shale and sandstones.

Geochemical analysis

A large number of geochemical publications, summarized by Tissot and Welte (1978) and Hunt (1979), document that hydrocarbon source rocks require minimum concentrations of finely disseminated organic matter, the composition and structure of which has to be of a certain type (kerogen quality, mainly determined by hydrogen content). For hydrocarbon generation to proceed, the source rocks have to experience favourable temperature conditions with burial during their geologic history (maturity of organic matter).

To classify a source rock unit, and evaluate which hydrocarbons the source rock can produce, it is necessary to use a number of techniques to evaluate the richness of the source rock, the type of kerogen in the source rock and the maturity. Such standard techniques used in this study were: measurements of total organic carbon; Rock—Eval pyrolysis; measurements of extractable organic matter; measurements of extractable hydrocarbons; gas-chromatographic analysis of saturated hydrocarbons; examination of kerogen in concentrates in transmitted light; and examination of polished samples in reflected light with measurements of vitrinite reflectance. Detailed descriptions of the experimental procedures have been given previously

(Bjorøy and Vigran, 1980; Bjorøy *et al.*, 1983a).

Most of our work was conducted on surface samples. In a study comparing shallow drilling cores (< 20-m deep) with adjacent surface samples from a cliff, Forsberg and Bjorøy (1983) concluded that the amount of organic carbon (TOC) and the Rock—Eval analytical values were only slightly affected by weathering, while gas-chromatograms might show stronger alteration. Consequently we have concentrated our conclusions on the first two techniques.

During the last 10 years IKU has analysed more than 1000 samples to evaluate the hydrocarbon source rock potential of the Mesozoic formations at Svalbard. The results are presented for the individual units in stratigraphic order.

Fig. 2. The Botneheia Mbr of the type locality of Botneheia in Isfjorden, central Spitsbergen. The dark cliff-forming unit is the organic-rich upper part of the Botneheia Mbr.

DATA PRESENTATION AND DISCUSSION

Sassendalen Group

Exposures of the Sassendalen Gp occur around the Tertiary basin on Spitsbergen, on Edgeøya, Barentsøya, the southern part of Nordaustlandet and far to the south on Bjørnøya (Fig. 1). The geology has been presented recently by Mørk *et al.* (1982), including sedimentary sequences from the whole area, and the same profiles have been utilized in the present study. The sediments show deposition in two regimes: a western facies belt, with input of sediment from a source area situated just west of the present western shoreline of Spitsbergen; and marine conditions towards the east. While the coarsening-upwards sequences seen in the west have led to the subdivision of the Sassendalen Gp into three formations, the more distal shaly facies to the east is represented by only one unit, the Barentsøya Fm. Three slight coarsening-upwards sequences are also seen here, and are represented by the subdivisions of the formation; these members being laterally equivalent to the formations in the western area.

The distribution of organic carbon (Figs. 3 and 4) in the Sassendalen Gp reflects the depositional environments of the individual units, with increasing

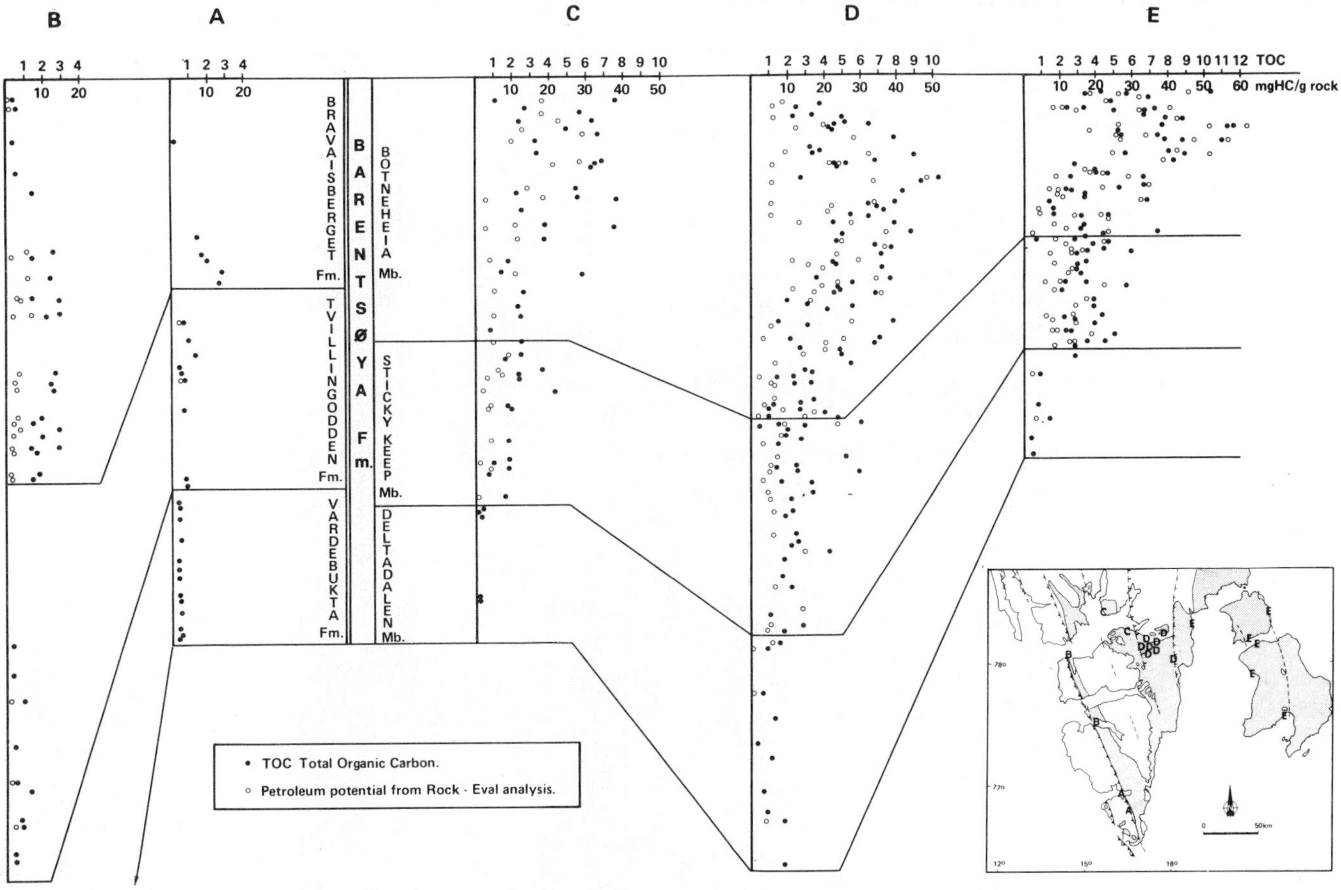

Fig. 3. Lateral variations in organic carbon content in the Sassendalen Gp. Both TOC and petroleum potential values show increasing values upwards and towards the east in the Sassendalen Gp. Vertical axes are adjusted for stratigraphical comparisons. Letters show locations for the individual profiles.

organic richness with increasing distance from the sediment source area in the west and with increasing depositional depth in the basin. The importance of the older major tectonic lineaments is also reflected as increased values are seen on the downthrown side of the lineaments (e.g. area D east of the Billefjorden lineament in Figs. 3 and 1).

Deltadalen Member and Vardebukta Formation

The sand/shale-dominated Vardebukta Fm deposited in marginal to shallow marine environments in the western area grades eastwards laterally into the Deltadalen Mbr which is dominated by silty shales. These mainly fine-grained clastic units were deposited after an early Triassic transgression. In the western area the Vardebukta Fm contains coarser intervals and also bioclastic limestones, representing local shoals and recurrent transgressive and regressive episodes. The development of the eastern Deltadalen Mbr is more uniform, with abundant thin siltstone beds and laminae in a grey-green silty shale.

A number of samples from these units were analysed. Most of the samples have organic carbon values less than 1%, i.e. rather lean samples (Fig. 4). Examination of the kerogen shows the samples to be of kerogen type III or IV, i.e. of a rather poor quality, which will only be capable of producing gas, at best. Therefore these samples will not be discussed further.

Sticky Keep Member and Tvillingodden Formation

The Tvillingodden Fm in the western area comprises one major coarsening upwards sequence, in part built up of several smaller coarsening upwards

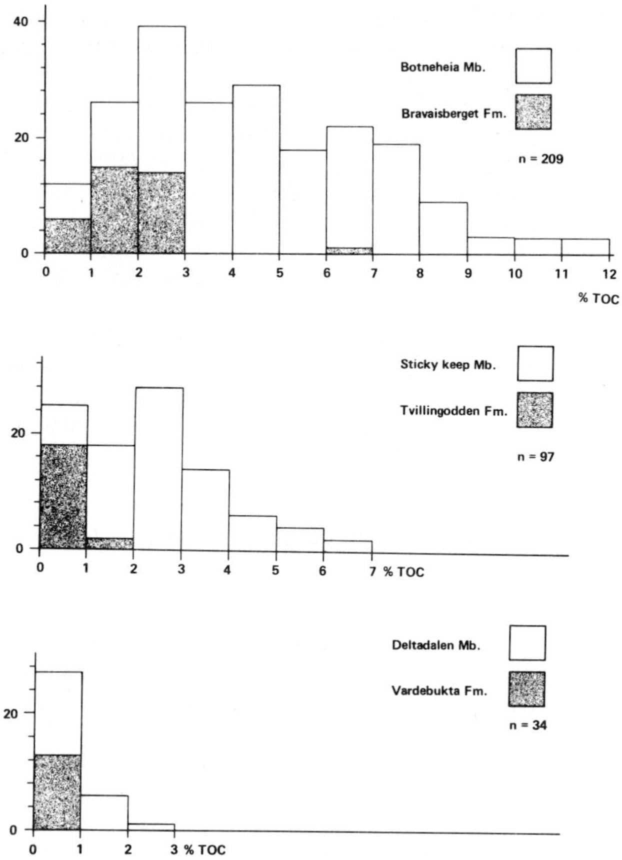

Fig. 4. Histograms showing frequency of the amount of total organic carbon (TOC) in the Sassendalen Gp.

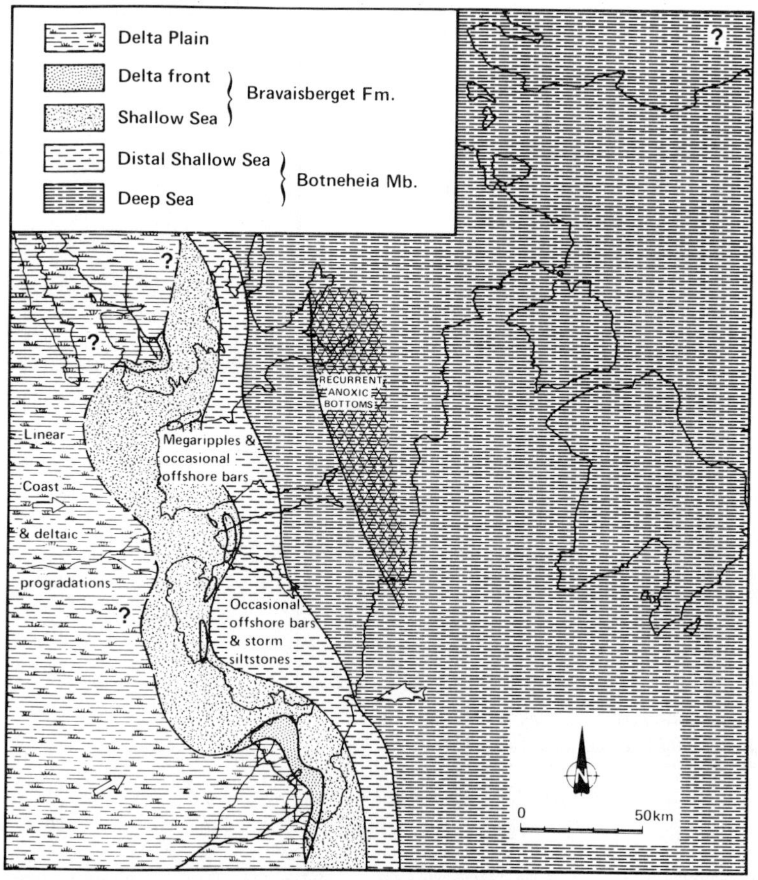

Fig. 5. Palaeographic reconstruction of the depositional environment in Middle Triassic time.

sequences. The development of the eastern equivalent, the Sticky Keep Mbr, is more uniform and is dominated by dark shales with an increasing content upwards of thin siltstones laminae and beds.

The samples analysed have organic carbon values in the 0.5—6% range (Fig. 4). Most of the samples have organic carbon values between 1.5 and 3%, with a mean at 2.5%. The regional variations of the organic carbon values (Fig. 3) basically show a trend similar to that in the overlying Middle Triassic units. The samples consist mainly of kerogen type III, but with an increasing content of kerogen type II in the northeastern part of the area, preferentially in the upper 10—30 m of the unit. On the whole, these units are dominated by kerogen type III, which is gas prone, but with some oil-prone kerogen in the eastern area. The regional trend indicates that the percentage of kerogen type II increases towards the northeast of the area investigated, but samples available for analysis from this part of the Svalbard Archipelago are scarce, and therefore it is not possible to make a complete assessment of this area.

Botneheia Member and Bravaisberget Formation

These units are of Middle Triassic age and comprise a coarsening upward sequence. As seen in the palaeogeographical reconstruction (Fig. 5) a land area was situated to the west with a deeper marine platform towards the east. There was a slow rate of sedimentation throughout the period in the eastern area, and consequently in part the sediments have a high organic content. Delta progradation from the west resulted in dark prodelta shales being overlain by delta front silts and sands, and in a few localities delta top sediments occur. This overall coarsening upwards sequence, the Bravaisberget Fm, can be traced all along the western coast of Spitsbergen.

The thinner development of the eastern equivalent, the Botneheia Mbr, reflects a more distal facies to the sediment source area. However, local thickening to the east of the N-trending lineaments (Fig. 1) shows the importance of synsedimentary subsidence.

Sediments with abundant trace fossils, as well as phosphatization and coquinas of small bivalves alternating with shales of high organic content, point towards an environment of fluctuating oxic and anoxic conditions. Storms probably ventilated a normally relatively anoxic sea-bottom, enabling fauna to develop. High biological productivity consequently led to utilization of the oxygen, resulting in the occurrence of anoxic conditions. Abundant siltstone beds, often with phosphate nodules at the base, were also deposited below normal wave base due to interruptions by occasional storms stirring up the sediment. This mechanism seems to have been responsible for deposition of a major part of the siltstones in the shale dominated unit in the lower part of the Bravaisberget Fm and in the Botneheia Mbr. The organic increase east of the N-trending lineaments (especially the Billefjorden Lineament, Figs. 1, 3 and 5), also points towards synsedimentary subsidence of this area, consequently with preservation of large amounts of organic content.

Maturity. The maturity of the samples from the Middle Triassic was measured by several methods. Due to the nature of the kerogen, very small amounts of vitrinite were found, and most of the maturity data are therefore based on the temperature of maximum pyrolytic degradation (T_{max}) and on visual examination of the colour of the kerogen, the thermal alteration index (TAI). The results of these analyses are transformed into vitrinite reflectance data and used together with reflectance measurements data to obtain reliable trends which can be compared with other formations. Although the Middle Triassic samples are of poor quality for maturation study, the calculated reflectance data are comparable to those measured on other Triassic units from the same or nearby locations, and a map showing the regional variation of maturity of the Middle Triassic sediments on Svalbard can be constructed (Fig. 6). The large similarities in the reflectance values found for the different units in Sassendalen Gp are as expected, since there is a relatively limited thickness of these sediments at most locations.

The regional variations in average vitrinite reflectance values and Rock—Eval T_{max} are very similar (Figs. 6 and 7), and allow the following conclusions: A general subdivision can be made into a narrow N-trending zone of highly mature Triassic sediments in the most westerly part of the study area, and to the east of this is a wide area, where maturity is less extreme and varies from moderate to early mature. The advanced maturity level in the west reflects the structural history of this area; higher thermal stress due to higher heat flux and/or deeper burial for the locations within the Tertiary deformation zone.

For all locations outside this deformation zone, iso-vitrinite reflectance and iso-T_{max} contours indicate a systematic and rapid decrease in maturity in the northeastern and northern parts of the area of exposure. The reason for this appears to be a rapid thinning and wedging out of overburden, mainly of Tertiary-age sediments, towards the eastern and northeastern edge of the Tertiary trough.

Except for the most northerly and easterly localities which were examined, the mid Triassic source rocks consist predominantly of amorphous

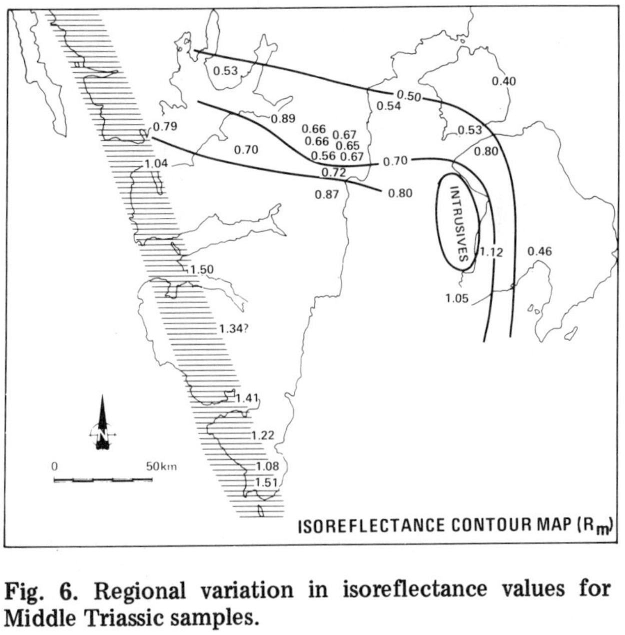

Fig. 6. Regional variation in isoreflectance values for Middle Triassic samples.

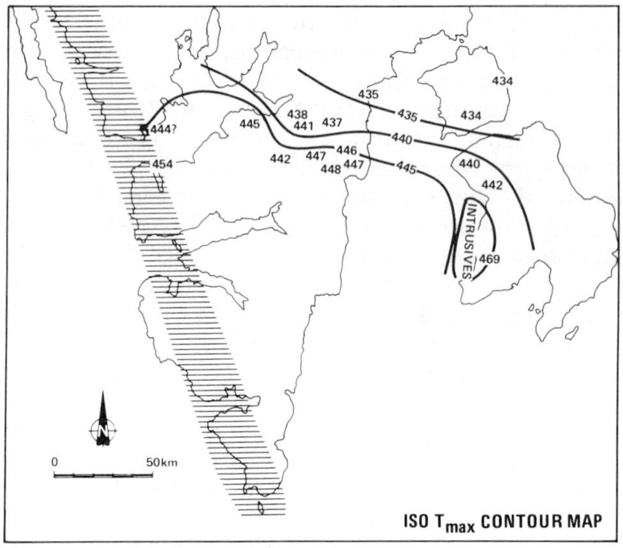

Fig. 7. Regional variation in iso-T_{max} values from Rock—Eval pyrolysis for Middle Triassic samples.

kerogen of type II-quality which have experienced favourable subsurface temperatures for effective petroleum generation (onset of maturity around 0.5% R_m, T_{max} 435°C). The northern and eastern localities are immature due to insufficient burial depth and temperatures.

Locally, advanced maturity levels are reached in the most westerly part of Edgeøya, probably due to dolerite intrusions. Thus, averaged mean reflectance-values of 1.2 and 1.05% and T_{max} values of 469°C and 465°C in two localities on southeastern Edgeøya have been recorded.

Richness. The evaluation of the organic richness of the Botneheia Mbr is based on the amount of organic carbon in the samples (TOC) (Fig. 3), the petroleum potential from the Rock—Eval pyrolyses (PP) (Fig. 3) and the extractability of both the total organic material and the hydrocarbons (EOM and EHC). Frequency study of the results from these analyses are shown in Table 1.

The following observations can be made from these analyses:

(1) A large number of the samples analysed were found to have organic richness values above those which are considered a minimum value for clastic source rocks, i.e. > 0.5% TOC, 2 mg hydrocarbon/g rock for the petroleum potential and > 20 and 10 mg/g organic carbon for extractable organic matter (EOM) and extractable hydrocarbons (EHC), respectively.

(2) Although there is no systematic trend with stratigraphic age, close relationships with the palaeogeographical development are seen and shales and siltstones of the Triassic Sassendalen Gp have the highest proportion of samples classified as rich by the source rock evaluation methods.

(3) Amongst all the stratigraphical units analysed, the Botneheia and the Sticky Keep Members are clearly the richest in organic carbon content, petroleum potential and extractable organic material. For these units, most samples contain in excess of 2.0% organic carbon, and they are classified as organic rich. These members also have a more uniform lithological development, with the analytical results being representative for the whole rock unit, whereas higher TOC values for other units may represent thinner and more locally enriched parts of the sedimentary sections.

Organic carbon contents of the Botneheia Mbr exhibit an unusually broad and bimodal frequency distribution. This is probably due to variation of organic carbon contents throughout deposition caused by fluctuating oxygen-rich and anoxic conditions during deposition and the overall palaeogeographic basin configuration.

The TOC values for the Middle Triassic units tend to increase towards the east and northeast. For the upper part of the Botneheia Mbr, an increase

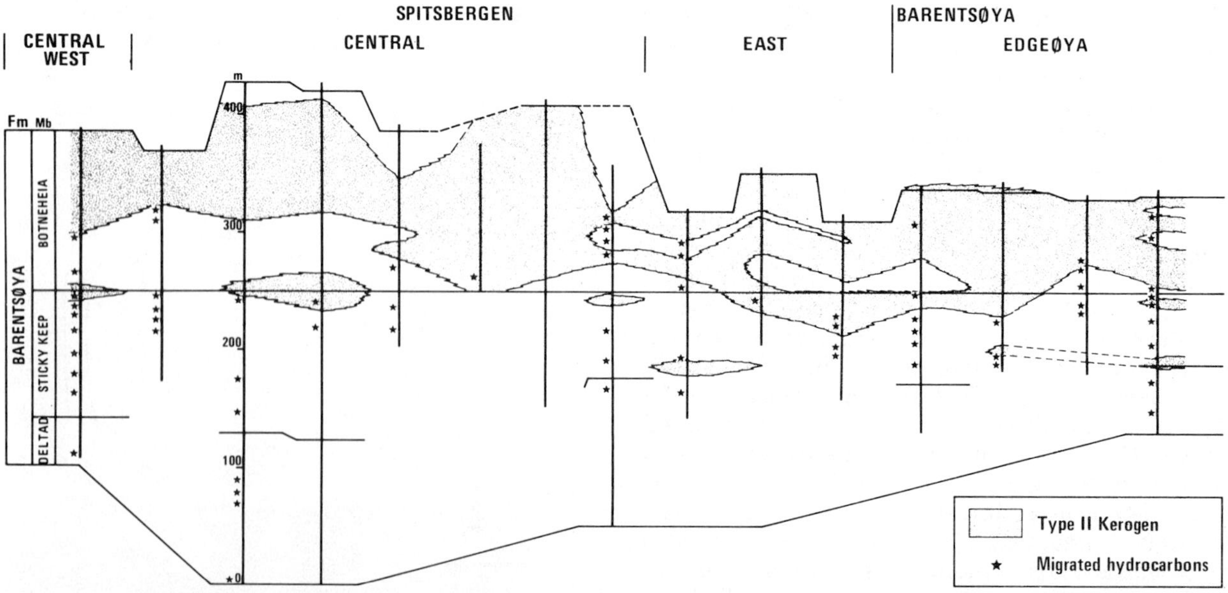

Fig. 8. A fence diagram showing the distribution of kerogen types based on Rock—Eval analysis. Shaded areas show dominance of type II, while white areas are dominated by type III, kerogens. Vertical lines refer to locations, triangles to Fig. 1, arranged in a W—E pattern. Stars indicate the presence of hydrocarbons measured by a high production index.

of mean organic carbon content is seen across the Billefjorden Lineament (Fig. 3, areas C and D) and reflects the influence of the lineament during deposition of these shales. The increased thickness of the Botneheia Mbr east of this lineament points towards synsedimentary subsidence of the basin. Anoxic conditions were probably more commonly developed in the sea bottom depression, resulting in enhanced preservation of organic matter than in the higher exposed area towards the west.

Similar observations can be made for the other richness parameters. These are, however, influenced more by local maturity increases due to intrusions and short distance to the Tertiary deformation zone, where the petroleum potential is almost nil. However, there are large quantities of extractable hydrocarbons in siltstone samples, indicating that migration out of the source rocks has occurred.

Kerogen. The kerogen-type for the Middle Triassic samples was determined on the basis of Rock—Eval pyrolysis and visual examination of kerogen concentrates in transmitted light. These analyses showed that most of the Botneheia samples are of kerogen type II (Fig. 8), i.e. oil-prone kerogen. The samples from the northeastern part of the study area, i.e. from the northeast of Spitsbergen, Edgeøya, Barentsøya and Nordaustlandet, contain a significant amount of kerogen type I. Examination in transmitted light showed that these samples contain the algae *Tasmanites.* Algal-rich kerogen will give the highest transformation ratio of kerogen to hydrocarbons, i.e., will produce more hydrocarbons per gram of kerogen than any other type of kerogen, and will produce a paraffinic oil.

For the type II kerogen source beds from the Middle Triassic units a mean hydrogen index (HI) was calculated for each sample location. The regional variation of these mean HI values reveals the following pattern (Fig. 9). High values (in excess of 450 mg HC/g Corg) occur in the north and the northeast, the highest value being reached at the northeasternmost locality on Spitsbergen (484). However, further northeast, on Nordaustlandet, HI values in excess of 600 mg HC/g Corg were found. To the southwest, a regular and fairly rapid reduction of mean HI-values is seen. The 250 HI iso-line follows the southwestern side of Sassendalen and extends into the Agardhbukta area.

This regional pattern reflects the above-established maturity trend: with increasing maturity the mean HI-value decreases due to conversion of hydrogen-rich moieties in the kerogen to liquid oil in the generation process.

Due to this predominant influence of maturity, the hydrogen index pattern shown in Fig. 9 cannot be expected to reflect the influence of the palaeoenvironment. This can, however, be reconstructed, if mean HI-values are corrected for the influence of maturity. The decrease in HI values due to increasing maturity can be calculated as shown by Tissot *et al.* (1974) and Espitalié *et al.* (1977) for kerogen type II in the Paris Basin. By undertaking similar calculations for the samples from the Botneheia Mbr, a pattern as shown in Fig. 10 is found. The pattern of the HI values reflects the facies conditions during deposition of the Botneheia shales: regional variations follow the type of organic matter deposited (Fig. 8) (the autochthonous

TABLE 1. Summary of geochemical data on Triassic rocks, Svalbard

Units	Total Organic Carbon %					Petroleum Potential (mg HC/g rock)				Extractable Organic Matter (EOM) (ppm)				Extractable Organic Matter (EOM) (mg/g C org)				Extractable hydrocarbons (EHC) (ppm)				Extractable hydrocarbons (EHC) (mg/g C org)			
	0–0.5	0.5–1.0	1.0–2	2–4	>4	<2	2–6	6–15	>15	<200	200–500	500–1000	>1000	<20	20–50	50–100	>100	<100	100–200	200–500	>500	<10	10–25	25–50	>50
De Geerdalen Fm.	14	10	23	3	0	18	2	—	—	7	15	2	—	9	10	1	—	15	8	1	—	15	8	1	—
Tschermakfjellet Fm.	8	6	10	2	0	13	1	1	1	4	12	4	2	4	18	3	—	9	9	2	2	11	9	2	—
Botneheia Mb.	3	15	19	75	147	7	22	37	75	—	—	3	100	—	25	77	11	2	5	15	97	20	49	37	12
Bravaisberget Fm.	3	3	15	13	1	9	7	—	—	3	5	1	19	8	9	1	—	6	—	3	16	5	4	10	—
Sticky Keep Mb.	4	16	22	44	15	6	19	28	8	—	2	1	68	—	—	26	19	1	2	5	41	1	5	18	23
Tvillingodden Fm.	10	8	2	0	0	6	—	—	—	5	—	2	1	5	4	4	1	5	—	3	—	6	2	3	—
Deltadalen Mb.	8	5	6	0	0	6	4	—	—	—	—	3	8	—	—	3	11	1	—	4	7	—	3	1	7
Vardebukta Fm.	12	1	0	0	0	—	—	—	—	1	1	—	2	1	—	1	12	1	1	2	—	—	1	1	2

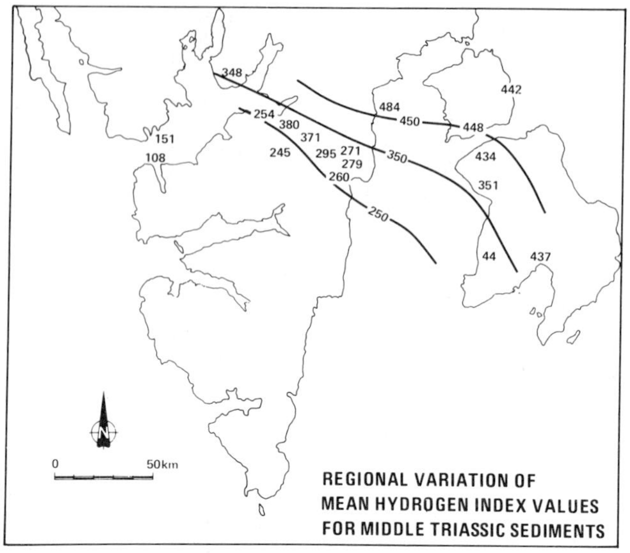

Fig. 9. Regional variation of the mean hydrogen index from Rock—Eval analysis for Middle Triassic sediments.

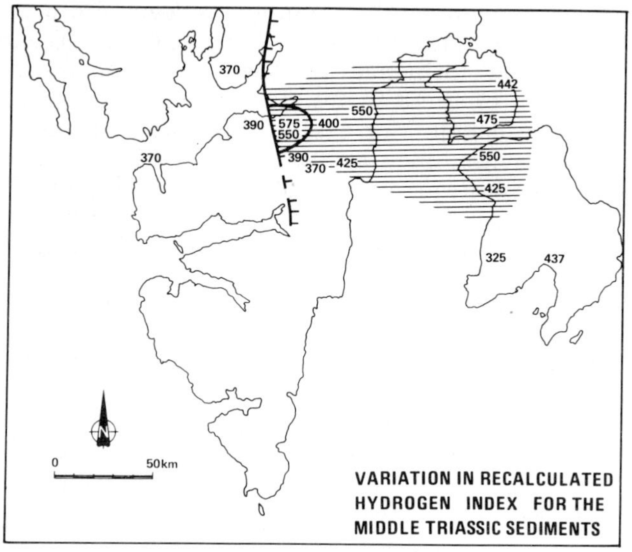

Fig. 10. Variation in recalculated hydrogen index from Rock—Eval analysis for Middle Triassic sediments. The shading reflects an area with high production of marine phytoplankton and deposition in reducing environments.

marine phytoplankton, type II, versus land plant material, type III) and depositional environment (oxygen-depleted euxinic conditions favouring preservation of the freshly deposited organic matter). Thus, a high productivity of marine phytoplankton and a reducing environment resulted in deposition of hydrogen-rich kerogens in the central and northeastern parts of the study area (shaded in Fig. 10). This area, with excellent petroleum source rock quality of the Botneheia shales, is believed to cover most of northeastern Spitsbergen, all of Barentsøya and the northern part of Edgeøya and extends into adjacent offshore areas.

Visual kerogen examination does not give a quantitative measure of source rock quality. It is, however, an excellent tool for the description of the kerogen, especially when the maturity is too high for pyrolysis techniques to distinguish the different types of kerogen from each other. By using this technique, most of the Botneheia and Sticky Keep shale deposits were found to consist mainly of marine, amorphous material. The percentage of terrestrial input varies for the samples analysed, but is mainly less than 50% of the total amount. A plot showing the kerogen composition based on visual kerogen examination shows approximately the same trends as that based on the pyrolysis data.

The Botneheia Mbr thus has the best potential as a source rock for hydrocarbons due to the dominance of kerogen type II, the high organic content and the thick development of the unit over a large area.

Generation of hydrocarbons

The petroleum potential calculated from the Rock—Eval analyses gives an indication of the total amount of hydrocarbons which may be produced from a given kerogen. A plot of the petroleum potential (Fig. 3) reveals a far higher potential for hydrocarbon production in the Botneheia Mbr than below, although significant amounts are also found in the upper part of the Sticky Keep Mbr.

The production index from Rock—Eval analysis gives an indication of the amount of hydrocarbons already generated in a sample or migrated into the analysed sample. Such migrated hydrocarbons (i.e. high production index values) occur in the samples from the Barentsøya Fm (marked on Fig. 8). The bulk of the Botneheia samples contain indigenous hydrocarbons, whereas adjacent units, mostly the Sticky Keep Mbr, contain migrated hydrocarbons.

The occurrence of abundant migrated hydrocarbons, especially in the siltstones of the Sticky Keep Mbr, points to short distance migration from the Botneheia Mbr source rocks. Gas chromatography analyses also show that the extractable hydrocarbons of these beds have smooth, front-end biased alkane distributions, more typically derived from type II kerogen. The regional dip of the beds varies throughout the area and there is generally a slight inclination of the units. Long-distance migration of petroleum along bedding planes, preferentially the extensive laterally developed siltstone beds, may have occurred from the Botneheia shales updip into underlying units. Bjorøy *et al.* (1983b) demonstrated that such migration has occurred by comparing gas-chromatography—mass-spectrometry analyses of hydrocarbons of shales and the adjacent siltstones from the Triassic sequence of Bjørnøya.

Migration of hydrocarbons into the sandstones of the overlying Kapp Toscana Gp was not investigated in the present study. These sandstones, which are above the shales with high source rock potential, have good possibilities as reservoirs if they are found in suitable geological settings in offshore areas further south.

Kapp Toscana Group

The group is of Ladinian to mid-Jurassic age and contains more sandstones than shales compared with the underlying group. The geological development has been presented by Mørk *et al.* (1982). They postulated a change in the general palaeogeographical pattern, from sediment supplied mainly

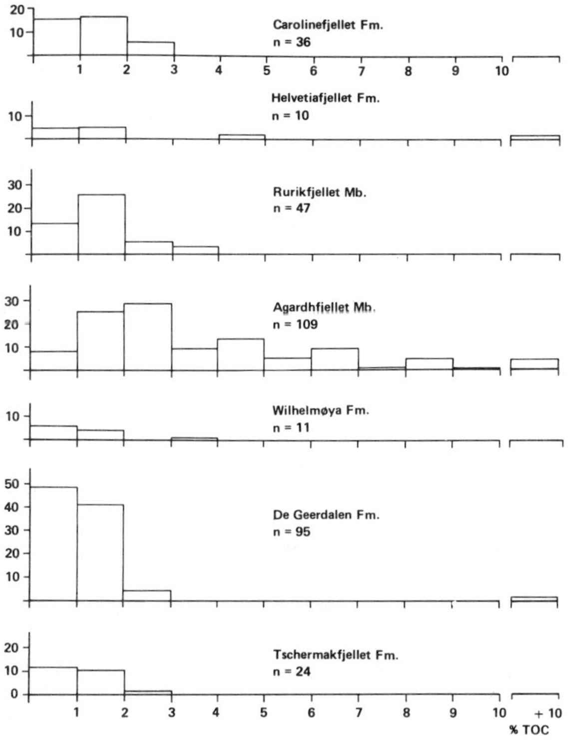

Fig. 11. Histograms showing frequency of the amount of total organic carbon (TOC) in the Kapp Toscana and Adventdalen Groups.

from the west in the lower part of the group to a deltaic input from the east and northeast in the upper part of the De Geerdalen Fm. While the sandstones of the De Geerdalen Fm are highly immature, and have low porosity, the sandstones of the Wilhelmøya Fm above are mature and were probably produced by extensive reworking. These latter sandstones have reservoir possibilities in the offshore area.

Organic content. A number of samples from the group were analysed (130), and most of the work was concentrated on the dark grey shale of the Tschermakfjellet Fm and the shaly intervals in the De Geerdalen Fm. As seen in Fig. 11, the amount of total organic carbon is in the range of 1—2%. The kerogen, however, shows a drastic lowering of the hydrogen index (from Rock—Eval analysis) compared with the underlying group. Visual kerogen examination shows that the insoluble organic matter is dominated by terrestrial debris. The dominance of type III kerogen means that the shales have only a fair, in some areas good, potential as source rocks for gas, although the content of organic matter is not poor. The lower values of extractable organic matter and hydrocarbons than in the unit below is also clearly seen in Table 1. The maturation of the group mainly follows the pattern for the underlying sediment, with a high level of maturity in the western Tertiary fold belt and also local intrusions, mainly on Edgeøya and Barentsøya.

Adventdalen Group

The group is subdivided into three formations: the Janusfjellet Fm, with up to 800 m of dark marine shales of Callovian to Hauterivian age; the Helvetiafjellet Fm, with up to 150 m of deltaic sandstones and shales of Barremian to ?Aptian age; and the Carolinefjellet Fm, with up to 850 m of mixed marine shales, siltstones and sandstones of Aptian to Albian age. The upper part of the Cretaceous is missing on Svalbard.

The maturation pattern mainly follows that described for the underlying units, and in the Agardhbukta area on eastern Spitsbergen a maturity gradient with depth could be established. An increase in mean vitrinite reflectance values from 0.66 to 0.87% R_m from the Helvetiafjellet Fm to the De Geerdalen Fm with an increase of overburden of 640 m gives a reflectance gradient of 0.33% R_m/km.

Janusfjellet Formation

The formation is composed of dark shales and mudstones with the thickest development (800 m) in western Spitsbergen. Drastic thinning of the unit is seen in the central Spitsbergen area at the tectonic lineaments, but in eastern Spitsbergen thicker sequences are again found.

The lower Agardhfjellet Mbr shows a distal shallow shelf facies in the eastern and northern part of Spitsbergen, with deeper shelf sediment towards the central area. A break in deposition occurred at the Jurassic—Cretaceous boundary, accompanied by local dolerite intrusions, before commencement of deposition of dark shales of the Rurikfjellet Mbr. This upper member has a higher content of silt and is bioturbated, showing that the unit was deposited under more oxic conditions than the Agardhfjellet Mbr which often developed anoxic conditions in the deeper shelf area.

Organic content. The content of organic matter clearly follows the depositional facies and high values are found in the Agardhfjellet Mbr (mean 2—3%, although a maximum of 12% was recorded) while lower values (1—2%) occur in the Rurikfjellet Mbr (Fig. 11).

Although we have a regional cover of samples to evaluate the organic richness of the formation, so far we have only conducted more detailed organic geochemical work in the Agardhbukta area in eastern Spitsbergen. These results were reported by Bjorøy and Vigran (1980). They found high TOC values in the lower Agardhfjellet Mbr and kerogens with relatively high contents of type II material, giving the unit a good source rock potential. The amount of extractable hydrocarbon increases upwards within this member.

In the Rurikfjellet Mbr, lower organic contents (1—2%) are due to more proximal and ventilated conditions during deposition. The hydrogen index values from Rock—Eval pyrolysis are lowered, showing that the kerogen is dominated by type III. The gas chromatograms of the saturated hydrocarbon fraction vary considerably, with some showing distribution patterns similar to that of the Agardhfjellet Mbr, while others have probably been affected by weathering. The Rurikfjellet Mbr is considered to have a fair to good potential for gas and oil.

Helvetiafjellet Formation

This formation has facies elements of deltaic origin as well as marine sandstones. The lower part of the unit contains a sandstone complex interpreted by Steel (1977), Steel *et al.* (1978) and Steel and Worsley (1984) to be diachronous with alluvial facies passing southeastwards into marine facies. The sandstone complex covers the whole of central and southern Spitsbergen and may have a considerable reservoir potential offshore (Edwards, 1979).

In the deltaic facies coals and coaly shales are present; however, their small local extent indicates a very limited potential as a source rock for gas.

Carolinefjellet Formation

The formation has up to 850 m of alternating shales, siltstones and sandstones. The units is interpreted by Steel and Worsley (1984) to be partly a lateral equivalent to the coal-bearing deltaic Helvetiafjellet sequence. They interpret the unit to represent shelf sedimentation while the sandier levels may be lower delta front facies.

Organic content. Although the amount of total organic carbon may rise to 2—3% (Fig. 11), Rock—Eval pyrolysis data show very low hydrogen index values and the kerogen appears to be dominated by terrestrial debris and inertinite. A large part of the kerogen is thus classified as kerogen type IV, which does not have any source rock potential.

Cretaceous uplift and Tertiary sedimentation

During late Cretaceous time, Spitsbergen was uplifted, preferentially in the north, and exposed for

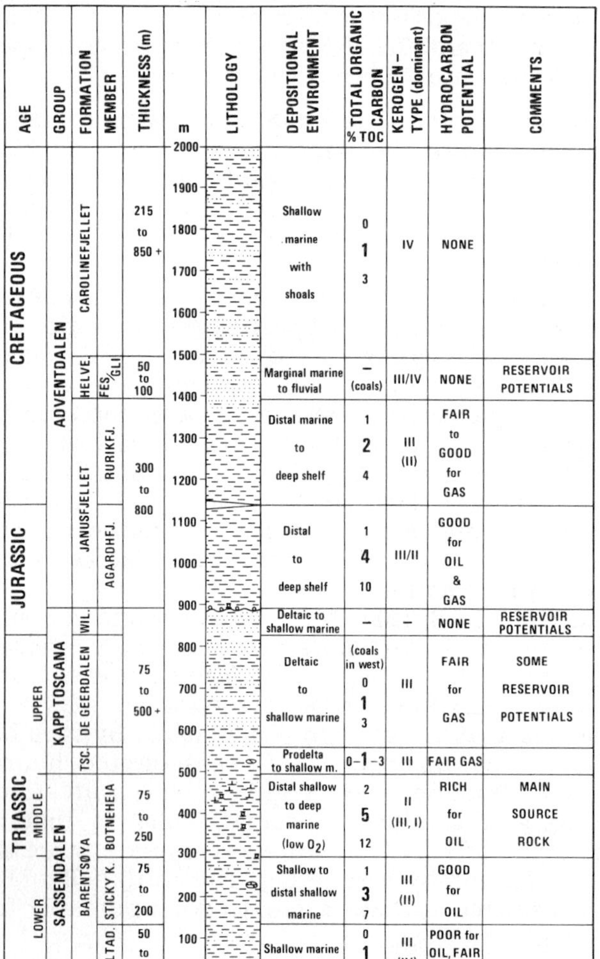

Fig. 12. Summary chart.

erosion, in an early phase before the later opening of the North Atlantic. During this opening of the North Atlantic, Svalbard was subjected to stress and compression, resulting in formation of Tertiary basins and in extensive folding along western Spitsbergen. Although an increased heat flow during folding may have taken place in the western fold belt, the maturity pattern of the underlying rocks mainly seems to be governed by the thickness of the Tertiary succession. Isopach maps for the Tertiary sediments of Spitsbergen have been produced by Kellogg (1975), who postulated that the depocentre of the early Tertiary basin was to the west of the present trough. He also indicated a rapid thinning of the early Tertiary sediment northwestward of the present area of exposure. An isoreflectance map, based on the coals at the base of the Tertiary sequence presented by Manum and Throndsen (1978), supports the conclusions of Kellogg. Their isopach map shows the thickest interpreted development (3.5 km) of the total Tertiary to be in the southwestern part of Spitsbergen. Although the Tertiary sediment present today thins rapidly eastwards, the rank of the coals shows that also in these northeasternmost exposed areas they have been subjected to overburden to the order of 2000 m. As seen from the isoreflectance map of the Triassic (Fig. 6), the Tertiary processes of sedimentation and folding may explain the maturity development of the Mesozoic sequence, with high values in the western area and progressively lower values towards the northeast.

SUMMARY AND CONCLUSIONS

Maturity

The whole Mesozoic sequence follows the trend shown in Figs. 6 and 7, with high values in the western fold belt, resulting in the sequence being mature to overmature. East of this zone maturation is governed by the thickness of the Tertiary overburden and while exposures in central Spitsbergen are in the oil window, immature Mesozoic sediments are found towards the north, east and south. Adding the thickness of the Tertiary succession as postulated by Manum and Throndsen (1978) and the maximum expected thickness of the Mesozoic succession (Fig. 12), the sediments of the lower part of the Mesozoic succession may be expected to be in the condensate or possibly even in the gas window below the Tertiary cover in southern Spitsbergen. On Bjørnøya the Triassic rocks have a maturity level within the lower part of the oil window.

Potential source rocks

Organic-rich shales of regional extent occur repeatedly in the Mesozoic succession of Svalbard; however, the type of kerogen and hence the hydrocarbon potential varies greatly between the individual units.

Of the Lower Triassic units, the Vardebukta Fm (< 1% TOC) and Deltadalen Mbr (1—2% TOC) have kerogen of mixed type III and IV, and thus have only a poor, although locally fair, potential for production of gas. The Tvillingodden Fm also follows this trend, whereas the eastern equivalent the Sticky Keep Mbr (< 7% TOC), although dominated by type III kerogen, also has some input of type II kerogen, especially in the upper part. As a whole this member has good potential for oil and gas. The Sticky Keep Mbr also contains hydrocarbons that have migrated into the unit, probably from the overlying source rock unit.

The Middle Triassic units have a high organic content (Bravaisberget Fm up to 3% TOC, and Botneheia Mbr up to 12% TOC) with good kerogen quality giving them a rich potential for hydrocarbons. The characteristics of this source rock will be summarised below in more detail.

The shales of the Kapp Toscana Gp have total organic carbon contents up to 2%; however, the organic matter is mostly of terrestrial origin and this kerogen type III means that the group has only a fair potential for gas production.

The regionally extensive dark shale unit, the Janusfjellet Fm, shows different characteristics for the hydrocarbon potential of the lower Jurassic and upper Cretaceous parts. The lower Agardhfjellet Mbr, with a mean TOC value of 3%, has a good potential for oil and gas due to a significant input of kerogen type II. The upper Rurikfjellet Mbr has a lower mean content of organic carbon (1—2%) and with a kerogen dominated by type III the potential is mainly for gas.

The Helvetiafjellet Fm with its local coals has only a very limited potential for gas. However, the unit may be of interest as a reservoir horizon, given possibilities for preservation of porosity in the sandstones during diagenesis.

The Carolinefjellet Fm, with its fine-grained clastic sediments, has only a very poor source rock potential due to the high content of kerogen type IV, although the content of organic matter varies between 1 and 3%.

The Middle Triassic source rock

The palaeogeographical setting of coastal to deltaic and shallow marine depositional environments in the western area, and progressively deeper marine conditions towards the east, results in a drastic increase in organic content in the central and eastern areas in the Lower and Middle Triassic. The quality of the source rock, i.e. more hydrogen rich, also improves eastwards and upwards in the Barentsøya Fm (Fig. 8). Local, very high values are found east of the Billefjorden lineament and probably reflect synsedimentary subsidence with the development of anoxic conditions. Towards the northeastern part of the exposed area an increasing content of algal-derived kerogen (containing *Tasmanites*) means that the sequence here has a very high hydrocarbon potential which would be mainly a source of paraffinic oil.

The most significant conclusions on source rock quality, hydrocarbon generation potential and maturity for the Middle Triassic deposits on Svalbard are summarized as follows:

(1) There is major generation of hydrocarbons in source rocks of the Botneheia and also the upper part of the Sticky Keep Members, in a wide area in central and eastern Spitsbergen. This area extends into the easternmost part of Edgeøya including a major part of the Storfjorden offshore area.

(2) Favourable maturity conditions are predicted, especially for a broad belt along the margin of the Tertiary trough. In the trough, Triassic source beds buried under Tertiary sediments should have experienced optimum temperature conditions for petroleum generation. Below the centre of the Tertiary trough and extending westwards towards the Tertiary deformation zone the maturity of Triassic source rocks is expected to have advanced to a level where condensate and gas may be the predominant products.

(3) In the northern exposure area, on Spitsbergen and also on Barentsøya and most of Edgeøya, petroleum generation processes are due to insufficient burial depths and temperatures and are in an early stage or have not been initiated at all. However, the petroleum generation potential of source rocks in the Botneheia and Sticky Keep Members is high and appears to improve continuously towards the northeast and east. Thus, major generation of hydrocarbons can be expected in the offshore areas to the northeast and east of Barentsøya and Edgeøya, provided there is sufficient burial of these source rock units.

(4) Triassic source rocks within the Tertiary deformation zone in the westernmost part of the study area have experienced severe temperature conditions in the past and contain a gas-prone kerogen type due to major influx of terrestrial-derived organic matter close to the ancient shoreline. Thus, this area is classified as a gas province.

ACKNOWLEDGEMENTS

Co-operation with the Svalbard Research Group of The University of Oslo, directed by Dr D. Worsley, has greatly improved our work, as has the extensive fieldwork support of Statoil in allowing us to participate on their Svalbard expeditions. At IKU, J.O. Vigran has conducted the kerogen analysis in transmitted light and we have also benefited from co-operation with Dr D. Leythaeuser, Institute for Petroleum and Organic Geochemistry, KFA, West-Germany, and A.W. Forsberg, University of Oslo. Samples from the Janusfjellet Fm were made available by H. Dypvik, E. Finnerud and M.B. Edwards, while M.B. Edwards and H.R. Moe supplied the samples from the Carolinefjellet Fm.

REFERENCES

Bjorøy, M. and Vigran, J.O. 1980. Geochemical study of the organic matter in outcrop samples from Agardhfjellet, Spitsbergen. In: A.G. Douglas and J.R. Maxwell (eds.), *Advances in Organic Geochemistry 1979*. Pergamon Press, Oxford, pp. 141—147.

Bjorøy, M., Bue, B. and Elvsborg, A. 1983a. Organic geochemical analysis of the first two wells in the Troms I area (Barents Sea). In: M. Bjorøy *et al.* (eds.), *Advances in Organic Geochemistry 1981*. John Wiley, London, pp. 16—27.

Bjorøy, M., Forsberg, A.W., Leythaeuser, D. and Mørk, A. Organic geochemical studies of the Mesozoic succession of Svalbard. (in preparation).

Bjorøy M., Mørk, A. and Vigran, J.O. 1983b. Organic geochemical studies of the Devonian to Triassic succession on Bjørnøya and the implications for the Barents Shelf.

In: M. Bjorøy *et al.* (eds.), *Advances in Organic Geochemistry 1981.* John Wiley, London, pp. 49—59.

Edwards, M.B. 1979. Sandstone in Lower Cretaceous Helvetiafjellet Formation, Svalbard: Bearing on reservoir potential of Barents Shelf. *Am. Assoc. Petrol. Geol. Bull.*, **63**, 2193—2203.

Espitalié, J., Laporte, J.L., Madec, M., Marquis, D., Leplat, P., Paulet, J. and Boutefeu, A. 1977. Methode rapide de charactérisation des roches méres de leur potential pétrolier et de leur degré d'évolution. *Rev. Inst. Fr. Pétrol.*, **33**, 23—42.

Forsberg, A. and Bjorøy, M. 1983. A sedimentological study of the Botneheia Formation, Svalbard, with special emphasis on the effects of weathering on the organic matter of shales. In: M. Bjorøy *et al.* (eds.). *Advances in Organic Geochemistry 1981,* John Wiley, London, pp. 60—68.

Hunt, J.M. 1979. *Petroleum Geochemistry and Geology.* W.H. Freeman, New York, NY, 617pp.

Kellogg, H.E. 1975. Tertiary stratigraphy and tectonism in Svalbard and continental drift. *Am. Assoc. Petrol. Geol. Bull.*, **59**, 465—485.

Manum, S.B. and Throndsen, T. 1978. Rank of coal and dispersed organic matter and its geological bearing in the Spitsbergen Tertiary. *Norsk Polarinst. Årbok,* **1977**, 159—177.

Mørk, A., Knarud, R. and Worsley, D. 1982. Depositional and diagenetic environments of the Triassic and Lower Jurassic succession of Svalbard. *Can. Soc. Petrol. Geol. Mem.*, **8**, 371—398.

Steel, R.J. 1977. Observations on some Cretaceous and Tertiary sandstone bodies in Nordedskiöld Land, Svalbard, *Norsk Polarinst. Årbok,* **1976**, 43—68.

Steel, R.J., Gjelberg, J. and Haarr, G. 1978. Helvetiafjellet Formation (Barremian) at Festningen, Spitsbergen — a field guide. *Norsk Polarinst. Årbok,* **1977**, 111—128.

Steel, R.J. and Worsley, D. 1984. Svalbard's post-Caledonian strata — an atlas of sedimentological patterns and palaeogeographical evaluation. This volume, pp. 109—136.

Tissot, B., Durand, B., Espitalié, J. and Combaz, A. 1974. Influence of the nature and diagenesis of organic matter in the formation of petroleum. *Am. Assoc. Petrol. Geol. Bull.*, **58**, 499—506.

Tissot, B.P. and Welte, D.H. 1978. *Petroleum Formation and Occurrence.* Springer-Verlag, Berlin, 538pp.

28 Geohistory and hydrocarbon evaluation of the Haltenbank area

N.B. Hollander
Saga Petroleum AS, Høvik, Norway

This chapter deals with a reconstruction of the evolutionary aspects of sedimentation, subsidence, structures, maturation and migration of hydrocarbons, as well as reserves.

The Triassic sediments consist of salt and red shales with sandstones which grade into carbonaceous shale of Triassic—Jurassic age. These are overlain by Jurassic sandstones and shales. The Cretaceous sediments are dominated by calcareous claystones and form the caprock in the area. The Palaeocene—Eocene ash-layers (Balder Fm) are a well defined sequence. The younger Tertiary sediments consist of claystones with minor sandstones and limestones. Interbedded clay, silt and sand deposited in Pliocene—Pleistocene time is very thick, reflecting the fast subsidence during this epoch.

The structural pattern of the area is characterized by the stable Trøndelag Platform in the east and a NE—SW-striking fault zone in the west. Between these there is a strongly faulted graben—horst area.

The occurrence of hydrocarbons in the area confirms the existence of mature source rocks. Analysis shows that the Upper Jurassic shales and the Triassic—Jurassic carbonaceous shales have a high potential for generation of hydrocarbons of types II and III respectively. Trapping of hydrocarbons occurs by a complex combination of structure and faulting. The only proven traps so far are a Jurassic horst structure drilled by the 6507/11-1 well and an anticline in block 6407/1. The structures in the study area probably contain 100—150 $\times$ $10^9 m^3$ while substantially larger reserves have to be searched for outside these six blocks.

INTRODUCTION

The purpose of this chapter is to review the stratigraphic and tectonic framework, as well as the reserves of hydrocarbons, in the Haltenbank area. This area is here defined as the six blocks opened so far for exploration northwest of the fishing bank (see Fig. 1). The review is based on the experience of 3 years' exploration with seven wells (of which Saga has drilled five) and a large amount of seismic data.

TECTONIC FRAMEWORK

The northern Atlantic and the Barents Sea developed in a tectonically active area between two stable plates: the Eurasian Plate and the North American Plate. The geological development was related to global plate tectonics with three main episodes:

(1) Caledonian mountain-building in Ordovician—Silurian with subsequent erosion and transgression in Devonian—Permian times.

(2) The formation of the Ural Mountains during Permian—Triassic, with subsequent tension and faulting during late Triassic to early Cretaceous.

(3) Opening of the Norwegian Sea during Tertiary time, involving faulting and subsidence of large areas, associated with strong volcanism.

STRATIGRAPHIC HISTORY

The stratigraphic history during Palaeozoic time, although interesting, is beyond the scope of this chapter and thus is not considered.

Triassic

The lowermost formations encountered in the first wells consist mainly of shale, and sand- and siltstone interbedded with minor shale (see Fig. 2). These are overlain by salt which is clear to translucent, occasionally light yellow. There are two salt sections, reaching a total thickness of nearly 800 m, separated by several hundred metres of

Petroleum Geology of the North European Margin © Norwegian Petroleum Society (Graham & Trotman, 1984) pp. 383—388.

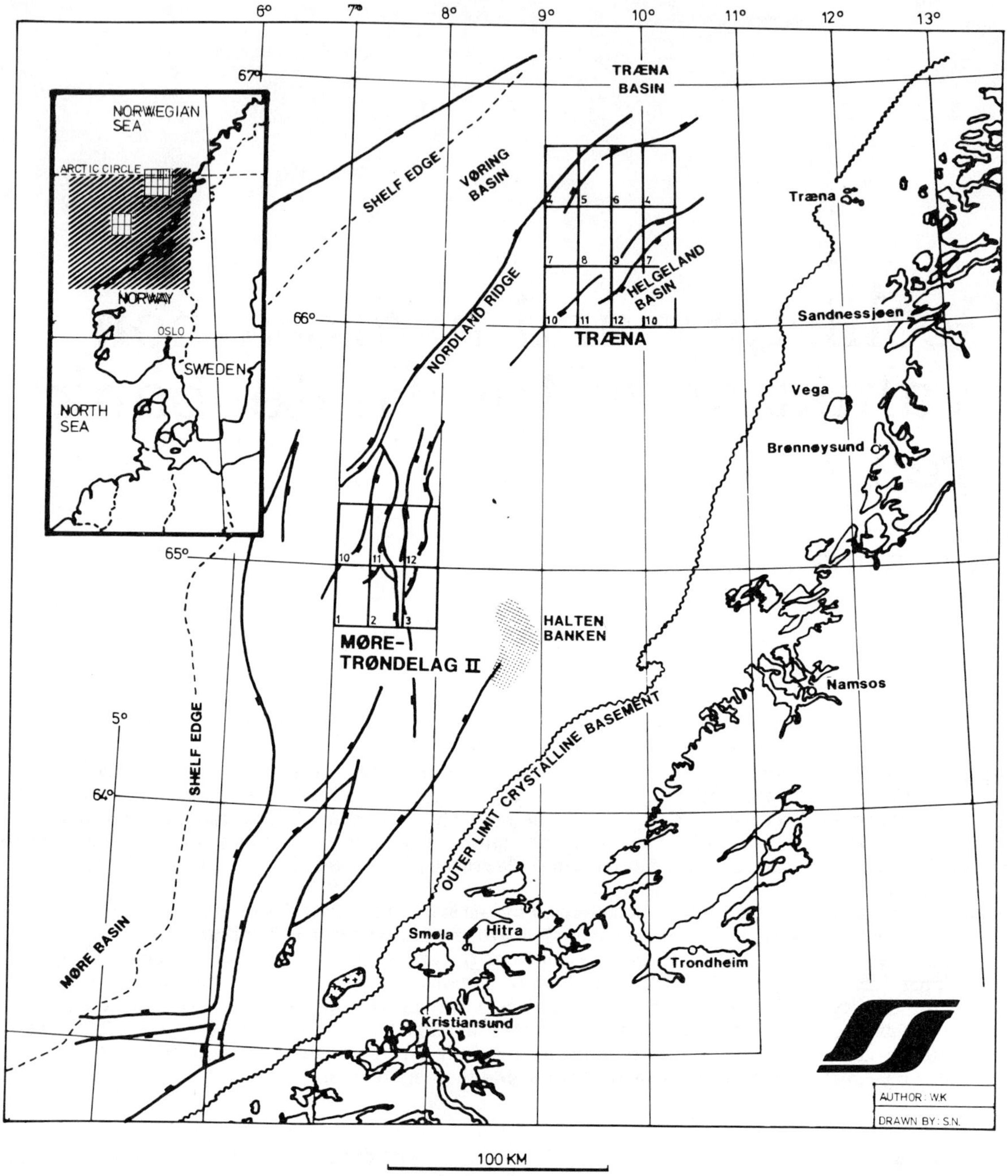

Fig. 1. Position of the study area: six blocks northwest of Trondheim, mid-Norway.

claystones/shales and siltstones with occasional beds of anhydrite. The siltstone is red to brown, hard and micaceous. Minor stringers of sandstone also occur. The sedimentary environment appears to have been an arid climate with repeated restrictions of seawater influx and depositions of halite.

The salt is overlain by alternating brown to red brown claystones and siltstones, with occasional stringers of grey to brown sandstone and limestone.

The Upper Triassic marks the start of a transgressive megacycle which culminated in Pliensbachian—Toarcian. In the first stage, grey friable sandstone and grey shale ('grey beds') were conformably deposited upon the redbeds, followed by an approximately 400-m thick unit of coal seams and carbonaceous shale interbedded with sandstone and shale — Formation I. The sedimentary environment was marginally marine to deltaic. These sediments have a large regional extension, continuing on to the Nordland Ridge in the north. It is, however, not clear whether they reappear west of the ridge. Eastward there is a facies change as the equivalent rocks outcropping on sea bottom consist of terrestrial sediments. South of the study area it is difficult to follow the corresponding seismic reflectors and therefore it is uncertain how far they continue. Apparently these sediments were deposited in one basin (Hollander, 1980; Jørgensen and Navrestad, 1980), extending from the Vøring and Helgeland Basins in the north to at least the area just south of block 6407/2. This coal sequence can be correlated with the Statfjord Fm in the North Sea and the Kap Stewart Fm in East Greenland.

The first growth faults become active at this stage, thus indicating the onset of an early Kimmerian tectonic phase. One of these is a deep westward dipping fault that marks the western limit of what later became the Trøndelag Platform (Hollander,

1980, 1982). These movements were the prelude to the generation of the horst and graben system of the Haltenbank area.

Jurassic

The Jurassic sediments are included here as an introduction only, as they are considered in another chapter (Karlsson, 1984). There is a gradual transition from the Triassic to the overlying Lower Jurassic sandstones (Formation II, a Cook Fm equivalent). A shale formation over the sandstone represents the culmination of a regional Lower Jurassic transgression. The shale (Formation III) is equivalent to the Drake Fm of the North Sea. The shale, which is a characteristic overall upward-coarsening sequence, terminates against a Lower—Middle Jurassic coastal sand complex (Formation IV) of large lateral extent, deposited at the same time as the Brent Group in the North Sea. The Upper Jurassic sequence (Formations V and VI) rests upon an unconformity of Middle Jurassic age and consists of sediments deposited in basins with restricted water circulation, resulting in a sequence of organic-rich shales with a good to excellent source rock potential, i.e. an equivalent to the Kimmeridge Clay Fm.

In early Jurassic times the faulting continued and a new system of growth faults was developed parallel to the first master faults, but with an easterly dip (Gabrielsen and Robinson, 1984). Thus, two small graben systems were subsiding on each side of a central stable horst.

In late Jurassic times, faulting of the central horst occurred at the same time as the graben areas were subjected to the generation of an axial fault system and subsidence. These movements continued into late Jurassic—early Cretaceous time, when most of the faults died out. It also appears that continued crustal extension caused rapid subsidence west of the area. During late Jurassic—early Cretaceous there was a period of erosion.

Cretaceous

The Jurassic—Cretaceous boundary is placed at a significant unconformity which marks the culmination of the late Kimmerian tectonic phase. This was formed by a significant eustatic sea-level drop that caused the emergence of large areas and a temporary restriction of marine sedimentation to the deeper parts of the basins. Thereby three separate basins developed in the area: the Vøring—Trœn Basin, the Helgeland Basin and the Haltenbank area. However, during the gradual rise of sea level towards late Albian times, claystones progressively onlapped the base Cretaceous unconformity. The Kimmeridge Clay equivalent was first overlain by early Cretaceous limestone, red brown marl and later by calcareous claystones (Cromer Knoll Gp equivalent), while the local highs were characterized by erosion or non-deposition. Later, Upper Cretaceous sediments were developed as predominantly grey claystones, variably calcareous, occasionally silty and with thin beds of sandstone. Thin beds of limestone occur. The late Cretaceous evolution was governed by a global transgression during which the basins were gradually filled and combined to form one main basin covering the former Vøring—Trœn, Helgeland and Halten Basins.

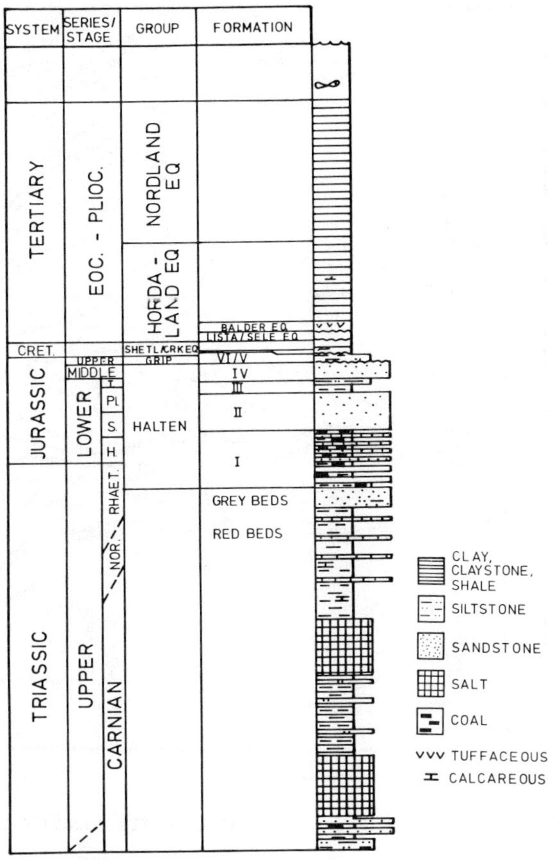

Fig. 2. A generalized lithologic column of the Haltenbank area.

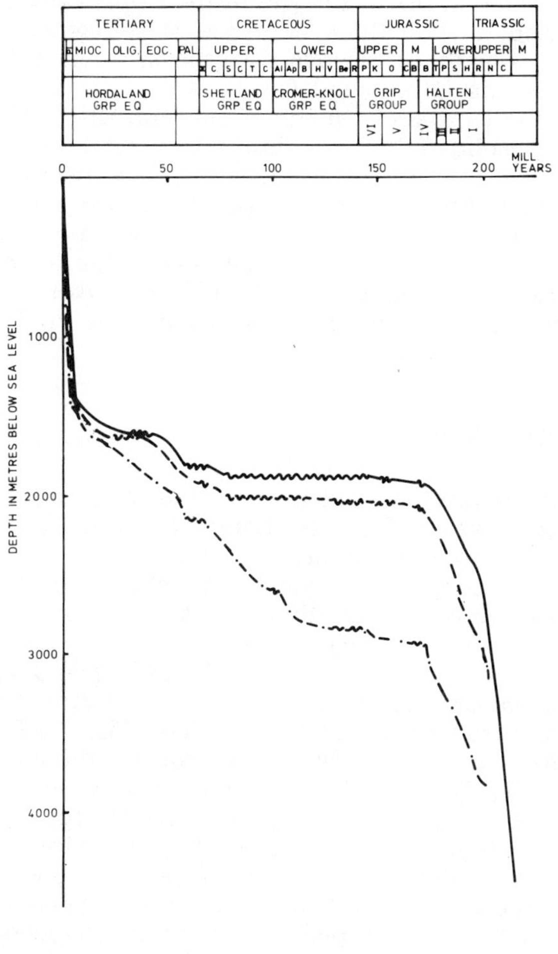

Fig. 3. Relative subsidence profiles of the Haltenbank area.

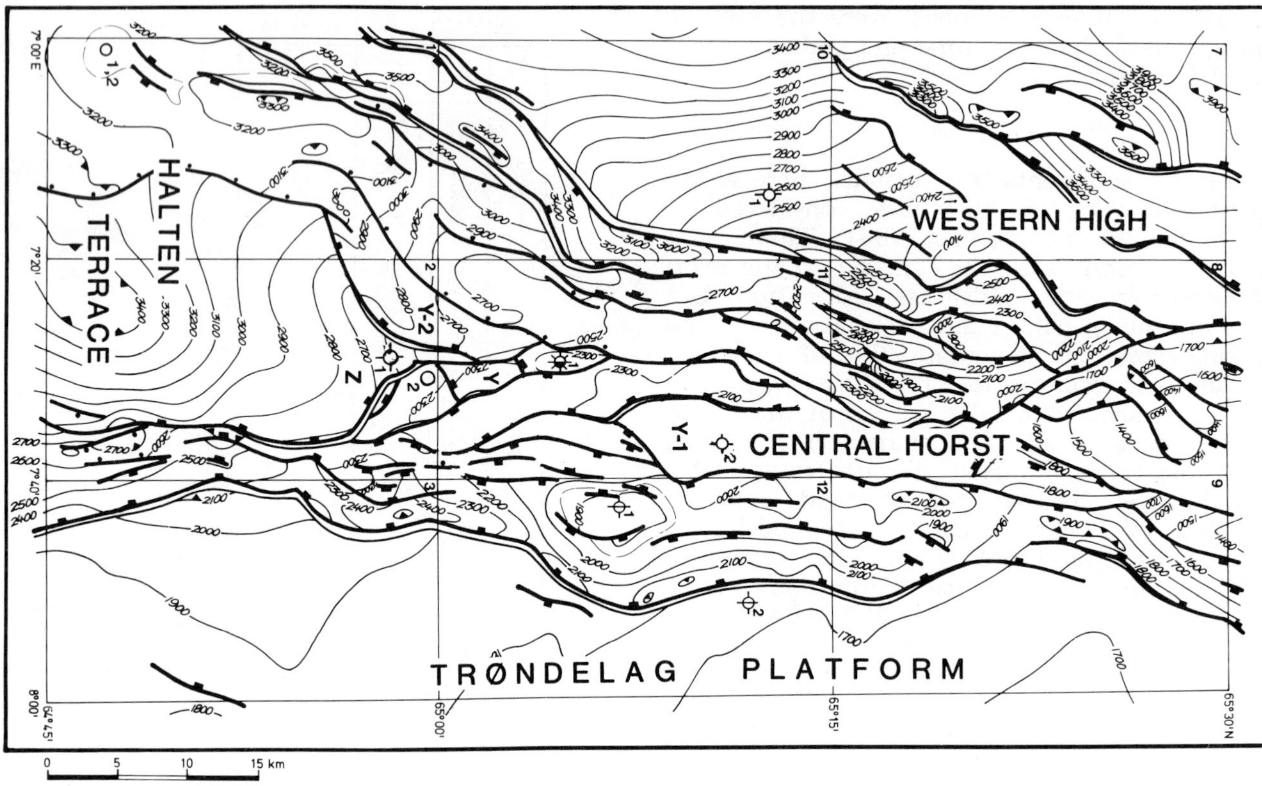

Fig. 4. A structural time map at base Cretaceous unconformity.

Tertiary

The Tertiary sequence rests on an unconformity surface that marks the end of the Upper Cretaceous period. It is a thick sequence of variegated claystones with tuffaceous claystones, sand, silt and limestone (Hordaland Gp equivalent) upon which lie the Nordland Gp equivalent, consisting of grey unconsolidated clays with minor sand and rock fragments. The tuffaceous claystones form a well-defined lithological unit (Balder Fm equivalent), which is easily correlated both on seismic sections and between the wells. The base of the Nordland Gp lies between 1300 and 1400 m below sea level. This reflects the rapid subsidence in Pliocene—Pleistocene times.

SUBSIDENCE

The relative subsidence of the area has been reconstructed from stratigraphic information obtained from the wells. The time scale is based on biostratigraphic data and hence is only approximate. Furthermore, the thicknesses have not been corrected for compaction.

All the wells in blocks 6507/10, 11 and 12 are located on structurally high areas and show similar reconstructed subsidence profiles (Fig. 3). During late Triassic to early Jurassic times sedimentation apparently took place in the same subsiding basin. The only tectonic activity was the development of the two growth faults. The rate of burial was high, but was abruptly reduced and became low during Middle Jurassic to Middle Cretaceous when strong erosion also took place. During late Cretaceous to Pliocene there was faster subsidence. However, strikingly greater subsidence rates occurred during the following 5 Ma, i.e. the Pliocene times.

The high rate of burial in 6407/2 in late Jurassic times shows that this area was detached both from the adjacent central horst and from the Trøndelag Platform. Later the gas structure (Y) also became separated from these areas. During late Cretaceous to Pliocene the basin in 6407/2 was subject to a higher rate of burial than the structurally higher areas. Through the Pliocene the rate of subsidence was equal in all the wells, indicating that the entire area was subsiding as one basin.

STRUCTURAL GEOLOGY

The Haltenbank area lies within the Kristiansund-Bodø Fault Complex. The eastern part of the study area covers the western edge of the Trøndelag Platform, which in its northern part is overlain by the Helgeland Basin. The northern limit of the platform is characterized by the complex faulting along the Nordland Ridge. North and west of the ridge the Cretaceous and younger sediments drop abruptly into the Træn and Vøring Basins. The western boundary of the platform is dominated by a zone of faulting. The southwestern part of the study area lies between the platform and the deep Møre Basin to the south.

The Haltenbank area is characterized by N—S to NNE—SSW-striking faults with modifying NNW—SSE faults, forming horsts and grabens. The structures at the base Cretaceous unconformity are dominated by the Trøndelag Platform in the east and a slightly deeper high in the west (Western High, Fig. 4). In the middle there is a central horst. The entire area rises gently to the north. West of the central horst there are two fault blocks strongly

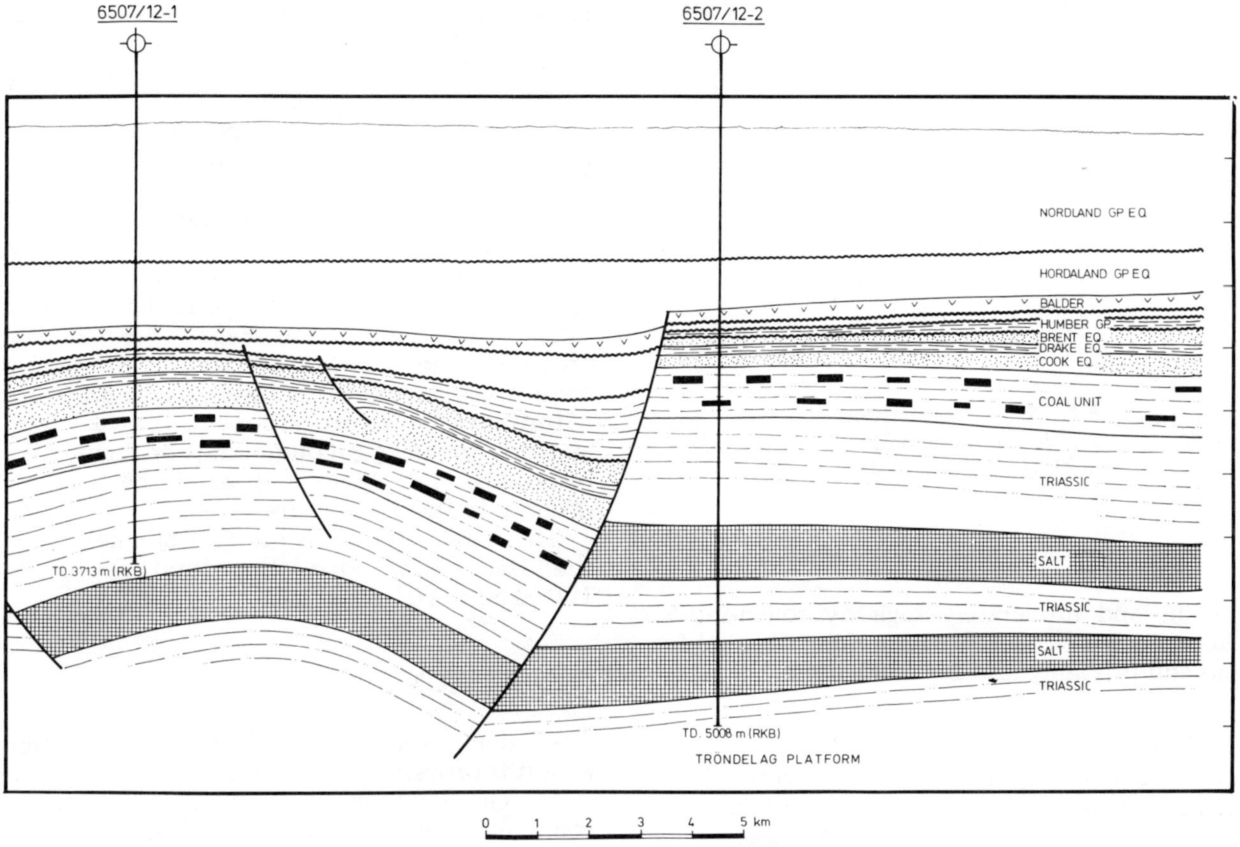

Fig. 5. A geological cross section from southwest to northeast.

tilted to the ESE. In blocks 6507/10—6407/1 there is a deep graben. The structures are illustrated in the geological cross-section (Fig. 5).

HYDROCARBON EVALUATION

In order to evaluate the possibility of finding hydrocarbons in the area, it is very important to identify and characterize the potential source rocks as well as analyse their maturity. Then it may be possible to predict what type of hydrocarbons can be generated and where the source rocks are buried deep enough for hydrocarbon generation.

Source rocks

The gas-condensate/light oil discoveries in wells 6507/11-1 and 6407/1-2 confirmed the existence of mature source rocks in this area. There are two types of rock with a high content as well as the right type of organic matter: the Kimmeridge Clay Fm equivalent and the coaly sequence (see Fig. 6). There are also some rocks with a poor potential as source rocks: Upper Cretaceous shale, Upper Jurassic Heather Fm equivalent and Middle Jurassic shale.

The Kimmeridge Clay Fm equivalent consists of dark radioactive shales with a high organic matter content and an amorphous, liptinic oil-generating kerogen. The content of organic carbon varies between 4 and 13%, averaging 10%. Pyrolysis shows a high potential for generation of hydrocarbons as well as high hydrogen indices. The kerogen can be classified as type II with a rich potential for oil generation. These properties confirm that the Kimmeridge Clay Fm equivalent is a good potential source rock for oil offshore mid-Norway.

All seven wells drilled in the area up to 1983 have encountered large amounts of coal and carbonaceous shale with interbedded sandstones. It is well known that both coal and carbonaceous shales are good source rocks for dry gas, especially at relatively high levels of maturity.

When moderately mature, condensate and some oil can also be generated. The amount of organic carbon varies in the formation, the best samples being very rich. The kerogen is, of course, composed of terrestrially derived organic material. However, pyrolysis shows a high potential of hydrocarbon

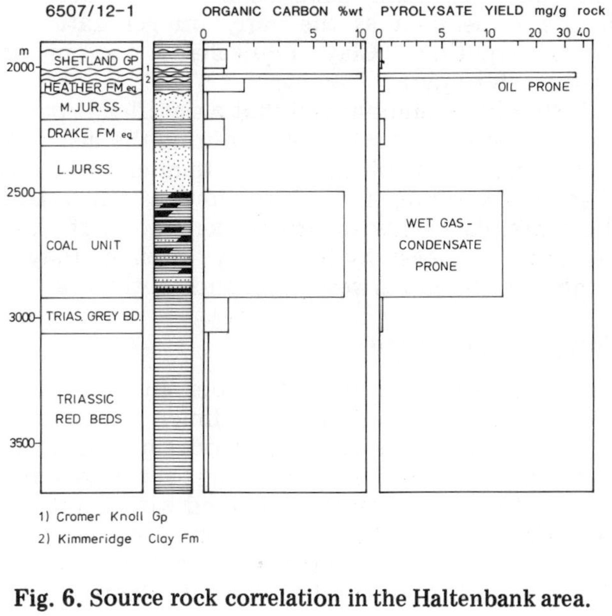

Fig. 6. Source rock correlation in the Haltenbank area.

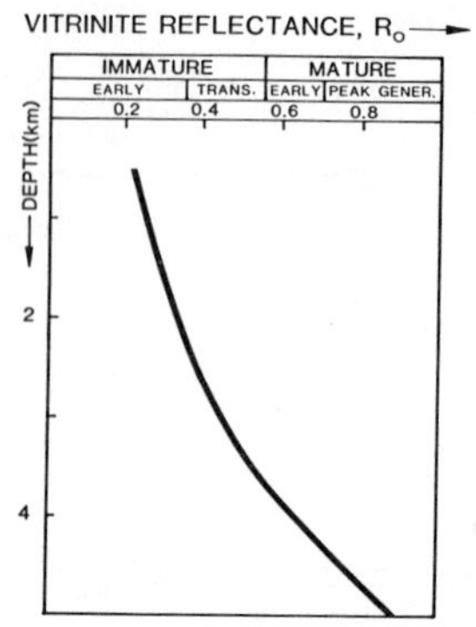

Fig. 7. A maturation profile for wells drilled in blocks 6407/2, 6507/11 and 12.

generation and moderately high hydrogen-index. The kerogen is of type III with a rich potential for wet gas and probably also for condensate.

Maturity

The vertical maturity has been interpreted from the vitrinite reflectivity measurements of samples from boreholes. The reflectivity increases continuously with depth (see Fig. 7). It appears that wet gas generation starts just below a depth of 2 km. The generation of oil does not begin until the sediments are buried well below 3 km. In order to evaluate the regional maturity, the seismic reflectors corresponding to base Cretaceous unconformity and near base Jurassic were chosen for reference. These markers are mapable and represent the top of the Kimmeridge Clay equivalent and an intra-coal unit reflector, respectively.

The base Cretaceous reflector reaches the zone of early oil generation in the southwestern part of the area, approximately SW of a line between the NW corner of block 6507/10 and SE corner of 6407/2. Maximum oil generation probably occurs west of the six blocks and possibly also in the graben area in NW part of block 6407/1.

The near base Jurassic reflector reaches the zone of maximum generation of hydrocarbons in the same area as the early oil generation of the Kimmeridge Clay equivalent, i.e. in the southwestern part.

It should be emphasized that a significant part of the subsidence has taken place during the last 5 Ma. This implies that oil generating temperatures have been achieved only for a short time, which restricts the amounts of hydrocarbons generated. If this is the case, the best hydrocarbon 'kitchen' may be located to the northwest, west and southwest.

Trapping and reserves

The well results show that Upper Jurassic, Cretaceous and Lower Tertiary sediments consist of impermeable shales forming good cap rocks. The main traps are defined by structures at top Reservoir or base Cretaceous level, combined with truncation of older rocks. These structures can be simple anticlines, or in combination with faulting, e.g. 6407/1-2 and 6507/12-1-structures. There are also tilted fault blocks and horsts, e.g. the Y-structure in block 6407/2.

In 1981 Saga made the first discovery offshore mid-Norway, when well 6507/11-1 proved a small gas/condensate field in an excellent reservoir. The well produced 764 000 Sm^3 (27MMSCF) gas and 138 m^3 (870Bbl) condensate per day through a 20.2-mm (51/64-inch) choke and the well flowed at a very small pressure drawdown. This flow rate was the maximum that the rig could handle. Probable reserves in the entire structure and other small structures in the Saga operated licences probably total 50—100 $\times$ $10^9 m^3$ of gas. Statoil made a discovery in block 6407/1 in 1983. The exact nature of the hydrocarbons is not known, but appears to be gas/condensate or light oil. The reserves in this block appear to be in the same order as in the Y-structure.

The total reserves in the five blocks awarded in the Haltenbank area thus total approximately 100—150 $\times$ $10^9 m^3$ of gas.

CONCLUSION

The conclusion is quite clear. In the Haltenbank area it is proven beyond doubt that:

(1) Good but immature to early mature source rocks exist in the study area. There are, however, strong indications that there are mature source rocks west of the Haltenbanken area.

(2) Reservoir rocks are of excellent quality and should be good even when these rocks are found at deeper levels.

(3) There are good seals for hydrocarbon traps.

So far only non-commercial amounts of hydrocarbons have been indicated, and it will be necessary to increase the area of exploration in order to find larger reserves.

ACKNOWLEDGEMENTS

The author wishes to express his gratitude to the partners in the Saga-operated blocks (Arco Norge AS, Deminex Norge AS, Elf Aquitaine Norge AS, Norsk Agip AS, Norske Shell, Norsk Hydro AS, Statoil and Volvo Petroleum Norge AS) for permission to publish this work and to R.H. Gabrielsen, T. Hagevang, W. Karlsson, C.I. Robinson and T.O. Trondsen for allowing use of their work.

REFERENCES

Gabrielsen, R.H. and Robinson, C.I. 1984. Tectonic inhomogeneities of the Kristiansund-Bodø complex, offshore mid-Norway. This volume.

Hollander, N.B. 1980. Evaluation of the oil/gas potential offshore mid-Norway. *26th. Int. Geol. Congr.*, Paris, Abstr.

Hollander, N.B. 1982. Evaluation of the hydrocarbon potential offshore mid-Norway. *Oil Gas J.*, Apr. 12 168—172.

Jørgensen, F. and Navrestad, T. 1980. Main structural elements and sedimentary succession on the shelf outside Nordland (Norway). *Norwegian Sea Symposium 1979.* (Norsk Petroleumsforening), Oslo, NSS.

Karlsson, W. 1984. Sedimentology and diagenesis of Jurassic sediments offshore mid-Norway. This volume, pp. 389—396.

29 Sedimentology and diagenesis of Jurassic sediments offshore mid-Norway

Werner Karlsson
Saga Petroleum AS, Høvik, Norway

The study area is located 130 km offshore mid-Norway at 65°N. The area was opened for hydrocarbon exploration in 1980. This chapter is based on data compiled from five wells, one of which was a gas/condensate discovery.

The Jurassic sequence is the main exploration target. The boundary between the Triassic and Jurassic succession is defined within a thick unit of coastal plain shales, sandstones and coal beds. This primarily continental facies was transgressed by shallow marine sands during the Sinemurian—Pliensbachian, followed by offshore clay deposition during the early Torcian. A major regressive cycle took place during late Toarcian—Bajocian, resulting in deposition of a continuous shallow marine sandstone. The Upper Jurassic is characterized by transgressive conditions with deposition of shales.

The mineralogical composition of the Lower Jurassic sandstones is immature, while the Middle Jurassic reveals sections of more mature quartz arenites. The diagenetic features include well developed K-feldspar overgrowth, weathering of plagioclase, early quartz overgrowth with subsequent etching, mica expansion, recrystallization of detrital clay and a late-stage ferroan calcite cementation.

None of the diagenetic processes have been detrimental to the reservoir properties. Detrital clay and kaolinite cement reduce permeability in the Lower Jurassic sandstones, while the Middle Jurassic sandstones generally have excellent porosity and permeability because of marine reworking/sorting, as well as minor grain cementation.

INTRODUCTION

The area studied was opened for commercial petroleum exploration in February 1980 and the first well, 6507/12-1, was drilled the following summer. To date, eight wells have been completed, of which three are gas/condensate discoveries.

The area is located in the Norwegian Sea offshore mid-Norway (Fig. 1) close to the Haltenbank fishing ground. The distance from the Norwegian coast is 130 km and the water depth is approximately 250 m. The well locations are marked on Fig. 2. Five blocks have been allocated, and the present study is based on well data from the three blocks where Saga Petroleum has been appointed operator. The following companies participate in these joint ventures: Statoil, Norsk Hydro, Elf, Shell, Arco, Agip, Deminex and Volvo Petroleum.

The purpose of this chapter is to present the lithology of the Jurassic succession, with emphasis on the potential reservoir units. No wells have been released and because of the confidential

Petroleum Geology of the North European Margin © Norwegian Petroleum Society (Graham & Trotman, 1984) pp. 389—396.

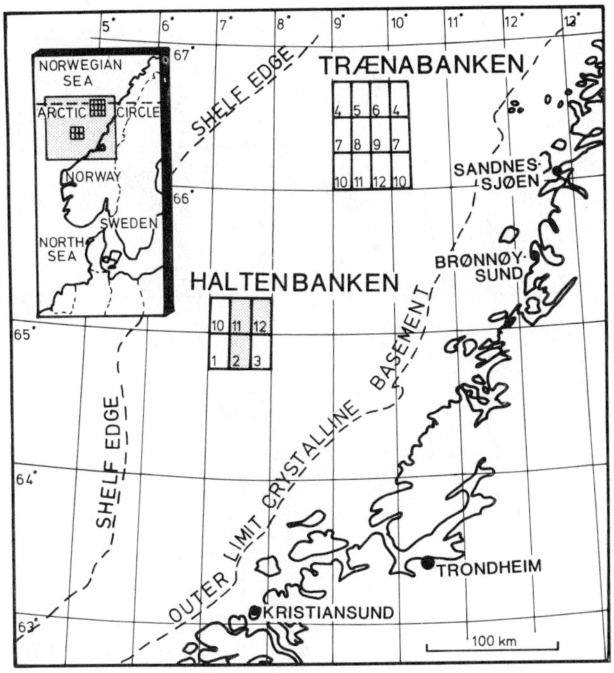

Fig. 1. Location of the Haltenbanken area. The study is based on wells from the three blocks 2, 11 and 12 where Saga Petroleum AS is operator.

nature of the data, a qualitative approach to data presentation has been necessary.

STRUCTURAL SETTING

The Haltenbank area is a part of a passive continental margin representing the predrift counterpart of the exposed East Greenland sedimentary basin (Talwani and Eldholm, 1977). The area is located at the boundary between the Møre Basin and the Trøndelag Platform and represents a southern extension of the Nordland Ridge (Jørgensen and Navrestad, 1981).

Basement is probably very deep in the area, so basement-related faulting has not been demonstrated from the seismic data. In the Triassic and Jurassic strata, normal faulting is pronounced, and has resulted in a generally N—S-trending series of rotated fault blocks as well as horst and graben structures.

The thickness distribution of the Jurassic and its relation to its structural position is shown on a cross-section based on seismic and well data (Fig. 3). Two prominent faults, one west of well 6507/12-2 and one west of well 6507/11-1, clearly demonstrate that growth faulting has been active, particularly in late Jurassic times. The structural highs demonstrate thinning of the Jurassic because of non-deposition or erosion during Middle Jurassic and particularly during late Jurassic to early Cretaceous times. The faulting continued locally throughout the Cretaceous and well into the Tertiary.

Most of the structuring of the Jurassic traps occurred during the Cretaceous. In addition to normal faulting, differential subsidence is a striking tectonic process during this period. Apart from some faults cutting the top Palaeocene reflector, the Tertiary was a tectonically quiet period of epeirogenic subsidence, as is normally seen in the adjacent North Sea Basin. The major part of the Tertiary comprises sediments of Pliocene—Pleistocene age.

STRATIGRAPHY

The Jurassic sedimentation in the area represents a period of sand-prone sedimentary environments in the Lower and Middle Jurassic, and a period of shale deposition in the Upper Jurassic. This has led to a lithostratigraphic subdivision of the Jurassic into two groups separated by a prominent Bathonian—Callovian hiatus (Fig. 4).

The lower group starts with a basal coal—coaly shale—sandstone sequence overlying Triassic red beds (Karlsson and Dypvik, 1981). This coal sequence ranges in age from the Middle Rhaetian to well into the Sinemurian. No major time gaps exist within the unit and the Jurassic seems to overlie, conformably, the Triassic part of the sequence. The coal unit is equivalent to the East Greenland Kap Stewart Formation (Birkelund, 1975; Clemmensen, 1976; Surlyk, 1975, 1977, 1978a,b; Raunsgaard-Pedersen and Lund, 1980) and to the North Sea Statfjord Fm and the lower part of the Dunlin Gp (Deegan and Scull, 1977).

The coaly sequence is followed by a Sinemurian—Toarcian shallow marine sandstone/siltstone unit of the same age as the upper part of the Amundsen and Cook Fms in the North Sea. The Neill Klinter Fm is a similar deposit in East Greenland (Sykes, 1974).

The Lower Jurassic sandstone sequence is overlain by a mudstone unit which is the time equivalent of the Drake Fm in the northern North Sea. The interval becomes sandier towards the top and the whole section is developed as an overall upward coarsening sequence.

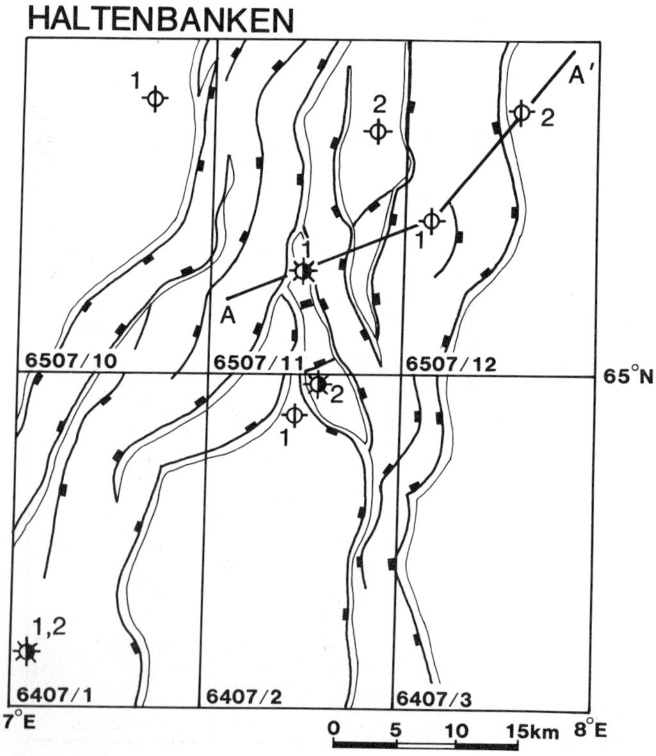

Fig. 2. Structural map and well locations for the six Halten blocks opened for exploration drilling in the Norwegian 5th round.

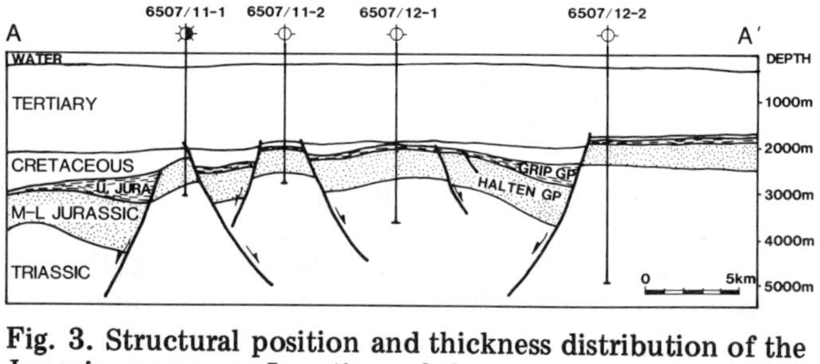

Fig. 3. Structural position and thickness distribution of the Jurassic sequence. Location of the cross-section is marked on Fig. 2.

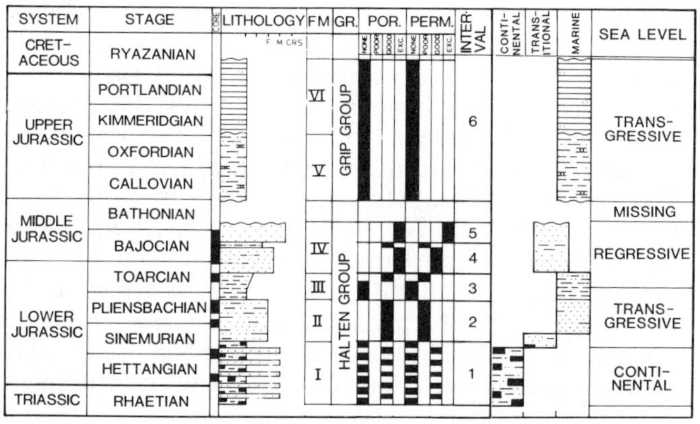

Fig. 4. Lithostratigraphy, reservoir characteristics and sedimentary settings for the Jurassic sediments in the Haltenbanken area offshore mid-Norway.

There is an abrupt change in lithology between this mudstone unit and the overlying Toarcian—Bathonian sandstone complex. This sandstone is bounded at its base by a carbonate cemented layer, while the top of the unit is a major unconformity often referred to as the pre-Callovian or main Kimmerian unconformity. Having a high organic content, it has some similarities to the Rannoch and Ness Fms in the Brent Province, but it is in general much more coarse grained and lacks continuous units of shales and coals. It also appears to be slightly older. At Andøya, off the Norwegian mainland, Dalland (1975) has described a sandstone — the Hesteberget Member — of similar age. This implies that the sandstone could be of large lateral extent.

Resting unconformably on top of the Toarcian—Bathonian sandstones is an interlaminated siltstone/shale sequence ranging in age from Callovian to the Ryazanian. The unit can be split into two formations, the lower containing frequent tight carbonate streaks, while the upper is a 'hot shale' rich in organic matter. The two units show from the wireline logs remarkable resemblance to the North Sea Heather and Kimmeridge Clay Fms. The typical anoxic, marine source beds of the uppermost Jurassic can also be correlated to the Ratjønna Member on Andøya as described by Dalland (1975). The rest of the Upper Jurassic, where exposed onshore, is sandier than that found offshore (Surlyk, 1975, 1978a,b; Clemmensen and Surlyk 1976). This is to be expected because in Upper Jurassic times the Haltenbank area is believed to have been more distal to the sand source than the Norwegian and East Greenland coastal areas.

SEDIMENTARY ENVIRONMENT

The main sedimentary settings for the Jurassic formations are shown in Fig. 4. There is no pronouced change in the sedimentary environments at the boundary between the Triassic and Jurassic sediments. A cyclic sequence of silty mudstones and medium to fine sands interbedded with coaly shales and coals was laid down as coastal plain and fluviodeltaic top-set sediments.

The mudstones represent overbank sediments, while the sands were laid down in fluviatile channels and as intra-channel flood stage sands. Pedogenic features, rootlets and sphaerosiderite nodules, have been documented from conventional cores (Fig. 5). The coals consist largely of *in situ* plant material accumulated in intra-channel swamps. The high proportion of carbonaceous beds clearly demonstrates a change in climate from dry to more humid conditions when going from the Upper Triassic to the Lower Jurassic. The coaly formation has a transitional upper boundary and grades from clearly continental deposits into shallow marine silts and sands deposited close to a terrestrial source. The top of the coaly formation is normally picked at the first coal bed. However, the marine sands above are very rich in reworked plant debris.

Continental conditions seem to have prevailed for a longer period in the Haltenbanken area in Lower Jurassic times, compared with the North Sea. While the main transgression took place during the Hettangian in the North Sea, continental deposits of early Sinemurian age are seen offshore mid-Norway.

The upper cored part of the Lower Jurassic marine sandstone (Fm II) consists of a heavily bioturbated tidal flat—shallow marine sequence with sand bar deposits and frequent storm surge sands with mud clasts and pebble sized rock fragments. Traces of glauconite have been found, supporting the other criteria for marine origin of these sands.

The Lower Jurassic sandstone sequence is followed by a mudstone unit grading upwards into bioturbated silts and fine sands (Fig. 5). The lower part of this unit (Fm III) is believed to be a shelf mud deposit. The more coarse grained upper part of this predominantly shaly unit is thought to be caused by a progressively shallowing water depth. Well-sorted glauconite sands at the top of the formation reveal a marine setting.

At the very top of the formation there is a tight bed with carbonate cement and pelecypod shell fragments. This is a hardground representing the base of a thick, persistent prograding, massive sand unit of Toarcian—Bajocian, possibly Bathonian, age (Fm IV).

The Middle—Lower Jurassic sandstone is believed to be the result of a major regressive phase with shallow marine coastal sands prograding a stable shelf platform. The paralic nature of the sediment is seen from the high content of coal debris in the formation, particularly in the middle part. The importance of wave action for the sediment texture is recognized by a well washed and sorted upper part. The Middle—Lower Jurassic sandstone terminates against a shale unit at the top. The boundary between these two units is a major unconformity, most of the Bathonian and the Callovian being absent in the area. The presence of a major mid-Jurassic unconformity is consistent with what is experienced from the northern North Sea sedimentary basin (Vail and Todd, 1981).

The Upper Jurassic shale sequence is a transgressive unit of suspension load material laid down in an offshore setting with moderate terrestrial influence. The uppermost part of the Upper Jurassic consists of black 'hot' shales of high organic content typical of an anoxic marine environment.

PETROGRAPHY AND DIAGENESIS

The sands of Fm I can be classified as feldspathic to quartz arenites, frequently very micaceous. The feldspar content is significantly higher in the fine overbank sands than in the medium-grained channel sands. Both the fine and medium sands have angular to subrounded quartz grains showing traces of overgrowths which are subsequently etched. Orthoclase and microcline feldspar show traces of clean overgrowths. Muscovites are partly expanded and illitized while biotites are expanded and sideritized. In the channel sands, detrital clay is very sparse but the pore space is filled partly with a well developed vermicular authigenic kaolinite lattice. The sands contain occasional rock fragments of polycrystalline quartz and granite which gives the sediment a rather immature composition. A granitic (Fiskenes granite) source for Jurassic sediments is also seen onshore Norway (Dypvik, 1979). The heavy minerals are dominated by garnet, but minor

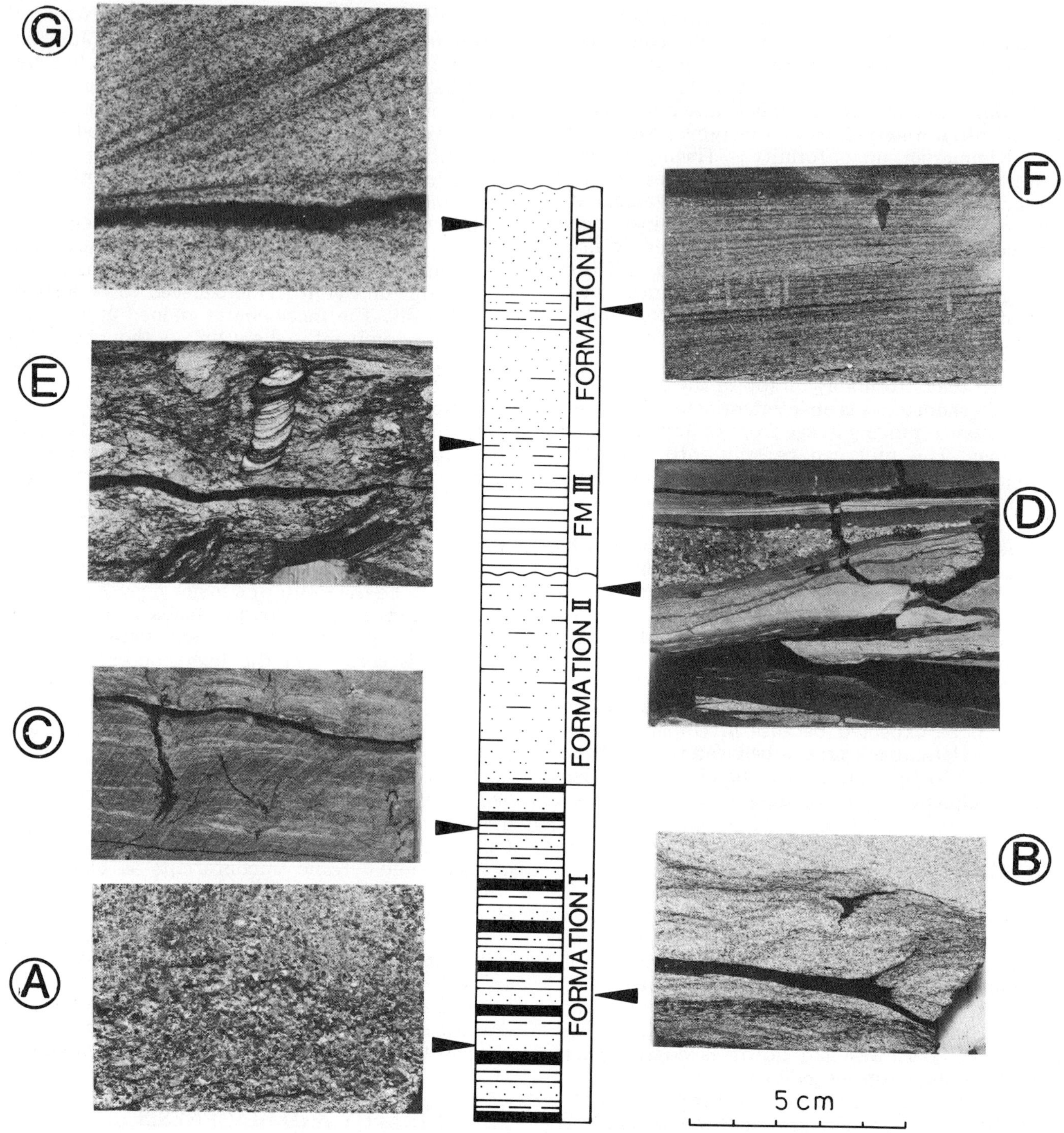

Fig. 5. Conventional cores from Jurassic offshore mid-Norway. (A) Fluviatile channel sand; near base of channel. (B) Fluviatile channel sand; near top of channel. Notice fossil plant remains. (C) Overbank silt with rootlets. (D) Wavy and lenticular bedded shallow marine tidal flat deposit. (E) Bioturbated clay and fine sand. (F) Mica sand. (G) Shallow marine massive and cross-bedded clean sand.

amounts of tourmaline and zircon are also observed. Traces of pyrite are seen as framboids associated with organic detritus. The sands contain some tight streaks. These are cemented by ferroan calcite which enclose all other diagenetic features.

In the wells examined, the coal has a dark brown streak indicating brown coal maturation stage. The coal grades into brown to dark grey, slightly laminated overbank claystones/siltstones.

The overlying transgressive sand formation (Fm II) consists of fine sands and silts with minor medium and coarse sand. The sands, particularly in the upper part, contain significant amounts of detrital clay and could be classified as wackes. The quartz grains are occasionally slightly overgrown but are later normally heavily etched by alkaline pore fluids. The K-feldspars demonstrate well-developed euhedral overgrowth while plagioclase is very rare. The muscovite is only slightly altered while the biotite is expanded and replaced by microgranular siderite. The detrital clay is partly recrystallized to illite and partly altered to siderite. Kaolinite is common and a loosely packed network partly fills the pore space. Pyrite is present through-

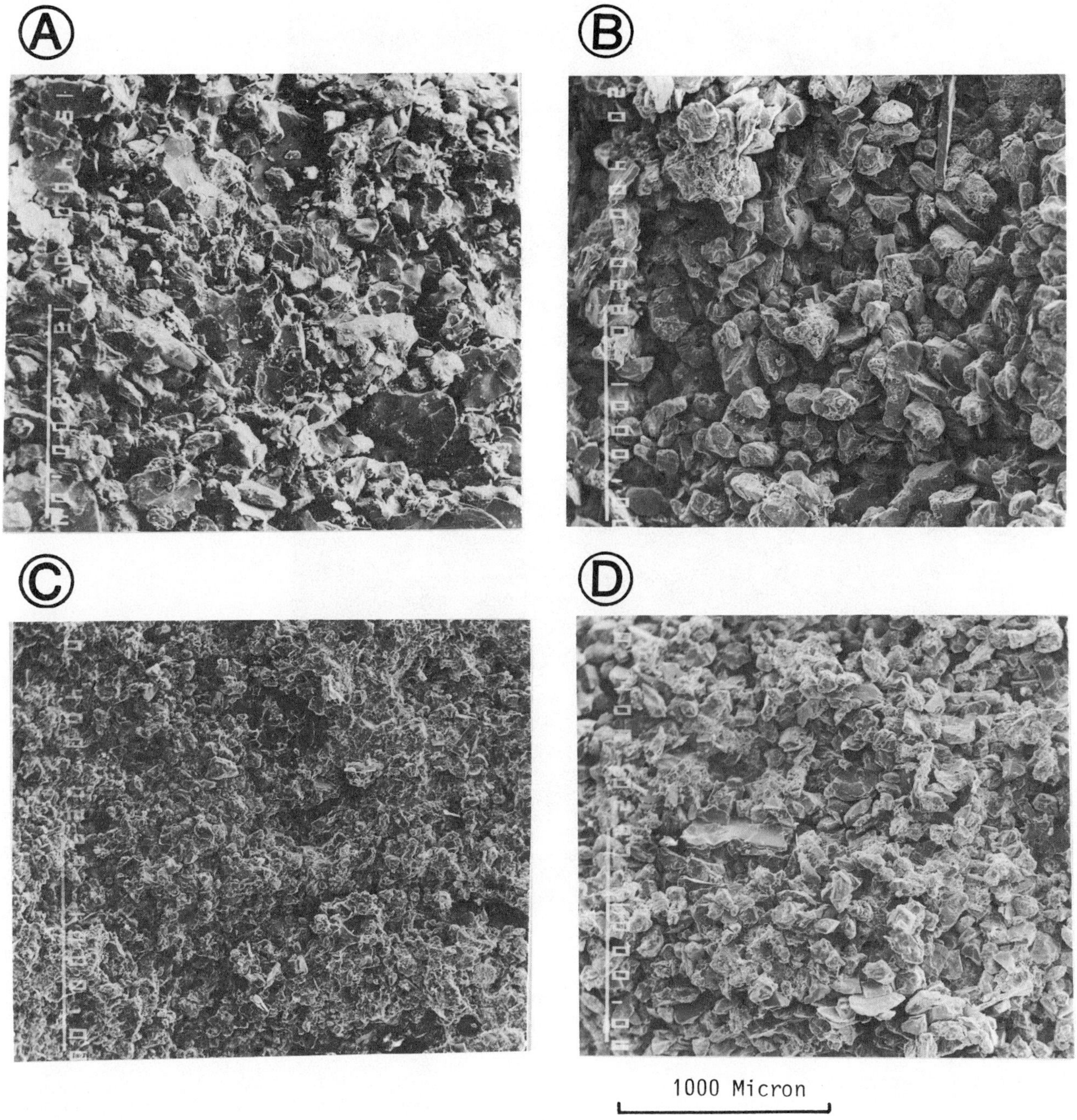

Fig. 6. Scanning electron micrographs of core samples. (A) Mica sand of poor reservoir quality from upper part of reservoir interval 4. (B) 'Clean' sand of excellent reservoir quality from reservoir interval 5. (C) 'Dirty' sand of poor reservoir quality from reservoir interval 3. (D) Sandstone of good reservoir quality of reservoir interval 4.

out the formation and is normally associated with the organic detritus. Traces of glauconite have been recorded.

The open marine deposit resting on top of the Lower Jurassic sands is a light grey, firm, predominantly non-calcareous sediment with claystones at the base and interlaminated silts and fine sands at the top. The fine sands are feldspathic, glauconitic and slightly micaceous. They are moderately to well sorted with some biogenically admixed detrital clay. The sand grains have normally tangential contacts, but matrix-supported grains occur. At the top of the formation there is a tight horizon that can be correlated between wells. This unit has a siderite coating on all grains while the remaining pore space is filled with sparry ferroan calcite.

The sands of Fm IV are generally medium grained but fine and coarse beds occur. The sorting is good with little detrital clay and silt (Fig. 6). The sands are quartz arenites except for the fine sands which are feldspathic.

Quartz overgrowth is uncommon and considerable solution etching has formed secondary porosity (Fig. 7). The feldspars consist of orthoclase and microcline with subordinate plagioclase. The K-feldspars show well-developed overgrowths, while the plagioclase is typically chemically weathered. In general, the fine sands are rich in mica. Some beds are extremely micaceous and appear, from a superficial visual inspection, to consist almost entirely of mica. The muscovite ends are often splayed into kaolinite. The biotites are generally more altered and contain aggregates of granular

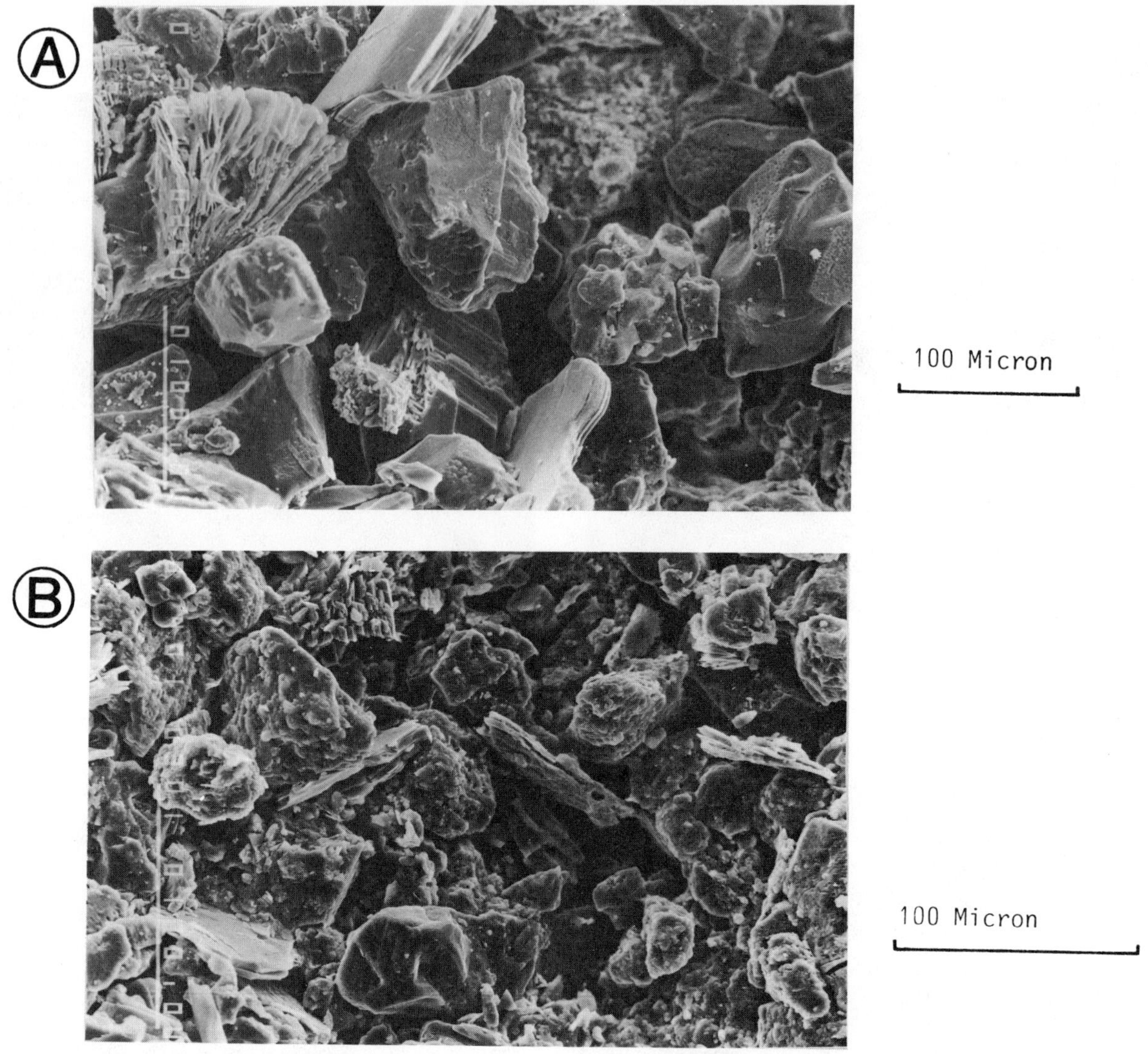

Fig. 7. Scanning electron micrographs from core samples. (A) Formation IV. Notice expansion of biotite, no expansion of muscovite, quartz etching and feldspar overgrowths. Almost no grain cementation occurs. (B) Formation III. Notice high content of detrital clay.

siderite with pyrite cubes. Kaolinite is found as cement, but is less common than in the underlying sands. Pyrite is commonly recorded, together with organic material which is found throughout the unit.

The Upper Jurassic contains shales only. The shale section can be split into two lithologies. The lower is a medium grey claystone while the upper is a black, subfissile to fissile 'hot' shale section. The lower (Fm V) contains some laminae of fine sand with pore spaces filled with detrital clay. The lithology of these fine sands is in general the same as that for the underlying sandstones. Tight carbonate bands, 1—3-m thick, are seen at irregular intervals throughout the formation.

The uppermost section (Fm VI) of black shales can be identified from its high uranium and organic carbon content which gives high radioactivities and low densities to the sediment. This hot shale is an excellent source rock which would yield predominantly oil when sufficient maturation is reached. XRD data indicate that the clay sized fraction of the hot shale consists entirely of smectites.

RESERVOIR QUALITY

Exploration drilling offshore mid-Norway so far has proved that the Jurassic sequence is of primary interest for finding hydrocarbon accumulations. The Jurassic demonstrates great variations in depositional environment as well as reservoir qualities. Both variation in grain size and sorting as well as diagenetic processes have affected the reservoir properties of the different stratigraphic units.

The Jurassic succession can be subdivided into six intervals related to reservoir properties (Fig. 4). These intervals are closely connected to the depositional environment and to the lithostratigraphy.

Interval 1. This interval corresponds to Fm I. Most of the sandstones are medium-grained and well-sorted channel sands giving good primary porosity. However, the permeabilities are only moderate because of a high content of authigenic pore-filling kaolinite. Some porosity has been produced by quartz etching, while feldspar overgrowth inhibits the reservoir quality improvement caused by this effect. Tight ferroan calcite cemented streaks occur throughout the formation. The reservoir quality of this interval is ranked as generally poor, because of a fairly low sand content (N/G less than 0.5) as well as expected lateral discontinuity of the channel sands,

Interval 2. This corresponds to Fm II and is a predominantly fine-grained sandstone with some medium to coarse, occasionally pebbly, layers. The original porosity and permeability were probably relatively poor because of a high proportion of biogenically admixed detrital clay. Permeabilities, in particular, are low because of the poor sorting. As seen in the other Jurassic sands, quartz etching and feldspar overgrowth is typical. The kaolinite content seems lower than that seen in interval 1. The reservoir quality of the interval varies from poor to good depending upon the amount of detrital clay present. The sand has been encountered in all wells and seems to be a continuous sand sheet in the Haltenbanken area.

Interval 3. Corresponding to Fm III, this is a mudstone sequence of non-reservoir quality except for the uppermost part. This part has bioturbated silts and fine sands. The permeabilities are generally very low and it is questionable whether hydrocarbons could be produced at commercial rates from this interval.

Interval 4 and 5. Corresponding to Fm IV, this is a shallow marine coastal sand with a generally fine-grained micaceous lower part, a silty/shaly middle part and a medium to coarse grained upper part. The clean sandstone at the top of the formation is labelled interval 5. Almost no grain cementation is developed in the wells studied and loose sand is recovered when coring. In contrast to the Lower Jurassic sand, the clastic clay content is low because of marine reworking, presumably by wave action. Because of good sorting and minor cementation, both porosities and permeabilities are good to excellent in the wells examined (see Fig. 6).

The primary porosity has been increased by quartz etching. This effect is to some extent balanced by well developed overgrowths on potassium feldspars. Some kaolin cement is present, but does not have severe effects on reservoir properties. Apart from a few ferroan carbonate cemented streaks and the mica-rich silty/shaly unit at the top of interval 4, this reservoir section is classified as good to excellent where tested. Marine reworking, little grain cementation and the expected large lateral extent of this sandstone, presumably will attract explorationists to continue their efforts to find petroleum reserves in this formation.

Interval 6. This consists of Upper Jurassic shales, silts and carbonates of non-reservoir properties and forms the cap rock in the area. Sands have not been identified in this interval.

CONCLUSIONS

The area offshore mid-Norway is in an early stage of exploration, but present drilling proves gas and condensate to be present. Three discoveries have been made, all in sandstones of Jurassic age. The Jurassic rocks are believed to form the major potential traps and are regarded as being of primary importance for further discoveries. The following conclusions have been reached about the Jurassic sediments offshore mid-Norway.

(1) The Jurassic sequence consists of sand-prone Middle and Lower Jurassic and a shale-prone Upper Jurassic, separated by a major Bathonian—Callovian unconformity.

(2) The sedimentary environment is varied with coastal plain sediments consisting of channels sands, overbank shales and coals at the Triassic—Jurassic boundary. This primarily continental facies was transgressed by shallow marine sands during the Sinemurian—Pliensbachian, which was followed by offshore clay deposition during the early Toarcian. A major regressive cycle took place during the late Toarcian—Bajocian, with nearshore reworked marine sands prograding the underlying shelf sediment. The Upper Jurassic is characterized by transgressive conditions with shale deposition. The uppermost part of the Jurassic consists of organic-rich black shales laid down under anoxic conditions.

(3) The petrography of the Jurassic sands reveals fairly immature mineral assemblages with a high content of potassium feldspars and biotite.

(4) The diagenetic features include well developed K-feldspar overgrowth, weathering of plagioclase, early quartz overgrowth with subsequent etching, mica expansion, recrystallization of detrital clay and biotite to siderite, and finally a late-stage ferroan calcite cementation.

(5) None of the diagenetic processes have been detrimental to the reservoir quality of the sands. Detrital clay and kaolinite cement reduce permeability in the Lower Jurassic sands, while the Middle Jurassic sands have excellent porosity—permeability characteristics because of marine reworking of the sands as well as minor grain cementation.

ACKNOWLEDGEMENTS

The author wishes to thank Saga Petroleum, Statoil, Norsk Hydro, Elf, Shell, Arco, Agip, Deminex and Volvo Petroleum for their support to the present study. I also acknowledge Robertson Research International for having done a fine job on a petrographic study on contract for Saga for the wells 6507/11-1 and 6507/12-2. The biostratigraphy presented is a synthesis of the reports from the consulting companies IKU and Paleoservices.

Particularly, I wish to thank the staff of the 'Seksjon N62' in Saga Petroleum for fruitful discussions and critical comments. Thanks are due especially to Terje Hagevang and Arvid Elvsborg for their professional contribution to these discussions. Charlotte Robinson kindly corrected the English manuscript.

REFERENCES

Birkelund, T. 1975. A review of the Jurassic of East-Greenland. In: *Jurassic Northern North Sea Symposium* (Norsk Petroleumsforening), pp. 1—27.

Clemmensen, L.B. 1976. Tidally influenced deltaic sequences from the Kap Stewart Formation (Rhaetic—Liassic), Scoresby Land, East Greenland. *Bull. Geol. Soc. Denmark*, **25**, 1—13.

Clemmensen, L.B. and Surlyk, F. 1976. Upper Jurassic coal bearing shoreline deposits, Hochstetter Forland, East Greenland. *Sediment. Geol.*, **15**, 193—211.

Dalland, A. 1975. The Mesozoic rocks of Andøy, Northern

Norway. *Norges Geol. Unders.,* **316**, 271—287.

Deegan, C.E. and Scull, B.J. 1977. *A Standard Lithostratigraphic Nomenclature for the Central and Northern North Sea.* H.M.S.O., London.

Dypvik, H. 1979. Mineralogy and geochemistry of the Mesozoic sediments of Andøya, northern Norway. *Sediment. Geol.,* 24, 45—67.

Jørgensen, F. and Navrestad T. 1981. The geology of the Norwegian shelf between 62°N and the Lofoten Islands. In: L.V. Illing and G.D. Hobson (eds.). *Petroleum Geology of the Continental Shelf of North-west Europe,* Heyden, London, 407—413.

Karlsson, W. and Dypvik H. 1981. *Corestudy, well 6507/12-1.* Saga Petroleum. Confidential Internal Report.

Raunsgaard-Pedersen, K. and Lund, J.J. 1980. Palynology of the plant-bearing Rhaetian to Hettangian Kap Stewart Formation, Scoresby Land, East Greenland. *Rev. Paleobot. Palynol.,* **31**, 1—69.

Surlyk, F. 1975. Block faulting and associated marine sedimentation at the Jurassic—Cretaceous boundary, East Greenland. In: *Jurassic Northern North Sea Symposium* (Norsk Petroleumsforening) pp. 1—31.

Surlyk, F. 1977. Stratigraphy, tectonics and palaeogeography of the Jurassic sediments of the areas north of Kong Oscars Fjord. East Greenland. *Bull. Grønlands Geol. Unders.,* **123**, 56pp.

Surlyk, F. 1978a. Submarine fan sedimentation along fault scarps on tilted fault blocks (Jurassic—Cretaceous boundary, East Greenland). *Bull. Grønlands Geol. Unders.,* **128**, 108pp.

Surlyk, F. 1978b. Jurassic basin evolution of East Greenland. *Nature, Lond.,* **274**, 130—133.

Surlyk, F. *et al.,* 1973a. Stratigraphy of the Jurassic—Lower Cretaceous sediments of Jameson Land and Scoresby Land, East Greenland. *Bull. Grønlands Geol. Unders.,* **105**, 76pp.

Surlyk, F., Callomon, J.H., Bromley, R.G. and Birkelund, T. 1973b. Stratigraphy of the Jurassic—Lower Cretaceous sediments of Jameson Land and Scoresby Land, East Greenland. *Bull. Grønland Geol. Unders.,* **105**, 1—75.

Talwani, M. and Eldholm O. 1977. Evolution of the Norwegian-Greenland Sea. *Bull. Geol. Soc. Am.* **88**, 969—999.

Sykes, R.M. 1974. Sedimentological studies in Southern Jameson Land East Greenland. II. Offshore estuarine regressive sequences in the Neill Klinter Formation (Pliensbachian—Toarcian). *Bull. Geol. Soc. Denmark,* **23**, 213—224.

Vail, P.R. and Todd, R.G. 1981. Northern North Sea Jurassic unconformities chronostratigraphy and sea-level changes from seismic stratigraphy. In: L.V. Illing and G.D. Hobson (eds.). *Petroleum Geology of the Continental Shelf of North-West Europe.* Heyden, London, pp. 216—235.

30 Tectonic inhomogeneities of the Kristiansund–Bodø Fault Complex, offshore mid-Norway

Roy H. Gabrielsen and Charlotte Robinson
Saga Petroleum AS, Høvik, Norway

The continental shelf offshore mid-Norway is dominated by a system of NE to ENE- and N-trending structures for which the name 'Kristiansund-Bodø Fault Complex' is proposed. On this system a less conspicuous regional NW structural trend is superimposed. The directional trends of these fracture systems find their counterparts in parallel structural trends onshore mid-Norway. Comparison of the different segments of the Kristiansund-Bodø Fault Complex reveals a complicated structural history, being different for the various parts of the fault complex. This is believed to be a consequence not only of difference in timing of tectonic events, but also of differences in stress situation for the segments.

Although the area was subject to faulting in pre-Jurassic times, the main deformation took place through middle to late Jurassic, with a major phase of subsidence of the adjacent basins in the Cretaceous. However, there are significant differences between the Træna and Haltenbank segments, the most conspicuous being strike-slip movements which have affected the northern (Træna) part. This system shows signs of Tertiary reactivation, while Haltenbanken was subject to basinwards gravity sliding at this time.

INTRODUCTION

The continental shelf off mid-Norway may be roughly divided into three megatectonic units: a landward platform area (the Trøndelag Platform), which on the oceanward side is bordered by large sedimentary basins (the Træn, Vøring and Møre Basins), and further west a complex transition zone between oceanic and continental crust. These mega-units are separated by two tectonized zones, namely the Vøring Plateau Escarpment and the Kristiansund-Bodø Fault Complex (Fig. 1) The main structural elements have been described in several papers (Grønlie and Ramberg, 1970; Talwani and Eldholm, 1972; Rønnevik *et al.*, 1975, 1979; Rønnevik and Navrestad, 1976, 1977; Jørgensen and Navrestad 1979, 1981a, b; Hinz *et al.*, 1982; Gabrielsen *et al.*, 1984; Price and Rattey, 1984).

THE KRISTIANSUND–BODØ FAULT COMPLEX

Rønnevik *et al.* (1975) recognized that a ridge-like feature, which they termed the Nordland Ridge, separated the Helgeland Basin, situated within the Trøndelag Platform, from the Vøring Basin to the west. Their map (Rønnevik *et al.*, 1975, fig. 10), however, indicated that the ridge may be followed southwards to about 63°20′N, thus also separating the southern part of the Trøndelag Platform from the Møre Basin. Further mapping in the area has confirmed that the Trøndelag Platform is bordered by a more or less continuous marginal high on its oceanward side, and that the high has been subject to erosion or non-deposition during Jurassic times (Price and Rattey, 1984).

The high is fault bounded on both sides, and is obviously a complex structural feature. Considering this, Gabrielsen *et al.* (1984) proposed to name the structure, which may be followed all the way from 63° to 65°20′N, the Kristiansund-Bodø Fault Complex. The fault complex may be subdivided into several segments, characterized by structural trend, fault geometry and structural history.

The northernmost segment, which is identical to the original definition of the Nordland Ridge (Rønnevik *et al.*, 1975) is found approximately between 65°20′N, 7°E and 66°50′N, 12°E (Fig. 2). The segment trends NE, and separates the Helgeland and Træn Basins. The Trænabank concession area is situated within this segment. The NE- to ENE-trending faults of this part of the fault complex have been associated with Cretaceous dextral

Petroleum Geology of the North European Margin © Norwegian Petroleum Society (Graham & Trotman, 1984) pp. 397—406.

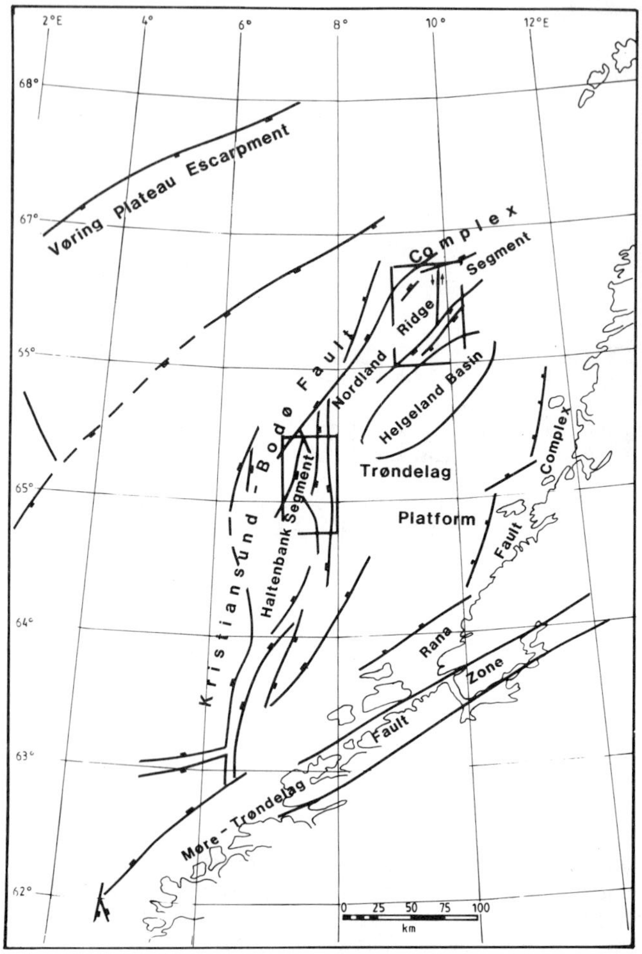

Fig. 1. Key map off shore mid-Norway. Black frames indicate the Trænabank (northern) and Haltenbank (southern) licence areas.

transcurrent movements (Price and Rattey, 1984). South of 65°20′N, the fault complex is dominated by a NNE-trend. This segment, which continues southwards, west of the Frøya High (Gabrielsen *et al.*, 1984) terminates close to the Møre coast, where it runs into the NE-trending Møre-Trøndelag Fault Zone.

Tectonic history of the Nordland Ridge segment

As stated by Rønnevik and Navrestad (1977), the Nordland Ridge changes character along its length. Its northern part has an ENE-trend, changing to NE west of 9°E (Fig. 2). Also, the structure is asymmetric, the western flank having the larger throw (Fig. 3). The stratigraphically oldest part of the structure is situated at the western edge, which is also associated with the maximum of a positive gravity anomaly (Talwani and Eldholm, 1972; Rønnevik *et al.*, 1975; Hollander, 1982). The Trænabank area is presently undrilled, and accordingly, dating of reflectors is uncertain. However, correlations to the Haltenbank area (Bleie *et al.*, 1982) give some control on the timing of the events. A schematic presentation of the geological history of the Nordland Ridge in the Trænabank area is shown in Fig. 4.

(?)Pre-Lower Triassic

A deep horst- and graben-type topography is seen in some of the sections. This fault system tentatively may be correlated to the Carboniferous–Permian tensional regime known to have affected large parts of Europe (e.g. Whiteman *et al.*, 1975; P.A. Ziegler, 1975, 1978, 1981, 1982; W.H. Ziegler, 1975; Eynon, 1981). This system was reactivated only to a minor degree in connection with later movements in the area.

Triassic

The relatively constant thickness of the assumed Triassic sequence indicates that the whole area acted as one sedimentary basin at this time. Only local thickness variations associated with a few main faults demonstrate that tectonic activity took place (Fig. 4.1). Similar movements are indicated in the Haltenbank area (see below) and in Greenland (Surlyk, 1977).

Early Jurassic

The fault activity continued into early Jurassic, with slight movements along the established faults. A series of small graben structures, which seem to be characterized by fast subsidence, probably mirrors an early Kimmerian tensional pulse (Fig. 4.2). Except for this, the entire area was a relatively quiet depositional basin (see also Jørgensen and Navrestad, 1981a).

Middle and early late Jurassic

This tectonic situation continued into middle Jurassic times with, however, a slight increase in activity along the major faults. The deformation corresponds to the main Kimmerian tectonic phase in the North Sea, and is, in the Nordland Ridge area, characterized by a brief subsidence of the Helgeland Basin and Træn Basin relative to the Nordland Ridge. As a consequence of the enlarged relief and maybe increasing tension, the fault activity spread into the Nordland Ridge (Fig. 4.3).

Late Jurassic

Movements within the established normal fault system continued into late Jurassic (Fig. 4.4). More important, however, was the establishment of a N-trending transpressional, probably sinistral, fault system dividing the Nordland Ridge into two segments. The movements along this shear system must have been significant, and probably affected the topography of the ridge. Although strike-slip faults are hard to detect on seismic sections, the following indications are seen: (1) Topographically, the Nordland Ridge shows a distinct sinistral off-set at base Cretaceous level. (2) The truncation line for the base Jurassic reflector shows the same type of off-set. (3) Local folding indicates transpressive type of movement (Fig. 5). (4) Reverse, steeply dipping faults are found at base Cretaceous and Palaeocene levels. (5) Thickness variations over the fault are indicated in the deeper parts of the sedimentary pile.

Deformation in the upper part of the section gives a Palaeocene minimum age. This deformation is moderate considering the relatively large-scale movements indicated from the off-set truncation line (in the order of some kilometres). Further, the fold structure associated with the shear zone is cut by the main Kimmerian unconformity (Fig. 4.4), suggesting that the shear movements were initiated

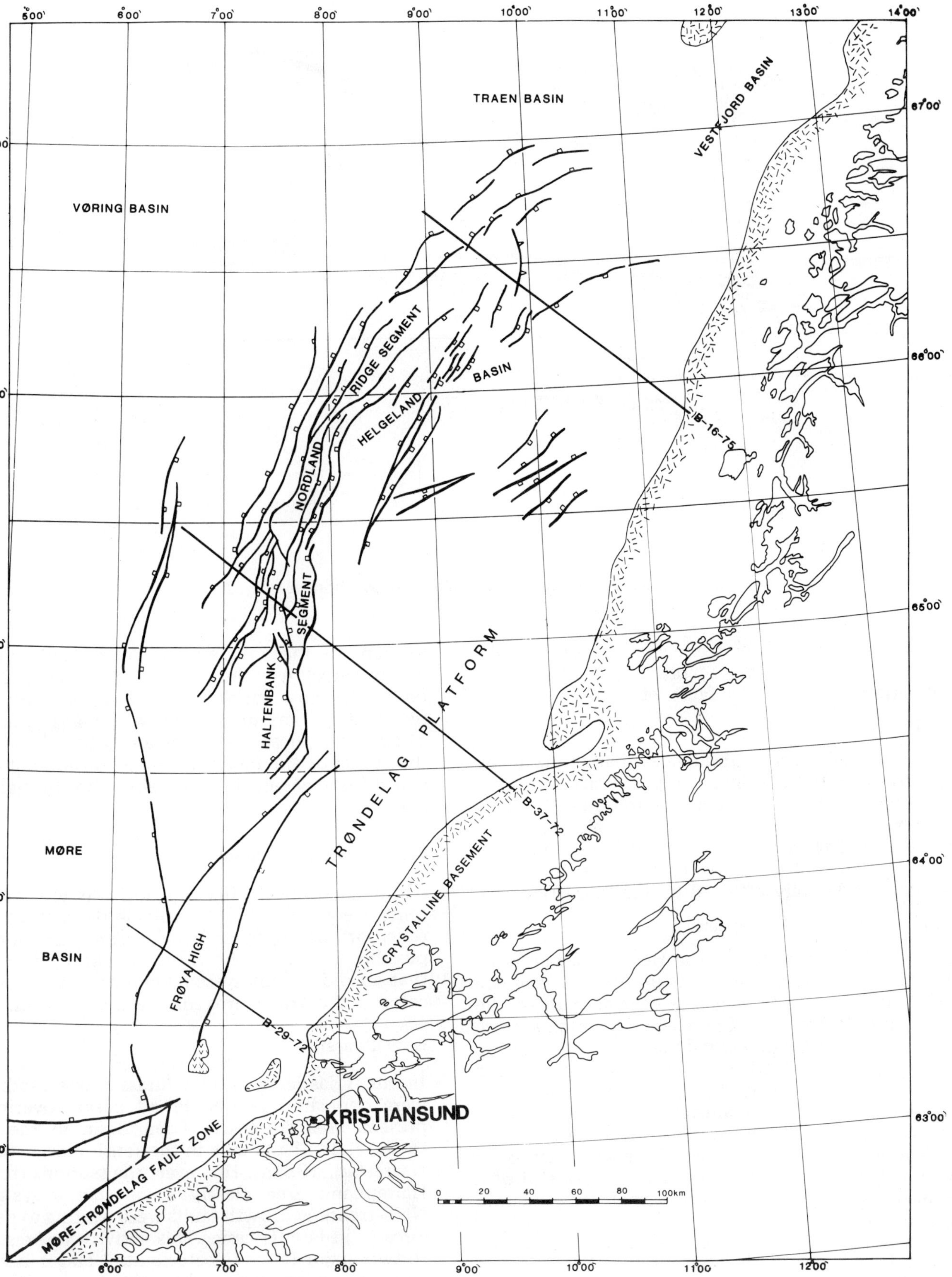

Fig. 2. The Kristiansund-Bodø Fault Complex with location of profiles of Fig. 3.

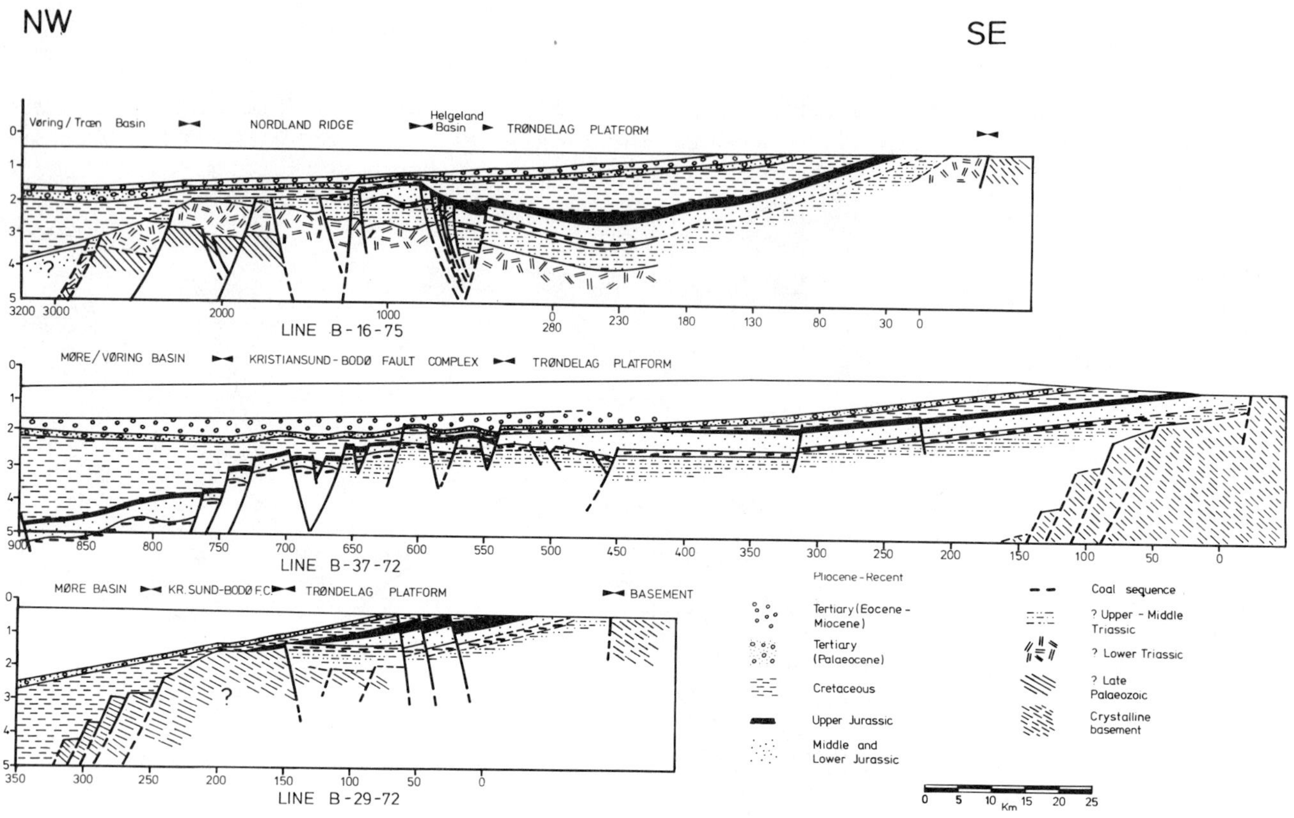

Fig. 3. Profiles across the fault complex. For locations see Fig. 2.

in late Jurassic, or even somewhat earlier. The offset at Palaeocene level accordingly is due to reactivation of the Jurassic fault set.

Early Cretaceous

The regional basin subsidence characterized by flexuring continued into early Cretaceous times with development of a considerable relief, leaving the Nordland Ridge as a positive element. The main crestal fault sets of the ridge became more or less inactive, and the lower part of the Cretaceous is only moderately affected by faulting (Fig. 4.5).

Late Cretaceous—Palaeocene

By late Cretaceous times the Helgeland and Træn Basins were nearly filled with sediments, and the relief was moderate. On the central part of the Nordland Ridge, late Cretaceous sediments were deposited directly on early Jurassic or even Triassic rocks. The Nordland Ridge may be considered a stable tectonic element during this time, and minor movements over the main faults may be traced as minor faults or flexures at base Tertiary level.

The Nordland Ridge Shear Zone was reactivated in the Palaeocene. The movements were still of a transpressive type, and must have been moderate (see above).

The subsidence of the Helgeland, Træn and Møre Basins caused instabilities in the Cretaceous and Tertiary sequence, generating basinward listric faulting above the major fault zones similar to those seen in the Haltenbank segment.

Tectonic history of the Haltenbank segment

The Haltenbank area contains fault sets of various trends, status, and tectonic history, although this segment of the Kristiansund-Bodø Fault Complex has an overall well defined NNE-trend (Fig. 2). The fault sets show variance with respect to geometry (orientation, amount of throw, dip-angle) and age of activity. These variations are believed to reflect the status of the faults (detached or non-detached) and differences in the stress situation through time (Fig. 6).

Pre-Triassic to Triassic

Although deeper reflecting horizons are indicated quite frequently in the available data, mapping of reflectors below ‘near base Jurassic’ has not been possible. It would, however, be natural to assume that an older fault system, similar to the one indicated in the Træna area, has been developed.

Late Triassic

In late Triassic to Middle Jurassic times, sediments were deposited in one broad basin covering the present area from the Træn Basin in the north to the Haltenbank area in the south (Jørgensen and Navrestad, 1981a). However, the sections from the Haltenbank area clearly show that a system of NNE-trending growth faults were starting to develop already in late Triassic. This activity was restricted to two major faults (Masterfaults 1 and 2 in Fig. 6). These are deep westward-dipping faults which may reflect basement structures and typically have a total throw of 0.9—0.5-seconds two-way travel time at the near base Jurassic level. (The total throw also may incorporate movements later than Jurassic.) The horst-and-graben system of the Haltenbank area started to develop from these two masterfaults, which continued to act as the most

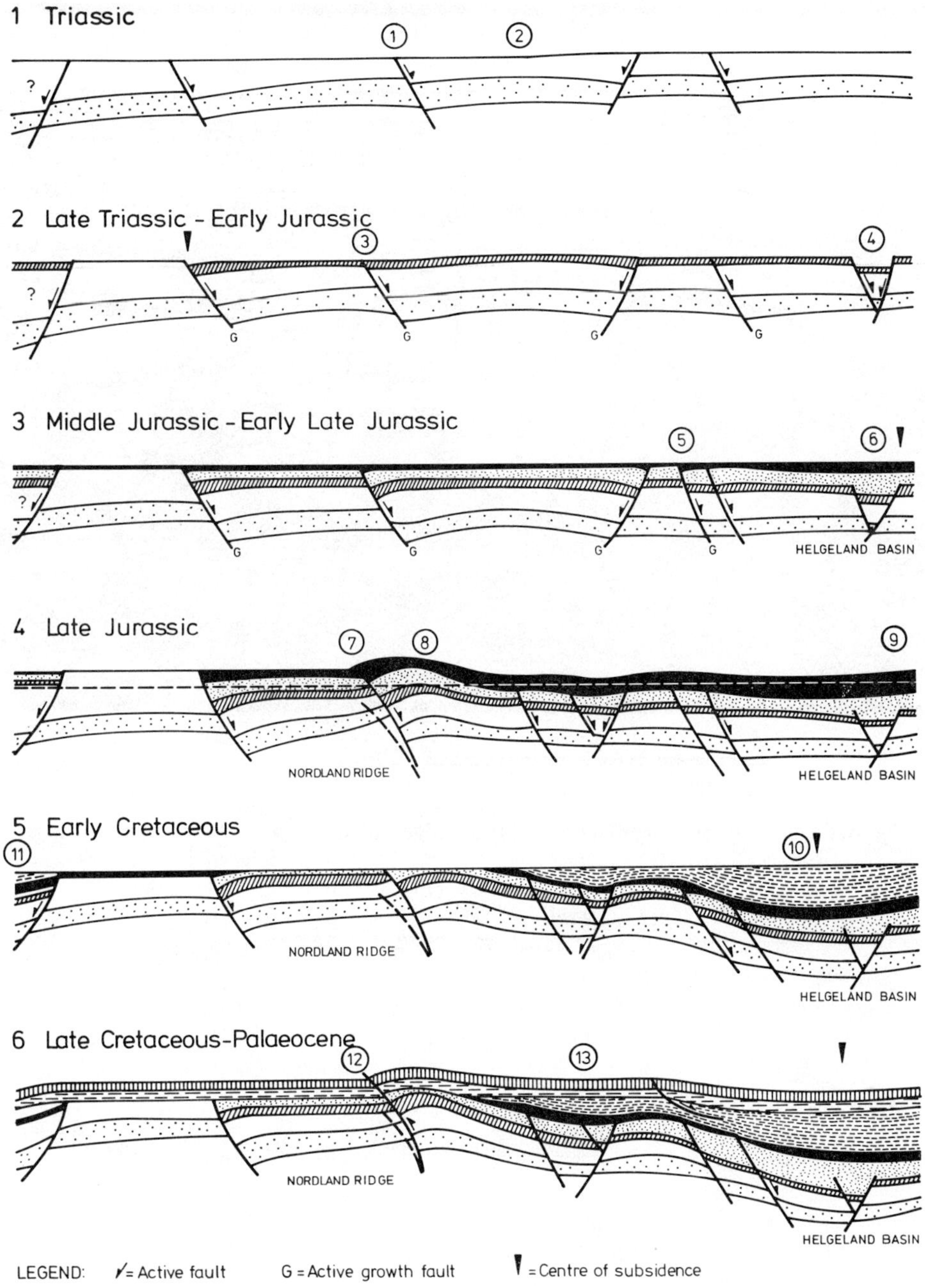

Fig. 4. Schematic tectonic development of the Nordland Ridge segment.

important structural elements through most of the tectonic history of the area.

A gentle dome structure with its crest situated at the central part of this future horst-and-graben system was developed, probably during late Triassic—early Jurassic times.

Early—middle Jurassic

As the faulting along the masterfaults continued, a new system of growth faults parallel to the masterfaults, but with an easterly dip, was developed (Fig. 6.2). This fault activity was initiated in early Jurassic, and resulted in the development of two small graben systems subsiding on each side of a stable horst. The two graben structures are characterized by local central doming and growth-fault activity along the margins. The doming mechanism is obscure, and may be related to either tectonic inversion or halokinesis. From the present data, neither of these models can be said to be preferred to the other, although the Lower Mesozoic sequence in the area is known to contain considerable amounts of salt.

Mechanically, these faults may be related to movements along the masterfaults (see above). Horsfield (1977) has shown that graben systems of this type may be generated in connection with basement faulting beneath a soft sedimentary cover. If this model is adopted, the geometry of the system would indicate basement faulting in the downward continuation of Masterfaults 1 and 2 where the basement fault planes are westward dipping at between 60° and 45°.

Late Jurassic

From late Jurassic the two small graben systems began to develop different deformation patterns. The central horst structure became less stable, and faulting took place along the shoulders, thus with a tendency to increase the width of the grabens. In

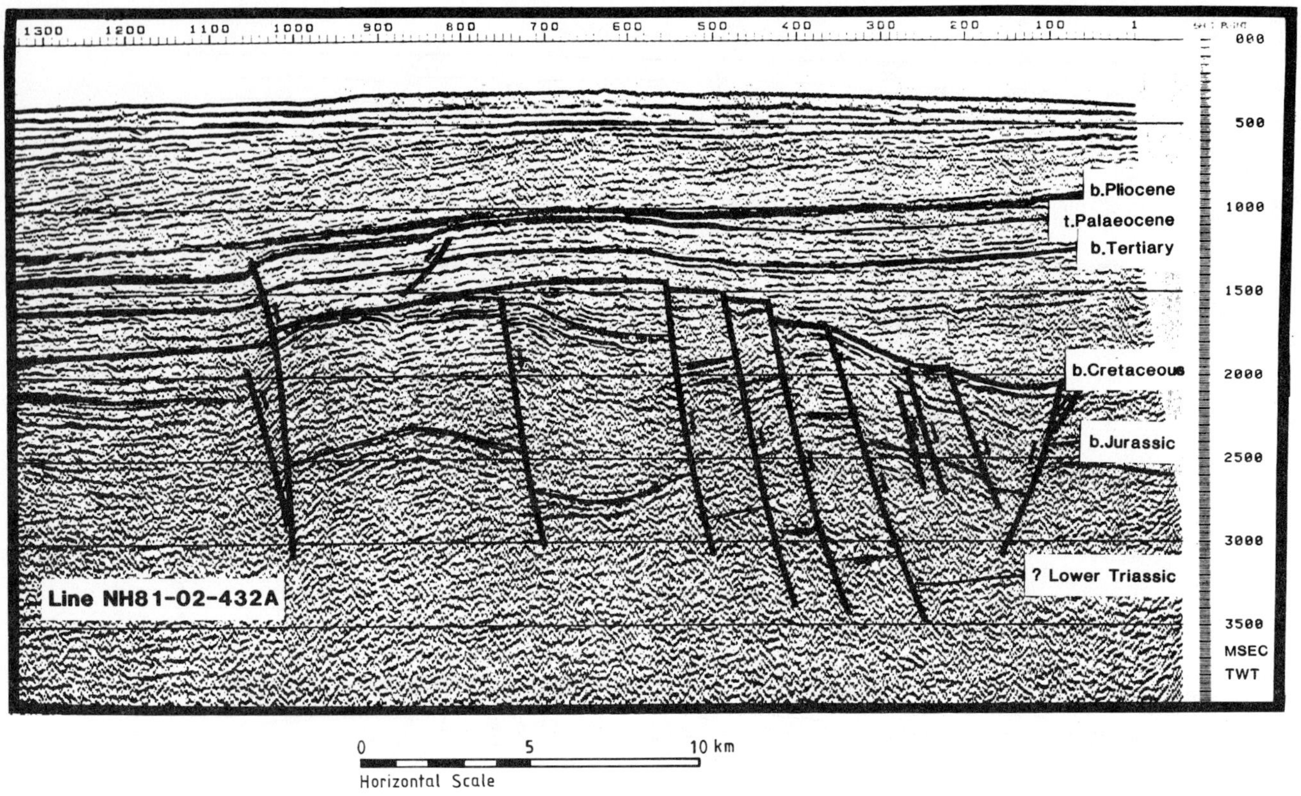

Fig. 5. Seismic section over the Nordland Ridge Shear Zone. Note reverse faulting.

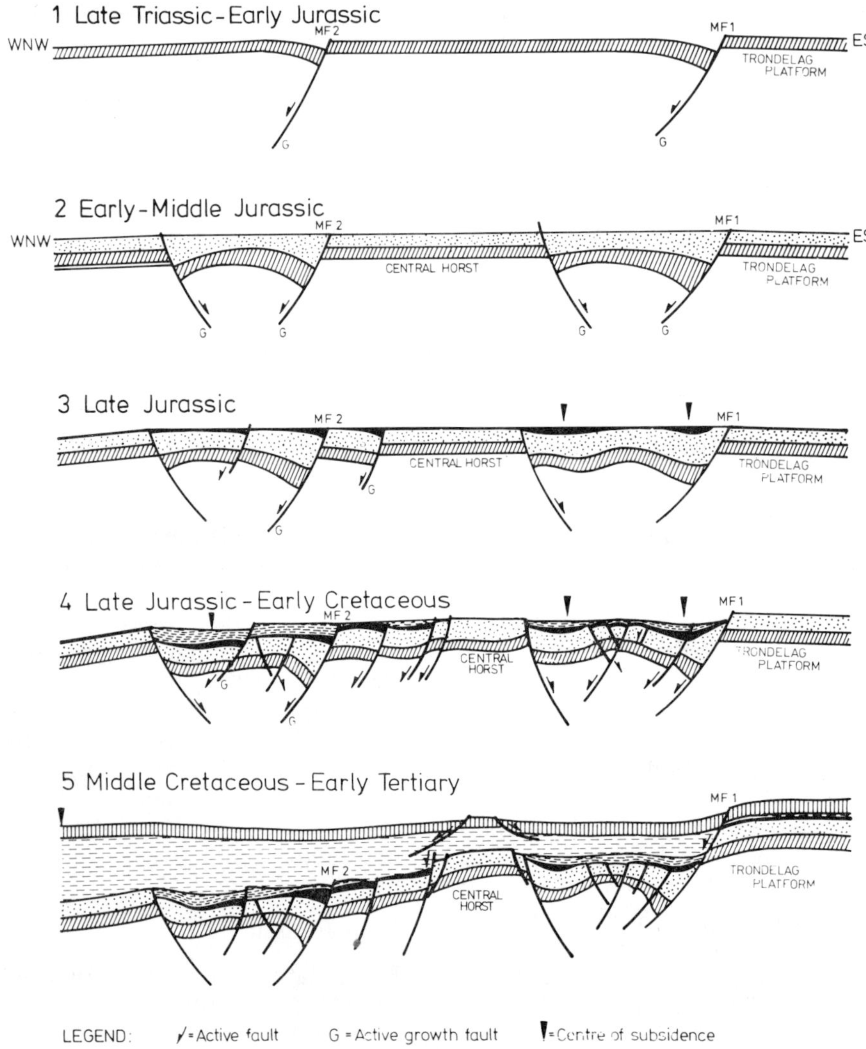

Fig. 6. Schematic tectonic development of the Haltenbank segment. MF 1 and MF 2 refer to Masterfaults 1 and 2 (see text).

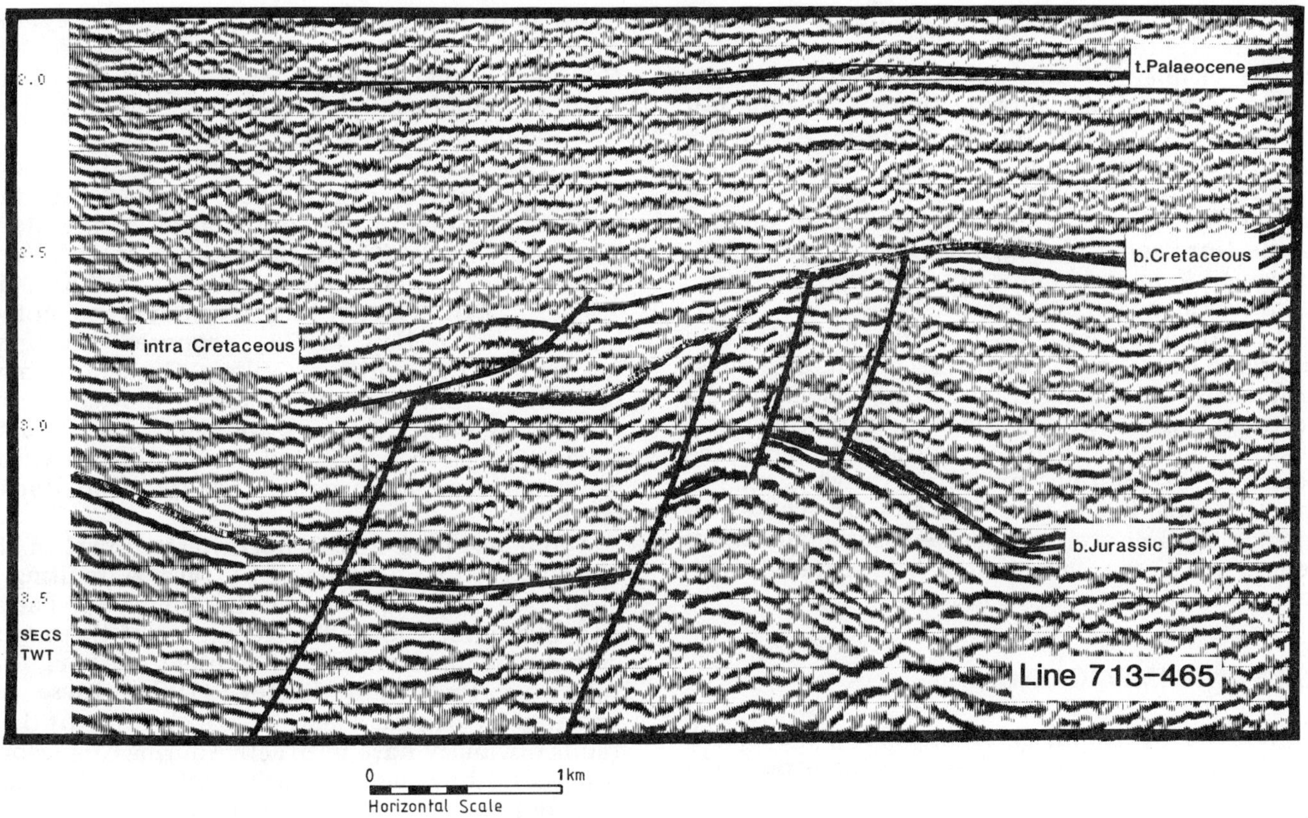

Fig. 7. Seismic section showing gravity sliding/listric faulting on intra-Cretaceous level.

the western graben, a collapse occurred with generation of an axial fault system, whereas faulting was less pronounced in the eastern graben system where subsidence resulted in generation of two minor sub-basins. The new active faults as well as some of the older, reactivated ones were generally characterized by growth faulting in the late Jurassic.

It should be noted that this 'double graben model' is valid for an area north of the northern part of block 6407/2 only. South of this area a large, stable platform-like structure is found. This platform, corresponding to the 'Halten Terrace' of Price and Rattey (1984), is bordered to the west by the western graben structure, and to the east by a step-like fault pattern up to the Trøndelag Platform.

Early Cretaceous

The movements along these faults continued in early Cretaceous times with generation of some new faults and extinction of some of the older ones. In general, the faulting affected increasing parts of the central horst structure. During this period, activity along most of the faults ceased, and movements were concentrated along a few major faults, namely the principal faults of the respective graben systems. A Lower Cretaceous sequence was deposited in the long-lived graben structures, and several of the major faults affect the Lower Cretaceous sequence.

Late Cretaceous—Tertiary

Although the bulk of the faults became inactive during the early Cretaceous, some fault activity persisted into the Tertiary. These movements may be ascribed to three types of activity:

(1) Activation of new structural elements. In the northern part of the eastern graben feature, a central horst was activated. This feature developed as a stable positive element in the region, and seems to have been so through most of the Cretaceous. The tectonic implication of this event is unclear, although the structure again emphasizes the difference between the eastern and western graben elements.

(2) Reactivation of older fault systems. Some of the major faults in the area show signs of activation also through most of Cretaceous times. The easternmost of the two masterfaults is the most conspicuous example of this process, which should be seen in connection with the general Cretaceous basin subsidence centred to the west of the study area.

(3) Gravity sliding. Minor faults are frequently seen to break the base Tertiary reflector in the area. These are usually situated above and on the upthrown side of the major faults, face into one of the local basins, and the fault planes are characterized by a gentle dip (Fig. 7). The geometry is characteristic of listric gravity faults. Accordingly, the pre-Cretaceous acted as a rigid basement relative to the younger sequence, giving rise to a new generation of 'basement detached' faults (Harding and Lowell, 1979) during this process.

THE SEDIMENTARY BASIN DEVELOPMENT

The patterns of subsidence and the activation of the different fault systems have, as already shown, affected the pattern of deposition of the sedimentary sequence in the Haltenbank area. During the late Triassic—early Jurassic, the entire area acted as one uniform sedimentary basin; however, with sediment accumulation along a few developing growth faults. The geographical position of an eventual regional depocentre is unknown.

In the Haltenbank area, sediments accumulated in the vicinity of the developing growth faults during early—middle Jurassic times, changing into graben systems during the late Jurassic. Within the graben systems, variance in the sedimentation pattern due to faulting and differential subsidence is seen. Subsidence of the Helgeland Basin relative to the Nordland Ridge segment started with slight movements in mid Jurassic, followed by more rapid subsidence in late Jurassic times.

Accordingly, from late Jurassic times the Nordland Ridge and Haltenbank segments developed differently. The transpressional movements in the northern parts activated the Nordland Ridge as a positive element, whereas subsidence and the beginning of rotation towards the west is seen in the Haltenbank area (Figs. 4 and 6). These differences became even more evident during the Cretaceous, until the entire area was covered with sediments in late Cretaceous times.

TECTONIC IMPLICATIONS

As noticed by Rønnevik *et al.* (1975), the overall development of the central and northern parts the Kristiansund-Bodø Fault Complex may show good similarities to the classical doming and rifting-type model, including a stretching episode (rift phase) and a subsequent episode of declining thermal subsidence and deposition of a more mature sediment blanket (e.g. McKenzie, 1978; Sengör and Burke, 1978). The major down-faulting of the Vøring and Møre Basins is believed to have taken place during Middle Cretaceous (Bott, 1978). This event has been associated with an early stage of continental rifting (Hinz and Moe, 1971; Hinz, 1972; Talwani and Eldholm, 1972), which accordingly predated the opening of the Norwegian Sea by approximately 55 Ma (Hinz *et al.*, 1982). However, considering the asymmetry of the Kristiansund-Bodø Fault Complex, and strong evidence of repeated shear movements in the Nordland Ridge, a more complicated series of events for the area might be suspected.

Rønnevik *et al.* (1975) suggested the possibility of shear movements having influenced the area, stating that "The Vøring and Møre Basins are connected by a tectonically complex zone where some compressional structures are indicated. This tectonism can be thought of as a result of movement along shear-fractures combining rift-axes" (pp. 124—125). Jørgensen and Navrestad (1981) suggested that the Nordland Ridge was subject to inversion at the end of the Jurassic. This may again point towards a compressional event at this time. On the other hand, Hinz *et al.* (1982) ascribed a series of Tertiary (middle Oligocene—late Miocene) highs west of the Nordland Ridge to either igneous intrusions, pre-Kimmerian evaporites or deep crustal movements.

Based on detailed mapping north and south of the Nordland Ridge, Price and Rattey (1984) proposed that this part of the Kristiansund-Bodø Fault Complex had been subject to dextral shear in Cretaceous times, and they found support for this theory from observations within the Møre-Trøndelag Fault Zone (Oftedal, 1972, 1975; Aanstad *et al.*, 1981; Aanstad, 1982). They also claimed that the shift in fault trend between the Nordland Ridge and Haltenbank segments of the Kristiansund-Møre Fault Complex resulted from rotation during this event.

Moving to the northern part of the Viking Graben, which is complicated by a NE-trending fault-set situated in the continuation of the Møre-Trøndelag Fault Zone, suggestions of strike-slip movements are not new (Halstead, 1975; W.H. Ziegler, 1975; Hay, 1978). It should be noted, however, that Hay (1978) proposed the fault pattern of the northern Viking Graben to have originated from sinistral strike-slip movements on the NE-trending faults in middle to late Jurassic times. This is in opposition to the dextral Cretaceous movements suggested by Price and Rattey (in press).

From mapping in the northwestern part of the Nordland Ridge, evidence of reverse faulting is clear (see also Gowers and Lunde, 1984; Larsen and Skarpnes, 1984). The geometry of these fault systems is unclear to the present authors, but following Price and Rattey (1984) these are initiated by dextral shear. Although some of these faults obviously have been active during Cretaceous, some also show signs of earlier activity. The strain ellipsoid associated with this type of movement is represented in Fig. 8. The most abundant faults developed a NE-oriented dextral shear system with synthetic, ENE-trending dextral structures. Also a set of NW-trending, sinistral faults would develop, although this antithetic set would be less abundant and rotate into a NNW-direction with increased deformation. Also, a series of folds with NE-striking axes would develop.

Can this simplistic model be fitted into the structural picture of the Nordland Ridge segment of the Kristiansund-Bodø Fault Complex? Figure 8 shows that geometrically, there is a fair degree of correlation between the strain diagram and structural trends of the region. The 'Røst Syncline' (Rønnevik *et al.*, in press) represents the NE-trending fold axes (Fig. 8(1)), the major fault sets of the Nordland Ridge itself the synthetic ENE-trending dextral faults (Fig. 8 (2)), and the NNW-trending sinistral shear zone crossing the Nordland Ridge has antithetic faults (Fig. 8(3)).

The major problem with this model seems to be the timing of the events in so far as the NNW-trending shear zone was initiated during late Jurassic (Fig. 4), whereas the major shear along the Nordland Ridge, according to Price and Rattey (1984), was a Cretaceous event. The NNW-trending sinistral strike-slip zone, however, was reactivated during late Cretaceous to Tertiary times, and two phases of movements are therefore established. The possibility of a late Jurassic initiation of the ENE-system should be held open, an argument that might be strengthened by the fact that middle to late Jurassic strike-slip movements have been proposed further south (Hay, 1978).

CONCLUSIONS

The asymmetry of the Kristiansund-Bodø Fault Complex and strong evidence of repeated shear

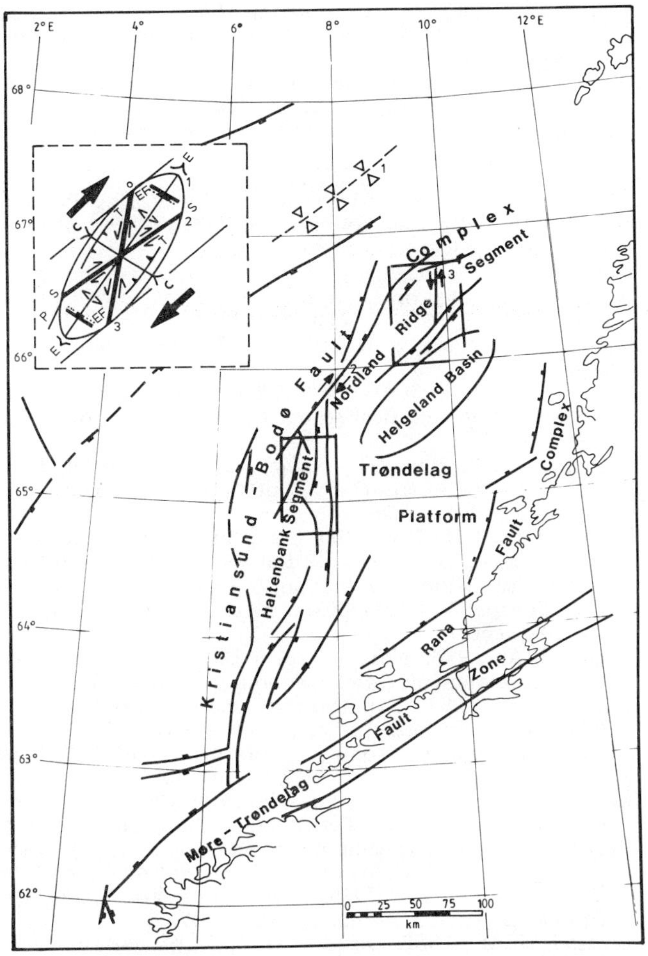

Fig. 8. Correlation between fault trends of the Kristiansund-Bodø Fault Complex and deformed strain ellipsoid for a dextral shear system. Numbers 1—3 refer to correlation of structural trends/features (see text).

movements in the Nordland Ridge indicate that the area has been subject to complex deformation. This impression is strengthened if comparisons are made between the different segments of the fault complex.

The area acted as a relatively uniform sedimentary basin in pre-Jurassic times (Rønnevik *et al.*, 1975; Hinz *et al.*, 1982). A few major faults were active, indicating that the nucleus of the Kristiansund-Bodø Fault Complex was already initiated.

In middle Jurassic times a slight subsidence east of the fault complex may be traced, contemporaneous with an increased activity within the fault complex itself. These events, which in time correspond to the main Kimmerian phase in the North Sea, introduced a longer period of instability in the Trøndelag Platform border zone.

From this time, a clear difference in tectonic development between the Nordland Ridge and Haltenbanken segments of the Kristiansund-Bodø Fault Complex became obvious. Whereas the Haltenbank segment was characterized by intense normal faulting and collapse, the Nordland Ridge segment was subject to inversion, strike-slip faulting and erosion.

In the Cretaceous, major basins in the area, namely the Møre and Vøring Basins to the west of the Kristiansund-Bodø Fault Complex and the Helgeland Basin east of the Nordland Ridge, subsided. Again this pattern resulted in different tectonic development along the Nordland Ridge and the Haltenbank segments, the latter rotating towards the major basins to the west, the former standing up as a positive element.

The differences between the segments are finally underlined by the reactivation of the Nordland Ridge Shear Zone in the Tertiary.

From the present state of knowledge, it is tempting to support the model of Price and Rattey (in press) indicating that considerable dextral shear took place in NE-trending fault zones in Cretaceous time. From our data, however, there are indications that these movements started in late Jurassic, and that the faults were activated in late Cretaceous to Tertiary.

Nevertheless, the model suffers from lack of more-detailed knowledge of the tectonic development in the vicinity of the Træna and Haltenbank licence areas, and a more comprehensive testing of the model must await further data from the region.

ACKNOWLEDGEMENTS

The authors wish to thank Saga Petroleum AS for giving us the opportunity to publish this work. The conclusions on the Haltenbank area are based on the interpretations of a number of people in Saga Petroleum AS. In particular we wish to thank Terje Hagevang and Werner Karlsson for discussions, and Benthe Englund, who typed the manuscript.

REFERENCES

Bleie, J., Oppebøen, K.A. and Nysæther, E. 1982. The hydrocarbon potential of the northern Norwegian shelf in the light of recent drilling. In: *The Geological Framework and Hydrocarbon Potential of Basins in Northern Seas. Offshore Northern Seas Conference* (Norsk Petroleumsforening), Article E/5.

Bott, M.H.P. 1978. The origin and development of the continental margins between the British Isles and south-eastern Greenland. In: D.R. Bowes and B.E. Leake (eds.), Crustal Evolution in Northwestern Britain and Adjacent Regions. Spec. Iss., *Geol. J.* **10**, 377—392.

Eynon, G. 1981. Basin development and sedimentation in the Middle Jurassic of the northern North Sea. In: L.V. Illing and G.D. Hobson (eds.), *Petroleum Geology of the Continental Shelf of North West Europe.* Heyden, London, pp. 196—204.

Gabrielsen, R.H., Færseth, R., Hamar, G. and Rønnevik, H. 1984. Nomenclature of the main structural features on the Norwegian Continental Shelf north of 62nd parallel. This volume, pp. 41—60.

Gowers, M. and Lunde, G. 1984. The geological history of Trænabanken. This volume, pp. 237—252.

Grønlie, G. and Ramberg, I.B. 1970. Gravity indications of deep sedimentary basins below the Norwegian continental shelf and the Vøring plateau. *Nor. Geol. Tidsskr.*, **50**, 375—391.

Halstead, P. 1975. Northern North Sea faulting. In: *Jurassic Northern North Sea Symposium.* Article 10. Norsk Petroleumsforening, Oslo.

Harding, T.P. and Lowell, J.D., 1979. Structural styles, their plate-tectonic habitats, and hydrocarbons traps in petroleum provinces. *Am. Assoc. Petrol. Geol. Bull.*, **63**, 1016—1058.

Hay, J.T.C. 1978. Structural development in the Northern North Sea. *J. Petrol. Geol.*, **1**, 65—77.

Hinz, K. 1972. The seismic crustal structure of the Norwegian continental margin in the Vøring Plateau, in the Norwegian deep sea and the eastern flanks of the Jan Mayen Ridge between 66° and 68° N. *Int. Geol. Congr. Montreal,* Sec. 8, 28—37.

Hinz, K. and Moe, A. 1971. Crustal structure in the Norwegian Sea. *Nat. Phys. Sci.*, **232**, 187—190.

Hinz, K., Dostmann, H.J. and Hanisch, J. 1982. Structural framework of the Norwegian Sea. In: *The Geological Framework and Hydrocarbon Potential of Basins in Northern Seas. Offshore Northern Seas Conference*, Article E/4.

Hollander, N.B. 1982. Evaluation of the hydrocarbon potential offshore mid-Norway. *Oil. Gas J.*, April 1982, 168—172.

Horsfield, W.T. 1977. An experimental approach to basement controlled faulting. *Geol. Mijnb.* **56**, 363—370.

Jørgensen, F. and Navrestad, T., 1979. Main structural elements and sedimentary succession on the shelf outside Nordland (Norway). In: *Norwegian Sea Symposium*. Article 11, Norsk Petroleumsforening, Oslo.

Jørgensen, F. and Navrestad, T. 1981a. The geology of the Norwegian Shelf between 62°N and the Lofoten Islands. In: L.V. Illing and G.D. Hobson (eds.). *Petroleum Geology of the Continental Shelf of Northwest Europe*, Heyden, London pp. 407—413.

Jørgensen, F. and Navrestad, T. 1981b. Main structural elements and sedimentary succession on the shelf outside Nordland (Norway). *Earth Evol. Sci.*, **1**, 43—49.

Larsen, R.M. and Skarpnes, O. 1983. Regional interpretation and hydrocarbon potential of the Træenabanken area. This volume, pp. 217—236.

McKenzie, D. 1978. Some remarks on the development of sedimentary basins. *Earth Planet Sci. Lett.*, **40**, 25—32.

Oftedal, C. 1972. A sideritic ironstone of Jurassic age in Beistadfjorden, Trøndelag. *Nor. Geol. Tidsskr.*, **52**, 123—134.

Oftedal, C. 1975. Middle Jurassic graben tectonics in mid-Norway. In: *Jurassic Northern North Sea Symposium*. (Norsk Petroleumsforening), Article 21.

Price, I. and Rattey, R.P. 1984. Cretaceous tectonics of mid-Norway: Implications for the Rockall-Faeroes Shetland Rift. *J. Geol. Soc. London* (in press).

Rønnevik, H. and Navrestad, H. 1976. Geology of the Norwegian shelf between 62°N and 69°N. In: *Exploration Geology and Geophysics. Offshore North Sea Conference*. Article G IV, p. 4.

Rønnevik, H. and Navrestad, T. 1977. Geology of the Norwegian Shelf between 62°N and 69°N. *Geojournal*, **1**, 33—45.

Rønnevik, H.C., Bergsager, E.I., Moe, A., Øvrebø, O., Navrestad, T. and Stangenes, J. 1975. The geology of the Norwegian continental shelf. In: A.W. Woodland (ed.). *Petroleum and the Continental Shelf of Northwest Europe*, Vol. 1, *Geology*, Applied Science, London. pp. 117—129.

Rønnevik, H., Eggen, S. and Vollset, J. Exploration of the Norwegian Shelf. In: J. Brooks (eds.), *Petroleum Geochemistry and Exploration of Europe*. Blackwell, Oxford, pp. 71—73.

Rønnevik, H., Jørgensen, G. and Motland, K. 1979. The geology of the northern part of the Vøring Plateau. In: *Norwegian Sea Symposium*. (Norsk Petroleumsforening), Article 12.

Sengör, A.M.C. and Burke, K. 1978. Relative timing of rifting volcanism on Earth and its tectonic implications. *Geophys. Res. Lett.*, **5**, 419—421.

Surlyk, F. 1977. Mesozoic faulting in East Greenland. *Geol. Mijnb.*, **56**, 311—327.

Talwani, M. and Eldholm, O. 1972. Continental margin off Norway: A geophysical study. *Geol. Soc. Am. Bull.*, **83**, 3575—3606.

Whiteman, A.J., Rees, G., Naylor, D. and Pegrum, R.M. 1975. North Sea troughs and plate tectonics. *Nor. Geol. Unders.*, **316**, 137—161.

Ziegler, P.A. 1975. North Sea basin history in the tectonic framework of north-western Europe. In: A.W. Woodland (eds.), *Petroleum and the Continental Shelf of Northwest Europe*, Vol 1, Geology. Applied Science, London, pp. 131—150.

Ziegler, P.A. 1978. North-Western Europe: tectonics and basin development. *Geol. Mijnbouw*, **57**, 589—626.

Ziegler, P.A. 1981. Evolution of sedimentary basins in North-West Europe. In: L.V. Illing and G.D. Hobson (eds.), *Petroleum Geology of the Continental Shelf of North-West Europe*. Heyden, London, pp. 3—39.

Ziegler, P.A. 1982. *Geological Atlas of Western and Central Europe*. Shell Internat. Petrol. Maatschappij B.V., 130pp.

Ziegler, W.H. 1975. Outline of the geological history of the North Sea. In: A.W. Woodland (ed.). *Petroleum and the Continental Shelf of North-West Europe*, Vol. 1, *Geology*, Applied Science, London, pp. 165—187.

Aanstad, K.M. 1982. Bruddtektonikk på fastlandet og kontinentalsokkelen mellom 62°N og 68°N Norge. Unpubl. cand. real. thesis, Univ. of Oslo, 195pp.

Aanstad, K.M., Gabrielsen, R.H., Hagevang, T., Ramberg, I.B. and Torvanger, O. 1981. Correlation of offshore and onshore structural features between 62°N and 68°N, Norway. In: *Norwegian Symposium on Exploration*. Article 11. Norsk Petroleumsforening, Oslo.

31 Structure and development of the mid-Norway Continental Margin

C. Bukovics, E.G. Cartier,** N.D. Shaw* and P.A. Ziegler***
*A/S Norske Shell, Stavanger, Norway
**Shell Internationale Petroleum Mij. B.V., The Hague, The Netherlands

The area reviewed covers the Mid-Norway Continental Margin between 62°N and 68°N. The main structural elements of the Mid-Norway Continental Shelf and slope basins, as defined at the base Cretaceous level, are the Trøndelag Platform, which underlies the inner shelf, the Møre and Vøring Basins, which are located beneath the outer shelf and slope, and the Møre Platform and the Outer Vøring Plateau, which form a base of slope trend of highs.

Sediments contained in the Mid-Norway Basin vary in age from late Palaeozoic to Cenozoic. The basement was consolidated during the Caledonian orogenic cycle. Devonian and early Carboniferous wrench movements along the axis of the Arctic/North Atlantic Caledonides are thought to have preceded the late Visean onset of crustal extension. Rifting processes were intermittently active for some 270 Ma until crustal separation between Greenland and Fennoscandia was achieved during the early Eocene.

In the course of the evolution of this rift system a stepwise concentration of the principal tectonic activity to its axial zone can be observed. During the late Palaeozoic to mid-Jurassic a broad zone was affected by faulting induced by crustal stretching. During the late Jurassic and Cretaceous the Trøndelag Platform was little affected by faulting whilst its seaward margin, the Nidaros Arch, became domed up and the Møre and Vøring Basins subsided rapidly. During the latest Cretaceous and early Palaeogene terminal rifting phase, only the areas of the western Møre and Vøring Basins were affected by intensive intrusive and extrusive igneous activity.

During the late Palaeogene and Neogene opening of the Norwegian—Greenland Sea, the Vøring segment of the Mid-Norway Marginal Basin subsided less rapidly than the Møre segment. During the Oligocene and Miocene, the Vøring Basin and to a lesser degree the Møre Basin were affected by compressional deformations.

INTRODUCTION

The area discussed covers the continental shelf and slope of mid-Norway between the northernmost North Sea (62°N) and the steep continental rise of the Lofoten area (68°N, Fig. 1). The Mid-Norway Basin covers an area of some 150 000 km^2 and contains a sedimentary prism, ranging in age from late Palaeozoic to Cenozoic, that attains a maximum thickness of some 10 km. The basin is highly structured. This chapter is based mainly on regional and detailed reflection seismic data which were integrated with exploration well data and information from core-holes (Fig. 2).

Petroleum Geology of the North European Margin © Norwegian Petroleum Society (Graham & Trotman, 1984) pp. 407—423.

GEOPHYSICAL DATA BASE AND PREVIOUS INTERPRETATIONS

The existence of a major sedimentary basin beneath the shelf and continental slope of mid-Norway gradually became evident during the years 1959—1969 (for summary see Rønnevik and Navrestad, 1977). The distribution and thickness of sediments was first estimated from magnetic and gravity data (Grønlie and Ramberg, 1970; Åm, 1970). These findings were subsequently confirmed by an increasing number of seismic refraction measurements (summarized by Sellevoll, 1975).

From 1970 onwards a rapidly increasing amount of reflection seismic data was acquired, mainly by the Norwegian Petroleum Directorate (NPD), but also by academic institutions, and during recent years by the oil industry (Fig. 2). Published accounts concentrated on the interpretation of reflection seismic data from the shelf areas (Rønnevik *et al.*,

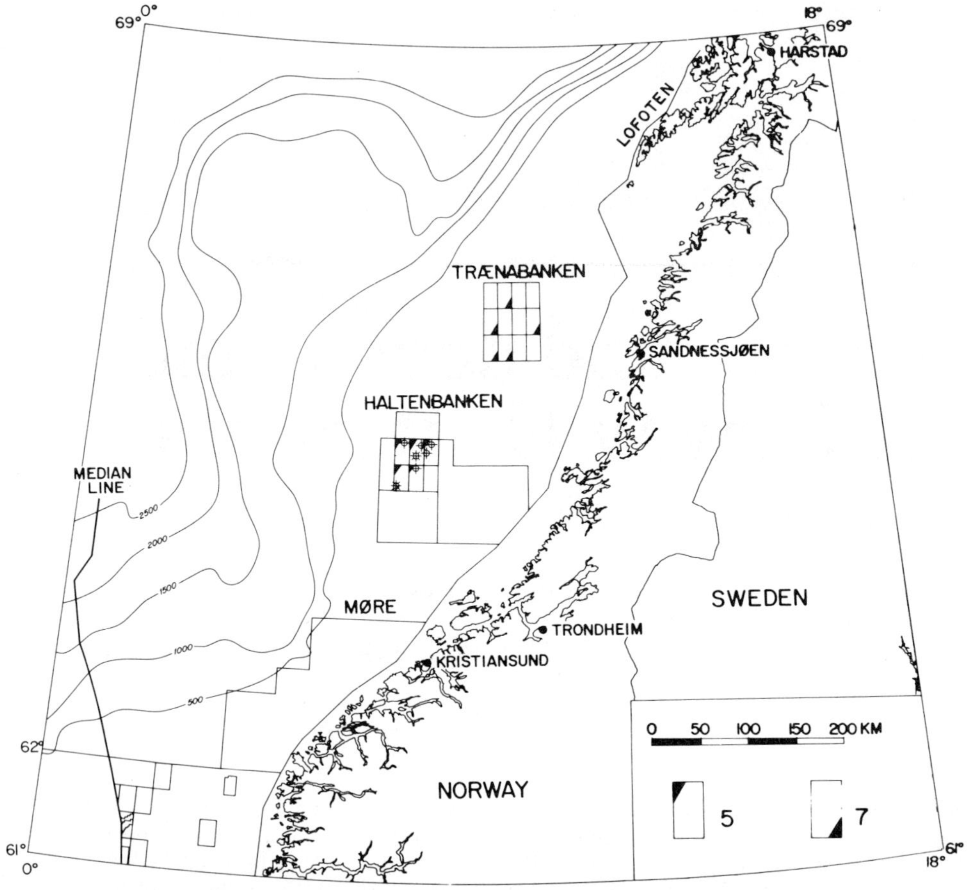

Fig. 1. Location map. Bathymetry in metres. 5, Block allocated in three phases of fifth licence round; 7, block allocated in seventh licence round.

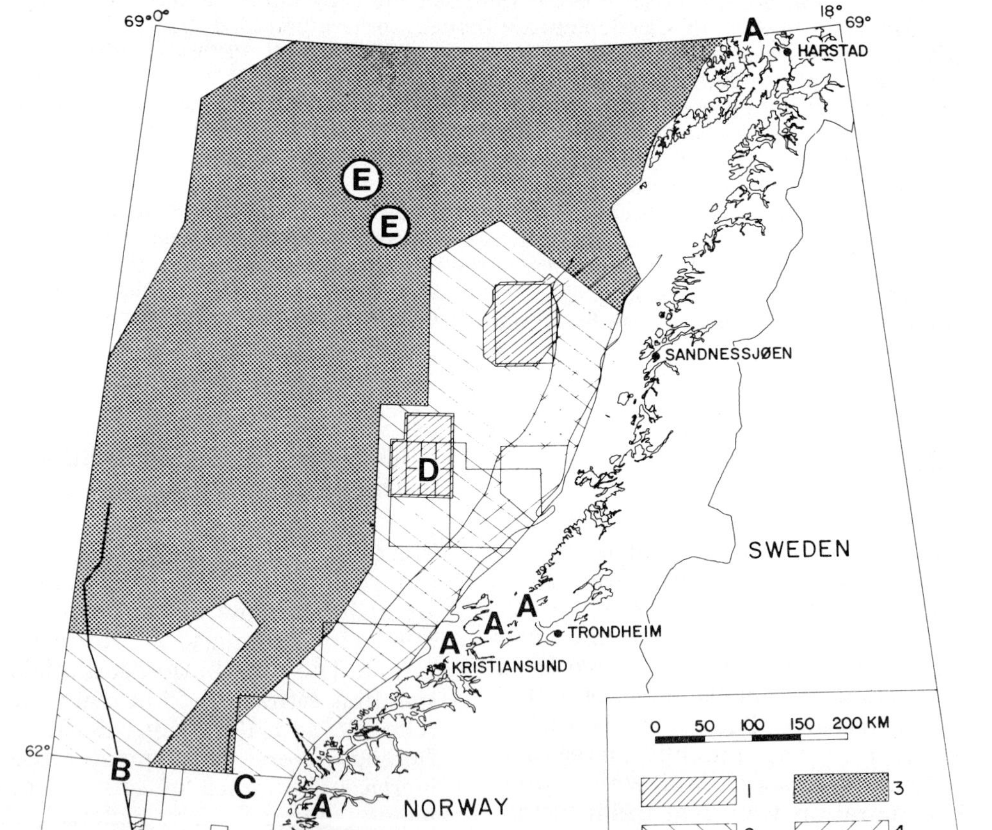

Fig. 2. Data base. Available multichannel reflection seismic: 1, 1 × 1 to 4 × 2 km grid; 2, 4 × 5 to 20 × 30 km grid; 3, 20 × 40 km and less dense grid. Geologic control: A, Onshore outcrops of sediments; B, UK wells NE of Shetland; C, wells in Norwegian blocks 35/3 and 36/1; D, Haltenbanken wells; E, DSDP core holes; 4, Offshore area of subcropping preQuaternary sediments, investigated by sparker seismic and sampling and coring.

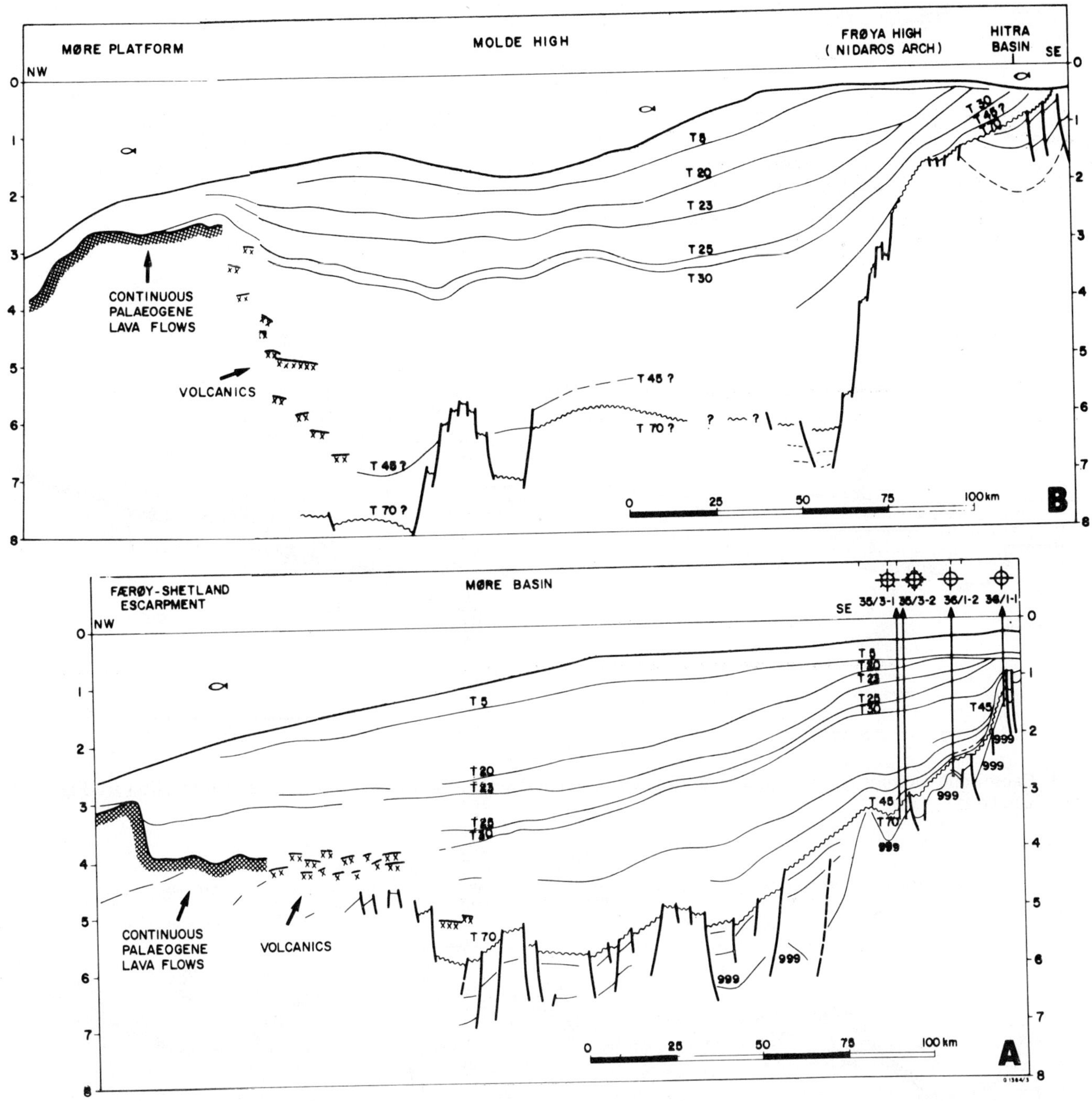

Fig. 3. Geoseismic sections of the Møre Basin. Vertical scale in seconds two-way time. For location see Fig. 8 and for horizon code see Fig. 6.

1975; Rønnevik and Navrestad, 1977; Jørgensen and Navrestad, 1979, 1981; Ånstad *et al.*, 1981) and on the interpretation of geophysical data from the Norwegian Sea as a whole, with special emphasis on areas west of the Vøring Plateau and the Faerøy-Shetland Escarpments. Gradually a detailed model for the Cenozoic separation of Europe and Greenland and the opening of the Norwegian-Greenland Sea was developed (Johnson *et al.*, 1970; Talwani and Eldholm, 1972; Eldholm and Talwani, 1977; Eldholm, 1979; Eldholm and Sundvor, 1979; Grønlie *et al.*, 1979; Eldholm and Thiede, 1980; Talwani and Mutter, 1980; Eldholm and Talwani, 1981).

However, relatively few publications deal with the area of the sedimentary basin underlying the continental slope of mid-Norway (e.g. Rønnevik *et al.*, 1979, 1982; Hinz *et al.* 1982). On a wider scale the geological history of the Norwegian continental margin has been considered in the context of the development of the Arctic areas by Nairn *et al.* (1981), and with reference to the evolution of northwest Europe by Ziegler (1982).

GEOLOGICAL DATA BASE AND STRATIGRAPHY

Stratigraphic control is provided by seven exploration wells drilled in the Haltenbanken blocks (Hollander, 1982; Rønnevik *et al.* 1982) (Figs. 2 and 4) and by sea-bottom samples and core-holes offshore Helgeland and Trøndelag (see Mørk, 1984; Fig. 5). The DSDP coreholes 338 to 343 that were drilled in 1974 on the Outer Vøring Plateau and in the Vøring Basin (Talwani *et al.*, 1976) (Fig. 5) yielded the only geological data from the western part of the study area. Furthermore, wells drilled

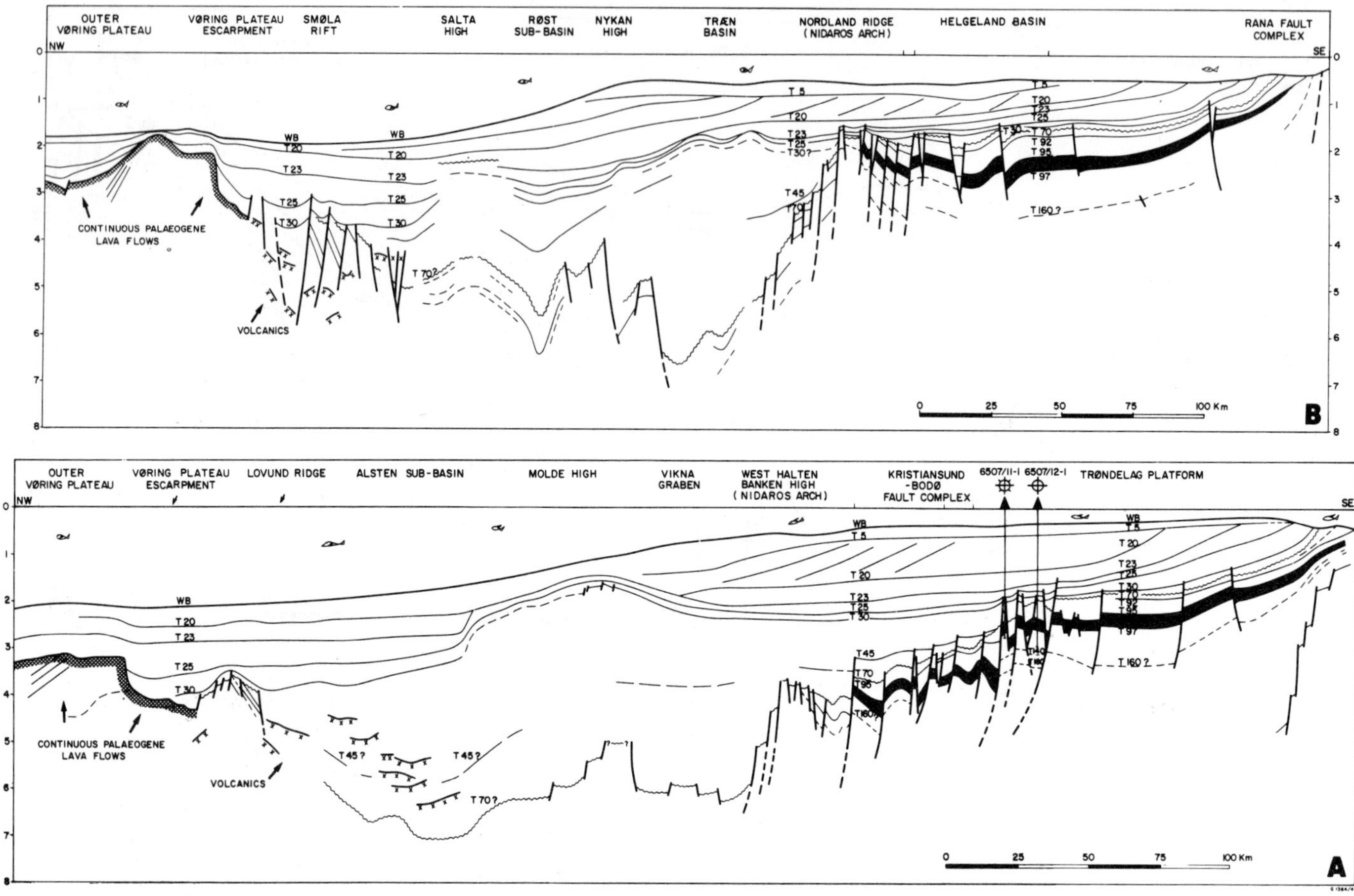

Fig. 4. Geoseismic sections of the southern Vϕring Basin. Vertical scale in seconds two-way time. For location see Fig. 8, and for horizon code see Fig. 6.

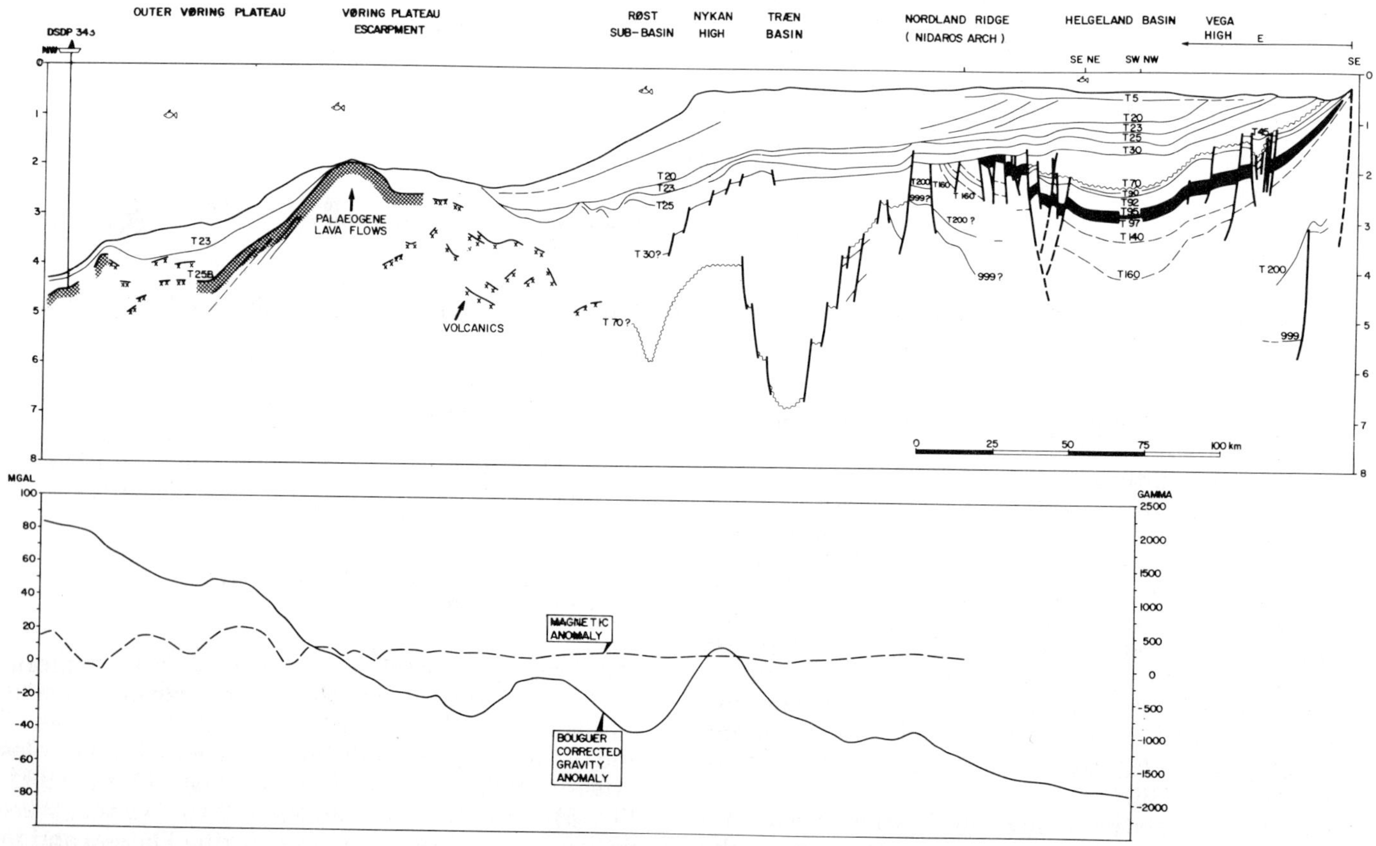

Fig. 5. Geoseismic sections of the northern Vϕring Basin. Vertical scale in seconds two-way time. For location see Fig. 8, and for horizon code see Fig. 6. E, Area of subcropping prequaternary sediments mapped from sparker data and investigated by sampling and coring (IKU).

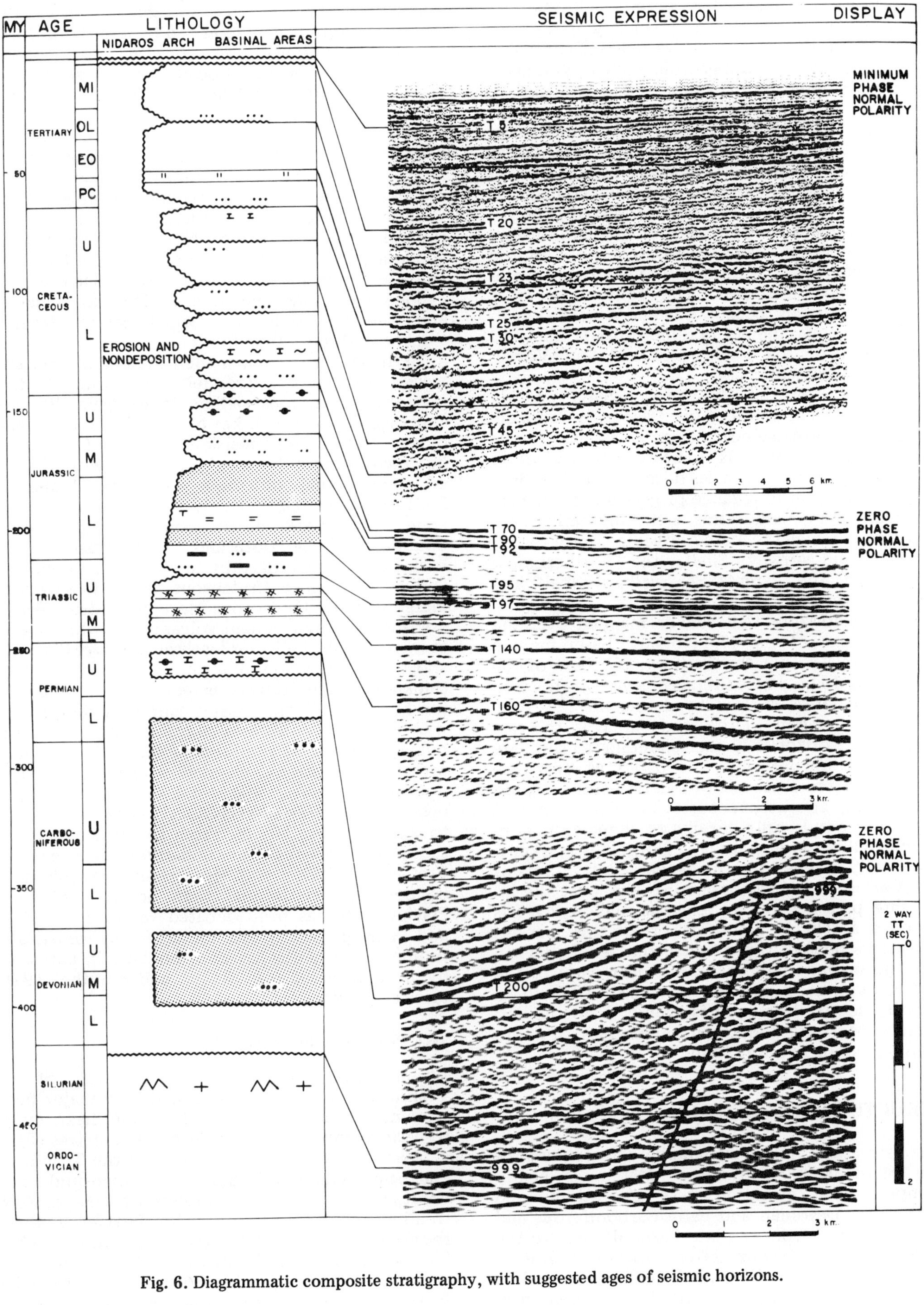

Fig. 6. Diagrammatic composite stratigraphy, with suggested ages of seismic horizons.

in the North Sea, for instance in the Norwegian blocks 35/3 and 36/1 (Myrland *et al.*, 1981) (Fig. 3) and also wells in UK waters, permit calibration of the seismically defined sedimentary sequences north of the 62° parallel. More indirect control is also provided by the stratigraphic sequences exposed on the east coast of Greenland and the limited outcrops of sediments on the Norwegian coast.

On the basis of this geological and geophysical data set, it is inferred that sediments contained in the Mid-Norway Basin range in age from Devonian to late Cenozoic and that a nearly continuous sedimentary record is probably preserved (Fig. 6). Sediments in the deepest parts of this basin attain thicknesses in excess of 10 km. The floor of the Mid-Norway Basin is formed by Caledonian metamorphics and intrusives. The continental crust thins progressively oceanwards and a transition to an Eocene and younger oceanic crust is likely to occur west of the Faerøy-Shetland and Vøring Plateau Escarpments.

Well data on the Mid-Norway Shelf provide control for the presence of the Triassic to Cenozoic series only. The existence of Permian, Carboniferous and Devonian strata is indicated on reflection seismic data, but their facies development can only be inferred from long-range projections from East Greenland. In general terms, a striking similarity is observed between the Mesozoic sedimentary record of the Mid-Norway Shelf, the northernmost North Sea, the Barents Sea-Svalbard Shelf and East Greenland.

Seismic horizons that can be mapped over the entire extent of the Mid-Norway Basin include several levels within the Tertiary, the base of the Tertiary and a near-base Cretaceous marker (horizon T70). The latter is used to define the Mesozoic structural elements of the study area. In the east and south, horizon T70 corresponding to the 'late Kimmerian unconformity' is dated as near-base Cretaceous (top Ryazanian). This reflector is related to a laterally consistent acoustic impedance contrast and corresponds only locally to a major unconformity (Rawson and Riley, 1982). This marker is less clearly expressed in the deeper, central parts of the Vøring and Møre Basins. To the west the increasing presence of volcanics at shallower levels hampers a reliable identification of this marker. Reflectors related to pre-Cretaceous sediments can only be mapped and correlated over parts of the Mid-Norway Basin and are particularly difficult to identify in its deeper parts where Cenozoic and Cretaceous strata can reach thicknesses of more than 6 seconds two-way time.

STRUCTURAL ELEMENTS

The structural and stratigraphic framework of the Mid-Norway Continental Margin is highly complex. Its evolution was shear-dominated during the Devonian and early Carboniferous and rift-dominated from the late Carboniferous to the late Palaeocene. Only after crustal separation between Fennoscandia and Greenland had taken place during the early Eocene did the area enter into a true post-rifting, passive continental margin setting.

The individual sub-basins on the Mid-Norway Continental Margin are defined by differences in structural style and in the age of their dominant sedimentary fill (Fig. 8). At shallow levels, the structural style is characterized by a fairly uniform seaward prograding clastic wedge that was deposited during and after the Eocene crustal separation between Greenland and Norway. This clastic wedge is little affected by faulting; there is, however, local evidence of mild compressional deformations. The nearshore limit of this Cenozoic sedimentary prism is erosional. Maximum thicknesses of some 3 km are attained near the shelf break in the area west of Trondheim and in the southern Møre Basin. Under the continental slope Cenozoic strata thin rapidly and onlap against and overstep the Møre Platform and the Outer Vøring Plateau. These platforms are capped by Palaeocene—Eocene Thulean volcanics and Cenozoic sediments, which rarely attain a thickness of more than 800 m.

In the area of the inner shelf the Cenozoic clastic wedge overlies the Trøndelag Platform. On this platform Cretaceous sediments are generally thin and overlie some 5—8 km of Jurassic, Triassic and older strata. The position of the Caledonian basement cannot be readily mapped with the aid of conventional reflection data. The Trøndelag Platform is limited to the west by a complex fault zone which marks at the same time the eastern boundary of the Møre and Vøring Basins. These basins are characterized by 2—3-km thick Cenozoic and 5—7-km thick Cretaceous strata. The western margins of the Cretaceous Møre and Vøring Basins are obscured by frequent volcanic sills which occur at increasingly shallower levels as the Møre Platform and the Outer Vøring Plateau are approached. The thickness of pre-Cretaceous sediments that underlie the Møre and Vøring Basins cannot be determined with the aid of conventional reflection seismic data, but appears to be rather variable. Published interpretations of gravity, magnetic and refraction data indicate, however, that total sedimentary thicknesses in the Vøring and Møre Basins are unlikely to exceed about 10 km. From this it is concluded that pre-Cretaceous strata underlying the essentially Cretaceous Vøring and Møre Basins, are likely to be considerably thinner than on the Trøndelag Platform. It has been debated whether the Møre and Vøring Basins are underlain by drastically thinned continental crust or possibly in part even by oceanic crust (e.g. Russel and Smythe, 1978; Price and Rattey, in press).

The Trøndelag Platform

At pre-Jurassic levels, the inner margin of the Trøndelag Platform is controlled by a major fault system, the Rana Fault Complex, which has a down-to-basin throw of possibly several kilometres (Figs. 4, 5 and 9). The western margin of the Trøndelag Platform is characterized by the Kristiansund-Bodø Fault Complex, which consists of NE- and N-trending fault segments. The heavily faulted, rhomb-shaped area of the Halten Terrace, which is located in the central part of this fault zone, takes in an intermediate position between the Trøndelag Platform and the westward adjacent Vøring Basin. The Kristiansund-Bodø Fault Complex is associated with a series of early Cretaceous palaeo-highs, which are here referred to as the Nidaros Arch.

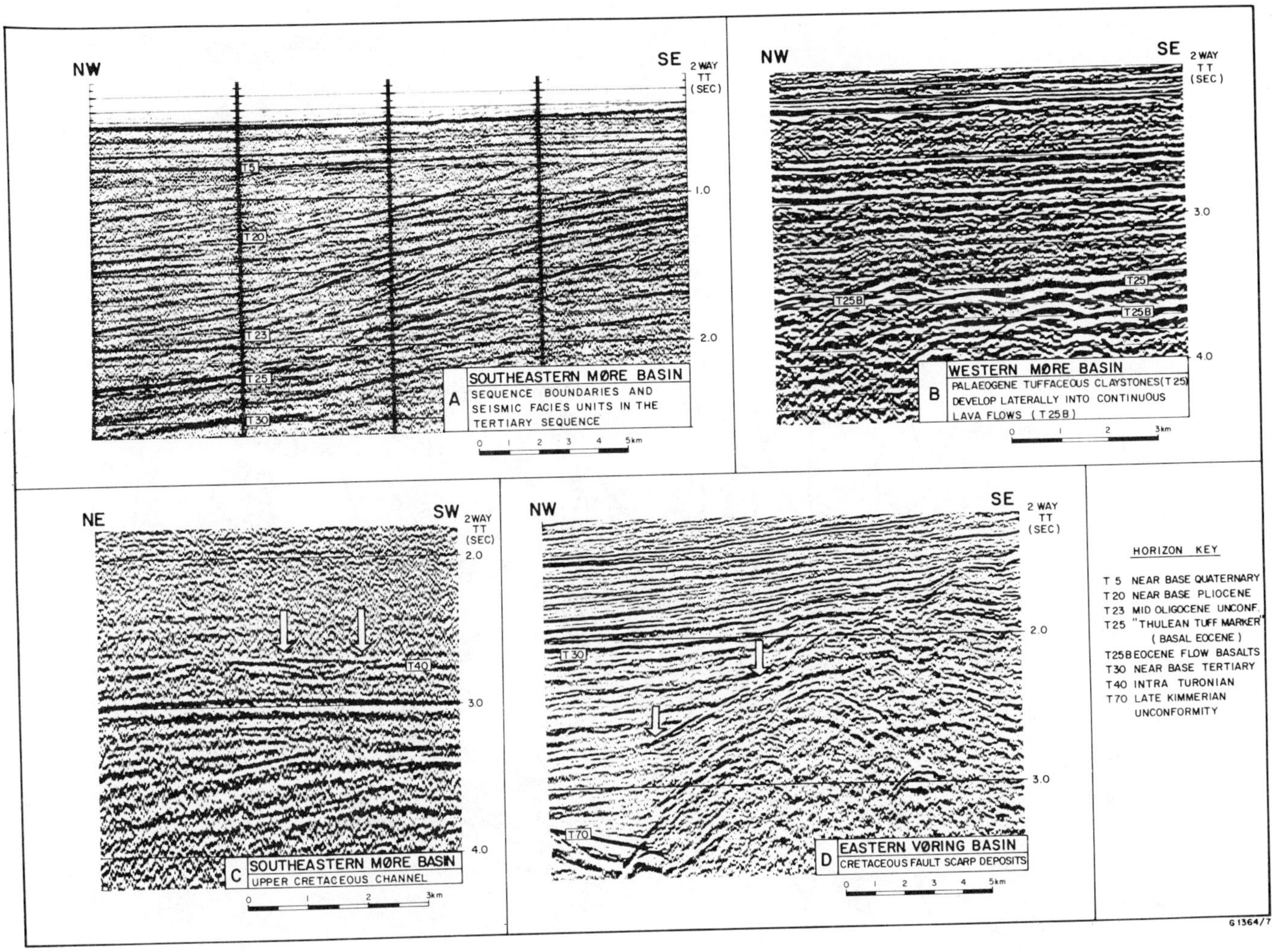

Fig. 7. Møre and Vøring Basins. Examples of the Cretaceous and Tertiary seismic stratigraphy.

At base Cretaceous level the Trøndelag Platform corresponds to a seaward dipping, relatively little faulted monocline, which grades northward into the gently southwest plunging synform of the Helgeland Basin. The Trøndelag Platform is underlain by thick Upper Palaeozoic strata. Locally, seismic data indicate that the base of this sequence is heavily blockfaulted. At a shallower level, a distinct reflector, tentatively identified as an intra-Permian marker, is comparatively less faulted and thus indicates that the early structural relief of this rift basin became infilled by presumably Carboniferous and Permian strata. The available seismic data are, however, of insufficient quality to reconstruct the details of the late Palaeozoic subsidence pattern of the Trøndelag Platform.

On the other hand, seismic data provide good control on the structural and stratigraphic configuration of the Trøndelag Platform at Triassic and Jurassic levels. The area of the Nidaros Arch subsided less rapidly during parts of the Permian, Triassic and Jurassic than the eastward adjacent part of the Trøndelag Platform. On the whole, the Triassic and Jurassic sequences increase in thickness towards the axis of the Trøndelag Platform, whereby syndepositional faulting often controlled local subsidence patterns. The thick Triassic evaporites that were encountered in a borehole have no equivalent in the northern North Sea and may indicate a restricted environment in a silled subbasin between the Nidaros Arch and the coast. Similarly, the Rhaeto-Liassic coal measures are considerably thicker on the Trøndelag Platform than in the northern North Sea. Middle and Upper Jurassic strata are best developed in the Helgeland Basin and show also a marked thickness increase towards their subcrop in nearshore areas. These strata are generally missing over the Nidaros Arch and are only thinly developed on the Halten Terrace.

The frequency of faults increases towards the western margin of the Trøndelag Platform. The Nidaros Arch consists of several asymmetrical palaeo-highs, which are bounded towards the west by steep fault zones with throws in the order of several kilometres. In contrast, on the eastern flanks of this arch, pre-Cretaceous sediments dip gently towards the east and thicken gradually in the same direction. In the crestal parts of the Nidaros Arch, Triassic and locally even older strata (e.g. in the area of the Rødøya High) are deeply truncated (Fig. 9). Most of this erosion probably took place during the early Cretaceous. This erosional phase correlates with the late Kimmerian unconformity in the North Sea. Also in the mid-Norway area, the late Kimmerian hiatus correlates to a major rifting phase that induced the rapid differential subsidence of the Møre and Vøring Basins and, to a lesser degree, also of the Halten Terrace.

Where mappable on detailed seismic grids, fault patterns of the Halten Terrace are characterized by the interference of NE- and N-striking fault segments. Fault patterns of the Kristiansund-Bodø

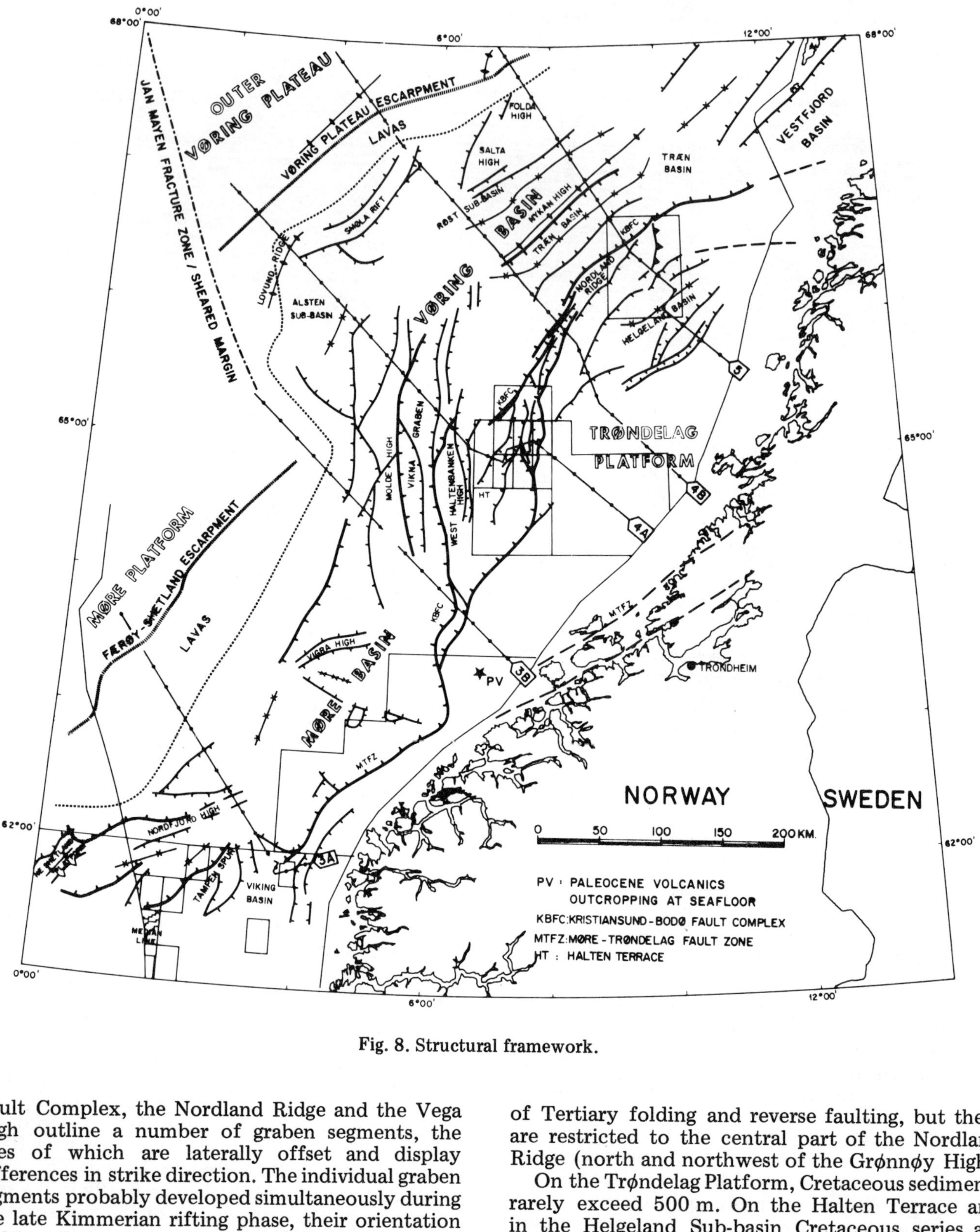

Fig. 8. Structural framework.

Fault Complex, the Nordland Ridge and the Vega High outline a number of graben segments, the axes of which are laterally offset and display differences in strike direction. The individual graben segments probably developed simultaneously during the late Kimmerian rifting phase, their orientation and location presumably being determined by pre-existing zones of weakness (see Illies, 1981), which are evident on Landsat images in the adjacent onshore areas (Ånstad *et al.*, 1981). The graben offsets resulting from this *en-echelon* faulting give rise to the zigzag configuration of the main graben boundary faults and local 'transform' zones (cf. the simulation of the fault pattern of the Rhine-Bresse Graben rift system in clay model studies by Elmohandes, 1981).

The interpretation is preferred to a two-stage rifting model whereby a major Cretaceous shearing phase would post-date earlier horst and graben formation. There are, however, clear indications of Tertiary folding and reverse faulting, but these are restricted to the central part of the Nordland Ridge (north and northwest of the Grønnøy High).

On the Trøndelag Platform, Cretaceous sediments rarely exceed 500 m. On the Halten Terrace and in the Helgeland Sub-basin Cretaceous series are somewhat thicker and more complete. Activity along most of the faults ceased by mid-Cretaceous times. However, downfaulting of the Halten Terrace is stratigraphically dated as mainly late Cretaceous. The uplift of the Vega High continued into the late Cretaceous, and border faults of the Nordland Ridge show reactivation during the Palaeogene and locally even during the Neogene. In the eastern Nordland Ridge, Cretaceous and Palaeogene strata transgress over a gently seaward dipping peneplain against which Jurassic and Triassic strata subcrop. The axial parts of the Nordland Ridge (North Haltenbanken High and Rødøy High) were only inundated during the Oligocene and Miocene/

Pliocene. In contrast, sedimentation had resumed on the West Haltenbanken High and on the Frøya High during the late early and late Cretaceous, respectively.

The Cenozoic subsidence pattern of the Trøndelag Platform was relatively uniform. Palaeogene and Neogene series reach a maximum thickness in the order of 2 km. The eastern parts of the Trøndelag Platform were strongly affected by the uplift of the Fennoscandian Shield.

Møre Basin

Seismic data indicate that in the Møre Basin the base of the Cretaceous is strongly affected by block faulting. Particularly in its southern part, there is seismic evidence that these fault blocks include thick Mesozoic and possibly older strata (Fig. 3). Early and late Cretaceous sediments onlap against discrete structural highs and ultimately bury a structural relief of several kilometres. Within the Tertiary prism several prograding units are observed (Fig. 7A). Cenozoic strata reach maximum thicknesses of about 3 km near the shelf break. Upper Cretaceous and Tertiary sediments are little affected by faulting.

The main structural elements are defined at the late Kimmerian unconformity level (T70). The eastern margin of the Møre Basin is formed by the Møre-Trøndelag Fault Zone and its western margin by the Faerøy-Shetland Escarpment (Fig. 8). Intrabasinal highs are the Nordfjord High, located to the northeast of the North East Shetland Platform, several basement highs west of the Møre-Trøndelag Fault Zone and the Molde and Vigra Highs. In the western parts of the Møre Basin, as the Faerøy-Shetland Escarpment is approached, the occurrence of volcanic flows and sills at increasingly shallower levels, progressively inhibits the definition of the structural configuration at deeper levels. Volcanics, dated from seismic as early Eocene (see Fig. 7B), form a 50–110-km wide continuous sheet to the east of these escarpments. Below this sheet only locally dipping events, interpreted to be related to faulted Mesozoic sediments, are observed.

To the south, the Møre Basin shallows somewhat and grades into the Viking Basin of the northernmost North Sea. The northern limit of the Møre Basin is defined by a zone to the north of which Cretaceous and Tertiary strata are much more affected by (partly steep reverse) faulting and folding. This border zone lies in the prolongation of the Jan Mayen Fracture Zone. South of it, the base of the Eocene (seismic marker T25) occurs at a depth of some 3.5 km, whilst to the north of it, in the Vøring Basin, the same marker occurs at a depth of approximately 2 km.

In the Møre Basin, Cretaceous sediments are thought to consist predominantly of deep-water clays. There is, however, on some seismic lines evidence of channelling that is probably associated with turbidite fans (Fig. 7C).

Seismic and well data from the northernmost North Sea indicate that the transition zone between the Møre Basin and the Viking Graben was temporarily domed up during the Cretaceous (Ziegler, 1982). This 'late Kimmerian' doming event was associated with a deep truncation of Jurassic and Triassic and locally even Permian strata. It is unknown to what extent contemporaneous doming affected also the main parts of the Møre Basin. Uplift of the Nidaros Arch is, however, dated as approximately latest Jurassic and earliest Cretaceous.

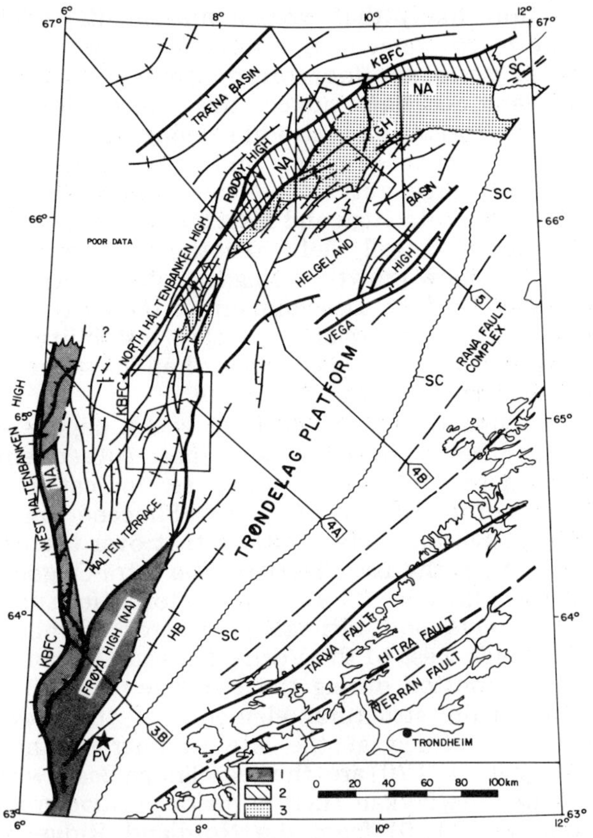

Fig. 9. Trøndelag Platform tectonic elements. 1, Pre-Jurassic overlain by Cretaceous; 2, Pre-Jurassic overlain by Tertiary; 3, Jurassic overlain by Tertiary. NA, Nidaros Arch; KBFC, Kristiansund-Bodø Fault Complex; HB, Hitra Basin; GH, Grønnøy High; SC, Subcrop of Jurassic and older sediments at sea-floor/base Quaternary; PV, Palaeocene volcanics.

Vøring Basin

Seismic data show that in the Vøring Basin the base of the Cretaceous is strongly affected by block faulting. A structural relief at the late Kimmerian unconformity level of some 7 km can be observed. Faults affecting this seismic marker horizon (T70) die out upwards within the Cretaceous strata. By the end of the late Cretaceous much of this block-faulted relief was apparently filled in, presumably by deeper water sediments. Along the Kristiansund-Bodø Fault Complex seismic data give evidence for the occurrence of local deep-water clastic fans (see Fig. 7D). The Cenozoic clastic wedge progrades into the Vøring Basin with the present-day shelf break being located in its eastern part.

In the axial parts of the Vøring Basin a complex NNE-striking anticlinal feature is observed at top Cretaceous and younger levels. This anticlinorium was formed during the Oligocene and early Miocene. DSDP core-holes 339 and 340 were drilled on this feature and recovered diapiric Oligocene and Eocene shales. Although it is possible that this anticlinorium was induced by the diapirism of the undercompacted Cretaceous and Palaeocene clays, it is more likely that wrench movements, related to the rearrangement of spreading axes during the early phases of sea-floor spreading in the Norwegian-Greenland Sea, have induced this deformation. The latter interpretation is also in keeping with Tertiary folding and reverse faulting on the Nordland Ridge. Moreover, large anticlinal and synclinal features, such as the Salta High and Røst Sub-basin (Fig. 4B), which are seismically defined at the base Cretaceous level, also show growth during the latest Cretaceous and early Tertiary. Development of these structures is unlikely to be related to clay instability, but rather points to the occurrence of an intra-Tertiary compressional deformation phase.

The area of the Folda High was a depocentre during the Cretaceous and has subsequently experienced repeated uplift, as indicated by the sea-crop of the Miocene—Pliocene unconformity. Seismic data illustrate that over the crest of this high, the base Tertiary, the intra-Oligocene and the Miocene—Pliocene unconformities successively merge and ultimately sea-crop. Similarly, high trends west of the Vøring Plateau Escarpment became uplifted during the Neogene (Fig. 4B, 5).

The main structural elements of the Vøring Basin as defined at the late Kimmerian unconformity level (T70) are: the deep Træna Basin which separates the Nykan High (a major positive gravity anomaly; Fig. 5) from the Nordland Ridge; the Vikna Graben which lies between the Molde High and the West Haltenbanken High and, in the west, the deep, but little structured, Alsten Sub-basin which is flanked towards the west by the Lovund Ridge. Seismic data suggest that the Lovund Ridge is upheld partly by steeply dipping, blockfaulted pre-Cretaceous strata and that it experienced uplift during the early Tertiary. East of the Vøring Plateau Escarpment lies the Mesozoic Smøla rift, which was also uplifted during the Palaeogene. The NE-trending synformal Røst Sub-basin is located to the west of the Nykan High and to the southeast of the Salta and Folda Highs.

Little is known about the thickness and age of the faulted and folded pre-Cretaceous strata of the Vøring Basin. The character of the late Kimmerian unconformity reflector, and the nature and the thickness of the reflective sequences beneath it, vary considerably. As in the Møre Basin, the structural definition of the western parts of the Vøring Basin is impeded by an increasing number of volcanic sills and flows that occur within the Cretaceous and Tertiary sediments. These isolated occurrences of igneous bodies merge westward into continuous flows near the Vøring Plateau Escarpment. On the Outer Vøring Plateau and along its western flank these flows are dated as Eocene (see Talwani *et al.*, 1976). It is uncertain whether some of the volcanics, which are located within the Cretaceous strata, were emplaced during the Cretaceous or whether they were also intruded during the early Tertiary (see Hinz *et al.*, 1982).

Møre Platform and Outer Vøring Plateau

On the platforms a moderately thick cover of Eocene and younger sediments overlies a continuous basalt layer. The Vøring Plateau, in particular, has been widely discussed in the literature. The principal controversy is whether these outer highs are underlain by thinned continental crust (Hinz *et al.*, 1982) or whether they are upheld by anomalously thick oceanic-type crust that was generated by subaerial sea-floor spreading (Talwani and Mutter, 1980).

EVOLUTION OF THE MID-NORWAY CONTINENTAL MARGIN

Passive margins, where crustal separation occurred at different geological times, exhibit a strikingly similar evolution (Kent, 1976, 1980). However, the length of the rifting stage, which precedes crustal separation, can vary considerably. The Mid-Norway Margin Basin is a classical continental margin basin. It is superimposed on the Caledonian suture between the Greenland and Fennoscandian Cratons. Subsidence of individual basins was initiated during the Devonian by wrench tectonics, which gave way during the late early Carboniferous to regional crustal extension. This rifting stage persisted until the early Eocene, by which time crustal separation between Greenland and Fennoscandia occurred. The evolution of the Mid-Norway Continental Margin differs from other passive margins in that it had an unusually long rifting stage which lasted some 270 Ma and that major tectonic activity continued also after crustal separation had been achieved Fig. 10.

Palaeozoic basin evolution

Collision of the Greenland-Laurentian Plate and the Fennoscandian-Russian Plate gave rise to the Ordovician—early Devonian Caledonian orogeny. Following the consolidation of the resulting Arctic—North Atlantic Caledonides, the area of the future Norwegian-Greenland Sea was affected from Middle Devonian to early Carboniferous times by major sinistral movements. These gave rise to the rapid subsidence of the intramontane Old Red Sandstone basins, in which many thousand metres of clastics accumulated. Subsidence of these basins was accompanied by transpressional deformations and a variable amount of volcanic and intrusive activity (Haller, 1971; Steel, 1976; Ziegler, 1982).

From late Carboniferous time onwards, the evolution of the Norwegian-Greenland Sea area, as well as the Svalbard-Barents Sea area, was dominated by regional crustal extension. In eastern Greenland, an extensive rift system developed in which Carboniferous and Permian clastics and carbonates accumulated in large half-grabens that are bounded by presumably listric, basement involving normal faults (Haller, 1971). It is surmised that similar half-grabens subsided also in areas of the Norwegian Shelf. This is supported by scarce seismic evidence, e.g. from the area east of the Vega High (Fig. 5). The eastern part of the Trøndelag Platform apparently subsided rapidly during the late Palaeozoic. The Palaeozoic history of the Møre and Vøring Basins is unknown.

Crustal extension persisted into late Permian times and led to the transgression of the Arctic Permian seas which advanced via the Norwegian-Greenland Sea Rift into the intra-cratonic northern and southern Permian basins of Northwest Europe (Ziegler, 1982). A presumed intra-Permian reflector is recognized on the Trøndelag Platform which is

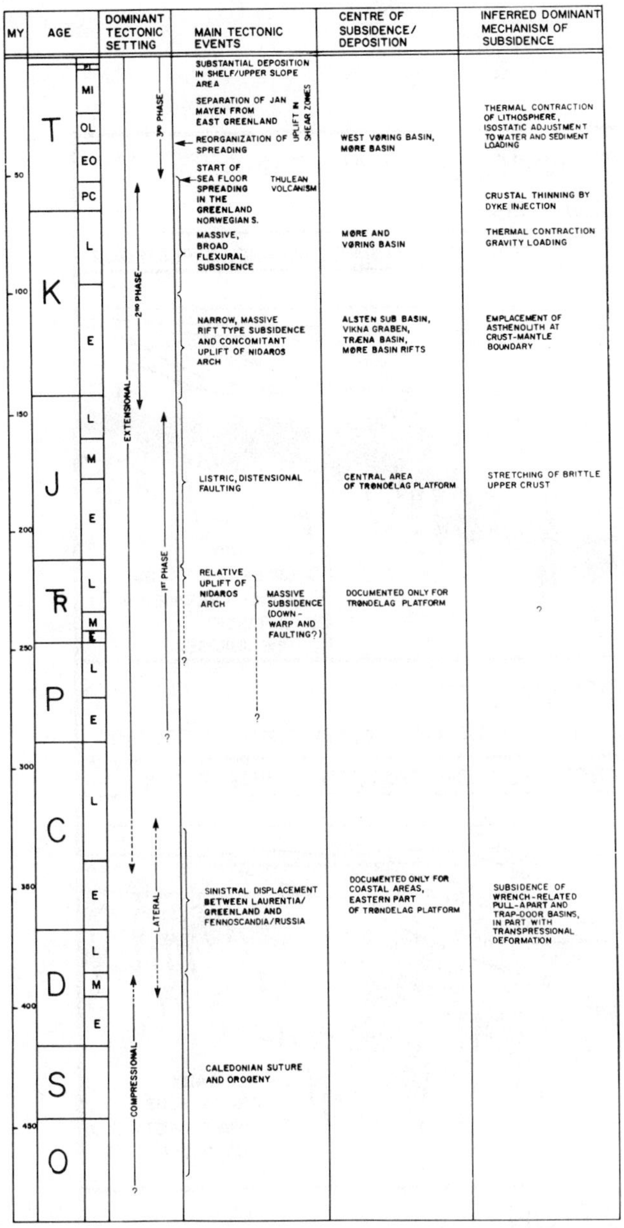

Fig. 10. Tectonic development of the mid-Norway Continental Margin.

affected by only minor syndepositional faulting. This suggests that Permo-Carboniferous strata had already infilled much of the pre-existing rift relief by the time the strata corresponding to this reflector were deposited. This high amplitude reflector, which has not yet been calibrated by well data, is characterized by great lateral continuity and could possibly be associated with Permian carbonates equivalent to those exposed in East Greenland and known from the Barents Sea area.

Early Mesozoic rifting stage

The late Permian—early Triassic regression, as known from the North Sea area, is thought to be related to a global sea-level low stand. In the Norwegian-Greenland Sea area crustal extension accelerated during the early Triassic and rifting propagated southward into the North Sea area. In East Greenland the subsidence of major fault blocks continued. On the Trøndelag Platform the accumulation of Triassic strata was accompanied by syndepositional tensional faulting. The bulk of the Triassic strata, as penetrated so far on the Mid-Norway Shelf, consists of continental red beds and only two halite layers give evidence for a mid-Triassic marine transgression. It is not known whether more marine Triassic series occur to the west of the Nidaros Arch. On the Trøndelag Platform, the Triassic series appear to thin towards the Nidaros Arch palaeo-high. On the latter, Cretaceous erosion has further modified the thickness of the Triassic strata. The observed overall thinning of the Triassic series toward the Nidaros Arch, whether depositional or partly due to late Triassic and Jurassic to early Cretaceous erosion, indicates that this arch probably had a long growth history.

On the Trøndelag Platform and the Halten Terrace seismic data give evidence for syndepositional Triassic and Jurassic synthetic listric faulting, with corresponding rotation of the downthrown blocks. Compensatory antithetic faults, which for geometric reasons can be expected within a set of extensional listric normal faults (Bally, 1981), are evident, e.g. on the Halten Terrace where they gave rise to typical horst and graben structures (Fig. 4). Some of the minor faults appear to sole out within the Triassic sediments, but the major ones are assumed to sole out within the basement. From this it is postulated that the late Palaeozoic to mid-Mesozoic subsidence of the Norwegian-Greenland Sea Rift was governed mainly by mechanical stretching of the crust (McKenzie 1978; and Fig. 11A). In the Helgeland and Hitra Sub-basins, Jurassic strata thicken towards the basin axis; this could indicate at least a gentle uplift of the Nidaros Arch. Åsheim and Larsen (1984) report that the mid-Jurassic reservoir sands encountered in the 'Tyrihans'-discovery well, 6407/1-2, were derived from western sources. This could be taken as evidence for mid-Jurassic uplift of parts of the Nidaros Arch. There is, however, no seismic evidence for Triassic or mid-Jurassic volcanic activity in the area.

From the seismic data available it is evident that during the Triassic—Jurassic rifting phases the entire Mid-Norway Shelf was affected ('phase 1' of Figs. 10 and 12). Already during the Jurassic, however, faulting in the western parts of the Trøndelag Platform appears to be more intense than in its eastern parts.

Late Mesozoic rifting stage

As in the North Sea area, the 'late Kimmerian' rifting pulse, at the transition from the Jurassic to the Cretaceous, strongly affected the mid-Norway area. Whilst the Nidaros Arch became sharply updomed the adjacent parts of the Møre and Vøring Basins (the Vikna Graben and the Træna Basin) began to subside rapidly. It is, however, uncertain whether parts of the Møre and Vøring Basins were also temporarily domed-up at this time. The Trøndelag Platform, on the other hand, was little affected by late Jurassic and early Cretaceous faulting and subsided only little during the Cretaceous.

This illustrates that rifting activity was now concentrated on the more axial parts of the East Greenland-Mid Norway Rift System (phase 2, Figs. 10 and 12). Doming of the rift flanks, in conjunction with the rapid subsidence of the rifts in the Møre-

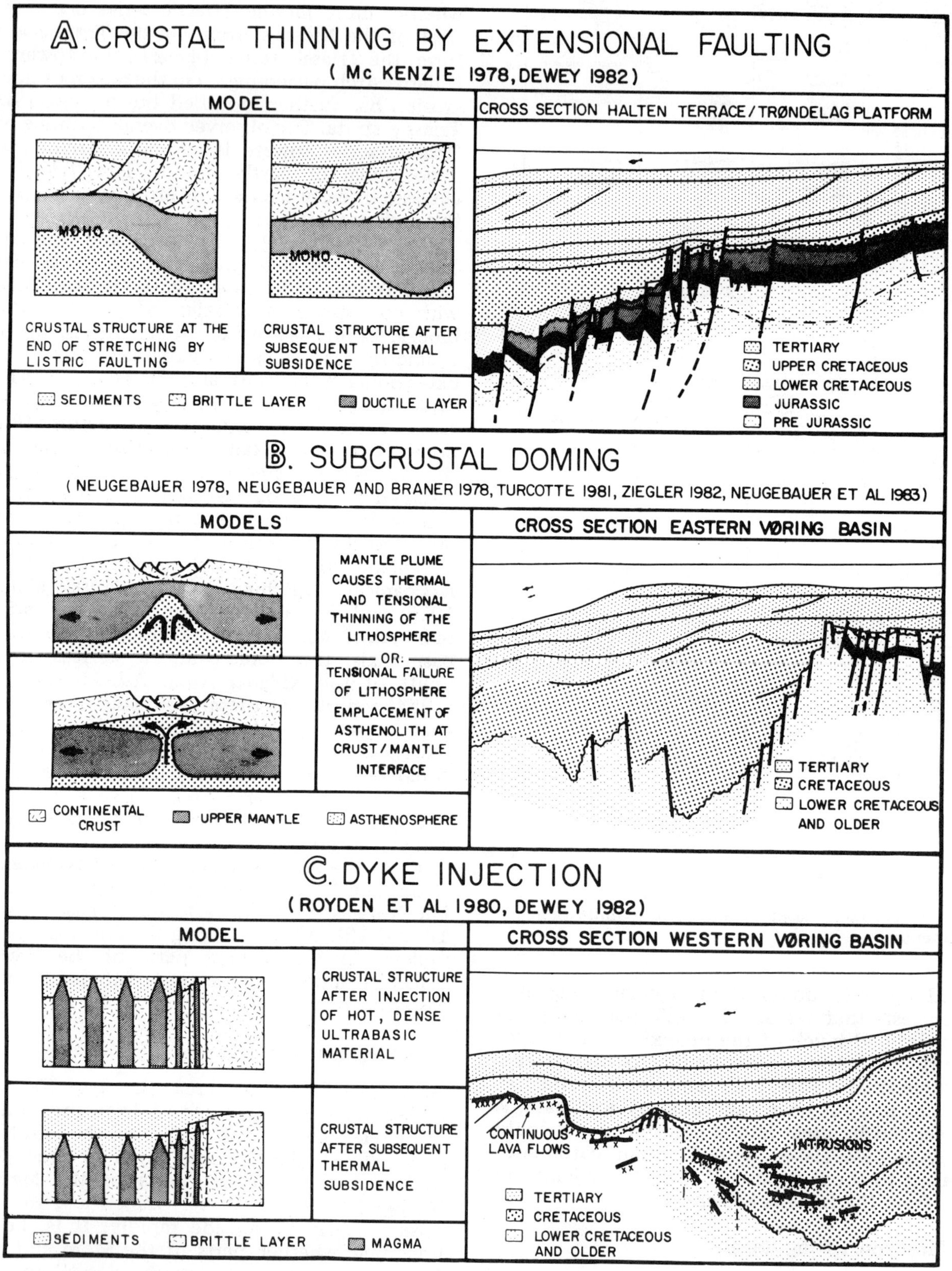

Fig. 11. Driving mechanisms of subsidence. Comparison of theoretical models with data from the mid-Norway area. Vertical exaggeration of cross-sections ca. 15:1.

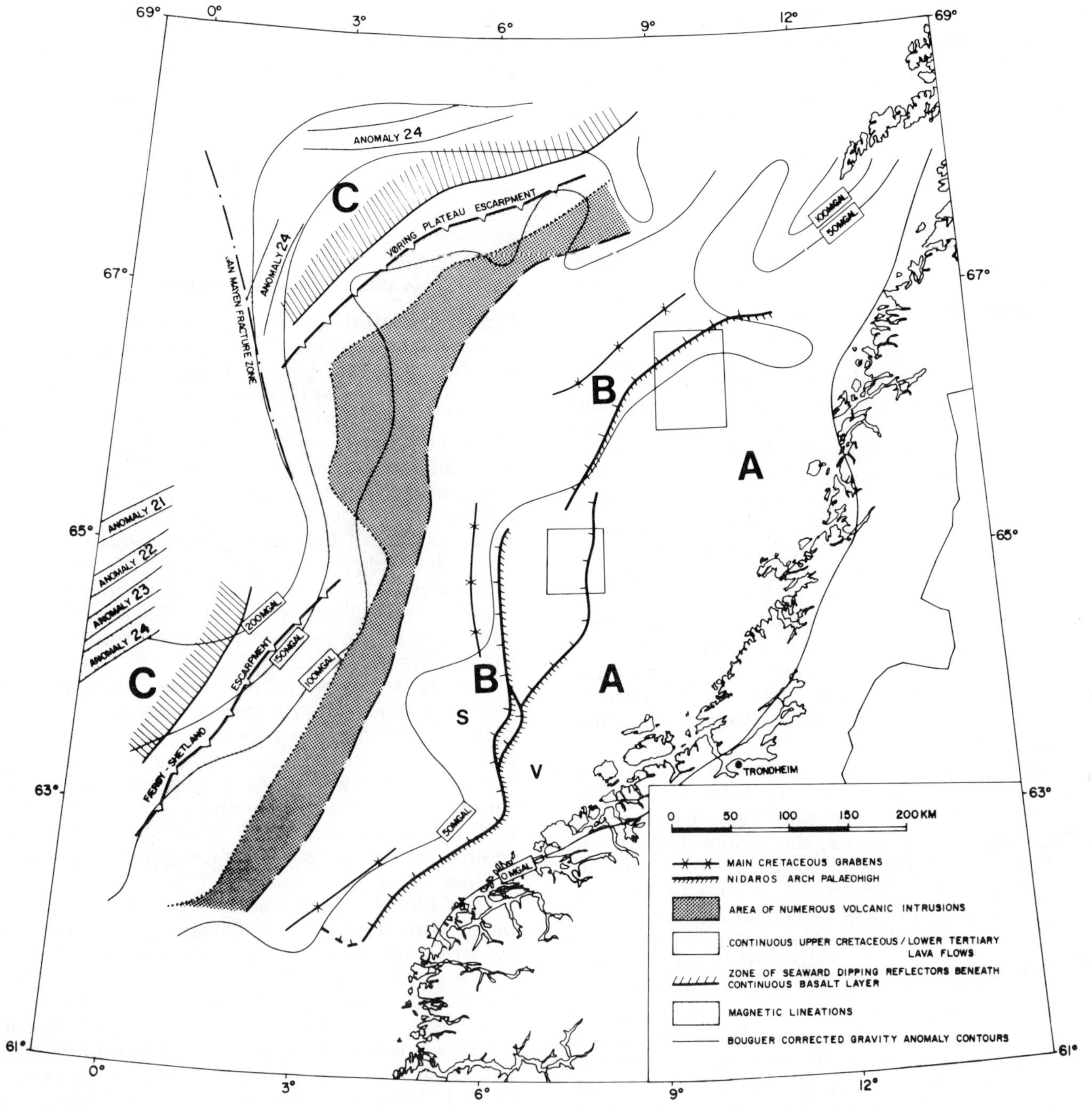

Fig. 12. Mid-Norway Continental Margin: main tectonic elements related to the continental separation. Contraction of area affected by tectonic activity towards its axis: A, First extensional phase. Carboniferous to Jurassic subsidence; B, Second extensional phase. Upper Jurassic—Lower Cretaceous rifting; C, Third extensional phase: Cenozoic continental separation; S, Storegga shelf slide; V, Palaeocene volcanics subcropping at seafloor.

Vøring Basin can be attributed to accelerated crustal extension and thinning.

The mechanism of doming of a rift can be either related to an ascent of the asthenosphere—lithosphere boundary caused by lower mantle convection (mantle plume model) or to failure of the lithosphere in response to a tensional stress system and the passive diapiric ascent of hot, lower mantle material to intra-upper mantle levels or to the crust—mantle interface where it spreads out to form an 'asthenolith' (tensional failure model, Turcotte, 1981; Neugebauer *et al.*, 1983). Both mechanisms cause doming of the crust—mantle interface. It has been shown in model calculations that this induces uplifting of the rift flanks and the concomitant rapid subsidence of the axial rift (Neugebauer, 1978; Neugebauer and Braner, 1978; Hellinger and Sclater, 1983; see Fig. 11B). The associated stress system induced by the combined effect of the surface load of the topography of the uplifted rift flanks and the upsurging low-density upper mantle enhances the horizontal deviatoric tension (Bott, 1979, 1981).

This model seems to be more applicable to the evolution of the Møre—Vøring Rifts than, for example, the concept that subaerial erosion over a major uplift causes the subsequent isostatic subsidence of the rift (Sleep, 1971).

A hypothetical mechanism that possibly contributed to the rapid Cretaceous subsidence of the Møre—Vøring area is crustal thinning by subcrustal erosion. These processes are probably triggered by the emplacement of hot, asthenospheric material at the crust—mantle interface (e.g. asthenolith) and are thought to involve intense dyke and sill intrusion of mantle material into the lower crust causing its gradual assimilation into the mantle. Subsequent cooling of the asthenolith itself can induce further subsidence that is not associated with major tensional faulting (Ziegler, 1982, 1983).

Geophysical data suggest the possible local occurrence of late Jurassic—early Cretaceous volcanics in the area NW of the Nordland Ridge and in the area of the Nykan High. From this and from the regional structural configuration of these areas, it can be construed that the late Kimmerian crustal thinning was even more intense in these areas than in other parts of the Møre and Vøring Basins.

It has been suggested that the rapid Cretaceous subsidence of the latter was associated with limited mid- and late Cretaceous sea-floor spreading (Hanisch, 1984; Price and Rattey, in press). Mid- and late Cretaceous limited sea-floor spreading is documented for the Rockall Trough and postulated for the Faerøy-Shetland Trough. However, gravity, magnetic and reflection seismic data lend no support for the hypothesis that parts of the Møre and Vøring Basins are underlain by oceanic crust.

Stratigraphic data from East Greenland and the Trøndelag Platform indicate that deep-water conditions were established during the late Jurassic and early Cretaceous in the Norwegian-Greenland Sea Rift System.

Whilst during the early Cretaceous the rifts in the Møre and Vøring Basins subsided rapidly and their rift flanks stayed generally high, the Møre and Vøring Basins as a whole, the Halten Terrace and the Trøndelag Platform all began to subside regionally during the late Cretaceous. In large parts of the study area, as in the Viking Graben, major faulting ceased in mid-Cretaceous times. This is indicative of a gradual decrease of the rifting activity. The late Cretaceous subsidence patterns of the mid-Norway area can be related to a gradual cooling of the thermal anomaly that was induced by the late Kimmerian rifting pulse. The resulting increase of the crustal rigidity probably led to a transition from narrow, fault controlled, 'Airy-type' subsidence to broad, flexural subsidence, driven by lithosphere cooling, that was enhanced by sediment and water loading (Watts, 1982; Karner and Watts, 1982; Watts *et al.*, 1982). This model, which explains elegantly the differences between syn- and post-rift subsidence and which provides a reason for the gradual post-rift enlargement of the subsiding area, has been successfully tested in the North Sea (Wood and Barton, 1983; Sclater and Christie, 1980).

In combination with this driving subsidence mechanism, the marked eustatic sea-level rise during the late Cretaceous contributed significantly to the inundation of the mid-Norway rift flanks. Limited well control and imprecise horizon identification, however, restricts our ability to determine quantitatively the Cretaceous subsidence history of the Mid-Norway Shelf.

From the above, it can be concluded that the Cretaceous evolution of the mid-Norway area started with a major rifting event that induced massive graben subsidence by crustal stretching and possibly also by subcrustal erosion. Rifting activity abated gradually in the course of the late Cretaceous. There is no tangible evidence that crustal separation between Greenland and Fennoscandia had already been achieved during the Cretaceous. The late Cretaceous subsidence pattern of the Mid-Norway Shelf can probably be attributed to lithospheric cooling and water and sedimentary loading of the increasingly more rigid crust whereby eustatically rising sea levels played an overprinting role.

Palaeogene crustal separation stage

During the latest Cretaceous and Palaeocene, rifting activity increased again but by now was centred on the future break-up axis between the Greenland and Fennoscandian Cratons (phase 3 of Figs. 10 and 12). This renewed phase of crustal extension was accompanied by the Palaeocene—early Eocene Thulean volcanic surge, which affected not only the Norwegian-Greenland Sea area, but also the Faerøy and Rockall Troughs. Crustal separation was achieved in the Norwegian-Greenland Sea and in the Iceland Sea during the early Eocene. On the Møre Shelf, the continental break-up phase was heralded by volcanic extrusions (Bugge *et al.*, 1978) which are located in line with the prolongation of the Jan Mayen transform fault.

During this late rifting phase, dyke intrusion into the crust and its gradual 'oceanization' probably played a major role (Hinz *et al.*, 1982). The intensity of igneous activity west of the Møre-Vøring Basin axis is very conspicuous on all relevant seismic lines (Figs. 3, 4, 5 and 12). Countless sills and dykes are emplaced in Jurassic to Eocene sediments and are lodged westwards at progressively younger stratigraphical levels (Fig. 11C).

The high level of volcanic activity can be compared with the extensive Palaeocene—Eocene dyke swarms that occur in southeastern Greenland. A model developed for this area proposes that massive invasion of the lighter crustal rocks by denser ultrabasic and basaltic mantle derived material induces isostatic subsidence and zones of weakness which provide suitable sites for the subsequent crustal breakup. Dewey (1982) proposed that during the rifting phase immediately preceding crustal separation, crustal thinning by massive dyke-invasion becomes progressively more important than crustal extension by faulting. In the end massive dyke invasion may become the driving force during the actual crustal separation phase. This model is compatible with the data from the mid-Norway area.

Oligocene and Neogene sea-floor spreading stage

Following crustal separation, the subsidence of the Mid-Norway Margin was governed by progressive lithospheric cooling and contraction and crustal loading by water and sediments. Locally it was also influenced by compressional deformations that can be related to the Oligocene rearrangement of the sea-floor spreading axes in the Norwegian-Greenland Sea. Pleistocene erosion of the coastal basin margin can be related to ice loading and unloading of the mainland areas.

The post crustal separation evolution of the Norwegian-Greenland Sea and of the mid-Norway area has been extensively described (e.g. Eldholm

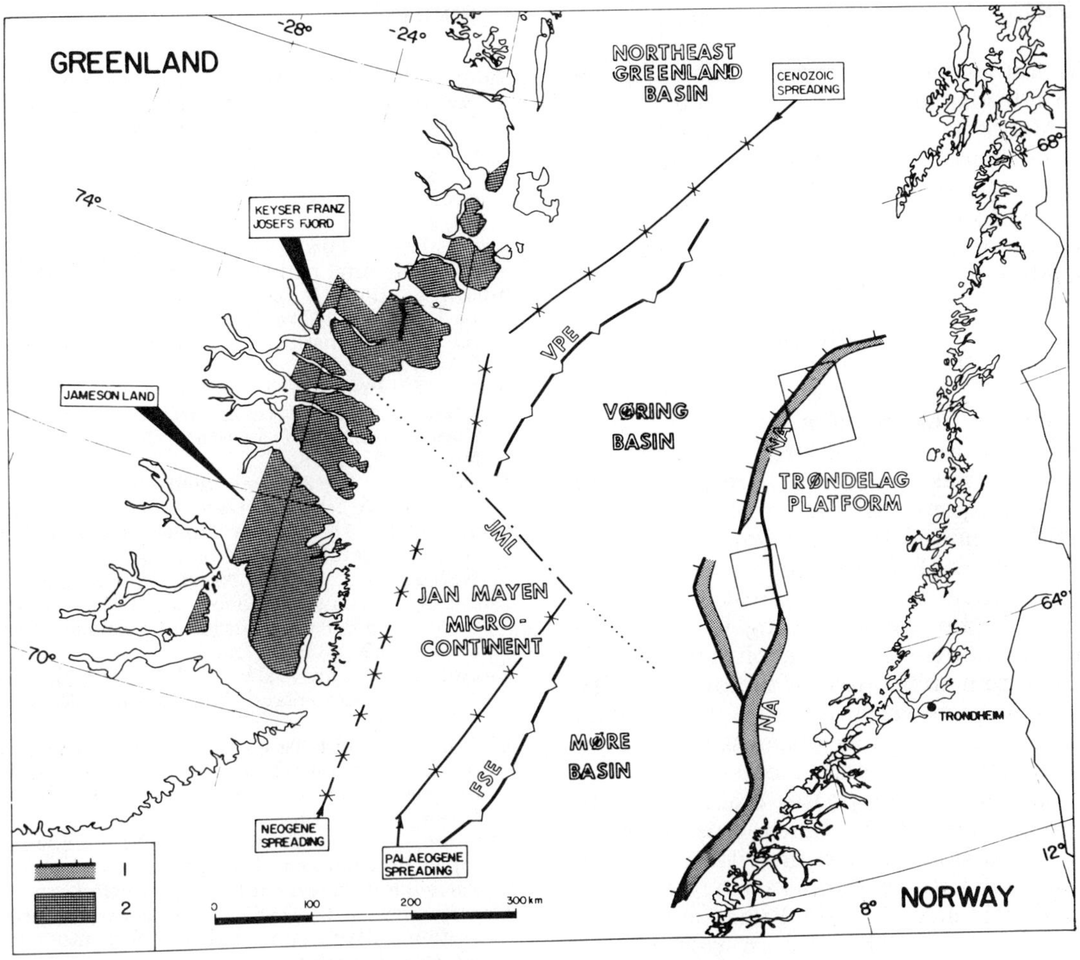

Fig. 13. Norwegian Sea continental separation.

and Thiede, 1980). A few additional points concerning the Cenozoic development of the Mid-Norway Shelf and especially the significance of the Jan Mayen Fracture Zone, can be highlighted:

(1) Pre-split reconstruction: A 'mirror image' rift model, assuming a roughly symmetrical geological structure on both sides of the rift system, has proved useful in understanding the North Sea area. For the mid-Norway area, a pre-split reconstruction has to take into account the fact that the Jan Mayen Fracture Zone developed from an offset of the break-up axis. The location of this offset was probably predetermined by a pre-break-up line of crustal weakness that possibly developed during the Devonian—early Carboniferous wrench movements. Accordingly, the Vøring Basin was juxtaposed to an area north of the Keyzer Franz Josef's Fjord of East Greenland. Thus, the real 'mirror image' of the Trænabanken area is to be found somewhere off-shore of the Northeast Greenland Basin (Fig. 13). Likewise, the stratigraphic framework of the northern parts of the Jan Mayen microcontinent may be inferred from its pre-drift position between Jameson Land of Greenland and the Møre Basin.

(2) The Cenozoic development of the Møre and Vøring Basins differs from a tectonic point of view, in so far as the latter was substantially more affected by transpressional deformations during the Oligo-Miocene.

(3) Large-scale Neogene slumping has been reported from sheared margin basins (Wilson and Williams, 1979). The well known giant Storegga slide in the Møre Basin was probably triggered by Holocene tectonic activity (Bugge *et al.*, 1979) in an area that is located in the eastward prolongation of the Jan Mayen Fracture Zone (Fig. 12).

GRAVITY DATA

Gravity data support, in a general way, the model proposed here for the development of the Mid-Norway Shelf Basin.

The main feature of the Bouguer anomaly map is a gradual increase of gravity towards the transition zone between continental and clearly oceanic crust (Fig. 12). This increase in gravity reflects the steady, progressive thinning of the continental crust toward the continent—ocean boundary which roughly corresponds to the 200-mGal contour. This boundary is located to the west of the Faerøe-Shetland and Vøring Escarpments in the zone where dipping sub-basalt reflectors can be observed and to the east of the first sea-floor magnetic anomalies. The 50-mGal contour tracks approximately the trend of the Nidaros Arch. This arch parallels the ocean—continent transition zone at a distance of some 100 km. As the Nidaros Arch became an outstanding structural feature only during the late Kimmerian rifting phase, the available gravity data suggest that the bulk of crustal thinning in areas to the west of it took place during the Cretaceous.

CONCLUSIONS

The information available on the crustal configuration and the subsidence history of the Mid-Norway Continental Margin is still inadequate for a detailed quantitative analysis. This study should therefore be seen as a first attempt to integrate several data sources in order to obtain a qualitative description of the developmental history of the Mid-Norway Continental Margin, a summary of which is given below:

Following the Caledonian suturing of the Laurentia-Greenland and Fennoscandian Cratons, Devonian and early Carboniferous wrench movements induced the development of a broad zone characterized by a deeply fractured and therefore weakened crust. These fractures were reactivated during the subsequent rifting phases whereby crustal extension caused the subsidence of a complex graben system. Rifting processes remained intermittently active from the latest early Carboniferous until the early Palaeogene or some 270 Ma. In the course of the evolution of this complex rift system the centre of tectonic activity contracted stepwise towards its axial zone, along which crustal separation was ultimately achieved during the late Palaeocene—early Eocene.

During the first rifting phase, embracing late Palaeozoic to Middle Jurassic times, crustal extension affected probably the entire width of the Norway-Greenland Rift System. On the Norwegian side, however, evidence for late Palaeozoic subsidence comes only from the eastern part of the Trøndelag Platform. Massive Triassic—Jurassic subsidence induced by crustal stretching giving rise to listric faulting is evident on the entire Trøndelag Platform. There is little evidence for thermal doming of the Norway-Greenland rift system during the initial 160 Ma of its evolution.

During the second rifting phase, embracing late Jurassic to Cretaceous times, crustal extension was concentrated on areas west of the Nidaros Arch. Uplift of the latter was caused by a thermally activated driving force. The pre-existing structural grain probably continued to influence the location of individual rift-segments.

During the third rifting phase, spanning latest Cretaceous—early Tertiary times, tectonic activity was concentrated along the western margins of the Møre—Vøring Basins, where dyke intrusion probably played a major role during crustal separation.

During the Oligocene to Recent sea-floor spreading stage the Møre Basin subsided more rapidly than the Vøring Basin; the latter was more affected by transpressional deformations which caused local uplift and reverse faulting.

ACKNOWLEDGEMENTS

The authors wish to thank B.M. Thomas of A/S Norske Shell for many invaluable discussions and suggestions during preparation of this work and are grateful to their colleagues in A/S Norske Shell and Shell Internationale Petroleum Mij. B.V. for their support and co-operation.

This paper is published with the permission of A/S Norske Shell and Shell Internationale Petroleum Mij. B.V.

REFERENCES

Bally, A.W. 1981. Atlantic-type margins. In: *Geology of Passive Continental Margins.* Education Course Note, Series No. 19. American Association of Petroleum Geology.

Bott, M.H.P. 1979. Subsidence mechanisms at passive continental margins. *Am. Assoc. Petrol. Geol. Mem.*, **29**, 3—9.

Bott, M.H.P. 1981. Crustal doming and the mechanism of continental rifting. *Tectonophysics*, **73**, 1—8.

Bugge, T., Lien, R.L. and Rokoengen, K. 1979. Store ras på kontinentalskråningen utenfor Midt-Norge. Continental Shelf Institute, Trondheim, Contribution No. P-155/12/79.

Bugge, T., Prestvik, T. and Rokoengen, K. 1978. Lower Tertiary volcanic rocks off Kristiansund-Mid Norway. Continental Shelf Institute, Trondheim, Contribution No. P-155/11/78.

Dewey, J.F. 1982. Plate tectonics and the evolution of the British Isles. *J. Geol. Soc. Lond.*, **139**, 371—412.

Eldholm, O. 1979. The Norwegian-Greenland Sea plate tectonic development. In: *Norwegian Sea Symposium* NSS/2, pp. 1—13. Norsk Petroleumsforening, Oslo.

Eldholm, O. and Sundvor, E. 1979. Geological events during the early formation of a passive margin. *Tectonophysics*, **59**, 233—237.

Eldholm, O. and Talwani, M. 1977. Evolution of the Norwegian-Greenland Sea. *Geol. Soc. Am. Bull.*, **88**, 969—999.

Eldholm, O. and Talwani, M. 1981. The passive margins of northern Europe and East Greenland. In: Dynamics of Passive Margins. Geodynamics Series, *Am. Geophys. Union*, **6**, 30—44.

Eldholm, O. and Thiede, J. 1980. Cenozoic continental separation between Europe and Greenland. *Palaeogeogr. Palaeoclim. Palaeoecol.*, **30**, 243—259.

Elmohandes, S.E. 1981. The central European graben system: rifting imitated by clay modelling. *Tectonophysics*, **73**, 69—78.

Grønlie, G. and Ramberg, I.B. 1970. Gravity indications of deep sedimentary basins below the Norwegian continental shelf and the Vøring Plateau. *Norsk Geologisk Tidsskr.* **50**, 356—391.

Grønlie, G., Chapman, M. and Talwani, M. 1979. Geophysical studies in the Norwegian-Greenland Sea. *Norsk Polarinst. Skr.*, No. 170.

Haller, J. 1971. *Geology of the East Greenland Caledonides.* Interscience, London.

Hanisch, J. 1984. The Cretaceous opening of the NE Atlantic and the west Spitsbergen orogeny — a plate tectonic model. This volume.

Hellinger, J.S. and Sclater, J.G. 1984. Some comments on two-layer extensional models for the evolution of sedimentary basins. *J. Geophys. Res.* (in press).

Hinz, K., Döstmann, H.J. Hanisch, J. 1982. Structural framework of the Norwegian Sea. *Offshore Northern Seas Conference*, Stavanger (Norsk Petroleumsforening) ONS-82/E/4, pp. 1—22.

Hollander, N.B. 1982. Evaluation of the hydrocarbon potential, offshore mid-Norway. *Oil Gas J.*, **15**, 168—172.

Illies, J.H. 1981. Mechanism of graben formation. *Tectonophysics*, **73**, 249—266.

Johnson, G.L., Vogt, P.R. and Avery, O.E. 1970. Evolution of the Norwegian Basin. In: F.M. Delany (ed.), *The Geology of the East Atlantic Continental Margin*, Vol. 2, *Europe.* Institute of Geological Sciences, Report No. 70/14, pp. 53—65.

Jørgensen, F. and Navrestad, T. 1979. Main structural elements and sedimentary succession on the shelf outside Nordland (Norway). *Norwegian Sea Symposium* (Norsk Petroleumsforening). NSS/11, pp. 1—20.

Jørgensen, F. and Navrestad, T. 1981. The geology of the Norwegian shelf between 62°N and the Lofoten Islands. In: L.V. Illing and G.D. Hobson (eds.). *Petroleum Geology of the Continental Shelf of North-West Europe*, Institute of Petroleum, London, pp. 407—413.

Karner, G.D. and Watts, A.B. 1982. On isostasy at Atlantic-type continental margins. *J. Geophys. Res.* **87**(B4), 2923—2948.

Kent, P.E. 1976. Major synchronous events in continental shelves. *Tectonophysics*, **36**, 87—91.

Kent, P.E. 1980. Vertical tectonics associated with rifting

and spreading. *Phil. Trans. R. Soc. Lond.*, **A294**, 125—135.

McKenzie, D. 1978. Some remarks on the development of sedimentary basins. *Earth Planet. Sci. Lett.*, **40**, 25—32.

Myrland, R., Messel, K. and Ræstad, N. 1981. Gas discovery in 35/3 — an exploration history. In: *Norwegian Symposium on Exploration.* (Norsk Petroleumsforening), Article 17.

Mørk, A. 1984. Geological mapping of Mesozoic rocks off mid-Norway (Møre-Lofoten). This volume.

Nairn, A.E.M., Churkin, M., Jr. and Stehl, F.G. 1981. *The Ocean Basins and Margins,* Vol. 5, *The Arctic Ocean.* Plenum Press, New York.

Neugebauer, H.J. 1978. Crustal doming and the mechanisms of rifting, part I: Rift formation. *Tectonophysics,* **45**, 159—186.

Neugebauer, H.J. and Braner, B. 1978. Crustal doming and the mechanisms of rifting, part II: Rift development of the upper Rhinegraben. *Tectonophysics,* **46**, 1—20.

Neugebauer, H.J., Woitt, W.D. and Wallner, H., 1983. Uplift volcanism and tectonics: evidence for mantle diapirs. In: *Plateau Uplift — the Rhenisch Massif. A Case of History.* Springer, Berlin.

Price, I. and Rattey, R.P. 1984. Cretaceous tectonics off mid-Norway: Implications for the Rockall-Faeroes-Shetland Rift. *J. Geol. Soc. Lond.* (in press).

Rawson, P.F. and Riley, L.A. 1982. Latest Jurassic—early Cretaceous events and the 'late Kimmerian unconformity' in North Sea area. *Am. Assoc. Petrol. Geol. Bull.*, **66**(12), 2628—2648.

Royden, L., Sclater, J.G. and von Herzen, R.P. 1980. Continental margin subsidence and heat flow. *AAPG Bull.*, **64**(2), 173—187.

Russell, M.J. and Smythe, D.K. 1978. Evidence of an early Permian oceanic rift in the northern North Atlantic. In: E.R. Neumann and I.B. Ramberg (eds.), *Petrology and Geochemistry of Continental Rifts.* Reidel, Dovdrecht, pp. 173—180.

Rønnevik, H.C. and Navrestad, T. 1977. Geology of the Norwegian Shelf between 62°N and 69°N. *GeoJournal,* **1**, 33—46.

Rønnevik, H.C., Bergsager, E.I., Moe, A. Øvrebø, O., Navrestad, T. and Stangenes, J. 1975. The Geology of the Norwegian continental shelf. In: A.W. Woodland (ed.). *Petroleum and the Continental Shelf of North-West Europe,* Vol. 1, Applied Science, London, pp. 117—129.

Rønnevik, H.C., Jørgensen, F. and Motland, K. 1979. The geology of the northern part of the Vøring Plateau. *Norwegian Sea Symposium* (Norsk Petroleumsforening), Oslo. NSS/12, pp. 1—21.

Rønnevik, H.C., Eggen, S. and Vollset, J. 1982. Exploration of the Norwegian Shelf. Norwegian Petroleum Directorate, Contribution No. 5.

Sclater, J.G. and Christie, P.A.F. 1980. Continental stretching: an explanation of the post-mid Cretaceous subsidence of the Central North Sea Basin. *J. Geophys. Res.*, **85**, 3711—3739.

Sellevoll, M.A. 1975. Seismic refraction measurements and continuous seismic profiling on the continental margin of Norway between 60° and 69°N. *Norges. Geol. Unders.*, **316**, 219—235.

Sleep, N.H. 1971. Thermal effects of the formation of Atlantic continental margins by continental break-up. *Geophys. J. R. Astron. Soc.*, **24**, 325—350.

Steel, R.J. 1976. Devonian basins of Western Norway: sedimentary response to tectonism and varying tectonic context. *Tectonophysics,* **36**, 207—224.

Talwani, M. and Eldholm, O. 1972. Continental margin off Norway: a geophysical study. *Geol. Soc. Am. Bull.*, **83**, 3575—3606.

Talwani, M. and Mutter, J. 1980. Initiation of opening of the Norwegian Sea. L-DGO Contribution No. 3000.

Talwani, M., Udintsev, G., *et al.* (eds.), 1976. *Initial Reports Deep Sea Drilling Project,* Vol. 38. U.S. Govt. Printing Office, Washington, D.C., 1256p.

Turcotte, D.L. 1981. Rifts — tensional failure of the lithosphere. In: *Conference on Processes of Planetary Rifting.* Lunar and Planet. Inst., Contrib. 451, pp. 5—8.

Watts, A.B. 1982. Tectonic subsidence, flexure and global changes of sea level. *Nature (London),* **297**, 469—474.

Watts, A.B., Karner, G.D. and Steckler, M.S. 1982. Lithospheric flexure and the evolution of sedimentary basins. *Phil. Trans. Roy. Soc. London.* **A305**, 249—281.

Wilson, R.C.L. and Williams, C.A. 1979. Oceanic transform structures and the development of Atlantic continental margin sedimentary basins — a review. *J. Geol. Soc. Lond.*, **136**(3), 311—320.

Wood, R. and Barton, P. 1983. Crustal thinning and subsidence in the North Sea. *Nature (London)*, **302**, 134—136.

Ziegler, P.A. 1982. *Geological Atlas of Western and Central Europe.* Elsevier, Amsterdam.

Ziegler, P.A. 1984. Comments to: Crustal thinning and subsidence in the North Sea. *Nature (London),* (in press).

Åm, K. 1970. Aeromagnetic investigations on the continental shelf of Norway, Stad-Lofoten (62—69°N). *Norges Geol. Unders.*, **266**.

Ånstad, M.K., Gabrielsen, R.H., Hagevang, T., Ramberg, I.B. and Torvanger, O. 1981. Correlation of offshore and onshore structural features between 62°N and 68°N, Norway. In: *Norwegian Symposium on Exploration.* NSE/11 (Norsk Petroleumsforening, Oslo), pp: 1—24.

Åsheim, S.M. and Larsen, V. 1984. Hydrocarbon finds at Haltenbanken. This volume, pp. 285—292.

Bibliography

B. Jacobsen and A.M. Spencer,* with contributions by M. Granviken,*† *G.H. Maisey*† *and A. Mørk*†
*BP Petroleum Development Ltd., Stavanger, Norway.
† Continental Shelf Institute (IKU), Trondheim, Norway

INTRODUCTION

In this bibliography we have aimed to include published references on offshore geology, geophysics and hydrocarbon exploration relevant to the Norwegian Continental Shelf from 62°N to Svalbard. Thus the geographical area included comprises the Norwegian Continental Shelf north of 62°, the Barents Shelf and all of the oceanic areas westwards as far as the continental shelf of Greenland. The references are restricted to articles describing the offshore areas; papers on onshore geology have been mostly excluded (e.g. for Norway, Svalbard, Bjørnøya), except for a few recent articles on onshore fracture lineament studies of possible direct relevance to the structure of the nearby offshore regions. Articles solely on plate tectonic theory and mechanisms have been excluded, as have those on the most superficial sea-floor sediments. We have included only full published articles and not abstracts and unpublished technical reports. A selection of references in the Russian language and of published translations from Russian are included. The 503 references here represent a first attempt at a bibliography for offshore earth-science research from the time of Nansen (1904) to mid-1983: the list will inevitably be incomplete.

The list is ordered according to the Norwegian alphabet which comprises 29 characters: all those of the English alphabet plus æ, ø, å/aa at the end.

BIBLIOGRAPHY

Ahlburg and Stoller. 1919. Gesteins- und Sedimentproben aus dem Barentsmeer. *Wissenschaftliche Meeresuntersuchungen, ABT. Helgoland* N.F. 13, H 1, S. XLV—XLVIII und 6—8.

Al-Kasim, F., Rønnevik, H.C. and Ulleberg, K. 1975. Review of the Jurassic offshore Norway. In: *Jurassic Northern North Sea Symposium* (Norsk Petroleumsforening), Article 3.

Anon. 1957. *Gravity Measurement in the Barents Sea in 1957. Free Air Anomalies.* Finnish Geodetic Inst., Publ. 55.

Anon. 1965. *Oceanographic Atlas of the North Atlantic Ocean,* Sect. V, *Marine Geology.* U.S. Naval Oceanographic Office, Publ. No. 700.

Anon. 1967. *Marine Geophysical Survey Program 1965—1967, North Atlantic Ocean, Norwegian Sea and Mediterranean Sea,* Sect. 5, *Geology and Geophysics.* U.S. Naval Oceanographic Office.

Anon. 1970. Anomalies de la pesanteur en Mer de Norvège. *Cahiers Oceanographiques,* **22** (5), 503—514.

Avery, O.E. and Burton, G.D. 1968. An aeromagnetic survey of the Norwegian Sea. *J. Geophys. Res.* **73**, 4583—4600.

Baker, E.W., Palmer, S.E. and Parrish, K.L. 1976. Tetrapyrrole pigments in DSDP Leg 38 sediments. In: M. Talwani *et al.* (eds). *Initial Reports Deep Sea Drilling Project,* Vol. 38, U.S. Govt. Printing Office, Washington, D.C., pp. 785—790.

Bakkelid, S. 1959. *Gravity Observations in a Submarine along the Norwegian Coast.* NGO (Geographical Survey of Norway), Geodetic Publ. No. 11.

Belderson, R.H. and Wilson, J.B. 1973. Iceberg plough marks in the vicinity of the Norwegian Trough. *Nor. Geol. Tidsskr.,* **53**(3), 323—328.

Bell, D.G., Vigran, J.O., Rønningsland, T.M., Edwards M.B. and Aarhus, N. 1981. *Study of Seabottom Samples from the NPD 1976 Barents Sea Cruise.* IKU Report, P-274/2/81, 43pp.

Berggren, W.A. and Schnitker, D. 1983. Cenozoic marine environments in the North Atlantic and Norwegian-Greenland Sea. In: M.H.P. Bott *et al.* (eds). *Structure and Development of the Greenland—Scotland Ridge,* Plenum Press, New York, pp.495—548.

Bergsager, E. 1981. First drilling in Norwegian Sea off Norway yeilds encouraging results. *Oil Gas J.,* **79**(23), 115—124.

Beskow, B., Moe, A. and Rønnevik, H.C. 1980. Status of exploration north of 62°N. In: *Exploration—Geological and Geophysical Highlights. Offshore North Sea Conference* (Norsk Petroleumsforening), Article E-III/9.

Bianchi, A., Daumas, R., Jacq, V., Laborde, P. and Romano, J.C. 1975. La géochimie organique des sédiments marins profonds, Mission Orgon 1, 1974 (Mer de Norvège). *Rev. Inst. Fr. Petrol.,* **30**(2), 197—212.

Birkenmajer, K. 1972. Tertiary history of Spitsbergen and continental drift. *Acta Geol. Pol.,* **22**, 193—218.

Birkenmajer, K. 1981. The geology of Svalbard, the western part of the Barents Sea, and the continental margin of Scandinavia. In: A.E.M. Nairn, M. Churkin Jr., and F.G.

Stehli, (eds.), *The Ocean Basins and Margins*, Vol. 5, *The Arctic Ocean*. Plenum Press. New York, pp. 265—329.

Bissett, C.B. 1930. Observations on material from the sea floor, Barents Sea and Arctic Ocean. (British Arctic Expedition of 1925). *Trans. Geol. Soc. Edin.* 12, 207—212.

Bjorφy, M. and Vigran, J.O. 1978. *Source Rock Studies on Outcrop Samples from the Norwegian Continental Shelf I.* IKU Report P-215/1/78, 63pp.

Bjorφy, M. and Vigran, J.O. 1979. *Organic Geochemistry of Mesozoic Shales on Andφy North Norway.* IKU Report, P-215/2/78, 48pp.

Bjorφy, M. and Vigran, J.O. 1979. Organic Geochemistry of the Lower Cretaceous of Andφya and Spitsbergen. In: *Norwegian Sea Symposium* (Norsk Petroleumsforening), Article 25.

Bjorφy, M. and Vigran, J.O. 1980. *Organic Geochemical Studies of Samples from Barents Sea.* IKU Report, P-160/3/80.

Bjorφy, M., Bue, B. and Elvsborg, A. 1983. Organic geochemical analysis of the two first wells in the Troms 1 area. In: M. Bjorφy, *et al.* (eds.), *Advances in Organic Geochemistry 1981*, Wiley, New York, pp.16—27.

Bjorφy, M., Hall, K. and Vigran, J.O. 1979. An organic geochemical study of Mesozoic shales from Andφya, North Norway. In: A.G. Douglas and J.R. Maxwell (eds.). *Advances in Organic Geochemistry.* Pergamon Press, Oxford, pp. 77—91.

Bjorφy, M., Mφrk, A. and Vigran, J.O. 1983. Organic geochemical studies of the Devonian to Triassic succession on Bjφrnφya and the implications for the Barents Shelf. In: M. Bjorφy, *et al.* (eds.), *Advances in Organic Geochemistry 1981.* Wiley, New York, pp.49—59.

Bjærke, T. and Manum, S.B. 1979. Geology of the Barents Sea Shelf: evidence from palynological studies of drift material. In: *Norwegian Sea Symposium* (Norsk Petroleumsforening), Article 17.

Bjærke, T. and Thusu, B. 1976. Cretaceous palynomorphs from Spitsbergenbanken, N.W. Barents Shelf. *Norsk Polarinst. Årbok.* 1974, 258—262.

Bjφrklund, K.R. 1976. Radiolaria from the Norwegian Sea, Leg 38 of the Deep Sea Drilling Project. In: M. Talwani *et al.* (eds.). *Initial Reports Deep Sea Drilling Project*, Vol. 38, U.S. Govt. Printing Office, Washington, D.C., pp. 1101—1168.

Bjφrklund, K.R. and Kellogg, D.E. 1972. Five new Eocene Radiolarian species from the Norwegian Sea. *Micropaleontology*, 18(3), 386—396.

Bjφrklund, K.R., Goll, R.M. and Meissner, R. 1979. Ice Age climates of the Norwegian-Greenland Sea. *GeoJournal*, 3(3), 273—286.

Bjφrklund, K.R., Thiede, J. and Holtedahl, H. 1979. Deglaciation of the Southeastern Norwegian Sea towards the end of the last glacial age. *Boreas*, 8, 133—135.

Bjφrlykke, K. and Elverhφi, A. 1975. Reworking of Mesozoic clayey material in the northwestern part of the Barents Sea. *Mar. Geol.*, 18(4), 29—34.

Bjφrlykke, K., Bue, B. and Elverhφi, A. 1978. Quaternary sediments in the northwestern part of the Barents Sea and their relation to the under-lying Mesozoic bedrock. *Sedimentology*, 25, 227—246.

Blanc, P.L., Rabussier, D. and Vergnaud-Grazzinic, 1980. North Atlantic deep water formed by the later Middle Miocene. *Nature (London)*, 283, 553—555.

Bleie, J., Nysæther, E. and Oppebφen, K.A. 1982. The hydrocarbon potential of the northern Norwegian Shelf in the light of recent drilling. In: *The Geological Framework and Hydrocarbon Potential of Basins in Northern Seas. Offshore Northern Seas Conference*, Article E/5.

Bogalyubova, L.J. and Timofeev, P.P. 1976. Plant organic matter in sediments from Hole 336, DSDP Leg 38. In: M. Talwani *et al.* (eds.), *Initial Reports Deep Sea Drilling Project*, Vol. 38, U.S. Govt. Printing Office, Washington, D.C., pp. 815—822.

Bott, M.H.P. 1973. The evolution of the Atlantic north of the Faeroe Islands. In: D.H. Tarling and S.K. Runcorn (eds.). *Implications of Continental Drift to the Earth Sciences.* Vol. 1, Academic Press, New York, pp. 175—189.

Bott, M.H.P. 1975. Structure and evolution of the North Scottish Shelf, the Faeroe Block and the intervening region. In: A.W. Woodland (ed.). *Petroleum and the Continental Shelf of North West Europe*, Vol. 1, *Geology*, Applied Science Publishers, pp. 105—115.

Bott, M.H.P. 1975. Sturcture and evolution of the Atlantic Floor between northern Scotland and Iceland. *Nor. Geol. Unders.*, **316**, 195—199.

Bott, M.H.P. 1978. The origin and development of the continental margins between the British Isles and southeastern Greenland. In: D.R. Bowes and B.E. Leake (eds.) Crustal Evolution in Northwestern Britain and Adjacent Regions. *Geol. J. (Liverpool)*, Spec. Iss., **10**, pp. 377.

Bott, M.H.P. 1983. The crust beneath the Iceland—Faeroe Ridge. In: M.H.P. Bott *et al.* (eds.), *Structure and Development of the Greenland-Scotland Ridge*, Plenum Press, New York, pp. 63—76.

Bott, M.H.P. 1983. Deep structure and geodynamics of the Greenland—Scotland Ridge: an introductory review. In: M.H.P. Bott *et al.* (eds.), *Structure and Development of the Greenland-Scotland Ridge*, Plenum Press, New York, pp. 3—10.

Bott, M.H.P. 1984. Deep structure and origin of the Faeroe-Shetland Channel. In: A.M. Spencer *et al.* (eds.), *Petroleum Geology of the North European Margin.* This volume pp. 341—348.

Bott, M.H.P. and Smith, P.J. 1984. Crustal structure of the Faeroe-Shetland Channel. *Geophys. J. R. Astron. Soc.* (in press).

Bott, M.H.P., Sunderland, J., Smith, P.J., Casten, U. and Saxov, S. 1974. Evidence for continental crust beneath the Faeroe Islands. *Nature (London)*, **248**, 202—204.

Bott, M.H.P., Nielsen, P.H. and Sunderland, J. 1976. Converted P-waves originating at the continental margin between the Iceland-Faeroe Ridge and the Faeroe Block. *Geophys. J. R. Astron. Soc.*, **44**, 229—238.

Briseid, E. and Mascle, J. 1975. Structure de la Marge Continentale Norvégienne au débouché de la Mer de Barentz. *Mar. Geophys. Res.*, **2** (2), 231—242.

Browitt, C.W.A. 1972. Seismic refraction investigation of deep sedimentary basin in the continental shelf West of Shetlands. *Nature (London)*, **236**, 161—163.

Buchardt, B. 1981. Tertiary deposits of the Norwegian-Greenland Sea region (Svalbard, Northeast and East Greenland, Iceland, the Faeroe Islands and the Norwegian-Greenland Sea) and their correlation to NW Europe. *Can. Soc. Petrol. Geol. Mem.* **7**, 585—610.

Bugge, T. 1975. *Kart med kystkontur og dybdekoter for den Norske Kontinentalsokkel. (Kartserier i UTM-projeksjon i Målestokk 1:100 000, 1.250 000 og 1,1 mill. med 25 og 50 m Ekvidistanse).* NTNF Continental Shelf, Publ. 55, 21pp.

Bugge, T. 1980. *Øvre Lags Geologi på Kontinentalsokkelen utenfor Mφre og Trφndelag (Shallow Geology on the Continental Shelf off Mφre and Trφndelag, Norway).* IKU Publ. 104.

Bugge, T. and Rokoengen, K. 1976. *Geologisk Kartlegging av de Øvre Lag på Kontinentalsokkelen utenfor Troms. (Geological Mapping Offshore Troms, Northern Norway).* IKU Publ. 85.

Bugge, T., Knarud, R. and Mφrk, A. 1984. Bedrock geology on the mid Norwegian Continental Shelf. In: A.M. Spencer *et al.* (eds.), *Petroleum Geology of the North European Margin.* This volume pp. 271—284.

Bugge, T., Lien, R. and Rokoengen, K. 1976. *Kartlegging av de Prekvartære Lag Utenfor Mφre/Trφndelag (63-65N) (Mapping of the Pre-quaternary Sediments Offshire Mφre-Trφndelag (63-65N).* NTNF Continental Shelf Project, Publ. 69, 47pp.

Bugge, T., Lien, R.L. and Rokoengen, K. 1978. *Kartlegging av Lφsmassene på Kontinentalsokkelen utenfor Mφre og Trφndelag: Seismisk Profilering. (Quaternary Deposits off Mφre and Trφndelag, Norway: Seismic Profiling).* IKU Publ. 99

Bugge, T., Lien, R.L. and Rokoengen, K. 1979. *Store Ras på Kontinentalskråningen utenfor Midt-Norge.* IKU Report P-155/12/79.

Bugge, T., Lφfaldli, H., Maisey, G.H., Rokoengen, K. and Skaar, F.E. 1975. Geological investigation of a Lower Tertiary/Quaternary core offshore Trφndelag, Norway. *Nor. Geol. Unders.*, **316**, 253—270.

Bugge, T., Maisey, G.H. and Rokoengen, K. 1974. *Refleksjonsseismiske undersφkelser med Sparker utenfor Troms 1972.* NTNF Continental Shelf Project Publ. 49, 28pp.

Bugge, T., Prestvik, T. and Rokoengen, K. 1980. Lower Tertiary volcanic rocks off Kristiansund — Mid Norway. *Mar. Geol.*, **35**(4), 277—286.

Bukovics, C., Cartier, E.G., Shaw, N.D. and Ziegler, P.A.

1984. Structure and development of the mid-Norway Continental Margin. In: A.M. Spencer *et al.* (eds.). *Petroleum Geology of the North European Margin.* This volume, pp. 407—424.

Bukry, D. 1976. Silicoflagellate and coccolith stratigraphy, Norwegian-Greenland Sea, Deep Sea Drilling Project Leg 38. In: M. Talwani *et al.* (eds.). *Initial Reports Deep Sea Drilling Project*, Vol. 38, U.S. Govt. Printing Office, Washington, D.C., pp. 843—856.

Bullard, E., Everett, J.E. and Smith, A.G. 1965. The Fit of the continents around the Atlantic. *Philos. Trans. R. Soc. London, A* **258**, 41—51.

Bungum, H. and Husebye, E.S. 1977. Seismicity of the Norwegian Sea; the Jan Mayen Fracture Zone. *Tectonophysics*, **40**, 351—360.

Bungum, H. and Husebye, E.S. 1979. The Meløy, N. Norway, Earthquake sequence — a unique intraplate phenomenon. *Norsar Contr.*, **262**, 189—193.

Bøen, F., Eggen, S. and Vollsæt, J. 1984. Structures and basins of the margin from 62° to 69°N and their development. In: A.M. Spencer (ed.). *Petroleum Geology of the North European Margin.* This volume, pp. 253—270.

Calcagnile, G. and Panza, G.F. 1978. Crust and upper mantle structure under the Baltic Shield and Barents Sea from the dispersion of Rayleigh waves. *Tectonophysics*, **47**, 59—71.

Cameron, D. 1978. Grain-size analyses, Leg 38. In: M. Talwani *et al.* (eds.). *Initial Reports Deep Sea Drilling Project.* Supplement to Vols. 38—41, U.S. Govt. Printing Office, Washington, D.C., pp. 429—432.

Casten, U. 1973. The crust beneath the Faeroe Islands. *Nat. Phys. Sci.*, **241**, 83—84.

Caston, V.N.D. 1976. Tertiary sediments of the Vøring Plateau, Norwegian Sea, recovered by Leg 38 of the Deep Sea Drilling Project. In: M. Talwani *et al.* (eds.). *Initial Reports, Deep Sea Drilling Project*, Vol. 38, U.S. Govt. Printing Office, Washington, D.C. pp. 761—782.

Chalmers, J.A. and Western, P.G. 1979. A Tertiary igneous centre north of the Shetland Islands. *Scott. J. Geol.*, **15**, 333—341.

Closs, H. 1972. Cruise to the Norwegian Sea with F.S. 'Planet', August 4—21, 1969; I. General report by the Chief Scientist; II, Preliminary contributions by the scientists. *'Meteor' Forschungsergeb.*, Reihe C. (MFRCBN), No. 8, 19pp.

Coles, R.L., Hannaford, W. and Haines, G.V. 1978. Magnetic anomalies and the evolution of the Arctic. In: J.F. Sweeney (ed.), *Arctic Geophysical Review Publ.* Vol. 45, No. 4, Earth Physics Branch, Ottawa, pp. 51—66.

Combaz, A., Bellet, J., Caratini, C. and Tissot, C. 1976. Étude microscopique de la matière organique des sédiments Quaternaires de la Mer de Norvège. *Cent Rech. Pau, Bull*, **10**(1), 159—160.

Corbel, J. 1960. Le soulèvement des terres autour de la Mer de Barentz. *Rev. Geogr.*, **35**, 253—274.

Crane, K., Eldholm, O., Myhre, A.M. and Sundvor, E. 1982. Thermal implications for the evolution of the Spitsbergen Transform Fault. *Tectonophysics*, **89**, 1—32.

Damuth, J.E. 1978. Echo character of the Norwegian-Greenland Sea; relationship to Quaternary sedimentation. *Mar. Geol.*, **28**(1—2), 1—36.

Danyushevskaya, A.I., Glebovskaya, Ye. A. and Rogozine, Ye. A. 1973. Geochemical investigations of organic materials in shelf sediments of the Barents Sea. *Geol. Nefti Gaza*, 1, 33—40. (In Russian.)

Daumas, R., Laborde, P., Romano, J.C., Bianchi, A. and Jacq, V. 1975. La géochimie organique des sédiments marins profonds: Mission Orgon 1, 1974. Généralités et résultats obtenus à la Mer, II. *Rev. Inst. Fr. Petrol.* **30**(2), 197—211.

Davies, M.L. 1980. Oil activities on the Norwegian Continental Shelf. *Geogr. Tijdschr.* (NLD), **14**(3), 208—218.

Dekko, T. 1975. *Refleksjonsseismiske Undersøkelser i Vestfjorden 1972.* NTNF Continental Shelf Project Publ. 77, 47pp.

Dekko, T. and Rokoengen, K. 1980. *Submerged Beaches off Troms, North Norway.* IKU Publ. 101, 34pp.

Demenitskaya, R.M. and Hunkins, K.L. 1970. Shape and structure of the Arctic Ocean. In: A.E. Maxwell (ed.). *The Sea*, Interscience, New York, pp. 223—249.

Demenitskaya, R.M., Ivanov, S.S. and Volk, V.E. 1973. Crust of the Arctic Seas of Eurasia. *Tectonophysics*, **20**, 97—104.

Demenitskaya, R.M. Karasik, A.M., Kiselev, Yu. G., Litvinenki, I.V. and Ushakov, S.A. 1968. The transition zone between the Eurasian continents and the Arctic Ocean. *Can. J. Earth Sci.*, **5**, 1125—1129.

Deutsch, E.R. 1969. Palaeomagnetism and North Atlantic paleogeography. *Mem. Am. Assoc. Petrol. Geol.*, **12**, 931—954.

Dibner, V.D., Basov, V.A., Gerke, M.F., Solovyesva, M.F. and Sosipatrova, G.P. 1970. The age of the pre-Quaternary deposits in the bottom of the Barents Sea. *Oceanology*, **10**(4), 520—529.

Dietrich, G. and Horn, W. 1973. Norwegian Sea Expedition 1969. *Meteor. Forsch. Ergeb.*, **A12**, 1—10.

Drivenes, G., Sellevoll, M.A., Renard, V., Avedik, F. and Pajchel, J. 1984. The continental margin/crustal structure off the Lofoten Islands, Northern Norway. In: A.M. Spencer *et al.* (eds.). *Petroleum Geology of the North European Margin.* This volume, pp. 211—216.

Duplessy, J.C., Gonthier, E., Latouche, C. and Moyes, J. 1977. Recent and Late Pleistocene deep sediments in the Northeast Atlantic and South of the Norwegian Sea *2nd Int. Exploit. Oceans Colloq. Proc.*, 4, 1—15pp. (In French.)

Dypvik, H. and Brunfelt, A.O. 1979. Distribution of rare earth elements in some North Atlantic Kimmeridgian black shales. *Nature, Lond.* **278**, 339—341.

Dypvik, H., Rueslåtten, H. and Throndsen, T. 1979. Composition of organic matter from North Atlantic Kimmeridgian shales. *Bull. Am. Assoc. Petrol. Geol.*, **63**(12), 2222—2226.

Dzinoridze, R.N., Jousé, A.P., Koroleva-Golikova, G., Kozlova, G.E. and Nagaeva, G.S. 1978. Diatom and Radiolarian Cenozoic stratigraphy Norwegian Basin: DSDP Leg 38. In: M. Talwani *et al.* (eds.). *Initial Reports Deep Sea Drilling Project.* Supplement to Vols. 38—41. U.S. Govt. Printing Office, Washington, D.C. pp. 289—428.

Edwards, M.B. 1975. Gravel fraction on the Spitsbergen Bank, N.W. Barents Shelf. *Nor. Geol. Unders.*, **316**, 205—217.

Edwards, M.B. 1979. Sandstone in Lower Cretaceous Helvetiafjellet Formation, Svalbard: bearing on reservoir potential of Barents Shelf. *Bull. Am. Assoc. Petrol. Geol.*, **63**, 2193—2203.

Eggvin, J. 1963. *Bathymetric Chart of the Norwegian Sea and Adjacent Areas*, Scale 1:10 000 000. Fiskeridirektoratets Havforskningsinst., Bergen, Norway.

Eldholm, O. 1970. Seismic refraction measurements on the Norwegian Continental Shelf between 62—65°N. *Nor. Geol. Tidsskr.*, **50**(3), 215—229.

Eldholm, O. 1976. The Sediment Distribution and Main Structural Elements of the Norwegian-Greenland Sea and Surrounding Areas Based on Marine Geophysical Data. D. Phil. Thesis, Univ. of Bergen.

Eldholm, O. 1978, Observations on the margin off Norway (66—70°) and the history of early Cenozoic rifting. In: I.B. Ramberg and E-R. Neumann (eds.). *Palaeorift Systems with Emphasis on the Permian Oslo Rift*, Reidel, Dordrecht, pp. 229—236.

Eldholm, O. 1979. Dannelsen av Norskehavet. *Naturen, Oslo*, **1**, 21—27.

Eldholm, O. 1979. The Norwegian-Greenland Sea—plate tectonic development. In: *Norwegian Sea Symposium* (Norsk Petroleumsforening), Article 2.

Eldholm, O. and Ewing, J. 1971. Marine geophysical survey in the South-Western Barents Sea. *J. Geophys. Res.*, **76**, 3832—3841.

Eldholm, O. and Myhre, A.M. 1977. Hovgaard Fracture Zone. *Norsk Polarinst. Årbok*, **1976**, 195—208.

Eldholm, O. and Nysæther, E. 1969. *Seismiske Undersøkelser på den Norske Kontinentalsokkel 1968*. Tekn. Rapp.No. 2C, Seismol. Obs. Univ. Bergen, 29pp.

Eldholm, O. and Sundvor, E. 1974. Barents Sea — geological and geophysical framework. In: *Exploration Geology and Geophysics. Offshore North Sea Conference*, Article G-IV/6.

Eldholm, O. and Sundvor, E. 1979. Geological events during the early formation of a passive margin. *Tectonophysics*, **59**, 233—237.

Eldholm, O. and Sundvor, E. 1980. The continental margins of the Norwegian-Greenland Sea: recent results and outstanding problems. *Philos. Trans. R. Soc. London, Ser. A*, 294, 77—86.

Eldholm, O. and Talwani, M. 1973. Structure and development of the Jan Mayen Ridge. *Eos, Trans. Am. Geophys. Union*, **54**, 324.

Eldhom, O. and Talwani, M. 1975. Barents Sea — geological and geophysical framework. *Oil Gas J.*, **73**(30), 176—177.

Eldholm, O. and Talwani, M. 1977. Sediment distribution and structural framework of the Barents Sea. *Bull. Geol. Soc. Am.*, **88**, 1015—1029.

Eldholm, O. and Talwani, M. 1982. The passive margins of Northern Europe and East-Greenland. In: R.A. Scrutton (ed.). *Dynamics of Passive Margins, Geodynamics Series*, Vol. 6,AGU/GSA, pp. 30—44.

Eldholm, O. and Thiede, J. 1980. Cenozoic continental separation between Europe and Greenland. *Palaeogeogr., Palaeoclimatol., Palaeoecol.*, **30**, 243—259.

Eldholm, O. and Windisch, C.C. 1974. Sediment distribution in the Norwegian-Greenland Sea. *Bull. Geol. Soc. Am.*, **85**, 1661—1676.

Eldholm, O. and Åm, K. 1976. Norwegian Sea — evolution of the continental margin. In: *Exploration Geology and Geophysics. Offshore North Sea Conference*, Article GIV/2.

Eldholm, O., Sundvor, E. and Myhre, A.M. 1979. Continental margin off Lofoten-Vesterålen, Northern Norway. *Mar. Geophys. Res.*, **4**, 3—35.

Eldholm, O., Sundvor, E., Myhre, A.M. and Faleide, J.I. 1984. Cenozoic evolution of the continental margin off Norway and Western Svalbard. In: A.M. Spencer *et al.* (eds.). *Petroleum Geology of the North European Margin.* This volume, pp. 3—18.

Elverhøi, A. 1977. Origin of framboidal pyrite in clayey sediments and in Jurassic black shale in the northwestern part of the Barents Sea. *Sedimentology*, **24**, 591—595.

Elverhøi, A. 1979. Sedimentological and mineralogical investigations of Quaternary bottom sediments of the Norwegian West Coast. *Nor. Geol. Tidsskr.*, **59**(3), 273—284.

Elverhøi, A. and Kristoffersen, Y. 1978. Holocene sedimentation on the shelf around Bjørnøya, northwestern part of the Barents Sea. *Norsk Polarinst. Årbok.*, **1977**, 199—207.

Elverhøi, A. and Kristoffersen, Y. 1978. Glacial deposits southeast of Bjørnøya, northwestern part of the Barents Sea. *Norsk Polarinst. Årbok.*, **1977**, 209—215.

Elvsborg, A. 1979. Late Quaternary sedimentation in a glacial trough on the continental shelf off Troms Northern Norway. *Nor. Geol. Tidsskr.*, **59**(4), 309—325.

Emelyanov, E.M., Blazchishin, A.I., Kharin, G.S., Lozovaya, N.G. and Zangalis, K.P. 1978. Mineral and chemical composition of sediments of the Vøring Plateau, DSDP Leg 38. In: M. Talwani *et al.* (eds.), *Initial Reports Deep Sea Drilling Project*, Supplement to Vols. 38—41. U.S. Govt. Printing Office, Washington, D.C. p. 31.

Emelyanov, E.M., Litvin, V.M., Levchenko, V.A. and Martynova, G.P. 1970. *The Geology of the Barents Sea.* Rep. Inst. Geol. Sci. London, 70/14, 5—15.

Erdman, J.G. and Schorno, K.S. 1976. Geochemistry of carbon, DSDP Leg. 38. In: M. Talwani *et al.* (eds.). *Initial Reports Deep Sea Drilling Project.* Vol. 38, U.S. Govt. Printing Office, Washington, D.C., pp. 791—800.

Eremenko, N.A., Malovitskiy, Ya. P. and Gramberg, I.S. 1973. Geologic structure of oil and gas project of USSR Continental Shelf. *Bull. Am. Assoc. Petrol. Geol.*, **57**(2), 235—243.

Evans, J.R. and Sacks, I.S. 1979. Deep structure of the Iceland Plateau. *J. Geophys. Res.*, **84**, 6859—6867.

Ewing, J. and Ewing, M. 1959. Seismic refraction measurements in the Atlantic Ocean basins, in the Mediterranean Sea, on the Mid-Atlantic Ridge and in the Norwegian Sea. *Bull. Geol. Soc. Am.*, **70**, 291—318.

Fairhall, D. 1977. North of Murmansk. *Geol. Mag.* **49**(6), 347—348, 350, 353.

Faleide, J.I. and Gudlaugsson, S.T. 1981. Geology of the Western Barents Sea — A Regional Study Based on Marine Geophysical Data. Cand. Real. Thesis in Applied Geophysics, Dept. of Geology, University of Oslo, 160 pp.

Faleide, J.I., Gudlaugsson, S.T. and Jacquart, G. Evolution of the Western Barents Sea. *J. Mar. Petrol. Geol.* (in press).

Faleide, J.I., Gudlaugsson, S.T., Johansen, B., Myhre, A.M. and Eldholm, O. Free-air gravity anomaly maps of the Greenland Sea and the Barents Sea. *Polar Res.* (in press).

Feden, R.H., Vogt, P.R. and Fleming, H.S. 1979. Magnetic and bathymetric evidence for the 'Yermak' Hot Spot northwest of Svalbard. *Earth Planet. Sci. Lett.*, **44**, 18—38.

Fleischer, U., Holzkamm, F., Vollbrecht, K. and Voppel, D. 1974. Die Struktur des Island-Färöer-Rückens aus Geophysikalischen Messungen. *Dt. Hydrogr. Z.*, **27**(3), 97—113.

Fønstelien, S. and Horvei, A. 1979. Thoughts and considerations arising from the study of a 'prospective' feature in the Troms 1 Area. In: *Norwegian Sea Symposium* (Norsk Petroleumsforening), Article 20.

Frebold, H. 1930. Die Mesozoische Entwicklung des Barentssee Schelfes. *Geol. Rundschau*, **21**, 343—345.

Frebold, H. 1934. Tatsachen und Deutungen zur Geologie der Arktis. *Medd. Dansk Geol. Foren.*, **6**, 301—326.

Frebold, H, 1951. *Geologie des Barentschelfes.* Abh. Deutsch. Akad. Wiss. Berlin Jahrg. 1950 M. No. 5 Klasse für Math. und Allgemeine Natur. Abh. zur. Geotektonik, No. 4. 51pp.

Gabrielsen, R.H. Long lived fault zones and their influence on tectonic development of the Southwestern Barents Sea. *J. Geol. Soc. Lond.* (in press).

Gabrielsen, R.H. and Ramberg, I.B. 1979. Fracture patterns in Norway from Landsat imagery: results and potential use. In: *Norwegian Sea Symposium* (Norsk Petroleumsforening), Article 23.

Gabrielsen, R.H. and Ramberg, I.B. 1979. Tectonic analysis of the Meløy Earthquake area based on Landsat lineament mapping. *Nor. Geol. Tidsskr.*, **59**, 183—187.

Gabrielsen, R.H. and Robinson, C. 1984. Tectonic inhomogeneities of the Kristiansund-Bodø Fault Complex, off-shore mid-Norway. In: A.M. Spencer *et al.* (eds.). *Petroleum Geology of the North European Margin.* This volume, pp. 397—406.

Gabrielsen, R.H., Færseth, R., Hamar, G. and Rønnevik, H. 1984. Nomenclature of the main structural features on the Norwegian Continental Shelf north of the 62nd parallel. In: A.M. Spencer *et al.* (eds.). *Petroleum Geology of the North European Margin.* This volume, pp. 41—60.

Gabrielsen, R.H., Ramberg, I.B., Mørk, M.B.E. and Tveiten, B. 1981. Regional geological, tectonic and geophysical features of Nordland, Norway. *Earth Evol. Sci.* **1**, 14—26.

Gairaud, H., Jacquart, G., Aubertin, F. and Beuzart, P. 1978. The Jan Mayen Ridge synthesis of geological knowledge and new data. *Ocean. Acta*, **1**, 335—358.

Gausland, I. 1973. Seismic exploration north of 62 degrees N. *Northern Offshore*, **2**(3), 61—68.

Gieskes, J.M., Lawrence, J.R. and Galleisky, G. 1978. Interstitial water studies Leg 38. In: M. Talwani *et al.* (eds). *Initial Reports Deep Sea Drilling Project*, Supplement to Vols. 38—41, U.S. Govt. Printing Office, Washington, D.C., pp. 121—134.

Gjelsvik, T. 1976. *Utforskningen av Kontinentalsokkelen i Barentshavet — Fra Fridtjof Nansens Tid til i dag* (English Summary). Nansen Minneforelesning XII, 10 Okt. 1975. Norske Vid. Ak., Oslo.

Gjelsvik, T. 1972. Norsk Polarinstitutts Virksomhet på Sokkelen. *Norsk Polarinst. Årbok.*, **1970**, 213—223.

Gloppen, T.G. and Westre, S. 1982. Petroleum potential big off Northern Norway. *Oil Gas J.*, **82**(23), 114—127.

Goll, R.M. 1978. Five Trissocyclid Radiolaria from Site 338. In: M. Talwani *et al.* (eds.). *Initial Reports Deep Sea Drilling Project*, Supplement to Vols. 38—41, U.S. Govt. Printing Office, Washington, D.C., pp. 177—192.

Goll, R.M. and Bjørklund, K.R. 1980. The evolution of *Eucoronis fridtjof nanseni*, n. sp. and its application to the Neogene biostratigraphy of the Norwegian-Greenland Sea. *Micropaleontology*, **26**(4), 356—371.

Gowers, M.B. and Lunde, G. 1984. The geological history of Traenabanken. In: A.M. Spencer *et al.* (eds.). *Petroleum Geology of the North European Margin.* This volume, pp. 237—252.

Gramberg, I.S. 1974. Geological premises for oil and gas possibilities of the West Arctic Seas of the USSR. In: *Exploration Geology and Geophysics. Offshore North Sea Conference*, Article G-IV.

Grønlie, G. 1979. Tertiary palaeogeography of the Norwegian-Greenland Sea. *Norsk Polarinst. Skr.*, **170**, 49—61.

Grønlie, G. and Ramberg, I.B. 1970. Gravity indications of deep sedimentary basins below the Norwegian Continental Shelf and the Vøring Plateau. *Nor. Geol. Tidsskr.*,

50, 375—391.

Grønlie, G. and Talwani, M. 1978. *Geophysical atlas; Norwegian-Greenland Sea,* Vema Research Series. Vol. 4. Geophysical Survey Maps, Lamont-Doherty Geological Observatory, Palisades, NY, 26 pp.

Grønlie, G. and Talwani, M. 1979. Bathymetry of the Norwegian-Greenland Sea. *Norsk Polarinst. Skr.*, **170**, 3—24.

Grønlie, G. and Talwani, M. 1982. The free-air gravity field of the Norwegian-Greenland Sea and adjacent areas. *Earth Evol. Sci.*, **2**, 79—103.

Grønlie, G., Chapman, M. and Talwani, M. 1979. Jan Mayen Ridge and Iceland Plateau: origin and evolution. *Norsk Polarinst. Skr.*, **170**, 25—48.

Grønlie, G., Elverhøi, A. and Kristoffersen, Y. 1980. A seismic velocity inversion on Bjørnøya — The Western Barents Shelf. *Mar. Geol.*, **35**(14), 17—26.

Gunleiksrud, T. and Rokoengen, K. 1979. *Regional Geological Mapping of the Norwegian Continental Shelf with Samples of Engineering Applications.* IKU Report P-155/1/79, 20pp.

Gunleiksrud, T. and Rokoengen, K. 1980. Regional geological mapping of the Norwegian Continental Shelf with examples of engineering applications. In: D.A. Aardus, (ed.). *Offshore Site Investigation,* Graham & Trotman, London, pp. 23—35.

Hagevang, T., Eldholm, O. and Aalstad, I. 1983. Pre-23 magnetic anomalies between Jan Mayen and Greenland-Senja Fracture Zones in the Norwegian Sea. *Mar. Geophys. Res.*, **5**, 345—363.

Haines, G.V., Hannaford, W. and Serson, P.H. 1970. Magnetic anomaly maps of the Nordic countries and the Greenland and Norwegian Seas. *Dominion Observ., Ottawa Pubs.*, **39**(5), 123—149.

Hallam, A. 1971. Mesozoic geology and the opening of the North Atlantic. *J. Geol.*, **79**, 129—157.

Halvorsen, S. 1974. Four sediment cores from the continental shelf outside Trøndelag. *Nor. Geol. Unders.*, **304**, 21—31.

Hamar, G.P. and Hjelle, K. 1984. Tectonic framework of the Møre Basin and the Northern North Sea. In: A.M. Spencer *et al.* (eds.). *Petroleum Geology of the North European Margin.* This volume, pp. 349—358.

Hanisch, J. 1983. The Structural Evolution of the NE Atlantic Region. *Geol. Jahrbuch, Reihe* B, 52, pp. 3—37.

Hanisch, J. The Cretaceous opening of the Northeast Atlantic. *Tectonophysics* (in press.)

Hanisch, J. 1984. West Spitsbergen foldbelt and Cretaceous opening of the NE Atlantic. In: A.M. Spencer *et al.* (ed.). *Petroleum Geology of the North European Margin.* This volume, pp. 187—198.

Hannaford, W. and Haines, G.V. 1968. A three component aeromagnetic survey of the Nordic countries and the Greenland Sea. *Can. Dominion Observ., Ottawa, Publ.*, **37**(5), 113—164.

Haq, B.U. and Lohmann, G.P. 1978. Remarks on the Oligocene calcareous mannoplankton biogeography of the Norwegian Sea, DSDP Leg 38. In: M. Talwani *et al.* (eds.). *Initial Reports Deep Sea Drilling Project,* Supplement to Vols. 38—41, U.S. Govt. Printing Office, Washington, D.C., pp. 141—146.

Harland, W.B. 1965. The tectonic evolution of the Arctic—North Atlantic Region. *Philos. Trans. R. Soc. London, Ser. A.* **258**, 59—75.

Harland, W.B. 1969. Contribution of Spitsbergen to understanding of tectonic evolution of North Atlantic region. *Mem. Am. Assoc. Petrol. Geol.*, **12**, 817—851.

Harland, W.B. 1969. Mantle changes beneath the Barents Shelf. *Trans. N.Y. Acad. Sci.*, Ser. II, **31**(1), 25—41.

Harland, W.B. 1973. Tectonic evolution of the Barents Shelf and related plates. *Mem. Am. Assoc. Petrol. Geol.*, **19**, 599—608.

Harland, W.B. and Gayer, R.A. 1972. The Arctic Caledonides and earlier oceans. *Geol. Mag.*, **109**(4), 289—314.

Harland, W.B., Gaskell, B.A., Heafford, A.P., Lind, E.K. and Perkins, P.J. 1984. Outline of Arctic post-Silurian continental displacements. In: A.M. Spencer *et al.* (eds.). *Petroleum Geology of the North European Margin.* This volume, pp. 137—148.

Hedberg, H.D. 1980. Discussion; continental margin off East Greenland. *10th World Petroleum Congress,* Vol. 2. Heyden, London, pp. 419—421.

Heezen, B.C. and Ewing, M. 1961. The mid-oceanic ridge and its extension through the Arctic Basin. In: G.O. Raasch (ed.). *Geology of the Arctic,* Univ. Toronto Press, pp. 522—643.

Heirtzler, J.R. 1973. The evolution of the North Atlantic Ocean. In: D.H. Tarling and S.K. Runcorn (eds.).*Implications of Continental Drift to the Earth Sciences,* Vol. 1, Academic Press, London, pp. 191—196.

Herron, D.A., Øvretveit, A. and Almaas, J. 1984. Anomalous seismic event in Lower Tertiary strata, Trænabanken. In: A.M. Spencer *et al.* (eds.). *Petroleum Geology of the North European Margin.* This volume, pp. 293—302.

Hinz, K. 1972. Der Krustenaufbau des Norwegischen Kontinentalrandes (Vøring Plateau) und der Norwegischen Tiefsee zwischen 66 und 68 N, nach seismischen Untersuchungen. *'Meteor' Forschungsergeb.*, Reihe, C, No. 10, pp. 1—16.

Hinz, K. 1972. The seismic crustal structure of the Norwegian Continental Margin in the Vøring Plateau, in the Norwegian deep sea and on the eastern flanks of the Jan Mayen Ridge between 66—68°N. *24th Int. Geol. Congr.*, Sect. 8, pp. 28—37.

Hinz, K. 1976. The Barents Sea. In: *Exploration Geology and Geophysics. Offshore North Sea Conference,* Article G-IV/5.

Hinz, K. and Moe, A. 1971. Crustal structure in the Norwegian Sea. *Nat. Phys. Sci.*, **232**, 187—190.

Hinz, K. and Schlüter, H.U. 1978. The geological structure of the Western Barents Sea. *Mar. Geol.*, **26**, 199—230.

Hinz, K. and Schlüter, H.U. 1978. Der Nordatlantik — Ergebnisse geophysikalischer Untersuchungen der Bundesanstalt für Geowissenschaften und Rohstoffe an Nordlantischen Kontinentalrändern. *Erdöl-Erdgas-Z.*, **94**, 271—280.

Hinz, K. and Schlüter, H.U. 1979. The North Atlantic; results of geophysical investigations by the Federal Institute for Geosciences and Natural Resources on North Atlantic continental margins. *Oil Gas Erdöl-Erdgas-Z.*, Int. Ed. (III), 5(3), 31—38.

Hinz, K. and Schlüter, H.U. 1980. Continental margin off East Greenland. *10th World Petroleum Congress,* Vol. 2. Heyden, London, pp. 405—418.

Hinz, K. and Weber, J. 1976. Zum Geologischen Aufbau des Norwegischen Kontinentalrandes und der Barents See nach Reflexionsseismischen Messungen. *Compendium 75/76 Erganzungsband der Zeitschrift Erdöl und Kohle, Erdgas, Petrochemie 3-29.*

Hinz, H.K., Dostman, H. and Hanisch, J. 1982. Structural Framework of the Norwegian Sea. In: *The Geological Framework and Hydrocarbon Potential of Basins in Northern Seas, Offshore Northern Seas Conference,* Article E/4.

Hollander, N.B. 1982. Evaluation of the hydrocarbon potential offshore mid-Norway. *Oil Gas. J.* **80**(15), 168—172.

Hollander, N.B. 1984. Geohistory and hydrocarbon evaluation of the Haltenbank area. In: A.M. Spencer *et al.* (eds.). *Petroleum Geology of the North European Margin.* This volume, pp. 383—388.

Holtedahl, H. 1955. On the Norwegian Continental Terrace, primarily outside Møre-Romsdal: its geomorphology and sediments. *Univ. Bergen Årbok,* **14**, 1—209.

Holtedahl, H. 1959. Some remarks on geomorphology of continental shelves off Norway, Labrador and Southeast Alaska. *J. Geol.*, **66**, 461—471.

Holtedahl, H. and Bjerkli, K. 1975. Pleistocene and Recent sediments of the Norwegian Continental Shelf (62°N—71°N) and the Norwegian Channel area. *Nor. Geol. Unders.*, **316**, 241—252.

Holtedahl, H. and Bjerkli, K. 1982. Late Quaternary sediments and stratigraphy on the continental shelf off Møre-Trøndelag, W. Norway. *Mar. Geol.*, 45 (3/4), 179—226.

Holtedahl, H. and Sellevoll, M.A. 1971. Notes on the *Continental Margin of the Eastern Norwegian Sea and of the Skagerrak.* Inst. Geol. Sci., Rep. 70/14, pp. 33—52.

Holtedahl, H. and Sellevoll, M.A. 1972. Notes on the influence of glaciation on the Norwegian Continental Shelf bordering on the Norwegian Sea. Special Report, *Ambio,* **2**, 31—38.

Holtedahl, H., Haldorsen, S. and Vigran, J.O. 1974. Two sediment cores from the Norwegian Continental Shelf between Haltenbanken and Frøyabanken 64°06′N 7°39′E. *Nor. Geol. Unders.*, **304**, 1—20.

Holtedahl, O. 1929. Some remarkable features of the

submarine relief on the north coast of the Varanger Peninsula Northern Norway. *Norske Vid. Ak., Oslo, 1, Mat. Nat. K1.*, (12), 1—14.

Holtedahl, O. 1933. Eine unterseeische Dislokations-Linie ausserhalb der Norwegischen Küste. *Z. Ges. Erkunde, Berlin*, 5(6), 198—204.

Holtedahl, O. 1937. On fault-lines indicated by the submarine relief in the shelf area west of Spitsbergen. *Norsk Geogr. Tidsskr.*, **6**, 214—221.

Holtedahl, O. 1940. The submarine relief off the Norwegian coast. With bathymetrical maps in seven sheets of the Norwegian coastal waters and adjoining areas. Spec. Publ., *Norske Vid. Ak., Oslo*, 1—43.

Holtedahl, O. 1950. Supposed marginal fault-lines in the shelf area off some High Northern Lands. *Bull. Geol. Soc. Am.*, **61**, 493—500.

Holtedahl, O. 1952. A comparison of a Scottish and a Norwegian Shelf area. *Trans. Edinburgh Geol. Soc.*, **15**, 214—220.

Holtedahl, O. and Holtedahl, H. 1961. On 'marginal channels' along continental borders and the problem of their origin. *Bull. Geol. Inst. Uppsala*, **40**, 183—187.

Hood, A., Castano, J.R. and Kendrick, J.W. 1976. Petroleum-generating potential and thermal history of DSDP Leg 38 sediments. In: M. Talwani *et al.* (eds.). *Initial Reports Deep Sea Drilling Project*, Vol. 38, U.S. Govt. Printing Office, Washington, D.C., pp. 801—804.

Horowitz, A. 1974. The geochemistry of sediments from the northern Reykjanes Ridge and the Iceland—Faeroes Ridge. *Mar. Geol.*, **17**, 103—122.

Houtz, R.E. 1980. Seafloor and near-surface sound velocities from Barents Sea sonobuoy data. *J. Geophys. Res.*, **85**, 4838—4844.

Houtz, R.E. 1980. Comparison of velocity—depth characteristics in Western North Atlantic and Norwegian Sea sediments. *J. Acoust. Soc. Am.*, **68**(5), 1409—1414.

Houtz, R.E. and Windisch, T.C. 1977. Barents Sea Continental Margin sonobuoy data. *Bull. Geol. Soc. Am.*, 8, 1030—1036.

Hunt, J.M. 1976. C4—C7 alkane yields. In: M. Talwani *et al.* (eds.). *Initial Reports Deep Sea Drilling Project*, Vol. 38, U.S. Govt. Printing Office, Washington, D.C., pp. 807—808.

Husebye, E.S., Gjøystad, H., Bungum, H. and Eldholm, O. 1975. The Seismicity of the Norwegian and Greenland Seas and adjacent continental shelf areas. *Tectonophysics*, **26**, 55—70.

Igiazarov, B.Kh. 1975. *Tectonic Map of the North Polar Regions of the Earth* (in Russian). Scientific Research Inst. Arctic Geol. (NIIGA), Ministry of Geology of USSR. 1:5 000 000 (6 sheets).

IKU, 1979. IKU with good background material for drilling north of the 62nd Parallel. *Scand. Oil-Gas Mag.*, 7, 18—22.

Ivanov, M.M. 1967. Some peculiarities of the magnetic field in the Eastern Norwegian Sea. *Oceanology*, **7**, 363—373.

Jackson, H.R. and Reid, T. 1980. Crustal refraction and seismic reflection results from Fram I. *Eos. Trans. Am. Geophys. Union*, **61**(17), 276.

Jacobsen, V.W. and Van Veen, P. 1984. The Triassic offshore Norway North of 62°N. In: A.M. Spencer *et al.* (eds.). *Petroleum Geology of the North European Margin.* This volume, pp. 317—328.

Johnson, G.L. 1968. Marine geology in the environs of Jan Mayen. *Norsk Polarinst. Årbok.*, **1966**, 105—111.

Johnson, G.L. 1972. The mid-oceanic ridge in the Greenland Sea. *Jøkull*, **22**, 65—68.

Johnson, G.L. 1976. The Jan Mayen Ridge. *Can Soc. Petrol. Geol. Mem.*, 4, 225—233.

Johnson, G.L. and Campsie, J. 1976. Morphology and structure of the western Jan Mayen Fracture Zone. *Norsk Polarinst. Årbok.*, **1974**, 69—81.

Johnson, G.L. and Eckhoff, O.B. 1966. Bathymetry of the North Greenland Sea. *Deep-Sea Res.*, **13**, 1161—1173.

Johnson, G.L. and Heezen, B.C. 1967. Morphology and evolution of the Norwegian-Greenland Sea. *Deep-Sea Res.*, **14**, 755—771.

Johnson, G.L. and Monahan, D. 1979. Geomorphology of the Arctic Basin and adjacent continental margins. *Eos. Trans. Am. Geophys. Union*, **60**(18), 372.

Johnson, G.L. and Vogt, P.R. 1973. Marine geology of the Atlantic Ocean north of the Arctic Circle. *Mem. Am. Assoc. Petrol. Geol.*, **19**, 161—170.

Johnson, G.L., Ballard, J.A. and Watson, J.A. 1968. Seismic studies of the Norwegian Continental Margin. *Norsk Polarinst. Årbok.*, **1966**, 112—119.

Johnson, G.L., Egloff, J. and McMillan, N.J. 1975. Continental margins of Eastern Greenland. *Can. Soc. Petrol. Geol. Mem.*, 4, 205—224.

Johnson, G.L., Freitag, J.S. and Pew, J.A. 1970. Structure of the Norwegian Basin. *Norsk Polarinst. Årbok.*, **1969**, 7—16.

Johnson, G.L., Sommerhoff, G. and Egloff, J. 1975. Structure and morphology of the West Reykjanes Basin and the Southeast Greenland Continental Margin. *Mar. Geol.*, **18**, 175—196.

Johnson, G.L., Taylor, P.T., Vogt., P.R. and Sweeney, J.F. 1978. Arctic Basin morphology. *Polarforschung*, **48**(1/2), 20—30.

Johnson, G.L., Vogt, P.R. and Avery, O.E. 1971. Evolution of the Norwegian Basin. In: F.M. Delany (ed.). *The Geology of the East Atlantic Continental Margin*, Vol. 2, Rep. Inst. Geol. Sci. London 70/14, pp. 53—65.

Jørgensen, F. and Navrestad, T. 1979. Main structural elements and sedimentary succession on the shelf outside Nordland. In: *Norwegian Sea Symposium* (Norsk Petroleumsforening), Article 11.

Jørgensen, F. and Navrestad, T. 1981. Main structural elements and sedimentary succession of the shelf outside Nordland (Norway). *Earth Evol. Sci.*, **1**, 43—49.

Jørgensen, F. and Navrestad, T. 1981. The geology of the Norwegian Shelf between 62°N and the Lofoten Islands. In: L.V. Illing and G.D. Hobson (eds.). *Petroleum Geology of the Continental Shelf of North-West Europe.* Heyden, London, pp.407—413.

Kanestrøm, R., Berge, A.M. and Pedersen, L.M. 1976. *Detailed Seismic Velocities along a 250 km Profile in the Barents Sea.* Rep. No. JRK—1, Seismol. Obs., Univ. Bergen.

Karlsson, W. 1984. Sedimentology and diagenesis of Jurassic sediments offshore mid-Norway. In: A.M. Spencer *et al.* (eds.). *Petroleum Geology of the North European Margin.* This volume, pp. 389—396.

Kay, M. 1969. Continental drift in North Atlantic Ocean. *Mem. Am. Assoc. Petrol. Geol.*, **12**, 965—973.

Kellogg, H.E. 1975. Tertiary stratigraphy and tectonism in Svalbard and continental drift. *Bull. Am. Assoc. Petrol. Geol.*, **59**, 465—485.

Kellogg, T.B. Duplessy, C.J. and Shackleton, N.J. 1978. Planktonic foraminiferal and oxygen isotopic stratigraphy and paleoclimatology of Norwegian Sea deep-sea cores. *Boreas*, **7**(1), 61—73.

Kent, D.V. and Opdyke, N.D. 1978. Paleomagnetism and magnetic properties of igneous rock samples, Leg. 38. In: M. Talwani *et al.* (eds.). *Initial Reports Deep Sea Drilling Project.* Supplement to Vols. 38—41. U.S. Govt. Printing Office, Washington, D.C., pp. 3—8.

Kharin, G.N. 1976. The petrology of magmatic rocks, DSDP Leg 38. In: M. Talwani *et al.* (eds.). *Initial Reports Deep Sea Drilling Project*, Vol. 38, U.S. Govt. Printing Office, Washington, D.C., pp.685—716.

Kharin, G.N., Udintsev, G.B., Bogatikov, O.A., Dmitriev, J.I., Raschka, H., Kreyzer, H., Mohr, M., Harre, W. and Eckhardt, F.J. 1976. K/Ar age of the basalts of Norwegian-Greenland Sea Glomar Challenger Leg 38 Deep Sea Drilling Project. In: M. Talwani *et al.* (eds.). *Initial Reports Deep Sea Drilling Project*, Vol. 38, U.S. Govt. Printing Office, Washington, D.C., pp.755—760.

Kihle, R. and Løfaldli, M. 1973. *Atlas of Foraminifera from Unconsolidated Sediments on the Norwegian Continental Shelf — Description and Stratigraphic Occurrence of 214 Species.* NTNF Continental Shelf Project, Publ. 35.

King, R.E. 1977. Norwegian and Barents Seas: oil for the 21st century? *World Oil*, **185**(7), 69—76.

Klenova, M.V. 1960. *The Barents Sea Geology* (in Russian). Akad. Nauk. SSSR, Moscow, 367pp. Translations available: Conclusions — P. Hagevoll, Norsk Polarinst.; Ch. 2, 4, 5 and 8 — US Navy Hydrogr. Office.

Klenova, M.V. 1966. Barents Sea and White Sea. In: R.W. Fairbridge (ed.). *Encyclopedia of Oceanography*, Vol. 1, Reinhold, New York, pp.95—101.

Kogan, L.I. and Milashin, A.P. 1970. Seismic investigations in the Greenland Sea. *Oceanology*, **10**(3), 358—362.

Koreneva, E.V., Zaklinskaya, E.D., Bratseva, G.M. and Kartashova, G.G. 1976. Palynology studies of Sites 336, 338, 345, 346 and 348, DSDP Leg 38. In: M. Talwani *et al.* (eds.). *Initial Reports Deep Sea Drilling Project*, Vol.38, U.S. Govt Printing Office, Washington,

D.C., pp.1169—1194.

Korsakov, O.D. 1975. Tectonics of the eastern part of the Norwegian Sea. *Akad. Sci. USSR, Dokl.Earth.Sci. Sect.*, **213**, 47—50.

Korsakov, O.D. 1975. Tectonics of the Faeroe-Shetland Trench. *Acad. Sci. USSR, Dokl., Earth Sci. Sect.*, **214**, 81—83.

Koshechkin, B.I. and Strelkov, S.A. 1976. Shoreline movements and shelf glaciations on the Norwegian, Barents and White Seas during the late Pleistocene and Holocene. In: *23rd International Geogr. Congress, Moscow, Section 1, Geomorphology and Paleogeography, M*, pp.328—330.

Kosminskaya, I.P. and Pavlenkova, N.I. 1979. Seismic models of inner parts of the Euro-Asian Continent and its margins. *Tectonophysics*, **59**, 307—320.

Kossovskaya, A.G. and Drits, V.A. 1978. Mineralogy, geochemistry, and petrography of sediments recovered at Site 345, DSDP Leg 38. In: M. Talwani *et al.* (eds.). *Initial Reports Deep Sea Drilling Project.* Supplement to Vols. 38—41, U.S. Govt. Printing Office, Washington, D.C., pp.31—54.

Kossovskaya, A.G., Timofeev, P.P. and Shutov, V.D. 1978. The lithology and genesis of the sedimentary deposits in the Norwegian Basin and western part of the Lofoten Basin. In: M. Talwani *et al.* (eds.). *Initial Reports Deep Sea Drilling Project.* Supplement to Vols. 38—41, U.S. Govt. Printing Office, Washington, D.C., pp.67—72.

Kovacs, L.C. and Vogt, P.R. 1979. Depth-to-magnetic source analysis of the Greenland-Norwegian Sea and Arctic Ocean. *Eos. Trans. Am. Geophys. Union*, **60**(18), 372.

Kristoffersen, Y. 1978. Sea-floor spreading and early opening of the North Atlantic. *Earth Planet. Sci. Lett.*, **38**, 273—290.

Kristoffersen, Y. 1977. Late Cretaceous sea floor spreading and the early opening of the North Atlantic. In: *Mesozoic Northern North Sea Symposium* (Norsk Petroleumsforening), Article 5.

Kristoffersen, Y. and Elverhøi, A. 1978. A diapir structure in Bjørnøyrenna. *Norsk Polarinst. Årbok.*, **1977**, 189—198.

Kvale, A., Sellevoll, M.A. and Gammelsæter, H. 1966. *Seismic Refraction Measurements on the Norwegian Continental Shelf at 63N 06°03'E.* Seismol. Obs., Univ. Bergen, 34pp.

Kvenvolden, K.A. 1976. Organic geochemistry, Leg 38: introduction to studies. In: M. Talwani *et al.* (eds.). *Initial Reports Deep Sea Drilling Project*, Vol. 38, U.S. Govt. Printing Office, Washington, D.C., pp.783—784.

Langseth, M.G. and Zielinski, G.W. 1974. Marine heat flow measurements in the Norwegian-Greenland Sea and in the vicinity of Iceland. In: Kristjansson (ed.). *Geodynamics of Iceland and the North Atlantic Area.* Reidel, Dordrecht, pp. 277—295.

Larese, R.E., Haskell, N.L., Prezbindowski, D.R. and Beju, D. 1984. Porosity development in selected Jurassic sandstones from the Norwegian and North Seas, Norway — an overview. In: A.M. Spencer *et al.* (eds.). *Petroleum Geology of the North European Margin.* This volume, pp. 81—96.

Larsen, H.C. 1978. Offshore continuation of East Greenland dyke swarm and North Atlantic Ocean formation. *Nature (London)*, **274**, 220—223.

Larsen, H.C. 1980. Geological perspectives of the East Greenland Continental Margin. *Geol. Soc. Denmark, Bull.*, **29**, 77—101.

Larsen, H.C. 1983. Marine geophysical investigations offshore East Greenland. *Rapp. Grønlands Geol. Unders.*, **115**, 93—100.

Larsen, H.C. 1984. Geology of the East Greenland Shelf. In: A.M. Spencer *et al.* (eds.). *Petroleum Geology of the North European Margin.* This volume, pp. 329—340.

Larsen, R.M. and Skarpnes, O. 1984. Regional interpretation and hydrocarbon potential of the Trænabanken area. In: A.M. Spencer *et al.* (eds.). *Petroleum Geology of the North European Margin.* This volume, pp. 217—236.

Lastochkin, A.N. 1979. Subaerial topography of North-Eurasian Shelf (in Russian). *Izve Vsesoya Obshch*, **111**(3), 208—211.

Latouche, C. and Parra, M. 1976. Mineralogy and geochemistry of Quaternary sediments from the north-eastern Atlantic Ocean. *Mar. Geol.*, **22**(1) 33—69.

Laughton, A.S. 1975. Tectonic evolution of the North-east Atlantic Ocean: a review. *Nor. Geol. Unders.*, **316**, 169—193.

Lebedev, L.I., Eremenko, N.A., Grdzelov, L.O. and Korsakov, O.D. 1975. Sedimentary basins of the submarine margins of North Europe and their oil and gas prospects. *Can. Soc. Petrol. Geol. Mem.*, **4**, 257—265.

Le Pichon, X., Sibuet, J.C. and Francheteau, J. 1977. The fit of the continents around the North Atlantic Ocean. *Tectonophysics*, **38**, 169—209.

Levshin, A. and Berteussen, K.A. 1979. Anomalous propagation of surface waves in the Barents Sea as inferred from Norsar recordings. *Geophys. J.R. Astron. Soc.*, **56**(1), 97—118.

Lien, R. 1983. Iceberg scouring on the Norwegian Continental Shelf. *Offshore Technology Conference, Houston*, No. 4585, Vol.3, pp.41—48.

Lien, R. and Myhre, L.A. 1977. *Undersøkelser av Overflatelaget på Kontinentalsokkelen utenfor Troms og Vest-Finnmark. IKU* Publ. 92.

Lien, R.L. 1976. *Ingeniørgeologisk Kartlegging på Kontinentalsokkelen utenfor Lofoten-Vesterålen*, IKU Publ. 78.

Litvin, V.M. 1964. Bottom relief of the Norwegian Seas (in Russian). *Trudy Polyar. Nauchnoissled. Inst. Morsk. Ryb. Khoz. Okeanogr.* (P.I.N.R.O.), **16**, 89—109.

Litvin, V.M. 1965. Origin of the bottom configuration of the Norwegian Sea. *Oceanol. Acad. Sci. USSR* **5**(4), 90—96.

Litvin, V.M. and Dibner, V.D. 1976. Main features of morphology of the bottom of Norway-Greenland Basin. Geology of the sea (in Russian). *Sbornik Naunych Trudov*, **5**, 69—85.

Litvin, V.M. and Treshnikov, A.F. 1976. Geomorphology of the floor of the Norwegian and Greenland Seas. In: *Problems of the Arctic and the Antarctic*, Vol. 42. Amerind, New Delhi, pp.10—17.

Litvinenko, I.V., Ankudinov, S.A. and Platonenkova, L.N. 1970. The structure of the Earths crust of the eastern part of the Baltic Shield and of the Barents Sea using deep seismic sounding (in Russian, English Summary.). In: *Yevropeyskaya Seismologicheskaya Komissiya, 10th Trudy*, Vol. 1, pp.324—336.

Lowell, J.D. 1972. Spitsbergen Tertiary orogenic belt and the Spitsbergen Fracture Zone. *Bull. Geol. Soc. Am.*, **83**, 3091—3102.

Løfaldli, M. and Rokoengen, K. 1980. *Late- and post-glacial foraminifera and sediments on Viknaryggen and Haltenbanken off mid-Norway.* IKU Publ., 103.

Maisey, G.H. 1974. Geologic properties of the Norwegian Continental Shelf relating to pipelines. *Northern Offshore*, **3**(8), 37—49.

Malod, J. and Mascle, J. 1975. Structures géologiques de la Marge Continental a l'ouest de Spitzberg. *Mar. Geophys. Res.*, **2**, 215—229.

Malod, J.A. 1972. Etude géologique en Mar de Barentz et de Norvège. *Univ. Rennes, Ser. C*, **81**, 1—250.

Manum, S.B. 1966. Deposits of probable Upper Cretaceous age offshore from Andøya, Northern Norway. *Nor. Geol. Tidsskr.*, **46**, 246—247.

Manum, S.B. 1976. Dinocysts in Tertiary Norwegian-Greenland Sea sediments with observations on palynomorphs and palynodebris in relation to environment. In: M. Talwani *et al.* (eds.). *Initial Reports Deep Sea Drilling Project*, Vol.38, U.S. Govt. Printing Office, Washington, D.C., pp.897—920.

Manum, S.B. 1979. Norskehavets Historie Belyst ved Mikropaleontologiske Undersøkelser. *Naturen, Oslo*, **1**, 28—34.

Markov, F.G., Igiazarov, B.Kh. and Lanina, N.N. 1975. Geological map of the Northern Polar area of the Earth. (In Russian). *Scientific Research Inst. Arctic Geol. (NIIGA), Ministry of Geology of USSR.* 1:5 000 000 (4 sheets).

Martini, E. and Müller, C. 1976. Eocene to Pleistocene silicoflagellates from the Norwegian-Greenland Sea, DSDP Leg 38. In: M. Talwani *et al.* (eds.). *Initial Reports Deep Sea Drilling Project*, Vol.38, U.S. Govt. Printing Office, Washington, D.C., pp.857—896.

Marty, A., Rabate, J., Cazes, Jacquart, G. and Auxietre, J.L. 1979. CEPM Barents Sea seismic survey. In: *Norwegian Sea Symposium* (Norsk Petroleumsforening), Article 16.

Matishov, G.G. 1977. Relief, morphotectonics and principal features of the evolution of the Barents Sea Shelf.

Oceanology, **17**(3), 322—326.

Matishov, G.G. 1978. Pleistocene glaciation of the Barents Sea Shelf. *Dokl. Acad. Sci. USSR., Earth Sci. Sect. (Washington D.C.)* **232**(1—6), pp. 47—50.

Matishov, G.G. 1978. The geomorphology of the sea floor and some distinctive features of glacial morphogenesis of the underwater margins of Western Spitsbergen. *Oceanology*, **18** (2), 168—172.

Meyerhoff, A.A. 1970. Continental drift II: high-latitude evaporite deposits and geologic history of Arctic and North Atlantic Oceans. *J. Geol.*, **78**, 406—444.

Moe, A. 1974. Continental Shelf north of 62°N Latitude. In: *Exploration Geology and Geophysics. Offshore North Sea Conference*, Article G-IV/5.

Mohr, M. 1976. Additional petrographic studies of basalts, DSDP, Leg 38. In: M. Talwani, *et al.* (eds.). *Initial Reports Deep Sea Drilling Project*, Vol.38, U.S. Govt. Printing Office, Washington, D.C., pp.717—718.

Morris, D.A. 1976. Organic diagenesis of Miocene sediments from Site 341, Vøring Plateau, Norway. In: M. Talwani *et al.* (eds.). *Initial Reports Deep Sea Drilling Project*, Vol. 38, U.S. Govt. Printing Office, Washington, D.C., pp.809—814.

Müller, C. 1976. Tertiary and Quaternary calcareous nannoplankton in the Norwegian-Greenland Sea, DSDP Leg 38. In: M. Talwani *et al.* (eds.). *Initial Reports Deep Sea Drilling Project*, Vol.38, U.S. Govt. Printing Office, Washington, D.C., pp.823—842.

Mutter, J.C. 1981. Layered oceanic basement complex of the Norwegian margin: a heuristic model. *Eos. Trans. Am. Geophys. Union*, **62**, 407.

Mutter, J.C. Cenozoic and Late Mesozoic stratigraphy and subsidence history of the Norwegian Margin. *Geol. Soc. Am. Bull.* (in press).

Mutter, J.C., Talwani, M. and Stoffa, P.L. 1982. Origin of seaward dipping reflectors in oceanic crust off the Norwegian Margin by 'subaerial sea-floor spreading'. *Geology* **10**(7), 353—357.

Mutter, J.C., Talwani, M. and Stoffa, P.L. 1984. Structure of the oceanic crust adjacent to the Norwegian Margin. *J. Geophys. Res.* **89**(B1), 483—502.

Myhre, A.M. and Eldholm, O. 1981. Sedimentary and crustal velocities in the Norwegian-Greenland Sea. *J. Geophys. Res.*, **86**, 5012—5022.

Myhre, A.M., Eldholm, O. and Sundvor, E. 1982. The margin between Senja and Spitsbergen Fracture Zones: implications from plate tectonics. *Tectonophysics*, **89**, 33—50.

Myhre, A.M., Eldholm, O. and Sundvor, E. The Jan Mayen Ridge: present status. *Polar Res.* (in press).

Myhre, L.A. and Scull, B.J. 1978. A combination of geochemical and geophysical prospecting for petroleum, offshore Norway. *Empresa Nacional Adaro de Investigaciones Mineral*, pp.561—569.

Nagy, J. 1973. *Fossilførende Blokker av Mesozoisk Alder fra Svalbardbanken.* NTNF Continental Shelf Project, Publ. 42, 26pp.

Nansen, F. 1904. The bathymetrical features of the North Polar Seas. With a discussion of the continental shelves and previous oscillations of the shoreline. *Sci. Res., Norwegian North Pole Expedition, 1893—1896*, 4(13), 1—232.

Navrestad, T. and Jørgensen, F. 1979. Aeromagnetic investigations on the Jan Mayen Ridge. In: *Norwegian Sea Symposium* (Norsk Petroleumsforening), Article 8.

Nesterova, M.P., Scherbokov, F.A., Schevchenko, A.J., Turanskaja, N.W. and Kazakova, W.P. 1978. Origin of the late Cenozoic sediments of the Icelandic Basin, DSDP Site 348, Leg 38. In: M. Talwani *et al.* (ed.). *Initial Reports Deep Sea Drilling Project.* Supplement to Vols. 38—41, U.S. Govt. Printing Office, Washington, D.C., pp.73—94.

Nielsen, P.H. 1976. Seismic refraction measurements around the Faeroe Islands. *Frooskaparrit*, **24**, 9—45.

Nilsen, T.H. 1978. Turbidites, redbeds, sedimentary structures, and trace fossils observed in DSDP Leg 38 Cores and the sedimentary history of the Norwegian-Greenland Sea. In: M. Talwani *et al.* (eds.). *Initial Reports Deep Sea Drilling Project.* Supplement to Vols. 38—41, U.S. Govt. Printing Office, Washington, D.C., pp.259—288.

Nilsen, T.H. 1978. Sedimentation in the Northeast Atlantic Ocean and Norwegian Sea. In: D.R. Bowes and B.E. Leake (eds.). Crustal Evolution in North-Western Britain and Adjacent Regions, Spec. Iss. *Geol. J. (Liverpool)*, **10**, 433—445.

Nilsen, T.H. 1983. Influence of the Greenland-Scotland Ridge on the geological history of the North Atlantic and Norwegian-Greenland Sea areas. In: M.H.P. Bott *et al.* (eds.). *Structure and Development of the Greenland-Scotland Ridge*, Plenum Press, New York, pp.457—478.

Nordhagen, R. 1921. Fosselførende Blokker fra Juratiden på Froøene utenfor Trondhjemsfjorden. *Naturen Oslo*, **45**, 110—115.

Nunns, A.G. 1983. Plate tectonic evolution of the Greenland-Scotland Ridge and surrounding regions. In: M.H.P. Bott *et al.* (eds.). *Structure and Development of the Greenland-Scotland Ridge*, Plenum Press, New York, pp.11—30.

Nunns, A.G. 1983. The structure and evolution of the Jan Mayen Ridge and surrounding regions. *Mem. Am. Assoc. Petrol. Geol.* **34**, 193—208.

Nunns, A.G. and Peacock, J.G. 1982. Correlation, identification and inversion of magnetic anomalies in the Norway Basin. *Earth Evol. Sci.*, 2(2), 130—138.

Nunns, A.G. *et al.* 1983. Magnetic anomalies over Iceland and surrounding seas. (Text to accompanying map.) In: M.H.P. Bott *et al.* (eds.). *Structure and Development of the Greenland-Scotland Ridge*, Plenum Press, New York, pp.661—678.

Nysæther, E. 1970. Notes on the Quaternary sediments of the Norwegian Continental Shelf between 62°N and 67°N (Abstr.). *Nor. Geol. Tidsskr.*, **50**(3), 275—276.

Nysæther, E., Eldholm, O. and Sundvor, E. 1969. *Seismiske Undersøkelser av den Norske Kontinentalsokkel, Sklinnabanken — Andøya.* Tekn. Rapp. No. 3, Seismol. Obs., Univ. Bergen, 23pp.

Oftedahl, C. 1972. A sideritic ironstone of Jurassic age in Beitstadfjorden, Trøndelag. *Nor. Geol. Tidsskr.*, **52**, 123—134.

Oftedahl, C. 1974. The possibilities of finding oil and gas north of latitude 62°N. *Northern Offshore*, **3**(1), 14—18.

Oftedahl, C. 1975. Middle Jurassic Graben tectonics in mid-Norway. In: *Jurassic Northern North Sea Symposium* (Norsk Petroleumsforening), Article 21.

Oftedahl, C. 1978. *Glacial Clay Basins in Trondheimsfjorden.* IKU Publ., 84.

Oftedahl, C. 1978. Oil geology north of 62. *Noroil*, **6**, 9, 89, 91 and 93.

Ohta, Y. 1982. Morpho-tectonic studies around Svalbard and the northernmost Atlantic. *Can. Soc. Petrol. Geol. Mem.*, **8**, 415—429.

Olaussen, S., Dalland, A., Gloppen, T.G. and Johannessen, E. 1984. Depositional environment and diagenesis of Jurassic reservoir sandstones in the eastern part of Troms I area. In: A.M. Spencer *et al.* (eds.). *Petroleum Geology of the North European Margin.* This volume, pp. 61—80.

Ostenso, N.A. and Wold, R.J. 1971. Aeromagnetic survey of the Arctic Ocean: techniques and interpretations. *Mar. Geophys. Res.*, **1**(2), 178—219.

Palmason, G. 1965. Seismic refraction measurements of the basalt lavas of the Faeroe Islands. *Tectonophysics*, **2**, 475—482.

Peach, B.N. 1912. Report on Rock Specimens dredged by the 'Michael Sars' in 1910, by H.M.S. 'Triton' in 1882 and by H.M.S. 'Knight Errant' in 1880. *Proc. R. Soc. Edinburgh*, **32**, 262—291.

Pedersen, G.S. 1979. Application of well data from Svalbard for paleotemperature analysis of source rocks for petroleum in the Barents Shelf. In: *Norwegian Sea Symposium* (Norsk Petroleumsforening), Article 14.

Pegrum, R.M. and Mounteney, N. 1978. Rift basins flanking North Atlantic Ocean and their relation to North Sea area. *Bull. Am. Assoc. Petrol. Geol.*, **62**, 419—441.

Pelet, R., Cauwet, G. and Combaz, A. 1975. La géochimie organique des sédiments marins profonds, Mission Orgon 1, 1974 (Mer de Norvège). *Rev. Inst. Fr. Petrol.*, **30**(1), 17—64.

Perch-Nielsen, K. 1978. Eocene to Pliocene archaeomonads, ebridians, and endoskeletal dinoflagellates from the Norwegian Sea, DSDP Leg 38. In: M. Talwani *et al.* (eds.). *Initial Reports Deep Sea Drilling Project.* Supplement to Vols. 38—41. U.S. Govt. Printing Office, Washington, D.C., pp.147—176.

Perry, R.K., Fleming, H.S. and Cherkis, N.Z. (compilers). 1978. Bathymetry of the Norwegian-Greenland and Western Barents Seas. Map Chart Series MC-21. *Geol. Soc. Am.*, Scale 1:2, **333**, 230.

Perry, E.A., Grady, S.J. and Kelly, W.M. 1978. Mineralogic

studies of sediments from the Norwegian-Greenland Sea (Sites 336, 343, 345 and 348). In: M. Talwani *et al.* (eds.). *Initial Reports Deep Sea Drilling Project.* Supplement to Vols. 38—41, U.S. Govt. Printing Office, Washington, D.C., pp.135—140.

Petrushevskaya, M. 1974. Radiolarians in the sediments of the Norwegian Sea and the North Atlantic Ocean. *Oceanology*, 14(6), 861—867.

Phillips, J.D., Fleming, H.S. and Feden, R. 1973. Aeromagnetic study of the Greenland and Norwegian Seas. *Geol. Soc. Am. Abstr.*, **5**, p. 767.

Pitman, W.C. and Talwani, M. 1972. Sea-floor spreading in the North Atlantic. *Bull. Geol. Soc. Am.*, **83**, 619—646.

Price, I. and Rattey, P.R. Cretaceous tectonics off mid-Norway. Implications for the Rockall—Faeroes—Shetland Rift. *J. Geol. Soc. London.* (in press).

Price, N.B. and Wright, P.L. 1971. *Variation in the Mineralogy and Chemistry of Sediments from the Southwestern Barents Sea.* Rep. Inst. Geol. Sci. London, 70/14, pp.17—31.

Price, N.B., Calvert S.E. and Jones, P.E.W. 1970. The distribution of iodine and bromine in the sediments of the southwestern Barents Sea. *J. Mar. Res.*, **28**(1), 22—34.

Ramberg, I.B. and Grønlie, G. 1970. Gravity Investigations on the Norwegian Continental Shelf between 62° and 69°N. *Nor. Geol. Tidsskr.*, **50**, 276—277.

Raschka, H. and Eckhardt, F.J. 1976. Geochemistry of basalts from the Norwegian-Greenland Sea, Leg 38, DSDP. In: M. Talwani *et al.* (eds.). *Initial Reports Deep Sea Drilling Project,* Vol. 38, U.S. Govt. Printing Office, Washington, D.C., pp.719—730.

Rateev, M.A., Renngarten, N.V., Shutov, V.D. and Drits, V.A. 1978. Lithology and clay mineralogy of sediments from Hole 346. In. M. Talwani *et al.* (eds.). *Initial Reports Deep Sea Drilling Project.* Supplement to Vols. 38—41, U.S. Govt. Printing Office, Washington, D.C., pp.56—66.

Renard, V. and Malod, J. 1974. Structure of the Barents Sea from seismic refraction. *Earth Planet. Sci. Lett.*, 24(1), 33—47.

Renngarten, N.V., Rateev, M.A., Shutov, V.D. and Drits, V.A. 1978. Lithology and clay mineralogy of sediments from site 337, DSDP Leg 38. In: M. Talwani *et al.* (eds.). *Initial Reports Deep Sea Drilling Project.* Supplement to Vols. 38—41, U.S. Govt. Printing Office, Washington, D.C. pp.21—30.

Ridd, M.F. 1983. Aspects of the Tertiary geology of the Faeroe-Shetland Channel. In: M.H.P. Bott *et al.* (eds.). *Structure and Development of the Greenland-Scotland Ridge,* Plenum Press, New York, pp.91—108.

Ridley, W.I. 1976. A synthesis of the magmatic evolution of the Brito-Arctic Province, including data from DSDP Leg 38. In: M. Talwani *et al.* (eds.). *Initial Reports Deep Sea Drilling Project.* Vol.38, U.S. Govt. Printing Office, Washington, D.C. pp.751—754.

Ridley, W.I., Perfit, M.R. and Adams, M.L. 1976. Petrology of basalts from Deep Sea Drilling Project, Leg 38. In: M. Talwani *et al.* (eds.). *Initial Reports Deep Sea Drilling Project.* Vol.38, U.S. Govt. Printing Office, Washington, D.C., pp.731—740.

Rokoengen, K. 1980. *Fjellgrunn og løsmassetykkelser utenfor Helgeland: Grunnseismisk Profilering.* IKU Report P-155/2/80.

Rokoengen, K. and Bugge, T. 1976. *Grunnforholdene på den norske kontinentalsokkel sør for 72°N.* IKU Publ. 81, 44pp.

Rokoengen, K., Bell, D.G., Bugge, T., Dekko, T. and Gunleiksrud, T. 1977. *Prøvetaking av fjellgrunn og løsmasser utenfor deler av Nord-Norge i 1976.* IKU, Publ. 91, 66pp.

Rokoengen, K., Bugge, T., Dekko, T., Gunleiksrud, T. and Lien, R.L. 1978. *Quaternary Deposits on the Continental Shelf off North-Norway.* IKU Report P-155/1/78.

Rokoengen, K., Bugge, T., Dekko, T., Gunleiksrud, T. and Lien, R.L. 1979. Shallow geology of the continental shelf off North Norway. *POAC 1979, Port and Ocean Engineering Under Arctic Conditions Conference Proc.* Vol. 2, pp.859—875.

Rokoengen, K., Bugge, T. and Løfaldli, M. 1979. Quaternary geology and deglaciation of the continental shelf off Troms, North Norway. *Boreas,* 8(2), 217—228.

Rozanov, L.N. and Nalivkin, V.D. 1969. *Tectonic Map of the Oil Bearing Regions of the USSR.* (In Russian). Ministry of Geology. All Union Oil Research Geol. Institute (VNIGRI). 1:2500 000.

Russell, M.J. 1976. A possible Lower Permian age for the onset of ocean floor spreading in the Northern North Atlantic. *Scott. J. Geol.*, **12**, 315—323.

Russell, M.J. and Smythe, D.K. 1978. Evidence for an early Permian oceanic rift in the Northern North Atlantic. In: E.R. Neuman, and I.B. Ramberg (eds.). *Petrology and Geochemistry of Continental Rifts,* Reidel, Dortrecht, pp.173—179.

Röser, H.A., Plaumann, S. and Forstner-Ballheim, E. 1975. The crustal structure of the Norwegian Continental Margin and the Norwegian Basin according to magnetic and gravimetric measurements. *Meteor. Forschungs Ergebnisse,* Reihe, C. 21, pp.1—14.

Rønnevik, H.C. 1981. Geology of the Barents Sea. In: L.V. Illing and G.D. Hobson (eds.). *Petroleum Geology of the Continental Shelf of North-West Europe,* Heyden, London, pp.395—406.

Rønnevik, H. and Jacobsen, H.-P. 1984. Structural highs and basins in the western Barents Sea. In: A.M. Spencer *et al.* (eds.). *Petroleum Geology of the North European Margin.* This volume, pp. 19—32.

Ronnevik, H.C. and Motland, K. 1979. Geology of the Barents Sea. In: *Norwegian Sea Symposium* (Norsk Petroleumsforening), Article 15.

Rønnevik, H.C. and Navrestad, H. 1976. Geology of the Norwegian Shelf between 62N and 69N. In: *Exploration Geology and Geophysics. Offshore North Sea Conference,* Article G-IV/4.

Rønnevik, H.C. and Navrestad, T. 1977. Geology of the Norwegian Shelf between 62—69°N. *GeoJournal,* 1, 33—46.

Rønnevik, H.C. Bergsager, E.I., Moe, A., Øvrebø, O. and Navrestad, T. 1975. The geology of the Norwegian Continental Shelf. In: A.W. Woodland (ed.). *Petroleum and the Continental Shelf of North-west Europe,* Vol. 1, *Geology,* Applied Science Publishers, London, pp. 117—129.

Rønnevik, H.C., Beskow, B. and Jacobsen, H.P. 1982. Structural and stratigraphic evolution of the Barents Sea. In: *The Geological Framework and Hydrocarbon Potential of Basins in Northern Seas. Offshore Northern Seas Conference,* Article E/3.

Rønnevik, H.C., Beskow, B. and Jacobsen, P. 1982. Structural and stratigraphic evolution of the Barents Sea. *Can. Soc. Petrol. Geol. Mem.*, **8**, 431—440.

Rønnevik, H.C., Eggen, S. and Vollset, J. 1983. Exploration of the Norwegian Shelf. *NPD-Contribution No. 5.* Publ. in: J. Brooks (ed.) Petroleum Geochemistry and Exploration of Europe. Blackwell, Oxford, pp. 71—94.

Rønnevik, H.C., Jørgensen, F. and Motland, K. 1979. The geology of the northern part of the Vøring Plateau. In: *Norwegian Sea Symposium* (Norsk Petroleumsforening), Article 12.

Sachs, V.N. and Strelkov, S.A. 1961. Mesozoic and Cenozoic of the Soviet Arctic. In: G.O. Raasch (ed.)., *Geology of the Arctic,* Vol. 1, Univ. Toronto Press, pp.48—67.

Schilling, J.G. 1976. Rare-earth, Sc, Cr, Fe, Co, and Na abundances in DSDP Leg 38 basement basalts: some additional evidence on the evolution of the Thulean Volcanic Province. In: M. Talwani *et al.* (eds.). *Initial Reports Deep Sea Drilling Project.* Vol. 38. U.S. Govt. Printing Office, Washington, D.C., pp. 741—750.

Schilling, J.G. 1979. Jan Mayen 'mantle plume'. *Eos. Trans. Am. Geophys. Union,* **60**(18), 372.

Schrader, H.J., Bjørklund, K., Manum, S., Martini, E. and Hinte, Van J. 1976. Cenozoic biostratigraphy, physical stratigraphy and paleoceanography in the Norwegian-Greenland Sea, DSDP Leg 38 paleontological synthesis. In: M. Talwani *et al.* (eds.). *Initial Reports Deep Sea Drilling Project,* Vol.38, U.S. Govt. Printing Office, Washington, D.C., pp.1197—1212.

Schrader, H.J. and Fenner, J. 1976. Norwegian Sea Cenozoic diatom biostratigraphy and taxonomy, parts 1 and 2. In: M. Talwani *et al.* (eds). *Initial Reports Deep Sea Drilling Project,* Vol.38, U.S. Govt. Printing Office, Washington, D.C., pp.921—1100.

Schüter, H.U. and Hinz, K. 1978. The Continental Margin of West Spitsbergen. *Polar Res.* **48**, 151—169.

Schüter, H.U. and Hinz, K. 1978. The geological structure of the western Barents Sea. *Mar. Geol.*, **26**, 199—230.

Schytt, V., Boström, K. and Hjort, C. 1981. Geoscience

during the Ymer-80 Expedition to the Arctic. *Forh. Geol. Foren. Stockholm*, **103**, 109—120.

Scull, B.J. 1977. Geochemical surveys for oil and gas, Norway. *9th Int. Sedimentology Congress Proc.*, 3, pp.79—84.

Sellevoll, M.A. 1970. Seismic refraction investigation on the Norwegian Continental Shelf north of 62°N. *Nor. Geol. Tidsskr.*, **50**(3), 278—279.

Sellevoll, M.A. 1973. Mohorovicic Discontinuity beneath Fennoscandia and adjacent parts of the Norwegian Sea and the North Sea. *Tectonophysics*, **20**, 359—366.

Sellevoll, M.A. 1973. A continuous seismic and magnetic profile across the Norwegian Continental Shelf and Vøring Plateau. *Nor. Geol. Unders.*, **300**, 1—10.

Sellevoll, M.A. 1975. Seismic refraction measurements and continuous seismic profiling on the continental margin off Norway between 62 and 69°N. *Nor. Geol. Unders.*, **316**, 219—235.

Sellevoll, M.A. and Sundvor, E. 1974. The origin of the Norwegian Channel — a discussion based on seismic measurements. *Can. J. Earth. Sci.*, **11**(2), 224—231.

Sellevoll, M.A. and Sundvor, E. (eds.). 1973. *Seismiske Undersøkelser av den Norske Kontinentalsokkel.* Tekn. Rapp. No. 7, Seismol. Obs., Univ. Bergen, 168pp.

Sellevoll, M.A., Eldholm, O. and Gammelsæter, H. 1967. *Refraksjonsseismiske undersøkelser av den Norske Kontinentalsokkelen mellom 62 og 65°N.* Tekn. Rapp. No. 1, Seismol. Obs., Univ. Bergen, 42pp.

Selnes, P.B. and Kleven, A. 1980. *Earthquake Analysis for Pipeline on the Norwegian Continental Shelf (Jordskjelvanalyse for rørledning på Norsk Kontinentalsokkel).* Nor. Geotek. Inst., Publ., No. 132, pp. 99—109.

Semonovich, V.N., Ramberg, I.S. and Nesterov, I.I. 1973. Oil and gas possibilities in the Soviet Arctic. *Mem. Am. Assoc. Petrol. Geol.*, **19**, 194—203.

Simoneit, B.R.T. 1976. Sources of the solvent-soluble organic matter in the glacial sequence of DSDP samples from the Norwegian-Greenland Sea, Leg 38. In: M. Talwani *et al.* (eds.). *Initial Reports Deep Sea Drilling Project,* Vol. 38, U.S. Govt. Printing Office, Washington, D.C., pp. 805—806.

Skaar, F.E. 1973. *Recent Scientific Investigations and Literature: The Norwegian Shelf and Adjacent Areas.* IKU Publ. 33.

Smythe, D.K. 1980. Identification of the continent—ocean boundary round the Faeroes by seismic reflection. *Geophys. J. R. Astron. Soc.*, **61**, 208.

Smythe, D.K. 1982. Early opening history of the North-west European Continental Margin. In: *The Geological Framework and Hydrocarbon Potential of Basins in Northern Seas. Offshore Northern Seas Conference,* Article E/6.

Smythe, D.K. 1983. Faeroe-Shetland Escarpment and continental margin north of the Faeroes. In: M.H.P. Bott *et al.* (eds.). *Structure and Development of the Greenland-Scotland Ridge,* Plenum Press, New York, pp. 109—120.

Smythe, D.K., Chalmers, J.A., Skuce, A.G., Dobinson, A. and Mould, A.S. 1983. Early opening history of the North Atlantic —1. Structure and origin of the Faeroe-Shetland Escarpment. *Geophys. J. R. Astron. Soc.* **72**, 373—399.

Sobczak, L.W. 1975. Gravity anomalies and passive continental margins, Canada and Norway. *Can. Soc. Petrol. Geol. Mem.*, **4**, 743—761.

Sokolov, V.N. 1973. Structural history of Spitsbergen and adjoining shelves. *Mem. Am. Assoc. Petrol. Geol.*, **19**, 269—274.

Sokolov, V.N., Krasiltschikov, A.A., Livshitz, J.J. and Semenevskij, D.V. 1972. A history of the formation of the structure of Spitsbergen and the contiguous shelves (in Russian). In: *Geologiya i perspektivy neftegazonosnosti Sovetskoy Arktiki,* NIIGA, Leningrad, pp. 60—65.

Spencer, A.M., Home, P.C. and Berglund, L.T. 1984. Tertiary structural development of the western Barents Shelf: Troms to Svalbard. In: A.M. Spencer *et al.* (eds.). *Petroleum Geology of the North European Margin.* This volume, pp. 199—210.

Spjeldnæs, N. 1971. Mesozoic (?) bedrock exposed on the bottom of the Barents Sea. *Mar. Geol.*, **11**, 47—50.

Stadum, C.J. and Ling, H.Y. 1969. Tripylean Radiolaria in deep sea sediments of the Norwegian Sea. *Micropaleontology*, **15**(4), 481—489.

Steven, R.R. 1981. North of 62nd seismic tests determine high risks. *Offshore*, **39**(13), 150—159.

Stewart, I.C.F. 1980. Anomalous travel times of teleseismic P-waves reflected under Arctic marine Areas. *Mar. Geophys. Res.*, **4**, 291—304.

Strass, I.F. 1978. Microtextures of quartz sand grains in coastal and shelf sediments, Møre, Western Norway. *Mar. Geol.*, **28**(1—2), 107—134.

Sundvor, E. 1971. Seismic refraction measurements on the Norwegian Continental Shelf between Andøya and Fugløybanken. *Mar. Geophys. Res.*, **1**(3), 303—313.

Sundvor, E. 1974. Seismic refraction and reflection measurements in the southern Barents Sea. *Mar. Geol.*, **16**(5), 255—273.

Sundvor, E. 1975. Thickness and distribution of sedimentary rocks in the southern Barents Sea. *Nor. Geol. Unders.*, **316**, 237—240.

Sundvor, E. and Eldholm, O. 1976. *Marine Geophysical Survey on the Continental Margin from Bear Island to Hornsund, Spitzbergen.* Sci. Rep. No. 3, Seismol. Obs., Univ. Bergen, 28pp.

Sundvor, E. and Eldholm, O. 1979. The western and northern margin off Svalbard. *Tectonophysics,* **59**, 239—250.

Sundvor, E. and Nysæther, E. 1975. Geological outline of the Norwegian Continental Margin between 60—68°N. *Can. Soc. Petrol. Geol. Mem.*, **4**, 267—281.

Sundvor, E. and Sellevoll, M.A. 1969. *Seismiske undersøkelser av den Norske Kontinentalsokkel. Andøya-Fugløybanken.* Tekn. Rapp. No. 4, Seismol. Obs., Univ. Bergen, 26pp.

Sundvor, E. and Sellevoll, M.A. 1971. *Seismiske undersøkelser av den Norske Kontinentalsokkel, Lofoten—Bjørnøya (68—75°N).* Tekn. Rapp. No. 5, Seismol. Obs., Univ. Bergen, 31pp.

Sundvor, E. and Sellevoll, M.A. 1972. *Seimiske undersøkelser av den Norske Kontinentalsokkel, Barentshavet.* Tekn. Rapp. No. 6, Seismol. Obs., Univ. Bergen, 24pp.

Sundvor, E., Eldholm, O., Gidskehaug, A. and Myhre, A. 1977. *Marine Geophysical Survey on the Western and Northern Continental Margin off Svalbard.* Sci Rep. No. 4, Seismol. Obs., Univ. Bergen, 35pp.

Sundvor, E., Eldholm, O., Haugland, K., Sellevoll, M.A. and Bruland, L. 1975. *Seismic Measurements on the Norwegian Continental Margin.* Sci. Rep. No. 2, Seismol. Obs., Univ. Bergen, 30pp.

Sundvor, E., Gidskehaug, A., Myhre, A. and Eldholm, O. 1978. *Marine Geophysical Survey on the Northern Svalbard Margin.* Sci. Rep. No. 5, Seismol. Obs., Univ. Bergen, 46pp.

Sundvor, E., Gidskehaug, A., Myhre, A. and Eldholm, O. 1979. *Marine Geophysical Survey on the Northern Jan Mayen Ridge.* Sci. Rep. No. 6, Seismol. Obs., Univ. Bergen, 18pp.

Sundvor. E., Myhre, A. and Eldholm, O. 1979. The Svalbard Continental Margin. In: *Norwegian Sea Symposium* (Norsk Petroleumsforening), Article 6.

Sundvor, E., Sellevoll, M.A. Gidskehaug, A. and Eldholm, O. 1978. Seismic investigation on the western and northern margin off Svalbard. *Polarforschung,* **48**, 41—43.

Sundvor, E., Sellevoll, M.A. and Haugland, K. 1974. *Seismic Measurements on the Norwegian Continental Margin.* Sci. Rep. No. 1, Seismol. Obs., Univ. Bergen, 34pp.

Sylvester, A.G. 1978. Petrography of volcanic ashes in deep-sea cores near Jan Mayen Island: Sites 338, 345—350 DSDP Leg 38. In: M. Talwani, *et al.* (eds.). *Initial Reports Deep Sea Drilling Project,* Supplement to Vols. 38—41, U.S. Govt. Printing Office, Washington, D.C., pp. 101—110.

Syrstad, E., Bergseth, S. and Navrestad, T. 1976. Gravity modelling offshore Troms, Northern Norway. In: *Exploration Geology and Geophysics. Offshore North Sea Conference,* Article T-1/4.

Talleraas, E. 1979. The Hammerfest Basin — an aulacogen? In: *Norwegian Sea Symposium* (Norsk Petroleumsforening), Article 18.

Talwani, M. 1976. Survey at Sites 346, 348, 349 and 350: the area of the Jan Mayen Ridge and the Icelandic Plateau. In: M. Talwani *et al.* (eds.). *Initial Reports Deep Sea Drilling Project,* Vol. 38, U.S. Govt. Printing Office, Washington, D.C., pp. 465—488.

Talwani, M. 1976. Sites 336 and 352. In: M. Talwani *et al.* (eds.). *Initial Reports Deep Sea Drilling Project,* Vol. 38, U.S. Govt. Printing Office, Washington, D.C., pp. 23—116.

Talwani, M. 1978. Geophysical surveys on the Iceland-Faeroe Ridge for selection of Sites 336 and 352. In: M. Talwani *et al.* (eds.). *Initial Reports Deep Sea Drilling Project.* Supplement to Vols. 38—41, U.S. Govt. Printing Office, Washington, D.C., pp. 445—450.

Talwani, M. 1978. Distribution of basement under the eastern North Atlantic Ocean and the Norwegian Sea. In: D.R. Bowes and B.E. Leake (eds.). Crustal Evolution in North-western Britain and Adjacent Regions. Spec. Iss. *Geol. J. (Liverpool)*, **10**, 347—376.

Talwani, M. 1978. Surveys and selection of Sites 338, 339, 340, 341, 342 and 343 on the Vøring Plateau. In: M. Talwani *et al.* (eds.). *Initial Reports Deep Sea Drilling Project.* Supplement to Vols. 38—41, U.S. Govt. Printing Office, Washington, D.C., pp. 451—454.

Talwani, M. 1978. Survey at site 344, rift mountains east of Knipovich Rift. In: M. Talwani *et al.* (eds.). *Initial Reports Deep Sea Drilling Project.* Supplement to Vols. 38—41, U.S. Govt. Printing Office, Washington, D.C., pp. 461—464.

Talwani, M. and Eldholm, O. 1972. Continental margin off Norway: a geophysical study. *Bull. Geol. Soc. Am.*, **83**, 3575—3606.

Talwani, M. and Eldholm, O. 1973. Boundary between continental and oceanic crust at the margin of rifted continents. *Nature (London)*, **241**, 325—330.

Talwani, M. and Eldholm, O. 1974. Margins of the Norwegian-Greenland Sea. In: C.A. Burke and C.L. Drake (eds.). *The Geology of Continental Margins*, Springer, New York, pp. 361—374.

Talwani, M. and Eldholm, O. 1977. Evolution of the Norwegian-Greenland Sea. *Bull. Geol. Soc. Am.*, **88**, 969—999.

Talwani, M. and Grønlie, G. 1976. *Free-air Gravity Field of the Norwegian-Greenland Seas.* Geol. Soc. Am., Map and Chart Series MC-15.

Talwani, M. and Sandal, S. 1978. Survey at Site 337, near the extinct axis in the Norway Basin. In: M. Talwani *et al.* (eds.). *Initial Reports Deep Sea Drilling Project.* Supplement to Vols. 38—41, U.S. Govt. Printing Office, Washington, D.C., pp. 455—460.

Talwani, M. and Udintsev, G. 1976. Tectonic synthesis. In: M. Talwani *et al.* (eds.). *Initial Reports Deep Sea Drilling Project.* Vol. 38, U.S. Govt. Printing Office, Washington, D.C., pp. 1213—1242.

Talwani, M. and Udintsev, G. (eds.). 1976. *Initial Reports of the Deep Sea Drilling Project,* Vol. 38. U.S. Govt. Printing Office, Washington, D.C., 1256pp.

Talwani, M., Mutter, J. and Eldholm, O. 1981. The initiation of opening of the Norwegian Sea. *26th Int. Geol. Congr.*, C3, pp. 23—30.

Talwani, M., Mutter, J. and Hinz, K. 1983. Ocean continent boundary under the Norwegian Continental Margin. In: M.H.P. Bott *et al.* (eds.). *Structure and Development of the Greenland-Scotland Ridge*, Plenum Press, New York, pp. 121—131.

Talwani, M., Udintsev, G., Bjørklund, K., Caston, V.N.D. and Faas, R.W. 1975. Leg 38; Deep Sea Drilling Project. *Geotimes*, **20**(2), 24—26.

Talwani, M., Udintsev, G.B. and White, S.M. 1976. Introduction and explanatory notes (and site reports), Leg 38, Deep Sea Drilling Project. In: M. Talwani *et al.* (eds.). *Initial Reports Deep Sea Drilling Project,* Vol. 38, U.S. Govt. Printing Office, Washington, D.C., pp. 3—19.

Talwani, M., Udintsev, G., Mirlin, E., Beresnev, A.F., Kanayev, V.F., Chapman, M., Grønlie, G. and Eldholm, O. 1978. Survey at sites 346, 347, 348, 349 and 350. The area of the Jan Mayen Ridge and the Icelandic Plateau. In: M. Talwani *et al.* (eds.). *Initial Reports Deep Sea Drilling Project.* Suppl. to Vols. 38—41, U.S. Govt. Printing Office, Washington, D.C., pp. 465—488.

Thiede, J. 1979. Paleophysiography of the North Atlantic Ocean and its surrounding epicontinental seas: implications for North Atlantic and Arctic palaeo-oceanography. In: *Norwegian Sea Symposium* (Norsk Petroleumsforening), Article 4.

Thiede, J. 1979. Paleogeography and paleobathymetry of the Mesozoic and Cenozoic North Atlantic Ocean. *GeoJournal,* **3**(3), 263—272.

Thiede, J. 1980. Palaeo-oceanography, margin stratigraphy, and palaeo-physiography of the Tertiary North Atlantic and Norwegian-Greenland Seas. *Philos. Trans. R. Soc. London, A,* **294**, 177—185.

Thiede, J. 1983. Outstanding geological problems of the Greenland-Scotland Ridge: an introduction. In: M.H.P. Bott *et al.* (eds.). *Structure and Development of the Greenland-Scotland Ridge.* Plenum Press, New York, pp. 313—318.

Thiede, J. and Eldholm, O. 1981. The growth of an ocean basin: The North Atlantic and its epicontinental seas. *Can. Soc. Petrol. Geol., Mem.* **7**, 185—196.

Thompson, K. 1978. Carbon and carbonate analysis, Leg 38. In: M. Talwani *et al.* (eds.). *Initial Reports Deep Sea Drilling Project.* Supplement to Vols. 38—41, U.S. Govt. Printing Office, Washington, D.C., pp. 433—436.

Thusu, B. 1978. *Aptian to Toarcian Dinoflagellate Cysts from Arctic Norway.* IKU Publ. 100, pp. 61—95.

Thusu, B. and Bjærke, T. 1976. Cretaceous palynomorphs from Spitsbergenbanken, NW Barents Shelf. *Norsk Polarinst. Årbok.*, **1974**, 258—262.

Timofeev, P.P., Renngarten, N.V. and Bogolyubova, L.J. 1978. Lithology and clay mineralogy of the sediments from Site 336, DSDP Leg 38. In: M. Talwani *et al.* (eds.). *Initial Reports Deep Sea Drilling Project.* Supplement to Vols. 38—41, U.S. Govt. Printing Office, Washington, D.C., pp. 9—20.

Torske, T. 1972. Tertiary oblique uplift of western Fennoscandia; crustal warping in connection with rifting and break-up of the Laurasian Continent. *Nor. Geol. Unders.*, **273**, 43—48.

Torske, T. 1975. Possible Mesozoic mantle plume activity beneath the continental margin of Norway. *Nor. Geol. Unders.*, **322**, 73—90.

Torvanger, O., Gausland, I. and Moe, A. 1979. Seismic status north of 62°N, 1959—1979. In: *Norwegian Sea Symposium* (Norsk Petroleumsforening), Article 5.

Tryti, J. 1979. Velocity study of seismic data in the Troms I Area. In: *Norwegian Sea Symposium* (Norsk Petroleumsforening), Article 19.

Udintsev, G.B. and Kharin, G.S. 1978. Sedimentary rocks of the Jan Mayen Ridge. In: M. Talwani *et al.* (eds.). *Initial Reports Deep Sea Drilling Project.* Supplement to Vols. 38—41, U.S. Govt. Printing Office, Washington, D.C., pp. 95—100.

Udintsev, G.B., Jemeljanow, E.M., Krüger, E.M., Litwin, W.M. and Charin, G.S. 1977. Geologische Untersuchungen im Gebiet der Insel Jan Mayen; Teil I. Vorläufige morphologishe und geologische Untersuchungsergebnisse aus dem Mohn-Rücken Nordlich Jan Mayen. *Z. Abgew. Geol.* **23**(10), 509—512.

United States Naval Oceanographic Office 1970. *Marine Geophysical Survey Program 1965—1967 North Atlantic Ocean, Norwegian Sea, and Mediterranean Sea, Area 3,* Vol. 3, *Geology and Geophysics.* Report No. NOO-SP-95-3-5, 110pp., 10/29/68, AD-845 252.

Vaage, S. 1980. Seismic evidence on complex tectonics in the Meløy earthquake area. In: *DP 1 Seismiske laster på Norsk Kontinentalsokkel.* Rep., Seismol. Obs., Univ. Bergen.

Vann, I.R. 1974. A modified predrift fit of Greenland and Western Europe. *Nature (London),* **251**, 209—211.

Varentsov, I.M. 1978. Lithologic—mineralogic studies of the sedimentary deposits from Hole 350, DSDP Leg 38. In: M. Talwani *et al.* (eds.). *Initial Reports Deep Sea Drilling Project.* Supplement to Vols. 38—41, U.S. Govt. Printing Office, Washington, D.C., pp. 111—120.

Vaschilov, Y.Y., Kucherinenko, V.A., Markov, G.S., Mirchink, I.M. and Popov, Y.I. 1974. New gravimetric data on the structure of the south-western Barents Sea Shelf (in Russian). *Dokl. Akad. Nauk, SSSR,* **217**, 94—97.

Vigran, J.O. 1979. Fragments of Middle Jurassic flora from the Northern Trøndelag, Norway. *Nor. Geol. Tidsskr.*, **50**, 193—214.

Vinogradova, P.S. and Litvin, V.M. 1960. Studies of bottom relief and sediments in the Barents and Norwegian Seas (In Russian, English summary). *Vniro Pinro,* 101—110.

Vogt, P.R. and Avery, O.E. 1974. Detailed magnetic surveys in the north-east Atlantic and Labrador Seas. *J. Geophys. Res.*, **79**, 363—390.

Vogt, P.R. and Östensö, N.A. 1973. Reconnaissance geophysical studies in the Barents and Kara Seas — summary. *Mem. Am. Assoc. Petrol. Geol.*, **19**, 588—598.

Vogt, P.R. and Perry, R.K. 1978. Post-rifting accretion of continental margins in the Norwegian-Greenland and Labrador Seas: morphological evidence. *Eos. Trans. Am. Geophys. Union,* **59**, p. 1024.

Vogt, P.R., Feden, R.H., Eldholm, O. and Sundvor, E. 1978. The ocean crust west and north of the Svalbard Archi-

pelago: synthesis and review of new results. *Polarforschung*, **48**, 1—19.
Vogt, P.R., Johnsen, G.L. and Kristjansson, L. 1980. Morphology and magnetic anomalies north of Iceland. *J. Geophys.* 47, 67—80.
Vogt, P.R., Kovacs, L.C., Johnson, G.L. and Feden, R.H. 1979. The Eurasia Basin. In: *Norwegian Sea Symposium* (Norsk Petroleumsforening), Article 3.
Vogt, P.R., Östenö, N.A. and Johnson, G.L. 1970. Magnetic and bathymetric data bearing on sea-floor spreading north of Iceland. *J. Geophys. Res.*, **75**, 903—920.
Vogt, P.R., Perry, R.K., Feden, R.H., Fleming, H.S. and Cherkis, N.Z. 1981. The Greenland-Norwegian Sea and Iceland environment; geology and geophysics. In: A.E.M. Nairn, M. Churkin, Jr. and F.G. Stehli, (eds.). *The Ocean Basins and Margins,* Vol. 5, *The Arctic Ocean,* Plenum Press, New York, pp. 493—598.
Vogt, P.R., Taylor, P. Kovacs, L.C. and Johnson, G.L. 1979. Detailed aeromagnetic investigation of the Arctic Basin. *J. Geophys. Res.*, **84**, 1071—1089.
Vorren, T. and Elvsborg, A. 1979. Late Weichselian deglaciation and paleoenvironment of the shelf and coastal areas of Troms, North Norway — a review. *Boreas,* **10**, 247—253.
Vorren, T.O. and Lind-Hansen, O.W. 1976. A Permo-Carboniferous fossil from Tromsøflaket, North Norway. *Nor. Geol. Tidsskr.*, **57**, 189—191.
Vorren, T.O. and Lind-Hansen, O.W. 1979. Quaternary sediments on the continental shelf off Troms and West Finnmark. In: *Norwegian Sea Symposium* (Norsk Petroleumsforening), Article 21.
Vorren, T.O., Strass, I. and Lind-Hansen, O.W. 1978. Late Quaternary sediments and stratigraphy on the continental shelf off Troms and West-Finnmark, Northern Norway. *Q. Res.*, **10**(3), 340—365.
Warnke, D.A. and Hansen, M.E. 1977. Sediments of glacial origins in the area of DSDP Leg 38 (Norwegian-Greenland Seas); preliminary results from Sites 336 and 344. *Naturforsch. Ges. Freib. Im Breisgau,* **67**, 371—392.
Westre, S. 1984. The Askeladden gas find—Troms I. In: A.M. Spencer *et al.* (eds.). *Petroleum Geology of the North European Margin.* This volume, pp. 33—40.
White, S.M. 1978. X-ray mineralogy of sediment, DSDP Leg 38. In: M. Talwani *et al.* (eds.). *Initial Reports Deep Sea Drilling Project.* Supplement to Vols. 38—41, U.S. Govt. Printing Office, Washington, D.C., pp. 437—444.
White, S.M. 1978. Sediments of the Norwegian-Greenland Sea, DSDP Leg 38. In: M. Talwani *et al.* (eds.). *Initial Reports Deep Sea Drilling Project.* Supplement to Vols. 38—41, U.S. Govt. Printing Office, Washington, D.C., pp. 193—258.
Whiteman, A., Roberts, D. and Sellevoll, M.A. 1975. Petroleum geology and geology of the North Sea and Northeast Atlantic Continental Margin. *Nor. Geol. Unders.*, **316**.
Williams, C.A. and McKenzie, D. 1971. The evolution of the North-east Atlantic. *Nature (London)*, **232**, 168—173.
Winsnes, T.S., Kristoffersen, Y and Elverhøi, A. 1977. *Barentshavprosjektet. Feltrapport 1977. Marin Geofysikk og Geologi.* Norsk Polarinstituttt.
Wright, P.L. 1974. The chemistry and mineralogy of the clay fraction of sediments from the southern Barents Sea. *Chem. Geology*, **13**(3), 197—216.
Wright, P.L. 1974. Recent sediment of the southwest Barents Sea. *Mar. Geol.*, **16**(2), 51—81.
Zielinski, G.W. 1979. On the thermal evolution of passive continental margins, thermal depth anomalies, and the Norwegian-Greenland Sea. *J. Geophys. Res.*, **84**(B13), 7577—7588.
Zorkin, L.M. and Bagirov, V.I. 1975. Gaseous hydrocarbons in bottom sediments of the Barents Sea. *Int. Geol. Rev.*, **17**, 631—634.
Østby, K.L. and Nagy, J. 1980. Foraminiferal Stratigraphy of Quaternary Sediments in the Western Barents Sea. In: J.W. Neale and M.D. Basier (eds.). *Microfossils from Recent and Fossil Shelf Seas.* Ellis Horwood, Chichester, pp. 261—273.
Østby, K.L. and Nagy, J. 1982. Foraminiferal distribution in the western Barents Sea Recent and Quaternary. *Polar Res.*, **1**, 53—96.
Øvrebø, O. and Talleraas, E. 1976. The structural geology of the Troms area. In: *Exploration Geology and Geophysics. Offshore North Sea Conference.* Article G-IV/6.
Øvrebø, O. and Talleraas, E. 1977. The structural geology of the Troms area (Barents Sea). *GeoJournal,* **1**, 47—54.
Aalstad, I. 1964. *Magnetiske flymålinger i Norskehavet, Molde, Bodø.* Nor. Geol. Unders., Report No. 534.
Aalstad, I. and Åm, K. 1970. Magnetic investigations on the continental shelf of Norway, Stadlandet—Lofoten. *Nor. Geol. Tidsskr.*, **50**, 280—281.
Aanstad, K.M., Gabrielsen, R.H., Hagevang, T., Ramberg, I.B. and Torvanger, O. 1981. Correlation of offshore and onshore structural features between 62°N and 68°N Norway. In: *Norwegian Symposium on Exploration* (Norsk Petroleumsforening), Article 11.
Aasheim, S.M. and Larsen, V. 1984. The Tyrihans discovery, preliminary results from well 6407/1-2. In: A.M. Spencer *et al.* (eds.). *Petroleum Geology of the North European Margin.* This volume, pp. 285—292.
Åm, K. 1970. Aeromagnetic investigations on the continental shelf of Norway Stad—Lofoten (62—60°). *Nor. Geol. Unders.*, **266**, 49—61.
Åm, K. 1971. Gravity indications of deep sedimentary basins below the Norwegian Continental Shelf and the Vøring Plateau. *Nor. Geol. Tidsskr.*, **51**(2), 199—202.
Åm, K. 1975. Aeromagnetic basement complex mapping north of latitude 62°N, Norway. *Nor. Geol. Unders.*, **316**, 351—374.
Åm, K. 1975. Magnetic profiling over Svalbard and surrounding shelf areas. *Norsk Polarinst. Årbok.* **1973**, 87—100.

ACKNOWLEDGEMENTS

We wish to thank C.A. Lawrance, BP Petroleum Dev. (Overseas) Ltd. and K.O. Kalseth, ex. NPD, for assistance with literature searches, and I. Gjermundsen, Universitetsbiblioteket i Oslo, S. Jensen Flo, Universitetsbiblioteket i Bergen, S. Strand, NTH, Trondheim, A. Gaare, NGU, Trondheim, O. Nordbø, NPD, A.M. Myhre, Universitetet i Oslo, E. Sundvor, Universitetet i Bergen, G. Grønlie, DNO, R. Lund, Norsk Polarinstitutt and Mrs E.R. Nutt, Geological Society of London for help with verification of references, and L. Ulvestad and R. Haaland for typing the reference list.